Sears and Zemansky's

UNIVERSITY PHYSICS

11th Edition

Volume 1

HUGH D. YOUNG
CARNEGIE MELLON UNIVERSITY

ROGER A. FREEDMAN
UNIVERSITY OF CALIFORNIA, SANTA BARBARA

CONTRIBUTING AUTHOR
A. LEWIS FORD
TEXAS A&M UNIVERSITY

PEARSON
Addison
Wesley

San Francisco Boston New York
Cape Town Hong Kong London Madrid Mexico City
Montreal Munich Paris Singapore Sydney Tokyo Toronto

Executive Editor: Adam Black, Ph.D.
Development Editors: David Chelton, Alice Houston
Project Editors: Nancy Benton, Richard Dal Porto
Production Editor, Art Coordinator, Copyeditor, Illustrator,
    Photo Editor, Permissions Editor, Proofreader, Compositor: Thompson Steele, Inc.
Media Producer: Claire Masson
Marketing Manager: Christy Lawrence
Director of Marketing: Stacy Treco
Art Director: Blake Kim
Manufacturing Supervisor: Vivian McDougal
Text and Cover Designer: Mark Ong, Side by Side Studios
Cover Printer: Phoenix Color Corporation
Printer and Binder: R.R. Donnelley & Sons

Cover Photo: Image of London's Millennium Bridge
created by Cimex Media, Ltd. http://www.cimexmedia.com
Photo Credits: see page A15

Library of Congress Cataloging-in-Publication Data

Young, Hugh D.
    Sears and Zemansky's university physics: with modern physics. — 11th ed. / Hugh D.
Young, Roger A. Freedman; contributing author, A. Lewis Ford.
      p. cm.
    Includes bibliographical references and index.
    ISBN 0-8053-8767-6 (alk, paper)
    1. Physics. I. Title: University physics. II. Freedman, Roger A. III. Sears, Francis
Weston, 1898–1975. University physics. IV. Title.

QC21.3.Y68 2004
530—dc21

2003049555

0-8053-8767-6   Volume 1 with Mastering Physics
0-8053-9186-X   Volume 1 without Mastering Physics

78910 — DOW —06 05
www.aw.com/bc

PREFACE

This book is the product of more than half a century of innovation in physics education. When the first edition of *University Physics* by Francis W. Sears and Mark W. Zemansky appeared in 1949, it was revolutionary among calculus-based physics textbooks in its emphasis on the fundamental principles of physics and how to apply them. The success of *University Physics* with generations of students and educators throughout the world is a testament to the merits of this approach. In preparing this new Eleventh Edition, we have further enhanced and developed *University Physics* to emphasize two key objectives: helping students develop conceptual understanding, and helping them build strong problem-solving skills.

New to This Edition

■ A **new arrangement of topics** places the chapters on mechanical waves (15 and 16) immediately after the chapters on oscillations and fluids and before the chapters on thermodynamics. The wave chapters have been rearranged to eliminate repetition of topics. The discussion of displacement current has been moved to Chapter 29 to follow the presentation of electromagnetic induction. The chapters on geometric optics and optical instruments have been streamlined into a single Chapter 34. Chapter 39 has an expanded discussion of the Schrödinger equation. Some of the optional sections from the Tenth Edition of *University Physics* have been moved to the companion web site.

■ A totally revised **art program** helps students better visualize key concepts. More discussion has been placed *on* the figures themselves, and consistent colors for each kind of vector quantity are used throughout the book. In the discussion of mechanics, color and design have been streamlined to help students identify the object of interest. New photographs throughout the book illustrate the applications of physics to the natural world and to today's technology.

■ A research-based, **step-by-step approach to problem solving** is used throughout the book. Called **ISEE**—an acronym for *Identify, Set up, Execute,* and *Evaluate*—this approach gives students a global strategy for solving physics problems of all kinds. The ISEE approach is used in *all* Problem-Solving Strategy boxes and in *all* worked examples. Educational research indicates that this consistent approach helps students build their confidence in attacking new problems.

■ To help students check their comprehension of the material, there are **Chapter Opening Questions** at the beginning of each chapter and **Test Your Understanding Questions** at the end of almost every chapter section. The answers to the questions are found at the end of the chapter.

■ New **visual chapter summaries** reinforce the key ideas in each chapter. They help students consolidate and connect important concepts by matching a descriptive explanation, a mathematical representation, and a thumbnail sketch in an easy-to-review grid.

■ **Careful use of subscripts** in kinematics, dynamics, and elsewhere makes it easier for students to distinguish between signed quantities such as velocity v_x and magnitudes such as speed v.

■ There are over **600 new end-of-chapter problems,** with an emphasis on context-rich problems, bringing the total problem count to more than 3700.

■ Accompanying *University Physics* are **two new web-based tools** for teaching and learning physics. **MasteringPhysics™** is a research-proven online tutorial and homework assignment system. MasteringPhysics™ includes an extensive library of entirely new and carefully tested problems that can be used for student assignments. These problems include *Skill Builders* (detailed worked examples that focus on conceptual understanding or key skills, and include multiple hint-giving options) and *Tutoring Problems* (standard multi-step homework problems that give on-demand hints and individualized feedback based on the student's particular answers). *End-of-Chapter Problems* (the majority of the end-of-chapter problems from *University Physics*) are also provided in the system with algorithmically generated variables and fewer hints to more accurately replicate a testing environment. **ActivPhysics OnLine** provides a comprehensive library of more than 220 applets and accompanying interactive activities that cover topics from mechanics through modern physics. It includes interactive conceptual exercises that encourage students to confront misconceptions, reason qualitatively, and think critically.

Key Features of *University Physics*

Standard, Extended, and Split Versions: This edition is available in three versions. The Standard version (ISBN 0-8053-8768-4) includes 37 chapters, ending with the special theory of relativity. The Extended version (ISBN 0-8053-8684-X) adds seven chapters on modern physics, including the physics of atoms, molecules, condensed matter, nuclei, and elementary particles. The Split version includes all 44 chapters in three volumes. Volume 1 (ISBN 0-8053-8767-6, Chapters 1–20) covers mechanics, waves, and thermodynamics; Volume 2 (ISBN 0-8053-8766-8, Chapters 21–37) covers electricity and magnetism, optics, and relativity; and Volume 3 (ISBN 0-8053-8765-X, Chapters 37–44) covers modern physics.

A Guide for the Student: Many physics students experience difficulty simply because they don't know how to use their textbook. A section entitled "How to Succeed in Physics by Really Trying," which follows this preface, is a "user's manual" to all the features of this book. This section, written by Professor Mark Hollabaugh (Normandale Community College), also gives a number of helpful study hints. *Every* student should read this section!

Chapter Organization: The first section of each chapter is an *Introduction* that gives specific examples of the chapter's content and connects it with what has come before. There is also a *Chapter Opening Question* to make the reader think about the subject matter of the chapter ahead. (To find the answer to the question, look for the ⟨ ? ⟩ icon.) Most sections end with a *Test Your Understanding Ques-*

tion, which can be conceptual or quantitative in nature. At the end of the last section of the chapter is a *Visual Chapter Summary* of the most important principles in the chapter, as well as a list of *Key Terms* with reference to the page number where each term is introduced. The answers to the Chapter Opening Question and Test Your Understanding Questions are on the page following the Key Terms.

Questions and Problems: At the end of each chapter is a collection of *Discussion Questions* that probe and extend the student's conceptual understanding. Following these are *Exercises,* which are single-concept problems keyed to specific sections of the text; *Problems,* usually requiring one or two nontrivial steps; and *Challenge Problems,* intended to challenge the strongest students. The problems include applications to such diverse fields as astrophysics, biology, and aerodynamics. Many problems have a conceptual part in which students must discuss and explain their results. The new questions, exercises, and problems for this edition were created and organized by Professor A. Lewis Ford (Texas A&M University) with contributions from Wayne Anderson (Sacramento City College) and Rich Gottfried (Frederick Community College).

Problem-Solving Strategies and Worked Examples: Throughout the book, *Problem-Solving Strategy* boxes provide students with specific tactics for solving particular types of problems. They address the needs of any student who's ever felt that they "understand the concepts but can't do the problems."

All Problem-Solving Strategy boxes follow the ISEE approach (Identify, Set up, Execute, and Evaluate) to solving problems. This approach helps students see how to begin with a seemingly complex situation, identify the relevant physical concepts, decide what tools are needed to solve the problem, carry out the solution, and then evaluate whether the result makes sense.

Each Problem-Solving Strategy box is followed by one or more worked-out *Examples* that illustrate the strategy. Many other worked-out Examples are found in each chapter. Like the Problem-Solving Strategy boxes, all of the quantitative Examples use the ISEE approach. Several of the examples are purely qualitative, and are labeled as *Conceptual Examples;* see, for instance, Conceptual Examples 6.6 (Comparing kinetic energies, p. 218), 8.1 (Momentum versus kinetic energy, p. 286) and 20.7 (A reversible adiabatic process, p. 775).

"Caution" paragraphs: Two decades of physics education research have revealed a number of conceptual pitfalls that commonly plague beginning physics students. These include the ideas that force is required for motion, that electric current is "used up" as it goes around a circuit, and that the product of an object's mass and its acceleration is itself a force. The "Caution" paragraphs alert students to these and other pitfalls, and explain why the wrong way to think about a certain situation (which may have occurred to the student first) is indeed wrong. (See, for example, pp. 132, 182, and 629.)

Notation and units: Students often have a hard time keeping track of which quantities are vectors and which are not. We use boldface italic symbols with an arrow on top for vector quantities, such as $\vec{v}$, $\vec{a}$, and $\vec{F}$; unit vectors such as $\hat{\imath}$, have a caret on top. Boldface $+$, $-$, $\times$, and $=$ signs are used in vector equations to emphasize the distinction between vector and scalar mathematical operations.

SI units are used exclusively (English unit conversions are included where appropriate). The joule is used as the standard unit of energy of all forms, including heat.

Flexibility: The book is adaptable to a wide variety of course outlines. There is plenty of material for a three-semester or a five-quarter course. Most instructors will find that there is too much material for a one-year course, but it is easy to tailor the book to a variety of one-year course plans by omitting certain chapters or sections. For example, any or all of the chapters on fluid mechanics, sound and hearing, electromagnetic waves, or relativity can be omitted without loss of continuity. In any case, no instructor should feel constrained to work straight through the entire book.

Online Supplements for Students and Instructors

MasteringPhysics™ (www.masteringphysics.com) is a sophisticated online tutoring and homework system developed specially for courses using *University Physics*. Originally developed by Professor David Pritchard of Effective Educational Technologies, MasteringPhysics™ provides students with individualized online tutoring by responding to their wrong answers and providing hints for solving multi-step problems when they get stuck. It gives them immediate and up-to-date assessment of their progress, and shows where they need to practice more. MasteringPhysics™ provides instructors with a fast and effective way to assign tried-and-tested online homework assignments that comprise a range of problem types. The powerful post-assignment diagnostics allow instructors to assess the progress of their class as a whole and individual students, and quickly identify areas of difficulty.

 ActivPhysics OnLine (www.aw.com/young11) was developed by Professors Alan Van Heuvelen and Paul D'Alessandris. It provides simulations and accompanying conceptual exercises from mechanics, waves, thermodynamics, electricity and magnetism, and modern physics. Icons throughout the book indicate which ActivPhysics simulations and exercises correspond to various sections of *University Physics*.

The Addison Wesley Tutor Center (www.aw.com/tutorcenter) provides one-on-one tutoring via telephone, fax, email, or interactive web site during evening hours and on weekends. Qualified college instructors answer questions and provide instruction regarding MasteringPhysics™ as well as examples, exercises, and other content from *University Physics*.

The Companion Web Site (www.aw.com/young11) includes additional study material as well as in-depth discussions of physics topics not included in *University Physics*.

Additional Student Supplements

The **Study Guide,** prepared by Professor James R. Gaines and William F. Palmer, reinforces the text's emphasis on problem-solving strategies and student misconceptions. *The Study Guide for Volume 1* (ISBN 0-8053-8778-1) covers Chapters 1–20, and the *Study Guide for Volumes 2 and 3* (ISBN 0-8053-8743-9) covers Chapters 21–44.

The **Student Solutions Manual,** prepared by Professor A. Lewis Ford, includes careful, step-by-step solutions for most of the odd-numbered exercises and problems in *University Physics*, following the same ISEE approach used in the Problem-Solving Strategy boxes and in the Examples. (The answers for all odd-numbered problems are found in *University Physics* following the Appendices.) The *Student Solutions Manual for Volume 1* (ISBN 0-8053-8777-3) covers Chapters 1–20, and the *Student Solutions Manual for Volumes 2 and 3* (ISBN 0-8053-8696-3) covers Chapters 21–44.

Additional Instructor Supplements

The **Instructor's Solutions Manual,** prepared by Professor A. Lewis Ford, contains worked-out solutions to all exercises, problems, and challenge problems. (It is available to qualifying instructors only.) The *Instructor's Solutions Manual for Volume 1* (ISBN 0-8053-8775-7) covers Chapters 1–20, and the *Instructor's Solutions Manual for Volumes 2 and 3* (ISBN 0-8053-8776-5) covers Chapters 21–44. The complete text of both printed manuals is also available in Word format on the cross-platform *Instructor's Solutions Manual* CD-ROM (ISBN 0-8053-8774-9).

The **Test Bank** is available in both print and electronic formats. The *Printed Test Bank* (ISBN 0-8053-8772-2) includes more than 1500 problems. All of these problems are included in the cross-platform *Computerized Test Bank* CD-ROM (ISBN 0-8053-8771-4); more than half of the problems on the CD-ROM now offer algorithmically generated functionality thanks to Benjamin Grinstein (University of California, San Diego).

Five Easy Lessons: Strategies for Successful Physics Teaching (ISBN 0-8053-8702-1), written by Professor Randall Knight (California Polytechnic State University, San Luis Obispo), is packed with creative ideas on how to enhance any physics course. It is an invaluable companion for both novice and veteran physics instructors.

The **Simulation and Image Presentation CD-ROM** (ISBN 0-8053-8769-2) includes a library of more than 220 simulation applets from ActivPhysics OnLine. These make it easy to illustrate challenging physics concepts in class in a dynamic way. The CD-ROM also includes all of the illustrations (excluding photographs) from *University Physics*, all chapter summaries, Problem-Solving Strategy boxes, and key equations. All these items can easily be exported into lecture-presentation programs such as PowerPoint®, or posted online. A set of **Transparency Acetates** (ISBN 0-8053-8773-0) with more than 200 illustrations from *University Physics* is also available.

Acknowledgments

We would like to thank the hundreds of reviewers and colleagues who have offered valuable comments and suggestions over the life of this textbook. The continuing success of *University Physics* is due in large measure to their contributions.

Edward Adelson (Ohio State University), Ralph Alexander (University of Missouri at Rolla), J. G. Anderson, R. S. Anderson, Wayne Anderson (Sacramento City College), Alex Azima (Lansing Community College), Dilip Balamore (Nassau Community College), Harold Bale (University of North Dakota), Arun Bansil (Northeastern University), John Barach (Vanderbilt University), J. D. Barnett, H. H. Barschall, Albert Bartlett (University of Colorado), Paul Baum (CUNY, Queens College), Frederick Becchetti (University of Michigan), B. Bederson, Lev I. Berger (San Diego State University),

Robert Boeke (William Rainey Harper College), S. Borowitz, A. C. Braden, James Brooks (Boston University), Nicholas E. Brown (California Polytechnic State University, San Luis Obispo), Tony Buffa (California Polytechnic State University, San Luis Obispo), A. Capecelatro, Michael Cardamone (Pennsylvania State University), Duane Carmony (Purdue University), Troy Carter (UCLA), P. Catranides, John Cerne (SUNY at Buffalo), Roger Clapp (University of South Florida), William M. Cloud (Eastern Illinois University), Leonard Cohen (Drexel University), W. R. Coker (University of Texas, Austin), Malcolm D. Cole (University of Missouri at Rolla), H. Conrad, David Cook (Lawrence University), Gayl Cook (University of Colorado), Hans Courant (University of Minnesota), Bruce A. Craver (University of Dayton), Larry Curtis (University of Toledo), Jai Dahiya (Southeast Missouri State University), Steve Detweiler (University of Florida), George Dixon (Oklahoma State University), Donald S. Duncan, Boyd Edwards (West Virginia University), Robert Eisenstein (Carnegie Mellon University), Amy Emerson Missouri (Virginia Institute of Technology), William Faissler (Northeastern University), William Fasnacht (U.S. Naval Academy), Paul Feldker (St. Louis Community College), Carlos Figueroa (Cabrillo College), L. H. Fisher, Neil Fletcher (Florida State University), Robert Folk, Peter Fong (Emory University), A. Lewis Ford (Texas A&M University), D. Frantszog, James R. Gaines (Ohio State University), Solomon Gartenhaus (Purdue University), Ron Gautreau (New Jersey Institute of Technology), J. David Gavenda (University of Texas, Austin), Dennis Gay (University of North Florida), James Gerhart (University of Washington), N. S. Gingrich, J. L. Glathart, S. Goodwin, Rich Gottfried (Frederick Community College), Walter S. Gray (University of Michigan), Paul Gresser (University of Maryland), Benjamin Grinstein (UC San Diego), Howard Grotch (Pennsylvania State University), John Gruber (San Jose State University), Graham D. Gutsche (U.S. Naval Academy), Michael J. Harrison (Michigan State University), Harold Hart (Western Illinois University), Howard Hayden (University of Connecticut), Carl Helrich (Goshen College), Laurent Hodges (Iowa State University), C. D. Hodgman, Michael Hones (Villanova University), Keith Honey (West Virginia Institute of Technology), Gregory Hood (Tidewater Community College), John Hubisz (North Carolina State University), M. Iona, John Jaszczak (Michigan Technical University), Alvin Jenkins (North Carolina State University), Robert P. Johnson (UC Santa Cruz), Lorella Jones (University of Illinois), John Karchek (GMI Engineering & Management Institute), Thomas Keil (Worcester Polytechnic Institute), Robert Kraemer (Carnegie Mellon University), Jean P. Krisch (University of Michigan), Robert A. Kromhout, Charles Lane (Berry College), Robert J. Lee, Alfred Leitner (Rensselaer Polytechnic University), Gerald P. Lietz (De Paul University), Gordon Lind (Utah State University), S. Livingston, Elihu Lubkin (University of Wisconsin, Milwaukee), Robert Luke (Boise State University), David Lynch (Iowa State University), Michael Lysak (San Bernardino Valley College), Jeffrey Mallow (Loyola University), Robert Mania (Kentucky State University), Robert Marchina (University of Memphis), David Markowitz (University of Connecticut), R. J. Maurer, Oren Maxwell (Florida International University), Joseph L. McCauley (University of Houston), T. K. McCubbin, Jr. (Pennsylvania State University), Charles McFarland (University of Missouri at Rolla), James Mcguire (Tulane University), Lawrence McIntyre (University of Arizona), Fredric Messing (Carnegie-Mellon University), Thomas Meyer (Texas A&M University), Andre Mirabelli (St. Peter's College, New Jersey), Herbert Muether (S.U.N.Y., Stony Brook), Jack Munsee (California State University, Long Beach), Lorenzo Narducci (Drexel University), Van E. Neie (Purdue University), David A. Nordling (U. S. Naval Academy), Benedict Oh (Pennsylvania State University), L. O. Olsen, Jim Pannell (DeVry Institute of Technology), W. F. Parks (University of Missouri), Jerry Peacher (University of Missouri at Rolla), Arnold Perlmutter (University of Miami), Lennart Peterson (University of Florida), R. J. Peterson (University of Colorado, Boulder), R. Pinkston, Ronald Poling (University of Minnesota), J. G. Potter, C. W. Price (Millersville University), Francis Prosser (University of Kansas), Shelden H. Radin, Michael Rapport (Anne Arundel Community College), R. Resnick, James A. Richards, Jr., John S. Risley (North Carolina State University), Francesc Roig (University of California, Santa Barbara), T. L. Rokoske, Richard Roth (Eastern Michigan University), Carl Rotter (University of West Virginia), S. Clark Rowland (Andrews University), Rajarshi Roy (Georgia Institute of Technology), Russell A. Roy (Santa Fe Community College), Bruce Schumm (UC Santa Cruz), Melvin Schwartz (St. John's University), F. A. Scott, L. W. Seagondollar, Stan Shepherd (Pennsylvania State University), Douglas Sherman (San Jose State), Bruce Sherwood (Carnegie Mellon University), Hugh Siefkin (Greenville College), C. P. Slichter, Charles W. Smith (University of Maine, Orono), Malcolm Smith (University of Lowell), Ross Spencer (Brigham Young University), Julien Sprott (University of Wisconsin), Victor Stanionis (Iona College), James Stith (American Institute of Physics), Chuck Stone (North Carolina A&T State University), Edward Strother (Florida Institute of Technology), Conley Stutz (Bradley University), Albert Stwertka (U.S. Merchant Marine Academy), Martin Tiersten (CUNY, City College), David Toot (Alfred University), Somdev Tyagi (Drexel University), F. Verbrugge, Helmut Vogel (Carnegie Mellon University), Robert Webb (Texas A & M), Thomas Weber (Iowa State University), M. Russell Wehr, (Pennsylvania State University), Robert Weidman (Michigan Technical University), Dan Whalen (UC San Diego), Lester V. Whitney, Thomas Wiggins (Pennsylvania State University), David Willey (University of Pittsburgh, Johnstown), George Williams (University of Utah), John Williams (Auburn University), Stanley Williams (Iowa State University), Jack Willis, Suzanne Willis (Northern Illinois University), Robert Wilson (San Bernardino Valley College), L. Wolfenstein, James Wood (Palm Beach Junior College), Lowell Wood (University of Houston), R. E. Worley, D. H. Ziebell (Manatee Community College), George O. Zimmerman (Boston University)

In addition, we both have individual acknowledgments we would like to make.

I want to extend my heartfelt thanks to my colleagues at Carnegie Mellon, especially Professors Robert Kraemer, Bruce Sherwood, Ruth Chabay, Helmut Vogel, and Brian Quinn, for many stimulating discussions about physics pedagogy and for their support and encouragement during the writing of several successive editions of this book. I am equally indebted to the many generations of Carnegie Mellon students who have helped me learn what good teaching and good writing are, by showing me what works and what doesn't. It is always a joy and a privilege to express my gratitude to my wife Alice and our children Gretchen and Rebecca for their love, support, and emotional sustenance during the writing of several successive editions of this book. May all men and women be blessed with love such as theirs. — H. D. Y.

I would like to thank my past and present colleagues at UCSB, including Francesc Roig, Elisabeth Nicol, Al Nash, and Carl Gwinn, for their wholehearted support and for many helpful discussions. I owe a special debt of gratitude to my early teachers Willa Ramsay, Peter Zimmerman, William Little, Alan Schwettman, and Dirk Walecka for showing me what clear and engaging physics teaching is all about, and to Stuart Johnson for inviting me to become a co-author of *University Physics* beginning with the 9th edition. I want to express special thanks to Adam Black and Nancy Benton at Addison Wesley for their superb editorial guidance and vision, to David Chelton for aiding in the development of this edition through his many valuable suggestions and constant good humor, and to Nancy Freihofer of Thompson Steele for shepherding this book through the Byzantine production process. I want to thank my father for his continued love and support and for keeping a space open on his bookshelf for this book. Most of all, I want to express my gratitude and love to my wife Caroline, to whom I dedicate my contributions to this book. Hey, Caroline, the new edition's done at last — let's go flying! — R. A. F.

PLEASE TELL US WHAT YOU THINK!

We welcome communications from students and professors, especially concerning errors or deficiencies that you find in this edition. We have devoted a lot of time and effort to writing the best book we know how to write, and we hope it will help you to teach and learn physics. In turn, you can help us by letting us know what still needs to be improved! Please feel free to contact us either electronically or by ordinary mail. Your comments will be greatly appreciated.

April 2003

Hugh D. Young
Department of Physics
Carnegie Mellon University
Pittsburgh, Pennsylvania 15213
hdy@andrew.cmu.edu

Roger A. Freedman
Department of Physics
University of California, Santa Barbara
Santa Barbara, California 93106-9530
airboy@physics.ucsb.edu
http://www.physics.ucsb.edu/~airboy/

HOW TO SUCCEED IN PHYSICS BY REALLY TRYING

Mark Hollabaugh, Normandale Community College

Physics encompasses the large and the small, the old and the new. From the atom to galaxies, from electrical circuitry to aerodynamics, physics is very much a part of the world around us. You probably are taking this introductory course in calculus-based physics because it is required for subsequent courses you plan to take in preparation for a career in science or engineering. Your professor wants you to learn physics and to enjoy the experience. He or she is very interested in helping you learn this fascinating subject. That is part of the reason your professor chose this textbook for your course. That is also the reason why Drs. Young and Freedman asked me to write this introductory section. We want you to succeed!

The purpose of this section of *University Physics* is to give you some ideas that will assist your learning. Specific suggestions on how to use the textbook will follow a brief discussion of general study habits and strategies.

Preparation for This Course

If you had high school physics, you will probably learn concepts faster than those who have not because you will be familiar with the language of physics. If English is a second language for you, keep a glossary of new terms that you encounter and make sure you understand how they are used in physics. Likewise, if you are farther along in your mathematics courses, you will pick up the mathematical aspects of physics faster. Even if your mathematics is adequate, you may find a book such as Arnold D. Pickar's *Preparing for General Physics: Math Skill Drills and Other Useful Help (Calculus Version)* to be useful. Your professor may actually assign sections of this math review to assist your learning.

Learning to Learn

Each of us has a different learning style and a preferred means of learning. Understanding your own learning style will help you to focus on aspects of physics that may give you difficulty and to use those components of your course that will help you overcome the difficulty. Obviously you will want to spend more time on those aspects that give you the most trouble. If you learn by hearing, lectures will be very important. If you learn by explaining, then working with other students will be useful to you. If solving problems is difficult for you, spend more time learning how to solve problems. Also, it is important to understand and develop good study habits. Perhaps the most important thing you can do for yourself is to set aside adequate, regularly scheduled, study time in a distraction-free environment.

Answer the following questions for yourself:
- Am I able to use fundamental mathematical concepts from algebra, geometry and trigonometry? (If not, plan a program of review with help from your professor.)
- In similar courses, what activity has given me the most trouble? (Spend more time on this.) What has been the easiest for me? (Do this first; it will help to build your confidence.)
- Do I understand the material better if I read the book before or after the lecture? (You may learn best by skimming the material, going to lecture, and then undertaking an in-depth reading.)

- Do I spend adequate time in studying physics? (A rule of thumb for a class like this is to devote, on the average, 2.5 hours out of class for each hour in class. For a course meeting 5 hours each week, that means you should spend about 10 to 15 hours per week studying physics.)
- Do I study physics every day? (Spread that 10 to 15 hours out over an entire week!) At what time of the day am I at my best for studying physics? (Pick a specific time of the day and stick to it.)
- Do I work in a quiet place where I can maintain my focus? (Distractions will break your routine and cause you to miss important points.)

Working with Others

Scientists or engineers seldom work in isolation from one another but rather work cooperatively. You will learn more physics and have more fun doing it if you work with other students. Some professors may formalize the use of cooperative learning or facilitate the formation of study groups. You may wish to form your own informal study group with members of your class who live in your neighborhood or dorm. If you have access to e-mail, use it to keep in touch with one another. Your study group is an excellent resource when reviewing for exams.

Lectures and Taking Notes

An important component of any college course is the lecture. In physics this is especially important because your professor will frequently do demonstrations of physical principles, run computer simulations, or show video clips. All of these are learning activities that will help you to understand the basic principles of physics. Don't miss lectures, and if for some reason you do, ask a friend or member of your study group to provide you with notes and let you know what happened.

Take your class notes in outline form, and fill in the details later. It can be very difficult to take word for word notes, so just write down key ideas. Your professor may use a diagram from the textbook. Leave a space in your notes and just add the diagram later. After class, edit your notes, filling in any gaps or omissions and noting things you need to study further. Make references to the textbook by page, equation number, or section number.

Make sure you ask questions in class, or see your professor during office hours. Remember the only "dumb" question is the one that is not asked. Your college may also have teaching assistants or peer tutors who are available to help you with difficulties you may have.

Examinations

Taking an examination is stressful. But if you feel adequately prepared and are well-rested, your stress will be lessened. Preparing for an exam is a continual process; it begins the moment the last exam is over. You should immediately go over the exam and understand any mistakes you made. If you worked a problem and made substantial errors, try this: Take a piece of paper and divide it down the middle with a line from top to bottom. In one column, write the proper solution to the problem. In the other column, write what you did and why, if you know, and why your solution was incorrect. If you are uncertain why you made your mistake, or how to avoid making it again, talk with your professor. Physics continually builds on fundamental ideas and it is important to correct any misunderstandings immediately. Warning: While cramming at the last minute may get you through the *present* exam, you will not adequately retain the concepts for use on the *next* exam.

Using Your Textbook

Now let's take a look at specific features of *University Physics* that will help you understand the concepts of physics. At its heart, physics is not equations and numbers. Physics is a way of looking at the universe and understanding how the universe works and how its various parts relate to each other. And although solving quantitative problems is an important part of physics, it is equally important for you to understand concepts qualitatively. Your textbook will help you in both areas.

First of all, don't be afraid to write in your book. It is more important for you to learn the concepts of physics than to keep your book in pristine condition. Write in the margins, make cross references. Take notes in your notebook as you read. *University Physics* is your primary "reference book" for this course. Refer to it often to help you understand the concepts you hear in lecture. Become familiar with the contents of the appendices and end papers.

Chapter Opening Questions and Test Your Understanding Questions

Test Your Understanding

You are driving your car on a country road when a mosquito splatters itself on the windshield. Which has the greater magnitude, the force that the car exerted on the mosquito or the force that the mosquito exerted on the car? Or are the magnitudes the same? If they are different, how can you reconcile this fact with Newton's third law? If they are equal, why is the mosquito splattered while the car is undamaged?

Each chapter of *University Physics* opens with a photograph, and next to that photograph is a question. As you read the chapter, try to figure out the answer to this question. You'll also find a *Test Your Understanding* question at the end of most sections of the book. These are designed to help you check if you've understood a key concept that's just been presented. If you can't see how to answer the question, you may need to read the section again with more care. (You'll find the answers to both kinds of questions toward the end of the chapter, but you'll get the most benefit if you figure out the answers yourself!)

Worked Examples

Your professor will work example problems in class to illustrate the application of the concepts of physics to real-world problems. You should work through all the examples in the textbook, filling in any missing steps, and making note of things you don't understand. Get help with the concepts that confuse you!

Example 5.8 Straight-line motion with friction

Suppose the wind is once again blowing steadily in the +*x*-direction as in Example 5.6, so that the iceboat has a constant acceleration $a_x = 1.5 \text{ m/s}^2$. Now, however, there is a constant horizontal friction force with magnitude 100 N that opposes the motion of the iceboat. In this case, what force F_W must the wind exert on the iceboat?

SOLUTION

IDENTIFY: Once again the target variable is F_W. We are given the acceleration, so to find F_W all we need is Newton's second law.

SET UP: Figure 5.9 shows the new free-body diagram. The only difference from Fig. 5.7b is the addition of the friction force $\vec{f}$, which points opposite to the motion. (Note that its *magnitude*, $f = 100$ N, is a positive quantity but that its *component* in the *x*-direction is negative, equal to $-f$ or -100 N.)

EXECUTE: There are now two forces (the force of the wind and the friction force) that have *x*-components. The *x*-component of Newton's second law gives

$$\sum F_x = F_W + (-f) = ma_x$$
$$F_W = ma_x + f = (200 \text{ kg})(1.5 \text{ m/s}^2) + (100 \text{ N}) = 400 \text{ N}$$

EVALUATE: Because there is friction, a greater force F_W is needed than in Example 5.6. We need 100 N to overcome friction and 300 N more to give the iceboat the necessary acceleration.

5.9 Free-body diagram for the iceboat and rider with a frictional force $\vec{f}$ opposing the motion.

Problem-Solving Strategies

One of the features of *University Physics* that first caught my eye as a teacher were the *Problem-Solving Strategy* boxes. This is the advice I would give to a student who came to me for help with a physics problem. Physics teachers approach a problem in a very systematic and logical manner. These boxes will help you as a beginning problem solver to do the same by following a systematic problem-solving strategy called ISEE—Identify, Set up, Execute, and Evaluate. Study these suggestions in great detail and implement them. In many cases these strategy boxes will tell you how to visualize an abstract concept. Finally, use the ISEE approach consistently. It will help you build confidence in tackling new problems.

Problem-Solving Strategy

Conservation of Momentum

IDENTIFY *the relevant concepts:* Before applying conservation of momentum to a problem, you must first decide whether momentum *is* conserved! This will be true *only* if the vector sum of the external forces acting on the system of particles is zero. If this is not the case, you can't use conservation of momentum.

SET UP *the problem using the following steps:*
1. Define a coordinate system. Make a sketch showing the coordinate axes, including the positive direction for each. Often it is easiest to choose the *x*-axis to have the direction of one of the initial velocities. Make sure you are using an inertial frame of reference. Most of the problems in this chapter deal with two-dimensional situations, in which the vectors have only *x*- and *y*-components; all of the following statements can be generalized to include *z*-components when necessary.
2. Treat each body as a particle. Draw "before" and "after" sketches, and include vectors on each to represent all known velocities. Label the vectors with magnitudes, angles, components, or whatever information is given, and give each unknown magnitude, angle, or component an algebraic symbol. You may find it helpful to use the subscripts 1 and 2 for velocities before and after the interaction, respectively; if you use these subscripts, use letters (not numbers) to label each particle.

3. As always, identify the target variable(s) from among the unknowns.

EXECUTE *the solution as follows:*
1. Write an equation in terms of symbols equating the total *initial x*-component of momentum (that is, before the interaction) to the total *final x*-component of momentum (that is, after the interaction), using $p_x = mv_x$ for each particle. Write another equation for the *y*-components, using $p_y = mv_y$ for each particle. Remember that the *x*- and *y*-components of velocity or momentum are *never* added together in the same equation! Even when all the velocities lie along a line (such as the *x*-axis), the components of velocity along this line can be positive or negative; be careful with signs!
2. Solve these equations to determine whatever results are required. In some problems you will have to convert from the *x*- and *y*-components of a velocity to its magnitude and direction, or the reverse.
3. In some problems, energy considerations give additional relationships among the various velocities, as we will see later in this chapter.

EVALUATE *your answer:* Does your answer make physical sense? If your target variable is a certain body's momentum, check that the direction of the momentum is reasonable.

Caution!

Educational research has identified misconceptions or misunderstandings that students frequently have when they study physics. Dr. Freedman has added Caution! paragraphs to warn you about these potential pitfalls and help you avoid them.

CAUTION Please note that the quantity $m\vec{a}$ is *not* a force. All that Eqs. (4.7) and (4.8) say is that the vector $m\vec{a}$ is equal in magnitude and direction to the vector sum $\Sigma\vec{F}$ of all the forces acting on the body. It's incorrect to think of acceleration as a force; rather, acceleration is a result of a nonzero net force. It's "common sense" to think that there is a "force of acceleration" that pushes you back into your seat when your car accelerates forward from rest. But *there is no such force;* instead, your inertia causes you to tend to stay at rest relative to the earth, and the car accelerates around you. The "common sense" confusion arises from trying to apply Newton's second law in a frame of reference where it isn't valid, like the non-inertial reference frame of an accelerating car. We will always examine motion relative to *inertial* frames of reference only.

Key Terms, Summary, and Problems

The most important concepts are listed in the *Key Terms*. Keep a glossary of terms in your notebook. Your professor may indicate through the use of course objectives which terms are important for you to know. The *Summary* will give you a quick review of the chapter's main ideas and how they can be represented descriptively (in text), visually (in a figure), and mathematically (with an equation). Use this to make sure you understand how the same concept can be represented in these different ways. Everything else can be derived from the general equations provided in the summary.

If your professor assigns *Problems* at the end of the chapter, make sure you work them carefully with other students. If solutions are available, do not look at the answer until you have struggled with the problem and compared your answer with someone else's. If the two of you agree on the answer, then look at the solution. If you have made a mistake, go back and rework the problem. You will note that the *Exercises* are keyed to specific sections of the chapter and are easier. Work on these before you attempt the *Problems* or *Challenge Problems* which typically use multiple concepts.

Well, there you have it. We hope these suggestions will benefit your study of physics. Strive for understanding and excellence, and be persistent in your learning.

CONTENTS

Note: Chapters 1–20 appear in Volume 1; Chapters 21–37 appear in Volume 2; and Chapters 37–44 appear in Volume 3.

MECHANICS

OPTICS

33 THE NATURE AND PROPAGATION OF LIGHT 1247

34 GEOMETRIC OPTICS AND OPTICAL INSTRUMENTS 1285

35 INTERFERENCE 1338

36 DIFFRACTION 1367

MODERN PHYSICS

37 RELATIVITY 1403

38 PHOTONS, ELECTRONS, AND ATOMS 1445

ABOUT THE AUTHORS

Hugh D. Young is Professor of Physics at Carnegie Mellon University in Pittsburgh, PA. He attended Carnegie Mellon for both undergraduate and graduate study and earned his Ph.D. in fundamental particle theory under the direction of the late Richard Cutkosky. He joined the faculty of Carnegie Mellon in 1956, and has also spent two years as a Visiting Professor at the University of California at Berkeley.

Prof. Young's career has centered entirely around undergraduate education. He has written several undergraduate-level textbooks, and in 1973 he became a co-author with Francis Sears and Mark Zemansky for their well-known introductory texts. With their deaths, he assumed full responsibility for new editions of these books until joined by Prof. Freedman for *University Physics*.

Prof. Young is an enthusiastic skier, climber, and hiker. He also served for several years as Associate Organist at St. Paul's Cathedral in Pittsburgh, and has played numerous organ recitals in the Pittsburgh area. Prof. Young and his wife Alice usually travel extensively in the summer, especially in Europe and in the desert canyon country of southern Utah.

Roger A. Freedman is a Lecturer in Physics at the University of California, Santa Barbara. Dr. Freedman was an undergraduate at the University of California campuses in San Diego and Los Angeles, and did his doctoral research in nuclear theory at Stanford University under the direction of Professor J. Dirk Walecka. He came to UCSB in 1981 after three years teaching and doing research at the University of Washington.

At UCSB, Dr. Freedman has taught in both the Department of Physics and the College of Creative Studies, a branch of the university intended for highly gifted and motivated undergraduates. He has published research in nuclear physics, elementary particle physics, and laser physics. In recent years, he has helped to develop computer-based tools for learning introductory physics and astronomy.

When not in the classroom or slaving over a computer, Dr. Freedman can be found either flying (he holds a commercial pilot's license) or driving with his wife, Caroline, in their 1960 Nash Metropolitan convertible.

A. Lewis Ford is Professor of Physics at Texas A&M University. He received a B.A. from Rice University in 1968 and a Ph.D. in chemical physics from the University of Texas at Austin in 1972. After a one-year postdoc at Harvard University, he joined the Texas A&M physics faculty in 1973 and has been there ever since. Professor Ford's research area is theoretical atomic physics, with a specialization in atomic collisions. At Texas A&M he has taught a variety of undergraduate and graduate courses, but primarily introductory physics.

UNITS, PHYSICAL QUANTITIES, AND VECTORS

Accurate measurement is essential in medical applications of physics. The laser beams shining on this cancer patient form a cross-hairs on the site of a tumor, which is then bombarded by a stream of high-energy neutrons coming from the square aperture on the right. The neutrons deposit their energy in the tumor, stopping its growth and, ideally, destroying it completely. Because the narrow neutron beam is very accurately targeted, little damage occurs to the healthy tissue surrounding the tumor.

? **Subatomic particles used in cancer therapy can be aimed at a tumor with an accuracy of 100 micrometers. How many human blood cells laid side by side would span this distance?**

Why study physics? For two reasons. First, physics is one of the most fundamental of the sciences. Scientists of all disciplines make use of the ideas of physics, from chemists who study the structure of molecules to paleontologists who try to reconstruct how dinosaurs walked. The principles of physics play an essential role in the scientific quest to understand how human activities affect the atmosphere and oceans, and in the search for alternative sources of energy. Physics is also the foundation of all engineering and technology. No engineer could design any kind of practical device without first understanding the basic principles involved. No engineer could design a DVD player, a flat-screen TV, an interplanetary spacecraft, or even a better mousetrap without first understanding the basic laws of physics.

But there's another reason. The study of physics is an adventure. You will find it challenging, sometimes frustrating, occasionally painful, and often richly rewarding and satisfying. It will appeal to your sense of beauty as well as to your rational intelligence. Our present understanding of the physical world has been built on the foundations laid by scientific giants such as Galileo, Newton, Maxwell, and Einstein, and their influence has extended far beyond science to affect profoundly the ways in which we live and think. You can share some of the excitement of their discoveries when you learn to use physics to solve practical problems and to gain insight into everyday phenomena. If you've ever wondered why the sky is blue, how radio waves can travel through empty space, or how a satellite stays in orbit, you can find the answers by using fundamental physics. Above all, you will come to see physics as a towering achievement of the human intellect in its quest to understand our world and ourselves.

In this opening chapter, we'll go over some important preliminaries that we'll need throughout our study. We'll discuss the nature of physical theory and the use of idealized models to represent physical systems. We'll introduce the systems of units used to describe physical quantities and discuss ways to describe the accuracy of a number. We'll look at examples of problems for which we can't (or don't want to) find a precise answer, but for which rough estimates can be useful and interesting. Finally, we'll study several aspects of vectors and vector algebra. Vectors will be needed throughout our study of physics to describe and analyze physical quantities, such as velocity and force, that have direction as well as magnitude.

1.1 | The Nature of Physics

Physics is an *experimental* science. Physicists observe the phenomena of nature and try to find patterns and principles that relate these phenomena. These patterns are called physical theories or, when they are very well established and of broad use, physical laws or principles.

> **CAUTION** Calling an idea a theory does *not* mean that it's just a random thought or an unproven concept. Rather, a theory is an explanation of natural phenomena based on observation and accepted fundamental principles. An example is the theory of biological evolution, which is the result of extensive research and observation by generations of biologists.

The development of physical theory requires creativity at every stage. The physicist has to learn to ask appropriate questions, design experiments to try to answer the questions, and draw appropriate conclusions from the results. Figure 1.1 shows two famous experimental facilities.

1.1 Two research laboratories. (a) The Leaning Tower in Pisa, Italy. According to legend, Galileo studied the motion of freely falling bodies by dropping them from the tower. He is also said to have studied pendulum motion by observing the swinging of the chandelier in the cathedral behind the tower. (b) The Hubble Space Telescope is the first major telescope to operate outside the earth's atmosphere. Its sensitive instruments measure visible, ultraviolet, and near-infrared light from astronomical objects. The Space Telescope has been used to study phenomena ranging from eruptions on the moons of Jupiter to the cores of distant galaxies. It is shown here in March 2002 while being repaired in orbit by the crew of the Space Shuttle *Columbia*.

(a)

(b)

According to legend, Galileo Galilei (1564–1642) dropped light and heavy objects from the top of the Leaning Tower of Pisa (Fig. 1.1a) to find out whether their rates of fall were the same or different. Galileo recognized that only experimental investigation could answer this question. From examining the results of his experiments (which were actually much more sophisticated than in the legend), he made the inductive leap to the principle, or theory, that the acceleration of a falling body is independent of its weight.

The development of physical theories such as Galileo's is always a two-way process that starts and ends with observations or experiments. This development often takes an indirect path, with blind alleys, wrong guesses, and the discarding of unsuccessful theories in favor of more promising ones. Physics is not simply a collection of facts and principles; it is also the *process* by which we arrive at general principles that describe how the physical universe behaves.

No theory is ever regarded as the final or ultimate truth. The possibility always exists that new observations will require that a theory be revised or discarded. It is in the nature of physical theory that we can disprove a theory by finding behavior that is inconsistent with it, but we can never prove that a theory is always correct.

Getting back to Galileo, suppose we drop a feather and a cannonball. They certainly do *not* fall at the same rate. This does not mean that Galileo was wrong; it means that his theory was incomplete. If we drop the feather and the cannonball *in a vacuum* to eliminate the effects of the air, then they do fall at the same rate. Galileo's theory has a **range of validity:** it applies only to objects for which the force exerted by the air (due to air resistance and buoyancy) is much less than the weight. Objects like feathers or parachutes are clearly outside this range.

Every physical theory has a range of validity outside of which it is not applicable. Often a new development in physics extends a principle's range of validity. Galileo's analysis of falling bodies was greatly extended half a century later by Newton's laws of motion and law of gravitation.

1.2 | Solving Physics Problems

At some point in their studies, almost all physics students find themselves thinking, "I understand the concepts, but I just can't solve the problems." But in physics, truly understanding a concept or principle is the same thing as being able to apply it to a variety of practical problems. Learning how to solve problems is absolutely essential; you don't *know* physics unless you can *do* physics.

How do you learn to solve physics problems? In every chapter of this book you will find *Problem-Solving Strategies* that offer techniques for setting up and solving problems efficiently and accurately. Following each *Problem-Solving Strategy* are one or more worked *Examples* that show these techniques in action. (The *Problem-Solving Strategies* will also help steer you away from some *incorrect* techniques that you may be tempted to use.) You'll also find additional examples that aren't associated with a particular *Problem-Solving Strategy*. Study these strategies and examples carefully, and work through each example for yourself on a piece of paper.

Different techniques are useful for solving different kinds of physics problems, which is why this book offers dozens of *Problem-Solving Strategies*. No matter what kind of problem you're dealing with, however, there are certain key steps that you'll always follow. (These same steps are equally useful for problems in math, engineering, chemistry, and many other fields.) In this book we've organized these steps into four stages of solving a problem.

Solving Physics Problems

IDENTIFY *the relevant concepts:* First, decide which physics ideas are relevant to the problem. Although this step doesn't involve any calculations, it's sometimes the most challenging part of solving the problem. Don't skip over this step, though; choosing the wrong approach at the beginning can make the problem more difficult than it has to be, or even lead you to an incorrect answer.

At this stage you'll also want to identify the **target variable** of the problem. The target variable is the quantity whose value you're trying to find. It could be the speed at which a projectile hits the ground, the intensity of a sound made by a siren, or the strength of a magnetic field generated by an electromagnet. (Sometimes the goal will be to find a mathematical expression for the target variable rather than a numerical value. Sometimes, too, the problem will have more than one target variable.) The target variable is the goal of the problem—solving process; make sure you don't lose sight of this goal as you work through the mathematics of the solution.

SET UP *the problem:* If appropriate, draw a sketch of the situation described in the problem. Based on the concepts you selected in the *Identify* step, choose the equations that you'll use to solve the problem and decide how you'll use them.

EXECUTE *the solution:* In this step, you "do the math." Before you launch into a flurry of calculations, make a list of all known and unknown quantities, and note which are the target variable or variables. Then solve the equations for the unknowns.

EVALUATE *your answer:* The goal of physics problem solving isn't just to get a number or a formula; it's to achieve better understanding. That means you must examine your answer to see what it's telling you. In particular, ask yourself, "Does this answer make sense?" If your target variable was the radius of the Earth and your answer is 6.38 centimeters (or if your answer turned out to be a negative number!), something went wrong in your problem-solving process. Go back and check your work, and revise your solution as necessary.

All of the *Problem-Solving Strategies* and *Examples* in this book will follow these four steps. (In some cases we will combine the first two or three steps.) We encourage you to follow these same steps when you solve problems yourself. You may find it useful to remember the acronym *I SEE*—short for *Identify, Set up, Execute,* and *Evaluate.*

Idealized Models

In everyday conversation we often use the word "model" to mean either a small-scale replica, such as a model railroad, or a person who displays articles of clothing (or the absence thereof). In physics a **model** is a simplified version of a physical system that would be too complicated to analyze in full detail.

For example, suppose we want to analyze the motion of a baseball thrown through the air. How complicated is this problem? The ball is neither perfectly spherical nor perfectly rigid; it has raised seams, and it spins as it moves through the air. Wind and air resistance influence its motion, the earth rotates beneath it, the ball's weight varies a little as its distance from the center of the earth changes, and so on. If we try to include all these things, the analysis gets hopelessly complicated. Instead, we invent a simplified version of the problem. We neglect the size and shape of the ball by representing it as a point object, or **particle.** We neglect air resistance by making the ball move in a vacuum, we ignore the earth's rotation, and we make the weight constant. Now we have a problem that is simple enough to deal with. We will analyze this model in detail in Chapter 3.

To make an idealized model of the system, we have to overlook quite a few minor effects to concentrate on the most important features of the system. Of course, we have to be careful not to neglect too much. If we ignore the effects of gravity completely, then our model predicts that when we throw the ball up, it goes in a straight line and disappears into space. We need to use some judgment and creativity to construct a model that simplifies a problem enough to make it manageable, yet keeps its essential features.

When we use a model to predict how a system will behave, the validity of our predictions is limited by the validity of the model. Going back to Galileo once more, we see that his prediction about falling bodies corresponds to an idealized model that does not include the effects of air resistance. This model works fairly well for a dropped cannonball, but not so well for a feather.

The concept of idealized models is extremely important in all physical science and technology. When we apply physical principles to complex systems, we always use idealized models, and we have to be aware of the assumptions we are making. In fact, the principles of physics themselves are stated in terms of idealized models; we speak about point masses, rigid bodies, ideal insulators, and so on. Idealized models play a crucial role throughout this book. Watch for them in discussions of physical theories and their applications to specific problems.

1.3 | Standards and Units

As we learned in Section 1.1, physics is an experimental science. Experiments require measurements, and we generally use numbers to describe the results of measurements. Any number that is used to describe a physical phenomenon quantitatively is called a **physical quantity.** For example, two physical quantities that describe you are your weight and your height. Some physical quantities are so fundamental that we can define them only by describing how to measure them. Such a definition is called an **operational definition.** Some examples are measuring a distance by using a ruler, and measuring a time interval by using a stopwatch. In other cases we define a physical quantity by describing how to calculate it from other quantities that we *can* measure. Thus we might define the average speed of a moving object as the distance traveled (measured with a ruler) divided by the time of travel (measured with a stopwatch).

When we measure a quantity, we always compare it with some reference standard. When we say that a Porsche Carrera GT is 4.56 meters long, we mean that it is 4.56 times as long as a meter stick, which we define to be 1 meter long. Such a standard defines a **unit** of the quantity. The meter is a unit of distance, and the second is a unit of time. When we use a number to describe a physical quantity, we must always specify the unit that we are using; to describe a distance simply as "4.56" wouldn't mean anything.

To make accurate, reliable measurements, we need units of measurement that do not change and that can be duplicated by observers in various locations. The system of units used by scientists and engineers around the world is commonly called "the metric system," but since 1960 it has been known officially as the **International System,** or **SI** (the abbreviation for its French name, *Système International*). A list of all SI units is given in Appendix A, as are definitions of the most fundamental units.

The definitions of the basic units of the metric system have evolved over the years. When the metric system was established in 1791 by the French Academy of Sciences, the meter was defined as one ten-millionth of the distance from the North Pole to the equator (Fig. 1.2). The second was defined as the time required

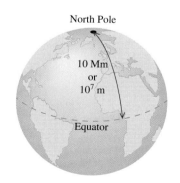

North Pole

10 Mm
or
10^7 m

Equator

1.2 In 1791 the distance from the North Pole to the equator was defined to be exactly 10^7 m. With the modern definition of the meter this distance is about 0.02% more than 10^7 m.

for a pendulum one meter long to swing from one side to the other. These definitions were cumbersome and hard to duplicate precisely, and by international agreement they have been replaced with more refined definitions.

Time

From 1889 until 1967, the unit of time was defined as a certain fraction of the mean solar day, the average time between successive arrivals of the sun at its highest point in the sky. The present standard, adopted in 1967, is much more precise. It is based on an atomic clock, which uses the energy difference between the two lowest energy states of the cesium atom. When bombarded by microwaves of precisely the proper frequency, cesium atoms undergo a transition from one of these states to the other. One **second** is defined as the time required for 9,192,631,770 cycles of this microwave radiation.

Length

In 1960 an atomic standard for the meter was also established, using the wavelength of the orange-red light emitted by atoms of krypton (^{86}Kr) in a glow discharge tube. Using this length standard, the speed of light in a vacuum was measured to be 299,792,458 m/s. In November 1983, the length standard was changed again so that the speed of light in a vacuum was *defined* to be precisely 299,792,458 m/s. The meter is defined to be consistent with this number and with the above definition of the second. Hence the new definition of the **meter** is the distance that light travels in a vacuum in 1/299,792,458 second. This provides a much more precise standard of length than the one based on a wavelength of light.

Mass

The standard of mass, the **kilogram,** is defined to be the mass of a particular cylinder of platinum-iridium alloy. That cylinder is kept at the International Bureau of Weights and Measures at Sèvres, near Paris. An atomic standard of mass would be more fundamental, but at present we cannot measure masses on an atomic scale with as much accuracy as on a macroscopic scale. The *gram* (which is not a fundamental unit) is 0.001 kilogram.

Unit Prefixes

Once we have defined the fundamental units, it is easy to introduce larger and smaller units for the same physical quantities. In the metric system these other units are related to the fundamental units (or, in the case of mass, to the gram) by multiples of 10 or $\frac{1}{10}$. Thus one kilometer (1 km) is 1000 meters, and one centimeter (1 cm) is $\frac{1}{100}$ meter. We usually express multiples of 10 or $\frac{1}{10}$ in exponential notation: $1000 = 10^3$, $\frac{1}{1000} = 10^{-3}$, and so on. With this notation, $1 \text{ km} = 10^3 \text{ m}$ and $1 \text{ cm} = 10^{-2} \text{ m}$.

The names of the additional units are derived by adding a **prefix** to the name of the fundamental unit. For example, the prefix "kilo-," abbreviated k, always means a unit larger by a factor of 1000; thus

$$1 \text{ kilometer} = 1 \text{ km} = 10^3 \text{ meters} = 10^3 \text{ m}$$
$$1 \text{ kilogram} = 1 \text{ kg} = 10^3 \text{ grams} = 10^3 \text{ g}$$
$$1 \text{ kilowatt} = 1 \text{ kW} = 10^3 \text{ watts} = 10^3 \text{ W}$$

? **1.3** Some typical lengths in the universe. (a) The most distant galaxies are about 10^{26} m, or 10^{23} km, away. (b) The sun is 1.50×10^{11} m, or 1.50×10^{8} km, from our earth. (c) The diameter of the earth is 1.28×10^{7} m, or 12,800 km. (d) A typical human is about 1.7 m, or 170 cm, tall. (e) Human red blood cells are about 8×10^{-6} m (0.008 mm, or 8 μm) in diameter. (f) These oxygen atoms, shown arrayed on the surface of a crystal, are about 10^{-10} m, or 10^{-4} μm, in radius. (g) Typical atomic nuclei (shown in an artist's impression) have radii of about 10^{-14} m, or 10^{-5} nm.

A table on the inside back cover of this book lists the standard SI prefixes, with their meanings and abbreviations.

Here are several examples of the use of multiples of 10 and their prefixes with the units of length, mass, and time. Figure 1.3 shows how these prefixes help describe both large and small distances.

Length

1 nanometer = 1 nm = 10^{-9} m (a few times the size of the largest atom)

1 micrometer = 1 μm = 10^{-6} m (size of some bacteria and living cells)

1 millimeter = 1 mm = 10^{-3} m (diameter of the point of a ballpoint pen)

1 centimeter = 1 cm = 10^{-2} m (diameter of your little finger)

1 kilometer = 1 km = 10^{3} m (a 10-minute walk)

Mass

1 microgram = 1 μg = 10^{-6} g = 10^{-9} kg (mass of a very small dust particle)

1 milligram = 1 mg = 10^{-3} g = 10^{-6} kg (mass of a grain of salt)

1 gram   = 1 g = 10^{-3} kg (mass of a paper clip)

Time

$$1 \text{ nanosecond } = 1 \text{ ns } = 10^{-9} \text{ s } \left(\text{time for light to travel } 0.3 \text{ m}\right)$$
$$1 \text{ microsecond} = 1 \text{ } \mu\text{s} = 10^{-6} \text{ s } \left(\text{time for an orbiting space shuttle to travel } 8 \text{ mm}\right)$$
$$1 \text{ millisecond } = 1 \text{ ms } = 10^{-3} \text{ s } \left(\text{time for sound to travel } 0.35 \text{ m}\right)$$

The British System

Finally, we mention the British system of units. These units are used only in the United States and a few other countries, and in most of these they are being replaced by SI units. British units are now officially defined in terms of SI units, as follows:

Length: 1 inch $= 2.54$ cm $\left(\text{exactly}\right)$
Force: 1 pound $= 4.448221615260$ newtons $\left(\text{exactly}\right)$

The newton, abbreviated N, is the SI unit of force. The British unit of time is the second, defined the same way as in SI. In physics, British units are used only in mechanics and thermodynamics; there is no British system of electrical units.

In this book we use SI units for all examples and problems, but we occasionally give approximate equivalents in British units. As you do problems using SI units, you may also wish to convert to the approximate British equivalents if they are more familiar to you (see Fig. 1.4). But you should try to think directly in SI as much as you can.

1.4 Many everyday items make use of both SI and British units. An example is this speedometer from a U.S.-built automobile, which shows the speed in both kilometers per hour and miles per hour.

1.4 | Unit Consistency and Conversions

We use equations to express relationships among physical quantities, represented by algebraic symbols. Each algebraic symbol always denotes both a number and a unit. For example, d might represent a distance of 10 m, t a time of 5 s, and v a speed of 2 m/s.

An equation must always be **dimensionally consistent.** You can't add apples and automobiles; two terms may be added or equated only if they have the same units. For example, if a body moving with constant speed v travels a distance d in a time t, these quantities are related by the equation

$$d = vt \tag{1.1}$$

If d is measured in meters, then the product vt must also be expressed in meters. Using the above numbers as an example, we may write

$$10 \text{ m} = \left(2 \frac{\text{m}}{\cancel{\text{s}}}\right)\left(5 \cancel{\text{s}}\right)$$

Because the unit 1/s on the right side of the equation cancels the unit s, the product vt has units of meters, as it must. In calculations, units are treated just like algebraic symbols with respect to multiplication and division.

CAUTION When a problem requires calculations using numbers with units, *always* write the numbers with the correct units and carry the units through the calculation as in the example above. This provides a very useful check for calculations. If at some stage in a calculation you find that an equation or an expression has inconsistent units, you know you have made an error somewhere. In this book we will always carry units through all calculations, and we strongly urge you to follow this practice when you solve problems.

Problem-Solving Strategy

Unit Conversions

IDENTIFY *the relevant concepts:* Unit conversion is important, but it's also important to recognize when it's needed. In most cases, you're best off using the fundamental SI units (lengths in meters, masses in kilograms, and time in seconds) within a problem. If you need the answer to be in a different set of units (such as kilometers, grams, or hours), wait until the end of the problem to make the conversion. In the following examples, we'll concentrate on unit conversion alone, so we'll skip the *Identify* step.

SET UP *the problem and* **EXECUTE** *the solution:* Units are multiplied and divided just like ordinary algebraic symbols. This gives us an easy way to convert a quantity from one set of units to another. The key idea is to express the same physical quantity in two different units and form an equality.

For example, when we say that 1 min = 60 s, we don't mean that the number 1 is equal to the number 60; rather, we mean that 1 min represents the same physical time interval as 60 s.

For this reason, the ratio (1 min)/(60 s) equals 1, as does its reciprocal (60 s)/(1 min). We may multiply a quantity by either of these factors without changing that quantity's physical meaning. For example, to find the number of seconds in 3 min, we write

$$3 \text{ min} = (3 \text{ min})\left(\frac{60 \text{ s}}{1 \text{ min}}\right) = 180 \text{ s}$$

EVALUATE *your answer:* If you do your unit conversions correctly, unwanted units will cancel, as in the example above. If instead you had multiplied 3 min by (1 min)/(60 s), your result would have been $\frac{1}{20}$ min²/s, which is a rather odd way of measuring time. To be sure you convert units properly, you must write down the units at *all* stages of the calculation.

Finally, check whether your answer is reasonable. Is the result 3 min = 180 s reasonable? The answer is yes; the second is a smaller unit than the minute, so there are more seconds than minutes in the same time interval.

Example 1.1

Converting speed units

The official world land speed record is 1228.0 km/h, set on October 15, 1997, by Andy Green in the jet engine car *Thrust SSC*. Express this speed in meters per second.

SOLUTION

IDENTIFY, SET UP and EXECUTE: The prefix k means 10^3, so the speed 1228.0 km/h = 1228.0×10^3 m/h. We also know that there are 3600 s in 1 h. So we must combine the speed of 1228.0×10^3 m/h and a factor of 3600. But should we multiply or divide by this factor? If we treat the factor as a pure number without units, we're forced to guess how to proceed.

The correct approach is to carry the units with each factor. We then arrange the factor so that the hour unit cancels:

$$1228.0 \text{ km/h} = \left(1228.0 \times 10^3 \frac{\text{m}}{\text{h}}\right)\left(\frac{1 \text{ h}}{3600 \text{ s}}\right) = 341.11 \text{ m/s}$$

If you multiplied by (3600 s)/(1 h) instead of (1 h)/(3600 s), the hour unit wouldn't cancel, and you would be able to easily recognize your error. Again, the *only* way to be sure that you correctly convert units is to carry the units throughout the calculation.

EVALUATE: While you probably have a good intuition for speeds in kilometers per hour or miles per hour, speeds in meters per second are likely to be a bit more mysterious. Notice that the length of an average person's stride is about one meter, and a good walking pace is about one stride per second. Thus a typical walking speed is about 1 m/s. By comparison, a speed of 341.11 m/s is rapid indeed!

Example
1.2   ## Converting volume units

The world's largest cut diamond is the First Star of Africa (mounted in the British Royal Sceptre and kept in the Tower of London). Its volume is 1.84 cubic inches. What is its volume in cubic centimeters? In cubic meters?

SOLUTION

IDENTIFY, SET UP and EXECUTE: To convert cubic inches to cubic centimeters, we multiply by $[(2.54 \text{ cm})/(1 \text{ in.})]^3$, not just $(2.54 \text{ cm})/(1 \text{ in.})$. We find

$$1.84 \text{ in.}^3 = (1.84 \text{ in.}^3)\left(\frac{2.54 \text{ cm}}{1 \text{ in.}}\right)^3$$

$$= (1.84)(2.54)^3 \frac{\cancel{\text{in.}^3} \text{ cm}^3}{\cancel{\text{in.}^3}} = 30.2 \text{ cm}^3$$

Also, $1 \text{ cm} = 10^{-2} \text{ m}$, and

$$30.2 \text{ cm}^3 = (30.2 \text{ cm}^3)\left(\frac{10^{-2} \text{ m}}{1 \text{ cm}}\right)^3$$

$$= (30.2)(10^{-2})^3 \frac{\cancel{\text{cm}^3} \text{ m}^3}{\cancel{\text{cm}^3}} = 30.2 \times 10^{-6} \text{ m}^3$$

EVALUATE: Our answer shows that while 1 centimeter is 10^{-2} of a meter (that is, $1 \text{ cm} = 10^{-2} \text{ m}$), a cubic centimeter (1 cm^3) is *not* 10^{-2} of a cubic meter. Rather, it is the volume of a cube whose sides are 1 cm long. So $1 \text{ cm}^3 = (1 \text{ cm})^3 = (10^{-2} \text{ m})^3 = (10^{-2})^3 \text{ m}^3$, or $1 \text{ cm}^3 = 10^{-6} \text{ m}^3$.

1.5 | Uncertainty and Significant Figures

Measurements always have uncertainties. If you measure the thickness of the cover of this book using an ordinary ruler, your measurement is reliable only to the nearest millimeter, and your result will be 3 mm. It would be *wrong* to state this result as 3.00 mm; given the limitations of the measuring device, you can't tell whether the actual thickness is 3.00 mm, 2.85 mm, or 3.11 mm. But if you use a micrometer caliper, a device that measures distances reliably to the nearest 0.01 mm, the result will be 2.91 mm. The distinction between these two measurements is in their **uncertainty.** The measurement using the micrometer caliper has a smaller uncertainty; it's a more accurate measurement. The uncertainty is also called the **error,** because it indicates the maximum difference there is likely to be between the measured value and the true value. The uncertainty or error of a measured value depends on the measurement technique used.

We often indicate the **accuracy** of a measured value—that is, how close it is likely to be to the true value—by writing the number, the symbol $\pm$, and a second number indicating the uncertainty of the measurement. If the diameter of a steel rod is given as 56.47 ± 0.02 mm, this means that the true value is unlikely to be less than 56.45 mm or greater than 56.49 mm. In a commonly used shorthand notation, the number 1.6454(21) means 1.6454 ± 0.0021. The numbers in parentheses show the uncertainty in the final digits of the main number.

We can also express accuracy in terms of the maximum likely **fractional error** or **percent error** (also called *fractional uncertainty* and *percent uncertainty*). A resistor labeled "47 ohms $\pm$ 10%" probably has a true resistance differing from 47 ohms by no more than 10% of 47 ohms—that is, about 5 ohms. The resistance is probably between 42 and 52 ohms. For the diameter of the steel rod given above, the fractional error is $(0.02 \text{ mm})/(56.47 \text{ mm})$, or about 0.0004; the percent error is $(0.0004)(100\%)$, or about 0.04%. Even small percent errors can sometimes be very significant (Fig. 1.5).

In many cases the uncertainty of a number is not stated explicitly. Instead, the uncertainty is indicated by the number of meaningful digits, or **significant figures,** in the measured value. We gave the thickness of the cover of this book

1.5 This spectacular mishap was the result of a very small percent error—traveling a few meters too far in a journey of hundreds of thousands of meters.

as 2.91 mm, which has three significant figures. By this we mean that the first two digits are known to be correct, while the third digit is uncertain. The last digit is in the hundredths place, so the uncertainty is about 0.01 mm. Two values with the *same* number of significant figures may have *different* uncertainties; a distance given as 137 km also has three significant figures, but the uncertainty is about 1 km.

When you use numbers having uncertainties to compute other numbers, the computed numbers are also uncertain. It is especially important to understand this when you compare a number obtained from measurements with a value obtained from a theoretical prediction. Suppose you want to verify the value of π, the ratio of the circumference of a circle to its diameter. The true value of this ratio to ten digits is 3.141592654. To make your own calculation, you draw a large circle and measure its diameter and circumference to the nearest millimeter, obtaining the values 135 mm and 424 mm. You punch these into your calculator and obtain the quotient 3.140740741. Does this agree with the true value or not?

First, the last seven digits in this answer are meaningless; they imply a smaller uncertainty than is possible with your measurements. When numbers are multiplied or divided, the number of significant figures in the result can be no greater than in the factor with the fewest significant figures. For example, $3.1416 \times 2.34 \times 0.58 = 4.3$. Your measurements each have three significant figures, so your measured value of π, equal to (424 mm)/(135 mm), can have only three significant figures. It should be stated simply as 3.14. Within the limit of three significant figures, your value does agree with the true value.

When we add and subtract numbers, it's the location of the decimal point that matters, not the number of significant figures. For example, $123.62 + 8.9 = 132.5$. Although 123.62 has an uncertainty of about 0.01, 8.9 has an uncertainty of about 0.1. So their sum has an uncertainty of about 0.1 and should be written as 132.5, not 132.52.

In the examples and problems in this book we usually give numerical values with three significant figures, so your answers should usually have no more than three significant figures. (Many numbers that you encounter in the real world have even less accuracy. An automobile speedometer, for example, usually gives only two significant figures.) You might do the arithmetic with a calculator having a display with ten digits. To give a ten-digit answer is not merely unnecessary, but genuinely wrong, because it misrepresents the accuracy of the results. Always round your final answer to keep only the correct number of significant figures or, in doubtful cases, one more at most. In Example 1.1 it would have been wrong to state the answer as 341.11111 m/s. Note that when you reduce such an answer to the appropriate number of significant figures, you must *round,* not *truncate.* Your calculator will tell you that the ratio of 525 m to 311 m is 1.688102894; to three significant figures, this is 1.69, not 1.68.

When we calculate with very large or very small numbers, we can show significant figures much more easily by using **scientific notation,** sometimes called **powers-of-10 notation.** The distance from the earth to the moon is about 384,000,000 m, but writing the number in this form gives no indication of the number of significant figures. Instead, we move the decimal point eight places to the left (corresponding to dividing by 10^8) and multiply by 10^8. That is,

$$384{,}000{,}000 \text{ m} = 3.84 \times 10^8 \text{ m}$$

In this form, it is clear that we have three significant figures. The number 4.00×10^{-7} also has three significant figures, even though two of them are zeros.

Note that in scientific notation the usual practice is to express the quantity as a number between 1 and 10 multiplied by the appropriate power of 10. Table 1.1 summarizes the rules for significant figures.

Table 1.1 Using Significant Figures

Mathematical operation	Significant figures in result
Multiplication or division	No more than in the number with the fewest significant figures *Example:* $(0.745 \times 2.2)/3.885 = 0.42$ *Example:* $(1.32578 \times 10^7) \times (4.11 \times 10^{-3}) = 5.45 \times 10^4$
Addition or subtraction	Determined by the number with the largest uncertainty (i.e., the fewest digits to the right of the decimal point) *Example:* $27.153 + 138.2 - 11.74 = 153.6$

Note: In this book we will usually give numerical values with three significant figures.

When an integer or a fraction occurs in a general equation, we treat that number as having no uncertainty at all. For example, in the equation $v_x^2 = v_{0x}^2 + 2a_x(x - x_0)$, which is Eq. (2.13) in Chapter 2, the coefficient 2 is *exactly* 2. We can consider this coefficient as having an infinite number of significant figures (2.000000 . . .). The same is true of the exponent 2 in v_x^2 and v_{0x}^2.

Finally, let's note that **precision** is not the same as *accuracy*. A cheap digital watch that says the time is 10:35:17 a.m. is very *precise* (the time is given to the second), but if the watch runs several minutes slow, then this value isn't very *accurate*. On the other hand, a grandfather clock might be very accurate (that is, display the correct time), but if the clock has no second hand, it isn't very precise. A high-quality measurement, like those used to define standards (Section 1.3), is both precise *and* accurate.

Example 1.3	**Significant figures in multiplication**

The rest energy E of an object with rest mass m is given by Einstein's equation

$$E = mc^2$$

where c is the speed of light in a vacuum. Find E for an object with $m = 9.11 \times 10^{-31}$ kg (to three significant figures, the mass of an electron). The SI unit for E is the joule (J); 1 J $= 1$ kg·m²/s².

SOLUTION

IDENTIFY and SET UP: Our target variable is the energy E. We are given the equation to use and the value of the mass m; from Section 1.3 the exact value of the speed of light is $c = 299,792,458$ m/s $= 2.99792458 \times 10^8$ m/s.

EXECUTE Substituting the values of m and c into Einstein's equation, we find

$$E = (9.11 \times 10^{-31} \text{ kg})(2.99792458 \times 10^8 \text{ m/s})^2$$
$$= (9.11)(2.99792458)^2(10^{-31})(10^8)^2 \text{ kg·m}^2/\text{s}^2$$
$$= (81.87659678)(10^{[-31 + (2 \times 8)]}) \text{ kg·m}^2/\text{s}^2$$
$$= 8.187659678 \times 10^{-14} \text{ kg·m}^2/\text{s}^2$$

Since the value of m was given to only three significant figures, we must round this to

$$E = 8.19 \times 10^{-14} \text{ kg·m}^2/\text{s}^2 = 8.19 \times 10^{-14} \text{ J}$$

Most calculators use scientific notation and add exponents automatically, but you should be able to do such calculations by hand when necessary.

EVALUATE: While the rest energy contained in an electron may seem ridiculously small, on the atomic scale it is tremendous. Compare our answer to 10^{-19} J, the energy gained or lost by a single atom during a typical chemical reaction; the rest energy of an electron is about 1,000,000 times larger! (We will discuss the significance of rest energy in Chapter 37.)

Test Your Understanding

The density of a material is equal to its mass divided by its volume. What is the density (in kg/m³) of a rock of mass 1.80 kg and volume 6.0×10^{-4} m³? Make sure your answer has the correct number of significant figures.

1.6 | Estimates and Orders of Magnitude

We have stressed the importance of knowing the accuracy of numbers that represent physical quantities. But even a very crude estimate of a quantity often gives us useful information. Sometimes we know how to calculate a certain quantity but have to guess at the data we need for the calculation. Or the calculation might be too complicated to carry out exactly, so we make some rough approximations. In either case our result is also a guess, but such a guess can be useful even if it is uncertain by a factor of two, ten, or more. Such calculations are often called **order-of-magnitude estimates.** The great Italian American nuclear physicist Enrico Fermi (1901–1954) called them "back-of-the-envelope calculations."

Exercises 1.18 through 1.29 at the end of this chapter are of the estimating, or "order-of-magnitude," variety. Some are silly, and most require guesswork for the needed input data. Don't try to look up a lot of data; make the best guesses you can. Even when they are off by a factor of ten, the results can be useful and interesting.

| Example 1.4 | **An order-of-magnitude estimate** |

You are writing an adventure novel in which the hero escapes across the border with a billion dollars worth of gold in his suitcase. Is this possible? Would that amount of gold fit in a suitcase? Would it be too heavy to carry?

SOLUTION

IDENTIFY, SET UP, and EXECUTE Gold sells for around $400 an ounce. On a particular day the price might be $200 or $600, but never mind. An ounce is about 30 grams. Actually, an ordinary (avoirdupois) ounce is 28.35 g; an ounce of gold is a troy ounce, which is 9.45% more. Again, never mind. Ten dollars' worth of gold has a mass somewhere around one gram, so a billion (10^9) dollars worth of gold is a hundred million (10^8) grams, or a hundred thousand (10^5) kilograms. This corresponds to a weight in British units of around 200,000 lb, or 100 tons. Whether the precise number is closer to 50 tons or 200 tons doesn't matter. Either way, the hero is not about to carry it across the border in a suitcase.

We can also estimate the *volume* of this gold. If its density were the same as that of water (1 g/cm^3), the volume would be 10^8 cm³, or 100 m³. But gold is a heavy metal; we might guess its density to be ten times that of water. Gold is actually 19.3 times as dense as water. But by guessing ten, we find a volume of 10 m³. Visualize ten cubical stacks of gold bricks, each 1 meter on a side, and ask yourself whether they would fit in a suitcase!

EVALUATE: Clearly, your novel needs rewriting. Try the calculation again with a suitcase full of five-carat (1-gram) diamonds, each worth $100,000. Would this work?

Test Your Understanding

Can you estimate the total number of teeth in all the mouths of everyone (students, staff, and faculty) on your campus? (*Hint:* How many teeth are in your mouth? Count them!)

1.7 | Vectors and Vector Addition

Some physical quantities, such as time, temperature, mass, density, and electric charge, can be described completely by a single number with a unit. But many other important quantities have a *direction* associated with them and cannot be described by a single number. Such quantities play an essential role in many of the central topics of physics, including motion and its causes and the phenomena of electricity and magnetism. A simple example of a quantity with direction is the motion of an airplane. To describe this motion completely, we must say not only how fast the plane is moving, but also in what direction. To fly from Chicago to New York, a plane has to head east, not south. The speed of the airplane combined with its direction of motion together constitute a quantity called *velocity.* Another example is *force,* which in physics means a push or pull exerted on a body. Giving a complete description of a force means describing both how hard the force pushes or pulls on the body and the direction of the push or pull.

When a physical quantity is described by a single number, we call it a **scalar quantity.** In contrast, a **vector quantity** has both a **magnitude** (the "how much" or "how big" part) and a direction in space. Calculations with scalar quantities use the operations of ordinary arithmetic. For example, 6 kg + 3 kg = 9 kg, or 4 × 2 s = 8 s. However, combining vectors requires a different set of operations.

To understand more about vectors and how they combine, we start with the simplest vector quantity, **displacement.** Displacement is simply a change in position of a point. (The point may represent a particle or a small body.) In Fig. 1.6a we represent the change of position from point P_1 to point P_2 by a line from P_1 to P_2, with an arrowhead at P_2 to represent the direction of motion. Displacement is a vector quantity because we must state not only how far the particle moves, but also in what direction. Walking 3 km north from your front door doesn't get you to the same place as walking 3 km southeast; these two displacements have the same magnitude, but different directions.

We usually represent a vector quantity such as displacement by a single letter, such as $\vec{A}$ in Fig. 1.6a. In this book we always print vector symbols in **boldface italic type with an arrow above them.** We do this to remind you that vector quantities have different properties from scalar quantities; the arrow is a reminder that vectors have direction. In handwriting, vector symbols are usually underlined or written with an arrow above them (Fig. 1.6). When you write a symbol for a vector, *always* write it in one of these ways. If you don't distinguish between scalar and vector quantities in your notation, you probably won't make the distinction in your thinking either, and hopeless confusion will result.

When drawing any vector, we always draw a line with an arrowhead at its tip. The length of the line shows the vector's magnitude, and the direction of the line shows the vector's direction. Displacement is always a straight-line segment, directed from the starting point to the end point, even though the actual path of the particle may be curved. In Fig. 1.6b the particle moves along the curved path shown from P_1 to P_2, but the displacement is still the vector $\vec{A}$. Note that displacement is not related directly to the total *distance* traveled. If the particle were to continue on to P_3 and then return to P_1, the displacement for the entire trip would be *zero.*

If two vectors have the same direction, they are **parallel.** If they have the same magnitude *and* the same direction, they are *equal,* no matter where they are located in space. The vector $\vec{A}'$ from point P_3 to point P_4 in Fig. 1.7 has the same length and direction as the vector $\vec{A}$ from P_1 to P_2. These two displacements are

Handwritten notation: $\underline{A}$ or $\vec{A}$

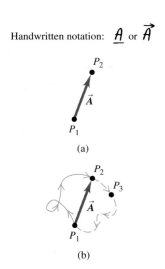

1.6 (a) Vector $\vec{A}$ is the displacement from point P_1 to point P_2. (b) A displacement is always a straight-line segment directed from the starting point to the end point, even if the actual path is curved. When a path ends at the same place where it started, the displacement is zero.

equal, even though they start at different points. We write this as $\vec{A} = \vec{A}'$ in Fig. 1.7, using a boldface equals sign to emphasize that equality of two vector quantities is not the same relationship as equality of two scalar quantities. Two vector quantities are equal only when they have the same magnitude *and* the same direction.

The vector $\vec{B}$ in Fig. 1.7, however, is not equal to $\vec{A}$ because its direction is *opposite* to that of $\vec{A}$. We define the **negative of a vector** as a vector having the same magnitude as the original vector but the *opposite* direction. The negative of vector quantity $\vec{A}$. is denoted as $-\vec{A}$, and we use a boldface minus sign to emphasize the vector nature of the quantities. If $\vec{A}$ is 87 m south, then $-\vec{A}$ is 87 m north. Thus the relation between $\vec{A}$ and $\vec{B}$ of Fig. 1.7 may be written as $\vec{A} = -\vec{B}$ or $\vec{B} = -\vec{A}$. When two vectors $\vec{A}$ and $\vec{B}$ have opposite directions, whether their magnitudes are the same or not, we say that they are **antiparallel.**

We usually represent the *magnitude* of a vector quantity (its length in the case of a displacement vector) by the same letter used for the vector, but in *light italic type* with *no* arrow on top, rather than boldface italic with an arrow (which is reserved for vectors). An alternative notation is the vector symbol with vertical bars on both sides:

$$\left(\text{Magnitude of } \vec{A}\right) = A = \left|\vec{A}\right| \tag{1.2}$$

By definition the magnitude of a vector quantity is a scalar quantity (a number) and is *always positive*. We also note that a vector can never be equal to a scalar because they are different kinds of quantities. The expression "$\vec{A} = 6$ m" is just as wrong as "2 oranges = 3 apples" or "6 lb = 7 km"!

When drawing diagrams with vectors, we'll generally use a scale similar to those used for maps, in which the distance on the diagram is proportional to the magnitude of the vector. For example, a displacement of 5 km might be represented in a diagram by a vector 1 cm long, since an actual-size diagram wouldn't be practical. When we work with vector quantities with units other than displacement, such as force or velocity, we *must* use a scale. In a diagram for force vectors we might use a scale in which a vector that is 1 cm long represents a force of magnitude 5 N. A 20-N force would then be represented by a vector 4 cm long, with the appropriate direction.

Vector Addition

Now suppose a particle undergoes a displacement $\vec{A}$, followed by a second displacement $\vec{B}$ (Fig. 1.8a). The final result is the same as if the particle had started at the same initial point and undergone a single displacement $\vec{C}$, as shown. We call displacement $\vec{C}$ the **vector sum,** or **resultant,** of displacements $\vec{A}$ and $\vec{B}$. We express this relationship symbolically as

$$\vec{C} = \vec{A} + \vec{B} \tag{1.3}$$

The boldface plus sign emphasizes that adding two vector quantities requires a geometrical process and is not the same operation as adding two scalar quantities such as 2 + 3 = 5. In vector addition we usually place the *tail* of the *second* vector at the *head,* or tip, of the *first* vector (Fig. 1.8a).

If we make the displacements $\vec{A}$ and $\vec{B}$ in reverse order, with $\vec{B}$ first and $\vec{A}$ second, the result is the same (Fig. 1.8b). Thus

$$\vec{C} = \vec{B} + \vec{A} \quad \text{and} \quad \vec{A} + \vec{B} = \vec{B} + \vec{A} \tag{1.4}$$

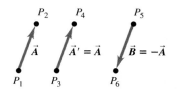

1.7 The displacement from P_3 to P_4 is equal to that from P_1 to P_2. The displacement $\vec{B}$ from P_5 to P_6 has the same magnitude as $\vec{A}$ and $\vec{A}'$ but opposite direction; displacement $\vec{B}$ is the negative of displacement $\vec{A}$.

1.8 Vector $\vec{C}$ is the vector sum of vectors $\vec{A}$ and $\vec{B}$. The order in vector addition doesn't matter; vector addition is commutative.

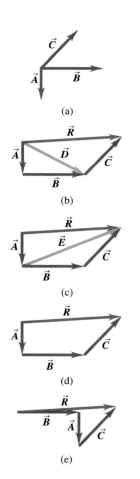

1.9 (a) In the special case when vectors $\vec{A}$ and $\vec{B}$ are parallel, and *only* in this case, the magnitude of their sum equals the sum of their magnitudes: $C = A + B$. (b) When $\vec{A}$ and $\vec{B}$ are antiparallel, the magnitude of their sum equals the *difference* of their magnitudes: $C = |A - B|$. Note that the vectors $\vec{A}$, $\vec{B}$, and $\vec{C}$ in part (a) are not the same as the vectors $\vec{A}$, $\vec{B}$, and $\vec{C}$ in part (b).

1.10 Several constructions for finding the vector sum $\vec{A} + \vec{B} + \vec{C}$.

This shows that the order of terms in a vector sum doesn't matter. In other words, vector addition obeys the commutative law.

Figure 1.8c shows an alternative representation of the vector sum. When vectors $\vec{A}$ and $\vec{B}$ are both drawn with their tails at the same point, vector $\vec{C}$ is the diagonal of a parallelogram constructed with $\vec{A}$ and $\vec{B}$ as two adjacent sides.

> **CAUTION** A word of warning is in order about vector addition. It's a common error to conclude that if $\vec{C} = \vec{A} + \vec{B}$, the magnitude C should just equal the magnitude A plus the magnitude B. Figure 1.8 shows that in general, this conclusion is *wrong;* you can see from the figure that $C < A + B$. The magnitude of the vector sum $\vec{A} + \vec{B}$ depends on the magnitudes of $\vec{A}$ and of $\vec{B}$ *and* on the angle between $\vec{A}$ and $\vec{B}$ (see Problem 1.88). Only in the special case in which $\vec{A}$ and $\vec{B}$ are *parallel* is the magnitude of $\vec{C} = \vec{A} + \vec{B}$ equal to the sum of the magnitudes of $\vec{A}$ and $\vec{B}$ (Fig. 1.9a). By contrast, when the vectors are *antiparallel* (Fig. 1.9b) the magnitude of $\vec{C}$ equals the *difference* of the magnitudes of $\vec{A}$ and $\vec{B}$. Students who aren't careful about distinguishing between scalar and vector quantities frequently make errors about the magnitude of a vector sum. Don't let this happen to you!

When we need to add more than two vectors, we may first find the vector sum of any two, add this vectorially to the third, and so on. Figure 1.10a shows three vectors $\vec{A}$, $\vec{B}$, and $\vec{C}$. In Fig. 1.10b, vectors $\vec{A}$ and $\vec{B}$ are added first, giving a vector sum $\vec{D}$; vectors $\vec{C}$ and $\vec{D}$ are then added by the same process to obtain the vector sum $\vec{R}$:

$$\vec{R} = (\vec{A} + \vec{B}) + \vec{C} = \vec{D} + \vec{C}$$

Alternatively, we can first add $\vec{B}$ and $\vec{C}$ (Fig. 1.10c) to obtain vector $\vec{E}$, and then add $\vec{A}$ and $\vec{E}$ to obtain $\vec{R}$:

$$\vec{R} = \vec{A} + (\vec{B} + \vec{C}) = \vec{A} + \vec{E}$$

We don't even need to draw vectors $\vec{D}$ and $\vec{E}$; all we need to do is draw the given vectors in succession, with the tail of each at the head of the one preceding it, and complete the polygon by a vector $\vec{R}$ from the tail of the first vector to the head of the last vector (Fig. 1.10d). The order makes no difference; Fig. 1.10e shows a different order, and we invite you to try others. We see that vector addition obeys the associative law.

We mentioned above that $-\vec{A}$ is a vector having the same magnitude as $\vec{A}$ but the opposite direction. This provides the basis for defining vector subtraction. We define the difference $\vec{A} - \vec{B}$ of two vectors $\vec{A}$ and $\vec{B}$ to be the vector sum of $\vec{A}$ and $-\vec{B}$:

$$\vec{A} - \vec{B} = \vec{A} + (-\vec{B}) \tag{1.5}$$

Figure 1.11 shows an example of vector subtraction. To construct the vector difference $\vec{A} - \vec{B}$, the tail of $-\vec{B}$ is placed at the head of $\vec{A}$.

A vector quantity such as a displacement can be multiplied by a scalar quantity (an ordinary number). The displacement $2\vec{A}$ is a displacement (vector quantity) in the same direction as the vector $\vec{A}$ but twice as long; this is the same as adding $\vec{A}$ to itself. In general, when a vector $\vec{A}$ is multiplied by a scalar c, the result $c\vec{A}$ has magnitude $|c|A$ (the absolute value of c multiplied by the magnitude of the vector

(a) (b) (c) (d)

1.11 (a) Vector $\vec{A}$ and vector $\vec{B}$. (b) Vector $\vec{A}$ and vector $-\vec{B}$. (c) The vector difference $\vec{A} - \vec{B}$ is the sum of vectors $\vec{A}$ and $-\vec{B}$. The tail of $-\vec{B}$ is placed at the head of $\vec{A}$. (d) To check: $(\vec{A} - \vec{B}) + \vec{B} = \vec{A}$.

$\vec{A}$). If c is positive, $c\vec{A}$ is in the same direction as $\vec{A}$; if c is negative, $c\vec{A}$ is in the direction opposite to $\vec{A}$. Thus $5\vec{A}$ is parallel to $\vec{A}$, while $-5\vec{A}$ is antiparallel to $\vec{A}$.

The scalar quantity used to multiply a vector may also be a physical quantity having units. For example, you may be familiar with the relationship $\vec{F} = m\vec{a}$; the net force $\vec{F}$ (a vector quantity) that acts on a body is equal to the product of the body's mass m (a positive scalar quantity) and its acceleration $\vec{a}$ (a vector quantity). The magnitude of the net force is equal to the mass (which is positive and equals its own absolute value) multiplied by the magnitude of the acceleration. The unit of the magnitude of force is the unit of mass multiplied by the unit of the magnitude of acceleration. The direction of $\vec{F}$ is the same as that of $\vec{a}$ because m is positive.

Example 1.5 Vector addition

A cross-country skier skis 1.00 km north and then 2.00 km east on a horizontal snow field. a) How far and in what direction is she from the starting point? b) What are the magnitude and direction of her resultant displacement?

SOLUTION

IDENTIFY: The problem involves combining displacements, so we can solve it using vector addition. The target variables in part (a) are the skier's total distance and direction from her starting point. Note that the target variables in part (b) are the *same* as those in part (a): the "magnitude of her resultant displacement" is just her final distance from the point of origin, and the "direction of her resultant displacement" is just the direction from the point of origin to where she stops.

1.12 The vector diagram, drawn to scale, for a cross-country ski trip.

SET UP: Figure 1.12 is a scale diagram of the skier's displacements. We describe the direction from the starting point by the angle ϕ (the Greek letter "phi"). By careful measurement we find that the distance from the starting point is about 2.2 km and that ϕ is about 63°. But we can *calculate* a much more accurate result by adding the 1.00-km and 2.00-km displacement vectors.

EXECUTE: a) The vectors in the diagram form a right triangle; the distance from the starting point is equal to the length of the hypotenuse. We find this length by using the Pythagorean theorem:

$$\sqrt{(1.00 \text{ km})^2 + (2.00 \text{ km})^2} = 2.24 \text{ km}$$

The angle ϕ can be found with a little simple trigonometry. If you need a review, the trigonometric functions and identities are summarized in Appendix B, along with other useful mathematical and geometrical relations. By the definition of the tangent function,

$$\tan \phi = \frac{\text{opposite side}}{\text{adjacent side}} = \frac{2.00 \text{ km}}{1.00 \text{ km}}$$

$$\phi = 63.4°$$

b) The magnitude of the resultant displacement is just the distance that we found in part (a), 2.24 km. We can describe the direction as 63.4° east of north or $90° - 63.4° = 26.6°$ north of east. Take your choice!

EVALUATE: It's good practice to check the results of a vector-addition problem by making measurements on a drawing of the situation. Happily, the answers we found by calculation (2.24 km and $\phi = 63.4°$) are very close to the cruder results we found by measurement (about 2.2 km and about 63°). If they were substantially different, we would have to go back and check for errors.

Test Your Understanding

If the skier had first gone 2.00 km east from her starting point, then 1.00 km north, what would have been the magnitude and direction of her resulting displacement?

1.8 | Components of Vectors

In Section 1.7 we added vectors by using a scale diagram and by using properties of right triangles. Measuring a diagram offers only very limited accuracy, and calculations with right triangles work only when the two vectors are perpendicular. So we need a simple but general method for adding vectors. This is called the method of *components*.

To define what we mean by the components of a vector, we begin with a rectangular (Cartesian) coordinate system of axes (Fig. 1.13). We then draw the vector we're considering with its tail at O, the origin of the coordinate system. We can represent any vector lying in the xy-plane as the sum of a vector parallel to the x-axis and a vector parallel to the y-axis. These two vectors are labeled $\vec{A}_x$ and $\vec{A}_y$ in the figure; they are called the **component vectors** of vector $\vec{A}$, and their vector sum is equal to $\vec{A}$. In symbols,

$$\vec{A} = \vec{A}_x + \vec{A}_y \qquad (1.6)$$

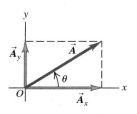

1.13 Vectors $\vec{A}_x$ and $\vec{A}_y$ are the rectangular component vectors of $\vec{A}$ in the directions of the x- and y-axes. For the vector $\vec{A}$ shown here, the components A_x and A_y are both positive.

By definition, each component vector lies along a coordinate-axis direction. Thus we need only a single number to describe each one. When the component vector $\vec{A}_x$ points in the positive x-direction, we define the number A_x to be equal to the magnitude of $\vec{A}_x$. When the component vector $\vec{A}_x$ points in the negative x-direction, we define the number A_x to be equal to the negative of that magnitude, keeping in mind that the magnitude of a vector quantity is itself never negative. We define the number A_y in the same way. The two numbers A_x and A_y are called the **components** of $\vec{A}$.

> **CAUTION** The components A_x and A_y of a vector $\vec{A}$ are just numbers; they are *not* vectors themselves. This is why we print the symbols for components in light italic type with no arrow on top instead of the boldface italic with an arrow reserved for vectors.

We can calculate the components of the vector $\vec{A}$ if we know its magnitude A and its direction. We'll describe the direction of a vector by its angle relative to some reference direction. In Fig. 1.13 this reference direction is the positive x-axis, and the angle between vector $\vec{A}$ and the positive x-axis is θ (the Greek letter "theta"). Imagine that the vector $\vec{A}$ originally lies along the $+x$-axis and that you then rotate it to its correct direction, as indicated by the arrow in Fig. 1.13 on the angle θ. If this rotation is from the $+x$-axis toward the $+y$-axis, as shown in Fig. 1.13, then θ is *positive;* if the rotation is from the $+x$-axis toward the $-y$-axis, θ is *negative*. Thus the $+y$-axis is at an angle of 90°, the $-x$-axis at 180°, and the $-y$-axis at 270° (or $-90°$). If θ is measured in this way, then from the definition of the trigonometric functions,

$$\frac{A_x}{A} = \cos\theta \qquad \text{and} \qquad \frac{A_y}{A} = \sin\theta$$

$$A_x = A\cos\theta \qquad \text{and} \qquad A_y = A\sin\theta \qquad (1.7)$$

$(\theta$ measured from the $+x$-axis, rotating toward the $+y$-axis$)$

> **CAUTION** Equations (1.7) are correct *only* when the angle θ is measured from the positive x-axis as described above. If the angle of the vector is given from a different reference direction or using a different sense of rotation, the relationships are different. Be careful! Example 1.6 illustrates this point.

In Fig. 1.13, A_x is positive because its direction is along the positive x-axis and A_y is positive because its direction is along the positive y-axis. This is consistent with Eqs. (1.7); θ is in the first quadrant (between $0°$ and $90°$), and both the cosine and the sine of an angle in this quadrant are positive. But in Fig. 1.14a the component B_x is negative; its direction is opposite to that of the positive x-axis. Again, this agrees with Eqs. (1.7); the cosine of an angle in the second quadrant is negative. The component B_y is positive ($\sin\theta$ is positive in the second quadrant). In Fig. 1.14b, both C_x and C_y are negative (both $\cos\theta$ and $\sin\theta$ are negative in the third quadrant).

(a)

(b)

1.14 The components of a vector may be positive or negative numbers.

| Example 1.6 | **Finding components** |

a) What are the x- and y-components of vector $\vec{D}$ in Fig. 1.15a? The magnitude of the vector is $D = 3.00$ m and the angle $\alpha = 45°$.
b) What are the x- and y-components of vector $\vec{E}$ in Fig. 1.15b? The magnitude of the vector is $E = 4.50$ m and the angle $\beta = 37.0°$.

SOLUTION

IDENTIFY and SET UP: The problem involves finding components, so it would seem that all we need is Eqs. (1.7). However, we need to be careful because the angles in Fig. 1.15 are *not* measured from the $+x$-axis toward the $+y$-axis.

EXECUTE: a) The angle between $\vec{D}$ and the positive x-axis is α (the Greek letter "alpha"), but this angle is measured toward the *negative* y-axis. So the angle we must use in Eqs. (1.7) is $\theta = -\alpha = -45°$. We find

$$D_x = D\cos\theta = (3.00\ \text{m})(\cos(-45°)) = +2.1\ \text{m}$$
$$D_y = D\sin\theta = (3.00\ \text{m})(\sin(-45°)) = -2.1\ \text{m}$$

The vector has a positive x-component and a negative y-component, as shown in the figure. Had you been careless and substituted $+45°$ for θ in Eqs. (1.7), you would have gotten the wrong sign for D_y.
b) The x-axis isn't horizontal in Fig. 1.15b, nor is the y-axis vertical. In general, *any* orientation of the x- and y-axes is permissible, just so the axes are mutually perpendicular. (Later we'll use axes like these to study the motion of an object sliding on a ramp or

incline; one axis will lie along the ramp and the other will be perpendicular to the ramp.)

Here the angle β (the Greek letter "beta") is the angle between $\vec{E}$ and the positive y-axis, *not* the positive x-axis, so we *cannot* use this angle in Eqs. (1.7). Instead, note that $\vec{E}$ defines the hypotenuse of a right triangle; the other two sides of the triangle are the magnitudes of E_x and E_y, the x- and y-components of $\vec{E}$. The sine of β is the opposite side (the magnitude of E_x) divided by the hypotenuse (the magnitude E), and the cosine of β is the adjacent side (the magnitude of E_y) divided by

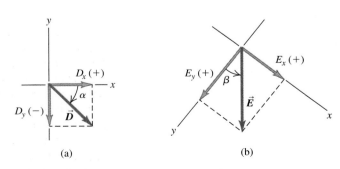

(a)   (b)

1.15 Calculating the x- and y-components of vectors.

the hypotenuse (again, the magnitude E). Both components of $\vec{E}$ are positive, so

$$E_x = E \sin \beta = (4.50 \text{ m})(\sin 37.0°) = +2.71 \text{ m}$$

$$E_y = E \cos \beta = (4.50 \text{ m})(\cos 37.0°) = +3.59 \text{ m}$$

Had you used Eqs. (1.7) directly and written $E_x = E \cos 37.0°$ and $E_y = E \sin 37.0°$, your answers for E_x and E_y would have been reversed!

If you insist on using Eqs. (1.7), you must first find the angle between $\vec{E}$ and the positive x-axis, measured toward the positive y-axis; this is $\theta = 90.0° - \beta = 90.0° - 37.0° = 53.0°$. Then $E_x = E \cos \theta$ and $E_y = E \sin \theta$. You can substitute the values of E and θ into Eqs. (1.7) to show that the results for E_x and E_y are the same as given above.

EVALUATE: Notice that the answers to part (b) have three significant figures, but the answers to part (a) have only two. Can you see why?

Using Components to Add Vectors

We can describe a vector completely by giving either its magnitude and direction or its x- and y-components. Equations (1.7) show how to find the components if we know the magnitude and direction. We can also reverse the process: we can find the magnitude and direction if we know the components. By applying the Pythagorean theorem to Fig. 1.13, we find that the magnitude of a vector $\vec{A}$ is

$$A = \sqrt{A_x^2 + A_y^2} \tag{1.8}$$

where we always take the positive root. Equation (1.8) is valid for any choice of x-axis and y-axis, as long as they are mutually perpendicular. The expression for the vector direction comes from the definition of the tangent of an angle. If θ is measured from the positive x-axis, and a positive angle is measured toward the positive y-axis (as in Fig. 1.13), then

$$\tan \theta = \frac{A_y}{A_x} \quad \text{and} \quad \theta = \arctan \frac{A_y}{A_x} \tag{1.9}$$

We will always use the notation arctan for the inverse tangent function. The notation $\tan^{-1}$ is also commonly used, and your calculator may have an INV button to be used with the TAN button. Microsoft Excel uses ATAN.

CAUTION There is one slight complication in using Eqs. (1.9) to find θ. Suppose $A_x = 2$ m and $A_y = -2$ m; then $\tan \theta = -1$. But there are two angles having tangents of -1, namely, 135° and 315° (or $-45°$). In general, any two angles that differ by 180° have the same tangent. To decide which is correct, we have to look at the individual components. Because A_x is positive and A_y is negative, the angle must be in the fourth quadrant; thus $\theta = 315°$ (or $-45°$) is the correct value. Most pocket calculators give $\arctan(-1) = -45°$. In this case that is correct; but if instead we have $A_x = -2$ m and $A_y = 2$ m, then the correct angle is 135°. Similarly, when A_x and A_y are both negative, the tangent is positive, but the angle is in the third quadrant. You should *always* draw a sketch to check which of the two possibilities is the correct one.

Here's how we use components to calculate the vector sum (resultant) of two or more vectors. Figure 1.16 shows two vectors $\vec{A}$ and $\vec{B}$ and their vector sum $\vec{R}$, along with the x- and y-components of all three vectors. You can see from the diagram that the x-component R_x of the vector sum is simply the sum $(A_x + B_x)$ of the x-components of the vectors being added. The same is true for the y-components. In symbols,

$$R_x = A_x + B_x, \quad R_y = A_y + B_y \quad (\text{components of } \vec{R} = \vec{A} + \vec{B}) \tag{1.10}$$

Figure 1.16 shows this result for the case in which the components A_x, A_y, B_x, and B_y are all positive. You should draw additional diagrams to verify for yourself that Eqs. (1.10) are valid for *any* signs of the components of $\vec{A}$ and $\vec{B}$.

If we know the components of any two vectors $\vec{A}$ and $\vec{B}$, perhaps by using Eqs. (1.7), we can compute the components of the vector sum $\vec{R}$. Then if we need the magnitude and direction of $\vec{R}$, we can obtain them from Eqs. (1.8) and (1.9) with the A's replaced by R's.

This procedure for finding the sum of two vectors can be extended to any number of vectors. Let $\vec{R}$ be the vector sum of $\vec{A}$, $\vec{B}$, $\vec{C}$, $\vec{D}$, $\vec{E}$, ... Then the components of $\vec{R}$ are

$$R_x = A_x + B_x + C_x + D_x + E_x + \cdots$$

$$R_y = A_y + B_y + C_y + D_y + E_y + \cdots \qquad (1.11)$$

We have talked only about vectors that lie in the xy-plane, but the component method works just as well for vectors having any direction in space. We introduce a z-axis perpendicular to the xy-plane; then in general a vector $\vec{A}$ has components A_x, A_y, and A_z in the three coordinate directions. The magnitude A is given by

$$A = \sqrt{A_x^2 + A_y^2 + A_z^2} \qquad (1.12)$$

Again, we always take the positive root. Also, Eqs. (1.11) for the components of the vector sum $\vec{R}$ have an additional member:

$$R_z = A_z + B_z + C_z + D_z + E_z + \cdots$$

Finally, while our discussion of vector addition has centered on combining *displacement* vectors, the method is applicable to all other vector quantities as well. When we study the concept of force in Chapter 4, we'll find that forces are vectors that obey the same rules of vector addition that we've used with displacement. Other vector quantities will make their appearance in later chapters.

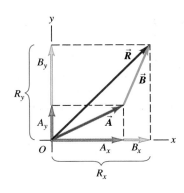

1.16 Vector $\vec{R}$ is the vector sum (resultant) of $\vec{A}$ and $\vec{B}$. Its x-component, R_x, equals the sum of the x-components of $\vec{A}$ and $\vec{B}$. The y-components are similarly related.

Problem-Solving Strategy **Vector Addition**

IDENTIFY *the relevant concepts and* **SET UP** *the problem:* Decide what your target variable is. It may be the magnitude of the vector sum, the direction, or both. Then draw the individual vectors being summed and the coordinate axes being used. In your drawing, place the tail of the first vector at the origin of coordinates; place the tail of the second vector at the head of the first vector; and so on. Draw the vector sum $\vec{R}$ from the tail of the first vector to the head of the last vector. By examining your drawing, make a rough estimate of the magnitude and direction of $\vec{R}$; you'll use these estimates later to check your calculations.

EXECUTE *the solution as follows:*
1. Find the x- and y-components of each individual vector and record your results in a table. If a vector is described by its magnitude A and its angle θ, measured from the

$+x$-axis toward the $+y$-axis, then the components are given by

$$A_x = A\cos\theta \qquad A_y = A\sin\theta$$

Some components may be positive and some may be negative, depending on how the vector is oriented (that is, what quadrant θ lies in). You can use this sign table as a check:

Quadrant	I	II	III	IV
A_x	+	−	−	+
A_y	+	+	−	−

If the angles of the vectors are given in some other way, perhaps using a different reference direction, convert them to angles measured from the $+x$-axis as described above. Be particularly careful with signs.

2. Add the individual x-components algebraically, including signs, to find R_x, the x-component of the vector sum. Do the same for the y-components to find R_y.
3. Then the magnitude R and direction θ of the vector sum are given by

$$R = \sqrt{R_x^2 + R_y^2}$$

$$\theta = \arctan\frac{R_y}{R_x}$$

EVALUATE *your answer:* Check your results for the magnitude and direction of the vector sum by comparing them with the rough estimates you made from your drawing. Remember that

the magnitude R is *always* positive and that θ is measured from the positive x-axis. The value of θ that you find with a calculator may be the correct one, or it may be off by 180°. You can decide by examining your drawing.

If your calculations disagree totally with the estimates from your drawing, check whether your calculator is set in "radians" or "degrees" mode. If it's in "radians" mode, entering angles in degrees will give nonsensical answers. Watch out for this problem if you use Microsoft Excel, in which all trigonometric functions use units of radians rather than degrees. To convert from one to the other, remember that 360 degrees is the same as 2π radians.

Example
1.7
Adding vectors with components

The three finalists in a contest are brought to the center of a large, flat field. Each is given a meter stick, a compass, a calculator, a shovel, and (in a different order for each contestant) the following three displacements:

72.4 m, 32.0° east of north;
57.3 m, 36.0° south of west;
17.8 m straight south.

The three displacements lead to the point where the keys to a new Porsche are buried. Two contestants start measuring immediately, but the winner first *calculates* where to go. What does she calculate?

SOLUTION

IDENTIFY and SET UP: The situation is shown in Fig. 1.17. We have chosen the $+x$-axis as east and the $+y$-axis as north, the usual choice for maps. Let $\vec{A}$ be the first displacement, $\vec{B}$ the second, and $\vec{C}$ the third. We can estimate from the diagram that the vector sum $\vec{R}$ is about 10 m, 40° west of north.

EXECUTE: The angles of the vectors, measured from the $+x$-axis toward the $+y$-axis, are $(90.0° - 32.0°) = 58.0°$, $(180.0° + 36.0°) = 216.0°$, and $270.0°$. We have to find the components of each. Because of our choice of axes, we may use Eqs. (1.7), and so the components of $\vec{A}$ are

$$A_x = A \cos\theta_A = (72.4 \text{ m})(\cos 58.0°) = 38.37 \text{ m}$$
$$A_y = A \sin\theta_A = (72.4 \text{ m})(\sin 58.0°) = 61.40 \text{ m}$$

Note that we have kept one too many significant figures in the components; we will wait until the end to round to the correct number of significant figures. The table below shows the components of all the displacements, the addition of components, and the other calculations. Always arrange your component calculations systematically like this.

Distance	Angle	x-component	y-component
$A = 72.4$ m	58.0°	38.37 m	61.40 m
$B = 57.3$ m	216.0°	−46.36 m	−33.68 m
$C = 17.8$ m	270.0°	0.00 m	−17.80 m
		$R_x = -7.99$ m	$R_y = 9.92$ m

$$R = \sqrt{(-7.99 \text{ m})^2 + (9.92 \text{ m})^2} = 12.7 \text{ m}$$

$$\theta = \arctan\frac{9.92 \text{ m}}{-7.99 \text{ m}} = 129° = 39° \text{ west of north}$$

The losers try to measure three angles and three distances totaling 147.5 m, one meter at a time. The winner measured only one angle and one much shorter distance.

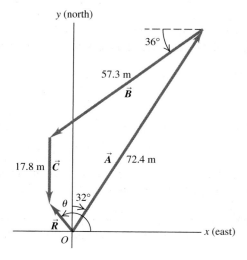

1.17 Three successive displacements $\vec{A}$, $\vec{B}$, and $\vec{C}$ and the resultant (vector sum) displacement $\vec{R} = \vec{A} + \vec{B} + \vec{C}$.

EVALUATE: Our calculated answers for R and θ are not too different from our estimates of 10 m and 40° west of north; that's good! Notice that $\theta = -51°$, or 51° south of east, also satisfies the equation for θ. But since the winner has made a drawing of the displacement vectors (Fig. 1.17), she knows that $\theta = 129°$ is the only correct solution for the angle.

| Example 1.8 | **A vector in three dimensions** |

After an airplane takes off, it travels 10.4 km west, 8.7 km north, and 2.1 km up. How far is it from the takeoff point?

SOLUTION

Let the $+x$-axis be east, the $+y$-axis north, and the $+z$-axis up. Then $A_x = -10.4$ km, $A_y = 8.7$ km, and $A_z = 2.1$ km; Eq. (1.12) gives

$$A = \sqrt{(-10.4\text{ km})^2 + (8.7\text{ km})^2 + (2.1\text{ km})^2} = 13.7\text{ km}$$

Test Your Understanding

How far are you from your starting point if you first traveled 4.00 km due west, then 4.00 km due south? What is the direction from where you started to where you ended up?

1.9 | Unit Vectors

A **unit vector** is a vector that has a magnitude of 1, with no units. Its only purpose is to *point;* that is, to describe a direction in space. Unit vectors provide a convenient notation for many expressions involving components of vectors. We will always include a caret or "hat" (^) in the symbol for a unit vector to distinguish it from ordinary vectors whose magnitude may or may not be equal to 1.

In an x-y coordinate system we can define a unit vector $\hat{\imath}$ that points in the direction of the positive x-axis and a unit vector $\hat{\jmath}$ that points in the direction of the positive y-axis. Then we can express the relationship between component vectors and components, described at the beginning of Section 1.8, as follows:

$$\vec{A}_x = A_x\hat{\imath}$$
$$\vec{A}_y = A_y\hat{\jmath} \tag{1.13}$$

Similarly, we can write a vector $\vec{A}$ in terms of its components as

$$\vec{A} = A_x\hat{\imath} + A_y\hat{\jmath} \tag{1.14}$$

Equations (1.13) and (1.14) are vector equations; each term, such as $A_x\hat{\imath}$, is a vector quantity (Fig. 1.18). The boldface equals and plus signs denote vector equality and addition.

When two vectors $\vec{A}$ and $\vec{B}$ are represented in terms of their components, we can express the vector sum $\vec{R}$ using unit vectors as follows:

$$\vec{A} = A_x\hat{\imath} + A_y\hat{\jmath}$$
$$\vec{B} = B_x\hat{\imath} + B_y\hat{\jmath}$$

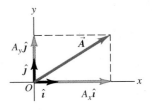

1.18 Using unit vectors, we can express a vector $\vec{A}$ in terms of its components A_x and A_y as $\vec{A} = A_x\hat{\imath} + A_y\hat{\jmath}$.

$$\vec{R} = \vec{A} + \vec{B}$$
$$= (A_x\hat{\imath} + A_y\hat{\jmath}) + (B_x\hat{\imath} + B_y\hat{\jmath})$$
$$= (A_x + B_x)\hat{\imath} + (A_y + B_y)\hat{\jmath}$$
$$= R_x\hat{\imath} + R_y\hat{\jmath} \qquad (1.15)$$

Equation (1.15) restates the content of Eqs. (1.10) in the form of a single vector equation rather than two component equations.

If the vectors do not all lie in the xy-plane, then we need a third component. We introduce a third unit vector $\hat{k}$ that points in the direction of the positive z-axis. The generalized forms of Eqs. (1.14) and (1.15) are

$$\vec{A} = A_x\hat{\imath} + A_y\hat{\jmath} + A_z\hat{k}$$
$$\vec{B} = B_x\hat{\imath} + B_y\hat{\jmath} + B_z\hat{k} \qquad (1.16)$$

$$\vec{R} = (A_x + B_x)\hat{\imath} + (A_y + B_y)\hat{\jmath} + (A_z + B_z)\hat{k}$$
$$= R_x\hat{\imath} + R_y\hat{\jmath} + R_z\hat{k} \qquad (1.17)$$

Example 1.9 Using unit vectors

Given the two displacements

$$\vec{D} = (6\hat{\imath} + 3\hat{\jmath} - \hat{k})\,\text{m} \quad \text{and} \quad \vec{E} = (4\hat{\imath} - 5\hat{\jmath} + 8\hat{k})\,\text{m}$$

find the magnitude of the displacement $2\vec{D} - \vec{E}$.

SOLUTION

IDENTIFY, SET UP, and EXECUTE: Letting $\vec{F} = 2\vec{D} - \vec{E}$, we have

$$\vec{F} = 2(6\hat{\imath} + 3\hat{\jmath} - \hat{k})\,\text{m} - (4\hat{\imath} - 5\hat{\jmath} + 8\hat{k})\,\text{m}$$
$$= [(12 - 4)\hat{\imath} + (6 + 5)\hat{\jmath} + (-2 - 8)\hat{k}]\,\text{m}$$
$$= (8\hat{\imath} + 11\hat{\jmath} - 10\hat{k})\,\text{m}$$

The units of the vectors $\vec{D}$, $\vec{E}$, and $\vec{F}$ are meters, so the components of these vectors are also in meters. From Eq. (1.12),

$$F = \sqrt{F_x^2 + F_y^2 + F_z^2}$$
$$= \sqrt{(8\,\text{m})^2 + (11\,\text{m})^2 + (-10\,\text{m})^2} = 17\,\text{m}$$

EVALUATE: Working with unit vectors makes vector addition and subtraction no more complicated than adding and subtracting ordinary numbers. Still, be sure to check for simple arithmetic errors.

Test Your Understanding

Express each of the vectors $\vec{A}$, $\vec{B}$, and $\vec{C}$ in Example 1.7 (Section 1.8) in terms of unit vectors.

1.10 | Products of Vectors

We have seen how addition of vectors develops naturally from the problem of combining displacements, and we will use vector addition for calculating many other vector quantities later. We can also express many physical relationships concisely by using *products* of vectors. Vectors are not ordinary numbers, so ordinary multiplication is not directly applicable to vectors. We will define two different kinds of products of vectors. The first, called the scalar product, yields a result that is a scalar quantity. The second, the vector product, yields another vector.

Scalar Product

The **scalar product** of two vectors $\vec{A}$ and $\vec{B}$ is denoted by $\vec{A} \cdot \vec{B}$. Because of this notation, the scalar product is also called the **dot product.**

To define the scalar product $\vec{A} \cdot \vec{B}$ of two vectors $\vec{A}$ and $\vec{B}$, we draw the two vectors with their tails at the same point (Fig. 1.19a). The angle between their directions is ϕ as shown; the angle ϕ always lies between 0° and 180°. (As usual, we use Greek letters for angles.) Figure 1.19b shows the projection of the vector $\vec{B}$ onto the direction of $\vec{A}$; this projection is the component of $\vec{B}$ parallel to $\vec{A}$ and is equal to $B\cos\phi$. (We can take components along any direction that's convenient, not just the x- and y-axes.) We define $\vec{A} \cdot \vec{B}$ to be the magnitude of $\vec{A}$ multiplied by the component of $\vec{B}$ parallel to $\vec{A}$. Expressed as an equation,

$$\vec{A} \cdot \vec{B} = AB\cos\phi = |\vec{A}||\vec{B}|\cos\phi \qquad (1.18)$$

(definition of the scalar (dot) product)

where ϕ ranges from 0° to 180°.

Alternatively, we can define $\vec{A} \cdot \vec{B}$ to be the magnitude of $\vec{B}$ multiplied by the component of $\vec{A}$ parallel to $\vec{B}$, as in Fig. 1.19c. Hence $\vec{A} \cdot \vec{B} = B(A\cos\phi) = AB\cos\phi$, which is the same as Eq. (1.18).

The scalar product is a scalar quantity, not a vector, and it may be positive, negative, or zero. When ϕ is between 0° and 90°, the scalar product is positive. When ϕ is between 90° and 180°, it is negative. You should draw a diagram like Fig. 1.19, but with ϕ between 90° and 180°, to confirm for yourself that the component of $\vec{B}$ parallel to $\vec{A}$ is negative in this case, as is the component of $\vec{A}$ parallel to $\vec{B}$. Finally, when $\phi = 90°$, $\vec{A} \cdot \vec{B} = 0$. *The scalar product of two perpendicular vectors is always zero.*

For any two vectors $\vec{A}$ and $\vec{B}$, $AB\cos\phi = BA\cos\phi$. This means that $\vec{A} \cdot \vec{B} = \vec{B} \cdot \vec{A}$. The scalar product obeys the commutative law of multiplication; the order of the two vectors does not matter.

We will use the scalar product in Chapter 6 to describe work done by a force. When a constant force $\vec{F}$ is applied to a body that undergoes a displacement $\vec{s}$, the work W (a scalar quantity) done by the force is given by

$$W = \vec{F} \cdot \vec{s}$$

The work done by the force is positive if the angle between $\vec{F}$ and $\vec{s}$ is between 0° and 90°, negative if this angle is between 90° and 180°, and zero if $\vec{F}$ and $\vec{s}$ are perpendicular. (This is another example of a term that has a special meaning in physics; in everyday language, "work" isn't something that can be positive or negative.) In later chapters we'll use the scalar product for a variety of purposes, from calculating electric potential to determining the effects that varying magnetic fields have on electric circuits.

We can calculate the scalar product $\vec{A} \cdot \vec{B}$ directly if we know the x-, y-, and z-components of $\vec{A}$ and $\vec{B}$. To see how this is done, let's first work out the scalar products of the unit vectors. This is easy, since $\hat{\imath}$, $\hat{\jmath}$, and $\hat{k}$ are all perpendicular to each other. Using Eq. (1.18), we find

$$\hat{\imath} \cdot \hat{\imath} = \hat{\jmath} \cdot \hat{\jmath} = \hat{k} \cdot \hat{k} = (1)(1)\cos 0 = 1$$

$$\hat{\imath} \cdot \hat{\jmath} = \hat{\imath} \cdot \hat{k} = \hat{\jmath} \cdot \hat{k} = (1)(1)\cos 90° = 0 \qquad (1.19)$$

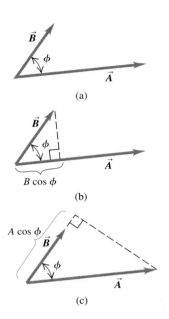

1.19 (a) Two vectors drawn from a common starting point to define their scalar product $\vec{A} \cdot \vec{B} = AB\cos\phi$. (b) $B\cos\phi$ is the component of $\vec{B}$ in the direction of $\vec{A}$, and $\vec{A} \cdot \vec{B}$ is the product of the magnitude of $\vec{A}$ and this component. (c) $\vec{A} \cdot \vec{B}$ is also the product of the magnitude of $\vec{B}$ and the component of $\vec{A}$ in the direction of $\vec{B}$.

Now we express $\vec{A}$ and $\vec{B}$ in terms of their components, expand the product, and use these products of unit vectors:

$$\vec{A} \cdot \vec{B} = (A_x\hat{\imath} + A_y\hat{\jmath} + A_z\hat{k}) \cdot (B_x\hat{\imath} + B_y\hat{\jmath} + B_z\hat{k})$$

$$= A_x\hat{\imath} \cdot B_x\hat{\imath} + A_x\hat{\imath} \cdot B_y\hat{\jmath} + A_x\hat{\imath} \cdot B_z\hat{k}$$

$$+ A_y\hat{\jmath} \cdot B_x\hat{\imath} + A_y\hat{\jmath} \cdot B_y\hat{\jmath} + A_y\hat{\jmath} \cdot B_z\hat{k}$$

$$+ A_z\hat{k} \cdot B_x\hat{\imath} + A_z\hat{k} \cdot B_y\hat{\jmath} + A_z\hat{k} \cdot B_z\hat{k}$$

$$= A_xB_x\hat{\imath} \cdot \hat{\imath} + A_xB_y\hat{\imath} \cdot \hat{\jmath} + A_xB_z\hat{\imath} \cdot \hat{k}$$

$$+ A_yB_x\hat{\jmath} \cdot \hat{\imath} + A_yB_y\hat{\jmath} \cdot \hat{\jmath} + A_yB_z\hat{\jmath} \cdot \hat{k}$$

$$+ A_zB_x\hat{k} \cdot \hat{\imath} + A_zB_y\hat{k} \cdot \hat{\jmath} + A_zB_z\hat{k} \cdot \hat{k} \qquad (1.20)$$

From Eqs. (1.19) we see that six of these nine terms are zero, and the three that survive give simply

$$\vec{A} \cdot \vec{B} = A_xB_x + A_yB_y + A_zB_z \qquad (1.21)$$

(scalar (dot) product in terms of components)

Thus the scalar product of two vectors is the sum of the products of their respective components.

The scalar product gives a straightforward way to find the angle ϕ between any two vectors $\vec{A}$ and $\vec{B}$ whose components are known. In this case, Eq. (1.21) can be used to find the scalar product of $\vec{A}$ and $\vec{B}$. From Eq. (1.18) the scalar product is also equal to $AB \cos \phi$. The vector magnitudes A and B can be found from the vector components with Eq. (1.12), so $\cos \phi$ and hence the angle ϕ can be determined (see Example 1.11).

Example 1.10	**Calculating a scalar product**

Find the scalar product $\vec{A} \cdot \vec{B}$ of the two vectors in Fig. 1.20. The magnitudes of the vectors are $A = 4.00$ and $B = 5.00$.

SOLUTION

IDENTIFY and SET UP: There are two ways to calculate the scalar product. The first way uses the magnitudes of the vectors and the angle between them (Eq. 1.18), and the second uses the components of the two vectors (Eq. 1.21).

EXECUTE: Using the first approach, the angle between the two vectors is $\phi = 130.0° - 53.0° = 77.0°$, so

$$\vec{A} \cdot \vec{B} = AB \cos \phi = (4.00)(5.00)\cos 77.0° = 4.50$$

This is positive because the angle between $\vec{A}$ and $\vec{B}$ is between $0°$ and $90°$.

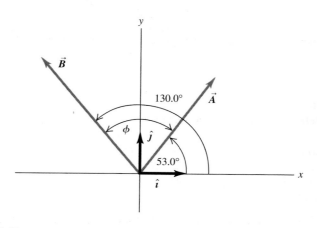

1.20 Two vectors in two dimensions.

To use the second approach, we first need to find the components of the two vectors. Since the angles of $\vec{A}$ and $\vec{B}$ are given with respect to the $+x$-axis, and these angles are measured in the sense from the $+x$-axis to the $+y$-axis, we can use Eqs. (1.7):

$$A_x = (4.00)\cos 53.0° = 2.407$$
$$A_y = (4.00)\sin 53.0° = 3.195$$
$$A_z = 0$$
$$B_x = (5.00)\cos 130.0° = -3.214$$
$$B_y = (5.00)\sin 130.0° = 3.830$$
$$B_z = 0$$

The z-components are zero because both vectors lie in the xy-plane. As in Example 1.7, we are keeping one too many significant figures in the components; we'll round to the correct number at the end. From Eq. (1.21) the scalar product is

$$\vec{A} \cdot \vec{B} = A_xB_x + A_yB_y + A_zB_z$$
$$= (2.407)(-3.214) + (3.195)(3.830)$$
$$+ (0)(0) = 4.50$$

EVALUATE: We get the same result for the scalar product with both methods, as we should.

| Example 1.11 | **Finding angles with the scalar product** |

Find the angle between the two vectors

$$\vec{A} = 2\hat{i} + 3\hat{j} + \hat{k} \quad \text{and} \quad \vec{B} = -4\hat{i} + 2\hat{j} - \hat{k}$$

SOLUTION

IDENTIFY: The scalar product of two vectors $\vec{A}$ and $\vec{B}$ is related to the angle ϕ between them and to the magnitudes A and B. The scalar product is also related to the components of the two vectors. If we are given the components of the vectors (as we are in this

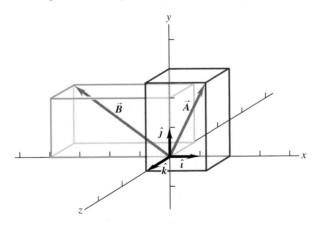

1.21 Two vectors in three dimensions.

example), we first determine the scalar product $\vec{A} \cdot \vec{B}$ and the values of A and B, and then determine the target variable ϕ.

SET UP and EXECUTE: The vectors are shown in Fig. 1.21. The scalar product of two vectors is given by either Eq. (1.18) or Eq. (1.21). Equating these two and rearranging, we obtain

$$\cos\phi = \frac{A_xB_x + A_yB_y + A_zB_z}{AB}$$

This formula can be used to find the angle between any two vectors $\vec{A}$ and $\vec{B}$. For our example the components of $\vec{A}$ are $A_x = 2$, $A_y = 3$, and $A_z = 1$, and the components of $\vec{B}$ are $B_x = -4$, $B_y = 2$, and $B_z = -1$. Thus

$$\vec{A} \cdot \vec{B} = A_xB_x + A_yB_y + A_zB_z$$
$$= (2)(-4) + (3)(2) + (1)(-1) = -3$$
$$A = \sqrt{A_x^2 + A_y^2 + A_z^2} = \sqrt{2^2 + 3^2 + 1^2} = \sqrt{14}$$
$$B = \sqrt{B_x^2 + B_y^2 + B_z^2} = \sqrt{(-4)^2 + 2^2 + (-1)^2} = \sqrt{21}$$
$$\cos\phi = \frac{A_xB_x + A_yB_y + A_zB_z}{AB} = \frac{-3}{\sqrt{14}\sqrt{21}} = -0.175$$
$$\phi = 100°$$

EVALUATE: As a check on this result, note that the scalar product $\vec{A} \cdot \vec{B}$ is negative. This means that ϕ is between $90°$ and $180°$, in agreement with our answer.

Vector Product

The **vector product** of two vectors $\vec{A}$ and $\vec{B}$, also called the **cross product,** is denoted by $\vec{A} \times \vec{B}$. We will use this product in Chapter 10 to describe torque

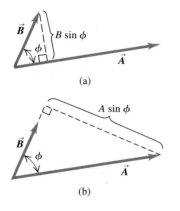

1.22 (a) Vectors $\vec{A}$ and $\vec{B}$ lie in a plane; the vector product $\vec{A} \times \vec{B}$ is perpendicular to this plane in a direction determined by the right-hand rule. (b) $\vec{B} \times \vec{A} = -\vec{A} \times \vec{B}$; the vector product is anticommutative.

1.23 (a) $B \sin \phi$ is the component of $\vec{B}$ perpendicular to the direction of $\vec{A}$, and the magnitude of $\vec{A} \times \vec{B}$ is the product of the magnitude of $\vec{A}$ and this component. (b) The magnitude of $\vec{A} \times \vec{B}$ is also the product of the magnitude of $\vec{B}$ and the component of $\vec{A}$ perpendicular to $\vec{B}$.

and angular momentum. Later we will also use it extensively for magnetic fields, where it will help us describe the relationships among the directions of several vector quantities.

To define the vector product $\vec{A} \times \vec{B}$ of two vectors $\vec{A}$ and $\vec{B}$, we again draw the two vectors with their tails at the same point (Fig. 1.22a). The two vectors then lie in a plane. We define the vector product to be a vector quantity with a direction perpendicular to this plane (that is, perpendicular to both $\vec{A}$ and $\vec{B}$) and a magnitude equal to $AB \sin \phi$. That is, if $\vec{C} = \vec{A} \times \vec{B}$, then

$$C = AB \sin \phi \qquad (1.22)$$

$$\left(\text{magnitude of the vector (cross) product of } \vec{A} \text{ and } \vec{B}\right)$$

We measure the angle ϕ from $\vec{A}$ toward $\vec{B}$ and take it to be the smaller of the two possible angles, so ϕ ranges from $0°$ to $180°$. Thus C in Eq. (1.22) is always positive, as a vector magnitude should be. Note also that when $\vec{A}$ and $\vec{B}$ are parallel or antiparallel, $\phi = 0$ or $180°$ and $C = 0$. That is, *the vector product of two parallel or antiparallel vectors is always zero.* In particular, *the vector product of any vector with itself is zero.* To see the contrast between the scalar product and the magnitude of the vector product, imagine that we vary the angle between $\vec{A}$ and $\vec{B}$ while keeping their magnitudes constant. When $\vec{A}$ and $\vec{B}$ are parallel, the scalar product will be maximum and the magnitude of the vector product will be zero. When $\vec{A}$ and $\vec{B}$ are perpendicular, the scalar product will be zero and the magnitude of the vector product will be maximum.

There are always *two* directions perpendicular to a given plane, one on each side of the plane. We choose which of these is the direction of $\vec{A} \times \vec{B}$ as follows. Imagine rotating vector $\vec{A}$ about the perpendicular line until it is aligned with $\vec{B}$, choosing the smaller of the two possible angles between $\vec{A}$ and $\vec{B}$. Curl the fingers of your right hand around the perpendicular line so that the fingertips point in the direction of rotation; your thumb will then point in the direction of $\vec{A} \times \vec{B}$. This **right-hand rule** is shown in Fig. 1.22a. The direction of the vector product is also the direction in which a right-hand screw advances if turned from $\vec{A}$ toward $\vec{B}$, as shown.

Similarly, we determine the direction of $\vec{B} \times \vec{A}$ by rotating $\vec{B}$ into $\vec{A}$ in Fig. 1.22b. The result is a vector that is *opposite* to the vector $\vec{A} \times \vec{B}$. The vector product is not commutative! In fact, for any two vectors $\vec{A}$ and $\vec{B}$,

$$\vec{A} \times \vec{B} = -\vec{B} \times \vec{A} \qquad (1.23)$$

Just as we did for the scalar product, we can give a geometrical interpretation of the magnitude of the vector product. In Fig. 1.23a, $B \sin \phi$ is the component of vector $\vec{B}$ that is *perpendicular* to the direction of vector $\vec{A}$. From Eq. (1.22) the magnitude of $\vec{A} \times \vec{B}$ equals the magnitude of $\vec{A}$ multiplied by the component of $\vec{B}$ perpendicular to $\vec{A}$. Figure 1.23b shows that the magnitude of $\vec{A} \times \vec{B}$ also equals the magnitude of $\vec{B}$ multiplied by the component of $\vec{A}$ perpendicular to $\vec{B}$. Note that Fig. 1.23 shows the case in which ϕ is between $0°$ and $90°$; you should draw a similar diagram for ϕ between $90°$ and $180°$ to show that the same geometrical interpretation of the magnitude of $\vec{A} \times \vec{B}$ still applies.

If we know the components of $\vec{A}$ and $\vec{B}$, we can calculate the components of the vector product, using a procedure similar to that for the scalar product. First we work out the multiplication table for the unit vectors $\hat{\imath}$, $\hat{\jmath}$, and $\hat{k}$. The vector product of any vector with itself is zero, so

$$\hat{\imath} \times \hat{\imath} = \hat{\jmath} \times \hat{\jmath} = \hat{k} \times \hat{k} = \mathbf{0}$$

The boldface zero is a reminder that each product is a zero *vector*—that is, one with all components equal to zero and an undefined direction. Using Eqs. (1.22) and (1.23) and the right-hand rule, we find

$$\hat{\imath} \times \hat{\jmath} = -\hat{\jmath} \times \hat{\imath} = \hat{k}$$

$$\hat{\jmath} \times \hat{k} = -\hat{k} \times \hat{\jmath} = \hat{\imath}$$

$$\hat{k} \times \hat{\imath} = -\hat{\imath} \times \hat{k} = \hat{\jmath} \qquad (1.24)$$

Next we express $\vec{A}$ and $\vec{B}$ in terms of their components and the corresponding unit vectors, and we expand the expression for the vector product:

$$\vec{A} \times \vec{B} = (A_x\hat{\imath} + A_y\hat{\jmath} + A_z\hat{k}) \times (B_x\hat{\imath} + B_y\hat{\jmath} + B_z\hat{k})$$

$$= A_x\hat{\imath} \times B_x\hat{\imath} + A_x\hat{\imath} \times B_y\hat{\jmath} + A_x\hat{\imath} \times B_z\hat{k}$$

$$+ A_y\hat{\jmath} \times B_x\hat{\imath} + A_y\hat{\jmath} \times B_y\hat{\jmath} + A_y\hat{\jmath} \times B_z\hat{k}$$

$$+ A_z\hat{k} \times B_x\hat{\imath} + A_z\hat{k} \times B_y\hat{\jmath} + A_z\hat{k} \times B_z\hat{k} \qquad (1.25)$$

We can also rewrite the individual terms as $A_x\hat{\imath} \times B_y\hat{\jmath} = (A_xB_y)\hat{\imath} \times \hat{\jmath}$, and so on. Evaluating these by using the multiplication table for the unit vectors and then grouping the terms, we find

$$\vec{A} \times \vec{B} = (A_yB_z - A_zB_y)\hat{\imath} + (A_zB_x - A_xB_z)\hat{\jmath} + (A_xB_y - A_yB_x)\hat{k} \quad (1.26)$$

Thus the components of $\vec{C} = \vec{A} \times \vec{B}$ are given by

$$C_x = A_yB_z - A_zB_y \qquad C_y = A_zB_x - A_xB_z \qquad C_z = A_xB_y - A_yB_x$$

$$\text{(components of } \vec{C} = \vec{A} \times \vec{B}) \qquad (1.27)$$

The vector product can also be expressed in determinant form as

$$\vec{A} \times \vec{B} = \begin{vmatrix} \hat{\imath} & \hat{\jmath} & \hat{k} \\ A_x & A_y & A_z \\ B_x & B_y & B_z \end{vmatrix}$$

If you aren't familiar with determinants, don't worry about this form.

   With the axis system of Fig. 1.24a, if we reverse the direction of the z-axis, we get the system shown in Fig. 1.24b. Then, as you may verify, the definition of the vector product gives $\hat{\imath} \times \hat{\jmath} = -\hat{k}$ instead of $\hat{\imath} \times \hat{\jmath} = \hat{k}$. In fact, all vector products of the unit vectors $\hat{\imath}$, $\hat{\jmath}$, and $\hat{k}$ would have signs opposite to those in Eqs. (1.24). We see that there are two kinds of coordinate systems, differing in the signs of the vector products of unit vectors. An axis system in which $\hat{\imath} \times \hat{\jmath} = \hat{k}$, as in Fig. 1.24a, is called a **right-handed system.** The usual practice is to use *only* right-handed systems, and we will follow that practice throughout this book.

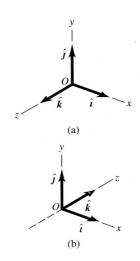

1.24 (a) A right-handed coordinate system, in which $\hat{\imath} \times \hat{\jmath} = \hat{k}$, $\hat{\jmath} \times \hat{k} = \hat{\imath}$, and $\hat{k} \times \hat{\imath} = \hat{\jmath}$. (b) A left-handed coordinate system, in which $\hat{\imath} \times \hat{\jmath} = -\hat{k}$, and so on. We'll use only right-handed systems.

Example 1.12	**Calculating a vector product**

Vector $\vec{A}$ has magnitude 6 units and is in the direction of the +x-axis. Vector $\vec{B}$ has magnitude 4 units and lies in the xy-plane, making an angle of 30° with the +x-axis (Fig. 1.25). Find the vector product $\vec{A} \times \vec{B}$.

SOLUTION

IDENTIFY and SET UP: We can find the vector product in one of two ways. The first way is to use Eq. (1.22) to determine the magnitude of $\vec{A} \times \vec{B}$ and then use the right-hand rule to find the direction of the vector product. The second way is to use the components of $\vec{A}$ and $\vec{B}$ to find the components of the vector product $\vec{C} = \vec{A} \times \vec{B}$ using Eqs. (1.27).

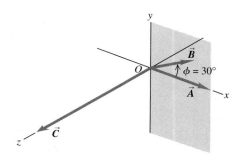

1.25 Vectors $\vec{A}$ and $\vec{B}$ and their vector product $\vec{C} = \vec{A} \times \vec{B}$. The vector $\vec{B}$ lies in the xy-plane.

EXECUTE: With the first approach, from Eq. (1.22) the magnitude of the vector product is

$$AB \sin \phi = (6)(4)(\sin 30°) = 12$$

From the right-hand rule, the direction of $\vec{A} \times \vec{B}$ is along the +z-axis, so we have $\vec{A} \times \vec{B} = 12\hat{k}$.

To use the second approach, we first write the components of $\vec{A}$ and $\vec{B}$:

$A_x = 6$	$A_y = 0$	$A_z = 0$
$B_x = 4 \cos 30° = 2\sqrt{3}$	$B_y = 4 \sin 30° = 2$	$B_z = 0$

Defining $\vec{C} = \vec{A} \times \vec{B}$, we have from Eqs. (1.27) that

$$C_x = (0)(0) - (0)(2) = 0$$
$$C_y = (0)(2\sqrt{3}) - (6)(0) = 0$$
$$C_z = (6)(2) - (0)(2\sqrt{3}) = 12$$

The vector product $\vec{C}$ has only a z-component, and it lies along the +z-axis. The magnitude agrees with the result we obtained with the first approach, as it should.

EVALUATE: For this example the first approach was more direct because we knew the magnitudes of each vector and the angle between them, and furthermore, both vectors lay in one of the planes of the coordinate system. But often you will need to find the vector product of two vectors that are not so conveniently oriented or for which only the components are given. In such a case the second approach, using components, is the more direct one.

Test Your Understanding

For the two vectors $\vec{A} = 3\hat{\imath} + 2\hat{\jmath}$ and $\vec{B} = 4\hat{\imath} + 5\hat{k}$, find the scalar product $\vec{A} \cdot \vec{B}$ and the vector product $\vec{A} \times \vec{B}$.

SUMMARY

The fundamental physical quantities of mechanics are mass, length, and time. The corresponding basic SI units are the kilogram, the meter, and the second. Other units for these quantities, related by powers of 10, are identified by adding prefixes to the basic unit. Derived units for other physical quantities are products or quotients of the basic units. Equations must be dimensionally consistent; two terms can be added only when they have the same units. (See Examples 1.1 and 1.2)

The accuracy of a measurement can be indicated by the number of significant figures or by a stated uncertainty. The result of a calculation usually has no more significant figures than the input data. When only crude estimates are available for input data, we can often make useful order-of-magnitude estimates. (See Examples 1.3 and 1.4)

Scalar quantities are numbers, and combine with the usual rules of arithmetic. Vector quantities have direction as well as magnitude, and combine according to the rules of vector addition. Graphically, two vectors $\vec{A}$ and $\vec{B}$ are added by placing the tail of $\vec{B}$ at the head, or tip, of $\vec{A}$. The vector sum $\vec{A} + \vec{B}$ then extends from the tail of $\vec{A}$ to the head of $\vec{B}$. (See Example 1.5)

Vector addition can be carried out using components of vectors. The x-component of $\vec{R} = \vec{A} + \vec{B}$ is the sum of the x-components of $\vec{A}$ and $\vec{B}$, and likewise for the y- and z-components. (See Examples 1.6 and 1.7)

$$R_x = A_x + B_x$$
$$R_y = A_y + B_y$$
$$R_z = A_z + B_z \qquad (1.10)$$

Unit vectors describe directions in space. A unit vector has a magnitude of one, with no units. The unit vectors $\hat{\imath}$, $\hat{\jmath}$, and $\hat{k}$, aligned with the x-, y-, and z-axes of a rectangular coordinate system, are especially useful. (See Example 1.9)

$$\vec{A} = A_x\hat{\imath} + A_y\hat{\jmath} + A_z\hat{k} \qquad (1.16)$$

The scalar product $C = \vec{A} \cdot \vec{B}$ of two vectors $\vec{A}$ and $\vec{B}$ is a scalar quantity. It can be expressed in two ways: in terms of the magnitudes of $\vec{A}$ and $\vec{B}$ and the angle ϕ between the two vectors, or in terms of the components of $\vec{A}$ and $\vec{B}$. The scalar product is commutative; for any two vectors $\vec{A}$ and $\vec{B}$, $\vec{A} \cdot \vec{B} = \vec{B} \cdot \vec{A}$. The scalar product of two perpendicular vectors is zero. (See Examples 1.10 and 1.11)

$$\vec{A} \cdot \vec{B} = AB \cos \phi = |\vec{A}||\vec{B}| \cos \phi \qquad (1.18)$$
$$\vec{A} \cdot \vec{B} = A_x B_x + A_y B_y + A_z B_z \qquad (1.21)$$

The vector product $\vec{C} = \vec{A} \times \vec{B}$ of two vectors $\vec{A}$ and $\vec{B}$ is another vector $\vec{C}$. The magnitude of $\vec{A} \times \vec{B}$ depends on the magnitudes of $\vec{A}$ and $\vec{B}$ and the angle ϕ between the two vectors. The direction of $\vec{A} \times \vec{B}$ is perpendicular to the plane of the two vectors being multiplied, as given by the right-hand rule. The components of $\vec{C} = \vec{A} \times \vec{B}$ can be expressed in terms of the components of $\vec{A}$ and $\vec{B}$. The vector product is not commutative; for any two vectors $\vec{A}$ and $\vec{B}$, $\vec{A} \times \vec{B} = -\vec{B} \times \vec{A}$. The vector product of two parallel or antiparallel vectors is zero. (See Example 1.12)

$$C = AB \sin\phi \qquad (1.22)$$

$$C_x = A_y B_z - A_z B_y$$

$$C_y = A_z B_x - A_x B_z$$

$$C_z = A_x B_y - A_y B_x \qquad (1.27)$$

Key Terms

Your Notes

Answer to Chapter Opening Question

Figure 1.3e shows that a typical human blood cell is about 8 μm in diameter. Twelve or thirteen such cells laid side by side would span a distance of 100 μm $= 10^{-4}$ m.

Answers to Test Your Understanding Questions

Section 1.5 Density $= (1.80 \text{ kg})/(6.0 \times 10^{-4} \text{ m}^3) =$ 3.0×10^3 kg/m^3. When multiplying or dividing, the number with the fewest significant figures controls the number of significant figures in the result.

Section 1.6 The answer depends on how many students are enrolled at your campus.

Section 1.7 When adding two vectors together, the order of the vectors doesn't matter. Hence the resultant displacement would be the same as in Example 1.5 (magnitude 2.24 km, direction 63.4° east of north).

Section 1.8 You are 5.66 km from your starting point in a direction 45.0° south of west.

Section 1.9
$\vec{A} = (38.37\hat{i} + 61.40\hat{j})$m, $\vec{B} = (-46.36\hat{i} - 33.68\hat{j})$m,
$\vec{C} = (-17.80\hat{j})$m

Section 1.10 $\vec{A} \cdot \vec{B} = (3)(4) + (2)(0) + (0)(5) = 12$
$$\vec{A} \times \vec{B} = [(2)(5) - (0)(0)]\hat{i} + [(0)(4) - (3)(5)]\hat{j}$$
$$+ [(3)(0) - (2)(4)]\hat{k}$$
$$= 10\hat{i} - 15\hat{j} - 8\hat{k}$$

Discussion Questions

Q1.1 How many correct experiments do we need to disprove a theory? How many to prove a theory? Explain.

Q1.2 A guidebook describes the rate of climb of a mountain trail as 120 meters per kilometer. How can you express this as a number with no units?

Q1.3 Suppose you are asked to compute the tangent of 5.00 meters. Is this possible? Why or why not?

Q1.4 A highway contractor stated that in building a bridge deck he poured 250 yards of concrete. What do you think he meant?

Q1.5 What is your height in centimeters? What is your weight in newtons?

Q1.6 The U.S. National Institute of Science and Technology (NIST) maintains several accurate copies of the international standard kilogram. Even after careful cleaning, these national standard kilograms are gaining mass at an average rate of about 1 μg/y (1 y $=$ 1 year) when compared every ten years or so to the standard international kilogram. Does this apparent change have any importance? Explain.

Q1.7 What physical phenomena (other than a pendulum or cesium clock) could you use to define a time standard?

Q1.8 Describe how you could measure the thickness of a sheet of paper with an ordinary ruler.

Q1.9 The quantity $\pi = 3.14159 \ldots$ is a number with no dimensions, since it is a ratio of two lengths. Describe two or three other geometrical or physical quantities that are dimensionless.

Q1.10 What are the units of volume? Suppose another student tells you that a cylinder of radius r and height h has volume given by $\pi r^3 h$. Explain why this cannot be right.

Q1.11 Three archers each fire four arrows at a target. Joe's four arrows hit at points 10 cm above, 10 cm below, 10 cm to the left, and 10 cm to the right of the center of the target. All four of Moe's arrows hit within 1 cm of a point 20 cm from the center, and Flo's four arrows all hit within 1 cm of the center. The contest judge says that one of the archers is precise but not accurate, another archer is accurate but not precise, and the third archer is both accurate and precise. Which description goes with which archer? Explain your reasoning.

Q1.12 A circular racetrack has a radius of 500 m. What is the displacement of a bicyclist when she travels around the track from the north side to the south side? When she makes one complete circle around the track? Explain your reasoning.

Q1.13 Can you find two vectors with different lengths that have a vector sum of zero? What length restrictions are required for three vectors to have a vector sum of zero? Explain your reasoning.

Q1.14 One sometimes speaks of the "direction of time," evolving from past to future. Does this mean that time is a vector quantity? Explain your reasoning.

Q1.15 Air traffic controllers give instructions to airline pilots telling them in which direction they are to fly. These instructions are called "vectors." If these are the only instructions given, is the name "vector" used correctly? Why or why not?

Q1.16 Can you find a vector quantity that has a magnitude of zero, but components that are different from zero? Explain. Can the magnitude of a vector be less than the magnitude of any of its components? Explain.

Q1.17 (a) Does it make sense to say that a vector is *negative*? Why? (b) Does it make sense to say that one vector is the negative of another? Why? Does your answer here contradict what you said in part (a)?

Q1.18 If $\vec{C}$ is the vector sum of $\vec{A}$ and $\vec{B}$, $\vec{C} = \vec{A} + \vec{B}$, what must be true if $C = A + B$? What must be true if $C = 0$?

Q1.19 If $\vec{A}$ and $\vec{B}$ are nonzero vectors, is it possible for $\vec{A} \cdot \vec{B}$ and $\vec{A} \times \vec{B}$ *both* to be zero? Explain.

Q1.20 What does $\vec{A} \cdot \vec{A}$, the scalar product of a vector with itself, give? What about $\vec{A} \times \vec{A}$, the vector product of a vector with itself?

Q1.21 Let $\vec{A}$ represent any nonzero vector. Why is $\vec{A}/A$ a unit vector and what is its direction? If θ is the angle that $\vec{A}$ makes with the $+x$-axis, explain why $(\vec{A}/A) \cdot \hat{i}$ is called the *direction cosine* for that axis.

Q1.22 Which of the following are legitimate mathematical operations: a) $\vec{A} \cdot (\vec{B} - \vec{C})$; b) $(\vec{A} - \vec{B}) \times \vec{C}$; c) $\vec{A} \cdot (\vec{B} \times \vec{C})$; d) $\vec{A} \times (\vec{B} \times \vec{C})$; e) $\vec{A} \times (\vec{B} \cdot \vec{C})$? In each case, give the reason for your answer.

Q1.23 Consider the two repeated vector products $\vec{A} \times (\vec{B} \times \vec{C})$ and $(\vec{A} \times \vec{B}) \times \vec{C}$. Give an example that illustrates the general rule that these two vector products do not have the same magnitude or direction. Can you choose the vectors $\vec{A}$, $\vec{B}$, and $\vec{C}$ such that these two vector products *are* equal to each other? If so, give an example.

Exercises

Section 1.3 Standards and Units
Section 1.4 Unit Consistency and Conversions

1.1 Starting with the definition 1 in. = 2.54 cm, find the number of kilometers in 1.00 mile.

1.2 According to the label on a bottle of salad dressing, the volume of the contents is 0.473 liter (L). Using only the conversions 1 L = 1000 cm^3 and 1 in. = 2.54 cm, express this volume in cubic inches.

1.3 How many nanoseconds does it take light to travel 1.00 km in vacuum?

1.4 The density of lead is 11.3 g/cm^3. What is this value in kilograms per cubic meter?

1.5 The most powerful engine available for the classic 1963 Chevrolet Corvette Sting Ray developed 360 horsepower and had a displacement of 327 cubic inches. Express this displacement in liters (L) by using only the conversions 1 L = 1000 cm^3 and 1 in. = 2.54 cm.

1.6 Told that he needs to set goals for himself, Billy Joe Bob decides to drink one cubic meter of his favorite beverage during the coming year. How many 16-fluid-ounce bottles should he drink each day? (Use Appendix E. A fluid ounce is a volume unit; 128 fluid ounces equals one gallon.)

1.7 The Concorde is the fastest airliner used for commercial service. It cruises at 1450 mi/h (about two times the speed of sound, or Mach 2). a) What is the cruise speed of the Concorde in km/h? b) What is it in m/s?

1.8 While driving in an exotic foreign land you see a speed limit sign on a highway that reads 180,000 furlongs per fortnight. How many miles per hour is this? (One furlong is $\frac{1}{8}$ mile, and a fortnight is 14 days. A furlong originally referred to the length of a plowed furrow.)

1.9 The gasoline consumption of a small car is advertised as 15.0 km/L (1 L = 1 liter). How many miles per gallon is this? Use the conversion factors in Appendix E.

1.10 The following conversions occur frequently in physics and are very useful. a) Use 1 mi = 5280 ft and 1 h = 3600 s to convert 60 mph to units of ft/s. b) The acceleration of a freely falling object is 32 ft/s^2. Use 1 ft = 30.48 cm to express this acceleration in units of m/s^2. c) The density of water is 1.0 g/cm^3. Convert this density to units of kg/m^3.

1.11 Neptunium. In the fall of 2002, a group of scientists at Los Alamos National Laboratory determined that the critical mass of neptunium–237 is about 60 kg. The critical mass of a fissionable material is the minimum amount that must be brought together to start a chain reaction. This element has a density of 19.5 g/cm^3. What would be the radius of a sphere of this material that has a critical mass?

Section 1.5 Uncertainty and Significant Figures

1.12 A useful and easy-to-remember approximate value for the number of seconds in a year is $\pi \times 10^7$. Determine the percent error in this approximate value. (There are 365.24 days in one year.)

1.13 Figure 1.5 shows the result of unacceptable error in the stopping position of a train. a) If a train travels 890 km from Berlin to Paris and then overshoots the end of the track by 10 m, what is the percentage error in the total distance covered? b) Would it be correct to write the total distance covered by the train as 890,010 m? Explain.

1.14 With a wooden ruler you measure the length of a rectangular piece of sheet metal to be 12 mm. You use micrometer calipers to measure the width of the rectangle and obtain the value 5.98 mm. Give your answers to the following questions to the correct number of significant figures. a) What is the area of the rectangle? b) What is the ratio of the rectangle's width to its length? c) What is the perimeter of the rectangle? d) What is the difference between the length and width? e) What is the ratio of the length to the width?

1.15 Estimate the percent error in measuring a) a distance of about 75 cm with a meter stick; b) a mass of about 12 g with a chemical balance; c) a time interval of about 6 min with a stopwatch.

1.16 A rectangular piece of aluminum is 5.10 ± 0.01 cm long and 1.90 ± 0.01 cm wide. a) Find the area of the rectangle and the uncertainty in the area. b) Verify that the fractional uncertainty in the area is equal to the sum of the fractional uncertainties in the length and in the width. (This is a general result; see Challenge Problem 1.94.)

1.17 As you eat your way through a bag of chocolate chip cookies, you observe that each cookie is a circular disk with a diameter of 8.50 ± 0.02 cm and a thickness of 0.050 ± 0.005 cm. a) Find the average volume of a cookie and the uncertainty in the volume. b) Find the ratio of the diameter to the thickness and the uncertainty in this ratio.

Section 1.6 Estimates and Orders of Magnitude

1.18 How many gallons of gasoline are used in the United States in one day?

1.19 A box of typewriter paper has dimensions 11 in. × 17 in. × 9 in.; it is marked "10 M." Does that mean it contains 10,000 sheets or 10 million?

1.20 How many kernels of corn does it take to fill a 2-L soft drink bottle?

1.21 How many words are there in this book?

1.22 What total volume of air does a person breathe in a lifetime? How does that compare with the volume of the Houston Astrodome? (Estimate that a person breathes about 500 cm^3 of air with each breath.)

1.23 How many hairs do you have on your head?

1.24 How many times does a human heart beat during a lifetime? How many gallons of blood does it pump? (Estimate that the heart pumps 50 cm^3 of blood with each beat.)

1.25 In Wagner's opera *Ring of the Nibelung*, the goddess Freia is ransomed for a pile of gold just tall enough and wide enough to hide her from sight. Estimate the monetary value of this pile. (Refer to Example 1.4 for information on the price per ounce and density of gold.)

1.26 How many drops of water are in all the oceans on earth?

1.27 How many pizzas are consumed each academic year by students at your school?

1.28 How many dollar bills would you have to stack to reach the moon? Would that be cheaper than building and launching a spacecraft?

1.29 How much would it cost to paper the entire United States (including Alaska and Hawaii) with dollar bills? What would be the cost to each person in the United States?

Section 1.7 Vectors and Vector Addition

1.30 Hearing rattles from a snake, you make two rapid displacements of magnitude 1.8 m and 2.4 m. In sketches (roughly to scale), show how your two displacements might add up to give a resultant of magnitude a) 4.2 m; b) 0.6 m; c) 3.0 m.

1.31 A postal employee drives a delivery truck along the route shown in Fig. 1.26. Determine the magnitude and direction of the resultant displacement by drawing a scale diagram. (See also Exercise 1.38 for a different approach to this same problem.)

Figure 1.26 Exercises 1.31 and 1.38.

1.32 For the vectors $\vec{A}$ and $\vec{B}$ in Fig. 1.27, use a scale drawing to find the magnitude and direction of a) the vector sum $\vec{A} + \vec{B}$; b) the vector difference $\vec{A} - \vec{B}$. From your answers to parts (a) and (b), find the magnitude and direction of c) $-\vec{A} - \vec{B}$; d) $\vec{B} - \vec{A}$. (See also Exercise 1.39 for a different approach to this problem.)

1.33 A spelunker is surveying a cave. She follows a passage 180 m straight west, then 210 m in a direction 45° east of south, and then 280 m at 30° east of north. After a fourth unmeasured displacement she finds herself back where she started. Use a scale drawing to determine the magnitude and direction of the fourth displacement. (See also Problem 1.69 for a different approach to this problem.)

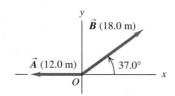

Figure 1.27 Exercises 1.32, 1.39, 1.44, and 1.54.

Section 1.8 Components of Vectors

1.34 Use a scale drawing to find the x- and y-components of the following vectors. For each vector the numbers given are i) the magnitude of the vector and ii) the angle, measured in the sense from the +x-axis toward the +y-axis, that it makes with the +x-axis. Find a) magnitude 9.30 m, angle 60.0°; b) magnitude 22.0 km, angle 135°; c) magnitude 6.35 cm, angle 307°.

1.35 Compute the x- and y-components of the vectors $\vec{A}, \vec{B}$, and $\vec{C}$ in Fig. 1.28.

1.36 Let the angle θ be the angle that the vector $\vec{A}$ makes with the +x-axis, measured counterclockwise from that axis. Find the angle θ for a vector that has the following components: a) $A_x = 2.00$ m, $A_y = -1.00$ m; b) $A_x = 2.00$ m, $A_y = 1.00$ m; c) $A_x = -2.00$ m, $A_y = 1.00$ m; d) $A_x = -2.00$ m, $A_y = -1.00$ m.

1.37 A rocket fires two engines simultaneously. One produces a thrust of 725 N directly forward while the other gives a 513 N thrust at 32.4° above the forward direction. Find the magnitude and direction (relative to the forward direction) of the resultant force which these engines exert on the rocket.

1.38 A postal employee drives a delivery truck over the route shown in Fig. 1.26. Use the method of components to determine the magnitude and direction of her resultant displacement. In a vector addition diagram (roughly to scale), show that the resultant displacement found from your diagram is in qualitative agreement with the result you obtained using the method of components.

1.39 For the vectors $\vec{A}$ and $\vec{B}$ in Fig. 1.27, use the method of components to find the magnitude and direction of a) the vector sum $\vec{A} + \vec{B}$; b) the vector sum $\vec{B} + \vec{A}$; c) the vector difference $\vec{A} - \vec{B}$; d) the vector difference $\vec{B} - \vec{A}$.

1.40 Find the magnitude and direction of the vector represented by the following pairs of components: a) $A_x = -8.60$ cm, $A_y = 5.20$ cm; b) $A_x = -9.70$ m, $A_y = -2.45$ m; c) $A_x = 7.75$ km, $A_y = -2.70$ km.

1.41 A disoriented physics professor drives 3.25 km north, then 4.75 km west, and then 1.50 km south. Find the magnitude and direction of the resultant displacement, using the method of components. In a vector addition diagram (roughly to scale), show that the resultant displacement found from your diagram is in qualitative agreement with the result you obtained using the method of components.

1.42 Vector $\vec{A}$ has components $A_x = 1.30$ cm, $A_y = 2.25$ cm; vector $\vec{B}$ has components $B_x = 4.10$ cm, $B_y = -3.75$ cm. Find a) the components of the vector sum $\vec{A} + \vec{B}$; b) the magnitude and direction of $\vec{A} + \vec{B}$; c) the components of the vector difference $\vec{B} - \vec{A}$; d) the magnitude and direction of $\vec{B} - \vec{A}$.

1.43 Vector $\vec{A}$ is 2.80 cm long and is 60.0° above the x-axis in the first quadrant. Vector $\vec{B}$ is 1.90 cm long and is 60.0° below the x-axis in the fourth quadrant (Fig. 1.29). Find the magnitude

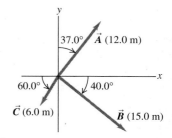

Figure 1.28 Exercises 1.35, 1.45, and 1.50, and Problem 1.68.

Figure 1.29 Exercises 1.43 and 1.56.

and direction of a) $\vec{A} + \vec{B}$; b) $\vec{A} - \vec{B}$; c) $\vec{B} - \vec{A}$. In each case, sketch the vector addition or subtraction and show that your numerical answers are in qualitative agreement with your sketch.

Section 1.9 Unit Vectors

1.44 Write each vector in Fig. 1.27 in terms of the unit vectors $\hat{\imath}$ and $\hat{\jmath}$.

1.45 Write each vector in Fig. 1.28 in terms of the unit vectors $\hat{\imath}$ and $\hat{\jmath}$.

1.46 a) Write each vector in Fig. 1.30 in terms of the unit vectors $\hat{\imath}$ and $\hat{\jmath}$. b) Use unit vectors to express the vector $\vec{C}$, where $\vec{C} = 3.00\vec{A} - 4.00\vec{B}$. c) Find the magnitude and direction of $\vec{C}$.

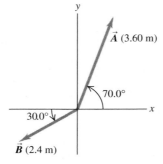

Figure 1.30 Exercise 1.46 and Problem 1.82.

1.47 Given two vectors $\vec{A} = 4.00\hat{\imath} + 3.00\hat{\jmath}$ and $\vec{B} = 5.00\hat{\imath} - 2.00\hat{\jmath}$, a) find the magnitude of each vector; b) write an expression for the vector difference $\vec{A} - \vec{B}$ using unit vectors; c) find the magnitude and direction of the vector difference $\vec{A} - \vec{B}$. d) In a vector diagram show $\vec{A}$, $\vec{B}$, and $\vec{A} - \vec{B}$, and also show that your diagram agrees qualitatively with your answer in part (c).

1.48 a) Is the vector $(\hat{\imath} + \hat{\jmath} + \hat{k})$ a unit vector? Justify your answer. b) Can a unit vector have any components with magnitude greater than unity? Can it have any negative components? In each case justify your answer. c) If $\vec{A} = a(3.0\hat{\imath} + 4.0\hat{\jmath})$, where a is a constant, determine the value of a that makes $\vec{A}$ a unit vector.

1.49 a) Use vector components to prove that two vectors commute for both addition and the scalar product. b) Use vector components to prove that two vectors anticommute for the vector product. That is, prove that $\vec{A} \times \vec{B} = -\vec{B} \times \vec{A}$.

Section 1.10 Products of Vectors

1.50 For the vectors $\vec{A}$, $\vec{B}$, and $\vec{C}$ in Fig. 1.28, find the scalar products a) $\vec{A} \cdot \vec{B}$; b) $\vec{B} \cdot \vec{C}$; c) $\vec{A} \cdot \vec{C}$.

1.51 a) Find the scalar product of the two vectors $\vec{A}$ and $\vec{B}$ given in Exercise 1.47. b) Find the angle between these two vectors.

1.52 Find the angle between each of the following pairs of vectors:

a) $\vec{A} = -2.00\hat{\imath} + 6.00\hat{\jmath}$ and $\vec{B} = 2.00\hat{\imath} - 3.00\hat{\jmath}$
b) $\vec{A} = 3.00\hat{\imath} + 5.00\hat{\jmath}$ and $\vec{B} = 10.00\hat{\imath} + 6.00\hat{\jmath}$
c) $\vec{A} = -4.00\hat{\imath} + 2.00\hat{\jmath}$ and $\vec{B} = 7.00\hat{\imath} + 14.00\hat{\jmath}$

1.53 Assuming a right-handed coordinate system, find the direction of the $+z$-axis in a) Fig. 1.15a; b) Fig. 1.15b.

1.54 For the two vectors in Fig. 1.27, a) find the magnitude and direction of the vector product $\vec{A} \times \vec{B}$; b) find the magnitude and direction of $\vec{B} \times \vec{A}$.

1.55 Find the vector product $\vec{A} \times \vec{B}$ (expressed in unit vectors) of the two vectors given in Exercise 1.47. What is the magnitude of the vector product?

1.56 For the two vectors in Fig. 1.29, a) find the magnitude and direction of the vector product $\vec{A} \times \vec{B}$; b) find the magnitude and direction of $\vec{B} \times \vec{A}$.

Problems

1.57 An acre, a unit of land measurement still in wide use, has a length of one furlong ($\frac{1}{8}$ mi) and a width one-tenth of its length. a) How many acres are in a square mile? b) How many square feet are in an acre? See Appendix E. c) An acre-foot is the volume of water that would cover one acre of flat land to a depth of one foot. How many gallons are in an acre-foot?

1.58 An estate on the California coast was offered for sale for $4,950,000. The total area of the estate was 102 acres (see Problem 1.57). a) Considering the price of the estate to be proportional to its area, what was the cost of one square meter of the estate? b) What would be the price of a portion of the estate the size of a postage stamp ($\frac{7}{8}$ in. by 1.0 in.)?

1.59 The Hydrogen Maser. You can use the radio waves generated by a hydrogen maser as a standard of frequency. The frequency of these waves is 1,420,405,751.786 hertz. (A hertz is another name for one cycle per second.) A clock controlled by a hydrogen maser is off by only 1 s in 100,000 years. For the following questions, use only three significant figures. (The large number of significant figures given for the frequency simply illustrates the remarkable accuracy to which it has been measured.) a) What is the time for one cycle of the radio wave? b) How many cycles occur in 1 h? c) How many cycles would have occurred during the age of the earth, which is estimated to be 4.6×10^9 years? d) By how many seconds would a hydrogen maser clock be off after a time interval equal to the age of the earth?

1.60 Estimate the number of atoms in your body. (*Hint:* Based on what you know about biology and chemistry, what are the most common types of atom in your body? What is the mass of each type of atom? Appendix D gives the atomic masses for different elements, measured in atomic mass units; you can find the value of an atomic mass unit, or 1 u, in Appendix F.)

1.61 Biological tissues are typically made up of 98% water. Given that the density of water is 1.0×10^3 kg/m^3, estimate the mass of a) the heart of an adult human; b) a cell with a diameter of 0.5 μm; c) a honey bee.

1.62 Iron has a property such that a 1.00 m^3 volume has a mass of 7.86×10^3 kg (density equals 7.86×10^3 kg/m^3). You want to manufacture iron into cubes and spheres. Find a) the length of the side of a cube of iron that has a mass of 200.0 g; b) the radius of a solid sphere of iron that has a mass of 200.0 g.

1.63 a) Estimate the number of dentists in your city. You will need to consider the number of people in your city, how often they need to go to the dentist, how often they actually go, how many hours a typical dental procedure (filling, root canal, and so on) takes, and how many hours a dentist works in a week. b) Using your local telephone directory, check to see whether your estimate was roughly correct.

1.64 Physicists, mathematicians, and others often deal with large numbers. The number 10^{100} has been given the whimsical name *googol* by mathematicians. Let us compare some large numbers in physics with the googol. (*Note:* This problem requires numerical values that you can find in the appendices of the book, with which you should become familiar.) a) Approximately how many atoms make up our planet? For simplicity, assume the average atomic mass of the atoms to be 14 g/mol. Avogadro's number gives the number of atoms in a mole. b) Approximately how many neutrons are in a neutron star? Neutron stars are composed almost entirely of neutrons and have approximately twice the mass of the sun. c) In the leading theory of the origin of the universe, the entire universe that we can now observe occupied, at a very early time, a sphere whose radius was approximately equal to the present distance of the earth to the sun. At that time the universe had a density (mass divided by volume) of 10^{15} g/cm^3. Assuming that $\frac{1}{3}$ of the particles were protons, $\frac{1}{3}$ of the particles were neutrons, and the remaining $\frac{1}{3}$ were electrons, how many particles then made up the universe?

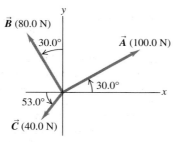

Figure 1.31 Problem 1.65.

1.65 Three horizontal ropes pull on a large stone stuck in the ground, producing the vector forces $\vec{A}$, $\vec{B}$ and $\vec{C}$ shown in Fig. 1.31. Find the magnitude and direction of a fourth force on the stone that will make the vector sum of the four forces zero.

1.66 Emergency Landing. A plane leaves the airport in Galisto and flies 170 km at 68° east of north and then changes direction to fly 230 km at 48° south of east, after which it makes an immediate emergency landing in a pasture. When the airport sends out a rescue crew, in which direction and how far should this crew fly to go directly to this plane?

1.67 You are to program a robotic arm on an assembly line to move in the xy-plane. Its first displacement is $\vec{A}$; its second displacement is $\vec{B}$, of magnitude 6.40 cm and direction 63.0° measured in the sense from the +x-axis toward the −y-axis. The resultant $\vec{C} = \vec{A} + \vec{B}$ of the two displacements should also have a magnitude of 6.40 cm, but a direction 22.0° measured in the sense from the +x-axis toward the +y-axis. a) Draw the vector addition diagram for these vectors, roughly to scale. b) Find the components of $\vec{A}$. c) Find the magnitude and direction of $\vec{A}$.

1.68 a) Find the magnitude and direction of the vector $\vec{R}$ that is the sum of the three vectors $\vec{A}$, $\vec{B}$, and $\vec{C}$ in Fig. 1.28. In a diagram, show how $\vec{R}$ is formed from the three vectors in Fig. 1.28. b) Find the magnitude and direction of the vector $\vec{S} = \vec{C} - \vec{A} - \vec{B}$. In a diagram, show how $\vec{S}$ is formed from the three vectors in Fig. 1.28.

1.69 As noted in Exercise 1.33, a spelunker is surveying a cave. She follows a passage 180 m straight west, then 210 m in a direction 45° east of south, then 280 m at 30° east of north. After a fourth unmeasured displacement she finds herself back where she started. Use the method of components to determine the magnitude and direction of the fourth displacement. Draw the vector addition diagram and show that it is in qualitative agreement with your numerical solution.

1.70 A sailor in a small sailboat encounters shifting winds. She sails 2.00 km east, then 3.50 km southeast, and then an additional distance in an unknown direction. Her final position is 5.80 km directly east of the starting point (Fig. 1.32). Find the magnitude and direction of the third leg of the journey. Draw the vector addi-

tion diagram and show that it is in qualitative agreement with your numerical solution.

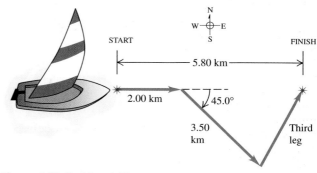

Figure 1.32 Problem 1.70.

1.71 A cross-country skier skis 2.80 km in the direction 45.0° west of south, then 7.40 km in the direction 30.0° north of east, and finally 3.30 km in the direction 22.0° south of west. a) Show these displacements in a diagram. b) How far is the skier from the starting point?

1.72 On a training flight, a student pilot flies from Lincoln, Nebraska to Clarinda, Iowa, then to St. Joseph, Missouri, and then to Manhattan, Kansas (Fig. 1.33). The directions are shown relative to north: 0° is north, 90° is east, 180° is south, and 270° is west. Use the method of components to find a) the distance she has to fly from Manhattan to get back to Lincoln; b) the direction (relative to north) she must

Figure 1.33 Problem 1.72.

fly to get there. Illustrate your solution with a vector diagram.

1.73 A graphic artist is creating a new logo for her company's Web site. In the graphics program she is using, each pixel in an image file has coordinates (x, y), where the origin $(0, 0)$ is at the upper-left corner of the image, the +x-axis points to the right, and the +y-axis points down. Distances are measured in pixels. a) The artist draws a line from the pixel location (10, 20) to the location (210, 200). She wishes to draw a second line that starts at (10, 20), is 250 pixels long, and is at angle of 30° measured clockwise from the first line. At which pixel location should this second line end? Give your answer to the nearest pixel. b) The artist now draws an arrow that connects the lower-right end of the first line to the lower right end of the second line. Find the length and direction of this arrow. Draw a diagram showing all three lines.

1.74 Getting Back. An explorer in the dense jungles of equatorial Africa leaves his hut. He takes 40 steps northeast, then 80 steps 60° north of west, then 50 steps due south. Assume his steps all have equal length. a) Sketch, roughly to scale, the three vectors and their resultant. b) Save him from becoming hopelessly lost in the jungle by giving him the displacement, calculated using the method of components, that will return him to his hut.

1.75 A ship leaves the island of Guam and sails 285 km at 40.0° north of west. In which direction must it now head and how far must it sail so that its resultant displacement will be 115 km directly east of Guam?

1.76 A boulder of weight w rests on a hillside that rises at a constant angle α above the horizontal, as shown in Fig. 1.34. Its weight is a force on the boulder that has direction vertically downward.
a) In terms of α and w, what is the component of the weight of the boulder in the direction parallel to the surface of the hill?
b) What is the component of the weight in the direction perpendicular to the surface of the hill?
c) An air conditioner unit is fastened to a roof that slopes upward at an angle of 35.0°. In order that the unit not slide down the roof, the component of the unit's weight parallel to the roof cannot exceed 550 N. What is the maximum allowed weight of the unit?

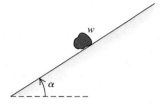

Figure 1.34 Problem 1.76.

1.77 Bones and Muscles. A patient in therapy has a forearm that weighs 20.5 N and that lifts a 112.0-N weight. These two forces have direction vertically downward. The only other significant forces on his forearm come from the biceps muscle (which acts perpendicularly to the forearm) and the force at the elbow. If the biceps produces a pull of 232 N when the forearm is raised 43° above the horizontal, find the magnitude and direction of the force that the elbow exerts on the forearm. (The sum of the elbow force and the biceps force must balance the weight of the arm and the weight it is carrying, so their vector sum must be 132.5 N, upward.)

1.78 You are hungry and decide to go to your favorite neighborhood fast-food restaurant. You leave your apartment and take the elevator 10 flights down (each flight is 3.0 m) and then go 15 m south to the apartment exit. You then proceed 0.2 km east, turn north, and go 0.1 km to the entrance of the restaurant. a) Determine the displacement from your apartment to the restaurant. Use unit-vector notation for your answer, being sure to make clear your choice of coordinates. b) How far did you travel along the path you took from your apartment to the restaurant and what is the magnitude of the displacement you calculated in part (a)?

1.79 You are canoeing on a lake. Starting at your camp on the shore, you travel 240 m in the direction 32° south of east to reach a store to purchase supplies. You know the distance because you have located both your camp and the store on a map of the lake. On the return trip you travel distance B in the direction 48° north of west, distance C in the direction 62° south of west, and then you are back at your camp. You measure the directions of travel with your compass, but you don't know the distances. Since you are curious to know the total distance you rowed, use vector methods to calculate the distances B and C.

1.80 You are camping with two friends, Joe and Karl. Since all three of you like your privacy, you don't pitch your tents close together. Joe's tent is 21.0 m from yours, in the direction 23.0° south of east. Karl's tent is 32.0 m from yours, in the direction 37.0° north of east. What is the distance between Karl's tent and Joe's tent?

1.81 Vectors $\vec{A}$ and $\vec{B}$ are drawn from a common point. Vector $\vec{A}$ has magnitude A and angle θ_A measured in the sense from the $+x$-axis to the $+y$-axis. The corresponding quantities for vector $\vec{B}$ are B and θ_B. Then $\vec{A} = A\cos\theta_A\hat{\imath} + A\sin\theta_A\hat{\jmath}$, $\vec{B} = B\cos\theta_B\hat{\imath} + B\sin\theta_B\hat{\jmath}$, and $\phi = |\theta_B - \theta_A|$ is the angle between $\vec{A}$ and $\vec{B}$.
a) Derive Eq. (1.18) from Eq. (1.21). b) Derive Eq. (1.22) from Eq. (1.27).

1.82 For the two vectors $\vec{A}$ and $\vec{B}$ in Fig. 1.30, a) find the scalar product $\vec{A} \cdot \vec{B}$; b) find the magnitude and direction of the vector product $\vec{A} \times \vec{B}$.

1.83 Figure 1.8c shows a parallelogram based on the two vectors $\vec{A}$ and $\vec{B}$. a) Show that the magnitude of the cross product of these two vectors is equal to the area of the parallelogram. (*Hint:* area = base × height.) b) What is the angle between the cross product and the plane of the parallelogram?

1.84 The vector $\vec{A}$ is 3.50 cm long and is directed into this page. Vector $\vec{B}$ points from the lower-right corner of this page to the upper-left corner of this page. Define an appropriate right-handed coordinate system and find the three components of the vector product $\vec{A} \times \vec{B}$, measured in cm². In a diagram, show your coordinate system and the vectors $\vec{A}$, $\vec{B}$, and $\vec{A} \times \vec{B}$.

1.85 Given two vectors $\vec{A} = -2.00\hat{\imath} + 3.00\hat{\jmath} + 4.00\hat{k}$ and $\vec{B} = 3.00\hat{\imath} + 1.00\hat{\jmath} - 3.00\hat{k}$, do the following. a) Find the magnitude of each vector. b) Write an expression for the vector difference $\vec{A} - \vec{B}$, using unit vectors. c) Find the magnitude of the vector difference $\vec{A} - \vec{B}$. Is this the same as the magnitude of $\vec{B} - \vec{A}$? Explain.

1.86 Bond Angle in Methane. In the methane molecule, CH_4, each hydrogen atom is at a corner of a regular tetrahedron with the carbon atom at the center. In coordinates where one of the C—H bonds is in the direction of $\hat{\imath} + \hat{\jmath} + \hat{k}$, an adjacent C—H bond is in the $\hat{\imath} - \hat{\jmath} - \hat{k}$ direction. Calculate the angle between these two bonds.

1.87 The two vectors $\vec{A}$ and $\vec{B}$ are drawn from a common point, and $\vec{C} = \vec{A} + \vec{B}$. a) Show that if $C^2 = A^2 + B^2$, the angle between the vectors $\vec{A}$ and $\vec{B}$ is 90°. b) Show that if $C^2 < A^2 + B^2$, the angle between the vectors $\vec{A}$ and $\vec{B}$ is greater than 90°. c) Show that if $C^2 > A^2 + B^2$, the angle between the vectors $\vec{A}$ and $\vec{B}$ is between 0° and 90°.

1.88 When two vectors $\vec{A}$ and $\vec{B}$ are drawn from a common point, the angle between them is ϕ. a) Using vector techniques, show that the magnitude of their vector sum is given by

$$\sqrt{A^2 + B^2 + 2AB\cos\phi}$$

b) If $\vec{A}$ and $\vec{B}$ have the same magnitude, for which value of ϕ will their vector sum have the same magnitude as $\vec{A}$ or $\vec{B}$? c) Derive a result analogous to that in part (a) for the magnitude of the vector difference $\vec{A} - \vec{B}$. d) If $\vec{A}$ and $\vec{B}$ have the same magnitude, for what value of ϕ will $\vec{A} - \vec{B}$ have this same magnitude?

1.89 A cube is placed so that one corner is at the origin and three edges are along the x-, y-, and z-axes of a coordinate system (Fig. 1.35). Use vectors to compute a) the angle between the edge along the z-axis (line ab) and the diagonal from the origin to the opposite corner (line ad); b) the angle between line ac (the diagonal of a face) and line ad.

Figure 1.35 Problem 1.89.

1.90 Obtain a *unit vector* perpendicular to the two vectors given in Problem 1.85.

1.91 You are given vectors $\vec{A} = 5.0\hat{\imath} - 6.5\hat{\jmath}$ and $\vec{B} = -3.5\hat{\imath} + 7.0\hat{\jmath}$. A third vector $\vec{C}$ lies in the xy-plane. Vector $\vec{C}$ is perpendicular to vector $\vec{A}$, and the scalar product of $\vec{C}$ with $\vec{B}$ is 15.0. From this information, find the components of vector $\vec{C}$.

1.92 Two vectors $\vec{A}$ and $\vec{B}$ have magnitude $A = 3.00$ and $B = 3.00$. Their vector product is $\vec{A} \times \vec{B} = -5.00\hat{k} + 2.00\hat{\imath}$. What is the angle between $\vec{A}$ and $\vec{B}$?

1.93 Later in our study of physics we will encounter quantities represented by $(\vec{A} \times \vec{B}) \cdot \vec{C}$. a) Prove that for any three vectors $\vec{A}$, $\vec{B}$, and $\vec{C}$, $\vec{A} \cdot (\vec{B} \times \vec{C}) = (\vec{A} \times \vec{B}) \cdot \vec{C}$. b) Calculate $(\vec{A} \times \vec{B}) \cdot \vec{C}$ for the three vectors $\vec{A}$ with magnitude $A = 5.00$ and angle $\theta_A = 26.0°$ measured in the sense from the $+x$-axis toward the $+y$-axis, $\vec{B}$ with $B = 4.00$ and $\theta_B = 63.0°$, and $\vec{C}$ with magnitude 6.00 and in the $+z$-direction. Vectors $\vec{A}$ and $\vec{B}$ are in the xy-plane.

Challenge Problems

1.94 The length of a rectangle is given as $L \pm l$, and its width as $W \pm w$. a) Show that the uncertainty in its area A is $a = Lw + lW$. Assume that the uncertainties l and w are small, so that the product lw is very small and you can ignore it. b) Show that the fractional uncertainty in the area is equal to the sum of the fractional uncertainty in length and the fractional uncertainty in width. c) A rectangular solid has dimensions $L \pm l$, $W \pm w$, and $H \pm h$. Find the fractional uncertainty in the volume, and show that it equals the sum of the fractional uncertainties in the length, width, and height.

1.95 Completed Pass. At Enormous State University (ESU), the football team records its plays using vector displacements, with the origin taken to be the position of the ball before the play starts. In a certain pass play, the receiver starts at $+1.0\hat{\imath} - 5.0\hat{\jmath}$, where the units are yards, $\hat{\imath}$ is to the right, and $\hat{\jmath}$ is downfield. Subsequent displacements of the receiver are $+9.0\hat{\imath}$ (in motion before the snap), $+11.0\hat{\jmath}$ (breaks downfield), $-6.0\hat{\imath} + 4.0\hat{\jmath}$ (zigs), and $+12.0\hat{\imath} + 18.0\hat{\jmath}$ (zags). Meanwhile, the quarterback has dropped straight back to a position $-7.0\hat{\jmath}$. How far and in which direction must the quarterback throw the ball? (Like the coach, you will be well advised to diagram the situation before solving it numerically.)

1.96 Navigating in the Solar System. The *Mars Polar Lander* spacecraft was launched on January 3, 1999. On December 3, 1999, the day that *Mars Polar Lander* touched down on the Martian surface, the positions of the earth and Mars were given by these coordinates:

	x	y	z
Earth	0.3182 AU	0.9329 AU	0.0000 AU
Mars	1.3087 AU	−0.4423 AU	−0.0414 AU

In these coordinates, the sun is at the origin and the plane of the earth's orbit is the xy-plane. The earth passes through the $+x$-axis once a year on the autumnal equinox, the first day of autumn in the northern hemisphere (on or about September 22). One AU, or *astronomical unit,* is equal to 1.496×10^8 km, the average distance from the earth to the sun. a) In a diagram, show the positions of the sun, the earth, and Mars on December 3, 1999. b) Find the following distances in AU on December 3, 1999: i) from the sun to the earth; ii) from the sun to Mars; iii) from the earth to Mars. c) As seen from the earth, what was the angle between the direction to the sun and the direction to Mars on December 3, 1999? d) Explain whether Mars was visible from your location at midnight on December 3, 1999. (When it is midnight at your location, the sun is on the opposite side of the earth from you.)

1.97 Navigating in the Big Dipper. All the stars of the Big Dipper (part of the constellation Ursa Major) may appear to be the same distance from the earth, but in fact they are very far from each other. Figure 1.36 shows the distances from the earth to each of these stars. The distances are given in light years (ly), the distance that light travels in one year. One light year equals 9.461×10^{15} m. a) Alkaid and Merak are 25.6° apart in the earth's sky. In a diagram, show the relative positions of Alkaid, Merak, and our sun. Find the distance in light years from Alkaid to Merak. b) To an inhabitant of a planet orbiting Merak, how many degrees apart in the sky would Alkaid and our sun be?

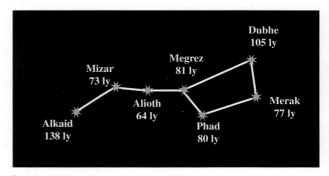

Figure 1.36 Challenge Problem 1.97.

1.98 The vector $\vec{r} = x\hat{\imath} + y\hat{\jmath} + z\hat{k}$, called the *position vector,* points from the origin (0, 0, 0) to an arbitrary point in space with coordinates (x, y, z). Use what you know about vectors to prove the following: All points (x, y, z) that satisfy the equation $Ax + By + Cz = 0$, where A, B, and C are constants, lie in a plane that passes through the origin and that is perpendicular to the vector $A\hat{\imath} + B\hat{\jmath} + C\hat{k}$. Sketch this vector and the plane.

2 MOTION ALONG A STRAIGHT LINE

Every successful space shuttle mission ends with a brief period of straight-line motion as it comes to a stop on the runway. No larger than a typical airliner, the space shuttle orbiter lands at more than 350 km/h (220 mi/h). Even with a drag parachute for braking, the fast-moving orbiter requires up to 3 km (about 2 mi) to stop.

? **Is it accurate to say that the space shuttle orbiter is *accelerating* as it slows to a stop?**

How do you describe the motion of a jet fighter being catapulted down the deck of an aircraft carrier? When you throw a baseball straight up in the air, how high does it go? When a glass slips from your hand, how much time do you have to catch it before it hits the floor? These are the kinds of questions you will learn to answer in this chapter. We are beginning our study of physics with *mechanics,* the study of the relationships among force, matter, and motion. The goal of this chapter and the next is to develop general methods for describing motion. The part of mechanics that deals with the description of motion is called *kinematics.* Later we will study *dynamics,* the relation of motion to its causes.

In this chapter we study the simplest kind of motion: a single particle moving along a straight line. We often use a particle as a model for a moving body when effects such as rotation or change of shape are not important. To describe the motion of a particle, we introduce the physical quantities *velocity* and *acceleration.* These quantities have simple definitions in physics; however, those definitions are more precise and slightly different than the ones used in everyday language. Paying careful attention to how velocity and acceleration are defined will help you work with these and other important physical quantities.

An important part of how a physicist defines velocity and acceleration is that these quantities are *vectors.* As you learned in Chapter 1, this means that they have both magnitude and direction. Our concern in this chapter is with motion along a straight line only, so we won't need the full mathematics of vectors just yet. But in Chapter 3 we'll extend our discussion to include motion in two or three dimensions, and using vectors will be essential.

An important special case of straight-line motion is when the acceleration is constant, a situation that we will encounter many times in our study of physics. An example is the motion of a freely falling body. We'll develop simple equations to describe motion with constant acceleration. We'll also consider situations in which the acceleration varies during the motion; in this case, it's necessary to use integration to describe the motion. (If you haven't studied integration yet, this section is optional.)

2.1 | Displacement, Time, and Average Velocity

Suppose a drag racer drives her AA-fuel dragster along a straight track. To study this motion, we need a coordinate system to describe the dragster's position. We choose the x-axis of our coordinate system to lie along the dragster's straight-line path, with the origin O at the starting line (Fig. 2.1). We will describe the dragster's position in terms of the position of a representative point, such as its front end. By doing this, we represent the entire dragster by that point and so treat the dragster as a **particle.**

A useful way to describe the motion of the front of the dragster—that is, the motion of the particle—is in terms of the change in the particle's position (in other words, the change in its coordinate x) over a time interval. Suppose that 1.0 s after the start the front of the dragster is at point P_1, 19 m from the origin, and 4.0 s after the start it is at point P_2, 277 m from the origin. The *displacement* of the particle is a vector that points from P_1 to P_2 (see Section 1.7). Figure 2.1 shows that this vector points along the x-axis. The x-component of the displacement is just the change in the value of x, $(277\ \text{m} - 19\ \text{m}) = 258\ \text{m}$, that took place during the time interval of $(4.0\ \text{s} - 1.0\ \text{s}) = 3.0$ s. We define the dragster's **average velocity** during this time interval as a *vector* quantity whose x-component is the change in x divided by the time interval: $(258\ \text{m})/(3.0\ \text{s}) = 86$ m/s. In general, the average velocity depends on the particular time interval chosen. For a 3.0-s time interval *before* the start of the race, the average velocity would be zero, because the dragster would be at rest at the starting line and would have zero displacement.

Let's generalize the concept of average velocity. At time t_1 the dragster is at point P_1, with coordinate x_1, and at time t_2 it is at point P_2, with coordinate x_2. The displacement of the dragster during the time interval from t_1 to t_2 is the vector from P_1 to P_2, with x-component $(x_2 - x_1)$ and with y- and z-components equal to zero. The x-component of the dragster's displacement is just the change in the coordinate x, which we write in a more compact way as

$$\Delta x = x_2 - x_1 \tag{2.1}$$

2.1 Positions of an AA-fuel dragster at two times during its run.

> **CAUTION** Note that Δx is *not* the product of Δ and x; it is a single symbol that means "the change in the quantity x." We will always use the Greek capital letter Δ ("delta") to represent a *change* in a quantity, calculated by taking the *final* value of the quantity minus the *initial* value—never the reverse. Likewise, the time interval from t_1 to t_2 is Δt, the change in the quantity t: $\Delta t = t_2 - t_1$ (final time minus initial time).

We can now define the x-component of average velocity more precisely: it is the x-component of displacement, Δx, divided by the time interval Δt during which the displacement occurs. We represent this quantity by the symbol $v_{\text{av-}x}$, where the subscript "av" signifies average value and the subscript x indicates that this is the x-component:

$$v_{\text{av-}x} = \frac{x_2 - x_1}{t_2 - t_1} = \frac{\Delta x}{\Delta t} \qquad \text{(average velocity, straight-line motion)} \quad (2.2)$$

For the above example we had $x_1 = 19$ m, $x_2 = 277$ m, $t_1 = 1.0$ s, and $t_2 = 4.0$ s, so Eq. (2.2) gives

$$v_{\text{av-}x} = \frac{277 \text{ m} - 19 \text{ m}}{4.0 \text{ s} - 1.0 \text{ s}} = \frac{258 \text{ m}}{3.0 \text{ s}} = 86 \text{ m/s}$$

The average velocity of the dragster is positive. This means that during the time interval, the coordinate x increased and the dragster moved in the positive x-direction (to the right in Fig. 2.1). If a particle moves in the *negative x-direction* during a time interval, its average velocity for that time interval is negative. For example, suppose an official's truck moves to the left along the track (Fig. 2.2). The truck is at $x_1 = 277$ m at $t_1 = 16.0$ s and is at $x_2 = 19$ m at $t_2 = 25.0$ s. Then $\Delta x = (19 \text{ m} - 277 \text{ m}) = -258$ m and $\Delta t = (25.0 \text{ s} - 16.0 \text{ s}) = 9.0$ s. The x-component of average velocity is $v_{\text{av-}x} = \Delta x/\Delta t = (-258 \text{ m})/(9.0 \text{ s}) = -29$ m/s. Whenever x is positive and increasing or is negative and becoming less negative, the particle is moving in the $+x$-direction and $v_{\text{av-}x}$ is positive (Fig. 2.1). Whenever x is positive and decreasing or is negative and becoming more negative, the particle is moving in the $-x$-direction and $v_{\text{av-}x}$ is negative (Fig. 2.2).

> **CAUTION** You might be tempted to conclude that positive average velocity necessarily implies motion to the right, as in Fig. 2.1, and that negative average velocity necessarily implies motion to the left, as in Fig. 2.2. But these conclusions are correct *only* if the positive x-direction is to the right, as we have cho-

START   FINISH

2.2 Positions of an official's truck at two times during its motion. The points P_1 and P_2 now refer to the motion of the truck, and so are different than in Fig. 2.1. The x-component of the truck's displacement is negative, so $v_{\text{av-}x}$ is negative.

sen it to be in Figs. 2.1 and 2.2. We could just as well have chosen the positive *x*-direction to be to the left, with the origin at the finish line. Then the dragster would have negative average velocity and the official's truck would have positive average velocity. In most problems the direction of the coordinate axis will be yours to choose. Once you've made your choice, you *must* take it into account when interpreting the signs of $v_{av\text{-}x}$ and other quantities that describe motion!

With straight-line motion we will usually call Δx simply the displacement and $v_{av\text{-}x}$ simply the average velocity. But be sure to remember that these are really the *x*-components of vector quantities that, in this special case, have *only* *x*-components. In Chapter 3, displacement, velocity, and acceleration vectors will have two or three nonzero components.

Figure 2.3 is a graph of the dragster's position as a function of time, that is, an ***x-t* graph.** The curve in the figure *does not* represent the dragster's path in space; as Fig. 2.1 shows, the path is a straight line. Rather, the graph is a pictorial way to represent how the dragster's position changes with time. The points on the graph labeled p_1 and p_2 correspond to the points P_1 and P_2 along the dragster's path. Line p_1p_2 is the hypotenuse of a right triangle with vertical side $\Delta x = x_2 - x_1$ and horizontal side $\Delta t = t_2 - t_1$. The average velocity $v_{av\text{-}x} = \Delta x/\Delta t$ of the dragster is then equal to the *slope* of the line p_1p_2, that is, the ratio of the vertical side Δx of the triangle to the horizontal side Δt.

Average velocity depends only on the total displacement $\Delta x = x_2 - x_1$ that occurs during the time interval $\Delta t = t_2 - t_1$, not on the details of what happens during the time interval. A second dragster might have passed through point P_1 in Fig. 2.1 at the same time t_1 as the first dragster, raced past the first dragster, then blown its engine and slowed down to pass through point P_2 at the same time t_2 as the first dragster. Both dragsters have the same displacement during the same time interval and so have the same average velocity.

If we express distance in meters and time in seconds, average velocity is measured in meters per second (m/s). Other common units of velocity are kilometers per hour (km/h), feet per second (ft/s), miles per hour (mi/h), and knots $(1 \text{ knot} = 1 \text{ nautical mile/h} = 6080 \text{ ft/h})$. Table 2.1 shows a few typical velocity magnitudes.

Table 2.1 Typical Velocity Magnitudes

A snail's pace	10^{-3} m/s	Random motion of air molecules	500 m/s
A brisk walk	2 m/s	Fastest airplane	1000 m/s
Fastest human	11 m/s	Orbiting communications satellite	3000 m/s
Running cheetah	35 m/s	Electron orbiting in a hydrogen atom	2×10^6 m/s
Fastest car	341 m/s	Light traveling in a vacuum	3×10^8 m/s

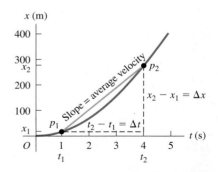

2.3 The position of an AA-fuel dragster as a function of time. The average velocity $v_{av\text{-}x}$ between points P_1 and P_2 in Fig. 2.1 is the slope of the line p_1p_2. This line slopes upward to the right, so the slope is positive and $v_{av\text{-}x}$ is positive.

Test Your Understanding

A truck drives due west from point *A* to point *B,* a distance of 60 km. A car drives due east from point *A* to point *C,* a distance of 30 km, then turns around and drives west to point *B.* The truck and car leave point *A* simultaneously and arrive at point *B* simultaneously. Explain why the car and truck have the *same* average velocity.

2.2 | Instantaneous Velocity

Sometimes the average velocity is all you need to know about a particle's motion. For example, a race along a straight line is really a competition to see whose average velocity, $v_{\text{av-}x}$, has the greatest magnitude. The prize goes to the competitor who can travel the displacement Δx from the start to the finish line in the briefest time interval, Δt (Fig. 2.4).

But the average velocity of a particle during a time interval can't tell us how fast, or in what direction, the particle was moving at any given time during the interval. To describe the motion in greater detail, we need to define the velocity at any specific instant of time or specific point along the path. Such a velocity is called **instantaneous velocity,** and it needs to be defined carefully.

Note that the word *instant* has a somewhat different definition in physics than in everyday language. You might use the phrase "It lasted just an instant" to refer to something that lasted for a very short time interval. But in physics an instant has *no* duration at all; it refers to a single value of time.

To find the instantaneous velocity of the dragster in Fig. 2.1 at the point P_1, we imagine moving the second point P_2 closer and closer to the first point P_1. We compute the average velocity $v_{\text{av-}x} = \Delta x/\Delta t$ over these shorter and shorter displacements and time intervals. Both Δx and Δt become very small, but their ratio does not necessarily become small. In the language of calculus, the limit of $\Delta x/\Delta t$ as Δt approaches zero is called the **derivative** of x with respect to t and is written dx/dt. *The instantaneous velocity is the limit of the average velocity as the time interval approaches zero; it equals the instantaneous rate of change of position with time.* We use the symbol v_x, with no "av" subscript, for the instantaneous velocity along the x-axis:

2.4 The winner of a 50-m swimming race is the swimmer whose average velocity has the greatest magnitude—that is, the swimmer who traverses a displacement Δx of 50 m in the shortest elapsed time Δt.

$$v_x = \lim_{\Delta t \to 0} \frac{\Delta x}{\Delta t} = \frac{dx}{dt} \qquad \text{(instantaneous velocity, straight-line motion)} \qquad (2.3)$$

We always assume that the time interval Δt is positive so that v_x has the same algebraic sign as Δx. If the positive x-axis points to the right, as in Fig. 2.1, a positive value of v_x means that x is increasing and the motion is toward the right; a negative value of v_x means that x is decreasing and the motion is toward the left. A body can have positive x and negative v_x, or the reverse; x tells us where the body is, while v_x tells us how it's moving (Fig 2.5).

Instantaneous velocity, like average velocity, is a vector quantity. Equation (2.3) defines its x-component, which can be positive or negative. In straight-line motion, all other components of instantaneous velocity are zero, and in this case we will often call v_x simply the instantaneous velocity. (In Chapter 3 we'll deal with the general case in which the instantaneous velocity can have nonzero x-, y-, and z-components.) When we use the term "velocity," we will always mean instantaneous rather than average velocity, unless we state otherwise.

2.5 Even when moving forward, this cyclist's instantaneous velocity can be negative—if he's traveling in the negative x-direction. In any problem, the choice of which direction is positive and which is negative is entirely up to you.

The terms "velocity" and "speed" are used interchangeably in everyday language, but they have distinct definitions in physics. We use the term **speed** to denote distance traveled divided by time, on either an average or an instantaneous basis. We use the symbol v with *no* subscripts to denote instantaneous speed. Instantaneous *speed* measures how fast a particle is moving; instantaneous *velocity* measures how fast *and* in what direction it's moving. For example, a particle with instantaneous velocity $v_x = 25$ m/s and a second particle with $v_x = -25$ m/s are moving in opposite directions at the same instantaneous speed 25 m/s. Instantaneous speed is the magnitude of instantaneous velocity, and so instantaneous speed can never be negative.

CAUTION Average speed, however, is *not* the magnitude of average velocity. When Alexander Popov set a world record in 1994 by swimming 100.0 m in 46.74 s, his average speed was $(100.0 \text{ m})/(46.74 \text{ s}) = 2.139$ m/s. But because he swam two lengths in a 50-m pool, he started and ended at the same point, giving him zero total displacement and zero average *velocity* for his effort! Both average speed and instantaneous speed are scalars, not vectors, because these quantities contain no information about direction.

| Example 2.1 | **Average and instantaneous velocities** |

A cheetah is crouched in ambush 20 m to the east of an observer's blind (Fig. 2.6). At time $t = 0$ the cheetah charges an antelope in a clearing 50 m east of the observer. The cheetah runs along a straight line. Later analysis of a videotape shows that during the first 2.0 s of the attack, the cheetah's coordinate x varies with time according to the equation $x = 20 \text{ m} + (5.0 \text{ m/s}^2)t^2$. (Note that the units for the numbers 20 and 5.0 *must* be as shown to make the expression dimensionally consistent.) a) Find the displacement of the cheetah during the interval between $t_1 = 1.0$ s and $t_2 = 2.0$ s. b) Find the average velocity during the same time interval. c) Find the instantaneous velocity at time $t_1 = 1.0$ s by taking $\Delta t = 0.1$ s, then $\Delta t = 0.01$ s, then $\Delta t = 0.001$ s. d) Derive a general expression for the instantaneous velocity as a function of time, and from it find v_x at $t = 1.0$ s and $= 2.0$ s.

SOLUTION

IDENTIFY: This problem requires us to use the definitions of displacement, average velocity, and instantaneous velocity. Using the first two of these involves algebra; the last one requires us to use calculus to take a derivative.

SET UP: Figure 2.6 depicts the cheetah's motion. To analyze this problem we use Eq. (2.1) for displacement, Eq. (2.2) for average velocity, and Eq. (2.3) for instantaneous velocity.

EXECUTE: a) At time $t_1 = 1.0$ s the cheetah's position x_1 is

$$x_1 = 20 \text{ m} + (5.0 \text{ m/s}^2)(1.0 \text{ s})^2 = 25 \text{ m}$$

At time $t_2 = 2.0$ s its position x_2 is

$$x_2 = 20 \text{ m} + (5.0 \text{ m/s}^2)(2.0 \text{ s})^2 = 40 \text{ m}$$

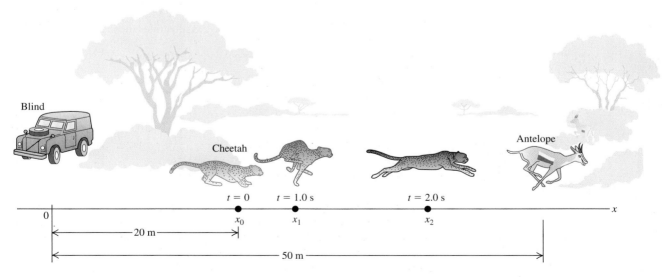

2.6 A cheetah attacking an antelope from ambush. The animals are not drawn to the same scale as the axis.

The displacement during this interval is

$$\Delta x = x_2 - x_1 = 40 \text{ m} - 25 \text{ m} = 15 \text{ m}$$

b) The average velocity during this time interval is

$$v_{\text{av-}x} = \frac{x_2 - x_1}{t_2 - t_1} = \frac{40 \text{ m} - 25 \text{ m}}{2.0 \text{ s} - 1.0 \text{ s}} = \frac{15 \text{ m}}{1.0 \text{ s}} = 15 \text{ m/s}$$

c) With $\Delta t = 0.1$ s, the time interval is from $t_1 = 1.0$ s to $t_2 = 1.1$ s. At time t_2, the position is

$$x_2 = 20 \text{ m} + (5.0 \text{ m/s}^2)(1.1 \text{ s})^2 = 26.05 \text{ m}$$

The average velocity during this interval is

$$v_{\text{av-}x} = \frac{26.05 \text{ m} - 25 \text{ m}}{1.1 \text{ s} - 1.0 \text{ s}} = 10.5 \text{ m/s}$$

You should follow this same pattern to work out the average velocities for the 0.01-s and 0.001-s intervals. The results are 10.05 m/s

and 10.005 m/s. As Δt gets smaller, the average velocity gets closer to 10.0 m/s, so we conclude that the instantaneous velocity at time $t = 1.0$ s is 10.0 m/s.

d) We find the instantaneous velocity as a function of time by taking the derivative of the expression for x with respect to t. For any n the derivative of t^n is nt^{n-1}, so the derivative of t^2 is $2t$. Therefore

$$v_x = \frac{dx}{dt} = (5.0 \text{ m/s}^2)(2t) = (10 \text{ m/s}^2)t$$

At time $t = 1.0$ s, $v_x = 10$ m/s as we found in part (c). At time $t = 2.0$ s, $v_x = 20$ m/s.

EVALUATE: Our results show that the cheetah picked up speed from $t = 0$ (when it was at rest) to $t = 1.0$ s ($v_x = 10$ m/s) to $t = 2.0$ s ($v_x = 20$ m/s). This makes sense; the cheetah covered only 5 m during the interval $t = 0$ to $t = 1.0$ s, but covered 15 m during the interval $t = 1.0$ s to $t = 2.0$ s.

Act|v
ONLINE
Physics

1.1   Analyzing Motion Using Diagrams

Finding Velocity on an *x-t* Graph

The velocity of a particle can also be found from the graph of the particle's position as a function of time. Suppose we want to find the velocity of the dragster in Fig. 2.1 at point P_1. As point P_2 in Fig. 2.1 approaches point P_1, point p_2 in the *x-t* graph of Fig. 2.3 approaches point p_1. This is shown in Figs. 2.7a and 2.7b, in which the average velocity is calculated over shorter time intervals Δt. In the limit that $\Delta t \to 0$, shown in Fig. 2.7c, the slope of the line $p_1 p_2$ equals the slope of the line tangent to the curve at point p_1. *On a graph of position as a function of time for straight-line motion, the instantaneous velocity at any point is equal to the slope of the tangent to the curve at that point.*

If the tangent to the *x-t* curve slopes upward to the right, as in Fig. 2.7c, then its slope is positive, the velocity is positive, and the motion is in the positive *x*-direction. If the tangent slopes downward to the right, the slope and velocity are negative and the motion is in the negative *x*-direction. When the tangent is horizontal, the slope is zero and the velocity is zero. Figure 2.8 illustrates these three possibilities.

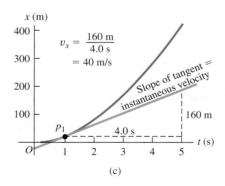

2.7 (a) and (b) As the average velocity $v_{\text{av-}x}$ is calculated over shorter and shorter time intervals, its value approaches the instantaneous velocity. (c) The instantaneous velocity v_x at a certain time equals the slope of the tangent to the *x-t* curve at that time. We find this slope by dividing any vertical interval (with distance units) along the tangent by the corresponding horizontal interval (with time units).

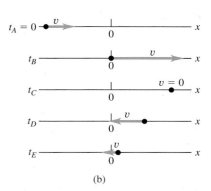

	x-t graph	Motion of particle
A	positive slope, so $v_x > 0$	moving in $+x$-direction
B	larger positive slope, so $v_x > 0$	moving in $+x$-direction faster than at A
C	zero slope, so $v_x = 0$	instantaneously at rest
D	negative slope, so $v_x < 0$	moving in $-x$-direction
E	smaller negative slope, so $v_x < 0$	moving in $-x$-direction more slowly than at D

(a)

(b)

2.8 (a) The x-t graph of the motion of a particular particle. The slope of the tangent at any point equals the velocity at that point. (b) A motion diagram showing the position and velocity of the particle at each of the five times labeled on the x-t graph. The particle speeds up from A to B, then slows down from B to C, where it comes to a momentary halt. It then moves in the negative x-direction, speeding up from C to D and slowing down from D to E.

Note that Fig. 2.8 depicts the motion of a particle in two ways. Figure 2.8a is an x-t graph, and Fig. 2.8b is an example of a **motion diagram.** A motion diagram shows the position of the particle at various times during the motion, like frames from a movie or video of the particle's motion, as well as arrows to represent the particle's velocity at each instant. Both x-t graphs and motion diagrams are helpful aids to understanding motion, and we will use both frequently in this chapter. You will find it worth your while to draw *both* an x-t graph and a motion diagram as part of solving any problem involving motion.

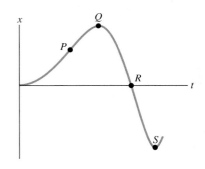

2.9 An x-t graph for a particle.

Test Your Understanding

Figure 2.9 is an x-t graph of the motion of a particle. At which of the points P, Q, R, and S is the velocity v_x positive? At which points is the velocity negative? At which points is it zero? At which point is the *speed* greatest?

2.3 | Average and Instantaneous Acceleration

When the velocity of a moving body changes with time, we say that the body has an *acceleration.* Just as velocity describes the rate of change of position with time, acceleration describes the rate of change of velocity with time. Like velocity, acceleration is a vector quantity. In straight-line motion its only nonzero component is along the axis along which the motion takes place.

Average Acceleration

Let's consider again the motion of a particle along the x-axis. Suppose that at time t_1 the particle is at point P_1 and has x-component of (instantaneous) velocity v_{1x}, and at a later time t_2 it is at point P_2 and has x-component of velocity v_{2x}. So the x-component of velocity changes by an amount $\Delta v_x = v_{2x} - v_{1x}$ during the time interval $\Delta t = t_2 - t_1$.

We define the **average acceleration** of the particle as it moves from P_1 to P_2 to be a vector quantity whose x-component is Δv_x, the change in the x-component of velocity, divided by the time interval Δt:

$$a_{\text{av-}x} = \frac{v_{2x} - v_{1x}}{t_2 - t_1} = \frac{\Delta v_x}{\Delta t} \qquad (2.4)$$

(average acceleration, straight-line motion)

For straight-line motion we will usually call $a_{\text{av-}x}$ simply the average acceleration, remembering that in fact it is the x-component of the average acceleration vector. (We'll encounter the other components of the average acceleration vector in Chapter 3.)

If we express velocity in meters per second and time in seconds, then average acceleration is in meters per second per second, or (m/s)/s. This is usually written as m/s² and is read "meters per second squared."

> **CAUTION** Be very careful not to confuse acceleration with velocity! Velocity describes how an object's position changes with time; it tells us how fast and in what direction the object moves. Acceleration describes how the velocity changes with time; it tells us how the speed and direction of motion are changing. It may help to remember the phrase "acceleration is to velocity as velocity is to position."

Example 2.2	**Average acceleration**

An astronaut has left an orbiting space shuttle to test a new personal maneuvering unit. As she moves along a straight line, her partner on board the shuttle measures her velocity every 2.0 s, starting at time $t = 1.0$ s, as follows:

t	v_x	t	v_x
1.0 s	0.8 m/s	9.0 s	−0.4 m/s
3.0 s	1.2 m/s	11.0 s	−1.0 m/s
5.0 s	1.6 m/s	13.0 s	−1.6 m/s
7.0 s	1.2 m/s	15.0 s	−0.8 m/s

Find the average acceleration, and describe whether the speed of the astronaut increases or decreases, for each of the following time intervals: a) $t_1 = 1.0$ s to $t_2 = 3.0$ s; b) $t_1 = 5.0$ s to $t_2 = 7.0$ s; c) $t_1 = 9.0$ s to $t_2 = 11.0$ s; d) $t_1 = 13.0$ s to $t_2 = 15.0$ s.

SOLUTION

IDENTIFY and SET UP: We use the definition of average acceleration, Eq. (2.4), to determine the value of $a_{\text{av-}x}$ from the change in *velocity* for each time interval. We determine the change in *speed* for each time interval by remembering that speed v is the magnitude of the instantaneous velocity v_x.

EXECUTE: The upper part of Fig. 2.10 graphs the velocity as a function of time. On this graph, the slope of the line connecting the

points at the beginning and end of each interval equals the average acceleration $a_{\text{av-}x} = \Delta v_x / \Delta t$ for that interval. The values of $a_{\text{av-}x}$ are

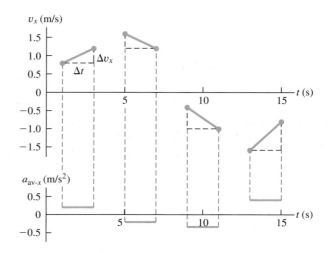

2.10 The slope of the line connecting two points on a graph of velocity versus time (upper graph) equals the average acceleration between those two points (lower graph).

graphed in the lower part of Fig. 2.10. For each time interval, we have

a) $a_{av-x} = (1.2 \text{ m/s} - 0.8 \text{ m/s})/(3.0 \text{ s} - 1.0 \text{ s}) = 0.2 \text{ m/s}^2$. The speed (magnitude of instantaneous velocity) increases from 0.8 m/s to 1.2 m/s.

b) $a_{av-x} = (1.2 \text{ m/s} - 1.6 \text{ m/s})/(7.0 \text{ s} - 5.0 \text{ s}) = -0.2 \text{ m/s}^2$. The speed decreases from 1.6 m/s to 1.2 m/s.

c) $a_{av-x} = [-1.0 \text{ m/s} - (-0.4 \text{ m/s})]/[11.0 \text{ s} - 9.0 \text{ s}] = -0.3 \text{ m/s}^2$. The speed increases from 0.4 m/s to 1.0 m/s.

d) $a_{av-x} = [-0.8 \text{ m/s} - (-1.6 \text{ m/s})]/[15.0 \text{ s} - 13.0 \text{ s}] = 0.4 \text{ m/s}^2$. The speed decreases from 1.6 m/s to 0.8 m/s.

EVALUATE: When the acceleration has the *same* direction (same algebraic sign) as the initial velocity, as in intervals a and c, the astronaut goes faster; when it has the *opposite* direction (opposite algebraic sign), as in intervals b and d, she slows down. When she moves in the negative direction with increasing speed (interval c), her velocity is algebraically decreasing (becoming more negative), and her acceleration is negative. But when she moves in the negative direction with decreasing speed (interval d), her velocity is algebraically increasing (becoming less negative) and her acceleration is positive.

Instantaneous Acceleration

We can now define **instantaneous acceleration,** following the same procedure that we used to define instantaneous velocity. Consider this situation: A race car driver has just entered the final straightaway at the Grand Prix. He reaches point P_1 at time t_1, moving with velocity v_{1x}. He passes point P_2, closer to the finish line, at time t_2 with velocity v_{2x} (Fig. 2.11).

2.11 A Grand Prix car at two points on the straightaway.

To define the instantaneous acceleration at point P_1, we take the second point P_2 in Fig. 2.11 to be closer and closer to the first point P_1 so that the average acceleration is computed over shorter and shorter time intervals. *The instantaneous acceleration is the limit of the average acceleration as the time interval approaches zero. In the language of calculus, instantaneous acceleration equals the instantaneous rate of change of velocity with time.* Thus

$$a_x = \lim_{\Delta t \to 0} \frac{\Delta v_x}{\Delta t} = \frac{dv_x}{dt} \qquad (2.5)$$

(instantaneous acceleration, straight-line motion)

Note that Eq. (2.5) is really the definition of the *x*-component of the acceleration vector; in straight-line motion, all other components of this vector are zero. Instantaneous acceleration plays an essential role in the laws of mechanics. From now on, when we use the term "acceleration," we will always mean instantaneous acceleration, not average acceleration.

Example 2.3	**Average and instantaneous accelerations**

Suppose the velocity v_x of the car in Fig. 2.11 at any time t is given by the equation

$$v_x = 60 \text{ m/s} + (0.50 \text{ m/s}^3)t^2$$

(Note that the units for the numbers 60 and 0.50 *must* be as shown to make the expression dimensionally consistent.) a) Find the change in velocity of the car in the time interval between $t_1 = 1.0 \text{ s}$ and $t_2 = 3.0 \text{ s}$. b) Find the average acceleration in this time interval.

c) Find the instantaneous acceleration at time $t_1 = 1.0$ s by taking Δt to be first 0.1 s, then 0.01 s, then 0.001 s. d) Derive an expression for the instantaneous acceleration at any time, and use it to find the acceleration at $t = 1.0$ s and $t = 3.0$ s.

SOLUTION

IDENTIFY: This example is analogous to Example 2.1 in Section 2.2. (Now would be a good time to review that example.) There we determined the average velocity over shorter and shorter time intervals by considering the change in displacement, and we determined the instantaneous velocity by differentiating the position as a function of time. In this example, we determine the average *acceleration* by considering changes in *velocity* over a time interval. Likewise, we find the *instantaneous* acceleration by differentiating the velocity as a function of time.

SET UP: We'll use Eq. (2.4) for average acceleration and Eq. (2.5) for instantaneous acceleration.

EXECUTE: a) We first find the velocity at each time by substituting each value of t into the equation. At time $t_1 = 1.0$ s,

$$v_{1x} = 60 \text{ m/s} + (0.50 \text{ m/s}^3)(1.0 \text{ s})^2 = 60.5 \text{ m/s}$$

At time $t_2 = 3.0$ s,

$$v_{2x} = 60 \text{ m/s} + (0.50 \text{ m/s}^3)(3.0 \text{ s})^2 = 64.5 \text{ m/s}$$

The change in velocity Δv_x is

$$\Delta v_x = v_{2x} - v_{1x} = 64.5 \text{ m/s} - 60.5 \text{ m/s} = 4.0 \text{ m/s}$$

The time interval is $\Delta t = 3.0 \text{ s} - 1.0 \text{ s} = 2.0$ s.
b) The average acceleration during this time interval is

$$a_{\text{av-}x} = \frac{v_{2x} - v_{1x}}{t_2 - t_1} = \frac{4.0 \text{ m/s}}{2.0 \text{ s}} = 2.0 \text{ m/s}^2$$

During the time interval from $t_1 = 1.0$ s to $t_2 = 3.0$ s, the velocity and average acceleration have the same algebraic sign (in this case, positive), and the car speeds up.
c) When $\Delta t = 0.1$ s, $t_2 = 1.1$ s and

$$v_{2x} = 60 \text{ m/s} + (0.50 \text{ m/s}^3)(1.1 \text{ s})^2 = 60.605 \text{ m/s}$$

$$\Delta v_x = 0.105 \text{ m/s}$$

$$a_{\text{av-}x} = \frac{\Delta v_x}{\Delta t} = \frac{0.105 \text{ m/s}}{0.1 \text{ s}} = 1.05 \text{ m/s}^2$$

We invite you to repeat this pattern for $\Delta t = 0.01$ s and $\Delta t = 0.001$ s; the results are $a_{\text{av-}x} = 1.005 \text{ m/s}^2$ and $a_{\text{av-}x} = 1.0005 \text{ m/s}^2$, respectively. As Δt gets smaller, the average acceleration gets closer to 1.0 m/s^2. We conclude that the instantaneous acceleration at $t = 1.0$ s is 1.0 m/s^2.
d) The instantaneous acceleration is $a_x = dv_x/dt$, the derivative of a constant is zero, and the derivative of t^2 is $2t$. Using these formulas, we obtain

$$a_x = \frac{dv_x}{dt} = \frac{d}{dt}[60 \text{ m/s} + (0.50 \text{ m/s}^3)t^2]$$

$$= (0.50 \text{ m/s}^3)(2t) = (1.0 \text{ m/s}^3)t$$

When $t = 1.0$ s,

$$a_x = (1.0 \text{ m/s}^3)(1.0 \text{ s}) = 1.0 \text{ m/s}^2$$

When $t = 3.0$ s,

$$a_x = (1.0 \text{ m/s}^3)(3.0 \text{ s}) = 3.0 \text{ m/s}^2$$

EVALUATE: Note that neither of the values we found in part (d) is equal to the average acceleration found in part (b). The instantaneous acceleration of this car varies with time. Automotive engineers sometimes call the rate of change of acceleration with time the "jerk."

Finding Acceleration on a v_x-t Graph or an x-t Graph

We interpreted average and instantaneous velocity in terms of the slope of a graph of position versus time. In the same way, we can get additional insight into the concepts of average and instantaneous acceleration from a graph with instantaneous velocity v on the vertical axis and time t on the horizontal axis; that is, a **v_x-t graph** (Fig. 2.12). The points on the graph labeled p_1 and p_2 correspond to points P_1 and P_2 in Fig. 2.11. The average acceleration $a_{\text{av-}x} = \Delta v_x/\Delta t$ during this interval is the slope of the line $p_1 p_2$. As point P_2 in Fig. 2.11 approaches point P_1, point p_2 in the v_x-t graph of Fig. 2.12 approaches point p_1, and the slope of the line $p_1 p_2$ approaches the slope of the line tangent to the curve at point p_1. Thus, *on a graph of velocity as a function of time, the instantaneous acceleration at any point is equal to the slope of the tangent to the curve at that point*. In Fig. 2.12 the instantaneous acceleration varies with time.

 Note that by itself, the algebraic sign of the acceleration doesn't tell you whether a body is speeding up or slowing down. You must compare the signs of the velocity and the acceleration. When v_x and a_x have the *same* sign, the body is

2.12 A v_x-t graph of the motion in Fig. 2.11. The average acceleration between t_1 and t_2 equals the slope of the line p_1p_2. The instantaneous acceleration at P_1 equals the slope of the tangent at p_1.

speeding up. If both are positive, the body is moving in the positive direction with increasing speed. If both are negative, the body is moving in the negative direction with a velocity that is becoming more and more negative, and again the speed is increasing. When v_x and a_x have *opposite* signs, the body is slowing down. If v_x is positive and a_x is negative, the body is moving in the positive direction with decreasing speed; if v_x is negative and a_x is positive, the body is moving in the negative direction with a velocity that is becoming less negative, and again the body is slowing down. Figure 2.13 illustrates these various possibilities.

The term *deceleration* is sometimes used for a decrease in speed. Because it may mean positive or negative a_x, depending on the sign of v_x, we avoid this term.

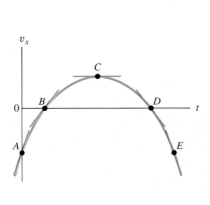

	v_x-t graph	Motion of particle
A	$v_x < 0$; positive slope, so $a_x > 0$	moving in $-x$-direction, slowing down
B	$v_x = 0$; positive slope, so $a_x > 0$	instantaneously at rest, about to move in $+x$-direction
C	$v_x > 0$; zero slope, so $a_x = 0$	moving in $+x$-direction at maximum speed
D	$v_x = 0$; negative slope, so $a_x < 0$	instantaneously at rest, about to move in $-x$-direction
E	$v_x < 0$; negative slope, so $a_x < 0$	moving in $-x$-direction, speeding up

(a)

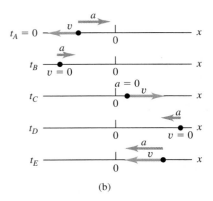

(b)

2.13 (a) A v_x-t graph of the motion of a particle (a different motion than in Fig. 2.8). The slope of the tangent at any point equals the acceleration at that point. (b) A motion diagram showing the position, velocity, and acceleration of the particle at each of the times labeled on the v_x-t graph. The positions are consistent with the v_x-t graph; for instance, from t_A to t_B the velocity is negative, so at t_B the particle is at a more negative value of x than at t_A.

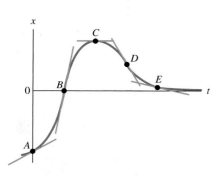

	x-t graph	Motion of particle
A	positive slope, upward curvature, so $v_x > 0$, $a_x > 0$	moving in +x-direction, speeding up
B	positive slope, zero curvature, so $v_x > 0$, $a_x = 0$	moving in +x-direction, speed not changing
C	zero slope, downward curvature, so $v_x = 0$, $a_x < 0$	instantaneously at rest, velocity changing from + to −
D	negative slope, zero curvature, so $v_x < 0$, $a_x = 0$	moving in −x-direction, speed not changing
E	negative slope, upward curvature, so $v_x < 0$, $a_x > 0$	moving in −x-direction, slowing down

(a)

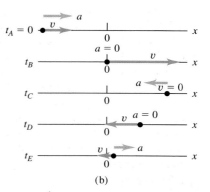

(b)

2.14 (a) The same x-t graph as shown in Fig. 2.8a. The velocity is equal to the *slope* of the graph, and the acceleration is given by the *concavity* or *curvature* of the graph. (b) A motion diagram showing the position, velocity, and acceleration of the particle at each of the times labeled on the x-t graph.

We can also learn about the acceleration of a body from a graph of its *position* versus time. Because $a_x = dv_x/dt$ and $v_x = dx/dt$, we can write

$$a_x = \frac{dv_x}{dt} = \frac{d}{dt}\left(\frac{dx}{dt}\right) = \frac{d^2x}{dt^2} \qquad (2.6)$$

That is, a_x is the second derivative of x with respect to t. The second derivative of any function is directly related to the *concavity* or *curvature* of the graph of that function. At a point where the x-t graph is concave up (curved upward), the acceleration is positive and v_x is increasing; at a point where the x-t graph is concave down (curved downward), the acceleration is negative and v is decreasing. At a point where the x-t graph has no curvature, such as an inflection point, the acceleration is zero and the velocity is constant. All three of these possibilities are shown in Fig. 2.14. Examining the curvature of an x-t graph is an easy way to decide what the *sign* of acceleration is. This technique is less helpful for determining numerical values of acceleration, because the curvature of a graph is hard to measure accurately.

Test Your Understanding

Look again at the x-t graph in Fig. 2.9 at the end of Section 2.2. At which of the points P, Q, R, and S is the acceleration a_x positive? At which points is the acceleration negative? At which points does the acceleration appear to be zero? Compare the acceleration at each point to the velocity v_x at that point, and decide whether the speed is increasing, decreasing, or not changing at that point.

2.4 | Motion with Constant Acceleration

The simplest accelerated motion is straight-line motion with *constant* acceleration. In this case the velocity changes at the same rate throughout the motion. This is a very special situation, yet one that occurs often in nature. As we will discuss in the next section, a falling body has a constant acceleration if the effects of the air are not important. The same is true for a body sliding on an incline or along a rough horizontal surface. Straight-line motion with nearly constant acceleration also occurs in technology, such as a jet fighter being catapulted from the deck of an aircraft carrier.

In this section we'll derive key equations for straight-line motion with constant acceleration. These equations will enable us to solve a wide range of problems.

Figure 2.15 is a motion diagram showing the position, velocity, and acceleration at five different times for a particle moving with constant acceleration. Figures 2.16 and 2.17 depict this same motion in the form of graphs. Since the acceleration a_x is constant, the a_x-t **graph** (graph of acceleration versus time) in Fig. 2.16 is a horizontal line. The graph of velocity versus time has a constant *slope* because the acceleration is constant, and so the v_x-t graph is a straight line (Fig. 2.17).

When the acceleration is constant, it's easy to derive equations for position x and velocity v_x as functions of time. Let's start with velocity. In Eq. (2.4) we can replace the average acceleration $a_{\text{av-}x}$ by the constant (instantaneous) acceleration a_x. We then have

$$a_x = \frac{v_{2x} - v_{1x}}{t_2 - t_1} \tag{2.7}$$

Now we let $t_1 = 0$ and let t_2 be any arbitrary later time t. We use the symbol v_{0x} for the x-component of velocity at the initial time $t = 0$; the x-component of velocity at the later time t is v_x. Then Eq. (2.7) becomes

$$a_x = \frac{v_x - v_{0x}}{t - 0} \quad \text{or}$$

$$v_x = v_{0x} + a_x t \quad \text{(constant acceleration only)} \tag{2.8}$$

We can interpret this equation as follows. The acceleration a_x is the constant rate of change of velocity; that is, the change in velocity per unit time. The term $a_x t$ is the product of the change in velocity per unit time, a_x, and the time interval t. Therefore it equals the *total* change in velocity from the initial time $t = 0$ to the later time t. The velocity v_x at any time t then equals the initial velocity v_{0x} (at $t = 0$) plus the change in velocity $a_x t$. Graphically, we can consider the height v_x of the graph in Fig. 2.17 at any time t to be the sum of two segments: one with length v_{0x} equal to the initial velocity, the other with length $a_x t$ equal to the change in velocity during time t. The graph of velocity as a function of time is a straight line with slope a_x that intercepts the vertical axis (the v-axis) at v_{0x}.

Another interpretation of Eq. (2.8) is that the change in velocity $v_x - v_{0x}$ of the particle between $t = 0$ and any later time t equals the *area* under the a_x-t graph between those two times. In Fig. 2.16, the area under the graph of acceleration versus time is shown as a green rectangle of vertical side a_x and horizontal side t. The area of this rectangle is equal to $a_x t$, which from Eq. (2.8) is just equal to the change in velocity $v_x - v_{0x}$. In Section 2.6 we'll show that even if the acceleration is not constant, the change in velocity during a time interval is still equal to the area under the a_x-t curve, although in that case Eq. (2.8) does not apply.

Next we want to derive an equation for the position x of a particle moving with constant acceleration. To do this, we make use of two different expressions for the average velocity $v_{\text{av-}x}$ during the interval from $t = 0$ to any later time t. The first expression comes from the definition of $v_{\text{av-}x}$, Eq. (2.2), which holds true whether or not the acceleration is constant. We call the position at time $t = 0$ the *initial position,*

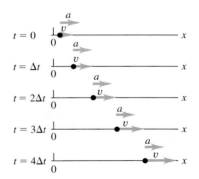

2.15 A motion diagram for a particle moving in a straight line in the positive x-direction with constant positive acceleration a_x. The position, velocity, and acceleration are shown at five equally spaced times. The velocity changes by equal amounts in equal time intervals because the acceleration is constant. The position changes by *different* amounts in equal time intervals because the velocity is changing.

2.16 An acceleration-time (a_x-t) graph for straight-line motion with constant positive acceleration a_x.

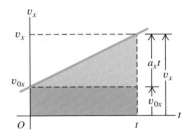

2.17 A velocity-time (v_x-t) graph for straight-line motion with constant positive acceleration a_x. The initial velocity v_{0x} is also positive.

denoted by x_0. The position at the later time t is simply x. Thus for the time interval $\Delta t = t - 0$ and the corresponding displacement $\Delta x = x - x_0$, Eq. (2.2) gives

$$v_{\text{av-}x} = \frac{x - x_0}{t} \tag{2.9}$$

We can also get a second expression for $v_{\text{av-}x}$ that is valid only when the acceleration is constant, so that the v_x-t graph is a straight line (as in Fig. 2.17) and the velocity changes at a constant rate. In this case the average velocity during any time interval is simply the arithmetic average of the velocities at the beginning and end of the interval. For the time interval 0 to t,

$$v_{\text{av-}x} = \frac{v_{0x} + v_x}{2} \qquad \text{(constant acceleration only)} \tag{2.10}$$

(This equation is *not* true if the acceleration varies and the v_x-t graph is a curve, as in Fig. 2.13.) We also know that with constant acceleration the velocity v_x at any time t is given by Eq. (2.8). Substituting that expression for v_x into Eq. (2.10), we find

$$v_{\text{av-}x} = \frac{1}{2}(v_{0x} + v_{0x} + a_x t)$$

$$= v_{0x} + \frac{1}{2}a_x t \qquad \text{(constant acceleration only)} \tag{2.11}$$

Finally, we equate Eqs. (2.9) and (2.11) and simplify the result:

$$v_{0x} + \frac{1}{2}a_x t = \frac{x - x_0}{t} \qquad \text{or}$$

$$x = x_0 + v_{0x}t + \frac{1}{2}a_x t^2 \qquad \text{(constant acceleration only)} \tag{2.12}$$

This equation states that if at the initial time $t = 0$, a particle is at position x_0 and has velocity v_{0x}, its new position x at any later time t is the sum of three terms: its initial position x_0, plus the distance $v_{0x}t$ that it would move if its velocity were constant, plus an additional distance $\frac{1}{2}a_x t^2$ caused by the change in velocity.

Just as the change in velocity of the particle equals the area under the a_x-t graph, the displacement—that is, the change in position—equals the area under the v_x-t graph. To be specific, the displacement $x - x_0$ of the particle between $t = 0$ and any later time t equals the area under the v_x-t graph between those two times. In Fig. 2.17 the area under the graph is divided into a dark-colored rectangle of vertical side v_{0x} and horizontal side t and a light-colored right triangle of vertical side $a_x t$ and horizontal side t. The area of the rectangle is $v_{0x}t$ and the area of the triangle is $\frac{1}{2}(a_x t)(t) = \frac{1}{2}a_x t^2$, so the total area under the curve is

$$x - x_0 = v_{0x}t + \frac{1}{2}a_x t^2$$

in agreement with Eq. (2.12).

The displacement during a time interval can always be found from the area under the v_x-t curve. This is true even if the acceleration is *not* constant, although in that case Eq. (2.12) does not apply. (We'll show this in Section 2.6.)

We can check whether Eqs. (2.8) and (2.12) are consistent with the assumption of constant acceleration by taking the derivative of Eq. (2.12). We find

$$v_x = \frac{dx}{dt} = v_{0x} + a_x t$$

which is Eq. (2.8). Differentiating again, we find simply

$$\frac{dv_x}{dt} = a_x$$

as we should expect.

In many problems, it's useful to have a relationship between position, velocity, and acceleration that does not involve the time. To obtain this, we first solve Eq. (2.8) for t, then substitute the resulting expression into Eq. (2.12) and simplify:

$$t = \frac{v_x - v_{0x}}{a_x}$$

$$x = x_0 + v_{0x}\left(\frac{v_x - v_{0x}}{a_x}\right) + \frac{1}{2}a_x\left(\frac{v_x - v_{0x}}{a_x}\right)^2$$

We transfer the term x_0 to the left side and multiply through by $2a_x$:

$$2a_x(x - x_0) = 2v_{0x}v_x - 2v_{0x}^2 + v_x^2 - 2v_{0x}v_x + v_{0x}^2$$

Finally, simplifying gives us

$$v_x^2 = v_{0x}^2 + 2a_x(x - x_0) \qquad \text{(constant acceleration only)} \qquad (2.13)$$

We can get one more useful relationship by equating the two expressions for $v_{\text{av-}x}$, Eqs. (2.9) and (2.10), and multiplying through by t. Doing this, we obtain

$$x - x_0 = \left(\frac{v_{0x} + v_x}{2}\right)t \qquad \text{(constant acceleration only)} \qquad (2.14)$$

Note that Eq. (2.14) does not contain the acceleration a_x. This is sometimes a useful equation when a_x is constant but its value is unknown.

Equations (2.8), (2.12), (2.13), and (2.14) are the *equations of motion with constant acceleration*. By using these equations, we can solve *any* kinematics problem involving straight-line motion of a particle with constant acceleration.

Figure 2.18 shows a graph of the coordinate x as a function of time for motion with constant acceleration. That is, it is a graph of Eq. (2.12); the x-t graph for constant acceleration is always a *parabola*. The curve intercepts the vertical axis (x-axis) at x_0, the position at $t = 0$. The slope of the tangent at $t = 0$ equals v_{0x}, the initial velocity, and the slope of the tangent at any time t equals the velocity v_x at that time. The graph in Fig. 2.18 is concave up (it curves upward). The slope and velocity are continuously increasing, so the acceleration a_x is positive. If a_x is negative, the x-t graph is a parabola that is concave down (has a downward curvature).

For the particular case of motion with constant acceleration depicted in Fig. 2.15 and graphed in Figs. 2.16, 2.17, and 2.18, the values of x_0, v_{0x}, and a_x are all positive. We invite you to redraw these figures for cases in which one, two, or all three of these quantities are negative.

A special case of motion with constant acceleration occurs when the acceleration is *zero*. The velocity is then constant, and the equations of motion become simply

$$v_x = v_{0x} = \text{constant}$$
$$x = x_0 + v_x t$$

2.18 A position-time (x-t) graph for straight-line motion with constant acceleration. This graph refers to the same motion as is shown in Figs. 2.15, 2.16, and 2.17. For this motion the initial position x_0, the initial velocity v_{0x}, and the acceleration a_x are all positive.

Motion with Constant Acceleration

IDENTIFY *the relevant concepts:* In most straight-line motion problems, you can use the constant-acceleration equations. Occasionally, however, you will encounter a situation in which the acceleration *isn't* constant. In such a case, you'll need a different approach (see Section 2.6).

SET UP *the problem using the following steps:*
1. You *must* decide at the beginning of a problem where the origin of coordinates is and which axis direction is positive. These choices are usually a matter of convenience. It is often easiest to place the particle at the origin at time $t = 0$; then $x_0 = 0$. It is always helpful to make a motion diagram showing these choices and some later positions of the particle.
2. Remember that your choice of the positive axis direction automatically determines the positive directions for velocity and acceleration. If x is positive to the right of the origin, then v_x and a_x are also positive toward the right.
3. Restate the problem in words first, and then translate this description into symbols and equations. *When* does the particle arrive at a certain point (that is, what is the value of t)? *Where* is the particle when its velocity has a speci-

fied value (that is, what is the value of x when v_x has the specified value)? Example 2.4 asks, "Where is the motorcyclist when his velocity is 25 m/s?" Translated into symbols, this becomes "What is the value of x when $v_x = 25$ m/s?"
4. Make a list of quantities such as x, x_0, v_x, v_{0x}, a_x, and t. In general, some of them will be known and some will be unknown. Write down the values of the known quantities, and decide which of the unknowns are the target variables. Be on the lookout for implicit information. For example, "A car sits at a stoplight" usually means $v_{0x} = 0$.

EXECUTE *the solution:* Choose an equation from Eqs. (2.8), (2.12), (2.13), and (2.14) that contains only one of the target variables. Solve this equation for the target variable, using symbols only. Then substitute the known values and compute the value of the target variable. Sometimes you will have to solve two simultaneous equations for two unknown quantities.

EVALUATE *your answer:* Take a hard look at your results to see whether they make sense. Are they within the general range of values you expected?

Example
2.4

Constant-acceleration calculations

A motorcyclist heading east through a small Iowa city accelerates after he passes the signpost marking the city limits (Fig. 2.19). His acceleration is a constant 4.0 m/s². At time $t = 0$ he is 5.0 m east of the signpost, moving east at 15 m/s. a) Find his position and velocity at time $t = 2.0$ s. b) Where is the motorcyclist when his velocity is 25 m/s?

SOLUTION

IDENTIFY: The statement of the problem tells us explicitly that the acceleration is constant, so we can use the constant-acceleration equations.

SET UP: We take the signpost as the origin of coordinates $(x = 0)$, and choose the positive x-axis to point east (Fig. 2.19). At the initial

time $t = 0$, the initial position is $x_0 = 5.0$ m and the initial velocity is $v_{0x} = 15$ m/s. The constant acceleration is $a_x = 4.0$ m/s². The unknown target variables in part (a) are the values of the position x and the velocity v_x at the later time $t = 2.0$ s; the target variable in part (b) is the value of x when $v_x = 25$ m/s.

EXECUTE: a) We can find the position at $t = 2.0$ s by using Eq. (2.12), which gives position x as a function of time t:

$$x = x_0 + v_{0x}t + \frac{1}{2}a_xt^2$$

$$= 5.0 \text{ m} + (15 \text{ m/s})(2.0 \text{ s}) + \frac{1}{2}(4.0 \text{ m/s}^2)(2.0 \text{ s})^2$$

$$= 43 \text{ m}$$

We can find the velocity at this same time by using Eq. (2.8), which gives velocity v_x as a function of time t:

$$v_x = v_{0x} + a_xt$$

$$= 15 \text{ m/s} + (4.0 \text{ m/s}^2)(2.0 \text{ s}) = 23 \text{ m/s}$$

b) From our solution to part (a), we can see that the velocity is $v_x = 25$ m/s at a time later than 2.0 s and at a point farther than 43 m from the signpost. From Eq. (2.13) we have

$$v_x^2 = v_{0x}^2 + 2a_x(x - x_0)$$

2.19 A motorcyclist traveling with constant acceleration.

Solving for x and substituting in the known values, we find

$$x = x_0 + \frac{v_x^2 - v_{0x}^2}{2a_x}$$

$$= 5.0 \text{ m} + \frac{(25 \text{ m/s})^2 - (15 \text{ m/s})^2}{2(4.0 \text{ m/s}^2)}$$

$$= 55 \text{ m}$$

Alternatively, we may use Eq. (2.8) to first find the time when $v_x = 25$ m/s:

$$v_x = v_{0x} + a_x t \quad \text{so}$$

$$t = \frac{v_x - v_{0x}}{a_x} = \frac{25 \text{ m/s} - 15 \text{ m/s}}{4.0 \text{ m/s}^2}$$

$$= 2.5 \text{ s}$$

Then from Eq. (2.12) we have

$$x = x_0 + v_{0x}t + \frac{1}{2}a_x t^2$$

$$= 5.0 \text{ m} + (15 \text{ m/s})(2.5 \text{ s}) + \frac{1}{2}(4.0 \text{ m/s}^2)(2.5 \text{ s})^2$$

$$= 55 \text{ m}$$

EVALUATE: Do these results make sense? According to our results in part (a), the motorcyclist accelerates from 15 m/s (about 34 mi/h, or 54 km/h) to 23 m/s (about 51 mi/h, or 83 km/h) in 2.0 s while traveling a distance of 38 m (about 125 ft). Our results in part (b) tell us that after an additional 0.5 s, the motorcyclist has moved an additional 12 m (about 39 ft) and has accelerated to 25 m/s (56 mi/h, or 90 km/h). This is pretty brisk acceleration, but well within the capabilities of a high-performance bike.

Example 2.5	**Two bodies with different accelerations**

A motorist traveling with constant velocity of 15 m/s (about 34 mi/h) passes a school-crossing corner, where the speed limit is 10 m/s (about 22 mi/h). Just as the motorist passes, a police officer on a motorcycle stopped at the corner starts off in pursuit with constant acceleration of 3.0 m/s² (Fig. 2.20a). a) How much time elapses before the officer catches up with the motorist? b) What is the officer's speed at that point? c) What is the total distance each vehicle has traveled at that point?

SOLUTION

IDENTIFY: The police officer and the motorist both move with constant acceleration (equal to zero for the motorist), so we can use the formulas we have developed.

SET UP: We take the origin at the corner, so $x_0 = 0$ for both, and take the positive direction to the right. Let x_P (for police) be the officer's position and x_M (for motorist) be the motorist's position at any time. The initial velocities are $v_{P0x} = 0$ for the officer and $v_{M0x} = 15$ m/s for the motorist; the constant accelerations are

$a_{Px} = 3.0$ m/s² for the officer and $a_{Mx} = 0$ for the motorist. Our target variable in part (a) is the time when the officer catches the motorist—that is, when the two vehicles are at the same position. In part (b) we're looking for the officer's speed v (the magnitude of his velocity) at the time found in part (a). In part (c) we want to find the position of either vehicle at this same time. Hence we use Eq. (2.12) (which relates position and time) in parts (a) and (c), and Eq. (2.8) (which relates velocity and time) in part (b).

EXECUTE: a) We want to find the value of the time t when the motorist and the police officer are at the same position: $x_M = x_P$. Applying Eq. (2.12), $x = x_0 + v_{0x}t + (1/2)a_x t^2$, to each vehicle, we get

$$x_M = 0 + v_{M0x}t + \frac{1}{2}(0)t^2 = v_{M0x}t$$

$$x_P = 0 + (0)t + \frac{1}{2}a_{Px}t^2 = \frac{1}{2}a_{Px}t^2$$

2.20 (a) Motion with constant acceleration overtaking motion with constant velocity. (b) Graph of x versus t for each vehicle.

Since $x_M = x_P$ at time t, we set these two expressions equal to each other and solve for t:

$$v_{M0x}t = \frac{1}{2}a_{Px}t^2$$

$$t = 0 \quad \text{or} \quad t = \frac{2v_{M0x}}{a_{Px}} = \frac{2(15 \text{ m/s})}{3.0 \text{ m/s}^2} = 10 \text{ s}$$

There are *two* times when both the vehicles have the same x-coordinate. The first, $t = 0$, is the time when the motorist passes the parked motorcycle at the corner. The second, $t = 10$ s, is the time when the officer catches up with the motorist.

b) We want the magnitude of the officer's velocity v_{Px} at the time t found in part (a). Her velocity at any time is given by Eq. (2.8):

$$v_{Px} = v_{P0x} + a_{Px}t = 0 + (3.0 \text{ m/s}^2)t$$

so when $t = 10$ s, we find $v_{Px} = 30$ m/s. When the officer overtakes the motorist, she is traveling twice as fast as the motorist is.

c) In 10 s the distance the motorist travels is

$$x_M = v_{M0x}t = (15 \text{ m/s})(10 \text{ s}) = 150 \text{ m}$$

and the distance the officer travels is

$$x_P = \frac{1}{2}a_{Px}t^2 = \frac{1}{2}(3.0 \text{ m/s}^2)(10 \text{ s})^2 = 150 \text{ m}$$

This verifies that at the time the officer catches the motorist, they have gone equal distances.

EVALUATE: Figure 2.20b shows graphs of x versus t for each vehicle. We see again that there are two times when the two positions are the same (where the two graphs cross). At neither of these times do the two vehicles have the same velocity (i.e., where the two graphs cross, their slopes are different). At $t = 0$, the officer is at rest; at $t = 10$ s, the officer has twice the speed of the motorist.

In a real pursuit the officer would accelerate to a speed faster than that of the motorist, then slow down in order to have the same velocity as the motorist when she catches him. We haven't treated this case here because it involves a changing acceleration (but see Problem 2.68).

> **Test Your Understanding**
>
> Figure 2.20b is an x-t graph for each vehicle in Example 2.5. Draw a v_x-t graph for each vehicle. Do the motorist and the police officer ever have the same velocity? At what time does this happen?

2.5 | Freely Falling Bodies

The most familiar example of motion with (nearly) constant acceleration is that of a body falling under the influence of the earth's gravitational attraction. Such motion has held the attention of philosophers and scientists since ancient times. In the 4th century B.C., Aristotle thought (erroneously) that heavy objects fall faster than light objects, in proportion to their weight. Nineteen centuries later, Galileo argued that a body should fall with a downward acceleration that is constant and independent of its weight. We mentioned in Section 1.1 that, according to legend, Galileo experimented by dropping bullets and cannonballs from the Leaning Tower of Pisa.

The motion of falling bodies has since been studied with great precision. When the effects of the air can be neglected, Galileo is right; all bodies at a particular location fall with the same downward acceleration, regardless of their size or weight. If the distance of the fall is small compared to the radius of the earth, the acceleration is constant. In the following discussion we use an idealized model in which we neglect the effects of the air, the earth's rotation, and the decrease of acceleration with increasing altitude. We call this idealized motion **free fall,** although it includes rising as well as falling motion. (In Chapter 3 we will extend the discussion of free fall to include the motion of projectiles, which move both vertically and horizontally.)

2.21 Multiflash photo of a freely falling ball.

Figure 2.21 is a photograph of a falling ball taken with a stroboscopic light source that produces a series of intense flashes at equal time intervals. Each flash is so short (a few millionths of a second) that there is little blur in the image of even a rapidly moving body. As each flash occurs, the position of the ball at that instant is recorded on the film. Because of the equal time intervals between flashes, the average velocity of the ball between any two flashes is proportional to the distance between corresponding images in the photograph. The increasing distances between images show that the velocity is continuously changing; the ball is accelerating downward. Careful measurement shows that the velocity change is the same in each time interval, so the acceleration of the freely falling ball is constant.

The constant acceleration of a freely falling body is called the **acceleration due to gravity,** and we denote its magnitude with the letter g. We will frequently use the approximate value of g at or near the earth's surface:

$$g = 9.8 \text{ m/s}^2 = 980 \text{ cm/s}^2$$

$$= 32 \text{ ft/s}^2 \quad \text{(approximate value near the earth's surface)}$$

The exact value varies with location, so we will often give the value of g at the earth's surface to only two significant figures. Because g is the magnitude of a vector quantity, it is always a *positive* number. On the surface of the moon the acceleration due to gravity is caused by the attractive force of the moon rather than the earth, and $g = 1.6 \text{ m/s}^2$. Near the surface of the sun, $g = 270 \text{ m/s}^2$.

In the following examples we use the constant-acceleration equations developed in Section 2.4. We suggest that you review the problem-solving strategies discussed in that section before you study the examples below.

| Example 2.6 | **A freely-falling coin** |

A one-euro coin is dropped from the Leaning Tower of Pisa. It starts from rest and falls freely. Compute its position and velocity after 1.0, 2.0, and 3.0 s.

SOLUTION

IDENTIFY: "Falls freely" means "has a constant acceleration due to gravity," so we can use the constant-acceleration equations to determine our target variables.

SET UP: We take the origin O at the starting point, the coordinate axis as vertical, and the upward direction as positive (Fig. 2.22). Because the coordinate axis is vertical, let's call the coordinate y instead of x. We replace all the x's in the constant-acceleration equations by y's. The initial coordinate y_0 and the initial velocity v_{0y} are both zero. The acceleration is downward (in the negative y-direction), so $a_y = -g = -9.8 \text{ m/s}^2$. (Remember that, by definition, g itself is *always* positive.) Hence our target variables are the values of y and v_y at the three given times. To find these, we use Eqs. (2.12) and (2.8) with x replaced by y.

EXECUTE: At an arbitrary time t, the position and velocity are

$$y = v_{0y}t + \frac{1}{2}a_yt^2 = 0 + \frac{1}{2}(-g)t^2 = (-4.9 \text{ m/s}^2)t^2$$

$$v_y = v_{0y} + a_yt = 0 + (-g)t = (-9.8 \text{ m/s}^2)t$$

2.22 Position and velocity of a coin freely falling from rest.

When $t = 1.0$ s, $y = (-4.9 \text{ m/s}^2)(1.0 \text{ s})^2 = -4.9$ m and $v_y = (-9.8 \text{ m/s}^2)(1.0 \text{ s}) = -9.8$ m/s; after one second, the coin is 4.9 m below the origin (y is negative) and has a downward velocity (v_y is negative) with magnitude 9.8 m/s.

The position and velocity at 2.0 and 3.0 s are found in the same way. The results are shown in Fig. 2.22; check the numerical values for yourself.

EVALUATE: All our answers for v_y are negative because we chose the positive y-axis to point upward. But we could just as well have chosen the positive y-axis to point downward. In that case the acceleration would have been $a_y = +g$ and all our answers for v_y would have been positive. Either choice of axis is fine; just make sure that you state your choice explicitly in your solution and confirm that the acceleration has the correct sign.

Example 2.7	**Up and down motion in free fall**

You throw a ball vertically upward from the roof of a tall building. The ball leaves your hand at a point even with the roof railing with an upward speed of 15.0 m/s; the ball is then in free fall. On its way back down, it just misses the railing. At the location of the building, $g = 9.80 \text{ m/s}^2$. Find a) the position and velocity of the ball 1.00 s and 4.00 s after leaving your hand; b) the velocity when the ball is 5.00 m above the railing; c) the maximum height reached and the time at which it is reached; and d) the acceleration of the ball when it is at its maximum height.

SOLUTION

IDENTIFY: The words "free fall" in the statement of the problem mean that the acceleration is constant and due to gravity. Our target variables are position [in parts (a) and (c)], velocity [in parts (a) and (b)], and acceleration [in part (d)].

SET UP: In Fig. 2.23 the downward path is displaced a little to the right of its actual position for clarity. Take the origin at the roof railing, at the point where the ball leaves your hand, and take the positive direction to be upward. First, let's collect our data. The initial position y_0 is zero. The initial velocity v_{0y} is $+15.0$ m/s, and the acceleration is $a_y = -g = -9.80 \text{ m/s}^2$. We'll again use Eqs. (2.12) and (2.8) to find the position and velocity, respectively, as functions of time. In part (b) we're asked to find the velocity at a certain *position* rather than at a certain *time,* so it will also be convenient to use Eq. (2.13) for that part.

EXECUTE: a) The position y and velocity v_y at any time t after the ball leaves your hand are given by Eqs. (2.12) and (2.8) with x's replaced by y's, so

$$y = y_0 = v_{0y}t + \frac{1}{2}a_y t^2 = y_0 + v_{0y}t + \frac{1}{2}(-g)t^2$$

$$= (0) + (15.0 \text{ m/s})t + \frac{1}{2}(-9.80 \text{ m/s}^2)t^2$$

$$v_y = v_{0y} + a_y t = v_{0y} + (-g)t$$

$$= 15.0 \text{ m/s} + (-9.80 \text{ m/s}^2)t$$

When $t = 1.00$ s, these equations give

$$y = +10.1 \text{ m} \qquad v_y = +5.2 \text{ m/s}$$

The ball is 10.1 m above the origin (y is positive), and it is moving upward (v_y is positive) with a speed of 5.2 m/s. This is less than the initial speed of 15.0 m/s, as expected.

When $t = 4.00$ s, the equations for y and v_y as functions of time t give

$$y = -18.4 \text{ m} \qquad v_y = -24.2 \text{ m/s}$$

The ball has passed its highest point and is 18.4 m *below* the origin (y is negative). It has a *downward* velocity (v_y is negative) with magnitude 24.2 m/s. This speed is greater than the initial speed, as we should expect for points below the ball's launching point. Note that to get these results, we don't need to find the highest point reached or the time at which it was reached. The equations of motion give the position and velocity at any time, whether the ball is on the way up or on the way down.

b) The velocity v_y at any position y is given by Eq. (2.13) with x's replaced by y's:

$$v_y^2 = v_{0y}^2 + 2a_y(y - y_0) = v_{0y}^2 + 2(-g)(y - 0)$$
$$= (15.0 \text{ m/s})^2 + 2(-9.80 \text{ m/s}^2)y$$

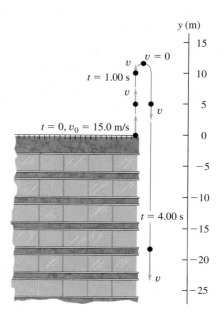

2.23 Position and velocity of a ball thrown vertically upward.

When the ball is 5.00 m above the origin, $y = +5.00$ m, so

$$v_y^2 = (15.0 \text{ m/s})^2 + 2(-9.80 \text{ m/s}^2)(5.00 \text{ m}) = 127 \text{ m}^2/\text{s}^2$$

$$v_y = \pm 11.3 \text{ m/s}$$

We get *two* values of v_y, one positive and one negative. As shown in Fig. 2.23, the ball passes this point twice, once on the way up and again on the way down. The velocity on the way up is $+11.3$ m/s, and on the way down it is -11.3 m/s.

c) At the highest point, the ball stops going up (positive v_y) and starts going down (negative v_y). Just at the instant when it reaches the highest point, $v_y = 0$. The maximum height y_1 can then be found in two ways. The first way is to use Eq. (2.13) and substitute $v_y = 0$, $y_0 = 0$, and $a_y = -g$:

$$0 = v_{0y}^2 + 2(-g)(y_1 - 0)$$

$$y_1 = \frac{v_{0y}^2}{2g} = \frac{(15.0 \text{ m/s})^2}{2(9.80 \text{ m/s}^2)} = +11.5 \text{ m}$$

The second way is find the time at which $v = 0$ using Eq. (2.8), $v_y = v_{0y} + a_y t$, and then substitute this value of t into Eq. (2.12) to find the position at this time. From Eq. (2.8), the time t_1 when the ball reaches the highest point is given by

$$v_y = 0 = v_{0y} + (-g)t_1$$

$$t_1 = \frac{v_{0y}}{g} = \frac{15.0 \text{ m/s}}{9.80 \text{ m/s}^2} = 1.53 \text{ s}$$

Substituting this value of t into Eq. (2.12), we find

$$y = y_0 + v_{0y}t + \frac{1}{2}a_y t^2 = (0) + (15 \text{ m/s})(1.53 \text{ s})$$

$$+ \frac{1}{2}(-9.8 \text{ m/s}^2)(1.53 \text{ s})^2 = +11.5 \text{ m}$$

Notice that with the first way of finding the maximum height, it's not necessary to find the time first.

d) **CAUTION** It's a common misconception that at the highest point of the motion the velocity is zero *and* the acceleration is zero. If this were so, once the ball reached the highest point it would hang there in midair forever! To see why, remember that acceleration is the rate of change of velocity. If the acceleration were zero at the highest point, the ball's velocity would no longer change, and once the ball was instantaneously at rest it would remain at rest forever.

The truth of the matter is that at the highest point, the acceleration is still $a_y = -g = -9.80$ m/s^2, the same value as when the ball is moving up and when it's moving down. The ball stops for an instant at the highest point, but its velocity is continuously changing, from positive values through zero to negative values.

EVALUATE: A useful way to check any motion problem is to draw the graphs of position versus time and velocity versus time. Figure 2.24 shows these graphs for the situation in this problem. Note that the v_y-t graph has a constant negative slope. This means that the acceleration is negative (downward) on the way up, at the highest point, and on the way down.

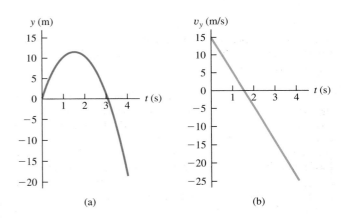

2.24 (a) Position and (b) velocity as functions of time for a ball thrown upward with an initial velocity of 15 m/s.

Example 2.8	**Two solutions or one?**

Find the time when the ball in Example 2.7 is 5.00 m below the roof railing.

SOLUTION

IDENTIFY AND SET UP: We again choose the y-axis as in Fig. 2.23, so y_0, v_{0y}, and $a_y = -g$ have the same values as in Example 2.7. The position y as a function of time t is again given by Eq. (2.12):

$$y = y_0 + v_{0y}t + \frac{1}{2}a_y t^2 = y_0 + v_{0y}t + \frac{1}{2}(-g)t^2$$

We want to solve this for the value of t when $y = -5.00$ m. Since this equation involves t^2, it is a *quadratic* equation for t.

EXECUTE: We first rearrange our equation into the same form as the standard quadratic equation for an unknown x, $Ax^2 + Bx + C = 0$:

$$\left(\frac{1}{2}g\right)t^2 + (-v_{0y})t + (y - y_0) = At^2 + Bt + C = 0$$

so $A = g/2$, $B = -v_{0y}$, and $C = y - y_0$. Using the quadratic formula (Appendix B), we find that this equation has *two* solutions:

$$t = \frac{-B \pm \sqrt{B^2 - 4AC}}{2A}$$

$$= \frac{-(-v_{0y}) \pm \sqrt{(-v_{0y})^2 - 4(g/2)(y - y_0)}}{2(g/2)}$$

$$t = \frac{v_{0y} \pm \sqrt{v_{0y}^2 - 2g(y - y_0)}}{g}$$

Substituting the values $y_0 = 0$, $v_{0y} = +15.0$ m/s, $g = 9.80$ m/s^2, and $y = -5.00$ m, we find

$$t = \frac{(15.0 \text{ m/s}) \pm \sqrt{(15.0 \text{ m/s})^2 - 2(9.80 \text{ m/s}^2)(-5.00 \text{ m} - 0)}}{9.80 \text{ m/s}^2}$$

$$t = +3.36 \text{ s} \quad \text{or} \quad t = -0.30 \text{ s}$$

To decide which of these is the right answer, the key question to ask is, "Are these answers reasonable?" The second answer, $t = -0.30$ s, is simply not reasonable; it refers to a time 0.30 s *before* the ball left your hand! The correct answer is $t = +3.36$ s. The ball is 5.00 m below the railing 3.36 s *after* it leaves your hand.

EVALUATE: Where did the erroneous "solution" $t = -0.30$ s come from? Remember that we began with Eq. (2.12) with $a_y = -g$, that

is, $y = y_0 + v_{0y}t + \frac{1}{2}(-g)t^2$. This equation has built into it the assumption that the acceleration is constant for *all* values of t, whether positive, negative, or zero. Taken at face value, this equation would tell us that the ball had been moving upward in free fall ever since the dawn of time; it eventually passes your hand at $y = 0$ at the special instant we chose to call $t = 0$, then continues in free fall. But anything that this equation describes happening before $t = 0$ is pure fiction, since the ball went into free fall only after leaving your hand at $t = 0$; the "solution" $t = -0.30$ s is part of this fiction.

You should repeat these calculations to find the times at which the ball is 5.00 m *above* the origin ($y = +5.00$ m). The two answers are $t = +0.38$ s and $t = +2.68$ s; these are both positive values of t, and both refer to the real motion of the ball after leaving your hand. The earlier time is when the ball passes through $y = +5.00$ m moving upward, and the later time is when the ball passes through this point moving downward. (Compare this with part (b) of Example 2.7.) You should also solve for the times at which $y = +15.0$ m. In this case, both solutions involve the square root of a negative number, so there are *no* real solutions. This makes sense; we found in part (c) of Example 2.7 that the ball's maximum height is only $y = +11.5$ m, so it never reaches $y = +15.0$ m. While a quadratic equation such as Eq. (2.12) always has two solutions, in some situations one or both of the solutions will not be physically reasonable.

Test Your Understanding

If you toss a ball upward with a certain velocity, it falls freely and reaches a maximum height h. What maximum height does the ball reach if you throw it upward with double the initial velocity?

*2.6 | Velocity and Position by Integration

Acceleration: Known
Velocity: To be determined
Position: To be determined

To London

From New York

2.25 The position and velocity of an airliner crossing the Atlantic are found by integrating its acceleration with respect to time.

This optional section is intended for students who have already learned a little integral calculus. In Section 2.4 we analyzed the special case of straight-line motion with constant acceleration. When a_x is not constant, as is frequently the case, the equations that we derived in that section are no longer valid. But even when a_x varies with time, we can still use the relation $v_x = dx/dt$ to find the velocity v_x as a function of time if the position x is a known function of time. And we can still use $a_x = dv_x/dt$ to find the acceleration a_x as a function of time if the velocity v_x is a known function of time.

In many physical situations, however, position and velocity are not known as functions of time, while the acceleration is. How can we find the position and velocity from the acceleration function $a_x(t)$? This problem arises in navigating an airliner between North America and Europe (Fig. 2.25). The crew of the airliner must know their position precisely at all times, because the airspace over the North Atlantic is very congested. But over the ocean an airliner is usually out of range of both radio navigation beacons on land and air traffic controllers' radar. To determine their position, airliners carry a device called an inertial navigation system (INS), which measures the acceleration of the moving airliner. This is done in much the same way that you can sense changes in the velocity of a car in which you're riding, even when your eyes are closed. (In Chapter 4 we'll discuss how your body detects acceleration.) Given this information, along with the airliner's initial position (say, a particular gate at Kennedy International Airport) and its initial velocity (zero when parked at the gate), the INS calculates and displays for the crew the current velocity and position of the airliner at all times during the flight. Our goal in the remainder of this section is to see how these calculations are done for the simpler case of motion in a straight line with time-varying acceleration. Figure 2.26 shows an everyday example of straight-line motion with varying acceleration.

We first consider a graphical approach. Figure 2.27 is a graph of acceleration versus time for a body whose acceleration is not constant but increases with time. We can divide the time interval between times t_1 and t_2 into many smaller intervals, calling a typical one Δt. Let the average acceleration during Δt be $a_{\text{av-}x}$. From Eq. (2.4) the change in velocity Δv_x during Δt is

$$\Delta v_x = a_{\text{av-}x}\,\Delta t$$

Graphically, Δv_x equals the area of the shaded strip with height $a_{\text{av-}x}$ and width Δt; that is, the area under the curve between the left and right sides of Δt. The total velocity change during any interval (say, t_1 to t_2) is the sum of the velocity changes Δv_x in the small subintervals. So the total velocity change is represented graphically by the *total* area under the a_x-t curve between the vertical lines t_1 and t_2. (In Section 2.4 we showed this to be true in the special case in which the acceleration is constant.)

In the limit that all the Δt's become very small and their number very large, the value of $a_{\text{av-}x}$ for the interval from any time t to $t + \Delta t$ approaches the instantaneous acceleration a_x at time t. In this limit, the area under the a_x-t curve is the *integral* of a_x (which is in general a function of t) from t_1 to t_2. If v_{1x} is the velocity of the body at time t_1 and v_{2x} is the velocity at time t_2, then

$$v_{2x} - v_{1x} = \int_{v_{1x}}^{v_{2x}} dv_x = \int_{t_1}^{t_2} a_x\,dt \qquad (2.15)$$

The change in velocity v_x is the integral of acceleration a_x with respect to time.

We can carry out exactly the same procedure with the curve of velocity versus time, where v_x is in general a function of t. If x_1 is a body's position at time t_1 and x_2 is its position at time t_2, from Eq. (2.2) the displacement Δx during a small time interval Δt is equal to $v_{\text{av-}x}\,\Delta t$, where $v_{\text{av-}x}$ is the average velocity during Δt. The total displacement $x_2 - x_1$ during the interval $t_2 - t_1$ is given by

$$x_2 - x_1 = \int_{x_1}^{x_2} dx = \int_{t_1}^{t_2} v_x\,dt \qquad (2.16)$$

The change in position x—that is, the displacement—is the time integral of velocity v_x. Graphically, the displacement between times t_1 and t_2 is the area under the v_x-t curve between those two times. (This is the same result that we obtained in Section 2.4 for the special case in which v_x is given by Eq. (2.8).)

If $t_1 = 0$ and t_2 is any later time t, and if x_0 and v_{0x} are the position and velocity, respectively, at time $t = 0$, then we can rewrite Eqs. (2.15) and (2.16) as follows:

2.26 When you push your car's accelerator pedal to the floorboard, the resulting acceleration is *not* constant: the greater the car's speed, the more slowly it gains additional speed. A typical car takes twice as long to accelerate from 50 km/h to 100 km/h as it does to accelerate from 0 to 50 km/h.

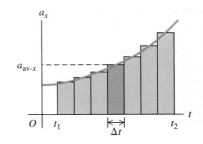

2.27 The area under an a_x-t graph between times t_1 and t_2 equals the change in velocity, $v_{2x} - v_{1x}$, that occurs between those times.

$$v_x = v_{0x} + \int_0^t a_x \, dt \qquad (2.17)$$

$$x = x_0 + \int_0^t v_x \, dt \qquad (2.18)$$

Here x and v_x are the position and velocity at time t. If we know the acceleration a_x as a function of time and we know the initial velocity v_{0x} we can use Eq. (2.17) to find the velocity v_x at any time; in other words, we can find v_x as a function of time. Once we know this function, and given the initial position x_0, we can use Eq. (2.18) to find the position x at any time.

| Example 2.9 | **Motion with changing acceleration** |

Sally is driving along a straight highway in her classic 1965 Mustang. At time $t = 0$, when Sally is moving at 10 m/s in the positive x-direction, she passes a signpost at $x = 50$ m. Her acceleration is a function of time:

$$a_x = 2.0 \text{ m/s}^2 - (0.10 \text{ m/s}^3)t$$

a) Derive expressions for her velocity and position as functions of time. b) At what time is her velocity greatest? c) What is the maximum velocity? d) Where is the car when it reaches maximum velocity?

SOLUTION

IDENTIFY AND SET UP: The acceleration is a function of time, so we cannot use the constant-acceleration formulas of Section 2.4. Instead we must use Eqs. (2.17) and (2.18) to find the velocity and position, respectively, as functions of time. Once we have those functions, we'll be able to answer a variety of questions about the motion.

EXECUTE: a) At time $t = 0$, Sally's position is $x_0 = 50$ m and her velocity is $v_{0x} = 10$ m/s. Since we are given the acceleration a_x as a function of time, we first use Eq. (2.17) to find the velocity v_x as a function of time t. The integral of t^n is $\int t^n dt = \frac{1}{n+1}t^{n+1}$ for $n \neq -1$, so

$$v_x = 10 \text{ m/s} + \int_0^t [2.0 \text{ m/s}^2 - (0.10 \text{ m/s}^3)t] dt$$

$$= 10 \text{ m/s} + (2.0 \text{ m/s}^2)t - \frac{1}{2}(0.10 \text{ m/s}^3)t^2$$

Then we use Eq. (2.18) to find x as a function of t:

$$x = 50 \text{ m} + \int_0^t \left[10 \text{ m/s} + (2.0 \text{ m/s}^2)t - \frac{1}{2}(0.10 \text{ m/s}^3)t^2 \right] dt$$

$$= 50 \text{ m} + (10 \text{ m/s})t + \frac{1}{2}(2.0 \text{ m/s}^2)t^2 - \frac{1}{6}(0.10 \text{ m/s}^3)t^3$$

Figure 2.28 shows graphs of x, v_x, and a as functions of time. Note that for any time t the slope of the x-t graph equals the value of v_x, and the slope of the v_x-t graph equals the value of a_x.

b) The maximum value of v_x occurs when v_x stops increasing and begins to decrease. At this instant, $dv_x/dt = a_x = 0$. Setting the expression for acceleration equal to zero, we obtain

$$0 = 2.0 \text{ m/s}^2 - (0.10 \text{ m/s}^3)t$$

$$t = \frac{2.0 \text{ m/s}^2}{0.10 \text{ m/s}^3} = 20 \text{ s}$$

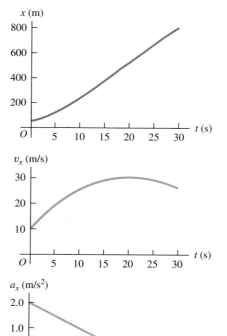

2.28 The position, velocity, and acceleration of the car in Example 2.9 as functions of time. Can you show that if this motion continues, the car will stop at $t = 44.5$ s?

c) We find the maximum velocity by substituting $t = 20$ s (when velocity is maximum) into the general velocity equation:

$$v_{max-x} = 10 \text{ m/s} + (2.0 \text{ m/s}^2)(20 \text{ s}) - \frac{1}{2}(0.10 \text{ m/s}^3)(20 \text{ s})^2$$

$$= 30 \text{ m/s}$$

d) The maximum value of v_x occurs at time $t = 20$ s. We obtain the position of the car (that is, the value of x) at that time by substituting $t = 20$ s into the general expression for x:

$$x = 50 \text{ m} + (10 \text{ m/s})(20 \text{ s}) + \frac{1}{2}(2.0 \text{ m/s}^2)(20 \text{ s})^2$$

$$- \frac{1}{6}(0.10 \text{ m/s}^3)(20 \text{ s})^3$$

$$= 517 \text{ m}$$

EVALUATE: We can use Fig. 2.28 to help us interpret our results. The bottom graph in this figure shows that a_x is positive between $t = 0$ and $t = 20$ s and negative after that. It is zero at $t = 20$ s, the time at which v_x is maximum (the high point in the middle graph). The car speeds up until $t = 20$ s (because v_x and a_x have the same sign) and slows down after $t = 20$ s (because v_x and a_x have opposite signs).

Since v_x is maximum at $t = 20$ s, the x-t graph (the top graph in Fig. 2.28) has its maximum positive slope at this time. Note that the x-t graph is concave up (curved upward) from $t = 0$ to $t = 20$ s, when a_x is positive. The graph is concave down (curved downward) after $t = 20$ s, when a_x is negative.

Example 2.10	**Constant-acceleration formulas via integration**

Use Eqs. (2.17) and (2.18) to find v_x and x as functions of time in the case in which the acceleration is constant. Compare the results to the constant-acceleration formulas $v_x = v_{0x} + a_x t$ (Eq. 2.8) and $x = x_0 + v_{0x}t + \frac{1}{2}a_x t^2$ (Eq. 2.12), which were derived in Section 2.4 without using integration.

SOLUTION

EXECUTE: From Eq. (2.17) the velocity is given by

$$v_x = v_{0x} + \int_0^t a_x \, dt = v_{0x} + a_x \int_0^t dt = v_{0x} + a_x t$$

We were able to take a_x outside the integral because it is constant. Our result is identical to Eq. (2.8), as it should be. Substituting this expression for v_x into Eq. (2.18), we get

$$x = x_0 + \int_0^t v_x \, dt = x_0 + \int_0^t (v_{0x} + a_x t) \, dt$$

Since v_{0x} and a_x are constants, they may be taken outside the integral:

$$x = x_0 + v_{0x} \int_0^t dt + a_x \int_0^t t \, dt = x_0 + v_{0x}t + \frac{1}{2}a_x t^2$$

EVALUATE: This result is the same as Eq. (2.12). Our expressions for v_x and x, Eqs. (2.17) and (2.18), which we developed to deal with cases in which acceleration depends on time, can be used just as well when the acceleration is constant.

Test Your Understanding

If the acceleration a_x is increasing with time, will the v_x-t graph be a straight line, concave up, or concave down?

SUMMARY

When a particle moves along a straight line, we describe its position with respect to an origin O by means of a coordinate such as x. The particle's average velocity $v_{av\text{-}x}$ during a time interval $\Delta t = t_2 - t_1$ is equal to its displacement $\Delta x = x_2 - x_1$ divided by Δt. (See Example 2.1)

$$v_{av\text{-}x} = \frac{x_2 - x_1}{t_2 - t_1} = \frac{\Delta x}{\Delta t} \quad (2.2)$$

The instantaneous velocity v_x at any time t is equal to the average velocity for the time interval from t to $t + \Delta t$ in the limit that Δt goes to zero. Equivalently, v_x is the derivative of the position function with respect to time. (See Example 2.1)

$$v_x = \lim_{\Delta t \to 0} \frac{\Delta x}{\Delta t} = \frac{dx}{dt} \quad (2.3)$$

The average acceleration $a_{av\text{-}x}$ during a time interval Δt is equal to the change in velocity $\Delta v_x = v_{2x} - v_{1x}$ during that time interval divided by Δt. The instantaneous acceleration a_x is the limit of $a_{av\text{-}x}$ as Δt goes to zero, or the derivative of v_x with respect to t. (See Examples 2.2 and 2.3)

$$a_{av\text{-}x} = \frac{v_{2x} - v_{1x}}{t_2 - t_1} = \frac{\Delta v_x}{\Delta t} \quad (2.4)$$

$$a_x = \lim_{\Delta t \to 0} \frac{\Delta v_x}{\Delta t} = \frac{dv_x}{dt} \quad (2.5)$$

When the acceleration is constant, four equations relate the position x and velocity v_x at any time t to the initial position x_0, the initial velocity v_{0x} (both measured at time $t = 0$), and the acceleration a_x. (See Examples 2.4 and 2.5)

constant acceleration only:

$$v_x = v_{0x} + a_x t \quad (2.8)$$

$$x = x_0 + v_{0x}t + \frac{1}{2}a_x t^2 \quad (2.12)$$

$$v_x^2 = v_{0x}^2 + 2a_x(x - x_0) \quad (2.13)$$

$$x - x_0 = \left(\frac{v_{0x} + v_x}{2}\right)t \quad (2.14)$$

Free fall is a case of motion with constant acceleration. The magnitude of the acceleration due to gravity is a positive quantity, g. The acceleration of a body in free fall is always downward. (See Examples 2.6 through 2.8)

When the acceleration is not constant but is a known function of time, we can find the velocity and position as functions of time by integrating the acceleration function. (See Examples 2.9 and 2.10)

$$v_x = v_{0x} + \int_0^t a_x \, dt \quad (2.17)$$

$$x = x_0 + \int_0^t v_x \, dt \quad (2.18)$$

Key Terms

particle, 41
average velocity, 41
x-t graph, 43
instantaneous velocity, 44
derivative, 44

speed, 44
motion diagram, 47
average acceleration, 48
instantaneous acceleration, 49

v_x-t graph, 50
a_x-t graph, 53
free fall, 58
acceleration due to gravity, 59

Your Notes

Answer to Chapter Opening Question

Yes. Acceleration refers to *any* change in velocity, including both speeding up and slowing down.

Answers to Test Your Understanding Questions

Section 2.1 Both the truck and the car undergo the same total displacement Δx (from point A to point B) during the same time interval Δt. Hence they have the same average velocity $v_{av\text{-}x} = \Delta x / \Delta t$. The details of what happens during the time interval don't matter.

Section 2.2 The velocity is positive when the slope of the x-t graph is positive (point P), negative when the slope is negative (point R), and zero when the slope is zero (points Q and S). The speed is greatest at times for which the slope of the x-t graph is steepest (either positive or negative), which is at point R.

Section 2.3 The acceleration is positive when the x-t graph is curved upward (concave up), as at point S, and negative when it is curved downward (concave down), as at point Q. The acceleration is zero when the x-t graph is not curved but is a straight line, as at points P and R. At P, $v_x > 0$ and $a_x = 0$ (speed is not changing); at Q, $v_x > 0$ and $a_x < 0$ (speed is decreasing); at R, $v_x < 0$ and $a_x = 0$ (speed is not changing); and at S, $v_x < 0$ and $a_x > 0$ (speed is decreasing).

Section 2.4 The motorist and the officer have the same velocity at $t = 5.0$ s.

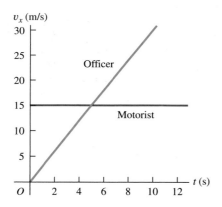

Section 2.5 Use Eq. (2.13) with x replaced by y and $a_y = -g$; $v_y^2 = v_{0y}^2 - 2g(y - y_0)$. The starting height is $y_0 = 0$ and the velocity at the maximum height $y = h$ is $v_y = 0$, so $0 = v_{0y}^2 - 2gh$ and $h = v_{0y}^2/2g$. If the initial velocity is increased by a factor of 2, the maximum height increases by a factor of $2^2 = 4$ and the ball goes to height $4h$.

Section 2.6 The acceleration a_x is equal to the slope of the v_x-t graph. If a_x is increasing, the slope of the v_x-t graph is also increasing and the graph is concave up.

Discussion Questions

Q2.1 Does the speedometer of a car measure speed or velocity? Explain.

Q2.2 In the same time interval, a car accelerates from 15 m/s to 20 m/s while a truck accelerates from 36 m/s to 40 m/s. Which vehicle has the greater average acceleration? Explain.

Q2.3 Can an object with constant acceleration reverse its direction of travel? Explain.

Q2.4 Under what conditions is average velocity equal to instantaneous velocity?

Q2.5 In a given time interval, is the total displacement of a particle equal to the product of the average velocity and the time interval, even when the velocity is not constant? Explain.

Q2.6 Under what conditions does the magnitude of the average velocity equal the average speed?

Q2.7 When a Dodge Viper is at Elwood's Car Wash, a BMW Z3 is at Elm and Main. Later, when the Dodge reaches Elm and Main, the BMW reaches Elwood's Car Wash. How are the cars' average velocities between these two times related?

Q2.8 A driver in Massachusetts was sent to traffic court for speeding. The evidence against the driver was that a policewoman observed the driver's car alongside a second car at a certain moment, and that the policewoman had already clocked the second car as going faster than the speed limit. The driver argued that "the second car was passing me. I was not speeding." The judge ruled against the driver because, in the judge's words, "if two cars were side by side, you were both speeding." If you were a lawyer representing the accused driver, how would you argue this case?

Q2.9 Can you have a zero displacement and a nonzero average velocity? A nonzero velocity? Illustrate your answers on a x-t graph.

Q2.10 Can you have zero acceleration and nonzero velocity? Explain using a v_x-t graph.

Q2.11 Can you have zero velocity and nonzero average acceleration? Zero velocity and nonzero acceleration? Explain using a v_x-t graph and give an example of such motion.

Q2.12 An automobile is traveling west. Can it have a velocity toward the west and at the same time have an acceleration toward the east? Under what circumstances?

Q2.13 The official's truck in Fig. 2.2 is at $x_1 = 277$ m at $t_1 = 16.0$ s, and is at $x_2 = 19$ m at $t_2 = 25.0$ s. a) Sketch *two* different possible x-t graphs for the motion of the truck. b) Does the average velocity $v_{av\text{-}x}$ during the time interval from t_1 to t_2 have the same value for both of your graphs? Why or why not?

Q2.14 Under constant acceleration the average velocity of a particle is half the sum of its initial and final velocities. Is this still true if the acceleration is *not* constant? Explain.

Q2.15 You throw a baseball straight up in the air so that it rises to a maximum height much greater than your height. Is the magnitude of the acceleration greater while it is being thrown or after it leaves your hand? Explain.

Q2.16 Prove these statements: i) As long as you can neglect the effects of the air, if you throw anything vertically upward, it will have the same speed when it returns to the release point as when it was released. ii) The time of flight will be twice the time it takes to get to its highest point.

Q2.17 In Example 2.7, substituting $y = -18.4$ m into Eq. (2.13) gives $v_y = \pm 24.2$ m/s. The negative root is the velocity at $t = 4.00$ s. Explain the meaning of the positive root.

Q2.18 If the initial position and initial velocity of a vehicle are known and a record is kept of the acceleration at each instant, can you compute the vehicle's position after a certain time from these data? If so, explain how this might be done.

Q2.19 You and your friend are leaning over the edge of the roof of a tall building. At the same instant you throw a baseball straight up with an initial speed v_0 and your friend throws a marble straight down with the same initial speed v_0. Assume that you can ignore air resistance. Which object will reach the ground first, the baseball or the marble? How will their speeds compare just before hitting the ground?

Q2.20 A ball is dropped from rest from the top of a building of height h. At the same instant, a second ball is projected vertically upward from ground level, such that it has zero speed when it reaches the top of the building. When the two balls pass each other, which ball has the greater speed, or do they have the same speed? Explain. Where will the two balls be when they are alongside each other: at height $h/2$ above the ground, below this height, or above this height? Explain.

Exercises

Section 2.1 Displacement, Time, and Average Velocity

2.1 A rocket carrying a satellite is accelerating straight up from the earth's surface. At 1.15 s after liftoff, the rocket clears the top of its launch platform, 63 m above the ground. After an additional 4.75 s it is 1.00 km above the ground. Calculate the magnitude of the average velocity of the rocket for a) the 4.75-s part of its flight; b) the first 5.90 s of its flight.

2.2 In an experiment, a shearwater (a seabird) was taken from its nest, flown 5150 km away, and released. It found its way back to its nest 13.5 days after release. If we place the origin in the nest and extend the $+x$-axis to the release point, what was the bird's average velocity in m/s a) for the return flight? b) For the whole episode, from leaving the nest to returning?

2.3 Trip Home. You normally drive on the freeway between San Diego and Los Angeles at an average speed of 105 km/h (65 mi/h), and the trip takes 2 h and 20 min. On a Friday afternoon, however, heavy traffic slows you down and you drive the same distance at an average speed of only 70 km/h (43 mi/h). How much longer does the trip take?

2.4 From Pillar to Post. Starting from a pillar, you run 200 m east (the $+x$-direction) at an average speed of 5.0 m/s, and then run 280 m west at an average speed of 4.0 m/s to a post. Calculate a) your average speed from pillar to post; b) your average velocity from pillar to post.

2.5 Two runners start simultaneously from the same point on a circular 200-m track and run in the same direction. One runs at a constant speed of 6.20 m/s and the other at a constant speed of 5.50 m/s. When will the fast one overtake ("lap") the slower one and how far from the starting point will each have run?

2.6 Geology. Earthquakes produce several types of shock waves. The most well-known are the P-waves (P for *primary* or *pressure*) and the S-waves (S for *secondary* or *shear*). In the earth's crust, the P-waves travel at around 6.5 km/s while the S-waves move at about 3.5 km/s. The actual speeds vary depending on the type of material they are going through. The time delay between the arrival of these two waves at a seismic recording station tells geologists how far away the earthquake occurred. If the time delay is 33 s, how far from the seismic station did the earthquake occur?

2.7 a) Your old VW van sputters along at an average speed of 8.0 m/s for 60 s, and then catches hold and zips along at an average speed of 20.0 m/s for another 60 s. Calculate your average speed for the entire 120 s. b) Suppose the speed of 8.0 m/s was maintained while you traveled 240 m, followed by the average speed of 20.0 m/s for another 240 m. Calculate the van's average speed for the entire distance. c) In which case is the average speed for the entire motion the numerical average of the two speeds?

2.8 A Honda Civic travels in a straight line along a road. Its distance x from a stop sign is given as a function of time t by the equation $x(t) = \alpha t^2 - \beta t^3$, where $\alpha = 1.50$ m/s^2 and $\beta = 0.0500$ m/s^3. Calculate the average velocity of the car for the following time intervals: a) $t = 0$ to $t = 2.00$ s; b) $t = 0$ to $t = 4.00$ s; c) $t = 2.00$ s to $t = 4.00$ s.

Section 2.2 Instantaneous Velocity

2.9 A car is stopped at a traffic light. It then travels along a straight road so that its distance from the light is given by $x(t) = bt^2 - ct^3$, where $b = 2.40$ m/s^2 and $c = 0.120$ m/s^3. a) Calculate the average velocity of the car for the time interval $t = 0$ to $t = 10.0$ s. b) Calculate the instantaneous velocity of the car at i) $t = 0$; ii) $t = 5.0$ s; iii) $t = 10.0$ s. c) How long after starting from rest is the car again at rest?

2.10 A physics professor leaves her house and walks along the sidewalk toward campus. After 5 min it starts to rain and she returns home. Her distance from her house as a function of time is shown in Fig. 2.29. At which of the labeled points is her velocity a) zero? b) constant and positive? c) constant and negative? d) increasing in magnitude? e) decreasing in magnitude?

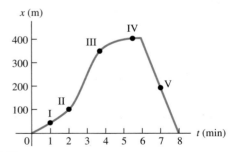

Figure 2.29 Exercise 2.10.

Section 2.3 Average and Instantaneous Acceleration

2.11 A test driver at Incredible Motors, Inc. is testing a new model car with a speedometer calibrated to read m/s rather than mi/h. The following series of speedometer readings was obtained during a test run along a long, straight road:

Time (s)	0	2	4	6	8	10	12	14	16
Speed (m/s)	0	0	2	6	10	16	19	22	22

a) Compute the average acceleration during each 2-s interval. Is the acceleration constant? Is it constant during any part of the test run? b) Make a v_x-t graph of the data above, using scales of 1 cm = 1 s horizontally and 1 cm = 2 m/s vertically. Draw a smooth curve through the plotted points. By measuring the slope of your curve, find the instantaneous acceleration at t = 9 s, 13 s, and 15 s.

2.12 Figure 2.30 shows the velocity of a solar-powered car as a function of time. The driver accelerates from a stop sign, cruises for 20 s at a constant speed of 60 km/h, and then brakes to come to a stop 40 s after leaving the stop sign. Compute the average acceleration for the following time intervals: a) t = 0 to t = 10 s; b) t = 30 s to t = 40 s; c) t = 10 s to t = 30 s; d) t = 0 to t = 40 s.

Figure 2.30 Exercises 2.12 and 2.13.

2.13 Refer to Exercise 2.12 and Fig. 2.30. a) During what time interval does the instantaneous acceleration a_x have its most positive value? b) During what time interval does the instantaneous acceleration a_x have its most negative value? c) What is the instantaneous acceleration at t = 20 s? d) What is the instantaneous acceleration at t = 35 s? e) In a motion diagram (like Fig. 2.13b or Fig. 2.14b), show the position, velocity, and acceleration of the car at the four times t = 5 s, t = 15 s, t = 25 s, and t = 35 s.

2.14 A person looking out the window of a tall office building observes what he suspects is a UFO. The person records the position of the object as a function of time and finds that it is given by $\vec{r}(t) = - (5.0 \text{ m/s})t\hat{i} + (10.0 \text{ m/s})t\hat{j} + [(7.0 \text{ m/s})t - (3.0 \text{ m/s}^2)t^2]\hat{k}$. a) Find the displacement, velocity, and acceleration vectors for the object at t = 5.0 s. b) Is there a time when the object has a velocity of zero? c) Is the acceleration of the object constant, or does it change in time?

2.15 A turtle crawls along a straight line, which we will call the x-axis with the positive direction to the right. The equation for the turtle's position as a function of time is $x(t)$ = 50.0 cm + (2.00 cm/s)t − (0.0625 cm/s^2)t^2. a) Find the turtle's initial velocity, initial position, and initial acceleration. b) At what time t is the velocity of the turtle zero? c) How long after starting does it take the turtle to return to its starting point? d) At what times t is the turtle a distance of 10.0 cm from its starting point? What is the velocity (magnitude and direction) of the turtle at each of these times? e) Sketch graphs of x versus t, v_x versus t, and a_x versus t, for the time interval t = 0 to t = 40 s.

2.16 An astronaut has left the International Space Station to test a new space scooter. Her partner measures the following velocity changes, each taking place in a 10-s interval. What are the magnitude, the algebraic sign, and the direction of the average acceleration in each interval? Assume that the positive direction is to the right. a) At the beginning of the interval the astronaut is moving toward the right along the x-axis at 15.0 m/s, and at the end of the interval she is moving toward the right at 5.0 m/s. b) At the beginning she is moving toward the left at 5.0 m/s, and at the end she is moving toward the left at 15.0 m/s. c) At the beginning she is moving toward the right at 15.0 m/s, and at the end she is moving toward the left at 15.0 m/s.

2.17 Auto Acceleration. a) Based on your experiences of riding in automobiles, estimate the magnitude of a car's average acceleration when it brakes from highway speeds to a sudden stop. b) Explain why this average acceleration could be regarded as either positive or negative.

2.18 A car's velocity as a function of time is given by $v_x(t) = \alpha + \beta t^2$, where α = 3.00 m/s and β = 0.100 m/s^3. a) Calculate the average acceleration for the time interval t = 0 to t = 5.00 s. b) Calculate the instantaneous acceleration for i) t = 0; ii) t = 5.00 s. c) Draw accurate v_x-t and a_x-t graphs for the car's motion between t = 0 and t = 5.00 s.

2.19 Figure 2.31 is a graph of the coordinate of a spider crawling along the x-axis. a) Graph its velocity and acceleration as functions of time. b) In a motion diagram (like Fig. 2.13b or Fig. 2.14b), show the position, velocity, and acceleration of the spider at the five times t = 2.5 s, t = 10 s, t = 20 s, t = 30 s, and t = 37.5 s.

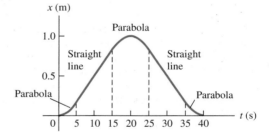

Figure 2.31 Exercise 2.19.

2.20 The position of the front bumper of a test car under microprocessor control is given by $x(t)$ = 2.17 m + (4.80 m/s^2)t^2 − (0.100 m/s^6)t^6. Find its a) position and acceleration at the instants when the car has zero velocity. b) Draw x-t, v_x-t, and a_x-t graphs for the motion of the bumper between t = 0 and t = 2.00 s.

Section 2.4 Motion with Constant Acceleration

2.21 An antelope moving with constant acceleration covers the distance between two points 70.0 m apart in 7.00 s. Its speed as it passes the second point is 15.0 m/s. a) What is its speed at the first point? b) What is the acceleration?

2.22 The catapult of the aircraft carrier *USS Abraham Lincoln* accelerates an F/A-18 Hornet jet fighter from rest to a takeoff speed of 173 mi/h in a distance of 307 ft. Assume constant acceleration. a) Calculate the acceleration of the fighter in m/s^2. b) Calculate the time required for the fighter to accelerate to takeoff speed.

2.23 Automobile Airbags. The human body can survive a negative acceleration trauma incident (sudden stop) if the magnitude of the acceleration is less than 250 m/s^2 (approximately 25g). If you are in an automobile accident with an initial speed of 105 km/h (65 mi/h) and are stopped by an airbag that inflates from the dashboard, over what distance must the airbag stop you for you to survive the crash?

2.24 An airplane travels 280 m down the runway before taking off. If it starts from rest, moves with constant acceleration, and becomes airborne in 8.00 s, what is its speed, in m/s, when it takes off?

2.25 Entering the Freeway. A car sits in an entrance ramp to a freeway, waiting for a break in the traffic. The driver sees a small gap between a van and an 18-wheel truck and accelerates with constant acceleration along the ramp and onto the freeway. The car starts from rest, moves in a straight line, and has a speed of 20 m/s (45 mi/h) when it reaches the end of the 120-m-long ramp. a) What is the acceleration of the car? b) How much time does it take the car to travel the length of the ramp? c) The traffic on the freeway is moving at a constant speed of 20 m/s. What distance does the traffic travel while the car is moving the length of the ramp?

2.26 Figures 2.15, 2.16, 2.17, and 2.18 were drawn for motion with constant acceleration with positive values of x_0, v_{0x}, and a_x. Redraw these four figures for the following cases: a) $x_0 < 0$, $v_{0x} < 0$, $a_x < 0$; b) $x_0 > 0$, $v_{0x} < 0$, $a_x > 0$; c) $x_0 > 0$, $v_{0x} > 0$, $a_x < 0$.

2.27 According to 1994 test data, a Ford Aspire automobile travels 0.250 miles in 19.9 s, starting from rest. The same car, when braking from 60.0 mph on dry pavement, stops in 146 ft. Assume constant acceleration in each part of the motion, but not necessarily the same acceleration when slowing down as when speeding up. a) Find the acceleration of this car when speeding up and when braking. b) If its acceleration is constant, how fast (in mph) should this car be traveling after 0.250 miles of acceleration? The actual measured speed is 70.0 mph; what does this tell you about the motion? c) How long does it take this car to stop while braking from 60.0 mph?

2.28 A cat walks in a straight line, which we shall call the x-axis with the positive direction to the right. As an observant physicist, you make measurements of this cat's motion and construct a graph of the feline's velocity as a function of time (Fig. 2.32). a) Find the cat's velocity at $t = 4.0$ s; at $t = 7.0$ s. b) What is the cat's acceleration at $t = 3.0$ s? At $t = 6.0$ s? At $t = 7.0$ s? c) What distance does

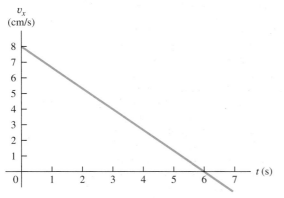

Figure 2.32 Exercise 2.28.

the cat move during the first 4.5 s? From $t = 0$ to $t = 7.5$ s? d) Sketch clear graphs of the cat's acceleration and position as functions of time, assuming that the cat started at the origin.

2.29 At time $t = 0$ a Corvette is traveling on a long, straight stretch of highway in Arizona at a constant speed of 30 m/s. This motion continues for 20 s. Then the driver, worried that she is behind schedule, accelerates at a constant rate for 5 s to bring the car's speed to 40 m/s. The car travels at this new speed for 10 s. But then the driver spots a police motorcycle parked behind a large cactus up ahead and slows with a constant acceleration of magnitude 4.0 m/s^2 until the car's speed is reduced once again to the legal limit of 30 m/s. She then maintains this speed, and waves to the officer as she passes him 5 s later. a) For the motion of the car from $t = 0$ until the moment the car passes the parked motorcycle, draw accurate a_x-t, v_x-t, and x-t graphs. b) In a motion diagram (like Fig. 2.13b or 2.14b), show the position, velocity, and acceleration of the car.

2.30 At $t = 0$ a car is stopped at a traffic light. When the light turns green, the car starts to speed up, and gains speed at a constant rate until it reaches a speed of 20 m/s 8 seconds after the light turns green. The car continues at a constant speed for 60 m. Then the driver sees a red light up ahead at the next intersection, and starts slowing down at a constant rate. The car stops at the red light, 180 m from where it was at $t = 0$. a) Draw accurate x-t, v_x-t, and a_x-t graphs for the motion of the car. b) In a motion diagram (like Fig. 2.13b or 2.14b), show the position, velocity, and acceleration of the car.

2.31 The graph in Fig. 2.33 shows the velocity of a motorcycle police officer plotted as a function of time. a) Find the instantaneous acceleration at $t = 3$ s, at $t = 7$ s, and at

Figure 2.33 Exercise 2.31.

$t = 11$ s. b) How far does the officer go in the first 5 s? The first 9 s? The first 13 s?

2.32 Figure 2.34 is a graph of the acceleration of a model railroad locomotive moving on the x-axis. Graph its velocity and coordinate as functions of time if $x = 0$ and $v_x = 0$ at $t = 0$.

Figure 2.34 Exercise 2.32.

2.33 A spaceship ferrying workers to Moon Base I takes a straight-line path from the earth to the moon, a distance of 384,000 km. Suppose it accelerates at 20.0 m/s^2 for the first 15.0 min of the trip, then travels at constant speed until the last 15.0 min, when it accelerates at -20.0 m/s^2, just coming to rest as it reaches the moon. a) What is the maximum speed attained? b) What fraction of the total distance is traveled at constant speed? c) What total time is required for the trip?

2.34 A subway train starts from rest at a station and accelerates at a rate of 1.60 m/s^2 for 14.0 s. It runs at constant speed for 70.0 s and slows down at a rate of 3.50 m/s^2 until it stops at the next station. Find the *total* distance covered.

2.35 Two cars, *A* and *B,* move along the *x*-axis. Figure 2.35 is a graph of the positions of *A* and *B* versus time. a) In motion diagrams (like Figure 2.13b or Figure 2.14b), show the position, velocity, and acceleration of each of the two cars at $t = 0$, $t = 1$ s, and $t = 3$ s. b) At what time(s), if any, do *A* and *B* have the same position? c) Graph velocity versus time for both *A* and *B*. d) At what time(s), if any, do *A* and *B* have the same velocity? e) At what time(s), if any, does car *A* pass car *B?* f) At what time(s), if any, does car *B* pass car *A?*

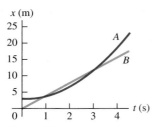

Figure 2.35 Exercise 2.35.

2.36 At the instant the traffic light turns green, a car that has been waiting at an intersection starts ahead with a constant acceleration of 3.20 m/s^2. At the same instant a truck, traveling with a constant speed of 20.0 m/s, overtakes and passes the car. a) How far beyond its starting point does the car overtake the truck? b) How fast is the car traveling when it overtakes the truck? c) Sketch an *x-t* graph of the motion of both vehicles. Take $x = 0$ at the intersection. d) Sketch a v_x-*t* graph of the motion of both vehicles.

2.37 As in Example 2.5, a car is traveling at a constant velocity with magnitude v_C. At the instant that the car passes a motorcycle officer, the motorcycle accelerates from rest with acceleration a_{Mx}. a) Sketch an *x-t* graph of the motion of both objects. Show that when the motorcycle overtakes the car, the motorcycle has a speed twice that of the car, no matter what the value of a_{Mx}. b) Let *d* be the distance the motorcycle travels before catching up with the car. In terms of *d,* how far has the motorcycle traveled when its velocity equals the velocity of the car?

Section 2.5 Freely Falling Bodies

2.38 Raindrops. If the effects of the air acting on falling raindrops is ignored, then we can treat raindrops as freely falling objects. a) Rain clouds are typically a few hundred meters above the ground. Estimate the speed with which raindrops would strike the ground if they were freely falling objects. Give your estimate in m/s, km/h, and mi/h. b) Estimate (from your own personal observations of rain) the speed with which raindrops actually strike the ground. c) Based on your answers to parts (a) and (b), is it a good approximation to neglect the effects of the air on falling raindrops? Explain.

2.39 a) If a flea can jump straight up to a height of 0.440 m, what is its initial speed as it leaves the ground? b) How long is it in the air?

2.40 Touchdown on the Moon. A lunar lander is making its descent to Moon Base I (Fig. 2.36). The lander descends slowly under the retro-thrust of its descent engine. The engine is cut off when the lander is 5.0 m above the surface and has a downward speed of 0.8 m/s. With the engine off, the lander is in free fall. What is the speed of the lander just before it touches the surface? The acceleration due to gravity on the moon is 1.6 m/s^2.

Figure 2.36 Exercise 2.40.

2.41 A Simple Reaction-Time Test. A meter stick is held vertically above your hand, with the lower end between your thumb and first finger. On seeing the meter stick released, you grab it with these two fingers. You can calculate your reaction time from the distance the meter stick falls, read directly from the point where your fingers grabbed it. a) Derive a relationship for your reaction time in terms of this measured distance, *d.* b) If the measured distance is 17.6 cm, what is the reaction time?

2.42 A brick is dropped (zero initial speed) from the roof of a building. The brick strikes the ground in 2.50 s. You may ignore air resistance, so the brick is in free fall. a) How tall, in meters, is the building? b) What is the magnitude of the brick's velocity just before it reaches the ground? c) Sketch a_y-*t*, v_y-*t*, and *y-t* graphs for the motion of the brick.

2.43 Jillian angrily throws her engagement ring straight up from the roof of a building, 12.0 m above the ground, with an initial speed of 5.00 m/s. You may ignore air resistance. For the motion from her hand to the ground, what are the magnitude and direction of a) the average velocity of the ring? b) the average acceleration of the ring? c) In how many seconds after being thrown does the ring strike the ground? d) What is the speed of the ring just before it strikes the ground? e) Sketch a_y-*t*, v_y-*t*, and *y-t* graphs for the motion.

2.44 A hot-air balloonist, rising vertically with a constant velocity of magnitude 5.00 m/s, releases a sandbag at an instant when the balloon is 40.0 m above the ground (Fig. 2.37). After it is released, the sandbag is in free fall. a) Compute the position and velocity of the sandbag at 0.250 s and 1.00 s after its release. b) How many seconds after its release will the bag strike the ground? c) With what magnitude of velocity does it strike? d) What is the greatest height above the ground that the sandbag reaches? e) Sketch a_y-*t*, v_y-*t*, and *y-t* graphs for the motion.

Figure 2.37 Exercise 2.44.

2.45 A student throws a water balloon vertically downward from the top of a building. The balloon leaves the thrower's hand with a speed of 6.00 m/s. Air resistance may be ignored, so the water balloon is in free fall after it leaves the thrower's hand. a) What is its speed after falling for 2.00 s? b) How far does it fall in 2.00 s? c) What is the magnitude of its velocity after falling 10.0 m? d) Sketch a_y-t, v_y-t, and y-t graphs for the motion.

2.46 An egg is thrown nearly vertically upward from a point near the cornice of a tall building. It just misses the cornice on the way down and passes a point 50.0 m below its starting point 5.00 s after it leaves the thrower's hand. Air resistance may be ignored. a) What is the initial speed of the egg? b) How high does it rise above its starting point? c) What is the magnitude of its velocity at the highest point? d) What are the magnitude and direction of its acceleration at the highest point? e) Sketch a_y-t, v_y-t, and y-t graphs for the motion of the egg.

2.47 The rocket-driven sled *Sonic Wind No. 2*, used for investigating the physiological effects of large accelerations, runs on a straight, level track 1070 m (3500 ft) long. Starting from rest, it can reach a speed of 224 m/s (500 mi/h) in 0.900 s. a) Compute the acceleration in m/s^2, assuming that it is constant. b) What is the ratio of this acceleration to that of a freely falling body (g)? c) What is the distance covered in 0.900 s? d) A magazine article states that at the end of a certain run, the speed of the sled decreased from 283 m/s (632 mi/h) to zero in 1.40 s and that during this time the magnitude of the acceleration was more than 40g. Are these figures consistent?

2.48 A large boulder is ejected vertically upward from a volcano with an initial speed of 40.0 m/s. Air resistance may be ignored. a) At what time after being ejected is the boulder moving at 20.0 m/s upward? b) At what time is it moving at 20.0 m/s downward? c) When is the displacement of the boulder from its initial position zero? d) When is the velocity of the boulder zero? e) What are the magnitude and direction of the acceleration while the boulder is i) moving upward? ii) moving downward? iii) at the highest point? f) Sketch a_y-t, v_y-t, and y-t graphs for the motion.

2.49 Suppose the acceleration of gravity were only 0.98 m/s^2 instead of 9.8 m/s^2, but the initial velocities with which you can jump upward or throw a ball are unchanged. a) Calculate the height to which you could jump vertically from a standing start if you can jump to 0.75 m when $g = 9.8$ m/s^2. b) How high could you throw a baseball if you can throw it 18 m up when $g = 9.8$ m/s^2? c) Calculate the maximum height of a window from which you would care to jump to a concrete sidewalk below if for $g = 9.8$ m/s^2 you are willing to jump from a height of 2.0 m, which is the typical height of a first-story window.

*Section 2.6 Velocity and Position by Integration

***2.50** The acceleration of a bus is given by $a_x(t) = \alpha t$, where $\alpha = 1.2$ m/s^3. a) If the bus's velocity at time $t = 1.0$ s is 5.0 m/s, what is its velocity at time $t = 2.0$ s? b) If the bus's position at time $t = 1.0$ s is 6.0 m, what is its position at time $t = 2.0$ s? c) Sketch a_x-t, v_x-t, and x-t graphs for the motion.

***2.51** The acceleration of a motorcycle is given by $a_x(t) = At - Bt^2$, where $A = 1.50$ m/s^3 and $B = 0.120$ m/s^4. The motorcycle is at rest at the origin at time $t = 0$. a) Find its position and velocity as functions of time. b) Calculate the maximum velocity it attains.

***2.52 Flying Leap of the Flea.** High speed motion pictures (3500 frames/second) of a jumping, 210-μg flea yielded the data used to plot the graph given in Fig. 2.38. (See "The Flying Leap of the Flea", by M. Rothschild, Y. Schlein, K. Parker, C. Neville, and S. Sternberg in the Nov. 1973 *Scientific American*.) This flea was about 2 mm long and jumped at a nearly vertical take-off angle. Use the graph to answer the following questions. a) Is the acceleration of the flea ever zero? If so, when? Justify your answer. b) Find the maximum height the flea reached in the first 2.5 ms. c) Find the flea's acceleration at 0.5 ms, 1.0 ms, and 1.5 ms. d) Find the flea's height at 0.5 ms, 1.0 ms, and 1.5 ms.

Figure 2.38 Exercise 2.52.

***2.53** The graph in Fig. 2.39 describes the acceleration as a function of time for a stone rolling down a hill starting from rest. a) Find the change in the stone's velocity between $t = 2.5$ s and $t = 7.5$ s. b) Sketch a graph of the stone's velocity as a function of time.

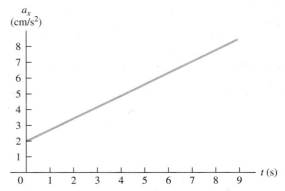

Figure 2.39 Exercise 2.53.

Problems

2.54 On a 20-mile bike ride, you ride the first 10 miles at an average speed of 8 mi/h. What must your average speed over the next 10 miles be to have your average speed for the total 20 miles be a) 4 mi/h? b) 12 mi/h? c) Given this average speed for the first 10 miles, can you possibly attain an average speed of 16 mi/h for the total 20-mile ride? Explain.

2.55 The position of a particle between $t = 0$ and $t = 2.00$ s is given by $x(t) = (3.00$ m/s$^3)t^3 - (10.0$ m/s$^2)t^2 + (9.00$ m/s$)t$. a) Draw the x-t, v_x-t, and a_x-t graphs of this particle. b) At what time(s) between $t = 0$ and $t = 2.00$ s is the particle instantaneously at rest? Does your numerical result agree with the v_x-t graph in part (a)? c) At each time calculated in part (b) is the acceleration of the particle positive or negative? Show that in each case the same answer is deduced from $a_x(t)$ and from the v_x-t graph. d) At what time(s) between $t = 0$ and $t = 2.00$ s is the velocity of the particle instantaneously not changing? Locate this point on the v_x-t and a_x-t graphs of part (a). e) What is the particle's greatest distance from the origin $(x = 0)$ between $t = 0$ and $t = 2.00$ s? f) At what time(s) between $t = 0$ and $t = 2.00$ s is the particle *speeding up* at the greatest rate? At what time(s) between $t = 0$ and $t = 2.00$ s is the particle *slowing down* at the greatest rate? Locate these points on the v_x-t and a_x-t graphs of part (a).

2.56 Relay Race. In a relay race, each contestant runs 25.0 m while carrying an egg balanced on a spoon, turns around, and comes back to the starting point. Edith runs the first 25.0 m in 20.0 s. On the return trip she is more confident and takes only 15.0 s. What is the magnitude of her average velocity for a) the first 25.0 m? b) the return trip? c) What is her average velocity for the entire round trip? d) What is her average speed for the round trip?

2.57 Dan gets on Interstate Highway I–80 at Seward, Nebraska, and drives due west in a straight line and at an average velocity of magnitude 88 km/h. After traveling 76 km, he reaches the Aurora exit (Fig. 2.40). Realizing he has gone too far, he turns around and drives due east 34 km back to the York exit at an average velocity of magnitude 72 km/h. For his whole trip from Seward to the York exit, what is a) his average speed? b) the magnitude of his average velocity?

Figure 2.40 Problem 2.57.

2.58 Freeway Traffic. According to a *Scientific American* article (May, 1990), current freeways can sustain about 2400 vehicles per lane per hour in smooth traffic flow at 96 km/h (60 mi/h). Above that figure the traffic flow becomes "turbulent" (stop-and-go). a) If a vehicle is 4.6 m (15 ft) long on the average, what is the average spacing between vehicles at the above traffic density? b) Collision-avoidance automated control systems, which operate by bouncing radar or sonar signals off surrounding vehicles and then accelerate or brake the car when necessary, could greatly reduce the required spacing between vehicles. If the average spacing is 9.2 m (two car lengths), how many vehicles per hour can a lane of traffic carry at 96 km/h?

2.59 A world-class sprinter accelerates to his maximum speed in 4.0 s. He then maintains this speed for the remainder of a 100-m race, finishing with a total time of 9.1 s. a) What is the runner's average acceleration during the first 4.0 s? b) What is his average acceleration during the last 5.1 s? c) What is his average acceleration for the entire race? d) Explain why your answer to part (c) is not the average of the answers to parts (a) and (b).

2.60 A sled starts from rest at the top of a hill and slides down with a constant acceleration. At some later time it is 14.4 m from the top; 2.00 s after that it is 25.6 m from the top, 2.00 s later 40.0 m from the top, and 2.00 s later it is 57.6 m from the top. a) What is the magnitude of the average velocity of the sled during each of the 2.00-s intervals after passing the 14.4-m point? b) What is the acceleration of the sled? c) What is the speed of the sled when it passes the 14.4-m point? d) How much time did it take to go from the top to the 14.4-m point? e) How far did the sled go during the first second after passing the 14.4-m point?

2.61 A car 3.5 m in length approaches an intersection that is 20 m wide (Fig. 2.41). The light turns yellow when the car's speed is 20 m/s and the front of the car is 50 m from the beginning of the intersection. If the driver steps on the brake, the car slows at -3.8 m/s². If he instead steps on the gas pedal, the car accelerates at 2.3 m/s². The light stays yellow for 3.0 s. Ignore the driver's reaction time. To avoid being in the intersection at the instant that the light turns red, should the driver step on the brake or the gas? Will the car remain clear of the intersection?

Figure 2.41 Problem 2.61.

2.62 In air or vacuum light travels at a constant speed of 3.0×10^8 m/s. To answer some of these questions you may need to look up astronomical data in Appendix F. a) The light year is defined as the distance light travels in one year. Use this information to determine how many meters there are in one light year. b) How far in meters does light travel in one nanosecond? c) When a solar flare occurs on our sun, how soon after its occurrence can we first observe it? d) By bouncing laser beams off a reflector placed on our moon by the Apollo astronauts, astronomers can make very accurate measurements of the earth-moon distance. How long after it is sent does it take such a laser beam (which is just a light beam) to return to earth? e) The Voyager probe, which passed by Neptune in Aug. 1989, was about 3.0 billion miles from earth at that time. Photographs and other information were sent to earth by radio waves, which travel at the speed of light. How long did it take these waves to reach earth from Voyager?

2.63 Use the information in Appendix F to answer each of the following questions. a) What is the speed of the Galapagos Islands, on the earth's equator, due to our planet's spin on its axis? b) What is the earth's speed due to its rotation around the sun? c) If light would bend around the curvature of the earth (which it does not), how many times would a light beam go around the equator in one second?

2.64 In a 350-m race, runner A starts from rest and accelerates at 1.6 m/s^2 for the first 30 m and then runs at constant speed. Runner B starts from rest and accelerates at 2.0 m/s^2 for the first 30 m and then runs at constant speed. Runner A begins running as soon as the race begins but B first takes a nap to rest up. What is the longest nap that B can take and still not lose the race?

2.65 A ball starts from rest and rolls down a hill with uniform acceleration, traveling 150 m during the second 5.0 s of its motion. How far did it roll during the first 5.0 s of motion?

2.66 Collision. The engineer of a passenger train traveling at 25.0 m/s sights a freight train whose caboose is 200 m ahead on the same track (Fig. 2.42). The freight train is traveling at 15.0 m/s in the same direction as the passenger train. The engineer of the passenger train immediately applies the brakes, causing a constant acceleration of -0.100 m/s^2, while the freight train continues with constant speed. Take $x = 0$ at the location of the front of the passenger train when the engineer applies the brakes. a) Will the cows nearby witness a collision? b) If so, where will it take place? c) On a single graph, sketch the positions of the front of the passenger train and the back of the freight train.

Figure 2.42 Problem 2.66.

2.67 Large cockroaches can run as fast as 1.50 m/s in short bursts. Suppose you turn on the light in a cheap motel and see one scurrying directly away from you at a constant 1.50 m/s as you move toward it at 0.80 m/s. If you start 0.90 m behind it, what minimum constant acceleration would you need to catch up with it when it has traveled 1.20 m, just short of safety under a counter?

2.68 Consider the situation described in Example 2.5. The example is somewhat unrealistic, because the constantly accelerating officer continues past the car. In a real pursuit the officer would accelerate to a speed faster than that of the car, and then slow down in order to have the same velocity as the car when she catches up with it. Suppose that the officer in Example 2.5 accelerates from rest at $a_x = 2.5$ m/s^2, until her speed is 20 m/s. She then slows down at a constant rate until she comes alongside the car at $x = 360$ m, traveling with the same speed of 15 m/s as the car. a) At what time does the officer catch up with the car? b) At what time does the officer change from speeding

up to slowing down? At this time, how far is the officer from the stop sign? How far behind the car is she? c) Find the acceleration of the officer while she is slowing down. d) Draw an x-t graph for the two vehicles. e) Draw a v_x-t graph for the two vehicles.

2.69 An automobile and a truck start from rest at the same instant, with the automobile initially at some distance behind the truck. The truck has a constant acceleration of 2.10 m/s^2, and the automobile an acceleration of 3.40 m/s^2. The automobile overtakes the truck after the truck has moved 40.0 m. a) How much time does it take the automobile to overtake the truck? b) How far was the automobile behind the truck initially? c) What is the speed of each when they are abreast? d) On a single graph, sketch the position of each vehicle as a function of time. Take $x = 0$ at the initial location of the truck.

2.70 Two stunt drivers drive directly toward each other. At time $t = 0$ the two cars are a distance D apart, car 1 is at rest, and car 2 is moving to the left with speed v_0. Car 1 begins to move at $t = 0$, speeding up with a constant acceleration a_x. Car 2 continues to move with a constant velocity. a) At what time do the two cars collide? b) Find the speed of car 1 just before it collides with car 2. c) Sketch x-t and v_x-t graphs for car 1 and car 2. Draw the curves for each car using the same axes.

2.71 In his Mustang, Juan rounds a curve onto a straight section of country road while traveling at 20 m/s (45 mi/h), only to find a fully loaded manure spreader completely blocking the road 37 m ahead of him. Startled, he travels at a constant velocity for 0.80 s of reaction time, then hits the brakes and slows at constant acceleration so that he manages to stop just short of the spreader. With the same reaction time and acceleration, if he had roared around the corner at 25 m/s (56 mi/h) instead of 20 m/s, a) how fast would he have been going when he hit the spreader? b) How much time would he have had for his life to flash before his eyes from the time he first sees the spreader until he hits it?

2.72 A police car is traveling in a straight line with constant speed v_p. A truck traveling in the same direction with speed $\frac{3}{2}v_p$ passes the police car. The truck driver realizes that she is speeding, and immediately begins to slow down at a constant rate until she comes to a stop. This is her lucky day, however, and the police car (still moving with the same constant speed) passes the truck driver without giving her a ticket. a) Show that the truck's speed at the instant that the police car passes the truck does *not* depend on the magnitude of the truck's acceleration as it slows down, and find the value of that speed. b) Sketch the x-t graph for the two vehicles.

2.73 Passing. The driver of a car wishes to pass a truck that is traveling at a constant speed of 20.0 m/s (about 45 mi/h). Initially, the car is also traveling at 20.0 m/s and its front bumper is 24.0 m behind the truck's rear bumper. The car accelerates at a constant 0.600 m/s^2, then pulls back into the truck's lane when the rear of the car is 26.0 m ahead of the front of the truck. The car is 4.5 m long and the truck is 21.0 m long. a) How much time is required for the car to pass the truck? b) What distance does the car travel during this time? c) What is the final speed of the car?

*2.74 An object's velocity is measured to be $v_x(t) = \alpha - \beta t^2$, where $\alpha = 4.00$ m/s and $\beta = 2.00$ m/s^3. At $t = 0$ the object is at $x = 0$. a) Calculate the object's position and acceleration as functions of time. b) What is the object's maximum *positive* displacement from the origin?

*2.75 The acceleration of a particle is given by $a_x(t) = -2.00 \text{ m/s}^2 + (3.00 \text{ m/s}^3)t$. a) Find the initial velocity v_{0x} such that the particle will have the same x-coordinate at $t = 4.00$ s as it had at $t = 0$. b) What will be the velocity at $t = 4.00$ s?

2.76 Egg Drop. You are on the roof of the physics building, 46.0 m above the ground (Fig. 2.43). Your physics professor, who is 1.80 m tall, is walking alongside the building at a constant speed of 1.20 m/s. If you wish to drop an egg on your professor's head, where should the professor be when you release the egg? Assume that the egg is in free fall.

2.77 A physics student with too much free time drops a watermelon from the roof of a building. He hears the sound of the watermelon going "splat" 2.50 s later. How high is the building? The speed of sound is 340 m/s. You may ignore air resistance.

Figure 2.43 Problem 2.76.

2.78 Elevators. Estimate the maximum speed and the magnitude of acceleration of an elevator. You will need to use your observations of roughly how long it takes the elevator to go from one floor to another, the approximate vertical distance from one floor to the next, and the distance over which an elevator accelerates to its maximum speed or slows to a stop.

2.79 Visitors at an amusement park watch divers step off a platform 21.3 m (70 ft) above a pool of water. According to the announcer, the divers enter the water at a speed of 56 mi/h (25 m/s). Air resistance may be ignored. a) Is the announcer correct in this claim? b) Is it possible for a diver to leap directly upward off the board so that, missing the board on the way down, she enters the water at 25.0 m/s? If so, what initial upward speed is required? Is the required initial speed physically attainable?

2.80 A flowerpot falls off a windowsill and falls past the window below. You may ignore air resistance. It takes 0.420 s to pass this window, which is 1.90 m high. How far is the top of the window below the windowsill from which the flowerpot fell?

2.81 A football is kicked vertically upward from the ground and a student gazing out of the window sees it moving upward past her at 5.00 m/s. The window is 12.0 m above the ground. You may ignore air resistance. a) How high does the football go above ground? b) How much time does it take to go from the ground to its highest point?

2.82 A model rocket has a constant upward acceleration of 40.0 m/s² while its engine is running. The rocket is fired vertically, and the engine runs for 2.50 s before it uses up the fuel. After the engine stops, the rocket is in free fall. The motion of the rocket is purely up and down. a) Sketch the a_x-t, v_x-t, and y-t graphs for the rocket. b) What is the maximum height that the rocket reaches? c) What will be the speed of the rocket just before it hits the ground? d) Is the total time of flight equal to twice the time the rocket takes to reach its highest point? Why or why not? (See Question Q2.16.)

2.83 Look Out Below. Sam heaves a 16-lb shot straight upward, giving it a constant upward acceleration from rest of 45.0 m/s² for 64.0 cm. He releases it 2.20 m above the ground. You may ignore air resistance. a) What is the speed of the shot when he releases it? b) How high above the ground does it go? c) How much time does he have to get out of its way before it returns to the height of the top of his head, 1.83 m above the ground?

2.84 A physics teacher performing an outdoor demonstration suddenly falls from rest off a high cliff and simultaneously shouts "Help." When she has fallen for 3.0 s, she hears the echo of her shout from the valley floor below. The speed of sound is 340 m/s. a) How tall is the cliff? b) If air resistance is neglected, how fast will she be moving just before she hits the ground? (Her actual speed will be less than this, due to air resistance.)

2.85 Juggling Act. A juggler performs in a room whose ceiling is 3.0 m above the level of his hands. He throws a ball upward so that it just reaches the ceiling. a) What is the initial velocity of the ball? b) What is the time required for the ball to reach the ceiling? At the instant when the first ball is at the ceiling, he throws a second ball upward with two-thirds the initial velocity of the first. c) How long after the second ball is thrown did the two balls pass each other? d) At what distance above the juggler's hand do they pass each other?

2.86 A helicopter carrying Dr. Evil takes off with a constant upward acceleration of 5.0 m/s². Secret agent Austin Powers jumps on just as the helicopter lifts off the ground. After the two men struggle for 10.0 s, Powers shuts off the engine and steps out of the helicopter. Assume that the helicopter is in free fall after its engine is shut off and ignore effects of air resistance. a) What is the maximum height above ground reached by the helicopter? b) Powers deploys a jet pack strapped on his back 7.0 s after leaving the helicopter, and then he has a constant downward acceleration with magnitude 2.0 m/s². How far is Powers above the ground when the helicopter crashes into the ground?

2.87 Building Height. Spiderman steps from the top of a tall building. He falls freely from rest to the ground a distance of h. He falls a distance of $h/4$ in the last 1.0 s of his fall. What is the height h of the building?

2.88 Cliff Height. You are climbing in the High Sierra where you suddenly find yourself at the edge of a fog-shrouded cliff. To find the height of this cliff, you drop a rock from the top and 10.0 s later hear the sound of it hitting the ground at the foot of the cliff. a) Ignoring air resistance, how high is the cliff if the speed of sound is 330 m/s? b) Suppose you had ignored the time it takes the sound to reach you. In that case, would you have overestimated or underestimated the height of the cliff? Explain your reasoning.

2.89 Falling Can. A painter is standing on scaffolding that is raised at constant speed. As he travels upward, he accidentally nudges a paint can off the scaffolding and it falls 15.0 m to the ground. You are watching, and measure with your stopwatch that it takes 3.25 s for the can to reach the ground. Ignore air resistance. a) What is the speed of the can just before it hits the ground? b) Another painter is standing on a ledge, with his hands 4.00 m above the can when it falls off. He has lightning-fast reflexes and if the can passes in front of him, he can catch it. Does he get the chance?

2.90 Determined to test the law of gravity for himself, a student walks off a skyscraper 180 m high, stopwatch in hand, and starts his free fall (zero initial velocity). Five seconds later, Superman arrives at the scene and dives off the roof to save the student. a) Superman leaves the roof with an initial speed v_0 that he produces by pushing himself downward from the edge of the roof with his legs of steel. He then falls with the same acceleration as any freely falling body. What must the value of v_0 be so that Superman catches the student just before they reach the ground? b) On the same graph, sketch the positions of the student and of Superman as functions of time. Take Superman's initial speed to have the value calculated in part (a). c) If the height of the skyscraper is less than some minimum value, even Superman can't reach the student before he hits the ground. What is this minimum height?

2.91 Another determined student (see Problem 2.90) walks off the top of the CN Tower in Toronto, which is 553 m high, and falls freely. His initial velocity is zero. The Rocketeer arrives at the scene 5.00 s later and dives off the top of the tower to save the student. The Rocketeer leaves the roof with an initial downward speed v_0. In order both to catch the student and to prevent injury to him, the Rocketeer should catch the student at a sufficiently great height above ground so that the Rocketeer and the student slow down and arrive at the ground with zero velocity. The upward acceleration that accomplishes this is provided by the Rocketeer's jet pack, which he turns on just as he catches the student; before then the Rocketeer is in free fall. To prevent discomfort to the student, the magnitude of the acceleration of the Rocketeer and the student as they move downward together should be no more than five times g. a) What is the minimum height above the ground at which the Rocketeer should catch the student? b) What must the Rocketeer's initial downward speed be so that he catches the student at the minimum height found in part (a)? c) Sketch the y-t, v_y-t, and a_y-t graphs for the student and for the Rocketeer. On each graph, draw the curves for the Rocketeer and for the student using the same axes.

2.92 A ball is thrown straight up from the ground with speed v_0. At the same instant, a second ball is dropped from rest from a height H, directly above the point where the first ball was thrown upward. There is no air resistance. a) Find the time at which the two balls collide. b) Find the value of H in terms of v_0 and g so that at the instant when the balls collide, the first ball is at the highest point of its motion.

2.93 Two cars, A and B, travel in a straight line. The distance of A from the starting point is given as a function of time by $x_A(t) = \alpha t + \beta t^2$, with $\alpha = 2.60$ m/s and $\beta = 1.20$ m/s^2. The distance of B from the starting point is $x_B(t) = \gamma t^2 - \delta t^3$, with $\gamma = 2.80$ m/s^2 and $\delta = 0.20$ m/s^3. a) Which car is ahead just after they leave the starting point? b) At what time(s) are the cars at the same point? c) At what time(s) is the distance from A to B neither increasing nor decreasing? d) At what time(s) do A and B have the same acceleration?

2.94 An apple drops from the tree and falls freely. The apple is originally at rest a height H above the top of the grass of a thick lawn, which is made of blades of grass of height h. When the apple enters the grass, it slows down at a constant rate so that its speed is

zero when it reaches ground level. a) Find the speed of the apple just before it enters the grass. b) Find the acceleration of the apple while it is in the grass. c) Sketch the y-t, v_y-t, and a_y-t graphs for the apple's motion.

Challenge Problems

2.95 Catching the Bus. A student is running at her top speed of 5.0 m/s to catch a bus, which is stopped at the bus stop. When the student is still 40.0 m from the bus, it starts to pull away, moving with a constant acceleration of 0.170 m/s^2. a) For how much time and what distance does the student have to run at 5.0 m/s before she overtakes the bus? b) When she reaches the bus, how fast is the bus traveling? c) Sketch an x-t graph for both the student and the bus. Take $x = 0$ at the initial position of the student. d) The equations you used in part (a) to find the time have a second solution, corresponding to a later time for which the student and bus are again at the same place if they continue their specified motions. Explain the significance of this second solution. How fast is the bus traveling at this point? e) If the student's top speed is 3.5 m/s, will she catch the bus? f) What is the *minimum* speed the student must have to just catch up with the bus? For what time and what distance does she have to run in that case?

2.96 In the vertical jump, an athlete starts from a crouch and jumps upward to reach as high as possible. Even the best athletes spend little more than 1.00 s in the air (their "hang time"). Treat the athlete as a particle and let y_{max} be his maximum height above the floor. To explain why he seems to hang in the air, calculate the ratio of the time he is above $y_{max}/2$ to the time it takes him to go from the floor to that height. You may ignore air resistance.

2.97 A ball is thrown straight up from the edge of the roof of a building. A second ball is dropped from the roof 1.00 s later. You may ignore air resistance. a) If the height of the building is 20.0 m, what must the initial speed be of the first ball if both are to hit the ground at the same time? On the same graph, sketch the position of each ball as a function of time, measured from when the first ball is thrown. Consider the same situation, but now let the initial speed v_0 of the first ball be given and treat the height h of the building as an unknown. b) What must the height of the building be for both balls to reach the ground at the same time for each of the following values of v_0: i) 6.0 m/s; ii) 9.5 m/s? c) If v_0 is greater than some value v_{max}, a value of h does not exist that allows both balls to hit the ground at the same time. Solve for v_{max}. The value v_{max} has a simple physical interpretation. What is it? d) If v_0 is less than some value v_{min}, a value of h does not exist that allows both balls to hit the ground at the same time. Solve for v_{min}. The value v_{min} also has a simple physical interpretation. What is it?

2.98 An alert hiker sees a boulder fall from the top of a distant cliff and notes that it takes 1.30 s for the boulder to fall the last third of the way to the ground. You may ignore air resistance. a) What is the height of the cliff in meters? b) If in part (a) you get two solutions of a quadratic equation and use one for your answer, what does the other solution represent?

3 MOTION IN TWO OR THREE DIMENSIONS

A spectacular example of motion occurs when a volcano ejects rock fragments into the air. Such eruptions are caused by the sudden release of high-pressure gas from within the magma (molten rock beneath the surface). The ejected fragments, called tephra, can be as large as a house and can fly out of the volcano at speeds of hundreds of meters per second.

? **If all the fragments are ejected at the same speed, for which ejection angle do the fragments land farthest from the volcano?**

When a baseball is hit by a bat, what determines where the baseball lands? How do you describe the motion of a roller coaster car along a curved track or the flight of a hawk circling over an open field? If you throw a water balloon horizontally from your window, will it take the same amount of time to hit the ground as a balloon that you simply drop?

We can't answer these kinds of questions using the techniques of Chapter 2, in which particles moved only along a straight line. Instead, we have to confront the reality that our world is three-dimensional. To understand the curved flight of a baseball, the orbital motion of a satellite, or the trajectory of a thrown projectile, we need to extend our descriptions of motion to two- and three-dimensional situations. We'll still use the vector quantities displacement, velocity, and acceleration, but now these quantities will have two or three components and will no longer lie along a single line. We'll find that several important and interesting kinds of motion take place in two dimensions only, that is, in a *plane*. These motions can be described with two coordinates and two components of velocity and acceleration.

We also need to consider how the motion of a particle is described by different observers who are moving relative to each other. The concept of *relative velocity* will play an important role later in the book when we study collisions, when we explore electromagnetic phenomena, and when we introduce Einstein's special theory of relativity.

This chapter merges the vector language we learned in Chapter 1 with the kinematic language of Chapter 2. As before, we are concerned with describing motion, not with analyzing its causes. But the language you learn here will be an essential tool in later chapters when you use Newton's laws of motion to study the relation between force and motion.

3.1 | Position and Velocity Vectors

To describe the *motion* of a particle in space, we first need to be able to describe the *position* of the particle. Consider a particle that is at a point P at a certain instant. The **position vector** $\vec{r}$ of the particle at this instant is a vector that goes from the origin of the coordinate system to the point P (Fig. 3.1). The figure also shows that the Cartesian coordinates x, y, and z of point P are the x-, y-, and z-components of vector $\vec{r}$. Using the unit vectors we introduced in Section 1.9, we can write

$$\vec{r} = x\hat{\imath} + y\hat{\jmath} + z\hat{k} \qquad (3.1)$$

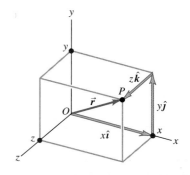

3.1 The position vector $\vec{r}$ from the origin to point P has components x, y, and z.

As the particle moves through space, the path that it follows is in general a curve (Fig. 3.2). During a time interval Δt the particle moves from P_1, where its position vector is $\vec{r}_1$, to P_2, where its position vector is $\vec{r}_2$. The change in position (the displacement) during this interval is $\Delta\vec{r} = \vec{r}_2 - \vec{r}_1 = (x_2 - x_1)\hat{\imath} + (y_2 - y_1)\hat{\jmath} + (z_2 - z_1)\hat{k}$. We define the **average velocity** $\vec{v}_{\text{av}}$ during this interval in the same way we did in Chapter 2 for straight-line motion, as the displacement divided by the time interval:

$$\vec{v}_{\text{av}} = \frac{\vec{r}_2 - \vec{r}_1}{t_2 - t_1} = \frac{\Delta\vec{r}}{\Delta t} \qquad \text{(average velocity vector)} \qquad (3.2)$$

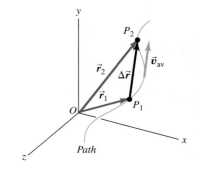

3.2 The average velocity $\vec{v}_{\text{av}}$ between points P_1 and P_2 has the same direction as the displacement $\Delta\vec{r}$.

Note that *dividing* a vector by a scalar is really a special case of *multiplying* a vector by a scalar, described in Section 1.7; the average velocity $\vec{v}_{\text{av}}$ is equal to the displacement vector $\Delta\vec{r}$ multiplied by $1/\Delta t$, the reciprocal of the time interval. Note also that the x-component of Eq. (3.2) is $v_{\text{av-}x} = (x_2 - x_1)/(t_2 - t_1) = \Delta x/\Delta t$. This is just Eq. (2.2), the expression for average velocity in one dimension that we found in Section 2.1.

We now define **instantaneous velocity** just as we did in Chapter 2; it is the limit of the average velocity as the time interval approaches zero, and it equals the instantaneous rate of change of position with time. The key difference is that position $\vec{r}$ and instantaneous velocity $\vec{v}$ are now both vectors:

$$\vec{v} = \lim_{\Delta t \to 0} \frac{\Delta\vec{r}}{\Delta t} = \frac{d\vec{r}}{dt} \qquad \text{(instantaneous velocity vector)} \qquad (3.3)$$

The *magnitude* of the vector $\vec{v}$ at any instant is the *speed* v of the particle at that instant. The *direction* of $\vec{v}$ at any instant is the same as the direction in which the particle is moving at that instant.

Note that as $\Delta t \to 0$, points P_1 and P_2 in Fig. 3.2 move closer and closer together. In this limit, the vector $\Delta\vec{r}$ becomes tangent to the curve. The direction of $\Delta\vec{r}$ in the limit is also the direction of the instantaneous velocity $\vec{v}$. This leads to an important conclusion: *at every point along the path, the instantaneous velocity vector is tangent to the path at that point* (Fig. 3.3).

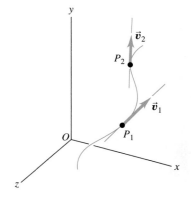

3.3 The instantaneous velocity $\vec{v}$ is tangent to the path at each point. Here $\vec{v}_1$ and $\vec{v}_2$ are the instantaneous velocities at the points P_1 and P_2 shown in Fig. 3.2.

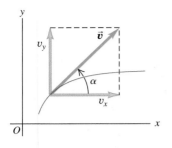

3.4 The two velocity components for motion in the *xy*-plane.

It's often easiest to calculate the instantaneous velocity vector using components. During any displacement $\Delta \vec{r}$, the changes Δx, Δy, and Δz in the three coordinates of the particle are the *components* of $\Delta \vec{r}$. It follows that the components v_x, v_y, and v_z of the instantaneous velocity $\vec{v}$ are simply the time derivatives of the coordinates x, y, and z. That is,

$$v_x = \frac{dx}{dt} \qquad v_y = \frac{dy}{dt} \qquad v_z = \frac{dz}{dt} \qquad (3.4)$$

(components of instantaneous velocity)

The *x*-component of $\vec{v}$ is $v_x = dx/dt$, which is the same as Eq. (2.3)—the expression for instantaneous velocity for straight-line motion that we obtained in Section 2.2. Hence Eq. (3.4) is a direct extension of the idea of instantaneous velocity to motion in three dimensions.

We can also get this result by taking the derivative of Eq. (3.1). The unit vectors $\hat{\imath}$, $\hat{\jmath}$, and $\hat{k}$ are constant in magnitude and direction, so their derivatives are zero, and we find

$$\vec{v} = \frac{d\vec{r}}{dt} = \frac{dx}{dt}\hat{\imath} + \frac{dy}{dt}\hat{\jmath} + \frac{dz}{dt}\hat{k} \qquad (3.5)$$

This shows again that the components of $\vec{v}$ are dx/dt, dy/dt, and dz/dt.

The magnitude of the instantaneous velocity vector $\vec{v}$—that is, the speed—is given in terms of the components v_x, v_y, and v_z by the Pythagorean relation

$$|\vec{v}| = v = \sqrt{v_x^2 + v_y^2 + v_z^2} \qquad (3.6)$$

Figure 3.4 shows the situation when the particle moves in the *xy*-plane. In this case, z and v_z are zero. Then the speed (the magnitude of $\vec{v}$) is

$$v = \sqrt{v_x^2 + v_y^2}$$

and the direction of the instantaneous velocity $\vec{v}$ is given by the angle α in the figure. We see that

$$\tan \alpha = \frac{v_y}{v_x} \qquad (3.7)$$

(We always use Greek letters for angles. We use α for the direction of the instantaneous velocity vector to avoid confusion with the direction θ of the *position* vector of the particle.)

The instantaneous velocity vector is usually more interesting and useful than the average velocity vector. From now on, when we use the word "velocity," we will always mean the instantaneous velocity vector $\vec{v}$ (rather than the average velocity vector). Usually, we won't even bother to call $\vec{v}$ a vector; it's up to you to remember that velocity is a vector quantity with both magnitude and direction.

Example
3.1

Calculating average and instantaneous velocity

A robotic vehicle, or rover, is being used to explore the surface of Mars. The landing craft is the origin of coordinates and the surrounding Martian surface lies in the *xy*-plane. The rover, which we represent as a point, has *x*- and *y*-coordinates that vary with time according to

$$x = 2.0 \text{ m} - (0.25 \text{ m/s}^2)t^2$$
$$y = (1.0 \text{ m/s})t + (0.025 \text{ m/s}^3)t^3$$

a) Find the rover's coordinates and its distance from the lander at time $t = 2.0$ s. b) Find the rover's displacement and average velocity vectors during the interval from $t = 0.0$ s to $t = 2.0$ s. c) Derive a general expression for the rover's instantaneous velocity vector, and find the instantaneous velocity at $t = 2.0$ s. Express the instantaneous velocity in component form and also in terms of magnitude and direction.

SOLUTION

IDENTIFY: This problem involves motion in a two-dimensional trajectory (i.e., in a plane). Hence we must use the expressions for the displacement, average velocity, and instantaneous velocity *vectors* obtained in this section. (The simpler expressions in Sections 2.1 and 2.2 don't involve vectors, but they apply only to motion along a straight line.)

SET UP: The rover's path is shown in Fig. 3.5. We'll use Eq. (3.1) for position $\vec{r}$, the expression $\Delta\vec{r} = \vec{r}_2 - \vec{r}_1$ for displacement, Eq. (3.2) for average velocity, and Eqs. (3.5) and (3.6) for instantaneous velocity and its direction. The target variables are as stated in the problem.

EXECUTE: a) At time $t = 2.0$ s the rover's coordinates are

$$x = 2.0 \text{ m} - (0.25 \text{ m/s}^2)(2.0 \text{ s})^2 = 1.0 \text{ m}$$
$$y = (1.0 \text{ m/s})(2.0 \text{ s}) + (0.025 \text{ m/s}^3)(2.0 \text{ s})^3 = 2.2 \text{ m}$$

The rover's distance from the origin at this time is

$$r = \sqrt{x^2 + y^2} = \sqrt{(1.0 \text{ m})^2 + (2.2 \text{ m})^2} = 2.4 \text{ m}$$

b) To find the displacement and average velocity, we express the position vector $\vec{r}$ as a function of time t. From Eq. (3.1), this is

$$\vec{r} = x\hat{i} + y\hat{j}$$
$$= [2.0 \text{ m} - (0.25 \text{ m/s}^2)t^2]\hat{i}$$
$$+ [(1.0 \text{ m/s})t + (0.025 \text{ m/s}^3)t^3]\hat{j}$$

At time $t = 0.0$ s the position vector is

$$\vec{r}_0 = (2.0 \text{ m})\hat{i} + (0.0 \text{ m})\hat{j}$$

From part (a) we find that at time $t = 2.0$ s the position vector $\vec{r}_2$ is

$$\vec{r}_2 = (1.0 \text{ m})\hat{i} + (2.2 \text{ m})\hat{j}$$

Therefore the displacement from $t = 0.0$ s to $t = 2.0$ s is

$$\Delta\vec{r} = \vec{r}_2 - \vec{r}_0 = (1.0 \text{ m})\hat{i} + (2.2 \text{ m})\hat{j} - (2.0 \text{ m})\hat{i}$$
$$= (-1.0 \text{ m})\hat{i} + (2.2 \text{ m})\hat{j}$$

During the time interval the rover has moved 1.0 m in the negative x-direction and 2.2 m in the positive y-direction. The average velocity during the interval from $t = 0.0$ s to $t = 2.0$ s is this displacement divided by the elapsed time, as in Eq. (3.2):

$$\vec{v}_{av} = \frac{\Delta\vec{r}}{\Delta t} = \frac{(-1.0 \text{ m})\hat{i} + (2.2 \text{ m})\hat{j}}{2.0 \text{ s} - 0.0 \text{ s}}$$
$$= (-0.50 \text{ m/s})\hat{i} + (1.1 \text{ m/s})\hat{j}$$

The components of this average velocity are

$$v_{av\text{-}x} = -0.50 \text{ m/s} \qquad v_{av\text{-}y} = 1.1 \text{ m/s}$$

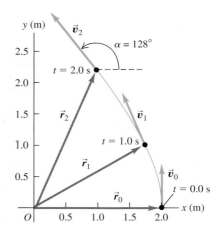

3.5 Path of a radio-controlled robotic vehicle. At $t = 0$ the rover has position vector $\vec{r}_0$ and instantaneous velocity vector $\vec{v}_0$. Likewise, $\vec{r}_1$ and $\vec{v}_1$ are the vectors at $t = 1.0$ s; $\vec{r}_2$ and $\vec{v}_2$ are the vectors at $t = 2.0$ s.

c) From Eq. (3.4), the components of instantaneous velocity are the time derivatives of the coordinates:

$$v_x = \frac{dx}{dt} = (-0.25 \text{ m/s}^2)(2t)$$

$$v_y = \frac{dy}{dt} = 1.0 \text{ m/s} + (0.025 \text{ m/s}^3)(3t^2)$$

Then we can write the instantaneous velocity vector $\vec{v}$ as

$$\vec{v} = v_x\hat{i} + v_y\hat{j} = (-0.50 \text{ m/s}^2)t\hat{i}$$
$$+ [1.0 \text{ m/s} + (0.075 \text{ m/s}^3)t^2]\hat{j}$$

At time $t = 2.0$ s, the components of instantaneous velocity are

$$v_x = (-0.50 \text{ m/s}^2)(2.0 \text{ s}) = -1.0 \text{ m/s}$$
$$v_y = 1.0 \text{ m/s} + (0.075 \text{ m/s}^3)(2.0 \text{ s})^2 = 1.3 \text{ m/s}$$

The magnitude of the instantaneous velocity (that is, the speed) at $t = 2.0$ s is

$$v = \sqrt{v_x^2 + v_y^2} = \sqrt{(-1.0 \text{ m/s})^2 + (1.3 \text{ m/s})^2}$$
$$= 1.6 \text{ m/s}$$

Its direction with respect to the positive x-axis is given by the angle α, where, from Eq. (3.7),

$$\tan\alpha = \frac{v_y}{v_x} = \frac{1.3 \text{ m/s}}{-1.0 \text{ m/s}} = -1.3 \quad \text{so} \quad \alpha = 128°$$

Your calculator will tell you that the inverse tangent of -1.3 is $-52°$. But as we learned in Section 1.8, you have to examine a sketch of a vector (Fig. 3.5) to decide on its direction. The correct answer for α is $-52° + 180° = 128°$.

EVALUATE: Take a moment to compare the components of *average* velocity that we found in part (b) for the interval from $t = 0.0$ s to $t = 2.0$ s $(v_{av\text{-}x} = -0.50 \text{ m/s}, v_{av\text{-}y} = 1.1 \text{ m/s})$ to the components of *instantaneous* velocity at $t = 2.0$ s that we found in part (c) $(v_x = -1.0$ m/s, $v_y = 1.3$ m/s$)$. The comparison shows that, just as in one dimension, the average velocity vector $\vec{v}_{av}$ over an interval is in general *not* equal to the instantaneous velocity $\vec{v}$ at the end of the interval (see Example 2.1).

We invite you to calculate the position, instantaneous velocity vector, speed, and direction of motion at $t = 0.0$ s and $t = 1.0$ s.

The position vectors and instantaneous velocity vectors at $t = 0.0$ s, 1.0 s, and 2.0 s are shown in Fig. 3.5. Notice that at every point, the instantaneous velocity vector $\vec{v}$ is tangent to the path. The magnitude of $\vec{v}$ increases as the rover moves, which shows that the rover's speed is increasing.

Test Your Understanding

Give an example of a situation in which the average velocity vector $\vec{v}_{av}$ over an interval *would* be equal to the instantaneous velocity $\vec{v}$ at the end of the interval.

3.2 | The Acceleration Vector

Now let's consider the *acceleration* of a particle moving in space. Just as for motion in a straight line, acceleration describes how the velocity of the particle changes. But now we'll generalize acceleration to describe changes in the magnitude of the velocity (that is, the speed) *and* changes in the direction of the velocity (that is, the direction in which the particle is moving in space).

In Fig. 3.6a, a particle is moving along a curved path. The vectors $\vec{v}_1$ and $\vec{v}_2$ represent the particle's instantaneous velocities at time t_1, when the particle is at point P_1, and at time t_2, when the particle is at point P_2. The two velocities may differ in both magnitude and direction. We define the **average acceleration** $\vec{a}_{av}$ of the particle as it moves from P_1 to P_2 as the *vector change in velocity*, $\vec{v}_2 - \vec{v}_1 = \Delta\vec{v}$, divided by the time interval $t_2 - t_1 = \Delta t$:

$$\vec{a}_{av} = \frac{\vec{v}_2 - \vec{v}_1}{t_2 - t_1} = \frac{\Delta\vec{v}}{\Delta t} \quad \text{(average acceleration vector)} \quad (3.8)$$

Average acceleration is a *vector* quantity in the same direction as the vector $\Delta\vec{v}$ (Fig. 3.6a). Note that $\vec{v}_2$ is the vector sum of the original velocity $\vec{v}_1$ and the change $\Delta\vec{v}$ (Fig. 3.6b). The x-component of Eq. (3.8) is $a_{av\text{-}x} = (v_{2x} - v_{1x})/(t_2 - t_1) = \Delta v_x/\Delta t$, which is just Eq. (2.4) for the average acceleration in straight-line motion.

As in Chapter 2, we define the **instantaneous acceleration** $\vec{a}$ at point P_1 as the limit of the average acceleration when point P_2 approaches point P_1 and $\Delta\vec{v}$ and Δt both approach zero; the instantaneous acceleration is also equal to the instantaneous rate of change of velocity with time. Because we are not restricted to straight-line motion, instantaneous acceleration is now a vector:

$$\vec{a} = \lim_{\Delta t \to 0} \frac{\Delta\vec{v}}{\Delta t} = \frac{d\vec{v}}{dt} \quad \text{(instantaneous acceleration vector)} \quad (3.9)$$

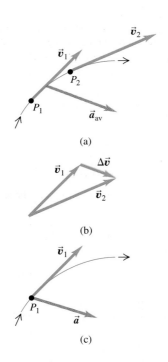

3.6 (a) The vector $\vec{a}_{av} = \Delta\vec{v}/\Delta t$ represents the average acceleration between P_1 and P_2. (b) Construction for obtaining $\Delta\vec{v} = \vec{v}_2 - \vec{v}_1$. (c) Instantaneous acceleration $\vec{a}$ at point P_1. Vector $\vec{v}$ is tangent to the path; vector $\vec{a}$ points toward the concave side of the path.

The velocity vector $\vec{v}$, as we have seen, is tangent to the path of the particle. But the construction in Fig. 3.6c shows that the instantaneous acceleration vector $\vec{a}$ of a moving particle always points toward the concave side of a curved path—that is, toward the inside of any turn that the particle is making. We can also see that when a particle is moving in a curved path, it *always* has nonzero acceleration, even when it moves with constant speed. This conclusion may seem contrary to your intuition, but it's really just contrary to the everyday use of the word "acceleration" to mean that speed is increasing. The more precise definition given in Eq. (3.9) shows that there is a nonzero acceleration whenever the velocity vector changes in *any* way, whether there is a change of speed, direction, or both.

To convince yourself that a particle has a nonzero acceleration when moving on a curved path with constant speed, think of your sensations when you ride in a car. When the car accelerates, you tend to move within the car in a direction *opposite* to the car's acceleration. (We'll discover the reason for this behavior in Chapter 4.) Thus you tend to slide toward the back of the car when it accelerates forward (speeds up) and toward the front of the car when it accelerates backward (slows down). If the car makes a turn on a level road, you tend to slide toward the outside of the turn; hence the car has an acceleration toward the inside of the turn.

We will usually be interested in the instantaneous acceleration, not the average acceleration. From now on, we will use the term "acceleration" to mean the instantaneous acceleration vector $\vec{a}$.

Each component of the acceleration vector is the derivative of the corresponding component of velocity:

$$a_x = \frac{dv_x}{dt} \qquad a_y = \frac{dv_y}{dt} \qquad a_z = \frac{dv_z}{dt} \qquad (3.10)$$

(components of instantaneous acceleration)

In terms of unit vectors,

$$\vec{a} = \frac{dv_x}{dt}\hat{i} + \frac{dv_y}{dt}\hat{j} + \frac{dv_z}{dt}\hat{k} \qquad (3.11)$$

The *x*-component of Eqs. (3.10) and (3.11), $a_x = dv_x/dt$, is a familiar result: it's the expression from Section 2.3 for instantaneous acceleration in one dimension, Eq. (2.5). Figure 3.7 shows an example of an acceleration vector that has both *x*- and *y*-components.

Also, because each component of velocity is the derivative of the corresponding coordinate, we can express the components a_x, a_y, and a_z of the acceleration vector $\vec{a}$ as

$$a_x = \frac{d^2x}{dt^2} \qquad a_y = \frac{d^2y}{dt^2} \qquad a_z = \frac{d^2z}{dt^2} \qquad (3.12)$$

and the acceleration vector $\vec{a}$ as

$$\vec{a} = \frac{d^2x}{dt^2}\hat{i} + \frac{d^2y}{dt^2}\hat{j} + \frac{d^2z}{dt^2}\hat{k} \qquad (3.13)$$

3.7 When a frog leaps, it accelerates both forward and upward. Thus its acceleration vector has both a horizontal component (a_x) and a vertical component (a_y).

| Example 3.2 | **Calculating average and instantaneous acceleration** |

Let's look again at the motions of the robotic rover in Example 3.1. We found that the components of instantaneous velocity at any time t are

$$v_x = \frac{dx}{dt} = (-0.25 \text{ m/s}^2)(2t)$$

$$v_y = \frac{dy}{dt} = 1.0 \text{ m/s} + (0.025 \text{ m/s}^3)(3t^2)$$

and that the velocity vector is

$$\vec{v} = v_x\hat{i} + v_y\hat{j} = (-0.50 \text{ m/s}^2)t\hat{i}$$
$$+ [1.0 \text{ m/s} + (0.075 \text{ m/s}^3)t^2]\hat{j}$$

a) Find the components of the average acceleration in the interval from $t = 0.0$ s to $t = 2.0$ s. b) Find the instantaneous acceleration at $t = 2.0$ s.

SOLUTION

IDENTIFY and SET UP: In part (a) we use Eq. (3.8) to calculate the components of the average acceleration. To do so, we first have to use the expressions given above to determine the values of v_x and v_y at the beginning and end of the interval. In part (b) we determine the instantaneous acceleration components at *any* time t by taking the time derivatives of the velocity components as in Eq. (3.10).

EXECUTE: a) If we substitute $t = 0.0$ s or $t = 2.0$ s into the above expressions for v_x and v_y, we find that at the beginning of the interval ($t = 0.0$ s) the velocity components are

$$v_x = 0.0 \text{ m/s} \qquad v_y = 1.0 \text{ m/s}$$

and that at the end of the interval ($t = 2.0$ s) the components are

$$v_x = -1.0 \text{ m/s} \qquad v_y = 1.3 \text{ m/s}$$

(Note that the values at $t = 2.0$ s are the same as those we found in Example 3.1.) Thus the components of average acceleration in this interval are

$$a_{\text{av-}x} = \frac{\Delta v_x}{\Delta t} = \frac{-1.0 \text{ m/s} - 0.0 \text{ m/s}}{2.0 \text{ s} - 0.0 \text{ s}} = -0.5 \text{ m/s}^2$$

$$a_{\text{av-}y} = \frac{\Delta v_y}{\Delta t} = \frac{1.3 \text{ m/s} - 1.0 \text{ m/s}}{2.0 \text{ s} - 0.0 \text{ s}} = 0.15 \text{ m/s}^2$$

b) Using Eq. (3.10), we find

$$a_x = \frac{dv_x}{dt} = -0.50 \text{ m/s}^2 \qquad a_y = \frac{dv_y}{dt} = (0.075 \text{ m/s}^3)(2t)$$

We can write the instantaneous acceleration vector $\vec{a}$ as

$$\vec{a} = a_x \hat{\imath} + a_y \hat{\jmath} = (-0.50 \text{ m/s}^2)\hat{\imath} + (0.15 \text{ m/s}^3)t\hat{\jmath}$$

At time $t = 2.0$ s, the components of instantaneous acceleration are

$$a_x = -0.50 \text{ m/s}^2 \qquad a_y = (0.15 \text{ m/s}^3)(2.0 \text{ s}) = 0.30 \text{ m/s}^2$$

The acceleration vector at this time is

$$\vec{a} = (-0.50 \text{ m/s}^2)\hat{\imath} + (0.30 \text{ m/s}^2)\hat{\jmath}$$

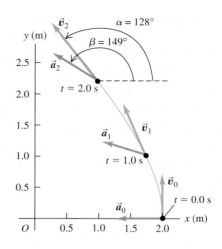

3.8 The path of the radio-controlled rover, showing the velocity and acceleration at $t = 0.0$ s ($\vec{v}_0$ and $\vec{a}_0$), $t = 1.0$ s ($\vec{v}_1$ and $\vec{a}_1$), and $t = 2.0$ s ($\vec{v}_2$ and $\vec{a}_2$).

The magnitude of acceleration at this time is

$$a = \sqrt{a_x^2 + a_y^2}$$
$$= \sqrt{(-0.50 \text{ m/s}^2)^2 + (0.30 \text{ m/s}^2)^2} = 0.58 \text{ m/s}^2$$

The direction of $\vec{a}$ with respect to the positive x-axis is given by the angle β, where

$$\tan \beta = \frac{a_y}{a_x} = \frac{0.30 \text{ m/s}^2}{-0.50 \text{ m/s}^2} = -0.60$$

$$\beta = 180° - 31° = 149°$$

EVALUATE: We invite you to calculate the instantaneous acceleration at $t = 0.0$ s and $t = 1.0$ s. Figure 3.8 shows the path of the car and the velocity and acceleration vectors at $t = 0.0$ s, 1.0 s, and 2.0 s. Note that the two vectors $\vec{v}$ and $\vec{a}$ are *not* in the same direction at any of these times. The velocity vector is tangent to the path at each point, and the acceleration vector points toward the concave side of the path.

Example 3.3	**Determining average acceleration graphically**

For the rover in Examples 3.1 and 3.2, use vector subtraction to find the approximate direction of the average acceleration in the interval from $t = 0.0$ s to 2.0 s.

SOLUTION

IDENTIFY: Figure 3.6 shows that the average acceleration for the interval from t_1 to t_2 is in the direction of the velocity change $\Delta\vec{v}$. Given the velocity vectors at the beginning and end of the interval, we can find the direction of $\Delta\vec{v}$, and hence of $\vec{a}_{\text{av}}$, graphically.

SET UP and EXECUTE: From Fig. 3.8, the velocity $\vec{v}_0$ at $t = 0.0$ s points in the positive y-direction and the velocity $\vec{v}_2$ at $t = 2.0$ s points at an angle between the negative x-direction and the positive y-direction. Furthermore, $\vec{v}_2$ has a larger magnitude than $\vec{v}_0$.

Figure 3.9 shows how to find the direction of $\Delta\vec{v} = \vec{v}_2 - \vec{v}_0$ by subtracting the two vectors graphically. The result points at a small angle above the negative x-direction; the average acceleration points in the same direction as $\Delta\vec{v}$.

EVALUATE: The direction of $\Delta \vec{v}$ in Fig. 3.9 is very close to the direction of $\vec{a}_1$ (the instantaneous acceleration at the midpoint of the interval) shown in Fig. 3.8. It's also intermediate between the directions of $\vec{a}_0$ and $\vec{a}_2$, the instantaneous acceleration at the beginning and end of the interval, respectively. This is just how we would expect the *average* acceleration to behave.

This graphical procedure usually won't help you determine numerical values for the acceleration; for that, you need to do calculations like those in Example 3.2. But the graphical approach does offer a way to check whether your calculations are reasonable. It also provides important insight into the character of the motion, as we describe below.

3.9 Using graphical vector subtraction to estimate the direction of average acceleration.

Parallel and Perpendicular Components of Acceleration

It's often useful to describe the acceleration of a particle moving along a curved path in terms of components parallel and perpendicular to the velocity at each point (Fig. 3.10). In Fig. 3.10 we label these components $a_\perp$ and $a_\parallel$. To see why these components are useful, we'll use the graphical interpretation of acceleration (as in Example 3.3) and consider two special cases.

In Fig. 3.11a the acceleration vector is *parallel* to the velocity $\vec{v}_1$. The change in $\vec{v}$ during a small time interval Δt is a vector $\Delta \vec{v}$ having the same direction as $\vec{a}$ and hence the same direction as $\vec{v}_1$. The velocity $\vec{v}_2$ at the end of Δt, given by $\vec{v}_2 = \vec{v}_1 + \Delta \vec{v}$, is a vector having the same direction as $\vec{v}_1$ but greater magnitude. In other words, during the time interval Δt the particle in Fig. 3.11a moved in a straight line with increasing speed.

In Fig. 3.11b the acceleration $\vec{a}$ is *perpendicular* to the velocity $\vec{v}$. In a small time interval Δt, the change $\Delta \vec{v}$ is a vector very nearly perpendicular to $\vec{v}_1$, as shown. Again $\vec{v}_2 = \vec{v}_1 + \Delta \vec{v}$, but in this case $\vec{v}_1$ and $\vec{v}_2$ have different directions. As the time interval Δt approaches zero, the angle ϕ in the figure also approaches zero, $\Delta \vec{v}$ becomes perpendicular to *both* $\vec{v}_1$ and $\vec{v}_2$, and $\vec{v}_1$ and $\vec{v}_2$ have the same magnitude. In other words, the speed of the particle stays the same, but the path of the particle curves.

Thus, when $\vec{a}$ is *parallel* (or antiparallel) to $\vec{v}$, its effect is to change the magnitude of $\vec{v}$ but not its direction; when $\vec{a}$ is *perpendicular* to $\vec{v}$, its effect is to change the direction of $\vec{v}$ but not its magnitude. In general, $\vec{a}$ may have components *both* parallel and perpendicular to $\vec{v}$, but the above statements are still valid for the individual components. In particular, when a particle travels along a curved path with constant speed, its acceleration is always perpendicular to $\vec{v}$ at each point.

Figure 3.12 shows a particle moving along a curved path for three different situations: constant speed, increasing speed, and decreasing speed. If the speed is constant, $\vec{a}$ is perpendicular, or *normal,* to the path and to $\vec{v}$ and points toward the concave side of the path (Fig. 3.12a). If the speed is increasing, there is still a perpendicular component of $\vec{a}$, but there is also a parallel component having the same direction as $\vec{v}$ (Fig. 3.12b). Then $\vec{a}$ points ahead of the normal to the path. (This was the case in Example 3.2.) If the speed is decreasing, the parallel component has the direction opposite to $\vec{v}$, and $\vec{a}$ points behind the normal to the path (Fig. 3.12c). These ideas help us describe the acceleration of a car going around a turn (Fig. 3.13). We will use these ideas again in Section 3.4 when we study the special case of motion in a circle.

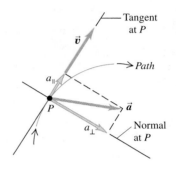

3.10 The acceleration can be resolved into a component $a_\parallel$ parallel to the path (and to the velocity) and a component $a_\perp$ perpendicular to the path (that is, along the normal to the path).

3.11 (a) When $\vec{a}$ is parallel to $\vec{v}$, the magnitude of $\vec{v}$ increases, but its direction does not change. The particle moves in a straight line with changing speed. (b) When $\vec{a}$ is perpendicular to $\vec{v}$, the direction of $\vec{v}$ changes, but its magnitude is constant. The particle moves along a curve with constant speed.

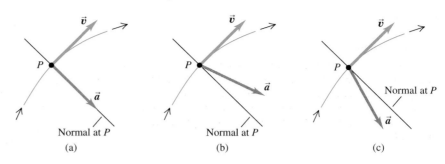

3.12 Velocity and acceleration vectors for a particle moving through a point P on a curved path with (a) constant speed, (b) increasing speed, and (c) decreasing speed.

3.13 A race car typically slows down when it enters a turn and speeds up as it exits the turn. Thus the car has a component of acceleration parallel or antiparallel to the direction of motion (which describes the change in speed) and a component perpendicular to the direction of motion (which describes the change in direction).

CAUTION Note that the two quantities

$$\frac{d|\vec{v}|}{dt} \quad \text{and} \quad \left|\frac{d\vec{v}}{dt}\right|$$

are not the same. The first is the rate of change of speed; it is zero whenever a particle moves with constant speed, even when its direction of motion changes. The second is the magnitude of the vector acceleration; it is zero only when the particle's acceleration is zero—that is, when the particle moves in a straight line with constant speed.

Example 3.4

Calculating parallel and perpendicular components of acceleration

For the rover of Examples 3.1 and 3.2, find the parallel and perpendicular components of the acceleration at $t = 2.0$ s.

SOLUTION

IDENTIFY and SET UP: To find the parallel and perpendicular components, we need to know the angle between the acceleration vector $\vec{a}$ and the velocity vector $\vec{v}$. We'll be able to determine this because we found the directions of $\vec{a}$ and $\vec{v}$ in Examples 3.2 and 3.1, respectively.

EXECUTE: In Example 3.2 we found that at $t = 2.0$ s the particle has an acceleration of magnitude 0.58 m/s^2 at an angle of 149° with respect to the positive x-axis. From Example 3.1, at this same time the velocity vector is at an angle of 128° with respect to the positive x-axis. So Fig. 3.8 shows that the angle between $\vec{a}$ and $\vec{v}$ is 149° − 128° = 21° (Fig. 3.14). The parallel and perpendicular components of acceleration are then

$$a_{\parallel} = a \cos 21° = (0.58 \text{ m/s}^2)\cos 21° = 0.54 \text{ m/s}^2$$
$$a_{\perp} = a \sin 21° = (0.58 \text{ m/s}^2)\sin 21° = 0.21 \text{ m/s}^2$$

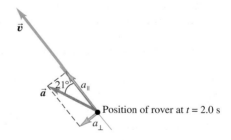

3.14 The parallel and perpendicular components of the acceleration of the rover at $t = 2.0$ s.

EVALUATE: The parallel component $a_{\parallel}$ is in the same direction as $\vec{v}$, which means that the speed is increasing at this instant. Because the perpendicular component $a_{\perp}$ is not zero, it follows that the rover's path is curved at this instant; in other words, the rover is turning.

Conceptual Example 3.5	**Acceleration of a skier**

A skier moves along a ski-jump ramp as shown in Fig. 3.15. The ramp is straight from point *A* to point *C* and curved from point *C* onward. The skier picks up speed as she moves downhill from point *A* to point *E*, where her speed is maximum. She slows down after passing point *E*. Draw the direction of the acceleration vector at points *B*, *D*, *E*, and *F*.

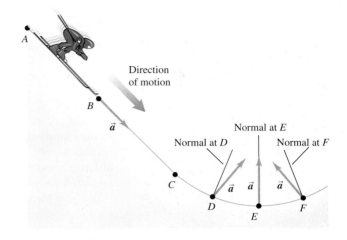

SOLUTION

At point *B* the skier is moving in a straight line with increasing speed, so her acceleration points downhill, in the same direction as her velocity.

At point *D* the skier is moving along a curved path, so her acceleration has a component perpendicular to the path. There is also a component in the direction of her motion because she is still speeding up at this point. So the acceleration vector points *ahead* of the normal to her path at point *D*.

The skier's speed is instantaneously not changing at point *E*; the speed is maximum at this point, so its derivative is zero. So there is no parallel component of $\vec{a}$, and the acceleration is perpendicular to her motion.

Finally, at point *F* the acceleration has a perpendicular component (because her path is curved at this point) and a parallel component *opposite* to the direction of her motion (because she's

3.15 The direction of a skier's acceleration is different at different points along the path.

slowing down). So at this point, the acceleration vector points *behind* the normal to her path.

In the next section we'll examine the skier's acceleration after she flies off the ramp.

Test Your Understanding

Consider the volcanic rock fragments shown in the photograph that opens this chapter. Each fragment slows down as it climbs toward the highest point of its path, then speeds up as it descends from this point. What is the direction of the fragment's acceleration at the highest point of its path?

3.3 | Projectile Motion

A **projectile** is any body that is given an initial velocity and then follows a path determined entirely by the effects of gravitational acceleration and air resistance. A batted baseball, a thrown football, a package dropped from an airplane, and a bullet shot from a rifle are all projectiles. The path followed by a projectile is called its **trajectory.**

To analyze this common type of motion, we'll start with an idealized model, representing the projectile as a single particle with an acceleration (due to gravity) that is constant in both magnitude and direction. We'll neglect the effects of air resistance and the curvature and rotation of the earth. Like all models, this one has limitations. Curvature of the earth has to be considered in the flight of long-range missiles, and air resistance is of crucial importance to a sky diver. Nevertheless, we can learn a lot from analysis of this simple model. For the remainder of this chapter

Activ Physics ONLINE

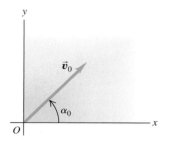

3.16 All projectile motion occurs in the vertical plane containing the initial velocity vector $\vec{v}_0$.

the phrase "projectile motion" will imply that we're ignoring air resistance. In Chapter 5 we will see what happens when air resistance cannot be ignored.

We first notice that projectile motion is always confined to a vertical plane determined by the direction of the initial velocity (Fig. 3.16). This is because the acceleration due to gravity is purely vertical; gravity can't move the projectile sideways. Thus projectile motion is *two-dimensional*. We will call the plane of motion the *xy*-coordinate plane, with the *x*-axis horizontal and the *y*-axis vertically upward.

The key to analyzing projectile motion is that we can treat the *x*- and *y*-coordinates separately. The *x*-component of acceleration is zero, and the *y*-component is constant and equal to $-g$. (Remember that by definition, g is always positive, while with our choice of coordinate directions, a_y is negative.) So *we can analyze projectile motion as a combination of horizontal motion with constant velocity and vertical motion with constant acceleration.* Figure 3.17 shows two projectiles with different *x*-motion but identical *y*-motion; one is dropped from rest and the other is projected horizontally, but both projectiles fall the same distance in the same time.

We can then express all the vector relationships for position, velocity, and acceleration by separate equations for the horizontal and vertical components. The actual motion of the projectile is the superposition of these separate motions. The components of $\vec{a}$ are

$$a_x = 0 \qquad a_y = -g \qquad \text{(projectile motion, no air resistance)} \qquad (3.14)$$

We will usually use $g = 9.8 \text{ m/s}^2$. Since the *x*-acceleration and *y*-acceleration are both constant, we can use Eqs. (2.8), (2.12), (2.13), and (2.14) directly. For example, suppose that at time $t = 0$ our particle is at the point (x_0, y_0) and that at this time its velocity components have the initial values v_{0x} and v_{0y}. The components of acceleration are $a_x = 0, a_y = -g$. Considering the *x*-motion first, we substitute 0 for a_x in Eqs. (2.8) and (2.12). We find

$$v_x = v_{0x} \qquad (3.15)$$

$$x = x_0 + v_{0x}t \qquad (3.16)$$

For the *y*-motion we substitute y for x, v_y for v_x, v_{0y} for v_{0x}, and $a_y = -g$ for a_x:

$$v_y = v_{0y} - gt \qquad (3.17)$$

$$y = y_0 + v_{0y}t - \frac{1}{2}gt^2 \qquad (3.18)$$

It's usually simplest to take the initial position (at time $t = 0$) as the origin; in this case, $x_0 = y_0 = 0$. This point might be, for example, the position of a ball at the instant it leaves the thrower's hand or the position of a bullet at the instant it leaves the gun barrel.

Figure 3.18 shows the path of a projectile that starts at (or passes through) the origin at time $t = 0$. The position, velocity, velocity components, and acceleration are shown at a series of times separated by equal intervals. The *x*-component of acceleration is zero, so v_x is constant. The *y*-component of acceleration is constant and not zero, so v_y changes by equal amounts in equal times. At the highest point in the trajectory, $v_y = 0$.

We can also represent the initial velocity $\vec{v}_0$ by its magnitude v_0 (the initial speed) and its angle α_0 with the positive *x*-axis. In terms of these quantities, the components v_{0x} and v_{0y} of the initial velocity are

$$v_{0x} = v_0 \cos \alpha_0 \qquad v_{0y} = v_0 \sin \alpha_0 \qquad (3.19)$$

3.17 Independence of horizontal and vertical motion. The red ball is dropped from rest, and the yellow ball is simultaneously projected horizontally; successive images in this stroboscopic photograph are separated by equal time intervals. At any given time, both balls have the same *y*-position, *y*-velocity, and *y*-acceleration, despite having different *x*-positions and *x*-velocities.

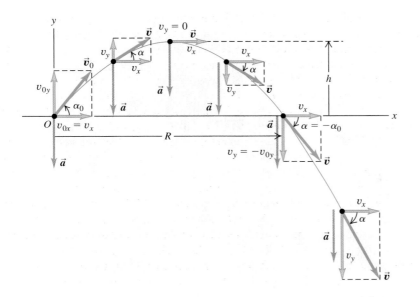

3.18 The trajectory of a body projected with an initial velocity $\vec{v}_0$ at an angle α_0 above the horizontal with negligible air resistance. The distance R is the horizontal range, and h is the maximum height.

Using these relations in Eqs. (3.15) through (3.18) and setting $x_0 = y_0 = 0$, we find

$$x = (v_0 \cos \alpha_0)t \qquad \text{(projectile motion)} \qquad (3.20)$$

$$y = (v_0 \sin \alpha_0)t - \frac{1}{2}gt^2 \qquad \text{(projectile motion)} \qquad (3.21)$$

$$v_x = v_0 \cos \alpha_0 \qquad \text{(projectile motion)} \qquad (3.22)$$

$$v_y = v_0 \sin \alpha_0 - gt \qquad \text{(projectile motion)} \qquad (3.23)$$

Act**i**v
ONLINE
Physics

3.7   Target Practice II

These equations describe the position and velocity of the projectile in Fig. 3.18 at any time t.

We can get a lot of information from these equations. For example, at any time the distance r of the projectile from the origin (the magnitude of the position vector $\vec{r}$) is given by

$$r = \sqrt{x^2 + y^2} \qquad (3.24)$$

The projectile's speed (the magnitude of its velocity) at any time is

$$v = \sqrt{v_x^2 + v_y^2} \qquad (3.25)$$

The *direction* of the velocity, in terms of the angle α it makes with the positive x-axis, is given by

$$\tan \alpha = \frac{v_y}{v_x} \qquad (3.26)$$

The velocity vector $\vec{v}$ is tangent to the trajectory at each point.

We can derive an equation for the trajectory's shape in terms of x and y by eliminating t. From Eqs. (3.20) and (3.21), which assume $x_0 = y_0 = 0$, we find $t = x/(v_0 \cos \alpha_0)$ and

$$y = (\tan \alpha_0)x - \frac{g}{2v_0^2 \cos^2 \alpha_0}x^2 \qquad (3.27)$$

3.19 Stroboscopic photograph of a bouncing ball, showing parabolic trajectories after each bounce. Successive images are separated by equal time intervals, as in Fig. 3.17. Each peak in the trajectory is lower than the preceding one because of energy loss during the "bounce," or collision with the horizontal surface.

3.20 Air resistance has a large cumulative effect on the motion of a baseball. In this simulation we allow the baseball to fall below the height from which it was thrown (for example, the baseball could have been thrown from a cliff).

Don't worry about the details of this equation; the important point is its general form. The quantities v_0, $\tan \alpha_0$, $\cos \alpha_0$, and g are constants, so the equation has the form

$$y = bx - cx^2$$

where b and c are constants. This is the equation of a *parabola*. In projectile motion, with our simple model, the trajectory is always a parabola (Fig. 3.19).

We mentioned at the beginning of this section that air resistance isn't always negligible. When it has to be included, the calculations become a lot more complicated; the effects of air resistance depend on velocity, so the acceleration is no longer constant. Figure 3.20 shows a computer simulation of the trajectory of a baseball with $v_0 = 50$ m/s and $\alpha_0 = 53.1°$, both without air resistance and with air resistance proportional to the square of the baseball's speed. We see that air resistance has a very large effect; the maximum height and range both decrease, and the trajectory is no longer a parabola.

| Conceptual Example 3.6 | **Acceleration of a skier, continued** |

Let's consider again the skier in Conceptual Example 3.5. What is her acceleration at points G, H, and I in Fig. 3.21 *after* she flies off the ramp? Neglect air resistance.

SOLUTION

The skier's acceleration changed from point to point while she was on the ramp. But as soon as she leaves the ramp, she becomes a pro-

jectile. So at points G, H, and I, and indeed at *all* points after she leaves the ramp, the skier's acceleration points vertically downward and has magnitude g. No matter how complicated the acceleration of a particle before it becomes a projectile, its acceleration as a projectile is given by $a_x = 0$, $a_y = -g$.

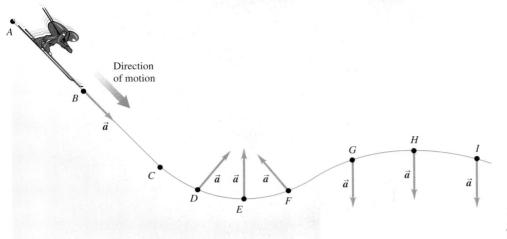

3.21 Once the skier is in projectile motion, her acceleration is constant and downward.

Projectile Motion

NOTE: The strategies we used in Sections 2.4 and 2.5 for straight-line, constant-acceleration problems are also useful here.

IDENTIFY *the relevant concepts:* The key concept to remember is that throughout projectile motion, the acceleration is downward and has a constant magnitude g. Be on the lookout for aspects of the problem that do *not* involve projectile motion. For example, the projectile-motion equations don't apply to throwing a ball, because during the throw the ball is acted on by both the thrower's hand and gravity. These equations come into play only after the ball leaves the thrower's hand.

SET UP *the problem using the following steps:*
1. Define your coordinate system and make a sketch showing your axes. Usually it's easiest to place the origin at the initial $(t = 0)$ position of the projectile. (If the projectile is a thrown ball or a dart shot from a gun, the initial position is where the projectile leaves the thrower's hand or exits the muzzle of the gun.) Also, it's usually best to take the x-axis as being horizontal and the y-axis as being upward. Then the initial position is $x_0 = 0$ and $y_0 = 0$, and the components of the (constant) acceleration are $a_x = 0$, $a_y = -g$.
2. List the unknown and known quantities, and decide which unknowns are your target variables. In some problems you'll be given the initial velocity (either in terms of components or in terms of magnitude and direction) and asked to find the coordinates and velocity components at some later time. In other problems you might be given two

points on the trajectory and asked to find the initial velocity. In any case, you'll be using Eqs. (3.20) through (3.23). Make sure that you have as many equations as there are target variables to be found.
3. It often helps to state the problem in words and then translate those words into symbols. For example, *when* does the particle arrive at a certain point? (That is, at what value of t?) *Where* is the particle when its velocity has a certain value? (That is, what are the values of x and y when v_x or v_y has the specified value?) At the highest point in a trajectory, $v_y = 0$. So the question "When does the particle reach its highest point?" translates into "What is the value of t when $v_y = 0$?" Similarly, "When does the projectile return to its initial elevation?" translates into "What is the value of t when $y = y_0$?"

EXECUTE *the solution:* Use Eqs. (3.20) through (3.23) to find the target variables. (Certain other equations given in Section 3.3 may be useful as well.) As you do so, resist the temptation to break the trajectory into segments and analyze each segment separately. You don't have to start all over, with a new axis and a new time scale, when the projectile reaches its highest point! It's almost always easier to set up Eqs. (3.20) through (3.23) at the start and continue to use the same axes and time scale throughout the problem.

EVALUATE *your answer:* As always, look at your results to see whether they make sense and whether the numerical values seem reasonable.

Example
3.7
A body projected horizontally

A motorcycle stunt rider rides off the edge of a cliff. Just at the edge his velocity is horizontal, with magnitude 9.0 m/s. Find the motorcycle's position, distance from the edge of the cliff, and velocity after 0.50 s.

SOLUTION

IDENTIFY: Once the rider leaves the cliff, he is in projectile motion. His velocity at the edge of the cliff is therefore his initial velocity.

SET UP: The coordinate system is shown in Fig. 3.22. We choose the origin at the edge of the cliff, where the motorcycle first becomes a projectile, so $x_0 = 0$ and $y_0 = 0$. The initial velocity is purely horizontal (that is, $\alpha_0 = 0$), so the initial velocity components are $v_{0x} = v_0 \cos \alpha_0 = 9.0$ m/s and $v_{0y} = v_0 \sin \alpha_0 = 0$. To find the motorcycle's position at time $t = 0.50$ s, we use Eqs. (3.20) and (3.21), which give x and y as functions of time. Given these values, we find the distance from the origin using Eq. (3.24). Finally, we determine the velocity at $t = 0.50$ s with Eqs. (3.22) and (3.23), which give v_x and v_y as functions of time.

EXECUTE: Where is the motorcycle at $t = 0.50$ s? From Eqs. (3.20) and (3.21), the x- and y-coordinates are

$$x = v_{0x}t = (9.0 \text{ m/s})(0.50 \text{ s}) = 4.5 \text{ m}$$

$$y = -\frac{1}{2}gt^2 = -\frac{1}{2}(9.8 \text{ m/s}^2)(0.50 \text{ s})^2 = -1.2 \text{ m}$$

The negative value of y shows that at this time the motorcycle is below its starting point.

What is the motorcycle's distance from the origin at this time? From Eq. (3.24),

$$r = \sqrt{x^2 + y^2} = \sqrt{(4.5 \text{ m})^2 + (-1.2 \text{ m})^2} = 4.7 \text{ m}$$

What is the velocity at time $t = 0.50$ s? From Eqs. (3.22) and (3.23), the components of velocity at this time are

$$v_x = v_{0x} = 9.0 \text{ m/s}$$

$$v_y = -gt = (-9.8 \text{ m/s}^2)(0.50 \text{ s}) = -4.9 \text{ m/s}$$

The motorcycle has the same horizontal velocity v_x as when it left the cliff at $t = 0$, but in addition there is a vertical velocity v_y

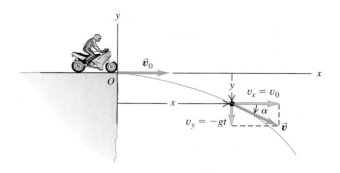

3.22 Trajectory of a body projected horizontally.

directed downward (in the negative y-direction). If we use unit vectors, the velocity at $t = 0.50$ s is

$$\vec{v} = v_x\hat{\imath} + v_y\hat{\jmath} = (9.0 \text{ m/s})\hat{\imath} + (-4.9 \text{ m/s})\hat{\jmath}$$

We can also express the velocity in terms of magnitude and direction. From Eq. (3.25), the speed (magnitude of the velocity) at this time is

$$v = \sqrt{v_x^2 + v_y^2}$$
$$= \sqrt{(9.0 \text{ m/s})^2 + (-4.9 \text{ m/s})^2} = 10.2 \text{ m/s}$$

From Eq. (3.26), the angle α of the velocity vector is

$$\alpha = \arctan \frac{v_y}{v_x}$$
$$= \arctan\left(\frac{-4.9 \text{ m/s}}{9.0 \text{ m/s}}\right) = -29°$$

At this time the velocity is 29° below the horizontal.

EVALUATE: Our answers illustrate that the horizontal aspect of the motion is unchanged by gravity; the motorcycle continues to move horizontally at 9.0 m/s, covering 4.5 m in 0.50 s. As the motorcycle initially has zero vertical velocity, it falls vertically just like an object released from rest, descending a distance $\frac{1}{2}gt^2 = 1.2$ m in 0.50 s.

Example
3.8
Height and range of a baseball

A batter hits a baseball so that it leaves the bat with an initial speed $v_0 = 37.0$ m/s at an initial angle $\alpha_0 = 53.1°$, at a location where $g = 9.80$ m/s² (Fig. 3.23). a) Find the position of the ball, and the magnitude and direction of its velocity, when $t = 2.00$ s. b) Find the time when the ball reaches the highest point of its

flight and find its height h at this point. c) Find the *horizontal range R*—that is, the horizontal distance from the starting point to the point at which the ball hits the ground. For each part, treat the baseball as a projectile.

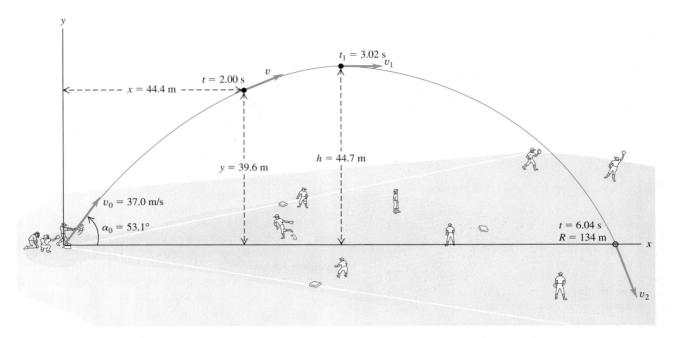

3.23 Trajectory of a batted baseball, treated as a projectile.

SOLUTION

IDENTIFY: As Fig. 3.20 shows, the effects of air resistance on the motion of a baseball aren't really negligible. For the sake of simplicity, however, we'll ignore air resistance for this example. Instead, we'll use the projectile-motion equations to describe the motion.

SET UP: We use the same coordinate system as in Fig. 3.18. Then we can use Eqs. (3.20) through (3.23) without any modifications. Our target variables are (1) the position and velocity of the ball 2.00 s after it leaves the bat, (2) the elapsed time after leaving the bat when the ball is at its maximum height—that is, when $v_y = 0$—and the y-coordinate at this time, and (3) the x-coordinate at the time when the y-coordinate is equal to the initial value y_0.

The ball is probably struck a meter or so above ground level, but we neglect this distance and assume that it starts at ground level $(y_0 = 0)$. The initial velocity of the ball has components

$$v_{0x} = v_0 \cos \alpha_0 = (37.0 \text{ m/s}) \cos 53.1° = 22.2 \text{ m/s}$$
$$v_{0y} = v_0 \sin \alpha_0 = (37.0 \text{ m/s}) \sin 53.1° = 29.6 \text{ m/s}$$

EXECUTE: a) We want to find x, y, v_x, and v_y at time $t = 2.00$ s. From Eqs. (3.20) through (3.23),

$$x = v_{0x}t = (22.2 \text{ m/s})(2.00 \text{ s}) = 44.4 \text{ m}$$
$$y = v_{0y}t - \frac{1}{2}gt^2$$
$$= (29.6 \text{ m/s})(2.00 \text{ s}) - \frac{1}{2}(9.80 \text{ m/s}^2)(2.00 \text{ s})^2$$
$$= 39.6 \text{ m}$$

$$v_x = v_{0x} = 22.2 \text{ m/s}$$
$$v_y = v_{0y} - gt = 29.6 \text{ m/s} - (9.80 \text{ m/s}^2)(2.00 \text{ s})$$
$$= 10.0 \text{ m/s}$$

The y-component of velocity is positive, which means that the ball is still moving upward at this time (Fig. 3.23). The magnitude and direction of the velocity are found from Eqs. (3.25) and (3.26):

$$v = \sqrt{v_x^2 + v_y^2} = \sqrt{(22.2 \text{ m/s})^2 + (10.0 \text{ m/s})^2}$$
$$= 24.3 \text{ m/s}$$
$$\alpha = \arctan\left(\frac{10.0 \text{ m/s}}{22.2 \text{ m/s}}\right) = \arctan 0.450 = 24.2°$$

The velocity is 24.2° above the horizontal.

b) At the highest point, the vertical velocity v_y is zero. When does this happen? Call the time t_1; then

$$v_y = 0 = v_{0y} - gt_1$$
$$t_1 = \frac{v_{0y}}{g} = \frac{29.6 \text{ m/s}}{9.80 \text{ m/s}^2} = 3.02 \text{ s}$$

The height h at this time is the value of y when $t = t_1 = 3.02$ s:

$$h = v_{0y}t_1 - \frac{1}{2}gt_1^2$$
$$= (29.6 \text{ m/s})(3.02 \text{ s}) - \frac{1}{2}(9.80 \text{ m/s}^2)(3.02 \text{ s})^2$$
$$= 44.7 \text{ m}$$

Alternatively, we can apply the constant-acceleration formula Eq. (2.13) to the y-motion:

$$v_y^2 = v_{0y}^2 + 2a_y(y - y_0) = v_{0y}^2 - 2g(y - y_0)$$

At the highest point, $v_y = 0$ and $y = h$. Substituting these, along with $y_0 = 0$, we find

$$0 = v_{0y}^2 - 2gh$$

$$h = \frac{v_{0y}^2}{2g} = \frac{(29.6 \text{ m/s})^2}{2(9.80 \text{ m/s}^2)} = 44.7 \text{ m}$$

c) We'll find the horizontal range in two steps. First, *when* does the ball hit the ground? This occurs when $y = 0$. Call this time t_2; then

$$y = 0 = v_{0y}t_2 - \frac{1}{2}gt_2^2 = t_2\left(v_{0y} - \frac{1}{2}gt_2\right)$$

This is a quadratic equation for t_2. It has two roots,

$$t_2 = 0 \quad \text{and} \quad t_2 = \frac{2v_{0y}}{g} = \frac{2(29.6 \text{ m/s})}{9.80 \text{ m/s}^2} = 6.04 \text{ s}$$

There are two times at which $y = 0$; $t_2 = 0$ is the time the ball *leaves* the ground, and $t_2 = 6.04$ s is the time of its return. This is exactly twice the time to reach the highest point, so the time of descent equals the time of ascent. (This is *always* true if the starting and end points are at the same elevation and air resistance can be neglected. We will prove this in Example 3.10.)

The horizontal range R is the value of x when the ball returns to the ground, that is, at $t = 6.04$ s:

$$R = v_{0x}t_2 = (22.2 \text{ m/s})(6.04 \text{ s}) = 134 \text{ m}$$

The vertical component of velocity when the ball hits the ground is

$$v_y = v_{0y} - gt_2 = 29.6 \text{ m/s} - (9.80 \text{ m/s}^2)(6.04 \text{ s})$$
$$= -29.6 \text{ m/s}$$

That is, v_y has the same magnitude as the initial vertical velocity v_{0y} but the opposite direction (down). Since v_x is constant, the angle $\alpha = -53.1°$ (below the horizontal) at this point is the negative of the initial angle $\alpha_0 = 53.1°$.

EVALUATE: Do our numerical results make sense? The maximum height $h = 44.7$ m in part (b) is roughly half the height above the playing field of the roof of Skydome in Toronto. The horizontal range $R = 134$ m in part (c) is greater than the 93.6-m distance from home plate to the right-field fence at Pacific Bell Park in San Francisco. Can you show that when the ball reaches this fence, it is 37.7 m above the ground? (This height is more than enough to clear the fence, so this ball is a home run.)

If the ball could continue to travel on *below* its original level (through an appropriately shaped hole in the ground), then negative values of y, corresponding to times greater than 6.04 s, would be possible. Can you compute the position and velocity at a time 7.55 s after the start, corresponding to the last position shown in Fig. 3.18? The results are

$$x = 168 \text{ m} \qquad y = -55.8 \text{ m}$$
$$v_x = 22.2 \text{ m/s} \qquad v_y = -44.4 \text{ m/s}$$

It's worth pointing out that in real life, a batted ball with the initial speed and angle we've used here won't go as high or as far as we've calculated. (If it did, home runs would be far more common and baseball would be a far less interesting game.) The reason is that air resistance, which we neglected in this example, is actually an important factor at the typical speeds of pitched and batted balls.

Example 3.9	**The zoo keeper and the monkey**

A clever monkey escapes from the zoo. The zoo keeper finds him in a tree. After failing to entice the monkey down, the zoo keeper points her tranquilizer gun directly at the monkey and shoots (Fig. 3.24). The clever monkey lets go at the same instant the dart leaves the gun barrel, intending to land on the ground and escape. Show that the dart *always* hits the monkey, regardless of the dart's muzzle velocity (provided that it gets to the monkey before he hits the ground).

SOLUTION

IDENTIFY: In this example we have *two* objects in projectile motion: the tranquilizer dart and the monkey. The dart and the monkey have different initial positions and initial velocities, but they go into projectile motion at the same time. To show that the dart hits the monkey, we have to prove that at some time the monkey and the dart have the same x-coordinate and the same y-coordinate.

SET UP: We make the usual choice for the x- and y-directions, and place the origin of coordinates at the end of the barrel of the tran-

quilizer gun (Fig. 3.24). We'll first use Eq. (3.20) to find the time t when the x-coordinates x_monkey and x_dart are the same. We'll then use Eq. (3.21) to check whether y_monkey and y_dart are also equal at this time; if they are, the dart hits the monkey.

EXECUTE: The monkey drops straight down, so $x_\text{monkey} = d$ at *all* times. For the dart, Eq. (3.20) tells us that $x_\text{dart} = (v_0 \cos \alpha_0)t$. When these x-coordinates are equal, $d = (v_0 \cos \alpha_0)t$, or

$$t = \frac{d}{v_0 \cos \alpha_0}$$

To have the dart hit the monkey, it must be true that $y_\text{monkey} = y_\text{dart}$ at this same time. The monkey is in one-dimensional free fall; his position at any time is given by Eq. (2.12), with appropriate symbol changes. Figure 3.24 shows that the monkey's initial height is $d \tan \alpha_0$ (the opposite side of a right triangle with angle α_0 and adjacent side d), and we find

$$y_\text{monkey} = d \tan \alpha_0 - \frac{1}{2}gt^2$$

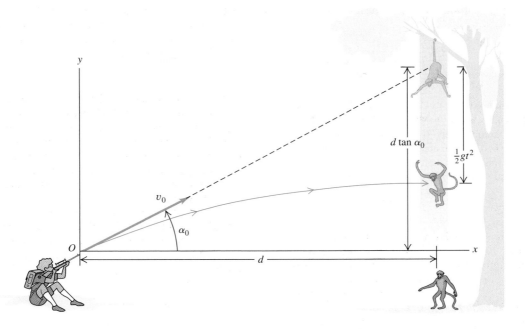

3.24 The tranquilizer dart hits the falling monkey.

For the dart we use Eq. (3.21):

$$y_{\text{dart}} = (v_0 \sin \alpha_0)t - \frac{1}{2}gt^2$$

So we see that if $d \tan \alpha_0 = (v_0 \sin \alpha_0)t$ at the time when the two x-coordinates are equal, then $y_{\text{monkey}} = y_{\text{dart}}$, and we have a hit. To prove that this happens, we replace t with $d/(v_0 \cos \alpha_0)$, the time when $x_{\text{monkey}} = x_{\text{dart}}$; then

$$(v_0 \sin \alpha_0)t = (v_0 \sin \alpha_0)\frac{d}{v_0 \cos \alpha_0} = d \tan \alpha_0$$

EVALUATE: We have proved that at the time the x-coordinates are equal, the y-coordinates are also equal; a dart aimed at the initial position of the monkey always hits it, no matter what v_0 is. This result is also independent of the value of g, the acceleration due to gravity. With no gravity $(g = 0)$, the monkey would remain motionless, and the dart would travel in a straight line to hit him. With gravity, both "fall" the same distance $(\frac{1}{2}gt^2)$ below their $g = 0$ positions, and the dart still hits the monkey.

Example 3.10	**Height and range of a projectile**

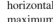

For a projectile launched with speed v_0 at initial angle α_0 (between 0 and 90°), derive general expressions for the maximum height h and horizontal range R (Fig. 3.18). For a given v_0, what value of α_0 gives maximum height? What value gives maximum horizontal range?

SOLUTION

IDENTIFY: This is really the same exercise as parts (b) and (c) of Example 3.8. The difference is that we are looking for general expressions for h and R. We'll also be looking for the values of α_0 that give the maximum values of h and R.

SET UP: As in Example 3.8, we use the coordinate system shown in Fig. 3.18. For convenience we choose the origin to be

at the initial position of the projectile. To determine the height h at the high point of the trajectory, we use Eq. (3.23) to find the time t_1 at which $v_y = 0$, then find the y-coordinate at this time with Eq. (3.21). To determine R, we first use Eq. (3.21) to find the time t_2 at which $y = 0$ (i. e., when the projectile returns to its initial height), then substitute t_2 into Eq. (3.20) to determine the x-coordinate at t_2.

EXECUTE: From Eq. (3.23), the time t_1 when $v_y = 0$ is given by the equation

$$v_y = v_0 \sin \alpha_0 - gt_1 = 0 \qquad t_1 = \frac{v_0 \sin \alpha_0}{g}$$

Then, from Eq. (3.21), the height at this time is

$$h = v_0 \sin \alpha_0 \left(\frac{v_0 \sin \alpha_0}{g} \right) - \frac{1}{2} g \left(\frac{v_0 \sin \alpha_0}{g} \right)^2$$
$$= \frac{v_0^2 \sin^2 \alpha_0}{2g}$$

If we vary α_0, the maximum value of h occurs when $\sin \alpha_0 = 1$ and $\alpha_0 = 90°$; that is, when the projectile is launched straight up. That's what we should expect. If it is launched horizontally, as in Example 3.7, $\alpha_0 = 0$ and the maximum height is zero!

To find a general expression for the horizontal range R, we first find the time t_2 at which $y = 0$. From Eq. (3.21), t_2 satisfies the equation

$$(v_0 \sin \alpha_0) t_2 - \frac{1}{2} g t_2^2 = t_2 \left(v_0 \sin \alpha_0 - \frac{1}{2} g t_2 \right) = 0$$

The two roots of this quadratic equation for t_2 are $t_2 = 0$ (the time of launch) and $t_2 = (2 v_0 \sin \alpha_0)/g$. The horizontal range R is the value of x at the second time. From Eq. (3.20),

$$R = (v_0 \cos \alpha_0) \frac{2 v_0 \sin \alpha_0}{g}$$

We can now use the trigonometric identity $2 \sin \alpha_0 \cos \alpha_0 = \sin 2\alpha_0$ to rewrite this as

$$R = \frac{v_0^2 \sin 2\alpha_0}{g}$$

The maximum value of $\sin 2\alpha_0$ is 1; this occurs when $2\alpha_0 = 90°$, or $\alpha_0 = 45°$. This angle gives the maximum range for a given initial speed.

EVALUATE: Figure 3.25 is based on a composite photograph of three trajectories of a ball projected from a spring gun at angles of 30°, 45°, and 60°. The initial speed v_0 is approximately the same in all three cases. The horizontal ranges are nearly the same for the 30° and 60° angles, and the range for 45° is greater than either. Can you prove that for a given value of v_0 the range is the same for an initial angle α_0 as for an initial angle $90° - \alpha_0$?

Comparing the expressions for t_1 and t_2, we see that $t_2 = 2t_1$; that is, the total flight time is twice the time up to the highest point. It follows that the time to reach the highest point equals the time to fall from there back to the initial elevation, as we asserted in Example 3.8.

CAUTION We don't recommend memorizing the above expressions for h and R. They are applicable only in the special circumstances we have described. In particular, the expression for the range R can be used *only* when launch and landing heights are equal. There are many end-of-chapter problems to which these equations are *not* applicable.

3.25 A firing angle of 45° (shown in purple) gives the maximum horizontal range. The range is less with firing angles of 30° (blue) or 60° (green).

| Example 3.11 | **Different initial and final heights** |

You toss a ball from your window 8.0 m above the ground. When the ball leaves your hand, it is moving at 10.0 m/s at an angle of 20° below the horizontal. How far horizontally from your window will the ball hit the ground? Ignore air resistance.

SOLUTION

IDENTIFY: As in our calculation of the horizontal range in Examples 3.8 and 3.10, we are trying to find the horizontal coordinate of a projectile when it is at a given value of y. The difference here is that this value of y is *not* equal to the initial y-coordinate.

SET UP: Once again we choose the x-axis to be horizontal and the y-axis to be upward. We place the origin of coordinates at the point where the ball leaves your hand, so that the ball's initial coordinates are $x_0 = 0$, $y_0 = 0$ (Fig. 3.26). We have $v_0 = 10.0$ m/s and $\alpha_0 = -20°$; the angle is negative because the initial velocity is below the horizontal. Our target variable is the value of x at the

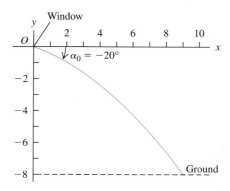

3.26 The trajectory of a ball launched at an angle below the horizontal.

point where the ball reaches the ground—that is, when $y = -8.0$ m. Because the initial and final heights of the ball are different, we can't simply use the expression for the horizontal range found in Example 3.10. Instead, we first use Eq. (3.21) to find the time t when the ball reaches $y = -8.0$ m, then calculate the value of x at this time, using Eq. (3.20).

EXECUTE: To determine t, we rewrite Eq. (3.21) in the standard form for a quadratic equation for t:

$$\frac{1}{2}gt^2 - (v_0 \sin \alpha_0)t + y = 0$$

The roots of this equation are

$$t = \frac{v_0 \sin \alpha_0 \pm \sqrt{(-v_0 \sin \alpha_0)^2 - 4\left(\frac{1}{2}g\right)y}}{2\left(\frac{1}{2}g\right)}$$

$$= \frac{v_0 \sin \alpha_0 \pm \sqrt{v_0^2 \sin^2 \alpha_0 - 2gy}}{g}$$

$$= \frac{\left[\begin{array}{c}(10.0 \text{ m/s}) \sin(-20°) \\ \pm\sqrt{(10.0 \text{ m/s})^2 \sin^2(-20°) - 2(9.8 \text{ m/s}^2)(-8.0 \text{ m})}\end{array}\right]}{9.8 \text{ m/s}^2}$$

We can discard the negative root, since it refers to a time before the ball left your hand. The positive root tells us that the ball takes 0.98 s to reach the ground. From Eq. (3.20), the ball's x-coordinate at that time is

$$x = (v_0 \cos \alpha_0)t = (10.0 \text{ m/s})[\cos(-20°)](0.98 \text{ s})$$

$$= 9.2 \text{ m}$$

The ball hits the ground a horizontal distance of 9.2 m from your window.

EVALUATE: The root $t = -1.7$ s is an example of a "fictional" solution to a quadratic equation. We discussed these in Example 2.8 in Section 2.5; you should review that discussion.

With our choice of origin we had initial and final heights $y_0 = 0$ and $y = -8.0$ m. Can you use Eqs. (3.16) and (3.18) to show that you get the same answers for t and x if you choose the origin to be at the point on the ground directly below where the ball leaves your hand?

Test Your Understanding

In Example 3.9, suppose the tranquilizer dart has a relatively low muzzle velocity so that the dart reaches a maximum height at a point P before striking the monkey (see Fig. 3.27). When the dart is at point P, will the monkey be at point A (higher than P), point B (at the same height as P), or at point C (lower than P)? Ignore air resistance.

3.27 The same situation as in Example 3.9, but with a low muzzle velocity.

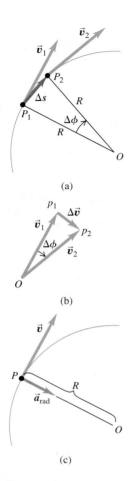

(a)

(b)

(c)

3.28 Finding the change in velocity, $\Delta\vec{v}$, of a particle moving in a circle with constant speed.

Actſv
ONLINE
Physics

4.1   Magnitude of Centripetal
      Acceleration

3.4 | Motion in a Circle

When a particle moves along a curved path, the direction of its velocity changes. As we saw in Section 3.2, this means that the particle *must* have a component of acceleration perpendicular to the path, even if its speed is constant. In this section we'll calculate the acceleration for the important special case of motion in a circle.

Uniform Circular Motion

When a particle moves in a circle with *constant speed,* the motion is called **uniform circular motion.** A car rounding a curve with constant radius at constant speed, a satellite moving in a circular orbit, and an ice skater skating in a circle with constant speed are all examples of uniform circular motion. There is no component of acceleration parallel (tangent) to the path; otherwise, the speed would change. The component of acceleration perpendicular (normal) to the path, which causes the direction of the velocity to change, is related in a simple way to the speed of the particle and the radius of the circle. Our next project is to derive this relation.

First we note that this is a different problem from the projectile motion considered in Section 3.3, in which the acceleration was constant in both its magnitude *(g)* and its direction (straight down). In uniform circular motion the acceleration is perpendicular to the velocity at each instant; as the direction of the velocity changes, the direction of the acceleration also changes. As we will see, the acceleration vector at each point in the circular path is directed toward the *center* of the circle.

Figure 3.28a shows a particle moving with constant speed in a circular path of radius R with center at O. The particle moves from P_1 to P_2 in a time Δt. The vector change in velocity $\Delta\vec{v}$ during this time is shown in Fig. 3.28b.

The angles labeled $\Delta\phi$ in Figs. 3.28a and 3.28b are the same because $\vec{v}_1$ is perpendicular to the line OP_1 and $\vec{v}_2$ is perpendicular to the line OP_2. Hence the triangles OP_1P_2 (Fig. 3.28a) and Op_1p_2 (Fig. 3.28b) are *similar.* Ratios of corresponding sides are equal, so

$$\frac{|\Delta\vec{v}|}{v_1} = \frac{\Delta s}{R} \quad \text{or} \quad |\Delta\vec{v}| = \frac{v_1}{R}\Delta s$$

The magnitude a_{av} of the average acceleration during Δt is therefore

$$a_{av} = \frac{|\Delta\vec{v}|}{\Delta t} = \frac{v_1}{R}\frac{\Delta s}{\Delta t}$$

The magnitude a of the *instantaneous* acceleration $\vec{a}$ at point P_1 is the limit of this expression as we take point P_2 closer and closer to point P_1:

$$a = \lim_{\Delta t \to 0}\frac{v_1}{R}\frac{\Delta s}{\Delta t} = \frac{v_1}{R}\lim_{\Delta t \to 0}\frac{\Delta s}{\Delta t}$$

But the limit of $\Delta s/\Delta t$ is the speed v_1 at point P_1. Also, P_1 can be any point on the path, so we can drop the subscript and let v represent the speed at any point. Then

$$a_{rad} = \frac{v^2}{R} \quad \text{(uniform circular motion)} \qquad (3.28)$$

We have added the subscript "rad" as a reminder that the direction of the instantaneous acceleration at each point is always along a radius of the circle, toward its center. This is consistent with our discussion in Section 3.2: the acceleration vector points toward the *concave* side of the particle's circular path, that is, toward the *inside* of the circle (never the outside). Because the speed is constant, the acceleration is always perpendicular to the instantaneous velocity. This is shown in Fig. 3.28c; compare with Fig. 3.12a.

We conclude: *In uniform circular motion, the magnitude a of the instantaneous acceleration is equal to the square of the speed v divided by the radius R of the circle. Its direction is perpendicular to $\vec{v}$ and inward along the radius.* Because the acceleration is always directed toward the center of the circle, it is sometimes called **centripetal acceleration.** The word *centripetal* is derived from two Greek words meaning "seeking the center." Figure 3.29 shows the directions of the velocity and acceleration vectors at several points for a particle moving with uniform circular motion. Compare this with the projectile motion shown in Fig. 3.18, in which the acceleration is always directed straight down and is *not* perpendicular to the path, except at one point.

We can also express the magnitude of the acceleration in uniform circular motion in terms of the **period** T of the motion, the time for one revolution (one complete trip around the circle). In a time T the particle travels a distance equal to the circumference $2\pi R$ of the circle, so its speed is

$$v = \frac{2\pi R}{T} \qquad (3.29)$$

When we substitute this into Eq. (3.28), we obtain the alternative expression

$$a_{\text{rad}} = \frac{4\pi^2 R}{T^2} \qquad (\text{uniform circular motion}) \qquad (3.30)$$

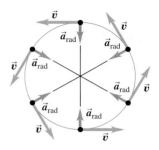

3.29 For a particle in uniform circular motion, the velocity at each point is tangent to the circle and the acceleration is directed toward the center.

| Example 3.12 | **Centripetal acceleration on a curved road** |

A BMW Z4 roadster has a "lateral acceleration" of $0.87g$, which is $(0.87)(9.8 \text{ m/s}^2) = 8.5 \text{ m/s}^2$. This represents the maximum centripetal acceleration that can be attained without skidding out of the circular path. If the car is traveling at a constant 40 m/s (about 89 mi/h, or 144 km/h), what is the minimum radius of curve it can negotiate? (Assume that the curve is unbanked.)

SOLUTION

IDENTIFY and SET UP: Because the car is moving along a curve—that is, a segment of a circle—at a constant speed, we can apply the ideas of uniform circular motion. In particular, we can use Eq. (3.28) to find the target variable R (the radius of the curve) in terms of the given centripetal acceleration a_{rad} and speed v.

EXECUTE: We are given a_{rad} and v, so we solve Eq. (3.28) for R:

$$R = \frac{v^2}{a_{\text{rad}}} = \frac{(40 \text{ m/s})^2}{8.5 \text{ m/s}^2} = 190 \text{ m} \qquad (\text{about } 600 \text{ ft})$$

EVALUATE: Our result shows that the required turning radius R is proportional to the *square* of the speed. Hence even a small reduction in speed can make R substantially smaller. For example, reducing v by 20% (from 40 to 32 m/s) would decrease R by 36% (from 190 m to 120 m).

Another way to make the required turning radius smaller is to *bank* the curve. We will investigate this option in Chapter 5.

Example
3.13   # Centripetal acceleration on a carnival ride

In a carnival ride, the passengers travel at constant speed in a circle of radius 5.0 m. They make one complete circle in 4.0 s. What is their acceleration?

SOLUTION

IDENTIFY and SET UP: The speed is constant, so this is a problem involving uniform circular motion. We are given the radius $R = 5.0$ m and the period $T = 4.0$ s, so we can use Eq. (3.30) to calculate the acceleration. Alternatively, we can first calculate the speed v using Eq. (3.29), then find the acceleration using Eq. (3.28).

EXECUTE: From Eq. (3.30),

$$a_{rad} = \frac{4\pi^2(5.0\text{ m})}{(4.0\text{ s})^2} = 12\text{ m/s}^2$$

We'll check this answer by using Eq. (3.28) after first determining the speed v. From Eq. (3.29), the speed is the circumference of the circle divided by the period T:

$$v = \frac{2\pi R}{T} = \frac{2\pi(5.0\text{ m})}{4.0\text{ s}} = 7.9\text{ m/s}$$

The centripetal acceleration is then

$$a_{rad} = \frac{v^2}{R} = \frac{(7.9\text{ m/s})^2}{5.0\text{ m}} = 12\text{ m/s}^2$$

Happily, we get the same answer for a_{rad} with both approaches.

EVALUATE: As in the preceding example, the direction of $\vec{a}$ is always toward the center of the circle. The magnitude of $\vec{a}$ is greater than g, the acceleration due to gravity, so this is not a ride for the faint-hearted. (Some roller coasters subject their passengers to accelerations as great as $4g$.)

Non-Uniform Circular Motion

We have assumed throughout this section that the particle's speed is constant. If the speed varies, we call the motion **non-uniform circular motion.** An example is a roller coaster car that slows down and speeds up as it moves around a vertical loop. In non-uniform circular motion, Eq. (3.28) still gives the *radial* component of acceleration $a_{rad} = v^2/R$, which is always *perpendicular* to the instantaneous velocity and directed toward the center of the circle. But since the speed v has different values at different points in the motion, the value of a_{rad} is not constant. The radial (centripetal) acceleration is greatest at the point in the circle where the speed is greatest.

In non-uniform circular motion there is also a component of acceleration that is *parallel* to the instantaneous velocity. This is the component $a_\parallel$ that we discussed in Section 3.2; here we call this component a_{tan} to emphasize that it is *tangent* to the circle. From the discussion at the end of Section 3.2 we see that the tangential component of acceleration a_{tan} is equal to the rate of change of *speed*. Thus

$$a_{rad} = \frac{v^2}{R} \quad \text{and} \quad a_{tan} = \frac{d|\vec{v}|}{dt} \quad \text{(non-uniform circular motion)} \quad (3.31)$$

The vector acceleration of a particle moving in a circle with varying speed is the vector sum of the radial and tangential components of accelerations. The tangential component is in the same direction as the velocity if the particle is speeding up, and is in the opposite direction if the particle is slowing down (Fig. 3.30).

In *uniform* circular motion there is no tangential component of acceleration, but the radial component is the magnitude of $d\vec{v}/dt$. We have mentioned before that the two quantities $|d\vec{v}/dt|$ and $d|\vec{v}|/dt$ are in general *not* equal. In uniform circular motion the first is constant and equal to v^2/R; the second is zero.

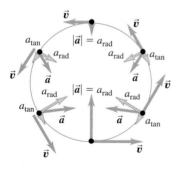

3.30 A particle moving in a vertical loop, like a roller coaster car, with a varying speed. The radial component of acceleration a_{rad} is largest where the speed is greatest (at the bottom of the loop) and smallest where the speed is least (at the top). The tangential component of acceleration a_{tan} is in the same direction as $\vec{v}$ when the particle is speeding up (going downhill) and opposite $\vec{v}$ when it is slowing down (going uphill).

Suppose that the roller coaster car in Fig. 3.30 experiences four times the acceleration at the bottom of the loop as it does at the top of the loop. How do the speeds at the top and bottom of the loop compare?

3.5 | Relative Velocity

You've no doubt observed how a car that is moving slowly forward appears to be moving backward when you pass it. In general, when two observers measure the velocity of a moving body, they get different results if one observer is moving relative to the other. The velocity seen by a particular observer is called the velocity *relative* to that observer, or simply **relative velocity.** Figure 3.31 shows a situation in which in which understanding relative velocity is extremely important.

We'll first consider relative velocity along a straight line, then generalize to relative velocity in a plane. Recall that for straight-line (one-dimensional) motion we use the term *velocity* to mean the component of the velocity vector along the line of motion; this can be positive, negative, or zero.

Relative Velocity in One Dimension

A woman walks with a velocity of 1.0 m/s along the aisle of a train that is moving with a velocity of 3.0 m/s (Fig. 3.32a). What is the woman's velocity? It's a simple enough question, but it has no single answer. As seen by a passenger sitting in the train, she is moving at 1.0 m/s. A person on a bicycle standing beside the train sees the woman moving at 1.0 m/s + 3.0 m/s = 4.0 m/s. An observer in another train going in the opposite direction would give still another answer. We have to specify which observer we mean, and we speak of the velocity *relative* to a particular observer. The woman's velocity relative to the train is 1.0 m/s, her velocity

3.31 Airshow pilots face a complicated problem involving relative velocities. They must keep track of their velocity relative to the air (to maintain enough airflow over the wings to sustain lift), their velocity relative to each other (to keep a tight formation without colliding), and their velocity relative to their audience (to remain in sight of the spectators).

(a)　　　　　(b)

3.32 (a) A woman walking in a train. (b) At the instant shown, the position of the woman (particle P) relative to frame of reference A is different from her position relative to frame of reference B.

relative to the cyclist is 4.0 m/s, and so on. Each observer, equipped in principle with a meter stick and a stopwatch, forms what we call a **frame of reference.** Thus a frame of reference is a coordinate system plus a time scale.

Let's call the cyclist's frame of reference (at rest with respect to the ground) A and the frame of reference of the moving train B (Fig. 3.32b). In straight-line motion the position of a point P relative to frame of reference A is given by the distance $x_{P/A}$ (the position of P with respect to A), and the position relative to frame B is given by $x_{P/B}$. The distance from the origin of A to the origin of B (position of B with respect to A) is $x_{B/A}$. We can see from the figure that

$$x_{P/A} = x_{P/B} + x_{B/A} \qquad (3.32)$$

This says that the total distance from the origin of A to point P is the distance from the origin of B to point P plus the distance from the origin of A to the origin of B.

The velocity of P relative to frame A, denoted by $v_{P/A}$, is the derivative of $x_{P/A}$ with respect to time. The other velocities are similarly obtained. So the time derivative of Eq. (3.32) gives us a relationship among the various velocities:

$$\frac{dx_{P/A}}{dt} = \frac{dx_{P/B}}{dt} + \frac{dx_{B/A}}{dt}$$

or

$$v_{P/A} = v_{P/B} + v_{B/A} \qquad \text{(relative velocity along a line)} \qquad (3.33)$$

Getting back to the woman on the train in Figure 3.32, A is the cyclist's frame of reference, B is the frame of reference of the train, and point P represents the woman. Using the above notation, we have

$$v_{P/B} = 1.0 \text{ m/s} \qquad v_{B/A} = 3.0 \text{ m/s}$$

From Eq. (3.33) the woman's velocity $v_{P/A}$ relative to the cyclist is

$$v_{P/A} = 1.0 \text{ m/s} + 3.0 \text{ m/s} = 4.0 \text{ m/s}$$

as we already knew.

In this example, both velocities are toward the right, and we have implicitly taken this as the positive direction. If the woman walks toward the *left* relative to the train, then $v_{P/B} = -1.0$ m/s, and her velocity relative to the cyclist is 2.0 m/s. The sum in Eq. (3.33) is always an algebraic sum, and any or all of the velocities may be negative.

When the woman looks out the window, the stationary cyclist on the ground appears to her to be moving backward; we can call the cyclist's velocity relative to her $v_{A/P}$. Clearly, this is just the negative of $v_{P/A}$. In general, if A and B are any two points or frames of reference,

$$v_{A/B} = -v_{B/A} \qquad (3.34)$$

Problem-Solving Strategy	**Relative Velocity**

IDENTIFY *the relevant concepts:* Whenever you see the phrase "velocity relative to" or "velocity with respect to," it's likely that the concepts of relative velocity will be helpful.

SET UP *the problem:* Label each frame of reference in the problem. Each moving object has its own frame of reference; in addi-

tion, you'll almost always have to include the frame of reference of the earth's surface. (Statements such as "The car is traveling north at 90 km/h" implicitly refer to the car's velocity relative to the surface of the earth.) Use the labels to help identify the target variable. For example, if you want to find the velocity of a car (C) with respect to a bus (B), your target variable is $v_{C/B}$.

EXECUTE *the solution:* Solve for the target variable using Eq. (3.33). (If the velocities are not along the same direction, you'll need to use the vector form of this equation, derived later in this section.) It's important to note the order of the double subscripts in Eq. (3.33): $v_{A/B}$ always means "velocity of A relative to B." These subscripts obey an interesting kind of algebra, as Eq. (3.33) shows. If we regard each one as a fraction, then the fraction on the left side is the *product* of the fractions on the right sides: $P/A = (P/B)(B/A)$. This is a handy rule you can use when applying Eq. (3.33) to any number of frames of refer-ence. For example, if there are three different frames of refer-ence A, B, and C, we can write immediately

$$v_{P/A} = v_{P/C} + v_{C/B} + v_{B/A}$$

EVALUATE *your answer:* Be on the lookout for stray minus signs in your answer. If the target variable is the velocity of a car relative to a bus $(v_{C/B})$, make sure that you haven't accidentally calculated the velocity of the *bus* relative to the *car* $(v_{B/C})$. If you have made this mistake, you can recover using Eq. (3.34).

Example 3.14	**Relative velocity on a straight road**

You are driving north on a straight two-lane road at a constant 88 km/h. A truck traveling at a constant 104 km/h approaches you (in the other lane, fortunately). a) What is the truck's velocity relative to you? b) What is your velocity with respect to the truck? c) How do the relative velocities change after you and the truck have passed each other?

SOLUTION

IDENTIFY and SET UP: Let you be Y, the truck be T, and the earth's surface be E, and let the positive direction be north (Fig. 3.33). Then your velocity relative to the earth is $v_{Y/E} = +88$ km/h. As the truck is initially approaching you, it must be moving south and its velocity with respect to the earth is $v_{T/E} = -104$ km/h. The target variable in part (a) is $v_{T/Y}$; the target variable in part (b) is $v_{Y/T}$. We'll find the answers for both parts of the problem by using the equation for rela-tive velocity, Eq. (3.33).

EXECUTE: a) To find $v_{T/Y}$, we first write Eq. (3.33) for the three frames Y, T, and E, and then rearrange:

$$v_{T/E} = v_{T/Y} + v_{Y/E}$$
$$v_{T/Y} = v_{T/E} - v_{Y/E}$$
$$= -104 \text{ km/h} - 88 \text{ km/h} = -192 \text{ km/h}$$

The truck is moving 192 km/h south relative to you.
b) From Eq. (3.34),

$$v_{Y/T} = -v_{T/Y} = -(-192 \text{ km/h}) = +192 \text{ km/h}$$

You are moving 192 km/h north relative to the truck.
c) The relative velocities don't change at all after you and the truck have passed each other. The relative positions of the bodies don't

3.33 Reference frames for you and the truck.

matter. The velocity of the truck relative to you is still -192 km/h, but it is now moving away from you instead of toward you.

EVALUATE: As a check on your answer in part (b), try using Eq. (3.33) directly in the form $v_{Y/T} = v_{Y/E} + v_{E/T}$. (Remember that the velocity of the earth with respect to the truck is the opposite of the velocity of the truck with respect to the earth: $v_{E/T} = -v_{T/E}$.) Do you get the same result?

Relative Velocity in Two or Three Dimensions

We can extend the concept of relative velocity to include motion in a plane or in space by using vector addition to combine velocities. Suppose that the woman in Fig. 3.32a is walking not down the aisle of the railroad car but from one side of the car to the other, with a speed of 1.0 m/s (Fig. 3.34a). We can again describe the

3.34 (a) A woman walking across a railroad car, seen from above. (b) The position vector depends on the frame of reference. (c) Vector diagram for the velocity of the woman relative to the ground.

woman's position P in two different frames of reference, A for the stationary ground observer and B for the moving train. But instead of coordinates x, we use position vectors $\vec{r}$ because the problem is now two-dimensional. Then, as Fig. 3.34b shows,

$$\vec{r}_{P/A} = \vec{r}_{P/B} + \vec{r}_{B/A} \tag{3.35}$$

Just as we did before, we take the time derivative of this equation to get a relation among the various velocities; the velocity of P relative to A is $\vec{v}_{P/A} = d\vec{r}_{P/A}/dt$, and so on for the other velocities. We get

$$\vec{v}_{P/A} = \vec{v}_{P/B} + \vec{v}_{B/A} \qquad \text{(relative velocity in space)} \tag{3.36}$$

If all three of these velocities lie along the same line, then Eq. (3.36) reduces to Eq. (3.33) for the components of the velocities along that line.

If the train's velocity relative to the ground has magnitude $v_{B/A} = 3.0$ m/s and the woman's velocity relative to the railroad car has magnitude $v_{P/B} = 1.0$ m/s, then her velocity vector $\vec{v}_{P/A}$ relative to the ground is as shown in the vector diagram of Fig. 3.34c. The Pythagorean theorem then gives us

$$v_{P/A} = \sqrt{(3.0 \text{ m/s})^2 + (1.0 \text{ m/s})^2} = \sqrt{10 \text{ m}^2/\text{s}^2} = 3.2 \text{ m/s}$$

We can also see from the diagram that the *direction* of her velocity vector relative to the ground makes an angle α with the train's velocity vector $\vec{v}_{B/A}$, where

$$\tan \alpha = \frac{v_{P/B}}{v_{B/A}} = \frac{1.0 \text{ m/s}}{3.0 \text{ m/s}} \quad \text{and} \quad \alpha = 18°$$

As in the case of motion along a straight line, we have the general rule that if A and B are *any* two points or frames of reference,

$$\vec{v}_{A/B} = -\vec{v}_{B/A} \tag{3.37}$$

The velocity of the woman relative to the train is the negative of the velocity of the train relative to the woman, and so on.

Equation (3.36) is known as the *Galilean velocity transformation*. It shows that the velocity of an object P with respect to frame A and its velocity with respect to frame B ($\vec{v}_{P/A}$ and $\vec{v}_{P/B}$ respectively) are related by the velocity of frame B with respect to frame A ($\vec{v}_{P/B}$).

Example 3.15   **Flying in a crosswind**

The compass of an airplane indicates that it is headed due north, and its airspeed indicator shows that it is moving through the air at 240 km/h. If there is a wind of 100 km/h from west to east, what is the velocity of the airplane relative to the earth?

SOLUTION

IDENTIFY and SET UP: This is clearly a relative velocity problem, and we will use Eq. (3.36) to find the velocity of the plane (P) relative to the earth (E). We are given the magnitude and direction of the velocity of the plane relative to the air (A). We are also given the magnitude and direction of the wind velocity, which is the velocity of the air with respect to the earth:

$$\vec{v}_{P/A} = 240 \text{ km/h} \quad \text{due north}$$
$$\vec{v}_{A/E} = 100 \text{ km/h} \quad \text{due east}$$

Our target variables are the magnitude and direction of $\vec{v}_{P/E}$.

EXECUTE: Using Eq. (3.36), we have

$$\vec{v}_{P/E} = \vec{v}_{P/A} + \vec{v}_{A/E}$$

The three relative velocities and their relationship are shown in Fig. 3.35; the unknowns are the speed $v_{P/E}$ and the angle ϕ. From this diagram we find

$$v_{P/E} = \sqrt{(240 \text{ km/h})^2 + (100 \text{ km/h})^2} = 260 \text{ km/h}$$

$$\phi = \arctan\left(\frac{100 \text{ km/h}}{240 \text{ km/h}}\right) = 23° \text{ E of N}$$

EVALUATE: The crosswind increases the speed of the airplane relative to the earth, but at the price of pushing the airplane off course.

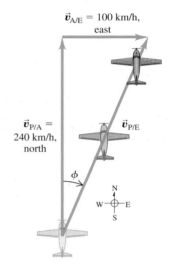

3.35 The plane is pointed north, but the wind blows east, giving the resultant velocity $\vec{v}_{P/E}$ relative to the earth.

Example 3.16   **Correcting for a crosswind**

In Example 3.15, in what direction should the pilot head to travel due north? What will then be her velocity relative to the earth? (Assume that her airspeed and the velocity of the wind are the same as in Example 3.15.)

SOLUTION

IDENTIFY: Figure 3.36 illustrates the situation. The vectors in Fig. 3.36 are arranged in accordance with the vector relative-velocity equation, Eq. (3.36):

$$\vec{v}_{P/E} = \vec{v}_{P/A} + \vec{v}_{A/E}$$

As Fig. 3.36 shows, the pilot points the nose of the airplane at an angle ϕ into the wind to compensate for the crosswind. This angle, which tells us the direction of the vector $\vec{v}_{P/A}$ (the velocity of the airplane relative to the air), is one of our target variables. The other target variable is the speed of the airplane over the ground, which is the magnitude of the vector $\vec{v}_{P/E}$ (the velocity of the airplane relative to the earth). Here are the known and unknown quantities:

$$\vec{v}_{P/E} = \text{magnitude unknown} \quad \text{due north}$$
$$\vec{v}_{P/A} = 240 \text{ km/h} \quad \text{direction unknown}$$
$$\vec{v}_{A/E} = 100 \text{ km/h} \quad \text{due east}$$

We can solve for the unknown target variables using Fig. 3.36 and trigonometry.

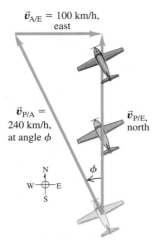

$\vec{v}_{A/E} = 100$ km/h, east

$\vec{v}_{P/A} = 240$ km/h, at angle ϕ

$\vec{v}_{P/E}$, north

N
W ⊕ E
S

ϕ

3.36 The pilot must point the plane in the direction of the vector $\vec{v}_{P/A}$ to travel due north relative to the earth.

EXECUTE: From the diagram, the speed $v_{P/E}$ and the angle ϕ are given by

$$v_{P/E} = \sqrt{(240 \text{ km/h})^2 - (100 \text{ km/h})^2} = 218 \text{ km/h}^2$$

$$\phi = \arcsin\left(\frac{100 \text{ km/h}}{240 \text{ km/h}}\right) = 25°$$

The pilot should point the airplane 25° west of north, and her ground speed is then 218 km/h.

EVALUATE: Note that there were two target variables—the magnitude of a vector and the direction of a vector—in both this example and Example 3.15. The difference is that in Example 3.15, the magnitude and direction referred to the *same* vector $(\vec{v}_{P/E})$, whereas in this example they referred to different vectors $(\vec{v}_{P/E}$ and $\vec{v}_{P/A})$.

It's no surprise that a headwind will reduce an airplane's speed relative to the ground. What this example shows is that a *crosswind* also slows an airplane down—an unfortunate fact of aeronautical life.

When we derived the relative-velocity relations, we assumed that all the observers use the same time scale. This is precisely the point at which Einstein's special theory of relativity departs from the physics of Galileo and Newton. When speeds approach the speed of light, denoted by c, the velocity-addition equation has to be modified. It turns out that if the woman in Fig. 3.32a could walk down the aisle at $0.30c$ and the train could move at $0.90c$, then her speed relative to the ground would be not $1.20c$ but $0.94c$; nothing can travel faster than light! We'll return to the special theory of relativity in Chapter 37.

Test Your Understanding

Suppose the wind in Example 3.16 dropped suddenly to zero, but the pilot kept the nose of the airplane pointed 25° to the west of north. How far off her desired south-to-north course would she be after one hour?

SUMMARY

The position vector $\vec{r}$ of a point P in space is the vector from the origin to P. Its components are the coordinates x, y, and z.

$$\vec{r} = x\hat{\imath} + y\hat{\jmath} + z\hat{k} \qquad (3.1)$$

The average velocity vector $\vec{v}_{av}$ during the time interval Δt is the displacement $\Delta\vec{r}$ (the change in the position vector $\vec{r}$) divided by Δt. The instantaneous velocity vector $\vec{v}$ is the time derivative of $\vec{r}$, and its components are the time derivatives of x, y, and z. The instantaneous speed is the magnitude of $\vec{v}$. The velocity $\vec{v}$ of a particle is always tangent to the particle's path. (See Example 3.1)

$$\vec{v}_{av} = \frac{\vec{r}_2 - \vec{r}_1}{t_2 - t_1} = \frac{\Delta\vec{r}}{\Delta t} \qquad (3.2)$$

$$\vec{v} = \lim_{\Delta t \to 0} \frac{\Delta\vec{r}}{\Delta t} = \frac{d\vec{r}}{dt} \qquad (3.3)$$

$$v_x = \frac{dx}{dt} \qquad v_y = \frac{dy}{dt} \qquad v_z = \frac{dz}{dt} \qquad (3.4)$$

The average acceleration vector $\vec{a}_{av}$ during the time interval Δt equals $\Delta\vec{v}$ (the change in the velocity vector $\vec{v}$) divided by Δt. The instantaneous acceleration vector $\vec{a}$ is the time derivative of $\vec{v}$, and its components are the time derivatives of v_x, v_y, and v_z. (See Examples 3.2 and 3.3)

$$\vec{a}_{av} = \frac{\vec{v}_2 - \vec{v}_1}{t_2 - t_1} = \frac{\Delta\vec{v}}{\Delta t} \qquad (3.8)$$

$$\vec{a} = \lim_{\Delta t \to 0} \frac{\Delta\vec{v}}{\Delta t} = \frac{d\vec{v}}{dt} \qquad (3.9)$$

$$a_x = \frac{dv_x}{dt} \qquad a_y = \frac{dv_y}{dt} \qquad a_z = \frac{dv_z}{dt} \qquad (3.10)$$

Acceleration can also be represented in terms of its components parallel and perpendicular to the direction of the instantaneous velocity. The parallel component of $\vec{a}$ affects the speed, while the perpendicular component of $\vec{a}$ affects the direction of motion. (See Examples 3.4 and 3.5)

In projectile motion with no air resistance, $a_x = 0$ and $a_y = -g$. The coordinates and velocity components are simple functions of time, and the shape of the path is always a parabola. We conventionally choose the origin to be at the initial position of the projectile. (See Examples 3.6 through 3.11)

$$x = (v_0 \cos \alpha_0)t \qquad (3.20)$$

$$y = (v_0 \sin \alpha_0)t - \frac{1}{2}gt^2 \qquad (3.21)$$

$$v_x = v_0 \cos \alpha_0 \qquad (3.22)$$

$$v_y = v_0 \sin \alpha_0 - gt \qquad (3.23)$$

When a particle moves in a circular path of radius R with constant speed v, its acceleration $\vec{a}$ is directed toward the center of the circle and perpendicular to $\vec{v}$. The magnitude a_{rad} of the acceleration can be expressed in terms of v and R or in terms of R and the period T (the time for one revolution), where $v = 2\pi R/T$. (See Examples 3.12 and 3.13)

$$a_{\text{rad}} = \frac{v^2}{R} \tag{3.28}$$

$$a_{\text{rad}} = \frac{4\pi^2 R}{T^2} \tag{3.30}$$

If the speed is not constant in circular motion, there is still a radial component of $\vec{a}$ given by Eq. (3.28) or (3.30), but there is also a component of $\vec{a}$ parallel to the path. This parallel component is equal to the rate of change of speed, dv/dt.

When a body P moves relative to a body (or reference frame) B, and B moves relative to A, we denote the velocity of P relative to B by $\vec{v}_{P/B}$, the velocity of P relative to A by $\vec{v}_{P/A}$, and the velocity of B relative to A by $\vec{v}_{B/A}$. If these velocities are all along the same line, their components along that line are related by Eq. (3.33). More generally, these velocities are related by Eq. (3.36). (See Examples 3.14 through 3.16)

$$v_{P/A} = v_{P/B} + v_{B/A}$$
(relative velocity along a line)
$$\tag{3.33}$$

$$\vec{v}_{P/A} = \vec{v}_{P/B} + \vec{v}_{B/A}$$
(relative velocity in space)
$$\tag{3.36}$$

Key Terms

position vector, 79
average velocity, 79
instantaneous velocity, 79
average acceleration, 82
instantaneous acceleration, 82

projectile, 87
trajectory, 87
uniform circular motion, 98
centripetal acceleration, 99

period, 99
non-uniform circular motion, 100
relative velocity, 101
frame of reference, 102

Your Notes

Answer to Chapter Opening Question

We show in Example 3.10 (Section 3.3) that a projectile has the maximum horizontal range when it is launched at a 45° angle.

Answers to Test Your Understanding Questions

Section 3.1 The average velocity vector $\vec{v}_{av}$ over an interval and the instantaneous velocity $\vec{v}$ at the end of the interval are equal if the velocity is constant over that interval.

Section 3.2 At the high point of the fragment's path, the speed is minimum. At that point the speed is neither increasing nor decreasing, and the parallel component of the acceleration (that is, the horizontal component) is zero. The acceleration has only a perpendicular component toward the inside of the fragment's curved path. In other words, the acceleration is downward. (We discuss the acceleration of projectiles in more depth in Section 3.3.)

Section 3.3 If there were no gravity $(g = 0)$, the monkey would not fall and the dart would follow a straight-line path (shown as a dashed line in Fig. 3.27). The effect of gravity is to make the monkey and dart both fall the same distance $\frac{1}{2}gt^2$ below their $g = 0$ positions. In Fig. 3.27, point A is the same distance below the monkey's initial position as point P is below the dashed straight line, so point A is where we would find the monkey at the time in question.

Section 3.4 At both the top and bottom of the loop, the acceleration is purely radial and is given by Eq. (3.28). The radius R is the same at both points, so the difference in acceleration is due purely to differences in speed. Since a_{rad} is proportional to the square of v, the speed must be twice as great at the bottom of the loop as at the top.

Section 3.5 By pointing the nose of the airplane to the west, the pilot has given the airplane's velocity relative to the air $(\vec{v}_{P/A})$ a westward component of 100 km/h—just enough to cancel the eastward wind. With no wind present, the airplane will drift westward by 100 km in an hour.

Discussion Questions

Q3.1 A simple pendulum (a mass swinging at the end of a string) swings back and forth in a circular arc. What is the direction of its acceleration at the ends of the swing? At the midpoint? In each case, explain how you obtain your answer.

Q3.2 Redraw Fig. 3.11a if $\vec{a}$ is antiparallel to $\vec{v}_1$. Does the particle move in a straight line? What happens to the speed?

Q3.3 A projectile moves in a parabolic path without air resistance. Is there any point at which $\vec{a}$ is parallel to $\vec{v}$? Perpendicular to $\vec{v}$? Explain.

Q3.4 When a rifle is fired at a distant target, the barrel is not lined up exactly on the target. Why not? Does the angle of correction depend on the distance of the target?

Q3.5 At the same instant that you fire a bullet horizontally from a gun, you drop a bullet from the height of the barrel. If there is no air resistance, which bullet hits the ground first? Explain.

Q3.6 A package falls out of an airplane that is flying in a straight line at a constant altitude and speed. If you could ignore air resistance, what would be the path of the package as observed by the pilot? As observed by a person on the ground?

Q3.7 Sketch the six graphs of the x- and y-components of position, velocity, and acceleration versus time for projectile motion with $x_0 = y_0 = 0$ and $0 < \alpha_0 < 90°$.

Q3.8 If $y_0 = 0$ and α_0 is negative, y can never be positive for a projectile. However, the expression for h in Example 3.10 seems to give a positive maximum height for a negative α_0. Explain this apparent contradiction.

Q3.9 If a jumping frog can give itself the same initial speed regardless of the direction in which it jumps (forward or straight up), how is the maximum vertical height to which it can jump related to its maximum horizontal range $R_{max} = v_0^2/g$?

Q3.10 When a projectile reaches its maximum height, is its speed zero? Explain.

Q3.11 In uniform circular motion, what are the *average* velocity and *average* acceleration for one revolution? Explain.

Q3.12 In uniform circular motion, how does the acceleration change when the speed is increased by a factor of 3? When the radius is decreased by a factor of 2?

Q3.13 In uniform circular motion, the acceleration is perpendicular to the velocity at every instant. Is this still true when the motion is not uniform, that is, when the speed is not constant?

Q3.14 Raindrops hitting the side windows of a car in motion often leave diagonal streaks. Why? Is the explanation the same or different for diagonal streaks on the windshield?

Q3.15 In a rainstorm with a strong wind, what determines the best position in which to hold an umbrella?

Q3.16 You are on the west bank of a river that is flowing north with a speed of 1.2 m/s. Your swimming speed relative to the water is 1.5 m/s and the river is 60 m wide. What is your path relative to earth that allows you to cross the river in the shortest time? Explain your reasoning.

Exercises

Section 3.1 Position and Velocity Vectors

3.1 A squirrel has x- and y-coordinates (1.1 m, 3.4 m) at time $t_1 = 0$ and coordinates (5.3 m, −0.5 m) at time $t_2 = 3.0$ s. For this time interval, find a) the components of the average velocity; b) the magnitude and direction of the average velocity.

3.2 A rhinoceros is at the origin of coordinates at time $t_1 = 0$. For the time interval from $t_1 = 0$ to $t_2 = 12.0$ s, the rhino's average velocity has x-component −3.8 m/s and y-component 4.9 m/s. At time $t_2 = 12.0$ s, a) what are the x- and y-coordinates of the rhino? b) How far is the rhino from the origin?

3.3 A web page designer creates an animation in which a dot on a computer screen has a position of $\vec{r} = [4.0\ \text{cm} + (2.5\ \text{cm/s}^2)t^2]\hat{i} + (5.0\ \text{cm/s})t\hat{j}$. a) Find the magnitude and direction of the dot's average velocity between $t = 0$ and $t = 2.0$ s. b) Find the magnitude and direction of the instantaneous velocity at $t = 0$, $t = 1.0$ s, and $t = 2.0$ s. c) Sketch the dot's trajectory from $t = 0$ to $t = 2.0$ s, and show the velocities calculated in part (b).

3.4 If $\vec{r} = bt^2\hat{\imath} + ct^3\hat{\jmath}$, where b and c are positive constants, when does the velocity vector make an angle of $45.0°$ with the x- and y-axes?

Section 3.2 The Acceleration Vector

3.5 A jet plane is flying at a constant altitude. At time $t_1 = 0$ it has components of velocity $v_x = 90$ m/s, $v_y = 110$ m/s. At time $t_2 = 30.0$ s the components are $v_x = -170$ m/s, $v_y = 40$ m/s. a) Sketch the velocity vectors at t_1 and t_2. How do these two vectors differ? For this time interval calculate b) the components of the average acceleration; c) the magnitude and direction of the average acceleration.

3.6 A dog running in an open field has components of velocity $v_x = 2.6$ m/s and $v_y = -1.8$ m/s at $t_1 = 10.0$ s. For the time interval from $t_1 = 10.0$ s to $t_2 = 20.0$ s, the average acceleration of the dog has magnitude 0.45 m/s^2 and direction $31.0°$ measured from the $+x$-axis toward the $+y$-axis. At $t_2 = 20.0$ s, a) what are the x- and y-components of the dog's velocity? b) What are the magnitude and direction of the dog's velocity? c) Sketch the velocity vectors at t_1 and t_2. How do these two vectors differ?

3.7 The coordinates of a bird flying in the xy-plane are given by $x(t) = \alpha t$ and $y(t) = 3.0$ m $- \beta t^2$, where $\alpha = 2.4$ m/s and $\beta = 1.2$ m/s^2. a) Sketch the path of the bird between $t = 0$ and $t = 2.0$ s. b) Calculate the velocity and acceleration vectors of the bird as functions of time. c) Calculate the magnitude and direction of the bird's velocity and acceleration at $t = 2.0$ s. d) Sketch the velocity and acceleration vectors at $t = 2.0$ s. At this instant, is the bird speeding up, is it slowing down, or is its speed instantaneously not changing? Is the bird turning? If so, in what direction?

3.8 A particle moves along a path as shown in Fig. 3.37. Between points B and D, the path is a straight line. Sketch the acceleration vectors at A, C, and E in the cases in which a) the particle moves with a constant speed; b) the particle moves with a steadily increasing speed; c) the particle moves with a steadily decreasing speed.

Figure 3.37 Exercise 3.8.

Section 3.3 Projectile Motion

3.9 A physics book slides off a horizontal table top with a speed of 1.10 m/s. It strikes the floor in 0.350 s. Ignore air resistance. Find a) the height of the table top above the floor; b) the horizontal distance from the edge of the table to the point where the book strikes the floor; c) the horizontal and vertical components of the book's velocity, and the magnitude and direction of its velocity, just before the book reaches the floor. d) Draw x-t, y-t, v_x-t, and v_y-t graphs for the motion.

3.10 A military helicopter on a training mission is flying horizontally at a speed of 60.0 m/s and accidentally drops a bomb (fortunately not armed) at an elevation of 300 m. You can ignore air resistance. a) How much time is required for the bomb to reach the earth? b) How far does it travel horizontally while falling? c) Find the horizontal and vertical components of its velocity just before it strikes the earth. d) Draw x-t, y-t, v_x-t, and v_y-t graphs for the bomb's motion. e) If the velocity of the helicopter remains constant, where is the helicopter when the bomb hits the ground?

3.11 Two crickets, Chirpy and Milada, jump from the top of a vertical cliff. Chirpy jumps horizontally and reaches the ground in 3.50 s. Milada jumps with an initial velocity of 95.0 cm/s at an angle of $32.0°$ above the horizontal. How far from the base of the cliff will Milada hit the ground?

3.12 A daring 510-N swimmer dives off a cliff with a running horizontal leap, as shown in Fig. 3.38. What must her minimum speed be just as she leaves the top of the cliff so that she will miss the ledge at the bottom, which is 1.75 m wide and 9.00 m below the top of the cliff?

Figure 3.38 Exercise 3.12

3.13 Leaping the River I. A $10,000$-N car comes to a bridge during a storm and finds the bridge washed out. The 650-N driver must get to the other side, so he decides to try leaping it with his car. The side the car is on is 21.3 m above the river, while the opposite side is a mere 1.8 m above the river. The river itself is a raging torrent 61.0 m wide. a) How fast should the car be traveling just as it leaves the cliff in order just to clear the river and land safely on the opposite side? b) What is the speed of the car just before it lands safely on the other side?

3.14 A student makes a bet with another student that she can roll a marble off the top of a table and have it land in a cup placed on the floor. She claims she can do this without letting the marble roll off the table before the actual attempt. If you were the student making the bet, what would you do to determine where to place the cup? Be sure to explain what measurements you need to make and how you would use them to determine the placement of the cup.

3.15 You throw a football with an initial speed $v_0 = 15.0$ m/s at an initial angle $\alpha_0 = 45.0°$. a) Find the time T when the football is at its maximum height. b) At the three times $t_1 = T - 0.50$ s, $t_2 = T$, and $t_3 = T + 0.50$ s, find the x- and y-components of the position vector. c) At the three times t_1, t_2, and t_3, find the magnitude and direction of the velocity vector. d) At the three times t_1, t_2, and t_3, find the component of the acceleration vector that is parallel (or antiparallel) to the velocity, and find the component of the acceleration vector that is perpendicular to the velocity. e) Sketch the trajectory of the football. On your sketch, label the position of the football at the three times t_1, t_2, and t_3. At each of these positions, draw the velocity vector and the parallel and perpendicular components of the acceleration vector. f) Discuss how the speed and the direction of motion of the football are changing at the three times t_1, t_2, and t_3, and explain how the vectors in your sketch describe these changes.

3.16 A tennis ball rolls off the edge of a tabletop 0.750 m above the floor and strikes the floor at a point 1.40 m horizontally from the edge of the table. You can ignore air resistance. a) Find the time of flight. b) Find the magnitude of the initial velocity. c) Find the magnitude and direction of the velocity of the ball just before it strikes the floor. d) Draw x-t, y-t, v_x-t, and v_y-t graphs for the motion.

3.17 A pistol that fires a signal flare gives the flare an initial speed (muzzle speed) of 120 m/s. a) If the flare is fired at an angle of 55° above the horizontal on the level salt flats of Utah, what is its horizontal range? You can ignore air resistance. b) If the flare is fired at the same angle over the flat Sea of Tranquility on the moon, where $g = 1.6$ m/s^2, what is its horizontal range?

3.18 A rookie quarterback throws a football with an initial upward velocity component of 16.0 m/s and a horizontal velocity component of 20.0 m/s. Ignore air resistance. a) How much time is required for the football to reach the highest point of the trajectory? b) How high is this point? c) How much time (after it is thrown) is required for the football to return to its original level? How does this compare with the time calculated in part (a)? d) How far has it traveled horizontally during this time? e) Draw x-t, y-t, v_x-t, and v_y-t graphs for the motion.

3.19 A major leaguer hits a baseball so that it leaves the bat at a speed of 30.0 m/s and at an angle of 36.9° above the horizontal. You can ignore air resistance. a) At what *two* times is the baseball at a height of 10.0 m above the point at which it left the bat? b) Calculate the horizontal and vertical components of the baseball's velocity at each of the two times calculated in part (a). c) What is the magnitude and direction of the baseball's velocity when it returns to the level at which it left the bat?

3.20 A shot putter releases the shot some distance above the level ground with a velocity of 12.0 m/s, 51.0° above the horizontal. The shot hits the ground 2.08 s later. You can ignore air resistance. a) What are the components of the shot's acceleration while in flight? b) What are the components of the shot's velocity at the beginning and at the end of its trajectory? c) How far did she throw the shot horizontally? d) Why does the expression for R in Example 3.10 *not* give the correct answer for part (c)? e) How high was the shot above the ground when she released it? f) Draw x-t, y-t, v_x-t, and v_y-t graphs for the motion.

3.21 Win the Prize. In a carnival booth, you win a stuffed giraffe if you toss a quarter into a small dish. The dish is on a shelf above

the point where the quarter leaves your hand and is a horizontal distance of 2.1 m from this point (Fig. 3.39). If you toss the coin with a velocity of 6.4 m/s at an angle of 60° above the horizontal, the coin lands in the dish. You can ignore air resistance. a) What is the height of the shelf above the point where the quarter leaves your hand? b) What is the vertical component of the velocity of the quarter just before it lands in the dish?

3.22 Suppose the departure angle α_0 in Fig. 3.24 is 42.0° and the distance d is 3.00 m. Where will the dart and monkey meet if the initial speed of the dart is a) 12.0 m/s? b) 8.0 m/s? c) What will happen if the initial speed of the dart is 4.0 m/s? Sketch the trajectory in each case.

3.23 A man stands on the roof of a 15.0-m-tall building and throws a rock with a velocity of magnitude 30.0 m/s at an angle of 33.0° above the horizontal. You can ignore air resistance. Calculate a) the maximum height above the roof reached by the rock; b) the magnitude of the velocity of the rock just before it strikes the ground; c) the horizontal distance from the base of the building to the point where the rock strikes the ground. d) Draw x-t, y-t, v_x-t, and v_y-t graphs for the motion.

3.24 Firemen are shooting a stream of water at a burning building using a high-pressure hose that shoots out the water with a speed of 25.0 m/s as it leaves the end of the hose. Once it leaves the hose, the water moves in projectile motion. The firemen adjust the angle of elevation α of the hose until the water takes 3.00 s to reach a building 45.0 m away. You can ignore air resistance; assume that the end of the hose is at ground level. a) Find the angle of elevation α. b) Find the speed and acceleration of the water at the highest point in its trajectory. c) How high above the ground does the water strike the building, and how fast is it moving just before it hits the building?

3.25 A 124-kg balloon carrying a 22-kg basket is descending with a constant downward velocity of 20.0 m/s. A 1.0-kg stone is thrown from the basket with an initial velocity of 15.0 m/s perpendicular to the path of the descending balloon, as measured relative to a person at rest in the basket. The person in the basket sees the stone hit the ground 6.00 s after being thrown. Assume that the balloon continues its downward descent with the same constant speed of 20.0 m/s. a) How high was the balloon when the rock was thrown out? b) How high is the balloon when the rock hits the ground? c) At the instant the rock hits the ground, how far is it from the basket? d) Just before the rock hits the ground, find its horizontal and vertical velocity components as measured by an observer: i) at rest in the basket; ii) at rest on the ground.

3.26 A cannon, located 60.0 m from the base of a vertical 25.0-m-tall cliff, shoots a 15-kg shell at 43.0° above the horizontal toward the cliff. a) What must the minimum muzzle velocity be for the shell to clear the top of the cliff? b) The ground at the top of the cliff is level, with a constant elevation of 25.0 m above the cannon. Under the conditions of part (a), how far does the shell land past the edge of the cliff?

3.27 An airplane is flying with a velocity of 90.0 m/s at an angle of 23.0° above the horizontal. When the plane is 114 m directly above a dog that is standing on level ground, a suitcase drops out of the luggage compartment. How far from the dog will the suitcase land? You can ignore air resistance.

Figure 3.39 Exercise 3.21.

$v = 6.4$ m/s

60°

2.1 m

Section 3.4 Motion in a Circle

3.28 On your first day at work for an appliance manufacturer, you are told to figure out what to do to the period of rotation during a washer spin cycle to triple the centripetal acceleration. You impress your boss by answering immediately. What do you tell her?

3.29 The earth has a radius of 6380 km and turns around once on its axis in 24 h. a) What is the radial acceleration of an object at the earth's equator? Give your answer in m/s^2 and as a fraction of g. b) If a_{rad} at the equator is greater than g, objects would fly off the earth's surface and into space. (We will see the reason for this in Chapter 5.) What would the period of the earth's rotation have to be for this to occur?

3.30 A model of a helicopter rotor has four blades, each 3.40 m in length from the central shaft to the blade tip. The model is rotated in a wind tunnel at 550 rev/min. a) What is the linear speed of the blade tip, in m/s? b) What is the radial acceleration of the blade tip expressed as a multiple of the acceleration of gravity, g?

3.31 In a test of a "g-suit," a volunteer is rotated in a horizontal circle of radius 7.0 m. What is the period of rotation at which the centripetal acceleration has a magnitude of a) $3.0g$? b) $10g$?

3.32 The radius of the earth's orbit around the sun (assumed to be circular) is 1.50×10^8 km, and the earth travels around this orbit in 365 days. a) What is the magnitude of the orbital velocity of the earth in m/s? b) What is the radial acceleration of the earth toward the sun in m/s^2? c) Repeat parts (a) and (b) for the motion of the planet Mercury (orbit radius $= 5.79 \times 10^7$ km, orbital period $= 88.0$ days).

3.33 A Ferris wheel with radius 14.0 m is turning about a horizontal axis through its center (Fig. 3.40). The linear speed of a passenger on the rim is constant and equal to 7.00 m/s. What are the magnitude and direction of the passenger's acceleration as she passes through a) the lowest point in her circular motion? b) the highest point in her circular motion? c) How much time does it take the Ferris wheel to make one revolution?

Figure 3.40 Exercises 3.33 and 3.34.

3.34 The Ferris wheel in Fig. 3.40, which rotates counterclockwise, is just starting up. At a given instant, a passenger on the rim of the wheel and passing through the lowest point of his circular motion is moving at 3.00 m/s and is gaining speed at a rate of $0.500 \ m/s^2$. a) Find the magnitude and the direction of the passenger's acceleration at this instant. b) Sketch the Ferris wheel and passenger showing his velocity and acceleration vectors.

3.35 **Elliptical Path.** A level racetrack has the shape of an ellipse. (See your math textbook, a dictionary, or an encyclopedia for an illustration of an ellipse.) A car drives around this track at a constant speed. a) Draw a picture of the track showing the car's velocity vector and acceleration vector at five or more different places around the track. b) Does the car's acceleration vector always point toward the geometrical center of the ellipse? Explain. c) At what point(s) on the ellipse does the car's acceleration have the greatest magnitude? Explain.

Section 3.5 Relative Velocity

3.36 A railroad flatcar is traveling to the right at a speed of 13.0 m/s relative to an observer standing on the ground. Someone is riding a motor scooter on the flatcar (Fig. 3.41). What is the velocity (magnitude and direction) of the motor scooter relative to the flatcar if its velocity relative to the observer on the ground is a) 18.0 m/s to the right? b) 3.0 m/s to the left? c) zero?

Figure 3.41 Exercise 3.36.

3.37 A "moving sidewalk" in an airport terminal building moves at 1.0 m/s and is 35.0 m long. If a woman steps on at one end and walks at 1.5 m/s relative to the moving sidewalk, how much time does she require to reach the opposite end if she walks a) in the same direction the sidewalk is moving? b) in the opposite direction?

3.38 Two piers, A and B, are located on a river: B is 1500 m downstream from A (Fig. 3.42). Two friends must make round trips from pier A to pier B and return. One rows a boat at a constant speed of 4.00 km/h relative to the water; the other walks on the shore at a constant speed of 4.00 km/h. The velocity of the river is 2.80 km/h in the direction from A to B. How much time does it take each person to make the round trip?

Figure 3.42 Exercise 3.38.

3.39 A canoe has a velocity of 0.40 m/s southeast relative to the earth. The canoe is on a river that is flowing 0.50 m/s east relative to the earth. Find the velocity (magnitude and direction) of the canoe relative to the river.

3.40 An airplane pilot wishes to fly due west. A wind of 80.0 km/h (about 50 mi/h) is blowing toward the south. a) If the airspeed of the plane (its speed in still air) is 320.0 km/h (about 200 mi/h), in which direction should the pilot head? b) What is the speed of the plane over the ground? Illustrate with a vector diagram.

3.41 Crossing the River I. A river flows due south with a speed of 2.0 m/s. A man steers a motorboat across the river; his velocity relative to the water is 4.2 m/s due east. The river is 800 m wide. a) What is his velocity (magnitude and direction) relative to the earth? b) How much time is required to cross the river? c) How far south of his starting point will he reach the opposite bank?

3.42 Crossing the River II. a) In which direction should the motorboat in Exercise 3.41 head in order to reach a point on the opposite bank directly east from the starting point? (The boat's speed relative to the water remains 4.2 m/s.) b) What is the velocity of the boat relative to the earth? c) How much time is required to cross the river?

3.43 The nose of an ultralight plane is pointed south, and its air speed indicator shows 35 m/s. The plane is in a 10 m/s wind blowing toward the southwest relative to the earth. a) In a vector addition diagram, show the relation of $\vec{v}_{P/E}$ (the velocity of the plane relative to the earth) to the two given vectors. b) Letting x be east and y be north, find the components of $\vec{v}_{P/E}$. c) Find the magnitude and direction of $\vec{v}_{P/E}$.

Problems

3.44 A faulty model rocket moves in the xy-plane (the positive y-direction is vertically upward). The rocket's acceleration has components $a_x(t) = \alpha t^2$ and $a_y(t) = \beta - \gamma t$, where $\alpha = 2.50$ m/s^4, $\beta = 9.00$ m/s^2, and $\gamma = 1.40$ m/s^3. At $t = 0$ the rocket is at the origin and has velocity $\vec{v}_0 = v_{0x}\hat{i} + v_{0y}\hat{j}$ with $v_{0x} = 1.00$ m/s and $v_{0y} = 7.00$ m/s. a) Calculate the velocity and position vectors as functions of time. b) What is the maximum height reached by the rocket? c) Sketch the path of the rocket. d) What is the horizontal displacement of the rocket when it returns to $y = 0$?

3.45 A student is moving in the xy-plane in a darkened room, trying to locate a lost $20 bill. The student's coordinates are given as functions of time by $x(t) = \alpha t$ and $y(t) = 15.0$ m $- \beta t^2$, where $\alpha = 1.20$ m/s and $\beta = 0.500$ m/s^2. Unknown to the student, the $20 bill is at the origin. a) At which times is the student's velocity perpendicular to his acceleration? b) At which times is the student's speed instantaneously not changing? c) At which times is the student's velocity perpendicular to his position vector? What are the locations of the student at these times? d) What is the minimum distance from the student to the $20 bill? At what time does this minimum occur? e) Sketch the path of the hapless student.

3.46 A bird flies in the xy-plane with a velocity vector given by $\vec{v} = (\alpha - \beta t^2)\hat{i} + \gamma t\hat{j}$, with $\alpha = 2.4$ m/s, $\beta = 1.6$ m/s^3, and $\gamma = 4.0$ m/s^2. The positive y-direction is vertically upward. At $t = 0$ the bird is at the origin. a) Calculate the position and acceleration vectors of the bird as functions of time. b) What is the bird's altitude (y-coordinate) as it flies over $x = 0$ for the first time after $t = 0$?

3.47 A Piper Warrior, a small four-seat airplane, requires 300 m of runway to take off. Its liftoff speed is 88 km/h. It then climbs with a constant speed of 88 km/h along a straight-line path, just clearing a power line 15 m high at a horizontal distance of 460 m from where the Piper began its takeoff roll. a) What was the acceleration of the Piper (assumed constant) during its takeoff roll? b) After the Piper takes off, what is the angle of its flight path above the horizontal? c) What is the rate of climb (in m/s) of the Piper? d) What is the elapsed time from the beginning of the takeoff roll to when the Piper just clears the power line?

3.48 A track and field coach (who is also a physics teacher) trains an athlete to throw a javelin so that it leaves the athlete's hand at a height h above the ground, moving at a speed of $\sqrt{25gh}/8$ and at an angle of 36.9° above the horizontal. The javelin continues in flight until it strikes the ground. The ground around the athlete is level, and air resistance is negligible. a) Graph the javelin's horizontal velocity versus time and the javelin's vertical velocity versus time. b) Find the maximum height reached by the javelin. c) Find the horizontal distance that the javelin travels from where it leaves the athlete's hand to where it strikes the ground.

3.49 Dynamite! A demolition crew uses dynamite to blow an old building apart. Debris from the explosion flies off in all directions, and is later found at distances up to 50 m from the explosion. Estimate the maximum speed at which debris was blown outward by the explosion. Describe any assumptions that you make.

3.50 Spiraling Up. It is common to see birds of prey rising upward on thermals. The paths they take may be spiral-like. You can model the spiral motion as uniform circular motion combined with a constant upward velocity. Assume a bird completes a circle of radius 8.00 m every 5.00 s and rises vertically at a rate of 3.00 m/s. Determine the following: a) the speed of the bird relative to the ground; b) the bird's acceleration (magnitude and direction); c) the angle between the bird's velocity vector and the horizontal.

3.51 A jungle veterinarian with a blow-gun loaded with a tranquilizer dart and a sly 1.5-kg monkey are each 25 m above the ground in trees 90 m apart. Just as the hunter shoots horizontally at the monkey, the monkey drops from the tree in a vain attempt to escape being hit. What must the minimum muzzle velocity of the dart have been for the hunter to hit the monkey before it reached the ground?

3.52 A movie stuntwoman drops from a helicopter that is 30.0 m above the ground and moving with a constant velocity whose components are 10.0 m/s upward and 15.0 m/s horizontal and toward the south. You can ignore air resistance. a) Where on the ground (relative to the position of the helicopter when she drops) should the stuntwoman have placed the foam mats that break her fall? b) Draw x-t, y-t, v_x-t, and v_y-t graphs of the motion.

3.53 In fighting forest fires, airplanes work in support of ground crews by dropping water on the fires. A pilot is practicing by dropping a canister of red dye, hoping to hit a target on the ground below. If the plane is flying in a horizontal path 90.0 m above the ground and with a speed of 64.0 m/s (143 mi/h), at what horizontal distance from the target should the pilot release the canister? Ignore air resistance.

3.54 Don't Do This I. A girl throws a water-filled balloon at an angle of 50.0° above the horizontal with a speed of 12.0 m/s. The

Figure 3.43 Problem 3.54.

horizontal component of the balloon's velocity is directed toward a car that is approaching the girl at a constant speed of 8.00 m/s (Fig. 3.43). If the balloon is to hit the car at the same height at which it leaves her hand, what is the maximum distance the car can be from the girl when the balloon is thrown? You can ignore air resistance.

3.55 The Longest Home Run. According to the *Guinness Book of World Records,* the longest home run ever measured was hit by Roy "Dizzy" Carlyle in a minor league game. The ball traveled 188 m (618 ft) before landing on the ground outside the ballpark. a) Assuming the ball's initial velocity was 45° above the horizontal and ignoring air resistance, what did the initial speed of the ball need to be to produce such a home run if the ball was hit at a point 0.9 m (3.0 ft) above ground level? Assume that the ground was perfectly flat. b) How far would the ball be above a fence 3.0 m (10 ft) in height if the fence was 116 m (380 ft) from home plate?

3.56 Don't Do This II. On the day after graduation, you decide to toss a burning match into the top of a cylindrical wastebasket (diameter D and height $2D$) full of old homework papers. To make this event more sporting, the bottom of the wastebasket is at the same height as the point where the match leaves your hand, and the near side of the wastebasket is a horizontal distance $6D$ from the point where the match leaves your hand. You launch the match at a 45.0° angle above the horizontal. Find the *minimum* and *maximum* values of the launch speed that will make the match enter the top of the wastebasket. You can ignore air resistance and give your answers in terms of g and D.

3.57 Hallway Catch. You are playing catch with a friend in the hallway of your dormitory. The distance from floor to ceiling is D, and you throw the ball with an initial speed $v_0 = \sqrt{6gD}$. What is the maximum horizontal distance (in terms of D) that the ball can travel without bouncing? (Assume that the ball is launched from the floor.)

3.58 A baseball thrown at an angle of 60.0° above the horizontal strikes a building 18.0 m away at a point 8.00 m above the point from which it is thrown. Ignore air resistance. a) Find the magnitude of the initial velocity of the baseball (the velocity with which the ball is thrown). b) Find the magnitude and direction of the velocity of the baseball just before it strikes the building. c) Draw x-t, y-t, v_x-t, and v_y-t graphs of the motion.

3.59 A projectile is launched with speed v_0 at an angle α_0 above the horizontal. The launch point is a height h above the ground. a) Show that if air resistance is ignored, the horizontal distance that the projectile travels before striking the ground is

$$x = \frac{v_0 \cos \alpha_0}{g} \left(v_0 \sin \alpha_0 + \sqrt{v_0^2 \sin^2 \alpha_0 + 2gh} \right)$$

Verify that if the launch point is at ground level so that $h = 0$, this is equal to the horizontal range R found in Example 3.10. b) For the case where $v_0 = 10$ m/s and $h = 5.0$ m, graph x as a function of launch angle α_0 for values of α_0 from 0° to 90°. Your graph should show that x is zero if $\alpha_0 = 90°$, but x is nonzero if $\alpha_0 = 0$; explain why this is so. c) We saw in Example 3.10 that for a projectile that lands at the same height from which it is launched, the horizontal range is maximum for $\alpha_0 = 45°$. For the case graphed in part (b), is the angle for maximum horizontal distance equal to, less than, or greater than 45°? (This is a general result for the situation where a projectile is launched from a point higher than where it lands.)

3.60 Look Out! A snowball rolls off a barn roof that slopes downward at an angle of 40° (Fig. 3.44). The edge of the roof is 14.0 m above the ground, and the snowball has a speed of 7.00 m/s as it rolls off the roof. Ignore air resistance. a) How far from the edge of the barn does the snowball strike the ground if it doesn't strike anything else while falling? b) Draw x-t, y-t, v_x-t, and v_y-t graphs for the

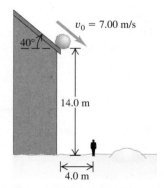

Figure 3.44 Problem 3.60.

motion in part (a). c) A man 1.9 m tall is standing 4.0 m from the edge of the barn. Will he be hit by the snowball?

3.61 a) Prove that a projectile launched at angle α_0 has the same horizontal range as one launched with the same speed at angle $(90° - \alpha_0)$. b) A frog jumps at a speed of 2.2 m/s and lands 25 cm from its starting point. At which angles above the horizontal could it have jumped?

3.62 On the Flying Trapeze. A new circus act is called the Texas Tumblers. Lovely Mary Belle swings from a trapeze, projects herself at an angle of 53°, and is supposed to be caught by Joe Bob, whose hands are 6.1 m above and 8.2 m horizontally from her launch point (Fig. 3.45). You can ignore air resistance. a) What initial speed v_0 must Mary Belle have just to reach Joe Bob? b) For the initial

Figure 3.45 Problem 3.62.

speed calculated in part (a), what are the magnitude and direction of her velocity when Mary Belle reaches Joe Bob? c) Assuming that Mary Belle has the initial speed calculated in part (a), draw x-t, y-t, v_x-t, and v_y-t graphs showing the motion of both tumblers. Your graphs should show the motion up until the point where Mary Belle reaches Joe Bob. d) The night of their debut performance, Joe Bob misses her completely as she flies past. How far horizontally does Mary Belle travel, from her initial launch point, before landing in the safety net 8.6 m below her starting point?

3.63 Leaping the River II. A physics professor did daredevil stunts in his spare time. His last stunt was an attempt to jump across a river on a motorcycle (Fig. 3.46). The takeoff ramp was inclined at 53.0°, the river was 40.0 m wide, and the far bank was 15.0 m lower than the top of the ramp. The river itself was 100 m below the ramp. You can ignore air resistance. a) What should his speed have been at the top of the ramp to have just made it to the edge of the far bank? b) If his speed was only half the value found in (a), where did he land?

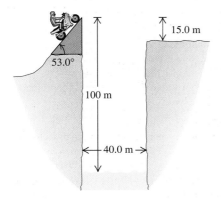

Figure 3.46 Problem 3.63.

3.64 A rock is thrown from the roof of a building with a velocity v_0 at an angle of α_0 from the horizontal. The building has height h. You can ignore air resistance. Calculate the magnitude of the velocity of the rock just before it strikes the ground, and show that this speed is independent of α_0.

3.65 A 5500-kg cart carrying a vertical rocket launcher moves to the right at a constant speed of 30.0 m/s along a horizontal track. It launches a 45.0-kg rocket vertically upward with an initial speed of 40.0 m/s relative to the cart. a) How high will the rocket go? b) Where, relative to the cart, will the rocket land? c) How far does the cart move while the rocket is in the air? d) At what angle, relative to the horizontal, is the rocket traveling just as it leaves the cart, as measured by an observer at rest on the ground? e) Sketch the rocket's trajectory as seen by an observer: i) stationary on the cart; ii) stationary on the ground.

3.66 A 2.7-kg ball is thrown upward with an initial speed of 20.0 m/s from the edge of a 45.0 m high cliff. At the instant the ball is thrown, a woman starts running away from the base of the cliff with a constant speed of 6.00 m/s. The woman runs in a straight line on level ground, and air resistance acting on the ball can be ignored. a) At what angle above the horizontal should the ball be thrown so that the runner will catch it just before it hits the ground,

and how far does the woman run before she catches the ball? b) Carefully sketch the ball's trajectory as viewed by: i) a person at rest on the ground; ii) the runner.

3.67 A 76.0-kg boulder is rolling horizontally at the top of a vertical cliff that is 20 m above the surface of a lake, as shown in Fig. 3.47. The top of the vertical face of a dam is located 100 m from the foot of the cliff, with the top of the dam level with the surface of the water in the lake. A level plain is 25 m below the top of the dam. a) What must the minimum speed of the rock be just as it leaves the cliff so it will travel to the plain without striking the dam? b) How far from the foot of the dam does the rock hit the plain?

Figure 3.47 Problem 3.67.

3.68 Tossing Your Lunch. Henrietta is going off to her physics class, jogging down the sidewalk at 3.05 m/s. Her husband Bruce suddenly realizes that she left in such a hurry that she forgot her lunch of bagels, so he runs to the window of their apartment, which is 43.9 m above the street level and directly above the sidewalk, to throw them to her. Bruce throws them horizontally 9.00 s after Henrietta has passed below the window, and she catches them on the run. You can ignore air resistance. a) With what initial speed must Bruce throw the bagels so Henrietta can catch them just before they hit the ground? b) Where is Henrietta when she catches the bagels?

3.69 Two tanks are engaged in a training exercise on level ground. The first tank fires a paint-filled training round with a muzzle speed of 250 m/s at 10.0° above the horizontal while advancing toward the second tank with a speed of 15.0 m/s relative to the ground. The second tank is retreating at 35.0 m/s relative to the ground, but is hit by the shell. You can ignore air resistance and assume the shell hits at the same height above ground from which it was fired. Find the distance between the tanks a) when the round was first fired; b) at the time of impact.

3.70 Bang! A student sits atop a platform a distance h above the ground. He throws a large firecracker horizontally with a speed v. However, a wind blowing parallel to the ground gives the firecracker a constant horizontal acceleration with magnitude a. This results in the firecracker reaching the ground directly under the student. Determine the height h in terms of v, a, and g. You can ignore the effect of air resistance on the vertical motion.

3.71 Free Throw. A basketball player is fouled and knocked to the floor during a layup attempt. The player is awarded two free throws. The center of the basket is a horizontal distance of 4.21 m (13.8 ft) from the foul line and is a height of 3.05 m (10.0 ft) above the floor (Fig. 3.48). On the first free-throw attempt, he shoots the ball at an angle of 35° above the horizontal and with a speed of

Figure 3.48 Problem 3.71.

$v_0 = 4.88$ m/s (16.0 ft/s). The ball is released 1.83 m (6.0 ft) above the floor. This shot misses badly. You can ignore air resistance. a) What is the maximum height reached by the ball? b) At what distance along the floor from the free-throw line does the ball land? c) For the second throw, the ball goes through the center of the basket. For this second free throw, the player again shoots the ball at 35° above the horizontal and releases it 1.83 m above the floor. What initial speed does the player give the ball on this second attempt? d) For the second throw, what is the maximum height reached by the ball? At this point, how far horizontally is the ball from the basket?

3.72 Romeo tosses a pebble at Juliet's window to wake her. Unfortunately, he throws too large a pebble too fast. Just before crashing through the glass, the pebble is moving horizontally, having traveled a horizontal distance x and a vertical distance y as a projectile. Find the magnitude and direction of the pebble's velocity as it leaves Romeo's hand.

3.73 A rocket is initially at rest on the ground. When its engines fire, the rocket flies off in a straight line at an angle of 53.1° above the horizontal with a constant acceleration of magnitude g. The engines stop at a time T after launch, after which the rocket is in projectile motion. You can ignore air resistance and assume g is independent of altitude. a) Draw the trajectory of the rocket from when its engines first fire until the rocket hits the ground. Indicate the direction of the velocity and acceleration vectors at various points along the trajectory. b) Sketch v_x-t and v_y-t graphs for the motion of the rocket from when its engines first fire until the rocket hits the ground. c) Find the maximum altitude reached by the rocket. Your answer will be in terms of g and T. d) Find the horizontal distance from the launch point to where the rocket hits the ground (that is, the range). Your answer will be in terms of g and T.

3.74 In an action-adventure film, the hero is supposed to throw a grenade from his car, which is going 90.0 km/h, to his enemy's car, which is going 110 km/h. The enemy's car is 15.8 m in front of the hero's when he lets go of the grenade. If the hero throws the grenade so its initial velocity relative to him is at an angle of 45° above the horizontal, what should the magnitude of the initial velocity be? The cars are both traveling in the same direction on a level road. You can ignore air resistance. Find the magnitude of the velocity both relative to the hero and relative to the earth.

3.75 A rock tied to a rope moves in the xy-plane. Its coordinates are given as functions of time by

$$x(t) = R \cos \omega t \qquad y(t) = R \sin \omega t$$

where R and ω are constants. a) Show that the rock's distance from the origin is constant and equal to R, that is, that its path is a circle of radius R. b) Show that at every point the rock's velocity is perpendicular to its position vector. c) Show that the rock's acceleration is always opposite in direction to its position vector and has magnitude $\omega^2 R$. d) Show that the magnitude of the rock's velocity is constant and equal to ωR. e) Combine the results of parts (c) and (d) to show that the rock's acceleration has constant magnitude v^2/R.

3.76 The speed of a particle moving in a plane is equal to the magnitude of its instantaneous velocity, $v = |\vec{v}| = \sqrt{v_x^2 + v_y^2}$. a) Show that the rate of change of the speed is $dv/dt = (v_x a_x + v_y a_y)/\sqrt{v_x^2 + v_y^2}$. b) Use this expression to find the value of dv/dt at time $t = 2.0$ s for the rover in Examples 3.1 through 3.4. Compare your answer to the components of acceleration found in Example 3.4. Explain why your answer is *not* equal to the magnitude of the acceleration found in part (b) of Example 3.2. c) Show that the rate of change of speed can be expressed as $dv/dt = \vec{v} \cdot \vec{a}/v$, and use this result to explain why dv/dt is equal to $a_\parallel$, the component of $\vec{a}$ that is parallel to $\vec{v}$.

3.77 A particle moves in the xy-plane. Its coordinates are given as functions of time by

$$x(t) = R(\omega t - \sin \omega t) \qquad y(t) = R(1 - \cos \omega t)$$

where R and ω are constants. a) Sketch the trajectory of the particle. (This is the trajectory of a point on the rim of a wheel that is rolling at a constant speed on a horizontal surface. The curve traced out by such a point as it moves through space is called a *cycloid*.) b) Determine the velocity components and the acceleration components of the particle at any time t. c) At which times is the particle momentarily at rest? What are the coordinates of the particle at these times? What are the magnitude and direction of the acceleration at these times? d) Does the magnitude of the acceleration depend on time? Compare to uniform circular motion.

3.78 You are flying in a light airplane spotting traffic for a radio station. Your flight carries you due east above a highway. Landmarks below tell you that your speed is 50 m/s relative to the ground and your air speed indicator also reads 50 m/s. However, the nose of your airplane is pointed somewhat south of east and the station's weather person tells you that a 20 m/s wind is blowing. In which direction is the wind blowing?

3.79 The Carrier Pigeon Problem. Larry is driving east at 40 km/h. His twin brother Harry is driving west at 30 km/h, toward Larry in an identical car on the same straight road. When they are 42 km apart, Larry sends out a carrier pigeon, which flies at a constant speed of 50 km/h. (All speeds are relative to the earth.) The pigeon flies to Harry, becomes confused and immediately returns to Larry, becomes more confused, and immediately flies back to Harry. This continues until the twins meet, at which time the dazed pigeon drops to the ground in exhaustion. Ignoring turnaround time, how far did the pigeon fly?

3.80 Raindrops. When a train's velocity is 12.0 m/s eastward, raindrops that are falling vertically with respect to the earth make traces that are inclined 30.0° to the vertical on the windows of the train. a) What is the horizontal component of a drop's velocity with respect to the earth? With respect to the train? b) What is the magnitude of the velocity of the raindrop with respect to the earth? With respect to the train?

3.81 An airplane pilot sets a compass course due west and maintains an airspeed of 220 km/h. After flying for 0.500 h, she finds herself over a town 120 km west and 20 km south of her starting point. a) Find the wind velocity (magnitude and direction). b) If the wind velocity is 40 km/h due south, in what direction should the pilot set her course to travel due west? Use the same airspeed of 220 km/h.

3.82 In an air race, an airplane flies from a point directly over Metropolis to a point directly over Gotham City, and then turns around and flies back to the starting point. The airspeed (that is, speed of the airplane relative to the air) is constant throughout the flight and equal to v. Gotham City lies a distance D due east of Metropolis. a) If there is no wind, how much time is required for the round trip? How much time is required for the round trip if a steady wind of speed w is blowing b) toward the east? c) toward the south? d) If $D = 3.00 \times 10^2$ km, $v = 4.00 \times 10^2$ km/h, and $w = 1.00 \times 10^2$ km/h, calculate the round-trip flight times for parts (a), (b), and (c). Which case gives the slowest round-trip?

3.83 An elevator is moving upward at a constant speed of 2.50 m/s. A bolt in the elevator ceiling 3.00 m above the elevator floor works loose and falls. a) How long does it take for the bolt to fall to the elevator floor? What is the speed of the bolt just as it hits the elevator floor b) according to an observer in the elevator? c) according to an observer standing on one of the floor landings of the building? d) According to the observer in part (c), what distance did the bolt travel between the ceiling and the floor of the elevator?

3.84 City A lies directly west of city B. When there is no wind, an airliner makes the 5310 km round trip flight between them in 6.60 h of flying time while traveling at the same speed in both directions. When a strong, steady wind is blowing from west to east and the airliner has the same air speed as before, the round trip takes 6.70 h of flying time. How fast is the wind blowing?

3.85 In a World Cup soccer match, Juan is running due north toward the goal with a speed of 8.00 m/s relative to the ground. A teammate passes the ball to him. The ball has a speed of 12.0 m/s and is moving in a direction of 37.0° east of north, relative to the ground. What are the magnitude and direction of the ball's velocity relative to Juan?

Challenge Problems

3.86 A man is riding on a flatcar traveling at a constant speed of 9.10 m/s (Fig. 3.49). He wishes to throw a ball through a stationary hoop 4.90 m above the height of his hands in such a manner that the ball will move horizontally as it passes through the hoop. He throws the ball with a speed of 10.8 m/s with respect to himself. a) What must the vertical component of the initial velocity of the ball be? b) How many seconds after he releases the ball will it pass through the hoop? c) At what horizontal distance in front of the hoop must he release the ball? d) When the ball leaves the man's

Figure 3.49 Challenge Problem 3.86.

hands, what is the direction of its velocity relative to the frame of reference of the flatcar? Relative to the frame of reference of an observer standing on the ground?

3.87 A shotgun fires a large number of pellets upward, with some pellets traveling very nearly vertically and others as much as 1.0° from the vertical. Assume that the initial speed of the pellets is uniformly 150 m/s, and ignore air resistance. a) Within what radius from the point of firing will the pellets land? b) If there are 1000 pellets, and they fall in a uniform distribution over a circle with the radius calculated in part (a), what is the probability that at least one pellet will fall on the head of the person who fires the shotgun? Assume that his head has a radius of 10 cm. c) Air resistance, in fact, has several effects. It slows down the rising pellets, decreases their horizontal component of velocity, and limits the speed with which they fall. Which of these effects will tend to make the radius greater than calculated in part (a), and which will tend to make it less? What do you think the overall effect of air resistance will be? (The effect of air resistance on a velocity component increases as the magnitude of the component increases.)

3.88 A projectile is thrown from a point P. It moves in such a way that its distance from P is always increasing. Find the maximum angle above the horizontal with which the projectile could have been thrown. You can ignore air resistance.

3.89 **Projectile Motion on an Incline I.** A baseball is given an initial velocity with magnitude v_0 at an angle ϕ above the surface of an incline, which is in turn inclined at an angle θ above the horizontal (Fig. 3.50). a) Calculate the distance, measured along the incline, from the launch point to where the baseball strikes the incline. Your answer will be in terms of v_0, g, θ, and ϕ. b) What angle ϕ gives the maximum range, measured along the incline? (*Note:* You might be interested in the three different methods of solution presented by I. R. Lapidus in *Amer. Jour. of Phys.,* Vol. 51 (1983), pp. 806 and 847. See also H. A. Buckmaster in *Amer. Jour. of Phys.,* Vol. 53 (1985), pp. 638–641, for a thorough study of this and some similar problems.)

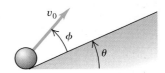

Figure 3.50 Challenge Problem 3.89.

3.90 **Projectile Motion on an Incline II.** Refer to Challenge Problem 3.89. a) An archer on ground that has a constant upward slope of 30.0° aims at a target 60.0 m farther up the incline. The arrow in the bow and the bull's-eye at the center of the target are each 1.50 m above the ground. The initial velocity of the arrow just after it leaves the bow has magnitude 32.0 m/s. At what angle above the *horizontal* should the archer aim to hit the bull's-eye? If there are two such angles, calculate the smaller of the two. You might have to solve the equation for the angle by iteration, that is, by trial and error. How does the angle compare to that required when the ground is level, with zero slope? b) Repeat the above for ground that has a constant *downward* slope of 30.0°.

3.91 For no apparent reason, a poodle is running at a constant speed of $v = 5.00$ m/s in a circle with radius $R = 2.50$ m. Let $\vec{v}_1$ be the velocity vector at time t_1, and let $\vec{v}_2$ be the velocity vector at time t_2. Consider $\Delta\vec{v} = \vec{v}_2 - \vec{v}_1$ and $\Delta t = t_2 - t_1$. Recall that $\vec{a}_{av} = \Delta\vec{v}/\Delta t$. For $\Delta t = 0.5$ s, 0.1 s, and 0.05 s, calculate the magnitude (to four significant figures) and direction (relative to $\vec{v}_1$) of the average acceleration $\vec{a}_{av}$. Compare your results to the general expression for the instantaneous acceleration $\vec{a}$ for uniform circular motion that is derived in the text.

3.92 A rocket designed to place small payloads into orbit is carried to an altitude of 12.0 km above sea level by a converted airliner. When the airliner is flying in a straight line at a constant speed of 850 km/h, the rocket is dropped. After the drop, the airliner maintains the same altitude and speed and continues to fly in a straight line. The rocket falls for a brief time, after which its rocket motor turns on. Once its rocket motor is on, the combined effects of thrust and gravity give the rocket a constant acceleration of magnitude 3.00g directed at an angle of 30.0° above the horizontal. For reasons of safety, the rocket should be at least 1.00 km in front of the airliner when it climbs through the airliner's altitude. Your job is to determine the minimum time that the rocket must fall before its engine starts. You can ignore air resistance. Your answer should include i) a diagram showing the flight paths of both the rocket and the airliner, labeled at several points with vectors for their velocities and accelerations; ii) an *x-t* graph showing the motions of both the rocket and the airliner; and iii) a *y-t* graph showing the motions of both the rocket and the airliner. In the diagram and the graphs, indicate when the rocket is dropped, when the rocket motor turns on, and when the rocket climbs through the altitude of the airliner.

3.93 Two students are canoeing on a river. While heading upstream they accidentally drop an empty bottle overboard. They then continue paddling for 60 minutes, reaching a point 2.0 km further upstream. At this point they realize that the bottle is missing and, driven by ecological awareness, they turn around and head downstream. They catch up with and retrieve the bottle (which has been moving along with the current) 5.0 km downstream from the turn-around point. a) Assuming a constant paddling effort throughout, how fast is the river flowing? b) What would be the canoe speed in a still lake for the same paddling effort?

NEWTON'S LAWS OF MOTION

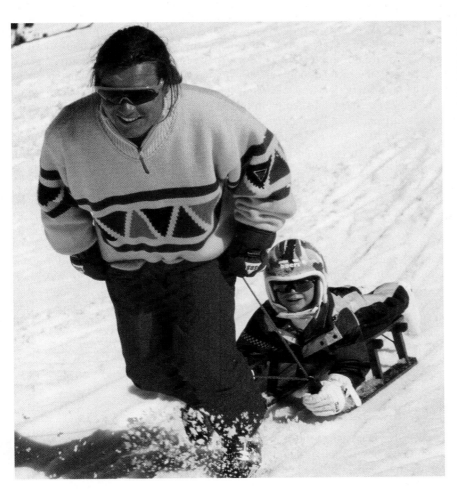

A man pulling his son's sled is illustrating Newton's third law—the law of action and reaction. To propel himself forward, the man pushes backward on the ground with his feet (the "action"), knowing that the ground will respond by pushing forward on him with an equal amount of force (the "reaction"). Similarly, the force that he exerts on the sled to pull it forward ("action") results in him feeling an equally strong force pulling him back ("reaction").

? **If the man pulls forward on the sled, and the sled pulls backward on the man with an equal amount of force, how can the man and sled move at all?**

We've seen in the last two chapters how to describe motion in one, two, or three dimensions. But what are the underlying *causes* of motion? For example, how can a tugboat push a cruise ship that's much heavier than the tug? Why does it take a long distance to stop the ship once it is in motion? Why is it harder to control a car on wet ice than on dry concrete? The answers to these and similar questions take us into the subject of **dynamics,** the relationship of motion to the forces that cause it. In the two preceding chapters we studied *kinematics,* the language for *describing* motion. Now we are ready to think about what makes bodies move the way they do.

In this chapter we will use the kinematic quantities displacement, velocity, and acceleration along with two new concepts, *force* and *mass,* to analyze the principles of dynamics. These principles can be wrapped up in a neat package of three statements called **Newton's laws of motion.** The first law states that when the net force on a body is zero, its motion doesn't change. The second law relates force to acceleration when the net force is *not* zero. The third law is a relation between the forces that two interacting bodies exert on each other.

Newton's laws are not the product of mathematical derivations, but rather a synthesis of what physicists have learned from a multitude of *experiments* about how objects move. These laws are truly fundamental, for they cannot be deduced or proved from other principles. The overarching importance of Newton's laws is that they make it possible to understand most familiar kinds of motion; they are the foundation of **classical mechanics** (also called **Newtonian mechanics**). Newton's laws are not universal, however; they require modification at very high speeds (near the speed of light) and for very small sizes (such as within the atom).

The laws of motion were clearly stated for the first time by Sir Isaac Newton (1642–1727), who published them in 1687 in his *Philosophiae Naturalis Principia Mathematica* ("Mathematical Principles of Natural Philosophy"). Many other scientists before Newton contributed to the foundations of mechanics, including Copernicus, Brahe, Kepler, and especially Galileo Galilei (1564–1642), who died the same year Newton was born. Indeed, Newton himself said, "If I have been able to see a little farther than other men, it is because I have stood on the shoulders of giants." Now it's your turn to stand on the shoulders of Newton and use his laws to understand how the physical world works.

Newton's laws are very simple to state, yet many students find these laws difficult to grasp and to work with. The reason is that before studying physics, you've spent years walking, throwing balls, pushing boxes, and doing dozens of things that involve motion. Along the way, you've developed a set of "common sense" ideas about motion and its causes. But many of these "common sense" ideas don't stand up to logical analysis. A big part of the job of this chapter—and of the rest of our study of physics—is helping you to recognize how "common sense" ideas can sometimes lead you astray, and how to adjust your understanding of the physical world to make it consistent with what experiments tell us.

4.1 | Force and Interactions

In everyday language, a **force** is a push or pull. The concept of force gives us a quantitative description of the interaction between two bodies or between a body and its environment. When you push on a car that is stuck in the snow, you exert a force on it. A locomotive exerts a force on the train it is pulling or pushing, a steel cable exerts a force on the beam it is hoisting at a construction site, and so on.

When a force involves direct contact between two bodies, we call it a **contact force.** Contact forces include the pushes or pulls you exert with your hand, the force of a rope pulling on a block to which it is tied, and the friction force that the ground exerts on a ball player sliding into home. There are also forces, called **long-range forces,** that act even when the bodies are separated by empty space. You've experienced long-range forces if you've ever played with a pair of magnets. Gravity, too, is a long-range force; the sun exerts a gravitational pull on the earth, even over a distance of 150 million kilometers, that keeps the earth in orbit. The force of gravitational attraction that the earth exerts on your body is called your **weight.**

Force is a vector quantity; you can push or pull a body in different directions. Thus, to describe a force, we need to describe the *direction* in which it acts as well

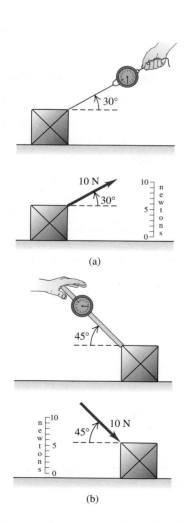

4.1 A force may be exerted on the box by either (a) pulling it or (b) pushing it. A force diagram illustrates each case.

as its *magnitude,* the quantity that describes "how much" or "how hard" the force pushes or pulls. The SI unit of the magnitude of force is the *newton,* abbreviated N. (We'll give a precise definition of the newton in Section 4.3.) Table 4.1 lists some typical force magnitudes.

Table 4.1 Typical Force Magnitudes

Sun's gravitational force on the earth	3.5×10^{22} N
Thrust of a space shuttle during launch	3.1×10^{7} N
Weight of a large blue whale	1.9×10^{6} N
Maximum pulling force of a locomotive	8.9×10^{5} N
Weight of a 250-lb linebacker	1.1×10^{3} N
Weight of a medium apple	1 N
Weight of smallest insect eggs	2×10^{-6} N
Electric attraction between the proton and the electron in a hydrogen atom	8.2×10^{-8} N
Weight of a very small bacterium	1×10^{-18} N
Weight of a hydrogen atom	1.6×10^{-26} N
Weight of an electron	8.9×10^{-30} N
Gravitational attraction between the proton and the electron in a hydrogen atom	3.6×10^{-47} N

A common instrument for measuring forces is the *spring balance.* It consists of a coil spring, enclosed in a case for protection, with a pointer attached to one end. When forces are applied to the ends of the spring, it stretches; the amount of stretch depends on the force. We can make a scale for the pointer and calibrate it by using a number of identical bodies with weights of exactly 1 N each. When two, three, or more of these are suspended simultaneously from the balance, the total force stretching the spring is 2 N, 3 N, and so on, and we can label the corresponding positions of the pointer 2 N, 3 N, and so on. Then we can use this instrument to measure the magnitude of an unknown force. We can also make a similar instrument that measures pushes instead of pulls.

Suppose we slide a box along the floor, applying a force to it by pulling it with a string or pushing it with a stick (Fig. 4.1). In each case we draw a vector to represent the force applied. The labels indicate the magnitude and direction of the force, and the length of the arrow (drawn to some scale, such as 1 cm = 10 N) also shows the magnitude.

When *two* forces $\vec{F}_1$ and $\vec{F}_2$ act at the same time at a point A of a body (Fig. 4.2), experiment shows that the effect on the body's motion is the same as the effect of a single force $\vec{R}$ equal to the *vector sum* of the original forces: $\vec{R} = \vec{F}_1 + \vec{F}_2$. More generally, the effect of any number of forces applied at a point on a body is the same as the effect of a single force equal to the vector sum of the forces. This important principle goes by the name **superposition of forces.**

The experimental discovery that forces combine according to vector addition is of the utmost importance. We will use this fact many times throughout our study of physics. It allows us to replace a force by its component vectors, as we did with displacements in Section 1.8. For example, in Fig. 4.3a, force $\vec{F}$ acts on a body at point O. The component vectors of $\vec{F}$ in the directions Ox and Oy are $\vec{F}_x$ and $\vec{F}_y$. When $\vec{F}_x$ and $\vec{F}_y$ are applied simultaneously, as in Fig. 4.3b, the effect is exactly the same as the effect of the original force $\vec{F}$. *Any force can be replaced by its component vectors, acting at the same point.*

It's frequently more convenient to describe a force $\vec{F}$ in terms of its x- and y-components F_x and F_y rather than by its component vectors (recall from

4.2 Two forces $\vec{F}_1$ and $\vec{F}_2$ acting simultaneously on a body have the same effect as a single force $\vec{R}$ equal to the vector sum (resultant) of $\vec{F}_1$ and $\vec{F}_2$.

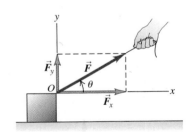

Component vectors: $\vec{F}_x$ and $\vec{F}_y$
Components: $F_x = F \cos \theta$ and $F_y = F \sin \theta$

(a)

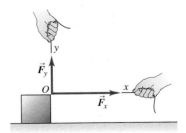

Component vectors $\vec{F}_x$ and $\vec{F}_y$ together have the same effect as original force $\vec{F}$.

(b)

4.3 The force $\vec{F}$, which acts at an angle θ from the x-axis, may be replaced by its rectangular component vectors $\vec{F}_x$ and $\vec{F}_y$.

4.4 F_x and F_y are the components of $\vec{F}$ parallel and perpendicular to the sloping surface of the inclined plane.

Section 1.8 that *component vectors* are vectors, but *components* are just numbers). For the case shown in Fig. 4.3, both F_x and F_y are positive; for other orientations of the force $\vec{F}$, either F_x or F_y can be negative or zero.

There is no law that says our coordinate axes have to be vertical and horizontal. Figure 4.4 shows a crate being pulled up a ramp by a force $\vec{F}$, represented by its components F_x and F_y parallel and perpendicular to the sloping surface of the ramp.

> **CAUTION** In Fig. 4.4 we draw a wiggly line through the force vector $\vec{F}$ to show that we have replaced it by its x- and y-components. Otherwise, the diagram would include the same force twice. We will draw such a wiggly line in any force diagram where a force is replaced by its components. Look for this wiggly line in other figures in this and subsequent chapters.

We will often need to find the vector sum (resultant) of *all* the forces acting on a body. We will call this the **net force** acting on the body. We will use the Greek letter Σ (capital "sigma," equivalent to the Roman S) as a shorthand notation for a sum. If the forces are labeled $\vec{F}_1$, $\vec{F}_2$, $\vec{F}_3$, and so on, we abbreviate the sum as

$$\vec{R} = \vec{F}_1 + \vec{F}_2 + \vec{F}_3 + \cdots = \sum \vec{F} \qquad (4.1)$$

where $\sum \vec{F}$ is read as "the vector sum of the forces" or "the net force." The component version of Eq. (4.1) is the pair of component equations

$$R_x = \sum F_x \qquad R_y = \sum F_y \qquad (4.2)$$

where $\sum F_x$ is the sum of the x-components, and so on. Each component may be positive or negative, so be careful with signs when you evaluate the sums in Eq. (4.2).

Once we have R_x and R_y, we can find the magnitude and direction of the net force $\vec{R} = \sum \vec{F}$ acting on the body. The magnitude is

$$R = \sqrt{R_x^2 + R_y^2}$$

and the angle θ between $\vec{R}$ and the $+x$-axis can be found from the relation $\tan \theta = R_y / R_x$. The components R_x and R_y may be positive, negative, or zero, and the angle θ may be in any of the four quadrants.

In three-dimensional problems, forces may also have z-components; then we add the equation $R_z = \Sigma F_z$ to Eq. (4.2). The magnitude of the net force is then

$$R = \sqrt{R_x^2 + R_y^2 + R_z^2}$$

Example 4.1	**Superposition of forces**

Three professional wrestlers are fighting over the same champion's belt. As viewed from above, they apply the three horizontal forces to the belt that are shown in Fig. 4.5a, where the belt is located at the origin. The magnitudes of the three forces are $F_1 = 250$ N, $F_2 = 50$ N, and $F_3 = 120$ N. Find the x- and y-components of the net force on the belt, and find the magnitude and direction of the net force.

SOLUTION

IDENTIFY AND SET UP: This example is just a problem in vector addition. We are asked to find the components of the net force, so we'll attack this problem using the component method. Our target variables are the magnitude and direction of the net force $\vec{R}$ as well as its x- and y-components; we'll find these using Eq. (4.2).

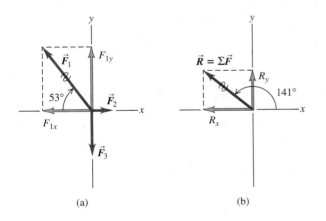

(a)   (b)

4.5 (a) Three forces acting on a belt. The x- and y-components of $\vec{F}_1$ are shown; $\vec{F}_2$ has zero y-component and $\vec{F}_3$ has zero x-component. (b) The net force $\vec{R}$ and its components.

EXECUTE: From Fig. 4.5a, the angles between the three forces $\vec{F}_1$, $\vec{F}_2$, and $\vec{F}_3$ and the $+x$-axis are $\theta_1 = 180° - 53° = 127°$, $\theta_2 = 0°$, and $\theta_3 = 270°$. The x- and y-components of the three forces are

$$F_{1x} = (250\text{ N})\cos 127° = -150\text{ N}$$
$$F_{1y} = (250\text{ N})\sin 127° = 200\text{ N}$$
$$F_{2x} = (50\text{ N})\cos 0° = 50\text{ N}$$
$$F_{2y} = (50\text{ N})\sin 0° = 0\text{ N}$$
$$F_{3x} = (120\text{ N})\cos 270° = 0\text{ N}$$
$$F_{3y} = (120\text{ N})\sin 270° = -120\text{ N}$$

From Eq. (4.2) the net force $\vec{R} = \Sigma \vec{F}$ has components

$$R_x = F_{1x} + F_{2x} + F_{3x} = (-150\text{ N}) + 50\text{ N} + 0\text{ N} = -100\text{ N}$$
$$R_y = F_{1y} + F_{2y} + F_{3y} = 200\text{ N} + 0\text{ N} + (-120\text{ N}) = 80\text{ N}$$

The net force has a negative x-component and a positive y-component, so it points to the left and toward the top of the page in Fig. 4.5b (that is, in the second quadrant).

The magnitude of the net force $\vec{R} = \Sigma \vec{F}$ is

$$R = \sqrt{R_x^2 + R_y^2} = \sqrt{(-100\text{ N})^2 + (80\text{ N})^2} = 128\text{ N}$$

To find the angle between the net force and the $+x$-axis, we use the relation $\tan\theta = R_y/R_x$, or

$$\theta = \arctan\frac{R_y}{R_x} = \arctan\left(\frac{80\text{ N}}{-100\text{ N}}\right) = \arctan(-0.80)$$

The two possible solutions are $\theta = -39°$ or $\theta = -39° + 180° = 141°$. Since the net force lies in the second quadrant, as mentioned earlier, the correct answer is $141°$ (see Fig. 4.5b).

EVALUATE: In this situation the net force is *not* zero, and you can see intuitively that wrestler 1 (who exerts the largest force, $\vec{F}_1$, on the belt) is likely to walk away with the belt at the end of the struggle. In Section 4.2 we will explore in detail what happens in situations in which the net force *is* zero.

Test Your Understanding

Suppose that in Figure 4.5a the forces $\vec{F}_1$ and $\vec{F}_2$ remain unchanged, but the magnitude and direction of $\vec{F}_3$ are changed so that the net force on the belt is zero. What are the x- and y-components of the new force $\vec{F}_3$? What are its magnitude and direction?

(a) Table:
puck stops short

(b) Waxed floor:
puck slides farther

(c) Air-hockey table:
puck slides even farther

4.6 A hockey puck is given an initial velocity. (a) It stops in a short distance on a table top. (b) A freshly waxed surface decreases the friction force, and the puck slides farther. (c) On an air-hockey table the friction force is practically zero, so the puck continues with almost constant velocity.

4.2 | Newton's First Law

We have discussed some of the properties of forces, but so far have said nothing about how forces affect motion. To begin with, let's consider what happens when the net force on a body is *zero*. You would almost certainly agree that if a body is at rest, and if no net force acts on it (that is, no net push or pull), that body will remain at rest. But what if there is zero net force acting on a body in *motion?*

To see what happens in this case, suppose you slide a hockey puck along a horizontal table top, applying a horizontal force to it with your hand (Fig. 4.6a). After you stop pushing, the puck *does not* continue to move indefinitely; it slows down and stops. To keep it moving, you have to keep pushing (that is, applying a force). You might come to the "common sense" conclusions that bodies in motion naturally come to rest and that a force is required to sustain motion.

But now imagine pushing the puck across the smooth surface of a freshly waxed floor (Fig. 4.6b). After you quit pushing, the puck will slide a lot farther before it stops. Put it on an air-hockey table, where it floats on a thin cushion of air, and it moves still farther (Fig. 4.6c). In each case, what slows the puck down is *friction,* an interaction between the lower surface of the puck and the surface on which it slides. Each surface exerts a frictional force on the puck that resists the puck's motion; the difference between the three cases is the magnitude of the frictional force. The slippery waxed floor exerts less friction than the table top, so the puck travels farther. The gas molecules of the air-hockey table exert the least friction of all. If we could eliminate friction completely, the puck would never slow down, and we would need no force at all to keep the puck moving once it had been started. Thus the "common sense" idea that a force is required to sustain motion is *incorrect.*

Experiments like the ones we've just described show that when no net force acts on a body, the body either remains at rest or moves with constant velocity in a straight line. Once a body has been set in motion, no net force is needed to keep it moving. In other words:

A body acted on by no net force moves with constant velocity (which may be zero) and zero acceleration.

This is **Newton's first law of motion.**

The tendency of a body to keep moving once it is set in motion results from a property called **inertia.** You use this when you try to get ketchup out of a bottle by shaking it. First you start the bottle (and the ketchup inside) moving forward; when you jerk the bottle back, the ketchup tends to keep moving forward and, you hope, ends up on your burger. The tendency of a body at rest to remain at rest is also due to inertia. You may have seen a tablecloth yanked out from under the china without breaking anything. The force on the china isn't great enough to make it move appreciably during the short time it takes to pull the tablecloth away.

It's important to note that the *net* force is what matters in Newton's first law. For example, a physics book at rest on a horizontal table top has two forces acting on it: the downward force of the earth's gravitational attraction (a long-range force that acts even if the table top is elevated above the ground) and an upward supporting force exerted by the table top (a contact force). The upward push of the surface is just as great as the downward pull of gravity, so the *net* force acting on the book (that is, the vector sum of the two forces) is zero. In agreement with Newton's first law, if the book is at rest on the table top, it remains at rest.

The same principle applies to a hockey puck sliding on a horizontal, frictionless surface: The vector sum of the upward push of the surface and the downward pull of gravity is zero. Once the puck is in motion, it continues to move with constant velocity because the *net* force acting on it is zero. (The upward supporting force of the surface is called a **normal force** because it is *normal,* or perpendicular, to the surface of contact. We'll discuss the normal force in more detail in Chapter 5.)

Here's another example. Suppose a hockey puck rests on a horizontal surface with negligible friction, such as an air-hockey table or a slab of wet ice. If the puck is initially at rest and a single horizontal force $\vec{F}_1$ acts on it (Fig. 4.7a), the puck starts to move. If the puck is in motion to begin with, the force changes its speed, its direction, or both, depending on the direction of the force. In this case the net force is equal to $\vec{F}_1$, which is *not* zero. (There are also two vertical forces: the earth's gravitational attraction and the upward contact force exerted by the surface. But as we mentioned earlier, these two forces cancel.)

Now suppose we apply a second force $\vec{F}_2$ (Fig. 4.7b), equal in magnitude to $\vec{F}_1$ but opposite in direction. The two forces are negatives of each other, $\vec{F}_2 = -\vec{F}_1$, and their vector sum is zero:

$$\sum \vec{F} = \vec{F}_1 + \vec{F}_2 = \vec{F}_1 + (-\vec{F}_1) = 0$$

Again, we find that if the body is at rest at the start, it remains at rest; if it is initially moving, it continues to move in the same direction with constant speed. These results show that in Newton's first law, *zero net force is equivalent to no force at all*. This is just the principle of superposition of forces that we saw in Section 4.1.

When a body is acted on by no forces, or by several forces such that their vector sum (resultant) is zero, we say that the body is in **equilibrium.** In equilibrium, a body is either at rest or moving in a straight line with constant velocity. For a body in equilibrium, the net force is zero:

$$\sum \vec{F} = 0 \qquad (\text{body in equilibrium}) \qquad (4.3)$$

For this to be true, each component of the net force must be zero, so

$$\sum F_x = 0 \qquad \sum F_y = 0 \qquad (\text{body in equilibrium}) \qquad (4.4)$$

When Eqs. (4.4) are satisfied, the body is in equilibrium.

We are assuming that the body can be represented adequately as a point particle. When the body has finite size, we also have to consider *where* on the body the forces are applied. We will return to this point in Chapter 11.

(a)

(b)

4.7 (a) A hockey puck accelerates in the direction of a net applied force $\vec{F}_1$. (b) When the net force is zero, the acceleration is zero, and the puck is in equilibrium.

| Conceptual Example 4.2 | **Zero net force means constant velocity** |

In the classic 1950 science fiction film *Rocketship X-M,* a spaceship is moving in the vacuum of outer space, far from any planet, when its engine dies. As a result, the spaceship slows down and stops. What does Newton's first law say about this event?

SOLUTION

In this situation there are no forces acting on the spaceship, so according to Newton's first law, it will *not* stop. It continues to move in a straight line with constant speed. Some science fiction movies have made use of very accurate science; this was not one of them.

Conceptual Example 4.3   **Constant velocity means zero net force**

You are driving a Porsche 911 Carrera on a straight testing track at a constant speed of 150 km/h. You pass a 1971 Volkswagen Beetle doing a constant 75 km/h. For which car is the net force greater?

SOLUTION

The key word in this question is "net." Both cars are in equilibrium because their velocities are both constant; therefore the *net* force on each car is *zero*.

This conclusion seems to contradict the "common sense" idea that the faster car must have a greater force pushing it. It's true that there is a forward force on both cars, and it's true that the forward force on your Porsche is much greater than that on the Volkswagen (thanks to your Porsche's high-power engine). But there is also a *backward* force acting on each car due to friction with the road and air resistance. The only reason these cars need their engines running at all is to counteract this backward force so that the vector sum of the forward and backward forces will be zero and the car will travel with constant velocity. The backward force on your Porsche is greater because of its greater speed, which is why your engine has to be more powerful than the Volkswagen's.

Inertial Frames of Reference

In discussing relative velocity in Section 3.5, we introduced the concept of *frame of reference*. This concept is central to Newton's laws of motion. Suppose you are in an airplane that is accelerating down the runway during takeoff. If you could stand in the aisle on roller skates, you would start moving *backward* relative to the plane as the pilot opens the throttle. If instead the airplane was landing, you would start moving forward down the aisle as the airplane slowed to a stop. In either case, it looks as though Newton's first law is not obeyed; there is no net force acting on you, yet your velocity changes. What's wrong?

The point is that the airplane is accelerating with respect to the earth and is *not* a suitable frame of reference for Newton's first law. This law is valid in some frames of reference, and is not valid in others. A frame of reference in which Newton's first law *is* valid is called an **inertial frame of reference.** The earth is at least approximately an inertial frame of reference, but the airplane is not. (The earth is not a completely inertial frame, owing to the acceleration associated with its rotation and its motion around the sun. These effects are quite small, however; see Exercises 3.29 and 3.32.) Because Newton's first law is used to define what we mean by an inertial frame of reference, it is sometimes called the *law of inertia*.

Figure 4.8 shows how we can use Newton's first law to understand what you experience when riding in a vehicle that's accelerating. In Fig. 4.8a a vehicle is initially at rest, then begins to accelerate to the right. A passenger on roller skates has virtually no net force acting on her, since the skates nearly eliminate the effects of friction; hence she tends to remain at rest relative to the inertial frame of the earth, in accordance with Newton's first law. As the vehicle accelerates around her, she moves backward relative to the vehicle. In the same way, a passenger in a vehicle that is slowing down tends to continue moving with constant velocity relative to the earth (Fig. 4.8b). This passenger moves forward relative to the vehicle. A vehicle is also accelerating if it moves at a constant speed but is turning (Fig. 4.8c). In this case a passenger tends to continue moving relative to the earth at constant speed in a straight line; relative to the vehicle, the passenger moves to the side of the vehicle on the outside of the turn.

In each case shown in Fig. 4.8, an observer in the vehicle's frame of reference might be tempted to conclude that there *is* a net force acting on the passenger, since the passenger's velocity *relative to the vehicle* changes in each case. This

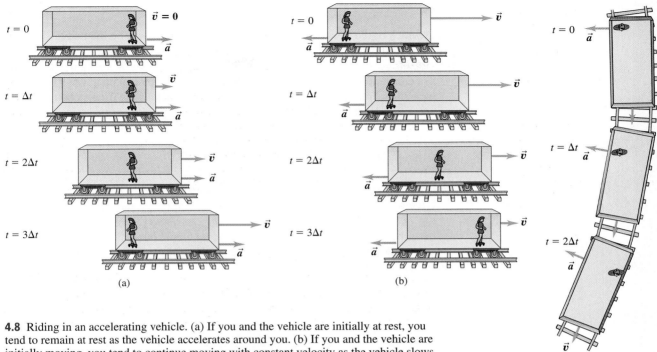

4.8 Riding in an accelerating vehicle. (a) If you and the vehicle are initially at rest, you tend to remain at rest as the vehicle accelerates around you. (b) If you and the vehicle are initially moving, you tend to continue moving with constant velocity as the vehicle slows down. (c) You tend to continue moving in a straight line as the vehicle turns.

conclusion is simply wrong; the net force on the passenger is indeed zero. The vehicle observer's mistake is in trying to apply Newton's first law in the vehicle's frame of reference, which is *not* an inertial frame and in which Newton's first law isn't valid (Fig. 4.9). In this book we will use *only* inertial frames of reference.

We've mentioned only one (approximately) inertial frame of reference: the earth's surface. But there are many inertial frames. If we have an inertial frame of reference A, in which Newton's first law is obeyed, then any second frame of reference B will also be inertial if it moves relative to A with constant velocity $\vec{v}_{B/A}$. To prove this, we use the relative velocity relation Eq. (3.36) from Section 3.5:

$$\vec{v}_{P/A} = \vec{v}_{P/B} + \vec{v}_{B/A}$$

Suppose that P is a body that moves with constant velocity $\vec{v}_{P/A}$ with respect to an inertial frame A. By Newton's first law the net force on this body is zero. The velocity of P relative to another frame B has a different value, $\vec{v}_{P/B} = \vec{v}_{P/A} - \vec{v}_{B/A}$. But if the relative velocity $\vec{v}_{B/A}$ of the two frames is constant, then $\vec{v}_{P/B}$ is constant as well. Thus B is also an inertial frame; the velocity of P in this frame is constant, and the net force on P is zero, so Newton's first law is obeyed in B. Observers in frames A and B will disagree about the velocity of P, but they will agree that P has a constant velocity (zero acceleration) and has zero net force acting on it.

There is no single inertial frame of reference that is preferred over all others for formulating Newton's laws. If one frame is inertial, then every other frame moving relative to it with constant velocity is also inertial. Viewed in this light, the state of rest and the state of motion with constant velocity are not very different; both occur when the vector sum of forces acting on the body is zero.

4.9 From the frame of reference of this car, it seems as though a force is pushing the crash test dummies forward as the car comes to a sudden stop. But there is really no such force: as the car stops, the dummies keep moving forward as a consequence of Newton's first law.

Test Your Understanding

In which of the following situations is there zero net force on the body: (i) an airplane flying due north at a steady 120 m/s and at a constant altitude; (ii) a car driving straight up a hill with a 3° slope at a constant 90 km/h; (iii) a hawk circling at a constant 20 km/h at a constant height of 15 m above an open field; (iv) a box with slick, frictionless surfaces in the back of a truck as the truck accelerates forward on a level road at 5 m/s²?

4.3 | Newton's Second Law

In discussing Newton's first law, we have seen that when a body is acted on by no force or zero net force, it moves with constant velocity and zero acceleration. In Fig. 4.10a a hockey puck is sliding to the right on wet ice, so there is negligible friction. There are no horizontal forces acting on the puck; the downward force of gravity and the upward contact force exerted by the ice surface sum to zero. So the net force $\sum \vec{F}$ acting on the puck is zero, the puck has zero acceleration, and its velocity is constant.

But what happens when the net force is *not* zero? In Fig. 4.10b we apply a constant horizontal force to a sliding puck in the same direction that the puck is moving. Then $\sum \vec{F}$ is constant and in the same horizontal direction as $\vec{v}$. We find that during the time the force is acting, the velocity of the puck changes at a constant rate; that is, the puck moves with constant acceleration. The speed of the puck increases, so the acceleration $\vec{a}$ is in the same direction as $\vec{v}$ and $\sum \vec{F}$.

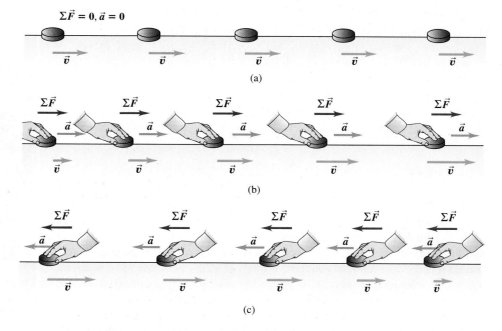

4.10 The acceleration of a body is in the same direction as the net force acting on the body (in this case, a hockey puck on a frictionless surface). (a) If $\sum \vec{F} = 0$, the puck is in equilibrium; the velocity is constant and the acceleration is zero. (b) If $\sum \vec{F}$ is to the right, the acceleration is to the right. (c) If $\sum \vec{F}$ is to the left, the acceleration is to the left.

Figure 4.10c shows another experiment, in which we reverse the direction of the force on the puck so that $\sum \vec{F}$ acts in the direction opposite to $\vec{v}$. In this case as well, the puck has an acceleration; the puck moves more and more slowly to the right. If the leftward force continues to act, the puck eventually stops and begins to move more and more rapidly to the left. The acceleration $\vec{a}$ in this experiment is to the left, in the same direction as $\sum \vec{F}$. As in the previous case, experiment shows that the acceleration is constant if $\sum \vec{F}$ is constant.

We conclude that the presence of a net force acting on a body causes the body to accelerate. The direction of the acceleration is the same as that of the net force. If the magnitude of the net force is constant, as in Figs. 4.10b and 4.10c, then so is the magnitude of the acceleration.

These conclusions about net force and acceleration also apply to a body moving along a curved path. For example, Fig. 4.11 shows a hockey puck moving in a horizontal circle on an ice surface of negligible friction. A string attaching the puck to the ice exerts a force of constant magnitude toward the center of the circle. The result is acceleration that is constant in magnitude and directed toward the center of the circle. The speed of the puck is constant, so this is uniform circular motion, as discussed in Section 3.4.

Figure 4.12a shows another experiment to explore the relationship between the acceleration of a body and the net force acting on that body. We apply a constant horizontal force to a puck on a frictionless horizontal surface, using the spring balance described in Section 4.1 with the spring stretched a constant amount. As in Figs. 4.10b and 4.10c, this horizontal force equals the net force on the puck. If we change the magnitude of the net force, the acceleration changes in the same proportion. Doubling the net force doubles the acceleration (Fig. 4.12b), halving the force halves the acceleration (Fig. 4.12c), and so on. Many such experiments show that for any given body the magnitude of the acceleration is directly proportional to the magnitude of the net force acting on the body.

For a given body the ratio of the magnitude $|\sum \vec{F}|$ of the net force to the magnitude $a = |\vec{a}|$ of the acceleration is constant, regardless of the magnitude of the net force. We call this ratio the inertial mass, or simply the **mass,** of the body and denote it by m. That is, $m = |\sum \vec{F}|/a$, or

$$\left| \sum \vec{F} \right| = ma \qquad (4.5)$$

This equation *defines* what we mean by mass, and it also tells us how to measure mass through the relation $m = |\sum \vec{F}|/a$. When you hold a piece of fruit in your hand at the supermarket and move it slightly up and down to estimate its heft, you're using this relation to estimate the fruit's mass: you're seeing how much force you must exert with your hand to make the fruit accelerate up and down.

Mass and Force

Mass is a quantitative measure of inertia, which we discussed in Section 4.2. The greater its mass, the more a body "resists" being accelerated. The concept is easy to relate to everyday experience. If you hit a table-tennis ball and then a basketball with the same force, the basketball has much smaller acceleration because it has much greater mass. If a force causes a large acceleration, the mass of the body is small; if the same force causes only a small acceleration, the mass of the body is large.

The SI unit of mass is the **kilogram.** We mentioned in Section 1.3 that the kilogram is officially defined to be the mass of a chunk of platinum-iridium alloy kept

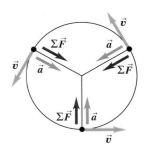

4.11 A top view of a hockey puck in uniform circular motion on a frictionless horizontal surface. At any point in the motion the acceleration $\vec{a}$ and the net force $\sum \vec{F}$ are in the same direction, toward the center of the circle.

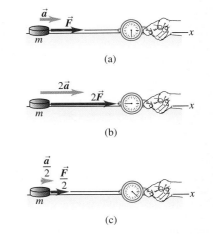

4.12 (a) Acceleration $\vec{a}$ is proportional to the net force $\sum \vec{F}$. (b) Doubling the net force doubles the acceleration. (c) Halving the net force halves the acceleration.

in a vault near Paris. We can use this standard kilogram, along with Eq. (4.5), to define the **newton:**

> **One newton is the amount of net force that gives an acceleration of one meter per second squared to a body with a mass of one kilogram.**

We can use this definition to calibrate the spring balances and other instruments used to measure forces. Because of the way we have defined the newton, it is related to the units of mass, length, and time. For Eq. (4.5) to be dimensionally consistent, it must be true that

$$1 \text{ newton} = (1 \text{ kilogram})(1 \text{ meter per second squared})$$

or

$$1 \text{ N} = 1 \text{ kg} \cdot \text{m/s}^2$$

We will use this relation many times in the next few chapters, so keep it in mind.

We can also use Eq. (4.5) to compare a mass with the standard mass and thus to *measure* masses. Suppose we apply a constant net force F to a body having a known mass m_1 and we find an acceleration of magnitude a_1. We then apply the same force to another body having an unknown mass m_2, and we find an acceleration of magnitude a_2. Then, according to Eq. (4.5),

$$m_1 a_1 = m_2 a_2$$

$$\frac{m_2}{m_1} = \frac{a_1}{a_2} \qquad (\text{same net force}) \qquad (4.6)$$

The ratio of the masses is the inverse of the ratio of the accelerations. Figure 4.13 shows the inverse proportionality between mass and acceleration. In principle we could use Eq. (4.6) to measure an unknown mass m_2, but it is usually easier to determine mass indirectly by measuring the body's *weight*. We'll return to this point in Section 4.4.

When two bodies with masses m_1 and m_2 are fastened together, we find that the mass of the composite body is always $m_1 + m_2$ (Fig. 4.13c). This additive property of mass may seem obvious, but it has to be verified experimentally. Ultimately, the mass of a body is related to the number of protons, electrons, and neutrons it contains. This wouldn't be a good way to *define* mass because there is no practical way to count these particles. But the concept of mass is the most fundamental way to characterize the quantity of matter in a body.

Newton's Second Law

We've been careful to state that the *net* force on a body is what causes that body to accelerate. Experiment shows that if a combination of forces $\vec{F}_1, \vec{F}_2, \vec{F}_3, \ldots$, is applied to a body, the body will have the same acceleration (magnitude and direction) as when only a single force is applied, if that single force is equal to the vector sum $\vec{F}_1 + \vec{F}_2 + \vec{F}_3 + \ldots$. In other words, the principle of superposition of forces also holds true when the net force is not zero and the body is accelerating.

Equation (4.5) relates the magnitude of the net force on a body to the magnitude of the acceleration that it produces. We have also seen that the direction of the net force is the same as the direction of the acceleration, whether the body's path is straight or curved. Newton wrapped up all these relationships and experi-

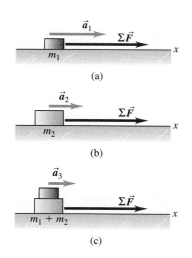

4.13 For a constant net force $\sum \vec{F}$, acceleration is inversely proportional to the mass of a body. Masses add like ordinary scalars. (a) $\vec{a}_1 = \sum \vec{F}/m_1$. (b) With a greater mass m_2, the acceleration has smaller magnitude: $\vec{a}_2 = \sum \vec{F}/m_2$. (c) If the two masses are combined, the acceleration is $\vec{a}_3 = \sum \vec{F}/(m_1 + m_2)$.

mental results in a single concise statement that we now call **Newton's second law of motion:**

If a net external force acts on a body, the body accelerates. The direction of acceleration is the same as the direction of the net force. The net force vector is equal to the mass of the body times the acceleration of the body.

In symbols,

$$\sum \vec{F} = m\vec{a} \qquad \text{(Newton's second law)} \qquad (4.7)$$

An alternative statement is that the acceleration of a body (the rate of change of its velocity) is equal to the vector sum (resultant) of all forces acting on the body, divided by its mass:

$$\vec{a} = \frac{\sum \vec{F}}{m}$$

The acceleration is in the same direction as the net force.

Newton's second law is a fundamental law of nature, the basic relation between force and motion. Most of the remainder of this chapter and all of the next are devoted to learning how to apply this principle in various situations.

Equation (4.7) has many practical applications (Fig. 4.14). In fact, you've actually been using it all your life to measure your body's acceleration. In your inner ear, microscopic hair cells sense the magnitude and direction of the force that they must exert to cause small membranes to accelerate along with the rest of your body. By Newton's second law, the acceleration of the membranes—and hence that of your body as a whole—is proportional to this force and has the same direction. In this way, you can sense the magnitude and direction of your acceleration even with your eyes closed!

There are at least four aspects of Newton's second law that deserve special attention. First, Eq. (4.7) is a *vector* equation. Usually, we will use it in component form, with a separate equation for each component of force and the corresponding acceleration:

$$\sum F_x = ma_x \qquad \sum F_y = ma_y \qquad \sum F_z = ma_z \qquad (4.8)$$
$$\text{(Newton's Second Law)}$$

This set of component equations is equivalent to the single vector equation (4.7). Each component of total force equals the mass times the corresponding component of acceleration.

Second, the statement of Newton's second law refers to *external* forces. By this we mean forces exerted on the body by other bodies in its environment. It's impossible for a body to affect its own motion by exerting a force on itself; if it were possible, you could lift yourself to the ceiling by pulling up on your belt! That's why only external forces are included in the sum $\sum \vec{F}$ in Eqs. (4.7) and (4.8).

Third, Eqs. (4.7) and (4.8) are valid only when the mass m is *constant*. It's easy to think of systems whose masses change, such as a leaking tank truck, a rocket ship, or a moving railroad car being loaded with coal. But such systems are better handled by using the concept of momentum; we'll get to that in Chapter 8.

Finally, Newton's second law is valid only in inertial frames of reference, just like the first law. Thus it is not valid in the reference frame of any of the accelerating

Lightweight body (small m)

Powerful engine (large F)

4.14 The design of high-performance motorcycles depends fundamentally on Newton's second law. To maximize the forward acceleration, the designer makes the motorcycle as light as possible (that is, minimizes the mass) and uses the most powerful engine possible (thus maximizing the forward force).

ActivPhysics ONLINE

2.1.3 Tension Change

2.1.4 Sliding on an Incline

vehicles in Fig. 4.8; relative to any of these frames, the passenger accelerates even though the net force on the passenger is zero. We will usually assume that the earth is an adequate approximation to an inertial frame, although because of its rotation and orbital motion it is not precisely inertial.

CAUTION Please note that the quantity $m\vec{a}$ is *not* a force. All that Eqs. (4.7) and (4.8) say is that the vector $m\vec{a}$ is equal in magnitude and direction to the vector sum $\Sigma\vec{F}$ of all the forces acting on the body. It's incorrect to think of acceleration as a force; rather, acceleration is a result of a nonzero net force. It's "common sense" to think that there is a "force of acceleration" that pushes you back into your seat when your car accelerates forward from rest. But *there is no such force;* instead, your inertia causes you to tend to stay at rest relative to the earth, and the car accelerates around you. The "common sense" confusion arises from trying to apply Newton's second law in a frame of reference where it isn't valid, like the non-inertial reference frame of an accelerating car. We will always examine motion relative to *inertial* frames of reference only.

In learning how to use Newton's second law, we will begin in this chapter with examples of straight-line motion. Then in Chapter 5 we will consider more general cases and develop more detailed problem-solving strategies for applying Newton's laws of motion.

| Example 4.4 | **Determining acceleration from force** |

A worker applies a constant horizontal force with magnitude 20 N to a box with mass 40 kg resting on a level floor with negligible friction. What is the acceleration of the box?

SOLUTION

IDENTIFY: This problem involves force and acceleration. Whenever you encounter a problem of this kind, you should approach it using Newton's second law.

SET UP: In *any* problem involving forces, the first steps are to (i) choose a coordinate system and (ii) identify all of the forces acting on the body in question.

It's usually convenient to take one axis either along or opposite to the direction of the body's acceleration, which in this case is horizontal (see Fig. 4.15a). Hence we take the $+x$-axis to be in the direction of the applied horizontal force (that is, the direction in

(a)   (b)

4.15 (a) A 20-N force accelerates the box to the right. (b) Vector diagram of forces acting on the box, considered as a particle. The tiles under the box are freshly waxed, so friction is negligible.

which the box accelerates) and the $+y$-axis to be upward (Fig. 4.15b). In most force problems that you'll encounter (including this one), the force vectors all lie in a plane, so the z-axis isn't used.

The forces acting on the box are (i) the horizontal force $\vec{F}$ exerted by the worker; (ii) the weight $\vec{w}$ of the box, that is, the downward force of the earth's gravitational attraction; and (iii) the upward supporting force $\vec{n}$ exerted by the floor. As in Section 4.2, we call the force $\vec{n}$ a *normal* force because it is normal (perpendicular) to the surface of contact. (We use an italic letter n to avoid confusion with the abbreviation N for newton.) We are told that friction is negligible, so no friction force is present.

Since the box doesn't move vertically at all, the y-acceleration is zero: $a_y = 0$. Our target variable is the x-component of acceleration, a_x. We'll find it using Newton's second law in component form as given by Eq. (4.8).

EXECUTE: From Fig. 4.15b, only the 20-N force has a nonzero x-component. Hence the first relation in Eqs. (4.8) tells us that

$$\sum F_x = F = 20 \text{ N}$$

Hence the x-component of acceleration is

$$a_x = \frac{\sum F_x}{m} = \frac{20 \text{ N}}{40 \text{ kg}} = \frac{20 \text{ kg} \cdot \text{m/s}^2}{40 \text{ kg}} = 0.50 \text{ m/s}^2$$

EVALUATE: The acceleration is in the $+x$-direction, the same direction as the net force. The net force is constant, so the acceleration is also constant. If we are given the initial position and velocity of the box, we can find the position and velocity at any later time from the equations of motion with constant acceleration we derived in Chapter 2.

Notice that to determine a_x, we didn't have to use the y-component of Newton's second law from Eq. (4.8), $\sum F_y = ma_y$. By using this equation, can you show that the magnitude n of the normal force in this situation is equal to the weight of the box?

Example 4.5	**Determining force from acceleration**

A waitress shoves a ketchup bottle with mass 0.45 kg toward the right along a smooth, level lunch counter. As the bottle leaves her hand, it has an initial velocity of 2.8 m/s. As it slides, it slows down because of the constant horizontal friction force exerted on it by the counter top. It slides a distance of 1.0 m before coming to rest. What are the magnitude and direction of the friction force acting on it?

SOLUTION

IDENTIFY: Like the previous example, this problem involves forces and acceleration (the slowing of the ketchup bottle), so we'll use Newton's second law to solve it.

SET UP: As in Example 4.4, our first steps are to choose a coordinate system and to identify the forces acting on the body (in this case, the ketchup bottle). We choose the $+x$-axis to be in the direction that the bottle slides, and we take the origin to be the point at which the bottle leaves the waitress's hand with the given initial velocity (Fig. 4.16a). The forces acting on the bottle are shown in Fig. 4.16b. The friction force $\vec{f}$ acts to slow the bottle down, so its direction must be opposite the direction of velocity (see Fig. 4.10c).

Our target variable is the magnitude f of the friction force. We'll find it using the x-component of Newton's second law from

(a)   (b)

4.16 (a) As the bottle slides to the right, the friction force slows it down. (b) Vector diagram of forces acting on the bottle, considered as a particle.

Eq. (4.8). To do so, we'll first need to know the x-component of the bottle's acceleration, a_x. We aren't told the value of a_x in the problem, but we are told that the friction force is constant. Hence the acceleration is constant as well, and we can calculate a_x by using one of the constant-acceleration formulas from Section 2.4. Since we know the bottle's initial x-coordinate and x-velocity ($x_0 = 0$, $v_{0x} = 2.8$ m/s) as well as its final x-coordinate and x-velocity ($x = 1.0$ m, $v_x = 0$), the easiest equation to use to determine a_x is Eq. (2.13).

EXECUTE: From Eq. (2.13),

$$v_x^2 = v_{0x}^2 + 2a_x(x - x_0)$$

$$a_x = \frac{v_x^2 - v_{0x}^2}{2(x - x_0)} = \frac{(0 \text{ m/s})^2 - (2.8 \text{ m/s})^2}{2(1.0 \text{ m} - 0 \text{ m})} = -3.9 \text{ m/s}^2$$

The negative sign means that the acceleration is toward the *left;* the velocity is in the opposite direction to acceleration, as it must be,

since the bottle is slowing down. The net force in the x-direction is the x-component $-f$ of the friction force, so

$$\sum F_x = -f = ma_x = (0.45 \text{ kg})(-3.9 \text{ m/s}^2)$$
$$= -1.8 \text{ kg} \cdot \text{m/s}^2 = -1.8 \text{ N}$$

Again the negative sign shows that the force on the bottle is directed toward the left. The magnitude of the friction force is $f = 1.8$ N. Remember that magnitudes are *always* positive!

EVALUATE: We chose the $+x$-axis to be in the direction of the bottle's motion, so that a_x was negative. As a check on the result, we invite you to repeat the calculation with the $+x$-axis directed *opposite* to the motion (to the left in Fig. 4.16b), so that a_x is positive. In this case you should find that $\sum F_x$ is equal to $+f$ (because the friction force is now in the $+x$-direction), which in turn is equal to $+1.8$ N. Your answers for the magnitudes of forces (which are always positive numbers) should never depend on your choice of coordinate axes!

Some Notes on Units

A few words about units are in order. In the cgs metric system (not used in this book), the unit of mass is the gram, equal to 10^{-3} kg, and the unit of distance is the centimeter, equal to 10^{-2} m. The corresponding unit of force is called the *dyne:*

$$1 \text{ dyne} = 1 \text{ g} \cdot \text{cm/s}^2$$

One dyne is equal to 10^{-5} N. In the British system, the unit of force is the *pound* (or pound-force) and the unit of mass is the *slug* (Fig. 4.17). The unit of acceleration is one foot per second squared, so

$$1 \text{ pound} = 1 \text{ slug} \cdot \text{ft/s}^2$$

The official definition of the pound is

$$1 \text{ pound} = 4.448221615260 \text{ newtons}$$

It is handy to remember that a pound is about 4.4 N and a newton is about 0.22 pound. Next time you want to order a "quarter-pounder," try asking for a "one-newtoner" and see what happens. Another useful fact: A body with a mass of 1 kg has a weight of about 2.2 lb at the earth's surface.

Table 4.2 summarizes the units of force, mass, and acceleration in the three systems.

4.17 Despite its name, the English unit of mass has nothing to do with the type of slug shown here. A common garden slug has a mass of about 15 grams, or about 10^{-3} slug.

Table 4.2 Units of Force, Mass, and Acceleration

System of Units	Force	Mass	Acceleration
SI	newton (N)	kilogram (kg)	m/s^2
cgs	dyne (dyn)	gram (g)	cm/s^2
British	pound (lb)	slug	ft/s^2

Suppose the championship belt in Example 4.1 (Section 4.1) has a mass of 8.0 kg. What are the magnitude and direction of the belt's acceleration? If the belt was initially at rest, how far has it moved after 0.50 s?

4.4 | Mass and Weight

The *weight* of a body is a familiar force. It is the force of the earth's gravitational attraction for the body. We will study gravitational interactions in detail in Chapter 12, but we need some preliminary discussion now. The terms *mass* and *weight* are often misused and interchanged in everyday conversation. It is absolutely essential for you to understand clearly the distinctions between these two physical quantities.

Mass characterizes the *inertial* properties of a body. Mass is what keeps the china on the table when you yank the tablecloth out from under it. The greater the mass, the greater the force needed to cause a given acceleration; this is reflected in Newton's second law, $\sum \vec{F} = m\vec{a}$. Weight, on the other hand, is a *force* exerted on a body by the pull of the earth or some other large body. Everyday experience shows us that bodies having large mass also have large weight. A large stone is hard to throw because of its large *mass,* and hard to lift off the ground because of its large *weight.* On the moon the stone would be just as hard to throw horizontally, but it would be easier to lift. So what exactly *is* the relationship between mass and weight?

The answer to this question, according to legend, came to Newton as he sat under an apple tree watching the apples fall. A freely falling body has an acceleration equal to g, and because of Newton's second law, a force must act to produce this acceleration. If a 1-kg body falls with an acceleration of 9.8 m/s², the required force has magnitude

$$F = ma = (1 \text{ kg})(9.8 \text{ m/s}^2) = 9.8 \text{ kg} \cdot \text{m/s}^2 = 9.8 \text{ N}$$

But the force that makes the body accelerate downward is the gravitational pull of the earth, that is, the *weight* of the body. Any body near the surface of the earth that has a mass of 1 kg *must* have a weight of 9.8 N to give it the acceleration we observe when it is in free fall. More generally, a body with mass m must have weight with magnitude w given by

$$w = mg \qquad (\text{magnitude of the weight of a body of mass } m) \qquad (4.9)$$

The weight of a body is a force, a vector quantity, and we can write Eq. (4.9) as a vector equation:

$$\vec{w} = m\vec{g} \qquad (4.10)$$

Remember that g is the *magnitude* of $\vec{g}$, the acceleration due to gravity, so g is always a positive number, by definition. Thus w, given by Eq. (4.9), is the *magnitude* of the weight and is also always positive.

CAUTION It is important to understand that the weight of a body acts on the body *all the time,* whether it is in free fall or not. When a 10-kg flowerpot hangs suspended from a chain, it is in equilibrium, and its acceleration is zero. But its weight, given by Eq. (4.10), is still pulling down on it. In this case the chain pulls up on the pot, applying an upward force. The *vector sum* of the forces is zero, and the pot is in equilibrium.

Act|v ONLINE
Physics

2.9   Pole-Vaulter Vaults

Conceptual Example 4.6	**Net force and acceleration in free fall**

In Example 2.6 (Section 2.5) a one-euro coin was dropped from rest from the Leaning Tower of Pisa. If the coin falls freely, so that the effects of the air are negligible, how does the net force on the coin vary as it falls?

SOLUTION

In free fall, the acceleration $\vec{a}$ of the coin is constant and equal to $\vec{g}$. Hence by Newton's second law the net force $\sum \vec{F} = m\vec{a}$ is also constant and equal to $m\vec{g}$, which is the weight $\vec{w}$ of the coin (Fig. 4.18). The velocity of the coin changes as it falls, but the net force acting on it remains constant. If this result surprises you, it's because you still believe in the erroneous "common sense" idea that greater speed implies greater force. If so, you should reread Conceptual Example 4.3.

The net force on a freely falling coin is constant even if you initially toss it upward. The force that your hand exerts on the coin is a contact force, and it disappears the instant that the coin loses contact with your hand. From then on, the only force acting on the coin is its weight $\vec{w}$.

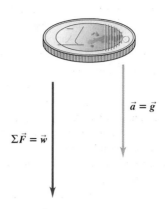

$\sum \vec{F} = \vec{w}$

$\vec{a} = \vec{g}$

4.18 The acceleration of a freely falling object is constant, and so is the net force acting on the object.

(a)

(b)

4.19 (a) A standard kilogram weighs about 9.8 N on earth. (b) The same kilogram weighs only about 1.6 N on the moon.

Variation of *g* with Location

We will use $g = 9.80 \text{ m/s}^2$ for problems on the earth (or, if the other data in the problem are given to only two significant figures, $g = 9.8 \text{ m/s}^2$). In fact, the value of g varies somewhat from point to point on the earth's surface, from about 9.78 to 9.82 m/s², because the earth is not perfectly spherical and because of effects due to its rotation and orbital motion. At a point where $g = 9.80 \text{ m/s}^2$, the weight of a standard kilogram is $w = 9.80 \text{ N}$. At a different point, where $g = 9.78 \text{ m/s}^2$, the weight is $w = 9.78 \text{ N}$ but the mass is still 1 kg. The weight of a body varies from one location to another; the mass does not. If we take a standard kilogram to the surface of the moon, where the acceleration of free fall (equal to the value of g at the moon's surface) is 1.62 m/s², its weight is 1.62 N, but its mass is still 1 kg (Fig. 4.19). An 80.0-kg astronaut has a weight on earth of $(80.0 \text{ kg})(9.80 \text{ m/s}^2) = 784 \text{ N}$, but on the moon the astronaut's weight would be only $(80.0 \text{ kg})(1.62 \text{ m/s}^2) = 130 \text{ N}$.

The following excerpt from Astronaut Buzz Aldrin's book *Men from Earth*, describing his experiences during the Apollo 11 mission to the moon in 1969, offers some interesting insights into the distinction between mass and weight:

> Our portable life-support system backpacks looked simple, but they were hard to put on and tricky to operate. On Earth the portable life-support system and space suit combination weighed 190 pounds, but on the moon it was only 30. Combined with my own body weight, that brought me to a total lunar-gravity weight of around 60 pounds.
>
> One of my tests was to jog away from the lunar module to see how maneuverable an astronaut was on the surface. I remembered what Isaac Newton had taught us two centuries before: Mass and weight are not the same. I weighed only 60 pounds, but my *mass* was the same as it was on Earth. Inertia was a problem. I had to plan ahead several steps to bring myself to a stop or to turn without falling.

Measuring Mass and Weight

In Section 4.3 we described a way to compare masses by comparing their accelerations when subjected to the same net force. Usually, however, the easiest way to measure the mass of a body is to measure its weight, often by comparing with a standard. Because of Eq. (4.9), two bodies that have the same weight at a particular location also have the same mass. We can compare weights very precisely; the familiar equal-arm balance (Fig. 4.20) can determine with great precision (up to 1 part in 10^6) when the weights of two bodies are equal and hence when their masses are equal. This method doesn't work in the apparent "zero-gravity" environment of outer space. Instead, we have to use Newton's second law directly. We apply a known force to the body, measure its acceleration, and compute the mass as the ratio of force to acceleration. This method, or a variation of it, is used to measure the masses of astronauts in orbiting space stations as well as the masses of atomic and subatomic particles.

4.20 An equal-arm balance determines the mass of a body by comparing its weight to a known weight.

The concept of mass plays two rather different roles in mechanics. The weight of a body (the gravitational force acting on it) is proportional to its mass; we may call the property related to gravitational interactions *gravitational mass*. On the other hand, we can call the inertial property that appears in Newton's second law the *inertial mass*. If these two quantities were different, the acceleration due to gravity might well be different for different bodies. However, extraordinarily precise experiments have established with a precision of better than one part in 10^{12} that in fact the two *are* the same.

> **CAUTION** The SI units for mass and weight are often misused in everyday life. Incorrect expressions such as "This box weighs 6 kg" are nearly universal. What is meant is that the *mass* of the box, probably determined indirectly by *weighing*, is 6 kg. This usage is so common that there is probably no hope of straightening things out, but be sure you recognize that the term *weight* is often used when *mass* is meant. Be careful to avoid this kind of mistake in your own work! In SI units, weight (a force) is measured in newtons, while mass is measured in kilograms.

Example 4.7	**Mass and weight**

A 1.96×10^4 N Lincoln Town Car traveling in the $+x$-direction makes a fast stop; the x-component of the net force acting on it is -1.50×10^4 N. What is its acceleration?

SOLUTION

IDENTIFY: Again we will use Newton's second law to relate force and acceleration. To use this relationship, we need to know the car's mass. However, because the newton is a unit for force, we know that 1.96×10^4 N is the car's *weight,* not its mass. So we'll also have to use the relationship between a body's mass and its weight.

SET UP: Our target variable is the x-component of acceleration of the car, a_x. (The motion is purely in the x-direction.) We use

Eq. (4.9) to determine the car's mass from its weight, then use the x-component of Newton's second law from Eq. (4.8) to determine a_x.

EXECUTE: The mass m of the car is

$$m = \frac{w}{g} = \frac{1.96 \times 10^4 \text{ N}}{9.80 \text{ m/s}^2} = \frac{1.96 \times 10^4 \text{ kg} \cdot \text{m/s}^2}{9.80 \text{ m/s}^2}$$
$$= 2000 \text{ kg}$$

Then $\Sigma F_x = ma_x$ gives

$$a_x = \frac{\Sigma F_x}{m} = \frac{-1.50 \times 10^4 \text{ N}}{2000 \text{ kg}} = \frac{-1.50 \times 10^4 \text{ kg} \cdot \text{m/s}^2}{2000 \text{ kg}}$$
$$= -7.5 \text{ m/s}^2$$

EVALUATE: The negative sign means that the acceleration vector points in the negative x-direction. This makes sense: the car is moving in the positive x-direction and is slowing down.

Note that the acceleration can alternatively be written as $-0.77g$. It's of interest that -0.77 is also the ratio of -1.50×10^4 N (the x-component of the net force) to 1.96×10^4 N (the weight). Indeed, the acceleration of a body expressed as a multiple of g is always equal to the ratio of the net force on the body to its weight. Can you see why?

Test Your Understanding

Suppose an astronaut landed on a planet where $g = 19.6 \text{ m/s}^2$. Compared to earth, would it be easier, harder, or just as easy for her to walk around? Would it be easier, harder, or just as easy for her to catch a ball that is moving horizontally at 12 m/s? (Assume that the astronaut's spacesuit is a lightweight model that doesn't impair her movements in any way.)

4.5 | Newton's Third Law

A force acting on a body is always the result of its interaction with another body, so forces always come in pairs. You can't pull on a doorknob without the doorknob pulling back on you. When you kick a football, the forward force that your foot exerts on the ball launches it into its trajectory, but you also feel the force the ball exerts back on your foot. If you kick a boulder, the pain you feel is due to the force that the boulder exerts on your foot.

In each of these cases, the force that you exert on the other body is in the opposite direction to the force that body exerts on you. Experiments show that whenever two bodies interact, the two forces that they exert on each other are always *equal in magnitude* and *opposite in direction*. This fact is called **Newton's third law of motion.** In Fig. 4.21, $\vec{F}_{A \text{ on } B}$ is the force applied *by* body A (first subscript) *on* body B (second subscript), and $\vec{F}_{B \text{ on } A}$ is the force applied *by* body B (first subscript) *on* body A (second subscript). The mathematical statement of Newton's third law is

$$\vec{F}_{A \text{ on } B} = -\vec{F}_{B \text{ on } A} \qquad (4.11)$$

Expressed in words,

> **If body A exerts a force on body B (an "action"), then body B exerts a force on body A (a "reaction"). These two forces have the same magnitude but are opposite in direction. These two forces act on *different* bodies.**

In this statement, "action" and "reaction" are the two opposite forces; we sometimes refer to them as an **action-reaction pair.** This is not meant to imply any cause-and-effect relationship; we can consider either force as the "action" and the other as the "reaction." We often say simply that the forces are "equal and opposite," meaning that they have equal magnitudes and opposite directions.

We stress that the two forces described in Newton's third law act on *different* bodies. This is important in problems involving Newton's first or second law, which involve the forces that act *on* a body. For instance, the net force acting on the football in Fig. 4.21 is the vector sum of the weight of the football and the

4.21 If body A exerts a force $\vec{F}_{A \text{ on } B}$ on body B, then body B exerts a force $\vec{F}_{B \text{ on } A}$ on body A that is equal in magnitude and opposite in direction: $\vec{F}_{A \text{ on } B} = -\vec{F}_{B \text{ on } A}$.

force $\vec{F}_{A \text{ on } B}$ exerted by the kicker. You would not include the force $\vec{F}_{B \text{ on } A}$ because this force acts on the *kicker,* not on the football.

In Fig. 4.21 the action and reaction forces are *contact* forces that are present only when the two bodies are touching. But Newton's third law also applies to *long-range* forces that do not require physical contact, such as the force of gravitational attraction. A table-tennis ball exerts an upward gravitational force on the earth that's equal in magnitude to the downward gravitational force the earth exerts on the ball. When you drop the ball, both the ball and the earth accelerate toward each other. The net force on each body has the same magnitude, but the earth's acceleration is microscopically small because its mass is so great. Nevertheless, it does move!

Conceptual Example 4.8   **Which force is greater?**

After your sports car breaks down, you start to push it to the nearest repair shop. While the car is starting to move, how does the force you exert on the car compare to the force the car exerts on you? How do these forces compare when you are pushing the car along at a constant speed?

SOLUTION

In *both* cases, the force you exert on the car is equal in magnitude and opposite in direction to the force the car exerts on you. It's true that you have to push harder to get the car going than to keep it going. But no matter how hard you push on the car, the car pushes just as hard back on you. Newton's third law gives the same result whether the two bodies are at rest, moving with constant velocity, or accelerating.

You may wonder how the car "knows" to push back on you with the same magnitude of force that you exert on it. It may help to remember that the forces you and the car exert on each other are really interactions between the atoms at the surface of your hand and the atoms at the surface of the car. These interactions are analogous to miniature springs between adjacent atoms, and a compressed spring exerts equally strong forces on either of its ends.

Fundamentally, though, the reason why we know that objects of different masses exert equally strong forces on each other is that *experiment tells us so.* Never forget that physics isn't merely a collection of rules and equations; rather, it's a systematic description of the natural world based on experiment and observation.

Conceptual Example 4.9   **Applying Newton's third law: Objects at rest**

An apple sits on a table in equilibrium. What forces act on it? What is the reaction force to each of the forces acting on the apple? What are the action-reaction pairs?

SOLUTION

Figure 4.22a shows the apple on the table, and Fig. 4.22b shows the forces acting on the apple. The apple is A, the table T, and the earth E. In the diagram, $\vec{F}_{E \text{ on } A}$ is the weight of the apple; that is, the downward gravitational force exerted *by* the earth E (first subscript) *on* the apple A (second subscript). Similarly, $\vec{F}_{T \text{ on } A}$ is the upward force exerted *by* the table T (first subscript) *on* the apple A (second subscript).

As the earth pulls down on the apple, the apple exerts an equally strong upward pull $\vec{F}_{A \text{ on } E}$ on the earth, as shown in Fig. 4.22d. $\vec{F}_{A \text{ on } E}$ and $\vec{F}_{E \text{ on } A}$ are an action-reaction pair, representing the mutual interaction of the apple and the earth, so

$$\vec{F}_{A \text{ on } E} = -\vec{F}_{E \text{ on } A}$$

Also, as the table pushes up on the apple with force $\vec{F}_{T \text{ on } A}$, the corresponding reaction is the downward force $\vec{F}_{A \text{ on } T}$ exerted by the apple on the table (Fig. 4.22c). So we have

$$\vec{F}_{A \text{ on } T} = -\vec{F}_{T \text{ on } A}$$

The two forces acting on the apple are $\vec{F}_{T \text{ on } A}$ and $\vec{F}_{E \text{ on } A}$. Are they an action-reaction pair? No, they aren't, despite being equal and opposite. They do not represent the mutual interaction of two bodies; they are two different forces acting on the *same* body. *The two forces in an action-reaction pair **never** act on the same body.* Here's another way to look at it. Suppose we suddenly yank the table out from under the apple (Fig. 4.22e). The two forces $\vec{F}_{A \text{ on } T}$ and $\vec{F}_{T \text{ on } A}$ then become zero, but $\vec{F}_{A \text{ on } E}$ and $\vec{F}_{E \text{ on } A}$ are still there (the gravitational interaction is still present). Since $\vec{F}_{T \text{ on } A}$ is now zero, it can't be the negative of $\vec{F}_{E \text{ on } A}$, and these two forces can't be an action-reaction pair.

4.22 The two forces in an action-reaction pair always act on different bodies.

| Conceptual Example 4.10 | **Applying Newton's third law: Objects in motion** |

A stonemason drags a marble block across a floor by pulling on a rope attached to the block (Fig. 4.23a). The block may or may not be in equilibrium. How are the various forces related? What are the action-reaction pairs?

SOLUTION

Figure 4.23b shows the horizontal forces acting on each body: the block (B), the rope (R), and the mason (M). We'll use subscripts on all the forces to help explain things. Vector $\vec{F}_{\text{M on R}}$ represents the force exerted *by* the mason *on* the rope. Its reaction is the equal and opposite force $\vec{F}_{\text{R on M}}$ exerted *by* the rope *on* the mason. Vector $\vec{F}_{\text{R on B}}$ represents the force exerted by the rope on the block. The reaction to it is the equal and opposite force $\vec{F}_{\text{B on R}}$ exerted by the block on the rope:

$$\vec{F}_{\text{R on M}} = -\vec{F}_{\text{M on R}} \quad \text{and} \quad \vec{F}_{\text{B on R}} = -\vec{F}_{\text{R on B}}$$

Be sure you understand that the forces $\vec{F}_{\text{M on R}}$ and $\vec{F}_{\text{B on R}}$ are *not* an action-reaction pair. Both of these forces act on the *same* body (the rope); an action and its reaction *must* always act on *different* bodies. Furthermore, the forces $\vec{F}_{\text{M on R}}$ and $\vec{F}_{\text{B on R}}$ are not necessarily equal in magnitude. Applying Newton's second law to the rope, we get

$$\sum \vec{F} = \vec{F}_{\text{M on R}} + \vec{F}_{\text{B on R}} = m_{\text{rope}}\vec{a}_{\text{rope}}$$

If the block and rope are accelerating (that is, speeding up or slowing down), the rope is not in equilibrium, and $\vec{F}_{\text{M on R}}$ must have a different magnitude than $\vec{F}_{\text{B on R}}$. By contrast, the action-reaction forces $\vec{F}_{\text{M on R}}$ and $\vec{F}_{\text{R on M}}$ are always equal in magnitude, as are

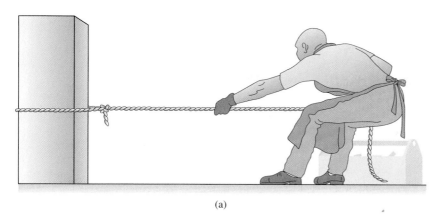

4.23 (a) A mason pulls on a rope attached to a block. (b) Separate diagrams showing the force of the rope on the block, the force of the rope on the mason, and the forces of the block and the mason on the rope. (c) If the rope is not accelerating or if its mass can be neglected, it can be considered to transmit an undiminished force from the mason to the block, and vice versa.

$\vec{F}_{\text{R on B}}$ and $\vec{F}_{\text{B on R}}$. Newton's third law holds whether or not the bodies are accelerating.

In the special case in which the rope is in equilibrium, the forces $\vec{F}_{\text{M on R}}$ and $\vec{F}_{\text{B on R}}$ are equal in magnitude. But this is an example of Newton's *first* law, not his *third*. Another way to look at this is that in equilibrium, $\vec{a}_{\text{rope}} = \mathbf{0}$ in the equation on the previous page. Then $\vec{F}_{\text{B on R}} = -\vec{F}_{\text{M on R}}$ because of Newton's first or second law.

This is also true if the rope is accelerating but has negligibly small mass compared to the block or the mason. In this case, $m_{\text{rope}} = 0$ in the equation on the previous page, so again $\vec{F}_{\text{B on R}} = -\vec{F}_{\text{M on R}}$. Since $\vec{F}_{\text{B on R}}$ *always* equals $-\vec{F}_{\text{R on B}}$ by New-

ton's third law (they are an action-reaction pair), in these same special cases, $\vec{F}_{\text{R on B}}$ also equals $\vec{F}_{\text{M on R}}$. In other words, in these cases the force of the rope on the block equals the force of the mason on the rope, and we can then think of the rope as "transmitting" to the block, without change, the force the person exerts on the rope (Fig. 4.23c). This is a useful point of view, but you have to remember that it is valid *only* when the rope has negligibly small mass or is in equilibrium.

If you feel as though you're drowning in subscripts at this point, take heart. Go over this discussion again, comparing the symbols with the vector diagrams, until you're sure you see what's going on.

| Conceptual Example 4.11 | **A Newton's third law paradox?** |

We saw in Conceptual Example 4.10 that the stonemason pulls as hard on the rope–block combination as that combination pulls back on him. Why, then, does the block move while the stonemason remains stationary?

SOLUTION

The way out of this seeming conundrum is to keep in mind the difference between Newton's *second* law and his *third* law. The only forces involved in Newton's second law are those that act *on* that body. The vector sum of these forces determines how the body accelerates (and whether it accelerates at all). By contrast, Newton's third law relates the forces that two *different* bodies exert on each other. The third law alone tells you nothing about the motion of either body.

The reason why the stonemason doesn't move is that the net force acting on him is zero. It is the vector sum of the upward normal force from the floor, his weight acting downward, the rope

pulling to the left, and the friction force of the floor pushing him to the right (Fig. 4.24). Because the stonemason has shoes with non-skid soles that don't slip on the floor, the friction force is strong enough to exactly balance the pull of the rope. If the floor were freshly waxed, so that there was little friction between the floor and the stonemason's shoes, he would indeed start sliding to the left.

Four forces also act on the block–rope combination: normal force, weight, friction, and the force of the stonemason pulling to the right (see Fig. 4.24). If the block is initially at rest, it begins to slide if the stonemason's force is *greater* than the friction force that the floor exerts on the block. (The marble block has a smooth underside, which helps to minimize friction.) Then the net force on the block is not zero, and the block accelerates to the right. Once the block is moving, the stonemason doesn't need to pull quite so hard; he need exert only enough force to exactly balance the friction force on the block. Then the net force on the moving block is zero, and the block continues to move at a constant velocity in accordance with Newton's first law.

The moral of this example is that when analyzing the motion of a body, remember that only forces acting *on* the body determine its motion. From this perspective, Newton's third law is merely a tool that can help you determine what those forces are.

4.24 Force diagrams for the stonemason (right) and the block–rope combination (left) in Conceptual Example 4.10. Both the block and the stonemason are acted on by normal forces exerted by the floor (n_B and n_M, respectively), weight forces (w_B and w_M, respectively), and friction forces exerted by the floor (f_B and f_M, respectively). (We ignore the relatively small mass of the rope.) The force of the rope on the stonemason has the same magnitude as the force of the stonemason on the rope ($F_{R \, on \, M} = F_{M \, on \, R}$).

A body, such as the rope in Fig. 4.23, that has pulling forces applied at its ends is said to be in **tension.** The tension at any point is the magnitude of force acting at that point. In Fig. 4.23b the tension at the right-hand end of the rope is the magnitude of $\vec{F}_{M \, on \, R}$ (or of $\vec{F}_{R \, on \, M}$), and the tension at the left-hand end equals the magnitude of $\vec{F}_{B \, on \, R}$ (or of $\vec{F}_{R \, on \, B}$). If the rope is in equilibrium and if no forces act except at its ends, the tension is the same at both ends and throughout the rope. Thus, if the magnitudes of $\vec{F}_{B \, on \, R}$ and $\vec{F}_{M \, on \, R}$ are 50 N each, the tension in the rope is 50 N (*not* 100 N). The *total* force vector $\vec{F}_{B \, on \, R} + \vec{F}_{M \, on \, R}$ acting on the rope in this case is zero!

CAUTION We emphasize once more a fundamental truth: The two forces in an action-reaction pair *never* act on the same body. Remembering this simple fact can often help you avoid confusion about action-reaction pairs and Newton's third law.

Test Your Understanding

You are driving your car on a country road when a mosquito splatters itself on the windshield. Which has the greater magnitude, the force that the car exerted on the mosquito or the force that the mosquito exerted on the car? Or are the magnitudes the same? If they are different, how can you reconcile this fact with Newton's third law? If they are equal, why is the mosquito splattered while the car is undamaged?

4.6 | Free-Body Diagrams

Newton's three laws of motion contain all the basic principles we need to solve a wide variety of problems in mechanics. These laws are very simple in form, but the process of applying them to specific situations can pose real challenges. In this brief section we'll point out some key ideas and techniques to use in any problems involving Newton's laws. You'll learn others in Chapter 5, which also extends the use of Newton's laws to cover more complex situations.

Newton's first and second laws apply to a specific body. Whenever you use Newton's first law, $\sum \vec{F} = \mathbf{0}$, for an equilibrium situation or Newton's second law, $\sum \vec{F} = m\vec{a}$, for a non-equilibrium situation, you must decide at the beginning to which body you are referring. This decision may sound trivial, but it isn't.

Only forces acting on the body matter. The sum $\sum \vec{F}$ includes all the forces that act *on* the body in question. Hence, once you've chosen the body to analyze, you have to identify all the forces acting on it. Don't get confused between the forces acting on a body and the forces exerted by that body on some other body. For example, to analyze a person walking, you would include in $\sum \vec{F}$ the force that the ground exerts on the person as he walks, but *not* the force that the person exerts on the ground (Fig. 4.25). These forces form an action-reaction pair and are related by Newton's third law, but only the member of the pair that acts on the body you're working with goes into $\sum \vec{F}$.

Free-body diagrams are essential to help identify the relevant forces. A **free-body diagram** is a diagram showing the chosen body by itself, "free" of its surroundings, with vectors drawn to show the magnitudes and directions of all the forces applied to the body by the various other bodies that interact with it. We have already shown some free-body diagrams in Figs. 4.15, 4.16, and 4.22. Be careful to include all the forces acting *on* the body, but be equally careful *not* to include any forces that the body exerts on any other body. In particular, the two forces in an action-reaction pair must *never* appear in the same free-body diagram because they never act on the same body. Furthermore, forces that a body exerts on itself are never included, since these can't affect the body's motion.

> **CAUTION** When you have a complete free-body diagram, you *must* be able to answer for each force the question, "What other body is applying this force?" If you can't answer that question, you may be dealing with a nonexistent force. Be especially on your guard to avoid nonexistent forces such as "the force of acceleration" or "the $m\vec{a}$ force," discussed in Section 4.3.

When a problem involves more than one body, you have to take the problem apart and draw a separate free-body diagram for each body. For example, Fig. 4.23b shows a separate free-body diagram for the rope in the case in which the rope is considered massless (so that no gravitational force acts on it). This figure also shows diagrams for the block and the mason, but these are *not* complete free-body diagrams because they don't show all the forces acting on each body. Which forces are missing in these diagrams? (Compare the complete free-body diagrams in Fig. 4.24.)

On the next page we present some real-life situations and the corresponding free-body diagrams. Note that while the situations depicted in the following figures occur in very different surroundings, the free-body diagrams are remarkably similar.

4.25 The simple act of walking depends crucially on Newton's third law. To start moving forward, you push backward on the ground with your foot. As a reaction, the ground pushes forward on your foot (and hence on your body as a whole) with a force of the same magnitude. This *external* force provided by the ground is what accelerates your body forward.

2.1.1 Force Magnitudes

A sprinter gains a large forward acceleration at the start of a race by kicking back hard against the angled starting blocks. The blocks exert an equally large reaction force $\vec{n}$ on the sprinter. This force has a large forward component n_x that springs her into motion, as well as a smaller upward component n_y. If this upward component has the same magnitude as the sprinter's weight w, the net vertical force acting on her is zero and she does not accelerate vertically.

Immersed in water, a person's body experiences an upward force $\vec{B}$ due to buoyancy. This is balanced by the downward force of the diver's weight $\vec{w}$. In this situation the diver's motion depends on the force with which the water pushes on her, due either to the water currents or to reaction forces to the swimmer's arm and leg movements.

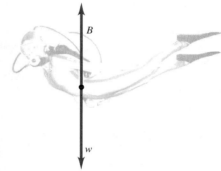

A basketball player jumps by pushing with his feet against the floor. The forces acting on him are the reaction force with which the floor pushes back up and his weight $\vec{w}$. Once the player is in the air, the only force acting on him is his weight; his acceleration is downward, even as he rises up. His stationary opponent is acted on by his own weight and by the upward normal force $\vec{n}$ exerted on him by the floor.

SUMMARY

Force is a quantitative measure of the interaction between two bodies. It is a vector quantity. When several forces act on a body, the effect on its motion is the same as when a single force, equal to the vector sum (resultant) of the forces, acts on the body. (See Example 4.1)

$$\vec{R} = \vec{F}_1 + \vec{F}_2 + \vec{F}_3 + \cdots = \sum \vec{F} \quad (4.1)$$

Newton's first law states that when the vector sum of all forces acting on a body (the *net force*) is zero, the body is in equilibrium. If the body is initially at rest, it remains at rest; if it is initially in motion, it continues to move with constant velocity. This law is valid only in inertial frames of reference. (See Examples 4.2 and 4.3)

$$\sum \vec{F} = 0 \quad (4.3)$$

The inertial properties of a body are characterized by its *mass*. The acceleration of a body under the action of a given set of forces is directly proportional to the vector sum of the forces (the *net force*) and inversely proportional to the mass of the body. This relationship is Newton's second law. Like Newton's first law, this law is valid only in inertial frames of reference. (See Examples 4.4 and 4.5)

$$\sum \vec{F} = m\vec{a} \quad (4.7)$$

$$\sum F_x = ma_x$$
$$\sum F_y = ma_y$$
$$\sum F_z = ma_z \quad (4.8)$$

The unit of force is defined in terms of the units of mass and acceleration. In SI units, the unit of force is the newton (N), equal to $1 \text{ kg} \cdot \text{m/s}^2$.

The weight of a body is the gravitational force exerted on it by the earth (or whatever other body exerts the gravitational force). Weight is a force and is therefore a vector quantity. The magnitude of the weight of a body at any specific location is equal to the product of its mass m and the magnitude of the acceleration due to gravity g at that location. While the weight of a body depends on its location, the mass is independent of location. (See Examples 4.6 and 4.7)

$$w = mg \quad (4.9)$$

Newton's third law states that "action equals reaction." That is, when two bodies interact, they exert forces on each other that at each instant are equal in magnitude and opposite in direction. Each force in an action–reaction pair acts on only one of the two bodies; the action and reaction forces never act on the same body. (See Examples 4.8 through 4.11)

$$\vec{F}_{A \text{ on } B} = -\vec{F}_{B \text{ on } A} \qquad (4.11)$$

Key Terms

Your Notes

Answer to Chapter Opening Question

What determines the motion of the man and sled together are the forces that *other* bodies exert on them. In particular, the man pushes backward on the ground with his feet, and the ground pushes forward on him with a force of equal magnitude (in accordance with Newton's third law). The ground also exerts a backward friction force on the sled. Thus, to pull the sled forward, the force that the man exerts on the ground must be at least as great as the friction force on the sled.

Answers to Test Your Understanding Questions

Section 4.1 We demand that the net force be zero, so

$$F_{1x} + F_{2x} + F_{3x} = 0 \quad \text{and} \quad F_{1y} + F_{2y} + F_{3y} = 0$$

Hence

$$F_{3x} = -(F_{1x} + F_{2x}) = -[(-150\text{ N}) + 50\text{ N}] = 100\text{ N}$$

$$F_{3y} = -(F_{1y} + F_{2y}) = -(200\text{ N} + 0\text{ N}) = -200\text{ N}$$

The magnitude is

$$F_3 = \sqrt{F_{3x}^2 + F_{3y}^2} = \sqrt{(100\text{ N})^2 + (-200\text{ N})^2} = 224\text{ N}$$

and the direction is

$$\theta = \arctan\left(\frac{F_{3y}}{F_{3x}}\right) = \arctan\left(\frac{-200\text{ N}}{100\text{ N}}\right) = \arctan(-2) = -63°$$

Section 4.2 There is zero net force in situations (i), (ii), and (iv). In situation (iii), the hawk is moving in a circle; hence it is accelerating and is not in equilibrium. In situation (iv), the box remains stationary (as seen in the inertial reference frame of the ground) as the truck accelerates forward, like the skater in Fig. 4.8a.

Section 4.3 The acceleration is equal to the net force divided by the mass. Hence the magnitude of the acceleration is $a = (128\text{ N})/(8.0\text{ kg}) = 16\text{ m/s}^2$. The acceleration is in the same direction as the net force, $\theta = 141°$. From Eq. (2.12), after 0.50 s the belt will have moved a distance $\frac{1}{2}at^2 = (1/2)(16\text{ m/s}^2)(0.50\text{ s})^2 = 2.0\text{ m}$.

Section 4.4 It would take twice the effort for the astronaut to walk around because her weight on the planet would be twice as much as on the earth. But it would be just as easy to catch the ball. The ball's *mass* remains the same, so the horizontal force the astronaut would have to exert to bring it to a stop (i.e., to give it the same acceleration) would be the same as on earth.

Section 4.5 By Newton's third law, the two forces have equal magnitudes. Because the car has much greater mass than the mosquito, it undergoes only a tiny, imperceptible acceleration in response to the force of the impact. By contrast, the mosquito, with its miniscule mass, undergoes a catastrophically large acceleration.

Discussion Questions

Q4.1 Can a body be in equilibrium when only one force acts on it? Explain.

Q4.2 A ball thrown straight up has zero velocity at its highest point. Is the ball in equilibrium at this point? Why or why not?

Q4.3 A helium balloon hovers in midair, neither ascending nor descending. Is it in equilibrium? What forces act on it?

Q4.4 When you fly in an airplane at night in smooth air, there is no sensation of motion, even though the plane may be moving at 800 km/h (500 mi/h). Why is this?

Q4.5 If the two ends of a rope in equilibrium are pulled with forces of equal magnitude and opposite direction, why is the total tension in the rope not zero?

Q4.6 You tie a brick to the end of a rope and whirl the brick around you in a horizontal circle. Describe the path of the brick after you suddenly let go of the rope.

Q4.7 When a car stops suddenly, the passengers tend to move forward relative to their seats. Why? When a car makes a sharp turn, the passengers tend to slide to one side of the car. Why?

Q4.8 Some people say that the "force of inertia" (or "force of momentum") throws the passengers forward when a car brakes sharply. What is wrong with this explanation?

Q4.9 A passenger in a moving bus with no windows notices that a ball that has been at rest in the aisle suddenly starts to move toward the rear of the bus. Think of two different possible explanations, and devise a way to decide which is correct.

Q4.10 Suppose you chose the fundamental SI units to be force, length, and time instead of mass, length, and time. What would be the units of mass in terms of those fundamental units?

Q4.11 Inertia is *not* a force that keeps things in place or that keeps them moving. How do we know this?

Q4.12 Why is the earth only approximately an inertial reference frame?

Q4.13 Does Newton's second law hold true for an observer in a van as it speeds up, slows down, or rounds a corner? Explain.

Q4.14 Some students refer to the quantity $m\vec{a}$ as "the force of acceleration." Is it correct to refer to this quantity as a force? If so, what exerts this force? If not, what is a better description of this quantity?

Q4.15 The acceleration of a falling body is measured in an elevator traveling upward at a constant speed of 9.8 m/s. What result is obtained?

Q4.16 You can play catch with a softball in a bus moving with constant speed on a straight road, just as though the bus were at rest. Is this still possible when the bus is making a turn at constant speed on a level road? Why or why not?

Q4.17 Is the belt of Example 4.1 in equilibrium? Explain.

Q4.18 The head of a hammer begins to come loose from its wooden handle. How should you strike the handle on a concrete sidewalk to reset the head? Why does this work?

Q4.19 Why can it hurt your foot more to kick a big rock than a small pebble? *Must* the big rock hurt more? Explain.

Q4.20 "It's not the fall that hurts you; it's the sudden stop at the bottom." Translate this saying into the language of Newton's laws of motion.

Q4.21 A person can dive into water from a height of 10 m without injury, but a person who jumps off the roof of a 10-m-tall building and lands on a concrete street is likely to be seriously injured. Why is there a difference?

Q4.22 Why are cars designed to crumple up in front and back for safety? Why not for side collisions and rollovers?

Q4.23 When a bullet is fired from a gun, what is the origin of the force that accelerates the bullet?

Q4.24 When a string barely strong enough lifts a heavy weight, it can lift the weight by a steady pull; but if you jerk the string, it will break. Explain in terms of Newton's laws of motion.

Q4.25 A large crate is suspended from the end of a vertical rope. Is the tension in the rope larger when the crate is at rest or when it is moving upward at constant speed? If the crate is traveling upward, is the tension in the rope greatest when the crate is speeding up or when it is slowing down? In each case explain in terms of Newton's laws of motion.

Q4.26 Which feels a greater pull due to the earth's gravity, a 10-kg stone or a 20-kg stone? If you drop them, why does the 20-kg stone not fall with twice the acceleration of the 10-kg stone? Explain your reasoning.

Q4.27 Why is it incorrect to say that 1.0 kg *equals* 2.2 lb?

Q4.28 A horse is hitched to a wagon. Since the wagon pulls back on the horse just as hard as the horse pulls on the wagon, why does the wagon not remain in equilibrium, no matter how hard the horse pulls?

Q4.29 A 450-N girl from the south slaps a 800-N boy from the north. Her fingers exert a force of 30 N to the west on his cheek. There may be other reactions, but by Newton's third law, what is the reaction *force* to the slap?

Q4.30 A large truck and a small compact car have a head-on collision. During the collision the truck exerts a force $\vec{F}_{\text{T on C}}$ on the car, and the car exerts a force $\vec{F}_{\text{C on T}}$ on the truck. Which force has the larger magnitude, or are they the same? Does your answer depend on how fast each vehicle was moving before the collision? Why or why not?

Q4.31 If you ask most people what force makes a car accelerate forward, they will say "the force of the engine." But what force is *directly* responsible for making the car accelerate?

Q4.32 A small compact car is pushing a large van that has broken down, and they travel along the road with equal velocities and accelerations. While the car is speeding up, is the force it exerts on the van larger than, smaller than, or the same magnitude as the force the van exerts on it? Which object, the car or the van, has the largest net force on it, or are the net forces the same? Explain.

Q4.33 Consider a tug-of-war between two people who pull in opposite directions on the ends of a rope. By Newton's third law, the force that *A* exerts on *B* is just as great as the force that *B* exerts on *A*. So what determines who wins? (*Hint:* Draw a free-body diagram showing all the forces that act on each person.)

Q4.34 On the moon, $g = 1.62 \text{ m/s}^2$. If a 2-kg brick drops on your foot from a height of 2 m, will this hurt more, or less, or the same if it happens on the moon instead of on the earth? Explain. If a 2-kg brick is thrown and hits you when it is moving horizontally at 6 m/s, will this hurt more, less, or the same if it happens on the moon instead of on the earth? Explain. (On the moon, assume that you are inside a pressurized structure, so you are not wearing a spacesuit.)

Q4.35 A manual for student pilots contains the following passage: "When an airplane flies at a steady altitude, neither climbing nor descending, the upward lift force from the wings equals the air-plane's weight. When the airplane is climbing at a steady rate, the upward lift is greater than the weight; when the airplane is descending at a steady rate, the upward lift is less than the weight." Are these statements correct? Explain.

Q4.36 If your hands are wet and no towel is handy, you can remove some of the excess water by shaking them. Why does this get rid of the water?

Q4.37 If you are squatting down (such as when you are examining the books on the bottom shelf in a library or bookstore) and suddenly get up, you can temporarily feel light-headed. What do Newton's laws of motion have to say about why this happens?

Q4.38 When a car is hit from behind, the passengers can receive a whiplash. Use Newton's laws of motion to explain what causes this to occur.

Q4.39 In a head-on auto collision, passengers not wearing seat belts can be thrown through the windshield. Use Newton's laws of motion to explain why this happens.

Q4.40 In a head-on collision between a compact 1000-kg car and a large 2500-kg car, which one experiences the greater force? Explain. Which one experiences the greater acceleration? Explain why. Now explain why passengers in the smaller car are more likely to be injured than those in the large car, even if the bodies of both cars are equally strong.

Q4.41 Suppose you are in a rocket with no windows, traveling in deep space far from any other objects. Without looking outside the rocket or making any contact with the outside world, explain how you could determine if the rocket is i) moving forward at a constant 80% of the speed of light; ii) accelerating in the forward direction.

Exercises

Section 4.1 Force and Interactions

4.1 Two forces have the same magnitude *F*. What is the angle between the two vectors if their sum has a magnitude of a) 2*F*? b) $\sqrt{2}F$? c) zero? Sketch the three vectors in each case.

4.2 Instead of using the *x*- and *y*-axes of Fig. 4.5 to analyze the situation of Example 4.1, use axes rotated 37.0° counterclockwise, so the *y*-axis is parallel to the 250-N force. a) For these axes find the *x*- and *y*-components of the net force on the belt. b) From the components computed in part (a) find the magnitude and direction of the net force. Compare your results to Example 4.1.

4.3 A warehouse worker pushes a crate along the floor, as shown in Fig. 4.1b, by a force of 10 N that points downward at an angle of 45° below the horizontal. Find the horizontal and vertical components of the force.

4.4 A man is dragging a trunk up the loading ramp of a mover's truck. The ramp has a slope angle of 20.0°, and the man pulls upward with a force $\vec{F}$ whose direction makes an angle of 30.0° with the ramp (Fig. 4.26). a) How large a force $\vec{F}$ is necessary for the component F_x parallel to the ramp to be 60.0

Figure 4.26 Exercise 4.4.

N? b) How large will the component F_y perpendicular to the ramp then be?

4.5 Two dogs pull horizontally on ropes attached to a post; the angle between the ropes is 60.0°. If dog A exerts a force of 270 N and dog B exerts a force of 300 N, find the magnitude of the resultant force and the angle it makes with dog A's rope.

4.6 Two forces, $\vec{F}_1$ and $\vec{F}_2$, act at a point. The magnitude of $\vec{F}_1$ is 9.00 N, and its direction is 60.0° above the x-axis in the second quadrant. The magnitude of $\vec{F}_2$ is 6.00 N, and its direction is 53.1° below the x-axis in the third quadrant. a) What are the x- and y-components of the resultant force? b) What is the magnitude of the resultant force?

Section 4.3 Newton's Second Law

4.7 If a net horizontal force of 132 N is applied to a person with mass 60 kg who is resting on the edge of a swimming pool, what horizontal acceleration is produced?

4.8 What magnitude of net force is required to give a 135-kg refrigerator an acceleration of magnitude 1.40 m/s²?

4.9 A box rests on a frozen pond, which serves as a frictionless horizontal surface. If a fisherman applies a horizontal force with magnitude 48.0 N to the box and produces an acceleration of magnitude 3.00 m/s², what is the mass of the box?

4.10 A dockworker applies a constant horizontal force of 80.0 N to a block of ice on a smooth horizontal floor. The frictional force is negligible. The block starts from rest and moves 11.0 m in 5.00 s. a) What is the mass of the block of ice? b) If the worker stops pushing at the end of 5.00 s, how far does the block move in the next 5.00 s?

4.11 A hockey puck with mass 0.160 kg is at rest at the origin $(x = 0)$ on the horizontal, frictionless surface of the rink. At time $t = 0$ a player applies a force of 0.250 N to the puck, parallel to the x-axis; he continues to apply this force until $t = 2.00$ s. a) What are the position and speed of the puck at $t = 2.00$ s? b) If the same force is again applied at $t = 5.00$ s, what are the position and speed of the puck at $t = 7.00$ s?

4.12 A crate with mass 32.5 kg initially at rest on a warehouse floor is acted on by a net horizontal force of 140 N. a) What acceleration is produced? b) How far does the crate travel in 10.0 s? c) What is its speed at the end of 10.0 s?

4.13 A hockey puck moves from point A to point B at constant velocity while under the influence of several forces. a) What can you say about the forces? b) Graph the puck's path from A to B. c) On the graph, continue the path to point C if a new constant force is added to the puck at point B, where the new force is perpendicular to the puck's velocity at point B. d) Continue the path to point D if at point C the constant force added at B is replaced by one of constant magnitude but in a direction always perpendicular to the path of the puck.

4.14 An electron (mass = 9.11×10^{-31} kg) leaves one end of a TV picture tube with zero initial speed and travels in a straight line to the accelerating grid, which is 1.80 cm away. It reaches the grid with a speed of 3.00×10^6 m/s. If the accelerating force is constant, compute a) the acceleration; b) the time to reach the grid; c) the net force, in newtons. (You can ignore the gravitational force on the electron.)

Section 4.4 Mass and Weight

4.15 Superman throws a 2400-N boulder at an adversary. What horizontal force must Superman apply to the boulder to give it a horizontal acceleration of 12.0 m/s²?

4.16 A bowling ball weighs 71.2 N (16.0 lb). The bowler applies a horizontal force of 160 N (36.0 lb) to the ball. What is the magnitude of the horizontal acceleration of the ball?

4.17 At the surface of Jupiter's moon Io, the acceleration due to gravity is $g = 1.81$ m/s². A watermelon weighs 44.0 N at the surface of the earth. a) What is its mass on the earth's surface? b) What are its mass and weight on the surface of Io?

4.18 a) What is the mass of a book that weighs 3.20 N at a point where $g = 9.80$ m/s²? b) At the same location, what is the weight of a dog whose mass is 14.0 kg?

Section 4.5 Newton's Third Law

4.19 World-class sprinters can accelerate out of the starting blocks with an acceleration that is nearly horizontal and has magnitude 15 m/s². How much horizontal force must a 55-kg sprinter exert on the starting blocks during a start to produce this acceleration? Which body exerts the force that propels the sprinter: the blocks or the sprinter herself?

4.20 Imagine that you are holding a book weighing 4 N at rest on the palm of your hand. Complete the following sentences: a) A downward force of magnitude 4 N is exerted on the book by _____. b) An upward force of magnitude _____ is exerted on _____ by the hand. c) Is the upward force in part (b) the reaction to the downward force in part (a)? d) The reaction to the force in part (a) is a force of magnitude _____, exerted on _____ by _____. Its direction is _____. e) The reaction to the force in part (b) is a force of magnitude _____, exerted on _____ by _____. Its direction is _____. f) The forces in parts (a) and (b) are equal and opposite because of Newton's _____ law. g) The forces in parts (b) and (e) are equal and opposite because of Newton's _____ law. Now suppose that you exert an upward force of magnitude 5 N on the book. h) Does the book remain in equilibrium? i) Is the force exerted on the book by your hand equal and opposite to the force exerted on the book by the earth? j) Is the force exerted on the book by the earth equal and opposite to the force exerted on the earth by the book? k) Is the force exerted on the book by your hand equal and opposite to the force exerted on your hand by the book? Finally, suppose that you snatch your hand away while the book is moving upward. l) How many forces then act on the book? m) Is the book in equilibrium?

4.21 A bottle is given a push along a tabletop and slides off the edge of the table. Do *not* ignore air resistance. a) What forces are exerted on the bottle while it is falling from the table to the floor? b) What is the reaction to each force; that is, on which body and by which body is the reaction exerted?

4.22 The upward normal force exerted by the floor is 620 N on an elevator passenger who weighs 650 N. What are the reaction forces to these two forces? Is the passenger accelerating? If so, what are the magnitude and direction of the acceleration?

4.23 A student with mass 45 kg jumps off a high diving board. Using 6.0×10^{24} kg for the mass of the earth, what is the acceleration of the earth toward her as she accelerates toward the earth with an acceleration of 9.8 m/s²? Assume that the net force on the earth is the force of gravity she exerts on it.

Section 4.6 Free-Body Diagrams

4.24 Two crates, A and B, sit at rest side-by-side on a frictionless horizontal surface, as shown in Fig. 4.27. The crates have masses m_A and m_B. A horizontal force $\vec{F}$ is applied to crate A and the two crates move off to the right. a) Draw clearly labeled free-body diagrams for crate A

Figure 4.27 Exercise 4.24.

and for crate B. Indicate which pairs of forces, if any, are third-law action-reaction pairs. b) If the magnitude of force $\vec{F}$ is less than the total weight of the two crates, will it cause the crates to move? Explain.

4.25 A ball is hanging from a long string that is tied to the ceiling of a train car that is traveling eastward on horizontal tracks and speeding up. An observer inside the train car sees the ball hang motionless. Draw a clearly labeled free-body diagram for the ball. Is the net force on the ball zero? Explain.

4.26 A large box containing your new computer sits on the bed of your pickup truck. You are stopped at a red light. The light turns green and you stomp on the gas and the truck accelerates. To your horror, the box starts to slide toward the back of the truck. Draw a clearly labeled free-body diagram for the truck and for the box. Indicate pairs of forces, if any, that are third-law action-reaction pairs.

4.27 A chair of mass 12.0 kg is sitting on the horizontal floor; the floor is not frictionless. You push on the chair with a force $F = 40.0$ N that is directed at an angle of 37.0° below the horizontal and the chair slides along the floor. a) Draw a clearly labeled free-body diagram for the chair. b) Use your diagram and Newton's laws to calculate the normal force that the floor exerts on the chair.

4.28 A skier of mass 65.0 kg is pulled up a snow-covered slope at constant speed by a tow rope that is parallel to the ground. The ground slopes upward at a constant angle of 26.0° above the horizontal and you can ignore friction. a) Draw a clearly labeled free-body diagram for the skier. b) Calculate the tension in the tow rope.

4.29 Your nephew Frank is riding his tricycle. The tricycle is connected by a light horizontal rope to a wagon in which his dog Spot is sitting. Treat the tricycle and Frank as one object and Spot and the wagon as one object. Draw a clearly labeled free-body diagram for each object. Indicate which pairs of forces, if any, form third-law action-reaction pairs.

Problems

4.30 A .22 rifle bullet, traveling at 350 m/s, strikes a block of soft wood, which it penetrates to a depth of 0.130 m. The block of wood is clamped in place and doesn't move. The mass of the bullet is 1.80 g. Assume a constant retarding force. a) How much time is required for the bullet to stop? b) What force, in newtons, does the wood exert on the bullet?

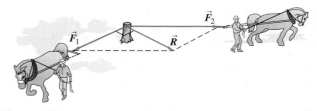

Figure 4.28 Problem 4.31.

4.31 Two horses pull horizontally on ropes attached to a stump. The two forces $\vec{F}_1$ and $\vec{F}_2$ that they apply to the stump are such that the net (resultant) force $\vec{R}$ has a magnitude equal to that of $\vec{F}_1$ and makes an angle of 90° with $\vec{F}_1$ (Fig. 4.28). Let $F_1 = 1300$ N and $R = 1300$ N also. Find the magnitude of $\vec{F}_2$ and its direction (relative to $\vec{F}_1$).

4.32 You have just landed on Planet X. You take out a 100-g ball, release it from rest from a height of 10.0 m and measure that it takes 2.2 s to reach the ground. You can ignore any force on the ball from the atmosphere of the planet. How much does the 100-g ball weigh on the surface of Planet X?

4.33 Two adults and a child want to push a wheeled cart in the direction marked x in Fig. 4.29. The two adults push with horizontal forces $\vec{F}_1$ and $\vec{F}_2$ as shown in the figure. a) Find the magnitude and direction of the *smallest* force that the child should exert. You can ignore the effects of friction. b) If the child

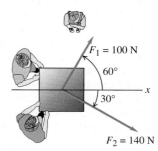

Figure 4.29 Problem 4.33.

exerts the minimum force found in part (a), the cart accelerates at 2.0 m/s² in the +x-direction. What is the weight of the cart?

4.34 An oil tanker's engines have broken down, and the wind is blowing the tanker straight toward a reef at a constant speed of 1.5 m/s (Fig. 4.30). When the tanker is 500 m from the reef, the wind dies down just as the engineer gets the engines going again. The rudder is stuck, so the only choice is to try to accelerate straight backward away from the reef. The mass of the tanker and cargo is 3.6×10^7 kg, and the engines produce a net horizontal force of 8.0×10^4 N on the tanker. Will the ship hit the reef? If it does, will the oil be safe? The hull can withstand an impact at a speed of 0.2 m/s or less. You can ignore the retarding force of the water on the tanker's hull.

Figure 4.30 Problem 4.34.

4.35 A Standing Vertical Jump. Basketball player Darrell Griffith is on record as attaining a standing vertical jump of 1.2 m (4 ft). (This means that he moved upward by 1.2 m after his feet left the floor.) Griffith weighed 890 N (200 lb). a) What is his speed as he leaves the floor? b) If the time of the part of the jump before his feet left the floor was 0.300 s, what was his average acceleration (magnitude and direction) while he was pushing against the floor? c) Draw his free-body diagram (see Section 4.6). In terms of the forces on the diagram, what is the net force on him? Use Newton's laws and the results of part (b) to calculate the average force he applied to the ground.

4.36 An advertisement claims that a particular automobile can "stop on a dime." What net force would actually be necessary to stop a 850-kg automobile traveling initially at 45.0 km/h in a distance equal to the diameter of a dime, which is 1.8 cm?

4.37 A 4.80-kg bucket of water is accelerated upward by a cord of negligible mass whose breaking strength is 75.0 N. a) Draw the free-body force diagram for the bucket. In terms of the forces on your diagram, what is the net force on the bucket? b) Apply Newton's second law to the bucket and find the maximum upward acceleration that can be given to the bucket without breaking the cord.

4.38 A parachutist relies on air resistance (mainly on her parachute) to decrease her downward velocity. She and her parachute have a mass of 55.0 kg and air resistance exerts a total upward force of 620 N on her and her parachute. a) What is the weight of the parachutist? b) Draw a free-body diagram for the parachutist (see Section 4.6). Use that diagram to calculate the net force on the parachutist. Is the net force upward or downward? c) What is the acceleration (magnitude and direction) of the parachutist?

4.39 Two crates, one with mass 4.00 kg and the other with mass 6.00 kg, sit on the frictionless surface of a frozen pond, connected by a light rope (Fig. 4.31). A woman wearing golf shoes (so she can get traction on the ice) pulls horizontally on the 6.00-kg crate with a force F that gives the crate an acceleration of 2.50 m/s^2. a) What is the acceleration of the 4.00-kg crate? b) Draw a free-body diagram for the 4.00-kg crate. Use that diagram and Newton's second law to find the tension T in the rope that connects the two crates. c) Draw a free-body diagram for the 6.00-kg crate. What is the direction of the net force on the 6.00-kg crate? Which is larger in magnitude, force T or force F? d) Use part (c) and Newton's second law to calculate the magnitude of the force F.

Figure 4.31 Problem 4.39.

4.40 An astronaut is tethered by a strong cable to a spacecraft. The astronaut and her spacesuit have a total mass of 105 kg, while the mass of the cable is negligible. The mass of the spacecraft is 9.05×10^4 kg. The spacecraft is far from any large astronomical bodies, so we can ignore the gravitational forces on it and the astronaut. We also assume that both the spacecraft and the astronaut are initially at rest in an inertial reference frame. The astronaut then pulls on the cable with a force of 80.0 N. a) What force does the cable exert on the astronaut? b) Since $\Sigma \vec{F} = m\vec{a}$, how can a "massless" ($m = 0$) cable exert a force? c) What is the astronaut's accel-

eration? d) What force does the cable exert on the spacecraft? e) What is the acceleration of the spacecraft?

4.41 To study damage to aircraft that collide with large birds, you design a test gun that will accelerate chicken-sized objects so that their displacement along the gun barrel is given by $x = (9.0 \times 10^3 \text{ m/s}^2)t^2 - (8.0 \times 10^4 \text{ m/s}^3)t^3$. The object leaves the end of the barrel at $t = 0.025$ s. a) How long must the gun barrel be? b) What will be the speed of the objects as they leave the end of the barrel? c) What net force must be exerted on a 1.50-kg object at i) $t = 0$? ii) $t = 0.025$ s?

4.42 A spacecraft descends vertically near the surface of Planet X. An upward thrust of 25.0 kN from its engines slows it down at a rate of 1.20 m/s^2, but it speeds up at a rate of 0.80 m/s^2 with an upward thrust of 10.0 kN. a) In each case, what is the direction of the acceleration of the spacecraft? b) Draw a free-body diagram for the spacecraft. In each case, speeding up or slowing down, what is the direction of the net force on the spacecraft? c) Apply Newton's second law to each case, slowing down or speeding up, and use this to find the spacecraft's weight near the surface of Planet X.

4.43 A train (an engine plus four cars) running on level tracks is speeding up with an acceleration of magnitude $|\vec{a}|$. If each car has a mass m and if each car has negligible frictional forces acting on it, what is a) the force of the engine on the first car? b) the force of the first car on the second car? c) the force of the second car on the third car? d) the force of the fourth car on the third car? e) What would these same four forces be if the train were slowing down with an acceleration of magnitude $|\vec{a}|$? Your answer to each question must include a clearly labeled free-body diagram. In each case indicate what body you are considering.

4.44 A gymnast of mass m climbs a vertical rope attached to the ceiling. You can ignore the weight of the rope. Draw a free-body diagram for the gymnast. Calculate the tension in the rope if the gymnast a) climbs at a constant rate; b) hangs motionless on the rope; c) accelerates up the rope with an acceleration of magnitude $|\vec{a}|$; d) slides down the rope with a downward acceleration of magnitude $|\vec{a}|$.

4.45 A loaded elevator with very worn cables has a total mass of 2200 kg, and the cables can withstand a maximum tension of 28,000 N. a) Draw the free-body force diagram for the elevator. In terms of the forces on your diagram, what is the net force on the elevator? Apply Newton's second law to the elevator and find the maximum upward acceleration for the elevator if the cables are not to break? b) What would be the answer to part (a) if the elevator were on the moon, where $g = 1.62$ m/s^2?

4.46 Jumping to the Ground. A 75.0-kg man steps off a platform 3.10 m above the ground. He keeps his legs straight as he falls, but at the moment his feet touch the ground his knees begin to bend, and, treated as a particle, he moves an additional 0.60 m before coming to rest. a) What is his speed at the instant his feet touch the ground? b) Treating him as a particle, what is his acceleration (magnitude and direction) as he slows down, if the acceleration is assumed to be constant? c) Draw his free-body diagram (see Section 4.6). In terms of the forces on the diagram, what is the net force on him? Use Newton's laws and the results of part (b) to calculate the average force his feet exert on the ground while he slows down. Express this force in newtons and also as a multiple of his weight.

4.47 A 4.9-N hammer head is stopped from an initial downward velocity of 3.2 m/s in a distance of 0.45 cm by a nail in a pine board. In addition to its weight, there is a 15-N downward force on the hammer head applied by the person using the hammer. Assume that the acceleration of the hammer head is constant while it is in contact with the nail and moving downward. a) Draw a free-body diagram for the hammer head. Identify the reaction force to each force in the diagram. b) Calculate the downward force $\vec{F}$ exerted by the hammer head on the nail while the hammer head is in contact with the nail and moving downward. c) Suppose the nail is in hardwood and the distance the hammer head travels in coming to rest is only 0.12 cm. The downward forces on the hammer head are the same as on part (b). What then is the force $\vec{F}$ exerted by the hammer head on the nail while the hammer head is in contact with the nail and moving downward?

4.48 A uniform cable of weight w hangs vertically downward, supported by an upward force of magnitude w at its top end. What is the tension in the cable a) at its top end? b) at its bottom end? c) at its middle? Your answer to each part must include a free-body diagram. (*Hint:* For each question choose the body to analyze to be a section of the cable or a point along the cable.) d) Graph the tension in the rope versus the distance from its top end.

4.49 The two blocks in Fig. 4.32 are connected by a heavy uniform rope with a mass of 4.00 kg. An upward force of 200 N is applied as shown. a) Draw three free-body diagrams, one for the 6.00-kg block, one for the 4.00-kg rope, and another one for the 5.00-kg block. For each force, indicate what body exerts that force. b) What is the acceleration of the system? c) What is the tension at the top of the heavy rope? d) What is the tension at the midpoint of the rope?

$F = 200$ N

6.00 kg

4.00 kg

5.00 kg

Figure 4.32 Problem 4.49.

4.50 An athlete whose mass is 90.0 kg is performing weight-lifting exercises. Starting from the rest position, he lifts, with constant acceleration, a barbell that weighs 490 N. He lifts the barbell a distance of 0.60 m in 1.6 s. a) Draw a clearly labeled free-body force diagram for the barbell and for the athlete. b) Use the diagrams in part (a) and Newton's laws to find the total force that his feet exert on the ground as he lifts the barbell.

4.51 A hot-air balloon consists of a basket, one passenger, and some cargo. Let the total mass be M. Even though there is an upward lift force on the balloon, the balloon is initially accelerating downward at a rate of $g/3$. a) Draw a free-body diagram for the descending balloon. b) Find the upward lift force in terms of the initial total weight Mg. c) The passenger notices that he is heading straight for a waterfall and decides he needs to go up. What fraction of the total weight must he drop overboard so that the balloon accelerates *upward* at a rate of $g/2$? Assume that the upward lift force remains the same.

4.52 A student tries to raise a chain consisting of three identical links. Each link has a mass of 300 g. The three-piece chain is connected to a string and then suspended vertically, with the student holding the upper end of the string and pulling upward. Because of the student's pull, an upward force of 12 N is applied to the chain by the string. a) Draw a free-body diagram for *each* of the links in the chain and also for the entire chain considered as a single body. b) Use the results of part (a) and Newton's laws to find (i) the acceleration of the chain; and (ii) the force exerted by the top link on the middle link.

4.53 The position of a 2.75×10^5 N training helicopter under test is given by $\vec{r} = (0.020 \text{ m/s}^3)t^3\,\hat{\imath} + (2.2 \text{ m/s})t\hat{\jmath} - (0.060 \text{ m/s}^2)t^2\hat{k}$. Find the net force on the helicopter at $t = 5.0$ s.

4.54 An object with mass m moves along the x-axis. Its position as a function of time is given by $x(t) = At - Bt^3$, where A and B are constants. Calculate the net force on the object as a function of time.

*4.55 An object with mass m initially at rest is acted on by a force $\vec{F} = k_1\hat{\imath} + k_2t^3\hat{\jmath}$, where k_1 and k_2 are constants. Calculate the velocity $\vec{v}(t)$ of the object as a function of time.

Challenge Problems

*4.56 If we know $F(t)$, the force as a function of time, for straight-line motion, Newton's second law gives us $a(t)$, the acceleration as a function of time. We can then integrate $a(t)$ to find $v(t)$ and $x(t)$. However, suppose we know $F(v)$ instead. a) The net force on a body moving along the x-axis equals $-Cv^2$. Use Newton's second law written as $\Sigma F = m\,dv/dt$ and two integrations to show that $x - x_0 = (m/C)\ln(v_0/v)$. b) Show that Newton's second law can be written as $\Sigma F = mv\,dv/dx$. Derive the same expression as in part (a) using this form of the second law and one integration.

*4.57 An object of mass m is at rest in equilibrium at the origin. At $t = 0$ a new force $\vec{F}(t)$ is applied that has components

$$F_x(t) = k_1 + k_2y \qquad F_y(t) = k_3t$$

where k_1, k_2, and k_3 are constants. Calculate the position $\vec{r}(t)$ and velocity $\vec{v}(t)$ vectors as functions of time.

APPLYING NEWTON'S LAWS

A bird glides by making use of Newton's first and third laws. The wings push downward on the air as it flows past the bird, and the air reacts by pushing upward on the wings with a force—called *lift*—of equal magnitude and opposite direction. In a steady glide the aerodynamic forces on the bird exactly balance the downward force of gravity, so that the net external force on the bird is zero.

? **Suppose the bird is caught in an updraft so that it is ascending at a steady rate. In this situation, which has a greater magnitude: the force of gravity or the aerodynamic forces?**

We saw in Chapter 4 that Newton's three laws of motion, the foundation of classical mechanics, can be stated very simply. But *applying* these laws to situations such as an iceboat skating across a frozen lake, a toboggan sliding down a hill, or a jet airplane making a steep turn requires analytical skills and problem-solving technique. In this chapter we'll help you extend the problem-solving skills you began to develop in Chapter 4.

We begin with equilibrium problems, in which a body is at rest or moving with constant velocity. Then we generalize our problem-solving techniques to include bodies that are not in equilibrium, for which we need to deal precisely with the relationships between forces and motion. We will learn how to describe and analyze the contact force acting on a body when it rests or slides on a surface. Finally, we study the important case of uniform circular motion, in which a body moves in a circle with constant speed.

All these situations involve the concept of force, a concept we'll use throughout our study of physics. We close the chapter with a brief look at the fundamental nature of force and the classes of forces found in our physical universe.

5.1 | Using Newton's First Law: Particles in Equilibrium

We learned in Chapter 4 that a body is in *equilibrium* when it is at rest or moving with constant velocity in an inertial frame of reference. A hanging lamp, a suspension bridge, an airplane flying straight and level at a constant speed—all are examples of equilibrium situations. In this section we consider only equilibrium of a body that can be modeled as a particle. (In Chapter 11, we'll consider the additional principles needed when the body can't be represented adequately as a particle.) The essential physical principle is Newton's first law: When a particle is at rest or is moving with constant velocity in an inertial frame of reference, the net force acting on it—that is, the vector sum of all the forces acting on it—must be zero:

$$\sum \vec{F} = 0 \qquad \text{(particle in equilibrium, vector form)} \qquad (5.1)$$

We most often use this equation in component form:

$$\sum F_x = 0 \qquad \sum F_y = 0 \qquad (5.2)$$
$$\text{(particle in equilibrium, component form)}$$

This section is about using Newton's first law to solve problems dealing with bodies in equilibrium. Some of these problems may seem complicated, but the important thing to remember is that *all* problems involving particles in equilibrium are done in the same way. The following strategy details the steps you need to follow for any and all such problems. (Like all Problem-Solving Strategies, this one follows the "I SEE" format—Identify, Set up, Execute, and Evaluate—that we introduced in Section 1.2.) Study this strategy carefully, look at how it's applied in the worked-out examples, and try to apply it yourself when you solve assigned problems.

Problem-Solving Strategy

Newton's First Law: Equilibrium of a Particle

IDENTIFY *the relevant concepts:* You must use Newton's first law for any problem that involves forces acting on a body in equilibrium. Remember that "equilibrium" means that the body either remains at rest or moves with constant velocity. For example, a car is in equilibrium when it's parked, but also when it's driving down a straight road at a steady speed.

If the problem involves more than one body, and the bodies interact with each other, you'll also need to use Newton's *third* law. This law allows you to relate the force that one body exerts on a second body to the force that the second body exerts on the first one.

Be certain that you identify the target variable(s). Common target variables in equilibrium problems include the magnitude of one of the forces, the components of a force, or the direction of a force.

SET UP *the problem using the following steps:*
1. Draw a very simple sketch of the physical situation, showing dimensions and angles. You don't have to be an artist!

2. For each body that is in equilibrium, draw a free-body diagram of this body. For the present, we consider the body as a particle, so a large dot will do to represent it. In your free-body diagram, *do not* include the other bodies that interact with it, such as a surface it may be resting on, or a rope pulling on it.

3. Now ask yourself what is interacting with the body by touching it or in any other way. On your free-body diagram, draw a force vector for each interaction. If you know the angle at which a force is directed, draw the angle accurately and label it. A surface in contact with the body exerts a normal force perpendicular to the surface and possibly a friction force parallel to the surface. Remember that a rope or chain can't push on a body, but can only pull in a direction along its length. Be sure to include the body's weight, except in cases where the body has negligible mass (and hence negligible weight). If the mass is given, use $w = mg$ to find the weight. Label each force with a symbol representing the *magnitude* of the force.

4. *Do not* show in the free-body diagram any forces exerted *by* the body on any other body. The sums in Eqs. (5.1) and (5.2) include only forces that act *on* the body. Make sure you can answer the question "What other body causes that force?" for each force. If you can't answer that question, you may be imagining a force that isn't there.

5. Choose a set of coordinate axes and include them in your free-body diagram. (If there is more than one body in the problem, you'll need to choose axes for each body separately.) Make sure you label the positive direction for each axis. This will be crucially important when you take components of the force vectors as part of your solution. Often you can simplify the problem by your choice of coordinate axes. For example, when a body rests or slides on a plane surface, it's usually simplest to take the axes in the directions parallel and perpendicular to this surface, even when the plane is tilted.

EXECUTE *the solution as follows:*

1. Find the components of each force along each of the body's coordinate axes. Draw a wiggly line through each force vector that has been replaced by its components, so you don't count it twice. Keep in mind that while the *magnitude* of a force is always positive, the *component* of a

force along a particular direction may be positive or negative.

2. Set the algebraic sum of all *x*-components of force equal to zero. In a separate equation, set the algebraic sum of all *y*-components equal to zero. (*Never* add *x*- and *y*-components in a single equation.) You can then solve these equations for up to two unknown quantities, which may be force magnitudes, components, or angles.

3. If there are two or more bodies, repeat all of the above steps for each body. If the bodies interact with each other, use Newton's third law to relate the forces they exert on each other.

4. Make sure that you have as many independent equations as the number of unknown quantities. Then solve these equations to obtain the target variables. This part is algebra, not physics, but it's an essential step.

EVALUATE *your answer:* Look at your results and ask whether they make sense. When the result is a symbolic expression or formula, try to think of special cases (particular values or extreme cases for the various quantities) for which you can guess what the results ought to be. Check to see that your formula works in these particular cases.

| Example 5.1 | **One-dimensional equilibrium: Tension in a massless rope** |

A gymnast with mass $m_G = 50.0$ kg suspends herself from the lower end of a hanging rope. The upper end of the rope is attached to the gymnasium ceiling (Fig. 5.1a). What is the gymnast's weight? What force (magnitude and direction) does the rope exert on her? What is the tension at the top of the rope? Assume that the mass of the rope itself is negligible.

SOLUTION

IDENTIFY: The gymnast and the rope are in equilibrium, so we can apply Newton's first law to both bodies. The gymnast and the rope exert forces on each other—that is, they interact—so we'll also use Newton's third law to relate these forces. The target variables are the weight of the gymnast, w_G; the force that the rope exerts on the gymnast (call it $T_{R \text{ on } G}$); and the tension that the ceiling exerts on the top of the rope (call it $T_{C \text{ on } R}$).

SET UP: We draw separate free-body diagrams for the gymnast (Fig. 5.1b) and the rope (Fig. 5.1c). We take the positive *y*-axis to be upward, as shown. Each force acts in the vertical direction and so has only a *y*-component.

The two forces $T_{R \text{ on } G}$ and $T_{G \text{ on } R}$ are the upward force of the rope on the gymnast (in Fig. 5.1b) and the downward force of the gymnast on the rope (in Fig. 5.1c). These forces form an action-reaction pair, so they must have the same magnitude.

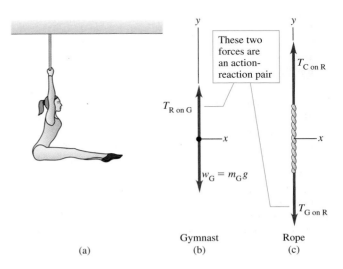

5.1 (a) A gymnast hanging at rest from the end of a vertical rope. (b) Free-body diagram for the gymnast. (c) Free-body diagram for the rope, assuming its weight to be negligible. The upward force of the rope on the gymnast and the downward force of the gymnast on the rope are an action-reaction pair.

Note also that the gymnast's weight w_G is the attractive (downward) force exerted on the gymnast by the earth. Its reaction force is the equal and opposite (upward) attractive force exerted on the *earth* by the *gymnast*. This force acts on the earth, not on the gymnast, so it doesn't appear in her free-body diagram. Compare the discussion of the apple in Conceptual Example 4.9 (Section 4.5).

EXECUTE: The magnitude of the weight of any object is the product of the object's mass and the acceleration due to gravity, g. For the gymnast this is

$$w_G = m_G g = (50.0 \text{ kg})(9.80 \text{ m/s}^2) = 490 \text{ N}$$

This force points in the negative y-direction, so its y-component is − 490 N. The upward force exerted by the rope has unknown magnitude $T_{R \text{ on } G}$. From Eq. (5.2), because the gymnast is in equilibrium, the sum of the y-components of force acting on her must be zero:

$$\text{Gymnast: } \sum F_y = T_{R \text{ on } G} + (-w_G) = 0 \quad \text{so}$$
$$T_{R \text{ on } G} = w_G = 490 \text{ N}$$

The rope pulls *up* on the gymnast with a force $T_{R \text{ on } G}$ of magnitude 490 N. By Newton's third law, the gymnast pulls *down* on the rope with a force of the same magnitude, $T_{G \text{ on } R} = 490$ N.

The rope is also in equilibrium. We have assumed that it is weightless, so the upward force of magnitude $T_{C \text{ on } R}$ that the ceiling exerts on its top end must make the *net* vertical force on the rope equal to zero. Expressed as an equation, this says

$$\text{Rope: } \sum F_y = T_{C \text{ on } R} + (-T_{G \text{ on } R}) = 0 \quad \text{so}$$
$$T_{C \text{ on } R} = T_{G \text{ on } R} = 490 \text{ N}$$

EVALUATE: The *tension* at any point in the rope is the force that acts at that point. For this weightless rope, the tension $T_{G \text{ on } R}$ at the lower end has the same value as the tension $T_{C \text{ on } R}$ at the upper end. Indeed, for an ideal weightless rope, the tension has the same value at any point along the rope's length. (Compare the discussion of Conceptual Example 4.10 in Section 4.5.)

Note that we have defined tension to be the *magnitude* of a force, so it is always positive. But the y-component of force acting on the rope at its lower end is $-T_{G \text{ on } R} = -490$ N.

Example 5.2	**One-dimensional equilibrium: Tension in a rope with mass**

Suppose that in Example 5.1, the weight of the rope is not negligible but is 120 N. Find the tension at each end of the rope.

SOLUTION

IDENTIFY: As in the previous example, we will apply Newton's first law to each of the two bodies in the problem (the gymnast and the rope) and use Newton's third law to relate the forces that the gymnast and rope exert on each other. The target variables are the magnitudes $T_{G \text{ on } R}$ and $T_{C \text{ on } R}$ of the forces that act at the bottom and top of the rope, respectively.

SET UP: Again we draw separate free-body diagrams for the gymnast (Fig. 5.2a) and the rope (Fig. 5.2b). The only difference in the diagrams from those in Example 5.1 is that there are now three forces acting on the rope: the downward force exerted by the gymnast ($T_{G \text{ on } R}$), the upward force exerted by the ceiling ($T_{C \text{ on } R}$), and the downward force of gravity (the weight of the rope, of magnitude $w_R = 120$ N).

EXECUTE: The gymnast's free-body diagram is unchanged from Example 5.1, so her equilibrium condition is also unchanged. From Newton's third law, $T_{R \text{ on } G} = T_{G \text{ on } R}$, and we have

$$\text{Gymnast: } \sum F_y = T_{R \text{ on } G} + (-w_G) = 0 \quad \text{so}$$
$$T_{R \text{ on } G} = T_{G \text{ on } R} = w_G = 490 \text{ N}$$

The equilibrium condition $\sum F_y = 0$ for the rope is

$$\text{Rope: } \sum F_y = T_{C \text{ on } R} + (-T_{G \text{ on } R}) + (-w_R) = 0$$

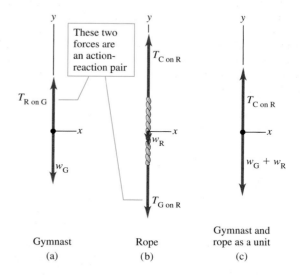

5.2 Free-body diagrams (a) for the gymnast (weight w_G) and (b) for the rope (weight w_R). (Compare Fig. 5.1.) (c) Free-body diagram for the gymnast and rope, considered as a single composite body.

Note that the y-component of $T_{C \text{ on } R}$ is positive because it points in the $+y$-direction, but the y-components of both $T_{G \text{ on } R}$ and w_R are negative. When we solve for $T_{C \text{ on } R}$ and substitute the values $T_{G \text{ on } R} = T_{R \text{ on } G} = 490$ N and $w_R = 120$ N, we find

$$T_{C \text{ on } R} = T_{G \text{ on } R} + w_R = 490 \text{ N} + 120 \text{ N} = 610 \text{ N}$$

EVALUATE: When we include the weight of the rope, we find that the tension is *different* at the rope's two ends. This makes sense: the force $T_{\text{C on R}}$ exerted by the ceiling has to hold up both the 490-N weight of the gymnast and the 120-N weight of the rope, so $T_{\text{C on R}} = 610$ N.

To see this more explicitly, draw a free-body diagram for a composite body consisting of the gymnast and rope considered as a unit (Fig. 5.2c). There are only two external forces acting on this composite body: the force $T_{\text{C on R}}$ exerted by the ceiling and the total weight $w_G + w_R = 490$ N $+ 120$ N $= 610$ N. (The forces $T_{\text{G on R}}$ and $T_{\text{R on G}}$ are *internal* to the composite body. Since Newton's first law involves only *external* forces, the internal forces play no role.) Hence Newton's first law applied to this composite body is

$$\text{Composite body: } \sum F_y = T_{\text{C on R}} + [-(w_G + w_R)] = 0$$

and so $T_{\text{C on R}} = w_G + w_R = 610$ N.

This method of treating the gymnast and rope as a composite body seems a lot simpler, and you may be wondering why we didn't use it first. The answer is that we can't find the tension $T_{\text{G on R}}$ at the bottom of the rope by this method. The moral is that whenever you have more than one body in a problem involving Newton's laws, the safest approach is to treat each body separately.

Example 5.3	**Two-dimensional equilibrium**

In Fig. 5.3a a car engine with weight w hangs from a chain that is linked at ring O to two other chains, one fastened to the ceiling and the other to the wall. Find the tensions in each of these three chains, assuming that w is given and the weights of the ring and chains are negligible.

SOLUTION

IDENTIFY: The target variables are the tensions T_1, T_2, and T_3 in the three chains (Fig. 5.3a). It may seem strange that we neglect the weight of the chains and ring in this example, while in Example 5.2 we did *not* neglect the weight of a mere rope. The reason is that the weight of the chains or ring is very small compared to the weight of the massive engine. By contrast, in Example 5.2 the rope weighed a reasonable fraction of the gymnast's weight (120 N compared to 490 N).

All the bodies in the example are in equilibrium, so we'll use Newton's first law to determine T_1, T_2, and T_3. We need three simultaneous equations, one for each target variable. However, applying Newton's first law to just one body gives us just *two* equations, as in Eq. (5.2). So to solve the problem, we'll have to consider more than one body in equilibrium. We'll look at the engine (which is acted on by T_1) and the ring (which is connected to all three chains and so is acted on by all three tensions).

SET UP: Figures 5.3b and 5.3c are free-body diagrams for the engine and the ring respectively. As in Examples 5.1 and 5.2, we've included an x-y coordinate system in each diagram.

The two forces acting on the engine are its weight w and the upward force T_1 exerted by the vertical chain; the three forces acting on the ring are the tensions from the vertical chain (T_1), the

(a)   (b)   (c)

5.3 (a) A car engine with weight w is suspended from a chain linked at O to two other chains. The chains are considered to be massless. (b) Free-body diagram for the engine. (c) Free-body diagram for the ring with $\vec{T}_3$ replaced by its components.

horizontal chain (T_2), and the slanted chain (T_3). Note that the vertical chain exerts forces of the same magnitude T_1 at both of its ends: upward on the engine in Fig. 5.3b and downward on the ring in Fig. 5.3c. This is because the weight of the chain is negligible (see Example 5.1). If the weight were not negligible, these two forces would have different magnitudes, as was the case for the rope in Example 5.2. Remember that we're also neglecting the weight of the ring, which is why it isn't included in the forces in Fig. 5.3c.

EXECUTE: The forces acting on the engine are along the y-axis only, so Newton's first law says

$$\text{Engine: } \sum F_y = T_1 + (-w) = 0 \quad \text{and} \quad T_1 = w$$

Note that the horizontal and slanted chains do not exert forces on the engine itself, because they are not attached to it. These forces do appear in the application of Newton's first law to the ring, however.

In the free-body diagram for the ring (Fig. 5.3c), remember that T_1, T_2, and T_3 are the *magnitudes* of the forces; their directions are shown by the vectors on the diagram. We first resolve the force with magnitude T_3 into its x- and y-components. We can then write the equilibrium conditions for the ring, writing separate equations for the x- and y-components. (Remember from the Problem-Solving Strategy that x- and y-components are *never* added together in a single equation.) We find

$$\text{Ring: } \sum F_x = T_3 \cos 60° + (-T_2) = 0$$
$$\text{Ring: } \sum F_y = T_3 \sin 60° + (-T_1) = 0$$

Because $T_1 = w$ (from the engine equation), we can rewrite the second ring equation as

$$T_3 = \frac{T_1}{\sin 60°} = \frac{w}{\sin 60°} = 1.155w$$

We can now use this result in the first ring equation:

$$T_2 = T_3 \cos 60° = w\frac{\cos 60°}{\sin 60°} = 0.577w$$

So we can express all three tensions as multiples of the weight w of the engine, which we assume is known. To summarize,

$$T_1 = w$$
$$T_2 = 0.577w$$
$$T_3 = 1.155w$$

EVALUATE: Our results show that the chain attached to the ceiling exerts a force on the ring of magnitude T_3, which is *greater* than the weight of the engine. If this seems strange, note that the vertical component of this force is equal to T_1, which in turn is equal to w. But since this force also has a horizontal component, its magnitude T_3 must be somewhat larger than w. Hence the chain attached to the ceiling is under the greatest tension, and is the one most susceptible to breaking.

It probably seemed to you at first glance that the most important body in this problem was the engine. But in order to get enough equations to solve the problem, we also had to consider the forces acting on a second body (in this case, the ring connecting the chains). Situations like this are fairly common in equilibrium problems, so keep this technique in mind.

Example 5.4	**An inclined plane**

A car rests on the slanted tracks of a ramp leading to a car-transporter trailer (Fig. 5.4a). Only a cable attached to the car and to the frame of the trailer prevents the car from rolling backward off the trailer. (The car's brakes and transmission lock are both released.) If the weight of the car is w, find the tension in the cable and the force with which the tracks push on the car's tires.

SOLUTION

IDENTIFY: The car is in equilibrium, so once again we use Newton's first law. One complication is that the ramp exerts *four* forces on the car, one on each of the car's tires. For simplicity, we lump all of these forces together into a single force. For a further simplification, we'll ignore the component of this force that acts *parallel* to the tracks. This is equivalent to saying that there's very little friction on the car. (If there were substantial friction, the tracks would exert a *friction force* acting parallel to the ramp that would tend to prevent the car from rolling down the ramp. We'll ignore this effect here, but will return to it in Section 5.3.) Hence we can say that the

ramp only exerts a force on the car that is *perpendicular* to the tracks. This force appears because the atoms on the surface of the track resist having the atoms of the tires squeezed into them. As in Section 4.2, we call this force *the normal force*.

The forces on the car do not all act along the same line—the force of gravity acts straight downward, but the normal force is perpendicular to the tilted surface of the tracks. Hence we have two separate equations for Newton's first law, one for the x-components of force and one for the y-components [see Eq. (5.2)]. This is just the right number of equations to solve for the two target variables, the magnitude T of the tension in the cable and the magnitude n of the normal force.

SET UP: Figure 5.4b shows a free-body diagram for the car. The three forces acting on the car are its weight (magnitude w), the tension in the cable (magnitude T), and the normal force (magnitude n). Note that the normal force acts up and to the left because it's preventing the car from penetrating into the solid tracks.

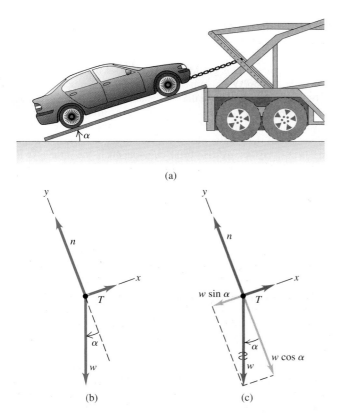

(a)

(b)   (c)

5.4 (a) A cable holds the car on the ramp. (b) Free-body diagram for the car. (c) Free-body diagram with the car's weight replaced by its components ($w \sin \alpha$ down the ramp, $w \cos \alpha$ perpendicular to the ramp).

We choose the x- and y-axes to be parallel and perpendicular to the ramp as shown. This choice makes the problem easier to analyze because only the weight force has both an x- and y-component. If we were to choose axes that were horizontal and vertical, our job would be harder because we'd need to find x- and y-components for *two* forces (the normal force and the tension).

Note that the angle α between the ramp and the horizontal is equal to the angle α between the weight vector $\vec{w}$ and the normal to the plane of the ramp.

EXECUTE: To write down the x- and y-components of Newton's first law, we need to find the components of the weight. One complication is that the angle α in Fig. 5.4b is *not* measured from the $+x$-axis toward the $+y$-axis. Hence we *cannot* use Eqs. (1.7) directly to find the components. (You may want to review Section 1.8 to make sure that you understand this important point.)

One approach to finding the components of $\vec{w}$ is to consider the right triangles in Fig. 5.4c. The sine of α is the magnitude of the x-component of $\vec{w}$ (that is, the side of the triangle opposite α)

divided by the magnitude w (the hypotenuse of the triangle). Similarly, the cosine of α is the magnitude of the y-component (the side of the triangle adjacent to α) divided by w. Both components are negative, so $w_x = -w \sin \alpha$ and $w_y = -w \cos \alpha$.

Another approach is to recognize that one component of $\vec{w}$ must involve $\sin \alpha$ while the other component involves $\cos \alpha$. To decide which is which, it helps to draw the free-body diagram so that the angle α is noticeably less than or greater than 45°. (You're free to do this, since the value of α isn't specified. You'll have to fight the natural tendency to draw such angles as being close to 45°.) We've drawn Figs. 5.4b and 5.4c so that α is less than 45°, which means that $\sin \alpha$ is less than $\cos \alpha$. Figure 5.4c shows that the x-component of $\vec{w}$ is smaller than the y-component, so the x-component must involve $\sin \alpha$ and the y-component must involve $\cos \alpha$. Since both components are negative, we again find $w_x = -w \sin \alpha$ and $w_y = -w \cos \alpha$.

We draw a wiggly line through the original vector representing the weight to remind us not to count it twice. The equilibrium conditions then give us

$$\sum F_x = T + (-w \sin \alpha) = 0$$
$$\sum F_y = n + (-w \cos \alpha) = 0$$

Be sure you understand how these signs are related to our choice of coordinates. Remember that, by definition, T, w, and n are all *magnitudes* of vectors and are therefore all positive.

Solving these equations for T and n, we find

$$T = w \sin \alpha$$
$$n = w \cos \alpha$$

EVALUATE: Our answers for T and n depend on the value of α. To check these answers to see whether they're reasonable, let's look at some special cases. If the angle α is zero, then $\sin \alpha = 0$ and $\cos \alpha = 1$. In this case, the tracks are horizontal; our answers tell us that no cable tension T is needed to hold the car, and the total normal force n is equal in magnitude to the weight. If the angle is 90°, then $\sin \alpha = 1$ and $\cos \alpha = 0$. Then the cable tension T equals the weight w, and the normal force n is zero. Are these the results you would expect for these particular cases?

> **CAUTION** It's a common error to automatically assume that the magnitude n of the normal force is equal to the weight w. But our result shows that this is *not* true in general. It's always best to treat n as a variable and solve for its value, as we have done here.

As a final note, ask yourself how the answers for T and n would be affected if the car were not stationary but were being pulled up the ramp at a constant speed. You should immediately realize that this, too, is an equilibrium situation, since the car's velocity is constant. So the calculation is exactly the same, and T and n have the same values as when the car is at rest. (It's true that T must be greater than $w \sin \alpha$ to *start* the car moving up the ramp, but that's not what we asked.)

Example 5.5	**Tension over a frictionless pulley**

Blocks of granite are being hauled up a $15°$ slope out of a quarry. For environmental reasons, dirt is also being dumped into the quarry to fill up old holes. You have been asked to find a way to use this dirt to move the granite out more easily. You design a system in which a granite block on a cart with steel wheels (weight w_1, including both the block and cart) is pulled uphill on steel rails by a dirt-filled bucket (weight w_2, including both the dirt and the bucket) dropping vertically into the quarry (Fig. 5.5a). Ignoring friction in the pulley and wheels and the weight of the cable, determine how the weights w_1 and w_2 must be related in order for the system to move with constant speed.

SOLUTION

IDENTIFY: The cart and bucket each move with a constant velocity (that is, in a straight line at constant speed). Hence each body is in equilibrium and we can apply Newton's first law to each.

Our two target variables are the weights w_1 and w_2. The forces that act on the bucket are its weight w_2 and an upward tension exerted by the cable; both these forces are purely vertical, so Newton's first law applied to the bucket gives us just one equation. The cart has *three* forces acting on it: its weight w_1, a normal force of magnitude n exerted by the rails, and a tension force from the cable. (We're ignoring friction, so we're assuming that the rails exert no force parallel to the incline.) This is exactly like the situation for the car on the ramp in Example 5.4. As in that example, the forces on the cart are not all along the same direction, so we'll need to use both components of Newton's first law in Eq. (5.2).

We're assuming that the cable has negligible weight, so the tension forces that the rope exerts on the cart and on the bucket have the same magnitude T.

SET UP: Figure 5.5b is our idealized model for the system. Figure 5.5c is the free-body diagram for the dirt and bucket and Fig. 5.5d is the free-body diagram for the granite block and cart. Note that we are at liberty to orient the axes differently for each body; the choices shown are the most convenient ones. We have represented the weight of the granite block in terms of its components in the chosen axis system; we find these components in the same way as in Example 5.4.

EXECUTE: Applying $\sum F_y = 0$ to the dirt-filled bucket in Fig. 5.5c, we find

$$\sum F_y = T + (-w_2) = 0 \quad \text{so} \quad T = w_2$$

Applying $\sum F_x = 0$ to the block and cart in Fig. 5.5d, we get

$$\sum F_x = T + (-w_1 \sin 15°) = 0 \quad \text{so} \quad T = w_1 \sin 15°$$

Equating the two expressions for T, we find

$$w_2 = w_1 \sin 15° = 0.26w_1$$

EVALUATE: If the weight of dirt and bucket totals 26% of the weight of the granite block and cart, the system can move with constant speed in *either* direction. This is because our analysis doesn't depend on the direction of motion, only on the velocity being constant. Can you see what would happen if w_2 were greater than $0.26w_1$? If it were less than $0.26w_1$?

Notice that we didn't need to apply the equation $\sum F_y = 0$ to the cart and block; this would be useful only if we wanted to find the value of n. Can you show that $n = w_1 \cos 15°$?

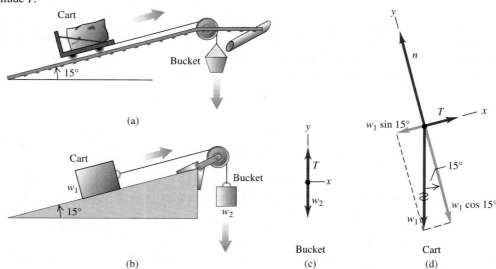

5.5 (a) A bucket of dirt pulls up a cart carrying a granite block. (b) Idealized model of the system. (c) Free-body diagram for the bucket of dirt. (d) Free-body diagram for the cart with block.

A traffic light of mass m hangs from two lightweight cables, one on each side of the light. Each cable hangs at a 45° angle from the horizontal. What is the tension in each cable?

5.2 | Using Newton's Second Law: Dynamics of Particles

We are now ready to discuss *dynamics* problems. In these problems, we apply Newton's second law to bodies that are accelerating and hence are *not* in equilibrium. In this case the net force on the body is not zero, but is equal to the mass of the body times its acceleration:

$$\sum \vec{F} = m\vec{a} \quad \text{(Newton's second law, vector form)} \quad (5.3)$$

We most often use this relation in component form:

$$\sum F_x = ma_x \qquad \sum F_y = ma_y \quad (5.4)$$
$$\text{(Newton's second law, component form)}$$

The following problem-solving strategy is very similar to our strategy for equilibrium problems in Section 5.1. We urge you to study it carefully, watch how we apply it in our examples, and use it when you tackle the end-of-chapter problems. Remember that you can solve *any* dynamics problem using this strategy.

CAUTION We emphasize again that the quantity $m\vec{a}$ is *not* a force; it's not a push or a pull exerted by anything in the body's environment. All that Eqs. (5.3) and (5.4) say is that the acceleration $\vec{a}$ is proportional to the net force $\sum \vec{F}$. When you draw the free-body diagram for an accelerating body (like the fruit in Fig. 5.6a), make sure you *never* include "the $m\vec{a}$ force," because *there is no such force* (Fig. 5.6b). You should review Section 4.3 if you're not clear on this point. Sometimes we will draw the acceleration vector $\vec{a}$ *alongside* a free-body diagram, as in Fig. 5.6c. But we will *never* draw the acceleration vector with its tail touching the body (a position reserved exclusively for the forces that act on the body).

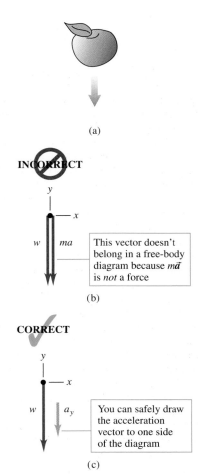

(a)

INCORRECT

This vector doesn't belong in a free-body diagram because $m\vec{a}$ is *not* a force

(b)

CORRECT

You can safely draw the acceleration vector to one side of the diagram

(c)

5.6 (a) The only force acting on this falling piece of fruit is the earth's gravitational pull. (b) An incorrect free-body diagram. (c) A correct free-body diagram.

Problem-Solving Strategy

Newton's Second Law: Dynamics of Particles

IDENTIFY *the relevant concepts:* You have to use Newton's second law for *any* problem that involves forces acting on an accelerating body.

As with any problem, identify the target variable—usually an acceleration or a force. If the target variable is something else, you'll need to identify another concept to use. For example, suppose the problem asks you to find how fast a sled is moving when it reaches the bottom of a hill. This means your target variable is the sled's final velocity. To find this, you'll first need to use Newton's second law to find the sled's acceleration. Then

you'll also have to use the constant-acceleration relationships from Section 2.4 to find velocity from acceleration.

SET UP *the problem using the following steps:*
1. Draw a simple sketch of the situation. Identify one or more moving bodies to which you'll apply Newton's second law.
2. For each body you identified, draw a free-body diagram that shows all the forces acting *on* the body. (Don't try to be fancy—just represent the object by a point.) Be careful

not to include any forces exerted *by* the object on some other object. Remember, the acceleration of a body is determined by the forces that act on it, not by the forces that it exerts on anything else. Make sure you can answer the question "What other body is applying this force?" for each force in your diagram. Furthermore, never include the quantity $m\vec{a}$ in your free-body diagram; it's not a force!

3. Label each force with an algebraic symbol for the force's *magnitude,* as well as a numerical value of the magnitude if it's given in the problem. (Remember that magnitudes are always positive. Minus signs show up later when you take components of the forces.) Usually, one of the forces will be the body's weight; it's usually best to label this as $w = mg$. If a numerical value of mass is given, you can compute the corresponding weight.

4. Choose your *x*- and *y*-coordinate axes for each object, and show them explicitly in each free-body diagram. Be sure to indicate which is the positive direction for each axis. If you know the direction of the acceleration, it usually simplifies things to take one positive axis along that direction. Note that if your problem involves more than one object and the objects accelerate in different directions, you can use a different set of axes for each object.

5. In addition to Newton's second law, $\sum \vec{F} = m\vec{a}$, identify any other equations you might need. (You need one equation for each target variable.) For example, you might need one or more of the equations for motion with constant acceleration. If more than one body is involved, there may be relationships among their motions; for example, they may be connected by a rope. Express any

such relationships as equations relating the accelerations of the various bodies.

EXECUTE *the solution as follows:*
1. For each object, determine the components of the forces along each of the object's coordinate axes. When you represent a force in terms of its components, draw a wiggly line through the original force vector to remind you not to include it twice.
2. For each object, write a separate equation for each component of Newton's second law, as in Eq. (5.4).
3. Make a list of all the known and unknown quantities. In your list, identify the target variable or variables.
4. Check that you have as many equations as there are unknowns. If you have too few equations, go back to step 5 of "Set up the problem." If you have too many equations, perhaps there is an unknown quantity that you haven't identified as such.
5. Do the easy part—the math! Solve the equations to find the target variable(s).

EVALUATE *your answer:* Does your answer have the correct units? (When appropriate, use the conversion $1 \text{ N} = 1 \text{ kg} \cdot \text{m/s}^2$.) Does it have the correct algebraic sign? (If the problem is about a sled sliding downhill, you probably took the positive *x*-axis to point down the hill. If you then find that the sled has a negative acceleration—that is, the acceleration is uphill—then something went wrong in your calculations.) When possible, consider particular values or extreme cases of quantities and compare the results with your intuitive expectations. Ask, "Does this result make sense?"

Example 5.6	**Straight-line motion with a constant force**

An iceboat is at rest on a perfectly frictionless horizontal surface (Fig. 5.7a). A steady wind is blowing (along the direction of the runners) so that 4.0 s after the iceboat is released, it attains a velocity of 6.0 m/s (about 22 km/h, or 13 mi/h). What constant horizontal force F_W does the wind exert on the iceboat? The mass of iceboat and rider is 200 kg.

SOLUTION

IDENTIFY: Our target variable is one of the forces (F_W) acting on the iceboat, so we'll need to use Newton's second law. This law involves forces and acceleration, but the acceleration isn't given; hence we'll need to find it. The wind is assumed to be steady, so the forces don't change with time and the resulting acceleration is constant. That means we'll be able to use one of the constant-acceleration formulas.

SET UP: Figure 5.7b shows the free-body diagram for the iceboat and rider considered as a unit. The forces acting on this object are

the weight w, the normal force n exerted by the surface, and the horizontal force F_W (our target variable). Note that the net force and hence the acceleration are to the right, so we chose the positive *x*-axis in this direction.

Since the acceleration isn't given, we'll need to find it from the other information given in the problem—namely, the final velocity $v_x = 6.0$ m/s and elapsed time $t = 4.0$ s. The iceboat starts at rest, so the initial velocity is $v_{0x} = 0$. From Section 2.4, an equation we can use to relate the *x*-acceleration a_x to these quantities is Eq. (2.8), $v_x = v_{0x} + a_x t$.

EXECUTE: The force F_W is in the positive *x*-direction, while the forces n and mg are in the positive and negative *y*-directions respectively. Hence the *x*- and *y*-equations for Newton's second law are

$$\sum F_x = F_W = ma_x$$
$$\sum F_y = n + (-mg) = 0$$

(a)

(b)

5.7 (a) An iceboat starting from rest. (b) Free-body diagram for the iceboat and rider with no friction.

(Note that there is zero acceleration in the y-direction; the iceboat doesn't accelerate up or down.) In addition, we have the constant-acceleration equation

$$v_x = v_{0x} + a_x t$$

The *known* quantities are the mass $m = 200$ kg, the final velocity $v_x = 6.0$ m/s, the initial velocity $v_{0x} = 0$, and the elapsed time $t = 4.0$ s. The *unknown* quantities are the acceleration a_x, the normal force n, and the horizontal force F_W (the target variable). There are as many unknowns (three) as equations, so we're in business.

To find F_W, first solve the constant-acceleration equation for a_x, then substitute this into the ΣF_x equation:

$$a_x = \frac{v_x - v_{0x}}{t} = \frac{6.0 \text{ m/s} - 0 \text{ m/s}}{4.0 \text{ s}} = 1.5 \text{ m/s}^2$$

$$F_W = ma_x = (200 \text{ kg})(1.5 \text{ m/s}^2) = 300 \text{ kg} \cdot \text{m/s}^2$$

One kg $\cdot$ m/s^2 is the same as one newton (N), so the final answer is

$$F_W = 300 \text{ N} \ (\text{about } 67 \text{ lb})$$

Note that we did not need the ΣF_y equation at all to find F_W. We would need this equation if we wanted to find the normal force n:

$$n - mg = 0$$
$$n = mg = (200 \text{ kg})(9.8 \text{ m/s}^2)$$
$$= 2.0 \times 10^3 \text{ N} \quad (\text{about } 440 \text{ lb})$$

The magnitude n of the normal force is equal to the combined weight of the iceboat and rider, because the surface is horizontal and these are the only vertical forces acting.

EVALUATE: Our answers for F_W and n have the correct units for a force, as they should. Does it seem reasonable that the force F_W is substantially less than the combined weight of the iceboat and rider?

Example 5.7

Straight-line motion with a time-varying force

Let's again consider the iceboat moving on a frictionless surface, as in Example 5.6. But now let's suppose that once the iceboat starts to move, its position as a function of time is

$$x = (1.2 \text{ m/s}^2)t^2 - (0.20 \text{ m/s}^3)t^3$$

Find the force F_W exerted by the wind as a function of time in this case. What is this force at time $t = 3.0$ s? For what times is the force zero? Positive? Negative?

SOLUTION

IDENTIFY: Just as in Example 5.6, F_W is our target variable, so we must once again use Newton's second law. We aren't given the acceleration, but we do know the position x as a function of time t; hence we can determine the acceleration by taking the second derivative of x with respect to t, as in Section 2.3.

Note that our expression for x includes a t^3 term, which doesn't appear in any of the constant-acceleration formulas. This tells us that the acceleration a_x is not constant, and hence neither is the force F_W. Hence the wind force is varying in this situation.

SET UP: The free-body diagram is exactly the same as in Fig. 5.7; the diagram is equally valid whether F_W is constant or not. To determine the acceleration a_x from the position x as a function of time, we use Eq. (2.6):

$$a_x = \frac{dv_x}{dt} = \frac{d^2x}{dt^2}$$

EXECUTE: Because the free-body diagram is unchanged from Example 5.6, the equation for the x-component of Newton's second law is also unchanged:

$$\sum F_x = F_W = ma_x.$$

To determine F_W, all that's left is to find a_x as a function of time. The second derivative of t^2 is 2, and the second derivative of t^3 is $6t$, so

$$a_x = \frac{d^2x}{dt^2} = \frac{d^2}{dt^2}\left[(1.2 \text{ m/s}^2)t^2 - (0.20 \text{ m/s}^3)t^3\right]$$

$$= 2.4 \text{ m/s}^2 - (1.2 \text{ m/s}^3)t$$

Then the force of the wind as a function of time is

$$F_W = ma_x = (200 \text{ kg})\left[2.4 \text{ m/s}^2 - (1.2 \text{ m/s}^3)t\right]$$

$$= 480 \text{ N} - (240 \text{ N/s})t$$

At time $t = 3.0$ s the value of F_W is 480 N − $(240 \text{ N/s})(3.0 \text{ s}) = -240$ N. The minus sign means that the direction of the wind force is actually opposite to what we assumed in Fig. 5.7. The wind has shifted and is now opposing the iceboat's motion! The force is zero when 480 N − $(240 \text{ N/s})t = 0$; this happens when $t = 2.0$ s, which is when the wind has momentarily died down. When $t < 2.0$ s, F_W is positive and the wind is pushing to the right in Fig. 5.7 (in the +x-direction). When $t > 2.0$ s, F_W is negative and the wind is pushing to the left.

EVALUATE: Figure 5.8 shows graphs of F_W and a_x as functions of time. Note that in this case F_W is the net horizontal force on the iceboat. You shouldn't be surprised that the net force and acceleration are directly proportional; according to Newton's second law, this is *always* true.

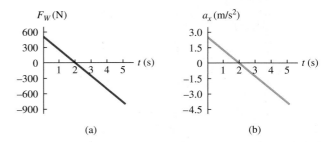

5.8 (a) The net force on the iceboat is directly proportional to (b) its acceleration.

<table>
<tr><td>Example
5.8</td><td>**Straight-line motion with friction**</td></tr>
</table>

Suppose the wind is once again blowing steadily in the +x-direction as in Example 5.6, so that the iceboat has a constant acceleration $a_x = 1.5 \text{ m/s}^2$. Now, however, there is a constant horizontal friction force with magnitude 100 N that opposes the motion of the iceboat. In this case, what force F_W must the wind exert on the iceboat?

SOLUTION

IDENTIFY: Once again the target variable is F_W. We are given the acceleration, so to find F_W all we need is Newton's second law.

SET UP: Figure 5.9 shows the new free-body diagram. The only difference from Fig. 5.7b is the addition of the friction force $\vec{f}$, which points opposite to the motion. (Note that its *magnitude*, $f = 100$ N, is a positive quantity but that its *component* in the x-direction is negative, equal to $-f$ or -100 N.)

EXECUTE: There are now two forces (the force of the wind and the friction force) that have x-components. The x-component of Newton's second law gives

$$\sum F_x = F_W + (-f) = ma_x$$

$$F_W = ma_x + f = (200 \text{ kg})(1.5 \text{ m/s}^2) + (100 \text{ N}) = 400 \text{ N}$$

EVALUATE: Because there is friction, a greater force F_W is needed than in Example 5.6. We need 100 N to overcome friction and 300 N more to give the iceboat the necessary acceleration.

5.9 Free-body diagram for the iceboat and rider with a frictional force $\vec{f}$ opposing the motion.

<table>
<tr><td>Example
5.9</td><td>**Tension in an elevator cable**</td></tr>
</table>

An elevator and its load have a total mass of 800 kg (Fig. 5.10a). The elevator is originally moving downward at 10.0 m/s; it slows to a stop with constant acceleration in a distance of 25.0 m. Find the tension T in the supporting cable while the elevator is being brought to rest.

5.10 (a) A loaded elevator moving downward and coming to rest. (b) Free-body diagram for the elevator. As in Fig. 5.6, we draw the acceleration vector to one side of the free-body diagram because acceleration is not a force.

SOLUTION

IDENTIFY: The target variable is the tension T, which we will find using Newton's second law. As in Example 5.6, we'll have to determine the acceleration using the constant-acceleration formulas.

SET UP: The free-body diagram in Figure 5.10b shows the only forces acting on the elevator: its weight w and the tension force T of the cable. The elevator is moving downward with decreasing speed,

so its acceleration is upward; we choose the positive y-axis to be in this direction.

The elevator is moving downward, in the negative y-direction. Hence its initial velocity v_{0y} and its displacement $y - y_0$ are both negative: $v_{0y} = -10.0$ m/s and $y - y_0 = -25.0$ m. The final velocity is $v_y = 0$. To find the acceleration a_y from this information, we'll use Eq. (2.13) in the form $v_y^2 = v_{0y}^2 + 2a_y(y - y_0)$. Once we have a_y, we'll substitute it into the y-component of Newton's second law from Eq. (5.4).

EXECUTE: First let's write out Newton's second law. The tension force acts upward and the weight acts downward, so

$$\sum F_y = T + (-w) = ma_y$$

We solve for the target variable T:

$$T = w + ma_y = mg + ma_y = m(g + a_y)$$

To determine a_y, we rewrite the constant-acceleration equation $v_y^2 = v_{0y}^2 + 2a_y(y - y_0)$:

$$a_y = \frac{v_y^2 - v_{0y}^2}{2(y - y_0)} = \frac{(0)^2 - (-10.0 \text{ m/s})^2}{2(-25.0 \text{ m})} = +2.00 \text{ m/s}^2$$

The acceleration is upward (positive), just as it should be for downward motion with decreasing speed.

Now we can substitute the acceleration into the equation for the tension:

$$T = m(g + a_y) = (800 \text{ kg})(9.80 \text{ m/s}^2 + 2.00 \text{ m/s}^2)$$
$$= 9440 \text{ N}$$

EVALUATE: Our answer shows that the tension is 1600 N *greater* than the weight ($w = mg = 7840$ N). This makes sense: there must be an upward net force to provide the upward acceleration that brings the elevator to a halt. Can you see that we would get the same answer for T if the elevator were moving *upward* and *gaining* speed at a rate of 2.00 m/s²?

<table>
<tr><td>Example
5.10</td><td>**Apparent weight in an accelerating elevator**</td></tr>
</table>

A 50.0-kg woman stands on a bathroom scale while riding in the elevator in Example 5.9 (Fig. 5.11a). What is the reading on the scale?

SOLUTION

IDENTIFY: The scale reads the magnitude of the downward force exerted *by* the woman *on* the scale. By Newton's third law, this equals the magnitude of the upward normal force exerted *by* the

scale *on* the woman. Hence our target variable is the magnitude n of the normal force.

We'll find n by applying Newton's second law. Happily, we already know the acceleration of the woman; it's the same as the acceleration of the elevator, which we calculated in Example 5.9.

SET UP: Figure 5.11b shows a free-body diagram for the woman. The forces acting on her are the normal force n exerted by the scale

Moving down with
decreasing speed

(a)   (b)

5.11 (a) A woman in the elevator as it slows down. (b) Free-body diagram for the woman.

and her weight $w = mg = (50.0 \text{ kg})(9.80 \text{ m/s}^2) = 490$ N. (The tension force, which played a major role in Example 5.9, doesn't appear here. That's because the tension doesn't act directly on the woman. What she feels pushing upward on her feet is the scale, *not* the elevator cable.) From Example 5.9, the acceleration of the elevator and of the woman is $a_y = +2.00 \text{ m/s}^2$.

EXECUTE: Newton's second law gives

$$\sum F_y = n + (-mg) = ma_y$$
$$n = mg + ma_y = m(g + a_y)$$
$$= (50.0 \text{ kg})(9.80 \text{ m/s}^2 + 2.00 \text{ m/s}^2) = 590 \text{ N}$$

EVALUATE: Our answer for n means that while the elevator is stopping, the scale pushes up on the woman with a force of 590 N. By Newton's third law, she pushes down on the scale with the same force; so the scale reads 590 N, which is 100 N more than her actual weight. The scale reading is called the passenger's **apparent weight.** The woman *feels* the floor pushing up harder on her feet than when the elevator is stationary or moving with constant velocity.

What would the woman feel if the elevator were accelerating *downward* so that $a_y = -2.00 \text{ m/s}^2$? This would be the case if the elevator were moving upward with decreasing speed or moving downward with increasing speed. To find the answer for this situation, we just insert the new value of a_y in our equation for n:

$$n = m(g + a_y) = (50.0 \text{ kg})[9.80 \text{ m/s}^2 + (-2.00 \text{ m/s}^2)]$$
$$= 390 \text{ N}$$

Now the woman feels as though she weighs only 390 N, or 100 N *less* than her actual weight.

You can feel these effects yourself; try taking a few steps in an elevator that is coming to a stop after descending (when your apparent weight is greater than your true weight w) or coming to a stop after ascending (when your apparent weight is less than w).

5.12 Astronauts in orbit feel "weightless" because they have the same acceleration as their spacecraft—*not* because they are "outside the pull of the earth's gravity." (If they were, the astronauts and their spacecraft wouldn't remain in orbit, but would fly off into deep space.)

Let's generalize the result of Example 5.10. When a passenger with mass m rides in an elevator with y-acceleration a_y, a scale shows the passenger's apparent weight to be

$$n = m(g + a_y)$$

When a_y is positive, the elevator is accelerating upward (moving upward with increasing speed, or moving downward with decreasing speed) and n is greater than the passenger's weight $w = mg$. When the elevator is accelerating downward (moving upward with decreasing speed, or moving downward with increasing speed), a_y is negative and n is less than the weight. If the passenger doesn't know the elevator is accelerating, she may feel as though her weight is changing; indeed, this is just what the scale shows.

The extreme case occurs when the elevator has a downward acceleration $a_y = -g$, that is, when it is in free fall. In that case, $n = 0$ and the passenger *seems* to be weightless. Similarly, an astronaut orbiting the Earth in a spacecraft experiences apparent weightlessness (Fig. 5.12). In each case, the person is not truly weightless, because there is still a gravitational force acting. But the effect of this free-fall condition is exactly the same as though the body were in outer space with no gravitational force at all. In both cases the person and vehicle (elevator or spacecraft) are falling together with the same acceleration g, so nothing pushes the person against the floor or walls of the vehicle.

Prolonged apparent weightlessness can have major physiological consequences for astronauts. The fluids of the body redistribute themselves, causing congested sinuses and shrinkage of the legs. Muscles and bones that normally carry weight on earth tend to deteriorate, and the resulting bone loss can cause kidney stones. Medical researchers are investigating ways to confront these and other effects of long-duration space flight.

Example 5.11	**Acceleration down a hill**

A toboggan loaded with vacationing students (total weight w) slides down a long, snow-covered slope (Fig. 5.13a). The hill slopes at a constant angle α and the toboggan is so well waxed that there is virtually no friction. What is its acceleration?

SOLUTION

IDENTIFY: Our target variable is the acceleration, which we'll find using Newton's second law. There is no friction, so just two forces act on the toboggan: its weight w and the normal force n exerted by the hill. The new feature in this example is that the normal force points at an angle from the vertical and is not opposite to the weight. Hence we must use both components of $\sum \vec{F} = m\vec{a}$ in Eq. (5.4).

SET UP: Figure 5.13b shows the free-body diagram. We take axes parallel and perpendicular to the surface of the hill, so that the acceleration (which is parallel to the hill) is along the positive x-direction.

EXECUTE: The normal force has only a y-component, but the weight has both x- and y-components: $w_x = w \sin \alpha$ and $w_y = -w \cos \alpha$. (Compare to Example 5.4, in which the x-component of weight was $-w \sin \alpha$. The difference is that the positive x-axis was uphill in Example 5.4, while in Fig. 5.13b it is down-

hill.) The wiggly line in Fig. 5.13b reminds us that we have resolved the weight into its components.

The acceleration is purely in the $+x$-direction, so $a_y = 0$. Newton's second law in component form then tells us that

$$\sum F_x = w \sin \alpha = ma_x$$
$$\sum F_y = n - w \cos \alpha = ma_y = 0$$

Since $w = mg$, the x-component equation tells us that $mg \sin \alpha = ma_x$, or

$$a_x = g \sin \alpha$$

Note that we didn't need the y-component equation to find the acceleration. That's the beauty of choosing the x-axis to lie along the acceleration direction! What the y-components tell us is the magnitude of the normal force that the surface of the hill exerts on the toboggan:

$$n = w \cos \alpha = mg \cos \alpha$$

EVALUATE: Notice that the mass does not appear in our answer for the acceleration. This means that *any* toboggan, regardless of its mass or number of passengers, slides down a frictionless hill with an acceleration of $g \sin \alpha$. In particular, if the plane is horizontal, $\alpha = 0$ and $a_x = 0$ (the toboggan does not accelerate); if the plane is vertical, $\alpha = 90°$ and $a_x = g$ (the toboggan is in free fall).

(a)   (b)

5.13 (a) A loaded toboggan slides down a frictionless hill. (b) The free-body diagram shows that the x-component of the weight, $w \sin \alpha$, accelerates the toboggan down the hill.

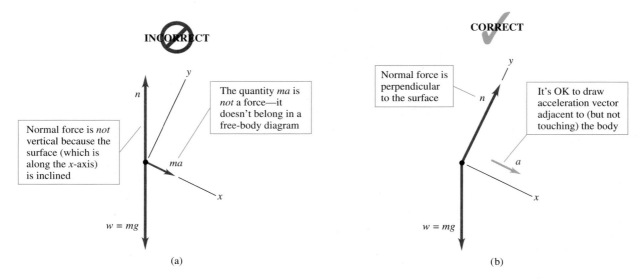

5.14 (a) An incorrect free-body diagram for the toboggan. (b) A correct free-body diagram.

Notice also that the normal force n is not equal to the toboggan's weight (compare Example 5.4 in Section 5.1). We don't need this result here, but it will be useful in a later example.

CAUTION   Figure 5.14a shows a common *incorrect* way to draw the free-body diagram for the toboggan. This diagram is wrong for two reasons: the normal force must be drawn perpendicular to the surface, and it's completely incorrect to include an "$m\vec{a}$ force." If you remember that "normal" means "perpendicular" and that the mass times the acceleration is not itself a force, you'll be well on your way to always drawing correct free-body diagrams (Fig. 5.14b)

<table>
<tr><td>Example
5.12</td><td>Two bodies with the same acceleration</td></tr>
</table>

You are pushing a 1.00-kg food tray through the cafeteria line with a constant 9.0-N force. As the tray moves, it pushes in turn on a 0.50-kg carton of milk (Fig. 5.15a). The tray and carton slide on a horizontal surface that is so greasy that friction can be neglected. Find the acceleration of the system and the horizontal force that the tray exerts on the carton.

SOLUTION

IDENTIFY: In this example we have *two* target variables, the acceleration a_x of the system of tray and carton and the force $F_{\text{T on C}}$ of the tray on the carton. Again we will use Newton's second law, but we'll have to apply it to two different bodies to get two equations (one for each target variable).

SET UP: We can set up the problem in either of two ways.

Method 1: We can treat the tray (mass m_{T}) and milk carton (mass m_{C}) as separate bodies, each with its own free-body diagram (Fig. 5.15b). Note that the force F that you exert on the tray doesn't appear in the free-body diagram for the milk carton. Instead, what's making the carton accelerate is the force of magnitude $F_{\text{T on C}}$ exerted on it by the tray. By Newton's third law, the carton exerts a force of equal magnitude on the tray: $F_{\text{C on T}} = F_{\text{T on C}}$. We take the acceleration to be in the positive x-direction; both the tray and milk carton move with the same acceleration a_x.

Method 2: We can treat the tray and milk carton as a composite body of mass $m = m_{\text{T}} + m_{\text{C}} = 1.50$ kg (Fig. 5.15c). The only horizontal force acting on this composite body is the force F that you exert. The forces $F_{\text{T on C}}$ and $F_{\text{C on T}}$ don't come into play because they're *internal* to this composite body, and Newton's second law tells us that only *external* forces affect a body's acceleration (see Section 4.3). That means we'll need an additional equation to find the magnitude $F_{\text{T on C}}$ using this method; we'll get this equation by also applying Newton's second law to the milk carton, as in Method 1.

EXECUTE: *Method 1:* The x-component equations of Newton's second law for the tray and for the carton are

$$\text{Tray:}\quad \sum F_x = F - F_{\text{C on T}} = F - F_{\text{T on C}} = m_{\text{T}}a_x$$
$$\text{Carton:}\quad \sum F_x = F_{\text{T on C}} = m_{\text{C}}a_x$$

These are two simultaneous equations for the two target variables a_x and $F_{\text{T on C}}$. (Two equations are all we need, which means that the y-components don't play a role in this example.) An easy way

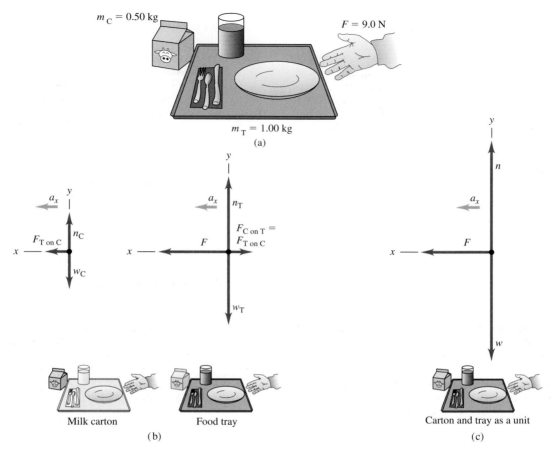

5.15 (a) Pushing a food tray and milk carton in the cafeteria line. (b) Individual free-body diagrams for the carton (left) and the tray (right). (c) Free-body diagram for the carton and the tray as a composite body.

to solve the two equations is to add them; this eliminates $F_{\text{T on C}}$, giving

$$F = m_T a_x + m_C a_x = (m_T + m_C) a_x$$

and

$$a_x = \frac{F}{m_T + m_C} = \frac{9.0 \text{ N}}{1.00 \text{ kg} + 0.50 \text{ kg}} = 6.0 \text{ m/s}^2$$

Substituting this back into the second equation (for the carton) gives

$$F_{\text{T on C}} = m_C a_x = (0.50 \text{ kg})(6.0 \text{ m/s}^2) = 3.0 \text{ N}$$

Method 2: The x-component of Newton's second law for the composite body of mass m is

$$\sum F_x = F = m a_x$$

and the acceleration of this composite body is

$$a_x = \frac{F}{m} = \frac{9.0 \text{ N}}{1.50 \text{ kg}} = 6.0 \text{ m/s}^2$$

Then, looking at the milk carton by itself, we see that to give it an acceleration of 6.0 m/s² requires that the tray exert a force

$$F_{\text{T on C}} = m_C a_x = (0.50 \text{ kg})(6.0 \text{ m/s}^2) = 3.0 \text{ N}$$

EVALUATE: The answers are the same with either method, as they should be. To check the answers, note that there are different forces on the two sides of the tray: $F = 9.0$ N on the right and $F_{\text{C on T}} = 3.0$ N on the left. Hence the net horizontal force on the tray is $F - F_{\text{C on T}} = 6.0$ N, exactly enough to accelerate a 1.00-kg tray at 6.0 m/s².

The method of treating the two bodies as a single composite body works *only* if the two bodies have the same magnitude *and* direction of acceleration. If the accelerations are different, we must treat the two bodies separately, as in the following example.

| Example 5.13 | **Two bodies with the same magnitude of acceleration** |

Figure 5.16a shows an air-track glider with mass m_1 moving on a level, frictionless air track in the physics lab. The glider is connected to a lab weight with mass m_2 by a light, flexible, nonstretching string that passes over a small frictionless pulley. Find the acceleration of each body and the tension in the string.

SOLUTION

IDENTIFY: We have objects that are accelerating, so again we must use Newton's second law. In this example there are *three* target variables: the tension T in the string and the accelerations of the two bodies. Hence we'll need to find three simultaneous equations that involve these variables.

SET UP: The two bodies have different motions—one horizontal, one vertical—so we can't consider them together as we did the bodies in Example 5.12. We need a separate free-body diagram for each. Figures 5.16b and 5.16c show a free-body diagram and a coordinate system for each body. It's convenient to have both bodies accelerate in positive axis directions, so we choose the positive y-direction for the lab weight to be downward. (It's perfectly all right to use different coordinate axes for the two bodies.)

There is no friction in the pulley and we consider the string to be massless, so the tension T in the string is the same throughout; it applies a force of the same magnitude T to each body. (You may want to review Conceptual Example 4.10 in Section 4.5, where we discussed the tension force exerted by a massless string.) The weights are m_1g and m_2g.

While the *directions* of the two accelerations are different, their *magnitudes* are the same. That's because the string doesn't stretch. Hence the two bodies must move equal distances in equal times, and so their speeds at any instant must be equal. When the speeds change, they change by equal amounts in a given time, so the accel-

erations of the two bodies must have the same magnitude a. We can express this relation as

$$a_{1x} = a_{2y} = a$$

Thanks to this relationship, we actually have only *two* target variables: a and the tension T.

EXECUTE: For the glider on the track, Newton's second law gives

$$\text{Glider: } \sum F_x = T = m_1a_{1x} = m_1a$$
$$\text{Glider: } \sum F_y = n + (-m_1g) = m_1a_{1y} = 0$$

For the lab weight, the only forces are in the y-direction, and

$$\text{Lab weight: } \sum F_y = m_2g + (-T) = m_2a_{2y} = m_2a$$

In these equations we've used the relationships $a_{1y} = 0$ (the glider doesn't accelerate vertically) and $a_{1x} = a_{2y} = a$ (the two objects have the same magnitude of acceleration).

The x-equation for the glider and the equation for the lab weight give us two simultaneous equations for the target variables T and a:

$$\text{Glider: } T = m_1a$$
$$\text{Lab weight: } m_2g - T = m_2a$$

We add the two equations to eliminate T, giving

$$m_2g = m_1a + m_2a = (m_1 + m_2)a$$

and so the magnitude of each body's acceleration is

$$a = \frac{m_2}{m_1 + m_2}g$$

Substituting this back into the first equation (for the glider), we get

$$T = \frac{m_1m_2}{m_1 + m_2}g$$

(a)

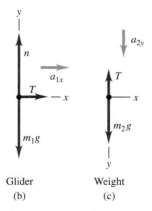

Glider
(b)

Weight
(c)

5.16 (a) A lab weight accelerates a glider along an air track. Free-body diagrams for (b) the glider (mass m_1) and for (c) the lab weight (mass m_2).

EVALUATE: The acceleration is less than g, as you might expect; the lab weight accelerates more slowly because the string tension pulls it back.

The tension T is *not* equal to the weight $m_2 g$ of the lab weight, but is *less* by a factor of $m_1/(m_1 + m_2)$. If T *were* equal to $m_2 g$, then the lab weight would be in equilibrium, and it isn't.

> **CAUTION** It's a common mistake to assume that if an object is attached to a vertical string, the string tension must be equal to the object's weight. That was the case in Example 5.5, where the acceleration was zero, but it would certainly be wrong in this

example! The only safe approach is to *always* treat the tension as a variable, as we have done here.

Finally, let's check some special cases. If $m_1 = 0$, then the lab weight would fall freely and there would be no tension in the string. The equations do give $T = 0$ and $a = g$ when $m_1 = 0$. Also, if $m_2 = 0$, we expect no tension and no acceleration; for this case the equations give $T = 0$ and $a = 0$. Thus in these two special cases the results agree with our intuitive expectations.

Test Your Understanding

You are riding a sled with frictionless runners down a hill with a 3.0° slope. The combined mass of you and the sled is 90 kg. As the sled moves, you drag your heels in the snow to keep your speed under control. What force must you exert parallel to the slope to make the sled slow down at a rate of 2.0 m/s²?

5.3 | Frictional Forces

We have seen several problems where a body rests or slides on a surface that exerts forces on the body, and we have used the terms *normal force* and *friction force* to describe these forces. Whenever two bodies interact by direct contact (touching) of their surfaces, we call the interaction forces *contact forces*. Normal forces and friction forces are both contact forces.

Our principal concern in this section is with friction, an important force in many aspects of everyday life. The oil in a car engine minimizes friction between moving parts, but without friction between the tires and the road we couldn't drive or turn the car. Air drag—the frictional force exerted by the air on a body moving through it—decreases automotive fuel economy but makes parachutes work. Without friction, nails would pull out, light bulbs and bottle caps would unscrew effortlessly, and sports such as bicycling and ice hockey would be hopeless (Fig. 5.17).

Kinetic and Static Friction

Let's consider a body sliding across a surface. When you try to slide a heavy box of books across the floor, the box doesn't move at all unless you push with a certain minimum force. Then the box starts moving, and you can usually keep it moving with less force than you needed to get it started. If you take some of the books out, you need less force than before to get it started or keep it moving. What general statements can we make about this behavior?

First, when a body rests or slides on a surface, we can always represent the contact force exerted by the surface on the body in terms of components of force perpendicular and parallel to the surface. As before, we call the perpendicular component vector the normal force, denoted by $\vec{n}$. The component vector parallel to the surface is the **friction force,** denoted by $\vec{f}$. By definition, $\vec{n}$ and $\vec{f}$ are always perpendicular to each other. If the surface is frictionless, then the contact force has only a normal component and $\vec{f}$ is zero. (Frictionless surfaces are an unattainable idealization, like a massless rope. But we can approximate a surface

5.17 The sport of ice hockey depends critically on having just the right amount of friction between a player's skates and the ice. If there were too much friction, the players would move too slowly; if there were too little friction, they could hardly keep from falling over.

as frictionless if the effects of friction are negligibly small.) The direction of the friction force is always such as to oppose relative motion of the two surfaces.

The kind of friction that acts when a body slides over a surface is called a **kinetic friction force** $\vec{f}_k$. The adjective "kinetic" and the subscript "k" remind us that the two surfaces are moving relative to each other. The *magnitude* of the kinetic friction force usually increases when the normal force increases. This is why it takes more force to slide a box full of books across the floor than to slide the same box when it is empty. This principle is also used in automotive braking systems: the harder the brake pads are squeezed against the rotating brake disks, the greater the braking effect. In many cases the magnitude of the kinetic friction force f_k is found experimentally to be approximately *proportional* to the magnitude n of the normal force. In such cases we represent the relation by the equation

$$f_k = \mu_k n \qquad (\text{magnitude of kinetic friction force}) \qquad (5.5)$$

where μ_k (pronounced "mu-sub-k") is a constant called the **coefficient of kinetic friction.** The more slippery the surface, the smaller the coefficient of friction. Because it is a quotient of two force magnitudes, μ_k is a pure number, without units.

> **CAUTION** Remember, the friction force and the normal force are always perpendicular. Equation (5.5) is not a vector equation, but a scalar relation between the magnitudes of the two perpendicular forces.

Equation (5.5) is only an approximate representation of a complex phenomenon. On a microscopic level, friction and normal forces result from the intermolecular forces (fundamentally electrical in nature) between two rough surfaces at points where they come into contact (Fig. 5.18). The actual area of contact is usually much smaller than the total surface area. As a box slides over the floor, bonds between the two surfaces form and break, and the total number of such bonds varies; hence the kinetic friction force is not perfectly constant. Smoothing the surfaces can actually increase friction, since more molecules are able to interact and bond; bringing two smooth surfaces of the same metal together can cause a "cold weld." Lubricating oils work because an oil film between two surfaces (such as the pistons and cylinder walls in a car engine) prevents them from coming into actual contact.

Table 5.1 shows a few representative values of μ_k. Although these values are given with two significant figures, they are only approximate since friction forces can also depend on the speed of the body relative to the surface. For now we'll ignore this effect and assume that μ_k and f_k are independent of speed, in order to concentrate on the simplest cases. Table 5.1 also lists coefficients of static friction; we'll define these shortly.

Friction forces may also act when there is *no* relative motion. If you try to slide that box of books across the floor, the box may not move at all because the floor exerts an equal and opposite friction force on the box. This is called a **static friction force** $\vec{f}_s$. In Fig. 5.19a, the box is at rest, in equilibrium, under the action of its weight $\vec{w}$ and the upward normal force $\vec{n}$. The normal force is equal in magnitude to the weight $(n = w)$ and is exerted on the box by the floor. Now we tie a rope to the box (Fig. 5.19b) and gradually increase the tension T in the rope. At first the box remains at rest because, as T increases, the force of static friction f_s also increases (staying equal in magnitude to T).

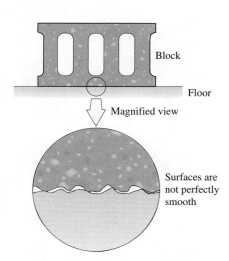

5.18 The normal and friction forces arise from interactions between molecules at high points on the surfaces of the block and the floor.

Table 5.1 Approximate Coefficients of Friction

Materials	Coefficient of static friction, μ_s	Coefficient of kinetic friction, μ_k
Steel on steel	0.74	0.57
Aluminum on steel	0.61	0.47
Copper on steel	0.53	0.36
Brass on steel	0.51	0.44
Zinc on cast iron	0.85	0.21
Copper on cast iron	1.05	0.29
Glass on glass	0.94	0.40
Copper on glass	0.68	0.53
Teflon on Teflon	0.04	0.04
Teflon on steel	0.04	0.04
Rubber on concrete (dry)	1.0	0.8
Rubber on concrete (wet)	0.30	0.25

No applied force, box at rest
$f_s = 0$

(a)

Weak applied force, box remains at rest
$f_s < \mu_s n$

(b)

Stronger applied force, box just about to slide
$f_s = \mu_s n$

(c)

Box sliding at constant speed
$f_k = \mu_k n$

(d)

5.19 (a), (b), (c) When there is no relative motion of the surfaces, the magnitude of the static friction force f_s is less than or equal to $\mu_s n$. (d) When there is relative motion, the magnitude of the kinetic friction force f_k equals $\mu_k n$.

At some point T becomes greater than the maximum static friction force f_s the surface can exert. Then the box "breaks loose" (the tension T is able to break the bonds between molecules in the surfaces of the box and floor) and starts to slide. Figure 5.19c is the force diagram when T is at this critical value. If T exceeds this value, the box is no longer in equilibrium. For a given pair of surfaces the maximum value of f_s depends on the normal force. Experiment shows that in many cases this maximum value, called $(f_s)_{max}$, is approximately *proportional* to n; we call the proportionality factor μ_s the **coefficient of static friction.** Table 5.1 lists some representative values of μ_s. In a particular situation, the actual force of static friction can have any magnitude between zero (when there is no other force parallel to the surface) and a maximum value given by $\mu_s n$. In symbols,

$$f_s \leq \mu_s n \quad (\text{magnitude of static friction force}) \quad (5.6)$$

Like Eq. (5.5), this is a relation between magnitudes, *not* a vector relation. The equality sign holds only when the applied force T, parallel to the surface, has reached the critical value at which motion is about to start (Fig. 5.19c). When T is less than this value (Fig. 5.19b), the inequality sign holds. In that case we have to use the equilibrium conditions $(\Sigma \vec{F} = \mathbf{0})$ to find f_s. If there is no applied force $(T = 0)$ as in Fig. 5.19a, then there is no static friction force either $(f_s = 0)$.

As soon as sliding starts (Fig. 5.19d), the friction force usually *decreases;* in other words, it's easier to keep the box moving than to start it moving. Hence the coefficient of kinetic friction is usually *less* than the coefficient of static friction for any given pair of surfaces, as shown in Table 5.1. If we start with no applied force $(T = 0)$ at time $t = 0$ and gradually increase the force, the friction force varies somewhat as shown in Fig. 5.20.

In some situations the surfaces will alternately stick (static friction) and slip (kinetic friction). This is what causes the horrible squeak made by chalk held at the wrong angle while writing on the blackboard. Another stick-slip phenomenon is the squeaky noise your windshield-wiper blades make when the glass is nearly dry; still another is the outraged shriek of tires sliding on asphalt pavement. A more positive example is the motion of a violin bow against the string.

When a body slides on a layer of gas, friction can be made very small. In the linear air track used in physics laboratories, the gliders are supported on a layer of air. The frictional force is velocity-dependent, but at typical speeds the effective coefficient of friction is of the order of 0.001. A similar device is the air hockey table, where the pucks are supported by an array of small air jets about 2 cm apart.

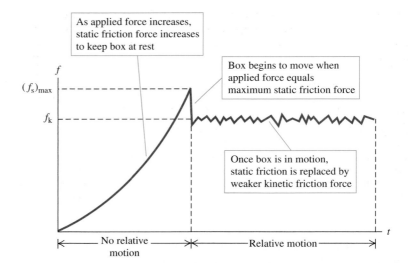

5.20 In response to an increasing externally applied force, the friction force increases to $(f_s)_{max}$. Then the surfaces begin to slide across one another, and the frictional force drops back to a nearly constant value f_k. The kinetic friction force varies somewhat as intermolecular bonds form and break.

As applied force increases, static friction force increases to keep box at rest

Box begins to move when applied force equals maximum static friction force

Once box is in motion, static friction is replaced by weaker kinetic friction force

No relative motion

Relative motion

| Example 5.14 | **Friction in horizontal motion** |

A delivery company has just unloaded a 500-N crate full of home exercise equipment in your driveway (Fig. 5.21a). You find that to get it started moving toward your garage, you have to pull with a horizontal force of magnitude 230 N. Once it "breaks loose" and starts to move, you can keep it moving at constant velocity with only 200 N. What are the coefficients of static and kinetic friction?

SOLUTION

IDENTIFY: The state of rest and the state of motion with constant velocity are both equilibrium conditions, so we use Newton's first law as expressed by Eq. (5.2). In either situation there are four

forces acting on the crate: the downward weight force (magnitude $w = 500$ N), the upward normal force (magnitude n) exerted by the ground, a tension force (magnitude T) to the right exerted by the rope, and a friction force to the left exerted by the ground. The target variables are μ_s and μ_k; to find these we'll need Eqs. (5.5) and (5.6), which involve the friction coefficients.

SET UP: Figure 5.21b shows the free-body diagram an instant before the crate starts to move, when the static friction force has its maximum possible value $(f_s)_{max} = \mu_s n$. Once the crate is moving to the right at constant velocity, the friction force changes to its

Crate as it starts to move
(b)

Crate moving at constant velocity
(c)

(a)

5.21 (a) Pulling a crate with a horizontal force. (b) Free-body diagram for the crate as it starts to move. (c) Free-body diagram for the crate moving at constant velocity.

kinetic form (Fig. 5.21c). Because the rope in Fig. 5.21a is in equilibrium, the tension is the same at both ends. Hence the tension force that the rope exerts on the crate has the same magnitude as the force you exert on the rope.

EXECUTE: Just before the crate starts to move, we have

$$\sum F_x = T + (-(f_s)_{max}) = 0 \quad \text{so} \quad (f_s)_{max} = T = 230 \text{ N}$$

$$\sum F_y = n + (-w) = 0 \quad \text{so} \quad n = w = 500 \text{ N}$$

To find the value of μ_s, we use the relationship $(f_s)_{max} = \mu_s n$. Hence

$$\mu_s = \frac{(f_s)_{max}}{n} = \frac{230 \text{ N}}{500 \text{ N}} = 0.46$$

After the crate starts to move, the forces are as shown in Fig. 5.21c, and we have

$$\sum F_x = T + (-f_k) = 0 \quad \text{so} \quad f_k = T = 200 \text{ N}$$

$$\sum F_y = n + (-w) = 0 \quad \text{so} \quad n = w = 500 \text{ N}$$

Then, using $f_k = \mu_k n$ from Eq. (5.5),

$$\mu_k = \frac{f_k}{n} = \frac{200 \text{ N}}{500 \text{ N}} = 0.40$$

EVALUATE: It's easier to keep the crate moving than to start it moving, and so the coefficient of kinetic friction is less than the coefficient of static friction.

Example 5.15	**Static friction can be less than the maximum**

In Example 5.14, what is the friction force if the crate is at rest on the surface and a horizontal force of 50 N is applied to it?

SOLUTION

IDENTIFY: The applied force is less than the maximum force of static friction, $(f_s)_{max} = 230$ N. Hence the crate remains at rest and the net force acting on it is zero. The target variable is the magnitude f_s of the friction force.

SET UP: The free-body diagram is the same as in Fig. 5.21b, but with $(f_s)_{max}$ replaced by f_s and $T = 230$ N replaced by $T = 50$ N.

EXECUTE: From the equilibrium conditions, Eq. (5.2), we have

$$\sum F_x = T + (-f_s) = 0 \quad \text{so} \quad f_s = T = 50 \text{ N}$$

EVALUATE: In this case, f_s is less than the maximum value $(f_s)_{max} = \mu_s n$. The frictional force can prevent motion for any horizontal applied force up to 230 N.

Example 5.16	**Minimizing kinetic friction**

In Example 5.14, suppose you try to move your crate full of exercise equipment by tying a rope around it and pulling upward on the rope at an angle of 30° above the horizontal (Fig. 5.22a). How hard do you have to pull to keep the crate moving with constant velocity? Is this easier or harder than pulling horizontally? Assume $w = 500$ N and $\mu_k = 0.40$.

SOLUTION

IDENTIFY: The crate is in equilibrium since its velocity is constant, so we again apply Newton's first law. Since the crate is in motion, the ground exerts a *kinetic* friction force. The target variable is the magnitude T of the tension force.

SET UP: Figure 5.22b is a free-body diagram showing the forces on the crate. The kinetic friction force f_k is still equal to $\mu_k n$, but now the normal force n is *not* equal in magnitude to the weight of the

crate. The force exerted by the rope has an additional vertical component that tends to lift the crate off the floor.

EXECUTE: From the equilibrium conditions and the equation $f_k = \mu_k n$, we have

$$\sum F_x = T \cos 30° + (-f_k) = 0 \quad \text{so} \quad T \cos 30° = \mu_k n$$

$$\sum F_y = T \sin 30° + n + (-w) = 0 \quad \text{so} \quad n = w - T \sin 30°$$

These are two simultaneous equations for the two unknown quantities T and n. To solve them, we can eliminate one unknown and solve for the other. There are many ways to do this; one way is to substitute the expression for n in the second equation back into the first equation:

$$T \cos 30° = \mu_k (w - T \sin 30°)$$

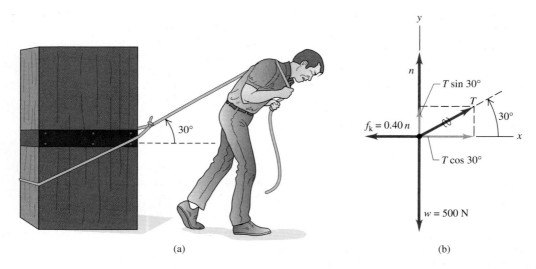

(a)   (b)

5.22 (a) Pulling a crate with a force applied at an upward angle. (b) Free-body diagram for the crate moving at constant velocity.

Then solve this equation for T, with the result

$$T = \frac{\mu_k w}{\cos 30° + \mu_k \sin 30°} = 188 \text{ N}$$

Finally, substitute this result back into either of the original equations to obtain n. If we use the second equation to do this, we get

$$n = w - T \sin 30° = (500 \text{ N}) - (188 \text{ N}) \sin 30° = 406 \text{ N}$$

EVALUATE: The normal force is *less* than the weight of the box ($w = 500 \text{ N}$) because the vertical component of tension pulls upward on the crate. Despite this, the tension required is a little less than the 200-N force needed when you pulled horizontally in Example 5.14. Try pulling at 22°; you'll find you need even less force (see Challenge Problem 5.123).

Example
5.17 **Toboggan ride with friction I**

Let's go back to the toboggan we studied in Example 5.11 (Section 5.2). The wax has worn off and there is now a coefficient of kinetic friction μ_k. The slope has just the right angle to make the toboggan slide with constant speed. Derive an expression for the slope angle in terms of w and μ_k.

SOLUTION

IDENTIFY: Our target variable is the slope angle α. The toboggan is in equilibrium because its velocity is constant, so once again we use Newton's first law. There are three forces acting on the toboggan: its weight, the normal force, and the kinetic friction force. Since the motion is downhill, the kinetic friction force (which opposes the motion of the toboggan over the hill) is directed uphill. The magnitude of the friction force is given by Eq. (5.5), $f_k = \mu_k n$.

SET UP: Figure 5.23 shows the free-body diagram. We take axes perpendicular and parallel to the surface and represent the weight in terms of its components in these two directions, as shown. (Compare Fig. 5.13b in Example 5.11.)

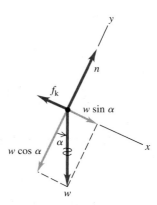

5.23 Free-body diagram for the toboggan with friction.

EXECUTE: The equilibrium conditions are

$$\sum F_x = w \sin \alpha + (-f_k) = w \sin \alpha - \mu_k n = 0$$
$$\sum F_y = n + (-w \cos \alpha) = 0$$

(We used the relationship $f_k = \mu_k n$ in the equation for the x-components.) Rearranging, we get

$$\mu_k n = w \sin \alpha \quad \text{and} \quad n = w \cos \alpha$$

Just as in Example 5.11, the normal force n is *not* equal to the weight w. When we divide the first of these equations by the second, we find

$$\mu_k = \frac{\sin \alpha}{\cos \alpha} = \tan \alpha \quad \text{so} \quad \alpha = \arctan \mu_k$$

EVALUATE: The weight w doesn't appear in this expression. *Any* toboggan, regardless of its weight, slides down an incline with constant speed if the coefficient of kinetic friction equals the tangent of the slope angle of the incline. The greater the coefficient of friction, the steeper the slope has to be for the toboggan to slide with constant velocity. This is just what we should expect.

Example 5.18	**Toboggan ride with friction II**

What if the same toboggan with the same coefficient of friction is sliding downhill as in Example 5.17, but on a steeper hill? This time the toboggan accelerates, although not as much as in Example 5.11, when there was no friction. Derive an expression for the acceleration in terms of g, α, μ_k, and w.

SOLUTION

IDENTIFY: The toboggan is no longer in equilibrium, but accelerates as it slides down the hill. Hence we must use Newton's second law, $\sum \vec{F} = m\vec{a}$, in its component form as given in Eq. (5.4). Our target variable is the downhill acceleration.

SET UP: The free-body diagram (Fig. 5.24) is almost the same as for Example 5.17. The toboggan's y-component of acceleration a_y is still zero, but the x-component a_x is not.

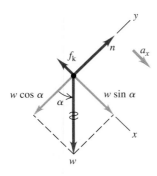

5.24 Free-body diagram for the toboggan with friction, going down a steeper hill.

EXECUTE: It's convenient to express the weight as $w = mg$. Then, using Newton's second law for the x- and y-components of force, we have

$$\sum F_x = mg \sin \alpha + (-f_k) = ma_x$$
$$\sum F_y = n + (-mg \cos \alpha) = 0$$

From the second equation and Eq. (5.5) we get an expression for f_k:

$$n = mg \cos \alpha$$
$$f_k = \mu_k n = \mu_k mg \cos \alpha$$

We substitute this back into the x-component equation. The result is

$$mg \sin \alpha + (-\mu_k mg \cos \alpha) = ma_x$$
$$a_x = g(\sin \alpha - \mu_k \cos \alpha)$$

EVALUATE: Does this result make sense? Here are some special cases we can check. First, if the hill is vertical, $\alpha = 90°$; then $\sin \alpha = 1$, $\cos \alpha = 0$, and $a_x = g$. This is free fall, just what we would expect. Second, on a hill at angle α with no friction, $\mu_k = 0$. Then $a_x = g \sin \alpha$. The situation is the same as in Example 5.11, and we get the same result; that's encouraging! Next, suppose that there is just enough friction to make the toboggan move with constant velocity. In that case, $a_x = 0$ and our result gives

$$\sin \alpha = \mu_k \cos \alpha \quad \text{and} \quad \mu_k = \tan \alpha$$

This agrees with our result from Example 5.17; good! Finally, note that there may be so much friction that $\mu_k \cos \alpha$ is actually greater than $\sin \alpha$. In that case, a_x is negative; if we give the toboggan an initial downhill push to start it moving, it will slow down and eventually stop.

We have pretty much beaten the toboggan problem to death, but there is an important lesson to be learned. We started out with a simple problem and then extended it to more and more general situations. The general result we found in this example includes *all* the previous ones as special cases, and that's a nice, neat package! Don't memorize this package; it is useful only for this one set of problems. But do try to understand how we obtained it and what it means.

One final variation that you may want to try out is the case in which we give the toboggan an initial push up the hill. The direction of the kinetic friction force is now reversed, so the acceleration is different from the downhill value. It turns out that the expression for a_x is the same as for downhill motion except that the minus sign becomes plus. Can you prove this?

Rolling Friction

It's a lot easier to move a loaded filing cabinet across a horizontal floor using a cart with wheels than to slide it. How much easier? We can define a **coefficient of rolling friction** μ_r, which is the horizontal force needed for constant speed on a flat surface divided by the upward normal force exerted by the surface. Transportation engineers call μ_r the *tractive resistance*. Typical values of μ_r are 0.002 to 0.003 for steel wheels on steel rails and 0.01 to 0.02 for rubber tires on concrete. These values show one reason why railroad trains are generally much more fuel-efficient than highway trucks.

Example 5.19	**Motion with rolling friction**

A typical car weighs about 12,000 N (about 2700 lb). If the coefficient of rolling friction is $\mu_r = 0.015$, what horizontal force is needed to make the car move with constant speed on a level road? Neglect air resistance.

SOLUTION

IDENTIFY: The car is moving with constant velocity, so this is an equilibrium problem that uses Newton's first law. The four forces on the car are the weight, the upward normal force, the backward force of rolling friction, and the unknown forward force F (our target variable).

SET UP: The free-body diagram is much like the one in Fig. 5.21c of Example 5.14, but with the kinetic friction force replaced by the rolling friction force f_r and with the tension force replaced by the unknown force F.

EXECUTE: As in Example 5.14, Newton's first law for the *vertical* components tells us that the normal force is equal in magnitude to the car's weight. Hence, from the definition of μ_r, the rolling friction force f_r is

$$f_r = \mu_r n = (0.015)(12{,}000 \text{ N}) = 180 \text{ N} \quad (\text{about 41 lb})$$

Newton's first law for the *horizontal* components tells us that a forward force with this magnitude is needed to keep the car moving with constant speed.

EVALUATE: The required force is rather small, which is why it's possible to push a car by hand. (As in the case of sliding, it's easier to keep a car rolling than it is to start it rolling.) We've ignored the effects of air resistance, which is a pretty good approximation if the car is moving slowly. But at typical highway speeds air resistance is a larger effect than rolling friction.

Try applying this analysis to your crate of exercise equipment (Example 5.14). If the delivery company brings it on a rubber-wheel dolly with $\mu_r = 0.02$, only a 10-N force is needed to keep it moving at constant velocity. Can you verify this?

Fluid Resistance and Terminal Speed

Sticking your hand out the window of a fast-moving car will convince you of the existence of **fluid resistance,** the force that a fluid (a gas or liquid) exerts on a body moving through it. The moving body exerts a force on the fluid to push it out of the way. By Newton's third law, the fluid pushes back on the body with an equal and opposite force.

The *direction* of the fluid resistance force acting on a body is always opposite the direction of the body's velocity relative to the fluid. The *magnitude* of the fluid resistance force usually increases with the speed of the body through the fluid; contrast this behavior with that of the kinetic friction force between two surfaces in contact, which we can usually regard as independent of speed. For low speeds, the magnitude f of the resisting force of the fluid is approximately proportional to the body's speed v:

$$f = kv \qquad (\text{fluid resistance at low speed}) \qquad (5.7)$$

where k is a proportionality constant that depends on the shape and size of the body and the properties of the fluid. In motion through air at the speed of a tossed tennis ball or faster, the resisting force is approximately proportional to v^2 rather than to v. It is then called **air drag** or simply *drag*. Airplanes, falling raindrops, and cars moving at high speed all experience air drag. In this case we replace Eq. (5.7) by

$$f = Dv^2 \qquad \text{(fluid resistance at high speed)} \qquad (5.8)$$

Because of the v^2 dependence, air drag increases rapidly with increasing speed. The air drag on a typical car is negligible at low speeds but comparable to or greater than tractive resistance at highway speeds. The value of D depends on the shape and size of the body and on the density of the air. You should verify that the units of the constant k in Eq. (5.7) are N·s/m or kg/s, and that the units of the constant D in Eq. (5.8) are N·s^2/m^2 or kg/m.

Because of the effects of fluid resistance, an object falling in a fluid does *not* have a constant acceleration. To describe its motion, we can't use the constant-acceleration relationships from Chapter 2; instead, we have to start over using Newton's second law. Let's consider the following situation. You release a rock at the surface of a deep pond and it falls to the bottom (Fig. 5.25a). The fluid resistance force in this situation is given by Eq. (5.7). What are the acceleration, velocity, and position of the rock as functions of time?

Figure 5.25b shows the free-body diagram. We take the positive y-direction to be downward and neglect any force associated with buoyancy in the water. Since the rock is moving downward, the speed v is equal to the y-velocity v_y and the fluid resistance force is in the $-y$-direction. There are no x-components, so Newton's second law gives

$$\sum F_y = mg + (-kv_y) = ma_y$$

When the rock first starts to move, $v_y = 0$, the resisting force is zero, and the initial acceleration is $a_y = g$. As the speed increases, the resisting force also increases, until finally it is equal in magnitude to the weight. At this time $mg - kv_y = 0$, the acceleration becomes zero, and there is no further increase in speed. The final speed v_t, called the **terminal speed,** is given by $mg - kv_t = 0$, or

$$v_t = \frac{mg}{k} \qquad \text{(terminal speed, fluid resistance } f = kv) \qquad (5.9)$$

Figure 5.26 shows how the acceleration, velocity, and position vary with time. As time goes by, the acceleration approaches zero and the velocity approaches v_t (remember that we chose the positive y-direction to be down). The slope of the graph of y versus t becomes constant as the velocity becomes constant.

To see how the graphs in Fig. 5.26 are derived, we must find the relation between speed and time during the interval before the terminal speed is reached. We go back to Newton's second law, which we rewrite as

$$m\frac{dv_y}{dt} = mg - kv_y$$

After rearranging terms and replacing mg/k by v_t, we integrate both sides, noting that $v_y = 0$ when $t = 0$:

$$\int_0^v \frac{dv_y}{v_y - v_t} = -\frac{k}{m}\int_0^t dt$$

5.25 (a) A rock falling downward through a fluid (water). (b) Free-body diagram for the rock.

5.26 Graphs of acceleration, velocity, and position versus time for a body falling with fluid resistance proportional to the speed (dark color curves) and without fluid resistance (light color curves).

which integrates to

$$\ln \frac{v_t - v_y}{v_t} = -\frac{k}{m}t \quad \text{or} \quad 1 - \frac{v_y}{v_t} = e^{-(k/m)t}$$

and finally

$$v_y = v_t \left[1 - e^{-(k/m)t} \right] \tag{5.10}$$

Note that v_y becomes equal to the terminal speed v_t only in the limit that $t \to \infty$; the rock cannot attain terminal speed in any finite length of time.

The derivative of v_y gives a_y as a function of time, and the integral of v_y gives y as a function of time. We leave the derivations for you to complete (see Exercise 5.40); the results are

$$a_y = ge^{-(k/m)t} \tag{5.11}$$

$$y = v_t \left[t - \frac{m}{k} \left(1 - e^{-(k/m)t} \right) \right] \tag{5.12}$$

Now look again at Fig. 5.26, which shows graphs of these three relations.

In deriving the terminal speed in Eq. (5.9), we assumed that the fluid resistance force is proportional to the speed. For an object falling through the air at high speeds, so that the fluid resistance is proportional to v^2 as in Eq. (5.8), you can show that the terminal speed v_t is given by

$$v_t = \sqrt{\frac{mg}{D}} \quad \left(\text{terminal speed, fluid resistance } f = Dv^2 \right) \tag{5.13}$$

This expression for terminal speed explains why heavy objects in air tend to fall faster than light objects. Two objects with the same physical size but different mass (say, a table-tennis ball and a lead ball with the same radius) have the same value of D but different values of m. The more massive object has a larger terminal speed and falls faster. The same idea explains why a sheet of paper falls faster if you first crumple it into a ball; the mass m is the same, but the smaller size makes D smaller (less air drag for a given speed) and v_t larger. Skydivers use the same principle to control their descent (Fig. 5.27).

Figure 5.28 shows the trajectory of a baseball with and without air drag, assuming a coefficient $D = 1.3 \times 10^{-3}$ kg/m (appropriate for a batted ball or

5.27 By changing the positions of their arms and legs while falling, skydivers can change the value of the constant D in Eq. (5.8) and hence adjust the terminal speed of their fall [Eq. (5.13)].

5.28 Computer-generated trajectories of a baseball (mass 0.145 kg, radius 0.0366 m) launched at a 35° angle above the horizontal at a speed of 50 m/s. The two curves show the trajectories with and without air drag. Note that the scales are different on the horizontal and vertical axes.

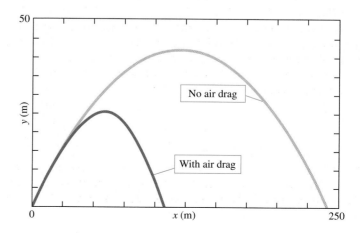

pitched fastball and a ballpark at sea level). In this example the baseball was given an initial velocity of 50 m/s at an angle of 35° above the +x-axis. You can see that both the range of the baseball and the maximum height reached are substantially less than the zero-drag calculation would lead you to believe. Hence the baseball trajectory we calculated in Example 3.8 (Section 3.3) by ignoring air drag is quite unrealistic. Air drag is an important part of the game of baseball!

Example 5.20	**Terminal speed of a skydiver**

For a human body falling though air in a spread-eagle position (Fig. 5.27), the numerical value of the constant D in Eq. (5.8) is about 0.25 kg/m. Find the terminal speed for an 80-kg skydiver.

SOLUTION

IDENTIFY AND SET UP: We use Eq. (5.13) to find the target variable v_t.

EXECUTE: We find

$$v_t = \sqrt{\frac{mg}{D}} = \sqrt{\frac{(80 \text{ kg})(9.8 \text{ m/s}^2)}{0.25 \text{ kg/m}}}$$

$$= 56 \text{ m/s (about 200 km/h, or 125 mi/h)}$$

EVALUATE: Because the terminal speed is quite high, skydives don't last very long. A drop from 2800 m (9200 ft) to the surface takes only $(2800 \text{ m})/(56 \text{ m/s}) = 50$ s.

When the sky diver deploys the parachute, the value of D increases greatly. Hence the terminal speed of the sky diver and parachute decreases dramatically to much less than 56 m/s.

Test Your Understanding

Once again you are riding a sled down a hill with a 3.0° slope. The combined mass of you and the sled is 90 kg. This time you don't drag your heels, and the sled speeds up at a rate of 0.25 m/s². What is the coefficient of kinetic friction between the snow and the sled's runners?

5.4 | Dynamics of Circular Motion

We talked about uniform circular motion in Section 3.4. We showed that when a particle moves in a circular path with constant speed, the particle's acceleration is always directed toward the center of the circle (perpendicular to the instantaneous velocity). The magnitude a_{rad} of the acceleration is constant and is given in terms of the speed v and the radius R of the circle by

$$a_{\text{rad}} = \frac{v^2}{R} \tag{5.14}$$

The subscript "rad" is a reminder that at each point the acceleration is radially inward toward the center of the circle, perpendicular to the instantaneous velocity. We explained in Section 3.4 why this acceleration is often called *centripetal acceleration*.

Act|v
ONLINE
Physics

4.2   Circular Motion Problem Solving

4.3   Cart Goes Over Circular Path

4.4   Ball Swings on a String

4.5   Car Circles a Track

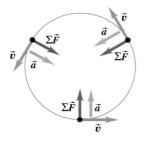

5.29 In uniform circular motion, both the acceleration and the net force are directed toward the center of the circle.

We can also express the centripetal acceleration a_{rad} in terms of the *period T*, the time for one revolution:

$$T = \frac{2\pi R}{v} \tag{5.15}$$

In terms of the period, a_{rad} is

$$a_{\text{rad}} = \frac{4\pi^2 R}{T^2} \tag{5.16}$$

Uniform circular motion, like all other motion of a particle, is governed by Newton's second law. The particle's acceleration toward the center of the circle must be caused by a force, or several forces, such that the vector sum $\sum \vec{F}$ is a vector that is always directed toward the center (Fig. 5.29). The magnitude of the acceleration is constant, so the magnitude F_{net} of the net inward radial force must also be constant. If an ice skater whirls his partner around in a circle on the ice, he must constantly pull on her inward toward the center of the circle. If he lets go, then the inward force no longer acts and his partner flies off in a straight line tangent to the circle (Fig. 5.30).

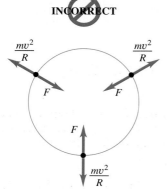

The quantity mv^2/R is *not* a force—it doesn't belong in a free-body diagram

(a)

CORRECT

If you include the acceleration, draw it to one side of the body to show that it's not a force

(b)

5.31 (a) Incorrect and (b) correct free-body diagrams for a body in uniform circular motion.

5.30 If the inward radial force suddenly ceases to act on a body in circular motion, the body flies off in a straight line with constant velocity (as it must, since the net force acting is zero).

The magnitude of the radial acceleration is given by $a_{\text{rad}} = v^2/R$, so the magnitude F_{net} of the net inward radial force on a particle with mass m must be

$$F_{\text{net}} = ma_{\text{rad}} = m\frac{v^2}{R} \tag{5.17}$$

Uniform circular motion can result from *any* combination of forces, just so the net force $\sum \vec{F}$ is always directed toward the center of the circle and has a constant magnitude.

CAUTION   In doing problems involving uniform circular motion, you may be tempted to include an extra outward force of magnitude $m(v^2/R)$ to "keep the body out there" or to "keep it in equilibrium"; this outward force is usually called *"centrifugal force."* ("Centrifugal" means "fleeing from the center.") Resist this temptation, because this approach is simply wrong! (See Fig. 5.31a.) Here's why: First, the body does *not* "stay out there." It is in constant motion around its circular path. Because its velocity is constantly changing in direction, the body accelerates and is *not* in equilibrium. Second, if there *were* an additional outward ("centrifugal") force to balance the inward force, there would then be no net inward force to cause the circular motion, and the body would move in a straight line, not a circle (Fig. 5.30). And third, the quantity $m(v^2/R)$ is *not* a force; it corresponds to the $m\vec{a}$ side of $\Sigma\vec{F} = m\vec{a}$, and does not appear in $\Sigma\vec{F}$ (Fig. 5.31b). It's certainly true that a passenger riding in a car going around a circular path on a level road tends to slide to the outside of the turn, as though responding to a "centrifugal force." But such a passenger is in an accelerating, noninertial frame of reference in which Newton's first and second laws don't apply. As we discussed in Section 4.2, what really happens is that the passenger tends to keep moving in a straight line, and the outer side of the car "runs into" the passenger as the car turns (Fig. 4.8c). *In an inertial frame of reference there is no such thing as "centrifugal force."* We promise not to mention this term again, and we strongly advise you to avoid using it as well.

Example 5.21	**Force in uniform circular motion**

A sled with a mass of 25.0 kg rests on a horizontal sheet of essentially frictionless ice. It is attached by a 5.00-m rope to a post set in the ice. Once given a push, the sled revolves uniformly in a circle around the post (Fig. 5.32a). If the sled makes five complete revolutions every minute, find the force F exerted on it by the rope.

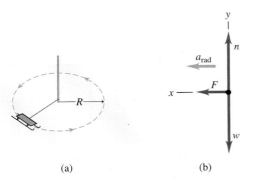

5.32 (a) A sled moving in uniform circular motion on a horizontal frictionless surface. (b) Free-body diagram for the sled. The $+x$-direction points toward the center of the circle.

SOLUTION

IDENTIFY: The sled is moving in uniform circular motion, so it has a radial acceleration. We will apply the component form of Newton's second law to the sled to determine the magnitude F of the force exerted by the rope, which is our target variable.

SET UP: Figure 5.32b shows the free-body diagram for the sled. The acceleration has only an x-component; this is toward the center of the circle, so we denote it as a_{rad}. The acceleration isn't given, so we'll need to determine its value using either Eq. (5.14) or Eq. (5.16).

EXECUTE: There is no acceleration in the y-direction, so the net force in that direction (the sum of the normal force and the weight) is zero. For the x-direction, Newton's second law gives

$$\Sigma F_x = F = ma_{\text{rad}}$$

We can find the centripetal acceleration a_{rad} using Eq. (5.16). The sled moves in a circle of radius $R = 5.00$ m with a period $T = (60.0\text{ s})/(5\text{ rev}) = 12.0$ s, so

$$a_{\text{rad}} = \frac{4\pi^2 R}{T^2} = \frac{4\pi^2(5.00\text{ m})}{(12.0\text{ s})^2} = 1.37\text{ m/s}^2$$

Alternatively, we can first use Eq. (5.15) to find the speed v:

$$v = \frac{2\pi R}{T} = \frac{2\pi(5.00\text{ m})}{12.0\text{ s}} = 2.62\text{ m/s}$$

Then, using Eq. (5.14),

$$a_{\text{rad}} = \frac{v^2}{R} = \frac{(2.62\text{ m/s})^2}{5.00\text{ m}} = 1.37\text{ m/s}^2$$

Hence the magnitude F of the force exerted by the rope is

$$F = ma_{\text{rad}} = (25.0\text{ kg})(1.37\text{ m/s}^2)$$
$$= 34.3\text{ kg}\cdot\text{m/s}^2 = 34.3\text{ N}$$

EVALUATE: A greater force would be needed if the sled moved around the circle at a higher speed v. In fact, if v were doubled while R remained the same, F would be four times greater. Can you show this? How would F change if v remained the same but the radius R were doubled?

Example
5.22   **The conical pendulum**

An inventor who dares to be different proposes to make a pendulum clock using a pendulum bob with mass m at the end of a thin wire of length L. Instead of swinging back and forth, the bob moves in a horizontal circle with constant speed v, with the wire making a constant angle β with the vertical direction (Fig. 5.33a). Assuming that the angle β is known, find the tension F in the wire and the period T (the time for one revolution of the bob).

SOLUTION

IDENTIFY: To find our two target variables—the tension F and period T—we need two simultaneous equations. These will be the horizontal and vertical components of Newton's second law applied to the bob. We'll find the acceleration of the bob toward the center of the circle using one of the circular motion equations.

SET UP: Figure 5.33b shows a free-body diagram for the bob as well as a coordinate system. The forces on the bob in the position shown are the weight mg and the tension F in the wire. Note that the center of the circular path is at the center of the shaded area, *not* at

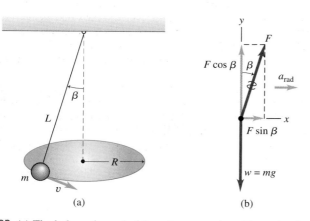

5.33 (a) The bob on the end of the wire moves in uniform circular motion. (b) Free-body diagram for the bob. The $+x$-direction points toward the center of the circle.

the top end of the wire. The horizontal component of tension is the force that produces the horizontal acceleration (toward the center of the circle).

EXECUTE: The tension has a horizontal component $F \sin \beta$ and a vertical component $F \cos \beta$. The system has zero vertical acceleration; the horizontal acceleration is toward the center of the circle, so we use the symbol a_{rad}. The $\sum \vec{F} = m\vec{a}$ equations are

$$\sum F_x = F \sin \beta = ma_{\text{rad}}$$
$$\sum F_y = F \cos \beta + (-mg) = 0$$

These are two simultaneous equations for the unknowns F and β. The equation for $\sum F_y$ gives $F = mg/\cos \beta$; substituting this result into the equation for $\sum F_x$ and using $(\sin \beta)/(\cos \beta) = \tan \beta$, we find

$$\tan \beta = \frac{a_{\text{rad}}}{g}$$

To relate β to the period T, we use Eq. (5.16) for a_{rad}. The radius of the circle is $R = L \sin \beta$, so

$$a_{\text{rad}} = \frac{4\pi^2 R}{T^2} = \frac{4\pi^2 L \sin \beta}{T^2}$$

Substituting this into $\tan \beta = a_{\text{rad}}/g$, we obtain

$$\tan \beta = \frac{4\pi^2 L \sin \beta}{gT^2}$$

which we can rewrite as

$$T = 2\pi \sqrt{\frac{L \cos \beta}{g}}$$

EVALUATE: For a given length L, as the angle β increases, the $\cos \beta$ decreases, the period T becomes smaller, and the tension $F = mg/\cos \beta$ increases. The angle can never be 90°, however; this would require that $T = 0$, $F = \infty$, and $v = \infty$. This system is called a *conical pendulum* because the suspending wire traces out a cone. It would not make a very good clock because the period depends on the angle β in such a direct way.

Example
5.23   **Rounding a flat curve**

The BMW Z4 roadster in Example 3.12 (Section 3.4) is rounding a flat, unbanked curve with radius R (Fig. 5.34a). If the coefficient of friction between tires and road is μ_s, what is the maximum speed v_{max} at which the driver can take the curve without sliding?

SOLUTION

IDENTIFY: The car's acceleration as it rounds the curve has magnitude $a_{\text{rad}} = v^2/R$, so the maximum speed corresponds to the maximum acceleration a_{rad}. This acceleration corresponds to the maximum horizontal force on the car toward the center of its circular path. The only horizontal force acting on the car is the friction force exerted by the road. So we'll approach this problem using Newton's second law and our knowledge of the friction force from Section 5.3.

SET UP: The free-body diagram in Fig. 5.34b includes the car's weight $w = mg$ and two contact forces exerted by the road: the normal force (magnitude n) and the horizontal friction force (magnitude f). The friction force must point in the positive x-direction (toward the center of the circular path) in order to cause the radial acceleration. Since the car doesn't move in the radial direction (that is, it doesn't slide toward or away from the center of the circle), the friction force is *static* friction with a maximum magnitude $f_{max} = \mu_s n$ [see Eq. (5.6)].

EXECUTE: The acceleration toward the center of the circular path is $a_{rad} = v^2/R$ and there is no vertical acceleration. Thus we have

$$\sum F_x = f = ma_{rad} = m\frac{v^2}{R}$$
$$\sum F_y = n + (-mg) = 0$$

The second equation shows that $n = mg$. The first equation shows that the friction force *needed* to keep the car moving in its circular path increases with the car's speed. But the maximum friction force *available* is $f_{max} = \mu_s n = \mu_s mg$, and this determines the car's maximum speed. Substituting f_{max} for f and v_{max} for v in the $\sum F_x$ equation, we find

$$\mu_s mg = m\frac{v_{max}^2}{R}$$

so the maximum speed is

$$v_{max} = \sqrt{\mu_s g R}$$

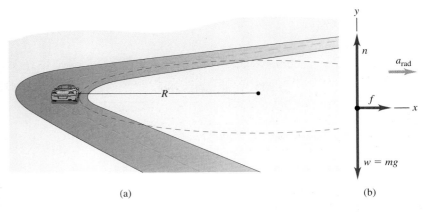

5.34 (a) A sports car rounding a curve on a level road. (b) Free-body diagram for the car. The +x-direction is toward the center of the car's circular path.

As an example, if $\mu_s = 0.87$ and $R = 230$ m, then

$$v_{max} = \sqrt{(0.87)(9.8 \text{ m/s}^2)(230 \text{ m})} = 44 \text{ m/s}$$

or about 160 km/h (100 mi/h). This is the maximum speed for this radius.

EVALUATE: If the car's speed is slower than $\sqrt{\mu_s g R}$, the required friction force is less than the maximum possible value $f_{max} = \mu_s mg$ and the car can easily make the curve. If we try to take the curve going *faster* than the maximum speed, the car can still go in a circle without skidding, but the radius will be larger and the car will run off the road.

Note that the maximum centripetal acceleration (called the "lateral acceleration" in Example 3.12) is equal to $\mu_s g$. If the coefficient of friction is reduced, the maximum centripetal acceleration and v_{max} are also reduced. That's why it's best to take curves at a lower speed if the road is wet or icy (either of which can reduce the value of μ_s).

Example 5.24	**Rounding a banked curve**

For a car traveling at a certain speed, it is possible to bank a curve at just the right angle so that no friction at all is needed to maintain the car's turning radius. Then a car can round the curve even on wet ice with Teflon tires. Bobsled racing depends on this same idea. An engineer proposes to rebuild the curve in Example 5.23 so that a car moving at speed v can safely make the turn even if there is no friction (Fig. 5.35a). At what angle β should the curve be banked?

SOLUTION

IDENTIFY: With no friction, the only two forces acting on the car are its weight and the normal force. Because the road is banked, the normal force (which acts perpendicular to the road surface) has a horizontal component. This component causes the car's centripetal

acceleration (which is in the horizontal direction, toward the center of the car's circular path). Since forces and acceleration are involved, we'll use Newton's second law to find the target variable β.

SET UP: The free-body diagram (Fig. 5.35b) is very similar to the diagram for the conical pendulum in Example 5.22 (Fig. 5.33b). The normal force acting on the car plays the role of the tension acting on the pendulum bob.

EXECUTE: The normal force $\vec{n}$ is perpendicular to the roadway at an angle β with the vertical. Thus it has a vertical component $n \cos \beta$ and a horizontal component $n \sin \beta$, as Fig. 5.35b shows. The acceleration in the x-direction is the centripetal acceleration,

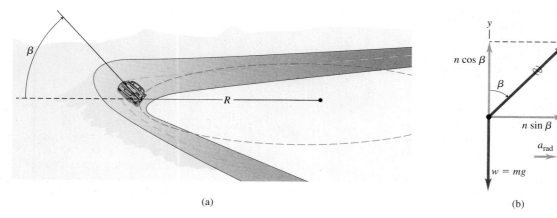

5.35 (a) A sports car rounding a banked curve. If the curved is banked at the correct angle β, no friction is needed to round the curve at speed v, (b) Free-body diagram for the car. The $+x$-direction points toward the center of the car's circular path.

there is no acceleration in the y-direction. Thus the equations of Newton's second law are

$$\sum F_x = n \sin \beta = ma_{\mathrm{rad}}$$

$$\sum F_y = n \cos \beta + (-mg) = 0$$

From the $\sum F_y$ equation, $n = mg/\cos \beta$. Substituting this into the $\sum F_x$ equation gives an expression for the banking angle:

$$\tan \beta = \frac{a_{\mathrm{rad}}}{g}$$

which is the same expression we found in Example 5.23. Finally, substituting the expression $a_{\mathrm{rad}} = v^2/R$, we have

$$\tan \beta = \frac{v^2}{gR}$$

EVALUATE: The banking angle depends on the speed and the radius. For a given radius, no one angle is correct for all speeds. In the design of highways and railroads, curves are often banked for the average speed of the traffic over them. If $R = 230$ m and $v = 25$ m/s (equal to a highway speed of 88 km/h, or 55 mi/h), then

$$\beta = \arctan \frac{(25 \text{ m/s})^2}{(9.8 \text{ m/s}^2)(230 \text{ m})} = 15°$$

This is within the range of banking angles actually used in highways. With the same radius and $v = 44$ m/s, as in Example 5.23, $\beta = 41°$; such steeply banked curves are found at automobile raceways.

The results of Example 5.24 also apply to an airplane when it makes a turn in level flight (Fig 5.36a). When an airplane is flying in a straight line at a constant speed and at a steady altitude, the airplane's weight is exactly balanced by the lift force $\vec{L}$ due to the wings. (The lift force is actually exerted by the air; as the wings move through the air, they push air downward, and so by Newton's third law, the air pushes upward on the wings.) To make the airplane turn, the pilot banks the airplane to one side (Fig. 5.36a) so that the lift force has a horizontal component as in Fig. 5.36b. (The pilot also changes the angle at which the wings "bite" into the air so that the vertical component of lift continues to balance the weight.) The bank angle is related to the airplane's speed v and the radius R of the turn by the same expression as in Example 5.24: $\tan \beta = v^2/gR$. For an airplane to make a tight turn (small R) at high speed (large v), $\tan \beta$ must be large and the required bank angle β must approach 90°.

We can also apply the results of Example 5.24 to the *pilot* of an airplane. The free-body diagram for the pilot of the airplane is exactly as shown in Fig. 5.35b; the normal force $n = mg/\cos \beta$ is exerted on the pilot by the seat. As in Example

(a)

(b)

5.36 (a) An airplane banks to one side in order to turn in that direction. (b) Free-body diagram for the airplane. The vertical component of the lift force $\vec{L}$ (provided by the air pushing on the wing) balances the force of gravity; the horizontal component of $\vec{L}$ causes the acceleration v^2/R.

5.10, n is equal to the apparent weight of the pilot, which is greater than the pilot's true weight mg. In a tight turn with a large bank angle β, the pilot's apparent weight can be tremendous: $n = 5.8mg$ at $\beta = 80°$ and $n = 9.6mg$ at $\beta = 84°$. Pilots black out in such tight turns because the apparent weight of their blood increases by the same factor, and the human heart isn't strong enough to pump such apparently "heavy" blood to the brain.

Motion in a Vertical Circle

In all the preceding examples the body moved in a horizontal circle. Motion in a vertical circle is no different in principle, but the weight of the body has to be treated carefully. The following example shows what we mean.

Example 5.25	**Uniform circular motion in a vertical circle**

A passenger on a carnival Ferris wheel moves in a vertical circle of radius R with constant speed v. Assuming that the seat remains upright during the motion, derive expressions for the force the seat exerts on the passenger at the top of the circle and at the bottom.

SOLUTION

IDENTIFY: At both the top and bottom of the circle, the target variable is the magnitude n of the normal force that the seat exerts on the passenger. We'll find this force at each position using Newton's second law and the equations of uniform circular motion.

SET UP: The diagram in Fig. 5.37a shows the passenger's velocity and acceleration at the two positions. Note that the acceleration points downward at the top of the circle but upward at the bottom of the circle. At each position the only forces acting are vertical: the

5.37 (a) The passenger's acceleration vector has the same magnitude at the top and bottom of the Ferris wheel, and in both cases points toward the center of the circle. (b) Free-body diagram for a passenger at the top of the circle, and (c) free-body diagram for a passenger at the bottom of the circle.

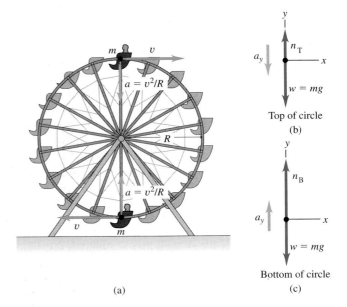

(a)

upward normal force and the downward force of gravity. Hence we only need the vertical component of Newton's second law.

EXECUTE: Figures 5.37b and 5.37c show free-body diagrams for the two positions. We take the positive y-direction as upward in both cases. Let n_T be the upward normal force the seat applies to the passenger at the top of the circle, and let n_B be the normal force at the bottom. At the top the acceleration has magnitude v^2/R, but its vertical component is negative because its direction is downward. Hence $a_y = -v^2/R$, and Newton's second law at the top tells us that

$$\text{Top: } \sum F_y = n_T + (-mg) = -m\frac{v^2}{R} \quad \text{or}$$

$$n_T = m\left(g - \frac{v^2}{R}\right)$$

At the bottom the acceleration is upward, so $a_y = +v^2/R$ and Newton's second law is

$$\text{Bottom: } \sum F_y = n_B + (-mg) = +m\frac{v^2}{R} \quad \text{or}$$

$$n_B = m\left(g + \frac{v^2}{R}\right)$$

EVALUATE: Our result for n_T tells us that at the top of the Ferris wheel, the upward force the seat applies to the passenger is *smaller* in magnitude than the passenger's weight. If the ride goes fast enough that $g - v^2/R$ becomes zero, the seat applies *no* force, and the passenger is about to become airborne. If v becomes still larger, n_T becomes negative; this means that a *downward* force (such as from a seat belt) is needed to keep the passenger in the seat. By contrast, the normal force n_B at the bottom is always *greater than* the passenger's weight. You feel the seat pushing up on you more firmly than when you are at rest.

We recognize that n_T and n_B are the values of the passenger's *apparent weight* at the top and bottom of the circle. As a specific example, suppose the passenger's mass is 60.0 kg, the radius of the circle is $R = 8.00$ m, and the wheel makes one revolution in 10.0 s. From Eqs. (5.14) and (5.16),

$$a_{rad} = \frac{v^2}{R} = \frac{4\pi^2 R}{T^2} = \frac{4\pi^2(8.00 \text{ m})}{(10.0 \text{ s})^2} = 3.16 \text{ m/s}^2$$

The two values of the apparent weight are

$$n_T = (60.0 \text{ kg})(9.80 \text{ m/s}^2 - 3.16 \text{ m/s}^2) = 398 \text{ N}$$
$$n_B = (60.0 \text{ kg})(9.80 \text{ m/s}^2 + 3.16 \text{ m/s}^2) = 778 \text{ N}$$

The passenger's weight is 588 N, and n_T and n_B are about 30% less than and greater than the weight, respectively.

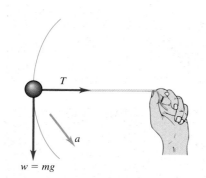

$$w = mg$$

5.38 A ball moving in a vertical circle. At most points around the circle, a component of the net force (and hence a component of acceleration) is tangent to the circle. Hence the ball is in nonuniform circular motion.

When we tie a string to an object and whirl it in a vertical circle, the analysis in Example 5.25 isn't directly applicable. The reason is that v is *not* constant in this case; at every point in the circle except the top and bottom, the net force (and hence the acceleration) does *not* point toward the center of the circle (Fig. 5.38). So both $\sum \vec{F}$ and $\vec{a}$ have a component tangent to the circle, and this is a case of *non-uniform* circular motion (see Section 3.5). Even worse, we can't use the constant-acceleration formulas to relate the speeds at various points because *neither* the magnitude nor the direction of the acceleration is constant. The speed relations we need are best obtained by using the concept of energy. We'll consider such problems in Chapter 7.

Test Your Understanding

Satellites are held in orbit by the force of our planet's gravitational attraction. A satellite in a small-radius orbit moves at a higher speed than a satellite in an orbit of large radius. Based on this information, explain how you can conclude that the Earth's gravitational attraction decreases with increasing distance from the planet.

*5.5 | The Fundamental Forces of Nature

We have discussed several kinds of forces—including weight, tension, friction, fluid resistance, and the normal force—and we will encounter others as we continue our study of physics. But just how many kinds of forces are there? Our current understanding is that all forces are expressions of just four distinct classes of *fundamental* forces, or interactions between particles (Fig. 5.39). Two are familiar in everyday experience. The other two involve interactions between subatomic particles that we cannot observe with the unaided senses.

Of the two familiar classes, **gravitational interactions** were the first to be studied in detail. The *weight* of a body results from the Earth's gravitational attraction acting on it. The sun's gravitational attraction for the Earth keeps the Earth in its nearly circular orbit around the sun. Newton recognized that the motions of the planets around the sun and the free fall of objects on Earth both result from gravitational forces. In Chapter 12 we will study gravitational interactions in greater detail, and we will analyze their vital role in the motions of planets and satellites.

The second familiar class of forces, **electromagnetic interactions,** includes electric and magnetic forces. If you run a comb through your hair, you can then use it to pick up bits of paper or fluff; this interaction is the result of electric charge on the comb. All atoms contain positive and negative electric charge, so atoms and molecules can exert electric forces on each other. Contact forces, including the normal force, friction, and fluid resistance, are the combination of all such forces exerted on the atoms of a body by atoms in its surroundings. *Magnetic* forces occur in interactions between magnets or between a magnet and a piece of iron. These may seem to form a different category, but magnetic interactions are actually the result of electric charges in motion. In an electromagnet an electric current in a coil of wire causes magnetic interactions. We will study electric and magnetic interactions in detail in the second half of this book.

These two interactions differ enormously in their strength. The electrical repulsion between two protons at a given distance is stronger than their gravitational attraction by a factor of the order of 10^{35}. Gravitational forces play no significant role in atomic or molecular structure. But in bodies of astronomical size, positive and negative charge are usually present in nearly equal amounts, and the resulting electrical interactions nearly cancel out. Gravitational interactions are thus the dominant influence in the motion of planets and in the internal structure of stars.

The other two classes of interactions are less familiar. One, the **strong interaction,** is responsible for holding the nucleus of an atom together. Nuclei contain electrically neutral neutrons and positively charged protons. The charged protons repel each other, and a nucleus could not be stable if it were not for the presence of an attractive force of a different kind that counteracts the repulsive electrical interactions. In this context the strong interaction is also called the *nuclear force.* It has much shorter range than electrical interactions, but within its range it is much stronger. The strong interaction is also responsible for the creation of unstable particles in high-energy particle collisions.

Finally, there is the **weak interaction.** It plays no direct role in the behavior of ordinary matter, but it is of vital importance in interactions among fundamental particles. The weak interaction is responsible for a common form of radioactivity called beta decay, in which a neutron in a radioactive nucleus is transformed into a proton while ejecting an electron and an essentially massless particle called an antineutrino. The weak interaction between the antineutrino and ordinary matter is so feeble that an antineutrino could easily penetrate a wall of lead a million kilometers thick!

During the past several decades a unified theory of the electromagnetic and weak interactions has been developed. We now speak of the *electroweak* interaction, and in a sense this reduces the number of classes of interactions from four to three. Similar attempts have been made to understand strong, electromagnetic, and weak interactions on the basis of a single unified theory called a *grand unified theory* (GUT), and the first tentative steps have been taken toward a possible unification of all interactions into a *theory of everything* (TOE). Such theories are still speculative, and there are many unanswered questions in this very active field of current research.

(a)

(b)

(c)

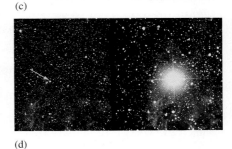
(d)

5.39 Examples of the fundamental interactions in nature. (a) Gravitational forces cause the motion of the moon around the earth. (b) Electromagnetic forces form atoms into molecules. This atomic force microscope image shows a molecule of bacterial plasmid DNA. (c) Strong forces between nuclear particles are responsible for thermonuclear reactions in the center of the sun. The released energy reaches us as sunlight. (d) Weak forces within atomic nuclei play a crucial role when a star (shown on the left by an arrow) explodes into a supernova (shown on the right).

SUMMARY

When a body is in equilibrium in an inertial frame of reference, the vector sum of forces acting on it must be zero. Free-body diagrams are essential in identifying the forces that act on the body being considered.

Newton's third law is also frequently needed in equilibrium problems. The two forces in an action-reaction pair *never* act on the same body. (See Examples 5.1 through 5.5)

Vector form:

$$\sum \vec{F} = 0 \qquad (5.1)$$

Component form:

$$\sum F_x = 0$$
$$\sum F_y = 0 \qquad (5.2)$$

If the vector sum of forces on a body is *not* zero, the body accelerates. Its acceleration is given by Newton's second law.

As they are for equilibrium problems, free-body diagrams are essential for solving problems involving Newton's second law. (See Examples 5.6 through 5.13)

Vector form:

$$\sum \vec{F} = m\vec{a} \qquad (5.3)$$

Component form:

$$\sum F_x = ma_x$$
$$\sum F_y = ma_y \qquad (5.4)$$

The contact force between two bodies can always be represented in terms of a normal force $\vec{n}$ perpendicular to the surface of contact and a friction force $\vec{f}$ parallel to the surface. The normal force exerted on a body by a surface is *not* always equal to the body's weight.
(See Examples 5.3, 5.10, and 5.11)

When a body is sliding over the surface, the friction force is called kinetic friction. Its magnitude f_k is approximately equal to the normal force magnitude n multiplied by the coefficient of kinetic friction μ_k.
(See Examples 5.14 and 5.16 through 5.18)

Magnitude of kinetic friction force:

$$f_k = \mu_k n \qquad (5.5)$$

When a body is not moving relative to a surface, the friction force is called static friction. The *maximum* possible static friction force is approximately equal to the magnitude n of the normal force multiplied by the coefficient of static friction μ_s. The *actual* static friction force may be anything from zero to this maximum value, depending on the situation. Usually μ_s is greater than μ_k for a given pair of surfaces in contact. (See Examples 5.14 and 5.15)

Magnitude of static friction force:

$$f_s \le \mu_s n \qquad (5.6)$$

In uniform circular motion, the acceleration vector is directed toward the center of the circle. Just as in any other dynamics problem, the motion is governed by Newton's second law, $\sum \vec{F} = m\vec{a}$. (See Examples 5.21 through 5.25)

Acceleration in uniform circular motion:

$$a_{\text{rad}} = \frac{v^2}{R} = \frac{4\pi^2 R}{T^2} \quad (5.14), (5.16)$$

Key Terms

apparent weight, 166
friction force, 171
kinetic friction force, 172
coefficient of kinetic friction, 172
static friction force, 172

coefficient of static friction, 173
coefficient of rolling friction, 178
fluid resistance, 178
air drag, 179
terminal speed, 179

gravitational interaction, 189
electromagnetic interaction, 189
strong interaction, 189
weak interaction, 189

Your Notes

Answer to Chapter Opening Question

Neither: the aerodynamic forces are still in balance with the force of gravity. Although the bird is ascending, its velocity is constant and so its acceleration is zero. Hence the net force on the bird must also be zero.

Answers to Test Your Understanding Questions

Section 5.1 The two cables are arranged symmetrically, so the tension in each cable has the same magnitude T. The vertical component of the tension from each cable is $T \sin 45°$ (or, equivalently, $T \cos 45°$), so Newton's first law applied to the vertical forces tells us that $2T \sin 45° - mg = 0$. Hence $T = mg/(2 \sin 45°) = mg/\sqrt{2} = 0.71mg$.

Section 5.2 The situation is the same as in Example 5.11, but with an additional force of unknown magnitude F_{heels} acting in the negative x-direction (uphill). Hence the x-component equation for Newton's second law is $w \sin \alpha - F_{heels} = ma_x$, and so $F_{heels} = w \sin \alpha - ma_x$. With $m = 90$ kg, $w = mg = (90 \text{ kg})(9.8 \text{ m/s}^2)$, $\alpha = 3.0°$, and $a_x = -2.0$ m/s^2 (the acceleration is uphill, opposite the motion), we get $F_{heels} = 2.3 \times 10^2$ N.

Section 5.3 This situation is the same as in Example 5.18. We found in that example that $a_x = g(\sin \alpha - \mu_k \cos \alpha)$; solving for μ_k, we get $\mu_k = (g \sin \alpha - a_x)/(g \cos \alpha) = [(9.8 \text{ m/s}^2)\sin 3.0° - (0.25 \text{ m/s}^2)]/[(9.8 \text{ m/s}^2)\cos 3.0°] = 0.027$.

Section 5.4 A satellite of mass m orbiting the earth at speed v in an orbit of radius r has an acceleration of magnitude v^2/r, so the net force acting on it from the earth's gravity has magnitude $F = mv^2/r$. The farther the satellite is from earth, the greater the value of r, the smaller the value of v, and hence the smaller the values of v^2/r and of F. In other words, the earth's gravitational force decreases with increasing distance.

Discussion Questions

Q5.1 A man sits in a seat that is suspended from a rope. The rope passes over a pulley suspended from the ceiling, and the man holds the other end of the rope in his hands. What is the tension in the rope, and what force does the seat exert on the man? Draw a free-body force diagram for the man.

Q5.2 "In general, the normal force is not equal to the weight." Give an example where the two forces are equal in magnitude, and at least two examples where they are not.

Q5.3 A clothesline hangs between two poles. No matter how tightly the line is stretched, it always sags a little at the center. Explain why.

Q5.4 A car is driven up a steep hill at constant speed. Discuss all the forces acting on the car. What pushes it up the hill?

Q5.5 For medical reasons it is important for astronauts in outer space to determine their body mass at regular intervals. Devise a scheme for measuring body mass in an apparently weightless environment.

Q5.6 To push a box up a ramp, is the force required smaller if you push horizontally or if you push parallel to the ramp? Why?

Q5.7 A woman in an elevator lets go of her briefcase but it does not fall to the floor. How is the elevator moving?

Q5.8 You can classify scales for weighing objects as those that use springs and those that use standard masses to balance unknown masses. Which group would be more accurate when you use it in an accelerating elevator? When you use it on the moon? Does it matter whether you are trying to determine mass or weight?

Q5.9 When you tighten a nut on a bolt, how are you increasing the frictional force? How does a lock washer work?

Q5.10 A block rests on an inclined plane with enough friction to prevent it from sliding down. To start the block moving, is it easier to push it up the plane, down the plane, or sideways? Why?

Q5.11 A crate of books rests on a level floor. To move it along the floor at a constant velocity, why do you exert a smaller force if you pull it at an angle θ above the horizontal than if you push it at the same angle below the horizontal?

Q5.12 To stop a car on an icy road, is it better to push the brake pedal hard enough to "lock" the wheels and make them slide or to push gently so that the wheels continue to roll? Why?

Q5.13 Can the force of kinetic friction acting on an object ever cause that object to *speed up*? If it cannot, explain why not. If it can, give at least one example. Repeat for the force of *static* friction.

Q5.14 When you stand with bare feet in a wet bathtub, the grip feels fairly secure, and yet a catastrophic slip is quite possible. Explain this in terms of the two coefficients of friction.

Q5.15 You are pushing a large crate from the back of a freight elevator to the front as the elevator is moving to the next floor. In which situation is the force you must apply to move the crate the smallest and in which is it the largest: when the elevator is accelerating upward, when it is accelerating downward, or when it is traveling at constant speed? Explain.

Q5.16 Roller Coaster Road near Tucson, Arizona, was so named because it resembled an amusement park ride. If you drove on it at constant speed, where was the normal force the greatest and the least? Why would it have been especially poor road design to put an unbanked right or left curve at the top of one of the hills?

Q5.17 An automotive magazine calls decreasing-radius curves "the bane of the Sunday driver." Explain.

Q5.18 If you hang fuzzy dice from your rear-view mirror and drive through a banked curve, how can you tell whether you are traveling less than, equal to, or greater than the speed used to calculate the banking angle?

Q5.19 If there is a net force on a particle in uniform circular motion, why does the particle's speed not change?

Q5.20 A curve in a road has the banking angle calculated for 80 km/h. However, the road is covered with ice, and you plan to creep around the highest lane at 20 km/h. What may happen to your car? Why?

Q5.21 If you swing a ball on the end of a lightweight string in a horizontal circle at constant speed, the string at any instant does not lie exactly along the radius vector from the center of the path to the ball's location. Does the string lie above or below the horizontal? In relation to the direction the ball is moving, is the string in front of or behind the radius vector? Draw a free-body force diagram for the ball and use it to explain your answers. (Note that air resistance may be a factor.)

Q5.22 The centrifugal force is not included in the free-body diagrams of Figs. 5.34b and 5.35b. Explain why not.

Q5.23 A professor swings a rubber stopper in a horizontal circle on the end of a string in front of his class. He tells Caroline, in the first row, that he is going to let the string go when the stopper is directly in front of her face. Should Caroline worry?

Q5.24 To keep the forces on the riders within allowable limits, loop-the-loop roller coaster rides are often designed so the loop, rather than being a perfect circle, has a larger radius of curvature at the bottom than at the top. Explain.

Q5.25 A tennis ball drops from rest at the top of a tall glass cylinder, first with the air pumped out of the cylinder so there is no air resistance, and then a second time after the air has been readmitted to the cylinder. You examine multiflash photographs (like Fig. 2.21) of the two drops. From these photos how can you tell which one is which, or can you?

Q5.26 If you throw a baseball straight upward with speed v_0, how does its speed, when it returns to the point from where you threw it, compare to v_0 a) in the absence of air resistance? b) in the presence of air resistance? Explain.

Q5.27 You take two identical tennis balls and fill one with water. You release both balls simultaneously from the top of a tall building. If air resistance is negligible, which ball strikes the ground first? Explain. What is the answer if air resistance is *not* negligible?

Q5.28 A sheet of paper is falling through the air at its terminal speed. An identical sheet of paper is crumpled tightly into a ball and then dropped from the top of a tall ladder. Which piece of paper, the flat sheet or the crumpled ball, falls with the greater terminal speed? Explain. How does the air resistance force on the flat sheet of paper compare with the air resistance force on the crumpled ball when each is falling at its terminal speed? Explain.

Q5.29 You throw a baseball straight upward. If air resistance is *not* ignored, how does the time required for the ball to go from the height at which it was thrown up to its maximum height compare to the time required for it to fall from its maximum height back down to the height from which it was thrown? Explain your answer.

Q5.30 When does a baseball in flight have an acceleration with a positive upward component? Explain in terms of the forces on the ball and also in terms of the velocity components compared to the terminal speed. Do *not* ignore air resistance.

Q5.31 When a batted baseball moves with air drag, does it spend more time climbing to its maximum height or descending from its maximum height back to the ground? Or is the time the same for both? Explain in terms of the forces acting on the ball.

Q5.32 When a batted baseball moves with air drag, does it travel a greater horizontal distance while climbing to its maximum height or while descending from its maximum height back to the ground? Or is the horizontal distance traveled the same for both? Explain in terms of the forces acting on the ball.

Q5.33 "A ball is thrown from the edge of a high cliff. No matter what the angle at which it is thrown, due to air resistance, the ball will eventually end up moving vertically downward." Justify this statement.

Exercises

Section 5.1 Using Newton's First Law: Particles In Equilibrium

5.1 Two 25.0-N weights are suspended at opposite ends of a rope that passes over a light, frictionless pulley. The pulley is attached to a chain that goes to the ceiling. a) What is the tension in the rope? b) What is the tension in the chain?

5.2 In Fig. 5.40 each of the suspended blocks has weight w. The pulleys are frictionless and the ropes have negligible weight. Calculate, in each case, the tension T in the rope in terms of the weight w. In each case, include the free-body diagram or diagrams you used to determine the answer.

(a)   (b)   (c)

Figure 5.40 Exercise 5.2.

5.3 An adventurous archaeologist crosses between two rock cliffs by slowly going hand-over-hand along a rope stretched between the cliffs. He stops to rest at the middle of the rope (Fig. 5.41). The rope will break if the tension in it exceeds 2.50×10^4 N, and our hero's mass is 90.0 kg. a) If the angle θ is 10.0°, find the tension in the rope. b) What is the smallest value the angle θ can have if the rope is not to break?

Figure 5.41 Exercise 5.3.

5.4 A picture frame hung against a wall is suspended by two wires attached to its upper corners. If the two wires make the same angle with the vertical, what must this angle be if the tension in each wire is equal to 0.75 of the weight of the frame? (Ignore any friction between the wall and the picture frame.)

5.5 Solve the problem in Example 5.4 using coordinate axes where the y-axis is vertical and the x-axis is horizontal. Do you get the same answers as obtained in the example using this different set of axes?

5.6 Certain streets in San Francisco make an angle of 17.5° with the horizontal. What force parallel to the street surface is required to keep a loaded 1967 Corvette of mass 1390 kg from rolling down such a street?

5.7 A large wrecking ball is held in place by two light steel cables (Fig. 5.42). If the mass m of the wrecking ball is 4090 kg, what is a) the tension T_B in the cable that makes an angle of 40° with the vertical? b) the tension T_A in the horizontal cable?

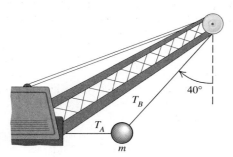

Figure 5.42 Exercise 5.7.

5.8 Find the tension in each cord in Fig. 5.43 if the weight of the suspended object is w.

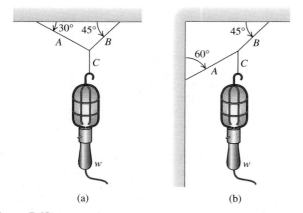

(a)   (b)

Figure 5.43 Exercise 5.8.

5.9 At a certain part of your drive from home to school, your car (mass 1600 kg) will coast in neutral at a constant speed of 72 km/h if there is no wind. Examination of a topographical map shows that for this stretch of straight highway the elevation decreases 200 m for each 6000 m of road. What is the total resistive force (friction plus air resistance) that acts on your car when it is traveling at 72 km/h?

5.10 A man pushes on a piano with mass 180 kg so that it slides at constant velocity down a ramp that is inclined at 11.0° above the horizontal. Neglect friction acting on the piano. If the force applied by the man is parallel to the incline, calculate the magnitude of this force.

5.11 In Fig. 5.44 the weight w is 60.0 N. a) What is the tension in the diagonal string? b) Find the magnitudes of the horizontal forces $\vec{F}_1$ and $\vec{F}_2$ that must be applied to hold the system in the position shown.

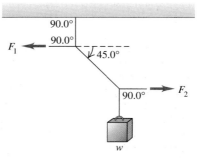

Figure 5.44 Exercise 5.11.

5.12 A tether ball leans against the post to which it is attached (Fig. 5.45). If the string to which the ball is attached is 1.40 m long and the ball has a radius of 0.110 m with mass 0.270 kg, what is the tension in the rope and what is the force the pole exerts on the ball? You can ignore any friction between the ball and the pole. (The string is attached to the ball such that a line along the string passes through the center of the ball.)

5.13 Two blocks, each with weight w, are held in place on a frictionless incline (Fig. 5.46). In terms of w and the angle α of the incline, calculate the tension in a) the rope connecting the

Figure 5.45 Exercise 5.12

blocks; b) the rope that connects block A to the wall. c) Calculate the magnitude of the force that the incline exerts on each block. d) Interpret your answers for the cases $\alpha = 0$ and $\alpha = 90°$.

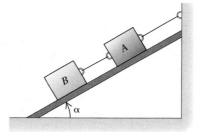

Figure 5.46 Exercise 5.13.

5.14 An airplane is in level flight at a constant speed. There are four forces acting on it: its weight $w = mg$; a forward force F provided by the engine (the *thrust*); the air resistance or *drag* force f, which acts opposite to the direction of flight; and an upward lift force L provided by the wings, which acts perpendicular to the plane of the wings and to the direction of flight. The drag force f is proportional to the square of the speed. a) Show that $F = f$ and $L = w$. b) Suppose the pilot pushes the throttle forward, doubling the value of F, while maintaining a constant altitude. The airplane will eventually attain a new, faster, constant speed. At this faster speed, how does the new value of f compare to the previous value? c) How much faster than the original speed is the airplane going now?

Section 5.2 Using Newton's Second Law: Dynamics of Particles

5.15 Atwood's Machine. A 15.0-kg load of bricks hangs from one end of a rope that passes over a small, frictionless pulley. A 28.0-kg counterweight is suspended from the other end of the rope, as shown in Fig. 5.47. The system is released from rest. a) Draw two free-body diagrams, one for the load of bricks and one for the counterweight. b) What is the magnitude of the upward acceleration of the load of bricks? c) What is the tension in the rope while the load is moving? How does the tension compare to the weight of the load of bricks? To the weight of the counterweight?

Figure 5.47 Exercise 5.15.

5.16 A 8.00-kg block of ice, released from rest at the top of a 1.50-m-long frictionless ramp, slides downhill, reaching a speed of 2.50 m/s at the bottom. What is the angle between the ramp and the horizontal?

5.17 A light rope is attached to a block with mass 4.00 kg that rests on a frictionless, horizontal surface. The horizontal rope passes over a frictionless, massless pulley and a block with mass m is suspended from the other end. When the blocks are released, the tension in the rope is 10.0 N. a) Draw two free-body diagrams, one for the 4.00-kg block and one for the block with mass m. What is b) the acceleration of either block? c) the mass m of the hanging block? d) How does the tension compare to the weight of the hanging block?

5.18 Runway Design. A transport plane takes off from a level landing field with two gliders in tow, one behind the other. The mass of each glider is 700 kg, and the total resistance (air drag plus friction with the runway) on each may be assumed constant and equal to 2500 N. The tension in the towrope between the transport plane and the first glider is not to exceed 12,000 N. a) If a speed of 40 m/s is required for takeoff, what minimum length of runway is needed? b) What is the tension in the towrope between the two gliders while they are accelerating for the takeoff?

5.19 Apparent Weight. A 550-N physics student stands on a bathroom scale in an elevator. As the elevator starts moving, the scale reads 450 N. a) Find the acceleration of the elevator (magnitude and direction). b) What is the acceleration if the scale reads 670 N? c) If the scale reads zero, should the student worry? Explain.

5.20 A physics student playing with an air hockey table (a frictionless surface) finds that if she gives the puck a velocity of 3.80 m/s along the length (1.75 m) of the table at one end, by the time it has reached the other end the puck has drifted 2.50 cm to the right but still has a velocity component along the length of 3.80 m/s. She correctly concludes that the table is not level and correctly calculates its inclination from the above information. What is the angle of inclination?

Section 5.3 Frictional Forces

5.21 Free-Body Diagrams. The first two steps in the solution of Newton's second law problems are to select an object for analysis and then to draw free-body diagrams for that object. Draw free-body diagrams for the following situations: a) a mass M sliding down a frictionless inclined plane of angle α; b) a mass M sliding up a frictionless inclined plane of angle α; c) a mass M sliding up an inclined plane of angle α with kinetic friction present; d) masses M and m sliding down an inclined plane of angle α with friction present, as shown in Fig. 5.48a. Here, draw free-body diagrams for both m and M. Identify the forces that are action-reaction pairs. e) Draw free-body diagrams for masses m and M shown in Fig. 5.48b. Identify all action-reaction pairs. There is a frictional force between all surfaces in contact. The pulley is frictionless and massless. In all cases be sure you have the correct direction of the forces and are completely clear on what object is causing each force in your free-body diagram.

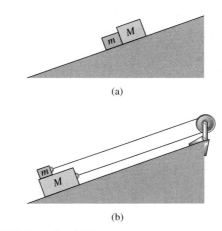

(a)

(b)

Figure 5.48 Exercise 5.21.

5.22 a) A large rock rests upon a rough horizontal surface. A bulldozer pushes on the rock with a horizontal force T that is slowly increased, starting from zero. In a graph with T along the x-axis and the friction force f along the y-axis, start at $T = 0$ and show the region of no motion, the point at which the rock is ready to move, and the region where the rock is moving. b) A block with weight w rests on a rough horizontal plank. The slope angle of the plank θ is gradually increased until the block starts to slip. Draw two graphs, both with θ along the x-axis. In one graph, show the ratio of the normal force to the weight, n/w, as a function of θ. In the second graph, show the ratio of the friction force to the weight, f/w. Indicate the region of no motion, the point at which the block is ready to move, and the region where the block is moving.

5.23 A stockroom worker pushes a box with mass 11.2 kg on a horizontal surface with a constant speed of 3.50 m/s. The coefficient of kinetic friction between the box and the surface is 0.20. a) What horizontal force must the worker apply to maintain the motion? b) If the force calculated in part (a) is removed, how far does the box slide before coming to rest?

5.24 A box of bananas weighing 40.0 N rests on a horizontal surface. The coefficient of static friction between the box and the surface is 0.40 and the coefficient of kinetic friction is 0.20. a) If no horizontal force is applied to the box and the box is at rest, how large is the friction force exerted on the box? b) What is the magnitude of the friction force if a monkey applies a horizontal force of 6.0 N to the box and the box is initially at rest? c) What minimum horizontal force must the monkey apply to start the box in motion? d) What minimum horizontal force must the monkey apply to keep the box moving at constant velocity once it has been started? e) If the monkey applies a horizontal force of 18.0 N, what is the magnitude of the friction force and what is the box's acceleration?

5.25 In a physics lab experiment, a 6.00-kg box is pushed across a flat table by a horizontal force $\vec{F}$. a) If the box is moving at a constant speed of 0.350 m/s and the coefficient of kinetic friction is 0.12, what is the magnitude of $\vec{F}$? b) What is the magnitude of $\vec{F}$ if the box is speeding up with a constant acceleration of 0.180 m/s²? c) How would your answers to parts (a) and (b) change if the experiments were performed on the moon, where $g = 1.62$ m/s²?

5.26 An 85-N box of oranges is being pushed across a horizontal floor. As it moves, it is slowing at a constant rate of 0.90 m/s each second. The push force has a horizontal component of 20 N and a vertical component of 25 N downward. Calculate the coefficient of kinetic friction between the box and floor.

5.27 You are to lower a safe with mass 260 kg at constant speed down skids 20.0 m long, from a truck 2.00 m high. a) If the coefficient of kinetic friction between safe and skids is 0.25, do you need to pull down or hold back the safe? b) How great a force parallel to the skids is needed?

5.28 Stopping Distance. a) If the coefficient of kinetic friction between tires and dry pavement is 0.80, what is the shortest distance in which you can stop an automobile by locking the brakes when traveling at 28.7 m/s (about 65 mi/h)? b) On wet pavement the coefficient of kinetic friction may be only 0.25. How fast should you drive on wet pavement in order to be able to stop in the same distance as in part (a)? (*Note:* Locking the brakes is *not* the safest way to stop.)

5.29 Coefficient of Friction. A clean brass washer slides along a horizontal clean steel surface until it stops. Using the values from Table 5.1, how many times farther would it slide with the same initial speed if the washer were Teflon-coated?

5.30 Consider the system shown in Fig. 5.49. Block *A* has weight w_A and block *B* has weight w_B. Once block *B* is set into downward motion, it descends at a constant speed. a) Calculate the coefficient of kinetic friction between block *A* and the tabletop. b) A cat, also of weight w_A, falls asleep on top of block *A*. If block *B* is now set into downward motion, what is its acceleration (magnitude and direction)?

Figure 5.49 Exercises 5.30, 5.35, and Problem 5.72.

5.31 Two crates connected by a rope lie on a horizontal surface (Fig. 5.50). Crate *A* has mass m_A and crate *B* has mass m_B. The coefficient of kinetic friction between each crate and the surface is μ_k. The crates are pulled to the right at constant velocity by a horizontal force $\vec{F}$. In terms of m_A, m_B, and μ_k, calculate a) the magnitude of the force $\vec{F}$ and b) the tension in the rope connecting the blocks. Include the free-body diagram or diagrams you used to determine each answer.

Figure 5.50 Exercise 5.31.

5.32 Rolling Friction. Two bicycle tires are set rolling with the same initial speed of 3.50 m/s on a long, straight road, and the distance each travels before its speed is reduced by half is measured. One tire is inflated to a pressure of 40 psi and goes 18.1 m; the other is at 105 psi and goes 92.9 m. What is the coefficient of rolling friction μ_r for each? Assume that the net horizontal force is due to rolling friction only.

5.33 Wheels. You find that it takes a horizontal force of 160 N to slide a box along the surface of a level floor at constant speed. The coefficient of static friction is 0.52 and the coefficient of kinetic friction is 0.47. If you place the box on a dolly of mass 5.3 kg and with coefficient of rolling friction 0.018, what horizontal acceleration would that 160-N force provide?

5.34 You find it takes 200 N of horizontal force to move an empty pickup truck along a level road at a speed of 2.4 m/s. You then load up the pickup and pump up its tires so that its total weight increases by 42% while the coefficient of rolling friction decreases by 19%. Now what horizontal force will you need to move the pickup along the same road at the same speed? The speed is low enough that you can ignore air resistance.

5.35 As shown in Fig. 5.49, block *A* (mass 2.25 kg) rests on a tabletop. It is connected by a horizontal cord passing over a light, frictionless pulley to a hanging block *B* (mass 1.30 kg). The coefficient of kinetic friction between block *A* and the table top is 0.450. After the blocks are released from rest, find a) the speed of each block after moving 3.00 cm and b) the tension in the cord. Include the free-body diagram or diagrams you used to determine the answers.

5.36 A 25.0-kg box of textbooks rests on a loading ramp that makes an angle α with the horizontal. The coefficient of kinetic friction is 0.25 and the coefficient of static friction is 0.35. a) As the angle α is increased, find the minimum angle at which the box starts to slip. b) At this angle, find the acceleration once the box has begun to move. c) At this angle, how fast will the box be moving after it has slid 5.0 m along the loading ramp?

5.37 A large crate with mass *m* rests on a horizontal floor. The coefficients of friction between the crate and the floor are μ_s and μ_k. A woman pushes downward at an angle θ below the horizontal on the crate with a force $\vec{F}$. a) What magnitude of force $\vec{F}$ is required to keep the crate moving at constant velocity? b) If μ_s is larger than some critical value, the woman cannot start the crate moving no matter how hard she pushes. Calculate this critical value of μ_s.

5.38 A box with mass m is dragged across a level floor having a coefficient of kinetic friction μ_k by a rope that is pulled upward at an angle θ above the horizontal with a force of magnitude F. a) In terms of m, μ_k, θ, and g, obtain an expression for the magnitude of force required to move the box with constant speed. b) Knowing that you are studying physics, a CPR instructor asks you how much force it would take to slide a 90-kg patient across a floor at constant speed by pulling on him at an angle of 25° above the horizontal. By dragging some weights wrapped in an old pair of pants down the hall with a spring balance, you find that $\mu_k = 0.35$. Use the result of part (a) to answer the instructor's question.

5.39 Blocks A, B, and C are placed as in Fig. 5.51 and connected by ropes of negligible mass. Both A and B weigh 25.0 N each, and the coefficient of kinetic friction between each block and the surface is 0.35. Block C descends with constant velocity. a) Draw two separate free-body diagrams showing the forces acting on A and on B. b) Find the tension in the rope connecting blocks A and B. c) What is the weight of block C? d) If the rope connecting A and B were cut, what would be the acceleration of C?

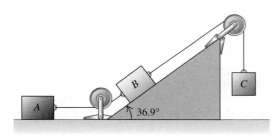

Figure 5.51 Exercise 5.39.

5.40 Starting from Eq. (5.10), derive Eqs. (5.11) and (5.12).

5.41 a) In Example 5.20, what value of D is required to make $v_t = 42$ m/s for this sky diver? b) If the sky diver's daughter, whose mass is 45 kg, is falling through the air and has the same D (0.25 kg/m) as her father, what is the daughter's terminal speed?

5.42 You throw a baseball straight up. The drag force is proportional to v^2. In terms of g, what is the y-component of the ball's acceleration when its speed is half its terminal speed and a) it is moving up? b) it is moving back down?

Section 5.4 Dynamics of Circular Motion

5.43 A stone with mass 0.80 kg is attached to one end of a string 0.90 m long. The string will break if its tension exceeds 600 N. (This is called the *breaking strength* of the string.) The stone is whirled in a horizontal circle on a frictionless tabletop; the other end of the string remains fixed. Find the maximum speed the stone can attain without breaking the string.

5.44 A flat (unbanked) curve on a highway has a radius of 220 m. A car rounds the curve at a speed of 25.0 m/s. What is the minimum coefficient of friction that will prevent sliding?

5.45 Aircraft experience a lift force (due to the air) that is perpendicular to the plane of the wings and to the direction of flight. Light

airplanes are designed so that their wings can safely provide a lift force of 3.8 times the weight of the airplane. A greater force can damage the structure of the wing. (Aerobatic airplanes and jet fighters are designed with much higher limits.) a) What is the maximum banking angle that a pilot can safely maintain in level flight without threatening the safety of the airplane (and her own safety)? b) Does your answer to part (a) depend on the speed of the airplane? Explain why or why not.

5.46 The "Giant Swing" at a county fair consists of a vertical central shaft with a number of horizontal arms attached at its upper end (Fig. 5.52). Each arm supports a seat suspended from a cable 5.00 m long, the upper end of the cable being fastened to the arm at a point 3.00 m from the central shaft. a) Find the time of one revolution of the swing if the cable supporting a seat makes an angle of 30.0° with the vertical. b) Does the angle depend on the weight of the passenger for a given rate of revolution?

Figure 5.52 Exercise 5.46.

5.47 Aircraft experience a lift force (due to the air) that is perpendicular to the plane of the wings and to the direction of flight. A small airplane is flying at a constant speed of 240 km/h. At what angle from the horizontal must the wings of the airplane be tilted for the plane to execute a horizontal turn from east to north with a turning radius of 1200 m?

5.48 A small button placed on a horizontal rotating platform with diameter 0.320 m will revolve with the platform when it is brought up to a speed of 40.0 rev/min, provided the button is no more than 0.150 m from the axis. a) What is the coefficient of static friction between the button and the platform? b) How far from the axis can the button be placed, without slipping, if the platform rotates at 60.0 rev/min?

5.49 **Rotating Space Stations.** One problem for humans living in outer space is that they are apparently weightless. One way around this problem is to design a space station that spins about its center at a constant rate. This creates "artificial gravity" at the outside rim of the station. a) If the diameter of the space station is 800 m, how many revolutions per minute are needed in order for the "artificial gravity" acceleration to be 9.80 m/s²? b) If the space station is a waiting area for travelers going to Mars, it might be desirable to simulate the acceleration due to gravity on the Martian surface (3.70 m/s^2). How many revolutions per minute are needed in this case?

5.50 The Cosmoclock 21 Ferris wheel in Yokohama City, Japan, has a diameter of 100 m. Its name comes from its 60 arms, each of which can function as a second hand (so that it makes one revolution every 60.0 s). a) Find the speed of the passengers when the Ferris wheel is rotating at this rate. b) A passenger weighs 882 N at the weight-guessing booth on the ground. What is his apparent weight at the highest and at the lowest point on the Ferris wheel? c) What would be the time for one revolution if the passenger's apparent weight at the highest point were zero? d) What then would be the passenger's apparent weight at the lowest point?

5.51 An airplane flies in a loop (a circular path in a vertical plane) of radius 150 m. The pilot's head always points toward the center of the loop. The speed of the airplane is not constant; the airplane goes slowest at the top of the loop and fastest at the bottom. a) At the top of the loop, the pilot feels weightless. What is the speed of the airplane at this point? b) At the bottom of the loop, the speed of the airplane is 280 km/h. What is the apparent weight of the pilot at this point? His true weight is 700 N.

5.52 A 50.0-kg stunt pilot who has been diving her airplane vertically pulls out of the dive by changing her course to a circle in a vertical plane. a) If the plane's speed at the lowest point of the circle is 95.0 m/s, what is the minimum radius of the circle for the acceleration at this point not to exceed 4.00g? b) What is the apparent weight of the pilot at the lowest point of the pullout?

5.53 Stay Dry! You tie a cord to a pail of water, and you swing the pail in a vertical circle of radius 0.600 m. What minimum speed must you give the pail at the highest point of the circle if no water is to spill from it?

5.54 A bowling ball weighing 71.2 N (16.0 lb) is attached to the ceiling by a 3.80-m rope. The ball is pulled to one side and released; it then swings back and forth as a pendulum. As the rope swings through the vertical, the speed of the bowling ball is 4.20 m/s. a) What is the acceleration of the bowling ball, in magnitude and direction, at this instant? b) What is the tension in the rope at this instant?

Problems

5.55 Two ropes are connected to a steel cable that supports a hanging weight as shown in Fig. 5.53. a) Draw a free-body diagram showing all of the forces acting at the knot connecting the two ropes to the steel cable. Based on your force diagram, which of the two ropes will have the greater tension? b) If the maximum tension either rope can sustain without breaking is 5000 N, determine the maximum value of the hanging weight that these ropes can safely support. You can ignore the weight of the ropes and the steel cable.

Figure 5.53 Problem 5.55.

5.56 In Fig. 5.54 a worker lifts a weight w by pulling down on a rope with a force $\vec{F}$. The upper pulley is attached to the ceiling by a chain, and the lower pulley is attached to the weight by another chain. In terms of w, find the tension in each chain and the magnitude of the force $\vec{F}$ if the weight is lifted at constant speed. Include the free-body diagram or diagrams you used to determine your answers. Assume that the rope, pulleys, and chains all have negligible weights.

Figure 5.54 Problem 5.56.

5.57 A man is pushing a refrigerator up a ramp at constant speed. The ramp is inclined at an angle α above the horizontal, but the man applies a *horizontal* force $\vec{F}$. Calculate the magnitude of $\vec{F}$ in terms of α and the refrigerator mass m. You can ignore friction on the refrigerator.

5.58 A Rope with Mass. In most problems in this book, the ropes, cords, or cables have so little mass compared to other objects in the problem that you can safely ignore their mass. But if the rope is the *only* object in the problem, then clearly you cannot ignore its mass. For example, suppose we have a clothesline attached to two poles (Fig. 5.55). The clothesline has a mass M, and each end makes an angle θ with the horizontal. What is a) the tension at the ends of the clothesline? b) the tension at the lowest point? c) Why can we not have $\theta = 0$? (See Discussion Question Q5.3.) d) Discuss your results for parts (a) and (b) in the limit that $\theta \to 90°$. The curve of the clothesline, or of any flexible cable hanging under its own weight, is called a *catenary*. For a more advanced treatment of this curve, see K. R. Symon, *Mechanics*, 3rd ed. (Addison–Wesley, Reading, MA, 1971), pp. 237–241.

Figure 5.55 Problem 5.58.

5.59 Another Rope with Mass. A block with mass M is attached to the lower end of a vertical, uniform rope with mass m and length L. A constant upward force $\vec{F}$ is applied to the top of the rope, causing the rope and block to accelerate upward. Find the tension in the rope at a distance x from the top end of the rope, where x can have any value from 0 to L.

5.60 A block with mass m_1 is placed on an inclined plane with slope angle α and is connected to a second hanging block with mass m_2 by a cord passing over a small, frictionless pulley (Fig. 5.56). The coefficient of static friction is μ_s and the coefficient of kinetic friction is μ_k. a) Find the mass m_2 for which block m_1 moves up the plane at constant speed once it is set in motion. b) Find the mass m_2 for which block m_1 moves down the plane at constant speed once it is set in motion. c) For what range of values of m_2 will the blocks remain at rest if they are released from rest?

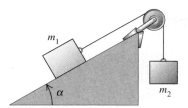

Figure 5.56 Problem 5.60.

5.61 a) Block A in Fig. 5.57 weighs 60.0 N. The coefficient of static friction between the block and the surface on which it rests is 0.25. The weight w is 12.0 N and the system is in equilibrium. Find the friction force exerted on block A. b) Find the maximum weight w for which the system will remain in equilibrium.

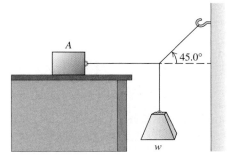

Figure 5.57 Problem 5.61.

5.62 Block A in Fig. 5.58 weighs 1.20 N and block B weighs 3.60 N. The coefficient of kinetic friction between all surfaces is 0.300. Find the magnitude of the horizontal force $\vec{F}$ necessary to drag block B to the left at constant speed a) if A rests on B and moves with it (Fig. 5.58a); b) if A is held at rest (Fig. 5.58b).

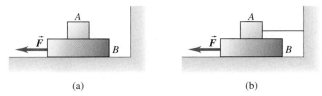

(a)   (b)

Figure 5.58 Problem 5.62.

5.63 A window washer pushes his scrub brush up a vertical window at constant speed by applying a force $\vec{F}$ as shown in Fig. 5.59. The brush weighs 12.0 N and the coefficient of kinetic friction is $\mu_k = 0.150$. Calculate a) the magnitude of the force $\vec{F}$ and b) the normal force exerted by the window on the brush.

5.64 The Flying Leap of a Flea. High-speed motion pictures (3500 frames/second) of a jumping 210 μg flea yielded the data

Figure 5.59 Problem 5.63.

to plot the flea's acceleration as a function of time as shown in Fig. 5.60. (See "The Flying Leap of the Flea," by M. Rothschild et al. in the Nov. 1973 *Scientific American*.) This flea was about 2 mm long and jumped at a nearly vertical takeoff angle. Use measurements made on the graph to answer the following questions. a) Find the *initial* net external force on the flea. How does it compare to the flea's weight? b) Find the *maximum* net external force on this jumping flea. When does this maximum force occur? c) Use the graph to find the flea's maximum speed.

Figure 5.60 Problem 5.64.

5.65 A 25,000-kg rocket blasts off vertically from the earth's surface with a constant acceleration. During the motion considered in the problem, assume that g remains constant (see Chapter 12). Inside the rocket, a 15.0-N instrument hangs from a wire that can support a maximum tension of 35.0 N. a) Find the minimum time for this rocket to reach the sound barrier (330 m/s) without breaking the inside wire and the maximum vertical thrust of the rocket engines under these conditions. b) How far is the rocket above the earth's surface when it breaks the sound barrier?

5.66 You are standing on a bathroom scale in an elevator in a tall building. Your mass is 72 kg. The elevator starts from rest and travels upward with a speed that varies with time according to $v(t) = (3.0 \text{ m/s}^2)t + (0.20 \text{ m/s}^3)t^2$. When $t = 4.0$ s, what is the reading of the bathroom scale?

5.67 Elevator Design. You are designing an elevator for a hospital. The force exerted on a passenger by the floor of the elevator is not to exceed 1.60 times the passenger's weight. The elevator accelerates upward with constant acceleration for a distance of 3.0 m and then starts to slow down. What is the maximum speed of the elevator?

5.68 You are working for a shipping company. Your job is to stand at the bottom of a 8.0-m-long ramp that is inclined at 37° above the horizontal. You grab packages off a conveyor belt and propel them up the ramp. The coefficient of kinetic friction between the packages and the ramp is $\mu_k = 0.30$. a) What speed do you need to give a package at the bottom of the ramp so that it has zero speed at the top of the ramp? b) Your coworker is supposed to grab the packages as they arrive at the top of the ramp, but she misses one and it slides back down. What is its speed when it returns to you?

5.69 A hammer is hanging by a light rope from the ceiling of a bus. The ceiling of the bus is parallel to the roadway. The bus is traveling in a straight line on a horizontal street. You observe that the hammer hangs at rest with respect to the bus when the angle between the rope and the ceiling of the bus is 74°. What is the acceleration of the bus?

5.70 A steel washer is suspended inside an empty shipping crate from a light string attached to the top of the crate. The crate slides down a long ramp that is inclined at an angle of 37° above the horizontal. The crate has mass 180 kg. You are sitting inside the crate (with a flashlight); your mass is 55 kg. As the crate is sliding down the ramp, you find the washer is at rest with respect to the crate when the string makes an angle of 68° with the top of the crate. What is the coefficient of kinetic friction between the ramp and the crate?

5.71 Lunch Time! You are riding your motorcycle one day down a wet street that slopes downward at an angle of 20° below the horizontal. As you start to ride down the hill, you notice a construction crew has dug a deep hole in the street at the bottom of the hill. A Siberian tiger, escaped from the City Zoo, has taken up residence in the hole. a) You apply the brakes and lock your wheels at the top of the hill, where you are moving with a speed of 20 m/s. The inclined street in front of you is 40 m long. a) Will you plunge into the hole and become the tiger's lunch, or do you skid to a stop before you reach the hole? (The coefficients of friction between your motorcycle tires and the wet pavement are $\mu_s = 0.90$ and $\mu_k = 0.70$.) b) What must your initial speed be if you are to stop just before reaching the hole?

5.72 In the system shown in Fig. 5.49, block A has mass m_A, block B has mass m_B, and the rope connecting them has a *nonzero* mass m_{rope}. The rope has a total length L and the pulley has a very small radius. You can ignore any sag in the horizontal part of the rope. a) If there is no friction between block A and the tabletop, find the acceleration of the blocks at an instant when a length d of rope hangs vertically between the pulley and block B. As block B falls, will the magnitude of the acceleration of the system increase, decrease, or remain constant? Explain. b) Let $m_A = 2.00$ kg, $m_B = 0.400$ kg, $m_{\text{rope}} = 0.160$ kg, and L = 1.00 m. If there is friction between block A and the tabletop, with $\mu_k = 0.200$ and $\mu_s = 0.250$, find the minimum value of the distance d such that the blocks will start to move if they are initially at rest. c) Repeat part (b) for the case $m_{\text{rope}} = 0.040$ kg. Will the blocks move in this case?

5.73 If the coefficient of static friction between a table and a uniform massive rope is μ_s, what fraction of the rope can hang over the edge of the table without the rope sliding?

5.74 A woman attempts to push a box of books that has mass m up a ramp inclined at an angle α above the horizontal. The coefficients of friction between the ramp and the box are μ_s and μ_k. The force $\vec{F}$ applied by the woman is *horizontal*. a) If μ_s is greater than some critical value, the woman cannot start the box moving up the ramp no matter how hard she pushes. Calculate this critical value of μ_s. b) Assume that μ_s is less than this critical value. What magnitude of force must the woman apply to keep the box moving up the ramp at constant speed?

5.75 A 30.0-kg packing case is initially at rest on the floor of a 1500-kg pickup truck. The coefficient of static friction between the case and the truck floor is 0.30 and the coefficient of kinetic friction is 0.20. Before each acceleration given below, the truck is traveling due north at constant speed. Find the magnitude and direction of the friction force acting on the case a) when the truck accelerates at 2.20 m/s² northward and b) when it accelerates at 3.40 m/s² southward.

5.76 Traffic Court. The truck in Problem 5.75 is traveling at a constant speed on a country road where the speed limit is 72 km/h (45 mi/h). Spotting a stop sign ahead, the driver applies the brakes and comes to a halt in a distance of 47.0 m. Without warning a policeman emerges from behind a bush and hands the driver a speeding ticket. When the driver protests that he was going less than the speed limit, the policeman replies that "I saw that packing case slide from the back to the front of your truck's cargo area. To make it slide like that, you must have had to brake very hard, which means you had to have been speeding." Will the policeman's argument hold up in traffic court? (Assume that the judge, like you, knows physics.)

5.77 Troy uses a frayed rope to pull a crate across a level floor. The maximum tension in the rope without breaking is T_{max} and the coefficient of kinetic friction is μ_k. a) Show that the maximum weight that he can pull at constant speed is $T_{\text{max}}/(\sin\theta)$, where $\theta = \arctan\mu_k$ is the angle of the rope above the horizontal. b) The answer to part (a) suggests that Troy can pull a weight that approaches infinity with a bit of spiderweb as the coefficient of kinetic friction approaches zero. Explain.

5.78 The airplane in Exercise 5.14 has an engine failure (so $F = 0$), and the airplane glides at constant speed toward a safe landing. The direction of flight is a constant angle α (called the *glide angle*) below the horizontal (Fig. 5.61). a) Find the magnitude of the lift force L (which acts perpendicular to the plane of the wings and to the direction of flight) and of the drag force f in terms of w and α. b) Show that $\alpha = \arctan(f/L)$. c) A fully loaded Cessna 182 (a single-engine airplane) weighs 12,900 N and has 1300 N of drag at a speed of 130 km/h. If this airplane has an engine failure at an altitude of 2500 m, what maximum distance over the ground can it glide while searching for a safe landing spot? d) Justify the statement that "it's drag, not gravity, that makes the airplane go down."

Figure 5.61 Problem 5.78.

5.79 Thrust. The pilot of the Cessna 182 airplane in part (c) of Problem 5.78 manages to get the engine restarted. He applies full throttle and the airplane climbs along a straight line angled above the horizontal. The airplane is flying at a constant speed of 130 km/h, weighs 12,900 N, and has 1300 N of drag. The rate-of-climb indicator on the instrument panel shows that the airplane is gaining altitude at a steady rate of 5.00 m/s (300 m/min, or about 980 feet/min). Determine the magnitude of the thrust (the forward force produced by the engine). (*Hint:* The thrust acts in the direction that the airplane is moving.)

5.80 Losing Cargo. A 12.0-kg box rests on the flat floor of a truck. The coefficients of friction between the box and floor are $\mu_s = 0.19$ and $\mu_k = 0.15$. The truck stops at a stop sign and then starts to move with an acceleration of 2.20 m/s². If the box is 1.80 m from the rear of the truck when the truck starts, how much time elapses before the box falls off the truck? How far does the truck travel in this time?

5.81 Block *A* in Fig. 5.62 weighs 1.40 N, and block *B* weighs 4.20 N. The coefficient of kinetic friction between all surfaces is 0.30. Find the magnitude of the horizontal force $\vec{F}$ necessary to drag block *B* to the left at constant speed if *A* and *B* are connected by a light, flexible cord passing around a fixed, frictionless pulley.

Figure 5.62 Problem 5.81.

5.82 You are part of a design team for future exploration of the planet Mars, where $g = 3.7$ m/s². An explorer is to step out of a survey vehicle traveling horizontally at 33 m/s when it is 1200 m above the surface and then fall freely for 20 s. At that time, a portable advanced propulsion system (PAPS) is to exert a constant force that will decrease the explorer's speed to zero at the instant she touches the surface. The total mass (explorer, suit, equipment, and PAPS) is 150 kg. Assume the change in mass of the PAPS to be negligible. Find the horizontal and vertical components of the force the PAPS must exert, and for what interval of time the PAPS must exert it. You can ignore air resistance.

5.83 Block *A* in Fig. 5.63 has a mass of 4.00 kg and block *B* has mass 12.0 kg. The coefficient of kinetic friction between block *B* and the horizontal surface is 0.25. a) What is the mass of block *C* if block *B* is moving to the right and speeding up with an acceleration 2.00 m/s²? b) What is the tension in each cord when block *B* has this acceleration?

Figure 5.63 Problem 5.83.

5.84 Two blocks connected by a cord passing over a small, frictionless pulley rest on frictionless planes (Fig. 5.64). a) Which way will the system move when the blocks are released from rest? b) What is the acceleration of the blocks? c) What is the tension in the cord?

Figure 5.64 Problem 5.84.

5.85 In terms of m_1, m_2, and g, find the accelerations of each block in Fig. 5.65. There is no friction anywhere in the system.

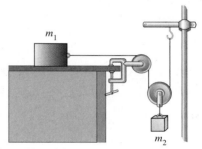

Figure 5.65 Problem 5.85.

5.86 Block *B*, with mass m_B, rests on block *A*, with mass m_A, which in turn is on a horizontal tabletop (Fig. 5.66). The coefficient of kinetic friction between block *A* and the tabletop is μ_k, and the coefficient of static friction between block *A* and block *B* is μ_s. A light string attached to block *A* passes over a frictionless, massless pulley and block *C* is suspended from the other end of the string. What is the largest mass m_C that block *C* can have so that blocks *A* and *B* still slide together when the system is released from rest?

Figure 5.66 Problem 5.86.

5.87 Two objects with masses 5.00 kg and 2.00 kg hang 0.600 m above the floor from the ends of a cord 6.00 m long passing over a frictionless pulley. Both objects start from rest. Find the maximum height reached by the 2.00-kg object.

5.88 Friction in an Elevator. You are riding in an elevator on the way to the eighteenth floor of your dormitory. The elevator is accelerating upward with $a = 1.90$ m/s². Beside you is the box containing your new computer; box and contents have a total mass 28.0 kg. While the elevator is accelerating upward, you push horizontally on the box to slide it at constant speed toward the elevator door. If the coefficient of kinetic friction between the box and elevator floor is $\mu_k = 0.32$, what magnitude of force must you apply?

5.89 A block is placed against the vertical front of a cart as shown in Fig. 5.67. What acceleration must the cart have in order that block A does not fall? The coefficient of static friction between the block and the cart is μ_s. How would an observer on the cart describe the behavior of the block?

Figure 5.67 Problem 5.89.

5.90 Two blocks with masses 4.00 kg and 8.00 kg are connected by a string and slide down a 30.0° inclined plane (Fig. 5.68). The coefficient of kinetic friction between the 4.00-kg block and the plane is 0.25; that between the 8.00-kg block and the plane is 0.35. a) Calculate the acceleration of each block. b) Calculate the tension in the string. c) What happens if the positions of the blocks are reversed, so the 4.00-kg block is above the 8.00-kg block?

Figure 5.68 Problem 5.90.

5.91 Block A, with weight $3w$, slides down an inclined plane S of slope angle 36.9° at a constant speed while plank B, with weight w, rests on top of A. The plank is attached by a cord to the wall (Fig. 5.69). a) Draw a diagram of all the forces acting on block A. b) If the coefficient of kinetic friction is the same between A and B and between S and A, determine its value.

Figure 5.69 Problem 5.91.

5.92 A man with mass 70.0 kg stands on a platform with mass 25.0 kg. He pulls on the free end of a rope that runs over a pulley on the ceiling and has its other end fastened to the platform. You can ignore the mass of the rope and the mass of the pulley, and the pulley is frictionless. The rope is vertical on either side of the pulley. a) With what force does he have to pull so that he and the platform have an upward acceleration of 1.80 m/s²? b) What is the acceleration of the rope relative to him?

5.93 Two blocks, with masses m_1 and m_2, are stacked as shown in Fig. 5.70 and placed on a frictionless horizontal surface. There is friction between the two blocks. An external force of magnitude F is applied to the top block at an angle α below the horizontal. a) If the two blocks move together, find their acceleration. b) Show that the two blocks will move together only if

Figure 5.70 Problem 5.93.

$$F \leq \frac{\mu_s m_1 (m_1 + m_2) g}{m_2 \cos \alpha - \mu_s (m_1 + m_2) \sin \alpha}$$

where μ_s is the coefficient of static friction between the two blocks.

5.94 Banked Curve I. A curve with a 120-m radius on a level road is banked at the correct angle for a speed of 20 m/s. If an automobile rounds this curve at 30 m/s, what is the minimum coefficient of static friction needed between tires and road to prevent skidding?

5.95 Banked Curve II. Consider a wet roadway banked as in Example 5.24 (Section 5.4), where there is a coefficient of static friction of 0.30 and a coefficient of kinetic friction of 0.25 between the tires and the roadway. The radius of the curve is $R = 50$ m. a) If the banking angle is $\beta = 25°$, what is the *maximum* speed the automobile can have before sliding *up* the banking? b) What is the *minimum* speed the automobile can have before sliding *down* the banking?

5.96 Maximum Safe Speed. As you travel every day to campus, the road makes a large turn that is approximately an arc of a circle. You notice the warning sign at the start of the turn, asking for a maximum speed of 55 mi/h. You also notice that in the curved portion the road is level; that is, not banked at all. On a dry day with very little traffic, you enter the turn at a constant speed of 80 mi/h and feel the car may skid if you do not slow down quickly. You conclude that your speed is at the limit of safety for this curve and you slow down. However, you remember reading that on dry pavement new tires have an average coefficient of static friction of about 0.76, while under the worst winter driving conditions, you may encounter wet ice for which the coefficient of static friction can be as low as 0.20. Wet ice is not unheard of on this road, so you ask yourself whether the speed limit for the turn on the roadside warning sign is for the worst-case scenario. a) Estimate the radius of the curve from your 80 mi/h experience in the dry turn. b) Use this estimate to find the maximum speed limit in the turn under the worst wet-ice conditions. How does this compare with the speed limit on the sign? Is the sign misleading drivers?

c) On a rainy day, the coefficient of static friction would be about 0.37. What is the maximum safe speed for the turn when the road is wet? Does your answer help you understand the maximum-speed sign?

5.97 On a rainy day, you drive on a wet one-way road. The road is level with two lanes and you drive in the second lane. You are on the straight part of the road, but you are aware that a turn having the shape of a circular arc will start in 0.80 km (about half a mile) from your current position. Your speed is 27 m/s (60 mi/h), and you know from past experience that on dry pavement this is the maximum speed at which you can round the coming curve safely. However, the road is wet and you also know that the rain makes the road slippery, reducing the coefficient of static friction to half its value for a dry road. Suddenly you notice a car 0.50 km (about one-third of a mile) behind you in the other lane and moving very fast, in your estimation at a speed of 36 m/s (80 mi/h). The driver of this car apparently does not notice the warning sign about the coming curve and continues to travel at constant speed. You realize that this car may catch up with you at the first section of the turn, skid in the turn, and involve you in a bad accident. a) On the wet road, what is the maximum safe speed for the turn? b) If you slow down with constant acceleration so that you have the speed calculated in (a), where will the second car be as you enter the turn? Is a collision likely?

5.98 You are riding in a school bus. As the bus rounds a flat curve at constant speed, a lunch box with mass 0.500 kg, suspended from the ceiling of the bus by a string 1.80 m long, is found to hang at rest relative to the bus when the string makes an angle of 30.0° with the vertical. In this position the lunch box is 50.0 m from the center of curvature of the curve. What is the speed v of the bus?

5.99 The Monkey and Bananas Problem. A 20-kg monkey has a firm hold on a light rope that passes over a frictionless pulley and is attached to a 20-kg bunch of bananas (Fig. 5.71). The monkey looks up, sees the bananas, and starts to climb the rope to get them. a) As the monkey climbs, do the bananas move up, down, or remain at rest? b) As the monkey climbs, does the distance between the monkey and the bananas decrease, increase, or remain constant? c) The monkey releases her hold on the rope. What happens to the distance between the monkey and the bananas while she is falling? d) Before reaching the ground, the monkey grabs the rope to stop her fall. What do the bananas do?

Figure 5.71 Problem 5.99.

5.100 You throw a rock downward into water with a speed of $3mg/k$, where k is the coefficient in Eq. (5.7). Assume that the rela-

tion between fluid resistance and speed is as given in Eq. (5.7) and calculate the speed of the rock as a function of time.

5.101 A rock with mass $m = 3.00$ kg falls from rest in a viscous medium. The rock is acted on by a net constant downward force of 18.0 N (a combination of gravity and the buoyant force exerted by the medium) and by a fluid resistance force $f = kv$, where v is the speed in m/s and $k = 2.20$ N·s/m. (See Section 5.3.) a) Find the initial acceleration, a_0. b) Find the acceleration when the speed is 3.00 m/s. c) Find the speed when the acceleration equals $0.1a_0$. d) Find the terminal speed, v_t. e) Find the coordinate, speed, and acceleration 2.00 s after the start of the motion. f) Find the time required to reach a speed $0.9v_t$.

5.102 A rock with mass m slides with initial velocity v_0 on a horizontal surface. A retarding force F_R that the surface exerts on the rock is proportional to the square root of the instantaneous velocity of the rock ($F_R = -kv^{1/2}$). a) Find expressions for the velocity and position of the rock as a function of time. b) In terms of m, k, and v_0, at what time will the rock come to rest? c) In terms of m, k, and v_0, what is the distance of the rock from its starting point when it comes to rest?

5.103 A fluid exerts an upward buoyancy force on an object immersed in it. In the derivation of Eq. (5.9) the buoyancy force exerted on an object by the fluid was ignored. But in some situations, where the density of the object is not much greater than the density of the fluid, you cannot ignore the buoyancy force. For a plastic sphere falling in water, you calculate the terminal speed to be 0.36 m/s when you ignore buoyancy, but you measure it to be 0.24 m/s. The buoyancy force is what fraction of the weight?

5.104 The 4.00-kg block in Fig. 5.72 is attached to a vertical rod by means of two strings. When the system rotates about the axis of the rod, the strings are extended as shown in the diagram and the tension in the upper string is 80.0 N. a) What is the tension in the lower cord? b) How many revolutions per minute does the system make? c) Find the number of revolutions per minute at which the lower cord just goes slack. d) Explain what happens if the number of revolutions per minute is less than in part (c).

Figure 5.72 Problem 5.104.

5.105 Equation (5.10) applies to the case where the initial velocity is zero. a) Derive the corresponding equation for $v_y(t)$ when the falling object has an initial downward velocity with magnitude v_0. b) For the case where $v_0 < v_t$, sketch a graph of v_y as a function of t and label v_t on your graph. c) Repeat part (b) for the case where $v_0 > v_t$. d) Discuss what your result says about $v_y(t)$ when $v_0 = v_t$.

5.106 A small rock moves in water and the force exerted on it by the water is given by Eq. (5.7). The terminal speed of the rock is measured and found to be 2.0 m/s. The rock is projected *upward* at an initial speed of 6.0 m/s. You can ignore the buoyancy force on the rock. a) In the absence of fluid resistance, how high will the rock rise and how long will it take to reach this maximum height? b) When the effects of fluid resistance are included, what are the answers to the questions in part (a)?

5.107 You observe a 1350-kg sports car rolling along flat pavement in a straight line. The only horizontal forces acting on it are a constant rolling friction and air resistance (proportional to the square of its speed). You take the following data during a time interval of 25 s: When its speed is 32 m/s, it slows down at a rate of −0.42 m/s², and when its speed is decreased to 24 m/s, it slows down at −0.30 m/s². a) Find the coefficient of rolling friction and the air drag constant D. b) At what constant speed will this car move down an incline that makes a 2.2° angle with the horizontal? c) How is the constant speed for an incline of angle β related to the terminal speed of this sports car if the car drops off a high cliff? Assume that in both cases the air resistance force is proportional to the square of the speed, and the air drag constant is the same.

5.108 A 70-kg person rides in a 30-kg cart moving at 12 m/s at the top of a hill that is in the shape of an arc of a circle with radius of 40 m. a) What is the apparent weight of the person as the cart passes over the top of the hill? b) Determine the maximum speed that the cart may travel at the top of the hill without losing contact with the surface. Does your answer depend on the mass of the cart or that of the person? Explain.

5.109 Merry-Go-Round. One December identical twins Jena and Jackie are playing on a large merry-go-round (a disk mounted parallel to the ground, on a vertical axle through its center) in their school playground in northern Minnesota. Each twin has mass 30.0 kg. The icy coating on the merry-go-round surface makes it frictionless. The merry-go-round revolves at a constant rate as the twins ride on it. Jena, sitting 1.80 m from the center of the merry-go-round, must hold on to one of the metal posts attached to the merry-go-round with a horizontal force of 60.0 N to keep from sliding off. Jackie is sitting at the edge, 3.60 m from the center. a) With what horizontal force must Jackie hold on to keep from falling off? b) If Jackie falls off, what will be her horizontal velocity when she becomes airborne?

5.110 A passenger with mass 85 kg rides in a Ferris wheel like that in Example 5.25. The seats travel in a circle of radius 35 m. The Ferris wheel rotates at constant speed and makes one complete revolution every 25 s. Calculate the magnitude and direction of the net force exerted on the passenger by the seat when she is a) one-quarter revolution past her lowest point and b) one-quarter revolution past her highest point.

5.111 On the ride "Spindletop" at the amusement park Six Flags Over Texas, people stood against the inner wall of a hollow vertical cylinder with radius 2.5 m. The cylinder started to rotate, and when it reached a constant rotation rate of 0.60 rev/s, the floor on which people were standing dropped about 0.5 m. The people remained pinned against the wall. a) Draw a force diagram for a person on this ride, after the floor has dropped. b) What minimum coefficient of static friction is required if the person on the ride is not to slide downward to the new position of the floor?

c) Does your answer in part (b) depend on the mass of the passenger? (*Note:* When the ride is over, the cylinder is slowly brought to rest. As it slows down, people slide down the walls to the floor.)

5.112 A physics major is working to pay his college tuition by performing in a traveling carnival. He rides a motorcycle inside a hollow transparent plastic sphere. After gaining sufficient speed, he travels in a vertical circle with a radius of 13.0 m. The physics major has mass 70.0 kg, and his motorcycle has mass 40.0 kg. a) What minimum speed must he have at the top of the circle if the tires of the motorcycle are not to lose contact with the sphere? b) At the bottom of the circle, his speed is twice the value calculated in part (a). What is the magnitude of the normal force exerted on the motorcycle by the sphere at this point?

5.113 Ulterior Motives. You are driving a classic 1954 Nash Ambassador with a friend who is sitting to your right on the passenger side of the front seat. The Ambassador has flat bench seats. You would like to be closer to your friend and decide to use physics to achieve your romantic goal by making a quick turn. a) Which way (to the left or to the right) should you turn the car to get your friend to slide closer to you? b) If the coefficient of static friction between your friend and the car seat is 0.35, and you keep driving at a constant speed of 20 m/s, what is the maximum radius you could make your turn and still have your friend slide your way?

5.114 A small block with mass m rests on a frictionless horizontal tabletop a distance r from a hole in the center of the table (Fig. 5.73). A string tied to the small block passes down through the hole, and a larger block with mass M is suspended from the free end of the string. The small block is set into uniform circular motion with radius r and speed v. What must v be if the large block is to remain motionless when released?

Figure 5.73 Problem 5.114.

5.115 A small bead can slide without friction on a circular hoop that is in a vertical plane and has a radius of 0.100 m. The hoop rotates at a constant rate of 4.00 rev/s about a vertical diameter (Fig. 5.74). a) Find the angle β at which the bead is in vertical equilibrium. (Of course, it has a radial acceleration toward the axis.) b) Is it possible for the bead to "ride" at the same elevation as the center of the hoop? c) What will happen if the hoop rotates at 1.00 rev/s?

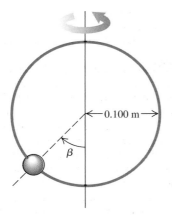

Figure 5.74 Problem 5.115.

5.116 A model airplane with mass 2.20 kg moves in the xy-plane such that its x- and y-coordinates vary in time according to $x(t) = \alpha - \beta t^3$ and $y(t) = \gamma t - \delta t^2$, where $\alpha = 1.50$ m, $\beta = 0.120$ m/s^3, $\gamma = 3.00$ m/s, and $\delta = 1.00$ m/s^2. a) Calculate the x- and y-components of the net force on the plane as functions of time. b) Sketch the trajectory of the airplane between $t = 0$ and $t = 3.00$ s, and draw on your sketch vectors showing the net force on the airplane at $t = 0$, $t = 1.00$ s, $t = 2.00$ s, and $t = 3.00$ s. For each of these times, relate the direction of the net force to the direction that the airplane is turning, and to whether the airplane is speeding up or slowing down (or neither). c) What are the magnitude and direction of the net force at $t = 3.00$ s?

5.117 A particle moves on a frictionless surface along a path as shown in Fig. 5.75. (The figure gives a view looking down on the surface.) The particle is initially at rest at point A, and then begins to move toward B as it gains speed at a constant rate. From B to C, the particle moves along a circular path at a constant speed. The speed remains constant along the straight-line path from C to D. From D to E, the particle moves along a circular path, but now its speed is decreasing at a constant rate. The speed continues to decrease at a constant rate as the particle moves from E to F; the particle comes to a halt at F. (The time intervals between the marked points are not equal.) At each point marked with a dot, draw arrows to represent the velocity, the acceleration, and the net force acting on the particle. Draw a longer or shorter arrow to represent vectors of larger or smaller magnitude.

Figure 5.75 Problem 5.117.

5.118 A small remote-control car with mass 1.60 kg moves at a constant speed of $v = 12.0$ m/s in a vertical circle inside a hollow

metal cylinder that has a radius of 5.00 m (Fig. 5.76). What is the magnitude of the normal force exerted on the car by the walls of the cylinder at a) point A (at the bottom of the vertical circle)? b) point B (at the top of the vertical circle)?

Figure 5.76 Problem 5.118.

5.119 A small block with mass m is placed inside an inverted cone that is rotating about a vertical axis such that the time for one revolution of the cone is T (Fig. 5.77). The walls of the cone make an angle β with the vertical. The coefficient of static friction between the block and the cone is μ_s. If the block is to remain at a constant height h above the apex of the cone, what are the maximum and minimum values of T?

Figure 5.77 Problem 5.119.

Challenge Problems

5.120 Moving Wedge. A wedge with mass M rests on a frictionless, horizontal table top. A block with mass m is placed on the wedge (Fig. 5.78a). There is no friction between the block and the wedge. The system is released from rest. a) Calculate the acceleration of the wedge and the horizontal and vertical components of the acceleration of the block. b) Do your answers to part (a) reduce to the correct results when M is very large? c) As seen by a stationary observer, what is the shape of the trajectory of the block?

5.121 A wedge with mass M rests on a frictionless horizontal tabletop. A block with mass m is placed on the wedge and a horizontal force $\vec{F}$ is applied to the wedge (Fig. 5.78b). What must the magnitude of $\vec{F}$ be if the block is to remain at a constant height above the tabletop?

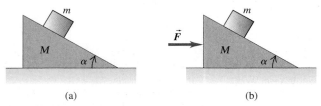

(a)   (b)

Figure 5.78 Challenge Problems 5.120 and 5.121.

5.122 A box of weight w is accelerated up a ramp by a rope that exerts a tension T. The ramp makes an angle α with the horizontal and the rope makes an angle θ above the ramp. The coefficient of kinetic friction between the box and the ramp is μ_k. Show that no matter what the value of α, the acceleration is maximum if $\theta = \arctan \mu_k$ (as long as the box remains in contact with the ramp).

5.123 Angle for Minimum Force. A box with weight w is pulled at constant speed along a level floor by a force $\vec{F}$ that is at an angle θ above the horizontal. The coefficient of kinetic friction between the floor and box is μ_k. a) In terms of θ, μ_k, and w, calculate F. b) For $w = 400$ N and $\mu_k = 0.25$, calculate F for θ ranging from $0°$ to $90°$ in increments of $10°$. Graph F versus θ. c) From the general expression in part (a), calculate the value of θ for which the value of F, required to maintain constant speed, is a minimum. (*Hint:* At a point where a function is minimum, what are the first and second derivatives of the function? Here F is a function of θ.) For the special case of $w = 400$ N and $\mu_k = 0.25$, evaluate this optimal θ and compare your result to the graph you constructed in part (b).

5.124 Falling Baseball. You drop a baseball from the roof of a tall building. As the ball falls, the air exerts a drag force proportional to the square of the ball's speed $(f = Dv^2)$. a) In a diagram, show the direction of motion and indicate, with the aid of vectors, all of the forces acting on the ball. b) Apply Newton's second law and infer from the resulting equation the general properties of the motion. c) Show that the ball acquires a terminal speed that is as given in Eq. (5.13). d) Derive the equation for the speed at any time. (*Note:*

$$\int \frac{dx}{a^2 - x^2} = \frac{1}{a} \operatorname{arctanh}\left(\frac{x}{a}\right)$$

where

$$\tanh(x) = \frac{e^x - e^{-x}}{e^x + e^{-x}} = \frac{e^{2x} - 1}{e^{2x} + 1}$$

defines the hyperbolic tangent.)

5.125 Double Atwood's Machine. In Fig. 5.79 masses m_1 and m_2 are connected by a light string A over a light frictionless pulley B.

The axle of pulley B is connected by a second light string C over a second light, frictionless pulley D to a mass m_3. Pulley D is suspended from the ceiling by an attachment to its axle. The system is released from rest. In terms of m_1, m_2, m_3, and g, what is a) the acceleration of block m_3? b) the acceleration of pulley B? c) the acceleration of block m_1? d) the acceleration of block m_2? e) the tension in string A? f) the tension in string C? g) What do your expressions give for the special case of $m_1 = m_2$ and $m_3 = m_1 + m_2$? Is this sensible?

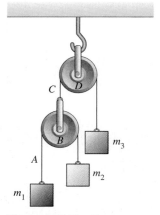

Figure 5.79 Challenge Problem 5.125.

5.126 The masses of blocks A and B in Fig. 5.80 are 20.0 kg and 10.0 kg, respectively. The blocks are initially at rest on the floor and are connected by a massless string passing over a massless and frictionless pulley. An upward force $\vec{F}$ is applied to the pulley. Find the accelerations $\vec{a}_A$ of block A and $\vec{a}_B$ of block B when F is a) 124 N; b) 294 N; c) 424 N.

5.127 A ball is held at rest at position A in Fig. 5.81 by two light strings. The horizontal string is cut and the ball starts swinging as a pendulum. Point B is the farthest to the right the ball goes as it swings back and forth. What is the ratio of the tension in the supporting string in position B to its value at A before the horizontal string was cut?

Figure 5.80 Challenge Problem 5.126.

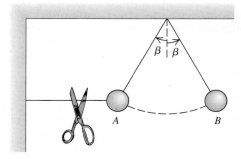

Figure 5.81 Challenge Problem 5.127.

WORK AND KINETIC ENERGY 6

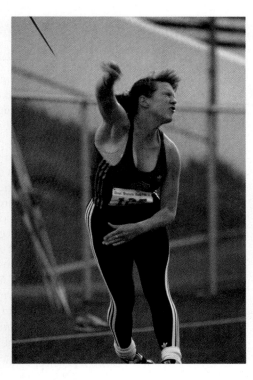

When an athlete throws the javelin in a track and field competition, she does work on the javelin—that is, she exerts a force over a distance. As a result the javelin gains kinetic energy, or energy of motion. At the end of the javelin's flight, this kinetic energy, in turn, does work on the ground as it penetrates into the surface.

? **According to Newton's third law, the javelin exerts as much force on the athlete as the athlete exerts on the javelin. Would it be correct to say that the *javelin* does work on the *athlete?***

Some problems are harder than they look. For example, suppose you try to find the speed of an arrow that has been shot from a bow. You apply Newton's laws and all the problem-solving techniques that we've learned, but you run across a major stumbling block: After the archer releases the arrow, the bow string exerts a *varying* force that depends on the arrow's position. As a result, the simple methods that we've learned aren't enough to calculate the speed. Never fear; we aren't by any means finished with mechanics, and there are other methods for dealing with such problems.

The new method that we're about to introduce uses the ideas of *work* and *energy*. The importance of the energy idea stems from the *principle of conservation of energy*: Energy is a quantity that can be converted from one form to another but cannot be created or destroyed. In an automobile engine, chemical energy stored in the fuel is converted partially to the energy of the automobile's motion and partially to thermal energy. In a microwave oven, electromagnetic energy obtained from your power company is converted to thermal energy of the food being cooked. In these and all other processes, the *total* energy—the sum of all energy present in all different forms—remains the same. No exception has ever been found.

We'll use the energy idea throughout the rest of this book to study a tremendous range of physical phenomena. This idea will help you understand why a sweater keeps you warm, how a camera's flash unit can produce a short burst of light, and the meaning of Einstein's famous equation $E = mc^2$.

In this chapter, though, our concentration will be on mechanics. We'll learn about one important form of energy called *kinetic energy,* or energy of motion, and how it relates to the concept of *work.* We'll also consider *power,* which is the time rate of doing work. In Chapter 7 we'll expand the ideas of work and kinetic energy into a deeper understanding of the concepts of energy and the conservation of energy.

6.1 | Work

You'd probably agree that it's hard work to pull a heavy sofa across the room, to lift a stack of encyclopedias from the floor to a high shelf, or to push a stalled car off the road. Indeed, all of these examples agree with the everyday meaning of *work*—any activity that requires muscular or mental effort.

In physics, work has a much more precise definition. By making use of this definition we'll find that in any motion, no matter how complicated, the total work done on a particle by all forces that act on it equals the change in its *kinetic energy*—a quantity that's related to the particle's speed. This relationship holds even when the forces acting on the particle aren't constant, a situation that can be difficult or impossible to handle with the techniques you learned in Chapters 4 and 5. The ideas of work and kinetic energy enable us to solve problems in mechanics that we could not have attempted before.

We'll develop the relationship between work and kinetic energy in Section 6.2, and see what to do with nonconstant forces in Section 6.3. Meanwhile, let's see how work is defined and how to calculate work in a variety of situations involving *constant* forces. Even though we already know how to solve problems in which the forces are constant, the idea of work is still useful in such problems.

The three examples of work described above—pulling a sofa, lifting encyclopedias, and pushing a car—have something in common. In each case you do work by exerting a *force* on a body while that body *moves* from one place to another, that is, undergoes a *displacement.* You do more work if the force is greater (you pull harder on the sofa) or if the displacement is greater (you pull the sofa farther across the room).

The physicist's definition of work is based on these observations. Consider a body that undergoes a displacement of magnitude s along a straight line. (For now, we'll assume that any body we discuss can be treated as a particle so that we can ignore any rotation or changes in shape of the body.) While the body moves, a constant force $\vec{F}$ acts on it in the same direction as the displacement $\vec{s}$ (Fig. 6.1). We define the **work** W done by this constant force under these circumstances as the product of the force magnitude F and the displacement magnitude s:

$$W = Fs \quad \text{(constant force in direction of straight-line displacement)} \quad (6.1)$$

The work done on the body is greater if either the force F or the displacement s is greater, in agreement with our observations above.

CAUTION Don't confuse W (work) with w (weight). Though the symbols may be almost the same, work and weight are different quantities.

6.1 When a constant force $\vec{F}$ acts in the same direction as the displacement $\vec{s}$, the work done by the force is $W = Fs$.

The SI unit of work is the **joule** (abbreviated J, pronounced "jewel," and named in honor of the 19th-century English physicist James Prescott Joule). From

Eq. (6.1) we see that in any system of units, the unit of work is the unit of force multiplied by the unit of distance. In SI units the unit of force is the newton and the unit of distance is the meter, so one joule is equivalent to one *newton-meter* $(N \cdot m)$:

$$1 \text{ joule} = (1 \text{ newton})(1 \text{ meter}) \quad \text{or } . \ 1 \text{ J} = 1 \text{ N} \cdot \text{m}$$

In the British system the unit of force is the pound (lb), the unit of distance is the foot, and the unit of work is the *foot-pound* $(ft \cdot lb)$. The following conversions are useful:

$$1 \text{ J} = 0.7376 \text{ ft} \cdot \text{lb} \qquad 1 \text{ ft} \cdot \text{lb} = 1.356 \text{ J}$$

As an illustration of Eq. (6.1), think of a person pushing a stalled car. If he pushes the car through a displacement $\vec{s}$ with a constant force $\vec{F}$ in the direction of motion, the amount of work he does on the car is given by Eq. (6.1): $W = Fs$. But what if the person pushes at an angle ϕ with the car's displacement (Fig. 6.2)? Only the component of force in the direction of the car's motion would be effective in moving the car. (Other forces must act on the car so that it moves along $\vec{s}$, not in the direction of $\vec{F}$. We're interested only in the work that the person does, however, so we'll consider only the force he exerts.) When the force $\vec{F}$ and the displacement $\vec{s}$ have different directions, we take the component of $\vec{F}$ in the direction of the displacement $\vec{s}$, and we define the work as the product of this component and the magnitude of the displacement. The component of $\vec{F}$ in the direction of $\vec{s}$ is $F \cos \phi$, so

$$W = Fs \cos \phi \qquad \text{(constant force, straight-line displacement)} \qquad (6.2)$$

We are assuming that F and ϕ are constant during the displacement. If $\phi = 0$, so that $\vec{F}$ and $\vec{s}$ are in the same direction, then $\cos \phi = 1$ and we are back to Eq. (6.1).

Equation (6.2) has the form of the *scalar product* of two vectors, introduced in Section 1.10: $\vec{A} \cdot \vec{B} = AB \cos \phi$. You may want to review that definition. Using this, we can write Eq. (6.2) more compactly as

$$W = \vec{F} \cdot \vec{s} \qquad \text{(constant force, straight-line displacement)} \qquad (6.3)$$

CAUTION Here's an essential point: work is a *scalar* quantity, even though it's calculated by using two vector quantities (force and displacement). A 5-N force toward the east acting on a body that moves 6 m to the east does exactly the same work as a 5-N force toward the north acting on a body that moves 6 m to the north.

6.2 When a constant force $\vec{F}$ acts at an angle ϕ to the displacement $\vec{s}$, the work done by the force is $(F \cos \phi) s = Fs \cos \phi$.

| Example 6.1 | **Work done by a constant force** |

a) Steve exerts a steady force of magnitude 210 N (about 47 lb) on the stalled car in Fig. 6.2 as he pushes it a distance of 18 m. The car also has a flat tire, so to make the car track straight Steve must push at an angle of 30° to the direction of motion. How much work does Steve do? b) In a helpful mood, Steve pushes a second stalled car with a steady force $\vec{F} = (160\,\text{N})\hat{\imath} - (40\,\text{N})\hat{\jmath}$. The displacement of the car is $\vec{s} = (14\,\text{m})\hat{\imath} + (11\,\text{m})\hat{\jmath}$. How much work does Steve do in this case?

SOLUTION

IDENTIFY and SET UP: In both parts (a) and (b), the target variable is the work W done by Steve. In each case the force is constant and the displacement is along a straight line, so we can use Eq. (6.2) or (6.3). The angle between $\vec{F}$ and $\vec{s}$ is given explicitly in part (a), so we can apply Eq. (6.2) directly. In part (b) the angle isn't given, so we're better off calculating the scalar product in Eq. (6.3) from the components of $\vec{F}$ and $\vec{s}$, as in Eq. (1.21).

EXECUTE: a) From Eq. (6.2),

$$W = Fs\cos\phi = (210\,\text{N})(18\,\text{m})\cos 30° = 3.3 \times 10^3\,\text{J}$$

b) The components of $\vec{F}$ are $F_x = 160\,\text{N}$ and $F_y = -40\,\text{N}$, and the components of $\vec{s}$ are $x = 14\,\text{m}$ and $y = 11\,\text{m}$. Hence, using Eqs. (1.21) and (6.3),

$$\begin{aligned}W = \vec{F}\cdot\vec{s} &= F_x x + F_y y \\ &= (160\,\text{N})(14\,\text{m}) + (-40\,\text{N})(11\,\text{m}) \\ &= 1.8 \times 10^3\,\text{J}\end{aligned}$$

EVALUATE: Our results show that 1 joule is a rather small amount of work.

In Example 6.1 the work done in pushing the cars was positive. But it's important to understand that work can also be negative or zero. This is the essential way in which work as defined in physics differs from the "everyday" definition of work. When the force has a component in the *same direction* as the displacement (ϕ between zero and 90°), $\cos\phi$ in Eq. (6.2) is positive and the work W is *positive* (Fig. 6.3a). When the force has a component *opposite* to the displacement (ϕ between 90° and 180°), $\cos\phi$ is negative and the work is *negative* (Fig. 6.3b). When the force is *perpendicular* to the displacement, $\phi = 90°$ and the work done by the force is *zero* (Fig. 6.3c). The cases of zero work and negative work bear closer examination, so let's look at some examples.

There are many situations in which forces act but do zero work. You might think it's "hard work" to hold a barbell motionless in the air for five minutes (Fig. 6.4). But in fact, you aren't doing any work at all on the barbell because there is no displacement. You get tired because the muscle fibers in your arm do work as they continually contract and relax. This is work done by one part of the arm exerting force on another part, however, *not* on the barbell. (We'll say more in Section 6.2 about work done by one part of a body on another part.) Even when

6.3 A constant force $\vec{F}$ can do positive, negative, or zero work depending on the angle between $\vec{F}$ and the displacement $\vec{s}$.

Force has component in direction of displacement: Work done is positive

(a)

Force has component opposite to displacement: Work done is negative

(b)

Force is perpendicular to displacement: Work done is zero

(c)

(a) Positive work

(b) Zero work

(c) Negative work

6.4 This weight lifter exerts an upward force $\vec{F}$ on the barbell at all times. (a) He does *positive* work on the barbell only when he is lifting it, so that the force is in the same direction as the displacement $\vec{s}$. (b) He does *zero* work while the barbell is stationary (the displacement is zero). (c) He does *negative* work while lowering the barbell ($\vec{F}$ is opposite to $\vec{s}$).

you walk with constant velocity on a level floor while carrying a book, you still do no work on it. The book has a displacement, but the (vertical) supporting force that you exert on the book has no component in the direction of the (horizontal) motion. Then $\phi = 90°$ in Eq. (6.2), and $\cos \phi = 0$. When a body slides along a surface, the work done on the body by the normal force is zero; and when a ball on a string moves in uniform circular motion, the work done on the ball by the tension in the string is also zero. In both cases the work is zero because the force has no component in the direction of motion.

What does it really mean to do *negative* work? The answer comes from Newton's third law of motion. When you catch a ball as in Fig. 6.5a, your hand and the ball move together with the same displacement $\vec{s}$ (Fig. 6.5b). The ball exerts a force $\vec{F}_{\text{B on H}}$ on your hand in the same direction as the hand's displacement, so the work done by the *ball* on your *hand* is positive. But by Newton's third law your hand exerts an equal and opposite force $\vec{F}_{\text{H on B}} = -\vec{F}_{\text{B on H}}$ on the ball (Fig. 6.5c). This force, which slows the ball to a stop, acts opposite to the ball's displacement. Thus the work done by your *hand* on the *ball* is negative. Because your hand and the ball have the same displacement, the work that your hand does on the ball is just the negative of the work that the ball does on your hand. In general, when one body does negative work on a second body, the second body does an equal amount of *positive* work on the first body.

CAUTION We always speak of work done *on* a particular body *by* a specific force. Always be sure to specify exactly what force is doing the work you are talking about. When you lift a book, you exert an upward force on the book and the book's displacement is upward, so the work done by the lifting force on the book is positive. But the work done by the *gravitational* force (weight) on a book being lifted is *negative* because the downward gravitational force is opposite to the upward displacement.

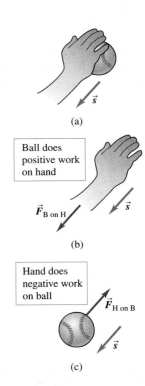
(a)

Ball does positive work on hand

$\vec{F}_{\text{B on H}}$

(b)

Hand does negative work on ball

$\vec{F}_{\text{H on B}}$

(c)

6.5 (a) When you catch a baseball, your hand and the ball have the same displacement $\vec{s}$. (b) The ball exerts a force $\vec{F}_{\text{B on H}}$ on your hand in the same direction as $\vec{s}$. (c) Your hand exerts an equal and opposite force $\vec{F}_{\text{H on B}} = -\vec{F}_{\text{B on H}}$ on the ball in the direction opposite to $\vec{s}$.

How do we calculate work when several forces act on a body? One way is to use Eq. (6.2) or (6.3) to compute the work done by each separate force. Then, because work is a scalar quantity, the *total* work W_{tot} done on the body by all the forces is the algebraic sum of the quantities of work done by the individual forces. An alternative way to find the total work W_{tot} is to compute the vector sum of the forces (that is, the net force) and then use this vector sum as $\vec{F}$ in Eq. (6.2) or (6.3).

Act|v
ONLINE
Physics

5.1   Work Calculations

<table>
<tr><td>Example
6.2</td><td>**Work done by several forces**</td></tr>
</table>

Farmer Benton hitches his tractor to a sled loaded with firewood and pulls it a distance of 20 m along level ground (Fig. 6.6a). The total weight of sled and load is 14,700 N. The tractor exerts a constant 5000-N force at an angle of 36.9° above the horizontal, as shown in Fig. 6.6b. There is a 3500-N friction force opposing the sled's motion. Find the work done by each force acting on the sled and the total work done by all the forces.

SOLUTION

IDENTIFY: All of the forces are constant and the displacement is along a straight line, so we can calculate the work using the formulas given in this section. We'll find the net work in two ways: (1) by adding together the work done on the sled by each force and (2) by finding the amount of work done by the net force on the sled. We'll find the net force using the techniques described in Chapter 5.

SET UP: As we are working with forces, our first steps are to draw a free-body diagram showing all of the forces acting on the sled and to choose a coordinate system (Fig. 6.6b). For each of the four forces—weight, normal force, force of the tractor, and friction force—we know the angle between the displacement (which is in the positive x-direction) and the force. Hence we can calculate the work done by each force using Eq. (6.2).

To find the net force, we'll add the components of the four forces. Newton's second law tells us that because the sled's motion is purely horizontal, the net force will have only a horizontal component.

EXECUTE: The work W_w done by the weight is zero because its direction is perpendicular to the displacement. (The angle between the force of gravity and the displacement is 90°, and the cosine of the angle is zero.) For the same reason, the work W_n done by the

normal force is also zero. So $W_w = W_n = 0$. (Incidentally, the normal force is *not* equal in magnitude to the weight; see Example 5.16 of Section 5.3, in which the free-body diagram is very similar to Fig. 6.6b.)

That leaves the force F_T exerted by the tractor and the friction force f. From Eq. (6.2) the work W_T done by the tractor is

$$W_T = F_T s \cos \phi = (5000 \text{ N})(20 \text{ m})(0.800) = 80,000 \text{ N} \cdot \text{m}$$
$$= 80 \text{ k J}$$

The friction force $\vec{f}$ is opposite to the displacement, so for this force $\phi = 180°$ and $\cos \phi = -1$. The work W_f done by the friction force is

$$W_f = fs \cos 180° = (3500 \text{ N})(20 \text{ m})(-1) = -70,000 \text{ N} \cdot \text{m}$$
$$= -70 \text{ k J}$$

The total work W_{tot} done on the sled by all forces is the *algebraic* sum of the work done by the individual forces:

$$W_{tot} = W_w + W_n + W_T + W_f = 0 + 0 + 80 \text{ k J} + (-70 \text{ k J})$$
$$= 10 \text{ k J}$$

In the alternative approach, we first find the *vector* sum of all the forces (the net force) and then use it to compute the total work. The vector sum is best found by using components. From Fig. 6.6b,

$$\sum F_x = F_T \cos \phi + (-f) = (5000 \text{ N}) \cos 36.9° - 3500 \text{ N}$$
$$= 500 \text{ N}$$
$$\sum F_y = F_T \sin \phi + n + (-w)$$
$$= (5000 \text{ N}) \sin 36.9° + n - 14,700 \text{ N}$$

We don't really need the second equation; we know that the y-component of force is perpendicular to the displacement, so it

(a)   (b)

6.6 (a) A tractor pulling a sled of firewood. (b) Free-body diagram for the sled and its load, treated as a particle.

does no work. Besides, there is no y-component of acceleration, so $\sum F_y$ has to be zero anyway. The total work is therefore the work done by the total x-component:

$$W_{tot} = (\textstyle\sum \vec{F}) \cdot \vec{s} = (\textstyle\sum F_x)s = (500 \text{ N})(20 \text{ m}) = 10,000 \text{ J}$$
$$= 10 \text{ kJ}$$

EVALUATE: We get the same result for W_{tot} with either method, as we should.

Note that the net force in the x-direction is *not* zero, and so the sled must accelerate as it moves. In Section 6.2 we'll return to this example and see how to use the concept of work to explore the sled's motion.

<table>
<tr><td>Conceptual
Example 6.3</td><td>**Total work when velocity is constant**</td></tr>
</table>

An electron moves in a straight line toward the east with a constant speed of 8×10^7 m/s. It has electric, magnetic, and gravitational forces acting on it. Calculate the total work done on the electron during a 1-m displacement.

SOLUTION

The electron moves with constant velocity, so its acceleration is zero. By Newton's second law, the vector sum of forces is also zero. Therefore the total work done by all the forces (equal to the work done by the vector sum of all the forces) must be *zero*. Individual forces may do nonzero work, but that's not what the problem asks for.

Test Your Understanding

Suppose the barbell in Fig. 6.4 weighs 500 N. Calculate how much work the weight lifter does on the barbell (a) to raise it at constant speed a distance of 1.5 m; (b) to hold it stationary for 3.0 s at a height 1.5 m above the ground; (c) to lower it at constant speed a distance of 1.5 m.

6.2 | Work and Kinetic Energy

The total work done on a body by external forces is related to the body's displacement—that is, to changes in its position. But the total work is also related to changes in the *speed* of the body. To see this, consider Fig. 6.7, which shows several exam-

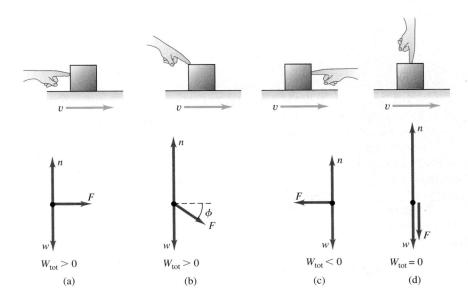

6.7 A block sliding on a frictionless table. (a) The net force causes the speed to increase and does positive work. (b) Again the net force causes the speed to increase and does positive work. (c) The net force opposes the displacement, causes the speed to decrease, and does negative work. (d) The net force is zero and does no work, and the speed is constant.

ples of a block sliding on a frictionless table. The forces acting on the block are its weight $\vec{w}$, the normal force $\vec{n}$, and the force $\vec{F}$ exerted on it by the hand.

In Fig. 6.7a the net force on the block is in the direction of its motion. From Newton's second law, this means that the block speeds up; from Eq. (6.1), this also means that the total work W_{tot} done on the block is positive. The total work is also positive in Fig. 6.7b, but only the component $F \cos \phi$ contributes to W_{tot}. The block again speeds up, and this same component $F \cos \phi$ is what causes the acceleration. The total work is *negative* in Fig. 6.7c because the net force opposes the displacement; in this case the block slows down. The net force is zero in Fig. 6.7d, so the speed of the block stays the same and the total work done on the block is zero. We can conclude that when a particle undergoes a displacement, it speeds up if $W_{tot} > 0$, slows down if $W_{tot} < 0$, and maintains the same speed if $W_{tot} = 0$.

Let's make these observations more quantitative. Consider a particle with mass m moving along the x-axis under the action of a constant net force with magnitude F directed along the positive x-axis (Fig. 6.1). The particle's acceleration is constant and given by Newton's second law, $F = ma_x$. Suppose the speed changes from v_1 to v_2 while the particle undergoes a displacement $s = x_2 - x_1$ from point x_1 to x_2. Using a constant-acceleration equation, Eq. (2.13), and replacing v_{0x} by v_1, v_x by v_2, and $(x - x_0)$ by s, we have

$$v_2^2 = v_1^2 + 2a_x s$$

$$a_x = \frac{v_2^2 - v_1^2}{2s}$$

When we multiply this equation by m and equate ma_x to the net force F, we find

$$F = ma_x = m \frac{v_2^2 - v_1^2}{2s}$$

and

$$Fs = \frac{1}{2} mv_2^2 - \frac{1}{2} mv_1^2 \tag{6.4}$$

The product Fs is the work done by the net force F and thus is equal to the total work W_{tot} done by all the forces acting on the particle. The quantity $\frac{1}{2}mv^2$ is called the **kinetic energy** K of the particle:

$$K = \frac{1}{2} mv^2 \quad \text{(definition of kinetic energy)} \tag{6.5}$$

Like work, the kinetic energy of a particle is a scalar quantity; it depends only on the particle's mass and speed, not its direction of motion. A car (viewed as a particle) has the same kinetic energy when going north at 10 m/s as when going east at 10 m/s. Kinetic energy can never be negative, and it is zero only when the particle is at rest.

We can now interpret Eq. (6.4) in terms of work and kinetic energy. The first term on the right side of Eq. (6.4) is $K_2 = \frac{1}{2}mv_2^2$, the final kinetic energy of the particle (that is, after the displacement). The second term is the initial kinetic energy, $K_1 = \frac{1}{2}mv_1^2$, and the difference between these terms is the *change* in kinetic energy. So Eq. (6.4) says that **the work done by the net force on a particle equals the change in the particle's kinetic energy:**

$$W_{tot} = K_2 - K_1 = \Delta K \quad \text{(work-energy theorem)} \tag{6.6}$$

This result is the **work-energy theorem.**

The work-energy theorem agrees with our observations about the block in Fig. 6.7. When W_{tot} is *positive;* K_2 is greater than K_1, the kinetic energy *increases,* and the particle is going faster at the end of the displacement than at the beginning. When W_{tot} is *negative,* the kinetic energy *decreases* and the speed is less after the displacement. When $W_{tot} = 0$, the initial and final kinetic energies K_1 and K_2 are the same and the speed is unchanged. We stress that the work-energy theorem by itself tells us only about changes in *speed,* not velocity, since the kinetic energy carries no information about the direction of motion.

From Eq. (6.4) or (6.6), kinetic energy and work must have the same units. Hence the joule is the SI unit of both work and kinetic energy (and, as we will see later, of all kinds of energy). To verify this, note that in SI units the quantity $K = \frac{1}{2}mv^2$ has units $\text{kg} \cdot (\text{m/s})^2$ or $\text{kg} \cdot \text{m}^2/\text{s}^2$; we recall that $1 \text{ N} = 1 \text{ kg} \cdot \text{m/s}^2$, so

$$1 \text{ J} = 1 \text{ N} \cdot \text{m} = 1 \, (\text{kg} \cdot \text{m/s}^2) \cdot \text{m} = 1 \text{ kg} \cdot \text{m}^2/\text{s}^2$$

In the British system the unit of kinetic energy and of work is

$$1 \text{ ft} \cdot \text{lb} = 1 \text{ ft} \cdot \text{slug} \cdot \text{ft/s}^2 = 1 \text{ slug} \cdot \text{ft}^2/\text{s}^2$$

Because we used Newton's laws in deriving the work-energy theorem, we can use it only in an inertial frame of reference. The speeds that we use to compute the kinetic energies and the distances that we use to compute work *must* be measured in an inertial frame. Note also that the work-energy theorem is valid in *any* inertial frame, but the values of W_{tot} and $K_2 - K_1$ may differ from one inertial frame to another (because the displacement and speed of a body may be different in different frames).

We have derived the work-energy theorem for the special case of straight-line motion with constant forces, and in the following examples we'll apply it to this special case only. We'll find in the next section that the theorem is valid in general, even when the forces are not constant and the particle's trajectory is curved.

| Problem Solving Strategy | **Work and Kinetic Energy** |

IDENTIFY *the relevant concepts:* The work-energy theorem is extremely useful in situations where you want to relate a body's speed v_1 at one point in its motion to its speed v_2 at a different point. This approach is less useful for problems that involve time, such as finding the time it takes a body to go from point 1 to point 2. The reason is that the work-energy theorem, $W_{tot} = K_2 - K_1$, doesn't involve time at all. For problems that involve time, it's usually best to use the relationships among time, position, velocity, and acceleration described in Chapters 2 and 3.

SET UP *the problem using the following steps:*
1. Choose the initial and final positions of the body, and draw a free-body diagram showing all the forces that act on the body.
2. Choose a coordinate system. (If the motion is along a straight line, it's usually easiest to have both the initial and final positions lie along the x-axis.)
3. List the unknown and known quantities, and decide which unknowns are your target variables. In some cases the tar-

get variable will be the body's initial or final speed; in other cases it will be the magnitude of one of the forces acting on the body.

EXECUTE *the solution:* Calculate the work done by each force. If the force is constant and the displacement is a straight line, you can use Eq. (6.2) or (6.3). (Later in this chapter we'll see how to handle varying forces and curved trajectories.) Be sure to check the sign of the work for each force; it must be positive if the force has a component in the direction of the displacement, negative if it has a component opposite the displacement, and zero if the force and displacement are perpendicular.

Add the amounts of work done by each force to find the total work W_{tot}. Be careful with signs! Sometimes it may be easier to calculate the vector sum of the forces (the net force), then find the work done by the net force; this value is also equal to W_{tot}.

Write expressions for the initial and final kinetic energies, K_1 and K_2. Note that kinetic energy involves *mass,* not *weight;* if you are given the body's weight, you'll need to use the relationship $w = mg$ to find the mass.

Finally, use the relationship $W_{tot} = K_2 - K_1$ to solve for the target variable. Remember that the right-hand side of this equation is the *final* kinetic energy minus the *initial* kinetic energy, never the other way around.

EVALUATE *your answer:* Check whether your answer makes physical sense. A key item to remember is that kinetic energy $K = \frac{1}{2}mv^2$ can never be negative. If you come up with a negative value of K, you've made a mistake. Perhaps you interchanged the initial and final kinetic energies in $W_{tot} = K_2 - K_1$ or made a sign error in one of the work calculations.

Example 6.4   Using work and energy to calculate speed

Let's look again at the sled in Fig. 6.6 and the numbers at the end of Example 6.2. Suppose the initial speed v_1 is 2.0 m/s. What is the speed of the sled after it moves 20 m?

SOLUTION

IDENTIFY and SET UP: We'll use the work-energy theorem, since we are given the initial speed $v_1 = 2.0$ m/s and want to find the final speed v_2. Figure 6.8 shows the free-body diagram from Example 6.2. The motion is in the positive x-direction.

EXECUTE: We already calculated the total work done by all the forces in Example 6.2, where we found $W_{tot} = 10$ kJ. Hence the kinetic energy of the sled and its load must increase by 10 kJ.

To write expressions for the initial and final kinetic energies, we need the mass of the sled and load. We are given that the *weight* is 14,700 N, so the mass is

$$m = w/g = (14{,}700 \text{ N})/(9.8 \text{ m/s}^2) = 1500 \text{ kg}$$

Then the initial kinetic energy K_1 is

$$K_1 = \frac{1}{2}mv_1^2 = \frac{1}{2}(1500 \text{ kg})(2.0 \text{ m/s})^2 = 3000 \text{ kg} \cdot \text{m}^2/\text{s}^2$$

$$= 3000 \text{ J}$$

The final kinetic energy K_2 is

$$K_2 = \frac{1}{2}mv_2^2 = \frac{1}{2}(1500 \text{ kg})v_2^2$$

where v_2 is the unknown speed we want to find. Equation (6.6) gives

$$K_2 = K_1 + W_{tot} = 3000 \text{ J} + 10{,}000 \text{ J} = 13{,}000 \text{ J}$$

Setting these two expressions for K_2 equal, substituting $1 \text{ J} = 1 \text{ kg} \cdot \text{m}^2/\text{s}^2$, and solving for v_2, we find

$$v_2 = 4.2 \text{ m/s}$$

EVALUATE: The total work is positive, so the kinetic energy increases $(K_2 > K_1)$ and the speed increases $(v_2 > v_1)$.

This problem can also be done without the work-energy theorem. We can find the acceleration from $\sum \vec{F} = m\vec{a}$ and then use the

equations of motion for constant acceleration to find v_2. Since the acceleration is along the x-axis,

$$a = a_x = \frac{\sum F_x}{m} = \frac{(5000 \text{ N}) \cos 36.9° - 3500 \text{ N}}{1500 \text{ kg}}$$

$$= 0.333 \text{ m/s}^2$$

then

$$v_2^2 = v_1^2 + 2as = (2.0 \text{ m/s})^2 + 2(0.333 \text{ m/s}^2)(20 \text{ m})$$

$$= 17.3 \text{ m}^2/\text{s}^2$$

$$v_2 = 4.2 \text{ m/s}$$

This is the same result that we obtained with the work-energy approach, but there we avoided the intermediate step of finding the acceleration. You will find several other examples in this chapter and the next that *can* be done without using energy considerations but that are easier when energy methods are used. When a problem can be done by two different methods, doing it by both methods (as we did in this example) is a very good way to check your work.

6.8 Free-body diagram for the sled and its load in Example 6.2.

Example 6.5	**Forces on a hammerhead**

In a pile driver, a steel hammerhead with mass 200 kg is lifted 3.00 m above the top of a vertical I-beam being driven into the ground (Fig. 6.9a). The hammer is then dropped, driving the I-beam 7.4 cm farther into the ground. The vertical rails that guide the hammerhead exert a constant 60-N friction force on the hammerhead. Use the work-energy theorem to find a) the speed of the hammerhead just as it hits the I-beam and b) the average force the hammerhead exerts on the I-beam. Ignore the effects of the air.

SOLUTION

IDENTIFY: This problem is an ideal candidate for the work-energy theorem, as it relates a body's speed at different locations and the forces acting on the body. In this problem there are *three* positions of interest: point 1, where the hammerhead starts from rest; point 2, where it first contacts the I-beam; and point 3, where the hammerhead comes to a halt (see Fig. 6.9a). We have two unknowns—the hammerhead's speed at point 2, and the force the I-beam exerts between points 2 and 3—we'll apply the work-energy theorem twice: once for the motion from 1 to 2 and once for the motion from 2 to 3.

SET UP: Figure 6.9b is a free-body diagram showing the vertical forces on the hammerhead as it falls from point 1 to point 2. (Because the displacement is vertical, we ignore any horizontal forces that may be present because they do no work.) For this part of the motion, our target variable is the hammerhead's speed v_2.

The free-body diagram in Fig. 6.9c shows the vertical forces on the hammerhead during the motion from point 2 to point 3. In addition to the forces shown in Fig. 6.9b, the I-beam exerts an upward normal force of magnitude n on the hammerhead. This force actually varies as the hammerhead comes to a halt, but for simplicity we'll treat n as a constant. Hence n represents the *average* value of this upward force during the motion. Our target variable for this part of the motion is the force that the *hammerhead* exerts on the I-beam; it is the reaction force to the normal force exerted by the I-beam, so by Newton's third law its magnitude is also n.

EXECUTE: a) From point 1 to point 2, the vertical forces are the downward weight $w = mg = (200 \text{ kg})(9.8 \text{ m/s}^2) = 1960 \text{ N}$ and the upward friction force $f = 60 \text{ N}$. Thus the net downward force is $w - f = 1900 \text{ N}$. The displacement of the hammerhead from

(a)

(b)

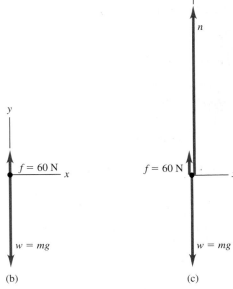

(c)

6.9 (a) A pile driver pounds an I-beam into the ground. (b) Free-body diagram for the hammerhead while falling. (c) Free-body diagram for the hammerhead while driving the I-beam. Vector lengths are not to scale.

point 1 to point 2 is downward and equal to $s_{12} = 3.00$ m. The total work done on the hammerhead as it moves from point 1 to point 2 is then

$$W_{tot} = (w - f)s_{12} = (1900 \text{ N})(3.00 \text{ m}) = 5700 \text{ J}$$

At point 1 the hammerhead is at rest, so its initial kinetic energy K_1 is zero. Equation (6.6) gives

$$W_{tot} = K_2 - K_1 = \frac{1}{2}mv_2^2 - 0$$

$$v_2 = \sqrt{\frac{2W_{tot}}{m}} = \sqrt{\frac{2(5700 \text{ J})}{200 \text{ kg}}} = 7.55 \text{ m/s}$$

This is the hammerhead's speed at point 2, just as it hits the beam. b) As the hammerhead moves downward between points 2 and 3, the net downward force acting on it is $w - f - n$ (see Fig. 6.9c). The total work done on the hammerhead during this displacement is

$$W_{tot} = (w - f - n)s_{23}$$

The initial kinetic energy for this part of the motion is K_2, which from part (a) equals 5700 J (the total work done on the hammerhead as it moved from point 1 to point 2). The final kinetic energy is

$K_3 = 0$, since the hammerhead ends at rest. Then, from the work-energy theorem,

$$W_{tot} = (w - f - n)s_{23} = K_3 - K_2$$

$$n = w - f - \frac{(K_3 - K_2)}{s_{23}}$$

$$= 1960 \text{ N} - 60 \text{ N} - \frac{(0 \text{ J} - 5700 \text{ J})}{0.074 \text{ m}}$$

$$= 79,000 \text{ N}$$

The downward force that the hammerhead exerts on the I-beam has this same magnitude, 79,000 N (about 9 tons)—more than 40 times the weight of the hammerhead.

EVALUATE: The total change in the hammerhead's kinetic energy during the whole process is zero; a relatively small net force does positive work over a large distance, and then a much larger net force does negative work over a much smaller distance. The same thing happens if you speed up your car gradually and then drive it into a brick wall. The very large force needed to reduce the kinetic energy to zero over a short distance is what does the damage to your car—and possibly to you.

The Meaning Of Kinetic Energy

Example 6.5 gives insight into the physical meaning of kinetic energy. The hammerhead is dropped from rest, and its kinetic energy when it hits the I-beam equals the total work done on it up to that point by the net force. This result is true in general: to accelerate a particle of mass m from rest (zero kinetic energy) up to a speed v, the total work done on it must equal the change in kinetic energy from zero to $K = \frac{1}{2}mv^2$;

$$W_{tot} = K - 0 = K$$

So *the kinetic energy of a particle is equal to the total work that was done to accelerate it from rest to its present speed* (Fig. 6.10). The definition $K = \frac{1}{2}mv^2$, Eq. (6.5), wasn't chosen at random; it's the *only* definition that agrees with this interpretation of kinetic energy.

In the second part of Example 6.5 the kinetic energy of the hammerhead did work on the I-beam and drove it into the ground. This gives us another interpretation of kinetic energy: *the kinetic energy of a particle is equal to the total work that particle can do in the process of being brought to rest.* This is why you pull your hand and arm backward when you catch a ball (Fig. 6.5). As the ball comes to rest, it does an amount of work (force times distance) on your hand equal to the ball's initial kinetic energy. By pulling your hand back, you maximize the distance over which the force acts and so minimize the force on your hand.

6.10 When a billiards player hits a cue ball at rest, the ball's kinetic energy after being hit is equal to the work that was done on it by the cue. The greater the force exerted by the cue and the greater the distance the ball moves while in contact with it, the greater the ball's kinetic energy.

| Conceptual Example 6.6 | **Comparing kinetic energies** |

Two iceboats like the one in Example 5.6 (Section 5.2) hold a race on a frictionless horizontal lake (Fig. 6.11). The two iceboats have masses m and $2m$, respectively. Each iceboat has an identical sail, so

the wind exerts the same constant force $\vec{F}$ on each iceboat. The two iceboats start from rest and cross the finish line a distance s away. Which iceboat crosses the finish line with greater kinetic energy?

SOLUTION

If you simply use the mathematical definition of kinetic energy, $K = \frac{1}{2}mv^2$, Eq. (6.5), the answer to this problem isn't immediately obvious. The iceboat of mass $2m$ has greater mass, so you might guess that the larger iceboat attains a greater kinetic energy at the finish line. But the smaller iceboat, of mass m, crosses the finish line with a greater speed, and you might guess that *this* iceboat has the greater kinetic energy. How can we decide?

The correct way to approach this problem is to remember that *the kinetic energy of a particle is just equal to the total work done to accelerate it from rest*. Both iceboats travel the same distance s, and only the horizontal force F in the direction of motion does work on either iceboat. Hence the total work done between the starting line and the finish line is the *same* for each iceboat, $W_{tot} = Fs$. At the finish line, each iceboat has a kinetic energy equal to the work W_{tot} done on it, because each iceboat started from rest. So both iceboats have the *same* kinetic energy at the finish line!

You might think this is a "trick" question, but it isn't. If you really understand the physical meanings of quantities such as kinetic energy, you can solve problems more easily and with better insight into the physics.

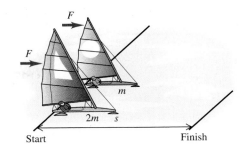

6.11 A race between iceboats.

Notice that we didn't need to say anything about how much time each iceboat took to reach the finish line. This is because the work-energy theorem makes no direct reference to time, only to displacement. In fact the iceboat of mass m takes less time to reach the finish line than does the larger iceboat of mass $2m$; we leave the calculation to you (Exercise 6.18).

Work and Kinetic Energy in Composite Systems

You may have noticed that in this section we've been careful to apply the work-energy theorem only to bodies that we can represent as *particles*—that is, as moving point masses. The reason is that new subtleties appear for more complex systems that have to be represented in terms of many particles with different motions. We can't go into these subtleties in detail in this chapter, but here's an example.

Consider a man standing on frictionless roller skates on a level surface, facing a rigid wall (Fig. 6.12). He pushes against the wall, setting himself in motion to the right. The forces acting on him are his weight $\vec{w}$, the upward normal forces $\vec{n}_1$ and $\vec{n}_2$ exerted by the ground on his skates, and the horizontal force $\vec{F}$ exerted on him by the wall. There is no vertical displacement, so $\vec{w}$, $\vec{n}_1$, and $\vec{n}_2$ do no work. The force $\vec{F}$ is the horizontal force that accelerates him to the right, but the parts of his body where that force is applied (the man's hands) do not move. Thus the force $\vec{F}$ also does no work. So where does the man's kinetic energy come from?

The difficulty is that it's simply not adequate to represent the man as a single point mass. For the motion to occur as we've described it, different parts of the man's body must have different motions; his hands are stationary against the wall while his torso is moving away from the wall. The various parts of his body interact with each other, and one part can exert forces and do work on another part. Therefore the *total* kinetic energy of this composite system of body parts can change, even though no work is done by forces applied by bodies (such as the wall) that are outside the system. This would not be possible with a system that can be represented as a single point particle. In Chapter 8 we'll consider further the motion of a collection of particles that interact with each other. We'll discover that just as for the man in this example, the total kinetic energy of such a system can change even when no work is done on any part of the system by anything outside it.

6.12 The external forces acting on a skater pushing off a wall. The work done by these forces is zero, but his kinetic energy changes nonetheless.

A baseball (mass 0.145 kg) leaves the pitcher's hand moving horizontally at a speed of 40 m/s (144 km/h, or 89 mi/h). If his throwing motion moves the ball forward over a distance of 2.0 m, what average horizontal force did the pitcher exert on the ball during the throw?

6.3 | Work and Energy with Varying Forces

So far in this chapter we've considered work done by *constant forces* only. But what happens when you stretch a spring? The more you stretch it, the harder you have to pull, so the force you exert is *not* constant as the spring is stretched. We've also restricted our discussion to *straight-line* motion. You can think of many situations in which a force that varies in magnitude, direction, or both acts on a body moving along a curved path. We need to be able to compute the work done by the force in these more general cases. Fortunately, we'll find that the work-energy theorem holds true even when varying forces are considered and when the body's path is not straight.

To add only one complication at a time, let's consider straight-line motion with a force that is directed along the line but with an *x*-component F_x that may change as the body moves. For example, imagine a train moving on a straight track with the engineer constantly changing the locomotive's throttle setting or applying the brakes. Suppose a particle moves along the *x*-axis from point x_1 to x_2. Figure 6.13a is a graph of the *x*-component of force as a function of the particle's coordinate *x*. To find the work done by this force, we divide the total displacement into small segments Δx_a, Δx_b, and so on (Fig. 6.13b). We approximate the work done by the force during segment Δx_a as the average force F_a in that segment multiplied by the displacement Δx_a. We do this for each segment and then add the results for all the segments. The work done by the force in the total displacement from x_1 to x_2 is approximately

$$W = F_a\Delta x_a + F_b\Delta x_b + \cdots$$

As the number of segments becomes very large and the width of each becomes very small, this sum becomes (in the limit) the *integral* of F_x from x_1 to x_2:

$$W = \int_{x_1}^{x_2} F_x\, dx \qquad (6.7)$$

(varying *x*-component of force, straight-line displacement)

Note that $F_a\,\Delta x_a$ represents the *area* of the first vertical strip in Fig. 6.13b and that the integral in Eq. (6.7) represents the area under the curve of Fig. 6.13a between x_1 and x_2. *On a graph of force as a function of position, the total work done by the force is represented by the area under the curve between the initial and final positions.* An alternative interpretation of Eq. (6.7) is that the work W equals the average force that acts over the entire displacement, multiplied by the displacement.

Equation (6.7) also applies if F_x, the *x*-component of the force, is constant. In that case, F_x may be taken outside the integral:

$$W = \int_{x_1}^{x_2} F_x\, dx = F_x \int_{x_1}^{x_2} dx = F_x(x_2 - x_1)$$

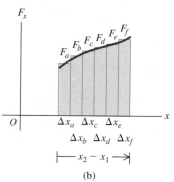

6.13 (a) Curve showing how a certain force F_x varies with *x*. (b) If the area is partitioned into small rectangles, the sum of their areas approximates the total work done during the displacement. The total area under the curve equals the work done by the force as the particle moves from x_1 to x_2.

But $x_2 - x_1 = s$, the total displacement of the particle. So in the case of a constant force F, Eq. (6.7) says that $W = Fs$, in agreement with Eq. (6.1). The interpretation of work as the area under the curve of F_x as a function of x also holds for a constant force; $W = Fs$ is the area of a rectangle of height F and width s (Fig. 6.14).

Now let's apply what we've learned to the stretched spring. To keep a spring stretched beyond its unstretched length by an amount x, we have to apply a force with magnitude F at each end (Fig. 6.15). If the elongation x is not too great, we find that the force we apply to the right-hand end has an x-component directly proportional to x:

$$F_x = kx \quad \text{(force required to stretch a spring)} \tag{6.8}$$

where k is a constant called the **force constant** (or spring constant) of the spring. Equation (6.8) shows that the units of k are force divided by distance, N/m in SI units and lb/ft in British units. A floppy toy spring such as a Slinky™ has a force constant of about 1 N/m; for the much stiffer springs in an automobile's suspension, k is about 10^5 N/m. The observation that elongation is directly proportional to force for elongations that are not too great was made by Robert Hooke in 1678 and is known as **Hooke's law.** It really shouldn't be called a "law," since it's a statement about a specific device and not a fundamental law of nature. Real springs don't always obey Eq. (6.8) precisely, but it's still a useful idealized model. We'll discuss Hooke's law more fully in Chapter 11.

To stretch a spring, we must do work. We apply equal and opposite forces to the ends of the spring and gradually increase the forces. We hold the left end stationary, so the force we apply at this end does no work. The force at the moving end *does* do work. Figure 6.16 is a graph of F_x as a function of x, the elongation of the spring. The work done by this force when the elongation goes from zero to a maximum value X is

$$W = \int_0^X F_x \, dx = \int_0^X kx \, dx = \frac{1}{2}kX^2 \tag{6.9}$$

We can also obtain this result graphically. The area of the shaded triangle in Fig. 6.16, representing the total work done by the force, is equal to half the product of the base and altitude, or

$$W = \frac{1}{2}(X)(kX) = \frac{1}{2}kX^2$$

This equation also says that the work is the *average* force $kX/2$ multiplied by the total displacement X. We see that the total work is proportional to the *square* of the final elongation X. To stretch an ideal spring by 2 cm, you must do four times as much work as is needed to stretch it by 1 cm.

Equation (6.9) assumes that the spring was originally unstretched. If initially the spring is already stretched a distance x_1, the work we must do to stretch it to a greater elongation x_2 is

$$W = \int_{x_1}^{x_2} F_x \, dx = \int_{x_1}^{x_2} kx \, dx = \frac{1}{2}kx_2^2 - \frac{1}{2}kx_1^2 \tag{6.10}$$

If the spring has spaces between the coils when it is unstretched, then it can also be compressed, and Hooke's law holds for compression as well as stretching. In this case the force F and displacement x are in the opposite directions from

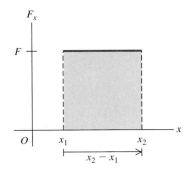

6.14 The work done by a constant force F in the x-direction as a particle moves from x_1 to x_2 is equal to the rectangular area under the graph of force versus displacement.

6.15 The force needed to stretch an ideal spring is proportional to its elongation: $F_x = kx$.

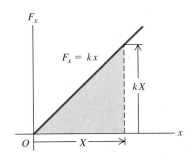

6.16 The work done to stretch a spring by a length X is equal to the triangular area under the graph of force versus displacement.

those shown in Fig. 6.15, and so F_x and x in Eq. (6.8) are both negative. Since both F_x and x are reversed, the force again is in the same direction as the displacement, and the work done by F_x is again positive. So the total work is still given by Eq. (6.9) or (6.10), even when X is negative or either or both of x_1 and x_2 are negative.

> **CAUTION** Note that the work given by Eq. (6.10) is the work that *you* must do *on* a spring to change its length. For example, if you stretch a spring that's originally relaxed, then $x_1 = 0$, $x_2 > 0$, and $W > 0$. That's because the force you apply to one end of the spring is in the same direction as the displacement and the work you do is positive. By contrast, the work that the *spring* does on whatever it's attached to is given by the *negative* of Eq. (6.10). Thus, as you pull on the spring, the spring does negative work on you. Paying careful attention to the sign of work will eliminate confusion later on!

Example 6.7	**Work done on a spring scale**

A woman weighing 600 N steps on a bathroom scale containing a stiff spring (Fig. 6.17). In equilibrium the spring is compressed 1.0 cm under her weight. Find the force constant of the spring and the total work done on it during the compression.

SOLUTION

IDENTIFY and SET UP: In equilibrium the upward force exerted by the spring balances the downward force of the woman's weight. We'll use this principle and Eq. (6.8) to determine the force con-

stant k, and we'll use Eq. (6.10) to calculate the work W that the woman does on the spring to compress it. We take positive values of x to correspond to elongation, so that the displacement of the spring and the force that the woman exerts on it are both negative.

EXECUTE: With our choice of coordinates, the top of the spring is displaced by $x = -1.0$ cm $= -0.010$ m and the force the woman exerts on the spring is $F_x = -600$ N. From Eq. (6.8) the force constant is

$$k = \frac{F_x}{x} = \frac{-600 \text{ N}}{-0.010 \text{ m}} = 6.0 \times 10^4 \text{ N/m}$$

Then, using $x_1 = 0$ and $x_2 = -0.010$ m in Eq. (6.10),

$$W = \frac{1}{2}kx_2^2 - \frac{1}{2}kx_1^2$$

$$= \frac{1}{2}(6.0 \times 10^4 \text{ N/m})(-0.010 \text{ m})^2 - 0 = 3.0 \text{ J}$$

EVALUATE: The applied force and the displacement of the end of the spring were in the same direction, so the work done must have been positive—which is just what we found by calculation. Our arbitrary choice of the positive direction has no effect on the answer for W. (You can test this by taking the positive x-direction to correspond to compression. You'll get the same values for k and for W.)

6.17 Compressing a spring in a bathroom scale. A negative force causes a negative displacement but does a positive amount of work on the spring.

Work-Energy Theorem for Straight-Line Motion, Varying Forces

In Section 6.2 we derived the work-energy theorem, $W_{\text{tot}} = K_2 - K_1$, for the special case of straight-line motion with a constant net force. We can now prove that this theorem is true even when the force varies with position. As in Section 6.2, let's consider a particle that undergoes a displacement x while being acted on by a net force with x-component F, which we now allow to vary. Just as in Fig. 6.13, we divide the total displacement x into a large number of small segments Δx. We can apply the work-energy theorem, Eq. (6.6), to each segment because the value of F in each small segment is approximately constant. The change in kinetic energy in segment Δx_a is equal to the work $F_a \Delta x_a$, and so on. The total change of kinetic energy is the sum of the changes in the individual segments, and thus is equal to the total work done on the particle during the entire displacement. So $W_{tot} = \Delta K$ holds for varying forces as well as for constant ones.

Here's an alternative derivation of the work-energy theorem for a force that may vary with position. It involves making a change of variable from x to v_x in the work integral. As a preliminary, we note that the acceleration a of the particle can be expressed in various ways, using $a_x = dv_x/dt$, $v_x = dx/dt$, and the chain rule for derivatives:

$$a_x = \frac{dv_x}{dt} = \frac{dv_x}{dx}\frac{dx}{dt} = v_x\frac{dv_x}{dx} \tag{6.11}$$

Using this result, Eq. (6.7) tells us that the total work done by the *net* force F_x is

$$W_{tot} = \int_{x_1}^{x_2} F_x\,dx = \int_{x_1}^{x_2} ma_x\,dx = \int_{x_1}^{x_2} mv_x\frac{dv_x}{dx}\,dx \tag{6.12}$$

Now $(dv_x/dx)\,dx$ is the change in velocity dv_x during the displacement dx, so in Eq. (6.12) we can substitute dv_x for $(dv_x/dx)\,dx$. This changes the integration variable from x to v_x, so we change the limits from x_1 and x_2 to the corresponding velocities v_1 and v_2 at these points. This gives us

$$W_{tot} = \int_{v_1}^{v_2} mv_x\,dv_x$$

The integral of $v_x\,dv_x$ is just $v_x^2/2$. Substituting the upper and lower limits, we finally find

$$W_{tot} = \frac{1}{2}mv_2^2 - \frac{1}{2}mv_1^2 \tag{6.13}$$

This is the same result as Eq. (6.6) but without the assumption that the net force F is constant. So the work-energy theorem is valid even when F varies during the displacement.

| Example 6.8 | **Motion with a varying force** |

An air-track glider of mass 0.100 kg is attached to the end of a horizontal air track by a spring with force constant 20.0 N/m (Fig. 6.18a). Initially the spring is unstretched and the glider is moving at 1.50 m/s to the right. Find the maximum distance d that the glider moves to the right a) if the air track is turned on so that there is no friction, and b) if the air is turned off so that there is kinetic friction with coefficient $\mu_k = 0.47$.

SOLUTION

IDENTIFY: The force exerted by the spring is not constant, so we *cannot* use the constant-acceleration formulas of Chapter 2 to solve

this problem. Instead, we'll use the work-energy theorem, which involves the distance moved (our target variable) through the formula for work.

SET UP: Figures 6.18b and 6.18c show the free-body diagrams for the glider without friction and with friction, respectively. We choose the positive x-direction to be to the right (in the direction of the glider's motion), with $x = 0$ at the glider's initial position (where the spring is relaxed) and $x = d$ (the target variable) at the position where the glider stops. In both cases the motion is purely horizontal, so only the horizontal forces do work. Note that

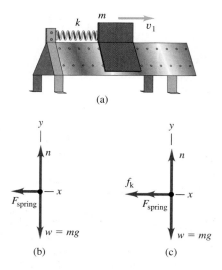

6.18 (a) A glider attached to an air track by a spring. (b) Free-body diagram for the glider without friction. (c) Free-body diagram for the glider with kinetic friction.

Eq. (6.10) gives the work done *on* the spring as it stretches, but to use the work-energy theorem we need the work done *by* the spring *on* the glider—which is the negative of Eq. (6.10).

EXECUTE: a) As the glider moves from $x_1 = 0$ to $x_2 = d$, it does an amount of work on the spring given by Eq. (6.10): $W = \frac{1}{2}kd^2 - \frac{1}{2}k(0)^2 = \frac{1}{2}kd^2$. The amount of work that the spring does on the glider is the negative of this value, or $-\frac{1}{2}kd^2$. The spring stretches until the glider comes instantaneously to rest, so the final kinetic energy K_2 is zero. The initial kinetic energy is $\frac{1}{2}mv_1^2$, where $v_1 = 1.50$ m/s is the glider's initial speed. Using the work-energy theorem, we find

$$-\frac{1}{2}kd^2 = 0 - \frac{1}{2}mv_1^2$$

so the distance the glider moves is

$$d = v_1\sqrt{\frac{m}{k}} = (1.50 \text{ m/s})\sqrt{\frac{0.100 \text{ kg}}{20.0 \text{ N/m}}}$$

$$= 0.106 \text{ m} = 10.6 \text{ cm}$$

The stretched spring subsequently pulls the glider back to the left, so the glider is at rest only instantaneously.

b) If the air is turned off, we must also include the work done by the constant force of kinetic friction. The normal force n is equal in magnitude to the weight of the glider, since the track is horizontal and there are no other vertical forces. The magnitude of the kinetic friction force is then $f_k = \mu_k n = \mu_k mg$. The friction force is directed opposite to the displacement, so the work done by friction is

$$W_{\text{fric}} = f_k d \cos 180° = -f_k d = -\mu_k mgd$$

The total work is the sum of W_{fric} and the work done by the spring, $-\frac{1}{2}kd^2$. Hence

$$-\mu_k mgd - \frac{1}{2}kd^2 = 0 - \frac{1}{2}mv_1^2$$

$$-(0.47)(0.100 \text{ kg})(9.8 \text{ m/s}^2)d - \frac{1}{2}(20.0 \text{ N/m})d^2$$

$$= -\frac{1}{2}(0.100 \text{ kg})(1.50 \text{ m/s})^2$$

$$(10.0 \text{ N/m})d^2 + (0.461 \text{ N})d - (0.113 \text{ N} \cdot \text{m}) = 0$$

This is a quadratic equation for d. The solutions are

$$d = \frac{-(0.461 \text{ N}) \pm \sqrt{(0.461 \text{ N})^2 - 4(10.0 \text{ N/m})(-0.113 \text{ N} \cdot \text{m})}}{2(10.0 \text{ N/m})}$$

$$= 0.086 \text{ m} \quad \text{or} \quad -0.132 \text{ m}$$

We have used d as the symbol for a positive displacement, so only the positive value of d makes sense. Thus with friction the glider moves a distance

$$d = 0.086 \text{ m} = 8.6 \text{ cm}$$

EVALUATE: With friction present, the glider goes a shorter distance and the spring stretches less, as you might expect. Again the glider stops instantaneously, and again the spring force pulls the glider to the left; whether it moves or not depends on how great the *static* friction force is. How large would the coefficient of static friction μ_s have to be to keep the glider from springing back to the left?

Work-Energy Theorem for Motion Along a Curve

We can generalize our definition of work further to include a force that varies in direction as well as magnitude, and a displacement that lies along a curved path. Suppose a particle moves from point P_1 to P_2 along a curve, as shown in Fig. 6.19a. We divide the portion of the curve between these points into many infinitesimal vector displacements, and we call a typical one of these $d\vec{l}$. Each $d\vec{l}$ is tangent to the path at its position. Let $\vec{F}$ be the force at a typical point along the path, and let ϕ be the angle between $\vec{F}$ and $d\vec{l}$ at this point. Then the small element of work dW done on the particle during the displacement $d\vec{l}$ may be written as

$$dW = F \cos \phi \, dl = F_\parallel \, dl = \vec{F} \cdot d\vec{l}$$

where $F_\parallel = F \cos \phi$ is the component of $\vec{F}$ in the direction parallel to $d\vec{l}$ (Fig. 6.19b). The total work done by $\vec{F}$ on the particle as it moves from P_1 to P_2 is then

$$W = \int_{P_1}^{P_2} F \cos \phi \, dl = \int_{P_1}^{P_2} F_\parallel \, dl = \int_{P_1}^{P_2} \vec{F} \cdot d\vec{l} \qquad (6.14)$$

(work done on a curved path)

We can now show that the work-energy theorem, Eq. (6.6), holds true even with varying forces and a displacement along a curved path. The force $\vec{F}$ is essentially constant over any given infinitesimal segment $d\vec{l}$ of the path, so we can apply the work-energy theorem for straight-line motion to that segment. Thus the change in the particle's kinetic energy K over that segment equals the work $dW = F_\parallel \, dl = \vec{F} \cdot d\vec{l}$ done on the particle. Adding up these infinitesimal quantities of work from all the segments along the whole path gives the total work done, Eq. (6.14), which equals the total change in kinetic energy over the whole path. So $W_{\text{tot}} = \Delta K = K_2 - K_1$ is true *in general,* no matter what the path and no matter what the character of the forces. This can be proved more rigorously by using steps like those in Eqs. (6.11) through (6.13) (see Challenge Problem 6.104).

Note that only the component of the net force parallel to the path, $F_\parallel$, does work on the particle, so only this component can change the speed and kinetic energy of the particle. The component perpendicular to the path, $F_\perp = F \sin \phi$, has no effect on the particle's speed; it only acts to change the particle's direction.

The integral in Eq. (6.14) is called a *line integral.* To evaluate this integral in a specific problem, we need some sort of detailed description of the path and of the way in which $\vec{F}$ varies along the path. We usually express the line integral in terms of some scalar variable, as in the following example.

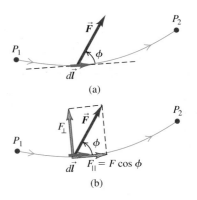

6.19 (a) A particle moves along a curved path from point P_1 to P_2, acted on by a force $\vec{F}$ that varies in magnitude and direction. During an infinitesimal displacement $d\vec{l}$ (a small segment of the path), the work dW done by the force is given by $dW = \vec{F} \cdot d\vec{l} = F \cos \phi \, dl$. (b) The force contributing to the work is the component of force parallel to the displacement, $F_\parallel = F \cos \phi$.

Example 6.9	**Motion on a curved path I**

At a family picnic you are appointed to push your obnoxious cousin Throckmorton in a swing (Fig. 6.20a). His weight is w, the length of the chains is R, and you push Throcky until the chains make an angle θ_0 with the vertical. To do this, you exert a varying horizontal force $\vec{F}$ that starts at zero and gradually increases just enough so that Throcky and the swing move very slowly and remain very nearly in equilibrium. What is the total work done on Throcky by all forces? What is the work done by the tension T in the chains? What is the work you do by exerting the force $\vec{F}$? (Neglect the weight of the chains and seat.)

SOLUTION

IDENTIFY and SET UP: The motion is along a curve, so we will use Eq. (6.14) to calculate the work done by the tension force, by the force $\vec{F}$, and by the net force. Figure 6.20b shows the free-body diagram as well as the coordinate system. We have replaced the tensions in the two chains with a single tension T.

EXECUTE: There are two ways to find the total work done during the motion: (1) by calculating the work done by each force and then

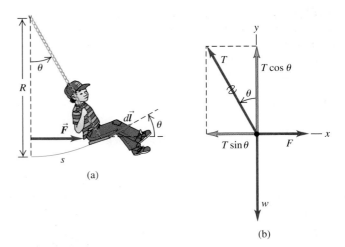

(a)

(b)

6.20 (a) Pushing cousin Throckmorton in a swing. (b) Free-body diagram for Throcky, considered as a particle and neglecting the small weight of chains and seat.

adding the quantities of work together, and (2) by calculating the work done by the net force. The second approach is far easier, because in this situation Throcky is in equilibrium at every point. Hence the net force on him is zero, so the integral of the net force in Eq. (6.14) is zero and the total work done on him by all forces is zero.

It's also easy to calculate the work done by the chain tension on Throcky, because the chain tension is perpendicular to the direction of motion at all points along the path. Hence at all points the angle between the chain tension and the displacement vector $d\vec{l}$ is 90° and the scalar product in Eq. (6.14) is zero. Thus the chain tension does zero work.

To compute the work done by $\vec{F}$, we need to know how this force varies with the angle θ. Throcky is in equilibrium at every point, so from $\sum F_x = 0$ we get

$$F + (-T \sin \theta) = 0$$

and from $\sum F_y = 0$ we find

$$T \cos \theta + (-w) = 0$$

By eliminating T from these two equations, we obtain

$$F = w \tan \theta$$

The point where $\vec{F}$ is applied swings through the arc s. The arc length s equals the radius R of the circular path multiplied by the length θ (in radians), so $s = R\theta$. Therefore the displacement $d\vec{l}$ corresponding to a small change of angle $d\theta$ has a magnitude $dl = ds = R\,d\theta$. The work done by $\vec{F}$ is

$$W = \int \vec{F} \cdot d\vec{l} = \int F \cos \theta \, ds$$

Now we express everything in terms of the varying angle θ:

$$W = \int_0^{\theta_0} (w \tan \theta) \cos \theta \, (R \, d\theta) = wR \int_0^{\theta_0} \sin \theta \, d\theta$$
$$= wR (1 - \cos \theta_0)$$

EVALUATE: If $\theta_0 = 0$, there is no displacement; in that case, $\cos \theta_0 = 1$ and $W = 0$, as we should expect. If $\theta_0 = 90°$, then $\cos \theta_0 = 0$ and $W = wR$. In that case the work you do is the same as if you had lifted Throcky straight up a distance R with a force equal to his weight w. In fact, the quantity $R(1 - \cos \theta_0)$ is the increase in his height above the ground during the displacement, so for any value of θ_0 the work done by force $\vec{F}$ is the change in height multiplied by the weight. This is an example of a more general result that we'll prove in Section 7.1.

Example 6.10	**Motion on a curved path II**

In Example 6.9 the infinitesimal vector displacement $d\vec{l}$ shown in Fig. 6.20a has a magnitude of ds, an x-component of $ds \cos \theta$, and a y-component of $ds \sin \theta$. Hence we can write $d\vec{l} = \hat{i}ds \cos \theta + \hat{j}ds \sin \theta$. Use this expression and Eq. (6.14) to calculate the work done during the motion by the chain tension, by the force of gravity, and by the force $\vec{F}$.

SOLUTION

IDENTIFY and SET UP: We use the same free-body diagram, Fig. 6.20b, as in Example 6.9. The only difference is that here we calculate the scalar product in Eq. (6.14) by using Eq. (1.21) for the scalar product in terms of components.

EXECUTE: From Fig. 6.20b, we can write the three forces in terms of unit vectors:

$$\vec{T} = \hat{i}(-T \sin \theta) + \hat{j}T \cos \theta$$
$$\vec{w} = \hat{j}(-w)$$
$$\vec{F} = \hat{j}F$$

To use Eq. (6.14), we must calculate the scalar product of each of these forces with $d\vec{l}$. Using Eq. (1.21),

$$\vec{T} \cdot d\vec{l} = (-T \sin \theta)(ds \cos \theta) + (T \cos \theta)(ds \sin \theta) = 0$$
$$\vec{w} \cdot d\vec{l} = (-w)(ds \sin \theta) = -w \sin \theta \, ds$$
$$\vec{F} \cdot d\vec{l} = F(ds \cos \theta) = F \cos \theta \, ds$$

Since $\vec{T} \cdot d\vec{l} = 0$, the integral of this quantity is zero and the work done by the chain tension is zero (just as we found in Example 6.9). Using $ds = R\,d\theta$ as in Example 6.9, the work done by the force of gravity is

$$\int \vec{w} \cdot d\vec{l} = \int (-w \sin \theta) R\,d\theta) = -wR \int_0^{\theta_0} \sin \theta \, d\theta$$
$$= -wR(1 - \cos \theta_0)$$

The work done by gravity is negative because gravity pulls down while Throcky moves upward. Finally, the work done by the force $\vec{F}$ is the integral $\int \vec{F} \cdot d\vec{l} = \int F \cos \theta \, ds$, which we calculated in Example 6.9; the answer is $+wR(1 - \cos \theta_0)$.

EVALUATE: As a check on our answers, note that the sum of all three quantities of work is zero. This is just what we concluded in Example 6.9 using the work-energy theorem.

The method of components is often the most convenient way to calculate scalar products. Use it when it will make your life easier!

A 1000-kg car being tested for crashworthiness is driven at 20 m/s into a giant spring of force constant 1.6×10^4 N/m. How much work does the car do on the spring as it comes to a halt? How much work does the spring do on the car? How much does the spring compress? Ignore friction.

6.4 | Power

The definition of work makes no reference to the passage of time. If you lift a barbell weighing 400 N through a vertical distance of 0.5 m at constant velocity, you do $(400 \text{ N})(0.5 \text{ m}) = 200$ J of work whether it takes you 1 second, 1 hour, or 1 year to do it. But often we need to know how quickly work is done. We describe this in terms of *power*. In ordinary conversation the word "power" is often synonymous with "energy" or "force." In physics we use a much more precise definition: **power** is the time *rate* at which work is done. Like work and energy, power is a scalar quantity.

When a quantity of work ΔW is done during a time interval Δt, the average work done per unit time or **average power** P_{av} is defined to be

$$P_{av} = \frac{\Delta W}{\Delta t} \qquad \text{(average power)} \qquad (6.15)$$

The rate at which work is done might not be constant. Even when it varies, we can define **instantaneous power** P as the limit of the quotient in Eq. (6.15) as Δt approaches zero:

$$P = \lim_{\Delta t \to 0} \frac{\Delta W}{\Delta t} = \frac{dW}{dt} \qquad \text{(instantaneous power)} \qquad (6.16)$$

The SI unit of power is the **watt** (W), named for the English inventor James Watt. One watt equals one joule per second $(1 \text{ W} = 1 \text{ J/s})$. The kilowatt $(1 \text{ kW} = 10^3 \text{ W})$ and the megawatt $(1 \text{ MW} = 10^6 \text{ W})$ are also commonly used. In the British system, work is expressed in foot-pounds, and the unit of power is the foot-pound per second. A larger unit called the *horsepower* (hp) is also used:

$$1 \text{ hp} = 550 \text{ ft} \cdot \text{lb/s} = 33,000 \text{ ft} \cdot \text{lb/min}$$

That is, a 1-hp motor running at full load does 33,000 ft·lb of work every minute. A useful conversion factor is

$$1 \text{ hp} = 746 \text{ W} = 0.746 \text{ kW}$$

or 1 horsepower equals about $\frac{3}{4}$ of a kilowatt.

The watt is a familiar unit of *electrical* power; a 100-W light bulb converts 100 J of electrical energy into light and heat each second. But there's nothing inherently electrical about a watt or a kilowatt. A light bulb could be rated in horsepower, and some automobile manufacturers rate their engines in kilowatts rather than horsepower.

The units of power can be used to define new units of work or energy. The *kilowatt-hour* (kWh) is the usual commercial unit of electrical energy. One kilowatt-hour is the total work done in 1 hour (3600 s) when the power is 1 kilowatt (10^3 J/s), so

$$1 \text{ kWh} = (10^3 \text{ J/s})(3600 \text{ s}) = 3.6 \times 10^6 \text{ J} = 3.6 \text{ MJ}$$

The kilowatt-hour is a unit of *work* or *energy,* not power.

In mechanics we can also express power in terms of force and velocity. Suppose that a force $\vec{F}$ acts on a body while it undergoes a vector displacement $\Delta\vec{s}$. If $F_{\parallel}$ is the component of $\vec{F}$ tangent to the path (parallel to $\Delta\vec{s}$), then the work done by the force is $\Delta W = F_{\parallel}\,\Delta s$, and the average power is

$$P_{\text{av}} = \frac{F_{\parallel}\Delta s}{\Delta t} = F_{\parallel}\frac{\Delta s}{\Delta t} = F_{\parallel}v_{\text{av}} \tag{6.17}$$

Instantaneous power P is the limit of this expression as $\Delta t \to 0$:

$$P = F_{\parallel}v \tag{6.18}$$

where v is the magnitude of the instantaneous velocity. We can also express Eq. (6.18) in terms of the scalar product:

$$P = \vec{F}\cdot\vec{v} \tag{6.19}$$
(instantaneous rate at which force $\vec{F}$ does work on a particle)

Example 6.11	**Force and power**

Each of the two jet engines on a Boeing 767 airliner develops a thrust (a forward force on the airplane) of 197,000 N (44,300 lb). When the airplane is flying at 250 m/s (900 km/h, or roughly 560 mi/h), what horsepower does each engine develop?

SOLUTION

IDENTIFY and SET UP: The thrust is in the direction of motion, so $F_{\parallel}$ in Eq. (6.18) is just equal to the thrust. Our target variable is the instantaneous power P.

EXECUTE: At $v = 250$ m/s, each engine develops power P given by Eq. (6.18):

$$P = F_{\parallel}v = (1.97 \times 10^5\,\text{N})(250\,\text{m/s}) = 4.93 \times 10^7\,\text{W}$$

$$= (4.93 \times 10^7\,\text{W})\frac{1\,\text{hp}}{746\,\text{W}} = 66{,}000\,\text{hp}$$

EVALUATE: The speed of modern airliners is directly related to the power of their engines (Fig. 6.21). The largest engines on propeller-driven airliners of the 1950s developed about 3400 hp $(2.5 \times 10^6\,\text{W})$, and these airliners had maximum speeds of about 600 km/h (370 mi/h). The power of each of the engines of a Boeing 767 is nearly 20 times greater, enabling it to fly at about 900 km/h (560 mi/h) and to carry a much heavier load.

What if the engines are running at maximum thrust while the airliner is at rest on the ground (with the brakes held)? Because $v = 0$, the engines develop *zero* power. Force and power are not the same thing!

(a)

(b)

6.21 (a) Propeller-driven and (b) jet airliners.

Example 6.12	**A "power climb"**

As part of a charity fund-raising drive, a Chicago marathon runner with mass 50.0 kg runs up the stairs to the top of the 443-m-tall Sears Tower, the tallest building in the United States (Fig. 6.22). To lift herself to the top in 15.0 minutes, what must be her average power output in watts? In kilowatts? In horsepower?

SOLUTION

IDENTIFY and SET UP: We'll treat the runner as a particle of mass m. We can calculate her average power output in two ways: (1) by first determining how much work she must do and then dividing by the elapsed time, as in Eq. (6.15), or (2) by calculating the average upward force she must exert (in the direction of the climb) and multiplying it by her upward velocity, as in Eq. (6.17).

EXECUTE: As in Example 6.9, lifting a mass m against gravity requires an amount of work equal to the weight mg multiplied by the height h it is lifted. Hence the work she must do is

$$W = mgh = (50.0 \text{ kg})(9.80 \text{ m/s}^2)(443 \text{ m})$$
$$= 2.17 \times 10^5 \text{ J}$$

The time is 15.0 min = 900 s, so from Eq. (6.15) the average power is

$$P_{av} = \frac{2.17 \times 10^5 \text{ J}}{900 \text{ s}} = 241 \text{ W} = 0.241 \text{ kW} = 0.323 \text{ hp}$$

Let's try the calculation again using the alternative approach with Eq. (6.17). The force exerted is vertical, and the average vertical component of velocity is $(443 \text{ m})/(900 \text{ s}) = 0.492$ m/s, so the average power is

6.22 How much power is required to run up the stairs of Chicago's Sears Tower in 15 minutes?

$$P_{av} = F_\parallel v_{av} = (mg)v_{av}$$
$$= (50.0 \text{ kg})(9.80 \text{ m/s}^2)(0.492 \text{ m/s}) = 241 \text{ W}$$

which is the same result as before.

EVALUATE: In fact the runner's *total* power output will be several times greater than we have calculated. The reason is that the runner isn't really a particle but a collection of parts that exert forces on each other and do work, such as the work done to inhale and exhale and to make her arms and legs swing. What we've calculated is only the part of her power output that goes into lifting her to the top of the building.

Test Your Understanding

In Example 6.2, what is the power associated with the force F_T when the sled is moving at 2.0 m/s? What is the power associated with the friction force at this speed?

SUMMARY

When a constant force $\vec{F}$ acts on a particle that undergoes a straight-line displacement $\vec{s}$, the work done by the force on the particle is defined to be the scalar product of $\vec{F}$ and $\vec{s}$. The unit of work in SI units is 1 joule = 1 newton-meter ($1\ \text{J} = 1\ \text{N} \cdot \text{m}$). Work is a scalar quantity; it has an algebraic sign (positive or negative) but no direction in space. (See Examples 6.1 through 6.3)

$$W = \vec{F} \cdot \vec{s} = Fs \cos \phi$$
$$\phi = \text{angle between } \vec{F} \text{ and } \vec{s}$$
(6.2), (6.3)

The kinetic energy K of a particle equals the amount of work required to accelerate the particle from rest to speed v. It is also equal to the amount of work the particle can do in the process of being brought to rest. Kinetic energy is a scalar quantity that has no direction in space; it is always positive or zero. Its units are the same as the units of work: $1\ \text{J} = 1\ \text{N} \cdot \text{m} = 1\ \text{kg} \cdot \text{m}^2/\text{s}^2$.

$$K = \frac{1}{2}mv^2$$
(6.5)

When forces act on a particle while it undergoes a displacement, the particle's kinetic energy changes by an amount equal to the total work done on the particle by all the forces. This relation, called the work-energy theorem, is valid whether the forces are constant or varying and whether the particle moves along a straight or curved path. It is applicable only to bodies that can be treated as a particle. (See Examples 6.4 through 6.6)

$$W_{\text{tot}} = K_2 - K_1 = \Delta K$$
(6.6)

When a force varies during a straight-line displacement, the work done by the force is given by an integral [Eq. (6.7)]. (See Examples 6.7 and 6.8)

$$W = \int_{x_1}^{x_2} F_x\, dx$$
(6.7)

If the particle follows a curved path, then the work done by a force $\vec{F}$ is given by an integral that involves the angle ϕ between the force and the displacement. This expression is valid even if the force magnitude and the angle ϕ vary during the displacement. (See Examples 6.9 and 6.10)

$$W = \int_{P_1}^{P_2} F \cos \phi\, dl = \int_{P_1}^{P_2} F_{\parallel}\, dl$$
$$= \int_{P_1}^{P_2} \vec{F} \cdot d\vec{l}$$
(6.14)

Power is the time rate of doing work. The average power P_{av} is the amount of work ΔW done in time Δt divided by that time. The instantaneous power is the limit of the average power as Δt goes to zero. When a force $\vec{F}$ acts on a particle moving with velocity $\vec{v}$, the instantaneous power (rate at which the force does work) is the scalar product of $\vec{F}$ and $\vec{v}$. Like work and kinetic energy, power is a scalar quantity. The SI unit of power is 1 watt = 1 joule/second $\left(1\ \text{W} = 1\ \text{J/s} \right)$. (See Examples 6.11 and 6.12)

$$P_{av} = \frac{\Delta W}{\Delta t} \tag{6.15}$$

$$P = \lim_{\Delta t \to 0} \frac{\Delta W}{\Delta t} = \frac{dW}{dt} \tag{6.16}$$

$$P = \vec{F} \cdot \vec{v} \tag{6.19}$$

Key Terms

work, 208
joule, 208
kinetic energy, 214
work-energy theorem, 214

force constant, 221
Hooke's law, 221
power, 227

average power, 227
instantaneous power, 227
watt, 227

Your Notes

Answer to Chapter Opening Question

It is indeed true that the javelin does work on the athlete. However, because the javelin exerts a backward force on the athlete's hand as her hand moves forward, the work done by the javelin is *negative* (see Section 6.1).

Answers to Test Your Understanding Questions

Section 6.1 In each case the barbell has zero acceleration, so the net force on it is zero. Hence the weight lifter must exert an upward force of the same magnitude as the barbell's weight, 500 N. (a) The force and displacement are in the same direction, so $W = Fs = (500 \text{ N})(1.5 \text{ m}) = 750 \text{ J}$. (b) The displacement is zero, so $W = 0$. (c) The displacement is opposite to the force, so $W = -Fs = -(500 \text{ N})(1.5 \text{ m}) = -750 \text{ J}$.

Section 6.2 The work that the pitcher did on the ball is equal to the kinetic energy change that he imparted to it. This work is equal to the product of the average horizontal force F that he applied and the displacement $s = 2.0$ m. The initial kinetic energy K_1 was zero (the ball began at rest), and the final kinetic energy was $K_2 = \frac{1}{2}mv_2^2 = \frac{1}{2}(0.145 \text{ kg})(40 \text{ m/s})^2 = 116$ J. Hence, by the work-energy theorem, $F(2.0 \text{ m}) = 116 \text{ J} - 0$ and $F = (116 \text{ J})/(2.0 \text{ m}) = 58$ N.

Section 6.3 As the car comes to a halt, it does an amount of work on the spring equal to its initial kinetic energy: $\frac{1}{2}mv_1^2 = \frac{1}{2}(1000 \text{ kg})(20 \text{ m/s})^2 = 2.0 \times 10^5$ J. The work done by the spring on the car is the negative of this value, -2.0×10^5 J. From Eq. (6.10), the work that the car does on the spring is $\frac{1}{2}kx_2^2 - \frac{1}{2}kx_1^2$; the spring is not compressed initially, so $x_1 = 0$ and $2.0 \times 10^5 \text{ J} = \frac{1}{2}kx_2^2 = \frac{1}{2}(1.6 \times 10^4 \text{ N/m})x_2^2$. Solving, we find $x_2 = 5.0$ m or -5.0 m; here the minus sign is appropriate because the spring is compressed. Hence the spring is compressed by 5.0 m.

Section 6.4 The component of the force F_T in the direction of the sled's motion is $F_{\parallel} = F_T \cos 36.9°$, with $F_T = 5000$ N. Hence the power is $P = F_{\parallel}v = (5000 \text{ N})(\cos 36.9°)(2.0 \text{ m/s}) = 8000$ W. For the friction force, $F_{\parallel} = -f = -3500$ N; the minus sign indicates that the force is opposite to the velocity. So the frictional power is $P = F_{\parallel}v = (-3500 \text{ N})(2.0 \text{ m/s}) = -7000$ W. In other words, the friction force does 7000 J of *negative* work on the sled each second.

Discussion Questions

Q6.1 For a constant force in the direction of the displacement, how can twice the work be done by a force of half the magnitude?

Q6.2 An elevator is hoisted by its cables at constant speed. Is the total work done on the elevator positive, negative, or zero? Explain.

Q6.3 A rope tied to a body is pulled, causing the body to accelerate. But according to Newton's third law, the body pulls back on the rope with an equal and opposite force. Is the total work done then zero? If so, how can the body's kinetic energy change? Explain.

Q6.4 When you use an automobile jack, you lift a car by exerting a force much smaller in magnitude than the weight of the car. Does this mean that you do less work on the car than if you had lifted the car directly? Explain.

Q6.5 If there is a net force on a moving object that is nonzero and constant in magnitude, is it possible for the total work done on the object to be zero? Explain, with an example that illustrates your answer.

Q6.6 In Example 5.5 (Section 5.1), how does the work done on the bucket by the tension in the cable compare to the work done on the cart by the tension in the cable?

Q6.7 In Example 5.21 (Section 5.4), does the force $\vec{F}$ do work on the sled? Do any of the forces on the sled do work on it? Is the speed of the sled constant? Is the velocity of the sled constant? Explain your answers.

Q6.8 In Example 5.11 (Section 5.2), gravity does work on the toboggan as it travels a distance d down the hill, where d is measured parallel to the slope. The magnitude of the gravity force depends on the mass of the toboggan and its contents but is independent of the hill's slope angle α. For a constant distance d, does the *work* done by gravity depend on α? Explain.

Q6.9 A force $\vec{F}$ is in the x-direction and has a magnitude that depends on x. Sketch a possible graph of F versus x such that the force does zero work on an object that moves from x_1 to x_2, even though the force magnitude is not zero at all x in this range.

Q6.10 Does the kinetic energy of a car change more when it speeds up from 10 to 15 m/s or from 15 to 20 m/s? Explain.

Q6.11 A falling brick has a mass of 1.5 kg and is moving straight downward with a speed of 5.0 m/s. A 1.5-kg physics book is sliding across the floor with a speed of 5.0 m/s. A 1.5-kg melon is traveling with a horizontal velocity component 3.0 m/s to the right and a vertical component 4.0 m/s upward. Do these objects all have the same velocity? Do these objects all have the same kinetic energy? For each question, give the reasoning behind your answer.

Q6.12 Can the *total* work done on an object during a displacement be negative? Explain. If the total work is negative, can its magnitude be larger than the initial kinetic energy of the object? Explain.

Q6.13 A net force acts on an object and accelerates it from rest to a speed v_1. In doing so, the force does an amount of work W_1. By what factor must the work done on the object be increased to produce three times the final speed, with the object again starting from rest?

Q6.14 A truck speeding down the highway has a lot of kinetic energy relative to a stopped state trooper, but no kinetic energy relative to the truck driver. In these two frames of reference, is the same amount of work required to stop the truck? Explain.

Q6.15 You are holding a briefcase by the handle, with your arm straight down by your side. Does the force your hand exerts do work on the briefcase when i) you walk at constant speed down a horizontal hallway; ii) you ride an escalator from the first to second floor of a building? In each case justify your answers.

Q6.16 When a book slides along a tabletop, the force of friction does negative work on it. Can friction ever do *positive* work? Explain. (*Hint:* Think of a box in the back of an accelerating truck, with friction between the box and the truck floor.)

Q6.17 Time yourself while running up a flight of steps and compute the average rate at which you do work against the force of gravity. Express your answer in watts and in horsepower.

Q6.18 When a constant force is applied to a body moving with constant acceleration, is the power of the force constant? If not, how would the force have to vary with speed for the power to be constant?

Q6.19 An advertisement for a portable electrical generating unit claims that the unit's diesel engine produces 28,000 hp to drive an electrical generator that produces 30 MW of electrical power. Is this possible? Explain.

Q6.20 A car speeds up while the engine delivers constant power. Is the acceleration greater at the beginning of this process or at the end? Explain.

Q6.21 Consider a graph of instantaneous power versus time, with the vertical P axis starting at $P = 0$. What is the physical significance of the area under the P versus t curve between vertical lines at t_1 and t_2? How could you find the average power from the graph? Draw a P versus t curve that consists of two straight-line sections and for which the peak power is equal to twice the average power.

Q6.22 A nonzero net force acts on an object. Is it possible for any of the following quantities to be constant: i) the particle's speed; ii) the particle's velocity; iii) the particle's kinetic energy?

Exercises

Section 6.1 Work

6.1 You push your physics book 1.50 m along a horizontal tabletop with a horizontal force of 2.40 N. The opposing force of friction is 0.600 N. a) How much work does your 2.40-N force do on the book? b) What is the work done on the book by the friction force? c) What is the total work done on the book?

6.2 An old oaken bucket of mass 6.75 kg hangs in a well at the end of a rope. The rope passes over a frictionless pulley at the top of the well, and you pull horizontally on the end of the rope to raise the bucket slowly a distance of 4.00 m. a) How much work do you do on the bucket in pulling it up? b) How much work is done by the gravitational force acting on the bucket? c) What is the total work done on the bucket?

6.3 A fisherman reels in 12.0 m of line while pulling in a fish that exerts a constant resisting force of 25.0 N. If the fish is pulled in at constant velocity, how much work is done on it by the tension in the line?

6.4 A factory worker pushes a 30.0-kg crate a distance of 4.5 m along a level floor at constant velocity by pushing horizontally on it. The coefficient of kinetic friction between the crate and floor is 0.25. a) What magnitude of force must the worker apply? b) How much work is done on the crate by this force? c) How much work is done on the crate by friction? d) How much work is done by the normal force? By gravity? e) What is the total work done on the crate?

6.5 Suppose the worker in Exercise 6.4 pushes downward at an angle of 30° below the horizontal. a) What magnitude of force must the worker apply to move the crate at constant velocity? b) How much work is done on the crate by this force when the crate is pushed a distance of 4.5 m? c) How much work is done on the crate by friction during this displacement? d) How much work is done by the normal force? By gravity? e) What is the total work done on the crate?

6.6 A boat with a horizontal tow rope pulls a water skier. She skis off to the side, so the rope makes an angle of 15.0° with her direction of motion, and then continues in a straight line. The tension in

the rope is 180 N. How much work is done on the skier by the rope during a displacement of 300 m?

6.7 Two tugboats pull a disabled supertanker. Each tug exerts a constant force of 1.80×10^6 N, one 14° west of north and the other 14° east of north, as they pull the tanker 0.75 km toward the north. What is the total work they do on the supertanker?

6.8 A loaded grocery cart is rolling across a parking lot in a strong wind. You apply a constant force $\vec{F} = (30\,\text{N})\hat{\imath} - (40\,\text{N})\hat{\jmath}$ to the cart as it undergoes a displacement $\vec{s} = (-9.0\,\text{m})\hat{\imath} - (3.0\,\text{m})\hat{\jmath}$. How much work does the force you apply do on the grocery cart?

6.9 A 0.800-kg ball is tied to the end of a string 1.60 m long and swung in a vertical circle. a) During one complete circle, starting anywhere, calculate the total work done on the ball by: (i) the tension in the string (ii) gravity. b) Repeat part (a) for motion along the semicircle from the lowest to the highest point on the path.

Section 6.2 Work and Kinetic Energy

6.10 a) Compute the kinetic energy, in joules, of a 1600-kg automobile traveling at 50.0 km/h. b) By what factor does the kinetic energy change if the speed is doubled?

6.11 **T. Rex.** The dinosaur *Tyrannosaurus rex* is thought to have had a mass of about 7000 kg. a) Treat the dinosaur as a particle and estimate its kinetic energy at a walking speed of 4.0 km/h. b) With what speed would a 70-kg person have to move to have the same kinetic energy as a walking *T. rex?*

6.12 a) How many joules of kinetic energy does a walking person have? What about a running person? b) How many joules of kinetic energy does a large car going at freeway speeds have? c) If you drop a 1-kg weight (about 2 lb) from shoulder height, how many joules of kinetic energy will it have when it reaches the ground?

6.13 You are a member of an Alpine Rescue Team and must project a box of supplies up an incline of constant slope angle α so that it reaches a stranded skier who is a vertical distance h above the bottom of the incline. The incline is slippery, but there is some friction present, with kinetic friction coefficient μ_k. Use the work-energy theorem to calculate the minimum speed you must give the box at the bottom of the incline so that it will reach the skier. Express your answer in terms of g, h, μ_k, and α.

6.14 You throw a 20-N rock vertically into the air from ground level. You observe that when it is 15.0 m above the ground, it is traveling at 25.0 m/s upward. Use the work-energy theorem to find a) its speed just as it left the ground; b) its maximum height.

6.15 A mass m slides down a smooth inclined plane from an initial vertical height h, making an angle α with the horizontal. a) The work done by a force is the sum of the work done by the components of the force. Consider the components of gravity parallel and perpendicular to the surface of the plane. Calculate the work done on the mass by each of the components and use these results to show that the work done by gravity is exactly the same as if the mass had fallen straight down through the air from a height h. b) Use the work-energy theorem to prove that the speed of the mass at the bottom of the incline is the same as if it had been dropped from height h, independent of the angle α of the incline. Explain how this speed can be independent of the slope angle. c) Use the results of part (b) to find the speed of a rock that slides down an icy frictionless hill, starting from rest 15.0 m above the bottom.

6.16 A car is stopped by a constant friction force that is independent of the car's speed. By what factor is the stopping distance changed if the car's initial speed is doubled? (Solve using work-energy methods.)

6.17 A baseball leaves a pitcher's hand at a speed of 32.0 m/s. The mass of the baseball is 0.145 kg. You can ignore air resistance. How much work has the pitcher done on the ball by throwing it?

6.18 Iceboat. In Conceptual Example 6.6 (Section 6.2), call the iceboat with mass m, boat A and the other iceboat, with mass $2m$, boat B. a) At the finish line what is the ratio v_A/v_B of the speeds of the two iceboats? b) Let t_A be the elapsed time it takes boat A to reach the finish line and t_B the elapsed time for boat B. What is the ratio t_B/t_A of these two times?

6.19 A moving electron has kinetic energy K_1. After a net amount of work W has been done on it, the electron is moving one-quarter as fast in the opposite direction. a) Find W in terms of K_1. b) Does your answer depend on the final direction of the electron's motion?

6.20 A sled with mass 8.00 kg moves in a straight line on a frictionless horizontal surface. At one point in its path, its speed is 4.00 m/s; after it has traveled 2.50 m beyond this point, its speed is 6.00 m/s. Use the work-energy theorem to find the force acting on the sled, assuming that this force is constant and that it acts in the direction of the sled's motion.

6.21 A soccer ball with mass 0.420 kg is initially moving with speed 2.00 m/s. A soccer player kicks the ball, exerting a constant force of magnitude 40.0 N in the same direction as the ball's motion. Over what distance must her foot be in contact with the ball to increase the ball's speed to 6.00 m/s?

6.22 A 12-pack of Omni-Cola (mass 4.30 kg) is initially at rest on a horizontal floor. It is then pushed in a straight line for 1.20 m by a trained dog who exerts a horizontal force with magnitude 36.0 N. Use the work-energy theorem to find the final speed of the 12-pack if a) there is no friction between the 12-pack and the floor; b) the coefficient of kinetic friction between the 12-pack and the floor is 0.30.

6.23 A pitcher throws a baseball with mass 0.145 kg straight upward with an initial speed of 25.0 m/s. a) How much work has gravity done on the baseball when it reaches a height of 20.0 m above the pitcher's hand? b) Use the work-energy theorem to calculate the speed of the baseball at a height of 20.0 m above the pitcher's hand. You can ignore air resistance. c) Does the answer to part (b) depend on whether the baseball is moving upward or downward at a height of 20.0 m? Explain.

6.24 A 4.80-kg watermelon is dropped (zero initial speed) from the roof of a 25.0-m-tall building. a) Calculate the work done by gravity on the watermelon during its displacement from the roof to the ground. b) What is the kinetic energy of the watermelon just before it strikes the ground? You can ignore air resistance.

6.25 A little red wagon with mass 7.00 kg moves in a straight line on a frictionless horizontal surface. It has an initial speed of 4.00 m/s and then is pushed 3.0 m in the direction of the initial velocity by a force with a magnitude of 10.0 N. a) Use the work-energy theorem to calculate the wagon's final speed. b) Calculate the acceleration produced by the force. Use this acceleration in the kinematic relations of Chapter 2 to calculate the wagon's final speed. Compare this result to that calculated in part (a).

6.26 A block of ice with mass 2.00 kg slides 0.750 m down an inclined plane that slopes downward at an angle of 36.9° below the horizontal. If the block of ice starts from rest, what is its final speed? You can ignore friction.

6.27 Stopping Distance. A car is traveling on a level road with speed v_0 at the instant when the brakes lock, so that the tires slide rather than roll. a) Use the work-energy theorem to calculate the minimum stopping distance of the car in terms of v_0, g, and the coefficient of kinetic friction μ_k between the tires and the road. b) The car stops in a distance of 91.2 m if $v_0 = 80.0$ km/h. What is the stopping distance if $v_0 = 60.0$ km/h? Assume that the value of μ_k remains the same.

Section 6.3 Work and Energy with Varying Forces

6.28 To stretch a spring 3.00 cm from its unstretched length, 12.0 J of work must be done. How much work must be done to compress this spring 4.00 cm from its unstretched length?

6.29 A force of 160 N stretches a spring 0.050 m beyond its unstretched length. a) What magnitude of force is required to stretch the spring 0.015 m beyond its unstretched length? To compress the spring 0.020 m? b) How much work must be done to stretch the spring 0.015 m beyond its unstretched length? How much work to compress the spring 0.020 m from its unstretched length?

6.30 A child applies a force $\vec{F}$ parallel to the x-axis to a 10.0-kg sled moving on the frozen surface of a small pond. As the child controls the speed of the sled, the x-component of the force she applies varies with the x-coordinate of the sled as shown in Fig. 6.23. Calculate the work done by the force $\vec{F}$ when the sled moves a) from $x = 0$ to $x = 8.0$ m; b) from $x = 8.0$ m to $x = 12.0$ m; c) from $x = 0$ to 12.0 m.

6.31 Suppose the sled in Exercise 6.30 is initially at rest at $x = 0$. Use the work-energy theorem to find the speed of the sled at a) $x = 8.0$ m; b) $x = 12.0$ m. You can ignore friction between the sled and the surface of the pond.

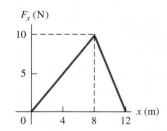

Figure 6.23 Exercises 6.30 and 6.31.

6.32 A balky cow is leaving the barn as you try harder and harder to push her back in. In coordinates with the origin at the barn door, the cow walks from $x = 0$ to $x = 6.9$ m as you apply a force with x-component $F_x = -[20.0 \text{ N} + (3.0 \text{ N/m})x]$. How much work does the force you apply do on the cow during this displacement?

6.33 A 6.0-kg box moving at 3.0 m/s on a horizontal, frictionless surface runs into a light spring of force constant 75 N/cm. Use the work-energy theorem to find the maximum compression of the spring.

6.34 Leg Presses. As part of your daily workout, you lie on your back and push with your feet against a platform attached to two stiff springs arranged side by side so that they are parallel to each other. When you push the platform, you compress the springs. You do 80.0 J of work when you compress the springs 0.200 m from their uncompressed length. a) What magnitude of force must you apply to hold the platform in this position? b) How much *additional* work must you do to move the platform 0.200 m *farther*, and what maximum force must you apply?

6.35 a) In Example 6.8 (Section 6.3) it is calculated that with the air track turned off, the glider travels 8.6 cm before it stops instantaneously. How large would the coefficient of static friction μ_s have to be to keep the glider from springing back to the left? b) If the coefficient of static friction between the glider and the track is $\mu_s = 0.60$, what is the maximum initial speed v_1 that the glider can be given and still remain at rest after it stops instantaneously? With the air track turned off, the coefficient of kinetic friction is $\mu_k = 0.47$.

6.36 A 4.00-kg block of ice is placed against a horizontal spring that has force constant $k = 200$ N/m and is compressed 0.025 m. The spring is released and accelerates the block along a horizontal surface. You can ignore friction and the mass of the spring. a) Calculate the work done on the block by the spring during the motion of the block from its initial position to where the spring has returned to its uncompressed length. b) What is the speed of the block after it leaves the spring?

6.37 A force $\vec{F}$ is applied to a 2.0-kg radio-controlled model car parallel to the x-axis as it moves along a straight track. The x-component of the force varies with the x-coordinate of the car as shown in Fig. 6.24. Calculate the work done by the force $\vec{F}$ when the car moves from a) $x = 0$ to $x = 3.0$ m; b)

Figure 6.24 Exercises 6.37 and 6.38.

$x = 3.0$ m to $x = 4.0$ m; c) $x = 4.0$ m to $x = 7.0$ m; d) $x = 0$ to $x = 7.0$ m; e) $x = 7.0$ m to $x = 2.0$ m.

6.38 Suppose the 2.0-kg model car in Exercise 6.37 is initially at rest at $x = 0$ and $\vec{F}$ is the net force acting on the car. Use the work-energy theorem to find the speed of the car at a) $x = 3.0$ m; b) $x = 4.0$ m; c) $x = 7.0$ m.

6.39 At a waterpark, sleds with riders are sent along a slippery, horizontal surface by the release of a large, compressed spring. The spring with a force constant $k = 4000$ N/m and negligible mass rests on the frictionless horizontal surface. One end is in contact with a stationary wall. A sled and rider with total mass 70.0 kg are pushed against the other end, compressing the spring 0.375 m. The sled is then released with zero initial velocity. What is the sled's speed when the spring a) returns to its uncompressed length? b) is still compressed 0.200 m?

6.40 In Example 6.9 (Section 6.3), instead of applying a varying horizontal force $\vec{F}$ that maintains Throcky very nearly in equilibrium, you apply a constant horizontal force with magnitude $F = 2w$, where w is Throcky's weight. As in Example 6.9, consider Throcky to be a particle and ignore the small weight of the chains and seat. You push Throcky until the chains make an angle θ_0 with the vertical. a) Use Eq. (6.14) to calculate the work done on Throcky by the force $\vec{F}$ that you apply. b) At the angle θ_0, how does the magnitude of $\vec{F}$ in this exercise compare to that in Example 6.9? c) How does the work done by $\vec{F}$ in this exercise compare to that in Example 6.7?

6.41 A small glider is placed against a compressed spring at the bottom of an air track that slopes upward at an angle of 40.0° above the horizontal. The glider has mass 0.0900 kg. The spring has $k = 640$ N/m and negligible mass. When the spring is released, the glider travels a maximum distance of 1.80 m along the air track before sliding back down. Before reaching this maximum distance, the glider loses contact with the spring. a) What distance was the spring originally compressed? b) When the glider has traveled along the air track 0.80 m from its initial position against the compressed spring, is it still in contact with the spring? What is the kinetic energy of the glider at this point?

6.42 Bricklaying. An ingenious bricklayer builds a device for shooting bricks up to the top of the wall where he is working. He places a brick on a vertical compressed spring with force constant $k = 450$ N/m and negligible mass. When the spring is released, the brick is propelled upward. If the brick has mass 1.80 kg and is to reach a maximum height of 3.6 m above its initial position on the compressed spring, what distance must the bricklayer compress the spring initially? (The brick loses contact with the spring when the spring returns to its uncompressed length. Why?)

Section 6.5 Power

6.43 How many joules of energy does a 100-watt light bulb use per hour? How fast would a 70-kg person have to run to have that amount of energy?

6.44 A 20.0-kg rock is sliding on a rough, horizontal surface at 8.00 m/s and eventually stops due to friction. The coefficient of kinetic friction between the rock and the surface is 0.200. What average thermal power is produced as the rock stops?

6.45 A tandem (two-person) bicycle team must overcome a force of 165 N to maintain a speed of 9.00 m/s. Find the power required per rider, assuming that each contributes equally. Express your answer in watts and in horsepower.

6.46 The total consumption of electrical energy in the United States is about 1.0×10^{19} joules per year. a) What is the average rate of electrical energy consumption in watts? b) If the population of the United States is 260 million, what is the average rate of electrical energy consumption per person? c) The sun transfers energy to the earth by radiation at a rate of approximately 1.0 kW per square meter of surface. If this energy could be collected and converted to electrical energy with 40% efficiency, how great an area (in square kilometers) would be required to collect the electrical energy used by the United States?

6.47 When its 75-kW (100-hp) engine is generating full power, a small single-engine airplane with mass 700 kg gains altitude at a rate of 2.5 m/s (150 m/min, or 500 ft/min). What fraction of the engine power is being used to make the airplane climb? (The remainder is used to overcome the effects of air resistance and of inefficiencies in the propeller and engine.)

6.48 Working Like a Horse. Your job is to lift 30-kg crates a vertical distance of 0.90 m from the ground onto the bed of a truck. a) How many crates would you have to load onto the truck in one minute for the average power output you use to lift the crates to equal 0.50 hp? b) How many crates for an average power output of 100 W?

6.49 An elevator has a mass 600 kg, not including passengers. The elevator is designed to ascend, at constant speed, a vertical distance of 20.0 m (five floors) in 16.0 s, and it is driven by a motor that can provide up to 40 hp to the elevator. What is the maximum number of passengers that can ride in the elevator? Assume that an average passenger has mass 65.0 kg.

6.50 The hammer of a pile driver weighs 3800 N and must be lifted 2.80 m vertically at constant speed in 4.00 s. What horsepower must the engine provide to the hammer?

6.51 The aircraft carrier *John F. Kennedy* has mass 7.4×10^7 kg. When its engines are developing their full power of 280,000 hp, the *John F. Kennedy* travels at its top speed of 35 knots (65 km/h). If 70% of the power output of the engines is applied to pushing the ship through the water, what is the magnitude of the force of water resistance that opposes the carrier's motion at this speed?

6.52 A ski tow operates on a 15.0° slope of length 300 m. The rope moves at 12.0 km/h and provides power for 50 riders at one time, with an average mass per rider of 70.0 kg. Estimate the power required to operate the tow.

6.53 A particle is accelerated from rest by a constant net force. a) Show that the instantaneous power supplied by the net force is ma^2t. b) To triple the acceleration at any given time, by what factor must the power be increased? c) At $t = 5.0$ s the power supplied by the net force is 36 W. What must the power be at $t = 15.0$ s to maintain constant acceleration?

6.54 Show that the instantaneous power P supplied by the net force acting on a particle is related to the particle's kinetic energy K by $P = dK/dt$.

6.55 A typical flying insect applies an average force equal to twice its weight during each downward stroke while hovering. Take the mass of the insect to be 10 g and assume the wings move an average downward distance of 1.0 cm during each stroke. Assuming 100 downward strokes per second, estimate the average power output of the insect.

Problems

6.56 Rotating Bar. A thin, uniform 12.0-kg bar that is 2.00 m long rotates uniformly about a pivot at one end, making 5.00 complete revolutions every 3.00 seconds. What is the kinetic energy of this bar? (*Hint:* Different points in the bar have different speeds. Break the bar up into infinitesimal segments of mass dm and integrate to add up the kinetic energy of all these segments.)

6.57 A luggage handler pulls a 20.0-kg suitcase up a ramp inclined at 25.0° above the horizontal by a force $\vec{F}$ of magnitude 140 N that acts parallel to the ramp. The coefficient of kinetic friction between the ramp and the incline is $\mu_k = 0.300$. If the suitcase travels 3.80 m along the ramp, calculate a) the work done on the suitcase by the force $\vec{F}$; b) the work done on the suitcase by the gravitational force; c) the work done on the suitcase by the normal force; d) the work done on the suitcase by the friction force; e) the total work done on the suitcase. f) If the speed of the suitcase is zero at the bottom of the ramp, what is its speed after it has traveled 3.80 m along the ramp?

6.58 Chin-Ups. While doing a chin-up, a man lifts his body 0.40 m. a) How much work must the man do per kilogram of body mass? b) The muscles involved in doing a chin-up can generate about 70 J of work per kilogram of muscle mass. If the man can just barely do a 0.40-m chin-up, what percentage of his body's mass do these muscles constitute? (For comparison, the *total* percentage of muscle in a typical 70-kg man with 14% body fat is about 43%.) c) Repeat part (b) for the man's young son, who has arms half as long as his father's but whose muscles can also generate 70 J of work per kilogram of muscle mass. d) Adults and children have about the same percentage of muscle in their bodies. Explain why children can commonly do chin-ups more easily than their fathers.

6.59 Simple Machines. Ramps for the disabled are used because a large weight w can be raised by relatively small force equal to $w \sin \alpha$ plus the small friction force. Such inclined planes are an example of a class of devices called *simple machines*. An input force F_{in} is applied to the system and results in an output force F_{out} applied to the object that is moved. For a simple machine the ratio of these forces, F_{out}/F_{in}, is called the actual mechanical advantage (AMA). The inverse ratio of the distances that the points of application of these forces move through during the motion of the object, s_{in}/s_{out}, is called the ideal mechanical advantage (IMA). a) Find the IMA for an inclined plane. b) What can we say about the relation between the work supplied to the machine, W_{in}, and the work output of the machine, W_{out}, if AMA = IMA? c) Sketch a single pulley arranged to give IMA = 2. d) We define the efficiency e of a simple machine to equal the ratio of the output work to the input work, $e = W_{out}/W_{in}$. Show that $e = $ AMA/IMA.

6.60 A woman stands in an elevator that has a constant upward acceleration while the elevator travels upward a distance of 18.0 m. During the 18.0-m displacement, the normal force exerted by the elevator floor does 8.25 kJ of work on her and gravity does -7.35 kJ of work on her. a) What is the mass of the woman? b) What is the normal force that the elevator floor exerts on her? c) What is the acceleration of the elevator?

6.61 The space shuttle *Endeavour,* with mass 86,400 kg, is in a circular orbit of radius 6.66×10^6 m around the earth. It takes 90.1 min for the shuttle to complete each orbit. On a repair mission, it is cautiously moving 1.00 m closer to a disabled satellite every 3.00 s. Calculate the shuttle's kinetic energy a) relative to the earth; b) relative to the satellite.

6.62 A 5.00-kg package slides 1.50 m down a long ramp that is inclined at 12.0° below the horizontal. The coefficient of kinetic friction between the package and the ramp is $\mu_k = 0.310$. Calculate a) the work done on the package by friction; b) the work done on the package by gravity; c) the work done on the package by the normal force; d) the total work done on the package. e) If the package has a speed of 2.20 m/s at the top of the ramp, what is its speed after sliding 1.50 m down the ramp?

6.63 The package in Problem 6.62 has a speed of 2.20 m/s at the top of the ramp. Use the work-energy theorem to find how far down the ramp the package slides before it stops.

6.64 An object is attracted toward the origin with a force given by $F_x = -k/x^2$. (Gravitational and electrical forces have this distance dependence.) a) Calculate the work done by the force F_x when the object moves in the x-direction from x_1 to x_2. If $x_2 > x_1$, is the work done by F_x positive or negative? b) The only other force acting on the object is a force that you exert with your hand to move the object slowly from x_1 to x_2. How much work do you do? If $x_2 > x_1$, is the work you do positive or negative? c) Explain the similarities and differences between your answers to parts (a) and (b).

6.65 The gravitational pull of the earth on an object is inversely proportional to the square of the distance of the object from the center of the earth. At the earth's surface this force is equal to the object's normal weight mg, where $g = 9.8$ m/s^2, and at large distances, the force is zero. If a 20,000-kg asteroid falls to earth from a very great distance away, what will be its minimum speed as it strikes the earth's surface, and how much kinetic energy will it impart to our planet? You can ignore the effects of the earth's atmosphere.

6.66 Varying Coefficient of Friction. A box is sliding with a speed of 4.50 m/s on a horizontal surface when, at point P, it encounters a rough section. On the rough section, the coefficient of friction is not constant, but starts at 0.100 at P and increases linearly with distance past P, reaching a value of 0.600 at 12.5 m past point P. a) Use the work-energy theorem to find how far this box slides before stopping. b) What is the coefficient of friction at the stopping point? c) How far would the box have slid if the friction coefficient didn't increase, but instead had the constant value of 0.100?

6.67 An object that can move along the x-axis is attracted toward the origin with a force of magnitude $F = \alpha x^3$, where $\alpha = 4.00$ N/m^3. What is the force F when the object is a) at $x = 1.00$ m? b) at $x = 2.00$ m? c) How much work is done by the force F when the object moves from $x = 1.00$ m to $x = 2.00$ m? Is this work positive or negative?

6.68 Consider a spring that does not obey Hooke's law very faithfully. One end of the spring is fixed. To keep the spring stretched or compressed an amount x, a force along the x-axis with x-component $F_x = kx - bx^2 + cx^3$ must be applied to the free end. Here $k = 100$ N/m, $b = 700$ N/m^2, and $c = 12,000$ N/m^3. Note that $x > 0$ when the spring is stretched and $x < 0$ when it is compressed. a) How much work must be done to stretch this spring by 0.050 m from its unstretched length? b) How much work must be done to *compress* this spring by 0.050 m from its unstretched length? c) Is it easier to stretch or compress this spring? Explain why in terms of the dependence of F_x on x. (Many real springs behave qualitatively in the same way.)

6.69 A small block with a mass of 0.120 kg is attached to a cord passing through a hole in a frictionless, horizontal surface (Fig. 6.25). The block is originally revolving at a distance of 0.40 m from the hole with a speed of 0.70 m/s. The cord is then pulled from below, shorten-

Figure 6.25 Problem 6.69.

ing the radius of the circle in which the block revolves to 0.10 m. At this new distance, the speed of the block is observed to be 2.80 m/s. a) What is the tension in the cord in the original situation when the block has speed $v = 0.70$ m/s? b) What is the tension in the cord in the final situation when the block has speed $v = 2.80$ m/s? c) How much work was done by the person who pulled on the cord?

6.70 Proton Bombardment. A proton with mass 1.67×10^{-27} kg is propelled at an initial speed of 3.00×10^5 m/s directly toward a uranium nucleus 5.00 m away. The proton is repelled by the uranium nucleus with a force of magnitude $F = \alpha/x^2$, where x is the separation between the two objects and $\alpha = 2.12 \times 10^{-26}$ N·m^2. Assume that the uranium nucleus remains at rest. a) What is the speed of the proton when it is 8.00×10^{-10} m from the uranium nucleus? b) As the proton approaches the uranium nucleus, the repulsive force slows down the proton until it comes momentarily to rest, after which the proton moves away from the uranium nucleus. How close to the uranium nucleus does the proton get? c) What is the speed of the proton when it is again 5.00 m away from the uranium nucleus?

6.71 A block of ice with mass 6.00 kg is initially at rest on a frictionless, horizontal surface. A worker then applies a horizontal force $\vec{F}$ to it. As a result, the block moves along the x-axis such that its position as a function of time is given by $x(t) = \alpha t^2 + \beta t^3$, where $\alpha = 0.200$ m/s^2 and $\beta = 0.0200$ m/s^3. a) Calculate the velocity of the object when $t = 4.00$ s. b) Calculate the magnitude of $\vec{F}$ when $t = 4.00$ s. c) Calculate the work done by the force $\vec{F}$ during the first 4.00 s of the motion.

6.72 A net force with magnitude $(5.00\ \text{N/m}^2)x^2$ and directed at a constant angle of 31.0° with the $+x$-axis acts on an object of mass 0.250 kg as the object moves parallel to the x-axis. How fast is the object moving at $x = 1.50$ m if it has a speed of 4.00 m/s at $x = 1.00$ m?

6.73 You and your bicycle have combined mass 80.0 kg. When you reach the base of a bridge, you are traveling along the road at 5.00 m/s (Fig. 6.26). At the top of the bridge, you have climbed a vertical distance of 5.20 m and have slowed to 1.50 m/s. You can ignore work done by friction and any inefficiency in the bike or your legs. a) What is the total work done on you and your bicycle when you go from the base to the top of the bridge? b) How much work have you done with the force you apply to the pedals?

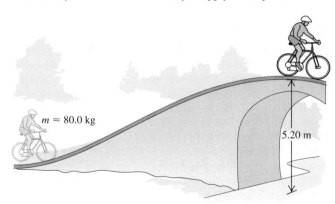

Figure 6.26 Problem 6.73.

6.74 A force in the $+x$-direction has magnitude $F = b/x^n$, where b and n are constants. a) For $n > 1$, calculate the work done on a particle by this force when the particle moves along the x-axis from $x = x_0$ to infinity. b) Show that for $0 < n < 1$, even though F becomes zero as x becomes very large, an infinite amount of work is done by F when the particle moves from $x = x_0$ to infinity.

6.75 You are asked to design spring bumpers for the walls of a parking garage. A freely rolling 1200-kg car moving at 0.65 m/s is to compress the spring no more than 0.070 m before stopping. What should be the force constant of the spring? Assume that the spring has negligible mass.

6.76 The spring of a spring gun has force constant $k = 400$ N/m and negligible mass. The spring is compressed 6.00 cm and a ball with mass 0.0300 kg is placed in the horizontal barrel against the compressed spring. The spring is then released, and the ball is propelled out the barrel of the gun. The barrel is 6.00 cm long, so the ball leaves the barrel at the same point that it loses contact with the spring. The gun is held so the barrel is horizontal. a) Calculate the speed with which the ball leaves the barrel if you can ignore friction. b) Calculate the speed of the ball as it leaves the barrel if a constant resisting force of 6.00 N acts on the ball as it moves along the barrel. c) For the situation in part (b), at what position along the barrel does the ball have the greatest speed, and what is that speed? (In this case, the maximum speed does not occur at the end of the barrel.)

6.77 A 2.50-kg textbook is forced against a horizontal spring of negligible mass and force constant 250 N/m, compressing the spring a distance of 0.250 m. When released, the textbook slides on a horizontal table top with coefficient of kinetic friction $\mu_k = 0.30$. Use the work-energy theorem to find how far the textbook moves from its initial position before coming to rest.

6.78 Pushing a Cat. Your cat "Ms." (mass 7.00 kg) is trying to make it to the top of a frictionless ramp 2.00 m long and inclined upward at 30.0° above the horizontal. Since the poor cat can't get any traction on the ramp, you push her up the entire length of the ramp by exerting a constant 100-N force parallel to the ramp. If Ms. takes a running start so that she is moving at 2.40 m/s at the bottom of the ramp, what is her speed when she reaches the top of the incline? Use the work-energy theorem.

6.79 Crash Barrier. A student proposes a design for an automobile crash barrier in which a 1700-kg sport utility vehicle moving at 20.0 m/s crashes into a spring of negligible mass that slows it to a stop. To avoid injuring the passengers, the acceleration of the vehicle as it slows can be no more than 5.00g. a) Find the required spring constant k, and find the distance the spring will compress in slowing the vehicle to a stop. In your calculation, disregard any deformation or crumpling of the vehicle and the friction between the vehicle and the ground. b) What disadvantages are there to this design?

6.80 A physics professor is pushed up a ramp inclined upward at 30.0° above the horizontal as he sits in his desk chair that slides on frictionless rollers. The combined mass of the professor and chair is 85.0 kg. He is pushed 2.50 m along the incline by a group of students who together exert a constant horizontal force of 600 N. The professor's speed at the bottom of the ramp is 2.00 m/s. Use the work-energy theorem to find his speed at the top of the ramp.

6.81 A 5.00-kg block is moving at $v_0 = 6.00$ m/s along a frictionless, horizontal surface toward a spring with force constant $k = 500$ N/m that is attached to a wall (Fig. 6.27). The spring

Figure 6.27 Problem 6.81.

has negligible mass. a) Find the maximum distance the spring will be compressed. b) If the spring is to compress by no more than 0.150 m, what should be the maximum value of v_0?

6.82 Consider the system shown in Fig. 6.28. The rope and pulley have negligible mass, and the pulley is frictionless. The coefficient of kinetic friction between the 8.00-kg block and the tabletop is $\mu_k = 0.250$. The blocks are released from rest. Use energy methods to calculate the speed of the 6.00-kg block after it has descended 1.50 m.

6.83 Consider the system shown in Fig. 6.28. The rope and pulley have negligible mass, and the pulley is frictionless. Initially the 6.00-kg block is moving downward and the 8.00-kg block is moving to the right, both with a speed of 0.900 m/s. The blocks come to rest after moving 2.00 m. Use the work-energy theorem to calculate the coefficient of kinetic friction between the 8.00-kg block and the tabletop.

Figure 6.28 Problems 6.82 and 6.83.

6.84 Bow and Arrow. Figure 6.29 shows how the force exerted by the string of a compound bow on an arrow varies as a function of how far back the arrow is pulled (*the draw length*). Assume that the same force is exerted on the arrow as it moves forward after being released. Full draw for this bow is at a draw length of 75.0 cm. If the bow shoots a 0.0250-kg arrow from full draw, what is the speed of the arrow as it leaves the bow?

Figure 6.29 Problem 6.84.

6.85 On an essentially frictionless, horizontal ice rink, a skater moving at 3.0 m/s encounters a rough patch that reduces her speed by 45% due to a friction force that is 25% of her weight. Use the work-energy theorem to find the length of this rough patch.

6.86 Rescue. Your friend (mass 65.0 kg) is standing on the ice in the middle of a frozen pond. There is very little friction between her feet and the ice, so she is unable to walk. Fortunately, there is a light rope tied around her waist and you stand on the bank holding the other end. You pull on the rope for 3.00 s and accelerate your friend from rest to a speed of 6.00 m/s while you remain at rest. What is the average power supplied by the force you applied?

6.87 A pump is required to lift 800 kg of water (about 210 gallons) per minute from a well 14.0 m deep and eject it with a speed of 18.0 m/s. a) How much work is done per minute in lifting the water? b) How much in giving the water the kinetic energy it has when ejected? c) What must be the power output of the pump?

6.88 Find the power output of the worker in Problem 6.71 as a function of time. What is the numerical value of the power (in watts) at $t = 4.00$ s?

6.89 A physics student spends part of her day walking between classes or for recreation, during which time she expends energy at an average rate of 280 W. The remainder of the day she is sitting in class, studying, or resting; during these activities, she expends energy at an average rate of 100 W. If she expends a total of 1.1×10^7 J of energy in a 24-hour day, how much of the day did she spend walking?

6.90 All birds, independent of their size, must maintain a power output of 10 to 25 watts per kilogram of body mass in order to fly by flapping their wings. (a) The Andean giant hummingbird *(Patagona gigas)* has mass 70 g and flaps its wings ten times per second while hovering. Estimate the amount of work done by such a hummingbird in each wingbeat. (b) A 70-kg athlete can maintain a power output of 1.4 kW for no more than a few seconds; the *steady* power output of a typical athlete is only 500 W or so. Is it possible for a human-powered aircraft to fly for extended periods by flapping its wings? Explain.

6.91 The Grand Coulee Dam is 1270 m long and 170 m high. The electrical power output from generators at its base is approximately 2000 MW. How many cubic meters of water must flow from the top of the dam per second to produce this amount of power if 92% of the work done on the water by gravity is converted to electrical energy? (Each cubic meter of water has a mass of 1000 kg.)

6.92 The engine of a car with mass m supplies a constant power P to the wheels to accelerate the car. You can ignore rolling friction and air resistance. The car is initially at rest. a) Show that the speed of the car is given as a function of time by $v = (2Pt/m)^{1/2}$. b) Show that the acceleration of the car is not constant, but is given as a function of time by $a = (P/2mt)^{1/2}$. c) Show that the displacement as a function of time is given by $x - x_0 = (8P/9m)^{1/2}t^{3/2}$.

6.93 Power of the Human Heart. The human heart is a powerful and extremely reliable pump. Each day it takes in and discharges about 7500 L of blood. Assume that the work done by the heart is equal to the work required to lift this amount of blood a height equal to that of the average American female (1.63 m). The density (mass per unit volume) of blood is 1.05×10^3 kg/m^3. a) How much work does the heart do in a day? b) What is its power output in watts?

6.94 Six diesel units in series can provide 13.4 MW of power to the lead car of a freight train. The diesel units have total mass 1.10×10^6 kg. The average car in the train has mass 8.2×10^4 kg and requires a horizontal pull of 2.8 kN to move at a constant 27 m/s on level tracks. a) How many cars can be in the train under these conditions? b) However, this would leave no power for accelerating or climbing hills. Show that the extra force needed to accelerate the train is about the same for a 0.10 m/s^2 acceleration or a 1.0% slope (slope angle $\alpha = \arctan 0.010$). c) With the 1.0% slope, show that an extra 2.9 MW of power is needed to maintain the 27 m/s speed of the diesel units. d) With 2.9 MW less power available, how many cars can the six diesel units pull up a 1.0% slope at a constant 27 m/s?

6.95 It takes a force of 53 kN on the lead car of a 16-car passenger train with mass 9.1×10^5 kg to pull it at a constant 45 m/s (101 mi/h) on level tracks. a) What power must the locomotive provide to the lead car? b) How much more power to the lead car than calculated in part (a) would be needed to give the train an acceleration of 1.5 m/s^2, at the instant that the train has a speed of 45 m/s on level tracks? c) How much more power to the lead car than that cal-

culated in part (a) would be needed to move the train up a 1.5% grade (slope angle $\alpha = \arctan 0.015$) at a constant 45 m/s?

6.96 An object has several forces acting on it. One of these forces is $\vec{F} = axy\hat{\imath}$, a force in the x-direction whose magnitude depends on the position of the object, with $\alpha = 2.50$ N/m^2. Calculate the work done on the object by this force for the following displacements of the object: a) The object starts at the point $x = 0$, $y = 3.00$ m and moves parallel to the x-axis to the point $x = 2.00$ m, $y = 3.00$ m. b) The object starts at the point $x = 2.00$ m, $y = 0$ and moves in the y-direction to the point $x = 2.00$ m, $y = 3.00$ m. c) The object starts at the origin and moves on the line $y = 1.5x$ to the point $x = 2.00$ m, $y = 3.00$ m.

6.97 Cycling. For a touring bicyclist the drag coefficient $C (f_{air} = \frac{1}{2} CA\rho v^2)$ is 1.00, the frontal area A is 0.463 m^2, and the coefficient of rolling friction is 0.0045. The rider has mass 50.0 kg, and her bike has mass 12.0 kg. a) To maintain a speed of 12.0 m/s (about 27 mi/h) on a level road, what must the rider's power output to the rear wheel be? b) For racing, the same rider uses a different bike with a coefficient of rolling friction 0.0030 and mass 9.00 kg. She also crouches down, reducing her drag coefficient to 0.88 and reducing her frontal area to 0.366 m^2. What must her power output to the rear wheel be then to maintain a speed of 12.0 m/s? c) For the situation in part (b), what power output is required to maintain a speed of 6.0 m/s? Note the great drop in power requirement when the speed is only halved. (For more on aerodynamic speed limitations for a wide variety of human-powered vehicles see "The Aerodynamics of Human-Powered Land Vehicles," *Scientific American,* December, 1983.)

6.98 Automotive Power I. A truck engine transmits 28.0 kW (37.5 hp) to the driving wheels when the truck is traveling at a constant velocity of magnitude 60.0 km/h (37.3 mi/h) on a level road. a) What is the resisting force acting on the truck? b) Assume that 65% of the resisting force is due to rolling friction, and that the remainder is due to air resistance. If the force of rolling friction is independent of speed, and the force of air resistance is proportional to the square of the speed, what power will drive the truck at 30.0 km/h? At 120.0 km/h? Give your answers in kilowatts and in horsepower.

6.99 Automotive Power II. a) If 8.00 hp are required to drive a 1800-kg automobile at 60.0 km/h on a level road, what is the total retarding force due to friction, air resistance, and so on? b) What power is necessary to drive the car at 60.0 km/h up a 10.0% grade (a hill rising 10.0 m vertically in 100.0 m horizontally)? c) What power is necessary to drive the car at 60.0 km/h *down* a 1.00% grade? d) Down what percent grade would the car coast at 60.0 km/h?

Challenge Problems

6.100 On a winter's day in Maine, a warehouse worker is shoving boxes up a rough plank inclined at an angle α above the horizontal. The plank is partially covered with ice, with more ice near the bottom of the plank than near the top, so that the coefficient of friction increases with the distance x along the plank: $\mu = Ax$, where A is a positive constant and the bottom of the plank is at $x = 0$. (For this plank the coefficients of kinetic and static friction are equal: $\mu_k = \mu_s = \mu$.) The worker shoves a box up the plank so that it leaves the bottom of the plank moving at speed v_0. Show that when the box first comes to rest, it will remain at rest if

$$v_0^2 \geq \frac{3g \sin^2 \alpha}{A \cos \alpha}$$

6.101 A Spring with Mass. We usually ignore the kinetic energy of the moving coils of a spring, but let's try to get a reasonable approximation to this. Consider a spring of mass M, equilibrium length L_0, and spring constant k. The work done to stretch or compress the spring by a distance L is $\frac{1}{2}kX^2$, where $X = L - L_0$. a) Consider a spring, as described above, that has one end fixed and the other end moving with speed v. Assume that the speed of points along the length of the spring varies linearly with distance l from the fixed end. Assume also that the mass M of the spring is distributed uniformly along the length of the spring. Calculate the kinetic energy of the spring in terms of M and v. (*Hint:* Divide the spring into pieces of length dl; find the speed of each piece in terms of l, v, and L; find the mass of each piece in terms of dl, M, and L; and integrate from 0 to L. The result is *not* $\frac{1}{2}Mv^2$, since not all of the spring moves with the same speed.) In a spring gun, a spring of mass 0.243 kg and force constant 3200 N/m is compressed 2.50 cm from its unstretched length. When the trigger is pulled, the spring pushes horizontally on a 0.053-kg ball. The work done by friction is negligible. Calculate the ball's speed when the spring reaches its uncompressed length b) ignoring the mass of the spring; c) including, using the results of part (a), the mass of the spring. d) In part (c), what is the final kinetic energy of the ball and of the spring?

6.102 An airplane in flight is subject to an air resistance force proportional to the square of its speed v. But there is an additional resistive force because the airplane has wings. Air flowing over the wings is pushed down and slightly forward, so from Newton's third law the air exerts a force on the wings and airplane that is up and slightly backward (Fig. 6.30). The upward force is the lift

Figure 6.30 Challenge Problem 6.102.

force that keeps the airplane aloft, and the backward force is called *induced drag*. At flying speeds, induced drag is inversely proportional to v^2, so that the total air resistance force can be expressed by $F_{air} = \alpha v^2 + \beta/v^2$, where α and β are positive constants that depend on the shape and size of the airplane and the density of the air. For a Cessna 150, a small single-engine airplane, $\alpha = 0.30\ \text{N} \cdot \text{s}^2/\text{m}^2$ and $\beta = 3.5 \times 10^5\ \text{N} \cdot \text{m}^2/\text{s}^2$. In steady flight, the engine must provide a forward force that exactly balances the air resistance force. a) Calculate the speed (in km/h) at which this airplane will have the maximum *range* (that is, travel the greatest distance) for a given quantity of fuel. b) Calculate the speed (in km/h) for which the airplane will have the maximum *endurance* (that is, will remain in the air the longest time).

6.103 Figure 6.31 shows the oxygen consumption rate of men walking and running at different speeds. The vertical axis shows the volume of oxygen (in cm^3) that a man consumes per kilogram of body mass per minute. Note the transition from walking to running that occurs naturally at about 9 km/h. The metabo-

Figure 6.31 Challenge Problem 6.103.

lism of 1 cm^3 of oxygen releases about 20 J of energy. Using the data in the graph, calculate the energy required for a 70-kg man to travel 1 km on foot at each of the following speeds a) 5 km/h (walking); b) 10 km/h (running); c) 15 km/h (running). d) Which speed is the most efficient, that is, requires the least energy to travel 1 km?

6.104 General Proof of the Work-Energy Theorem. Consider a particle that moves along a curved path in space from (x_1, y_1, z_1) to (x_2, y_2, z_2). At the initial point, the particle has velocity $\vec{v} = v_{1x}\hat{\imath} + v_{1y}\hat{\jmath} + v_{1z}\hat{k}$. The path that the particle follows may be divided into infinitesimal segments $d\vec{l} = dx\hat{\imath} + dy\hat{\jmath} + dz\hat{k}$. As the particle moves, it is acted on by a net force $\vec{F} = F_x\hat{\imath} + F_y\hat{\jmath} + F_z\hat{k}$. The force components F_x, F_y, and F_z are in general functions of position. By the same sequence of steps used in Eqs. (6.11) through (6.13), prove the work-energy theorem for this general case. That is, prove that

$$W_{tot} = K_2 - K_1$$

where

$$W_{tot} = \int_{(x_1, y_1, z_1)}^{(x_2, y_2, z_2)} \vec{F} \cdot d\vec{l} = \int_{(x_1, y_1, z_1)}^{(x_2, y_2, z_2)} \left(F_x\, dx + F_y\, dy + F_z\, dz \right)$$

POTENTIAL ENERGY AND ENERGY CONSERVATION

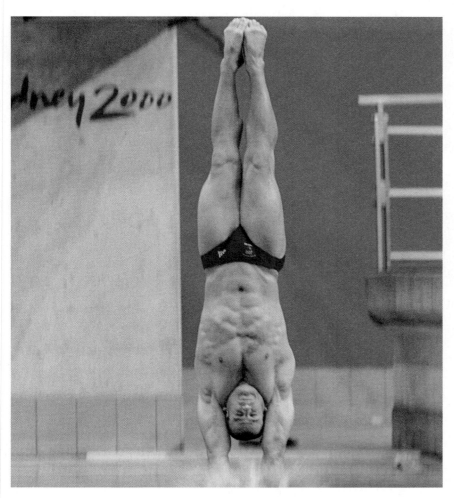

As a competitive diver plummets toward the water, gravitational potential energy is converted into the diver's kinetic energy. When the diver enters the water, he loses speed and his kinetic energy is converted into thermal energy of the water, raising its temperature slightly. Careful measurements show that no energy is lost in this process—instead, the energy merely changes form.

? **As the diver enters the water, is the water doing positive or negative work on him? Is the force of gravity doing positive or negative work on him?**

When a diver jumps off a high board into a swimming pool, he hits the water moving pretty fast, with a lot of kinetic energy. Where does that energy come from? The answer we learned in Chapter 6 was that the gravitational force (his weight) does work on the diver as he falls. The diver's kinetic energy—energy associated with his *motion*—increases by an amount equal to the work done.

However, there is a very useful alternative way to think about work and kinetic energy. This new approach is based on the concept of *potential energy,* which is energy associated with the *position* of a system rather than its motion. In this approach, there is *gravitational potential energy* even while the diver is standing on the high board. Energy is not added to the earth–diver system as the diver falls, but rather a storehouse of energy is *transformed* from one form (potential energy) to another (kinetic energy) as he falls. In this chapter we'll see how the work-energy theorem explains this transformation.

If the diver bounces on the end of the board before he jumps, the bent board stores a second kind of potential energy called *elastic potential energy.* We'll discuss elastic potential energy of simple systems such as a stretched or compressed spring. (An important third kind of potential energy is associated with the positions of electrically charged particles relative to each other. We'll encounter this potential energy in Chapter 23.)

We will prove that in some cases the sum of a system's kinetic and potential energy, called the *total mechanical energy* of the system, is constant during the motion of the system. This will lead us to the general statement of the *law of conservation of energy,* one of the most fundamental and far-reaching principles in all of science.

7.1 | Gravitational Potential Energy

A particle gains or loses kinetic energy because it interacts with other objects that exert forces on it. We learned in Chapter 6 that during any interaction the change in a particle's kinetic energy is equal to the total work done on the particle by the forces that act on it.

In many situations it seems as though energy has been stored in a system, to be recovered later. It's like a savings account in a bank; you deposit money, and then you can draw it out later. For example, you must do work on the hammerhead of a pile driver to lift it. It seems reasonable that in hoisting the hammerhead into the air you are storing energy in the system, energy that is later converted into kinetic energy as the hammerhead falls. Or consider your cousin Throckmorton on a swing. Suppose you give him a shove, which gives him some kinetic energy, and then you let him swing freely back and forth. He stops momentarily when he reaches the front and back end points of his arc, so at these points he has no kinetic energy. But he regains his kinetic energy as he passes through the low point in the arc. It seems as though at the high points the energy is stored in some other form, related to his height above the ground, and is converted back to kinetic energy as he swings toward the low point.

Both these examples point to the idea of an energy associated with the *position* of bodies in a system. This kind of energy is a measure of the *potential* or *possibility* for work to be done; when a hammerhead is raised into the air, there is a potential for work to be done on it by the gravitational force, but only if the hammerhead is allowed to fall to the ground. For this reason, energy associated with position is called **potential energy.** Our discussion suggests that there is potential energy associated with a body's weight and its height above the ground. We call this *gravitational potential energy.*

When a body falls without air resistance, gravitational potential energy decreases and the body's kinetic energy increases. But in Chapter 6 we used the work-energy theorem to conclude that the kinetic energy of a falling body increases because the force of the earth's gravity (the body's weight) does work on the body. Let's use the work-energy theorem to show that these two descriptions of a falling body are equivalent and to derive the expression for gravitational potential energy.

Let's consider a body with mass m that moves along the (vertical) y-axis, as in Fig. 7.1. The forces acting on it are its weight, with magnitude $w = mg$, and possibly some other forces; we call the vector sum (resultant) of all the other forces $\vec{F}_{\text{other}}$.

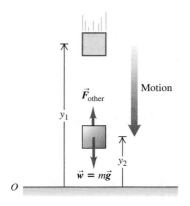

(a) Gravitational potential energy U decreases

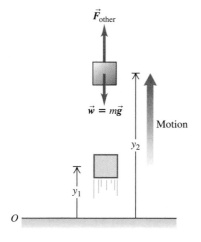

(b) Gravitational potential energy U increases

7.1 Work is done by the gravitational force $\vec{w}$ during the vertical motion of a body from an initial height y_1 to a final height y_2. The gravitational potential energy (a) decreases if the body moves downward and (b) increases if it moves upward.

We'll assume that the body stays close enough to the earth's surface that the weight is constant. (We'll find in Chapter 12 that weight decreases with altitude.) We want to find the work done by the weight when the body drops from a height y_1 above the origin to a lower height y_2 (Fig. 7.1a). The weight and displacement are in the same direction, so the work W_{grav} done on the body by its weight is positive;

$$W_{grav} = Fs = w(y_1 - y_2) = mgy_1 - mgy_2 \qquad (7.1)$$

This expression also gives the correct work when the body moves upward and y_2 is greater than y_1 (Fig. 7.1b). In that case the quantity $(y_1 - y_2)$ is negative, and W_{grav} is negative because the weight and displacement are opposite in direction.

Equation (7.1) shows that we can express W_{grav} in terms of the values of the quantity mgy at the beginning and end of the displacement. This quantity, the product of the weight mg and the height y above the origin of coordinates, is called the **gravitational potential energy,** U:

$$U = mgy \qquad \text{(gravitational potential energy)} \qquad (7.2)$$

Its initial value is $U_1 = mgy_1$ and its final value is $U_2 = mgy_2$. The change in U is the final value minus the initial value, or $\Delta U = U_2 - U_1$. We can express the work W_{grav} done by the gravitational force during the displacement from y_1 to y_2 as

$$W_{grav} = U_1 - U_2 = -(U_2 - U_1) = -\Delta U \qquad (7.3)$$

The negative sign in front of ΔU is *essential*. When the body moves up, y increases, the work done by the gravitational force is negative, and the gravitational potential energy increases $(\Delta U > 0)$. When the body moves down, y decreases, the gravitational force does positive work, and the gravitational potential energy decreases $(\Delta U < 0)$. It's like drawing money out of the bank (decreasing U) and spending it (doing positive work). As Eq. (7.3) shows, the unit of potential energy is the joule (J), the same unit as is used for work.

> **CAUTION** Although you might be tempted to do so, it is not correct to call $U = mgy$ the "gravitational potential energy of the body." The reason is that gravitational potential energy is a *shared* property of the body and the earth. Gravitational potential energy increases if the earth stays fixed and the height of the body increases; it also increases if the body is fixed in space and the earth is moved away from the body. Notice that the formula $U = mgy$ involves characteristics of both the body (its mass m) and the earth (the value of g).

Conservation of Mechanical Energy (Gravitational Forces Only)

To see what gravitational potential energy is good for, suppose the body's weight is the *only* force acting on it, so $\vec{F}_{other} = 0$. The body is then falling freely with no air resistance, and can be moving either up or down. Let its speed at point y_1 be v_1 and let its speed at y_2 be v_2. The work-energy theorem, Eq. (6.6), says that the total work done on the body equals the change in the body's kinetic energy; $W_{tot} = \Delta K = K_2 - K_1$. If gravity is the only force that acts, then from Eq. (7.3), $W_{tot} = W_{grav} = -\Delta U = U_1 - U_2$. Putting these together, we get

$$\Delta K = -\Delta U \quad \text{or} \quad K_2 - K_1 = U_1 - U_2$$

which we can rewrite as

$$K_1 + U_1 = K_2 + U_2 \qquad \text{(if only gravity does work)} \qquad (7.4)$$

or

$$\frac{1}{2}mv_1^{\,2} + mgy_1 = \frac{1}{2}mv_2^{\,2} + mgy_2 \qquad \text{(if only gravity does work)} \qquad (7.5)$$

We now define the sum $K + U$ of kinetic and potential energy to be E, the **total mechanical energy of the system.** By "system" we mean the body of mass m and the earth considered together, because gravitational potential energy U is a shared property of both bodies. Then $E_1 = K_1 + U_1$ is the total mechanical energy at y_1 and $E_2 = K_2 + U_2$ is the total mechanical energy at y_2. Equation (7.4) says that when the body's weight is the only force doing work on it, $E_1 = E_2$. That is, E is constant; it has the same value at y_1 and y_2. But since the positions y_1 and y_2 are arbitrary points in the motion of the body, the total mechanical energy E has the same value at *all* points during the motion;

$$E = K + U = \text{constant} \qquad \text{(if only gravity does work)}$$

A quantity that always has the same value is called a *conserved* quantity. *When only the force of gravity does work, the total mechanical energy is constant, that is, is conserved* (Fig. 7.2). This is our first example of the **conservation of mechanical energy.**

When we throw a ball into the air, its speed decreases on the way up as kinetic energy is converted to potential energy; $\Delta K < 0$ and $\Delta U > 0$. On the way back down, potential energy is converted back to kinetic energy and the ball's speed increases; $\Delta K > 0$ and $\Delta U < 0$. But the *total* mechanical energy (kinetic plus potential) is the same at every point in the motion, provided that no force other than gravity does work on the ball (that is, air resistance must be negligible). It's still true that the gravitational force does work on the body as it moves up or down, but we no longer have to calculate work directly; keeping track of changes in the value of U takes care of this completely.

CAUTION An important point about gravitational potential energy is that it doesn't matter what height we choose to be $y = 0$, the origin of coordinates. If we shift the origin for y, the values of y_1 and y_2 both change, but the *difference*

Moving up
K decreases
U increases

Moving down
K increases
U decreases

$\vec{w} = m\vec{g}$

7.2 During the time this athlete is in midair, only gravity does work on him (if we neglect the minor effects of air resistance). Mechanical energy—the sum of kinetic and gravitational potential energy—is conserved.

$y_2 - y_1$ stays the same. It follows that although U_1 and U_2 depend on where we place the origin, the difference $U_2 - U_1 = mg(y_2 - y_1)$ does not. The physically significant quantity is not the value of U at a particular point, but only the *difference* in U between two points. So we can define U to be zero at whatever point we choose without affecting the physics of the situation. This is shown in the following example.

| Example 7.1 | **Height of a baseball from energy conservation** |

You throw a 0.145-kg baseball straight up in the air, giving it an initial upward velocity of magnitude 20.0 m/s. Use conservation of energy to find how high it goes, ignoring air resistance.

SOLUTION

IDENTIFY: After the ball leaves your hand, the only force doing work on the ball is its weight. Hence we can use conservation of mechanical energy.

SET UP: We'll use Eqs. (7.4) and (7.5), taking point 1 to be where the ball leaves your hand and point 2 to be where the ball reaches its maximum height. As in Fig. 7.1, we take the y-axis to point vertically upward. The ball's speed at point 1 is $v_1 = 20.0$ m/s. The ball is instantaneously at rest at the high point of its motion (point 2), so $v_2 = 0$.

Our target variable is how far the ball moves vertically between these two points, so our target variable is the displacement $y_2 - y_1$. For simplicity, let's take the origin to be at point 1, where the ball leaves your hand. Then $y_1 = 0$ (Fig. 7.3) and the target variable is just y_2.

7.3 After a baseball leaves your hand, the only force acting on it is its weight (neglecting air resistance), so mechanical energy $E = K + U$ is conserved. The energy bar graphs show the values of E, K, and U at $y_1 = 0$ and y_2.

EXECUTE: Since $y_1 = 0$, the potential energy at point 1 is $U_1 = mgy_1 = 0$. Furthermore, since the ball is at rest at point 2, the kinetic energy at that point is $K_2 = \frac{1}{2}mv_2^2 = 0$. Hence Eq. (7.4), which says that $K_1 + U_1 = K_2 + U_2$, becomes

$$K_1 = U_2$$

As the energy bar graphs in Fig. 7.3 show, the kinetic energy of the ball at point 1 is completely converted into gravitational potential energy at point 2. At point 1 the kinetic energy is

$$K_1 = \frac{1}{2}mv_1^2 = \frac{1}{2}(0.145 \text{ kg})(20.0 \text{ m/s})^2 = 29.0 \text{ J}$$

This equals the gravitational potential energy $U_2 = mgy_2$ at point 2, so

$$y_2 = \frac{U_2}{mg} = \frac{29.0 \text{ J}}{(0.145 \text{ kg})(9.80 \text{ m/s}^2)} = 20.4 \text{ m}$$

We can also solve the equation $K_1 = U_2$ algebraically for y_2:

$$\frac{1}{2}mv_1^2 = mgy_2$$

$$y_2 = \frac{v_1^2}{2g} = \frac{(20.0 \text{ m/s})^2}{2(9.80 \text{ m/s}^2)} = 20.4 \text{ m}$$

EVALUATE: The mass divides out, as we should expect; we learned in Chapter 2 that the motion of a body in free fall doesn't depend on its mass. Indeed, we could have derived the result $y_2 = v_1^2/2g$ using Eq. (2.13).

In carrying out the calculation above, we chose the origin to be at point 1, so $y_1 = 0$ and $U_1 = 0$. What happens if we make a different choice? As an example, suppose we choose the origin to be 5.0 m below point 1, so $y_1 = 5.0$ m. With this choice the total mechanical energy at point 1 is part kinetic and part potential, while at point 2 it's purely potential energy. If you work through the calculation again with this choice of origin, you'll find $y_2 = 25.4$ m; this is 20.4 m above point 1, just as with the first choice of origin. In every problem, it's up to you to choose the height at which $U = 0$; don't agonize over the choice, however, because the physics of the answer doesn't depend on your choice.

Effect of Other Forces

When other forces act on the body in addition to its weight, then $\vec{F}_{\text{other}}$ in Fig. 7.1 is *not* zero. For the pile driver described in Example 6.5 (Section 6.2), the force applied by the hoisting cable and the friction with the vertical guide rails are examples of forces that might be included in $\vec{F}_{\text{other}}$. The gravitational work W_{grav} is still given by Eq. (7.3), but the total work W_{tot} is then the sum of W_{grav} and the work done by $\vec{F}_{\text{other}}$. We will call this additional work W_{other}, so the total work done by all forces is $W_{\text{tot}} = W_{\text{grav}} + W_{\text{other}}$. Equating this to the change in kinetic energy, we have

$$W_{\text{other}} + W_{\text{grav}} = K_2 - K_1 \qquad (7.6)$$

Also, from Eq. (7.3), $W_{\text{grav}} = U_1 - U_2$, so

$$W_{\text{other}} + U_1 - U_2 = K_2 - K_1$$

which we can rearrange in the form

$$K_1 + U_1 + W_{\text{other}} = K_2 + U_2 \qquad (7.7)$$

(if forces other than gravity do work)

Finally, using the appropriate expressions for the various energy terms, we obtain

$$\frac{1}{2}mv_1^2 + mgy_1 + W_{\text{other}} = \frac{1}{2}mv_2^2 + mgy_2 \qquad (7.8)$$

(if forces other than gravity do work)

The meaning of Eqs. (7.7) and (7.8) is this: *The work done by all forces* other than the gravitational force *equals the change in the total mechanical energy* $E = K + U$ *of the system, where U is the gravitational potential energy.* When W_{other} is positive, E increases, and $K_2 + U_2$ is greater than $K_1 + U_1$. When W_{other} is negative, E decreases. In the special case in which no forces other than the body's weight do work, $W_{\text{other}} = 0$. The total mechanical energy is then constant, and we are back to Eq. (7.4) or (7.5).

Problem-Solving Strategy

Problems Using Mechanical Energy

IDENTIFY *the relevant concepts:* First decide whether the problem should be solved by energy methods, by using $\sum \vec{F} = m\vec{a}$ directly, or by a combination of these strategies. The energy approach is particularly useful when the problem involves motion with varying forces, along a curved path (discussed later in this section), or both. But if the problem involves elapsed time, the energy approach is usually *not* the best choice, because this approach doesn't involve time directly.

SET UP *the problem using the following steps:*
1.   When using the energy approach, first decide what the initial and final states (the positions and velocities) of the system are. Use the subscript 1 for the initial state and the

subscript 2 for the final state. It helps to draw sketches showing the initial and final states.

2.   Define your coordinate system, particularly the level at which $y = 0$. You will use it to compute gravitational potential energies. Equation (7.2) assumes that the positive direction for y is upward; we suggest that you use this choice consistently.

3.   Identify all nongravitational forces that do work. A free-body diagram is always helpful. If some of the quantities you need are unknown, represent them by algebraic symbols.

4.   List the unknown and known quantities, including the coordinates and velocities at each point. Decide which unknowns are your target variables.

EXECUTE *the solution:* Write expressions for the initial and final kinetic and potential energies—that is, K_1, K_2, U_1, and U_2. In general some of these quantities will be known and some will be unknown. Then relate the kinetic and potential energies and the nongravitational work W_{other} using Eq. (7.7). (You will have to calculate W_{other} in terms of the nongravitational forces.) If there is no nongravitational work, this expression becomes Eq. (7.5). It's helpful to draw bar graphs showing the initial and

final values of K, U, and $E = K + U$. Then solve to find whatever unknown quantity is required.

EVALUATE *your answer:* Check whether your answer makes physical sense. Keep in mind, here and in later sections, that the work done by each force must be represented either in $U_1 - U_2 = -\Delta U$ or as W_{other}, but *never* in both places. The gravitational work is included in ΔU, so make sure you did not include it again in W_{other}.

| Example 7.2 | **Work and energy throwing a baseball** |

In Example 7.1, suppose your hand moves up 0.50 m while you are throwing the ball, which leaves your hand with an upward velocity of 20.0 m/s. Again ignore air resistance. a) Assuming that your hand exerts a constant upward force on the ball, find the magnitude of that force. b) Find the speed of the ball at a point 15.0 m above the point where it leaves your hand.

SOLUTION

IDENTIFY: In Example 7.1 we used conservation of mechanical energy because only gravity did work. In this example, however, we must also include the nongravitational work done by your hand.

SET UP: Figure 7.4 shows a diagram of the situation, including a free-body diagram for the ball while it is being thrown. The ball's motion occurs in two stages: while it is in contact with your hand and after it leaves your hand. To keep track of these stages, we let point 1 be where your hand first starts to move, point 2 be where the ball leaves your hand, and point 3 be where the ball is 15.0 m above point 2. The nongravitational force $\vec{F}$ of your hand acts only between points 1 and 2. Using the same coordinate system as in Example 7.1, we have $y_1 = -0.50$ m, $y_2 = 0$, and $y_3 = 15.0$ m. The ball starts at rest at point 1, so $v_1 = 0$, and we are given that the ball's speed as it leaves your hand is $v_2 = 20.0$ m/s. Our target variables are (a) the magnitude F of the force of your hand and (b) the speed v_3 at point 3.

EXECUTE:

a) To determine the magnitude of $\vec{F}$, we'll first use Eq. (7.7) to calculate the work W_{other} done by this force. We have

$$K_1 = 0$$

$$U_1 = mgy_1 = (0.145 \text{ kg})(9.80 \text{ m/s}^2)(-0.50 \text{ m}) = -0.71 \text{ J}$$

$$K_2 = \frac{1}{2}mv_2^2 = \frac{1}{2}(0.145 \text{ kg})(20.0 \text{ m/s})^2 = 29.0 \text{ J}$$

$$U_2 = mgy_2 = (0.145 \text{ kg})(9.80 \text{ m/s}^2)(0) = 0$$

The initial potential energy U_1 is negative because the ball was initially below the origin. According to Eq. (7.7), $K_1 + U_1 + W_{\text{other}} = K_2 + U_2$, so

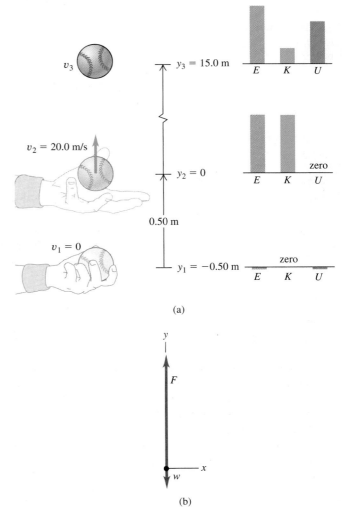

7.4 (a) Throwing a ball vertically upward. (b) The free-body diagram for the ball as the force $\vec{F}$ applied by the hand does the work W_{other} on the ball. The force $\vec{F}$ and gravity both act between y_1 and y_2. From y_2 to y_3, only gravity acts on the ball.

$$W_{other} = (K_2 - K_1) + (U_2 - U_1)$$
$$= (29.0 \text{ J} - 0) + (0 - (-0.71 \text{ J})) = 29.7 \text{ J}$$

The kinetic energy of the ball increases by $K_2 - K_1 = 29.0$ J, and the potential energy increases by $U_2 - U_1 = 0.71$ J; the sum is $E_2 - E_1$, the change in total mechanical energy, which is equal to W_{other}.

Assuming that the upward force $\vec{F}$ that your hand applies is constant, the work W_{other} done by this force is equal to the magnitude F of the force multiplied by the upward displacement $y_2 - y_1$ over which it acts:

$$W_{other} = F(y_2 - y_1)$$
$$F = \frac{W_{other}}{y_2 - y_1} = \frac{29.7 \text{ J}}{0.50 \text{ m}} = 59 \text{ N}$$

This is about 40 times greater than the weight of the ball.

b) To find the speed at point 3, note that between points 2 and 3, total mechanical energy is conserved; the force of your hand no longer acts, so $W_{other} = 0$. We can then find the kinetic energy at point 3 using Eq. (7.4):

$$K_2 + U_2 = K_3 + U_3$$
$$U_3 = mgy_3 = (0.145 \text{ kg})(9.80 \text{ m/s}^2)(15.0 \text{ m}) = 21.3 \text{ J}$$
$$K_3 = (K_2 + U_2) - U_3 = (29.0 \text{ J} + 0 \text{ J}) - 21.3 \text{ J} = 7.7 \text{ J}$$

Since $K_3 = \frac{1}{2}mv_{3y}^2$, where v_{3y} is the y-component of the ball's velocity at point 3, we have

$$v_{3y} = \pm\sqrt{\frac{2K_3}{m}} = \pm\sqrt{\frac{2(7.7 \text{ J})}{0.145 \text{ kg}}} = \pm10 \text{ m/s}$$

The significance of the plus-or-minus sign is that the ball passes point 3 *twice*, once on the way up and again on the way down. The total mechanical energy E is constant and equal to 29.0 J while the ball is in free fall, and the potential energy at point 3 is $U_3 = 21.3$ J whether the ball is moving up or down. So at point 3, the ball's kinetic energy K_3 and *speed* don't depend on the direction the ball is moving. The velocity v_{3y} is positive ($+10$ m/s) when the ball is moving up and negative (-10 m/s) when it is moving down; the speed v_3 is 10 m/s in either case.

EVALUATE: As a check on our result, recall from Example 7.1 that the ball reaches a maximum height $y = 20.4$ m. At that point all of the kinetic energy that the ball had when it left your hand at $y = 0$ has been converted to gravitational potential energy. At $y = 15.0$ m, the ball is about three-fourths of the way to its maximum height, so about three-fourths of its mechanical energy should be in the form of potential energy. Can you show that this is true from our results for K_3 and U_3?

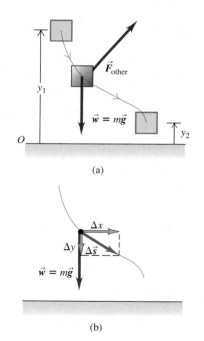

7.5 (a) A displacement along a curved path. (b) The work done by the gravitational force $\vec{w} = m\vec{g}$ depends only on the vertical component of displacement Δy (in this figure, Δy is negative).

Gravitational Potential Energy for Motion Along a Curved Path

In our first two examples the body moved along a straight vertical line. What happens when the path is slanted or curved (Fig. 7.5a)? The body is acted on by the gravitational force $\vec{w} = m\vec{g}$ and possibly by other forces whose resultant we call $\vec{F}_{other}$. To find the work done by the gravitational force during this displacement, we divide the path up into small segments $\Delta\vec{s}$; a typical segment is shown in Fig. 7.5b. The work done by the gravitational force over this segment is the scalar product of the force and the displacement. In terms of unit vectors, the force is $\vec{w} = m\vec{g} = -mg\hat{j}$ and the displacement is $\Delta\vec{s} = \Delta x\hat{i} + \Delta y\hat{j}$, so the work done by the gravitational force is

$$\vec{w} \cdot \Delta\vec{s} = -mg\hat{j} \cdot (\Delta x\hat{i} + \Delta y\hat{j}) = -mg\Delta y$$

The work done by gravity is the same as though the body had been displaced vertically a distance Δy, with no horizontal displacement. This is true for every segment, so the *total* work done by the gravitational force is $-mg$ multiplied by the *total* vertical displacement $(y_2 - y_1)$:

$$W_{grav} = -mg(y_2 - y_1) = mgy_1 - mgy_2 = U_1 - U_2$$

This is the same as Eq. (7.1) or (7.3), in which we assumed a purely vertical path. So even if the path a body follows between two points is curved, the total work done by the gravitational force depends only on the difference in height between the two points of the path. This work is unaffected by any horizontal motion that may occur. So *we can use the same expression for gravitational potential energy whether the body's path is curved or straight.*

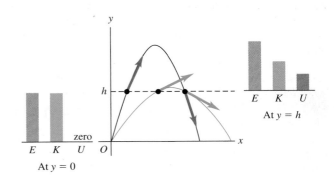

| | Conceptual Example 7.3 | **Energy in projectile motion** |

A batter hits two identical baseballs with the same initial speed and height but different initial angles. Prove that at a given height h, both balls have the same speed if air resistance can be neglected.

SOLUTION

If there is no air resistance, the only force acting on each ball after it is hit is its weight. Hence the total mechanical energy for each ball is constant. Figure 7.6 shows the trajectories of two balls batted at the same height with the same initial speed, and thus the same total mechanical energy, but with different initial angles. At all points at the same height the potential energy is the same. Thus the kinetic energy at this height must be the same for both balls, and the speeds are the same.

7.6 For the same initial speed and initial height, the speed of a projectile at a given elevation h is always the same, neglecting air resistance.

| | Example 7.4 | **Maximum height of a projectile, using energy methods** |

In Example 3.10 (Section 3.3) we derived an expression for the maximum height h of a projectile launched with initial speed v_0 at initial angle α_0:

$$h = \frac{v_0^2 \sin^2 \alpha_0}{2g}$$

Derive this expression using energy considerations.

SOLUTION

IDENTIFY and SET UP: We neglect air resistance, so just as in Conceptual Example 7.3, total mechanical energy is conserved. Let

point 1 at $y = 0$ be the launch point, where the speed is $v_1 = v_0$, and let point 2 at $y = h$ be the highest point on the trajectory (Fig. 7.7). Our target variable is the maximum height h, at which the kinetic energy is minimum and the gravitational potential energy is maximum.

This appears to be an easy problem: the potential energy at point 2 is $U_2 = mgh$, so it may seem that all we need to do is solve the energy-conservation equation $K_1 + U_1 = K_2 + U_2$ for U_2. However, while we know the initial kinetic and potential energies ($K_1 = \frac{1}{2}mv_1^2 = \frac{1}{2}mv_0^2$ and $U_1 = 0$), we don't know the speed or kinetic energy at point 2. To get around this issue, we'll use two results from our study of projectile motion in Chapter 3: (1) the x-component of acceleration is zero, so the x-component of velocity is constant, and (2) the y-component of velocity is zero at point 2 (the highest point of the trajectory).

EXECUTE: We can express the kinetic energy at each point in terms of the components of velocity, using $v^2 = v_x^2 + v_y^2$:

$$K_1 = \frac{1}{2}m(v_{1x}^2 + v_{1y}^2)$$

$$K_2 = \frac{1}{2}m(v_{2x}^2 + v_{2y}^2)$$

Conservation of energy then gives $K_1 + U_1 = K_2 + U_2$, so

$$\frac{1}{2}m(v_{1x}^2 + v_{1y}^2) + 0 = \frac{1}{2}m(v_{2x}^2 + v_{2y}^2) + mgh$$

To simplify this, we multiply through by $2/m$ to obtain

$$v_{1x}^2 + v_{1y}^2 = v_{2x}^2 + v_{2y}^2 + 2gh$$

7.7 Trajectory of a projectile.

Now we use our results from projectile motion. Since the x-component of velocity doesn't change, $v_{1x} = v_{2x}$ and we can cancel the v_x^2 terms from both sides of the equation given above. Furthermore, because the projectile has zero vertical velocity at the highest point of its motion, $v_{2y} = 0$. Hence we get

$$v_{1y}^2 = 2gh$$

But v_{1y} is just the y-component of initial velocity, which is equal to $v_0 \sin \alpha_0$. Making this substitution and solving for h, we find

$$h = \frac{v_0^2 \sin^2 \alpha_0}{2g}$$

EVALUATE: This agrees with the result of Example 3.10, as it must.

| Example 7.5 | **Calculating speed along a vertical circle** |

Your cousin Throckmorton skateboards down a curved playground ramp. Treating Throcky and his skateboard as a particle, he moves through a quarter-circle with radius R (Fig. 7.8). The total mass of Throcky and his skateboard is 25.0 kg. He starts from rest and there is no friction. (a) Find his speed at the bottom of the ramp. (b) Find the normal force that acts on him at the bottom of the curve.

SOLUTION

IDENTIFY: We can't use the equations of motion with constant acceleration; the acceleration isn't constant because the slope decreases as Throcky descends. Instead, we'll use the energy

approach. Since Throcky moves along a circular arc, we'll also use what we learned about circular motion in Section 5.4.

SET UP: Since there is no friction, the only force other than Throcky's weight is the normal force $\vec{n}$ exerted by the ramp (Fig. 7.8b). Although this force acts all along the path, it does *zero* work because $\vec{n}$ is perpendicular to Throcky's velocity at every point. Hence $W_{\text{other}} = 0$ and mechanical energy is conserved.

Take point 1 at the starting point and point 2 at the bottom of the curved ramp, and let $y = 0$ be at the bottom of the ramp (Fig. 7.8a). Then $y_1 = R$ and $y_2 = 0$. (We are treating Throcky as if his entire mass were concentrated at his center.) Throcky starts at rest at the top, so $v_1 = 0$. Our target variable in part (a) is his speed at the bottom, v_2. In part (b) we want to find the magnitude n of the normal force at point 2. As this force does no work, it doesn't appear in the energy equation, so we'll use Newton's second law instead.

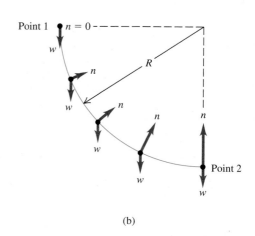

7.8 (a) Throcky skateboarding down a frictionless circular ramp. The total mechanical energy is constant. (b) Free-body diagrams for Throcky and his skateboard at various points on the ramp.

EXECUTE:
a) The various energy quantities are

$$K_1 = 0 \qquad U_1 = mgR$$

$$K_2 = \frac{1}{2}mv_2^2 \qquad U_2 = 0$$

From conservation of energy,

$$K_1 + U_1 = K_2 + U_2$$

$$0 + mgR = \frac{1}{2}mv_2^2 + 0$$

$$v_2 = \sqrt{2gR}$$

The speed is the same as if Throcky had fallen vertically through a height R, and it is independent of his mass.
As a numerical example, let $R = 3.00$ m. Then

$$v_2 = \sqrt{2(9.80 \text{ m/s}^2)(3.00 \text{ m})} = 7.67 \text{ m/s}$$

Notice that this answer doesn't depend on the ramp being circular; no matter what the shape of the ramp, Throcky will have the same speed $v_2 = \sqrt{2gR}$ at the bottom. This would be true even if the wheels of his skateboard lost contact with the ramp during the ride, because only the gravitational force would still do work.
b) To find n at point 2 using Newton's second law, we need the free-body diagram at that point (Fig. 7.8b). At point 2, Throcky is moving at speed $v_2 = \sqrt{2gR}$ in a circle of radius R; his acceleration is toward the center of the circle and has magnitude

$$a_{\text{rad}} = \frac{v_2^2}{R} = \frac{2gR}{R} = 2g$$

If we take the positive y-direction to be upward, the y-component of Newton's second law is

$$\sum F_y = n + (-w) = ma_{\text{rad}} = 2mg$$

$$n = w + 2mg = 3mg$$

At point 2 the normal force is three times Throcky's weight. This result is independent of the radius of the circular ramp. We learned in Example 5.10 (Section 5.2) and Example 5.25 (Section 5.4) that the magnitude of n is the *apparent weight,* so Throcky feels as though he weighs three times his true weight mg. But as soon as he reaches the horizontal part of the ramp to the right of point 2, the normal force decreases to $w = mg$ and Throcky feels normal again. Can you see why?

EVALUATE: This example shows a general rule about the role of forces in problems in which we use energy techniques: What matters is not simply whether a force acts, but whether that force does work. If the force does no work, like the normal force $\vec{n}$ in this example, then it does not appear at all in Eq. (7.7), $K_1 + U_1 + W_{\text{other}} = K_2 + U_2$.
Notice we had to use *both* the energy approach and Newton's second law to solve this problem; energy conservation gave us the speed and $\sum \vec{F} = m\vec{a}$ gave us the normal force. For each part of the problem we used the technique that most easily led to the answer.

Example 7.6	**A vertical circle with friction**

In Example 7.5, suppose that the ramp is not frictionless and that Throcky's speed at the bottom is only 6.00 m/s. What work was done by the friction force acting on him? Use $R = 3.00$ m.

SOLUTION

IDENTIFY and SET UP: We use the same coordinate system and the same initial and final points as in Example 7.5 (Fig. 7.9). Again the normal force does no work, but now there is a friction force $\vec{f}$ that *does* do work. Hence the nongravitational work done on Throcky between points 1 and 2, W_{other}, is just equal to the work done by friction, W_f. This is our target variable, which we'll find using Eq. (7.7).

EXECUTE: The energy quantities are

$$K_1 = 0$$

$$U_1 = mgR = (25.0 \text{ kg})(9.80 \text{ m/s}^2)(3.00 \text{ m}) = 735 \text{ J}$$

$$K_2 = \frac{1}{2}mv_2^2 = \frac{1}{2}(25.0 \text{ kg})(6.00 \text{ m/s})^2 = 450 \text{ J}$$

$$U_2 = 0$$

From Eq. (7.7),

$$W_f = K_2 + U_2 - K_1 - U_1$$

$$= 450 \text{ J} + 0 - 0 - 735 \text{ J} = -285 \text{ J}$$

The work done by the friction force is -285 J, and the total mechanical energy *decreases* by 285 J. Do you see why W_f has to be negative?

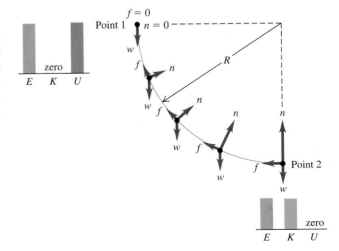

7.9 Free-body diagram for Throcky skateboarding down a ramp with friction. The total mechanical energy decreases as Throcky moves down the ramp.

EVALUATE: Throcky's motion is determined by Newton's second law, $\sum \vec{F} = m\vec{a}$. But it would be very difficult to apply the second law directly to this problem because the normal and friction forces and the acceleration are continuously changing in both magnitude and direction as Throcky skates down. The energy approach, by contrast, relates the motions at the top and bottom of the ramp without involving the details of what happens in between. Many problems are easy if energy considerations are used but very complex if we try to use Newton's laws directly.

Example 7.7 | **An inclined plane with friction**

We want to load a 12-kg crate into a truck by sliding it up a ramp 2.5 m long, inclined at 30°. A worker, giving no thought to friction, calculates that he can get the crate up the ramp by giving it an initial speed of 5.0 m/s at the bottom and letting it go. But friction is *not* negligible; the crate slides 1.6 m up the ramp, stops, and slides back down (Fig. 7.10). a) Assuming that the friction force acting on the crate is constant, find its magnitude. b) How fast is the crate moving when it reaches the bottom of the ramp?

The friction force of 35 N, acting over 1.6 m, causes the mechanical energy of the crate to decrease from 150 J to 94 J (Fig. 7.10c). b) The crate returns to point 3 at the bottom of the ramp; $y_3 = 0$ and

SOLUTION

IDENTIFY: The friction force does work on the crate as it slides. As in Example 7.2, we'll use the energy approach in part (a) to find the magnitude of the nongravitational force that does work (in this case, friction). Once we know the magnitude of the friction force, we can calculate how much nongravitational work this force does as the crate slides back down. We can then use the energy approach again to find the crate's speed at the bottom of the ramp.

SET UP: The first part of the motion is from point 1, at the bottom of the ramp, to point 2, where the crate stops instantaneously (Fig. 7.10a). In the second part of the motion, the crate returns to the bottom of the ramp, which we'll call point 3 (Fig. 7.10b). We'll take $y = 0$ (and hence $U = 0$) to be at ground level, so $y_1 = 0$, $y_2 = (1.6 \text{ m}) \sin 30° = 0.80$ m, and $y_3 = 0$. We are given that $v_1 = 5.0$ m/s and $v_2 = 0$ (the crate is instantaneously at rest at point 2). Our target variable in part (a) is f, the magnitude of the friction force, which we'll find using Eq. (7.7). In part (b) our target variable is v_3, the speed at the bottom of the ramp.

EXECUTE:
a) The energy quantities are

$$K_1 = \frac{1}{2}(12 \text{ kg})(5.0 \text{ m/s})^2 = 150 \text{ J} \qquad U_1 = 0$$

$$K_2 = 0 \qquad U_2 = (12 \text{ kg})(9.8 \text{ m/s}^2)(0.80 \text{ m}) = 94 \text{ J}$$

$$W_{\text{other}} = -fs$$

Here f is the unknown magnitude of the friction force and $s = 1.6$ m. Using Eq. (7.7), we find

$$K_1 + U_1 + W_{\text{other}} = K_2 + U_2$$

$$W_{\text{other}} = -fs = (K_2 + U_2) - (K_1 + U_1)$$

$$f = -\frac{(K_2 + U_2) - (K_1 + U_1)}{s}$$

$$= -\frac{[(0 + 94 \text{ J}) - (150 \text{ J} + 0)]}{1.6 \text{ m}} = 35 \text{ N}$$

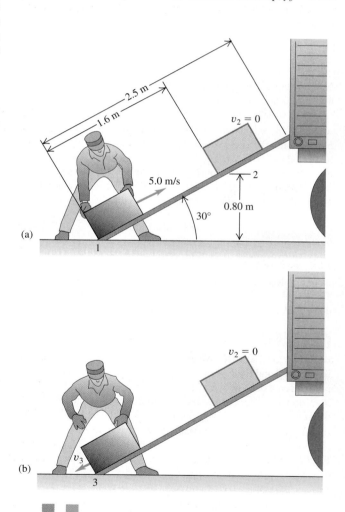

7.10 (a) A crate slides part way up the ramp, stops, and (b) slides back down. (c) Energy bar graphs for points 1, 2, and 3.

$U_3 = 0$ (Fig. 7.10b). On the way down, the friction force and the displacement both reverse direction but have the same magnitudes, so the frictional work has the same negative value for each half of the trip. The total work done by friction between points 1 and 3 is

$$W_{\text{other}} = W_{\text{fric}} = -2fs = -2(35 \text{ N})(1.6 \text{ m}) = -112 \text{ J}$$

From part (a), $K_1 = 150$ J and $U_1 = 0$. Equation (7.7) then gives

$$K_1 + U_1 + W_{\text{other}} = K_3 + U_3$$
$$K_3 = K_1 + U_1 - U_3 + W_{\text{other}}$$
$$= 150 \text{ J} + 0 - 0 + (-112 \text{ J}) = 38 \text{ J}$$

The crate returns to the bottom of the ramp with only 38 J of the original 150 J of mechanical energy (Fig. 7.10c). Using $K_3 = \frac{1}{2}mv_3^2$, we get

$$v_3 = \sqrt{\frac{2K_3}{m}} = \sqrt{\frac{2(38 \text{ J})}{12 \text{ kg}}} = 2.5 \text{ m/s}$$

EVALUATE: Notice that the crate's speed when it returns to the bottom of the ramp, $v_3 = 2.5$ m/s, is less than the speed $v_1 = 5.0$ m/s at which it left that point. That's good—energy was lost due to friction.

Notice also that in part (b) we applied Eq. (7.7) to points 1 and 3, considering the entire round trip as a whole. Alternatively, we could have considered the second part of the motion by itself and applied Eq. (7.7) to points 2 and 3. Try it and see whether you get the same result for v_3.

Test Your Understanding

Figure 7.11 shows two different frictionless ramps. The heights y_1 and y_2 are the same for each ramp. If a block of mass m is released from rest at the left-hand end of each ramp, which block arrives at the right-hand end with the greater speed?

7.11 Two ramps with the same y_1 and y_2.

7.2 | Elastic Potential Energy

When a railroad car runs into a spring bumper at the end of the track, the spring is compressed as the car is brought to a stop. If there is no friction, the bumper springs back and the car moves away in the opposite direction with its original speed. During the interaction with the spring, the car's kinetic energy has been "stored" in the elastic deformation of the spring. Something very similar happens in a rubber-band slingshot. Work is done on the rubber band by the force that stretches it, and that work is stored in the rubber band until you let it go. Then the rubber band gives kinetic energy to the projectile.

This is the same pattern we saw with the pile-driver in Section 7.1: do work on the system to store energy, which can later be converted to kinetic energy. We'll describe the process of storing energy in a deformable body such as a spring or rubber band in terms of *elastic potential energy* (Fig. 7.12). A body is called *elastic* if it returns to its original shape and size after being deformed. To be specific, we'll consider storing energy in an ideal spring, like the ones we discussed in Section 6.3. To keep such an ideal spring stretched by a distance x, we must exert a force $F = kx$, where k is the force constant of the spring. The ideal

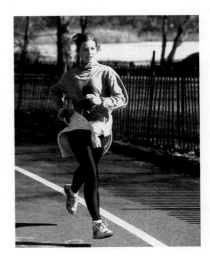

7.12 The Achilles tendon, which runs along the back of the ankle to the heel bone, acts like a natural spring. When it stretches and then relaxes, this tendon stores and then releases elastic potential energy. This spring action reduces the amount of work your leg muscles must do as you run.

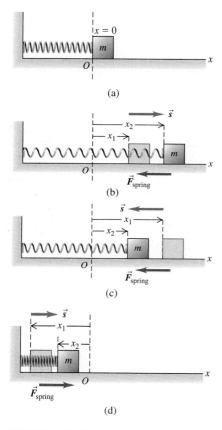

7.13 (a) A block attached to a spring in equilibrium ($x = 0$) on a horizontal surface. (b) As the spring is stretched, it does negative work on the block. (c) As the spring relaxes, it does positive work on the block. (d) A compressed spring also does positive work on the block as it relaxes.

spring is a useful idealization because many other elastic bodies show this same direct proportionality between force $\vec{F}$ and displacement x, provided that x is sufficiently small.

We proceed just as we did for gravitational potential energy. We begin with the work done by the elastic (spring) force and then combine this with the work-energy theorem. The difference is that gravitational potential energy is a shared property of a body and the earth, but elastic potential energy is stored just in the spring (or other deformable body).

Figure 7.13 shows the ideal spring from Fig. 6.15, with its left end held stationary and its right end attached to a block with mass m that can move along the x-axis. In Fig. 7.13a the body is at $x = 0$ when the spring is neither stretched nor compressed. We move the block to one side, thereby stretching or compressing the spring, and then let it go. As the block moves from one position x_1 to another position x_2, how much work does the elastic (spring) force do on the block?

We found in Section 6.3 that the work we must do *on* the spring to move one end from an elongation x_1 to a different elongation x_2 is

$$W = \frac{1}{2}kx_2^2 - \frac{1}{2}kx_1^2 \qquad (\text{work done } on \text{ a spring})$$

where k is the force constant of the spring. If we stretch the spring farther, we do positive work on the spring; if we let the spring relax while holding one end, we do negative work on it. We also saw that this expression for work is still correct if the spring is compressed, not stretched, so that x_1 or x_2 or both are negative. Now we need to find the work done *by* the spring. From Newton's third law the two quantities of work are just negatives of each other. Changing the signs in this equation, we find that in a displacement from x_1 to x_2 the spring does an amount of work W_{el} given by

$$W_{el} = \frac{1}{2}kx_1^2 - \frac{1}{2}kx_2^2 \qquad (\text{work done } by \text{ a spring})$$

The subscript "el" stands for *elastic*. When x_1 and x_2 are both positive and $x_2 > x_1$ (Fig. 7.13b), the spring does negative work on the block, which moves in the $+x$-direction while the spring pulls on it in the $-x$-direction. The spring stretches farther, and the block slows down. When x_1 and x_2 are both positive and $x_2 < x_1$ (Fig. 7.13c), the spring does positive work as it relaxes and the block speeds up. If the spring can be compressed as well as stretched, x_1 or x_2 or both may be negative, but the expression for W_{el} is still valid. In Fig. 7.13d, both x_1 and x_2 are negative, but x_2 is less negative than x_1; the compressed spring does positive work as it relaxes, speeding the block up.

Just as for gravitational work, we can express the work done by the spring in terms of a given quantity at the beginning and end of the displacement. This quantity is $\frac{1}{2}kx^2$, and we define it to be the **elastic potential energy:**

$$U = \frac{1}{2}kx^2 \qquad (\text{elastic potential energy}) \qquad (7.9)$$

Figure 7.14 is a graph of Eq. (7.9). The unit of U is the joule (J), the unit used for *all* energy and work quantities; to see this from Eq. (7.9), recall that the units of k are N/m and that $1\ \text{N} \cdot \text{m} = 1\ \text{J}$.

We can use Eq. (7.9) to express the work W_{el} done on the block by the elastic force in terms of the change in potential energy:

$$W_{el} = \frac{1}{2}kx_1^2 - \frac{1}{2}kx_2^2 = U_1 - U_2 = -\Delta U \qquad (7.10)$$

When a stretched spring is stretched farther, as in Fig. 7.13b, W_{el} is negative and *U increases;* a greater amount of elastic potential energy is stored in the spring. When a stretched spring relaxes, as in Fig. 7.13c, x decreases, W_{el} is positive, and *U decreases;* the spring loses elastic potential energy. Negative values of x refer to a compressed spring. But, as Fig. 7.14 shows, U is positive for both positive and negative x, and Eqs. (7.9) and (7.10) are valid for both cases. Thus, when a compressed spring is compressed farther, $W_{el} < 0$ and U increases; when a compressed spring relaxes (Fig. 7.13d), $W_{el} > 0$ and U decreases. The more a spring is compressed *or* stretched, the greater its elastic potential energy.

> **CAUTION** An important difference between gravitational potential energy $U = mgy$ and elastic potential energy $U = \frac{1}{2}kx^2$ is that we do *not* have the freedom to choose $x = 0$ to be wherever we wish. To be consistent with Eq. (7.9), $x = 0$ *must* be the position at which the spring is neither stretched nor compressed. At that position, its elastic potential energy and the force that it exerts are both zero.

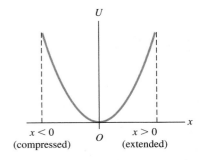

7.14 The graph of elastic potential energy for an ideal spring is a parabola: $U = \frac{1}{2}kx^2$, where x is the extension or compression of the spring. For extension (stretching), x is positive. For compression (when that is possible), x is negative. Elastic potential energy U is never negative.

The work-energy theorem says that $W_{tot} = K_2 - K_1$, no matter what kind of forces are acting on a body. If the elastic force is the *only* force that does work on the body, then

$$W_{tot} = W_{el} = U_1 - U_2$$

The work-energy theorem $W_{tot} = K_2 - K_1$ then gives us

$$K_1 + U_1 = K_2 + U_2 \qquad \text{(if only the elastic force does work)} \quad (7.11)$$

Here U is given by Eq. (7.9), so

$$\frac{1}{2}mv_1^2 + \frac{1}{2}kx_1^2 = \frac{1}{2}mv_2^2 + \frac{1}{2}kx_2^2 \qquad (7.12)$$
$$\text{(if only the elastic force does work)}$$

In this case the total mechanical energy $E = K + U$ (the sum of kinetic and elastic potential energy) is *conserved.* An example of this is the motion of the block in Fig. 7.13, provided that the horizontal surface is frictionless so that no force does work other than that exerted by the spring.

For Eq. (7.12) to be strictly correct, the ideal spring that we've been discussing must also be *massless.* If the spring has a mass, it also has kinetic energy as the coils of the spring move back and forth. We can neglect the kinetic energy of the spring if its mass is much less than the mass m of the body attached to the spring. For instance, a typical automobile has a mass of 1200 kg or more. The springs in its suspension have masses of only a few kilograms, so their mass can be neglected if we want to study how a car bounces on its suspension.

If forces other than the elastic force also do work on the body, we call their work W_{other}, as before. Then the total work is $W_{tot} = W_{el} + W_{other}$, and the work-energy theorem gives

$$W_{el} + W_{other} = K_2 - K_1$$

The work done by the spring is still $W_{el} = U_1 - U_2$, so again

$$K_1 + U_1 + W_{other} = K_2 + U_2 \qquad (7.13)$$
$$\text{(if forces other than the elastic force do work)}$$

and

$$\frac{1}{2}mv_1^2 + \frac{1}{2}kx_1^2 + W_{other} = \frac{1}{2}mv_2^2 + \frac{1}{2}kx_2^2 \qquad (7.14)$$
$$\text{(if forces other than the elastic force do work)}$$

This equation shows that *the work done by all forces* other than the elastic force *equals the change in the total mechanical energy $E = K + U$ of the system, where U is the elastic potential energy.* The "system" is made up of the body of mass m and the spring of force constant k. When W_{other} is positive, E increases; when W_{other} is negative, E decreases. You should compare Eq. (7.14) to Eq. (7.8), which describes situations in which there is gravitational potential energy but no elastic potential energy.

Situations with Both Gravitational and Elastic Potential Energy

Equations (7.11), (7.12), (7.13), and (7.14) are valid when the only potential energy in the system is elastic potential energy. What happens when we have *both* gravitational and elastic forces, such as a block attached to the lower end of a vertically hanging spring? We can still use Eq. (7.13), but now U_1 and U_2 are the initial and final values of the *total* potential energy, including both gravitational and elastic potential energies. That is, $U = U_{grav} + U_{el}$. Thus *the most general statement* of the relationship between kinetic energy, potential energy, and work done by other forces is

$$K_1 + U_{grav,\,1} + U_{el,\,1} + W_{other} = K_2 + U_{grav,\,2} + U_{el,\,2} \qquad (7.15)$$
$$\text{(valid in general)}$$

That is, **the work done by all forces other than the gravitational force or elastic force equals the change in the total mechanical energy $E = K + U$ of the system, where U is the sum of the gravitational potential energy and the elastic potential energy.** If the gravitational and elastic forces are the *only* forces that do work on the body, then $W_{other} = 0$ and the total mechanical energy (including both gravitational and elastic potential energy) is conserved.

Bungee jumping (Fig. 7.15) is an example of transformations among kinetic energy, elastic potential energy, and gravitational potential energy. As the jumper falls, gravitational potential energy decreases and is converted into the kinetic energy of the jumper and the elastic potential energy of the bungee cord. Beyond a certain point in the fall, the jumper's speed decreases so that both gravitational potential energy and kinetic energy are converted into elastic potential energy.

7.15 The fall of a bungee jumper involves an interplay among kinetic energy, gravitational potential energy, and elastic potential energy. Mechanical energy is not conserved, however, because work is also done by frictional forces within the bungee cord and by air resistance. This is a good thing—if mechanical energy were conserved, the bungee jumper would keep bouncing up and down forever!

The strategy outlined in Section 7.1 is equally useful in solving problems that involve elastic forces as well as gravitational forces. The only new wrinkle is that the potential energy U now includes the elastic potential energy $U_{el} = \frac{1}{2}kx^2$, where x is the displacement of the spring *from its unstretched length*. The work done by the gravitational and elastic forces is accounted for by their potential energies; the work of the other forces, W_{other}, has to be included separately.

Activ
ONLINE
Physics

5.4   Inverse Bungee Jumper

5.5   Spring-Launched Bowler

Example 7.8	**Motion with elastic potential energy**

In Fig. 7.16a a glider with mass $m = 0.200$ kg sits on a frictionless horizontal air track, connected to a spring with force constant $k = 5.00$ N/m. You pull on the glider, stretching the spring 0.100 m, and then release it with no initial velocity (Fig. 7.16b). The glider begins to move back toward its equilibrium position $(x = 0)$. What is its x-velocity when $x = 0.080$ m?

SET UP: The spring force is the only force doing work on the glider, so $W_{other} = 0$ and we may use Eq. (7.11). Let point 1 be where the glider is released (Fig. 7.16b), and let point 2 be at $x = 0.080$ m (Fig. 7.16c). We know the velocity at point 1 $(v_{1x} = 0)$; our target variable is the x-velocity at point 2, v_{2x}.

EXECUTE: The energy quantities are

$$K_1 = \frac{1}{2}(0.200 \text{ kg})(0)^2 = 0$$

$$U_1 = \frac{1}{2}(5.00 \text{ n/m})(0.100 \text{ m})^2 = 0.0250 \text{ J}$$

$$K_2 = \frac{1}{2}mv_{2x}^2$$

$$U_2 = \frac{1}{2}(5.00 \text{ N/m})(0.080 \text{ m})^2 = 0.0160 \text{ J}$$

SOLUTION

IDENTIFY: Because the spring force varies with position, this problem can't be solved with the equations for motion with constant acceleration. Instead, we'll use the energy method to find a simple solution. In particular, we'll use the idea that as the glider starts to move, elastic potential energy is converted into kinetic energy. (The glider remains at the same height throughout the motion, so gravitational potential energy is not a factor.)

7.16 (a) An air-track glider attached to a spring. (b) Elastic potential energy is added to the system by stretching the spring. (c) Elastic potential energy is transformed to kinetic energy as the glider moves back toward its equilibrium position.

Then from Eq. (7.11),

$$K_2 = K_1 + U_1 - U_2 = 0 + 0.0250 \text{ J} - 0.0160 \text{ J} = 0.0090 \text{ J}$$

$$v_{2x} = \pm \sqrt{\frac{2K_2}{m}} = \pm \sqrt{\frac{2(0.0090 \text{ J})}{0.200 \text{ kg}}} = \pm 0.30 \text{ m/s}$$

We choose the negative root because the glider is moving in the $-x$-direction; the answer we want is $v_{2x} = -0.30$ m/s.

EVALUATE: What is the meaning of the second solution, $v_{2x} = +0.30$ m/s? Eventually the spring will compress and push the glider back to the right in the positive x-direction (see Fig. 7.13d). The second solution tells us that when the glider passes through $x = 0.080$ m while moving to the right, its speed will be 0.30 m/s—the same speed as when it passed through this point while moving to the left.

When the glider passes through the point $x = 0$, the spring is relaxed and all of the mechanical energy is in the form of kinetic energy. Can you show that the speed of the glider at this point is 0.50 m/s?

| Example 7.9 | Motion with elastic potential energy and work done by other forces |

For the system of Example 7.8, suppose the glider is initially at rest at $x = 0$, with the spring unstretched. You then apply a constant force $\vec{F}$ in the $+x$-direction with magnitude 0.610 N to the glider. What is the glider's velocity when it has moved to $x = 0.100$ m?

SOLUTION

IDENTIFY: Although the force $\vec{F}$ you apply is constant, the spring force isn't, so the acceleration of the glider won't be constant. Total mechanical energy is not conserved because of the work done by the force $\vec{F}$, but we can still use the energy relation as given by Eq. (7.13).

SET UP: Let point 1 be at $x = 0$, where the velocity is $v_{1x} = 0$, and let point 2 be at $x = 0.100$ m. (These points are different from the ones labeled in Fig. 7.16.) Our target variable is v_{2x}, the velocity at point 2.

EXECUTE: The energy quantities are

$$K_1 = 0 \qquad U_1 = 0$$

$$K_2 = \frac{1}{2}mv_{2x}^2 \qquad U_2 = \frac{1}{2}(5.00 \text{ N/m})(0.100 \text{ m})^2$$

$$= 0.0250 \text{ J}$$

$$W_{\text{other}} = (0.610 \text{ N})(0.100 \text{ m}) = 0.0610 \text{ J}$$

(To calculate W_{other} we multiplied the magnitude of the force by the displacement, since both are in the $+x$-direction.) Initially, the total mechanical energy is zero; the work done by the force $\vec{F}$ increases the total mechanical energy to 0.0610 J, of which 0.0250 J is elastic potential energy. The remainder is kinetic energy. From Eq. (7.13),

$$K_1 + U_1 + W_{\text{other}} = K_2 + U_2$$

$$K_2 = K_1 + U_1 + W_{\text{other}} - U_2$$

$$= 0 + 0 + 0.0610 \text{ J} - 0.0250 \text{ J} = 0.0360 \text{ J}$$

$$v_{2x} = \sqrt{\frac{2K_2}{m}} = \sqrt{\frac{2(0.0360 \text{ J})}{0.200 \text{ kg}}} = 0.60 \text{ m/s}$$

We choose the positive square root because the glider is moving in the $+x$-direction.

EVALUATE: To test our answer, think what would be different if we disconnected the glider from the spring. Then $\vec{F}$ would be the only force doing work, there would be zero potential energy at all times, and Eq. (7.13) would give us

$$K_2 = K_1 + W_{\text{other}} = 0 + 0.0610 \text{ J}$$

$$v_{2x} = \sqrt{\frac{2K_2}{m}} = \sqrt{\frac{2(0.0610 \text{ J})}{0.200 \text{ kg}}} = 0.78 \text{ m/s}$$

We found a lower velocity than this value because the spring does negative work on the glider as it stretches (see Fig. 7.13b).

| Example 7.10 | Motion with elastic potential energy after other forces have ceased |

In Example 7.9, suppose the force $\vec{F}$ is removed when the glider reaches the point $x = 0.100$ m. How much farther does the glider move before coming to rest?

SOLUTION

IDENTIFY: After $\vec{F}$ is removed, the spring force is the only force doing work. Hence for this part of the motion the mechanical energy $E = K + U$ is conserved.

SET UP: We'll let point 2 be at $x = 0.100$ m, as in Example 7.9, and let point 3 be where the glider comes instantaneously to rest. Our target variable is the coordinate x_3 of this point. We'll find its value using the conservation of energy expressions, Eq. (7.11), along with the relation $U = \frac{1}{2}kx^2$ for elastic potential energy.

EXECUTE: We saw in Example 7.9 that the kinetic and potential energies at point 2 are $K_2 = 0.0360$ J and $U_2 = 0.0250$ J. The total mechanical energy at and beyond this point is therefore

$K_2 + U_2 = 0.0610$ J. When the glider comes to rest at $x = x_3$, the kinetic energy K_3 is zero and the elastic potential energy U_3 is equal to the total mechanical energy 0.0610 J. We can also see this from $K_2 + U_2 = K_3 + U_3$:

$$U_3 = K_2 + U_2 - K_3 = 0.0360 \text{ J} + 0.0250 \text{ J} - 0 = 0.0610 \text{ J}$$

But $U_3 = \frac{1}{2}kx_3^2$, so

$$x_3 = \sqrt{\frac{2U_3}{k}} = \sqrt{\frac{2(0.0610 \text{ J})}{5.00 \text{ N/m}}} = 0.156 \text{ m}$$

The body moves an additional 0.056 m after the force is removed at $x = 0.100$ m.

EVALUATE: The total mechanical energy for the motion from point 2 to point 3 is 0.0610 J, the same as the work W_{other} done by the force $\vec{F}$ in Example 7.8. Is this just coincidence? Not at all; the system of glider and spring had zero mechanical energy initially (at point 1 in Example 7.9), so all the mechanical energy it has came from the work done by $\vec{F}$.

Example 7.11	Motion with gravitational, elastic, and friction forces

In a "worst-case" design scenario, a 2000-kg elevator with broken cables is falling at 25 m/s when it first contacts a cushioning spring at the bottom of the shaft. The spring is supposed to stop the elevator, compressing 3.00 m as it does so (Fig. 7.17). During the motion a safety clamp applies a constant 17,000-N frictional force to the elevator. As a design consultant, you are asked to determine what the force constant of the spring should be.

SOLUTION

Identify: We'll use the energy approach to determine the force constant, which appears in the expression for elastic potential energy. Note that this problem involves *both* gravitational and elastic potential energy. Furthermore, total mechanical energy is not conserved because the friction force does negative work W_{other} on the elevator.

SET UP: Since mechanical energy isn't conserved and more than one kind of potential energy is involved, we'll use the most general form of the energy relation, Eq. (7.15). We take point 1 as the position of the bottom of the elevator when it initially contacts the

7.17 The fall of an elevator is stopped by a spring and by a constant friction force.

spring, and take point 2 as its position when it is at rest. We choose the origin to be at point 1, so $y_1 = 0$ and $y_2 = -3.00$ m. With this choice the coordinate of the upper end of the spring is the same as the coordinate of the elevator, so the elastic potential energy at any point between point 1 and point 2 is $U_{\text{el}} = \frac{1}{2}ky^2$. (The gravitational potential energy is $U_{\text{grav}} = mgy$ as usual.) We know the initial and final speeds of the elevator and the magnitude of the friction force, so the only unknown is the force constant k (our target variable).

EXECUTE: The elevator's initial speed is $v_1 = 25$ m/s, so the initial kinetic energy is

$$K_1 = \frac{1}{2}mv_1^2 = \frac{1}{2}(2000 \text{ kg})(25 \text{ m/s})^2 = 625,000 \text{ J}$$

The elevator stops at point 2, so $K_2 = 0$. The potential energy at point 1, U_1, is zero; U_{grav} is zero because $y_1 = 0$, and $U_{\text{el}} = 0$ because the spring is not yet compressed. At point 2 there is both gravitational and elastic potential energy, so

$$U_2 = mgy_2 + \frac{1}{2}ky_2^2$$

The gravitational potential energy at point 2 is

$$mgy_2 = (2000 \text{ kg})(9.80 \text{ m/s}^2)(-3.00 \text{ m}) = -58,800 \text{ J}$$

The other force is the 17,000-N friction force, acting opposite to the 3.00-m displacement, so

$$W_{\text{other}} = -(17,000 \text{ N})(3.00 \text{ m}) = -51,000 \text{ J}$$

Putting these terms into $K_1 + U_1 + W_{\text{other}} = K_2 + U_2$, we have

$$K_1 + 0 + W_{\text{other}} = 0 + \left(mgy_2 + \frac{1}{2}ky_2^2\right)$$

so the force constant of the spring is

$$k = \frac{2(K_1 + W_{\text{other}} - mgy_2)}{y_2^2}$$

$$= \frac{2[625,000 \text{ J} + (-51,000 \text{ J}) - (-58,800 \text{ J})]}{(-3.00 \text{ m})^2}$$

$$= 1.41 \times 10^5 \text{ N/m}$$

This is comparable to the springs in an automobile suspension.

EVALUATE: Let's note what might seem to be a paradox in this problem. The elastic potential energy in the spring at point 2 is

$$\frac{1}{2}ky_2^2 = \frac{1}{2}(1.41 \times 10^5 \text{ N/m})(-3.00 \text{ m})^2 = 632,800 \text{ J}$$

This is *more* than the total mechanical energy at point 1,

$$E_1 = K_1 + U_1 = 625,000 \text{ J} + 0 = 625,000 \text{ J}$$

But the friction force caused the mechanical energy of the system to *decrease* by 51,000 J between point 1 and point 2. Does this mean that energy appeared from nowhere? Don't panic; there is no paradox. At point 2 there is also *negative* gravitational potential energy $mgy_2 = -58,800$ J because point 2 is below the origin. The total mechanical energy at point 2 is

$$E_2 = K_2 + U_2 = 0 + \frac{1}{2}ky_2^2 + mgy_2$$

$$= 632,800 \text{ J} + (-58,800 \text{ J}) = 574,000 \text{ J}$$

This is just the initial mechanical energy of 625,000 J, minus 51,000 J lost to friction.

Your next job as design consultant would be to tell your client that the elevator won't stay at the bottom of the shaft. Instead, it will bounce back up. The reason is that at point 2 the compressed spring exerts an upward force of magnitude $F_{\text{spring}} = (1.41 \times 10^5 \text{ N/m})(3.00 \text{ m}) = 422,000$ N. The weight of the elevator is only $w = mg = (2000 \text{ kg})(9.80 \text{ m/s}^2) = 19,600$ N, so the net force will be upward. The elevator bounces back upward even if the safety clamp now exerts a *downward* friction force of magnitude $f = 17,000$ N; the spring force is greater than the sum of f and mg. The elevator will bounce again and again until enough mechanical energy has been removed by friction for it to stop.

Can you also show that the acceleration of the falling elevator at point 2 is unacceptably high?

Test Your Understanding

First the value of y when the elevator in Example 7.11 finally comes to a stop, and find the values of K, U_{grav}, U_{el}, and E at this point. Assume that the safety clamp exerts an upward force of 17,000 N. (*Hint:* What is the relationship between the compression of the spring and the force it exerts on the elevator?) How does the mechanical energy at this point compare to the value at point 1?

7.3 | Conservative and Nonconservative Forces

In our discussions of potential energy we have talked about "storing" kinetic energy by converting it to potential energy. We always have in mind that later we may retrieve it again as kinetic energy. When you throw a ball up in the air, it slows down as kinetic energy is converted into potential energy. But on the way down, the conversion is reversed, and the ball speeds up as potential energy is converted back to kinetic energy. If there is no air resistance, the ball is moving just as fast when you catch it as when you threw it.

If a glider moving on a frictionless horizontal air track runs into a spring bumper at the end of the track, the spring compresses and the glider stops. But then it bounces back, and if there is no friction, the glider has the same speed and kinetic energy it had before the collision. Again, there is a two-way conversion from kinetic to potential energy and back. In both cases we find that we can define a potential-energy function so that the total mechanical energy, kinetic plus potential, is constant or *conserved* during the motion.

A force that offers this opportunity of two-way conversion between kinetic and potential energies is called a **conservative force.** We have seen two examples of conservative forces: the gravitational force and the spring force. An essential feature of conservative forces is that their work is always *reversible*. Anything that we deposit in the energy "bank" can later be withdrawn without loss. Another important aspect of conservative forces is that a body may move from point 1 to point 2 by various paths, but the work done by a conservative force is the same for

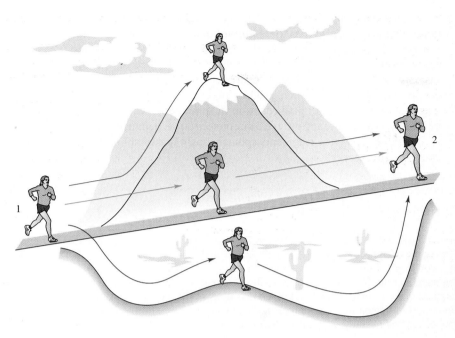

7.18 For any conservative force, the work done by that force depends only on the end points, not on the path taken. Thus the gravitational force, which is conservative, does the same work on the runner no matter what path he takes from point 1 to point 2.

all of these paths (Fig. 7.18). Thus, if a body stays close to the surface of the earth, the gravitational force $m\vec{g}$ is independent of height, and the work done by this force depends only on the change in height. If the body moves around a closed path, ending at the same point where it started, the *total* work done by the gravitational force is always zero.

The work done by a conservative force *always* has these properties:

1. It can always be expressed as the difference between the initial and final values of a *potential energy* function.
2. It is reversible.
3. It is independent of the path of the body and depends only on the starting and ending points.
4. When the starting and ending points are the same, the total work is zero.

When the only forces that do work are conservative forces, the total mechanical energy $E = K + U$ is constant.

Not all forces are conservative. Consider the friction force acting on the crate sliding on a ramp in Example 7.7 (Section 7.1). When the body slides up and then back down to the starting point, the total work done on it by the friction force is *not* zero. When the direction of motion reverses, so does the friction force, and friction does *negative* work in *both* directions. When a car with its brakes locked skids across the pavement with decreasing speed (and decreasing kinetic energy), the lost kinetic energy cannot be recovered by reversing the motion or in any other way, and mechanical energy is *not* conserved. There is no potential-energy function for the friction force.

In the same way, the force of fluid resistance (Section 5.3) is not conservative. If you throw a ball up in the air, air resistance does negative work on the ball while it's rising *and* while it's descending. The ball returns to your hand with less speed and less kinetic energy than when it left, and there is no way to get back the lost mechanical energy.

A force that is not conservative is called a **nonconservative force.** The work done by a nonconservative force *cannot* be represented by a potential-energy function. Some nonconservative forces, like kinetic friction or fluid resistance, cause mechanical energy to be lost or dissipated; a force of this kind is called a **dissipative force.** There are also nonconservative forces that *increase* mechanical energy. The fragments of an exploding firecracker fly off with very large kinetic energy, thanks to a chemical reaction of gunpowder with oxygen. The forces unleashed by this reaction are nonconservative because the process is not reversible. Imagine the fragments spontaneously reassembling themselves into a complete firecracker!

Example 7.12	**Frictional work depends on the path**

You are rearranging your furniture and wish to move a 40.0-kg futon 2.50 m across the room (Fig. 7.19). However, the straight-line path is blocked by a heavy coffee table that you don't want to move. Instead, you slide the futon in a dogleg path over the floor; the doglegs are 2.00 m and 1.50 m long. Compared to the straight-line path, how much more work must you do to push the futon in the dogleg path? The coefficient of kinetic friction is 0.200.

SOLUTION

IDENTIFY: Here work is done both by you and by the force of friction, so we must use the energy relation that includes forces other than elastic or gravitational forces. We'll use this relation to find a connection between the work that *you* do and the work done by *friction*.

7.19 A view from above of furniture being moved. How much more work must you do to move the futon on a dogleg path?

SET UP: Figure 7.19 shows the initial and final points. The futon is at rest at both point 1 and point 2, so $K_1 = K_2 = 0$. The gravitational potential energy does not change because the futon moves only horizontally; to be specific, we'll say $U_1 = U_2 = 0$. From Eq. (7.7) it follows that $W_{other} = 0$. The other work done on the futon is the sum of the positive work you do, W_{you}, and the negative work W_{fric} done by the kinetic friction force. Since the sum of these is zero, we have

$$W_{you} = -W_{fric}$$

Thus to determine W_{you}, we'll calculate the work done by friction.

EXECUTE: Because the floor is horizontal, the normal force on the futon equals its weight mg, and the magnitude of the friction force is $f_k = \mu_k n = \mu_k mg$. The work you must do over each path is then

$$\begin{aligned} W_{you} &= -W_{fric} = -(-f_k s) = +\mu_k mgs \\ &= (0.200)(40.0 \text{ kg})(9.80 \text{ m/s}^2)(2.50 \text{ m}) \\ &= 196 \text{ J} \quad (\text{straight-line path}) \end{aligned}$$

$$\begin{aligned} W_{you} &= -W_{fric} \\ &= (0.200)(40.0 \text{ kg})(9.80 \text{ m/s}^2)(2.00 \text{ m} + 1.50 \text{ m}) \\ &= 274 \text{ J} \quad (\text{dogleg path}) \end{aligned}$$

The extra work you must do is $274 \text{ J} - 196 \text{ J} = 78 \text{ J}$.

EVALUATE: The work done by friction is $W_{fric} = -W_{you} = -196 \text{ J}$ on the straight-line path, and -274 J on the dogleg. The work done by friction depends on the path taken, which illustrates that friction is a nonconservative force.

Example 7.13	**Conservative or nonconservative?**

In a certain region of space the force on an electron is $\vec{F} = Cx\hat{j}$, where C is a positive constant. The electron moves in a counterclockwise direction around a square loop in the xy-plane (Fig. 7.20). The corners of the square are at $(x, y) = (0, 0)$, $(L, 0)$, (L, L) and $(0, L)$. Calculate the work done on the electron by the force $\vec{F}$ during one complete trip around the square. Is this force conservative or nonconservative?

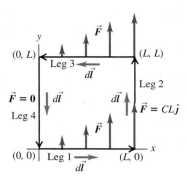

7.20 An electron moving around a square loop while being acted on by the force $\vec{F} = Cx\hat{j}$.

SOLUTION

IDENTIFY AND SET UP: In Example 7.12 the force of friction was constant in magnitude and always opposite to the displacement, so it was easy to calculate the work done. Here, however, the force $\vec{F}$ is not constant and in general is not in the same direction as the displacement. So we'll use the more general expression for work, Eq. (6.14):

$$W = \int_{P_1}^{P_2} \vec{F} \cdot d\vec{l}$$

where $d\vec{l}$ is an infinitesimal displacement. Let's calculate the work done by the force $\vec{F}$ on each leg of the square and then add the results to find the work done on the round trip.

EXECUTE: On the first leg, from $(0, 0)$ to $(L, 0)$, the force varies but is everywhere perpendicular to the displacement. So $\vec{F} \cdot d\vec{l} = 0$, and the work done on the first leg is $W_1 = 0$. The force has the same value $\vec{F} = CL\hat{j}$ everywhere on the second leg from $(L, 0)$ to (L, L). The displacement on this leg is in the $+y$-direction, so $d\vec{l} = dy\hat{j}$ and

$$\vec{F} \cdot d\vec{l} = CL\hat{j} \cdot dy\hat{j} = CL\,dy$$

The work done on the second leg is then

$$W_2 = \int_{(L, 0)}^{(L, L)} \vec{F} \cdot d\vec{l} = \int_{y=0}^{y=L} CL\,dy = CL \int_0^L dy = CL^2$$

On the third leg, from (L, L) to $(0, L)$, $\vec{F}$ is again perpendicular to the displacement so $W_3 = 0$. The force is zero on the final leg, from $(0, L)$ to $(0, 0)$, so no work is done and $W_4 = 0$. The work done by the force $\vec{F}$ on the round trip is

$$W = W_1 + W_2 + W_3 + W_4 = 0 + CL^2 + 0 + 0 = CL^2$$

The starting and ending points are the same, but the total work done by $\vec{F}$ is not zero. This is a nonconservative force; it cannot be represented by a potential energy function.

EVALUATE: Because W is positive, the mechanical energy of the electron *increases* as it goes around the loop. This is not a mathematical curiosity; it's a description of what happens in an electrical generating plant. A loop of wire is moved through a magnetic field, which gives rise to a nonconservative force similar to the one in this example. Electrons in the wire gain energy as they move around the loop, and this energy is carried via transmission lines to the consumer. (We'll discuss how this works in detail in Chapter 29.) All the electrical energy used in the home and in industry comes from work done by nonconservative forces!

How would the value of W change if the electron went around the loop clockwise instead of counterclockwise? The force $\vec{F}$ would be unaffected, but the direction of each infinitesimal displacement $d\vec{l}$ would reverse. Thus the sign of work would also reverse, and the work for a clockwise round trip would be $W = -CL^2$. This is a different behavior than the nonconservative friction force. When a body slides over a stationary surface with friction, the work done by friction is always negative, no matter what the direction of motion (see Example 7.7 in Section 7.1).

The Law of Conservation of Energy

Nonconservative forces cannot be represented in terms of potential energy. But we can describe the effects of these forces in terms of kinds of energy other than kinetic and potential energy. When a car with locked brakes skids to a stop, the tires and the road surface both become hotter. The energy associated with this change in the state of the materials is called **internal energy.** Raising the temperature of a body increases its internal energy; lowering the body's temperature decreases its internal energy.

To see the significance of internal energy, let's consider a block sliding on a rough surface. Friction does *negative* work on the block as it slides, and the change in internal energy of the block and surface (both of which get hotter) is *positive*. Careful experiments show that the increase in the internal energy is *exactly* equal to the absolute value of the work done by friction. In other words,

$$\Delta U_{\text{int}} = -W_{\text{other}}$$

7.21 When a liter of gasoline is burned in an automotive engine, it releases 3.3×10^7 J of internal energy. Hence $\Delta U_{\text{int}} = -3.3 \times 10^7$ J, where the minus sign means that the amount of energy stored in the gasoline has decreased. This energy can be converted into kinetic energy (making the car go faster) or into potential energy (enabling the car to climb uphill).

Activ
ONLINE
Physics

5.7   Modified Atwood Machine

where ΔU_{int} is the change in internal energy. If we substitute this into Eq. (7.7), (7.13), or (7.15), we find

$$K_1 + U_1 - \Delta U_{\text{int}} = K_2 + U_2$$

Writing $\Delta K = K_2 - K_1$ and $\Delta U = U_2 - U_1$, we can finally express this as

$$\Delta K + \Delta U + \Delta U_{\text{int}} = 0 \qquad \text{(law of conservation of energy)} \qquad (7.16)$$

This remarkable statement is the general form of the **law of conservation of energy.** In a given process, the kinetic energy, potential energy, and internal energy of a system may all change. But the *sum* of those changes is always zero. If there is a decrease in one form of energy, it is made up for by an increase in the other forms (Fig. 7.21). When we expand our definition of energy to include internal energy, Eq. (7.16) says that *energy is never created or destroyed; it only changes form.* No exception to this rule has ever been found.

Notice that the concept of work has been banished from Eq. (7.16). Instead, this equation invites us to think purely in terms of the conversion of energy from one form to another. When you throw a baseball straight up, you convert a portion of the internal energy of your molecules into kinetic energy of the baseball. This is converted into gravitational potential energy as the ball climbs and back to kinetic energy as the ball falls. If there is air resistance, part of the energy is used to heat up the air and the ball and increase their internal energy. Energy is converted back into the kinetic form as the ball falls. If you catch the ball in your hand, whatever energy was not lost to the air once again becomes internal energy; the ball and your hand are now warmer than they were at the beginning.

In a hydroelectric generating station, falling water is used to drive turbines ("water wheels"), which in turn run electric generators. Fundamentally, gravitational potential energy is released as the water falls, and the generating station converts this into electrical energy. Even if we know nothing about the details of how this is done, we can still use the law of conservation of energy, Eq. (7.16), to draw an important conclusion: the amount of electrical energy produced can be no greater than the amount of gravitational potential energy lost. (Because of friction, some of the potential energy goes into raising the temperatures of the water and the mechanism.)

In later chapters we will study the relation of internal energy to temperature changes, heat, and work. This is the heart of the area of physics called *thermodynamics.*

Example 7.14	**Work done by friction**

Let's look again at Example 7.6 in Section 7.1, in which your cousin Throcky skateboards down a curved ramp. He starts with zero kinetic energy and 735 J of potential energy, and at the bottom he has 450 J of kinetic energy and zero potential energy. So $\Delta K = +450$ J and $\Delta U = -735$ J. The work $W_{\text{other}} = W_{\text{fric}}$ done by the nonconservative friction forces is -285 J, so the change in internal energy is $\Delta U_{\text{int}} = -W_{\text{other}} = +285$ J. The wheels, the

bearings, and the ramp all get a little warmer as Throcky rolls down. In accordance with Eq. (7.16), the sum of the energy changes equals zero:

$$\Delta K + \Delta U + \Delta U_{\text{int}} = +450 \text{ J} + (-735 \text{ J}) + 285 \text{ J} = 0$$

The total energy of the system (including nonmechanical forms of energy) is conserved.

For the situation in Example 7.12, calculate the total amount of work you must do to move the futon from point 1 to point 2 along the 2.00-m and 1.50-m doglegs, then back to point 1 via the straight-line path shown as a dashed line in Fig. 7.19. (Assume that the coffee table was first moved out of the way.) Since the futon ended up back where it started, why isn't the total work you do on the futon equal to zero?

7.4 | Force and Potential Energy

For the two kinds of conservative forces (gravitational and elastic) we have studied, we started with a description of the behavior of the force and derived from that an expression for the potential energy. For example, for a body with mass m in a uniform gravitational field, the gravitational force is $F_y = -mg$. We found that the corresponding potential energy is $U(y) = mgy$. To stretch an ideal spring by a distance x, we exert a force equal to $+kx$. By Newton's third law the force that an ideal spring exerts on a body is opposite this, or $F_x = -kx$. The corresponding potential energy function is $U(x) = \frac{1}{2}kx^2$.

In studying physics, however, you'll encounter situations in which you are given an expression for the *potential energy* as a function of position and have to find the corresponding *force*. We'll see several examples of this kind when we study electric forces later in this book: it's often far easier to calculate the electric potential energy first, then determine the corresponding electric force afterward.

Here's how we find the force that corresponds to a given potential-energy expression. First let's consider motion along a straight line, with coordinate x. We denote the x-component of force, a function of x, by $F_x(x)$, and the potential energy as $U(x)$. This notation reminds us that both F_x and U are *functions* of x. Now we recall that in any displacement, the work W done by a conservative force equals the negative of the change ΔU in potential energy:

$$W = -\Delta U$$

Let's apply this to a small displacement Δx. The work done by the force $F_x(x)$ during this displacement is approximately equal to $F_x(x)\Delta x$. We have to say "approximately" because $F_x(x)$ may vary a little over the interval Δx. But it is at least approximately true that

$$F_x(x)\Delta x = -\Delta U \quad \text{and} \quad F_x(x) = -\frac{\Delta U}{\Delta x}$$

You can probably see what's coming. We take the limit as $\Delta x \to 0$; in this limit, the variation of F_x becomes negligible, and we have the exact relation

$$F_x(x) = -\frac{dU(x)}{dx} \quad \text{(force from potential energy, one dimension)} \quad (7.17)$$

This result makes sense; in regions where $U(x)$ changes most rapidly with x (that is, where $dU(x)/dx$ is large), the greatest amount of work is done during a given displacement, and this corresponds to a large force magnitude. Also, when $F_x(x)$ is in the positive x-direction, $U(x)$ *decreases* with increasing x. So $F_x(x)$ and $dU(x)/dx$

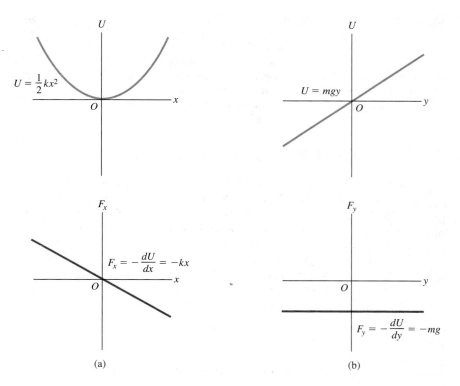

7.22 Graphs of potential energy and force versus position for (a) the spring force and (b) the gravitational force. In each case the force is the negative derivative of the potential energy.

should indeed have opposite signs. The physical meaning of Eq. (7.17) is that *a conservative force always acts to push the system toward lower potential energy.*

As a check, let's consider the function for elastic potential energy, $U(x) = \frac{1}{2}kx^2$. Substituting this into Eq. (7.17) yields

$$F_x(x) = -\frac{d}{dx}\left(\frac{1}{2}kx^2\right) = -kx$$

which is the correct expression for the force exerted by an ideal spring (Fig. 7.22a). Similarly, for gravitational potential energy we have $U(y) = mgy$; taking care to change x to y for the choice of axis, we get $F_y = -dU/dy = -d(mgy)/dy = -mg$, which is the correct expression for gravitational force (Fig. 7.22b).

Example 7.15   An electric force and its potential energy

An electrically charged particle is held at rest at the point $x = 0$, while a second particle with equal charge is free to move along the positive x-axis. The potential energy of the system is

$$U(x) = \frac{C}{x}$$

where C is a positive constant that depends on the magnitude of the charges. Derive an expression for the x-component of force acting on the movable charge, as a function of its position.

SOLUTION

IDENTIFY AND SET UP: We are given the potential-energy function $U(x)$. Hence we can use Eq. (7.17) to find our target, the function $F_x(x)$.

EXECUTE: The derivative with respect to x of the function $1/x$ is $-1/x^2$. So the force on the movable charge for $x > 0$ is

$$F_x(x) = -\frac{dU(x)}{dx} = -C\left(-\frac{1}{x^2}\right) = \frac{C}{x^2}$$

EVALUATE: The x-component of force is positive, corresponding to a repulsive interaction between like electric charges. The potential energy is very large for small x and approaches zero as x becomes large; the force pushes the movable charge toward large positive values of x, for which the potential energy is less. The force varies as $1/x^2$; it is small when the particles are far apart (large x) but becomes large when the particles are close together (small x). This is an example of Coulomb's law for electric interactions, which we will study in greater detail in Chapter 21.

Force and Potential Energy in Three Dimensions

We can extend this analysis to three dimensions, where the particle may move in the x-, y-, or z-direction, or all at once, under the action of a conservative force that has components F_x, F_y, and F_z. Each component of force may be a function of the coordinates x, y, and z. The potential-energy function U is also a function of all three space coordinates. We can now use Eq. (7.17) to find each component of force. The potential-energy change ΔU when the particle moves a small distance Δx in the x-direction is again given by $-F_x \Delta x$; it doesn't depend on F_y and F_z, which represent force components that are perpendicular to the displacement and do no work. So we again have the approximate relation

$$F_x = -\frac{\Delta U}{\Delta x}$$

The y- and z-components of force are determined in exactly the same way:

$$F_y = -\frac{\Delta U}{\Delta y} \qquad F_z = -\frac{\Delta U}{\Delta z}$$

To make these relations exact, we need to take the limits $\Delta x \to 0$, $\Delta y \to 0$, and $\Delta z \to 0$ so that these ratios become derivatives. Because U may be a function of all three coordinates, we need to remember that when we calculate each of these derivatives, only one coordinate changes at a time. We compute the derivative of U with respect to x by assuming that y and z are constant and only x varies, and so on. Such a derivative is called a *partial derivative*. The usual notation is $\partial U / \partial x$ and so on; the symbol ∂ is a modified d to remind us of the nature of this operation. So we write

$$F_x = -\frac{\partial U}{\partial x} \qquad F_y = -\frac{\partial U}{\partial y} \qquad F_z = -\frac{\partial U}{\partial z} \qquad (7.18)$$
$$\text{(force from potential energy)}$$

We can use unit vectors to write a single compact vector expression for the force $\vec{F}$:

$$\vec{F} = -\left(\frac{\partial U}{\partial x}\hat{\imath} + \frac{\partial U}{\partial y}\hat{\jmath} + \frac{\partial U}{\partial z}\hat{k}\right) \quad \text{(force from potential energy)} \quad (7.19)$$

The expression inside the parentheses represents a particular operation on the function U, in which we take the partial derivative of U with respect to each coordinate, multiply by the corresponding unit vector, and take the vector sum. This operation is called the **gradient** of U and is often abbreviated as $\vec{\nabla}U$. Thus the force is the negative of the gradient of the potential-energy function:

$$\vec{F} = -\vec{\nabla}U \qquad (7.20)$$

As a check, let's substitute into Eq. (7.20) the function $U = mgy$ for gravitational potential energy:

$$\vec{F} = -\vec{\nabla}(mgy) = -\left(\frac{\partial(mgy)}{\partial x}\hat{i} + \frac{\partial(mgy)}{\partial y}\hat{j} + \frac{\partial(mgy)}{\partial z}\hat{k}\right) = (-mg)\hat{j}$$

This is just the familiar expression for the gravitational force.

Example 7.16

Force and potential energy in two dimensions

A puck slides on a level, frictionless air-hockey table. The coordinates of the puck are x and y. It is acted on by a conservative force described by the potential-energy function

$$U(x, y) = \frac{1}{2}k(x^2 + y^2)$$

Derive an expression for the force acting on the puck, and find an expression for the magnitude of the force as a function of position.

SOLUTION

IDENTIFY AND SET UP: We'll find the components of the force from the function $U(x, y)$ using Eq. (7.18), then determine the magnitude of the force using the formula for the magnitude of a vector: $F = \sqrt{F_x^2 + F_y^2}$.

EXECUTE: The components of the force are

$$F_x = -\frac{\partial U}{\partial x} = -kx \qquad F_y = -\frac{\partial U}{\partial y} = -ky$$

From Eq. (7.19) this corresponds to the vector expression

$$\vec{F} = -k(x\hat{i} + y\hat{j})$$

Now $x\hat{i} + y\hat{j}$ is just the position vector $\vec{r}$ of the particle, so we can rewrite this expression as $\vec{F} = -k\vec{r}$. This represents a force that at

each point is opposite in direction to the position vector of the point—that is, a force that at each point is directed toward the origin. The potential energy is minimum at the origin, so again the force pushes in the direction of decreasing potential energy.

The *magnitude* of the force at any point is

$$F = \sqrt{(-kx)^2 + (-ky)^2} = k\sqrt{x^2 + y^2} = kr$$

where r is the particle's distance from the origin. This is the force exerted by a spring that obeys Hooke's law and has a negligibly small length (compared to the other distances in the problem) when it is not stretched. So the motion of the puck is the same as if it were attached to one end of an ideal spring of negligible unstretched length; the other end is attached to the air-hockey table at the origin.

EVALUATE: We can check our result by noting that the potential-energy function can also be expressed as $U = \frac{1}{2}kr^2$. Written this way, U is a function of a single coordinate r, so we can find the force using Eq. (7.17) with x replaced by r:

$$F_r = -\frac{dU}{dr} = -\frac{d}{dr}\left(\frac{1}{2}kr^2\right) = -kr$$

Just as we calculated above, the force has magnitude kr; the minus sign indicates that the force is radially inward (toward the origin).

Test Your Understanding

A particle moving along the x-axis is acted on by a conservative force F_x. At a certain point, the force is zero. What does this tell you about the value of the potential-energy function $U(x)$ at that point? What about the value of the derivative of $U(x)$ at that point?

7.5 | Energy Diagrams

When a particle moves along a straight line under the action of a conservative force, we can get a lot of insight into the possible motions by looking at the graph of the potential-energy function $U(x)$. Figure 7.23a shows a glider with mass m that moves along the x-axis on an air track. The spring exerts on the glider a force with x-component $F_x = -kx$. Figure 7.23b is a graph of the corresponding potential-energy function $U(x) = \frac{1}{2}kx^2$. If the elastic force of the spring is the *only* horizontal force acting on the glider, the total mechanical energy $E = K + U$ is constant, independent of x. A graph of E as a function of x is thus a straight horizontal line.

The vertical distance between the U and E graphs at each point represents the difference $E - U$, equal to the kinetic energy K at that point. We see that K is greatest at $x = 0$. It is zero at the values of x where the two graphs cross, labeled A and $-A$ in the diagram. Thus the speed v is greatest at $x = 0$, and it is zero at $x = \pm A$, the points of *maximum* possible displacement from $x = 0$ for a given value of the total energy E. The potential energy U can never be greater than the total energy E; if it were, K would be negative, and that's impossible. The motion is a back-and-forth oscillation between the points $x = A$ and $x = -A$.

At each point, the force F_x on the glider is equal to the negative of the slope of the $U(x)$ curve: $F_x = -dU/dx$. When the particle is at $x = 0$, the slope and the force are zero, so this is an *equilibrium* position. When x is positive, the slope of the $U(x)$ curve is positive and the force F_x is negative, directed toward the origin. When x is negative, the slope is negative and F_x is positive, again toward the origin. Such a force is sometimes called a *restoring force;* when the glider is displaced to either side of $x = 0$, the resulting force tends to "restore" it back to $x = 0$. An analogous situation is a marble rolling around in a round-bottomed salad bowl. We say that $x = 0$ is a point of **stable equilibrium.** More generally, *any minimum in a potential-energy curve is a stable equilibrium position.*

Figure 7.24a shows a hypothetical but more general potential-energy function $U(x)$. Figure 7.24b shows the corresponding force $F_x = -dU/dx$. Points x_1 and x_3 are stable equilibrium points. At each of these points, F_x is zero because the slope

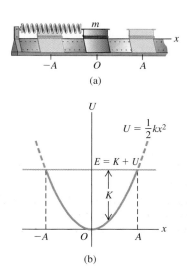

7.23 (a) A glider on an air track. The spring exerts a force $F_x = -kx$. (b) The potential-energy function. The limits of the motion are the points where the U curve intersects the horizontal line representing the total mechanical energy E.

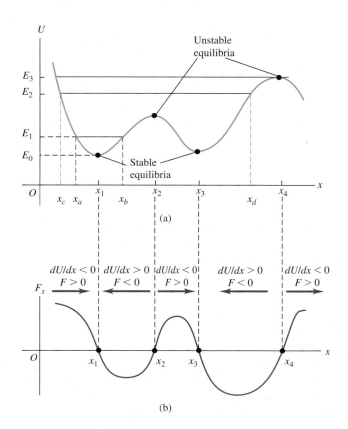

7.24 (a) A hypothetical potential-energy function $U(x)$. (b) The corresponding force $F_x = -dU/dx$. Maxima and minima of $U(x)$ correspond to points where $F_x = 0$.

of the $U(x)$ curve is zero. When the particle is displaced to either side, the force pushes back toward the equilibrium point. The slope of the $U(x)$ curve is also zero at points x_2 and x_4, and these are also equilibrium points. But when the particle is displaced a little to the right of either point, the slope of the $U(x)$ curve becomes negative, corresponding to a positive F_x that tends to push the particle still farther from the point. When the particle is displaced a little to the left, F_x is negative, again pushing away from equilibrium. This is analogous to a marble rolling on the top of a bowling ball. Points x_2 and x_4 are called **unstable equilibrium** points; *any maximum in a potential-energy curve is an unstable equilibrium position.*

> **CAUTION** The direction of the force on a body is *not* determined by the sign of the potential energy U. Rather, it's the sign of $F_x = -dU/dx$ that matters. As we discussed in Section 7.1, the physically significant quantity is the *difference* in the value of U between two points, which is just what the derivative $F_x = -dU/dx$ measures. This means that you can always add a constant to the potential-energy function without changing the physics of the situation.

If the total energy is E_1 and the particle is initially near x_1, it can move only in the region between x_a and x_b determined by the intersection of the E_1 and U graphs (Fig. 7.24a). Again, U cannot be greater than E_1 because K can't be negative. We speak of the particle as moving in a *potential well*, and x_a and x_b are the *turning points* of the particle's motion (since at these points, the particle stops and reverses direction). If we increase the total energy to the level E_2, the particle can move over a wider range, from x_c to x_d. If the total energy is greater than E_3, the particle can "escape" and move to indefinitely large values of x. At the other extreme, E_0 represents the least possible total energy the system can have.

Test Your Understanding

The potential-energy function for a particle moving along the positive x-axis is $U(x) = (0.600 \text{ N})x + (2.40 \text{ N} \cdot \text{m}^2)/x$. Where is the point of equilibrium? (Just consider positive values of x.) What is the potential energy at that point? What is the force at that point?

SUMMARY

The work done on a particle by a constant gravitational force can be represented as a change in the gravitational potential energy $U = mgy$. This energy is a shared property of the particle and the earth.

$$W_{grav} = mgy_1 - mgy_2$$
$$= U_1 - U_2 = -\Delta U$$
(7.1), (7.3)

An ideal stretched or compressed spring exerts an elastic force $F_x = -kx$ on a particle, where x is the amount of stretch or compression. The work done by this force can be represented as a change in the elastic potential energy of the spring, $U = \frac{1}{2}kx^2$.

$$W_{el} = \frac{1}{2}kx_1^2 - \frac{1}{2}kx_2^2$$
$$= U_1 - U_2 = -\Delta U$$
(7.10)

The total potential energy is the sum of the gravitational and elastic potential energy. If no forces other than the gravitational and elastic forces do work on a particle, the sum of kinetic and potential energy is conserved. This sum $E = K + U$ is called the total mechanical energy. (See Examples 7.1, 7.3, 7.4, 7.5, 7.8, and 7.10)

$$K_1 + U_1 = K_2 + U_2$$
(7.11)

When forces other than the gravitational and elastic forces do work on a particle, the work W_{other} done by these other forces equals the change in total mechanical energy (kinetic energy plus total potential energy) (See Examples 7.2, 7.6, 7.7, 7.9, and 7.11)

$$K_1 + U_1 + W_{other} = K_2 + U_2$$
(7.13)

All forces are either conservative or nonconservative. A conservative force is one for which the work-kinetic energy relation is completely reversible. The work of a conservative force can always be represented by a potential-energy function, but the work of a nonconservative force cannot.

The work done by nonconservative forces manifests itself as changes in the internal energy of bodies. The sum of kinetic, potential, and internal energy is always conserved.
(See Examples 7.12 through 7.14)

$$\Delta K + \Delta U + \Delta U_{\text{int}} = 0 \qquad (7.16)$$

For motion along a straight line, a conservative force $F_x(x)$ is the negative derivative of its associated potential-energy function U. In three dimensions, the components of a conservative force are negative partial derivatives of U.
(See Examples 7.15 and 7.16)

$$F_x(x) = -\frac{dU(x)}{dx} \qquad (7.17)$$

$$F_x = -\frac{\partial U}{\partial x}, \quad F_y = -\frac{\partial U}{\partial y} \quad F_z = -\frac{\partial U}{\partial y} \qquad (7.18)$$

$$\vec{F} = -\left(\frac{\partial U}{\partial x}\hat{i} + \frac{\partial U}{\partial y}\hat{j} + \frac{\partial U}{\partial z}\hat{k}\right) \qquad (7.19)$$

Key Terms

Your Notes

Answer to Chapter Opening Question

The water is doing negative work on the diver; it exerts an upward force of fluid resistance as he moves downward. Gravity is doing positive work, since this force is in the same downward direction as the diver's displacement.

Answers to Test Your Understanding Questions

Section 7.1 The initial kinetic energy $K_1 = 0$, the initial potential energy $U_1 = mgy_1$, and the final potential energy $U_2 = mgy_2$ are the same for both ramps. Mechanical energy is conserved in both cases, so the final kinetic energy $K = \frac{1}{2}mv_2^2$ is also the same for both ramps. Hence the speed at the right-hand end is the *same* in both cases!

Section 7.2 When the elevator finally comes to a stop, the net force on it is zero. Its weight is $(2000 \text{ kg})(9.8 \text{ m/s}^2) = 19{,}600 \text{ N}$, and friction provides an upward force of 17,000 N. Hence the spring must provide an additional 2600 N of upward force. Using the relationship $(\text{force}) = (\text{force constant})(\text{compression})$ for a spring, we find that the spring is compressed by $(2600 \text{ N})/(1.41 \times 10^5 \text{ N/m}) = 0.018 \text{ m}$, or 1.8 cm. Hence $y = -0.018$ m. At this point $K = 0$, $U_{\text{grav}} = mgy = (2000 \text{ kg})(9.8 \text{ m/s}^2)(-0.018 \text{ m}) = -360 \text{ J}$, $U_{\text{el}} = \frac{1}{2}ky^2 = \frac{1}{2}(1.41 \times 10^5 \text{ N/m})(-0.018 \text{ m})^2 = 23 \text{ J}$, and $E = K + U_{\text{grav}} + U_{\text{el}} = -337 \text{ J}$. The initial mechanical energy of 625,000 J was completely lost to friction, plus another 337 J.

Section 7.3 The total work *you* would do is 274 J + 196 J = 470 J. The *friction* force would do -470 J of work, so the total work done by *all* forces is, in fact, zero: 470 J $-$ 470 J = 0. The moral is that it's important to include *all* work done by forces other than gravity or spring forces.

Section 7.4 If $F_x = 0$ at a point, then the derivative of $U(x)$ is zero at that point. However, this tells us absolutely nothing about the *value* of $U(x)$ at that point.

Section 7.5 Equilibrium occurs where $F_x = 0$. The force is

$$F_x = -dU/dx = -0.600 \text{ N} + (2.40 \text{ N} \cdot \text{m}^2)/x^2$$

Solving for the value of x where $F_x = 0$ (the equilibrium position) gives $x^2 = (2.40 \text{ N} \cdot \text{m}^2)/(0.600 \text{ N}) = 4.00 \text{ m}^2$, or $x = 2.00$ m. At that point $U = (0.600 \text{ N})(2.00 \text{ m}) + (2.40 \text{ N} \cdot \text{m}^2)/(2.00 \text{ m}) = 2.40 \text{ N} \cdot \text{m} = 2.40 \text{ J}$; the force is zero.

Discussion Questions

Q7.1 A baseball is thrown straight up with initial speed v_1, as in Example 7.1 (Section 7.1). If air resistance cannot be ignored when the ball returns to its initial height its speed is less than v_1. Explain, using energy concepts.

Q7.2 In Fig. 7.6 the projectile has the same initial kinetic energy in each case. Why does it not rise to the same maximum height in each case?

Q7.3 In Example 7.5 (Section 7.1), does Throckmorton's speed at the bottom depend on the shape of the ramp or just the difference in height of points 1 and 2? Explain. Answer this same question if the ramp is not frictionless, as in Example 7.6 (Section 7.1).

Q7.4 An egg is released from rest at the roof of a building and falls to the ground. Its fall is observed by a student on the roof of the building, who uses coordinates with origin at the roof and by a student on the ground, who uses coordinates with origin at the ground. Do the two students assign the same or different values to the initial gravitational potential energy, the final gravitational potential energy, the change in gravitational potential energy, and the kinetic energy of the egg just before it strikes the ground? Explain.

Q7.5 A physics teacher had a bowling ball suspended from a very long rope attached to the high ceiling of a large lecture hall. To illustrate his faith in conservation of energy, he would back up to one side of the stage, pull the ball far to one side until the taut rope brought it just to the end of his nose, and then release it. The massive ball would swing in a mighty arc across the stage and then return to stop momentarily just in front of the nose of the stationary, unflinching teacher. However, one day after the demonstration he looked up just in time to see a student at the other side of the stage *push* the ball away from his nose as he tried to duplicate the demonstration. Tell the rest of the story and explain the reason for the potentially tragic outcome.

Q7.6 When a returning space shuttle touches down on the runway, it has lost almost all the kinetic energy it had in orbit. The gravitational potential energy has also decreased considerably. Where did all that energy go?

Q7.7 Is it possible for a frictional force to *increase* the mechanical energy of a system? If so, give examples.

Q7.8 A woman bounces on a trampoline, going a little higher with each bounce. Explain how she increases the total mechanical energy.

Q7.9 A box slides down a ramp and work is done on the box by the forces of gravity and friction. Can the work of each of these forces be expressed in terms of the change in a potential energy function? For each force explain why or why not.

Q7.10 A compressed spring is clamped in its compressed position and then is dissolved in acid. What becomes of its potential energy?

Q7.11 When an object moves away from the earth the gravitational potential energy increases; when it moves closer the potential energy decreases. But the potential energy for a spring increases both when the spring is stretched and when it is compressed. Explain the reason for the difference in behavior of these two potential energies.

Q7.12 Since only changes in potential energy are important in any problem, a student decides to let the elastic potential energy of a spring be zero when the spring is stretched a distance x_1. The student decides, therefore, to let $U = \frac{1}{2}k(x - x_1)^2$. Is this correct? Explain.

Q7.13 Figure 7.22a shows the potential energy function for the force $F_x = -kx$. Sketch the potential energy function for the force $F_x = +kx$. For this force, is $x = 0$ a point of equilibrium? Is this equilibrium stable or unstable? Explain.

Q7.14 Figure 7.22b shows the potential energy function associated with the gravitational force between an object and the earth. Use this graph to explain why objects always fall toward the earth when they are released.

Q7.15 For a system of two particles we often let the potential energy for the force between the particles approach zero as the separation of the particles approaches infinity. If this choice is made, explain why the potential energy at noninfinite separation is positive if the particles repel one another and negative if they attract.

Q7.16 Explain why the points $x = A$ and $x = -A$ in Fig. 7.23b are called *turning points*. How are the values of E and U related at a turning point?

Q7.17 A particle is in *neutral equilibrium* if the net force on it is zero and remains zero if the particle is displaced slightly in any direction. Sketch the potential energy function near a point of neutral equilibrium, for the case of one-dimensional motion. Give an example of an object in neutral equilibrium.

Q7.18 The net force on a particle of mass m has the potential-energy function graphed in Figure 7.24a. If the total energy is E_1, graph the speed v of the particle versus its position x. At what value of x is the speed greatest? Sketch v versus x if the total energy is E_2.

Q7.19 A ball moves from point 1 to point 2. The gravitational potential energy at point 2 is *greater* than at point 1. When the ball moves from point 1 to point 2, does gravity do positive or negative work?

Q7.20 The potential energy function for a force $\vec{F}$ is $U = \alpha x^3$, where α is a positive constant. What is the direction of $\vec{F}$?

Exercises

Section 7.1 Gravitational Potential Energy

7.1 What is the potential energy for an 800-kg elevator at the top of Chicago's Sears Tower, 440 m above street level? Let the potential energy be zero at street level.

7.2 A 5.00-kg sack of flour is lifted vertically at a constant speed of 3.50 m/s through a height of 15.0 m. a) How great a force is required? b) How much work is done on the sack by the lifting force? What becomes of this work?

7.3 Repeat part (a) of Example 6.5 (Section 6.2) using Eq. (7.7).

7.4 A mail bag with a mass of 120 kg is suspended by a vertical rope 6.0 m long. a) What horizontal force is necessary to hold the bag in a position displaced sideways 3.0 m from its initial position? b) How much work is done by the worker in moving the bag to this position?

7.5 A baseball is thrown from the roof of a 22.0-m-tall building with an initial velocity of magnitude 12.0 m/s and directed at an angle of 53.1° above the horizontal. a) What is the speed of the ball just before it strikes the ground? Use energy methods and ignore air resistance. b) What is the answer for part (a) if the initial velocity is at an angle of 53.1° *below* the horizontal? c) If the effects of air resistance are included, will part (a) or (b) give the higher speed?

7.6 a) In Example 7.7 (Section 7.1), calculate the minimum initial speed required for the crate to reach the top of the ramp. b) If the initial speed of the crate in Example 7.7 is 11.0 m/s, what is its speed at the top of the ramp?

7.7 Answer part (b) of Example 7.7 (Section 7.1) by applying Eq. (7.7) to points 2 and 3, rather than to points 1 and 3 as was done in the example.

7.8 An empty crate slides down a ramp, starting with a speed v_0 and reaching the bottom with a speed v and kinetic energy K. Some books are placed in the box, so that the total mass is quadrupled. The coefficient of kinetic friction is constant and air resistance is negligible. Starting again with v_0 at the top of the ramp, what are the speed and kinetic energy at the bottom? Explain the reasoning behind your answers.

7.9 A small rock with mass 0.20 kg is released from rest at point A, which is at the top edge of a large, hemispherical bowl with radius $R = 0.50$ m (Fig. 7.25). Assume that the size of the rock is small compared to R, so that the rock can be treated as a particle, and assume that the rock slides rather than rolling. The work done by friction on the rock when it moves from point A to point B at the bottom of the bowl is -0.22 J. What is the speed of the rock when it reaches point B?

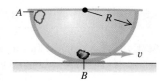

Figure 7.25 Exercise 7.9.

7.10 **Jumping Flea.** High-speed motion pictures (3500 frames/second) of a jumping 210 μg flea yielded the data to plot the graph shown in Fig. 7.26 (see "The Flying Leap of the Flea," by M. Rothschild, Y. Schlein, K. Parker, C. Neville, and S. Sternberg in the Nov. 1973 *Scientific American*). This flea was about 2 mm long and jumped at a nearly vertical take-off angle. a) Find the maximum height above the ground reached by the flea. (Can you ignore the distance traveled by the flea before its speed becomes constant? Explain.) b) How much work was done by the flea in launching itself into the air?

Figure 7.26 Exercise 7.10.

7.11 You are testing a new amusement park roller coaster with an empty car with mass 120 kg. One part of the track is a vertical loop with radius 12.0 m. At the bottom of the loop (point A) the car has speed 25.0 m/s and at the top of the loop (point B) has speed 8.0 m/s. As the car rolls from point A to point B, how much work is done by friction?

7.12 **Tarzan and Jane.** Tarzan, in one tree, sights Jane in another tree. He grabs the end of a vine with length 20 m that makes an angle of 45° with the vertical, steps off his tree limb, and swings down and then up to Jane's open arms. When he arrives, his vine makes an angle of 30° with the vertical. Determine whether he gives her a tender embrace or knocks her off her limb by calculating Tarzan's speed just before he reaches Jane. You can ignore air resistance and the mass of the vine.

7.13 A 10.0-kg microwave oven is pushed 8.00 m up the sloping surface of a loading ramp inclined at an angle of 36.9° above the horizontal, by a constant force $\vec{F}$ with a magnitude 110 N and acting parallel to the ramp. The coefficient of kinetic friction between the oven and the ramp is 0.250. a) What is the work done on the oven by the force $\vec{F}$? b) What is the work done on the oven by the friction force? c) Compute the increase in potential energy for the oven. d) Use your answers to parts (a), (b), and (c) to calculate the increase in the oven's kinetic energy. e) Use $\sum \vec{F} = m\vec{a}$ to calculate the acceleration of the oven. Assuming that the oven is initially at rest, use the acceleration to calculate the oven's speed after traveling 8.00 m. From this, compute the increase in the oven's kinetic energy, and compare it to the answer you got in part (d).

7.14 Pendulum. A small rock with mass 0.12 kg is fastened to a massless string with length 0.80 m to form a pendulum. The pendulum is swinging so as to make a maximum angle of 45° with the vertical. Air resistance is negligible. a) What is the speed of the rock when the string passes through the vertical position? b) What is the tension in the string when it makes an angle of 45° with the vertical? c) What is the tension in the string as it passes through the vertical?

Section 7.2 Elastic Potential Energy

7.15 A force of 800 N stretches a certain spring a distance of 0.200 m. a) What is the potential energy of the spring when it is stretched 0.200 m? b) What is its potential energy when it is compressed 5.00 cm?

7.16 A force of 720 N stretches a certain spring a distance of 0.150 m. What is the potential energy of the spring when a 60.0-kg mass hangs vertically from it?

7.17 A spring of negligible mass has force constant $k = 1600$ N/m. a) How far must the spring be compressed for 3.20 J of potential energy to be stored in it? b) You place the spring vertically with one end on the floor. You then drop a 1.20-kg book onto it from a height of 0.80 m above the top of the spring. Find the maximum distance the spring will be compressed.

7.18 A slingshot will shoot a 10-g pebble 22.0 m straight up. a) How much potential energy is stored in the slingshot's rubber band? b) With the same potential energy stored in the rubber band, how high can the slingshot shoot a 25-g pebble? c) What physical effects did you ignore in solving this problem?

7.19 A 1.20-kg piece of cheese is placed on a vertical spring of negligible mass and force constant $k = 1800$ N/m that is compressed 15.0 cm. When the spring is released, how high does the cheese rise from this initial position? (The cheese and the spring are *not* attached.)

7.20 Consider the glider of Example 7.8 (Section 7.2) and Fig. 7.16. As in the example, the glider is released from rest with the spring stretched 0.100 m. What is the displacement x of the glider from its equilibrium position when its speed is 0.20 m/s? (You should get more than one answer. Explain why.)

7.21 Consider the glider of Example 7.8 (Section 7.2) and Fig. 7.16. a) As in the example, the glider is released from rest with the spring stretched 0.100 m. What is the speed of the glider when it returns to $x = 0$? b) What must the initial displacement of the glider be if its maximum speed in the subsequent motion is to be 2.50 m/s?

7.22 Consider the glider of Example 7.8 (Section 7.2) and Fig. 7.16. As in the example, the glider is released from rest with the spring stretched 0.100 m. But now the air track is turned off so that there is a kinetic friction force acting on the glider. a) If the coefficient of kinetic friction between the air track and the glider is $\mu_k = 0.050$, what is the speed of the glider when it has traveled 0.020 m, so that $x = 0.080$ m? b) If $\mu_k = 0.050$, what is the speed

of the glider when it has traveled 0.100 m, so that $x = 0$? c) What must the value of μ_k be so that the glider reaches the $x = 0$ position with zero speed?

7.23 a) For the elevator of Example 7.11 (Section 7.2), what is the speed of the elevator after it has moved downward 1.00 m from point 1 in Fig. 7.17? b) When the elevator is 1.00 m below point 1 in Fig. 7.17, what is its acceleration?

7.24 You are asked to design a spring that will give a 1160-kg satellite a speed of 2.50 m/s relative to an orbiting space shuttle. Your spring is to give the satellite a maximum acceleration of $5.00g$. The spring's mass, the recoil kinetic energy of the shuttle, and changes in gravitational potential energy will all be negligible. a) What must the force constant of the spring be? b) What distance must the spring be compressed?

Section 7.3 Conservative and Nonconservative Forces

7.25 A 0.75-kg book moves vertically upward a distance of 16 m, and then vertically downward 16 m, returning to its initial position. a) How much work is done by gravity during the upward motion of the book? b) How much work is done by gravity during the downward motion of the book? c) What is the total work done on the book by gravity during the complete round trip? d) On the basis of your answer to part (c), would you say that the gravitational force is conservative or nonconservative? Explain.

7.26 A 0.050-kg rock moves from the origin to the point (3.0 m, 5.0 m) in a coordinate system in which the positive y-direction is upward. a) The rock first moves horizontally from the origin to the point (3.0 m, 0). Then it moves vertically from (3.0 m, 0) to (3.0 m, 5.0 m). Sketch the path of the rock in the xy-plane. How much work is done on the rock by gravity during the displacement? b) Instead of the path in part (a), let the rock first move vertically from the origin to (0, 5.0 m) and then horizontally from (0, 5.0 m) to (3.0 m, 5.0 m). Sketch the path of the rock in the xy-plane. How much work is done on the rock by gravity during the displacement? c) Compare the answers to parts (a) and (b). From your results would you say that the gravitational force is conservative or nonconservative? Explain.

7.27 In an experiment, one of the forces exerted on a proton is $\vec{F} = -\alpha x^2 \hat{\imath}$, where $\alpha = 12$ N/m^2. a) How much work does $\vec{F}$ do when the proton moves along the straight-line path from the point (0.10 m, 0) to the point (0.10 m, 0.40 m)? b) Along the straight-line path from the point (0.10 m, 0) to the point (0.30 m, 0)? c) Along the straight-line path from the point (0.30 m, 0) to the point (0.10 m, 0)? d) Is the force $\vec{F}$ conservative? Explain. If $\vec{F}$ is conservative, what is the potential energy function for it? Let $U = 0$ when $x = 0$.

7.28 Consider the electron and force $\vec{F}$ of Example 7.13 (Section 7.3). a) The electron travels from the point (0, 0) to the point (L, L) along the straight line from (0, 0) to (0, L) and then along the straight line from (0, L) to (L, L). What is the work done by the force $\vec{F}$ during this displacement? b) The electron travels from the point (0, 0) to the point (L, L) along the straight line from (0, 0) to (L, 0) and then along the straight line from (L, 0) to (L, L). What is the work done by the force $\vec{F}$ during this displacement? c) The electron travels from the point (0, 0) to the point (L, L) along the straight line from (0, 0) to (L, L). What is the work done by the force $\vec{F}$ during this displacement? d) Compare the answers in parts (a), (b), and (c) and explain the results of this comparison.

7.29 A 0.60-kg book slides on a horizontal table. The kinetic friction force on the book has magnitude 1.2 N. a) How much work is done on the book by friction during a displacement of 3.0 m to the left? b) The book now slides 3.0 m to the right, returning to its starting point. During this second 3.0-m displacement, how much work is done on the book by friction? c) What is the total work done on the book by friction during the complete round trip? d) On the basis of your answer to part (c), would you say that the friction force is conservative or nonconservative? Explain.

7.30 A 30.0-kg packing crate in a warehouse is pushed to the loading dock by a worker who applies a horizontal force. The coefficient of kinetic friction between the crate and the floor is 0.20. The loading dock is 15.0 m southwest of the initial position of the crate. a) If the crate is pushed 10.6 m south and then 10.6 m west, what is the total work done on the crate by friction? b) If the crate is pushed along a straight-line path to the dock, so that it travels 15.0 m southwest, what is the work done on the crate by friction? c) Sketch the paths along which the crate moves in parts (a) and (b). Based on your answers to parts (a) and (b), would you say that the friction force is conservative or nonconservative? Explain.

7.31 You and three friends stand at the corners of a square whose sides are 8.0 m long in the middle of the gym floor, as shown in Fig. 7.27. You take your physics book and push it from one person to the other. The book has a mass of 1.5 kg, and the coefficient of kinetic friction between the book and the floor is $\mu_k = 0.25$. a) The book slides from you to Beth and then from Beth to Carlos, along the lines connecting these people. What is the work done by friction during this displacement? b) You slide the book from you to Carlos along the diagonal of the square. What is the work done by friction during this displacement? c) You slide the book to Kim who then slides it back to you. What is the total work done by friction during this motion of the book? d) Is the friction force on the book conservative or nonconservative? Explain.

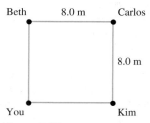

Figure 7.27 Exercise 7.31.

7.32 A block with mass m is attached to an ideal spring that has force constant k. a) The block moves from x_1 to x_2, where $x_2 > x_1$. How much work does the spring force do during this displacement? b) The block moves from x_1 to x_2 and then from x_2 to x_1. How much work does the spring force do during the displacement from x_2 to x_1? What is the total work done by the spring during the entire $x_1 \rightarrow x_2 \rightarrow x_1$ displacement? Explain why you obtained the answer you did. c) The block moves from x_1 to x_3, where $x_3 > x_2$. How much work does the spring force do during this displacement? The block then moves from x_3 to x_2. How much work does the spring force do during this displacement? What is the total work done by the spring force during the $x_1 \rightarrow x_3 \rightarrow x_2$ displacement? Compare your answer to the answer to part (a), where the starting and ending points are the same but the path is different.

Section 7.4 Force and Potential Energy

7.33 The potential energy of a pair of hydrogen atoms separated by a large distance x is given by $U(x) = -C_6/x^6$, where C_6 is a positive constant. What is the force that one atom exerts on the other? Is this force attractive or repulsive?

7.34 A force parallel to the x-axis acts on a particle moving along the x-axis. This force produces a potential energy $U(x)$ given by $U(x) = \alpha x^4$, where $\alpha = 1.20$ J/m^4. What is the force (magnitude and direction) when the particle is at $x = -0.800$ m?

7.35 An experimental device moving in the xy-plane is acted on by a conservative force described by the potential-energy function $U(x, y) = k(x^2 + y^2) + k'xy$, where k and k' are positive constants. Derive an expression for the force expressed in terms of the unit vectors $\hat{\imath}$ and $\hat{\jmath}$.

7.36 An object moving in the xy-plane is acted on by a conservative force described by the potential-energy function $U(x, y) = \alpha(1/x^2 + 1/y^2)$, where α is a positive constant. Derive an expression for the force expressed in terms of the unit vectors $\hat{\imath}$ and $\hat{\jmath}$.

Section 7.5 Energy Diagrams

7.37 The potential energy of two atoms in a diatomic molecule is approximated by $U(r) = a/r^{12} - b/r^6$, where r is the spacing between atoms and a and b are positive constants. a) Find the force $F(r)$ on one atom as a function of r. Make two graphs, one of $U(r)$ versus r and one of $F(r)$ versus r. b) Find the equilibrium distance between the two atoms. Is this equilibrium stable? c) Suppose the distance between the two atoms is equal to the equilibrium distance found in part (b). What minimum energy must be added to the molecule to *dissociate* it, that is, to separate the two atoms to an infinite distance apart? This is called the *dissociation energy* of the molecule. d) For the molecule CO, the equilibrium distance between the carbon and oxygen atoms is 1.13×10^{-10} m and the dissociation energy is 1.54×10^{-18} J per molecule. Find the values of the constants a and b.

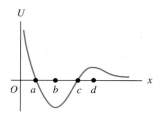

Figure 7.28 Exercise 7.38.

7.38 A marble moves along the x-axis. The potential-energy function is shown in Fig. 7.28. a) At which of the labeled x-coordinates is the force on the marble zero? b) Which of the labeled x-coordinates is a position of stable equilibrium? c) Which of the labeled x-coordinates is a position of unstable equilibrium?

Problems

7.39 A man with mass 70.0 kg sits on a platform suspended from a movable pulley, as shown in Fig. 7.29, and raises himself at constant speed by a rope passing over a fixed pulley. The platform and the pulleys have

Figure 7.29 Problem 7.39.

negligible mass. Assume that there are no friction losses. a) Find the force he must exert. b) Find the increase in the energy of the system when he raises himself 1.20 m. (Answer by calculating the increase in potential energy and also by computing the product of the force on the rope and the length of the rope passing through his hands.)

7.40 Two blocks with different mass are attached to either end of a light rope that passes over a light, frictionless pulley that is suspended from the ceiling. The masses are released from rest, and the more massive one starts to descend. After this block has descended 1.20 m, its speed is 3.00 m/s. If the total mass of the two blocks is 15.0 kg, what is the mass of each block?

7.41 Legal Physics. In an auto accident, a car hit a pedestrian and the driver then slammed on the brakes to stop the car. During the subsequent trial, the driver's lawyer claimed that he was obeying the posted 35 mph speed limit, but that the legal speed was too high to allow him to see and react to the pedestrian in time. You have been called in as the state's expert witness. Your investigation of the accident yields the following measurements: the skid marks made while the brakes were applied were 280 ft long, and the tread on the tires produced a coefficient of kinetic friction of 0.30 with the road. a) In your testimony in court, will you say that the driver was obeying the posted speed? You must be able to back up your conclusion with clear reasoning, because one of the lawyers will surely cross-examine you. b) If the driver's speeding ticket were $10 for each mph he was driving above the posted speed limit, would he have to pay a ticket, and if so, how much would it be?

7.42 A 2.00-kg block is pushed against a spring with negligible mass and force constant $k = 400$ N/m, compressing it 0.220 m. When the block is released, it moves along a frictionless, horizontal surface and then up a frictionless incline with slope 37.0° (Fig. 7.30). a) What is the speed of the block as it slides along the horizontal surface after having left the spring? b) How far does the block travel up the incline before starting to slide back down?

Figure 7.30 Problem 7.42.

7.43 A block with mass 0.50 kg is forced against a horizontal spring of negligible mass, compressing the spring a distance of

Figure 7.31 Problem 7.43.

0.20 m (Fig. 7.31). When released, the block moves on a horizontal table top for 1.00 m before coming to rest. The spring constant k is 100 N/m. What is the coefficient of kinetic friction, μ_k, between the block and the table?

7.44 On a horizontal surface, a crate with mass 50.0 kg is placed against a spring that stores 360 J of energy. The spring is released and the crate slides 5.60 m before coming to rest. What is the speed of the crate when it is 2.00 m from its initial position?

7.45 Bouncing Ball. A 650-gram rubber ball is dropped from an initial height of 2.50 m and on each bounce it returns to 75% of its previous height. a) What is the initial mechanical energy of the ball, just after it is released from its initial height? b) How much mechanical energy does the ball lose during its first bounce? What happens to this energy? c) How much mechanical energy is lost during the second bounce?

7.46 Riding a Loop-the-loop. A car in an amusement park ride rolls without friction around the track shown in Fig. 7.32. It starts from rest at point A at a height h above the bottom of the loop. Treat the car as a particle. a) What is the minimum value of h

Figure 7.32 Problem 7.46.

(in terms of R) such that the car moves around the loop without falling off at the top (point B)? b) If $h = 3.50R$ and $R = 20.0$ m, compute the speed, radial acceleration, and tangential acceleration of the passengers when the car is at point C, which is at the end of a horizontal diameter. Show these acceleration components in a diagram, approximately to scale.

7.47 A 2.0-kg piece of wood slides on the surface shown in Fig. 7.33. The curved sides are perfectly smooth, but the rough horizontal bottom is 30 m long and has a kinetic friction coefficient of 0.20 with the wood. The piece of wood starts from rest 4.0 m above the rough bottom. a) Where will this wood eventually come to rest? b) For the

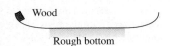

Figure 7.33 Problem 7.47.

motion from the initial release until the piece of wood comes to rest, what is the total amount of work done by friction?

7.48 Up and Down the Hill. A 28-kg rock approaches the foot of a hill with a speed of 15 m/s. This hill slopes upward at a constant angle of 40.0° above the horizontal. The coefficients of static and kinetic friction between the hill and the rock are 0.75 and 0.20, respectively. a) Use energy conservation to find the maximum height above the foot of the hill reached by the rock. b) Will the rock remain at rest at its highest point, or will it slide back down the hill? c) If the rock does slide back down, find its speed when it returns to the bottom of the hill.

7.49 A 15.0-kg stone slides down a snow-covered hill (Fig. 7.34), leaving point A with a speed of 10.0 m/s. There is no friction on the hill between points A and B, but there is friction on the level ground at the

Figure 7.34 Problem 7.49.

bottom of the hill, between B and the wall. After entering the rough horizontal region, the stone travels 100 m and then runs into a very long, light spring with force constant 2.00 N/m. The coefficients of kinetic and static friction between the stone and the horizontal ground are 0.20 and 0.80, respectively. a) What is the speed of the stone when it reaches point B? b) How far will the stone compress the spring? c) Will the stone move again after it has been stopped by the spring?

7.50 A 2.8-kg block slides over the smooth, icy hill shown in Fig. 7.35. The top of the hill is horizontal and 70 m higher than its base. What minimum speed must the block have at the base of the hill so that it will not fall into the pit on the far side of the hill?

Figure 7.35 Problem 7.50.

7.51 Bungee Jump. A bungee cord is 30.0 m long and when stretched a distance x, it exerts a restoring force of magnitude kx. Your father-in-law (mass 95.0 kg) stands on a platform 45.0 m above the ground, and one end of the cord is tied securely to his ankle and the other end to the platform. You have promised him that when he steps off the platform he will fall a maximum distance of only 41.0 m before the cord stops him. You had several bungee cords to select from and tested them by stretching them out, tying one end to a tree, and pulling on the other end with a force of 380.0 N. When you do this, what distance will the bungee cord that you should select have stretched?

7.52 Ski Jump Ramp. You are designing a ski jump ramp for the next Winter Olympics. You need to calculate the vertical height h from the starting gate to the bottom of the ramp. The skiers push off hard with their ski poles at the start, just above the starting gate, so they typically have a speed of 2.0 m/s as they reach the gate. For safety, the skiers should have a speed of no more than 30.0 m/s when they reach the bottom of the ramp. You determine that for a 85.0-kg skier with good form, friction and air resistance will do a total of -4000 J of work on him during his run down the slope. What is the maximum height h for which the maximum safe speed will not be exceeded?

7.53 The Great Sandini is a 60-kg circus performer who is shot from a cannon (actually a spring gun). You don't find many men of his caliber, so you help him design a new gun. This new gun has a very large spring with a very small mass and a force constant of 1100 N/m that he will compress with a force of 4400 N. The inside of the gun barrel is coated with Teflon, so the average friction force will be only 40 N during the 4.0 m he moves in the barrel. At what speed will he emerge from the end of the barrel, 2.5 m above his initial rest position?

7.54 You are designing a delivery ramp for crates containing exercise equipment. The 1470-N crates will move at 1.8 m/s at the top of a ramp that slopes downward at 22.0°. The ramp exerts a 550-N kinetic friction force on each crate, and the maximum static friction force also has this value. Each crate will compress a spring at the bottom of the ramp and will come to rest after traveling a total distance of 8.0 m along the ramp. Once stopped, a crate must not rebound back up the ramp. Calculate the force constant of the spring that will be needed in order to meet the design criteria.

7.55 A system of two paint buckets connected by a light-weight rope is released from rest with the 12.0-kg bucket 2.00 m above the floor (Fig. 7.36). Use the principle of conservation of energy to find the speed with which this bucket strikes the floor. You can ignore friction and the inertia of the pulley.

7.56 a) For the elevator of Example 7.11 (Section 7.2), how much potential energy is stored in the spring when the elevator is at point 2 in Fig. 7.17? b) What maximum distance does the elevator travel upward from point 2

Figure 7.36 Problem 7.55.

before it starts to slide down the shaft again? c) As the elevator slides back down, what is its speed just before it reaches the spring? d) When the elevator compresses the spring the second time, what is the maximum energy that is stored in the spring and what is the force that the spring exerts on the elevator?

7.57 Redesign the elevator safety system of Example 7.11 (Section 7.2) so that the elevator does not bounce but stays at rest the first time its speed becomes zero. The mass of the elevator is 2000 kg and its speed when it first touches the spring is 25 m/s. There is a kinetic friction force of 17,000 N, and the maximum static friction force on the elevator is also 17,000 N. You can ignore the mass of the spring. a) What spring constant is required, and what distance is the spring compressed when the elevator is stopped? Do you think this design is practical? Explain. b) What is the maximum magnitude of the acceleration of the elevator?

7.58 A wooden rod of negligible mass and length 80.0 cm is pivoted about a horizontal axis through its center. A white rat with mass 0.500 kg clings to one end of the stick, and a mouse with mass 0.200 kg clings to the other end. The system is released from rest with the rod horizontal. If the animals can manage to hold on, what are their speeds as the rod swings through a vertical position?

7.59 A 0.100-kg potato is tied to a string with length 2.50 m, and the other end of the string is tied to a rigid support. The potato is held straight out horizontally from the point of support, with the string pulled taut, and is then released. a) What is the speed of the potato at the lowest point of its motion? b) What is the tension in the string at this point?

7.60 The following is data from a computer simulation for a batted baseball with mass 0.145 kg, including air resistance:

t	x	y	v_x	v_y
0	0	0	30.0 m/s	40.0 m/s
3.05 s	70.2 m	53.6 m	18.6 m/s	0
6.59 s	124.4 m	0	11.9 m/s	-28.7 m/s

a) How much work was done by the air on the baseball as it moved from its initial position to its maximum height? b) How much work was done by the air on the baseball as it moved from its maximum height back to the starting elevation? c) Explain why the magnitude of the answer to part (b) is smaller than the magnitude of the answer in part (a).

7.61 Down the Pole. A fireman of mass m slides a distance d down a pole. He starts from rest. He moves as fast at the bottom as if he had stepped off a platform a distance $h \le d$ above the ground and descended with negligible air resistance. a) What average friction force did the fireman exert on the pole? Does your answer make sense in the special cases of $h = d$ and $h = 0$? b) Find a numerical value for the average friction force a 75-kg fireman exerts, for $d = 2.5$ m and $h = 1.0$ m. c) In terms of g, h, and d, what is the speed of the fireman when he is a distance y above the bottom of the pole?

7.62 A 60.0-kg skier starts from rest at the top of a ski slope 65.0 m high. a) If frictional forces do -10.5 kJ of work on her as she descends, how fast is she going at the bottom of the slope? b) Now moving horizontally, the skier crosses a patch of soft snow, where $\mu_k = 0.20$. If the patch is 82.0 m wide and the average force of air resistance on the skier is 160 N, how fast is she going after crossing the patch? c) The skier hits a snowdrift and penetrates 2.5 m into it before coming to a stop. What is the average force exerted on her by the snowdrift as it stops her?

7.63 A skier starts at the top of a very large, frictionless snowball, with a very small, initial speed, and skis straight down the side (Fig. 7.37). At what point does she lose contact with the snowball and fly off at a tangent? That is, at the instant she loses contact with the snowball, what angle α does a radial line from the center of the snowball to the skier make with the vertical?

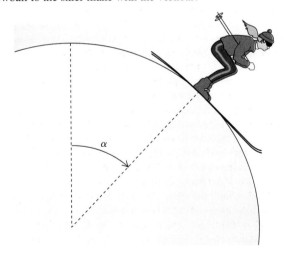

Figure 7.37 Problem 7.63.

7.64 A rock is tied to a cord and the other end of the cord is held fixed. The rock is given an initial tangential velocity that causes it to rotate in a vertical circle. Prove that the tension in the cord at the lowest point exceeds that at the highest point by six times the weight of the rock.

7.65 In a truck-loading station at a post office, a small 0.200-kg package is released from rest at point A on a track that is one-quarter of a circle with radius 1.60 m (Fig. 7.38). The size of the package is much less than 1.60 m, so the package can be treated as a particle. It slides down the track and reaches point B with a speed

Figure 7.38 Problem 7.65.

of 4.80 m/s. From point B, it slides on a level surface a distance of 3.00 m to point C, where it comes to rest. a) What is the coefficient of kinetic friction on the horizontal surface? b) How much work is done on the package by friction as it slides down the circular arc from A to B?

7.66 A truck with mass m has a brake failure while going down an icy mountain road of constant downward slope angle α (Fig. 7.39). Initially the truck is moving downhill at speed v_0. After careening downhill a distance L with negligible friction, the truck driver steers the runaway vehicle onto a runaway truck ramp of constant upward slope angle β. The truck ramp has a soft sand surface for which the coefficient of rolling friction is μ_r. What is the distance that the truck moves up the ramp before coming to a halt? Use energy methods.

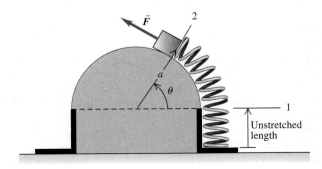

Figure 7.39 Problem 7.66.

7.67 A certain spring is found *not* to obey Hooke's law; it exerts a restoring force $F_x(x) = -\alpha x - \beta x^2$ if it is stretched or compressed, where $\alpha = 60.0$ N/m and $\beta = 18.0$ N/m². The mass of the spring is negligible. a) Calculate the potential energy function $U(x)$ for this spring. Let $U = 0$ when $x = 0$. b) An object with mass 0.900 kg on a frictionless, horizontal surface is attached to this spring, pulled a distance 1.00 m to the right (the $+x$-direction) to stretch the spring, and released. What is the speed of the object when it is 0.50 m to the right of the $x = 0$ equilibrium position?

7.68 A variable force $\vec{F}$ is maintained tangent to a frictionless, semicircular surface (Fig. 7.40). By slowly varying the force, a

Figure 7.40 Problem 7.68.

block with weight w is moved, and the spring to which it is attached is stretched from position 1 to position 2. The spring has negligible mass and force constant k. The end of the spring moves in an arc of radius a. Calculate the work done by the force $\vec{F}$.

7.69 A 0.150-kg block of ice is placed against a horizontal, compressed spring mounted on a horizontal tabletop that is 1.20 m above the floor. The spring has force constant 1900 N/m and is initially compressed 0.045 m. The mass of the spring is negligible. The spring is released, and the block slides along the table, goes off the edge, and travels to the floor. If there is negligible friction between the block of ice and the table, what is the speed of the block of ice when it reaches the floor?

7.70 An 80.0-kg man jumps from a height of 2.50 m onto a platform mounted on springs. As the springs compress, he pushes the platform down a maximum distance of 0.240 m below its initial position, and then it rebounds. The platform and springs have negligible mass. a) What is the man's speed at the instant he depresses the platform 0.120 m? b) If the man just steps gently onto the platform, what maximum distance would he push it down?

7.71 An experimental apparatus with mass m is placed on a vertical spring of negligible mass and pushed down until the spring is compressed a distance x. The apparatus is then released and reaches its maximum height at a distance h above the point where it is released. The apparatus is not attached to the spring, and at its maximum height it is no longer in contact with the spring. The maximum magnitude of acceleration the apparatus can have without being damaged is a, where $a > g$. a) What should the force constant of the spring be? b) What distance x must the spring be compressed initially?

7.72 If a fish is attached to a vertical spring and slowly lowered to its equilibrium position, it is found to stretch the spring by an amount d. If the same fish is attached to the end of the unstretched spring and then allowed to fall from rest, through what maximum distance does it stretch the spring? (*Hint:* Calculate the force constant of the spring in terms of the distance d and the mass m of the fish.)

7.73 A wooden block with mass 1.50 kg is placed against a compressed spring at the bottom of an incline of slope 30.0° (point A). When the spring is released, it projects the block up the incline. At point B, a distance of 6.00 m up the incline from A, the block is moving up the incline at 7.00 m/s and is no longer in contact with the spring. The coefficient of kinetic friction between the block and incline is $\mu_k = 0.50$. The mass of the spring is negligible. Calculate the amount of potential energy that was initially stored in the spring.

7.74 A 2.00-kg package is released on a 53.1° incline, 4.00 m from a long spring with force constant 120 N/m that is attached at the bottom of the incline (Fig. 7.41). The coefficients of friction between the package and the incline are $\mu_s = 0.40$ and $\mu_k = 0.20$. The mass of the spring is negligible. a) What is the speed of the pack-

Figure 7.41 Problem 7.74.

age just before it reaches the spring? b) What is the maximum compression of the spring? c) The package rebounds back up the incline. How close does it get to its initial position?

7.75 A 0.500-kg block, attached to a spring with length 0.60 m and force constant 40.0 N/m is at rest with the back of the block at point A on a frictionless, horizontal air table (Fig. 7.42). The mass of the spring is negligible. You move the block to the right along the surface by pulling with a constant 20.0-N horizontal force. a) What is the block's speed when the back of the block reaches point B, which is 0.25 m to the right of point A? b) When the back of the block reaches point B, you let go of the block. In the subsequent motion, how close does the block get to the wall where the left end of the spring is attached?

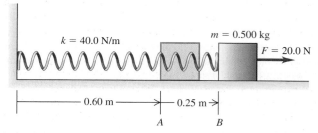

Figure 7.42 Problem 7.75.

7.76 A box of mass m is pushed against a spring of negligible mass and force constant k, compressing it a distance x. The box is then released and travels up a ramp that is at an angle α above the horizontal. The coefficient of kinetic friction between the box and the ramp is μ_k, where $\mu_k < 1$. The box is still moving up the ramp after traveling a distance $s > |x|$ along the ramp. Calculate the angle α for which the speed of the box after traveling distance s is a minimum. Explain why the minimum speed doesn't occur when $\alpha = 90°$, even though for this α there is the maximum increase in gravitational potential energy.

7.77 Fraternity Physics. The brothers of Iota Eta Pi fraternity build a platform, supported at all four corners by vertical springs, in the basement of their frat house. A brave fraternity brother wearing a football helmet stands in the middle of the platform; his weight compresses the springs by 0.18 m. Then four of his fraternity brothers, pushing down at the corners of the platform, compress the springs another 0.53 m until the top of the brave brother's helmet is 0.90 m below the basement ceiling. They then simultaneously release the platform. You can ignore the masses of the springs and platform. a) When the dust clears, the fraternity asks you to calculate their fraternity brother's speed just before his helmet hit the flimsy ceiling. b) Without the ceiling, how high would he have gone? c) In discussing their probation, the dean of students suggests that the next time they try this, they do it outdoors on another planet. Would the answer to part (b) be the same if this stunt were performed on a planet with a different value of g? Assume that the fraternity brothers push the platform down 0.53 m as before. Explain your reasoning.

7.78 A particle with mass m is acted on by a conservative force and moves along a path given by $x = x_0 \cos \omega_0 t$ and $y = y_0 \sin \omega_0 t$, where x_0, y_0 and ω_0 are constants. a) Find the components of the force that acts on the particle. b) Find the potential energy of the particle as a function of x and y. Take $U = 0$ when $x = 0$ and $y = 0$.

c) Find the total energy of the particle when: (i) $x = x_0$, $y = 0$ (ii) $x = 0$, $y = y_0$.

7.79 When it is burned, one gallon of gasoline produces 1.3×10^8 J of energy. A 1500-kg car accelerates from rest to 37 m/s in 10 s. The engine of this car is only 15% efficient (which is typical), meaning that only 15% of the energy from the combustion of the gasoline is used to accelerate the car. The rest goes into things like the internal kinetic energy of the engine parts, as well as heating of the exhaust air and engine. a) How many gallons of gasoline does this car use during the acceleration? b) How many such accelerations will it take to burn up one gallon of gas?

7.80 A hydroelectric dam holds back a lake of surface area 3.0×10^6 m^2 that has vertical sides below the water level. The water level in the lake is 150 m above the base of the dam. When the water passes through turbines at the base of the dam, its mechanical energy is converted into electrical energy with 90% efficiency. a) If gravitational potential energy is taken to be zero at the base of the dam, how much energy is stored in the top meter of the water in the lake? The density of water is 1000 kg/m^3. b) What volume of water must pass through the dam to produce 1000 kilowatt-hours of electrical energy? What distance does the level of water in the lake fall when this much water passes through the dam?

7.81 How much total energy is stored in the lake in Problem 7.80? As in Problem 7.80, take the gravitational potential energy to be zero at the base of the dam. Express your answer in joules and in kilowatt-hours. (*Hint:* Break the lake up into infinitesimal, horizontal layers of thickness dy and integrate to find the total potential energy.)

7.82 a) Is the force $\vec{F} = Cy^2\hat{j}$, where C is a negative constant with units of N/m^2, conservative or nonconservative? Justify your answer. b) Is the force $\vec{F} = Cy^2\hat{i}$, where C is a negative constant with units of N/m^2, conservative or nonconservative? Justify your answer.

7.83 A cutting tool under microprocessor control has several forces acting on it. One force is $\vec{F} = -\alpha xy^2\hat{j}$, a force in the negative y-direction whose magnitude depends on the position of the tool. The constant is $\alpha = 2.50$ N/m^3. Consider the displacement of the tool from the origin to the point $x = 3.00$ m, $y = 3.00$ m. a) Calculate the work done on the tool by $\vec{F}$ if this displacement is along the straight line $y = x$ that connects these two points. b) Calculate the work done on the tool by $\vec{F}$ if the tool is first moved out along the x-axis to the point $x = 3.00$ m, $y = 0$ and then moved parallel to the y-axis to $x = 3.00$ m, $y = 3.00$ m. c) Compare the work done by $\vec{F}$ along these two paths. Is $\vec{F}$ conservative or nonconservative? Explain.

7.84 An object has several forces acting on it. One force is $\vec{F} = \alpha xy\hat{i}$, a force in the x-direction whose magnitude depends on the position of the object. (See Problem 6.96.) The constant is $\alpha = 2.00$ N/m^2. The object moves along the following path: (1) it starts at the origin and moves along the y-axis to the point $x = 0$, $y = 1.50$ m; (2) it moves parallel to the x-axis to the point $x = 1.50$ m, $y = 1.50$ m; (3) it moves parallel to the y-axis to the

point $x = 1.50$ m, $y = 0$; (4) it moves parallel to the x-axis back to the origin. a) Sketch this path in the xy-plane. b) Calculate the work done on the object by $\vec{F}$ for each leg of the path and for the complete round trip. c) Is $\vec{F}$ conservative or nonconservative? Explain.

7.85 A Hooke's law force $-kx$ and a constant conservative force F in the $+x$-direction act on an atomic ion. a) Show that a possible potential-energy function for this combination of forces is $U(x) = \frac{1}{2}kx^2 - Fx - F^2/2k$. Is this the *only* possible function? Explain. b) Find the stable equilibrium position. c) Graph $U(x)$ (in units of F^2/k) versus x (in units of F/k) for values of x between $-5F/k$ to $5F/k$. d) Are there any unstable equilibrium positions? e) If the total energy is $E = F^2/k$, what are the maximum and minimum values of x that the ion reaches in its motion? f) If the ion has mass m, find its maximum speed if the total energy is $E = F^2/k$. For what value of x is the speed maximum?

7.86 A particle moves along the x-axis while acted on by a single conservative force parallel to the x-axis. The force corresponds to the potential-energy function graphed in Fig. 7.43. The particle is released from rest at point A. a) What is the direction of the force on the particle when it is at point A? b) At point B? c) At what value of x is the kinetic energy of the particle a maximum? d) What is the force on the particle when it is at point C? e) What is the largest value of x reached by the particle during its motion? f) What value or values of x correspond to points of stable equilibrium? g) Of unstable equilibrium?

Figure 7.43 Problem 7.86.

Challenge Problem

7.87 A proton with mass m moves in one dimension. The potential-energy function is $U(x) = \alpha/x^2 - \beta/x$, where α and β are positive constants. The proton is released from rest at $x_0 = \alpha/\beta$. a) Show that $U(x)$ can be written as

$$U(x) = \frac{\alpha}{x_0^2}\left[\left(\frac{x_0}{x}\right)^2 - \frac{x_0}{x}\right]$$

Graph $U(x)$. Calculate $U(x_0)$ and thereby locate the point x_0 on the graph. b) Calculate $v(x)$ the speed of the proton as a function of position. Graph $v(x)$ and give a qualitative description of the motion. c) For what value of x is the speed of the proton a maximum? What is the value of that maximum speed? d) What is the force on the proton at the point in part (c)? e) Let the proton be released instead at $x_1 = 3\alpha/\beta$. Locate the point x_1 on the graph of $U(x)$. Calculate $v(x)$ and give a qualitative description of the motion. f) For each release point ($x = x_0$ and $x = x_1$), what are the maximum and minimum values of x reached during the motion?

8 MOMENTUM, IMPULSE, AND COLLISIONS

When one football player tackles another, the dynamics of the collision between their bodies can be quite complex. Yet the collision is governed by a simple physical principle: if you calculate the product of mass and vector velocity for each player, and then add together the results for each player, the sum has the same value just before the collision as it does just after the collision. This is an example of the law of conservation of momentum.

? **Which could potentially cause you the greater injury: being tackled by a lightweight, fast-moving football player, or being tackled by a player with double the mass but moving at half the speed?**

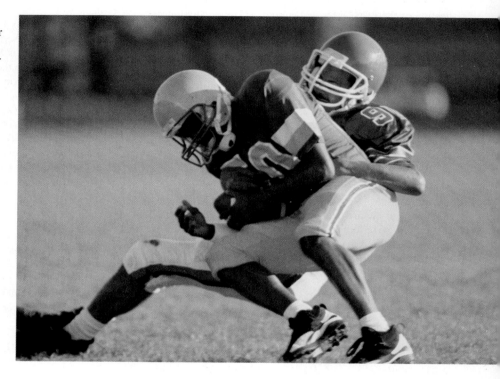

There are many questions involving forces that cannot be answered by directly applying Newton's second law, $\sum \vec{F} = m\vec{a}$. For example, when an eighteen-wheeler collides head-on with a compact car, what determines which way the wreckage goes after the collision? In playing pool, how do you decide how to aim the cue ball in order to knock the eight ball into the pocket? And when a meteorite collides with the earth, how much of the meteorite's kinetic energy is released in the impact?

A common theme of all of these questions is that they involve forces about which we know very little: the forces between the car and the eighteen-wheeler, between the two pool balls, or between the meteorite and the earth. Remarkably, we will find in this chapter that we don't have to know *anything* about these forces to answer questions of this kind!

Our approach uses two new concepts, *momentum* and *impulse*, and a new conservation law, *conservation of momentum*. This conservation law is every bit as important as that of conservation of energy. The law of conservation of momentum is valid even in situations in which Newton's laws are inadequate, such as bodies moving at very high speeds (near the speed of light) or objects on a very small scale (such as the constituents of atoms). Within the domain of Newtonian mechanics, conservation of momentum enables us to analyze many situations that would be very difficult if we tried to use Newton's laws directly. Among these are *collision* problems, in which two bodies collide and can exert very large forces on each other for a short time.

8.1 | Momentum and Impulse

In Chapter 6 we re-expressed Newton's second law for a particle, $\sum \vec{F} = m\vec{a}$, in terms of the work-energy theorem. This theorem helped us tackle a great number of physics problems and led us to the law of conservation of energy. Let's now return to $\sum \vec{F} = m\vec{a}$ and see yet another useful way in which this fundamental law can be restated.

Consider a particle of constant mass m. (Later in this chapter we'll see how to deal with situations in which the mass of a body changes.) Because $\vec{a} = d\vec{v}/dt$, we can write Newton's second law for this particle as

$$\sum \vec{F} = m\frac{d\vec{v}}{dt} = \frac{d}{dt}(m\vec{v}) \qquad (8.1)$$

We can take the mass m inside the derivative because it is constant. Thus, Newton's second law says that the net force $\sum \vec{F}$ acting on a particle equals the time rate of change of the combination $m\vec{v}$, the product of the particle's mass and velocity. We'll call this combination the **momentum,** or **linear momentum,** of the particle. Using the symbol $\vec{p}$ for momentum, we have

$$\vec{p} = m\vec{v} \qquad (\text{definition of momentum}) \qquad (8.2)$$

Momentum is a vector quantity that has a magnitude (mv) and a direction (the same as the velocity vector $\vec{v}$). The momentum of a car driving north at 20 m/s is different from the momentum of the same car driving east at the same speed. A fastball thrown by a major-league pitcher has greater magnitude of momentum than the same ball thrown by a child because the speed is greater. An eighteen-wheeler going 65 mi/h has greater magnitude of momentum than a Saturn automobile with the same speed because the truck's mass is greater.

The units of the magnitude of momentum are units of mass times speed; the SI units of momentum are kg·m/s. The plural of momentum is "momenta."

Substituting Eq. (8.2) into Eq. (8.1), we have

$$\sum \vec{F} = \frac{d\vec{p}}{dt} \qquad (\text{Newton's second law in terms of momentum}) \quad (8.3)$$

The net force (vector sum of all forces) acting on a particle equals the time rate of change of momentum of the particle. This, not $\sum \vec{F} = m\vec{a}$, is the form in which Newton originally stated his second law (though he called momentum the "quantity of motion"). This law is valid only in inertial frames of reference.

According to Eq. (8.3), a rapid change in momentum requires a large net force, while a gradual change in momentum requires less net force. This principle is used in the design of automobile safety devices such as air bags. The driver of a fast-moving automobile has a large momentum (the product of the driver's mass and velocity). If the car stops suddenly in a collision, the driver's momentum becomes zero. An air bag causes the driver to lose momentum more gradually than would an abrupt collision with the steering wheel, reducing the force exerted on the driver (and the possibility for injury). The same principle applies to the padding used to package fragile objects for shipping.

We often express the momentum of a particle in terms of its components. If the particle has velocity components v_x, v_y, and v_z, its momentum components p_x, p_y, and p_z are given by

$$p_x = mv_x \qquad p_y = mv_y \qquad p_z = mv_z \qquad (8.4)$$

These three component equations are equivalent to Eq. (8.2).

A particle's momentum $\vec{p} = m\vec{v}$ and its kinetic energy $K = \frac{1}{2}mv^2$ both depend on the mass and velocity of the particle. What is the fundamental difference between these two quantities? A purely mathematical answer is that momentum is a vector whose magnitude is proportional to speed, while kinetic energy is a scalar proportional to the speed squared. But to see the *physical* difference between momentum and kinetic energy, we must first define a quantity closely related to momentum called *impulse*.

Let's first consider a particle acted on by a *constant* net force $\sum\vec{F}$ during a time interval Δt from t_1 to t_2. (We'll look at the case of varying forces shortly.) The **impulse** of the net force, denoted by $\vec{J}$, is defined to be the product of the net force and the time interval:

$$\vec{J} = \sum\vec{F}(t_2 - t_1) = \sum\vec{F}\Delta t \qquad \text{(assuming constant net force)} \quad (8.5)$$

Impulse is a vector quantity; its direction is the same as the net force $\sum\vec{F}$. Its magnitude is the product of the magnitude of the net force and the length of time that the net force acts. The SI unit of impulse is the newton-second $(\text{N}\cdot\text{s})$. Because $1\,\text{N} = 1\,\text{kg}\cdot\text{m/s}^2$, an alternative set of units for impulse is $\text{kg}\cdot\text{m/s}$, the same as the units of momentum.

To see what impulse is good for, let's go back to Newton's second law as restated in terms of momentum, Eq. (8.3). If the net force $\sum\vec{F}$ is constant, then $d\vec{p}/dt$ is also constant. In that case, $d\vec{p}/dt$ is equal to the *total* change in momentum $\vec{p}_2 - \vec{p}_1$ during the time interval $t_2 - t_1$, divided by the interval:

$$\sum\vec{F} = \frac{\vec{p}_2 - \vec{p}_1}{t_2 - t_1}$$

Multiplying this equation by $(t_2 - t_1)$, we have

$$\sum\vec{F}(t_2 - t_1) = \vec{p}_2 - \vec{p}_1$$

Comparing to Eq. (8.5), we end up with a result called the **impulse-momentum theorem:**

$$\vec{J} = \vec{p}_2 - \vec{p}_1 \qquad \text{(impulse-momentum theorem)} \qquad (8.6)$$

The change in momentum of a particle during a time interval equals the impulse of the net force that acts on the particle during that interval.

The impulse-momentum theorem also holds when forces are not constant. To see this, we integrate both sides of Newton's second law $\sum\vec{F} = d\vec{p}/dt$ over time between the limits t_1 and t_2:

$$\int_{t_1}^{t_2}\sum\vec{F}\,dt = \int_{t_1}^{t_2}\frac{d\vec{p}}{dt}\,dt = \int_{\vec{p}_1}^{\vec{p}_2}d\vec{p} = \vec{p}_2 - \vec{p}_1$$

The integral on the left is defined to be the impulse $\vec{J}$ of the net force $\sum \vec{F}$ during this interval:

$$\vec{J} = \int_{t_1}^{t_2} \sum \vec{F}\, dt \qquad \text{(general definition of impulse)} \qquad (8.7)$$

With this definition the impulse-momentum theorem $\vec{J} = \vec{p}_2 - \vec{p}_1$, Eq. (8.6), is valid even when the net force $\sum \vec{F}$ varies with time.

We can define an *average* net force $\vec{F}_{av}$ such that even when $\sum \vec{F}$ is not constant, the impulse $\vec{J}$ is given by

$$\vec{J} = \vec{F}_{av}(t_2 - t_1) \qquad (8.8)$$

When $\sum \vec{F}$ is constant, $\sum \vec{F} = \vec{F}_{av}$ and Eq. (8.8) reduces to Eq. (8.5).

Figure 8.1 shows a graph of the *x*-component of net force $\sum F_x$ as a function of time during a collision. This might represent the force on a soccer ball that is in contact with a player's foot from time t_1 to t_2. The *x*-component of impulse during this interval is represented by the red area under the curve between t_1 and t_2. This area is equal to the rectangular area bounded by t_1, t_2, and $(F_{av})_x$, so $(F_{av})_x(t_2 - t_1)$ is equal to the impulse of the actual time-varying force during the same interval.

Impulse and momentum are both vector quantities, and Eqs. (8.5) through (8.8) are all vector equations. In specific problems it is often easiest to use them in component form:

$$J_x = \int_{t_1}^{t_2} \sum F_x\, dt = (F_{av})_x(t_2 - t_1) = p_{2x} - p_{1x} = mv_{2x} - mv_{1x}$$

$$J_y = \int_{t_1}^{t_2} \sum F_y\, dt = (F_{av})_y(t_2 - t_1) = p_{2y} - p_{1y} = mv_{2y} - mv_{1y} \qquad (8.9)$$

and similarly for the *z*-component.

Momentum and Kinetic Energy Compared

We can now see the fundamental difference between momentum and kinetic energy. The impulse-momentum theorem $\vec{J} = \vec{p}_2 - \vec{p}_1$ says that changes in a particle's momentum are due to impulse, which depends on the *time* over which the net force acts. By contrast, the work-energy theorem $W_{tot} = K_2 - K_1$ tells us that kinetic energy changes when work is done on a particle; the total work depends on the *distance* over which the net force acts. Consider a particle that starts from rest at t_1 so that $\vec{v} = 0$. Its initial momentum is $\vec{p}_1 = m\vec{v}_1 = 0$, and its initial kinetic energy is $K_1 = \frac{1}{2}mv_1^2 = 0$. Now let a constant net force equal to $\vec{F}$ act on that particle from time t_1 until time t_2. During this interval the particle moves a distance s in the direction of the force. From Eq. (8.6), the particle's momentum at time t_2 is

$$\vec{p}_2 = \vec{p}_1 + \vec{J} = \vec{J}$$

where $\vec{J} = \vec{F}(t_2 - t_1)$ is the impulse that acts on the particle. So *the momentum of a particle equals the impulse that accelerated it from rest to its present speed;* impulse is the product of the net force that accelerated the particle and the *time* required for the acceleration. By comparison, the kinetic energy of the particle at t_2 is $K_2 = W_{tot} = Fs$, the total *work* done on the particle to accelerate it from rest. The total work is the product of the net force and the *distance* required to accelerate the particle (Fig. 8.2).

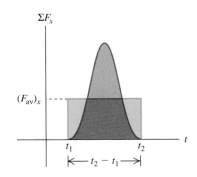

8.1 The *x*-component of the impulse of $\sum F_x$ between t_1 and t_2 equals the red area under the $\sum F_x$–*t* curve, which also equals the area under the rectangle with height $(F_{av})_x$.

8.2 The kinetic energy of a pitched baseball is equal to the work the pitcher does on it (force multiplied by the distance the ball moves during the throw). The momentum of the ball is equal to the impulse the pitcher imparts to it (force multiplied by the time it took to bring the ball up to speed).

6.1   Momentum and Energy Change

Here's an application of the distinction between momentum and kinetic energy. Suppose you have a choice between catching a 0.50-kg ball moving at 4.0 m/s or a 0.10-kg ball moving at 20 m/s. Which will be easier to catch? Both have the same magnitude of momentum, $p = mv = (0.50 \text{ kg})(4.0 \text{ m/s}) = (0.10 \text{ kg})(20 \text{ m/s}) = 2.0 \text{ kg·m/s}$. However, the two balls have very different values of kinetic energy $K = \frac{1}{2}mv^2$; the large, slow-moving ball has $K = 4.0$ J, while the small, fast-moving ball has $K = 20$ J. Since the momentum is the same for both balls, both require the same *impulse* to be brought to rest. But stopping the 0.10-kg ball with your hand requires five times more *work* than stopping the 0.50-kg ball, because the smaller ball has five times as much kinetic energy. Hence for a given force that you exert with your hand, it takes the same amount of time (the duration of the catch) to stop either ball, but your hand and arm will be pushed back five times farther if you choose to catch the small, fast-moving ball. To minimize arm strain, you should choose to catch the 0.50-kg ball with its lower kinetic energy. (Kinetic energy may not be the only factor to consider, however. A pitched baseball and a .22 bullet fired from a rifle have roughly the same kinetic energy, and the bullet has less momentum than does the baseball. Nonetheless, you'd probably prefer to catch the baseball rather than the bullet. Why?)

Both the impulse-momentum and work-energy theorems are relationships between force and motion, and both rest on the foundation of Newton's laws. They are *integral* principles, relating the motion at two different times separated by a finite interval. By contrast, Newton's second law itself (in either of the forms $\Sigma \vec{F} = m\vec{a}$ or $\Sigma \vec{F} = d\vec{p}/dt$) is a *differential* principle, relating the forces to the rate of change of velocity or momentum at each instant.

<table>
<tr><td>Conceptual
Example 8.1</td><td>**Momentum versus kinetic energy**</td></tr>
</table>

Consider again the race described in Conceptual Example 6.6 (Section 6.2) between two iceboats on a frictionless frozen lake. The iceboats have masses m and $2m$, respectively, and the wind exerts the same constant horizontal force $\vec{F}$ on each iceboat (Fig. 6.10). The two iceboats start from rest and cross the finish line a distance s away. Which iceboat crosses the finish line with greater momentum?

SOLUTION

In Conceptual Example 6.6 we asked how the kinetic energies of the iceboats compare when they cross the finish line. The way to determine this was not by using the formula $K = \frac{1}{2}mv^2$, but rather by remembering that a body's kinetic energy is equal to the total work done to accelerate it from rest. Both iceboats started from rest, and the total work done between the starting and finish lines was the same for both iceboats (because the net force and displacement were the same for both). Hence both iceboats cross the finish line with the same kinetic energy.

Similarly, the best way to compare the momenta of the iceboats is *not* to use the formula for momentum, $\vec{p} = m\vec{v}$. By itself this formula isn't enough to decide which iceboat has greater momentum at the finish line. The iceboat of mass $2m$ has greater mass, which

suggests greater momentum; but this iceboat crosses the finish line going slower than the other one, which suggests less momentum.

Instead, we use the idea that the momentum of each iceboat equals the impulse that accelerated it from rest. For each iceboat the downward force of gravity and the upward normal force add to zero, so the net force equals the constant horizontal wind force $\vec{F}$. Let Δt be the time an iceboat takes to reach the finish line, so that the impulse on the iceboat during that time is $\vec{J} = \vec{F} \Delta t$. Since the iceboat starts from rest, this just equals the iceboat's momentum $\vec{p}$ at the finish line:

$$\vec{p} = \vec{F} \Delta t$$

Both iceboats are subjected to the same force $\vec{F}$, but they do not take the same amount of time Δt to reach the finish line. The more massive iceboat accelerates more slowly and takes a longer time to travel the distance s; thus there is a greater impulse on this iceboat between the starting and finish lines. So the iceboat of mass $2m$ crosses the finish line with a greater magnitude of momentum than the iceboat of mass m (but with the same kinetic energy). Can you show that the iceboat of mass $2m$ has $\sqrt{2}$ times as much momentum at the finish line as the iceboat of mass m?

Example
8.2

A ball hits a wall

Suppose you throw a ball with a mass of 0.40 kg against a brick wall. It hits the wall moving horizontally to the left at 30 m/s and rebounds horizontally to the right at 20 m/s. a) Find the impulse of the net force on the ball during its collision with the wall. b) If the ball is in contact with the wall for 0.010 s, find the average horizontal force that the wall exerts on the ball during the impact.

SOLUTION

IDENTIFY: We're given enough information to determine the initial and final values of the ball's momentum, so we can use the impulse-momentum theorem to find the impulse. We'll then use the definition of impulse to determine the average force.

SET UP: The motion is purely horizontal, so we just need a single axis; we'll take the x-axis to be horizontal and the positive direction to be to the right (Fig. 8.3). Our target variable in part (a) is the x-component of impulse, J_x, which we'll find from the x-components of momentum before and after the impact, using Eqs. (8.9). In part (b), our target variable is the average x-component of force $(F_{av})_x$; once we know J_x, we can also find this force by using Eqs. (8.9).

EXECUTE: a) With our choice of x-axis, the initial and final x-components of momentum of the ball are

$$p_{1x} = mv_{1x} = (0.40 \text{ kg})(-30 \text{ m/s}) = -12 \text{ kg} \cdot \text{m/s}$$
$$p_{2x} = mv_{2x} = (0.40 \text{ kg})(+20 \text{ m/s}) = +8.0 \text{ kg} \cdot \text{m/s}$$

From the x-equation in Eqs. (8.9), the x-component of impulse equals the *change* in the x-component of momentum:

$$J_x = p_{2x} - p_{1x}$$
$$= 8.0 \text{ kg} \cdot \text{m/s} - (-12 \text{ kg} \cdot \text{m/s}) = 20 \text{ kg} \cdot \text{m/s} = 20 \text{ N} \cdot \text{s}$$

b) The collision time is $t_2 - t_1 = \Delta t = 0.010$ s. From the x-equation in Eqs. (8.9), $J_x = (F_{av})_x (t_2 - t_1) = (F_{av})_x \Delta t$, so

$$(F_{av})_x = \frac{J_x}{\Delta t} = \frac{20 \text{ N} \cdot \text{s}}{0.010 \text{ s}} = 2000 \text{ N}$$

$v_1 = -30 \text{ m/s}$

$+x$ direction

$v_2 = 20 \text{ m/s}$

8.3 A ball moving to the left, hitting a wall, and rebounding to the right.

8.4 Typically, a tennis ball is in contact with the racket for approximately 0.01 s. The ball flattens noticeably due to the tremendous force exerted by the racket.

EVALUATE: The x-component of impulse is positive—that is, to the right in Fig. 8.3. This is as it should be: the impulse represents the "kick" that the wall imparts to the ball, and this "kick" is certainly to the right.

CAUTION Because momentum is a vector, we had to include the negative sign in p_{1x}. Had we been careless and omitted it, we would have calculated the impulse to be $8.0 \text{ kg} \cdot \text{m/s} - (12 \text{ kg} \cdot \text{m/s}) = -4 \text{ kg} \cdot \text{m/s}$. This answer is clearly nonsensical, because it says that the wall would somehow have given the ball a kick to the *left!* Make sure that you account for the direction of momentum while carrying out your calculations.

The average force that the wall exerts on the ball is quite substantial, 2000 N (equal to 450 lb, or the weight of a 200-kg object). The magnitude of this force has to be large to change the ball's momentum in such a short time interval. Other forces that act on the ball during the collision are very weak by comparison; for instance, the gravitational force is only 3.9 N. Thus, during the brief time that the collision lasts, we can ignore all other forces on the ball to a very good approximation. Figure 8.4 is a photograph showing the impact of a tennis ball and racket.

Note that the 2000-N value we calculated is just the *average* horizontal force that the wall exerts on the ball during the impact. It corresponds to the horizontal line $(F_{av})_x$ in Fig. 8.1. The horizontal force is zero before impact, rises to a maximum, and then decreases to zero when the ball loses contact with the wall. If the ball is relatively rigid, like a baseball or golf ball, the collision lasts a short time and the maximum force is large, as in curve (a) in Fig. 8.5. If

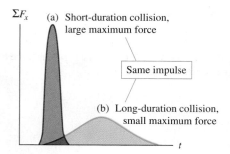

8.5 (a) A "hard" collision, such as that of a golf ball. (b) A "soft" collision, such as that of a tennis ball.

the ball is softer, like a tennis ball, the collision time is longer and the maximum force is less, as in curve (b). In either case, the impulse is the same: it equals the change in the ball's momentum. From Eq. (8.7), we know that impulse is the integral of force with respect to time, which means that the *area* under each curve in Fig. 8.5 is the same.

Example 8.3 — Kicking a soccer ball

A soccer ball has a mass of 0.40 kg. Initially, it is moving to the left at 20 m/s, but then it is kicked and given a velocity at 45° upward and to the right, with a magnitude of 30 m/s (Fig. 8.6a). Find the impulse of the net force and the average net force, assuming a collision time $\Delta t = 0.010$ s.

SOLUTION

IDENTIFY: This example uses the same principles as Example 8.2. The key difference is that the initial and final velocities are not along the same line, so we have to be careful to treat momentum and impulse as vector quantities, using their x- and y-components.

SET UP: We take the x-axis to be horizontally to the right and the y-axis to be vertically upward. Our target variables are the components of the net impulse on the ball, J_x and J_y, and the components of the average net force on the ball, $(F_{av})_x$ and $(F_{av})_y$. We'll find them using the x- and y-components of Eqs. (8.9).

EXECUTE: With our choice of axes, we find the following velocity components:

$$v_{1x} = -20 \text{ m/s} \qquad v_{1y} = 0$$
$$v_{2x} = v_{2y} = (0.707)(30 \text{ m/s}) = 21.2 \text{ m/s}$$
$$(\text{since } \cos 45° = \sin 45° = 0.707)$$

The x-component of impulse is equal to the x-component of momentum change, and the same is true for the y-components:

$$J_x = p_{2x} - p_{1x} = m(v_{2x} - v_{1x})$$
$$= (0.40 \text{ kg})[21.2 \text{ m/s} - (-20 \text{ m/s})] = 16.5 \text{ kg} \cdot \text{m/s}$$
$$J_y = p_{2y} - p_{1y} = m(v_{2y} - v_{1y})$$
$$= (0.40 \text{ kg})(21.2 \text{ m/s} - 0) = 8.5 \text{ kg} \cdot \text{m/s}$$

The components of the average net force on the ball are

$$(F_{av})_x = \frac{J_x}{\Delta t} = 1650 \text{ N} \qquad (F_{av})_y = \frac{J_y}{\Delta t} = 850 \text{ N}$$

(a)   (b)

8.6 (a) A soccer ball (1) before and (2) after being kicked. (b) Finding the average force from its components.

The magnitude and direction of the average force are

$$F_{av} = \sqrt{(1650 \text{ N})^2 + (850 \text{ N})^2} = 1.9 \times 10^3 \text{ N}$$

$$\theta = \arctan \frac{850 \text{ N}}{1650 \text{ N}} = 27°$$

where θ is measured upward from the $+x$-axis (Fig. 8.6b). Note that because the ball was not initially at rest, the ball's final velocity does *not* have the same direction as the average force acting on it.

EVALUATE: The average net force $\vec{F}_{av}$ includes the effects of the force of gravity, though these are small; the weight of the ball is only 3.9 N. As in Example 8.2, the average force acting during the collision is exerted almost entirely by the object that the ball hit (in this case, the soccer player's foot).

Test Your Understanding

Consider the falling hammerhead in Example 6.4 (Section 6.2). (a) Find the hammerhead's y-component of momentum just before it hit the I-beam and the impulse of the net force that acted on the hammerhead to bring it to a stop. (b) Use impulse ideas to find the time it took for this net force to bring the hammerhead to a stop.

8.2 Conservation of Momentum

The concept of momentum is particularly important in situations in which we have two or more interacting bodies. Let's consider first an idealized system consisting of two bodies that interact with each other but not with anything else—for example, two astronauts who touch each other as they float freely in the zero-gravity environment of outer space (Fig. 8.7a). Think of the astronauts as particles. Each particle exerts a force on the other; according to Newton's third law, the

(a)

(b)   (c)

8.7 (a) Two astronauts push each other as they float freely in the zero-gravity environment of space. (b) Free-body diagram for astronaut A. (c) Free-body diagram for astronaut B. The only forces that act are internal forces; they don't change the total momentum of the two astronauts.

two forces are always equal in magnitude and opposite in direction. Hence, the *impulses* that act on the two particles are equal and opposite, and the changes in momentum of the two particles are equal and opposite.

Let's go over that again with some new terminology. For any system, the forces that the particles of the system exert on each other are called **internal forces.** Forces exerted on any part of the system by some object outside it are called **external forces.** For the system we have described, the internal forces are $\vec{F}_{B \text{ on } A}$, exerted by particle B on particle A, and $\vec{F}_{A \text{ on } B}$ exerted by particle A on particle B (Figs. 8.7b, c). There are *no* external forces; when this is the case, we have an **isolated system.**

The net force on particle A is $\vec{F}_{B \text{ on } A}$ and the net force on particle B is $\vec{F}_{A \text{ on } B}$, so from Eq. (8.3) the rates of change of the momenta of the two particles are

$$\vec{F}_{B \text{ on } A} = \frac{d\vec{p}_A}{dt} \qquad \vec{F}_{A \text{ on } B} = \frac{d\vec{p}_B}{dt} \tag{8.10}$$

The momentum of each particle changes, but these changes are not independent; according to Newton's third law, the two forces $\vec{F}_{B \text{ on } A}$ and $\vec{F}_{A \text{ on } B}$ are always equal in magnitude and opposite in direction. That is, $\vec{F}_{B \text{ on } A} = -\vec{F}_{A \text{ on } B}$, so $\vec{F}_{B \text{ on } A} + \vec{F}_{A \text{ on } B} = \mathbf{0}$. Adding together the two equations in Eq. (8.10), we have

$$\vec{F}_{B \text{ on } A} + \vec{F}_{A \text{ on } B} = \frac{d\vec{p}_A}{dt} + \frac{d\vec{p}_B}{dt} = \frac{d(\vec{p}_A + \vec{p}_B)}{dt} = \mathbf{0} \tag{8.11}$$

The rates of change of the two momenta are equal and opposite, so that the rate of change of the vector sum $\vec{p}_A + \vec{p}_B$ is zero. We now define the **total momentum $\vec{P}$** of the system of two particles as the vector sum of the momenta of the individual particles. That is,

$$\vec{P} = \vec{p}_A + \vec{p}_B \tag{8.12}$$

Then Eq. (8.11) becomes, finally,

$$\vec{F}_{B \text{ on } A} + \vec{F}_{A \text{ on } B} = \frac{d\vec{P}}{dt} = \mathbf{0} \tag{8.13}$$

The time rate of change of the *total* momentum $\vec{P}$ is zero. Hence the total momentum of the system is constant, even though the individual momenta of the particles that make up the system can change.

If external forces are also present, they must be included on the left side of Eq. (8.13) along with the internal forces. Then the total momentum is, in general, not constant. But if the vector sum of the external forces is zero, as in Fig. 8.8, these forces don't contribute to the sum, and $d\vec{P}/dt$ is again zero. Thus we have the following general result:

If the vector sum of the external forces on a system is zero, the total momentum of the system is constant.

This is the simplest form of the **principle of conservation of momentum.** This principle is a direct consequence of Newton's third law. What makes this principle useful is that it doesn't depend on the detailed nature of the internal forces that act between members of the system. This means that we can apply conservation of momentum even if (as is often the case) we know very little about the internal forces. We have used Newton's second law to derive this principle, so we have to be careful to use it only in inertial frames of reference.

(a)

(b)   (c)

8.8 (a) Two ice skaters push each other as they skate on a frictionless, horizontal surface. (b) Free-body diagram for skater A. (c) Free-body diagram for skater B. The normal forces and gravitational forces are external forces, but the vector sum of the external forces is zero and total momentum is conserved. (Compare to Fig. 8.7.)

We can generalize this principle for a system containing any number of particles A, B, C, ... interacting only with each other. The total momentum of such a system is

$$\vec{P} = \vec{p}_A + \vec{p}_B + \cdots = m_A\vec{v}_A + m_B\vec{v}_B + \cdots \quad \begin{array}{l}\text{(total momentum of a}\\ \text{system of particles)}\end{array} \quad (8.14)$$

We make the same argument as before; the total rate of change of momentum of the system due to each action-reaction pair of internal forces is zero. Thus the total rate of change of momentum of the entire system is zero whenever the vector sum of the external forces acting on it is zero. The internal forces can change the momenta of individual particles in the system but not the *total* momentum of the system.

CAUTION When you apply the conservation of momentum to a system, it's absolutely essential to remember that momentum is a *vector* quantity. Hence you must use vector addition to compute the total momentum of a system (Fig. 8.9). Using components is usually the simplest method. If p_{Ax}, p_{Ay}, and p_{Az} are the components of momentum of particle A, and similarly for the other particles, then Eq. (8.14) is equivalent to the component equations

$$P_x = p_{Ax} + p_{Bx} + \cdots$$
$$P_y = p_{Ay} + p_{By} + \cdots$$
$$P_z = p_{Az} + p_{Bz} + \cdots \quad (8.15)$$

If the vector sum of the external forces on the system is zero, then P_x, P_y, and P_z are all constant.

$p_A = 18\ \text{kg} \cdot \text{m/s}$
$p_B = 24\ \text{kg} \cdot \text{m/s}$

(a)

INCORRECT $P = p_A + p_B = 42\ \text{kg} \cdot \text{m/s}$

(b)

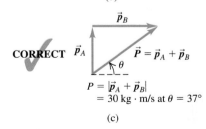

CORRECT

$$P = |\vec{p}_A + \vec{p}_B|$$
$$= 30\ \text{kg} \cdot \text{m/s at } \theta = 37°$$

(c)

8.9 (a) A system of two particles. Note that their momenta do *not* point in the same direction. (b) The magnitude of the total momentum is *not* the sum of the individual momentum magnitudes. (c) The correct way to find the total momentum is by vector addition.

In some ways the principle of conservation of momentum is more general than the principle of conservation of mechanical energy. For example, mechanical energy is conserved only when the internal forces are *conservative*—that is, when the forces allow two-way conversion between kinetic and potential energy—but conservation of momentum is valid even when the internal forces are *not* conservative. In this chapter we will analyze situations in which both momentum and mechanical energy are conserved, and others in which only momentum is conserved. These two principles play a fundamental role in all areas of physics, and we will encounter them throughout our study of physics.

Problem-Solving Strategy	**Conservation of Momentum**

IDENTIFY *the relevant concepts:* Before applying conservation of momentum to a problem, you must first decide whether momentum *is* conserved! This will be true *only* if the vector sum of the external forces acting on the system of particles is zero. If this is not the case, you can't use conservation of momentum.

SET UP *the problem using the following steps:*
1. Define a coordinate system. Make a sketch showing the coordinate axes, including the positive direction for each. Often it is easiest to choose the *x*-axis to have the direction of one of the initial velocities. Make sure you are using an inertial frame of reference. Most of the problems in this chapter deal with two-dimensional situations, in which the vectors have only *x*- and *y*-components; all of the following statements can be generalized to include *z*-components when necessary.
2. Treat each body as a particle. Draw "before" and "after" sketches, and include vectors on each to represent all known velocities. Label the vectors with magnitudes, angles, components, or whatever information is given, and give each unknown magnitude, angle, or component an algebraic symbol. You may find it helpful to use the subscripts 1 and 2 for velocities before and after the interaction, respectively; if you use these subscripts, use letters (not numbers) to label each particle.

3. As always, identify the target variable(s) from among the unknowns.

EXECUTE *the solution as follows:*
1. Write an equation in terms of symbols equating the total *initial x*-component of momentum (that is, before the interaction) to the total *final x*-component of momentum (that is, after the interaction), using $p_x = mv_x$ for each particle. Write another equation for the *y*-components, using $p_y = mv_y$ for each particle. Remember that the *x*- and *y*-components of velocity or momentum are *never* added together in the same equation! Even when all the velocities lie along a line (such as the *x*-axis), the components of velocity along this line can be positive or negative; be careful with signs!
2. Solve these equations to determine whatever results are required. In some problems you will have to convert from the *x*- and *y*-components of a velocity to its magnitude and direction, or the reverse.
3. In some problems, energy considerations give additional relationships among the various velocities, as we will see later in this chapter.

EVALUATE *your answer:* Does your answer make physical sense? If your target variable is a certain body's momentum, check that the direction of the momentum is reasonable.

Example 8.4	**Recoil of a rifle**

A marksman holds a rifle of mass $m_R = 3.00$ kg loosely in his hands, so as to let it recoil freely when fired. He fires a bullet of mass $m_B = 5.00$ g horizontally with a velocity relative to the ground of $v_{Bx} = 300$ m/s (Fig. 8.10). What is the recoil velocity v_{Rx} of the rifle? What are the final momentum and kinetic energy of the bullet? Of the rifle?

SOLUTION

IDENTIFY: We consider an idealized model in which the horizontal forces the marksman exerts on the rifle are negligible. Then there is no net horizontal force on the system (the bullet and rifle) during the firing of the rifle, and so the total horizontal momentum of the system is the same before and after the rifle is fired (i.e., is conserved).

SET UP: Take the positive x-axis to be the direction the rifle is aimed. Initially, both the rifle and bullet are at rest, so the initial x-component of total momentum is zero. After the shot is fired, the x-component of the bullet's momentum is $p_{Bx} = m_B v_{Bx}$ and the x-component of the rifle's momentum is $p_{Rx} = m_R v_{Rx}$. Our target variables are v_R, p_{Bx}, p_{Rx}, and $K_B = \frac{1}{2} m_B v_{Bx}^2$ and, $K_R = \frac{1}{2} m_R v_{Rx}^2$ (the final kinetic energies of the bullet and rifle, respectively).

EXECUTE: Conservation of the x-component of total momentum gives

$$P_x = 0 = m_B v_{Bx} + m_R v_{Rx}$$

$$v_{Rx} = -\frac{m_B}{m_R} v_{Bx} = -\left(\frac{0.00500 \text{ kg}}{3.00 \text{ kg}}\right)(300 \text{ m/s}) = -0.500 \text{ m/s}$$

The negative sign means that the recoil is in the direction opposite to that of the bullet. If the butt of a rifle were to hit your shoulder traveling at this speed, you'd feel it. That's what the "kick" of a rifle is all about. It's more comfortable to hold the rifle tightly against your shoulder when you fire it; then m_R is replaced by the sum of your mass and the rifle's mass, and the recoil speed is much less.

The final momentum and kinetic energy of the bullet are

$$p_{Bx} = m_B v_{Bx} = (0.00500 \text{ kg})(300 \text{ m/s}) = 1.50 \text{ kg} \cdot \text{m/s}$$

$$K_B = \frac{1}{2} m_B v_{Bx}^2 = \frac{1}{2}(0.00500 \text{ kg})(300 \text{ m/s})^2 = 225 \text{ J}$$

For the rifle, the momentum and final kinetic energy are

$$p_{Rx} = m_R v_{Rx} = (3.00 \text{ kg})(-0.500 \text{ m/s}) = -1.50 \text{ kg} \cdot \text{m/s}$$

$$K_R = \frac{1}{2} m_R v_{Rx}^2 = \frac{1}{2}(3.00 \text{ kg})(-0.500 \text{ m/s})^2 = 0.375 \text{ J}$$

8.10 The ratio of bullet speed to recoil speed is the inverse of the ratio of bullet mass to rifle mass.

EVALUATE: The bullet and the rifle have equal and opposite momenta after the interaction because they were subjected to equal and opposite interaction forces for the same amount of *time* (i.e., equal and opposite impulses). But the bullet acquires much greater kinetic energy than the rifle because the bullet travels a much greater *distance* than the rifle during the interaction. Thus the force on the bullet does more work than the force on the rifle. The ratio of the two kinetic energies, 600:1, is equal to the inverse ratio of the masses; in fact, it can be shown that this always happens in recoil situations. We leave the proof as a problem (see Exercise 8.21).

Notice that our calculation doesn't depend at all on the details of how the rifle works. In a real rifle, the bullet is propelled forward by an explosive charge; if instead a very stiff spring were used to give the same bullet velocity, the answers would have been exactly the same.

Example 8.5	**Collision along a straight line**

Two gliders move toward each other on a frictionless linear air track (Figure 8.11a). After they collide, glider B moves away with a final velocity of $+2.0$ m/s (Fig. 8.11b). What is the final velocity of glider A? How do the changes in momentum and in velocity compare for the two gliders?

SOLUTION

IDENTIFY: The total vertical force on each glider is zero; the net force on each glider is the horizontal force exerted on it by the other glider. The net *external* force on the two gliders together is zero, so the total momentum is conserved.

SET UP: We take the x-axis to lie along the air track, with the positive direction to the right. We are given the masses and initial velocities of both gliders, as well as the final velocity of glider B (see Fig. 8.11). Our target variables are v_{A2x}, the final x-component of velocity of glider A, and the changes in momentum and in velocity of the two gliders (that is, the value after the collision minus the value before the collision).

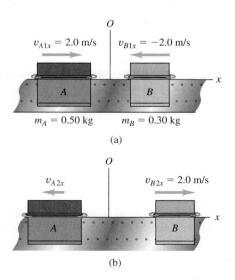

8.11 Two gliders (a) before and (b) after their collision.

EXECUTE: The x-component of total momentum before the collision is

$$P_x = m_A v_{A1x} + m_B v_{B1x}$$
$$= (0.50 \text{ kg})(2.0 \text{ m/s}) + (0.30 \text{ kg})(-2.0 \text{ m/s})$$
$$= 0.40 \text{ kg} \cdot \text{m/s}$$

This is positive (to the right in Fig. 8.11) because glider A has a greater magnitude of momentum before the collision than does glider B. The x-component of total momentum has the same value after the collision, so

$$P_x = m_A v_{A2x} + m_B v_{B2x}$$

Solving this equation for v_{A2x}, the final velocity of A, we find

$$v_{A2x} = \frac{P_x - m_B v_{B2x}}{m_A} = \frac{0.40 \text{ kg} \cdot \text{m/s} - (0.30 \text{ kg})(2.0 \text{ m/s})}{0.50 \text{ kg}}$$
$$= -0.40 \text{ m/s}$$

The change in momentum of glider A is

$$m_A v_{A2x} - m_A v_{A1x} = (0.50 \text{ kg})(-0.40 \text{ m/s})$$
$$- (0.50 \text{ kg})(2.0 \text{ m/s}) = -1.2 \text{ kg} \cdot \text{m/s}$$

and the change in momentum of glider B is

$$m_B v_{B2x} - m_B v_{B1x} = (0.30 \text{ kg})(2.0 \text{ m/s})$$
$$- (0.30 \text{ kg})(-2.0 \text{ m/s}) = +1.2 \text{ kg} \cdot \text{m/s}$$

The two interacting gliders undergo changes in momentum that are equal in magnitude and opposite in direction. The same is not true of their changes in velocity, however. For A, $v_{A2x} - v_{A1x} = (-0.40 \text{ m/s}) - 2.0 \text{ m/s} = -2.4 \text{ m/s}$; for B, $v_{B2x} - v_{B1x} = 2.0 \text{ m/s} - (-2.0 \text{ m/s}) = +4.0 \text{ m/s}$.

EVALUATE: Why do the momentum changes have the same magnitude for the two gliders, while the velocity changes do not? By Newton's third law, both gliders were acted on for equal amounts of time by an interaction force of the same magnitude. Hence both gliders experienced impulses of the same magnitude, as well as equal-magnitude changes in momentum. But by Newton's second law, the less massive glider (B) had a greater magnitude of acceleration and hence a greater velocity change.

When a sports utility vehicle (SUV) collides with a car of normal size, both cars undergo equal changes in momentum. The occupants of the normal-size car, however, are subjected to significantly greater acceleration (and significantly greater chance of injury) than the occupants of the SUV. An even more extreme example is what happens when an SUV collides with an insect: the SUV driver won't notice the resulting acceleration at all, but the insect surely will!

Example 8.6   Collision in a horizontal plane

Figure 8.12a shows two battling robots sliding on a frictionless surface. Robot A, with mass 20 kg, initially moves at 2.0 m/s parallel to the x-axis. It collides with robot B, which has mass 12 kg and is initially at rest. After the collision, robot A is moving at 1.0 m/s in a direction that makes an angle $\alpha = 30°$ with its initial direction (Fig. 8.12b). What is the final velocity of robot B?

SOLUTION

IDENTIFY: There are no horizontal (x or y) external forces, so both the x-component and the y-component of the total momentum of the system are conserved in the collision.

SET UP: The coordinate axes are shown in Fig. 8.12. The velocities are not all along a single line, so we have to treat momentum as a vector quantity, using components of each momentum in the x- and y-directions. Then momentum conservation requires that the sum of the x-components *before* the collision (subscript 1) must equal the sum *after* the collision (subscript 2), and similarly for the y-components. Just as with force equilibrium problems, we write a separate equation for each component. Our target variable is $\vec{v}_{B2}$, the final velocity of robot B. We'll express our answer in terms of the x- and y-components as well as in terms of magnitude and direction.

(a)

(b)

8.12 Views from above of the velocities (a) before and (b) after the collision.

EXECUTE: For the x-components (before and after) we have

$$m_A v_{A1x} + m_B v_{B1x} = m_A v_{A2x} + m_B v_{B2x}$$

$$v_{B2x} = \frac{m_A v_{A1x} + m_B v_{B1x} - m_A v_{A2x}}{m_B}$$

$$= \frac{\left[\begin{array}{c} (20 \text{ kg})(2.0 \text{ m/s}) + (12 \text{ kg})(0) \\ -(20 \text{ kg})(1.0 \text{ m/s})(\cos 30°) \end{array} \right]}{12 \text{ kg}}$$

$$= 1.89 \text{ m/s.}$$

Conservation of the y-component of total momentum gives

$$m_A v_{A1y} + m_B v_{B1y} = m_A v_{A2y} + m_B v_{B2y}$$

$$v_{B2y} = \frac{m_A v_{A1y} + m_B v_{B1y} - m_A v_{A2y}}{m_B}$$

$$= \frac{\left[\begin{array}{c} (20 \text{ kg})(0) + (12 \text{ kg})(0) \\ -(20 \text{ kg})(1.0 \text{ m/s})(\sin 30°) \end{array} \right]}{12 \text{ kg}}$$

$$= -0.83 \text{ m/s}$$

After the collision, robot B moves in the positive x-direction and the negative y-direction (Figure 8.12b). The magnitude of $\vec{v}_{B2}$ is

$$v_{B2} = \sqrt{(1.89 \text{ m/s})^2 + (-0.83 \text{ m/s})^2} = 2.1 \text{ m/s}$$

and the angle of its direction from the positive x-axis is

$$\beta = \arctan \frac{-0.83 \text{ m/s}}{1.89 \text{ m/s}} = -24°$$

EVALUATE: A useful way to check our answer is to look at the values of the momentum components before and after the collision. Initially all of the momentum is in robot A, which has x-momentum $m_A v_{A1x} = (20 \text{ kg})(2.0 \text{ m/s}) = 40 \text{ kg} \cdot \text{m/s}$ and zero y-momentum. After the collision, robot A has x-momentum $m_A v_{A2x} = (20 \text{ kg})(1.0 \text{ m/s}) \times (\cos 30°) = 17 \text{ kg} \cdot \text{m/s}$, while robot B has x-momentum $m_B v_{B2x} = (12 \text{ kg})(1.89 \text{ m/s}) = 23 \text{ kg} \cdot \text{m/s}$; the total x-momentum is 40 $\text{kg} \cdot \text{m/s}$, the same as before the collision (as it should be). In the y-direction, the momentum of robot A after the collision is $m_A v_{A2y} = (20 \text{ kg})(1.0 \text{ m/s})(\sin 30°) = 10 \text{ kg} \cdot \text{m/s}$, while robot B has y-momentum $m_B v_{B2y} = (12 \text{ kg})(-0.83 \text{ m/s}) = -10 \text{ kg} \cdot \text{m/s}$. Note that while robot A acquires a positive y-momentum in the collision, robot B acquires a negative y-momentum of the same magnitude. Thus the *total* y-component of momentum after the collision has the same value (zero) as before the collision.

Test Your Understanding

A spring-loaded toy of mass 0.60 kg sits at rest on a horizontal frictionless surface. When the spring releases, the toy breaks into three equal-mass pieces, A, B, and C, which slide along the surface. Piece A moves off in the negative x-direction at 3.0 m/s, while piece B moves off in the negative y-direction at 4.0 m/s. What is the velocity (magnitude and direction) of piece C?

8.3 | Inelastic Collisions

To the average person on the street the term *collision* is likely to mean some sort of automotive disaster. We'll use it in that sense, but we'll also broaden the meaning to include any strong interaction between bodies that lasts a relatively short time. So we include not only car accidents but also balls colliding on a billiard table, neutrons hitting atomic nuclei in a nuclear reactor, a bowling ball striking the pins, the impact of a meteor on the Arizona desert, and a close encounter of a spacecraft with the planet Saturn.

If the forces between the bodies are much larger than any external forces, as is the case in most collisions, we can neglect the external forces entirely and treat the bodies as an *isolated* system. Then momentum is conserved and the total momentum of the system has the same value before and after the collision. Two cars colliding at an icy intersection provide a good example. Even two cars colliding on dry pavement can be treated as an isolated system during the collision if the forces between the cars are much larger than the friction forces of pavement against tires.

6.4   Collision Problems

6.8   Skier and Cart

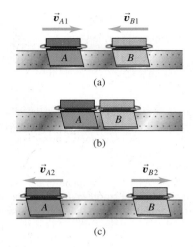

8.13 Model of an elastic collision: Glider A and glider B approach each other on a frictionless surface. Each glider has a steel spring bumper on the end to ensure an elastic collision.

If the forces between the bodies are also *conservative,* so that no mechanical energy is lost or gained in the collision, the total *kinetic* energy of the system is the same after the collision as before. Such a collision is called an **elastic collision.** A collision between two marbles or two billiard balls is almost completely elastic. Figure 8.13 shows a model for an elastic collision. When the bodies collide, the springs are momentarily compressed and some of the original kinetic energy is momentarily converted to elastic potential energy. Then the bodies bounce apart, the springs expand, and this potential energy is converted back to kinetic energy.

A collision in which the total kinetic energy after the collision is *less* than that before the collision is called an **inelastic collision.** A meatball landing on a plate of spaghetti and a bullet embedding itself in a block of wood are examples of inelastic collisions. An inelastic collision in which the colliding bodies stick together and move as one body after the collision is often called a **completely inelastic collision.** An example is shown in Fig. 8.14; we have replaced the spring bumper in Fig. 8.13 with a ball of putty that squashes and sticks the two bodies together.

> **CAUTION** It's a common misconception that the *only* inelastic collisions are those in which the colliding bodies stick together. In fact, inelastic collisions include many situations in which the bodies do *not* stick. If two cars bounce off each other in a "fender-bender," the work done to deform the fenders cannot be recovered as kinetic energy of the cars, and so the collision is inelastic (Fig. 8.15).

Remember the following rule: **In any collision in which external forces can be neglected, momentum is conserved and the total momentum before equals the total momentum after; in elastic collisions *only*, the total kinetic energy before equals the total kinetic energy after.**

Completely Inelastic Collisions

Let's look at what happens to momentum and kinetic energy in a *completely* inelastic collision of two bodies (A and B), as in Fig. 8.14. Because the two bodies stick together after the collision, their final velocities must be equal:

$$\vec{v}_{A2} = \vec{v}_{B2} = \vec{v}_2$$

Conservation of momentum gives the relation

$$m_A\vec{v}_{A1} + m_B\vec{v}_{B1} = (m_A + m_B)\vec{v}_2 \quad \text{(completely inelastic collision)} \quad (8.16)$$

If we know the masses and initial velocities, we can compute the common final velocity $\vec{v}_2$.

Suppose, for example, that a body with mass m_A and initial x-component of velocity v_1 collides inelastically with a body with mass m_B that is initially at rest $(v_{B1} = 0)$. From Eq. (8.16) the common x-component of velocity v_2 of both bodies after the collision is

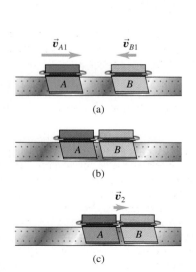

8.14 Model of a completely inelastic collision. The spring bumpers on the gliders are replaced by putty, so the gliders stick together after collision.

$$v_2 = \frac{m_A}{m_A + m_B}v_1 \quad \text{(completely inelastic collision, } B \text{ initially at rest)} \quad (8.17)$$

Let's verify that the total kinetic energy after this completely inelastic collision is less than before the collision. The kinetic energies K_1 and K_2 before and after the collision, respectively, are

$$K_1 = \frac{1}{2}m_A v_1^2$$

$$K_2 = \frac{1}{2}(m_A + m_B)v_2^2 = \frac{1}{2}(m_A + m_B)\left(\frac{m_A}{m_A + m_B}\right)^2 v_1^2$$

The ratio of final to initial kinetic energy is

$$\frac{K_2}{K_1} = \frac{m_A}{m_A + m_B} \quad \text{(completely inelastic collision, } B \text{ initially at rest)} \quad (8.18)$$

The right side is always less than unity because the denominator is always greater than the numerator. Even when the initial velocity of m_B is not zero, it is not hard to verify that the kinetic energy after a completely inelastic collision is always less than before.

Please note: We don't recommend memorizing Eqs. (8.17) or (8.18). We derived them only to prove that kinetic energy is always lost in a completely inelastic collision.

8.15 Automobile collisions are intended to be inelastic, so that the structure of the car absorbs as much of the energy of the collision as possible. This absorbed energy cannot be recovered, since it goes into a permanent deformation of the car.

| Example 8.7 | **A completely inelastic collision** |

Suppose that in the collision described in Example 8.5 (Section 8.2), the two gliders do not bounce but stick together after the collision. The masses and initial velocities are as shown in Figure 8.11a. Find the common final velocity v_{2x}, and compare the initial and final kinetic energies.

SOLUTION

IDENTIFY and SET UP: As in Example 8.5, we take the positive x-axis to point to the right in Fig. 8.11a. There are no external forces in the x-direction, so the x-component of momentum is conserved. Our target variables are the final velocity v_{2x} and the initial and final kinetic energies of the system.

EXECUTE: From conservation of the x-component of momentum,

$$m_A v_{A1x} + m_B v_{B1x} = (m_A + m_B)v_{2x}$$

$$v_{2x} = \frac{m_A v_{A1x} + m_B v_{B1x}}{m_A + m_B}$$

$$= \frac{(0.50 \text{ kg})(2.0 \text{ m/s}) + (0.30 \text{ kg})(-2.0 \text{ m/s})}{(0.50 \text{ kg} + 0.30 \text{ kg})}$$

$$= 0.50 \text{ m/s}$$

Because v_{2x} is positive, the gliders move together to the right (the $+x$-direction) after the collision.

Before the collision, the kinetic energies of gliders A and B are

$$K_A = \frac{1}{2}m_A v_{A1x}^2 = \frac{1}{2}(0.50 \text{ kg})(2.0 \text{ m/s})^2 = 1.0 \text{ J}$$

$$K_B = \frac{1}{2}m_B v_{B1x}^2 = \frac{1}{2}(0.30 \text{ kg})(-2.0 \text{ m/s})^2 = 0.60 \text{ J}$$

(Note that the kinetic energy of glider B is positive, even though the x-components of its velocity v_{B1x} and momentum $m_B v_{B1x}$ are both negative.) The *total* kinetic energy before the collision is 1.6 J. The kinetic energy after the collision is

$$\frac{1}{2}(m_A + m_B)v_{2x}^2 = \frac{1}{2}(0.50 \text{ kg} + 0.30 \text{ kg})(0.50 \text{ m/s})^2 = 0.10 \text{ J}$$

EVALUATE: The final kinetic energy is only $\frac{1}{16}$ of the original; $\frac{15}{16}$ is converted from mechanical energy to various other forms. If there is a ball of chewing gum between the gliders, it squashes and becomes warmer. If the gliders couple together like two freight cars, the energy goes into elastic vibrations that eventually dissipate. If there is a spring between the gliders that is compressed as they lock together, then the energy is stored as potential energy of the spring. In all of these cases the *total* energy of the system is conserved, although *kinetic* energy is not. However, in an isolated system, momentum is *always* conserved, whether the collision is elastic or not.

Example
8.8   **The ballistic pendulum**

Figure 8.16 shows a ballistic pendulum, a system for measuring the speed of a bullet. The bullet, with mass m, is fired into a block of wood with mass M, suspended like a pendulum, and makes a completely inelastic collision with it. After the impact of the bullet, the block swings up to a maximum height y. Given the values of y, m, and M, what is the initial velocity v_x of the bullet?

Another unknown quantity is the x-component of velocity of the block and bullet as a unit just after the collision (that is, just at the end of the first stage); we'll call it V_x. We'll use momentum conservation in the first stage to relate v_x to V_x, and we'll use energy conservation in the second stage to relate V_x to the (given) maximum height y.

SOLUTION

IDENTIFY: We'll analyze this event in two stages: (1) the embedding of the bullet in the block and (2) the subsequent swinging of the block on its strings. Let's first find out which physical quantities are conserved in each stage.

During the first stage, the bullet embeds itself in the block so quickly that the block has no time to swing appreciably away from its straight-down position. During this short-duration impact the supporting strings remain very nearly vertical, and negligible external horizontal force acts on the system of bullet plus block. Thus, in this first stage, the *horizontal component of momentum* is conserved. Note that mechanical energy is *not* conserved in this stage because a nonconservative force acts on the system (the force of friction between the bullet and the block).

In the second stage, after the collision, the block and bullet move as a unit. The only forces acting on this unit are gravity (a conservative force) and the string tensions (which do no work). Thus, as the block swings upward and to the right, *mechanical energy* is conserved. Momentum is *not* conserved during this stage because there is a net external force (the sum of gravity and the string tensions, which don't cancel when the strings are inclined).

SET UP: We take the positive x-axis to be to the right and the positive y-axis to be upward in Fig. 8.16. Our target variable is v_y.

EXECUTE: In the first stage, momentum conservation gives

$$mv_x = (m + M)V_x \qquad v_x = \frac{m + M}{m}V_x$$

At the beginning of the second stage, the block–bullet unit has kinetic energy $K = \frac{1}{2}(m + M)V_x^2$. (As in Eq. (8.18), this is less than the kinetic energy before the collision; the collision is inelastic!) The block–bullet unit swings up and comes to rest for an instant at a height y, where its kinetic energy is zero and its potential energy is $(m + M)gy$; it then swings back down. Energy conservation gives

$$\frac{1}{2}(m + M)V_x^2 = (m + M)gy \qquad V_x = \sqrt{2gy}$$

Now we substitute this expression into the momentum equation to find an expression for our target variable v_x:

$$v_x = \frac{m + M}{m}\sqrt{2gy}$$

By measuring m, M, and y, we can compute the original speed of the bullet.

EVALUATE: Let's check our answers by plugging in some realistic numbers. If $m = 5.00$ g $= 0.00500$ kg, $M = 2.00$ kg, and $y = 3.00$ cm $= 0.0300$ m, the initial velocity of the bullet is

$$v_x = \frac{2.00 \text{ kg} + 0.00500 \text{ kg}}{0.00500 \text{ kg}}\sqrt{2(9.80 \text{ m/s}^2)(0.0300 \text{ m})}$$

$$= 307 \text{ m/s}$$

The velocity V_x of the block just after impact is

$$V_x = \sqrt{2gy} = \sqrt{2(9.80 \text{ m/s}^2)(0.0300 \text{ m})}$$

$$= 0.767 \text{ m/s}$$

The kinetic energy of the bullet just before impact is $\frac{1}{2}(0.00500 \text{ kg})$ $\times (307 \text{ m/s})^2 = 236$ J. Just after impact the kinetic energy of the bullet and block is $\frac{1}{2}(2.005 \text{ kg})(0.767 \text{ m/s})^2 = 0.589$ J. Nearly all the kinetic energy disappears as the wood splinters and the bullet and block become hotter.

8.16 A ballistic pendulum.

<table>
<tr><td>Example
8.9</td><td>**Analysis of an automobile collision**</td></tr>
</table>

A small compact car with a mass of 1000 kg is traveling north on Morewood Avenue at 15 m/s. At the intersection of Morewood and Fifth Avenue it collides with an enormous SUV with a mass of 2000 kg, traveling east on Fifth Avenue at 10 m/s (Fig. 8.17a). Fortunately, all occupants are wearing seat belts and there are no injuries, but the two cars are thoroughly tangled and move away from the impact point as one mass. a) Treating each car as a particle, find the total momentum just before the collision. b) The insurance adjustor has asked you to find the velocity of the wreckage just after impact. What do you tell her?

SOLUTION

IDENTIFY: We are given the masses and initial velocities of the cars, so it's a straightforward procedure to find the total momentum before the collision. To find out what happens *after* the collision, we'll assume that we can treat the cars as an isolated system during the collision. We can do so because the horizontal forces that the cars exert on each other during the collision have very large magnitudes, great enough to crumple the cars' metal skins. Compared to these forces, we can neglect any external forces such as friction. (We'll justify this assumption later.) Hence the momentum of the system of two cars has the same value just before and just after the collision.

SET UP: In Fig. 8.17, we have drawn coordinate axes and have labeled the small car A and the sports utility vehicle B. Our first target variable is the total momentum before the collision, $\vec{P}$, which we'll find using Eqs. (8.15). The momentum has the same value just after the collision; hence, once we've found $\vec{P}$, we'll be able to find the velocity $\vec{V}$ just after the collision (our second target variable) using the relationship $\vec{P} = M\vec{V}$, where M is the combined mass of the wreckage.

EXECUTE:

a) From Eqs. (8.15) the components of the total momentum $\vec{P}$ are

$$P_x = p_{Ax} + p_{Bx} = m_A v_{Ax} + m_B v_{Bx}$$
$$= (1000 \text{ kg})(0) + (2000 \text{ kg})(10 \text{ m/s})$$
$$= 2.0 \times 10^4 \text{ kg} \cdot \text{m/s}$$
$$P_y = p_{Ay} + p_{By} = m_A v_{Ay} + m_B v_{By}$$
$$= (1000 \text{ kg})(15 \text{ m/s}) + (2000 \text{ kg})(0)$$
$$= 1.5 \times 10^4 \text{ kg} \cdot \text{m/s}$$

The magnitude of $\vec{P}$ is

$$P = \sqrt{(2.0 \times 10^4 \text{ kg} \cdot \text{m/s})^2 + (1.5 \times 10^4 \text{ kg} \cdot \text{m/s})^2}$$
$$= 2.5 \times 10^4 \text{ kg} \cdot \text{m/s}$$

and its direction is given by the angle θ shown in Fig. 8.17b, where

$$\tan \theta = \frac{P_y}{P_x} = \frac{1.5 \times 10^4 \text{ kg} \cdot \text{m/s}}{2.0 \times 10^4 \text{ kg} \cdot \text{m/s}} = 0.75 \qquad \theta = 37°$$

b) The total momentum just after the collision is the same as just before, which we found to be 2.5×10^4 kg·m/s in a direction 37° north of due east. Assuming that no parts fall off, the total mass of wreckage is $M = m_A + m_B = 3000$ kg. Using $\vec{P} = M\vec{V}$, the direction of the velocity $\vec{V}$ just after the collision is the same as that of the momentum, and its magnitude is

$$V = \frac{P}{M} = \frac{2.5 \times 10^4 \text{ kg} \cdot \text{m/s}}{3000 \text{ kg}} = 8.3 \text{ m/s}$$

EVALUATE: This is an inelastic collision, so we expect the total kinetic energy to be less after the collision than before. We invite you to carry out the calculations; you will find that the initial

(a)

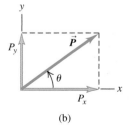

(b)

8.17 (a) View from a traffic helicopter overhead of the start of the collision at Morewood and Fifth. (b) Finding the total momentum just before the collision.

kinetic energy is 2.1×10^5 J and the final value is 1.0×10^5 J. Over half of the initial kinetic energy is converted to other forms.

We still need to justify our assumption that we can neglect the external forces on the vehicles during the collision. To do so, let's look at some numbers. The mass of the SUV is 2000 kg, its weight is about 20,000 N, and, if the coefficient of friction is about 0.5, the friction force when it slides across the pavement is about 10,000 N. The SUV's kinetic energy just before the impact is $\frac{1}{2}(2000 \text{ kg}) \times (10 \text{ m/s})^2 = 1.0 \times 10^5$ J. The car may crumple 0.2 m or so; to do -1.0×10^5 J of work on the car (required to stop it) in a distance of 0.2 m would require a force of 5.0×10^5 N, which is 50 times greater than the friction force. So it's quite reasonable to ignore the external force of friction as compared to the internal forces that the vehicles exert on each other.

CAUTION You may have been tempted to try to find the final velocity by taking the vector sum of the initial velocities. But there is *no* "law of conservation of velocities." The conserved quantity is the total momentum of the system, and the above analysis is the only correct way to do the problem.

Test Your Understanding

Find the kinetic energies of the robots in Example 8.6 (Section 8.2) before and after the collision. Was the collision inelastic? Was it totally inelastic?

Activ
ONLINE
Physics

6.2   Collisions and Elasticity
6.5   Car Collisions: Two Dimensions
6.9   Pendulum Bashes Box

8.18 Billiard balls deform very little when they collide, and quickly spring back from any deformation they do undergo. Hence the force of interaction between the balls is almost perfectly conservative, and the collision is almost perfectly elastic.

8.4 | Elastic Collisions

As discussed in Section 8.3, an *elastic collision* in an isolated system is a collision in which kinetic energy (as well as momentum) is conserved. Elastic collisions occur when the forces between the colliding bodies are *conservative*. When two steel balls collide, they squash a little near the surface of contact, but then they spring back. Some of the kinetic energy is stored temporarily as elastic potential energy, but at the end it is reconverted to kinetic energy (Fig. 8.18).

Let's look at an elastic collision between two bodies A and B. We start with a one-dimensional collision, in which all the velocities lie along the same line; we choose this line to be the x-axis. Each momentum and velocity then has only an x-component. We call the x-components of velocity before the collision v_{A1x} and v_{B1x}, and those after the collision v_{A2x} and v_{B2x}. From conservation of kinetic energy we have

$$\frac{1}{2}m_A v_{A1x}^2 + \frac{1}{2}m_B v_{B1x}^2 = \frac{1}{2}m_A v_{A2x}^2 + \frac{1}{2}m_B v_{B2x}^2$$

and conservation of momentum gives

$$m_A v_{A1x} + m_B v_{B1x} = m_A v_{A2x} + m_B v_{B2x}$$

If the masses m_A and m_B and the initial velocities v_{A1x} and v_{B1x} are known, these two equations can be solved simultaneously to find the two final velocities v_{A2x} and v_{B2x}.

The general solution is a little complicated, so we will concentrate on the particular case in which body B is at rest before the collision. Think of it as a target for body A to hit. We can then simplify the velocity notation; we let v_x be the initial x-component of velocity of A, and v_{Ax} and v_{Bx} be the final x-components for A and B. Then the kinetic energy and momentum conservation equations are, respectively,

$$\frac{1}{2}m_A v_x^2 = \frac{1}{2}m_A v_{Ax}^2 + \frac{1}{2}m_B v_{Bx}^2 \tag{8.19}$$

$$m_A v_x = m_A v_{Ax} + m_B v_{Bx} \tag{8.20}$$

We may solve for v_{Ax} and v_{Bx} in terms of the masses and the initial velocity v_x. This involves some fairly strenuous algebra, but it's worth it. No pain, no gain! The simplest approach is somewhat indirect, but along the way it uncovers an additional interesting feature of elastic collisions.

First rearrange Eqs. (8.19) and (8.20) as follows:

$$m_B v_{Bx}^2 = m_A (v_x^2 - v_{Ax}^2) = m_A (v_x - v_{Ax})(v_x + v_{Ax}) \tag{8.21}$$

$$m_B v_{Bx} = m_A (v_x - v_{Ax}) \tag{8.22}$$

Now divide Eq. (8.21) by Eq. (8.22) to obtain

$$v_{Bx} = v_x + v_{Ax} \tag{8.23}$$

Substitute this expression back into Eq. (8.22) to eliminate v_{Bx} and then solve for v_{Ax}:

$$m_B(v_x + v_{Ax}) = m_A(v_x - v_{Ax})$$

$$v_{Ax} = \frac{m_A - m_B}{m_A + m_B}v_x \tag{8.24}$$

Finally, substitute this result back into Eq. (8.23) to obtain

$$v_{Bx} = \frac{2m_A}{m_A + m_B}v_x \tag{8.25}$$

Now we can interpret the results. Suppose body A is a ping-pong ball and body B is a bowling ball. Then we expect A to bounce off after the collision with a velocity nearly equal to its original value but in the opposite direction (Fig. 8.19a), and we expect B's velocity to be much smaller. That's just what the equations predict. When m_A is much smaller than m_B, the fraction in Eq. (8.24) is approximately equal to (-1), so v_{Ax} is approximately equal to $-v_x$. The fraction in Eq. (8.25) is much smaller than unity, so v_{Bx} is much less than v_x. Figure 8.19b shows the opposite case, in which A is the bowling ball and B the ping-pong ball and m_A is much larger than m_B. What do you expect to happen then? Check your predictions against Eqs. (8.24) and (8.25).

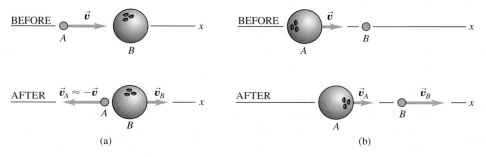

BEFORE $\vec{v}$ A B x BEFORE $\vec{v}$ A B x

AFTER $\vec{v}_A \approx -\vec{v}$ A $\vec{v}_B$ B x AFTER $\vec{v}_A$ A $\vec{v}_B$ B x

(a)   (b)

8.19 Collisions between a ping-pong ball and a bowling ball. (a) Initially stationary bowling ball. (b) Initially stationary ping-pong ball.

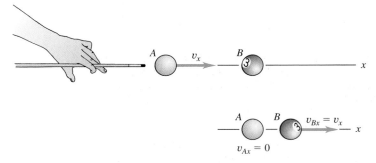

8.20 In a one-dimensional elastic collision between bodies of equal mass, if one body is at rest, it receives all the momentum and kinetic energy of the moving body.

Another interesting case occurs when the masses are equal (Fig. 8.20). If $m_A = m_B$, then Eqs. (8.24) and (8.25) give $v_{Ax} = 0$ and $v_{Bx} = v_x$. That is, the body that was moving stops dead; it gives all its momentum and kinetic energy to the body that was at rest. This behavior is familiar to all pool players and marbles shooters.

Let's return to the general case in which A and B have different masses. Equation (8.23) can be rewritten as

$$v_x = v_{Bx} - v_{Ax} \qquad (8.26)$$

Here $v_{Bx} - v_{Ax}$ is just the velocity of B relative to A *after* the collision; from Eq. (8.26), this equals v_x, which is the *negative* of the velocity of B relative to A *before* the collision. (We discussed relative velocity in Section 3.5.) The relative velocity has the same magnitude, but opposite sign, before and after the collision. The sign changes because A and B are approaching each other before the collision but receding from each other after the collision. If we view this collision from a second coordinate system moving with constant velocity relative to the first, the velocities of the bodies are different but the *relative* velocities are the same. Hence our statement about relative velocities holds for *any* straight-line elastic collision, even when neither body is at rest initially. *In a straight-line elastic collision of two bodies, the relative velocities before and after the collision have the same magnitude but opposite sign.* This can be expressed using our original notation as

$$v_{B2x} - v_{A2x} = -(v_{B1x} - v_{A1x}) \qquad (8.27)$$

It turns out that a *vector* relationship similar to Eq. (8.27) is a general property of *all* elastic collisions, even when both bodies are moving initially and the velocities do not all lie along the same line. This result provides an alternative and equivalent definition of an elastic collision: *In an elastic collision the relative velocity of the two bodies has the same magnitude before and after the collision.* Whenever this condition is satisfied, the total kinetic energy is also conserved.

When an elastic two-body collision isn't head-on, the velocities don't all lie along a single line. If they all lie in a plane, each final velocity has two unknown components, and there are four unknowns in all. Conservation of energy and conservation of the x- and y-components of momentum give only three equations. To determine the final velocities uniquely, we need additional information, such as the direction or magnitude of one of the final velocities.

Example
8.10

An elastic straight-line collision

The situation in Fig. 8.21 is the same as for Example 8.5 in Section 8.2, but now we've added ideal spring bumpers to the gliders so that the collision is elastic. What are the velocities of A and B after the collision?

SOLUTION

IDENTIFY and SET UP: As in Example 8.5, the net external force on the system of two gliders is zero and the momentum of the system is conserved. We again choose the positive x-axis to point to the right. We'll find our target variables, v_{A2x} and v_{B2x}, using Eq. (8.27) and the equation of momentum conservation.

EXECUTE: From conservation of momentum,

$$m_A v_{A1} + m_B v_{B1} = m_A v_{A2} + m_B v_{B12}$$
$$(0.50 \text{ kg})(2.0 \text{ m/s}) + (0.30 \text{ kg})(-2.0 \text{ m/s})$$
$$= (0.50 \text{ kg})v_{A2x} + (0.30 \text{ kg})v_{B2x}$$
$$0.50v_{A2x} + 0.30v_{B2x} = 0.40 \text{ m/s}$$

(In the last equation we have divided through by the unit "kg.") From Eq. (8.27), the relative velocity relation for an elastic collision,

$$v_{B2x} - v_{A2x} = -(v_{B1x} - v_{A1x})$$
$$= -(-2.0 \text{ m/s} - 2.0 \text{ m/s}) = 4.0 \text{ m/s}$$

Before the collision, the velocity of B relative to A is to the left at 4.0 m/s; after the collision, the velocity of B relative to A is to the right at 4.0 m/s. Solving these equations simultaneously, we find

$$v_{A2x} = -1.0 \text{ m/s} \qquad v_{B2x} = 3.0 \text{ m/s}$$

EVALUATE: Both bodies reverse their directions of motion; A moves to the left at 1.0 m/s and B moves to the right at 3.0 m/s. This

$v_{A1x} = 2.0 \text{ m/s}$ $v_{B1x} = -2.0 \text{ m/s}$

$m_A = 0.50 \text{ kg}$ $m_B = 0.30 \text{ kg}$

(a)

v_{A2x} v_{B2x}

(b)

8.21 Air-track gliders (a) before and (b) after an elastic collision.

is different from the result of Example 8.5 because that collision was *not* elastic. The total kinetic energy after the elastic collision is

$$\frac{1}{2}(0.50 \text{ kg})(-1.0 \text{ m/s})^2 + \frac{1}{2}(0.30 \text{ kg})(3.0 \text{ m/s})^2 = 1.6 \text{ J}$$

As expected, this equals the total kinetic energy before the collision (which we calculated in Example 8.7, Section 8.3).

CAUTION You might have been tempted to solve this problem using Eqs. (8.24) and (8.25). These equations apply *only* to situations in which body B is initially at rest, which isn't the case here. When in doubt, always solve the problem at hand using equations that are applicable to a broad variety of cases.

Example
8.11

Moderator in a nuclear reactor

High-speed neutrons are produced in a nuclear reactor during nuclear fission processes. Before a neutron can trigger additional fissions, it has to be slowed down by collisions with nuclei in the *moderator* of the reactor. The first nuclear reactor (built in 1942 at the University of Chicago) and the reactor involved in the 1986 Chernobyl accident both used carbon (graphite) as the moderator material. A neutron (mass 1.0 u) traveling at 2.6×10^7 m/s undergoes a head-on elastic collision with a carbon nucleus (mass 12 u) initially at rest. The external forces during the collision are negligible. What are the velocities after the collision? (1 u is the *atomic mass unit,* equal to 1.66×10^{-27} kg.)

SOLUTION

IDENTIFY and SET UP: We are given that the external forces can be neglected (so momentum is conserved in the collision) and that the collision is elastic. We take the x-axis to be in the direction that the neutron is moving initially. Because the collision is head-on, both the neutron and the carbon nucleus move along this same axis after the collision. Furthermore, because one body is initially at rest, we can use Eqs. (8.24) and (8.25) with

$$m_A = 1.0 \text{ u} \qquad m_B = 12 \text{ u} \quad \text{and} \quad v_x = 2.6 \times 10^7 \text{ m/s}$$

to solve for the target variables v_{Ax} and v_{Bx} (the final velocities of the neutron and the carbon nucleus, respectively).

EXECUTE: We'll let you do the arithmetic; the results are

$$v_{Ax} = -2.2 \times 10^7 \text{ m/s} \qquad v_{Bx} = 0.4 \times 10^7 \text{ m/s}$$

EVALUATE: The neutron ends up with $\frac{11}{13}$ of its initial speed, and the speed of the recoiling carbon nucleus is $\frac{2}{13}$ of the neutron's initial speed. [These ratios are just the factors $(m_A - m_B)/(m_A + m_B)$

and $2m_A/(m_A + m_B)$ that appear in Eqs. (8.24) and (8.25).] Kinetic energy is proportional to speed squared, so the neutron's final kinetic energy is $\left(\frac{11}{13}\right)^2$, or about 0.72 of its original value. If it makes a second such collision, its kinetic energy is $(0.72)^2$, or about half its original value, and so on. After several collisions the neutron will be moving quite slowly and will be able to trigger a fission reaction in a uranium nucleus.

Example 8.12	**The gravitational slingshot effect**

Figure 8.22 shows the planet Saturn moving in the negative x-direction at its orbital speed (with respect to the sun) of 9.6 km/s. The mass of Saturn is 5.69×10^{26} kg. A spacecraft with mass 2150 kg approaches Saturn, moving initially in the $+x$-direction at 10.4 km/s. The gravitational attraction of Saturn (a conservative force) causes the spacecraft to swing around it and head off in the opposite direction. Find the final speed of the spacecraft after it is far enough away to be nearly free of Saturn's gravitational pull.

SOLUTION

IDENTIFY: Here the "collision" is not an impact but rather a gravitational interaction. Nonetheless, we can treat it as a one-dimensional elastic collision because the interaction force is conservative and the spacecraft departs in the direction opposite to that in which it arrives.

SET UP: Let the spacecraft be body A and let Saturn be body B. We can assume that Saturn's speed is essentially constant during the interaction because its mass is so much greater than that of the spacecraft. If we take the positive x-direction to be the direction in

which the spacecraft is moving initially, we have $v_{A1x} = 10.4$ km/s and $v_{B1x} = v_{B2x} = -9.6$ km/s. To find our target variable, the magnitude of v_{A2x}, we'll use the result given by Eq. (8.27): that the relative velocity of the spacecraft and Saturn has the same magnitude before and after the interaction.

EXECUTE: Being very careful with signs, we find from Eq. (8.27) that

$$
\begin{aligned}
v_{A2x} &= v_{B2x} + v_{B1x} - v_{A1x} \\
&= (-9.6 \text{ km/s}) + (-9.6 \text{ km/s}) - 10.4 \text{ km/s} \\
&= -29.6 \text{ km/s}
\end{aligned}
$$

The spacecraft's final speed (relative to the sun) is 29.6 km/s, nearly three times as great as before the encounter with Saturn. Not only does the spacecraft get a close look at Saturn, but its kinetic energy increases by a factor of $(29.6/10.4)^2 = 8.1$ in the process.

EVALUATE: How can the spacecraft acquire so much additional speed and kinetic energy? The reason is that Saturn is *moving* in its orbit around the sun. If Saturn were initially at rest, the spacecraft would reverse direction but would undergo hardly any change in

8.22 The gravitational slingshot effect: a simplified model of a spacecraft making a non-impact "collision" with the planet Saturn.

speed; a baseball behaves similarly when it bounces off a bat that's held stationary (a "bunt"). But when a baseball collides with a bat that's swinging toward it, the baseball leaves the bat with a much greater speed, perhaps enough to make a home run. In our example, Saturn takes the place of the swinging bat, and the spacecraft gains speed much like a batted baseball.

This is a simplified version of the *gravitational slingshot effect,* used to give an extra boost to spacecraft. (The actual motion of a spacecraft is never purely one-dimensional as we have assumed here. We'll discuss the trajectories of spacecraft and planets in detail in Chapter 12.) The *Voyager 2* spacecraft, launched in 1977, used the slingshot effect in flying past Jupiter, Saturn, and Uranus. Thanks to the kinetic energy gained in this way, *Voyager 2* reached the planet Neptune in 1989; without the slingshot effect, *Voyager 2* would not have arrived at Neptune until 2008. The *Cassini* spacecraft, launched in 1997, used an even more elaborate trajectory that involved slingshot flybys of Venus, Earth, and Jupiter to help propel it toward a rendezvous with Saturn in 2004.

In this bit of celestial gymnastics, does the spacecraft get something for nothing? Where does its extra energy come from?

| Example 8.13 | **A two-dimensional elastic collision** |

Figure 8.23 shows an elastic collision of two pucks on a frictionless air-hockey table. Puck A has mass $m_A = 0.500$ kg and puck B has mass $m_B = 0.300$ kg. Puck A has an initial velocity of 4.00 m/s in the positive x-direction and a final velocity of 2.00 m/s in an unknown direction. Puck B is initially at rest. Find the final speed v_{B2} of puck B and the angles α and β in the figure.

SOLUTION

IDENTIFY and SET UP: The target variables are given in the statement of the problem. Although the collision is elastic, it is *not* one-dimensional, so we can't use any of the one-dimensional formulas derived in this section. Instead, we'll use the equations for conservation of energy, conservation of x-momentum, and conservation of y-momentum. These three equations should be enough to solve for our three target variables.

EXECUTE: Because the collision is elastic, the initial and final kinetic energies are equal:

$$\frac{1}{2}m_A v_{A1}^2 = \frac{1}{2}m_A v_{A2}^2 + \frac{1}{2}m_B v_{B2}^2$$

$$v_{B2}^2 = \frac{m_A v_{A1}^2 - m_A v_{A2}^2}{m_B}$$

$$= \frac{(0.500 \text{ kg})(4.00 \text{ m/s})^2 - (0.500 \text{ kg})(2.00 \text{ m/s})^2}{0.300 \text{ kg}}$$

$$v_{B2} = 4.47 \text{ m/s}$$

Conservation of the x-component of total momentum gives

$$m_A v_{A1x} = m_A v_{A2x} + m_B v_{B2x}$$

$$(0.500 \text{ kg})(4.00 \text{ m/s}) = (0.500 \text{ kg})(2.00 \text{ m/s})(\cos \alpha)$$
$$+ (0.300 \text{ kg})(4.47 \text{ m/s})(\cos \beta)$$

and conservation of the y-component gives

$$0 = m_A v_{A2y} + m_B v_{B2y}$$
$$0 = (0.500 \text{ kg})(2.00 \text{ m/s})(\sin \alpha)$$
$$- (0.300 \text{ kg})(4.47 \text{ m/s})(\sin \beta)$$

These are two simultaneous equations for α and β. The simplest solution is to eliminate β as follows: We solve the first equation for

$\cos \beta$ and the second for $\sin \beta$; we then square each equation and add. Since $\sin^2 \beta + \cos^2 \beta = 1$, this eliminates β and leaves an equation that we can solve for $\cos \alpha$ and hence for α. We can then substitute this value back into either of the two equations and solve the result for β. We leave the details for you to work out as a problem; the results are

$$\alpha = 36.9° \qquad \beta = 26.6°$$

EVALUATE: A quick way to check the answers is to make sure that the y-momentum, which was zero before the collision, is still zero after the collision. The y-momenta of the pucks are

$$p_{A2y} = (0.500 \text{ kg})(2.00 \text{ m/s})(\sin 36.9°) = +0.600 \text{ kg} \cdot \text{m/s}$$
$$p_{B2y} = -(0.300 \text{ kg})(4.47 \text{ m/s})(\sin 26.6°) = -0.600 \text{ kg} \cdot \text{m/s}$$

The sum of these values is zero, as it should be.

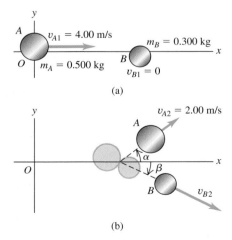

8.23 (a) Before and (b) after an elastic collision that isn't head-on.

The examples in this and the preceding sections show that we can classify collisions according to energy considerations. A collision in which kinetic energy is conserved is called *elastic*. A collision in which the total kinetic energy decreases is called *inelastic*. When the two bodies have a common final velocity, we say that the collision is *completely inelastic*. There are also cases in which the final kinetic energy is *greater* than the initial value. Rifle recoil, discussed in Example 8.4 (Section 8.2), is an example.

Finally, we emphasize again that we can sometimes use momentum conservation even when there are external forces acting on the system, if the net external force acting on the colliding bodies is small in comparison to the internal forces during the collision.

Test Your Understanding

Most present-day nuclear reactors use water as a moderator (see Example 8.11). After a neutron collides with a water molecule (mass 18 u), what fraction of its initial speed does the neutron have? What fraction of its initial kinetic energy? Are water molecules a better or worse moderator than carbon atoms? (One advantage of water is that it also acts as a coolant for the reactor's radioactive core.)

8.5 | Center of Mass

We can restate the principle of conservation of momentum in a useful way by using the concept of **center of mass.** Suppose we have several particles, with masses m_1, m_2, and so on. Let the coordinates of m_1 be (x_1, y_1), those of m_2 be (x_2, y_2), and so on. We define the center of mass of the system as the point having coordinates (x_{cm}, y_{cm}) given by

$$x_{cm} = \frac{m_1 x_1 + m_2 x_2 + m_3 x_3 + \cdots}{m_1 + m_2 + m_3 + \cdots} = \frac{\sum_i m_i x_i}{\sum_i m_i}$$

$$y_{cm} = \frac{m_1 y_1 + m_2 y_2 + m_3 y_3 + \cdots}{m_1 + m_2 + m_3 + \cdots} = \frac{\sum_i m_i y_i}{\sum_i m_i} \qquad \text{(center of mass)} \qquad (8.28)$$

The position vector $\vec{r}_{cm}$ of the center of mass can be expressed in terms of the position vectors $\vec{r}_1$, $\vec{r}_2$, ... of the particles as

$$\vec{r}_{cm} = \frac{m_1 \vec{r}_1 + m_2 \vec{r}_2 + m_3 \vec{r}_3 + \cdots}{m_1 + m_2 + m_3 + \cdots} = \frac{\sum_i m_i \vec{r}_i}{\sum_i m_i} \qquad \text{(center of mass)} \qquad (8.29)$$

In statistical language, the center of mass is a *mass-weighted average* position of the particles.

| Example 8.14 | **Center of mass of a water molecule** |

Figure 8.24 shows a simple model of the structure of a water molecule. The separation between atoms is $d = 9.57 \times 10^{-11}$ m. Each hydrogen atom has mass 1.0 u and the oxygen atom has mass 16.0 u. Find the position of the center of mass.

SOLUTION

IDENTIFY and SET UP: The coordinate system is shown in Fig. 8.24. Nearly all the mass of each atom is concentrated in its nucleus, which is only about 10^{-5} times the overall radius of the atom, so we can safely represent each atom as a point. We'll use Eqs. (8.28) to determine the coordinates x_{cm} and y_{cm}.

EXECUTE: The x-coordinate of each hydrogen atom is $d \cos$ (105°/2); the y-coordinates of the upper and lower hydrogen atoms are $+d \sin$ (105°/2) and $-d \sin$ (105°/2), respectively. The coordinates of the oxygen atom are $x = 0$, $y = 0$. From Eqs. (8.28) the x-coordinate of the center of mass is

$$x_{cm} = \frac{\left[\begin{array}{c} (1.0\,u)\,(d \cos 52.5°) + (1.0\,u) \\ \times\,(d \cos 52.5°) + (16.0\,u)\,(0) \end{array} \right]}{1.0\,u + 1.0\,u + 16.0\,u}$$

$$= 0.068\,d$$

and the y-coordinate is

$$y_{cm} = \frac{\left[\begin{array}{c} (1.0\,u)\,(d \sin 52.5°) + (1.0\,u) \\ \times\,(-d \sin 52.5°) + (16.0\,u)\,(0) \end{array} \right]}{1.0\,u + 1.0\,u + 16.0\,u}$$

$$= 0$$

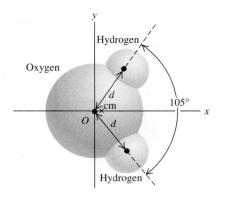

8.24 Where is the center of mass of a water molecule?

The actual value of d for the water molecule is 9.57×10^{-11} m, so we find

$$x_{cm} = (0.068)\,(9.57 \times 10^{-11}\,\text{m}) = 6.5 \times 10^{-12}\,\text{m}$$

EVALUATE: The center of mass is much closer to the oxygen atom than to either hydrogen atom because the oxygen atom is much more massive. Notice that the center of mass lies along the x-axis, which is the *axis of symmetry* of this molecule. If the molecule is rotated by 180° around this axis, it looks exactly the same as before. The position of the center of mass can't be affected by this rotation, so it must lie on the axis of symmetry.

For solid bodies, in which we have (at least on a macroscopic level) a continuous distribution of matter, the sums in Eqs. (8.28) have to be replaced by integrals. The calculations can get quite involved, but we can say some general things about such problems. First, whenever a homogeneous body has a geometric center, such as a billiard ball, a sugar cube, or a can of frozen orange juice, the center of mass is at the geometric center. Second, whenever a body has an axis of symmetry, such as a wheel or a pulley, the center of mass always lies on that axis. Third, there is no law that says the center of mass has to be within the body. For example, the center of mass of a doughnut is right in the middle of the hole. You can probably think of other examples.

We'll talk a little more about locating the center of mass in Chapter 11 in connection with the related concept *center of gravity.*

Motion of the Center of Mass

To see the significance of the center of mass of a collection of particles, we must ask what happens to the center of mass when the particles move. The x- and y-components of velocity of the center of mass, $v_{cm\text{-}x}$ and $v_{cm\text{-}y}$, are the time deriva-

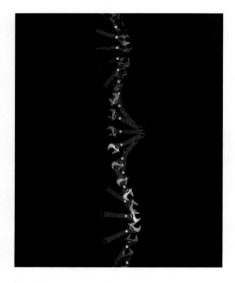

8.25 The center of mass of this wrench is marked with a white dot. The total external force acting on the wrench is almost zero. As it spins on a smooth horizontal surface, the center of mass moves in a straight line with nearly constant velocity.

tives of x_{cm} and y_{cm}. Also, dx_1/dt is the x-component of velocity of particle 1, and so on, so $dx_1/dt = v_{1x}$, and so on. Taking time derivatives of Eqs. (8.28), we get

$$v_{cm\text{-}x} = \frac{m_1 v_{1x} + m_2 v_{2x} + m_3 v_{3x} + \cdots}{m_1 + m_2 + m_3 + \cdots}$$

$$v_{cm\text{-}y} = \frac{m_1 v_{1y} + m_2 v_{2y} + m_3 v_{3y} + \cdots}{m_1 + m_2 + m_3 + \cdots} \qquad (8.30)$$

These equations are equivalent to the single vector equation obtained by taking the time derivative of Eq. (8.29):

$$\vec{v}_{cm} = \frac{m_1 \vec{v}_1 + m_2 \vec{v}_2 + m_3 \vec{v}_3 + \cdots}{m_1 + m_2 + m_3 + \cdots} \qquad (8.31)$$

We denote the *total* mass $m_1 + m_2 + \cdots$ by M. We can then rewrite Eq. (8.31) as

$$M\vec{v}_{cm} = m_1 \vec{v}_1 + m_2 \vec{v}_2 + m_3 \vec{v}_3 + \cdots = \vec{P} \qquad (8.32)$$

The right side is simply the total momentum $\vec{P}$ of the system. Thus we have proved that *the total momentum is equal to the total mass times the velocity of the center of mass.* When you catch a baseball, you are really catching a collection of a very large number of molecules of masses $m_1, m_2, m_3, \ldots$. The impulse you feel is due to the total momentum of this entire collection. But this impulse is the same as if you were catching a single particle of mass $M = m_1 + m_2 + m_3 + \cdots$ moving with velocity $\vec{v}_{cm}$, the velocity of the collection's center of mass. So Eq. (8.32) helps to justify representing an extended body as a particle.

For a system of particles on which the net external force is zero, so that the total momentum $\vec{P}$ is constant, the velocity of the center of mass $\vec{v}_{cm} = \vec{P}/M$ is also constant. Suppose we mark the center of mass of an adjustable wrench, which lies at a point partway down the handle. We then slide the wrench with a spinning motion across a smooth, horizontal table top (Fig. 8.25). The overall motion appears complicated, but the center of mass follows a straight line, as though all the mass were concentrated at that point.

<table>
<tr><td>Example
8.15</td><td>**A tug-of-war on the ice**</td></tr>
</table>

James and Ramon are standing 20.0 m apart on the slippery surface of a frozen pond. Ramon has mass 60.0 kg and James has mass 90.0 kg. Midway between the two men a mug of their favorite beverage sits on the ice (Fig. 8.26). They pull on the ends of a light rope that is stretched between them. When James has moved 6.0 m toward the mug, how far and in what direction has Ramon moved?

SOLUTION

IDENTIFY: The frozen surface is horizontal and essentially frictionless, so the net external force on the system of James, Ramon, and the rope is zero. Hence their total momentum is conserved. Initially, there is no motion, so the total momentum is zero; thus the velocity of the center of mass is zero, and the center of mass remains at rest. We can use this to relate the positions of James and Ramon.

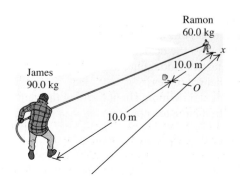

8.26 No matter how hard James or Ramon pulls, their center of mass will not move.

SET UP: Let's take the origin at the position of the mug, and let the $+x$-axis extend from the mug toward Ramon. Since the rope is light, we can neglect its mass in calculating the position of the center of mass with Eq (8.28).

EXECUTE: The initial x-coordinates of James and Ramon are -10.0 m and $+10.0$ m, respectively, so the x-coordinate of the center of mass is

$$x_{cm} = \frac{(90.0 \text{ kg})(-10.0 \text{ m}) + (60.0 \text{ kg})(10.0 \text{ m})}{90.0 \text{ kg} + 60.0 \text{ kg}} = -2.0 \text{ m}$$

When James moves 6.0 m toward the mug, his new x-coordinate is -4.0 m; we'll call Ramon's new x-coordinate x_2. The center of mass doesn't move, so

$$x_{cm} = \frac{(90.0 \text{ kg})(-4.0 \text{ m}) + (60.0 \text{ kg})x_2}{90.0 \text{ kg} + 60.0 \text{ kg}} = -2.0 \text{ m}$$

$$x_2 = 1.0 \text{ m}$$

James has moved 6.0 m in the positive x-direction and is still 4.0 m from the mug, but Ramon has moved 9.0 m in the negative x-direction and is only 1.0 m from it.

EVALUATE: The ratio of how far each man moved, $(6.0 \text{ m})/(9.0 \text{ m}) = \frac{2}{3}$, equals the inverse ratio of their masses. Can you see why? If the two men keep moving (and if the surface is frictionless, they will!), Ramon will reach the mug first. This result is completely independent of how hard either person pulls; pulling harder just helps Ramon quench his thirst sooner.

External Forces and Center-of-Mass Motion

If the net external force on a system of particles is not zero, then total momentum is not conserved and the velocity of the center of mass changes. Let's look at the relation between the motion of the center of mass and the forces acting on the system.

Equations (8.31) and (8.32) give the *velocity* of the center of mass in terms of the velocities of the individual particles. Proceeding one additional step, we take the time derivatives of these equations to show that the accelerations are related in the same way. Let $\vec{a}_{cm} = d\vec{v}_{cm}/dt$ be the acceleration of the center of mass; then we find

$$M\vec{a}_{cm} = m_1\vec{a}_1 + m_2\vec{a}_2 + m_3\vec{a}_3 + \cdots \qquad (8.33)$$

Now $m_1\vec{a}_1$ is equal to the vector sum of forces on the first particle, and so on, so the right side of Eq. (8.33) is equal to the vector sum $\sum \vec{F}$ of *all* the forces on *all* the particles. Just as we did in Section 8.2, we may classify each force as *internal* or *external.* The sum of all forces on all the particles is then

$$\sum \vec{F} = \sum \vec{F}_{ext} + \sum \vec{F}_{int} = M\vec{a}_{cm}$$

Because of Newton's third law, the internal forces all cancel in pairs, and $\sum \vec{F}_{int} = 0$. What survives on the left side is only the sum of the *external* forces, and we have

$$\sum \vec{F}_{ext} = M\vec{a}_{cm} \quad \text{(body or collection of particles)} \qquad (8.34)$$

When a body or a collection of particles is acted on by external forces, the center of mass moves just as though all the mass were concentrated at that point and it were acted on by a net force equal to the sum of the external forces on the system.

This result may not sound very impressive, but in fact it is central to the whole subject of mechanics. In fact, we've been using this result all along; without it, we would not be able to represent an extended body as a point particle when we apply Newton's laws. It explains why only *external* forces can affect the motion of an extended body. If you pull upward on your belt, your belt exerts an equal downward force on your hands; these are *internal* forces that cancel and have no effect on the overall motion of your body.

(a)

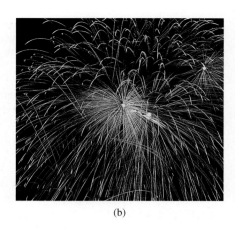

(b)

8.27 (a) A shell explodes into two parts in flight. If air resistance is ignored, the center of mass continues on the same trajectory as the shell's path before exploding—until one of the fragments hits the ground. (b) The same effect occurs with exploding fireworks.

Suppose a cannon shell traveling in a parabolic trajectory (neglecting air resistance) explodes in flight, splitting into two fragments with equal mass (Fig. 8.27a). The fragments follow new parabolic paths, but the center of mass continues on the original parabolic trajectory, just as though all the mass were still concentrated at that point. A skyrocket exploding in air (Fig. 8.27b) is a spectacular example of this effect.

This property of the center of mass is important when we analyze the motion of rigid bodies. We describe the motion of an extended body as a combination of translational motion of the center of mass and rotational motion about an axis through the center of mass. We will return to this topic in Chapter 10. This property also plays an important role in the motion of astronomical objects. It's not correct to say that the moon orbits the earth; rather, the earth and moon both move in orbits around their center of mass.

There's one more useful way to describe the motion of a system of particles. Using $\vec{a}_{cm} = d\vec{v}_{cm}/dt$, we can rewrite Eq. (8.33) as

$$M\vec{a}_{cm} = M\frac{d\vec{v}_{cm}}{dt} = \frac{d(M\vec{v}_{cm})}{dt} = \frac{d\vec{P}}{dt} \qquad (8.35)$$

The total system mass M is constant, so we're allowed to take it inside the derivative. Substituting Eq. (8.35) into Eq. (8.34), we find

$$\sum \vec{F}_{ext} = \frac{d\vec{P}}{dt} \qquad \text{(extended body or system of particles)} \qquad (8.36)$$

This equation looks like Eq. (8.3). The difference is that Eq. (8.36) describes a *system* of particles, such as an extended body, while Eq. (8.3) describes a single particle. The interactions between the particles that make up the system can change the individual momenta of the particles, but the *total* momentum $\vec{P}$ of the system can be changed only by external forces acting from outside the system.

Finally, we note that if the net external force is zero, Eq. (8.34) shows that the acceleration $\vec{a}_{cm}$ of the center of mass is zero. So the center-of-mass velocity $\vec{v}_{cm}$ is constant, as for the wrench in Figure 8.25. From Eq. (8.36) the total momentum $\vec{P}$ is also constant. This reaffirms our statement in Section 8.3 of the principle of conservation of momentum.

Test Your Understanding

Find the location of the center of mass of the earth–moon system using information from Appendix F. Where is the center of mass relative to the earth's surface?

*8.6 | Rocket Propulsion

Momentum considerations are particularly useful for analyzing a system in which the masses of parts of the system change with time. In such cases we can't use Newton's second law $\sum \vec{F} = m\vec{a}$ directly because m changes. Rocket propulsion offers a typical and interesting example of this kind of analysis. A rocket is propelled forward by rearward ejection of burned fuel that initially was in the rocket. The forward force on the rocket is the reaction to the backward force on the ejected material. The total mass of the system is constant, but the mass of the rocket itself decreases as material is ejected.

As a simple example, we consider a rocket fired in outer space, where there is no gravitational force and no air resistance. We let m denote the mass of the rocket, which will change as it expends fuel. We choose our x-axis to be along the rocket's direction of motion. Figure 8.28a shows the rocket at a time t, when its mass is m and the x-component of its velocity relative to our coordinate system is v. (For simplicity, we will drop the subscript x in this discussion.) The x-component of total momentum at this instant is $P_1 = mv$. In a short time interval dt the mass of the rocket changes by an amount dm. This is an inherently negative quantity because the rocket's mass m *decreases* with time. During dt a *positive* mass $-dm$ of burned fuel is ejected from the rocket. Let v_{ex} be the exhaust *speed* of this material *relative to the rocket;* the burned fuel is ejected opposite the direction of motion, so its x-component of *velocity* relative to the rocket is $-v_{ex}$. The x-component of velocity v_{fuel} of the burned fuel relative to our coordinate system is then

$$v_{fuel} = v + (-v_{ex}) = v - v_{ex}$$

and the x-component of momentum of the ejected mass $(-dm)$ is

$$(-dm)v_{fuel} = (-dm)(v - v_{ex})$$

As shown in Fig. 8.28b, at the end of the time interval dt, the x-component of velocity of the rocket and unburned fuel has increased to $v + dv$, and its mass has decreased to $m + dm$ (remember that dm is negative). The rocket's momentum at this time is

$$(m + dm)(v + dv)$$

Thus the *total x*-component of momentum P_2 of rocket plus ejected fuel at time $t + dt$ is

$$P_2 = (m + dm)(v + dv) + (-dm)(v - v_{ex})$$

Activ
Physics
ONLINE

6.6   Saving an Astronaut

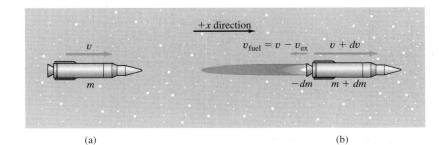

(a)

(b)

8.28 (a) A rocket moving in gravity-free outer space with mass m and x-component of velocity v at time t. (b) At time $t + dt$ the rocket's mass has decreased, so dm is negative. The mass of the rocket (including the unburned fuel) is $m + dm$, and its velocity component is $v + dv$. The mass of burned fuel in the exhaust is $-dm$, and its velocity relative to the coordinate system shown is $v_{fuel} = v - v_{ex}$.

According to our initial assumption, the rocket and fuel are an isolated system. Thus momentum is conserved, and the total x-component of momentum of the system must be the same at time t and at time $t + dt$: $P_1 = P_2$. Hence

$$mv = (m + dm)(v + dv) + (-dm)(v - v_{ex})$$

This can be simplified to

$$m\,dv = -dm\,v_{ex} - dm\,dv$$

We can neglect the term $(-dm\,dv)$ because it is a product of two small quantities and thus is much smaller than the other terms. Dropping this term, dividing by dt, and rearranging, we find

$$m\frac{dv}{dt} = -v_{ex}\frac{dm}{dt} \qquad (8.37)$$

Now dv/dt is the acceleration of the rocket, so the left side of this equation (mass times acceleration) equals the net force F, or *thrust*, on the rocket,

$$F = -v_{ex}\frac{dm}{dt} \qquad (8.38)$$

The thrust is proportional both to the relative speed v_{ex} of the ejected fuel and to the mass of fuel ejected per unit time, $-dm/dt$. (Remember that dm/dt is negative because it is the rate of change of the rocket's mass.)

The x-component of acceleration of the rocket is

$$a = \frac{dv}{dt} = -\frac{v_{ex}}{m}\frac{dm}{dt} \qquad (8.39)$$

This is positive because v_{ex} is positive (remember, it's the exhaust *speed*) and dm/dt is negative. The rocket's mass m decreases continuously while the fuel is being consumed. If v_{ex} and dm/dt are constant, the acceleration increases until all the fuel is gone.

Equation (8.38) tells us that an effective rocket burns fuel at a rapid rate (large $-dm/dt$) and ejects the burned fuel at a high relative speed (large v_{ex}), as in Fig. 8.29. In the early days of rocket propulsion, people who didn't understand conservation of momentum thought that a rocket couldn't function in outer space because "it doesn't have anything to push against." On the contrary, rockets work *best* in outer space, where there is no air resistance! The rocket in Fig. 8.29 is *not* "pushing against the ground" to get into the air.

If the exhaust speed v_{ex} is constant, we can integrate Eq. (8.39) to find a relation between the velocity v at any time and the remaining mass m. At time $t = 0$, let the mass be m_0 and the velocity v_0. Then we rewrite Eq. (8.39) as

$$dv = -v_{ex}\frac{dm}{m}$$

We change the integration variables to v' and m', so we can use v and m as the upper limits (the final speed and mass). Then we integrate both sides, using limits v_0 to v and m_0 to m, and take the constant v_{ex} outside the integral:

$$\int_{v_0}^{v} dv' = -\int_{m_0}^{m} v_{ex}\frac{dm'}{m'} = -v_{ex}\int_{m_0}^{m}\frac{dm'}{m'}$$

$$v - v_0 = -v_{ex}\ln\frac{m}{m_0} = v_{ex}\ln\frac{m_0}{m} \qquad (8.40)$$

8.29 Launch of a space shuttle, a dramatic example of rocket propulsion. To provide enough upward thrust to overcome gravity, space launch vehicles can consume over 10,000 kg/s of fuel and eject the burned fuel at speeds of over 4000 m/s.

The ratio m_0/m is the original mass divided by the mass after the fuel has been exhausted. In practical spacecraft this ratio is made as large as possible to maximize the speed gain, which means that the initial mass of the rocket is almost all fuel. The final velocity of the rocket will be greater in magnitude (and is often *much* greater) than the relative speed v_{ex} if $\ln(m_0/m) > 1$, that is, if $m_0/m > e = 2.71828\ldots$.

We've assumed throughout this analysis that the rocket is in gravity-free outer space. However, gravity must be taken into account when a rocket is launched from the surface of a planet, as in Fig. 8.29 (see Problem 8.104).

Example 8.16   Acceleration of a rocket

A rocket is in outer space, far from any planet, when the rocket engine is turned on. In the first second of firing, the rocket ejects $\frac{1}{120}$ of its mass with a relative speed of 2400 m/s. What is the rocket's initial acceleration?

SOLUTION

IDENTIFY and SETUP: Equation (8.39) gives the acceleration of the rocket—our target variable—in terms of the exhaust speed v_{ex}, the current value m of the rocket's mass, and the rate of change of the mass dm/dt. While we are not given the mass or the rate of change, we are told what fraction of the initial mass is lost during a given time interval; this information should be all we need.

EXECUTE: The initial rate of change of mass is

$$\frac{dm}{dt} = -\frac{m_0/120}{1\text{ s}} = -\frac{m_0}{120\text{ s}}$$

where m_0 is the initial $(t = 0)$ mass of the rocket. From Eq. (8.39) the initial acceleration is

$$a = -\frac{v_{ex}}{m_0}\frac{dm}{dt} = -\frac{2400\text{ m/s}}{m_0}\left(-\frac{m_0}{120\text{ s}}\right) = 20\text{ m/s}^2$$

EVALUATE: Note that the answer didn't depend on the value of m_0. If v_{ex} is the same, the initial acceleration is the same for a 120,000-kg spacecraft that ejects 1000 kg/s as for a 60-kg astronaut equipped with a small rocket that ejects 0.5 kg/s.

Example 8.17   Speed of a rocket

Suppose that $\frac{3}{4}$ of the initial mass m_0 of the rocket in Example 8.16 is fuel, so the final mass is $m = m_0/4$, and that the fuel is completely consumed at a constant rate in a total time $t = 90$ s. If the rocket starts from rest in our coordinate system, find its speed at the end of this time.

SOLUTION

IDENTIFY and SET UP: We are given the initial velocity v_0 (equal to zero), the exhaust speed v_{ex}, and the final mass m in terms of the initial mass m_0. Hence we can use Eq. (8.40) directly to find the final speed v.

EXECUTE: We have $m_0/m = 4$, so from Eq. (8.40),

$$v = v_0 + v_{ex}\ln\frac{m_0}{m} = 0 + (2400\text{ m/s})(\ln 4) = 3327\text{ m/s}$$

EVALUATE: Let's examine what happens as the rocket gains speed. At the start of the flight, when the velocity of the rocket is zero, the ejected fuel is moving to the left, relative to our coordinate system, at 2400 m/s. At the end of the first second ($t = 1$ s) the rocket is moving at 20 m/s, and the fuel's speed relative to our system is 2380 m/s. During the next second the acceleration, given by Eq. (8.39), is a little greater. At $t = 2$ s the rocket is moving at a little over 40 m/s, and the fuel's speed is a little less than 2360 m/s. Detailed calculation shows that at about $t = 75.6$ s, the rocket's velocity v in our coordinate system equals 2400 m/s. The burned fuel ejected after this time moves *forward*, not backward, in our system. Since the final velocity of the rocket is 3327 m/s and the relative velocity is 2400 m/s, the last portion of the ejected fuel has a forward velocity (relative to our frame of reference) of $(3327 - 2400)$ m/s $= 927$ m/s. (To illustrate our point, we are using more figures than are significant.)

Test Your Understanding

If a rocket in gravity-free outer space has the same thrust at all times, is its acceleration constant, increasing, or decreasing? If the rocket has the same acceleration at all times, is the thrust constant, increasing, or decreasing?

SUMMARY

The momentum $\vec{p}$ of a particle is a vector quantity equal to the product of the particle's mass m and velocity $\vec{v}$. Newton's second law says that the net force on a particle is equal to the rate of change of the particle's momentum.

$$\vec{p} = m\vec{v} \qquad (8.2)$$

$$\sum \vec{F} = \frac{d\vec{p}}{dt} \qquad (8.3)$$

$p_A = 18 \text{ kg} \cdot \text{m/s}$
$p_B = 24 \text{ kg} \cdot \text{m/s}$

If a constant net force $\sum \vec{F}$ acts on a particle for a time interval Δt from t_1 to t_2, the impulse $\vec{J}$ of the net force is the product of the net force and the time interval. If $\sum \vec{F}$ varies with time, $\vec{J}$ is the integral of the net force over the time interval. In any case, the change in a particle's momentum during a time interval equals the impulse of the net force that acted on it during that interval. The momentum of a particle equals the impulse that accelerated it from rest to its present speed.
(See Examples 8.1 through 8.3)

$$\vec{J} = \sum \vec{F}(t_2 - t_1) = \sum \vec{F}\Delta t \quad (8.5)$$

$$\vec{J} = \int_{t_1}^{t_2} \sum \vec{F} \, dt \qquad (8.7)$$

$$\vec{J} = \vec{p}_2 - \vec{p}_1 \qquad (8.6)$$

Net force $\sum \vec{F}$
Displacement $\vec{s}$ in time Δt
Kinetic energy gained by ball = $\sum \vec{F} \cdot \vec{s}$
Momentum gained by ball = $\sum \vec{F} \Delta t$

The total momentum of a system of particles A, B, C, ... is the vector sum of the momenta of the individual particles.

$$\vec{P} = \vec{p}_A + \vec{p}_B = \cdots$$
$$= m_A\vec{v}_A + m_B\vec{v}_B + \cdots \quad (8.14)$$

An internal force is a force exerted by one part of a system on another. An external force is a force exerted on any part of a system by something outside the system. If the net external force on a system is zero, the total momentum of the system is constant (conserved); each component of total momentum is separately conserved.
(See Examples 8.4 through 8.6)

If $\sum \vec{F} = 0$ $\vec{P} = $ constant

v_{Rx} $v_{Bx} = 300 \text{ m/s}$
                $-+x$
$m_R = 3.00 \text{ kg}$ $m_B = 5.00 \text{ g}$

In collisions of all kinds, the initial and final total momenta are equal. In an elastic collision between two bodies, the initial and final total kinetic energies are also equal, and the initial and final relative velocities have the same magnitude. In an inelastic two-body collision, the total kinetic energy is less after the collision than before. If the two bodies have the same final velocity, the collision is completely inelastic.
(See Examples 8.7 through 8.13)

$v_{A1x} = 2.0 \text{ m/s}$ $v_{B1x} = -2.0 \text{ m/s}$ v_{A2x} $v_{B2x} = 2.0 \text{ m/s}$

$m_A = 0.50 \text{ kg}$ $m_B = 0.30 \text{ kg}$

The position vector of the center of mass of a system of particles, $\vec{r}_{cm}$, is a weighted average of the positions $\vec{r}_1$, $\vec{r}_2$, ... of the individual particles.
(See Example 8.14)

$$\vec{r}_{cm} = \frac{m_1\vec{r}_1 + m_2\vec{r}_2 + m_3\vec{r}_3 + \cdots}{m_1 + m_2 + m_3 + \cdots}$$

$$= \frac{\sum_i m_i\vec{r}_i}{\sum_i m_i} \qquad (8.29)$$

Oxygen   Hydrogen
cm $105°$
Hydrogen

The total momentum $\vec{P}$ of a system equals its total mass M multiplied by the velocity of its center of mass, $\vec{v}_{cm}$. The center of mass moves as through all the mass M was concentrated at that point. If the net external force on the system is zero, the center-of-mass velocity $\vec{v}_{cm}$ is constant. If the net external force is not zero, the center of mass accelerates as though it was a particle of mass M being acted on by the same net external force. (See Example 8.15)

$$\vec{P} = m_1\vec{v}_1 + m_2\vec{v}_2 + m_3\vec{v}_3 + \cdots$$

$$= M\vec{v}_{cm} \qquad (8.32)$$

$$\sum \vec{F}_{ext} = M\vec{a}_{cm} \qquad (8.34)$$

Shell explodes into two parts

Fragments follow individual parabolic paths

Center of mass continues on shell's original trajectory

In rocket propulsion, the mass of a rocket changes as the fuel is used up and ejected from the rocket. Analysis of the motion of the rocket must include the momentum carried away by the spent fuel as well as the momentum of the rocket itself. (See Examples 8.16 and 8.17)

$+x$ direction

$v_{fuel} = v - v_{ex} \qquad v + dv$

$-dm \qquad m + dm$

Key Terms

momentum (linear momentum), 283
impulse, 284
impulse-momentum theorem, 284
internal force, 290
external force, 290

isolated system, 290
total momentum, 290
principle of conservation
  of momentum, 290

elastic collision, 296
inelastic collision, 296
completely inelastic collision, 296
center of mass, 306

Your Notes

Answer to Chapter Opening Question

The two players have the same magnitude of momentum $p = mv$ (the product of mass and speed), but the lightweight player has twice as much kinetic energy $K = \frac{1}{2}mv^2$. Hence, the lightweight player can do twice as much work on you (and twice as much damage) in the process of coming to a halt.

Answers to Test Your Understanding Questions

Section 8.1 (a) The velocity was $v_y = -7.55$ m/s (downward), so $p_y = mv_y = (200 \text{ kg})(-7.55 \text{ m/s}) = -1510$ kg·m/s. The final momentum was zero, so $J_y = 0 - (-1510 \text{ kg·m/s}) = +1510$ kg·m/s $= +1510$ N·s. (b) The net upward force was $F_y = n + f - w = 79,000 \text{ N} + 60 \text{ N} - 1960 \text{ N} = 77,100$ N. From the definition of impulse, $J_y = F_y \Delta t$, so the elapsed time was $\Delta t = J_y/F_y = (1510 \text{ N·s})/(77,100 \text{ N}) = 0.0196$ s.

Section 8.2 There are no external horizontal forces, so the x- and y-components of the total momentum of the system are both conserved. Both components of the total momentum are zero before the spring releases, so they must be zero after the spring releases. Hence

$$P_x = 0 = m_A v_{A2x} + m_B v_{B2x} + m_C v_{C2x}$$

$$P_y = 0 = m_A v_{A2y} + m_B v_{B2y} + m_C v_{C2y}$$

We are given that $m_A = m_B = m_C = 0.20$ kg, $v_{A2x} = -3.0$ m/s, $v_{A2y} = 0$, $v_{B2x} = 0$, and $v_{B2y} = -4.0$ m/s. You can solve the above equations to show that $v_{C2x} = +3.0$ m/s and $v_{C2y} = +4.0$ m/s. Hence the velocity of piece C has magnitude $\sqrt{(3.0 \text{ m/s})^2 + (4.0 \text{ m/s})^2} = 5.0$ m/s and is at an angle from the positive x-axis of $\arctan[(4.0 \text{ m/s})/(3.0 \text{ m/s})] = 53°$.

Section 8.3 The kinetic energy before the collision is $K_1 = \frac{1}{2}m_A v_{A1}^2 = \frac{1}{2}(20 \text{ kg})(2.0 \text{ m/s})^2 = 40$ J. After the collision, the kinetic energy is $K_2 = \frac{1}{2}m_A v_{A2}^2 + \frac{1}{2}m_B v_{B2}^2 = \frac{1}{2}(20 \text{ kg})(1.0 \text{ m/s})^2 + \frac{1}{2}(12 \text{ kg}) \times (2.1 \text{ m/s})^2 = 36$ J, which is less than K_1. Because kinetic energy was lost, the collision was inelastic. It was not, however, *totally* inelastic, because the two robots did not stick together.

Section 8.4 After the collision, the speed of the neutron is $|(m_A - m_B)/(m_A + m_B)| = |(18 \text{ u} - 1.0 \text{ u})/(18 \text{ u} + 1.0 \text{ u})| = (\frac{17}{19})$ of its initial speed, and its kinetic energy is $(\frac{17}{19})^2 = 0.80$ of the initial value. Hence a water molecule is a poorer moderator than a carbon atom.

Section 8.5 Let the subscripts 1 and 2 represent the earth and the moon, respectively, and take the origin to be at the center of the earth. Then, using data from Appendix F,

$$x_{cm} = (m_1 x_1 + m_2 x_2)/(m_1 + m_2)$$

$$= \frac{\left[\begin{array}{c} (5.97 \times 10^{24} \text{ kg})(0) + (7.35 \times 10^{22} \text{ kg}) \\ \times (3.84 \times 10^8 \text{ m}) \end{array} \right]}{(5.97 \times 10^{24} \text{ kg} + 7.35 \times 10^{22} \text{ kg})}$$

$$= 4.67 \times 10^6 \text{ m} = 4670 \text{ km}$$

The radius of the earth is 6.38×10^6 m $= 6380$ km. Hence the center of mass of the earth–moon system is actually within the earth's interior, 1710 km below the surface.

Section 8.6 From Eqs. (8.37) and (8.38), the thrust F is equal to $m \, dv/dt$, where m is the rocket's mass and dv/dt is its acceleration. Because m decreases with time, if the thrust F is constant, then the acceleration must increase with time (the same force acts on a smaller mass); if the acceleration dv/dt is constant, then the thrust must decrease with time (a smaller force is all that's needed to accelerate a smaller mass).

Discussion Questions

Q8.1 In splitting logs with a hammer and wedge, is a heavy hammer more effective than a lighter hammer? Why?

Q8.2 Suppose you catch a baseball and then someone invites you to catch a bowling ball with either the same momentum or with the same kinetic energy as the baseball. Which would you choose? Explain.

Q8.3 When rain falls from the sky, what happens to its momentum as it hits the ground? Is your answer also valid for Newton's famous apple?

Q8.4 A car has the same kinetic energy when it is traveling south at 30 m/s as when it is traveling northwest at 30 m/s. Is the momentum of the car the same in both cases? Explain.

Q8.5 A truck is accelerating as it speeds down the highway. One inertial frame of reference is attached to the ground with its origin at a fence post. A second frame of reference is attached to a police car that is traveling down the highway at constant velocity. Is the momentum of the truck the same in these two reference frames? Explain. Is the rate of change of the truck's momentum the same in these two frames? Explain.

Q8.6 When a large, heavy truck collides with a passenger car, the occupants of the car are more likely to be hurt than the truck driver. Why?

Q8.7 A woman holding a large rock stands on a frictionless, horizontal sheet of ice. She throws the rock with speed v_0 at an angle α above the horizontal. Consider the system consisting of the woman plus the rock. Is the momentum of the system conserved? Why or why not? Is any component of the momentum of the system conserved? Again, why or why not?

Q8.8 In Example 8.7 (Section 8.3), where the two gliders in Fig. 8.11a stick together after the collision, the collision is inelastic since $K_2 < K_1$. In Example 8.5 (Section 8.2), is the collision inelastic? Explain.

Q8.9 In a completely inelastic collision between two objects, where the objects stick together after the collision, is it possible for the final kinetic energy of the system to be zero? If so, give an example in which this would occur. If the final kinetic energy is zero, what must the initial momentum of the system be? Is the initial kinetic energy of the system zero? Explain.

Q8.10 Since for a particle the kinetic energy is given by $K = \frac{1}{2}mv^2$ and the momentum by $\vec{p} = m\vec{v}$, it is easy to show that $K = p^2/2m$. How, then, is it possible to have an event during which the total momentum of the system is constant, but the total kinetic energy changes?

Q8.11 In each of Examples 8.10, 8.11, 8.12, and 8.13 (Section 8.4), verify that the relative velocity vector of the two bodies has the same magnitude before and after the collision. In each case what happens to the *direction* of the relative velocity vector?

Q8.12 A glass dropped on the floor is more likely to break if the floor is concrete than if it is wood. Why? (Refer to Fig. 8.5.)

Q8.13 In Fig. 8.22 the kinetic energy of the spacecraft is larger after its interaction with Saturn than before. From where does the extra energy come? Describe the event in terms of conservation of energy.

Q8.14 A machine gun is fired at a steel plate. Is the average force on the plate from the bullet impact greater if the bullets bounce off or if they are squashed and stick to the plate? Explain.

Q8.15 A net force of 4 N acts on an object initially at rest for 0.25 s and gives it a final speed of 5 m/s. How could a net force of 2 N produce the same final speed?

Q8.16 A net force with x-component ΣF_x acts on an object from time t_1 to time t_2. The x-component of the momentum of the object is the same at t_1 as it is at t_2, but ΣF_x is not zero at all times between t_1 and t_2. What can you say about the graph of ΣF_x versus t?

Q8.17 A tennis player hits a tennis ball with a racket. Consider the system made up of the ball and the racket. Is the total momentum of the system the same just before and just after the hit? Is the total momentum just after the hit the same as two seconds later, when the ball is in midair at the high point of its trajectory? Explain any differences between the two cases.

Q8.18 In Example 8.4 (Section 8.2), consider the system consisting of the rifle plus the bullet. What is the speed of the system's center of mass after the rifle is fired? Explain.

Q8.19 An egg is released from rest from the roof of a building and falls to the ground. As the egg falls, what happens to the momentum of the system of the egg plus the earth?

Q8.20 A woman stands in the middle of a perfectly smooth, frictionless, frozen lake. She can set herself in motion by throwing things, but suppose she has nothing to throw. Can she propel herself to shore *without* throwing anything?

Q8.21 In a zero-gravity environment, can a rocket-propelled spaceship ever attain a speed greater than the relative speed with which the burnt fuel is exhausted?

Q8.22 When an object breaks into two pieces (explosion, radioactive decay, recoil, etc.), the lighter fragment gets more kinetic energy than the heavier one. This is a consequence of momentum conservation, but can you also explain it using Newton's laws of motion?

Exercises

Section 8.1 Momentum and Impulse

8.1 a) What is the magnitude of the momentum of a 10,000-kg truck whose speed is 12.0 m/s? b) What speed would a 2,000-kg sport utility vehicle have to attain in order to have i) the same momentum? ii) the same kinetic energy?

8.2 In Conceptual Example 8.1 (Section 8.1), show that the iceboat with mass $2m$ has $\sqrt{2}$ times as much momentum at the finish line as does the iceboat with mass m.

8.3 a) Show that the kinetic energy K and the momentum magnitude p of a particle with mass m are related by $K = p^2/2m$. b) A 0.040-kg cardinal (*Richmondena cardinalis*) and a 0.145-kg baseball have the same kinetic energy. Which has the greater magnitude of momentum? What is the ratio of the cardinal's magnitude of momentum to the baseball's? c) A 700-N man and a 450-N woman have the same momentum. Who has the greater kinetic energy? What is the ratio of the man's kinetic energy to that of the woman?

8.4 A soccer ball with mass 0.420 kg is traveling with a speed of 4.50 m/s at an angle of 20.0° counterclockwise from the +x-axis (Fig. 8.30). What are the ball's x- and y-components of momentum?

Figure 8.30 Exercise 8.4.

8.5 A baseball with mass 0.145 kg is moving in the +y-direction with a speed of 1.30 m/s, and a tennis ball with mass 0.0570 kg is moving in the −y-direction with a speed of 7.80 m/s. What are the magnitude and direction of the total momentum of the system consisting of the two balls?

8.6 A golf ball with mass 0.045 kg is moving in the +x-direction with a speed of 9.00 m/s, and a baseball with mass 0.145 kg is moving in the −y-direction with a speed of 7.00 m/s. What are the magnitude and direction of the total momentum of the system that consists of the two balls?

8.7 **Force of a Golf Swing.** A 0.0450-kg golf ball initially at rest is given a speed of 25.0 m/s when a club strikes. If the club and ball are in contact for 2.00 ms, what average force acts on the ball? Is the effect of the ball's weight during the time of contact significant? Why or why not?

8.8 **Force of a Baseball Swing.** A baseball has mass 0.145 kg. a) If the velocity of a pitched ball has a magnitude of 45.0 m/s and the batted ball's velocity is 55.0 m/s in the opposite direction, find the magnitude of the change in momentum of the ball and of the impulse applied to it by the bat. b) If the ball remains in contact with the bat for 2.00 ms, find the magnitude of the average force applied by the bat.

8.9 A 0.160-kg hockey puck is moving on an icy, frictionless, horizontal surface. At $t = 0$ the puck is moving to the right at 3.00 m/s. a) Calculate the velocity of the puck (magnitude and direction) after a force of 25.0 N directed to the right has been applied for 0.050 s. b) If instead, a force of 12.0 N directed to the left is applied from $t = 0$ to $t = 0.050$ s, what is the final velocity of the puck?

8.10 An engine of the orbital maneuvering system (OMS) on a space shuttle exerts a force of $(26{,}700 \text{ N})\hat{\jmath}$ for 3.90 s, exhausting a negligible mass of fuel relative to the 95,000-kg mass of the shuttle. a) What is the impulse of the force for this 3.90 s? b) What is the shuttle's change in momentum from this impulse? c) What is the shuttle's change in velocity from this impulse? d) Why can't we find the resulting change in kinetic energy of the shuttle?

8.11 Coach Johnson's bat exerts a horizontal force on a 0.145-kg baseball of $\vec{F} = [(1.60 \times 10^7 \text{ N/s})t - (6.00 \times 10^9 \text{ N/s}^2)t^2]\hat{\imath}$

between $t = 0$ and $t = 2.50$ ms. At $t = 0$ the baseball's velocity is $-(40.0\hat{\imath} + 5.0\hat{\jmath})$ m/s. a) Calculate the impulse exerted by the bat on the ball during the 2.50 ms that they are in contact. b) Calculate the impulse exerted by gravity on the ball during this time interval. c) Calculate the average force exerted by the bat on the ball during this time interval. d) Calculate the momentum and the velocity of the baseball at $t = 2.50$ ms.

8.12 A bat strikes a 0.145-kg baseball. Just before impact, the ball is traveling horizontally to the right at 50.0 m/s, and it leaves the bat traveling to the left at an angle of 30° above horizontal with a speed of 65.0 m/s. If the ball and bat are in contact for 1.75 ms, find the horizontal and vertical components of the average force on the ball.

8.13 A net force $\sum F_x(t) = A + Bt^2$ in the $+x$-direction is applied to a girl on roller skates. The girl has mass m. The force starts at $t_1 = 0$ and continues until $t = t_2$. a) What is the impulse J_x of the force? b) If the girl is initially at rest, what is her speed at time t_2?

Section 8.2 Conservation of Momentum

8.14 Frustrated by having the goalie block his shots, a 75.0-kg hockey player standing on ice throws a 0.160-kg puck horizontally at the net with a speed of 20.0 m/s. With what speed and in what direction will the hockey player begin to move if there is no friction between his feet and the ice?

8.15 You are standing on a sheet of ice that covers the football stadium parking lot in Buffalo; there is negligible friction between your feet and the ice. A friend throws you a 0.400-kg ball that is traveling horizontally at 10.0 m/s. Your mass is 70.0 kg. a) If you catch the ball, with what speed do you and the ball move afterwards? b) If the ball hits you and bounces off your chest, so afterwards it is moving horizontally at 8.0 m/s in the opposite direction, what is your speed after the collision?

8.16 On a frictionless, horizontal air table, puck A (with mass 0.250 kg) is moving toward puck B (with mass 0.350 kg), that is initially at rest. After the collision, puck A has a velocity of 0.120 m/s to the left, and puck B has velocity 0.650 m/s to the right. a) What was the speed of puck A before the collision? b) Calculate the change in the total kinetic energy of the system that occurs during the collision.

8.17 Energy Change During a Hip Check. Ice-hockey star Wayne Gretzky is skating at 13.0 m/s toward a defender, who in turn is skating at 5.00 m/s toward Gretzky (Fig. 8.31). Gretzky's weight is 756 N; that of the defender is 900 N. Immediately after the collision, Gretzky is moving at 1.50 m/s in his original direction. You can ignore external horizontal forces applied by the ice to the skaters during the collision. a) What is the velocity of the defender immediately after the collision? b) Calculate the change in the combined kinetic energy of the two players.

13.0 m/s

5.0 m/s

$w = 756$ N $w = 900$ N

Figure 8.31 Exercise 8.17.

8.18 The expanding gases that leave the muzzle of a rifle also contribute to the recoil. A .30 caliber bullet has mass 0.00720 kg and a speed of 601 m/s relative to the muzzle when fired from a rifle that has mass 2.80 kg. The loosely held rifle recoils at a speed of 1.85 m/s relative to the earth. Find the momentum of the propellant gases in a coordinate system attached to the earth as they leave the muzzle of the rifle.

8.19 Block A in Fig. 8.32 has mass 1.00 kg, and block B has mass 3.00 kg. The blocks are forced together, compressing a spring S between them; then the system is released from rest on a level, frictionless surface. The spring, that has negligible mass, is not fastened to either block and drops to the surface after it has expanded. Block B acquires a speed of 1.20 m/s. a) What is the final speed of block A? b) How much potential energy was stored in the compressed spring?

$m_A = 1.00$ kg $m_B = 3.00$ kg

S

Figure 8.32 Exercise 8.19.

8.20 One of James Bond's adversaries is standing on a frozen lake; there is no friction between his feet and the ice. He throws his steel-lined hat with a velocity of 22.0 m/s at 36.9° above the horizontal, hoping to hit James. If his mass is 120 kg and that of his hat is 4.50 kg, what is the magnitude of his horizontal recoil velocity?

8.21 A ceramic penguin on top of your television set suddenly bursts apart into two pieces. One piece, mass m_A, travels off to the left with speed v_A. The other piece, mass m_B, travels off to the right with speed v_B. a) Use conservation of momentum to solve for v_B in terms of m_A, m_B, and v_A. b) Use the results of part (a) to show that $K_A/K_B = m_B/m_A$, where K_A and K_B are the kinetic energies of the two pieces.

8.22 The nucleus of ^{214}Po decays radioactively by emitting an alpha particle (mass 6.65×10^{-27} kg) with kinetic energy 1.23×10^{-12} J, as measured in the laboratory reference frame. Assuming that the Po was initially at rest in this frame, find the recoil velocity of the nucleus that remains after the decay.

8.23 You are standing on a large sheet of frictionless ice and are holding a large rock. In order to get off the ice, you throw the rock so it has velocity relative to the earth of 12.0 m/s at an angle of 35.0° above the horizontal. If your mass is 70.0 kg and the rock's mass is 15.0 kg, what is your speed after you throw the rock (see Discussion Question Q8.7)?

8.24 Two ice skaters, Daniel (mass 65.0 kg) and Rebecca (mass 45.0 kg), are practicing. Daniel stops to tie his shoelace and, while at rest, is struck by Rebecca, who is moving at 13.0 m/s before she collides with him. After the collision, Rebecca has a velocity of magnitude 8.00 m/s at an angle of 53.1° from her initial direction. Both skaters move on the frictionless, horizontal surface of the rink. a) What are the magnitude and direction of Daniel's velocity after the collision? b) What is the change in total kinetic energy of the two skaters as a result of the collision?

8.25 Ken and Kim are skating together on a rink at 3.00 m/s. Ken keeps asking Kim how much she weighs. Annoyed, Kim pushes away from Ken so that she speeds up to 4.00 m/s and he slows down to 2.25 m/s in the same direction. Friction, in the physics sense, is negligible in this drama. If Ken weighs 700 N, what does Kim weigh?

8.26 Changing Mass. An open-topped freight car with mass 24,000 kg is coasting without friction along a level track. It is raining very hard, and the rain is falling vertically downward. Originally, the car is empty and moving with a speed of 4.00 m/s. What is the speed of the car after it has collected 3000 kg of rainwater?

8.27 A hockey puck B rests on a smooth ice surface and is struck by a second puck A, which was originally traveling at 40.0 m/s and which is deflected 30.0° from its original direction (Fig. 8.33). Puck B acquires a velocity at a 45.0° angle to the original direction of A. The pucks have the same mass. a) Compute the speed of each puck after the collision. b) What fraction of the original kinetic energy of puck A dissipates during the collision?

Figure 8.33 Exercise 8.27.

Section 8.3 Inelastic Collisions

8.28 On a greasy, essentially frictionless lunch counter, a 0.500-kg submarine sandwich, moving 3.00 m/s to the left, collides with an 0.250-kg grilled cheese sandwich moving 1.20 m/s to the right. a) If the two sandwiches stick together, what is the final velocity? b) How much mechanical energy dissipates in the collision?

8.29 Your 1050-kg sports car, parked on a hill without the parking brake set, has rolled to the bottom of the hill and is moving westbound on a level road at a speed of 15.0 m/s. A truck driver driving east on the same road sees the sports car coming, and decides to stop the car by running his 6320-kg truck into it. The two vehicles remain locked together after the collision. a) If the truck is moving at 10.0 m/s when it hits your sports car head-on, what is the velocity (magnitude and direction) of the two vehicles just after the collision? b) At what speed should the truck be moving so that the truck and car are both stopped in the collision? c) Find the change in kinetic energy of the system of two vehicles for the situation of part (a) and the situation of part (b). For which situation is the change in kinetic energy greater in magnitude?

8.30 On a very muddy football field, a 110-kg linebacker tackles an 85-kg halfback. Immediately before the collision, the linebacker is slipping with a velocity of 8.8 m/s north and the halfback is sliding with a velocity of 7.2 m/s east. What is the velocity (magnitude and direction) at which the two players move together immediately after the collision?

8.31 Two frisky grasshoppers collide in midair at the top of their respective trajectories and grab onto each other, holding tight thereafter. One is a robust 250-g beast initially moving south at 20.0 cm/s, while the other is a svelte 150-g creature initially moving north at 60.0 cm/s. Calculate the decrease in kinetic energy that results from the collision. What happens to the "lost" kinetic energy?

8.32 Two cars, one a compact with mass 1200 kg and the other a large gas-guzzler with mass 3000 kg, collide head-on at typical freeway speeds. a) Which car has a greater magnitude of momentum change? Which car has a greater velocity change? Calculate the change in the velocity of the small car relative to that of the large car. b) Which car's occupants would you expect to sustain greater injuries? Explain.

8.33 In Dallas, the morning after a winter ice storm, a 1400-kg automobile going west on Chestnut Street at 35.0 km/h collides with a 2800-kg truck going south *across* Chestnut Street at 50.0 km/h. If they become coupled on collision, what are the magnitude and direction of their velocity after colliding? You can ignore forces between the vehicles and the icy road.

8.34 At the intersection of Texas Avenue and University Drive, a blue, subcompact car with mass 950 kg traveling east on University collides with a maroon pickup truck with mass 1900 kg that is traveling north on Texas and ran a red light (Fig. 8.34). The two vehicles stick together as a result of the collision and, after the collision, the wreckage is sliding at 16.0 m/s in the direction 24.0° east of north. Calculate the speed of each vehicle before the collision. The collision occurs during a heavy rainstorm; you can ignore friction forces between the vehicles and the wet road.

Figure 8.34 Exercise 8.34.

8.35 A 5.00-g bullet is fired horizontally into a 1.20-kg wooden block resting on a horizontal surface. The coefficient of kinetic friction between block and surface is 0.20. The bullet remains embedded in the block, which is observed to slide 0.230 m along the surface before stopping. What was the initial speed of the bullet?

8.36 A Ballistic Pendulum. A 12.0-g rifle bullet is fired with a speed of 380 m/s into a ballistic pendulum with mass 6.00 kg, suspended from a cord 70.0 cm long (see Example 8.8 in Section 8.3). Compute a) the vertical height through which the pendulum rises; b) the initial kinetic energy of the bullet; c) the kinetic energy of the bullet and pendulum immediately after the bullet becomes embedded in the pendulum.

8.37 You and your friends are doing physics experiments on a frozen pond that serves as a frictionless, horizontal surface. Sam, with mass 80.0 kg, is given a push and slides eastward. Abigail, with mass 50.0 kg, is sent sliding northward. They collide and, after the collision, Sam is moving at 37.0° north of east with a speed of 6.00 m/s and Abigail is moving at 23.0° south of east with a speed of 9.00 m/s. a) What was the speed of each person before the collision? b) By how much did the total kinetic energy of the two people decrease during the collision?

Section 8.4 Elastic Collisions

8.38 Blocks *A* (mass 2.00 kg) and *B* (mass 10.00 kg) move on a frictionless, horizontal surface. Initially, block *B* is at rest and block *A* is moving toward it at 2.00 m/s. The blocks are equipped with ideal spring bumpers, as in Example 8.10. The collision is head-on, so all motion before and after the collision is along a straight line. a) Find the maximum energy stored in the spring bumpers and the velocity of each block at that time. b) Find the velocity of each block after they have moved apart.

8.39 A 0.150-kg glider is moving to the right on a frictionless, horizontal air track with a speed of 0.80 m/s. It has a head-on collision with a 0.300-kg glider that is moving to the left with a speed of 2.20 m/s. Find the final velocity (magnitude and direction) of each glider if the collision is elastic.

8.40 A 10.0-g marble slides to the left with a velocity of magnitude 0.400 m/s on the frictionless, horizontal surface of an icy, New York sidewalk and has a head-on elastic collision with a larger 30.0-g marble sliding to the right with a velocity of magnitude 0.200 m/s (Fig. 8.35). a) Find the velocity of each marble (magnitude and direction) after the collision. (Since the collision is head-on, all the motion is along a line.) b) Calculate the *change in momentum* (that is, the momentum after the collision minus the momentum before the collision) for each marble. Compare the values you get for each marble. c) Calculate the *change in kinetic energy* (that is, the kinetic energy after the collision minus the kinetic energy before the collision) for each marble. Compare the values you get for each marble.

Figure 8.35 Exercise 8.40.

8.41 Supply the details of the calculation of α and β in Example 8.13 (Section 8.4).

8.42 Moderators. Canadian nuclear reactors use *heavy water* moderators in which elastic collisions occur between the neutrons and deuterons of mass 2.0 u (see Example 8.11 in Section 8.4). a) What is the speed of a neutron, expressed as a fraction of its original speed, after a head-on, elastic collision with a deuteron which is initially at rest? b) What is its kinetic energy, expressed as a fraction of its original kinetic energy? c) How many such successive collisions will reduce the speed of a neutron to 1/59,000 of its original value?

8.43 You are at the controls of a particle accelerator, sending a beam of 1.50×10^7 m/s protons (mass *m*) at a gas target of an unknown element. Your detector tells you that some protons bounce straight back after a collision with one of the nuclei of the unknown element. All such protons rebound with a speed of 1.20×10^7 m/s. Assume that the initial speed of the target nucleus is negligible and that the collision is elastic. a) Find the mass of one nucleus of the unknown element. Express your answer in terms of the proton mass *m*. b) What is the speed of the unknown nucleus immediately after such a collision?

Section 8.5 Center of Mass

8.44 Three odd-shaped blocks of chocolate have the following masses and center-of-mass coordinates: (1) 0.300 kg, (0.200 m, 0.300 m); (2) 0.400 kg, (0.100 m, −0.400 m); (3) 0.200 kg, (−0.300 m, 0.600 m). Find the coordinates of the center of mass of the system of three chocolate blocks.

8.45 Find the position of the center of mass of the system of the sun and Jupiter. (Since Jupiter is more massive than the rest of the planets put together, this is essentially the position of the center of mass of the solar system.) Does the center of mass lie inside or outside the sun? Use the data in Appendix F.

8.46 A 1200-kg station wagon is moving along a straight highway at 12.0 m/s. Another car, with mass 1800 kg and speed 20.0 m/s, has its center of mass 40.0 m ahead of the center of mass of the station wagon (Fig. 8.36). a) Find the position of the center of mass of the system consisting of the two automobiles. b) Find the magnitude of the total momentum of the system from the above data. c) Find the speed of the center of mass of the system. d) Find the total momentum of the system, using the speed of the center of mass. Compare your result with that of part (b).

Figure 8.36 Exercise 8.46.

8.47 At one instant, the center of mass of a system of two particles is located on the *x*-axis at $x = 2.0$ m and has a velocity of $(5.0 \text{ m/s})\hat{\imath}$. One of the particles is at the origin. The other particle has a mass of 0.10 kg and is at rest on the *x*-axis at $x = 8.0$ m. a) What is the mass of the particle at the origin? b) Calculate the total momentum of this system. c) What is the velocity of the particle at the origin?

8.48 In Example 8.15 (Section 8.5) Ramon pulls on the rope to give himself a speed of 0.70 m/s. What is the speed of James?

8.49 A system consists of two particles. At $t = 0$ one particle is at the origin; the other, which has a mass of 0.50 kg, is on the *y*-axis at $y = 6.0$ m. At $t = 0$ the center of mass of the system is on the *y*-axis at $y = 2.4$ m. The velocity of the center of mass is given by $(0.75 \text{ m/s}^3)t^2\hat{\imath}$. a) Find the total mass of the system. b) Find the acceleration of the center of mass at any time *t*. c) Find the net external force acting on the system at $t = 3.0$ s.

8.50 A radio-controlled model airplane has a momentum given by $[(-0.75 \text{ kg} \cdot \text{m/s}^3)t^2 + (3.0 \text{ kg} \cdot \text{m/s})]\hat{\imath} + (0.25 \text{ kg} \cdot \text{m/s}^2)t\hat{\jmath}$. What are the *x*-, *y*-, and *z*-components of the net force on the airplane?

*Section 8.6 Rocket Propulsion

8.51 A small rocket burns 0.0500 kg of fuel per second, ejecting it as a gas with a velocity relative to the rocket of magnitude 1600 m/s. a) What is the thrust of the rocket? b) Would the rocket operate in outer space where there is no atmosphere? If so, how would you steer it? Could you brake it?

8.52 A 70-kg astronaut floating in space in a 110-kg MMU (manned maneuvering unit) experiences an acceleration of 0.029 m/s^2 when he fires one of the MMU's thrusters. a) If the speed of the escaping N_2 gas relative to the astronaut is 490 m/s, how much gas is used by the thruster in 5.0 s? b) What is the thrust of the thruster?

8.53 A rocket is fired in deep space, where gravity is negligible. If the rocket has an initial mass of 6000 kg and ejects gas at a relative velocity of magnitude 2000 m/s, how much gas must it eject in the first second to have an initial acceleration of 25.0 m/s^2?

8.54 A rocket is fired in deep space, where gravity is negligible. In the first second it ejects $\frac{1}{160}$ of its mass as exhaust gas and has an acceleration of 15.0 m/s^2. What is the speed of the exhaust gas relative to the rocket?

8.55 A C 6-5 model rocket engine has an impulse of 10.0 N·s for 1.70 s, while burning 0.0125 kg of propellant. It has a maximum thrust of 13.3 N. The initial mass of the engine plus propellant is 0.0258 kg. a) What fraction of the maximum thrust is the average thrust? b) Calculate the relative speed of the exhaust gases, assuming it is constant. c) Assuming that the relative speed of the exhaust gases is constant, find the final speed of the engine if it was attached to a very light frame and fired from rest in gravity-free outer space.

8.56 A single-stage rocket is fired from rest from a deep-space platform, where gravity is negligible. If the rocket burns its fuel in a time of 50.0 s and the relative speed of the exhaust gas is $v_{ex} = 2100$ m/s, what must the mass ratio m_0/m be for a final speed v of 8.00 km/s (about equal to the orbital speed of an earth satellite)?

8.57 Obviously, we can make rockets to go very fast, but what is a reasonable top speed? Assume that a rocket is fired from rest at a space station in deep space, where gravity is negligible. a) If the rocket ejects gas at a relative speed of 2000 m/s and you want the rocket's speed eventually to be $1.00 \times 10^{-3} c$, where c is the speed of light, what fraction of the initial mass of the rocket and fuel is *not* fuel? b) What is this fraction if the final speed is to be 3000 m/s?

Problems

8.58 A steel ball with mass 40.0 g is dropped from a height of 2.00 m onto a horizontal steel slab. The ball rebounds to a height of 1.60 m. a) Calculate the impulse delivered to the ball during impact. b) If the ball is in contact with the slab for 2.00 ms, find the average force on the ball during impact.

8.59 The net force acting on a 2.00-kg discus while it is being thrown is $(\alpha t^2)\hat{\imath} + (\beta + \gamma t)\hat{\jmath}$, where $\alpha = 25.0$ N/s^2, $\beta = 30.0$ N, and $\gamma = 5.0$ N/s. If the discus was originally at rest, what is its

velocity after the net force has acted for 0.500 s? Express your answer in terms of the unit vectors $\hat{\imath}$ and $\hat{\jmath}$.

8.60 Just before it is struck by a racket, a tennis ball weighing 0.560 N has a velocity of $(20.0 \text{ m/s})\hat{\imath} - (4.0 \text{ m/s})\hat{\jmath}$. During the 3.00 ms that the racket and ball are in contact, the net force on the ball is constant and equal to $-(380 \text{ N})\hat{\imath} + (110 \text{ N})\hat{\jmath}$. a) What are the x- and y-components of the impulse of the net force applied to the ball? b) What are the x- and y-components of the final velocity of the ball?

8.61 Three coupled railroad cars roll along and couple with a fourth car, which is initially at rest. These four roll along and couple with a fifth car initially at rest. This process continues until the speed of the final collection of railroad cars is one-fifth the speed of the initial three railroad cars. All the cars are identical. Ignoring friction, how many cars are in the final collection?

8.62 A 1500-kg, blue convertible is traveling south, and a 2000-kg red sport utility vehicle is traveling west. If the total momentum of the system consisting of the two cars is 8000 kg·m/s directed at 60.0° west of south, what is the speed of each car?

8.63 Three identical pucks on a horizontal air table have repelling magnets. They are held together and then released simultaneously. Each has the same speed at any instant. One puck moves due west. What is the direction of the velocity of each of the other two pucks?

8.64 Spheres A (mass 0.020 kg), B (mass 0.030 kg), and C (mass 0.050 kg), are each approaching the origin as they slide on a frictionless air table (Fig. 8.37). The initial velocities of A and B are given in the figure. All three spheres arrive at the origin at the same time and stick together. a) What must the x- and y-components of the initial velocity of C be if all three objects are to end up moving at 0.50 m/s in the $+x$-direction after the collision? b) If C has the velocity found in part (a), what is the change in the kinetic energy of the system of three spheres as a result of the collision?

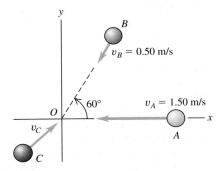

Figure 8.37 Problem 8.64.

8.65 A railroad handcar is moving along straight, frictionless tracks with negligible air resistance. In the following cases, the car initially has a total mass (car and contents) 200 kg and is traveling east with a velocity of magnitude 5.00 m/s. Find the *final velocity* of the car in each case, assuming that the handcar does not leave the tracks. a) A 25.0-kg mass is thrown sideways out of the car with a velocity of magnitude 2.00 m/s relative to the car's initial velocity.

b) A 25.0-kg mass is thrown backward out of the car with a velocity of 5.00 m/s relative to the initial motion of the car. c) A 25.0-kg mass is thrown into the car with a velocity of 6.00 m/s relative to the ground and opposite in direction to the initial velocity of the car.

8.66 Changing Mass. A railroad hopper car filled with sand is rolling with an initial speed of 15.0 m/s on straight, horizontal tracks. You can ignore frictional forces on the railroad car. The total mass of the car plus sand is 85,000 kg. The hopper door is not fully closed so sand leaks out the bottom. After 20 minutes, 13,000 kg of sand has leaked out. Then what is the speed of the railroad car? (Compare your analysis to that used to solve Exercise 8.26.)

8.67 At a classic auto show, a 840-kg 1955 Nash Metropolitan motors by at 9.0 m/s, followed by a 1620-kg 1957 Packard Clipper purring past at 5.0 m/s. a) Which car has the greater kinetic energy? What is the ratio of the kinetic energy of the Nash to that of the Packard? b) Which car has the greater magnitude of momentum? What is the ratio of the magnitude of momentum of the Nash to that of the Packard? c) Let F_N be the net force required to stop the Nash in time t, and let F_P be the net force required to stop the Packard in the same time. Which is larger, F_N or F_P? What is the ratio F_N/F_P of these two forces? d) Now let F_N be the net force required to stop the Nash in a distance d, and let F_P be the net force required to stop the Packard in the same distance. Which is larger, F_N or F_P? What is the ratio F_N/F_P?

8.68 A soldier on a firing range fires an 8-shot burst from an assault weapon at a full automatic rate of 1000 rounds per minute. Each bullet has a mass of 7.45 g and a speed of 293 m/s relative to the ground as it leaves the barrel of the weapon. Calculate the average recoil force exerted on the weapon during that burst.

8.69 A 0.150-kg frame, when suspended from a coil spring, stretches the spring 0.050 m. A 0.200-kg lump of putty is dropped from rest onto the frame from a height of 30.0 cm (Fig. 8.38). Find the maximum distance the frame moves downward from its initial position.

Figure 8.38 Problem 8.69.

8.70 A rifle bullet with mass 8.00 g strikes and embeds itself in a block with mass 0.992 kg that rests on a frictionless, horizontal surface and is attached to a coil spring (Fig. 8.39). The impact compresses the spring 15.0 cm. Calibration of the spring shows that a force of 0.750 N is required to compress the spring 0.250 cm.

Figure 8.39 Problem 8.70.

a) Find the magnitude of the block's velocity just after impact.
b) What was the initial speed of the bullet?

8.71 A Ricocheting Bullet. A 0.100-kg stone rests on a frictionless, horizontal surface. A bullet of mass 6.00 g, traveling horizontally at 350 m/s, strikes the stone and rebounds horizontally at right angles to its original direction with a speed of 250 m/s. a) Compute the magnitude and direction of the velocity of the stone after it is struck. b) Is the collision perfectly elastic?

8.72 A movie stuntman (mass 80.0 kg) stands on a window ledge 5.0 m above the floor (Fig. 8.40). Grabbing a rope attached to a chandelier, he swings down to grapple with the movie's villain (mass 70.0 kg), who is standing directly under the chandelier. (Assume that the stuntman's center of mass moves downward 5.0 m.

Figure 8.40 Problem 8.72.

He releases the rope just as he reaches the villain.) a) With what speed do the entwined foes start to slide across the floor? b) If the coefficient of kinetic friction of their bodies with the floor is $\mu_k = 0.250$, how far do they slide?

8.73 Two identical masses are released from rest in a smooth hemispherical bowl of radius R, from the positions shown in Fig. 8.41. You can ignore friction between the masses and the surface of the bowl. If they stick together when they collide, how high above the bottom of the bowl will the masses go after colliding?

Figure 8.41 Problem 8.73.

8.74 A ball with mass M, moving horizontally at 5.00 m/s, collides elastically with a block with mass $3M$ that is initially hanging at rest from the ceiling on the end of a 50.0-cm wire. Find the maximum angle through which the block swings after it is hit.

8.75 A 20.00-kg lead sphere is hanging from a hook by a thin wire 3.50 m long, and is free to swing in a complete circle. Suddenly it is struck horizontally by a 5.00-kg steel dart that embeds itself in the lead sphere. What must be the minimum initial speed of the dart, so that the combination makes a complete circular loop after the collision?

8.76 A 8.00-kg ball, hanging from the ceiling by a wire 135 cm long, is struck in an elastic collision by a 2.00-kg ball moving with speed v_0 just before the collision. If the maximum tension that the wire can have without breaking is 1600 N, what is the largest that v_0 can be without the wire breaking?

8.77 A 4.00-g bullet, traveling horizontally with a velocity of magnitude 400 m/s, is fired into a wooden block with mass 0.800 kg, initially at rest on a level surface. The bullet passes through the block and emerges with its speed reduced to 120 m/s. The block slides a distance of 45.0 cm along the surface from its initial position. a) What is the coefficient of kinetic friction between block and surface? b) What is the decrease in kinetic energy of the bullet? c) What is the kinetic energy of the block at the instant after the bullet passes through it?

8.78 A 5.00-g bullet is shot *through* a 1.00-kg wood block suspended on a string 2.000 m long. The center of mass of the block rises a distance of 0.45 cm. Find the speed of the bullet as it emerges from the block if its initial speed is 450 m/s.

8.79 A neutron with mass m makes a head-on, elastic collision with a nucleus of mass M, which is initially at rest. a) Show that if the neutron's initial kinetic energy is K_0, the kinetic energy that it loses during the collision is $4mMK_0/(M + m)^2$. b) For what value of M does the incident neutron lose the most energy? c) When M has the value calculated in part (b), what is the speed of the neutron after the collision?

8.80 Energy Sharing in Elastic Collisions. A stationary object with mass m_B is struck head-on by an object with mass m_A that is moving initially at speed v_0. a) If the collision is elastic, what percentage of the original energy does each object have after the collision? b) What does your answer in part (a) give for the following special cases: i) $m_A = m_B$; ii) $m_A = 5m_B$. c) For what values, if any, of the mass ratio m_A/m_B is the original kinetic energy shared equally by the two objects after the collision?

8.81 In a shipping company distribution center, an open cart of mass 50.0 kg is rolling to the left at a speed of 5.00 m/s (Fig. 8.42). You can ignore friction between the cart and the floor. A 15.0-kg package slides down a chute that is inclined at 37.0° from the horizontal and leaves the end of the chute with a speed of 3.00 m/s. The package lands in the cart and they roll off together. If the lower end of the chute is a vertical distance of 4.00 m above the bottom of the cart, what is a) the speed of the package just before it lands in the cart; b) the final speed of the cart?

Figure 8.42 Problem 8.81.

8.82 A blue puck with mass 0.0400 kg, sliding with a velocity of magnitude 0.200 m/s on a frictionless, horizontal air table, makes a perfectly elastic, head-on collision with a red puck with mass m, initially at rest. After the collision, the velocity of the blue puck is 0.050 m/s in the same direction as its initial velocity. Find a) the velocity (magnitude and direction) of the red puck after the collision; b) the mass m of the red puck.

8.83 Two asteroids with masses m_A and m_B are moving with velocities $\vec{v}_A$ and $\vec{v}_B$ with respect to an astronomer in a space vehicle. a) Show that the total kinetic energy as measured by the astronomer is

$$K = \frac{1}{2}Mv_{cm}^2 + \frac{1}{2}(m_A v_A'^2 + m_B v_B'^2)$$

with $\vec{v}_{cm}$ and M defined as in Section 8.5, $\vec{v}_A' = \vec{v}_A - \vec{v}_{cm}$ and $\vec{v}_B' = \vec{v}_B - \vec{v}_{cm}$. In this expression the total kinetic energy of the two asteroids is the energy associated with their center of mass plus that associated with the internal motion relative to the center of mass. b) If the asteroids collide with each other, what is the *mini-*

mum possible kinetic energy they can have after the collision, as measured by the astronomer? Explain.

8.84 Suppose you hold a small ball in contact with, and directly over, the center of a large ball. If you then drop the small ball a short time after dropping the large ball, the small ball rebounds with surprising speed. To show the extreme case, ignore air resistance and suppose the large ball makes an elastic collision with the floor, and then rebounds to make an elastic collision with the still-descending small ball. Just before the collision between the two balls, the large ball is moving upward with velocity $\vec{v}$, and the small ball has velocity $-\vec{v}$. (Do you see why?) Assume the large ball has a much greater mass than the small ball. a) What is the velocity of the small ball immediately after its collision with the large ball? b) From the answer to part (a), what is the ratio of the small ball's rebound distance to the distance it fell before the collision?

8.85 Jack and Jill are standing on a crate at rest on the frictionless, horizontal surface of a frozen pond. Jack has mass 75.0 kg, Jill has mass 45.0 kg, and the crate has mass 15.0 kg. They remember that they must fetch a pail of water, so each jumps horizontally from the top of the crate. Just after each jumps, that person is moving away from the crate with a speed of 4.00 m/s relative to the crate. a) What is the final speed of the crate if both Jack and Jill jump simultaneously and in the same direction? (*Hint:* Use an inertial coordinate system attached to the ground.) b) What is the final speed of the crate if Jack jumps first and then a few seconds later Jill jumps in the same direction? c) What is the final speed of the crate if Jill jumps first and then Jack, again in the same direction?

8.86 Energy Sharing. An object with mass m, initially at rest, explodes into two fragments, one with mass m_A and the other with mass m_B, where $m_A + m_B = m$. a) If energy Q is released in the explosion, how much kinetic energy does each fragment have immediately after the collision? b) What percentage of the total energy released does each fragment get when one fragment has four times the mass of the other?

8.87 Neutron Decay. A neutron at rest decays (breaks up) into a proton and an electron. Energy is released in the decay, and appears as kinetic energy of the proton and electron. The mass of a proton is 1836 times the mass of an electron. What fraction of the total energy released goes into the kinetic energy of the proton?

8.88 A ^{232}Th (thorium) nucleus at rest decays to a ^{228}Ra (radon) nucleus with the emission of an alpha particle. The total kinetic energy of the decay fragments is 6.54×10^{-13} J. An alpha particle has 1.76% of the mass of a ^{228}Ra nucleus. Calculate the kinetic energy of a) the recoiling ^{228}Ra nucleus; b) the alpha particle.

8.89 Antineutrino. In beta decay, a nucleus emits an electron. A ^{210}Bi (bismuth) nucleus at rest undergoes beta decay to ^{210}Po (polonium). Suppose that the emitted electron moves to the right with a momentum of 5.60×10^{-22} kg·m/s. The ^{210}Po nucleus, with mass 3.50×10^{-25} kg, recoils to the left at a speed of 1.14×10^{-3} m/s. Momentum conservation requires that a second particle, called an antineutrino, must also be emitted. Calculate the magnitude and direction of the momentum of the antineutrino that is emitted in this decay.

8.90 A proton moving with speed v_{A1} in the $+x$-direction makes an elastic, off-center collision with an identical proton originally at rest. After impact, the first proton moves with speed v_{A2} in the first quadrant at an angle α with the x-axis, and the second moves with speed v_{B2} in the fourth quadrant at an angle β with the x-axis (Fig. 8.12). a) Write the equations expressing conservation of linear momentum in the x- and y-directions. b) Square the equations from part (a) and add them. c) At this point, introduce the fact that the collision is elastic. d) Prove that $\alpha + \beta = \pi/2$. (You have shown that this equation is obeyed in any elastic, off-center collision between objects of equal mass when one object is initially at rest.)

8.91 Hockey puck B rests on a smooth ice surface and is struck by a second puck A, which has the same mass. Puck A is initially traveling at 15.0 m/s and is deflected 25.0° from its initial direction. Assume that the collision is perfectly elastic. Find the final speed of each puck and the direction of B's velocity after the collision. (*Hint:* Use the relation derived in part (d) of Problem 8.90.)

8.92 Jonathan and Jane are sitting in a sleigh that is at rest on frictionless ice. Jonathan's weight is 800 N, Jane's weight is 600 N, and that of the sleigh is 1000 N. They see a poisonous spider on the floor of the sleigh and immediately jump off. Jonathan jumps to the left with a velocity (relative to the ice) of 5.00 m/s at 30.0° above the horizontal, and Jane jumps to the right at 7.00 m/s at 36.9° above the horizontal (relative to the ice). Calculate the sleigh's horizontal velocity (magnitude and direction) after they jump out.

8.93 The objects in Fig. 8.43 are constructed of uniform wire bent into the shapes shown. Find the position of the center of mass of each.

(a)   (b)   (c)   (d)

Figure 8.43 Problem 8.93.

8.94 A 45.0-kg woman stands up in a 60.0-kg canoe of length 5.00 m. She walks from a point 1.00 m from one end to a point 1.00 m from the other end (Fig. 8.44). If you ignore resistance to motion of the canoe in the water, how far does the canoe move during this process?

Figure 8.44 Problem 8.94.

8.95 You are standing on a concrete slab, that in turn is resting on a frozen lake. Assume that there is no friction between the slab and the ice. The slab has a weight five times your weight. If you begin walking forward at 2.00 m/s, relative to the ice, with what speed, relative to the ice, does the slab move?

8.96 A 20.0-kg projectile is fired at an angle of 60.0° above the horizontal and with a speed of 80.0 m/s. At the highest point of its trajectory the projectile explodes into two fragments with equal mass, one of which falls vertically with zero initial speed. You can ignore air resistance. a) How far from the point of firing does the other fragment strike if the terrain is level? b) How much energy is released during the explosion?

8.97 A fireworks rocket is fired vertically upward. At its maximum height of 80.0 m it explodes and breaks into two pieces, one with mass 1.40 kg and the other with mass 0.28 kg. In the explosion, 860 J of chemical energy is converted into kinetic energy of the two fragments. a) What is the speed of each fragment just after the explosion? b) It is observed that the two fragments hit the ground at the same time. What is the distance between the points on the ground where they land? Assume that the ground is level and that air resistance can be ignored.

8.98 A 12.0-kg shell is launched at an angle of 55.0° above the horizontal with an initial speed of 150 m/s. When it is at its highest point, the shell exploded into two fragments, one three times as heavy as the other. The two fragments reach the ground at the same time. Assume that air resistance can be ignored. If the heavier fragment lands back at the same point from which the shell was launched, where will the lighter fragment land and how much energy was released in the explosion?

8.99 A Nuclear Reaction. Fission, the process that supplies energy in nuclear power plants, occurs when a heavy nucleus is split into two medium-sized nuclei. One such reaction occurs when a neutron colliding with a ^{235}U (uranium) nucleus splits that nucleus into a ^{141}Ba (barium) nucleus and a ^{92}Kr (krypton) nucleus. In this reaction, two neutrons also are split off from the original ^{235}U. Before the collision, we have the arrangement in Fig. 8.45a. After the collision, we have the ^{141}Ba nucleus moving in the $+z$-direction and the ^{92}Kr nucleus in the $-z$-direction. The three neutrons are moving in the xy-plane, as shown in Fig. 8.45b. If the incoming neutron has an initial velocity of magnitude 3.0×10^3 m/s and a final velocity of magnitude 2.0×10^3 m/s in the directions shown, what are the speeds of the other two neutrons, and what can you say about the speeds of the ^{141}Ba and ^{92}Kr nuclei?

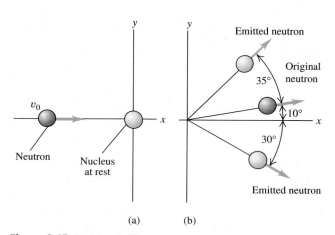

(a)   (b)

Figure 8.45 Problem 8.99.

(The mass of the ^{141}Ba nucleus is approximately 2.3×10^{-25} kg, and that of ^{92}Kr is about 1.5×10^{-25} kg.)

8.100 Center-of-Mass Coordinate System. Puck A (mass m_A) is moving on a frictionless, horizontal air table in the $+x$-direction with velocity $\vec{v}_{A1}$ and makes an elastic, head-on collision with puck B (mass m_B) that is initially at rest. After the collision, both pucks are moving along the x-axis. a) Calculate the velocity of the center of mass of the two-puck system before the collision. b) Consider a coordinate system whose origin is at the center of mass and moves with it. Is this an inertial reference frame? c) What are the initial velocities $\vec{u}_{A1}$ and $\vec{u}_{B1}$ of the two pucks in this center-of-mass reference frame? What is the total momentum in this frame? d) Use conservation of momentum and energy, applied in the center-of-mass reference frame, to relate the final momentum of each puck to its initial momentum and thus the final velocity of each puck to its initial velocity. Your results should show that a one-dimensional elastic collision has a very simple description in center-of-mass coordinates. e) Let $m_A = 0.400$ kg, $m_B = 0.200$ kg, and $v_{A1} = 6.00$ m/s. Find the center-of-mass velocities $\vec{u}_{A1}$ and $\vec{u}_{B1}$, apply the simple result found in part (d), and transform back to velocities in a stationary frame to find the final velocities of the pucks. Does your result agree with Eqs. (8.24) and (8.25)?

8.101 The *coefficient of restitution* ϵ for a head-on collision is defined as the ratio of the relative speed after the collision to the relative speed before. a) What is ϵ for a completely inelastic collision? b) What is ϵ for an elastic collision? c) A ball is dropped from a height h onto a stationary surface and rebounds back to a height H_1. Show that $\epsilon = \sqrt{H_1/h}$. d) A properly inflated basketball should have a coefficient of restitution 0.85. When dropped from a height of 1.2 m above a solid wood floor, to what height should a properly inflated basketball bounce? e) The height of the first bounce is H_1. If ϵ is constant, show that the height of the nth bounce is $H_n = \epsilon^{2n}h$. f) If ϵ is constant, what is the height of the eighth bounce of a properly inflated basketball dropped from 1.2 m?

8.102 Binding Energy of the Hydrogen Molecule. When two hydrogen atoms of mass m combine to form a diatomic hydrogen molecule (H_2), the potential energy of the system after they combine is $-\Delta$, where Δ is a positive quantity called the *binding energy* of the molecule. a) Show that in a collision that only involves two hydrogen atoms, it is *impossible* to form an H_2 molecule because momentum and energy cannot simultaneously be conserved. (*Hint:* If you can show this to be true in one frame of reference, then it is true in all frames of reference. Can you see why?) b) An H_2 molecule can be formed in a collision that involves *three* hydrogen atoms. Suppose that before such a collision, each of the three atoms has speed 1.00×10^3 m/s, and they are approaching at $120°$ angles so that at any instant, the atoms lie at the corners of an equilateral triangle. Find the speeds of the H_2 molecule and of the single hydrogen atom that remains after the collision. The binding energy of H_2 is $\Delta = 7.23 \times 10^{-19}$ J, and the mass of the hydrogen atom is 1.67×10^{-27} kg.

8.103 A wagon with two boxes of gold, having total mass 300 kg, is cut loose from the horses by an outlaw when the wagon is at rest 50 m up a $6.0°$ slope (Fig. 8.46). The outlaw plans to have the wagon roll down the slope and across the level ground, and then fall into a canyon where his confederates wait. But in a tree 40 m

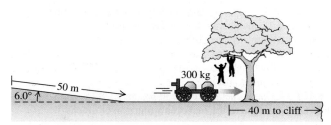

Figure 8.46 Problem 8.103.

from the canyon edge wait the Lone Ranger (mass 75.0 kg) and Tonto (mass 60.0 kg). They drop vertically into the wagon as it passes beneath them. a) If they require 5.0 s to grab the gold and jump out, will they make it before the wagon goes over the edge? The wagon rolls with negligible friction. b) When the two heroes drop into the wagon, is the kinetic energy of the system of the heroes plus the wagon conserved? If not, does it increase or decrease, and by how much?

*8.104 In Section 8.6 we considered a rocket fired in outer space where there is no air resistance and where gravity is negligible. Suppose instead that the rocket is accelerating vertically upward from rest on the earth's surface. Continue to ignore air resistance and consider only that part of the motion where the altitude of the rocket is small so that g may be assumed to be constant. a) How is Eq. (8.37) modified by the presence of the gravity force? b) Derive an expression for the acceleration a of the rocket, analogous to Eq. (8.39). c) What is the acceleration of the rocket in Example 8.16 (Section 8.6) if it is near the earth's surface rather than in outer space? You can ignore air resistance. d) Find the speed of the rocket in Example 8.17 (Section 8.6) after 90 s if the rocket is fired from the earth's surface rather than in outer space. You can ignore air resistance. How does your answer compare to the rocket speed calculated in Example 8.17?

*8.105 A Multistage Rocket. Suppose the first stage of a two-stage rocket has a total mass 12,000 kg, of which 9000 kg is fuel. The total mass of the second stage is 1000 kg, of which 700 kg is fuel. Assume that the relative speed v_{ex} of ejected material is constant, and ignore any effect of gravity. (The latter effect is small during the firing period if the rate of fuel consumption is large.) a) Suppose the entire fuel supply carried by the two-stage rocket is utilized in a single-stage rocket with the same total mass of 13,000 kg. In terms of v_{ex}, what is the speed of the rocket, starting from rest, when its fuel is exhausted? b) For the two-stage rocket, what is the speed when the fuel of the first stage is exhausted if the first stage carries the second stage with it to this point? This speed then becomes the initial speed of the second stage. At this point, the second stage separates from the first stage. c) What is the final speed of the second stage? d) What value of v_{ex} is required to give the second stage of the rocket a speed of 7.00 km/s?

*8.106 The equation $F = -v_{ex}(dm/dt)$ for the thrust on a rocket, Eq. (8.38), can also be applied to an airplane propeller. In fact, there are *two* contributions to the thrust: one positive and one negative. The positive contribution comes from air pushed backward, away from the propeller (so $dm/dt < 0$), at a speed v_{ex} relative to the propeller. The negative contribution comes from this same quantity of air flowing into the front of the propeller (so $dm/dt > 0$) at speed v, equal to the speed of the airplane through the air. a) Write an equation for the *net* thrust developed by an aircraft propeller in terms of v, v_{ex}, and the absolute value $|dm/dt|$. b) For a Cessna 182 (a single-engine airplane) flying at 130 km/h, 150 kg of air flows through the propeller each second and the propeller develops a net thrust of 1300 N. Determine the speed increase (in km/h) that the propeller imparts to the air.

*8.107 For the rocket described in Examples 8.16 and 8.17 (Section 8.6), the mass of the rocket as a function of time is

$$
m(t) = \begin{cases} m_0 & \text{for } t < 0 \\ m_0\left(1 - \dfrac{t}{120 \text{ s}}\right) & \text{for } 0 \le t \le 90 \text{ s} \\ m_0/4 & \text{for } t \ge 90 \text{ s} \end{cases}
$$

a) Calculate and graph the velocity of the rocket as a function of time from $t = 0$ s to $t = 100$ s. b) Calculate and graph the acceleration of the rocket as a function of time from $t = 0$ s to $t = 100$ s. c) A 75-kg astronaut lies on a reclined chair during the firing of the rocket. What is the maximum net force exerted by the chair on the astronaut during the firing? How does your answer compare to her weight on earth?

Challenge Problems

8.108 You are going to entertain your Aunt Emma by pulling a tablecloth out from under the dishes at her birthday party. The birthday cake is resting on a tablecloth at the center of a round table of radius $r = 0.90$ m. The tablecloth is the same size as the table top. You grab the edge of the tablecloth and pull sharply. The table-cloth and cake are in contact for time t after you start pulling. Then the sliding cake is stopped (you hope) by the friction between the cake and table top. The coefficient of kinetic friction between the cake and tablecloth is $\mu_{\text{k1}} = 0.30$, and that between the cake and table top is $\mu_{\text{k2}} = 0.40$. Apply the impulse-momentum relation (Eq. 8.9) and the work-energy theorem (Eq. 6.6) to calculate the maximum value of t if the cake is not to end up on the floor. (*Hint:* Assume that the cake moves a distance d while still on the table-cloth and therefore a distance $r - d$ while sliding on the table top. Assume that the friction forces are independent of the relative speed of the sliding surfaces. You can easily try this trick yourself by pulling a sheet of paper out from under a glass of water, but have a mop handy just in case!)

8.109 In Section 8.5 we calculated the center of mass by considering objects composed of a *finite* number of point masses or objects that, by symmetry, could be represented by a finite number of point masses. For a solid object whose mass distribution does not allow for a simple determination of the center of mass by symmetry, the sums of Eqs. (8.28) must be generalized to integrals

$$
x_{\text{cm}} = \frac{1}{M} \int x \, dm \qquad y_{\text{cm}} = \frac{1}{M} \int y \, dm
$$

where x and y are the coordinates of the small piece of the object that has mass dm. The integration is over the whole of the object. Consider a thin rod of length L, mass M, and cross-sectional area A. Let the origin of coordinates be at the left end of the rod and the

positive x-axis lie along the rod. a) If the density $\rho = M/V$ of the object is uniform, perform the integration described above to show that the x-coordinate of the center of mass of the rod is at its geometrical center. b) If the density of the object varies linearly with x—that is, $\rho = \alpha x$, where α is a positive constant—calculate the x-coordinate of the rod's center of mass.

8.110 Use the methods of Challenge Problem 8.109 to calculate the x- and y-coordinates of the center of mass of a semicircular metal plate with uniform density ρ and thickness t. Let the radius of the plate be a. The mass of the plate is thus $M = \frac{1}{2}\rho \pi a^2 t$. Use the coordinate system indicated in Fig. 8.47.

Figure 8.47 Challenge Problem 8.110.

8.111 One-fourth of a rope of length l is hanging down over the edge of a frictionless table. The rope has a uniform, linear density (mass per unit length) λ ("lambda"), and the end already on the table is held by a person. How much work does the person do when she pulls on the rope to raise the rest of the rope slowly onto the table? Do the problem in two ways as follows. a) Find the force that the person must exert to raise the rope and from this the work done. Note that this is a variable force because at different times different amounts of rope are hanging over the edge. b) Suppose the segment of the rope initially hanging over the edge of the table has all of its mass concentrated at its center of mass. Find the work necessary to raise this to table height. You will probably find this approach simpler than that of part (a). How do the answers compare, and why is this so?

*8.112 A Variable-Mass Raindrop. In a rocket-propulsion problem the mass is variable. Another such problem is a raindrop falling through a cloud of small water droplets. Some of these small droplets adhere to the raindrop, thereby *increasing* its mass as it falls. The force on the raindrop is

$$
F_{\text{ext}} = \frac{dp}{dt} = m\frac{dv}{dt} + v\frac{dm}{dt}
$$

Suppose the mass of the raindrop depends on the distance x that it has fallen. Then $m = kx$, where k is a constant, and $dm/dt = kv$. This gives, since $F_{\text{ext}} = mg$

$$
mg = m\frac{dv}{dt} + v(kv)
$$

Or, dividing by k

$$
xg = x\frac{dv}{dt} + v^2
$$

This is a differential equation that has a solution of the form $v = at$, where a is the acceleration and is constant. Take the initial velocity of the raindrop to be zero. a) Using the proposed solution for v, find the acceleration a. b) Find the distance the raindrop has fallen in $t = 3.00$ s. c) Given that $k = 2.00$ g/m, find the mass of the raindrop at $t = 3.00$ s. For many more intriguing aspects of this problem see K. S. Krane, *Amer. Jour. Phys.*, Vol. 49 (1981), pp. 113–117.

ROTATION OF RIGID BODIES

9

No matter where you sit on a rotating amusement park ride, you have the same angular velocity and angular acceleration. But your *linear* velocity and *linear* acceleration depend on how far you sit from the axis of rotation. In general, the farther you are from the axis, the greater the linear velocity and acceleration and the wilder the ride.

? **If you double your distance from the ride's axis of rotation, how many times greater is your speed? How many times greater is your acceleration toward the axis of rotation?**

What do the motions of a compact disc, a Ferris wheel, a circular saw blade, and a ceiling fan have in common? None of these can be represented adequately as a moving *point;* each involves a body that *rotates* about an axis that is stationary in some inertial frame of reference.

Rotation occurs at all scales, from the motion of electrons in atoms to the motions of entire galaxies. We need to develop some general methods for analyzing the motion of a rotating body. In this chapter and the next we consider bodies that have definite size and definite shape, and that in general can have rotational as well as translational motion.

Real-world bodies can be very complicated; the forces that act on them can deform them—stretching, twisting, and squeezing them. We'll neglect these deformations for now and assume that the body has a perfectly definite and unchanging shape and size. We call this idealized model a **rigid body.** This chapter and the next are mostly about rotational motion of a rigid body.

We begin with kinematic language for *describing* rotational motion. Next we look at the kinetic energy of rotation, the key to using energy methods for rotational motion. Then in Chapter 10 we'll develop dynamic principles that relate the forces on a body to its rotational motion.

9.1 A speedometer needle (an example of a rigid body) rotating counterclockwise about a fixed axis. The axis passes through O and is perpendicular to the plane of the diagram.

9.1 | Angular Velocity and Acceleration

In analyzing rotational motion, let's think first about a rigid body that rotates about a fixed axis. By *fixed axis* we mean an axis that is at rest in some inertial frame of reference and does not change direction relative to that frame. The rotating rigid body might be a motor shaft, a chunk of roast beef on a barbecue skewer, or a merry-go-round.

Figure 9.1 shows a rigid body (in this case, the indicator needle of a speedometer) rotating about a fixed axis. The axis passes through point O and is perpendicular to the plane of the diagram, which we choose to call the *xy*-plane. One way to describe the rotation of this body would be to choose a particular point P on the body and to keep track of the x- and y-coordinates of this point. This isn't a terribly convenient method, since it takes two numbers (the two coordinates x and y) to specify the rotational position of the body. Instead, we notice that the line OP is fixed in the body and rotates with it. The angle θ that this line makes with the +x-axis describes the rotational position of the body; we will use this single quantity θ as a *coordinate* for rotation.

The angular coordinate θ of a rigid body rotating around a fixed axis can be positive or negative. If we choose positive angles to be measured counterclockwise from the positive x-axis, then the angle θ in Fig. 9.1 is positive. If we instead choose the positive rotation direction to be clockwise, then θ in Fig. 9.1 is negative. When we considered the motion of a particle along a straight line, it was essential to specify the direction of positive displacement along that line; in discussing rotation around a fixed axis, it's just as essential to specify the direction of positive rotation.

In describing rotational motion, the most natural way to measure the angle θ is not in degrees, but in **radians.** As shown in Fig. 9.2a, one radian (1 rad) is the angle subtended at the center of a circle by an arc with a length equal to the radius of the circle. In Fig. 9.2b an angle θ is subtended by an arc of length s on a circle of radius r. The value of θ (in radians) is equal to s divided by r:

$$\theta = \frac{s}{r} \quad \text{or} \quad s = r\theta \tag{9.1}$$

An angle in radians is the ratio of two lengths, so it is a pure number, without dimensions. If s = 3.0 m and r = 2.0 m, then θ = 1.5, but we will often write this as 1.5 rad to distinguish it from an angle measured in degrees or revolutions.

The circumference of a circle (that is, the arc length all the way around the circle) is 2π times the radius, so there are 2π (about 6.283) radians in one complete revolution (360°). Therefore

$$1 \text{ rad} = \frac{360°}{2\pi} = 57.3°$$

Similarly, 180° = π rad, 90° = π/2 rad, and so on. If we had insisted on measuring the angle θ in degrees, we would have needed to include an extra factor of (2π/360) on the right-hand side of s = rθ in Eq. (9.1). By measuring angles in radians, we keep the relationship between angle and distance along an arc as simple as possible.

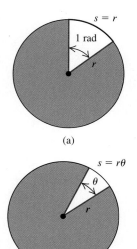

9.2 An angle θ in radians is defined as the ratio of the arc length s to the radius r.

Angular Velocity

The coordinate θ shown in Fig. 9.1 specifies the rotational position of a rigid body at a given instant. We can describe the rotational *motion* of such a rigid body in terms of the rate of change of θ. We'll do this in an analogous way to our descrip-

9.3 (a) Angular displacement $\Delta\theta$ of a rotating body. (b) Every part of a rigid body has the same angular velocity.

tion of straight-line motion in Chapter 2. In Fig. 9.3a a reference line OP in a rotating body makes an angle θ_1 with the $+x$-axis at time t_1. At a later time t_2 the angle has changed to θ_2. We define the **average angular velocity** $\omega_{\text{av-}z}$ (the Greek letter "omega") of the body in the time interval $\Delta t = t_2 - t_1$ as the ratio of the **angular displacement** $\Delta\theta = \theta_2 - \theta_1$ to Δt:

$$\omega_{\text{av-}z} = \frac{\theta_2 - \theta_1}{t_2 - t_1} = \frac{\Delta\theta}{\Delta t} \qquad (9.2)$$

The subscript z indicates that the body in Fig. 9.3a is rotating about the z-axis, which is perpendicular to the plane of the diagram. The **instantaneous angular velocity** ω_z is the limit of $\omega_{\text{av-}z}$ as Δt approaches zero, that is, the derivative of θ with respect to t:

$$\omega_z = \lim_{\Delta t \to 0} \frac{\Delta\theta}{\Delta t} = \frac{d\theta}{dt} \qquad (\text{definition of angular velocity}) \qquad (9.3)$$

When we refer simply to "angular velocity," we mean the instantaneous angular velocity, not the average angular velocity.

The angular velocity ω_z can be positive or negative, depending on the direction in which the rigid body is rotating. The angular *speed* ω, which we will use extensively in Sections 9.3 and 9.4, is the magnitude of angular velocity. Like ordinary (linear) speed v, the angular speed is never negative.

CAUTION Keep in mind the distinction between angular velocity ω_z and ordinary velocity v_x (Section 2.2). If an object has a velocity v_x, the object as a whole is *moving* along the x-axis. By contrast, if an object has an angular velocity ω_z, then it is *rotating* around the z-axis. We do *not* mean that the object is moving along the z-axis.

Different points on a rotating rigid body move different distances in a given time interval, depending on how far the point lies from the rotation axis. But because the body is rigid, *all* points rotate through the same angle in the same time (Fig. 9.3b). Hence *at any instant, every part of a rotating rigid body has the same angular velocity.* The angular velocity is positive if the body is rotating in the direction of increasing θ and negative if it is rotating in the direction of decreasing θ.

If the angle θ is in radians, the unit of angular velocity is the radian per second (rad/s). Other units, such as the revolution per minute (rev/min or rpm), are often used. Since 1 rev = 2π rad, two useful conversions are

$$1\text{rev/s} = 2\pi \text{ rad/s} \quad \text{and} \quad 1 \text{ rev/min} = 1 \text{ rpm} = \frac{2\pi}{60}\text{rad/s}$$

That is, 1 rad/s is about 10 rpm.

Example 9.1 Calculating angular velocity

Figure 9.4 shows the flywheel in a car engine under test. The angular position θ of the flywheel is given by

$$\theta = (2.0 \text{ rad/s}^3)t^3$$

The diameter of the flywheel is 0.36 m. a) Find the angle θ, in radians and in degrees, at times $t_1 = 2.0$ s and $t_2 = 5.0$ s. b) Find the distance that a particle on the rim moves during that time interval. c) Find the average angular velocity, in rad/s and in rev/min (rpm), between $t_1 = 2.0$ s and $t_2 = 5.0$ s. d) Find the instantaneous angular velocity at time $t = t_2 = 5.0$ s.

SOLUTION

IDENTIFY and SET UP: We're given the angular position θ as a function of time. Hence we can easily find our first two target variables, the values θ_1 and θ_2 of the angular position at times t_1 and t_2. Once we have these values, we can calculate the angular displacement $\Delta\theta$ between t_1 and t_2, and from this result find the distance and the average angular velocity using Eqs. (9.1) and (9.2), respectively. To find the instantaneous angular velocity, we'll take the derivative of θ with respect to time, as in Eq. (9.3).

Flywheel
diameter = 0.36 m

9.4 Rotating flywheel in a car engine.

EXECUTE: a) We substitute the values of t into the given equation:

$$\theta_1 = (2.0 \text{ rad/s}^3)(2.0 \text{ s})^3 = 16 \text{ rad}$$
$$= (16 \text{ rad})\frac{360°}{2\pi \text{ rad}} = 920°$$
$$\theta_2 = (2.0 \text{ rad/s}^3)(5.0 \text{ s})^3 = 250 \text{ rad}$$
$$= (250 \text{ rad})\frac{360°}{2\pi \text{ rad}} = 14,000°$$

b) The flywheel turns through an angular displacement of $\Delta\theta = \theta_2 - \theta_1 = 250 \text{ rad} - 16 \text{ rad} = 234 \text{ rad}$. The radius r is half the diameter, or 0.18 m. Equation (9.1) gives

$$s = r\theta = (0.18 \text{ m})(234 \text{ rad}) = 42 \text{ m}$$

To use Eq. (9.1), the angle *must* be expressed in radians. We drop "radians" from the unit for s because θ is really a dimensionless pure number; s is a distance and is measured in meters, the same unit as r.

c) In Eq. (9.2) we have

$$\omega_{\text{av-}z} = \frac{\theta_2 - \theta_1}{t_2 - t_1} = \frac{250 \text{ rad} - 16 \text{ rad}}{5.0 \text{ s} - 2.0 \text{ s}} = 78 \text{ rad/s}$$
$$= \left(78\frac{\text{rad}}{\text{s}}\right)\left(\frac{1 \text{ rev}}{2\pi \text{ rad}}\right)\left(\frac{60 \text{ s}}{1 \text{ min}}\right) = 740 \text{ rev/min}$$

d) We use Eq. (9.3):

$$\omega_z = \frac{d\theta}{dt} = \frac{d}{dt}[(2.0 \text{ rad/s}^3)t^3] = (2.0 \text{ rad/s}^3)(3t^2)$$
$$= (6.0 \text{ rad/s}^3)t^2$$

At time $t = 5.0$ s,

$$\omega_z = (6.0 \text{ rad/s}^3)(5.0 \text{ s})^2 = 150 \text{ rad/s}$$

EVALUATE: Our result in part (d) shows that ω_z is proportional to t^2 and hence increases with time. Our numerical results are consistent with this result: the 150-rad/s instantaneous angular velocity at $t = 5.0$ s is greater than the 78-rad/s average angular velocity for the 2.0-s interval leading up to that time (from $t_1 = 3.0$ s to $t_2 = 5.0$ s).

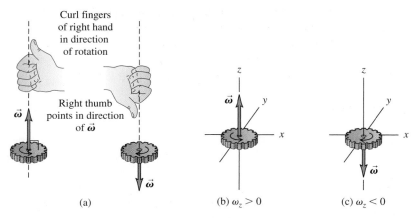

9.5 (a) The right-hand rule for the direction of the angular velocity vector $\vec{\omega}$. Reversing the direction of rotation reverses the direction of $\vec{\omega}$. If the rotation is along the z-axis, the sign of ω_z depends on whether $\vec{\omega}$ (b) points along or (c) points opposite to the positive z-direction.

As we have seen, our notation for the angular velocity ω_z about the z-axis is reminiscent of the notation v_x for the ordinary velocity along the x-axis (see Section 2.2). Just as v_x is the x-component of the velocity vector $\vec{v}$, ω_z is the z-component of an angular velocity *vector* $\vec{\omega}$ directed along the axis of rotation. As Fig. 9.5a shows, the direction of $\vec{\omega}$ is given by the right-hand rule that we used to define the vector product in Section 1.10. If the rotation is about the z-axis, then $\vec{\omega}$ has only a z-component; this component is positive if $\vec{\omega}$ is along the positive z-axis (Fig. 9.5b) and negative if $\vec{\omega}$ is along the negative z-axis (Fig. 9.5c).

The vector formulation is especially useful in situations in which the direction of the rotation axis *changes*. We'll examine such situations briefly at the end of Chapter 10. In this chapter, however, we'll consider only situations in which the rotation axis is fixed. Hence throughout this chapter we'll use "angular velocity" to refer to ω_z, the component of the angular velocity vector $\vec{\omega}$ along the axis.

Angular Acceleration

When the angular velocity of a rigid body changes, it has an *angular acceleration*. When you pedal your bicycle harder to make the wheels turn faster or apply the brakes to bring the wheels to a stop, you're giving the wheels an angular acceleration. You also impart an angular acceleration whenever you change the rotation speed of a piece of spinning machinery, such as an automobile engine's crankshaft or the rotor of a helicopter (Fig. 9.6).

If ω_{1z} and ω_{2z} are the instantaneous angular velocities at times t_1 and t_2, we define the **average angular acceleration** $\alpha_{\text{av-}z}$ over the interval $\Delta t = t_2 - t_1$ as the change in angular velocity divided by Δt:

$$\alpha_{\text{av-}z} = \frac{\omega_{2z} - \omega_{1z}}{t_2 - t_1} = \frac{\Delta\omega_z}{\Delta t} \tag{9.4}$$

The **instantaneous angular acceleration** α_z is the limit of $\alpha_{\text{av-}z}$ as $\Delta t \to 0$:

$$\alpha_z = \lim_{\Delta t \to 0} \frac{\Delta\omega_z}{\Delta t} = \frac{d\omega_z}{dt} \quad \text{(definition of angular acceleration)} \tag{9.5}$$

9.6 The pilot of a helicopter can change the angular velocity of the rotor—that is, give it an angular acceleration—by adjusting a control in the cockpit.

The usual unit of angular acceleration is the radian per second per second, or rad/s^2. Henceforth we will use the term "angular acceleration" to mean the instantaneous angular acceleration rather than the average angular acceleration.

Because $\omega_z = d\theta/dt$, we can also express angular acceleration as the second derivative of the angular coordinate:

$$\alpha_z = \frac{d}{dt}\frac{d\theta}{dt} = \frac{d^2\theta}{dt^2} \tag{9.6}$$

You have probably noticed that we are using Greek letters for angular kinematic quantities: θ for angular position, ω_z for angular velocity, and α_z for angular acceleration. These are analogous to x for position, v_x for velocity, and a_x for acceleration, respectively, in straight-line motion. In each case, velocity is the rate of change of position with respect to time and acceleration is the rate of change of velocity with respect to time. We will sometimes use the terms *linear* velocity and *linear* acceleration for the familiar quantities we defined in Chapters 2 and 3 to make sure we distinguish clearly between these and the *angular* quantities introduced in this chapter.

In rotational motion, if the angular acceleration α_z is positive, then the angular velocity ω_z is increasing; if α_z is negative, then ω_z is decreasing. The rotation is speeding up if α_z and ω_z have the same sign and slowing down if α_z and ω_z have opposite signs. (These are exactly the same relationships as those between *linear* acceleration a_x and *linear* velocity v_x for straight-line motion; see Section 2.3.)

| Example 9.2 | **Calculating angular acceleration** |

In Example 9.1 we found that the instantaneous angular velocity ω_z of the flywheel at ány time t is given by

$$\omega_z = (6.0 \text{ rad/s}^3)t^2$$

a) Find the average angular acceleration between $t_1 = 2.0$ s and $t_2 = 5.0$ s. b) Find the instantaneous angular acceleration at time $t_2 = 5.0$ s.

SOLUTION

IDENTIFY and SET UP: We'll use Eqs. (9.4) and (9.5) to find our two target variables, the average angular acceleration $\alpha_{av\text{-}z}$ between t_1 and t_2 and the instantaneous angular acceleration α_z at $t = t_2$.

EXECUTE: a) The values of ω_z at the two times are

$$\omega_{1z} = (6.0 \text{ rad/s}^3)(2.0 \text{ s})^2 = 24 \text{ rad/s}$$
$$\omega_{2z} = (6.0 \text{ rad/s}^3)(5.0 \text{ s})^2 = 150 \text{ rad/s}$$

From Eq. (9.4) the average angular acceleration is

$$\alpha_{av\text{-}z} = \frac{150 \text{ rad/s} - 24 \text{ rad/s}}{5.0 \text{ s} - 2.0 \text{ s}} = 42 \text{ rad/s}^2$$

b) From Eq. (9.5) the instantaneous angular acceleration at any time t is

$$\alpha_z = \frac{d\omega_z}{dt} = \frac{d}{dt}[(6.0 \text{ rad/s}^3)(t^2)] = (6.0 \text{ rad/s}^3)(2t)$$
$$= (12 \text{ rad/s}^3)t$$

At time $t = 5.0$ s,

$$\alpha_z = (12 \text{ rad/s}^3)(5.0 \text{ s}) = 60 \text{ rad/s}^2$$

EVALUATE: Note that the angular acceleration is *not* constant in this situation. The angular velocity ω_z is always increasing because α_z is always positive. Furthermore, the rate at which the angular velocity increases is itself increasing, since α_z increases with time.

Just as we did for angular velocity, it's useful to define an angular acceleration vector $\vec{\alpha}$. Mathematically, $\vec{\alpha}$ is the time derivative of the angular velocity vector $\vec{\omega}$. If the object rotates around the fixed z-axis, then $\vec{\alpha}$ has only a z-component; the quantity α_z is just that component. In this case, $\vec{\alpha}$ points in the same direction as

$\vec{\omega}$ if the rotation is speeding up and opposite to $\vec{\omega}$ if the rotation is slowing down (Fig. 9.7).

The angular acceleration vector will be particularly useful in Chapter 10 when we discuss what happens when the rotation axis can change direction. In this chapter, however, the rotation axis will always be fixed and we need use only the z-component α_z.

Speeding up   Slowing down

9.7 When the rotation axis is fixed, the angular acceleration and angular velocity vectors both lie along that axis.

Test Your Understanding

Figure 9.8 is a graph of ω_z and α_z versus time for a particular rotating body. During which time intervals is the rotation speeding up? During which time intervals is the rotation slowing down?

9.2 | Rotation with Constant Angular Acceleration

In Chapter 2 we found that straight-line motion is particularly simple when the acceleration is constant. This is also true of rotational motion about a fixed axis. When the angular acceleration is constant, we can derive equations for angular velocity and angular position using exactly the same procedure that we used for straight-line motion in Section 2.4. In fact, the equations we are about to derive are identical to Eqs. (2.8), (2.12), (2.13), and (2.14) if we replace x with θ, v_x with ω_z, and a_x with α_z. We suggest that you review Section 2.4 before continuing.

Let ω_{0z} be the angular velocity of a rigid body at time $t = 0$ and let ω_z be its angular velocity at any later time t. The angular acceleration α_z is constant and equal to the average value for any interval. Using Eq. (9.4) with the interval from 0 to t, we find

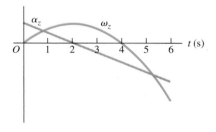

9.8 Graph of angular velocity ω_z and angular acceleration α_z versus time for a rotating body.

$$\alpha_z = \frac{\omega_z - \omega_{0z}}{t - 0} \quad \text{or}$$

$$\omega_z = \omega_{0z} + \alpha_z t \quad (\text{constant angular acceleration only}) \quad (9.7)$$

The product $\alpha_z t$ is the total change in ω_z between $t = 0$ and the later time t; the angular velocity ω_z at time t is the sum of the initial value ω_{0z} and this total change.

With constant angular acceleration, the angular velocity changes at a uniform rate, so its average value between 0 and t is the average of the initial and final values:

$$\omega_{\text{av-}z} = \frac{\omega_{0z} + \omega_z}{2} \quad (9.8)$$

We also know that $\omega_{\text{av-}z}$ is the total angular displacement $(\theta - \theta_0)$ divided by the time interval $(t - 0)$;

$$\omega_{\text{av-}z} = \frac{\theta - \theta_0}{t - 0} \quad (9.9)$$

When we equate Eqs. (9.8) and (9.9) and multiply the result by t, we get

Act|v
ONLINE
Physics

7.7   Rotational Kinematics

$$\theta - \theta_0 = \frac{1}{2}(\omega_{0z} + \omega_z)t \quad (\text{constant angular acceleration only}) \quad (9.10)$$

To obtain a relation between θ and t that doesn't contain ω_z, we substitute Eq. (9.7) into Eq. (9.10):

$$\theta - \theta_0 = \frac{1}{2}[\omega_{0z} + (\omega_{0z} + \alpha_z t)]t$$

$$\theta = \theta_0 + \omega_{0z}t + \frac{1}{2}\alpha_z t^2 \quad \text{(constant angular acceleration only)} \quad (9.11)$$

That is, if at the initial time $t = 0$ the body is at angular position θ_0 and has angular velocity ω_{0z}, then its angular position θ at any later time t is the sum of three terms: its initial angular position θ_0, plus the rotation $\omega_{0z}t$ it would have if the angular velocity were constant, plus an additional rotation $\frac{1}{2}\alpha_z t^2$ caused by the changing angular velocity.

Following the same procedure as for straight-line motion in Section 2.4, we can combine Eqs. (9.7) and (9.11) to obtain a relation between θ and ω_z that does not contain t. We invite you to work out the details, following the same procedure we used to get Eq. (2.13). (See Exercise 9.12.) In fact, because of the perfect analogy between straight-line and rotational quantities, we can simply take Eq. (2.13) and replace each straight-line quantity by its rotational analog. We get

$$\omega_z^2 = \omega_{0z}^2 + 2\alpha_z(\theta - \theta_0) \quad \text{(constant angular acceleration only)} \quad (9.12)$$

CAUTION Keep in mind that all of these results are valid *only* when the angular acceleration α_z is *constant;* be careful not to try to apply them to problems in which α_z is *not* constant. Table 9.1 shows the analogy between Eqs. (9.7), (9.10), (9.11), and (9.12) for fixed-axis rotation with constant angular acceleration and the corresponding equations for straight-line motion with constant linear acceleration.

Table 9.1 Comparison of Linear and Angular Motion with Constant Acceleration

Straight-Line Motion with Constant Linear Acceleration	Fixed-Axis Rotation with Constant Angular Acceleration
$a_x = \text{constant}$	$\alpha_z = \text{constant}$
$v_x = v_{0x} + a_x t$	$\omega_z = \omega_{0z} + \alpha_z t$
$x = x_0 + v_{0x}t + \frac{1}{2}a_x t^2$	$\theta = \theta_0 + \omega_{0z}t + \frac{1}{2}\alpha_z t^2$
$v_x^2 = v_{0x}^2 + 2a_x(x - x_0)$	$\omega_z^2 = \omega_{0z}^2 + 2\alpha_z(\theta - \theta_0)$
$x - x_0 = \frac{1}{2}(v_x + v_{0x})t$	$\theta - \theta_0 = \frac{1}{2}(\omega_z + \omega_{0z})t$

Example 9.3

Rotation with constant angular acceleration

You have just finished watching a movie on DVD and the disc is slowing to a stop. The angular velocity of the disc at $t = 0$ is 27.5 rad/s and its angular acceleration is a constant -10.0 rad/s^2. A line PQ on the surface of the disc lies along the $+x$-axis at $t = 0$ (Fig. 9.9). a) What is the disc's angular velocity at $t = 0.300$ s? b) What angle does the line PQ make with the $+x$-axis at this time?

SOLUTION

IDENTIFY: The angular acceleration of the disc is constant, so we can use any of the equations derived in this section. Our target variables are the angular velocity and the angular displacement at $t = 0.300$ s.

SET UP: We are given the initial angular velocity $\omega_{0z} = 27.5$ rad/s, the initial angle $\theta_0 = 0$ between the line PQ and the $+x$-axis, the angular acceleration $\alpha_z = -10.0$ rad/s^2, and the time $t = 0.300$ s of interest. With this information it's easiest to use Eqs. (9.7) and (9.11) to find the target variables ω_z and θ, respectively.

EXECUTE: a) From Eq. (9.7), at $t = 0.300$ s we have

$$\omega_z = \omega_{0z} + \alpha_z t = 27.5 \text{ rad/s} + (-10.0 \text{ rad/s}^2)(0.300 \text{ s})$$

$$= 24.5 \text{ rad/s}$$

b) From Eq. (9.11),

$$\theta = \theta_0 + \omega_{0z} t + \frac{1}{2}\alpha_z t^2$$

$$= 0 + (27.5 \text{ rad/s})(0.300 \text{ s}) + \frac{1}{2}(-10.0 \text{ rad/s}^2)(0.300 \text{ s})^2$$

$$= 7.80 \text{ rad} = 7.80 \text{ rad}\left(\frac{1 \text{ rev}}{2\pi \text{ rad}}\right) = 1.24 \text{ rev}$$

The DVD has turned through one complete revolution plus an additional 0.24 rev—that is, through an additional angle of $(0.24 \text{ rev})(360°/\text{rev}) = 87°$. Hence the line PQ is at an angle of $87°$ with the $+x$-axis.

EVALUATE: Our answer to part (a) tells us that the angular velocity has decreased. This is as it should be, since α_z is negative. We can also use our answer for ω_z in part (a) to check our result for θ in part

9.9 A line PQ on a rotating DVD at $t = 0$. What angle does this line make with the $+x$-axis at $t = 0.300$ s?

(b). To do so, we solve Eq. (9.12), $\omega_z^2 = \omega_{0z}^2 + 2\alpha_z(\theta - \theta_0)$, for the angle θ:

$$\theta = \theta_0 + \left(\frac{\omega_z^2 - \omega_{0z}^2}{2\alpha_z}\right)$$

$$= 0 + \frac{(24.5 \text{ rad/s})^2 - (27.5 \text{ rad/s})^2}{2(-10.0 \text{ rad/s}^2)} = 7.80 \text{ rad}$$

which agrees with the result we found earlier.

Test Your Understanding

At what time t does the DVD in Example 9.3 come to rest? Through how many revolutions does it rotate between $t = 0$ and coming to rest?

9.3 | Relating Linear and Angular Kinematics

How do we find the linear velocity and acceleration of a particular point in a rotating rigid body? We need to answer this question to proceed with our study of rotation. For example, to find the kinetic energy of a rotating body, we have to start from $K = \frac{1}{2}mv^2$ for a particle, and this requires knowing v for each particle in the body. So it's worthwhile to develop general relations between the *angular* velocity and acceleration of a rigid body rotating about a fixed axis and the *linear* velocity and acceleration of a specific point or particle in the body.

When a rigid body rotates about a fixed axis, every particle in the body moves in a circular path. The circle lies in a plane perpendicular to the axis and is centered on the axis. The speed of a particle is directly proportional to the body's angular velocity; the faster the body rotates, the greater the speed of each particle. In Fig. 9.10, point P is a constant distance r from the axis of rotation, so it moves in a circle of radius r. At any time the angle θ (in radians) and the arc length s are related by

$$s = r\theta$$

9.10 A rigid body rotating about a fixed axis through point O. The distance s that point P on the body moves through equals $r\theta$ if θ is measured in radians. The linear speed v of point P equals $r\omega$ if the angular speed ω is measured in rad/s.

We take the time derivative of this, noting that r is constant for any specific particle, and take the absolute value of both sides:

$$\left| \frac{ds}{dt} \right| = r \left| \frac{d\theta}{dt} \right|$$

Now $|ds/dt|$ is the absolute value of the rate of change of arc length, which is equal to the instantaneous *linear* speed v of the particle. Analogously, $|d\theta/dt|$, the absolute value of the rate of change of the angle, is the instantaneous **angular speed** ω—that is, the magnitude of the instantaneous angular velocity in rad/s. Thus

$$v = r\omega \qquad \text{(relation between linear and angular speed)} \qquad (9.13)$$

The farther a point is from the axis, the greater its linear speed. The *direction* of the linear velocity *vector* is tangent to its circular path at each point (Fig. 9.10).

CAUTION   Keep in mind the distinction between the linear and angular *speeds* v and ω (which appear in Eq. (9.13)) and the linear and angular *velocities* v_x and ω_z. The quantities without subscripts, v and ω, are never negative; they are the magnitudes of the vectors $\vec{v}$ and $\vec{\omega}$, respectively, and their values tell you only how fast a particle is moving (v) or how fast a body is rotating (ω). The corresponding quantities with subscripts, v_x and ω_z, can be either positive or negative; their signs tell you the direction of the motion.

9.11 The component of acceleration of point P parallel to $\vec{v}$ is $a_{\text{tan}} = r\alpha$; the component perpendicular to $\vec{v}$, radially inward toward the rotation axis, is $a_{\text{rad}} = \omega^2 r$. The rigid body shown here is speeding up.

We can represent the acceleration of a particle moving in a circle in terms of its centripetal and tangential components, a_{rad} and a_{tan} (Fig. 9.11), as we did in Section 3.4. It would be a good idea to review that section now. We found that the **tangential component of acceleration** a_{tan}, the component parallel to the instantaneous velocity, acts to change the *magnitude* of the particle's velocity (i.e., the speed) and is equal to the rate of change of speed. Taking the derivative of Eq. (9.13), we find

$$a_{\text{tan}} = \frac{dv}{dt} = r\frac{d\omega}{dt} = r\alpha \qquad (9.14)$$

(tangential acceleration of a point on a rotating body)

This component of a particle's acceleration is always tangent to the circular path of the particle.

The quantity $\alpha = d\omega/dt$ in Eq. (9.14) is the rate of change of the angular *speed*. It is not quite the same as $\alpha_z = d\omega_z/dt$, which is the rate of change of the angular *velocity*. For example, consider a body rotating so that its angular velocity vector points in the $-z$-direction (Fig. 9.5c). If the body is gaining angular speed at a rate of 10 rad/s per second, then $\alpha = 10$ rad/s^2. But ω_z is negative and becoming more negative as the rotation gains speed, so $\alpha_z = -10$ rad/s^2. The rule for rotation about a fixed axis is that α is equal to α_z if ω_z is positive but equal to $-\alpha_z$ if ω_z is negative.

The component of the particle's acceleration directed toward the rotation axis, the **centripetal component of acceleration** a_{rad}, is associated with the change of *direction* of the particle's velocity. In Section 3.4 we worked out the relation $a_{\text{rad}} = v^2/r$. We can express this in terms of ω by using Eq. (9.13):

$$a_{rad} = \frac{v^2}{r} = \omega^2 r \qquad (9.15)$$

(centripetal acceleration of a point on a rotating body)

This is true at each instant *even when ω and v are not constant*. The centripetal component always points toward the axis of rotation.

The vector sum of the centripetal and tangential components of acceleration of a particle in a rotating body is the linear acceleration $\vec{a}$ (Fig. 9.11).

CAUTION It's important to remember that Eq. (9.1), $s = r\theta$, is valid *only* when θ is measured in radians. The same is true of any equation derived from this, including Eqs. (9.13), (9.14), and (9.15). When you use these equations, you *must* express the angular quantities in radians, not revolutions or degrees (Fig. 9.12).

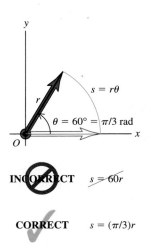

9.12 In any equation that relates linear quantities to angular quantities, the angles must be expressed in radians—never in degrees or revolutions.

Equations (9.1), (9.13), and (9.14) also apply to any particle that has the same tangential velocity as a point in a rotating rigid body. For example, when a rope wound around a circular cylinder unwraps without stretching or slipping, its speed and acceleration at any instant are equal to the speed and tangential acceleration of the point at which it is tangent to the cylinder. The same principle holds for situations such as bicycle chains and sprockets, belts and pulleys that turn without slipping, and so on. We will have several opportunities to use these relations later in this chapter and in Chapter 10. Note that Eq. (9.15) for the centripetal component a_{rad} is applicable to the rope or chain *only* at points that are in contact with the cylinder or sprocket. Other points do not have the same acceleration toward the center of the circle that points on the cylinder or sprocket have.

Example 9.4	**Throwing a discus**

A discus thrower moves the discus in a circle of radius 80.0 cm. At a certain instant, the thrower is spinning at an angular speed of 10.0 rad/s and the angular speed is increasing at 50.0 rad/s². At this instant, find the tangential and centripetal components of the acceleration of the discus and the magnitude of the acceleration.

9.13 Whirling a discus in a circle.

SOLUTION

IDENTIFY and SET UP: We model the discus as a particle traveling on a circular path (Fig. 9.13), so we can use the ideas developed in this section. We are given the radius $r = 0.800$ m, the angular speed $\omega = 10.0$ rad/s, and the rate of change of angular speed $\alpha = 50.0$ rad/s². The first two target variables are the acceleration components a_{tan} and a_{rad}, which we'll find with Eqs. (9.14) and (9.15), respectively. Given these components of the acceleration vector, we'll find its magnitude (the third target variable) using the Pythagorean theorem (see Fig. 9.11).

EXECUTE: From Eqs. (9.14) and (9.15),

$$a_{tan} = r\alpha = (0.800 \text{ m})(50.0 \text{ rad/s}^2) = 40.0 \text{ m/s}^2$$
$$a_{rad} = \omega^2 r = (10.0 \text{ rad/s})^2(0.800 \text{ m}) = 80.0 \text{ m/s}^2$$

The magnitude of the acceleration vector is

$$a = \sqrt{a_{tan}^2 + a_{rad}^2} = 89.4 \text{ m/s}^2$$

EVALUATE: Note that we have dropped the unit "radian" from our results for a_{tan}, a_{rad}, and a. We can do this because "radian" is a dimensionless quantity.

The magnitude a is about nine times g, the acceleration due to gravity. Can you show that if the angular speed doubles to 20.0 rad/s while α remains the same, the acceleration magnitude a increases to 322 m/s², or almost 33g?

Example 9.5	**Designing a propeller**

You are asked to design an airplane propeller to turn at 2400 rpm. The forward air speed of the plane is to be 75.0 m/s (270 km/h, or about 168 mi/h), and the speed of the tips of the propeller blades through the air must not exceed 270 m/s (Fig. 9.14a). (This is about 0.80 times the speed of sound in air. If the propeller tips were to move at the speed of sound, they would produce a tremendous amount of noise. Keeping the speed comfortably below this keeps the noise level acceptably low.) a) What is the maximum radius the propeller can have? b) With this radius, what is the acceleration of the propeller tip?

SOLUTION

IDENTIFY: The object of interest in this example is a particle at the tip of the propeller; our target variables are the particle's distance from the axis and its acceleration. Note that the speed of this particle through the air (which cannot exceed 270 m/s) is due to both the propeller's rotation *and* the forward motion of the airplane.

SET UP: As Fig. 9.14b shows, the velocity $\vec{v}_{\text{total}}$ of a particle at the propeller tip is the vector sum of its tangential velocity due to the propeller's rotation (magnitude v_P, given by Eq. (9.13)) and the forward velocity of the airplane (magnitude $v_A = 75.0$ m/s). The rotation plane of the propeller is perpendicular to the direction of flight, so these two vectors are perpendicular and we can use the Pythagorean theorem to relate v_P and v_A to v_{total}. We will then set $v_{\text{total}} = 270$ m/s and solve for the radius r. Note that the angular speed of the propeller is constant, so the acceleration of the propeller tip has only a radial component; we'll find it using Eq. (9.15).

EXECUTE: We first convert ω to rad/s (see Fig. 9.12):

$$\omega = 2400 \text{ rpm} = \left(2400\frac{\text{rev}}{\text{min}}\right)\left(\frac{2\pi \text{ rad}}{1 \text{ rev}}\right)\left(\frac{1 \text{ min}}{60 \text{ s}}\right)$$

$$= 251 \text{ rad/s}$$

a) From Fig. 9.14b and Eq. (9.13), the velocity magnitude v_{total} is given by

$$v_{\text{total}}^2 = v_A^2 + v_P^2 = v_A^2 + r^2\omega^2$$

so

$$r^2 = \frac{v_{\text{total}}^2 - v_A^2}{\omega^2} \quad \text{and} \quad r = \frac{\sqrt{v_{\text{total}}^2 - v_A^2}}{\omega}$$

If $v_{\text{total}} = 270$ m/s, the propeller radius is

$$r = \frac{\sqrt{(270 \text{ m/s})^2 - (75.0 \text{ m/s})^2}}{251 \text{ rad/s}} = 1.03 \text{ m}$$

b) The purely centripetal acceleration is

$$a_{\text{rad}} = \omega^2 r$$

$$= (251 \text{ rad/s})^2(1.03 \text{ m}) = 6.5 \times 10^4 \text{ m/s}^2$$

EVALUATE: From $\sum\vec{F} = m\vec{a}$, the propeller must exert a force of 6.5×10^4 N on each kilogram of material at its tip! This is why propellers are made out of tough material (usually aluminum alloy).

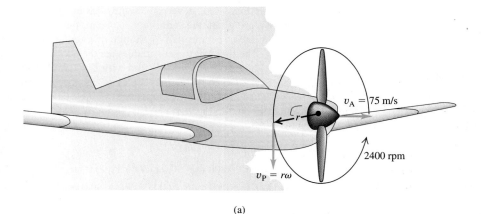

$v_A = 75$ m/s

2400 rpm

$v_P = r\omega$

(a)

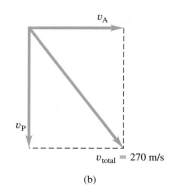

v_A

v_P

$v_{\text{total}} = 270$ m/s

(b)

9.14 Designing an airplane propeller.

Conceptual Example 9.6	**Bicycle gears**

How are the angular speeds of the two bicycle sprockets in Fig. 9.15 related to the number of teeth on each sprocket?

SOLUTION

The chain does not slip or stretch, so it moves at the same tangential speed v on both sprockets. From Eq. (9.13),

$$v = r_1\omega_1 = r_2\omega_2 \quad \text{so} \quad \frac{\omega_2}{\omega_1} = \frac{r_1}{r_2}$$

The angular speed is inversely proportional to the radius. This relationship also holds for pulleys connected by a belt, provided that the belt doesn't slip. For chain sprockets the teeth must be equally spaced on the circumferences of both sprockets for the chain to mesh properly with both. Let N_1 and N_2 be the numbers of teeth; the condition that the tooth spacing is the same on both sprockets is

$$\frac{2\pi r_1}{N_1} = \frac{2\pi r_2}{N_2} \quad \text{or} \quad \frac{r_1}{r_2} = \frac{N_1}{N_2}$$

Combining this with the other equation, we get

$$\frac{\omega_2}{\omega_1} = \frac{N_1}{N_2}$$

The angular speed of each sprocket is inversely proportional to the number of teeth. On a multi-speed bike, you get the highest

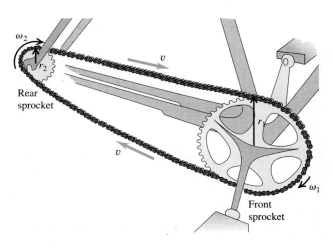

9.15 How are the angular velocities of the two sprockets related to their number of teeth?

angular speed ω_2 of the rear wheel for a given pedaling rate ω_1 when the ratio N_1/N_2 is maximum; this means using the largest-radius front sprocket (largest N_1) and the smallest-radius rear sprocket (smallest N_2).

Test Your Understanding

Information is stored on a CD or DVD (Fig. 9.9) in a coded pattern of tiny pits. The pits are arranged in a track that spirals outward toward the rim of the disc. As the disc spins inside a player, the track is scanned at a constant *linear* speed. How must the rotation speed of the disc change as the player's scanning head moves over the track?

9.4 | Energy in Rotational Motion

A rotating rigid body consists of mass in motion, so it has kinetic energy. We can express this kinetic energy in terms of the body's angular velocity and a new quantity called *moment of inertia* that we define in this section. To develop this relationship, we think of the body as being made up of a large number of particles, with masses $m_1, m_2, \ldots$, at distances $r_1, r_2, \ldots$ from the axis of rotation. We label the particles with the index i: The mass of the ith particle is m_i and its distance from the axis of rotation is r_i. The particles don't necessarily all lie in the same plane, so we specify that r_i is the *perpendicular* distance from the axis to the ith particle.

When a rigid body rotates about a fixed axis, the speed v_i of the ith particle is given by Eq. (9.13), $v_i = r_i\omega$, where ω is the body's angular speed. Different particles have different values of r, but ω is the same for all (otherwise, the body wouldn't be rigid). The kinetic energy of the ith particle can be expressed as

$$\frac{1}{2}m_i v_i^2 = \frac{1}{2}m_i r_i^2 \omega^2$$

The *total* kinetic energy of the body is the sum of the kinetic energies of all its particles:

$$K = \frac{1}{2}m_1 r_1^2 \omega^2 + \frac{1}{2}m_2 r_2^2 \omega^2 + \cdots = \sum_i \frac{1}{2}m_i r_i^2 \omega^2$$

Taking the common factor $\omega^2/2$ out of this expression, we get

$$K = \frac{1}{2}(m_1 r_1^2 + m_2 r_2^2 + \cdots)\omega^2 = \frac{1}{2}\left(\sum_i m_i r_i^2\right)\omega^2$$

The quantity in parentheses, obtained by multiplying the mass of each particle by the square of its distance from the axis of rotation and adding these products, is denoted by I and is called the **moment of inertia** of the body for this rotation axis:

$$I = m_1 r_1^2 + m_2 r_2^2 + \cdots = \sum_i m_i r_i^2 \qquad (9.16)$$

(definition of moment of inertia)

The word "moment" means that I depends on how the body's mass is distributed in space; it has nothing to do with a "moment" of time. For a body with a given rotation axis and a given total mass, the greater the distance from the axis to the particles that make up the body, the greater the moment of inertia. In a rigid body, the distances r_i are all constant and I is independent of how the body is rotating around the given axis. The SI unit of moment of inertia is the kilogram-meter^2 $(\text{kg}\cdot\text{m}^2)$.

In terms of moment of inertia I, the **rotational kinetic energy** K of a rigid body is

$$K = \frac{1}{2}I\omega^2 \qquad \text{(rotational kinetic energy of a rigid body)} \qquad (9.17)$$

The kinetic energy given by Eq. (9.17) is not a new form of energy; it's the sum of the kinetic energies of the individual particles that make up the rigid body, written in a compact and convenient form in terms of the moment of inertia. When using Eq. (9.17), ω *must* be measured in radians per second, not revolutions or degrees per second, to give K in joules; this is because we used $v_i = r_i\omega$ in our derivation.

Equation (9.17) gives a simple physical interpretation of moment of inertia: the greater the moment of inertia, the greater the kinetic energy of a rigid body rotating with a given angular speed ω. We learned in Chapter 6 that the kinetic energy of a body equals the amount of work done to accelerate that body from rest. So the greater a body's moment of inertia, the harder it is to start the body rotating if it's at rest and the harder it is to stop its rotation if it's already rotating (Fig. 9.16). For this reason, I is also called the *rotational inertia*.

Example 9.7 shows how *changing* the rotation axis can affect the value of I.

• Mass close to axis
• Small moment of inertia
• Easy to start apparatus rotating

• Mass farther from axis
• Greater moment of inertia
• Harder to start apparatus rotating

9.16 An apparatus free to rotate around a vertical axis. The moment of inertia can be varied by locking the two equal-mass cylinders into different positions on the horizontal shaft.

Example
9.7

Moments of inertia for different rotation axes

An engineer is designing a one-piece machine part consisting of three heavy connectors linked by light molded struts (Fig. 9.17). a) What is the moment of inertia of this body about an axis through point A, perpendicular to the plane of the diagram? b) What is the moment of inertia about an axis coinciding with rod BC? c) If the body rotates about an axis through A perpendicular to the plane of the diagram, with angular speed $\omega = 4.0$ rad/s, what is its kinetic energy?

9.17 A strangely shaped machine part.

SOLUTION

IDENTIFY: We'll consider the connectors as massive particles and the lightweight struts as massless rods. Then we can use the ideas of this section to calculate the moment of inertia of this collection of three particles.

SET UP: In parts (a) and (b), we'll use Eq. (9.16) to find the moments of inertia for each of the two axes. Given the moment of inertia for axis A, we'll use Eq. (9.17) in part (c) to find the rotational kinetic energy.

EXECUTE: a) The particle at point A lies *on* the axis. Its distance r from the axis is zero, so it contributes nothing to the moment of inertia. Equation (9.16) gives

$$I = \sum m_i r_i^2 = (0.10 \text{ kg})(0.50 \text{ m})^2 + (0.20 \text{ kg})(0.40 \text{ m})^2$$

$$= 0.057 \text{ kg} \cdot \text{m}^2$$

b) The particles at B and C both lie *on* the axis, so for them $r = 0$ and neither contributes to the moment of inertia. Only A contributes, and we have

$$I = \sum m_i r_i^2 = (0.30 \text{ kg})(0.40 \text{ m})^2 = 0.048 \text{ kg} \cdot \text{m}^2$$

c) From Eq. (9.17),

$$K = \frac{1}{2} I \omega^2 = \frac{1}{2}(0.057 \text{ kg} \cdot \text{m}^2)(4.0 \text{ rad/s})^2 = 0.46 \text{ J}$$

EVALUATE: Our results show that the moment of inertia for the axis through A is greater than that for the axis through B and C. Hence, of the two axes, it's easier to make the machine part rotate about axis BC.

CAUTION The results of parts (a) and (b) of Example 9.7 show that the moment of inertia of a body depends on the location and orientation of the axis. It's not enough to just say, "The moment of inertia of this body is 0.048 kg $\cdot$ m^2." We have to be specific and say, "The moment of inertia of this body *about axis BC* is 0.048 kg $\cdot$ m^2."

7.6   Rotational Inertia

In Example 9.7 we represented the body as several point masses, and we evaluated the sum in Eq. (9.16) directly. When the body is a *continuous* distribution of matter, such as a solid cylinder or plate, the sum becomes an integral, and we need to use calculus to calculate the moment of inertia. We will give several examples of such calculations in Section 9.6; meanwhile, Table 9.2 gives moments of inertia for several familiar shapes in terms of the masses and dimensions. Each body shown in Table 9.2 is *uniform;* that is, the density has the same value at all points within the solid parts of the body.

CAUTION You may be tempted to try to compute the moment of inertia of a body by assuming that all the mass is concentrated at the center of mass and then multiplying the total mass by the square of the distance from the center of mass to the axis. Resist that temptation; it doesn't work! For example, when a

Table 9.2 Moments of Inertia of Various Bodies

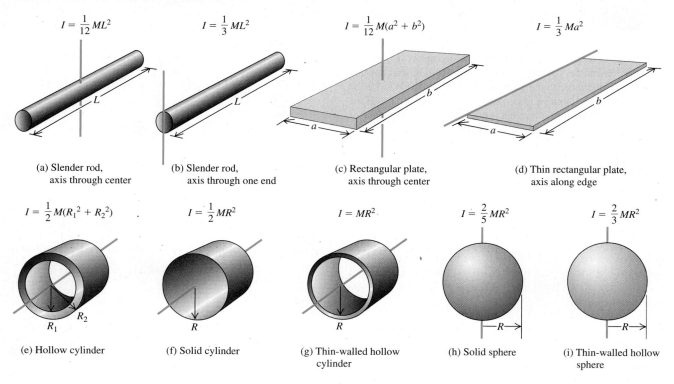

$I = \frac{1}{12} ML^2$

(a) Slender rod,
axis through center

$I = \frac{1}{3} ML^2$

(b) Slender rod,
axis through one end

$I = \frac{1}{12} M(a^2 + b^2)$

(c) Rectangular plate,
axis through center

$I = \frac{1}{3} Ma^2$

(d) Thin rectangular plate,
axis along edge

$I = \frac{1}{2} M(R_1^2 + R_2^2)$

(e) Hollow cylinder

$I = \frac{1}{2} MR^2$

(f) Solid cylinder

$I = MR^2$

(g) Thin-walled hollow
cylinder

$I = \frac{2}{5} MR^2$

(h) Solid sphere

$I = \frac{2}{3} MR^2$

(i) Thin-walled hollow
sphere

uniform thin rod of length L and mass M is pivoted about an axis through one end, perpendicular to the rod, the moment of inertia is $I = ML^2/3$ (case (b) in Table 9.2). If we took the mass as concentrated at the center, a distance $L/2$ from the axis, we would obtain the *incorrect* result $I = M(L/2)^2 = ML^2/4$.

Now that we know how to calculate the kinetic energy of a rotating rigid body, we can apply the energy principles of Chapter 7 to rotational motion. Here are some points of strategy and some examples.

Problem-Solving Strategy   **Rotational Energy**

IDENTIFY *the relevant concepts:* You can use work–energy relations and conservation of energy to find relations involving position and motion of a rigid body rotating around a fixed axis. As we saw in Chapter 7, the energy method is usually not helpful for problems that involve elapsed time. In Chapter 10 we'll see how to approach rotational problems of this kind.

SET UP *the problem* using the same steps as in the Problem-Solving Strategy in Section 7.1, with the following addition:

5. Many problems involve a rope or cable wrapped around a rotating rigid body, which functions as a pulley. In these situations, remember that the point on the pulley that contacts the rope has the same linear speed as the rope, provided the rope doesn't slip on the pulley. You can then take advantage of Eqs. (9.13) and (9.14), which relate the linear speed and tangential acceleration of a point on a rigid body to the angular velocity and angular acceleration of the body. Examples 9.8 and 9.9 illustrate this point.

EXECUTE *the solution:* As in Chapter 7, write expressions for the initial and final kinetic and potential energies $(K_1, K_2, U_1,$ and $U_2)$ and the nonconservative work W_{other} (if any). The new feature is rotational kinetic energy, which is expressed in terms of the body's moment of inertia I for the given axis and its angular speed ω $(K = \frac{1}{2}I\omega^2)$ instead of its mass m and speed v. Substitute these expressions into $K_1 + U_1 + W_{other} = K_2 + U_2$ (if nonconservative work is done) or $K_1 + U_1 = K_2 + U_2$ (if only conservative work is done) and solve for the target variable(s). As in Chapter 7, it's helpful to draw bar graphs showing the initial and final values of K, U, and $E = K + U$.

EVALUATE *your answer:* As always, check whether your answer makes physical sense.

Example 9.8 An unwinding cable I

A light, flexible, nonstretching cable is wrapped several times around a winch drum, a solid cylinder of mass 50 kg and diameter 0.120 m, which rotates about a stationary horizontal axis held by frictionless bearings (Fig. 9.18). The free end of the cable is pulled with a constant force of magnitude 9.0 N for a distance of 2.0 m. It unwinds without slipping, turning the cylinder as it does so. If the cylinder is initially at rest, find its final angular speed and the final speed of the cable.

SOLUTION

IDENTIFY: We will solve this problem using energy methods. Point 1 is when the cylinder first begins to move and point 2 is when the cable has moved 2.0 m. Because the cable is light, we'll assume that it's massless, so that only the cylinder has kinetic energy. The vertical position of the cylinder doesn't change, so there are no changes in gravitational potential energy. There is friction between the cable and the cylinder, which is what makes the cylinder rotate when the cable is pulled. But because the cable doesn't slip, there is no sliding of the cable relative to the cylinder and no mechanical energy is lost in friction. Because the cable is massless, the force that the cable exerts on the cylinder rim is equal to the applied force F.

9.18 A cable unwinds from a cylinder. The force F does work and increases the kinetic energy of the cylinder.

SET UP: The cylinder is initially at rest, so the initial kinetic energy is $K_1 = 0$. Between points 1 and 2, the force F does work on the cylinder over a distance $s = 2.0$ m. As a result, the kinetic energy at point 2 is $K_2 = \frac{1}{2}I\omega^2$. One of our target variables is ω; the other is the speed of the cable at point 2, which is equal to the tangential speed v of the cylinder at that point. We'll find v from ω by using Eq. (9.13).

EXECUTE: The work done on the cylinder is $W_{other} = Fs = (9.0\text{ N})(2.0\text{ m}) = 18$ J. From Table 9.2 the moment of inertia is

$$I = \frac{1}{2}MR^2 = \frac{1}{2}(50\text{ kg})(0.060\text{ m})^2 = 0.090\text{ kg} \cdot \text{m}^2$$

(The radius R is just half the diameter of the cylinder.) The relationship $K_1 + U_1 + W_{other} = K_2 + U_2$ then gives

$$0 + 0 + W_{other} = \frac{1}{2}I\omega^2 + 0$$

$$\omega = \sqrt{\frac{2W_{other}}{I}} = \sqrt{\frac{2(18\text{ J})}{0.090\text{ kg} \cdot \text{m}^2}}$$

$$= 20\text{ rad/s}$$

The final tangential speed of the cylinder, and hence the final speed of the cable, is

$$v = R\omega = (0.060\text{ m})(20\text{ rad/s}) = 1.2\text{ m/s}$$

EVALUATE: We assumed that the cable has negligible mass. If the mass of the cable can't be neglected, then some of the work done would go into the kinetic energy of the cable. Hence the cylinder would end up with less kinetic energy and a smaller angular speed than we calculated here.

Example 9.9 An unwinding cable II

In a lab experiment to test conservation of energy in rotational motion, we wrap a light, flexible cable around a solid cylinder with mass M and radius R. The cylinder rotates with negligible friction about a stationary horizontal axis (Fig. 9.19). We tie the free end of the cable to an object of mass m and release the object with no initial velocity at a distance h above the floor. As the object falls, the

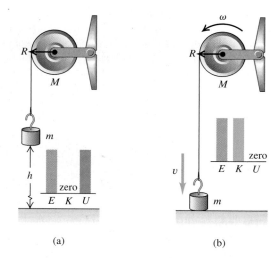

9.19 (a) A cable wrapped around a cylinder is attached to a hanging object. (b) As the cable unwinds, gravitational potential energy is converted into kinetic energy of the falling object and the rotating cylinder.

cable unwinds without stretching or slipping, turning the cylinder. Find the speed of the falling object and the angular speed of the cylinder just as the object strikes the floor.

SOLUTION

IDENTIFY: As in Example 9.8, the cable doesn't slip and friction does no work. (The cable does no net work; at its upper end the force and displacement are in the same direction, and at its lower end they are in opposite directions. Thus the total work done by both ends of the cable is zero.) The only work done is by gravity, so mechanical energy is conserved.

SET UP: Initially the system has no kinetic energy (Fig. 9.19a), so $K_1 = 0$. We take the potential energy to be zero when the object is at floor level; then $U_1 = mgh$ and $U_2 = 0$. (We can ignore the gravitational potential energy for the rotating cylinder, since its height doesn't change.) Just before the falling object hits the floor, both this object and the cylinder have kinetic energy (Fig. 9.19b). The total kinetic energy K_2 at that instant is

$$K_2 = \frac{1}{2}mv^2 + \frac{1}{2}I\omega^2$$

From Table 9.2 the moment of inertia of the cylinder is $I = \frac{1}{2}MR^2$. Also, v and ω are related by $v = R\omega$, since the speed of the falling mass must be equal to the tangential speed at the outer surface of the cylinder. We'll use these relationships to solve for the target variables v and ω at point 2 (Fig. 9.19b).

EXECUTE: Using our expressions for K_1, U_1, K_2, and U_2 and the relation $\omega = v/R$ in the energy-conservation equation $K_1 + U_1 = K_2 + U_2$, we find

$$0 + mgh = \frac{1}{2}mv^2 + \frac{1}{2}\left(\frac{1}{2}MR^2\right)\left(\frac{v}{R}\right)^2 + 0 = \frac{1}{2}\left(m + \frac{1}{2}M\right)v^2$$

$$v = \sqrt{\frac{2gh}{1 + M/2m}}$$

The final angular speed ω is obtained from $\omega = v/R$.

EVALUATE: Let's check some particular cases. When M is much larger than m, v is very small, as we would expect. When M is much smaller than m, v is nearly equal to $\sqrt{2gh}$, which is the speed of a body that falls freely from an initial height h. Does it surprise you that v doesn't depend on the radius of the cylinder?

Act**i**v
Phys**i**cs
ONLINE

7.12 Woman and Flywheel Elevator—
      Energy Approach

7.13 Rotoride—Energy Approach

Gravitational Potential Energy for an Extended Body

In Example 9.9 the cable was of negligible mass, so we could ignore its kinetic energy as well as the gravitational potential energy associated with it. If the mass is *not* negligible, we need to know how to calculate the *gravitational potential energy* associated with such an extended body. If the acceleration of gravity g is the same at all points on the body, the gravitational potential energy is the same as though all the mass were concentrated at the center of mass of the body. Suppose we take the y-axis vertically upward. Then for a body with total mass M, the gravitational potential energy U is simply

$$U = Mgy_{cm} \tag{9.18}$$

(gravitational potential energy for an extended body)

where y_{cm} is the y-coordinate of the center of mass. This expression applies to any extended body, whether it is rigid or not (Fig. 9.20).

To prove Eq. (9.18), we again represent the body as a collection of mass elements m_i. The potential energy for element m_i is $m_i g y_i$, so the total potential energy is

$$U = m_1 g y_1 + m_2 g y_2 + \cdots = (m_1 y_1 + m_2 y_2 + \cdots)g$$

But from Eq. (8.28), which defines the coordinates of the center of mass,

$$m_1 y_1 + m_2 y_2 + \cdots = (m_1 + m_2 + \cdots)y_{cm} = M y_{cm}$$

where $M = m_1 + m_2 + \cdots$ is the total mass. Combining this with the above expression for U, we find $U = Mg y_{cm}$, in agreement with Eq. (9.18).

We leave the application of Eq. (9.18) to the problems. We'll make use of this relationship in Chapter 10 in the analysis of rigid-body problems where the axis of rotation moves.

Test Your Understanding

Suppose the cylinder and hanging object in Example 9.9 have the same mass, so $m = M$. Just before the object strikes the floor, what fraction of the total kinetic energy of the system is in the motion of the object and what fraction is in the rotation of the cylinder?

9.5 | Parallel-Axis Theorem

We pointed out in Section 9.4 that a body doesn't have just one moment of inertia. In fact, it has infinitely many, because there are infinitely many axes about which it might rotate. But there is a simple relationship between the moment of inertia I_{cm} of a body of mass M about an axis through its center of mass and the moment of inertia I_P about any other axis parallel to the original one but displaced from it by a distance d. This relationship, called the **parallel-axis theorem**, states that

$$I_P = I_{cm} + Md^2 \quad \text{(parallel-axis theorem)} \quad (9.19)$$

To prove this theorem, we consider two axes, both parallel to the z-axis, one through the center of mass and the other through a point P (Fig. 9.21). First we take a very thin slice of the body, parallel to the xy-plane and perpendicular to the z-axis. We take the origin of our coordinate system to be at the center of mass of the body; the coordinates of the center of mass are then $x_{cm} = y_{cm} = z_{cm} = 0$. The axis through the center of mass passes through this thin slice at point O, and the parallel axis passes through point P, whose x- and y-coordinates are (a, b). The distance of this axis from the axis through the center of mass is d, where $d^2 = a^2 + b^2$.

We can write an expression for the moment of inertia I_P about the axis through point P. Let m_i be a mass element in our slice, with coordinates (x_i, y_i, z_i). Then the moment of inertia I_{cm} of the slice about the axis through the center of mass (at O) is

$$I_{cm} = \sum_i m_i (x_i^2 + y_i^2)$$

The moment of inertia of the slice about the axis through P is

$$I_P = \sum_i m_i [(x_i - a)^2 + (y_i - b)^2]$$

These expressions don't involve the coordinates z_i measured perpendicular to the slices, so we can extend the sums to include *all* particles in *all* slices. Then I_P

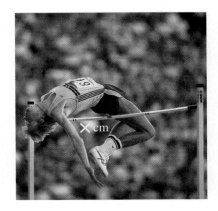

9.20 In a technique called the "Fosbury flop" after its innovator, this athlete arches his body as he passes over the bar in the high jump. As a result, his center of mass actually passes *under* the bar. This technique requires a smaller increase in gravitational potential energy [Eq. (9.18)] than the older method of straddling the bar.

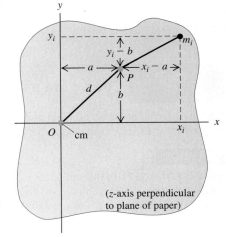

9.21 The mass element m_i has coordinates (x_i, y_i) with respect to an axis of rotation through the center of mass and perpendicular to the plane of the figure. The mass element has coordinates $(x_i - a, y_i - b)$ with respect to the parallel axis through point P.

becomes the moment of inertia of the *entire* body for an axis through P. We then expand the squared terms and regroup, and obtain

$$I_P = \sum_i m_i(x_i^2 + y_i^2) - 2a \sum_i m_i x_i - 2b \sum_i m_i y_i + (a^2 + b^2) \sum_i m_i$$

The first sum is I_{cm}. From Eq. (8.28), the definition of the center of mass, the second and third sums are proportional to x_{cm} and y_{cm}; these are zero because we have taken our origin to be the center of mass. The final term is d^2 multiplied by the total mass, or Md^2. This completes our proof that $I_P = I_{cm} + Md^2$.

As Eq. (9.19) shows, a rigid body has a lower moment of inertia about an axis through its center of mass than about any other parallel axis. Thus it's easier to start a body rotating if the rotation axis passes through the center of mass. This suggests that it's somehow most natural for a rotating body to rotate about an axis through its center of mass; we'll make this idea more quantitative in Chapter 10.

Example 9.10   Using the parallel-axis theorem I

A part of a mechanical linkage (Fig. 9.22) has a mass of 3.6 kg. We measure its moment of inertia about an axis 0.15 m from its center of mass to be $I_P = 0.132$ kg·m². What is the moment of inertia I_{cm} about a parallel axis through the center of mass?

SOLUTION

IDENTIFY AND SET UP: We'll use Eq. (9.19) to determine the target variable I_{cm}.

EXECUTE: Rearranging the equation and substituting the values,

$$I_{cm} = I_P - Md^2 = 0.132 \text{ kg·m}^2 - (3.6 \text{ kg})(0.15 \text{ m})^2$$
$$= 0.051 \text{ kg·m}^2$$

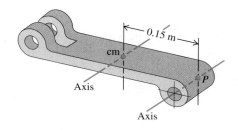

9.22 Calculating I_{cm} from a measurement of I_P.

EVALUATE: Our result shows that I_{cm} is less than I_P. This is as it should be: as we saw earlier, the moment of inertia for an axis through the center of mass is lower than for any other parallel axis.

Example 9.11   Using the parallel-axis theorem II

Find the moment of inertia of a thin uniform disk of mass M and radius R about an axis perpendicular to its plane at the edge.

SOLUTION

IDENTIFY AND SET UP: Here the target variable in Eq. (9.19) is I_P. The distance from P (at the rim of the disk) to the disk's center of mass is the radius R, and the moment of inertia through the center of mass is given in Table 9.2: $I_{cm} = MR^2/2$.

EXECUTE: The parallel-axis theorem gives

$$I_P = I_{cm} + Md^2 = \frac{MR^2}{2} + MR^2 = \frac{3MR^2}{2}$$

EVALUATE: As in the previous example, $I_{cm} < I_P$. This is a very simple solution of a problem that would be fairly complicated if we tried to solve it using the integration techniques to be discussed in the next section.

Test Your Understanding

The discus in Example 9.4 (Section 9.3) is a uniform disk of mass 2.00 kg and radius 11.0 cm. Find its moment of inertia about the thrower's shoulder, 80.0 cm from the center of the disk (Fig. 9.13).

*9.6 | Moment of Inertia Calculations

When a rigid body cannot be represented by a few point masses but is a continuous distribution of mass, the sum of masses and distances that defines the moment of inertia [Eq. (9.16)] becomes an integral. Imagine dividing the body into small mass elements dm so that all points in a particular element are at essentially the same perpendicular distance from the axis of rotation. We call this distance r, as before. Then the moment of inertia is

$$I = \int r^2 \, dm \tag{9.20}$$

To evaluate the integral, we have to represent r and dm in terms of the same integration variable. When we have an effectively one-dimensional object, such as the slender rods (a) and (b) in Table 9.2, we can use a coordinate x along the length and relate dm to an increment dx. For a three-dimensional object it is usually easiest to express dm in terms of an element of volume dV and the *density* ρ of the body. Density is mass per unit volume, $\rho = dm/dV$, so we may also write Eq. (9.20) as

$$I = \int r^2 \rho \, dV$$

This expression tells us that a body's moment of inertia depends on how its density varies within its volume (Fig. 9.23). If the body is uniform in density, then we may take ρ outside the integral:

$$I = \rho \int r^2 \, dV \tag{9.21}$$

To use this equation, we have to express the volume element dV in terms of the differentials of the integration variables, such as $dV = dx \, dy \, dz$. The element dV must always be chosen so that all points within it are at very nearly the same distance from the axis of rotation. The limits on the integral are determined by the shape and dimensions of the body. For regularly shaped bodies, this integration can often be carried out quite easily.

9.23 By measuring small variations in the orbits of satellites, geophysicists can measure the earth's moment of inertia. This tells us how our planet's mass is distributed within its interior. The data show that the earth is far denser at the core than in its outer layers.

Example 9.12	Uniform thin rod, axis perpendicular to length

Figure 9.24 shows a slender uniform rod with mass M and length L. It might be a baton held by a twirler in a marching band (less the rubber end caps). Compute its moment of inertia about an axis through O, at an arbitrary distance h from one end.

We'll determine I from Eq. (9.20) with r replaced by x (see Fig. 9.24).

SOLUTION

IDENTIFY AND SET UP: We choose as an element of mass a short section of rod with length dx at a distance x from point O. The ratio of the mass dm of this element to the total mass M is equal to the ratio of its length dx to the total length L:

$$\frac{dm}{M} = \frac{dx}{L} \quad \text{so} \quad dm = \frac{M}{L} dx$$

9.24 Finding the moment of inertia of a thin rod about an axis through O. The mass element is a segment with length dx.

EXECUTE: Figure 9.24 shows that the integration limits on x are from $-h$ to $(L - h)$. Hence we obtain

$$I = \int x^2 \, dm = \frac{m}{L} \int_{-h}^{L-h} x^2 \, dx$$

$$= \left[\frac{M}{L} \left(\frac{x^3}{3} \right) \right]_{-h}^{L-h} = \frac{1}{3} M (L^2 - 3Lh + 3h^2)$$

EVALUATE: From this general expression we can find the moment of inertia about an axis through any point on the rod. For example, if the axis is at the left end, $h = 0$ and

$$I = \frac{1}{3} ML^2$$

If the axis is at the right end, we should get the same result. Putting $h = L$, we again get

$$I = \frac{1}{3} ML^2$$

If the axis passes through the center, the usual place for a twirled baton, then $h = L/2$ and

$$I = \frac{1}{12} ML^2$$

These results agree with the expressions in Table 9.2.

<table>
<tr><td>Example
9.13</td><td>**Hollow or solid cylinder, rotating about axis of symmetry**</td></tr>
</table>

Figure 9.25 shows a hollow, uniform cylinder with length L, inner radius R_1, and outer radius R_2. It might be a steel cylinder in a printing press or a sheet-steel rolling mill. Find the moment of inertia about the axis of symmetry of the cylinder.

SOLUTION

IDENTIFY AND SET UP: We choose as a volume element a thin cylindrical shell of radius r, thickness dr, and length L. All parts of this element are at very nearly the same distance from the axis. The volume of the element is very nearly equal to that of a flat sheet with thickness dr, length L, and width $2\pi r$ (the circumference of the shell). Then

$$dm = \rho \, dV = \rho (2\pi r L \, dr)$$

Axis
r
dr
L
R_1
R_2

9.25 Finding the moment of inertia of a hollow cylinder about its symmetry axis. The mass element is a cylindrical shell with radius r and thickness dr.

We will use this expression in Eq. (9.20) and integrate from $r = R_1$ to $r = R_2$.

EXECUTE: The moment of inertia is given by

$$I = \int r^2 \, dm = \int_{R_1}^{R_2} r^2 \rho (2\pi r L \, dr)$$

$$= 2\pi \rho L \int_{R_1}^{R_2} r^3 \, dr$$

$$= \frac{2\pi \rho L}{4} (R_2^4 - R_1^4)$$

$$= \frac{\pi \rho L}{2} (R_2^2 - R_1^2)(R_2^2 + R_1^2)$$

It is usually more convenient to express the moment of inertia in terms of the total mass M of the body, which is its density ρ multiplied by the total volume V. The volume is

$$V = \pi L (R_2^2 - R_1^2)$$

so the total mass M is

$$M = \rho V = \pi L \rho (R_2^2 - R_1^2)$$

and the moment of inertia is

$$I = \frac{1}{2} M (R_1^2 + R_2^2)$$

EVALUATE: This agrees with Table 9.2, case (e). If the cylinder is solid, such as a lawn roller, $R_1 = 0$. Calling the outer radius R_2 simply R, we find that the moment of inertia of a solid cylinder of radius R is

$$I = \frac{1}{2} MR^2$$

If the cylinder has a very thin wall (like a stovepipe), R_1 and R_2 are very nearly equal; if R represents this common radius,

$$I = MR^2$$

We could have predicted this last result; in a thin-walled cylinder, all the mass is the same distance $r = R$ from the axis, so $I = \int r^2 \, dm = R^2 \int dm = MR^2$.

Note that the moment of inertia of a cylinder about an axis coinciding with its axis of symmetry depends on its mass and radii but not on its length L. Two hollow cylinders with the same inner and outer radii, one of wood and one of brass, but having the same mass M, have equal moments of inertia even though the length of the wood cylinder is much greater. The moment of inertia depends only on the *radial* distribution of mass, not on its distribution along the axis. Thus the above results hold also for a thin disk, such as a washer or a compact disc (CD).

| Example 9.14 | Uniform sphere with radius R, axis through center |

The object in Fig. 9.26 might be a billiard ball, a steel ball bearing, or the world's largest ball of string. Find the moment of inertia about an axis through the center of the sphere.

SOLUTION

IDENTIFY AND SET UP: We divide the sphere into thin disks of thickness dx, whose moment of inertia we know from Example 9.13. The radius r of the disk shown is

$$r = \sqrt{R^2 - x^2}$$

Its volume is

$$dV = \pi r^2 \, dx = \pi (R^2 - x^2) \, dx$$

and its mass is

$$dm = \rho \, dV = \pi \rho (R^2 - x^2) \, dx$$

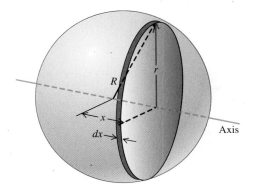

9.26 Finding the moment of inertia of a sphere about an axis through its center. The mass element is a disk with thickness dx.

EXECUTE: From Example 9.13, the moment of inertia of a disk of radius r and mass dm is

$$dI = \frac{1}{2} r^2 \, dm = \frac{1}{2} (\sqrt{R^2 - x^2})^2 [\pi \rho (R^2 - x^2) \, dx]$$
$$= \frac{\pi \rho}{2} (R^2 - x^2)^2 \, dx$$

Integrating this expression from $x = 0$ to $x = R$ gives the moment of inertia of the right hemisphere. From symmetry, the total I for the entire sphere is just twice this:

$$I = (2) \frac{\pi \rho}{2} \int_0^R (R^2 - x^2)^2 \, dx$$

Carrying out the integration, we obtain

$$I = \frac{8\pi \rho}{15} R^5$$

The mass M of the sphere of volume $V = 4\pi R^3 / 3$ is

$$M = \rho V = \frac{4 \pi \rho R^3}{3}$$

By comparing the expressions for I and M, we find

$$I = \frac{2}{5} MR^2$$

EVALUATE: This result agrees with the expression in Table 9.2, case (h). Note that the moment of inertia of a solid sphere of mass M and R is less than the moment of inertia of a solid *cylinder* of the same mass and radius, $I = \frac{1}{2} MR^2$. The reason is that more of the sphere's mass is located close to the axis.

Test Your Understanding

Consider the rod in Example 9.12. Use calculus to show that the value of h that gives the smallest moment of inertia is $h = L/2$.

SUMMARY

When a rigid body rotates about a stationary axis (usually called the z-axis), its position is described by an angular coordinate θ. The angular velocity ω_z is the time derivative of the angular coordinate θ: The angular acceleration α_z is the time derivative of the angular velocity ω_z or the second derivative of the angular coordinate θ. (See Examples 9.1 and 9.2)

$$\omega_z = \lim_{\Delta t \to 0} \frac{\Delta \theta}{\Delta t} = \frac{d\theta}{dt} \qquad (9.3)$$

$$\alpha_z = \lim_{\Delta t \to 0} \frac{\Delta \omega_z}{\Delta t} = \frac{d\omega_z}{dt} = \frac{d^2\theta}{dt^2} \qquad (9.5, 9.6)$$

When a rigid body rotates about a fixed axis with constant angular acceleration, the angular coordinate, velocity, and acceleration are related by simple kinematic equations. These involve θ_0 and ω_{0z}, the initial values of the angular coordinate and angular velocity, respectively. (See Example 9.3)

$$\theta = \theta_0 + \omega_{0z}t + \frac{1}{2}\alpha_z t^2 \qquad (9.11)$$

$$\theta - \theta_0 = \frac{1}{2}(\omega_z + \omega_{0z})t \qquad (9.10)$$

$$\omega_z = \omega_{0z} + \alpha_z t \qquad (9.7)$$

$$\omega_z^2 = \omega_{0z}^2 + 2\alpha_z(\theta - \theta_0) \qquad (9.12)$$

The angular speed ω of a rigid body is the magnitude of its angular velocity. The rate of change of ω is $\alpha = d\omega/dt$. For a particle in the body a distance r from the rotation axis, the speed v and the components of the acceleration $\vec{a}$ are related to ω and α. (See Examples 9.4 through 9.6)

$$v = r\omega \qquad (9.13)$$

$$a_{\text{tan}} = \frac{dv}{dt} = r\frac{d\omega}{dt} = r\alpha \qquad (9.14)$$

$$a_{\text{rad}} = \frac{v^2}{r} = \omega^2 r \qquad (9.15)$$

The moment of inertia I of a body about a given axis is a measure of its rotational inertia: the greater the value of I, the more difficult it is to change the state of the body's rotation. The moment of inertia can be expressed as a sum over the particles m_i that make up the body, each of which is at its own perpendicular distance r_i from the axis. (See Examples 9.7 and 9.12 through 9.14)

$$I = m_1 r_1^2 + m_2 r_2^2 + \cdots = \sum_i m_i r_i^2 \qquad (9.16)$$

- Mass close to axis
- Small moment of inertia
- Easy to start apparatus rotating

- Mass farther from axis
- Greater moment of inertia
- Harder to start apparatus rotating

The rotational kinetic energy of a rigid body rotating about a fixed axis depends on the angular speed ω and the moment of inertia I for that rotation axis. (See Examples 9.8 and 9.9)

$$K = \frac{1}{2}I\omega^2 \qquad (9.17)$$

The moment of inertia I_{cm} of a body of mass M about an axis through the center of mass is related to the moment of inertia I_P about a parallel axis at a distance d from the first axis. (See Examples 9.10 and 9.11)

$$I_P = I_{cm} + Md^2 \qquad (9.19)$$

Key Terms

rigid body, 327
radian, 328
average angular velocity, 329
angular displacement, 329
instantaneous angular velocity, 329

average angular acceleration, 331
instantaneous angular acceleration, 331
angular speed, 336
tangential component of acceleration, 336

centripetal component of acceleration, 336
moment of inertia, 340
rotational kinetic energy, 340
parallel-axis theorem, 345

Your Notes

Answer to Chapter Opening Question

From Eqs. (9.13) and (9.15), doubling the distance r for the same angular speed ω doubles the linear speed $v = r\omega$ and doubles the radial acceleration $a_{\text{rad}} = \omega^2 r$.

Answers to Test Your Understanding Questions

Section 9.1 The rotation is speeding up between $t = 0$ and $t = 2$ s (ω_z and α_z are both positive) and between $t = 4$ s and $t = 6$ s (ω_z and α_z are both negative). The rotation is slowing down between $t = 2$ s and $t = 4$ s (ω_z is positive and α_z is negative).

Section 9.2 When the DVD comes to rest, $\omega_z = 0$. From Eq. (9.7), the time when this occurs is

$$t = \frac{\omega_z - \omega_{0z}}{\alpha_z} = \frac{0 - 27.5 \text{ rad/s}}{-10.0 \text{ rad/s}^2} = 2.75 \text{ s}$$

Either Eq. (9.10), (9.11), or (9.12) can be used to find the angle θ. Using Eq. (9.10),

$$\theta = \theta_0 + \frac{1}{2}(\omega_z + \omega_{0z})t = 0 + \frac{1}{2}(0 + 27.5 \text{ rad/s})(2.75 \text{ s})$$

$$= 37.8 \text{ rad} = 37.8 \text{ rad}\left(\frac{1 \text{ rev}}{2\pi \text{ rad}}\right) = 6.02 \text{ rev}$$

Section 9.3 From Eq. (9.13), $v = r\omega$. To maintain a constant linear speed v, the angular speed ω must decrease as the scanning head moves outward (greater r).

Section 9.4 The kinetic energy in the falling object is $\frac{1}{2}mv^2$ and the kinetic energy in the rotating cylinder is $\frac{1}{2}I\omega^2 = \frac{1}{2}(\frac{1}{2}mR^2)(\frac{v}{R})^2 = \frac{1}{4}mv^2$. Hence the total kinetic energy of the system is $\frac{3}{4}mv^2$, of which two-thirds is in the object and one-third in the cylinder.

Section 9.5 From Table 9.2, the moment of inertia of the discus around an axis through its center of mass and perpendicular to the plane of the discus is $I_{\text{cm}} = MR^2/2 = (2.00 \text{ kg})(0.110 \text{ m})^2/2 = 0.0121 \text{ kg} \cdot \text{m}^2$. The discus is rotating about an axis parallel to this one, so we use the parallel-axis theorem: the moment of inertia around the thrower's shoulder is $I_P = I_{\text{cm}} + Md^2 = 0.0121 \text{ kg} \cdot \text{m}^2 + (2.00 \text{ kg})(0.800 \text{ m})^2 = 1.29 \text{ kg} \cdot \text{m}^2$.

Section 9.6 To find the minimum of I, take its derivative with respect to h and set it equal to zero:

$$\frac{dI}{dh} = \frac{d}{dh}\frac{1}{3}M(L^2 - 3Lh + 3h^2) = M(-L + 2h) = 0$$

This has solution $h = L/2$. The second derivative $d^2I/dh^2 = 2M$ is positive, as it must be at a minimum.

Discussion Questions

Q9.1 When rewinding an audio or video tape, why does the tape wind up fastest at the end?

Q9.2 Does a body rotating about a fixed axis have to be *perfectly* rigid for all points on the body to have the same angular velocity and the same angular acceleration? Explain.

Q9.3 What is the difference between tangential and radial acceleration for a point on a rotating body?

Q9.4 In Fig. 9.15, all points on the chain have the same linear speed v. Is the magnitude a of the linear acceleration also the same for all points on the chain? How are the angular accelerations of the two sprockets related? Explain.

Q9.5 In Fig. 9.15, how are the radial accelerations of points at the teeth of the two sprockets related? Explain the reasoning behind your answer.

Q9.6 A flywheel rotates with constant angular velocity. Does a point on its rim have a tangential acceleration? A radial acceleration? Are these accelerations constant in magnitude? In direction? In each case give the reasoning behind your answer.

Q9.7 What is the purpose of the spin cycle of a washing machine? Explain in terms of acceleration components.

Q9.8 Although angular velocity and angular acceleration can be treated as vectors, the angular displacement θ, despite having a magnitude and a direction, cannot. This is because θ does not follow the commutative law of vector addition (Eq. 1.4). Prove this to yourself in the following way. Lay your physics textbook flat on the desk in front of you with the cover side up so you can read the writing on it. Rotate the farthest edge toward you by 90° about a horizontal axis. Call this angular displacement θ_1. Then rotate the left edge toward you by 90° about a vertical axis. Call this angular displacement θ_2. The spine of the book should now face you, with the writing on it oriented so that you can read. Now start over again but carry out the two rotations in the reverse order. Do you get a different result? That is, is $\theta_1 + \theta_2$ equal to $\theta_2 + \theta_1$? Now repeat this experiment but this time with an angle of 1° rather than 90°. Do you think that the infinitesimal displacement $d\vec{\theta}$ obeys the commutative law of addition and hence qualifies as a vector? If so, how is the direction of $d\vec{\theta}$ related to the direction of $\vec{\omega}$?

Q9.9 Can you think of a body that has the same moment of inertia for all possible axes? If so, give an example, and if not, explain why this is not possible. Can you think of a body that has the same moment of inertia for all axes passing through a certain point? If so, give an example and indicate where the point is located.

Q9.10 To maximize the moment of inertia of a flywheel while minimizing its weight, what shape and distribution of mass should it have? Explain.

Q9.11 How might you determine experimentally the moment of inertia of an irregularly shaped body about a given axis?

Q9.12 A cylindrical body has mass M and radius R. Can the mass be distributed within the body in such a way that its moment of inertia about its axis of symmetry is greater than MR^2? Explain.

Q9.13 Describe how you could use part (b) of Table 9.2 to derive the result in part (d).

Q9.14 The moment of inertia I of a rigid body for an axis through its center of mass is I_{cm}. Is there any other axis parallel to this axis for which I is less than I_{cm}? Explain.

Q9.15 For the equations for I given in parts (a) and (b) of Table 9.2 to be valid, must the rod have a circular cross section? Is there any restriction on the size of the cross section in order for these equations to apply? Explain.

Q9.16 In part (d) of Table 9.2, the thickness of the plate must be much less than a for the expression given for I to apply. But in part (c), the expression given for I applies no matter how thick the plate is. Explain.

Q9.17 In Fig. 5.32a, use the expressions $K = \frac{1}{2}I\omega^2$ and $K = \frac{1}{2}mv^2$ to calculate the kinetic energy of the sled (treated as a single particle). How do these two results compare? Explain.

Q9.18 Equation (9.18) says to use y_{cm} to calculate U for an extended rigid body. But in Example 9.9 (Section 9.4), y is measured to the bottom of the hanging mass rather than to the center of mass. Is this a mistake? Explain.

Q9.19 You can use any angular measure—radians, degrees, or revolutions—in some of the equations in Chapter 9, but you can use only radian measure in others. Identify those for which using radians is necessary and those for which it is not, and in each case give the reasoning behind your answer.

Q9.20 When calculating the moment of inertia of an object, can we treat all its mass as if it were concentrated at the center of mass of the object? Justify your answer.

Q9.21 A wheel is rotating about an axis perpendicular to the plane of the wheel and passing through the center of the wheel. The angular speed of the wheel is increasing at a constant rate. Point A is on the rim of the wheel and point B is midway between the rim and center of the wheel. For each of the following quantities, is its magnitude larger at point A, at point B, or is it the same at both points: angular speed, tangential speed, angular acceleration, tangential acceleration, and radial acceleration. Justify each of your answers.

Exercises

Section 9.1 Angular Velocity and Acceleration

9.1 a) What angle in radians is subtended by an arc 1.50 m in length on the circumference of a circle of radius 2.50 m? What is this angle in degrees? b) An arc 14.0 cm in length on the circumference of a circle subtends an angle of 128°. What is the radius of the circle? c) The angle between two radii of a circle with radius 1.50 m is 0.700 rad. What length of arc is intercepted on the circumference of the circle by the two radii?

9.2 An airplane propeller is rotating at 1900 rev/min. a) Compute the propeller's angular velocity in rad/s. b) How long in seconds does it take for the propeller to turn through 35°?

9.3 Consider the flywheel in Examples 9.1 and 9.2 (Section 9.1). a) Calculate the instantaneous angular acceleration at $t = 3.5$ s. Explain why your result is equal to the average angular acceleration for the time interval from 2.0 s to 5.0 s. b) Calculate the instantaneous angular velocity at $t = 3.5$ s. Explain why your result is *not* equal to the average angular velocity for the 2.0 s to 5.0 s time interval, even though 3.5 s is at the middle of this time interval.

9.4 A fan blade rotates with angular velocity given by $\omega_z(t) = \gamma - \beta t^2$, where $\gamma = 5.00$ rad/s and $\beta = 0.800$ rad/s^3. a) Calculate the angular acceleration as a function of time. b) Calculate the

instantaneous angular acceleration α_z at $t = 3.00$ s and the average angular acceleration α_{av-z} for the time interval $t = 0$ to $t = 3.00$ s. How do these two quantities compare? If they are different, why are they different?

9.5 A child is pushing a merry-go-round. The angle through which the merry-go-round has turned varies with time according to $\theta(t) = \gamma t + \beta t^3$, where $\gamma = 0.400$ rad/s and $\beta = 0.0120$ rad/s^3. a) Calculate the angular velocity of the merry-go-round as a function of time. b) What is the initial value of the angular velocity? c) Calculate the instantaneous value of the angular velocity ω_z at $t = 5.00$ s and the average angular velocity ω_{av-z} for the time interval $t = 0$ to $t = 5.00$ s. Show that ω_{av-z} is *not* equal to the average of the instantaneous angular velocities at $t = 0$ and $t = 5.00$ s, and explain why it is not.

9.6 At $t = 0$ the current to a dc electric motor is reversed, resulting in an angular displacement of the motor shaft given by $\theta(t) = (250 \text{ rad/s})t - (20.0 \text{ rad/s}^2)t^2 - (1.50 \text{ rad/s}^3)t^3$. a) At what time is the angular velocity of the motor shaft zero? b) Calculate the angular acceleration at the instant that the motor shaft has zero angular velocity. c) How many revolutions does the motor shaft turn through between the time when the current is reversed and the instant when the angular velocity is zero? d) How fast was the motor shaft rotating at $t = 0$, when the current was reversed? e) Calculate the average angular velocity for the time period from $t = 0$ to the time calculated in part (a).

9.7 The angle θ through which a bicycle wheel turns is given by $\theta(t) = a + bt^2 - ct^3$, where a, b, and c are positive constants such that for t in seconds, θ will be in radians. a) Calculate the angular acceleration of the wheel as a function of time. b) At what time is the angular velocity of the wheel instantaneously not changing?

9.8 A wheel is rotating about an axis that is in the z-direction. The angular velocity ω_z is -6.00 rad/s at $t = 0$, increases linearly with time and is $+8.00$ m/s at $t = 7.00$ s. We have taken counterclockwise rotation to be positive. a) Is the angular acceleration during this time interval positive or negative? b) During what time interval is the speed of the wheel increasing? Decreasing? c) What is the angular displacement of the wheel at $t = 7.00$ s?

Section 9.2 Rotation with Constant Angular Acceleration

9.9 A high-speed flywheel in a motor is spinning at 500 rpm when a power failure suddenly occurs. The flywheel has mass 40.0 kg and diameter 75.0 cm. The power is off for 30.0 s and during this time the flywheel slows due to friction in its axle bearings. During the time the power is off, the flywheel makes 200 complete revolutions. a) At what rate is the flywheel spinning when the power comes back on? b) How long after the beginning of the power failure would it have taken the flywheel to stop if the power had not come back on, and how many revolutions would the wheel have made during this time?

9.10 A bicycle wheel has an initial angular velocity of 1.50 rad/s. a) If its angular acceleration is constant and equal to 0.300 rad/s^2, what is its angular velocity at $t = 2.50$ s? b) Through what angle has the wheel turned between $t = 0$ and $t = 2.50$ s?

9.11 An electric fan is turned off, and its angular velocity decreases uniformly from 500 rev/min to 200 rev/min in 4.00 s.

a) Find the angular acceleration in rev/s² and the number of revolutions made by the motor in the 4.00-s interval. b) How many more seconds are required for the fan to come to rest if the angular acceleration remains constant at the value calculated in part (a)?

9.12 a) Derive Eq. (9.12) by combining Eqs. (9.7) and (9.11) to eliminate t. b) The angular velocity of an airplane propeller increases from 12.0 rad/s to 16.0 rad/s while turning through 7.00 rad. What is the angular acceleration in rad/s²?

9.13 The rotating blade of a blender turns with constant angular acceleration 1.50 rad/s². a) How much time does it take to reach an angular velocity of 36.0 rad/s, starting from rest? b) Through how many revolutions does the blade turn in this time interval?

9.14 A flywheel requires 4.00 s to rotate through 162 rad. Its angular velocity at the end of this time is 108 rad/s. Find a) the angular velocity at the beginning of the 4.00-s interval; b) the constant angular acceleration.

9.15 Emilie's potter's wheel rotates with a constant 2.25 rad/s² angular acceleration. After 4.00 s the wheel has rotated through an angle of 60.0 rad. What was the angular velocity of the wheel at the beginning of the 4.00 s interval?

9.16 A circular saw blade 0.200 m in diameter starts from rest. In 6.00 s it accelerates with constant angular acceleration to an angular velocity of 140 rad/s. Find the angular acceleration and the angle through which the blade has turned.

9.17 A safety device brings the blade of a power mower from an initial angular speed of ω_1 to rest in 1.00 revolution. At the same constant acceleration, how many revolutions would it take the blade to come to rest from an initial angular speed ω_3 that was three times as great, $\omega_3 = 3\omega_1$?

9.18 A straight piece of reflecting tape extends from the center of a wheel to its rim. You darken the room and use a camera and strobe unit that flashes once every 0.050 s to take pictures of the wheel as it rotates counterclockwise. You trigger the strobe so that the first flash ($t = 0$) occurs when the tape is horizontal to the right at an angular displacement of zero. For the following situations draw a sketch of the photo you will get for the time exposure over five flashes (at $t = 0, 0.050$ s, 0.100 s, 0.150 s, and 0.200 s) and graph θ versus t and ω versus t for $t = 0$ to $t = 0.200$ s. a) The angular velocity is constant at 10.0 rev/s. b) The wheel starts from rest with a constant angular acceleration of 25.0 rev/s². c) The wheel is rotating at 10.0 rev/s at $t = 0$ and changes angular velocity at a constant rate of -50.0 rev/s².

9.19 At $t = 0$ a grinding wheel has an angular velocity of 24.0 rad/s. It has a constant angular acceleration of 30.0 rad/s² until a circuit breaker trips at $t = 2.00$ s. From then on, it turns through 432 rad as it coasts to a stop at constant angular acceleration. a) Through what total angle did the wheel turn between $t = 0$ and the time it stopped? b) At what time did it stop? c) What was its acceleration as it slowed down?

9.20 a) Derive a constant-angular-acceleration equation that gives $\theta - \theta_0$ in terms of ω_z, α_z, and t (no ω_{0z} in the equation). b) At $t = 8.0$ s a gear is rotating about a fixed axis at 4.50 rad/s. During the preceding 8.0 s, it has turned through 40.0 rad. Use the result of part (a) to calculate the gear's constant angular acceleration. c) What was the angular velocity of the gear at $t = 0$?

Section 9.3 Relating Linear and Angular Kinematics

9.21 The main rotor of a helicopter is turning in a horizontal plane at 90.0 rev/min. The distance from the center of the rotor shaft to each blade tip is 5.00 m. Calculate the speed through the air of the blade tip if a) the helicopter is sitting on the ground; b) the helicopter is rising vertically at 4.00 m/s.

9.22 Compact Disc. A compact disc (CD) stores music in a coded pattern of tiny pits 10^{-7} m deep. The pits are arranged in a track that spirals outward toward the rim of the disc; the inner and outer radii of this spiral are 25.0 mm and 58.0 mm, respectively. As the disc spins inside a CD player, the track is scanned at a constant *linear* speed of 1.25 m/s. a) What is the angular speed of the CD when scanning the innermost part of the track? The outermost part of the track? b) The maximum playing time of a CD is 74.0 min. What would be the length of the track on such a maximum-duration CD if it were stretched out in a straight line? c) What is the average angular acceleration of a maximum-duration CD during its 74.0-min playing time? Take the direction of rotation of the disc to be positive.

9.23 A wheel rotates with a constant angular velocity of 6.00 rad/s. a) Compute the radial acceleration of a point 0.500 m from the axis, using the relation $a_{\text{rad}} = \omega^2 r$. b) Find the tangential speed of the point and compute its radial acceleration from the relation $a_{\text{rad}} = v^2/r$.

9.24 Find the required angular speed (in rev/min) of an ultracentrifuge for the radial acceleration of a point 2.50 cm from the axis to equal 400,000g (that is, 400,000 times the acceleration due to gravity).

9.25 A flywheel with a radius of 0.300 m starts from rest and accelerates with a constant angular acceleration of 0.600 rad/s². Compute the magnitude of the tangential acceleration, the radial acceleration and the resultant acceleration of a point on its rim a) at the start; b) after it has turned through 60.0°; c) after it has turned through 120.0°.

9.26 An electric ceiling fan with blades 0.750 m in diameter is rotating about a fixed axis with an initial angular velocity of 0.250 rev/s. The angular acceleration is 0.900 rev/s². a) Compute the angular velocity after 0.200 s. b) Through how many revolutions has the blade turned in this time interval? c) What is the tangential speed of a point on the tip of the blade at $t = 0.200$ s? d) What is the magnitude of the resultant acceleration of a point on the tip of the blade at $t = 0.200$ s?

9.27 Centrifuge. An advertisement claims that a centrifuge takes up only 0.127 m of bench space but can produce a radial acceleration of 3000g at 5000 rev/min. Calculate the required radius of the centrifuge. Is the claim a realistic one?

9.28 a) Derive an equation for the radial acceleration that includes v and ω, but not r. b) You are designing a merry-go-round for which a point on the rim will have a radial acceleration of 0.500 m/s² when the tangential velocity of that point has magnitude 2.00 m/s. What angular velocity is required to achieve these values?

9.29 A Drill Problem. When drilling a 12.7-mm-diameter hole in wood, plastic, or aluminum, a shop manual recommends a drill speed of 1250 rev/min. For a 12.7-mm-diameter drill bit turning at a constant 1250 rev/min, find a) the maximum linear speed of any part of the bit; b) the maximum radial acceleration of any part of the bit.

9.30 At $t = 3.00$ s a point on the rim of a 0.200-m-radius wheel has a tangential speed of 50.0 m/s as the wheel slows down with a tangential acceleration of constant magnitude 10.0 m/s². a) Calculate the wheel's constant angular acceleration. b) Calculate the angular velocities at $t = 3.00$ s and $t = 0$. c) Through what angle did the wheel turn between $t = 0$ and $t = 3.00$ s? d) At what time will the radial acceleration equal g?

9.31 The spin cycles of a washing machine have two angular speeds, 423 rev/min and 640 rev/min. The internal diameter of the drum is 0.470 m. a) What is the ratio of the maximum radial force on the laundry for the higher angular speed to that for the lower speed? b) What is the ratio of the maximum tangential speed of the laundry for the higher angular speed to that for the lower speed? c) Find the laundry's maximum tangential speed and the maximum radial acceleration, in terms of g.

9.32 You are to design a rotating cylindrical axle to lift 800-N buckets of cement from the ground to a rooftop 78.0 m above the ground. The buckets will be attached to a hook on the free end of a cable that wraps around the rim of the axle; as the axle turns, the buckets will rise. a) What should the diameter of the axle be in order to raise the buckets at a steady 2.00 cm/s when it is turning at 7.5 rpm? b) If instead the axle must give the buckets an upward acceleration of 0.400 m/s², what should the angular acceleration of the axle be?

9.33 While riding a multi-speed bicycle, the rider can select the radius of the rear sprocket, that is fixed to the rear axle. The front sprocket of a bicycle has radius 12.0 cm. If the angular speed of the front sprocket is 0.600 rev/s, what is the radius of the rear sprocket for which the tangential speed of a point on the rim of the rear wheel will be 5.00 m/s? The rear wheel has radius 0.330 m.

Section 9.4 Energy in Rotational Motion

9.34 Small blocks, each with mass m, are clamped at the ends and at the center of a light rod of length L. Compute the moment of inertia of the system about an axis perpendicular to the rod and passing through a point $\frac{1}{4}$ of the length from one end. You can ignore the moment of inertia of the light rod.

9.35 A twirler's baton is made of a slender metal cylinder of mass M and length L. Each end has a rubber cap of mass m and you can accurately treat each cap as a particle in this problem. Find the total moment of inertia of the baton about the usual twirling axis (perpendicular to the baton through its center).

9.36 Find the moment of inertia about each of the following axes for a rod that is 0.300 cm in diameter and 1.50 m long, and with mass 0.0420 kg. Use the formulas of Table 9.2. a) About an axis perpendicular to the rod and passing through its center. b) About an axis perpendicular to the rod and passing through one end. c) About a longitudinal axis passing through the center of the rod.

9.37 Four small spheres, each of which you can regard as a point of mass 0.200 kg, are arranged in a square 0.400 m on a side and connected by light rods (Fig. 9.27). Find the moment of inertia of the system about an

Figure 9.27 Exercise 9.37.

axis a) through the center of the square, perpendicular to its plane (an axis through point O in the figure); b) bisecting two opposite sides of the square (an axis along the line AB in the figure); c) that passes through the centers of the upper left and lower right spheres and through point O.

9.38 A uniform bar has two small balls glued to its ends. The bar is 2.00 m long and with mass 4.00 kg, while the balls each have mass 0.500 kg and can be treated as point masses. Find the moment of inertia of this combination about each of the following axes: a) an axis perpendicular to the bar through its center; b) an axis perpendicular to the bar through one of the balls; c) an axis parallel to the bar through both balls; d) an axis parallel to the bar and 0.500 m from it.

9.39 A compound disk of outside diameter 140.0 cm is made up of a uniform solid disk of radius 50.0 cm and area density 3.00 g/cm² surrounded by a concentric ring of inner radius 50.0 cm, outer radius 70.0 cm and area density 2.00 g/cm². Find the moment of inertia of this object about an axis perpendicular to the plane of the object and passing through its center.

9.40 How I Scales. If we multiply all design dimensions of an object by a scaling factor f, its volume and mass will be multiplied by f^3. a) By what factor will its moment of inertia be multiplied? b) If a $\frac{1}{48}$-scale model has a rotational kinetic energy of 2.5 J, what will be the kinetic energy for the full-scale object of the same material rotating at the same angular velocity?

9.41 A wagon wheel is constructed as shown in Fig. 9.28. The radius of the wheel is 0.300 m, and the rim has mass 1.40 kg. Each of the eight spokes, that lie along a diameter and are 0.300 m long, has mass 0.280 kg. What is the moment of inertia of the wheel about an axis through its center and perpendicular to the plane of the wheel? (Use the formulas given in Table 9.2.)

Figure 9.28 Exercise 9.41.

9.42 An airplane propeller is 2.08 m in length (from tip to tip) with mass 117 kg. The propeller is rotating at 2400 rev/min about an axis through its center. a) What is its rotational kinetic energy? Treat the propeller as a slender rod. b) If it were not rotating, how far would it have to drop in free fall to acquire the same kinetic energy?

9.43 a) Show that $\frac{1}{2}I\omega^2$ has units equivalent to joules. Explain why you need not include "rads" in the units. b) Often ω is expressed in rev/min rather than rad/s. Write an expression for rotational kinetic energy such that if I is in kg · m² and ω is in rev/min, the energy will be in joules.

9.44 An antique phonograph turntable has a kinetic energy of 0.0250 J when turning at 45.0 rev/min. What is the moment of inertia of the turntable about the rotation axis?

9.45 The flywheel of a gasoline engine is required to give up 500 J of kinetic energy while its angular velocity decreases from 650 rev/min to 520 rev/min. What moment of inertia is required?

9.46 A light, flexible rope is wrapped several times around a *hollow* cylinder with a weight of 40.0 N and a radius of 0.25 m, that rotates without friction about a fixed horizontal axis. The cylinder is attached to the axle by spokes of a negligible moment of inertia. The cylinder is initially at rest. The free end of the rope is pulled with a constant force P for a distance of 5.00 m, at which point the end of the rope is moving at 6.00 m/s. If the rope does not slip on the cylinder, what is the value of P?

9.47 Energy is to be stored in a 70.0-kg flywheel in the shape of a uniform solid disk with radius $R = 1.20$ m. To prevent structural failure of the flywheel, the maximum allowed radial acceleration of a point on its rim is 3500 m/s^2. What is the maximum kinetic energy that can be stored in the flywheel?

9.48 Suppose the solid cylinder in the apparatus described in Example 9.9 (Section 9.4) is replaced by a thin-walled, hollow cylinder with the same mass M and radius R. The cylinder is attached to the axle by spokes of a negligible moment of inertia. a) Find the speed of the hanging mass m just as it strikes the floor. b) Is the answer to part (a) greater than, equal to, or less than the speed found in Example 9.9? Explain why, using energy concepts.

9.49 Rate of Kinetic Energy Loss. A rigid body with a moment of inertia I spins once every T seconds. The rotation is slowing down, so that $dT/dt > 0$. a) Express the kinetic energy of rotation of the body in terms of I and T. b) Express the rate of change of the kinetic energy of rotation in terms of I, T, and dT/dt. c) A large flywheel has $I = 8.0$ kg $\cdot$ m^2. What is the kinetic energy of the flywheel when the period of rotation is 1.5 s? d) What is the rate of change of the kinetic energy of the flywheel in part (c) at the instant when the period is 1.5 s and is changing at the rate of $dT/dt = 0.0060$?

9.50 A uniform rope 10.0 m long and with mass 3.00 kg is hanging with one end attached to a gymnasium ceiling and the other end just touching the floor. The upper end of the rope is released, and the rope falls to the floor. What is the change in the gravitational potential energy if the rope ends up flat on the floor (not coiled up)?

9.51 Center of Mass of an Extended Object. How much work must a wrestler do to raise the center of mass of his 120-kg opponent a vertical distance of 0.700 m?

Section 9.5 Parallel-Axis Theorem

9.52 Find the moment of inertia of a hoop (a thin-walled, hollow ring) with mass M and radius R about an axis perpendicular to the hoop's plane at an edge.

9.53 About what axis will a uniform, balsa-wood sphere have the same moment of inertia as for a thin-walled, hollow, lead sphere of the same mass and radius, with the axis along a diameter?

9.54 Use the parallel-axis theorem to show that the moments of inertia given in parts (a) and (b) of Table 9.2 are consistent.

9.55 A thin, rectangular sheet of metal has mass M and sides of length a and b. Use the parallel-axis theorem to calculate the moment of inertia of the sheet for an axis that is perpendicular to the plane of the sheet and that passes through one corner of the sheet.

9.56 a) For the thin rectangular plate shown in part (d) of Table 9.2, find the moment of inertia about an axis that lies in the plane of the plate, passes through the center of the plate, and is parallel to the axis shown in the figure. b) Find the moment of inertia of the plate for an axis that lies in the plane of the plate, passes through the center of the plate, and is perpendicular to the axis in part (a).

*Section 9.6 Moment of Inertia Calculations

*9.57 Using the information in Table 9.2 and the parallel-axis theorem, find the moment of inertia of the slender rod with mass M and length L shown in Fig. 9.24 about an axis through O, at an arbitrary distance h from one end. Compare your result to that found by integration in Example 9.12 (Section 9.6).

*9.58 Use Eq. (9.20) to calculate the moment of inertia of a uniform, solid disk with mass M and radius R for an axis perpendicular to the plane of the disk and passing through its center.

*9.59 Use Eq. (9.20) to calculate the moment of inertia of a slender, uniform rod with mass M and length L about an axis at one end, perpendicular to the rod.

*9.60 A slender rod with length L has a mass per unit length that varies with distance from the left-hand end, where $x = 0$, according to $dm/dx = \gamma x$, where γ has units of kg/m^2. a) Calculate the total mass of the rod in terms of γ and L. b) Use Eq. (9.20) to calculate the moment of inertia of the rod for an axis at the left-hand end, perpendicular to the rod. Use the expression you derived in part (a) to express I in terms of M and L. How does your result compare to that for a uniform rod? Explain this comparison. c) Repeat part (b) for an axis at the right-hand end of the rod. How do the results for parts (b) and (c) compare? Explain this result.

Problems

9.61 Sketch a wheel lying in the plane of your paper and rotating counterclockwise. Choose a point on the rim and draw a vector $\vec{r}$ from the center of the wheel to that point. a) What is the direction of $\vec{\omega}$? b) Show that the velocity of the point is $\vec{v} = \vec{\omega} \times \vec{r}$. c) Show that the radial acceleration of the point is $\vec{a}_{rad} = \vec{\omega} \times \vec{v} = \vec{\omega} \times (\vec{\omega} \times \vec{r})$ (see Exercise 9.28).

9.62 Trip to Mars. You are working on a project with NASA to launch a rocket to Mars, with the rocket blasting off from earth when earth and Mars are just aligned along a straight line from the sun. As a first step in doing the calculation, assume circular orbits for both planets. If Mars is now 60° ahead of earth in its orbit around the sun, when should you launch the rocket? (*Note:* All the planets orbit the sun in the same direction, and the year on Mars is 1.9 earth-years.)

9.63 Speedometer. Your car's speedometer converts the angular speed of the wheels to the linear speed of the car, assuming standard-size tires and no slipping on the pavement. a) If your car's standard tires are 24 inches in diameter, at what rate (in rpm) are your wheels rotating when you are driving at a freeway speed of 60 mph? b) Suppose you put oversize, 30-inch diameter tires on your car. How fast are you really going when your speedometer reads 60 mph? c) If you now put on undersize, 20-inch diameter tires, what will the speedometer read when you are actually traveling at 50 mph?

9.64 a) Prove that when an object starts from rest and rotates about a fixed axis with constant angular acceleration, the radial acceleration of a point in the object is directly proportional to its angular displacement. b) Through what angle has the object turned at the instant when the resultant acceleration of a point makes an angle of 36.9° with the radial direction?

9.65 A roller in a printing press turns through an angle $\theta(t)$ given by $\theta(t) = \gamma t^2 - \beta t^3$, where $\gamma = 3.20$ rad/s^2 and $\beta = 0.500$ rad/s^3. a) Calculate the angular velocity of the roller as a function of time. b) Calculate the angular acceleration of the roller as a function of time. c) What is the maximum positive angular velocity, and at what value of t does it occur?

*9.66 A bicycle wheel with a radius of 0.33 m turns with angular acceleration $\alpha(t) = \gamma - \beta t$, where $\gamma = 1.80$ rad/s^2 and $\beta = 0.25$ rad/s^3. It is at rest at $t = 0$. a) Calculate the angular velocity and angular displacement as functions of time. b) Calculate the maximum positive angular velocity and maximum positive angular displacement of the wheel. (*Hint:* See Section 2.6.)

9.67 When a toy car is rapidly scooted across the floor, it stores energy in a flywheel. The car has mass 0.180 kg, and its flywheel has a moment of inertia of 4.00×10^{-5} kg $\cdot$ m^2. The car is 15.0 cm long. An advertisement claims that the car can travel at a scale speed of up to 700 km/h (440 mi/h). The scale speed is the speed of the toy car multiplied by the ratio of the length of an actual car to the length of the toy. Assume a length of 3.0 m for a real car. a) For a scale speed of 700 km/h, what is the actual translational speed of the car? b) If all the kinetic energy that is initially in the flywheel is converted to the translational kinetic energy of the toy, how much energy is originally stored in the flywheel? c) What initial angular velocity of the flywheel was needed to store the amount of energy calculated in part (b)?

9.68 A classic 1957 Chevrolet Corvette mass 1240 kg starts from rest and speeds up with a constant tangential acceleration of 3.00 m/s^2 on a circular test track of radius 60.0 m. Treat the car as a particle. a) What is its angular acceleration? b) What is its angular speed 6.00 s after it starts? c) What is its radial acceleration at this time? d) Sketch a view from above showing the circular track, the car, the velocity vector, and the acceleration component vectors 6.00 s after the car starts. e) What are the magnitudes of the total acceleration and net force for the car at this time? f) What angle do the total acceleration and net force make with the car's velocity at this time?

9.69 The flywheel of a punch press has a moment of inertia of 16.0 kg $\cdot$ m^2 and runs at 300 rev/min. The flywheel supplies all the energy needed in a quick punching operation. a) Find the speed in rev/min to which the flywheel will be reduced by a sudden punching operation requiring 4000 J of work. b) What must the constant power supply to the flywheel (in watts) be to bring it back to its initial speed in a time of 5.00 s?

9.70 A totally inedible cafeteria meatball with mass 40.0 g is attached to the free end of a 2.50-m piece of string that is attached to the ceiling. The meatball is pulled to the left so that the string makes a 36.9° angle with the vertical and is then released. a) What is the magnitude and direction of the angular velocity of the meatball the first time the angular acceleration is zero? b) What is it the second time that $\alpha_z = 0$? c) At the times described in parts (a) and (b), what is the magnitude and direction of the meatball's radial acceleration? d) Show that the answer to part (c) is independent of the length of the string.

9.71 A vacuum cleaner belt is looped over a shaft of radius 0.45 cm and a wheel of radius 2.00 cm. The arrangement of the belt, shaft, and wheel is similar to that of the chain and sprockets in Fig. 9.15. The motor turns the shaft at 60.0 rev/s and the moving belt turns the wheel, that in turn is connected by another shaft to the roller that beats the dirt out of the rug being vacuumed. Assume that the belt doesn't slip on either the shaft or the wheel. a) What is the speed of a point on the belt? b) What is the angular velocity of the wheel in rad/s?

9.72 The motor of a table saw is rotating at 3450 rev/min. A pulley attached to the motor shaft drives a second pulley of half the diameter by means of a V-belt. A circular saw blade of diameter 0.208 m is mounted on the same rotating shaft as the second pulley. a) The operator is careless and the blade catches and throws back a small piece of wood. This piece of wood moves with linear speed equal to the tangential speed of the rim of the blade. What is this speed? b) Calculate the radial acceleration of points on the outer edge of the blade to see why sawdust doesn't stick to its teeth.

9.73 A wheel changes its angular velocity with a constant angular acceleration while rotating about a fixed axis through its center. a) Show that the change in the magnitude of the radial acceleration during any time interval of a point on the wheel is twice the product of the angular acceleration, the angular displacement, and the perpendicular distance of the point from the axis. b) The radial acceleration of a point on the wheel that is 0.250 m from the axis changes from 25.0 m/s^2 to 85.0 m/s^2 as the wheel rotates through 15.0 rad. Calculate the tangential acceleration of this point. c) Show that the change in the wheel's kinetic energy during any time interval is the product of the moment of inertia about the axis, the angular acceleration, and the angular displacement. d) During the 15.0-rad angular displacement of part (b), the kinetic energy of the wheel increases from 20.0 J to 45.0 J. What is the moment of inertia of the wheel about the rotation axis?

9.74 A sphere consists of a solid wooden ball of uniform density 800 kg/m^3 and radius 0.20 m and is covered with a thin coating of lead foil with area density 20 kg/m^2. Calculate the moment of inertia of this sphere about an axis passing through its center.

9.75 Estimate your own moment of inertia about a vertical axis through the center of the top of your head when you are standing up straight with your arms outstretched. Make reasonable approximations and measure or estimate necessary quantities.

9.76 A thin uniform rod 50.0 cm long and with mass 0.320 kg is bent at its center into a V shape, with a 70.0° angle at its vertex. Find the moment of inertia of this V-shaped object about an axis perpendicular to the plane of the V at its vertex.

9.77 It has been argued that power plants should make use of off-peak hours (such as late at night) to generate mechanical energy and store it until it is needed during peak load times, such as the middle of the day. One suggestion has been to store the energy in large flywheels spinning on nearly frictionless ball-bearings. Consider a flywheel made of iron (density 7800 kg/m^3) and in the shape of a 10.0 cm thick uniform disk. a) What would the diameter of such a disk need to be if it is to store 10.0 megajoules of kinetic energy when spinning at 90.0 rpm about an axis perpendicular to the disk at its center? b) What would be the centripetal acceleration of a point on its rim when spinning at this rate?

9.78 The three uniform objects shown in Fig. 9.29 have the same mass *m*. Object *A* is a solid cylinder with radius *R*. Object *B* is a hollow, thin cylinder with radius *R*. Object *C* is a solid cube whose sides are of length 2*R*. The axis of rotation of each object is perpendicular to the page and through the object's center of mass. a) Which object has the smallest moment of inertia? Explain. b) Which object has the largest moment of inertia? Explain. c) Where would the moment of inertia of a uniform solid sphere rank if its radius is *R*, its mass is *m,* and the axis of rotation is along a diameter of the sphere? Explain.

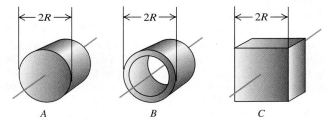

Figure 9.29 Problem 9.78.

9.79 The earth, which is not a uniform sphere, has a moment of inertia of $0.3308MR^2$ about an axis through its north and south poles. It takes the earth 86,164 s to spin once about this axis. Use Appendix F to calculate a) the earth's kinetic energy due to its rotation about this axis and b) the earth's kinetic energy due to its orbital motion around the sun. c) Explain how the value of the earth's moment of inertia tells us that the mass of the earth is concentrated toward the planet's center.

9.80 A uniform, solid disk with mass *m* and radius *R* is pivoted about a horizontal axis through its center. A small object of the same mass *m* is glued to the rim of the disk. If the disk is released from rest with the small object at the end of a horizontal radius, find the angular speed when the small object is directly below the axis.

9.81 A metal sign for a car dealership is a thin, uniform, right triangle with base length *b* and height *h*. The sign has mass *M*. a) What is the moment of inertia of the sign for rotation about the side of length *h*? b) If *M* = 5.40 kg, *b* = 1.60 m, and *h* = 1.20 m, what is the kinetic energy of the sign when it is rotating about an axis along the 1.20-m side at 2.00 rev/s?

9.82 *Measuring I.* As an intern with an engineering firm, you are asked to measure the moment of inertia of a large wheel, for rotation about an axis through its center. Since you were a good physics student, you know what to do. You measure the diameter of the wheel to be 0.740 m and find that it weighs 280 N. You mount the wheel, using frictionless bearings, on a horizontal axis through the wheel's center. You wrap a light rope around the wheel and hang a 8.00-kg mass from the free end of the rope, as shown in Fig. 9.19. You release the mass from rest; the mass descends and the wheel turns as the rope unwinds. You find that the mass has speed 5.00 m/s after it has descended 2.00 m. a) What is the moment of inertia of the wheel for an axis perpendicular to the wheel at its center? b) Your boss tells

you that a larger *I* is needed. He asks you to design a wheel of the same mass and radius that has $I = 19.0$ kg · m². How do you reply?

9.83 A meter stick with a mass of 0.160 kg is pivoted about one end so it can rotate without friction about a horizontal axis. The meter stick is held in a horizontal position and released. As it swings through the vertical, calculate a) the change in gravitational potential energy that has occurred; b) the angular speed of the stick; c) the linear speed of the end of the stick opposite the axis. d) Compare the answer in part (c) to the speed of a particle that has fallen 1.00 m, starting from rest.

9.84 Exactly one turn of a flexible rope with mass *m* is wrapped around a uniform cylinder with mass *M* and radius *R*. The cylinder rotates without friction about a horizontal axle along the cylinder axis. One end of the rope is attached to the cylinder. The cylinder starts with an angular speed ω_0. After one revolution of the cylinder the rope has unwrapped and, at this instant, hangs vertically down, tangent to the cylinder. Find the angular speed of the cylinder and the linear speed of the lower end of the rope at this time. You can ignore the thickness of the rope. (*Hint:* Use Eq. (9.18).)

9.85 The pulley in Fig. 9.30 has radius *R* and a moment of inertia *I*. The rope does not slip over the pulley, and the pulley spins on a frictionless axle. The coefficient of kinetic friction between block *A* and the tabletop is μ_k. The system is released from rest, and block *B* descends. Block *A* has mass m_A and block *B* has mass m_B. Use energy methods to calculate the speed of block *B* as a function of the distance *d* that it has descended.

Figure 9.30 Problem 9.85.

9.86 The pulley in Fig. 9.31 has radius 0.160 m and a moment of inertia 0.480 kg · m². The rope does not slip on the pulley rim. Use energy methods to calculate the speed of the 4.00-kg block just before it strikes the floor.

9.87 You hang a thin hoop with radius *R* over a nail at the rim of the hoop. You displace it to the side through an angle *β* from its equilibrium position and let it go. What is its angular speed when it returns to its equilibrium position? (*Hint:* Use Eq. (9.18).)

Figure 9.31 Problem 9.86.

9.88 A passenger bus in Zurich, Switzerland derived its motive power from the energy stored in a large flywheel. The wheel was brought up to speed periodically, when the bus stopped at a station, by an electric motor, which could then be attached to the electric power lines. The flywheel was a solid cylinder with mass 1000 kg and diameter 1.80 m; its top angular speed was 3000 rev/min. a) At this angular speed, what is the kinetic energy of the flywheel? b) If the average power required to operate the bus is 1.86×10^4 W, how long could it operate between stops?

9.89 Two metal disks, one with radius $R_1 = 2.50$ cm and mass $M_1 = 0.80$ kg and the other with radius $R_2 = 5.00$ cm and mass $M_2 = 1.60$ kg, are welded together and mounted on a frictionless axis through their common center (Fig. 9.32). a) What is the total moment of inertia of the two disks? b) A light string is wrapped around the edge of the smaller disk, and a 1.50-kg block, suspended from the free end of the string. If the block is released from rest at a distance of 2.00 m above the floor, what is its speed just before it strikes the floor? c) Repeat the calculation of part (b), this time with the string wrapped around the edge of the larger disk. In which case is the final speed of the block the greatest? Explain why this is so.

1.50 kg

Figure 9.32 Problem 9.89.

9.90 In the cylinder and mass combination described in Example 9.9 (Section 9.4), suppose the falling mass m is made of ideal rubber, so that no mechanical energy is lost when the mass hits the ground. a) If the cylinder is originally not rotating and the mass m is released from rest at a height h above the ground, to what height will this mass rebound if it bounces straight back up from the floor? b) Explain, in terms of energy, why the answer to (a) is *less* than h.

9.91 In the system shown in Fig. 9.19, a 12.0-kg mass is released from rest and falls, causing the uniform 10.0-kg cylinder of diameter 30.0 cm to turn about a frictionless axle through its center. How far will the mass have to descend to give the cylinder 250 J of kinetic energy?

9.92 In Fig. 9.33, the cylinder and pulley turn without friction about stationary horizontal axles that pass through their centers. A light rope is wrapped around the cylinder, passes over the pulley, and has a 3.00-kg box suspended from its free end. There is no slipping between the rope and the pulley surface. The uniform cylinder has mass 5.00 kg and radius 40.0 cm. The pulley is a uniform disk with mass 2.00 kg and radius 20.0 cm. The box is released from rest and descends as the rope unwraps from the cylinder. Find the speed of the box when it has fallen 1.50 m.

Figure 9.33 Problem 9.92.

9.93 A thin, flat, uniform disk has mass M and radius R. A circular hole of radius $R/4$, centered at a point $R/2$ from the disk's center, is then punched in the disk. a) Find the moment of inertia of the disk

with the hole about an axis through the original center of the disk, perpendicular to the plane of the disk. (*Hint:* Find the moment of inertia of the piece punched from the disk.) b) Find the moment of inertia of the disk with the hole about an axis through the center of the hole, perpendicular to the plane of the disk.

9.94 A pendulum is made of a uniform solid sphere with mass M and radius R suspended from the end of a light rod. The distance from the pivot at the upper end of the rod to the center of the sphere is L. The pendulum's moment of inertia I_P for rotation about the pivot is usually approximated as ML^2. a) Use the parallel-axis theorem to show that if R is 5% of L and the mass of the rod is ignored, I_P is only 0.1% larger than ML^2. b) If the mass of the rod is 1% of M and R is much less than L, what is the ratio of I_{rod} for an axis at the pivot to ML^2?

9.95 Perpendicular-Axis Theorem. Consider a rigid body that is a thin, plane sheet of arbitrary shape. Take the body to lie in the xy-plane and let the origin O of coordinates be located at any point within or outside the body. Let I_x and I_y be the moments of inertia about the x- and y-axes and let I_O be the moment of inertia about an axis through O perpendicular to the plane. a) By considering mass elements m_i with coordinates (x_i, y_i), show that $I_x + I_y = I_O$. This is called the perpendicular-axis theorem. Note that point O does not have to be the center of mass. b) For a thin washer with mass M and with inner and outer radii R_1 and R_2, use the perpendicular-axis theorem to find the moment of inertia about an axis that is in the plane of the washer and that passes through its center. You may use the information in Table 9.2. c) Use the perpendicular-axis theorem to show that for a thin, square sheet with mass M and side L, the moment of inertia about *any* axis in the plane of the sheet that passes through the center of the sheet is $\frac{1}{2}ML^2$. You may use the information in Table 9.2.

9.96 A thin, uniform rod is bent into a square of side length a. If the total mass is M, find the moment of inertia about an axis through the center and perpendicular to the plane of the square. (*Hint:* Use the parallel-axis theorem.)

*9.97 A cylinder with radius R and mass M has density that increases linearly with distance r from the cylinder axis, $\rho = \alpha r$, where α is a positive constant. a) Calculate the moment of inertia of the cylinder about a longitudinal axis through its center in terms of M and R. b) Is your answer greater or smaller than the moment of inertia of a cylinder of the same mass and radius but of uniform density? Explain why this result makes qualitative sense.

9.98 Neutron Stars and Supernova Remnants. The Crab Nebula is a cloud of glowing gas about 10 light years across, located about 6500 light years from the earth (Fig. 9.34). It is the remnant of a star that underwent a *supernova explosion,* seen on earth in 1054 A.D. Energy is released by the Crab

Figure 9.34 Problem 9.98.

Nebula at a rate of about 5×10^{31} W, about 10^5 times the rate at which the sun radiates energy. The Crab Nebula obtains its energy from the rotational kinetic energy of a rapidly spinning *neutron star* at its center. This object rotates once every 0.0331 s, and this period is increasing by 4.22×10^{-13} s for each second of time that elapses. a) If the rate at which energy is lost by the neutron star is equal to the rate at which energy is released by the nebula, find the moment of inertia of the neutron star. (Use the expression derived in Exercise 9.49.) b) Theories of supernovae predict that the neutron star in the Crab Nebula has a mass of about 1.4 times that of the sun. Modeling the neutron star as a solid uniform sphere, calculate its radius in kilometers. c) What is the linear speed of a point on the equator of the neutron star? Compare to the speed of light. d) Assume that the neutron star is uniform and calculate its density. Compare to the density of ordinary rock (3000 kg/m³) and to the density of an atomic nucleus (about 10^{17} kg/m³). Justify the statement that a neutron star is essentially a large atomic nucleus.

Challenge Problems

9.99 The moment of inertia of a sphere with uniform density about an axis through its center is $\frac{2}{5}MR^2 = 0.400MR^2$. Satellite observations show that the earth's moment of inertia is $0.3308MR^2$. Geophysical data suggests the earth consists of five main regions: the inner core ($r = 0$ to $r = 1220$ km) of average density 12,900 kg/m³, the outer core ($r = 1220$ km to $r = 3480$ km) of average density 10,900 kg/m³, the lower mantle ($r = 3480$ km to $r = 5700$ km) of average density 4900 kg/m³, the upper mantle ($r = 5700$ km to $r = 6350$ km) of average density 3600 kg/m³, and the outer crust and oceans ($r = 6350$ km to $r = 6370$ km) of average density 2400 kg/m³. a) Show that the moment of inertia about a diameter of a uniform spherical shell of inner radius R_1, outer radius R_2, and density ρ is $I = \rho(8\pi/15)(R_2^5 - R_1^5)$. (*Hint:* Form the shell by superposition of a sphere of density ρ and a smaller sphere of density $-\rho$.) b) Check the given data by using them to calculate the mass of the earth. c) Use the given data to calculate the earth's moment of inertia in terms of MR^2.

*9.100 Calculate the moment of inertia of a uniform solid cone about an axis through its center (Fig. 9.35). The cone has mass M and altitude h. The radius of its circular base is R.

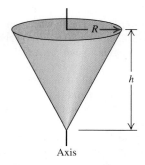

Figure 9.35 Challenge Problem 9.100.

9.101 On a compact disc (CD), music is coded in a pattern of tiny pits arranged in a track that spirals outward toward the rim of the disc. As the disc spins inside a CD player, the track is scanned at a constant *linear* speed of $v = 1.25$ m/s. Because the radius of the track varies as it spirals outward, the *angular* speed of the disc must change as the CD is played. (See Exercise 9.22.) Let's see what angular acceleration is required to keep v constant. The equation of a spiral is $r(\theta) = r_0 + \beta\theta$, where r_0 is the radius of the spiral at $\theta = 0$ and β is a constant. On a CD, r_0 is the inner radius of the spiral track. If we take the rotation direction of the CD to be positive, β must be positive so that r increases as the disc turns and θ increases. a) When the disc rotates through a small angle $d\theta$, the distance scanned along the track is $ds = r \, d\theta$. Using the above expression for $r(\theta)$, integrate ds to find the total distance s scanned along the track as a function of the total angle θ through which the disc has rotated. b) Since the track is scanned at a constant linear speed v, the distance s found in part (a) is equal to vt. Use this to find θ as a function of time. There will be two solutions for θ; choose the positive one, and explain why this is the solution to choose. c) Use your expression for $\theta(t)$ to find the angular velocity ω_z and angular acceleration α_z as functions of time. Is α_z constant? d) On a CD, the inner radius of the track is 25.0 mm, the track radius increases by 1.55 μm per revolution, and the playing time is 74.0 min. Find the values of r_0 and β, and find the total number of revolutions made during the playing time. e) Using your results from parts (c) and (d), make graphs of ω_z (in rad/s) versus t and α_z (in rad/s²) versus t between $t = 0$ and $t = 74.0$ min.

DYNAMICS OF ROTATIONAL MOTION 10

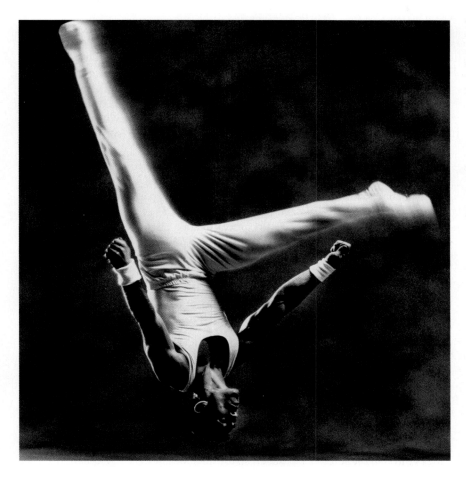

This tumbling acrobat is not a rigid body, which enables him to vary his rotational speed in mid-air. Moving his arms and legs outward makes his rotation slow down, while pulling his arms and legs inward makes him rotate more rapidly.

? **If the acrobat isn't touching the ground, how can he change his rotation speed? What physical principle is at work here?**

We learned in Chapters 4 and 5 that a net force applied to a body gives that body an acceleration. But what does it take to give a body an *angular* acceleration? That is, what does it take to start a stationary body rotating or to bring a spinning body to a halt? A force is required, but it must be applied in a way that gives a twisting or turning action.

In this chapter we will define a new physical quantity, *torque,* that describes the twisting or turning effort of a force. We'll find that the net torque acting on a rigid body determines its angular acceleration, in the same way that the net force on a body determines its linear acceleration. We'll also look at work and power in rotational motion so as to understand such problems as how energy is transmitted by the rotating drive shaft in a car. Finally, we will develop a new conservation principle, *conservation of angular momentum,* that is tremendously useful for understanding the rotational motion of both rigid and nonrigid bodies.

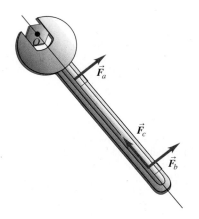

10.1 Which of these three equal-magnitude forces is most likely to loosen the tight bolt?

Act|v
ONLINE
Physics

7.1   Calculating Torques

Torque =
(Force magnitude) × (lever arm)

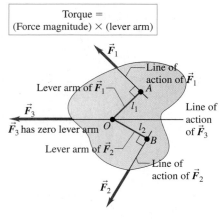

10.2 The torque or moment of a force about a point is the product of the force magnitude and the lever arm of the force.

We'll finish this chapter by studying *gyroscopes,* rotating devices that seemingly defy common sense and don't fall over when you might think they should—but that actually behave in perfect accordance with the dynamics of rotational motion.

10.1 | Torque

What is it about a force that determines how effective it is in causing or changing rotational motion? The magnitude and direction of the force are important, but so is the position of the point where the force is applied. When you try to swing a heavy door open, it's a lot more effective to push farther away from the axis of rotation (near the knob) than close to the axis (near the hinges). In Fig. 10.1 a wrench is being used to loosen a tight bolt. Force $\vec{F}_b$, applied near the end of the handle, is more effective than an equal force $\vec{F}_a$ applied near the bolt. Force $\vec{F}_c$ doesn't do any good at all; it's applied at the same point and has the same magnitude as $\vec{F}_b$, but it's directed along the length of the handle.

The quantitative measure of the tendency of a force to cause or change the rotational motion of a body is called *torque.* Figure 10.2 shows a body that can rotate about an axis that passes through point O and is perpendicular to the plane of the figure. The body is acted on by three forces, $\vec{F}_1$, $\vec{F}_2$, and $\vec{F}_3$, in the plane of the figure. The tendency of $\vec{F}_1$ to cause a rotation about point O depends on its magnitude F_1. It also depends on the *perpendicular* distance l_1 between the **line of action** of the force (that is, the line along which the force vector lies) and point O. We call the distance l_1 the **lever arm** (or **moment arm**) of force $\vec{F}_1$ about O. The twisting effort is directly proportional to both F_1 and l_1. We define the **torque** (or *moment*) of the force $\vec{F}_1$ with respect to point O as the product $F_1 l_1$. We use the Greek letter τ ("tau") for torque. For a force of magnitude F whose line of action is a perpendicular distance l from the point O, the torque is

$$\tau = Fl \qquad (10.1)$$

Physicists usually use the term "torque," while engineers usually use "moment" (unless they are talking about a rotating shaft). Both groups use the term "lever arm" or "moment arm" for the distance l.

The lever arm of $\vec{F}_1$ in Fig. 10.2 is the perpendicular distance OA or l_1, and the lever arm of $\vec{F}_2$ is the perpendicular distance OB or l_2. The line of action of $\vec{F}_3$ passes through the reference point O, so the lever arm for $\vec{F}_3$ is zero and its torque with respect to point O is zero. In the same way, force $\vec{F}_c$ in Fig. 10.1 has zero torque with respect to point O. On the other hand, $\vec{F}_b$ has a greater torque than $\vec{F}_a$ because its lever arm is greater.

CAUTION Note that torque is always defined with reference to a specific point, often (but not always) the origin of a coordinate system. If we shift the position of this point, the torque of each force may also change. For example, the torque of force $\vec{F}_3$ in Fig. 10.2 is zero with respect to point O, but is *not* zero about point A or B. When describing the torque of a certain force, it's not enough to call it "the torque of $\vec{F}$"; you must say "the torque of $\vec{F}$ with respect to point X" or "the torque of $\vec{F}$ about point X."

Force $\vec{F}_1$ in Fig. 10.2 tends to cause *counterclockwise* rotation about O, while $\vec{F}_2$ tends to cause *clockwise* rotation. To distinguish between these two possibilities, we need to choose a positive sense of rotation. With the choice that *counter-clockwise torques are positive and clockwise torques are negative*, the torques of $\vec{F}_1$ and $\vec{F}_2$ about O are

$$\tau_1 = +F_1 l_1 \qquad \tau_2 = -F_2 l_2$$

We will often use the symbol

to indicate our choice of the positive sense of rotation.

The SI unit of torque is the newton-meter. In our discussion of work and energy we called this combination the joule. But torque is *not* work or energy, and torque should be expressed in newton-meters, *not* joules.

Figure 10.3 shows a force $\vec{F}$ applied at a point P described by a position vector $\vec{r}$ with respect to the chosen point O. There are several ways to calculate the torque of this force. One is to find the lever arm l and use $\tau = Fl$. Or we can determine the angle ϕ between the vectors $\vec{r}$ and $\vec{F}$; the lever arm is $r \sin \phi$, so $\tau = rF \sin \phi$. A third method is to represent $\vec{F}$ in terms of a radial component F_{rad} along the direction of $\vec{r}$ and a tangential component F_{tan} at right angles, perpendicular to $\vec{r}$. (We call this a tangential component because if the body rotates, the point where the force acts moves in a circle, and this component is tangent to that circle.) Then $F_{\text{tan}} = F \sin \phi$ and $\tau = r(F \sin \phi) = F_{\text{tan}} r$. The component F_{rad} has no torque with respect to O because its lever arm with respect to that point is zero (compare to forces $\vec{F}_c$ in Fig. 10.1 and $\vec{F}_3$ in Fig. 10.2). Summarizing these three expressions for torque, we have

$$\tau = Fl = rF \sin \phi = F_{\text{tan}} r \qquad \text{(magnitude of torque)} \qquad (10.2)$$

We saw in Section 9.1 that angular velocity and angular acceleration can be represented as vectors; the same is true for torque. To see how to do this, note that the quantity $rF \sin \phi$ in Eq. (10.2) is the magnitude of the *vector product* $\vec{r} \times \vec{F}$ that we defined in Section 1.10. You should review that definition. We now generalize the definition of torque as follows: When a force $\vec{F}$ acts at a point having a position vector $\vec{r}$ with respect to an origin O, as in Fig. 10.3, the torque $\vec{\tau}$ of the force with respect to O is the *vector* quantity

$$\vec{\tau} = \vec{r} \times \vec{F} \qquad \text{(definition of torque vector)} \qquad (10.3)$$

The torque as defined in Eq. (10.2) is just the magnitude of the torque vector $\vec{r} \times \vec{F}$. The direction of $\vec{\tau}$ is perpendicular to both $\vec{r}$ and $\vec{F}$. In particular, if both $\vec{r}$ and $\vec{F}$ lie in a plane perpendicular to the axis of rotation, as in Fig. 10.3, then the torque vector $\vec{\tau} = \vec{r} \times \vec{F}$ is directed along the axis of rotation, with a sense given by the right-hand rule (Fig. 1.22). The direction relationships are shown in Fig. 10.4.

In diagrams that involve $\vec{r}$, $\vec{F}$, and $\vec{\tau}$, it's common to have one of the vectors oriented perpendicular to the page. (Indeed, by the very nature of the cross product, $\vec{\tau} = \vec{r} \times \vec{F}$ *must* be perpendicular to the plane of the vectors $\vec{r}$ and $\vec{F}$.) We use a dot (•) to represent a vector that points out of the page (see Fig. 10.4) and a cross (✖) to represent a vector that points into the page.

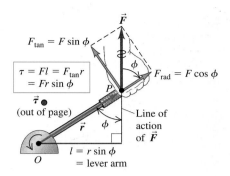

10.3 The torque of the force $\vec{F}$ about the point O is defined by $\vec{\tau} = \vec{r} \times \vec{F}$. The magnitude of $\vec{\tau}$ is $Fr \sin \phi$. In this figure, $\vec{r}$ and $\vec{F}$ are in the plane of the page and $\vec{\tau}$ points out of the page toward you.

10.4 The torque vector $\vec{\tau} = \vec{r} \times \vec{F}$ is directed along the axis of the bolt, perpendicular to both $\vec{r}$ and $\vec{F}$. The direction of $\vec{\tau}$ is given by the right-hand rule. Note that the fingers of the right hand curl in the direction of the rotation that the torque tends to cause.

In the following sections we will usually be concerned with rotation of a body about an axis oriented in a specified constant direction. In that case, only the component of torque along that axis is of interest, and we often call that component the torque with respect to the specified axis.

<table>
<tr><td>Example
10.1</td><td>**Applying a torque**</td></tr>
</table>

A weekend plumber, unable to loosen a pipe fitting, slips a piece of scrap pipe (a "cheater") over his wrench handle. He then applies his full weight of 900 N to the end of the cheater by standing on it. The distance from the center of the fitting to the point where the weight acts is 0.80 m, and the wrench handle and cheater make an angle of 19° with the horizontal (Fig. 10.5a). Find the magnitude and direction of the torque he applies about the center of the pipe fitting.

SOLUTION

IDENTIFY and SET UP: We'll use Eq. (10.1) or (10.2) to find the magnitude of the torque and the right-hand rule with Eq. (10.3) to find the torque direction. Figure 10.5b shows the vectors $\vec{r}$ and $\vec{F}$ and the angle between them ($\phi = 109°$).

EXECUTE: To use Eq. (10.1), we first calculate the lever arm l. As Fig. 10.5b shows, it is the perpendicular distance from O to the line of action of the force:

$$l = (0.80 \text{ m}) \sin 109° = (0.80 \text{ m}) \sin 71° = 0.76 \text{ m}$$

Then Eq. (10.1) tells us that the magnitude of the torque is

$$\tau = Fl = (900 \text{ N})(0.76 \text{ m}) = 680 \text{ N} \cdot \text{m}$$

Or, from Eq. (10.2),

$$\tau = rF \sin \phi = (0.80 \text{ m})(900 \text{ N})(\sin 109°) = 680 \text{ N} \cdot \text{m}$$

Alternatively, we can find F_{tan}, the tangential component of $\vec{F}$. This is the component that acts perpendicular to $\vec{r}$ (that is, perpendicular to the "cheater"). The vector $\vec{r}$ is oriented 19° from the horizontal, so the perpendicular to $\vec{r}$ is oriented 19° from the vertical. Since $\vec{F}$ is vertical, this means $F_{\text{tan}} = F(\cos 19°) = (900 \text{ N})(\cos 19°) = 851 \text{ N}$. Then the torque is

$$\tau = F_{\text{tan}} r = (851 \text{ N})(0.80 \text{ m}) = 680 \text{ N} \cdot \text{m}$$

If you curl the fingers of your right hand from the direction of $\vec{r}$ (in the plane of Fig. 10.5b, to the right and up) into the direction of $\vec{F}$ (straight down), your right thumb points into the plane of the figure. This is the direction of the torque $\vec{\tau}$.

EVALUATE: We've already checked our answer for the magnitude τ by calculating it in three different ways. To check our result for the direction of the torque, note that the force in Fig. 10.5 tends to produce a clockwise rotation about O. If you curl the fingers of your right hand in a clockwise direction, the thumb points into the plane of Fig. 10.5, which is indeed the direction of the torque.

(a)

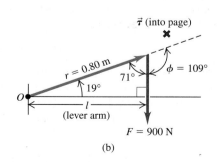

(b)

10.5 (a) A weekend plumber tries to loosen a pipe fitting by standing on a "cheater." (b) Vector diagram to find the torque about O.

How great a downward force would the weekend plumber in Example 10.1 have to exert to produce the same torque without the "cheater"? The wrench by itself is 25 cm long.

10.2 | Torque and Angular Acceleration for a Rigid Body

We are now ready to develop the fundamental relation for the rotational dynamics of a rigid body. We will show that the angular acceleration of a rotating rigid body is directly proportional to the sum of the torque components along the axis of rotation. The proportionality factor is the moment of inertia.

To develop this relation, we again imagine the body as being made up of a large number of particles. We choose the axis of rotation to be the z-axis; the first particle has mass m_1 and distance r_1 from this axis (Fig. 10.6). The *net force* acting on this particle has a component $F_{1,\text{rad}}$ along the radial direction, a component $F_{1,\text{tan}}$ that is tangent to the circle of radius r_1 in which the particle moves as the body rotates, and a component F_{1y} along the axis of rotation. Newton's second law for the tangential component is

$$F_{1,\text{tan}} = m_1 a_{1,\text{tan}} \tag{10.4}$$

We can express the tangential acceleration of the first particle in terms of the angular acceleration α_z, using Eq. (9.14): $a_{1,\text{tan}} = r_1 \alpha_z$. Using this relation and multiplying both sides of Eq. (10.4) by r_1, we obtain

$$F_{1,\text{tan}} r_1 = m_1 r_1{}^2 \alpha_z \tag{10.5}$$

From Eq. (10.2), $F_{1,\text{tan}} r_1$ is just the *torque* of the net force with respect to the rotation axis, equal to the component τ_{1z} of the torque vector along the rotation axis. The subscript z is a reminder that the torque affects rotation around the z-axis, in the same way that the subscript on F_{1z} is a reminder that this force affects the motion of particle 1 along the z-axis.

Neither of the components $F_{1,\text{rad}}$ or F_{1z} contributes to the torque about the z-axis, since neither tends to change the particle's rotation about that axis. So $\tau_{1z} = F_{1,\text{tan}} r_1$ is the total torque acting on the particle with respect to the rotation axis. Also, $m_1 r_1{}^2$ is I_1, the moment of inertia of the particle about the rotation axis. With this in mind, we rewrite Eq. (10.5) as

$$\tau_{1z} = I_1 \alpha_z = m_1 r_1{}^2 \alpha_z$$

We write an equation like this for every particle in the body and then add all these equations:

$$\tau_{1z} + \tau_{2z} + \cdots = I_1 \alpha_z + I_2 \alpha_z + \cdots = m_1 r_1{}^2 \alpha_z + m_2 r_2{}^2 \alpha_z + \cdots$$

or

$$\sum \tau_{iz} = \left(\sum m_i r_i{}^2 \right) \alpha_z$$

The left side of this equation is the sum of all the torques about the rotation axis that act on all the particles. The right side is $I = \sum m_i r_i{}^2$, the total moment of inertia about the rotation axis, multiplied by the angular acceleration α_z. Note that α_z

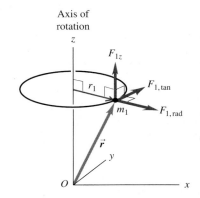

10.6 Three components of the net force acting on one of the particles of a rigid body. Only $F_{1,\text{tan}}$ has a z-component of torque about O.

10.7 Loosening or tightening a screw requires giving it an angular acceleration and hence applying a torque. This is made easier by using a screwdriver with a large-radius handle, which provides a large lever arm for the force you apply with your hand.

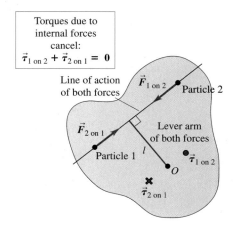

Torques due to
internal forces
cancel:
$$\vec{\tau}_{1 \text{ on } 2} + \vec{\tau}_{2 \text{ on } 1} = 0$$

10.8 Two particles in a rigid body exert equal and opposite forces on each other. If the forces act along the line joining the particles, the lever arms of the forces are the same and the torques due to the two forces are equal and opposite. Only *external* torques affect the body's rotation.

Act|v
Phys|cs ONLINE

7.8　Rotoride—Dynamics Approach

7.9　Falling Ladder

7.10　Woman and Flywheel Elevator—Dynamics Approach

is the same for every particle because this is a *rigid* body. Thus for the entire rigid body we have the *rotational analog of Newton's second law:*

$$\sum \tau_z = I\alpha_z \qquad (10.6)$$

(rotational analog of Newton's second law for a rigid body)

Just as Newton's second law says that the net force on a particle equals the particle's mass times its acceleration, Eq. (10.6) says that the net torque on a rigid body equals the body's moment of inertia about the rotation axis times its angular acceleration (Fig. 10.7).

We emphasize that Eq. (10.6) is valid *only* for *rigid* bodies. If the body is not rigid, like a rotating tank of water or a swirling tornado of air, the angular acceleration α_z is different for different particles in the body, and the derivation of Eq. (10.6) isn't valid. Also note that since our derivation used Eq. (9.14), $a_{\text{tan}} = r\alpha_z$, α_z must be measured in rad/s².

The torque on each particle is due to the net force on that particle, which is the vector sum of external and internal forces (defined in Section 8.2). According to Newton's third law, the *internal* forces that any pair of particles in the rigid body exert on each other are equal and opposite (Fig. 10.8). If these forces act along the line joining the two particles, their lever arms with respect to any axis are also equal. So the torques for each such pair are equal and opposite, and add to zero. Indeed, *all* the internal torques add to zero, so the sum $\sum \tau_z$ in Eq. (10.6) includes only the torques of the *external* forces.

Often, one of the important external forces acting on a body is its *weight*. This force is not concentrated at a single point; it acts on every particle in the entire body. Nevertheless, it turns out that if the value of $\vec{g}$ is the same at all points, we always get the correct torque (about any specified axis) if we assume that all the weight is concentrated at the *center of mass* of the body. We will prove this statement in Chapter 11, but meanwhile we will use it for some of the problems in this chapter.

Problem-Solving Strategy　　**Rotational Dynamics for Rigid Bodies**

Our strategy for solving problems in rotational dynamics is very similar to the strategy presented in Section 5.1 for solving problems that involve Newton's second law.

IDENTIFY *the relevant concepts:* The equation $\sum \tau_z = I\alpha_z$ is useful whenever torques act on a rigid body—that is, whenever forces act on a rigid body in such a way as to change the state of the body's rotation.

In some cases you may be able to use an energy approach instead, as we did in Section 9.4. However, if the target variable is a force, a torque, an acceleration, an angular acceleration, or an elapsed time, the approach using $\sum \tau_z = I\alpha_z$ is almost always the most efficient one.

SET UP *the problem using the following steps:*

1. Draw a sketch of the situation and select the body or bodies to be analyzed.
2. For each body, draw a free-body diagram isolating the body and including all the forces (and *only* those forces) that act on the body, including its weight. Label unknown quantities with algebraic symbols. A new consideration is that you must show the *shape* of the body accurately, including all dimensions and angles you will need for torque calculations.
3. Choose coordinate axes for each body and indicate a positive sense of rotation for each rotating body. If there is a linear acceleration, it's usually simplest to pick a positive axis in its direction. If you know the sense of α_z in advance, picking it as the positive sense of rotation simplifies the calculations. When you represent a force in

terms of its components, cross out the original force to avoid including it twice.

EXECUTE *the solution as follows:*

1. For each body in the problem, decide whether it undergoes translational motion, rotational motion, or both. Depending on the behavior of the body in question, apply $\Sigma \vec{F} = m\vec{a}$ (as in Section 5.1), $\Sigma \tau_z = I\alpha_z$, or both to the body. Be careful to write separate equations of motion for each body.

2. There may be *geometrical* relations between the motions of two or more bodies, as with a string that unwinds from a pulley while turning it or a wheel that rolls without slipping (to be discussed in Section 10.3). Express these relations in algebraic form, usually as relations between two

linear accelerations or between a linear acceleration and an angular acceleration.

3. Check that the number of equations matches the number of unknown quantities. Then solve the equations to find the target variable(s).

EVALUATE *your answer:* Check that the algebraic signs of your results make sense. As an example, suppose the problem is about a spool of thread. If you are pulling thread off the spool, your answers should *not* tell you that the spool is turning in the direction that rolls the thread back on the spool! Whenever possible, check the results for special cases or extreme values of quantities and compare them with your intuitive expectations. Ask yourself: "Does this result make sense?"

| Example 10.2 | **An unwinding cable** |

Figure 10.9a shows the same situation that we analyzed in Example 9.8 (Section 9.4) using energy methods. A cable is wrapped several times around a uniform solid cylinder that can rotate about its axis. The cylinder has diameter 0.120 m and mass 50 kg. The cable is pulled with a force of 9.0 N. Assuming that the cable unwinds without stretching or slipping, what is its acceleration?

SOLUTION

IDENTIFY: Our target variable is the acceleration of the cable, which we cannot find directly using the energy method of Section 9.4 (which does not involve acceleration). Instead, we'll apply rotational dynamics to the cylinder. To obtain the acceleration of the cable, we'll find a relationship between the motion of the cable and the motion of the rim of the cylinder.

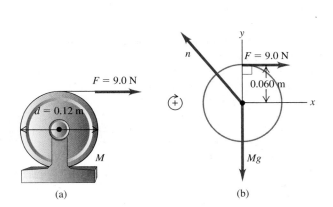

10.9 (a) Cylinder and cable. (b) Free-body diagram for the cylinder.

SET UP: Figure 10.9b shows the free-body diagram for the cylinder of mass $M = 50$ kg. The cylinder rotates clockwise when the cable is pulled, so we take the clockwise rotation to be positive. The net force on the cylinder must be zero because its center of mass remains at rest. Note that the weight (magnitude Mg) and the normal force (magnitude n) exerted by the cylinder bearings act along lines through the rotation axis. Hence these forces produce no torque with respect to that axis.

EXECUTE: The only torque about the rotation axis is due to the force F, which has a lever arm equal to the radius R of the cylinder: $l = R = 0.060$ m, so $\tau_z = FR$. (This torque is positive as it tends to cause a clockwise rotation.) From Example 9.8, the moment of inertia of the cylinder about the rotation axis is $I = \frac{1}{2}MR^2$. Hence Eq. (10.6) gives us the *angular* acceleration of the cylinder:

$$\alpha_z = \frac{\tau_z}{I} = \frac{FR}{MR^2/2} = \frac{2F}{MR} = \frac{2(9.0 \text{ N})}{(50 \text{ kg})(0.060 \text{ m})} = 6.0 \text{ rad/s}^2$$

(Be certain to check that these units are correct. We can add the "rad" to our result because a radian is a dimensionless quantity.)

To get the *linear* acceleration of the cable, we need a kinematic relation. We remarked in Section 9.3 that the acceleration of a cable unwinding from a cylinder is the same as the tangential component of acceleration of a point on the surface of the cylinder where the cable is tangent to it. This tangential acceleration is given by Eq. (9.14):

$$a_x = R\alpha = (0.060 \text{ m})(6.0 \text{ rad/s}^2) = 0.36 \text{ m/s}^2$$

EVALUATE: Can you use this result, together with an equation from Chapter 2, to determine the speed of the cable after it has been pulled 2.0 m? Try it, and compare your result with Example 9.8, in which we found this speed using work and energy considerations.

Example
10.3

An unwinding cable II

Figure 10.10a shows the same situation that we analyzed in Example 9.9 (Section 9.4) using energy methods. Find the acceleration of the object of mass m.

SOLUTION

IDENTIFY: We'll apply translational dynamics to the hanging object and rotational dynamics to the cylinder. Because the cable doesn't slip on the cylinder, there is a relationship between the linear acceleration of the hanging object (our target variable) and the angular acceleration of the cylinder.

SET UP: We have to treat the two bodies separately. Figure 10.10b shows a free-body diagram for each body. We take the positive sense of rotation for the cylinder to be counterclockwise and the positive direction of the y-coordinate for the object to be downward.

EXECUTE: For the object, Newton's second law gives

$$\sum F_y = mg + (-T) = ma_y$$

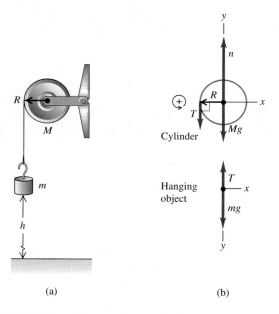

10.10 (a) Cylinder, object, and cable. (b) Free-body diagrams for the cylinder and the hanging object. The cable is assumed to have negligible mass.

For the cylinder, the weight Mg and the normal force n (exerted by the bearing) have no torques with respect to the rotation axis because they act along lines through that axis, just as in Example 10.2. The only torque is that due to the cable tension T. Applying Eq. (10.6) to the cylinder gives

$$\sum \tau_z = RT = I\alpha_z = \frac{1}{2}MR^2\alpha_z$$

As in Example 10.2, the acceleration of the cable is the same as the tangential acceleration of a point on the cylinder rim. According to Eq. (9.14), this acceleration is given by $a_y = a_{\text{tan}} = R\alpha_z$. We use this to replace $(R\alpha_z)$ with a_y in the cylinder equation above, and then divide by R; the result is

$$T = \frac{1}{2}Ma_y$$

Now we substitute this expression for T into Newton's second law for the object and solve for the acceleration a_y:

$$mg - \frac{1}{2}Ma_y = ma_y$$

$$a_y = \frac{g}{1 + M/2m}$$

EVALUATE: The acceleration is positive (in the downward direction) and less than g, as it should be since the cable is holding the object back. To see how much force the cable exerts, substitute our expression for a_y back into Newton's second law for the object to find T:

$$T = mg - ma_y = mg - m\left(\frac{g}{1 + M/2m}\right) = \frac{mg}{1 + 2m/M}$$

The tension in the cable is *not* equal to the weight mg of the object; if it were, the object could not accelerate.

Let's check some particular cases. When M is much larger than m, the tension is nearly equal to mg, and the acceleration is correspondingly much less than g. When M is zero, $T = 0$ and $a_y = g$; the object then falls freely. If the object starts from a height h above the floor with an initial speed v_0, its speed v when it strikes the ground is given by $v^2 = v_0^2 + 2a_yh$. If it starts from rest, $v_0 = 0$ and

$$v = \sqrt{2a_yh} = \sqrt{\frac{2gh}{1 + M/2m}}$$

This is the same result we obtained from energy considerations in Example 9.9.

Example
10.4

Two masses and a rotating pulley

In Fig. 10.11a a glider of mass m_1 slides without friction on a horizontal air track. It is attached to an object of mass m_2 by a massless string. The pulley is a thin cylindrical shell (with massless spokes) with mass M and radius R, and the string turns the pulley without slipping or stretching. Find the acceleration of each body, the angular acceleration of the pulley, and the tension in each part of the string.

SOLUTION

IDENTIFY: We will use translational dynamics to describe the motion of the glider and hanging object, and rotational dynamics to describe the motion of the pulley. Because the string doesn't stretch, both the glider and the object have the same magnitude of acceleration; the rim of the pulley has the same magnitude of *tangential* acceleration because the string doesn't slip.

SET UP: Figure 10.11b shows the free-body diagrams and coordinate systems for the three bodies. With our choices of coordinates the glider and object accelerate in their positive x- and y-directions, respectively. We likewise choose the positive sense of rotation to be clockwise (the same sense as the pulley's actual angular acceleration). We have five target variables: the acceleration of the glider (a_{1x}), the acceleration of the object (a_{2y}), the pulley's angular acceleration α_z, and the two tensions (T_1 and T_2). This may sound like an awesome problem, but we will have as many equations as we have unknowns, and solving them will be a lot easier than you might imagine.

CAUTION We considered a similar situation in Example 5.13 (Section 5.2). In that example the string slid without friction over a fixed pulley, and the tension was the same throughout the massless string. But with a rotating pulley, and with friction between the pulley and string to prevent slipping, the two tensions T_1 and T_2 *cannot* be equal. If they were, the pulley could not have an angular acceleration. Labeling the tension in both parts of the string as simply T would be a serious error. Be on your guard against this error in any problem that involves a rotating pulley.

EXECUTE: The equations of motion for the glider and the object are

$$\text{Glider:} \quad \sum F_x = T_1 = m_1 a_{1x} \quad (10.7)$$

$$\text{Object:} \quad \sum F_y = m_2 g + (-T_2) = m_2 a_{2y} \quad (10.8)$$

The unknown normal force n_2 acts on a line through the pulley's axis of rotation, so it has no lever arm or torque with respect to that axis. From Table 9.2 the moment of inertia of the pulley about this axis is $I = MR^2$. Then the pulley's equation of motion is

$$\text{Pulley:} \quad \sum \tau_z = T_2 R + (-T_1 R) = I\alpha_z = (MR^2)\alpha_z \quad (10.9)$$

Because the string does not stretch or slip, we have the additional *kinematic* relations

$$a_{1x} = a_{2y} = R\alpha_z \quad (10.10)$$

(The accelerations of the glider and the object have different directions but the same magnitude.)

Equations (10.7) through (10.10) are five simultaneous equations for the five unknowns a_{1x}, a_{2y}, α, T_1, and T_2. (Equation (10.10) is actually two equations.) First we use Eqs. (10.10) to eliminate a_{2y}

and α_z from Eqs. (10.7) through (10.9). We then have three equations for the three unknowns T_1, T_2, and a_{1x}:

$$T_1 = m_1 a_{1x}$$
$$m_2 g - T_2 = m_2 a_{1x}$$
$$T_2 - T_1 = Ma_{1x}$$

The easiest way to solve these is simply to *add* the three equations, which eliminates T_1 and T_2, and then solve for a_{1x}:

$$a_{1x} = \frac{m_2 g}{m_1 + m_2 + M}$$

From Eq. (10.10), the acceleration a_{2y} of the hanging object is equal to a_{1x} and the angular acceleration α_z of the pulley is equal to a_{1x} divided by R. We can then substitute a_{1x} back into Eqs. (10.7) and (10.8) to find the tensions. The results are

$$T_1 = \frac{m_1 m_2 g}{m_1 + m_2 + M} \qquad T_2 = \frac{(m_1 + M)m_2 g}{m_1 + m_2 + M}$$

EVALUATE: Let's check some special cases to see whether these results make sense. First, if either m_1 or M is much larger than m_2, the accelerations are very small and T_2 is approximately $m_2 g$. But if m_2 is much *larger* than either m_1 or M, the acceleration is approximately g. Both of these results are what we should expect. When $M = 0$, do we get the same result as Example 5.13 (Section 5.2)? Why or why not? Can you think of other special cases to check?

10.11 (a) An air-track glider pulled by an object hanging over a pulley. (b) Free-body diagrams for all three bodies.

10.12 The motion of a rigid body like this thrown hammer is a combination of translational motion of the center of mass and rotation around the center of mass.

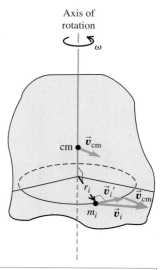

Velocity $\vec{v}_i$ of particle in rotating, translating rigid body = (velocity $\vec{v}_{cm}$ of center of mass) plus (particle's velocity $\vec{v}_i'$ relative to center of mass)

10.13 A rigid body with both translation and rotation.

Test Your Understanding

Suppose the system in Example 10.4 is initially in motion, with the glider moving to the left, the hanging object rising, and the pulley rotating counterclockwise. In this situation, what are the linear accelerations of the glider and the object and the angular acceleration of the pulley?

10.3 | Rigid-Body Rotation About a Moving Axis

We can extend our analysis of the dynamics of rotational motion to some cases in which the axis of rotation moves. When that happens, the motion of the body is **combined translation and rotation.** The key to understanding such situations is this: every possible motion of a rigid body can be represented as a combination of *translational motion of the center of mass* and *rotation about an axis through the center of mass.* This is true even when the center of mass accelerates, so that it is not at rest in any inertial frame. A graphic example is the motion of a hammer thrown upward (Fig. 10.12). The center of mass of the hammer follows a parabolic curve, as though the hammer were a particle located at the center of mass. At the same time, the hammer rotates with constant angular velocity about the center of mass (compare the motion of the wrench in Fig. 8.25). The translation of the center of mass and the rotation about the center of mass can be treated as separate but related problems. Other examples of such motion are a ball rolling down a hill and a yo-yo unwinding at the end of a string.

Combined Translation and Rotation: Energy Relations

It's beyond the scope of this book to prove that the motion of a rigid body can always be divided into translation of the center of mass and rotation about the center of mass. But we can show that this is true for the *kinetic energy* of a rigid body that has both translational and rotational motion. In this case, the body's kinetic energy is the sum of a part $\frac{1}{2}Mv_{cm}^2$ associated with motion of the center of mass and a part $\frac{1}{2}I_{cm}\omega^2$ associated with rotation about an axis through the center of mass:

$$K = \frac{1}{2}Mv_{cm}^2 + \frac{1}{2}I_{cm}\omega^2$$

$$\text{(rigid body with both translation and rotation)} \qquad (10.11)$$

To prove this relation, we again imagine the rigid body to be made up of particles. Consider a typical particle with mass m_i as shown in Fig. 10.13. The velocity $\vec{v}_i$ of this particle relative to an inertial frame is the vector sum of the velocity $\vec{v}_{cm}$ of the center of mass and the velocity $\vec{v}_i'$ of the particle *relative to* the center of mass:

$$\vec{v}_i = \vec{v}_{cm} + \vec{v}_i' \qquad (10.12)$$

The kinetic energy K_i of this particle in the inertial frame is $\frac{1}{2}m_iv_i^2$, which we can also express as $\frac{1}{2}m_i(\vec{v}_i \cdot \vec{v}_i)$. Substituting Eq. (10.12) into this, we get

$$K_i = \frac{1}{2}m_i(\vec{v}_{cm} + \vec{v}_i') \cdot (\vec{v}_{cm} + \vec{v}_i')$$
$$= \frac{1}{2}m_i(\vec{v}_{cm} \cdot \vec{v}_{cm} + 2\vec{v}_{cm} \cdot \vec{v}_i' + \vec{v}_i' \cdot \vec{v}_i')$$
$$= \frac{1}{2}m_i(v_{cm}^2 + 2\vec{v}_{cm} \cdot \vec{v}_i' + v_i'^2)$$

The total kinetic energy is the sum $\sum K_i$ for all the particles making up the body. Expressing the three terms in this equation as separate sums, we get

$$K = \sum K_i = \sum \left(\frac{1}{2} m_i v_{\text{cm}}^2\right) + \sum (m_i \vec{v}_{\text{cm}} \cdot \vec{v}_i') + \sum \left(\frac{1}{2} m_i v_i'^2\right)$$

The first and second terms have common factors that can be taken outside the sum:

$$K = \frac{1}{2}\left(\sum m_i\right) v_{\text{cm}}^2 + \vec{v}_{\text{cm}} \cdot \left(\sum m_i \vec{v}_i'\right) + \sum \left(\frac{1}{2} m_i v_i'^2\right) \qquad (10.13)$$

Now comes the reward for our effort. In the first term, $\sum m_i$ is the total mass M. The second term is zero because $\sum m_i \vec{v}_i'$ is M times the velocity of the center of mass *relative to the center of mass,* and this is zero by definition. The last term is the sum of kinetic energies of the particles computed by using their speeds with respect to the center of mass; this is just the kinetic energy of rotation around the center of mass. Using the same steps that led to Eq. (9.17) for the rotational kinetic energy of a rigid body, we can write this last term as $\frac{1}{2} I_{\text{cm}} \omega^2$, where I_{cm} is the moment of inertia with respect to the axis through the center of mass and ω is the angular speed. So Eq. (10.13) becomes Eq. (10.11):

$$K = \frac{1}{2} M v_{\text{cm}}^2 + \frac{1}{2} I_{\text{cm}} \omega^2$$

An important case of combined translation and rotation is **rolling without slipping,** such as the motion of the wheel shown in Fig. 10.14. The wheel is symmetrical, so its center of mass is at its geometric center. We view the motion in an inertial frame of reference in which the surface on which the wheel rolls is at rest. In this frame, the point on the wheel that contacts the surface must be instantaneously *at rest* so that it does not slip. Hence the velocity $\vec{v}_1'$ of the point of contact relative to the center of mass must have the same magnitude but opposite direction as the center-of-mass velocity $\vec{v}_{\text{cm}}$. If the radius of the wheel is R and its angular speed about the center of mass is ω, then the magnitude of $\vec{v}_1'$ is $R\omega$; hence we must have

$$v_{\text{cm}} = R\omega \qquad (\text{condition for rolling without slipping}) \qquad (10.14)$$

As Fig. 10.14 shows, the velocity of a point on the wheel is the vector sum of the velocity of the center of mass and the velocity of the point relative to the center of mass. Thus while point 1, the point of contact, is instantaneously at rest,

Act|v
ONLINE
Physics

7.11 Race Between a Block and a Disk

Wheel as a whole translates with velocity $\vec{v}_{\text{cm}}$ Wheel rotates around center of mass, speed at rim $= v_{\text{cm}}$ Rolling without slipping

10.14 The motion of a rolling wheel is the sum of the translational motion of the center of mass plus the rotational motion of the wheel around the center of mass. For rolling without slipping, the speed of the rim relative to the center of mass must be equal to the magnitude of $\vec{v}_{\text{cm}}$. Note that the wheel is instantaneously at rest at the point where it contacts the ground.

10.15 The smoke rising from this drag racer's rear tires shows that the tires are slipping on the road, so v_{cm} is *not* equal to $R\omega$.

point 3 at the top of the wheel is moving forward *twice as fast* as the center of mass, and points 2 and 4 at the sides have velocities at 45° to the horizontal.

At any instant we can think of the wheel as rotating about an "instantaneous axis" of rotation that passes through the point of contact with the ground. The angular velocity ω is the same for this axis as for an axis through the center of mass; an observer at the center of mass sees the rim make the same number of revolutions per second as does an observer at the rim watching the center of mass spin around him. If we think of the motion of the rolling wheel in Fig. 10.14 in this way, the kinetic energy of the wheel is $K = \frac{1}{2} I_1 \omega^2$, where I_1 is the moment of inertia of the wheel about an axis through point 1. But by the parallel-axis theorem, Eq. (9.19), $I_1 = I_{cm} + MR^2$, where M is the total mass of the wheel and I_{cm} is the moment of inertia with respect to an axis through the center of mass. Using Eq. (10.14), the kinetic energy of the wheel is

$$K = \frac{1}{2} I_1 \omega^2 = \frac{1}{2} I_{cm}\omega^2 + \frac{1}{2} MR^2\omega^2 = \frac{1}{2} I_{cm}\omega^2 + \frac{1}{2} M v_{cm}^2$$

which is the same as Eq. (10.11).

CAUTION It's important to note that the relationship $v_{cm} = R\omega$ holds only if there is rolling without slipping. When a drag racer first starts to move, the rear tires are spinning very fast while the racer is hardly moving, so $R\omega$ is greater than v_{cm} (Fig. 10.15). If a driver applies the brakes too heavily so that the car skids, the tires will spin hardly at all and $R\omega$ is less than v_{cm}.

If a rigid body changes height as it moves, we must also consider gravitational potential energy. As we discussed in Section 9.4, the gravitational potential energy associated with any extended body of mass M, rigid or not, is the same as if we replace the body by a particle of mass M located at the body's center of mass. That is,

$$U = Mgy_{cm}$$

Example 10.5	**A rolling cylindrical shell**

A hollow cylindrical shell with mass M and radius R rolls without slipping with speed v_{cm} on a flat surface. What is its kinetic energy?

SOLUTION

IDENTIFY and SET UP: We'll use Eq. (10.11) to find the kinetic energy. The moment of inertia is $I = MR^2$ from Table 9.2, and the angular speed is $\omega = v_{cm}/R$ since there is rolling without slipping.

EXECUTE: Substituting these expressions into Eq. (10.11), we get

$$K = \frac{1}{2} M v_{cm}^2 + \frac{1}{2}(MR^2)\left(\frac{v_{cm}}{R}\right)^2$$
$$= M v_{cm}^2$$

EVALUATE: The kinetic energy is twice as great as it would be if the cylindrical shell were sliding at speed v_{cm} without rolling. Half of the total kinetic energy is translational and half is rotational.

Example 10.6	**Speed of a primitive yo-yo**

A primitive yo-yo is made by wrapping a string several times around a solid cylinder with mass M and radius R (Fig. 10.16). You hold the end of the string stationary while releasing the cylinder with no initial motion. The string unwinds but does not slip or stretch as the cylinder drops and rotates. Use energy considerations to find the speed v_{cm} of the center of mass of the solid cylinder after it has dropped a distance h.

10.16 Calculating the speed of a primitive yo-yo.

SOLUTION

IDENTIFY: The upper end of the string is held fixed, not pulled upward, so the hand in Fig. 10.16 does no work on the system of string and cylinder. As in Example 9.8 (Section 9.4), there is friction between the string and the cylinder, but because the string never slips on the surface of the cylinder, no mechanical energy is lost. Thus we can use conservation of mechanical energy.

SET UP: The potential energies are $U_1 = Mgh$ and $U_2 = 0$. The string has no kinetic energy because it's massless. The initial kinetic energy of the cylinder is $K_1 = 0$, and its final kinetic energy K_2 is given by Eq. (10.11). The moment of inertia is $I = \frac{1}{2}MR^2$, and $\omega = v_{cm}/R$ because the cylinder does not slip on the string.

EXECUTE: Using Eq. (10.11), the kinetic energy at point 2 is

$$K_2 = \frac{1}{2}Mv_{cm}^2 + \frac{1}{2}\left(\frac{1}{2}MR^2\right)\left(\frac{v_{cm}}{R}\right)^2$$

$$= \frac{3}{4}Mv_{cm}^2$$

Then, conservation of energy gives

$$K_1 + U_1 = K_2 + U_2$$

$$0 + Mgh = \frac{3}{4}Mv_{cm}^2 + 0$$

and

$$v_{cm} = \sqrt{\frac{4}{3}gh}$$

EVALUATE: This is less than the speed $\sqrt{2gh}$ that a dropped object would have, because one-third of the potential energy released as the cylinder falls appears as rotational kinetic energy.

Example 10.7	**Race of the rolling bodies**

In a physics lecture demonstration, an instructor "races" various round rigid bodies by releasing them from rest at the top of an inclined plane (Fig. 10.17). What shape should a body have to reach the bottom of the incline first?

SOLUTION

IDENTIFY: We can again use conservation of energy because there is no sliding of the rigid bodies over the inclined plane. No work is done by kinetic friction if the bodies roll without slipping. We can also ignore the effects of *rolling friction,* introduced in Section 5.3, provided that the bodies and the surface on which they roll are perfectly rigid. (Later in this section we'll explain why this is so.)

10.17 Which body rolls down the incline fastest, and why?

SET UP: Each body starts from rest at the top of an incline with height h, so $K_1 = 0$, $U_1 = Mgh$, and $U_2 = 0$. The kinetic energy at the bottom of the incline is given by Eq. (10.11). If the bodies roll without slipping, $\omega = v_{cm}/R$. The moments of inertia of all the round bodies in Table 9.2 (about axes through their centers of mass) can be expressed as $I_{cm} = cMR^2$, where c is a pure number less than or equal to 1 that depends on the shape of the body. Our goal is to find the value of c that gives the body the greatest speed at the bottom of the incline.

EXECUTE: From conservation of energy,

$$K_1 + U_1 = K_2 + U_2$$

$$0 + Mgh = \frac{1}{2}Mv_{cm}^2 + \frac{1}{2}cMR^2\left(\frac{v_{cm}}{R}\right)^2$$

$$= \frac{1}{2}(1 + c)Mv_{cm}^2$$

so the speed at the bottom of the incline is

$$v_{cm} = \sqrt{\frac{2gh}{1 + c}}$$

EVALUATE: This is a fairly amazing result; the speed doesn't depend on either the mass M of the body or its radius R. All uniform solid cylinders have the same speed at the bottom, even if their masses and radii are different, because they have the same c. All solid spheres have the same speed, and so on. The smaller the value of c, the faster the body is moving at the bottom (and at any point

on the way down). Small-c bodies always beat large-c bodies, because they have less of their kinetic energy tied up in rotation and have more available for translation. Reading the values of c from Table 9.2, we see that the order of finish is as follows: any solid sphere, any solid cylinder, any thin-walled hollow sphere, and any thin-walled hollow cylinder.

Combined Translation and Rotation: Dynamics

We can also analyze the combined translational and rotational motion of a rigid body from the standpoint of dynamics. We showed in Section 8.5 that for a body with total mass M, the acceleration $\vec{a}_{cm}$ of the center of mass is the same as that of a point mass M acted on by all the external forces on the actual body:

$$\sum \vec{F}_{ext} = M\vec{a}_{cm} \qquad (10.15)$$

The rotational motion about the center of mass is described by the rotational analog of Newton's second law, Eq. (10.6):

$$\sum \tau_z = I_{cm}\alpha_z \qquad (10.16)$$

where I_{cm} is the moment of inertia with respect to an axis through the center of mass and the sum $\sum \tau_z$ includes all external torques with respect to this axis. It's not immediately obvious that Eq. (10.16) should apply to the motion of a translating rigid body; after all, our derivation of $\sum \tau_z = I\alpha_z$ in Section 10.2 assumed that the axis of rotation was stationary. But in fact, Eq. (10.16) is valid *even when the axis of rotation moves,* provided the following two conditions are met:

1. The axis through the center of mass must be an axis of symmetry.
2. The axis must not change direction.

These conditions are satisfied for many types of rotation (Fig. 10.18). Note that in general this moving axis of rotation is *not* at rest in an inertial frame of reference.

We can now solve dynamics problems involving a rigid body that undergoes translational and rotational motion at the same time, provided that the rotation axis satisfies the two conditions just mentioned. The problem-solving strategy outlined in Section 10.2 is equally useful here, and you should review it now. Keep in mind that when a body undergoes translational and rotational motion at the same time, we need two separate equations of motion *for the same body.* One of these, Eq. (10.15), describes the translational motion of the center of mass. The other equation of motion, Eq. (10.16), describes the rotational motion about the axis through the center of mass.

10.18 The axle of a bicycle wheel passes through the wheel's center of mass and is an axis of symmetry. Hence the rotation of the wheel is described by Eq. (10.16), provided the bicycle doesn't turn or tilt to one side (which would change the orientation of the axle).

Example 10.8	Acceleration of a primitive yo-yo

For the primitive yo-yo in Example 10.6, find the downward acceleration of the cylinder and the tension in the string.

SOLUTION

IDENTIFY and SET UP: We'll use Eqs. (10.15) and (10.16), along with the condition that the string does not slip on the cylinder. Figure 10.19 shows a free-body diagram for the yo-yo, including the

choice of positive coordinate directions. With these coordinates, our target variable is $a_{cm\text{-}y}$.

EXECUTE: The equation for the translational motion of the center of mass is

$$\sum F_y = Mg + (-T) = Ma_{cm\text{-}y} \qquad (10.17)$$

10.19 Free-body diagram for a primitive yo-yo (see Fig. 10.16).

The moment of inertia for an axis through the center of mass is $I_{cm} = \frac{1}{2}MR^2$. Only the tension force has a torque with respect to the axis through the center of mass, so the equation for rotational motion about this axis is

$$\sum \tau_z = TR = I_{cm}\alpha_z = \frac{1}{2}MR^2\alpha_z \qquad (10.18)$$

The string unwinds without slipping, so $v_{cm} = R\omega$ from Eq. (10.14); the derivative of this relation with respect to time is

$$a_{cm\text{-}y} = R\alpha_z \qquad (10.19)$$

We now use Eq. (10.19) to eliminate α_z from Eq. (10.18) and then solve Eqs. (10.17) and (10.18) simultaneously for T and $a_{cm\text{-}y}$. The results are amazingly simple:

$$a_{cm\text{-}y} = \frac{2}{3}g \qquad T = \frac{1}{3}Mg$$

Using the constant-acceleration formula $v_{cm\text{-}y}^2 = v_{cm\text{-}0y}^2 + 2a_{cm\text{-}y}h$, you can show that the speed of the yo-yo after it has fallen a distance h is $v_{cm} = \sqrt{\frac{4}{3}gh}$, just as we found in Example 10.6.

EVALUATE: From the standpoint of dynamics, the tension force is essential; it causes the yo-yo's acceleration to be less than g, and its torque is what causes the yo-yo to turn. Yet when we analyzed this situation using energy methods in Example 10.6, we didn't have to consider the tension force at all! Because no mechanical energy was lost or gained, from the energy standpoint the string is significant only as a way of helping to convert some of the gravitational potential energy into rotational kinetic energy.

Example 10.9	**Acceleration of a rolling sphere**

A solid bowling ball rolls without slipping down the return ramp at the side of the alley (Fig. 10.20a). The ramp is inclined at an angle β to the horizontal. What is the ball's acceleration? Treat the ball as a uniform solid sphere, ignoring the finger holes.

SOLUTION:

IDENTIFY: As in Example 10.8, we use Eq. (10.15) to describe the translational motion and Eq. (10.16) to describe the rotational motion. Our target variable is the acceleration of the ball's center of mass.

SET UP: Figure 10.20b shows a free-body diagram, with positive coordinate directions indicated. From Table 9.2 the moment of inertia of a solid sphere is $I_{cm} = \frac{2}{5}MR^2$. The equations of motion for translation and for rotation about the axis through the center of mass, respectively, are

$$\sum F_x = Mg \sin \beta + (-f) = Ma_{cm\text{-}x} \qquad (10.20)$$

$$\sum \tau_z = fR = I_{cm}\alpha_z = \left(\frac{2}{5}MR^2\right)\alpha_z \qquad (10.21)$$

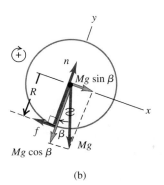

(a)   (b)

10.20 (a) A bowling ball rolling down a ramp. (b) Free-body diagram for the bowling ball.

Note that only the friction force f has a torque with respect to the axis through the center of mass. If the ball rolls without slipping, we have the same kinematic relation $a_{\text{cm-}x} = R\alpha_z$ as in Example 10.8. We use this to eliminate α_z from Eq. (10.21):

$$fR = \frac{2}{5}MRa_{\text{cm-}x}$$

This equation and Eq. (10.20) are two equations for two unknowns, $a_{\text{cm-}x}$ and f. We solve Eq. (10.20) for f, substitute the expression into the above equation to eliminate f, and then solve for $a_{\text{cm-}x}$ to obtain

$$a_{\text{cm-}x} = \frac{5}{7}g \sin \beta$$

The acceleration is just $\frac{5}{7}$ as large as it would be if the ball could *slide* without friction down the slope, like the toboggan in Example 5.9 (Section 5.2). Finally, we substitute this back into Eq. (10.20) and solve for f:

$$f = \frac{2}{7}Mg \sin \beta$$

EVALUATE: Because the ball does not slip at the instantaneous point of contact with the ramp, the friction force f is a *static* friction force; it prevents slipping and gives the ball its angular acceleration. We can derive an expression for the minimum coefficient of static friction μ_s needed to prevent slipping. The normal force is $n = Mg \cos \beta$. The maximum force of static friction equals $\mu_s n$, so the coefficient of friction must be at least as great as

$$\mu_s = \frac{f}{n} = \frac{\frac{2}{7}Mg \sin \beta}{Mg \cos \beta} = \frac{2}{7} \tan \beta$$

If the plane is tilted only a little, β is small, and only a small value of μ_s is needed to prevent slipping. But as the angle increases, the required value of μ_s increases, as we might expect intuitively. If the ball begins to slip, Eqs. (10.20) and (10.21) are both still valid, but it's no longer true that $v_{\text{cm-}x} = R\omega_z$ or $a_{\text{cm-}x} = R\alpha_z$; we have only two equations for three unknowns $(a_{\text{cm-}x}, \alpha_z, \text{and } f)$. To solve the problem of rolling *with* slipping requires taking *kinetic* friction into account (see Challenge Problem 10.101).

If the bowling ball descends a vertical distance h as it moves down the ramp, the displacement along the ramp is $h/\sin \beta$. You should be able to show that the speed of the ball at the bottom of the ramp would be $v_{\text{cm}} = \sqrt{\frac{10}{7}gh}$, which is just the result you found in Example 10.7 with $c = \frac{2}{5}$.

If the ball were rolling *uphill*, the force of friction would still be directed uphill as in Fig. 10.20b. Can you see why?

Rolling Friction

In Example 10.7 we said that we can ignore rolling friction if both the rolling body and the surface over which it rolls are perfectly rigid. In Fig. 10.21a a perfectly rigid sphere is rolling down a perfectly rigid incline. The line of action of the normal force passes through the center of the sphere, so its torque is zero; there is no sliding at the point of contact, so the friction force does no work. Fig. 10.21b shows a more realistic situation, in which the surface "piles up" in front of the sphere and the sphere rides in a shallow trench. Because of these deformations, the contact forces on the sphere no longer act along a single point, but over an area; the forces are concentrated on the front of the sphere as shown. As a

10.21 (a) Forces on a perfectly rigid sphere rolling down a perfectly rigid incline. (b) If the sphere or incline is deformable, the contact forces act at different positions. The normal force produces a counterclockwise torque that opposes the clockwise rotation. The deformation is shown greatly exaggerated.

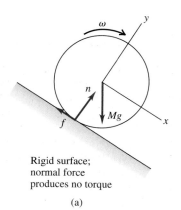

Rigid surface;
normal force
produces no torque

(a)

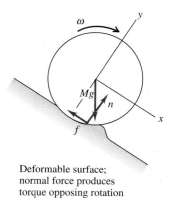

Deformable surface;
normal force produces
torque opposing rotation

(b)

result, the normal force now exerts a torque that opposes the rotation. In addition, there is some sliding of the sphere over the surface due to the deformation, causing mechanical energy to be lost. The combination of these two effects is the phenomenon of *rolling friction*. Rolling friction also occurs if the rolling body is deformable, such as an automobile tire. Often the rolling body and surface are rigid enough that rolling friction can be ignored, as we have assumed in all the examples in this section.

Test Your Understanding

In Example 10.9, what would be the acceleration and the static friction force if the ball were a hollow sphere?

10.4 | Work and Power in Rotational Motion

When you pedal a bicycle, you apply forces to a rotating body and do work on it. Similar things happen in many other real-life situations, such as a rotating motor shaft driving a power tool or a car engine propelling the vehicle. We can express this work in terms of torque and angular displacement.

Suppose a tangential force $\vec{F}_{\text{tan}}$ acts at the rim of a pivoted disk—for example, a child running while pushing on a playground merry-go-round (Fig. 10.22a). The disk rotates through an infinitesimal angle $d\theta$ about a fixed axis during an infinitesimal time interval dt (Fig. 10.22b). The work dW done by the force $\vec{F}_{\text{tan}}$ while a point on the rim moves a distance ds is $dW = F_{\text{tan}}\, ds$. If $d\theta$ is measured in radians, then $ds = R\, d\theta$ and

$$dW = F_{\text{tan}}R\, d\theta$$

Now $F_{\text{tan}}R$ is the *torque* τ_z due to the force $\vec{F}_{\text{tan}}$, so

$$dW = \tau_z\, d\theta \tag{10.22}$$

The total work W done by the torque during an angular displacement from θ_1 to θ_2 is

$$W = \int_{\theta_1}^{\theta_2} \tau_z\, d\theta \quad \text{(work done by a torque)} \tag{10.23}$$

If the torque is *constant* while the angle changes by a finite amount $\Delta\theta = \theta_2 - \theta_1$, then

$$W = \tau_z(\theta_2 - \theta_1) = \tau_z\Delta\theta \quad \text{(work done by a constant torque)} \tag{10.24}$$

The work done by a *constant* torque is the product of torque and the angular displacement. If torque is expressed in newton-meters ($\text{N}\cdot\text{m}$) and angular displacement in radians, the work is in joules. Equation (10.24) is the rotational analog of Eq. (6.1), $W = Fs$, and Eq. (10.23) is the analog of Eq. (6.7), $W = \int F_x\, dx$, for the work done by a force in a straight-line displacement.

If the force in Fig. 10.22 had an axial or radial component, that component would do no work because the displacement of the point of application has only a tangential component. An axial or radial component of force would also make no contribution to the torque about the axis of rotation, so Eqs. (10.23) and (10.24) are correct for any force, no matter what its components.

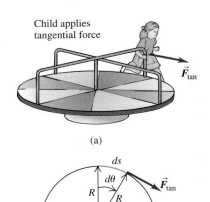

Child applies tangential force

$\vec{F}_{\text{tan}}$

(a)

Overhead view of merry-go-round

(b)

10.22 A tangential force applied to a rotating body does work.

When a torque does work on a rotating rigid body, the kinetic energy changes by an amount equal to the work done. We can prove this by using exactly the same procedure that we used in Eqs. (6.11) through (6.13) for a particle. We first let τ_z represent the *net* torque on the body so that from Eq. (10.6), $\tau_z = I\alpha_z$. By using this equation, we are assuming that the body is rigid so that the moment of inertia I is constant. We then transform the integrand in Eq. (10.23) into an integral on ω_z as follows:

$$\tau_z \, d\theta = (I\alpha_z) \, d\theta = I\frac{d\omega_z}{dt} \, d\theta = I\frac{d\theta}{dt} \, d\omega_z = I\omega_z \, d\omega_z$$

Since τ_z is the net torque, the integral in Eq. (10.23) is the *total* work done on the rotating rigid body. This equation then becomes

$$W_{\text{tot}} = \int_{\omega_1}^{\omega_2} I\omega_z \, d\omega_z = \frac{1}{2}I\omega_2^2 - \frac{1}{2}I\omega_1^2 \qquad (10.25)$$

The change in the rotational kinetic energy of a *rigid* body equals the work done by forces exerted from outside the body. This equation is analogous to Eq. (6.13), the work-energy theorem for a particle.

What about the *power* associated with work done by a torque acting on a rotating body? When we divide both sides of Eq. (10.22) by the time interval dt during which the angular displacement occurs, we find

$$\frac{dW}{dt} = \tau_z \frac{d\theta}{dt}$$

But dW/dt is the rate of doing work, or *power P*, and $d\theta/dt$ is angular velocity ω_z, so

$$P = \tau_z \omega_z \qquad (10.26)$$

When a torque τ_z (with respect to the axis of rotation) acts on a body that rotates with angular velocity ω_z, its power (rate of doing work) is the product of τ_z and ω_z. This is the analog of the relation $P = \vec{F} \cdot \vec{v}$ that we developed in Section 6.4 for particle motion.

| Example 10.10 | **Engine power and torque** |

The power output of an automobile engine is advertised to be 200 hp at 6000 rpm. What is the corresponding torque?

SOLUTION

IDENTIFY and SET UP: We are given the power output P and the angular velocity ω_z, so we can find the torque using Eq. (10.26).

EXECUTE: First we have to convert the power to watts and the angular velocity to rad/s:

$$P = 200 \text{ hp} = 200 \text{ hp}\left(\frac{746 \text{ W}}{1 \text{ hp}}\right) = 1.49 \times 10^5 \text{ W}$$

$$\omega_z = 6000 \text{ rev/min} = \left(\frac{6000 \text{ rev}}{1 \text{ min}}\right)\left(\frac{2\pi \text{ rad}}{1 \text{ rev}}\right)\left(\frac{1 \text{ min}}{60 \text{ s}}\right)$$
$$= 628 \text{ rad/s}$$

From Eq. (10.26),

$$\tau_z = \frac{P}{\omega_z} = \frac{1.49 \times 10^5 \text{ N} \cdot \text{m/s}}{628 \text{ rad/s}} = 237 \text{ N} \cdot \text{m}$$

EVALUATE: You could apply this much torque by using a wrench 0.25 m long and applying a force of 948 N (213 lb) to the end of its handle. Could you do it?

| Example 10.11 | **Calculating power from torque** |

An electric motor exerts a constant torque of 10 N·m on a grindstone mounted on its shaft. The moment of inertia of the grindstone about the shaft is 2.0 kg·m². If the system starts from rest, find the work done by the motor in 8.0 seconds and the kinetic energy at the end of this time. What was the average power delivered by the motor?

SOLUTION

IDENTIFY and SET UP: We use the rotational version of Newton's second law, $\Sigma\tau_z = I\alpha_z$, to find the angular acceleration of the grindstone. Given this we use the kinematics equations from Section 9.2 to find the angle through which the grindstone turns in 8.0 s (which, through Eq. (10.24), tells us the work done) and the angular velocity at that time (which tells us the kinetic energy). We find the average power by dividing the work done by the time interval.

EXECUTE: We have $\Sigma\tau_z = 10$ N·m (the only torque acting is that due to the motor) and $I = 2.0$ kg·m², so from $\Sigma\tau_z = I\alpha_z$ the angular acceleration is 5.0 rad/s². From Eq. (9.11) the total angle through which the system turns in 8.0 s is

$$\Delta\theta = \frac{1}{2}\alpha_z t^2 = \frac{1}{2}(5.0 \text{ rad/s}^2)(8.0 \text{ s})^2 = 160 \text{ rad}$$

and the total work done by the torque is

$$W = \tau_z\Delta\theta = (10 \text{ N}\cdot\text{m})(160 \text{ rad}) = 1600 \text{ J}$$

From Eqs. (9.7) and (9.17), the angular velocity and kinetic energy at $t = 8.0$ s are

$$\omega_z = \alpha_z t = (5.0 \text{ rad/s}^2)(8.0 \text{ s}) = 40 \text{ rad/s}$$

$$K = \frac{1}{2}I\omega_z^2 = \frac{1}{2}(2.0 \text{ kg}\cdot\text{m}^2)(40 \text{ rad/s})^2 = 1600 \text{ J}$$

The initial kinetic energy was zero, so the work done equals the increase in kinetic energy [see Eq. (10.25)].

The average power is

$$P_{av} = \frac{1600 \text{ J}}{8.0 \text{ s}} = 200 \text{ J/s} = 200 \text{ W}$$

EVALUATE: We can check our answer for *average* power by considering the *instantaneous* power $P = \tau_z\omega_z$. Note that because ω_z increases continuously, P increases continuously as well; its value is zero at $t = 0$ and increases to $(10 \text{ N}\cdot\text{m})(40 \text{ rad/s}) = 400$ W at $t = 8.0$ s. The angular velocity and the power increase uniformly with time, so the *average* power is just half this maximum value, or 200 W.

Test Your Understanding

You apply equal torques to two different cylinders, one of which has a moment of inertia twice as large as the other cylinder. Each cylinder is initially at rest. After one complete rotation, which cylinder has the greater kinetic energy?

10.5 | Angular Momentum

Every rotational quantity that we have encountered in Chapters 9 and 10 is the analog of some quantity in the translational motion of a particle. The analog of *momentum* of a particle is **angular momentum,** a vector quantity denoted as $\vec{L}$. Its relation to momentum $\vec{p}$ (which we will sometimes call *linear momentum* for clarity) is exactly the same as the relation of torque to force, $\vec{\tau} = \vec{r} \times \vec{F}$. For a particle with constant mass m, velocity $\vec{v}$, momentum $\vec{p} = m\vec{v}$, and position vector $\vec{r}$ relative to the origin O of an inertial frame, we define angular momentum $\vec{L}$ as

$$\vec{L} = \vec{r} \times \vec{p} = \vec{r} \times m\vec{v} \quad \text{(angular momentum of a particle)} \quad (10.27)$$

The value of $\vec{L}$ depends on the choice of origin O, since it involves the position vector of the particle relative to the origin. The units of angular momentum are kg·m²/s.

In Fig. 10.23 a particle moves in the *xy*-plane; its position vector $\vec{r}$ and momentum $\vec{p} = m\vec{v}$ are shown. The angular momentum vector $\vec{L}$ is perpendicular to the

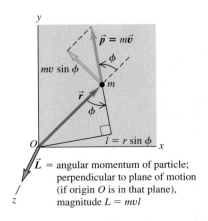

$\vec{L}$ = angular momentum of particle; perpendicular to plane of motion (if origin O is in that plane), magnitude $L = mvl$

10.23 Calculating the angular momentum $\vec{L} = \vec{r} \times m\vec{v} = \vec{r} \times \vec{p}$ of a particle with mass m moving in the *xy*-plane.

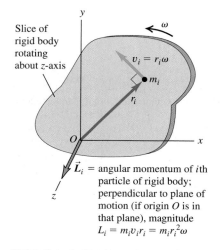

$\vec{L}_i$ = angular momentum of ith particle of rigid body; perpendicular to plane of motion (if origin O is in that plane), magnitude $L_i = m_i v_i r_i = m_i r_i^2 \omega$

10.24 Calculating the angular momentum of a particle of mass m_i in a rotating rigid body. (Compare Fig. 10.23.) Every particle in the body moves in a circle about the rotation axis with the same angular speed ω.

xy-plane. The right-hand rule for vector products shows that its direction is along the $+z$-axis, and its magnitude is

$$L = mvr \sin \phi = mvl \qquad (10.28)$$

where l is the perpendicular distance from the line of $\vec{v}$ to O. This distance plays the role of "lever arm" for the momentum vector.

When a net force $\vec{F}$ acts on a particle, its velocity and momentum change, so its angular momentum may also change. We can show that the *rate of change* of angular momentum is equal to the torque of the net force. We take the time derivative of Eq. (10.27), using the rule for the derivative of a product:

$$\frac{d\vec{L}}{dt} = \left(\frac{d\vec{r}}{dt} \times m\vec{v}\right) + \left(\vec{r} \times m\frac{d\vec{v}}{dt}\right) = (\vec{v} \times m\vec{v}) + (\vec{r} \times m\vec{a})$$

The first term is zero because it contains the vector product of the vector $\vec{v} = d\vec{r}/dt$ with itself. In the second term we replace $m\vec{a}$ with the net force $\vec{F}$, obtaining

$$\frac{d\vec{L}}{dt} = \vec{r} \times \vec{F} = \vec{\tau} \qquad \text{(for a particle acted on by net force } \vec{F}) \qquad (10.29)$$

The rate of change of angular momentum of a particle equals the torque of the net force acting on it. Compare this result to Eq. (8.3), which states that the rate of change $d\vec{p}/dt$ of the *linear* momentum of a particle equals the net force that acts on it.

We can use Eq. (10.28) to find the total angular momentum of a rigid body rotating about the z-axis with angular speed ω. First consider a thin slice of the body lying in the xy-plane (Fig. 10.24). Each particle in the slice moves in a circle centered at the origin, and at each instant its velocity $\vec{v}_i$ is perpendicular to its position vector $\vec{r}_i$, as shown. Hence in Eq. (10.28), $\phi = 90°$ for every particle. A particle with mass m_i at a distance r_i from O has a speed v_i equal to $r_i\omega$. From Eq. (10.28) the magnitude L_i of its angular momentum is

$$L_i = m_i(r_i\omega)r_i = m_i r_i^2 \omega \qquad (10.30)$$

The direction of each particle's angular momentum, as given by the right-hand rule for the vector product, is along the $+z$-axis.

The *total* angular momentum of the slice of the body lying in the xy-plane is the sum $\sum L_i$ of the angular momenta L_i of the particles. Summing Eq. (10.30), we have

$$L = \sum L_i = \left(\sum m_i r_i^2\right)\omega = I\omega$$

where I is the moment of inertia of the slice about the z-axis.

We can do this same calculation for the other slices of the body, all parallel to the xy-plane. For points that do not lie in the xy-plane, a complication arises because the $\vec{r}$ vectors have components in the z-direction as well as the x- and y-directions; this gives the angular momentum of each particle a component perpendicular to the z-axis. But *if the z-axis is an axis of symmetry,* the perpendicular components for particles on opposite sides of this axis add up to zero (Fig. 10.25). So when a body rotates about an axis of symmetry, its angular momentum vector $\vec{L}$ lies along the symmetry axis, and its magnitude is $L = I\omega$.

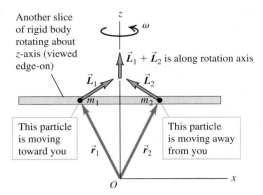

Another slice of rigid body rotating about z-axis (viewed edge-on)

$\vec{L}_1 + \vec{L}_2$ is along rotation axis

This particle is moving toward you

This particle is moving away from you

10.25 Catching the angular momentum of two particles in a rotating rigid body. The two particles have the same mass and are located symmetrically on either side of the rotation axis. While the angular momentum vectors $\vec{L}_1$ and $\vec{L}_2$ of the two particles do not lie along the rotation axis, their vector sum $\vec{L}_1 + \vec{L}_2$ does.

The angular velocity vector $\vec{\omega}$ also lies along the rotation axis, as we discussed at the end of Section 9.1. Hence for a rigid body rotating around an axis of symmetry, $\vec{L}$ and $\vec{\omega}$ are in the same direction (Fig. 10.26). So we have the *vector* relationship

$$\vec{L} = I\vec{\omega} \quad \text{(for a rigid body rotating around a symmetry axis)} \quad (10.31)$$

From Eq. (10.29) the rate of change of angular momentum of a particle equals the torque of the net force acting on the particle. For any system of particles (including both rigid and nonrigid bodies), the rate of change of the *total* angular momentum equals the sum of the torques of all forces acting on all the particles. The torques of the *internal* forces add to zero if these forces act along the line from one particle to another, as in Fig. 10.8, and so the sum of torques includes only the torques of the *external* forces. (A similar cancellation occurred in our discussion of center-of-mass motion in Section 8.5.) If the total angular momentum of the system of particles is $\vec{L}$ and the sum of external torques is $\sum \vec{\tau}$, then

$$\sum \vec{\tau} = \frac{d\vec{L}}{dt} \quad \text{(for any system of particles)} \quad (10.32)$$

Curl fingers of right hand in direction of rotation

Right thumb points in direction of $\vec{\omega}$: if rotation axis is axis of symmetry, this is also direction of $\vec{L}$

10.26 For rotation about an axis of symmetry, $\vec{\omega}$ and $\vec{L}$ are parallel and along the axis. The directions of both vectors are given by the right-hand rule (compare Fig. 9.5).

Finally, if the system of particles is a rigid body rotating about a symmetry axis (the z-axis), then $L_z = I\omega_z$ and I is constant. If this axis has a fixed direction in space, then the vectors $\vec{L}$ and $\vec{\omega}$ change only in magnitude, not in direction. In that case, $dL_z/dt = I\,d\omega_z/dt = I\alpha_z$, or

$$\sum \tau_z = I\alpha_z$$

which is again our basic relation for the dynamics of rigid-body rotation. If the body is *not* rigid, I may change, and in that case, L changes even when ω is constant. For a non-rigid body, Eq. (10.32) is still valid, even though Eq. (10.6) is not.

When the axis of rotation is *not* a symmetry axis, the angular momentum is in general *not* parallel to the axis (Fig. 10.27). As the body turns, the angular momentum vector $\vec{L}$ traces out a cone around the rotation axis. Because $\vec{L}$ changes, there must be a net external torque acting on the body even though the angular velocity magnitude ω may be constant. If the body is an unbalanced wheel on a car, this torque is provided by friction in the bearings, which causes the bearings to wear out. "Balancing" a wheel means distributing the mass so that the rotation axis is an axis of symmetry; then $\vec{L}$ points along the rotation axis, and no net torque is required to keep the wheel turning.

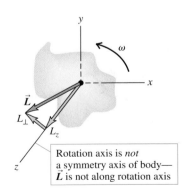

Rotation axis is *not* a symmetry axis of body— $\vec{L}$ is not along rotation axis

10.27 If the rotation axis of a rigid body is not a symmetry axis, $\vec{L}$ does not in general lie along the rotation axis. Even if $\vec{\omega}$ is constant, the direction of $\vec{L}$ changes and a net torque is required to maintain rotation.

In fixed-axis rotation we often use the term "angular momentum of the body" to refer just to the *component* of $\vec{L}$ along the rotation axis of the body (the *z*-axis in Fig. 10.27), with a sign to indicate the sense of rotation just as with angular velocity.

Example 10.12	Angular momentum and torque

A turbine fan in a jet engine (Fig. 10.28) has a moment of inertia of $2.5 \text{ kg} \cdot \text{m}^2$ about its axis of rotation. As the turbine is starting up, its angular velocity as a function of time is

$$\omega_z = (40 \text{ rad/s}^3)t^2$$

a) Find the fan's angular momentum as a function of time, and find its value at time $t = 3.0$ s. b) Find the net torque acting on the fan as a function of time, and find the torque at time $t = 3.0$ s.

SOLUTION

IDENTIFY and SET UP: Just like an electric fan, the turbine fan rotates about an axis of symmetry. Hence we can use Eq. (10.31) to find L_z from ω_z and use Eq. (10.32) to relate the net torque to the time derivative of L_z.

EXECUTE:
a) The only component of angular momentum is along the rotation (*z*) axis:

$$L_z = I\omega_z = (2.5 \text{ kg} \cdot \text{m}^2)(40 \text{ rad/s}^3)t^2 = (100 \text{ kg} \cdot \text{m}^2/\text{s}^3)t^2$$

(We dropped "rad" from the answer because a radian is a dimensionless quantity.) At time $t = 3.0$ s, $L_z = 900 \text{ kg} \cdot \text{m}^2/\text{s}$.
b) The direction of the angular momentum does not change, so the torque is also along the rotation axis. From Eq. (10.32), its component along that axis is

$$\tau_z = \frac{dL_z}{dt} = (100 \text{ kg} \cdot \text{m}^2/\text{s}^3)(2t) = (200 \text{ kg} \cdot \text{m}^2/\text{s}^3)t$$

At time $t = 3.0$ s,

$$\tau_z = (200 \text{ kg} \cdot \text{m}^2/\text{s}^3)(3.0 \text{ s}) = 600 \text{ kg} \cdot \text{m}^2/\text{s}^2 = 600 \text{ N} \cdot \text{m}$$

EVALUATE: As a check on our result, note that the angular acceleration of the turbine fan is $\alpha_z = d\omega_z/dt = (40 \text{ rad/s}^2)(2t) = (80 \text{ rad/s}^2)t$. From the rotational equivalent of Newton's second law, the torque on the fan is $\tau_z = I\alpha_z = (2.5 \text{ kg} \cdot \text{m}^2)(80 \text{ rad/s}^2)t = (200 \text{ kg} \cdot \text{m}^2/\text{s}^3)t$, just as we calculated above.

10.28 A turbine fan is used to force air into a turbojet engine.

Test Your Understanding

A ball is attached to one end of a piece of string. You hold the other end of the string and whirl the ball in a circle around your hand. If the ball moves at a constant speed, is its linear momentum $\vec{p}$ constant? Is its angular momentum $\vec{L}$ constant? Why is there a difference?

10.6 | Conservation of Angular Momentum

We have just seen that angular momentum can be used for an alternative statement of the basic dynamic principle for rotational motion. It also forms the basis for the **principle of conservation of angular momentum.** Like conservation of energy and of linear momentum, this principle is a universal conservation law, valid at all scales from atomic and nuclear systems to the motions of galaxies. This principle follows directly from Eq. (10.32): $\sum \vec{\tau} = d\vec{L}/dt$. If $\sum \vec{\tau} = 0$, then $d\vec{L}/dt = 0$, and $\vec{L}$ is constant. **When the net external torque acting on a system is zero, the total angular momentum of the system is constant (conserved).**

7.14 Ball Hits Bat

A circus acrobat, a diver, and an ice skater pirouetting on the toe of one skate all take advantage of this principle. Suppose an acrobat has just left a swing with arms and legs extended and rotating counterclockwise about her center of mass. When she pulls her arms and legs in, her moment of inertia I_{cm} with respect to her center of mass changes from a large value I_1 to a much smaller value I_2. The only external force acting on her is her weight, which has no torque with respect to an axis through her center of mass. So her angular momentum $L_z = I_{cm}\omega_z$ remains constant, and her angular velocity ω_z increases as I_{cm} decreases. That is,

$$I_1\omega_{1z} = I_2\omega_{2z} \qquad \text{(zero net external torque)} \qquad (10.33)$$

When a skater or ballerina spins with arms outstretched and then pulls her arms in, her angular velocity increases as her moment of inertia decreases. In each case there is conservation of angular momentum in a system in which the net external torque is zero.

When a system has several parts, the internal forces that the parts exert on each other cause changes in the angular momenta of the parts, but the *total* angular momentum doesn't change. Here's an example. Consider two bodies A and B that interact with each other but not with anything else, such as the astronauts we discussed in Section 8.2 (Fig. 8.7). Suppose body A exerts a force $\vec{F}_{A \text{ on } B}$ on body B; the corresponding torque (with respect to whatever point we choose) is $\vec{\tau}_{A \text{ on } B}$. According to Eq. (10.32), this torque is equal to the rate of change of angular momentum of B:

$$\vec{\tau}_{A \text{ on } B} = \frac{d\vec{L}_B}{dt}$$

At the same time, body B exerts a force $\vec{F}_{B \text{ on } A}$ on body A, with a corresponding torque $\vec{\tau}_{B \text{ on } A}$, and

$$\vec{\tau}_{B \text{ on } A} = \frac{d\vec{L}_A}{dt}$$

From Newton's third law, $\vec{F}_{B \text{ on } A} = -\vec{F}_{A \text{ on } B}$. Furthermore, if the forces act along the same line, as in Fig. 10.8, their lever arms with respect to the chosen axis are equal. Thus the *torques* of these two forces are equal and opposite, and $\vec{\tau}_{B \text{ on } A} = -\vec{\tau}_{A \text{ on } B}$. So if we add the two previous equations, we find

$$\frac{d\vec{L}_A}{dt} + \frac{d\vec{L}_B}{dt} = 0$$

or, because $\vec{L}_A + \vec{L}_B$ is the *total* angular momentum $\vec{L}$ of the system,

$$\frac{d\vec{L}}{dt} = 0 \quad \text{(zero net external torque)} \qquad (10.34)$$

That is, the total angular momentum of the system is constant. The torques of the internal forces can transfer angular momentum from one body to the other, but they can't change the *total* angular momentum of the system (Fig. 10.29).

10.29 A falling cat twists different parts of its body in different directions so that it lands feet first. At all times during this process the angular momentum of the cat as a whole remains zero.

Example
10.13
Anyone can be a ballerina

An acrobatic physics professor stands at the center of a turntable, holding his arms extended horizontally with a 5.0-kg dumbbell in each hand (Fig. 10.30). He is set rotating about a vertical axis, making one revolution in 2.0 s. Find the prof's new angular velocity if he pulls the dumbbells in to his stomach, and discuss how this affects the kinetic energy. His moment of inertia (without the dumbbells) is 3.0 kg·m² when his arms are outstretched, dropping to 2.2 kg·m² when his hands are at his stomach. The dumbbells are 1.0 m from the axis initially and 0.20 m from it at the end. Treat the dumbbells as particles.

SOLUTION

IDENTIFY and SET UP: If we neglect friction in the turntable, no external torques act about the vertical (z) axis. Hence the angular momentum about this axis is constant and we can use Eq. (10.33) to find our target variable, the final angular velocity ω_{2z}.

EXECUTE: The moment of inertia of the system is $I = I_{prof} + I_{dumbbells}$. Each dumbbell of mass m contributes mr^2 to $I_{dumbbells}$, where r is the perpendicular distance from the rotation axis to the dumbbell. Initially we have

$$I_1 = 3.0 \text{ kg·m}^2 + 2(5.0 \text{ kg})(1.0 \text{ m})^2 = 13 \text{ kg·m}^2$$

$$\omega_{1z} = \frac{1 \text{ rev}}{2.0 \text{ s}} = 0.50 \text{ rev/s}$$

The final moment of inertia is

$$I_2 = 2.2 \text{ kg·m}^2 + 2(5.0 \text{ kg})(0.20 \text{ m})^2 = 2.6 \text{ kg·m}^2$$

From Eq. (10.33), the final angular velocity is

$$\omega_{2z} = \frac{I_1}{I_2}\omega_{1z} = \frac{13 \text{ kg·m}^2}{2.6 \text{ kg·m}^2}(0.50 \text{ rev/s}) = 2.5 \text{ rev/s}$$

That is, the angular velocity increases by a factor of 5 while the angular momentum remains constant. Note that we didn't have to change "revolutions" to "radians" in this calculation. Why not?

EVALUATE: To calculate the kinetic energy, we must express ω_1 and ω_2 in rad/s. (Why?) We have $\omega_{1z} = (0.50 \text{ rev/s})(2\pi \text{ rad/rev}) = 3.14 \text{ rad/s}$ and $\omega_{2z} = (2.5 \text{ rev/s})(2\pi \text{ rad/rev}) = 15.7 \text{ rad/s}$. The initial kinetic energy is

$$K_1 = \frac{1}{2}I_1\omega_{1z}^2 = \frac{1}{2}(13 \text{ kg·m}^2)(3.14 \text{ rad/s})^2 = 64 \text{ J}$$

and the final kinetic energy is

$$K_2 = \frac{1}{2}I_2\omega_{2z}^2 = \frac{1}{2}(2.6 \text{ kg·m}^2)(15.7 \text{ rad/s})^2 = 320 \text{ J}$$

Where did the extra energy come from?

10.30 Fun with conservation of angular momentum—if you don't get dizzy.

Example
10.14
A rotational "collision" I

Figure 10.31 shows two disks: one an engine flywheel, and the other a clutch plate attached to a transmission shaft. Their moments of inertia are I_A and I_B; initially, they are rotating with constant angular speeds ω_A and ω_B, respectively. We then push the disks together with forces acting along the axis, so as not to apply any torque on either disk. The disks rub against each other and eventually reach a common final angular speed ω. Derive an expression for ω.

SOLUTION

IDENTIFY: The only torque acting on either disk is the torque applied by the other disk; there are no external torques. Thus the total angular momentum of the system of two disks is the same before and after they are pushed together. At the end they rotate together as one body with total moment of inertia $I = I_A + I_B$ and angular speed ω, which is our target variable.

10.31 When the net external torque is zero, angular momentum is conserved. The forces shown are along the axis of rotation and thus exert no torque about any point on it.

EXECUTE: Figure 10.31 shows that all of the angular velocities are in the same direction, so we can regard ω_A, ω_B, and ω as the components of angular velocity along the rotation axis. Conservation of angular momentum then gives

$$I_A\omega_A + I_B\omega_B = (I_A + I_B)\omega$$

$$\omega = \frac{I_A\omega_A + I_B\omega_B}{I_A + I_B}$$

EVALUATE: This "collision" between two disks is analogous to a completely inelastic collision (Section 8.3). When two objects in translational motion along the same axis come together and stick, the linear momentum of the system is conserved. In the situation shown in Fig. 10.31, two objects in *rotational* motion along the same axis come together and stick, and the *angular* momentum is conserved. The kinetic energy of the system decreases in a completely inelastic collision; in Example 10.15 we'll see what becomes of the kinetic energy in the "collision" of two rotating disks.

Example 10.15	**A rotational "collision" II**

In Example 10.14, suppose flywheel A has a mass of 2.0 kg, a radius of 0.20 m, and an initial angular speed of 50 rad/s (about 500 rpm) and that clutch plate B has a mass of 4.0 kg, a radius of 0.10 m, and an initial angular speed of 200 rad/s. Find the common final angular speed ω after the disks are pushed into contact. What happens to the kinetic energy during this process?

SOLUTION

IDENTIFY and SET UP: We'll use the result of Example 10.14 and the expression $K = \frac{1}{2}I\omega^2$ for rotational kinetic energy.

EXECUTE: The moments of inertia of the two disks are

$$I_A = \frac{1}{2}m_Ar_A^2 = \frac{1}{2}(2.0\text{ kg})(0.20\text{ m})^2 = 0.040\text{ kg}\cdot\text{m}^2$$

$$I_B = \frac{1}{2}m_Br_B^2 = \frac{1}{2}(4.0\text{ kg})(0.10\text{ m})^2 = 0.020\text{ kg}\cdot\text{m}^2$$

From Example 10.14 we have

$$\omega = \frac{I_A\omega_A + I_B\omega_B}{I_A + I_B}$$

$$= \frac{(0.040\text{ kg}\cdot\text{m}^2)(50\text{ rad/s}) + (0.020\text{ kg}\cdot\text{m}^2)(200\text{ rad/s})}{0.040\text{ kg}\cdot\text{m}^2 + 0.020\text{ kg}\cdot\text{m}^2}$$

$$= 100\text{ rad/s}$$

The initial kinetic energy is

$$K_1 = \frac{1}{2}I_A\omega_A^2 + \frac{1}{2}I_B\omega_B^2$$

$$= \frac{1}{2}(0.040\text{ kg}\cdot\text{m}^2)(50\text{ rad/s})^2$$

$$+ \frac{1}{2}(0.020\text{ kg}\cdot\text{m}^2)(200\text{ rad/s})^2$$

$$= 450\text{ J}$$

The final kinetic energy is

$$K_2 = \frac{1}{2}(I_A + I_B)\omega^2$$

$$= \frac{1}{2}(0.040\text{ kg}\cdot\text{m}^2 + 0.020\text{ kg}\cdot\text{m}^2)(100\text{ rad/s})^2 = 300\text{ J}$$

EVALUATE: One-third of the initial kinetic energy was lost during this "angular collision," the rotational analog of a completely inelastic collision. We shouldn't expect kinetic energy to be conserved, even though the resultant external force and torque are zero, because nonconservative (frictional) internal forces act while the two disks rub together and gradually approach a common angular velocity.

Example
10.16 **Angular momentum in a crime bust**

A door 1.00 m wide, of mass 15 kg, is hinged at one side so that it can rotate without friction about a vertical axis. It is unlatched. A police officer fires a bullet with a mass of 10 g and a speed of 400 m/s into the exact center of the door, in a direction perpendicular to the plane of the door (Fig. 10.32). Find the angular speed of the door just after the bullet embeds itself in the door. Is kinetic energy conserved?

SOLUTION

IDENTIFY: We consider the door and bullet together as a system. There is no external torque about the axis defined by the hinges, so angular momentum about this axis is conserved.

SET UP: The initial angular momentum is wholly in the bullet and is given by Eq. (10.28). The final angular momentum is that of a rigid body composed of the door and embedded bullet. We'll set these two equal to each other and solve for the angular speed ω of the door and bullet just after the collision.

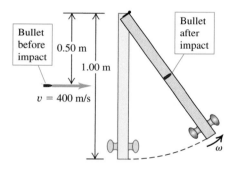

10.32 Shooting open a door (top view). The bullet embeds itself in the center of the door.

EXECUTE: The initial angular momentum of the bullet is:

$$L = mvl = (0.010 \text{ kg})(400 \text{ m/s})(0.50 \text{ m}) = 2.0 \text{ kg} \cdot \text{m}^2/\text{s}$$

The final angular momentum is $I\omega$, where $I = I_{door} + I_{bullet}$. From Table 9.2, for a door of width d,

$$I_{door} = \frac{Md^2}{3} = \frac{(15 \text{ kg})(1.0 \text{ m})^2}{3} = 5.0 \text{ kg} \cdot \text{m}^2$$

The moment of inertia of the bullet (with respect to the axis along the hinges) is

$$I_{bullet} = ml^2 = (0.010 \text{ kg})(0.50 \text{ m})^2 = 0.0025 \text{ kg} \cdot \text{m}^2$$

Conservation of angular momentum requires that $mvl = I\omega$, or

$$\omega = \frac{mvl}{I} = \frac{2.0 \text{ kg} \cdot \text{m}^2/\text{s}}{5.0 \text{ kg} \cdot \text{m}^2 + 0.0025 \text{ kg} \cdot \text{m}^2} = 0.40 \text{ rad/s}$$

The collision of bullet and door is inelastic because nonconservative forces act during the impact. Thus we do not expect kinetic energy to be conserved. To check, we calculate initial and final kinetic energies:

$$K_1 = \frac{1}{2}mv^2 = \frac{1}{2}(0.010 \text{ kg})(400 \text{ m/s})^2 = 800 \text{ J}$$

$$K_2 = \frac{1}{2}I\omega^2 = \frac{1}{2}(5.0025 \text{ kg} \cdot \text{m}^2)(0.40 \text{ rad/s})^2$$

$$= 0.40 \text{ J}$$

The final kinetic energy is only 1/2000 of the initial value!

EVALUATE: The final angular speed of the door is quite slow: at 0.40 rad/s, the door takes 3.9 s to swing through 90° ($\pi/2$ radians). Can you see that the speed would double if the bullet were shot into the edge of the door near the doorknob?

If the polar ice caps were to completely melt due to global warming, the melted ice would redistribute itself over the earth. Use angular momentum ideas to explain how this change would affect the length of the day (the time needed for the earth to rotate once on its axis). Assume that the sun, moon, and planets exert negligibly small torques on the earth.

10.7 | Gyroscopes and Precession

In all of the situations we've looked at so far in this chapter, the axis of rotation either has stayed fixed or has moved and kept the same direction (such as rolling without slipping). But a variety of new physical phenomena, some quite unexpected, can occur when the axis of rotation can change direction. For example, consider a toy gyroscope that's supported at one end (Fig. 10.33). If we hold it with the flywheel axis horizontal and let go, the free end of the axis simply drops owing to gravity—*if* the flywheel isn't spinning. But if the flywheel *is* spinning, what happens is quite different. One possible motion is a steady circular motion of the axis in a horizontal plane, combined with the spin motion of the flywheel about the axis. This surprising, non-intuitive motion of the axis is called **precession.** Precession is found in nature, as well as in rotating machines such as gyroscopes. As you read these words, the earth itself is precessing; its spin axis (through the north and south poles) slowly changes direction, going through a complete cycle of precession every 26,000 years.

To study this strange phenomenon of precession, we must remember that angular velocity, angular momentum, and torque are all *vector* quantities. In particular, we need the general relation between the net torque $\sum \vec{\tau}$ that acts on a body and the rate of change of the body's angular momentum $\vec{L}$, given by Eq. (10.32), $\sum \vec{\tau} = d\vec{L}/dt$. Let's first apply this equation to the case in which the flywheel is *not* spinning (Fig. 10.34a). We take the origin O at the pivot and assume that the flywheel is symmetrical, with mass M and moment of inertia I about the flywheel axis. The flywheel axis is initially along the x-axis. The only external forces on the gyroscope are the normal force $\vec{n}$ acting at the pivot (assumed to be frictionless) and the weight $\vec{w}$ of the flywheel that acts at its center of mass, a distance r from the pivot. The normal force has zero torque with respect to the pivot and the weight has a torque $\vec{\tau}$ in the y-direction, as shown in Fig. 10.34a. Initially, there is no rotation, and the initial angular momentum $\vec{L}_i$ is zero. From Eq. (10.32) the *change $d\vec{L}$* in angular momentum in a short time interval dt following this is

$$d\vec{L} = \vec{\tau} \, dt \qquad (10.35)$$

This change is in the y-direction because $\vec{\tau}$ is. As each additional time interval dt elapses, the angular momentum changes by additional increments $d\vec{L}$ in the y-direction because the direction of the torque is constant (Fig. 10.34b). The steadily increasing horizontal angular momentum means that the gyroscope rotates downward faster and faster around the y-axis until it hits either the stand or the table on which it sits.

Now let's see what happens if the flywheel *is* spinning initially, so the initial angular momentum $\vec{L}_i$ is not zero (Fig. 10.35a). Since the flywheel rotates around its symmetry axis, $\vec{L}_i$ lies along the axis. But each change in angular momentum $d\vec{L}$ is perpendicular to the axis, because the torque $\vec{\tau} = \vec{r} \times \vec{w}$ is perpendicular to the axis (Fig. 10.35b). This causes the *direction* of $\vec{L}$ to change, but not its magni-

10.33 A gyroscope supported at one end. Because the flywheel is spinning at angular speed ω, the flywheel and axis do not drop but instead move in a horizontal circular motion called precession. The angular speed of precession is Ω.

10.34 (a) The flywheel is initially not spinning. The torque $\vec{\tau}$ is due to the weight $\vec{w}$. (b) Looking down from directly over the gyroscope. In each successive time interval dt, the torque produces a change $d\vec{L} = \vec{\tau} dt$ in the angular momentum. In this case the final angular momentum $\vec{L}_f$ is in the same direction as $\vec{\tau}$, and the flywheel axis falls.

Flywheel initially at rest: torque makes it rotate about y-axis (flywheel axis falls)

Zero initial angular momentum ($\vec{L}_i = \mathbf{0}$), torque $\vec{\tau}$ always in same direction, all $d\vec{L}$ vectors in same direction

View from above

(a)

(b)

10.35 (a) The flywheel is spinning initially with angular momentum $\vec{L}_i$. The forces (not shown) are the same as those in Fig. 10.34a. (b) Because the initial angular momentum is not zero, each change $d\vec{L} = \vec{\tau}\,dt$ in angular momentum is perpendicular to $\vec{L}$. As a result, the magnitude of $\vec{L}$ remains the same but its direction changes continuously.

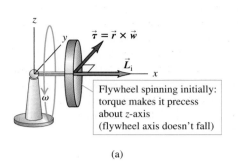

Flywheel spinning initially: torque makes it precess about z-axis (flywheel axis doesn't fall)

(a)

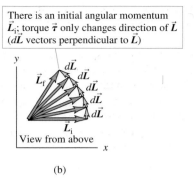

There is an initial angular momentum $\vec{L}_i$; torque $\vec{\tau}$ only changes direction of $\vec{L}$ ($d\vec{L}$ vectors perpendicular to $\vec{L}$)

View from above

(b)

tude. The changes $d\vec{L}$ are always in the horizontal xy-plane, so the angular momentum vector and the flywheel axis with which it moves are always horizontal. In other words, the axis doesn't fall—it just precesses.

If this still seems mystifying to you, think about a ball attached to a string. If the ball is initially at rest and you pull the string toward you, the ball moves toward you also. But if the ball is initially moving and you continuously pull the string in a direction perpendicular to the ball's motion, the ball moves in a circle around your hand; it does not approach your hand at all. In the first case the ball has zero linear momentum $\vec{p}$ to start with; when you apply a force $\vec{F}$ toward you for a time dt, the ball acquires a momentum $d\vec{p} = \vec{F}\,dt$, which is also toward you. But if the ball already has linear momentum $\vec{p}$, a change in momentum $d\vec{p}$ that's perpendicular to $\vec{p}$ changes the direction of motion, not the speed. Replace $\vec{p}$ with $\vec{L}$ and $\vec{F}$ with $\vec{\tau}$ in this argument, and you'll see that precession is simply the rotational analog of uniform circular motion.

At the instant shown in Fig. 10.35a, the gyroscope has angular momentum $\vec{L}$. A short time interval dt later, the angular momentum is $\vec{L} + d\vec{L}$; the infinitesimal change in angular momentum is $d\vec{L} = \vec{\tau}\,dt$, which is perpendicular to $\vec{L}$. As the vector diagram in Fig. 10.36 shows, this means that the flywheel axis of the gyroscope has turned through a small angle $d\phi$ given by $d\phi = |d\vec{L}|/|\vec{L}|$. The rate at which the axis moves, $d\phi/dt$, is called the **precession angular speed;** denoting this quantity by Ω, we find

$$\Omega = \frac{d\phi}{dt} = \frac{|d\vec{L}|/|\vec{L}|}{dt} = \frac{\tau_z}{L_z} = \frac{wr}{I\omega} \tag{10.36}$$

10.36 Detailed view of part of Fig. 10.35b. In a time dt, the angular momentum vector and the flywheel axis precess together through an angle $d\phi$.

Thus the precession angular speed is *inversely* proportional to the angular speed of spin about the axis. A rapidly spinning gyroscope precesses slowly; if friction in its bearings causes the flywheel to slow down, the precession angular speed *increases!* The precession angular speed of the earth is very slow (1 rev/26,000 yr) because its spin angular momentum L_z is large and the torque τ_z, due to the gravitational influences of the moon and sun, is relatively small.

As a gyroscope precesses, its center of mass moves in a circle with radius r in a horizontal plane. Its vertical component of acceleration is zero, so the upward normal force $\vec{n}$ exerted by the pivot must be just equal in magnitude to the weight. The circular motion of the center of mass with angular speed Ω requires a force $\vec{F}$ directed toward the center of the circle, with magnitude $F = M\Omega^2 r$. This force must also be supplied by the pivot.

One key assumption that we made in our analysis of the gyroscope was that the angular momentum vector $\vec{L}$ is associated only with the spin of the flywheel and is purely horizontal. But there will also be a vertical component of angular momentum associated with the precessional motion of the gyroscope. By ignoring this, we've tacitly assumed that the precession is *slow*—that is, that the precession angular speed Ω is very much less than the spin angular speed ω. As Eq. (10.36) shows, a large value of ω automatically gives a small value of Ω, so this approximation is reasonable. When the precession is not slow, additional effects show up, including an up-and-down wobble or *nutation* of the flywheel axis that's superimposed on the precessional motion. You can see nutation occurring in a gyroscope as its spin slows down, so that Ω increases and the vertical component of $\vec{L}$ can no longer be ignored.

Example 10.17	**A precessing gyroscope**

Figure 10.37a shows a top view of a cylindrical gyroscope wheel that has been set spinning by an electric motor. The pivot is at O, and the mass of the axle is negligible. a) As seen from above, is the precession clockwise or counterclockwise? b) If the gyro takes 4.0 s for one revolution of precession, at what angular speed does the wheel spin?

SOLUTION

IDENTIFY and SET UP: We'll determine the direction of precession using the right-hand rule as in Fig. 10.35, which shows the same kind of gyroscope as Fig. 10.37. We'll use the relationship between precession angular speed Ω and the spin angular speed ω, Eq. (10.36), to find the value of ω.

EXECUTE: a) The right-hand rule shows that $\vec{\omega}$ and $\vec{L}$ are to the left (Fig. 10.37b). The weight $\vec{w}$ points into the page in this top view and acts at the center of mass (denoted by an ✕); the torque $\vec{\tau} = \vec{r} \times \vec{w}$ is toward the top of the page; and $d\vec{L}/dt$ is also toward

the top of the page. Adding a small $d\vec{L}$ to the $\vec{L}$ that we have initially changes the direction of $\vec{L}$ as shown, so the precession is clockwise as seen from above.

b) Be careful not to confuse ω and Ω! We are given $\Omega = (1 \text{ rev})/(4.0 \text{ s}) = (2\pi \text{ rad})/(4.0 \text{ s}) = 1.57 \text{ rad/s}$. The weight is equal to mg, and the moment of inertia about its symmetry axis of a solid cylinder with radius R is $I = \frac{1}{2}mR^2$. Solving Eq. (10.36) for ω, we find

$$\omega = \frac{wr}{I\Omega} = \frac{mgr}{(mR^2/2)\Omega} = \frac{2gr}{R^2\Omega}$$

$$= \frac{2(9.8 \text{ m/s}^2)(2.0 \times 10^{-2} \text{ m})}{(3.0 \times 10^{-2} \text{ m})^2(1.57 \text{ rad/s})} = 280 \text{ rad/s} = 2600 \text{ rev/min}$$

EVALUATE: The precession angular speed Ω is very much less than the spin angular speed ω, so this is an example of slow precession.

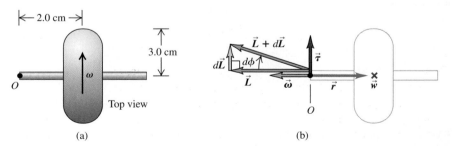

(a)   Top view

(b)

10.37 In which direction does this gyroscope precess?

Test Your Understanding

Suppose the mass of the flywheel in Fig. 10.35 were doubled but all other dimensions and the spin angular speed remained the same. What effect would this change have on the precession angular speed?

SUMMARY

When a force $\vec{F}$ acts on a body, the torque of that force with respect to a point O has a magnitude given by the product of the force magnitude F and the lever arm l. More generally, torque is a vector $\vec{\tau}$ equal to the vector product of $\vec{r}$ (the position vector of the point at which the force acts) and $\vec{F}$. (See Example 10.1)

$$\tau = Fl \qquad (10.2)$$

$$\vec{\tau} = \vec{r} \times \vec{F} \qquad (10.3)$$

Curl fingers of right hand from direction of $\vec{r}$ into direction of $\vec{F}$; outstretched thumb points in direction of $\vec{\tau}$

The rotational analog of Newton's second law says that the net torque acting on a body equals the product of the body's moment of inertia and its angular acceleration. (See Examples 10.2 through 10.4)

$$\sum \tau_z = I\alpha_z \qquad (10.6)$$

If a rigid body is both moving through space and rotating, its motion may be regarded as a superposition of translational motion of the center of mass and rotational motion about an axis through the center of mass. This approach can be used for the kinetic energy, which is a sum of a translational and rotational kinetic energies. It can also be applied to dynamics: Newton's second law describes the motion of the center of mass, and the rotational equivalent of Newton's second law describes rotation about the center of mass. In the case of rolling without slipping, there is a special relationship between the motion of the center of mass and the rotational motion. (See Example 10.5 through 10.9)

$$K = \frac{1}{2}Mv_{cm}^2 + \frac{1}{2}I_{cm}\omega^2 \quad (10.11)$$

$$\sum \vec{F}_{ext} = M\vec{a}_{cm} \qquad (10.15)$$

$$\sum \tau_z = I_{cm}\alpha_z \qquad (10.16)$$

$$v_{cm} = R\omega \qquad (10.14)$$
(rolling without slipping)

When a torque acts on a rigid body that undergoes an angular displacement, the torque does work on the body. The work can be expressed as an integral of the torque or, if the torque is constant, the product of the torque and the angular displacement. The work—energy theorem for rotational motion of a rigid body—says that the total rotational work done on a body is equal to the change in rotational kinetic energy. The power, or rate at which the torque does work, is the product of the torque and the angular velocity. (See Examples 10.10 and 10.11)

$$W = \int_{\theta_1}^{\theta_2} \tau_z d\theta \qquad (10.23)$$

$$W = \tau_z(\theta_2 - \theta_1) = \tau_z \Delta\theta \qquad (10.24)$$

$$W_{tot} = \frac{1}{2}I\omega_2^2 - \frac{1}{2}I\omega_1^2 \qquad (10.25)$$

$$P = \tau_z\omega_z \qquad (10.26)$$

Child applies tangential force

Overhead view of merry-go-round

The angular momentum of a particle with respect to point O is the vector product of the particle's position vector $\vec{r}$ relative to O and its momentum $\vec{p} = m\vec{v}$. When a symmetric body rotates about a stationary axis of symmetry, its angular momentum is the product of its moment of inertia and its angular velocity vector $\vec{\omega}$. If the body is not symmetric or the rotation (z) axis is not an axis of symmetry, the component of angular momentum along the rotation axis is $I\omega_z$. (See Example 10.12)

$$\vec{L} = \vec{r} \times \vec{p} = \vec{r} \times m\vec{v} \qquad (10.27)$$
(particle)

$$\vec{L} = I\vec{\omega} \qquad (10.31)$$
(rigid body rotating about axis of symmetry)

Curl fingers of right hand in direction of rotation

$\vec{L}$

$\vec{\omega}$

Right thumb points in direction of $\vec{\omega}$: if rotation axis is axis of symmetry, this is also direction of $\vec{L}$

The basic dynamic relation for the rotational motion of any system is that the net external torque is equal to the rate of change of angular momentum. If there is zero net external torque on a system, the total angular momentum of the system is constant (conserved). (See Examples 10.13 through 10.16)

$$\sum \vec{\tau} = \frac{d\vec{L}}{dt} \qquad (10.32)$$

Key Terms

Your Notes

Answer to Chapter Opening Question

When the gymnast is in mid-air, no net torque acts about his center of mass. Hence the angular momentum of his body (the product of the moment of inertia I and the angular speed ω) around the center of mass remains constant. By moving his limbs outward, he increases I and hence ω decreases; if he pulls his limbs in, I decreases and ω increases.

Answers to Test Your Understanding Questions

Section 10.1 The torque is proportional to the product of the distance r and the force magnitude F. Without the "cheater," the distance r is smaller by a factor of $(0.40)/(0.80) = 0.50$. To produce the same torque, the force must therefore be increased by a factor of $(0.80)/(0.40) = 2.0$. Hence the required force is $(2.0)(900\ \mathrm{N}) = 1800\ \mathrm{N}$.

Section 10.2 The answers would be unchanged. The reason is that none of the forces acting on the glider, pulley, or hanging object depends on the direction of motion, or on whether there is any motion at all.

Section 10.3 With a hollow sphere rather than a solid one, the moment of inertia increases from $\frac{2}{5}MR^2$ to $\frac{2}{3}MR^2$ (see Table 9.2). Repeating the calculation of Example 10.9, we find $a_{\mathrm{cm}\text{-}x} = \frac{3}{5}g \sin \beta$. Since $3/5 = 0.600$ is less than $5/7 = 0.714$, the acceleration must be less for the hollow ball than for the solid ball. We would expect this result, since the hollow ball has a greater moment of inertia. Following the steps in Example 10.9 also tells us that the friction force on a hollow ball is $f = \frac{2}{5}Mg \sin \beta$. This is greater than for a solid ball $(2/5 = 0.400$ is greater than $2/7 = 0.286)$, which gives us another way to understand why the hollow ball accelerates more slowly down the incline.

Section 10.4 You apply the same torque over the same angular displacement to both cylinders. Hence, by Eq. (10.24), you do the same amount of work to both cylinders and impart the same kinetic energy to both. (The one with the smaller moment of inertia ends up with a greater angular speed, but that isn't what was asked for. Compare Conceptual Example 6.3 in Section 6.2.)

Section 10.5 As the ball goes around the circle, the magnitude of $\vec{p} = m\vec{v}$ remains the same (the speed is constant) but its direction changes, so the linear momentum vector isn't constant. But $\vec{L} = \vec{r} \times \vec{p}$ *is* constant: it maintains a constant magnitude (the speed and the perpendicular distance from your hand to the ball are both constant) and a constant direction (along the rotation axis, perpendicular to the plane of the ball's motion). The linear momentum changes because there is a net force $\vec{F}$ on the ball (toward the center of the circle). The angular momentum remains constant because there is no net torque; the vector $\vec{r}$ points from your hand to the ball and the force $\vec{F}$ on the ball is directed toward your hand, so the vector product $\vec{\tau} = \vec{r} \times \vec{F}$ is zero.

Section 10.6 In the absence of any external torques, the earth's angular momentum $L_z = I\omega_z$ would remain constant. The melted ice would move from the poles toward the equator—that is, away from our planet's rotation axis—and the earth's moment of inertia I would increase slightly. Hence the angular velocity ω_z would

decrease slightly and the length of the day would be slightly longer.

Section 10.7 Doubling the flywheel mass would double both its moment of inertia I and its weight w, so the ratio I/w would be unchanged. Equation (10.36) shows that the precession angular speed depends on this ratio, so there would be *no* effect on the value of Ω.

Discussion Questions

Q10.1 When tightening cylinder-head bolts in an automobile engine, the critical quantity is the *torque* applied to the bolts. Why is this more important than the actual *force* applied to the wrench handle?

Q10.2 Can a single force applied to a body change both its translational and rotational motion? Explain.

Q10.3 Suppose you could use wheels of any type in the design of a soapbox-derby racer (an unpowered, four-wheel vehicle that coasts from rest down a hill). Conforming to the rules on the total weight of the vehicle and rider, should you design with large massive wheels or small light wheels? Should you use solid wheels or wheels with most of the mass at the rim? Explain.

Q10.4 Unless $\vec{r}$ and $\vec{F}$ are perpendicular, there are always two angles between their directions that give the same torque for given magnitudes of $\vec{r}$ and $\vec{F}$. Explain why. Illustrate your answer with a sketch.

Q10.5 The crankshaft of an automobile engine has a flywheel to increase the moment of inertia about the rotation axis. Why is this desirable?

Q10.6 The harder you hit the brakes while driving forward, the more the front end of your car will move down (and the rear end move up). Why? What happens when accelerating? Why do drag racers not use front-wheel drive only?

Q10.7 When an acrobat walks on a tightrope, she extends her arms straight out from her sides. She does this to make it easier for her to catch herself if she should tip to one side or the other. Explain how this works. (*Hint:* Think about Eq. (10.6).)

Q10.8 When you turn on an electric motor, it takes longer to come up to final speed if there is a grinding wheel attached to the shaft. Why?

Q10.9 Experienced cooks can tell whether an egg is raw or hard-boiled by rolling it down a slope (taking care to catch it at the bottom). How is this possible? What are they looking for?

Q10.10 The work done by a force is the product of force and distance. The torque due to a force is a product of force and distance. Does this mean that torque and work are equivalent? Explain.

Q10.11 A valued client brings a treasured ball to your engineering firm, wanting to know if the ball is solid or hollow. He has tried tapping on it, but that has given insufficient information. Design a simple, inexpensive experiment that you could perform quickly, without injuring the precious ball, to find out if it is solid or hollow.

Q10.12 Force A has a magnitude twice that of force B. Is it possible for force B to exert a greater torque on an object than force A?

Q10.13 An electric grinding wheel coasts for a minute or more after the power is turned off, but an electric drill coasts for only a few seconds. Why is there a difference?

Q10.14 The force of gravity acts on the hammer in Fig. 10.12 and forces produce torques that cause a body's angular velocity to change. Why, then, is the angular velocity of the hammer in the figure constant?

Q10.15 In Example 10.6 (Section 10.3) suppose you pulled your hand and the end of the string upward. Would you conserve mechanical energy in this situation? Why or why not?

Q10.16 A wheel is rolling without slipping on a horizontal surface. In an inertial frame of reference in which the surface is at rest, is there any point on the wheel that has a velocity that is purely vertical? Is there any point that has a horizontal velocity component opposite to the velocity of the center of mass? Explain. Do your answers change if the wheel is slipping as it rolls? Why or why not?

Q10.17 Part of the kinetic energy of a moving automobile is in the rotational motion of its wheels. When the brakes are applied hard on an icy street, the wheels "lock," and the car starts to slide. What becomes of the rotational kinetic energy?

Q10.18 A hoop, a uniform solid cylinder, a spherical shell, and a uniform solid sphere are released from rest at the top of an incline. What is the order in which they arrive at the bottom of the incline? Does it matter whether or not the masses and radii of the objects are all the same? Explain.

Q10.19 A ball is rolling along at speed v without slipping on a horizontal surface when it comes to a hill that rises at a constant angle above the horizontal. In which case will it go higher up the hill: if the hill has enough friction to prevent slipping or if the hill is perfectly smooth? Justify your answers in both cases in terms of energy conservation and in terms of Newton's second law.

Q10.20 You are standing at the center of a large, horizontal turntable in a carnival funhouse. The turntable is set rotating on frictionless bearings, and rotates freely (that is, there is no motor driving the turntable). As you walk toward the edge of the turntable, what happens to the combined angular momentum of you and the turntable? What happens to the rotation speed of the turntable? Explain your answer.

Q10.21 A point particle travels in a circle at constant speed. With respect to the origin at the center of the circle, is there a net torque acting on the particle? A net force? What if the speed of the particle is changing? In each case, explain your answers.

Q10.22 A point particle travels in a straight line at constant speed and the closest distance it comes to the origin of coordinates is a distance l. With respect to this origin, does the particle have nonzero angular momentum? As the particle moves along its straight-line path, does its angular momentum with respect to origin change?

Q10.23 In Example 10.13 (Section 10.6) the angular speed ω changes and this must mean that there is nonzero angular acceleration. But there is no torque about the rotation axis if the forces the professor applies to the weights are directly, radially inward. Then by Eq. (10.6), α_z must be zero. Explain what is wrong with this reasoning that leads to this apparent contradiction.

Q10.24 In Example 10.13 (Section 10.6) the rotational kinetic energy of the professor and dumbbells increases. But since there are no external torques, no work is being done to change the rotational kinetic energy. Then, by Eq. (10.25) the kinetic energy must remain the same! Explain what is wrong with this reasoning that leads to this apparent contradiction. Where *does* the extra kinetic energy come from?

Q10.25 As discussed in Section 10.6, the angular momentum of a circus acrobat is conserved as she tumbles through the air. Is her *linear* momentum conserved? Why or why not?

Q10.26 If you stop a spinning raw egg for the shortest instant you can and then release it, the egg will start spinning again. If you do the same to a hard-boiled egg, it will remain stopped. Try it. Explain it.

Q10.27 A helicopter has a large main rotor that rotates in a horizontal plane and provides lift. There is also a small rotor on the tail that rotates in a vertical plane. What is the purpose of the tail rotor? (*Hint:* if there were no tail rotor, what would happen when the pilot changed the angular speed of the main rotor?) Some helicopters have no tail rotor, but instead have two large main rotors that rotate in a horizontal plane. Why is it important that the two main rotors rotate in opposite directions?

Q10.28 In a common design for a gyroscope, the flywheel and flywheel axis are enclosed in a light, spherical frame with the flywheel at the center of the frame. The gyroscope is then balanced on top of a pivot so that the flywheel is directly above the pivot. Does the gyroscope precess if it is released while the flywheel is spinning? Explain.

Q10.29 A gyroscope takes 3.8 s to precess 1.0 revolution about a vertical axis. Two minutes later, it takes only 1.9 s to precess 1.0 revolution. No one has touched the gyroscope. Explain.

Q10.30 A gyroscope is precessing as in Fig. 10.33. What happens if you gently add some weight to the end of the flywheel axis farthest from the pivot?

Q10.31 A bullet emerges from a rifle spinning on its axis. Explain how this prevents the bullet from tumbling and keeps the streamlined end pointed forward.

Exercises

Section 10.1 Torque

10.1 Calculate the torque (magnitude and direction) about point O due to the force $\vec{F}$ in each of the situations sketched in Fig. 10.38. In each case, the force $\vec{F}$ and the rod both lie in the plane of the

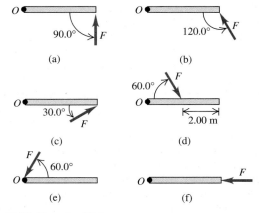

Figure 10.38 Exercise 10.1.

page, the rod has length 4.00 m, and the force has magnitude $F = 10.0$ N.

10.2 Calculate the net torque about point O for the two forces applied as in Fig. 10.39. The rod and both forces are in the plane of the page.

Figure 10.39 Exercise 10.2.

10.3 A square metal plate 0.180 m on each side is pivoted about an axis through point O at its center and perpendicular to the plate (Fig. 10.40). Calculate the net torque about this axis due to the three forces shown in the figure if the magnitudes of the forces are $F_1 = 18.0$ N, $F_2 = 26.0$ N, and $F_3 = 14.0$ N. The plate and all forces are in the plane of the page.

Figure 10.40 Exercise 10.3.

10.4 Forces $F_1 = 7.50$ N and $F_2 = 5.30$ N are applied tangentially to a wheel with radius 0.330 m, as shown in Fig. 10.41. What is the net torque on the wheel due to these two forces for an axis perpendicular to the wheel and passing through its center?

10.5 One force acting on a machine part is $\vec{F} = (-5.00\text{ N})\hat{\imath} + (4.00\text{ N})\hat{\jmath}$. The vector from the origin to the point where the force is applied is $\vec{r} = (-0.450\text{ m})\hat{\imath} + (0.150\text{ m})\hat{\jmath}$. a) In a sketch, show $\vec{r}$, $\vec{F}$, and the origin. b) Use the right-hand rule to determine the direction of the torque. c) Calculate the vector torque produced by this force. Verify that the direction of the torque is the same as you obtained in (b).

Figure 10.41 Exercise 10.4.

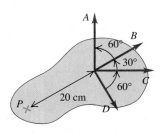

Figure 10.42 Exercise 10.6.

10.6 In Fig. 10.42, the forces $\vec{A}$, $\vec{B}$, $\vec{C}$, and $\vec{D}$ each have magnitude 50 N and act at the same point on the object. a) What torque (magnitude and direction) does each of these forces exert on the object about point P? b) What is the total torque about point P?

Section 10.2 Torque and Angular Acceleration for a Rigid Body

10.7 A uniform, 8.40-kg, spherical shell 50.0 cm in diameter has four small 2.00 kg masses attached to its outer surface and equally spaced around it. This combination is spinning about an axis running through the center of the sphere and two of the small masses (Fig. 10.43). What friction torque is needed to reduce its angular speed from 75.0 rpm to 50.0 rpm in 30.0 s?

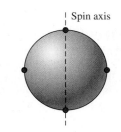

Figure 10.43 Exercise 10.7.

10.8 The flywheel of an engine has a moment of inertia 2.50 kg·m² about its rotation axis. a) What constant torque is required to bring it up to an angular speed of 400 rev/min in 8.00 s, starting from rest? b) What is its final kinetic energy?

10.9 Using the value of a_x calculated in Example 10.2 (Section 10.2), what is the speed of the cable after it has been pulled 2.0 m? Compare your result with Example 9.8 (Section 9.4).

10.10 A cord is wrapped around the rim of a wheel 0.250 m in radius, and a steady pull of 40.0 N is exerted on the cord. The wheel is mounted on frictionless bearings on a horizontal shaft through its center. The moment of inertia of the wheel about this shaft is 5.00 kg·m². Compute the angular acceleration of the wheel.

10.11 a) Find the magnitude n of the normal force for the situation described in Example 10.3 (Section 10.2). b) Is your answer to part (a) less than, equal to, or greater than the total weight $(M + m)g$ of the cylinder and mass? Explain why this should be so. c) Suppose the cylinder is initially rotating clockwise so that the hanging mass m is initially moving upward at speed v_0 (the cable remains taut). What effect does this have on the tension T and normal force n? Explain.

10.12 a) In the situation described in Example 10.2 of Section 10.2 (Fig. 10.9), the normal force $\vec{n}$ exerted on the cylinder by the bearing is directed up and to the left. Why must the normal force be in this direction? b) Calculate the magnitude and direction of $\vec{n}$.

10.13 A grindstone in the shape of a solid disk with diameter 0.520 m and a mass of 50.0 kg is rotating at 850 rev/min. You press an ax against the rim with a normal force of 160 N (Fig. 10.44), and the grindstone comes to rest in 7.50 s. Find the coefficient of friction

Figure 10.44 Exercise 10.13 and Problem 10.52.

between the ax and the grindstone. You can ignore friction in the bearings.

10.14 A stone is suspended from the free end of a wire that is wrapped around the outer rim of a pulley, as shown in Fig. 10.10. The pulley is a uniform disk with mass 10.0 kg and radius 50.0 cm and turns on frictionless bearings. You measure that the stone travels 12.6 m in the first 3.00 s starting from rest. Find a) the mass of the stone; b) the tension in the wire.

10.15 A solid, uniform cylinder with mass 8.25 kg and diameter 15.0 cm is spinning at 220 rpm on a thin, frictionless axle that passes along the cylinder axis. You design a simple friction-brake to stop the cylinder by pressing the brake against the outer rim with a normal force. The coefficient of kinetic friction between the brake and rim is 0.333. What must the applied normal force be to bring the cylinder to rest after it has turned through 5.25 revolutions?

10.16 A 15.0-kg bucket of water is suspended by a rope wrapped around a windlass, that is a solid cylinder 0.300 m in diameter with mass 12.0 kg. The cylinder pivots on a frictionless axle through its center. The bucket is released from rest at the top of a well and falls 10.0 m to the water. You can ignore the weight of the rope. a) What is the tension in the rope while the bucket is falling? b) With what speed does the bucket strike the water? c) What is the time of fall? d) While the bucket is falling, what is the force exerted on the cylinder by the axle?

10.17 A 2.00-kg textbook rests on a frictionless, horizontal surface. A cord attached to the book passes over a pulley whose diameter is 0.150 m, to a hanging book with mass 3.00 kg. The system is released from rest, and the books are observed to move 1.20 m in 0.800 s. a) What is the tension in each part of the cord? b) What is the moment of inertia of the pulley about its rotation axis?

10.18 A thin, horizontal rod with length l and mass M pivots about a vertical axis at one end. A force with constant magnitude F is applied to the other end, causing the rod to rotate in a horizontal plane. The force is maintained perpendicular to the rod and to the axis of rotation. Calculate the magnitude of the angular acceleration of the rod.

Section 10.3 Rigid-Body Rotation about a Moving Axis

10.19 A string is wrapped several times around the rim of a small hoop with radius 0.0800 m and mass 0.180 kg. If the free end of the string is held in place and the hoop is released from rest (Fig. 10.45), calculate a) the tension in the string while the hoop descends as the string unwinds; b) the time it takes the hoop to descend 0.750 m; c) the angular speed of the rotating hoop after it has descended 0.750 m.

0.0800 m

Figure 10.45 Exercise 10.19 and Problem 10.71.

10.20 Repeat part (c) of Exercise 10.19, this time using energy considerations.

10.21 In Example 10.5 (Section 10.3) we found that for a hollow, cylindrical shell rolling without slipping on a horizontal surface, half of the total kinetic energy is translational and half is rotational. What

fraction of the total kinetic energy is rotational for the following objects rolling without slipping on a horizontal surface? a) A uniform solid cylinder. b) A uniform sphere. c) A thin-walled hollow sphere. d) A hollow, cylinder with outer radius R and inner radius $R/2$.

10.22 A hollow, spherical shell with mass 2.00 kg rolls without slipping down a 38.0° slope. a) Find the acceleration, the friction force, and the minimum coefficient of friction needed to prevent slipping. b) How would your answers to part (a) change if the mass were doubled to 4.00 kg?

10.23 A solid ball is released from rest and slides down a hillside that slopes downward at 65.0° from the horizontal. a) What minimum value must the coefficient of static friction between the hill and ball surfaces have for no slipping to occur? b) Would the coefficient of friction calculated in part (a) be sufficient to prevent a hollow ball (such as a soccer ball) from slipping? Justify your answer. c) In part (a), why did we use the coefficient of static friction and not the coefficient of kinetic friction?

10.24 A uniform marble rolls down a symmetric bowl, starting from rest at the top of the left side. The top of each side is a distance h above the bottom of the bowl. The left half of the bowl is rough enough to cause the marble to roll without slipping, but the right half has no friction because it is coated with oil. a) How far up the smooth side will the marble go, measured vertically from the bottom? b) How high would the marble go if both sides were as rough as the left side? c) How do you account for the fact that the marble goes *higher* with friction on the right side than without friction?

10.25 A 392-N wheel comes off a moving truck and rolls without slipping along a highway. At the bottom of a hill it is rotating at 25.0 rad/s. The radius of the wheel is 0.600 m and its moment of inertia about its rotation axis is $0.800MR^2$. Friction does work on the wheel as it rolls up the hill to a stop, a height h above the bottom of the hill; this work has absolute value 3500 J. Calculate h.

10.26 A Ball Rolling Uphill. A bowling ball rolls without slipping up a ramp that slopes upward at an angle β to the horizontal (see Example 10.9 in Section 10.3). Treat the ball as a uniform, solid sphere, ignoring the finger holes. a) Draw the free-body diagram for the ball. Explain why the friction force must be directed *uphill*. b) What is the acceleration of the center of mass of the ball? c) What minimum coefficient of static friction is needed to prevent slipping?

Section 10.4 Work and Power in Rotational Motion

10.27 A playground merry-go-round has radius 2.40 m and a moment of inertia 2100 kg·m² about a vertical axle through its center, and turns with negligible friction. a) A child applies an 18.0-N force tangentially to the edge of the merry-go-round for 15.0 s. If the merry-go-round is initially at rest, what is its angular speed after this 15.0-s interval? b) How much work did the child do on the merry-go-round? c) What is the average power supplied by the child?

10.28 Example 9.5 (Section 9.3) described the design of an aircraft propeller. The engine delivers 175 hp to the propeller at 2400 rev/min. a) How much torque does the aircraft engine provide? b) How much work does the engine do in one revolution of the propeller?

10.29 A 1.50-kg grinding wheel is in the form of a solid cylinder of radius 0.100 m. a) What constant torque will bring it from rest to an angular speed of 1200 rev/min in 2.5 s? b) Through what angle has it turned during that time? c) Use Eq. (10.24) to calculate the work done by the torque. d) What is the grinding wheel's kinetic energy when it is rotating at 1200 rev/min? Compare your answer to the result in part (c).

10.30 What is the power output in horsepower of an electric motor turning at 4800 rev/min and developing a torque of 4.30 N·m?

10.31 The carbide tips of the cutting teeth of a circular saw are 8.6 cm from the axis of rotation. a) The no-load speed of the saw, when it is not cutting anything, is 4800 rev/min. Why is its no-load power output negligible? b) While cutting lumber, the angular speed of the saw slows to 2400 rev/min and the power output is 1.9 hp. What is the tangential force that the wood exerts on the carbide tips?

10.32 An airplane propeller is 2.08 m in length (from tip to tip) and has a mass of 117 kg. When the airplane's engine is first started, it applies a constant torque of 1950 N·m to the propeller, which starts from rest. a) What is the angular acceleration of the propeller? Treat the propeller as a slender rod. (*Hint:* See Table 9.2.) b) What is the propeller's angular speed after making 5.00 rev? c) How much work is done by the engine during the first 5.00 rev? d) What is the average power output of the engine during the first 5.00 rev? e) What is the instantaneous power output of the motor at the instant that the propeller has turned through 5.00 rev?

10.33 a) Compute the torque developed by an industrial motor whose output is 150 kW at an angular speed of 4000 rev/min. b) A drum with negligible mass, 0.400 m in diameter, is attached to the motor shaft and the power output of the motor is used to raise a weight hanging from a rope wrapped around the drum. How heavy a weight can the motor lift at constant speed? c) At what constant speed will the weight rise?

Section 10.5 Angular Momentum

10.34 A woman with mass 50 kg is standing on the rim of a large disk that is rotating at 0.50 rev/s about an axis through its center. The disk has mass 110 kg and a radius 4.0 m. Calculate the magnitude of the total angular momentum of the woman-plus-disk system. (Assume that you can treat the woman as a point.)

10.35 A 2.00-kg rock has a horizontal velocity of magnitude 12.0 m/s when it is at point P in Fig. 10.46. a) At this instant, what is the magnitude and direction of its angular momentum relative to point O? b) If the only force acting on the rock is its weight, what is the rate of change (magnitude and direction) of its angular momentum at this instant?

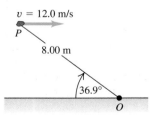

Figure 10.46 Exercise 10.35.

10.36 a) Calculate the magnitude of the angular momentum of the earth considered as a particle orbiting the sun. The mass of the earth is 5.97×10^{24} kg. Treat it as moving in a circular orbit of radius 1.50×10^{11} m at a speed of 2.98×10^4 m/s. b) Calculate the magnitude of the angular momentum of the earth due to its rotation around an axis through the north and south poles. Treat the earth as a uniform sphere of radius 6.38×10^6 m that makes one revolution in 24.0 hours.

10.37 Find the magnitude of the angular momentum of the second hand on a clock about an axis through the center of the clock face. The clock hand has a length of 15.0 cm and a mass of 6.00 g. Take the second hand to be a slender rod rotating with constant angular velocity about one end.

Section 10.6 Conservation of Angular Momentum

10.38 Under some circumstances, a star can collapse into an extremely dense object made mostly of neutrons and called a *neutron star*. The density of a neutron star is roughly 10^{14} times as great as that of ordinary solid matter. Suppose we represent the star as a uniform, solid, rigid sphere, both before and after the collapse. The star's initial radius was 7.0×10^5 km (comparable to our sun); its final radius is 16 km. If the original star rotated once in 30 days, find the angular speed of the neutron star.

10.39 A small block on a frictionless, horizontal surface has a mass of 0.0250 kg. It is attached to a massless cord passing through a hole in the surface (Fig. 10.47). The block is originally revolving at a distance of 0.300 m from the hole with an angular speed of 1.75 rad/s. The cord is then pulled from below, shortening the radius of the circle in which the block revolves to 0.150 m. You may treat the block as a particle. a) Is angular momentum conserved? Why or why not? b) What is the new angular speed? c) Find the change in kinetic energy of the block. d) How much work was done in pulling the cord?

Figure 10.47 Exercise 10.39, Problem 10.92, and Challenge Problem 10.103.

10.40 The Spinning Figure Skater. The outstretched hands and arms of a figure skater preparing for a spin can be considered a slender rod pivoting about an axis through its center (Fig. 10.48). When his hands and arms are brought in and wrapped around his

Figure 10.48 Exercise 10.40.

body to execute the spin, the hands and arms can be considered a thin-walled hollow cylinder. His hands and arms have a combined mass 8.0 kg. When outstretched, they span 1.8 m; when wrapped, they form a cylinder of radius 25 cm. The moment of inertia about the rotation axis of the remainder of his body is constant and equal to $0.40 \text{ kg} \cdot \text{m}^2$. If his original angular speed is 0.40 rev/s, what is his final angular speed?

10.41 A diver comes off a board with arms straight up and legs straight down, giving her a moment of inertia about her rotation axis of $18 \text{ kg} \cdot \text{m}^2$. She then tucks into a small ball, decreasing this moment of inertia to $3.6 \text{ kg} \cdot \text{m}^2$. While tucked, she makes two complete revolutions in 1.0 s. If she hadn't tucked at all, how many revolutions would she have made in the 1.5 s from board to water?

10.42 A large turntable rotates about a fixed vertical axis, making one revolution in 6.00 s. The moment of inertia of the turntable about this axis is $1200 \text{ kg} \cdot \text{m}^2$. A child of mass 40.0 kg, initially standing at the center of the turntable, runs out along a radius. What is the angular speed of the turntable when the child is 2.00 m from the center? (Assume that you can treat the child as a particle.)

10.43 A large, wooden turntable in the shape of a flat disk has a radius of 2.00 m and a total mass of 120 kg. The turntable is initially rotating at 3.00 rad/s about a vertical axis through its center. Suddenly, a 70.0-kg parachutist makes a soft landing on the turntable at a point near the outer edge. a) Find the angular speed of the turntable after the parachutist lands. (Assume that you can treat the parachutist as a particle.) b) Compute the kinetic energy of the system before and after the parachutist lands. Why are these kinetic energies not equal?

10.44 A solid wood door 1.00 m wide and 2.00 m high is hinged along one side and has a total mass 40.0 kg. Initially open and at rest, the door is struck at its center by a handful of sticky mud with mass 0.500 kg, traveling perpendicular to the door at 12.0 m/s just before impact. Find the final angular speed of the door. Does the mud make a significant contribution to the moment of inertia?

10.45 A small 10.0-g bug stands at one end of a thin, uniform bar which is initially at rest on a smooth, horizontal table. The other end of the bar pivots about a nail driven into the table and can rotate freely, without friction. The bar has mass 50.0 g and is 100 cm in length. The bug jumps off in the horizontal direction, perpendicular to the bar, with a speed of 20.0 cm/s relative to the table. a) What is the angular speed of the bar just after the frisky insect leaps? b) What is the total kinetic energy of the system just after the bug leaps? c) Where does this energy come from?

10.46 A thin, uniform, metal bar, 2.00 m long and weighing 90.0 N, is hanging vertically from the ceiling by a frictionless pivot. Suddenly it is struck 1.50 m below the ceiling by a small 3.00-kg ball, initially traveling horizontally at 10.0 m/s. The ball rebounds in the opposite direction with a speed of 6.00 m/s. a) Find the angular speed of the bar just after the collision. b) During the collision, why is the angular momentum conserved, but not the linear momentum?

Section 10.7 Gyroscopes and Precession

10.47 Draw a top view of the gyroscope shown in Fig. 10.33. a) Draw labeled arrows on your sketch for $\vec{\omega}$, $\vec{L}$, and $\vec{\tau}$. Draw $d\vec{L}$ produced by $\vec{\tau}$. Draw $\vec{L} + d\vec{L}$. Determine the sense of precession by

examining the directions of $\vec{L}$ and $\vec{L} + d\vec{L}$. b) Reverse the direction of the spin angular velocity of the rotor and repeat all steps in part (a). c) Move the pivot to the other end of the shaft, with the same direction of spin angular velocity as in part (b), and repeat all the steps. d) Keeping the pivot as in part (c), reverse the spin angular velocity of the rotor and repeat all steps.

10.48 The rotor (flywheel) of a toy gyroscope has mass 0.140 kg. Its moment of inertia about its axis is $1.20 \times 10^{-4} \text{ kg} \cdot \text{m}^2$. The mass of the frame is 0.0250 kg. The gyroscope is supported on a single pivot (Fig. 10.49) with its center of mass a horizontal distance of 4.00 cm from the pivot. The gyroscope is precessing in a horizontal plane at the rate of one revolution in 2.20 s. a) Find the upward force exerted by the pivot. b) Find the angular speed

Figure 10.49 Exercise 10.48.

with which the rotor is spinning about its axis, expressed in rev/min. c) Copy the diagram and draw vectors to show the angular momentum of the rotor and the torque acting on it.

10.49 A Gyro Stabilizer. The stabilizing gyroscope of a ship is a solid disk with mass 60,000 kg; its radius is 2.00 m, and it rotates about a vertical axis with an angular speed of 500 rev/min. a) How much time is required to bring it up to speed, starting from rest, with a constant power input of $7.46 \times 10^4 \text{ W}$? b) Find the torque needed to cause the axis to precess in a vertical fore-and-aft plane at the rate of $1.00°/\text{s}$.

10.50 A gyroscope is precessing about a vertical axis. Describe what happens to the precession angular speed if the following changes in the variables are made, with all other variables remaining the same: a) the angular speed of the spinning flywheel is doubled; b) the total weight is doubled; c) the moment of inertia about the axis of the spinning flywheel is doubled; d) the distance from the pivot to the center of gravity is doubled. e) What happens if all four of the variables in parts (a) through (d) are doubled?

10.51 The earth precesses once every 26,000 years, and spins on its axis once a day. Estimate the magnitude of the torque that causes the precession of the earth. You may need some data from Appendix F. Make the estimate by assuming i) the earth is a uniform sphere and ii) the precession of the earth is like that of the gyroscope shown in Fig. 10.34. In this model, the precession axis and rotation axis are perpendicular. Actually, the angle between these two axes for the earth is only $23\frac{1}{2}°$; this affects the calculated torque by about a factor of 2.

Problems

10.52 A 55.0-kg grindstone is a solid disk 0.520 m in diameter. You press an ax down on the rim with a normal force of 160 N (Fig. 10.44). The coefficient of kinetic friction between the blade and the stone is 0.60, and there is a constant friction torque of

6.50 N·m between the axle of the stone and its bearings. a) How much force must be applied tangentially at the end of a crank handle 0.500 m long to bring the stone from rest to 120 rev/min in 9.00 s? b) After the grindstone attains an angular speed of 120 rev/min, what tangential force at the end of the handle is needed to maintain a constant angular speed of 120 rev/min? c) How much time does it take the grindstone to come from 120 rev/min to rest if it is acted on by the axle friction alone?

10.53 An experimental bicycle wheel is placed on a test stand so that it is free to turn on its axle. If a constant net torque of 5.00 N·m is applied to the tire for 2.00 s, the angular speed of the tire increases from zero to 100 rev/min. The external torque is then removed, and the wheel is brought to rest by friction in its bearings in 125 s. Compute a) the moment of inertia of the wheel about the rotation axis; b) the friction torque; c) the total number of revolutions made by the wheel in the 125-s time interval.

10.54 A flywheel 0.600 m in diameter pivots on a horizontal axis. A rope is wrapped around the outside of the flywheel, and a steady pull of 40.0 N is exerted on the rope. The flywheel starts from rest, and 5.00 m of rope are unwound in 2.00 s. a) What is the angular acceleration of the flywheel? b) What is its final angular speed? c) What is its final kinetic energy? d) What is its moment of inertia about the rotation axis?

10.55 A wheel starts from rest and rotates with constant angular acceleration about a fixed axis. a) Prove that the power at any given time is proportional to the square of the net torque about the axis. b) If the power at $t = 3.00$ s is 500 W with a constant net torque of 20.0 N·m, what would the power have been at $t = 3.00$ s with a constant net torque of 60.0 N·m? c) Prove that the power at any given angular displacement is proportional to the $\frac{3}{2}$ power of the total torque about the axis at that angular displacement. d) If the power after rotating 37.5 rad with a torque of 20.0 N·m is 500 W, what would the power have been after rotating 37.5 rad with a torque of 60.0 N·m? e) Do the answers to parts (a) and (b) contradict the answers to parts (c) and (d)? Why or why not?

10.56 A beam of length l lies on the $+x$-axis with its left end at the origin. A cable pulls on the beam in the $+y$-direction with a force $\vec{F}$ with a magnitude that depends on where along the beam it is exerted, $F = F_0(1 - x/l)$, where F_0 is a constant equal to the force magnitude when it is applied at the left end of the beam. a) What is the direction of the torque due to $\vec{F}$? The rotation axis is perpendicular to the beam and passes through the origin. b) Graph F versus x from $x = 0$ to $x = l$. Express F in terms of F_0 and x in terms of l. c) Graph the torque versus x from $x = 0$ to $x = l$. Express the torque in terms of $F_0 l$, and express x in terms of l. d) Where along the beam should the force be applied to produce maximum torque and what is that maximum torque?

10.57 While exploring a castle, Exena the Exterminator is spotted by a dragon who chases her down a hallway. Exena runs into a room and attempts to swing the heavy door shut before the dragon gets her. The door is initially perpendicular to the wall, so it must be turned through 90° to close. The door is 3.00 m tall and 1.25 m wide, and it weighs 750 N. You can ignore the friction at the hinges. If Exena applies a force of 220 N at the edge of the door and perpendicular to it, how much time does it take her to close the door?

10.58 A thin rod of length l lies on the $+x$-axis with its left end at the origin. A string pulls on the rod with a force $\vec{F}$ directed toward a point P a distance h above the rod. Where along the rod should you attach the string to get the greatest torque about the origin if point P is a) above the right end of the rod? b) above the left end of the rod? c) above the center of the rod?

10.59 Balancing Act. Attached to one end of a long, thin, uniform rod of length L and mass M is a small blob of clay of the same mass M. a) Locate the position of the center of mass of the system of rod and clay. Note this position on a drawing of the rod. b) You carefully balance the rod on a frictionless tabletop so that it is standing vertically, with the end without the clay touching the table. If the rod is now tipped so that it is a small angle θ away from the vertical, determine its angular acceleration at this instant. Assume that the end without the clay remains in contact with the tabletop. (*Hint:* See Table 9.2.) c) You again balance the rod on the frictionless tabletop so that it is standing vertically, but now the end of the rod *with* the clay is touching the table. If the rod is again tipped so that it is a small angle θ away from the vertical, determine its angular acceleration at this instant. Assume that the end with the clay remains in contact with the tabletop. How does this compare to the angular acceleration in part (b)? d) A pool cue is a tapered wooden rod that is thick at one end and thin at the other. You can easily balance a pool cue vertically on one finger if the thin end is in contact with your finger; this is quite a bit harder to do if the thick end is in contact with your finger. Explain why there is a difference.

10.60 You connect a light string to a point on the edge of a uniform vertical disk with radius R and mass M. The disk is free to rotate without friction about a stationary horizontal axis through its center. Initially, the disk is at rest with the string connection at the highest point on the disk. You pull the string with a constant horizontal force $\vec{F}$ until the wheel has made exactly one-quarter revolution about a horizontal axis through its center, and then let go. a) Use Eq. (10.23) to find the work done by the string. b) Use Eq. (6.14) to find the work done by the string. Do you obtain the same result as in part (a)? c) Find the final angular speed of the disk. d) Find the maximum tangential acceleration of a point on the disk. e) Find the maximum radial (centripetal) acceleration of a point on the disk.

10.61 The mechanism shown in Fig. 10.50 is used to raise a crate of supplies from a ship's hold. The crate has total mass 50 kg. A rope is wrapped around a wooden cylinder that turns on a metal axle. The cylinder has radius 0.25 m and a moment of inertia $I = 2.9$ kg·m^2 about the axle. The crate is suspended from the free end of the rope. One end of the axle pivots on frictionless bearings; a crank handle is attached to the other end. When the crank is turned, the end of the handle rotates about the axle

Figure 10.50 Problem 10.61.

in a vertical circle of radius 0.12 m, the cylinder turns, and the crate is raised. What magnitude of the force $\vec{F}$ applied tangentially to the rotating crank is required to raise the crate with an acceleration of 0.80 m/s^2? (You can ignore the mass of the rope as well as the moments of inertia of the axle and the crank.)

10.62 A large 16.0-kg roll of paper with radius $R = 18.0$ cm rests against the wall and is held in place by a bracket attached to a rod through the center of the roll (Fig. 10.51). The rod turns without friction in the bracket, and the moment of inertia of the paper and rod about the axis is 0.260 kg·m^2. The other end of the bracket is attached by a frictionless hinge to the wall such that the bracket makes an angle of 30.0° with the wall. The weight of the bracket is negligible. The coefficient of kinetic friction between the paper and the wall is $\mu_k = 0.25$. A constant vertical force $F = 40.0$ N is applied to the paper, and the paper unrolls. a) What is the magnitude of the force that the rod exerts on the paper as it unrolls? b) What is the magnitude of the angular acceleration of the roll?

30.0°

R

40.0 N

Figure 10.51 Problem 10.62.

10.63 A block with mass $m = 5.00$ kg slides down a surface inclined 36.9° to the horizontal (Fig. 10.52). The coefficient of kinetic friction is 0.25. A string attached to the block is wrapped around a flywheel on a fixed axis at O. The flywheel has mass 25.0 kg and moment of inertia 0.500 kg·m^2 with respect to the axis of rotation. The string pulls without slipping at a perpendicular distance of 0.200 m from that axis. a) What is the acceleration of the block down the plane? b) What is the tension in the string?

O

5.00 kg

36.9°

Figure 10.52 Problem 10.63.

10.64 Two metal disks, one with radius $R_1 = 2.50$ cm and mass $M_1 = 0.80$ kg and the other with radius $R_2 = 5.00$ cm and mass $M_2 = 1.60$ kg, are welded together and mounted on a frictionless axis through their common center, as in Problem 9.89. a) A light string is wrapped around the edge of the smaller disk and a 1.50-kg block is suspended from the free end of the string. What is the magnitude of the downward acceleration of the block after it is released? b) Repeat the calculation of part (a), this time with the string wrapped around the edge of the larger disk. In which case is the acceleration of the block the greatest? Does your answer make sense?

10.65 A lawn roller in the form of a thin-walled, hollow cylinder with mass M is pulled horizontally with a constant horizontal force F applied by a handle attached to the axle. If it rolls without slipping, find the acceleration and the friction force.

10.66 In Fig. 10.53, find the magnitudes of the linear accelerations of blocks A and B, the magnitude of the angular acceleration of the wheel C, and the tension in each side of the cord if there is no slipping between the cord and the surface of the wheel. The masses of blocks A and B are m_A and m_B, respectively, with $m_A > m_B$.

The wheel has movement of inertia I and radius R.

10.67 A solid disk is rolling without slipping on a level surface at a constant speed of 2.50 m/s. a) If the disk rolls up a 30.0° ramp, how far along the ramp will it move before it stops? b) Explain why your answer to (a) does not depend on either the mass or the radius of the disk.

10.68 The Yo-yo. A yo-yo is made from two uniform disks, each with mass m and radius R, connected by a light axle of radius b. A light, thin string is wound several times around the axle and then held stationary while the yo-yo is released from rest, dropping as the string unwinds. Find the linear acceleration and angular acceleration of the yo-yo and the tension in the string.

10.69 A uniform marble of radius r starts from rest with its center of mass a height h above the lowest point of a loop-the-loop track of radius R, a track shaped like the one shown in Fig. 7.32. It rolls without slipping. Rolling friction and air resistance are negligible. a) What is the minimum value of h such that the marble will not leave the track at the top of the loop? (*Note:* The radius r is *not* negligible compared to the radius R.) b) What is the answer to part (a) if the track is well lubricated, making friction negligible?

10.70 Fig. 10.54 shows three identical yo-yos initially at rest on a horizontal surface. For each yo-yo, the string is pulled in the direction shown. In each case, there is sufficient friction for the yo-yo to roll without slipping. Draw the free-body diagram for each yo-yo. In what direction will each yo-yo rotate? (Try it!) Explain your answers.

C

A

B

Figure 10.53 Problem 10.66.

F

F

F

Figure 10.54 Problem 10.70.

10.71 As shown in Fig. 10.45, a string is wrapped several times around the rim of a small hoop with radius 0.0800 m and mass 0.180 kg. The free end of the string is pulled upward in just the right way so that the hoop does not move vertically as the string unwinds. a) Find the tension in the string as the string unwinds. b) Find the angular acceleration of the hoop as the string unwinds. c) Find the upward acceleration of the hand that pulls on the free end of the string. d) How would your answers be different if the hoop were replaced by a solid disk of the same mass and radius?

10.72 Starting from rest, a constant force $F = 100$ N is applied to the free end of a 50-m cable wrapped around the outer rim of a uniform solid cylinder, similar to the situation shown in Fig. 10.9(a).

The cylinder has mass 4.00 kg and diameter 30.0 cm and is free to turn about a fixed, frictionless axle through its center. a) How long does it take to unwrap all the cable, and how fast is the cable moving just as the last bit comes off? b) Now suppose that the cylinder is replaced by a uniform hoop, with all other quantities remaining unchanged. In this case, would the answers to the questions in part (a) be larger or smaller? Explain.

10.73 A uniform marble rolls without slipping down the path shown in Fig. 10.55, starting from rest. a) Find the minimum height h required for the marble not to fall into the pit. b) The moment of inertia of the marble depends on its radius. Explain why the answer to part (a) does not depend on the radius of the marble. c) Solve part (a) for a block that slides without friction instead of the rolling marble. How does the minimum h in this case compare to the answer in part (a)?

Figure 10.55 Exercise 10.73.

10.74 Rolling Stones. A solid, uniform spherical boulder starts from rest and rolls down a 50.0-m high hill, as shown in Fig. 10.56. The top half of the hill is rough enough to cause the boulder to roll without slipping, but the lower half is covered with ice and there is no friction. What is the translational speed of the boulder when it reaches the bottom of the hill?

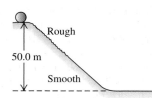

Figure 10.56 Problem 10.74.

10.75 A solid, uniform ball rolls without slipping up a hill, as shown in Fig. 10.57. At the top of the hill, it is moving horizontally, and then it goes over the vertical cliff. a) How far from the foot of the cliff does the ball land and how fast is it moving just before it lands? b) Notice that when the balls lands, it has a larger translational speed than when it was at the bottom of the hill. Does

Figure 10.57 Problem 10.75.

this mean that the ball somehow gained energy? Explain!

10.76 A 42.0-cm diameter wheel, consisting of a rim and six spokes, is constructed from a thin rigid plastic material having a linear mass density of 25.0 g/cm. This wheel is released from rest at the top of a hill 58.0 m high. a) How fast is it rolling when it reaches the bottom of the hill? b) How would your answer change if the linear mass density and the diameter of the wheel were each doubled?

10.77 A high-wheel antique bicycle has a large front wheel with the foot-powered crank mounted on its axle and a small rear wheel turning independently of the front wheel; there is no chain connecting the wheels. The radius of the front wheel is 65.5 cm and the

radius of the rear wheel is 22.0 cm. Your modern bike has a wheel diameter of 66.0 cm (26 inches) and front and rear sprockets with radii of 11.0 cm and 5.5 cm, respectively. The rear sprocket is rigidly attached to the axle of the rear wheel. You ride your modern bike and turn the front sprocket at 1.00 rev/s. The wheels of both bikes roll along the ground without slipping. a) What is your linear speed when you ride your modern bike? b) At what rate must you turn the crank of the antique bike in order to travel at the same speed as in part (a)? c) What then is the angular speed (in rev/s) of the small rear wheel of the antique bike?

10.78 In a lab experiment you let a uniform ball roll down a curved track. The ball starts from rest and rolls without slipping. While on the track, the ball descends a vertical distance h. The lower end of the track is horizontal and extends over the edge of the lab table; the ball leaves the track traveling horizontally. While free falling after leaving the track, the ball moves a horizontal distance x and a vertical distance y. a) Calculate x in terms of h and y, ignoring the work done by friction. b) Would the answer to part (a) be any different on the moon? c) In doing the experiment very carefully, your measured value of x is consistently a bit smaller than that calculated in part (a). Why? d) What would x be for the same h and y as in part (a) if you let a silver dollar roll down the track? You can ignore the work done by friction.

10.79 In a spring gun, a spring of force constant 400 N/m is compressed 0.15 m. When fired, 80.0% of the elastic potential energy stored in the spring is eventually converted into kinetic energy of a 0.0590-kg uniform ball that is rolling without slipping at the base of a ramp. The ball continues to roll without slipping up the ramp with 90.0% of the kinetic energy at the bottom converted into an increase in gravitational potential energy at the instant it stops. a) What is the speed of the ball's center of mass at the base of the ramp? b) At this position, what is the speed of a point at the top of the ball? c) At this position, what is the speed of a point at the bottom of the ball? d) What maximum vertical height up the ramp does the ball move?

10.80 If a wheel rolls along a horizontal surface at constant speed, the coordinates of a certain point on the rim of the wheel are $x(t) = R[(2\pi t/T) - \sin(2\pi t/T)]$ and $y(t) = R[1 - \cos(2\pi t/T)]$, where R and T are constants. a) Sketch the trajectory of the point from $t = 0$ to $t = 2T$. A curve with this shape is called a *cycloid*. b) What are the meanings of the constants R and T? c) Find the x- and y-components of the velocity and of the acceleration of the point at any time t. d) Find the times at which the point is instantaneously at rest. What are the x- and y-components of the acceleration at these times? e) Find the magnitude of the acceleration of the point. Does it depend on time? Compare to the magnitude of the acceleration of a particle in uniform circular motion, $a_{rad} = 4\pi^2 R/T^2$. Explain your result for the magnitude of the acceleration of the point on the rolling wheel, using the idea that rolling is a combination of rotational and translational motion.

10.81 A child rolls a 0.600-kg basketball up a long ramp. The basketball can be considered a thin-walled hollow sphere. When the child releases the basketball at the bottom of the ramp, it has a speed of 8.0 m/s. When the ball returns to her after rolling up the ramp and then rolling back down, it has a speed of 4.0 m/s. Assume the work done by friction on the basketball is the same when the ball moves up or down the ramp and that the basketball rolls without slipping. Find the maximum vertical height increase of the ball as it rolls up the ramp.

10.82 A wheel rotates from rest about a fixed axis through its center of mass such that $\theta = bt^3$, where b is a positive constant with units rad/s^3. a) Use Eq. (10.23) to show that the work done by the net torque on the wheel when it has turned through an angle θ is $\frac{9}{2}I_{cm}b^{2/3}\theta^{4/3}$. b) Use Eq. (9.3) to calculate the angular speed of the wheel when it has turned through an angle θ. c) Use the result of part (b) to calculate the kinetic energy of the wheel when it has turned through an angle θ. Is the work-energy theorem, Eq. (10.25), obeyed? Explain.

10.83 A uniform, solid cylinder with mass M and radius $2R$ rests on a horizontal tabletop. A string is attached by a yoke to a frictionless axle through the center of the cylinder so that the cylinder can rotate about the axle. The string runs over a disk-shaped pulley mass M and radius R that is mounted on a frictionless axle through its center. A block of mass M is suspended from the free end of the string (Fig. 10.58). The string doesn't slip over the pulley surface, and the cylinder rolls without slipping on the table top. Find the magnitude of the acceleration of the block after the system is released from rest.

Figure 10.58 Problem 10.83.

10.84 A uniform drawbridge 8.00 m long is attached to the roadway by a frictionless hinge at one end and can be raised by a cable attached to the other end. The bridge is at rest, suspended at 60.0° above the horizontal, when the cable suddenly breaks. a) Find the angular acceleration of the drawbridge just after the cable breaks. b) Could you use Eq.(9.7) to calculate the angular speed of the drawbridge at a later time? Explain. c) What is the angular speed of the drawbridge as it becomes horizontal?

10.85 A 5.00-kg ball is dropped from a height of 12.0 m above one end of a uniform bar that pivots at its center. The bar has mass 8.00 kg and is 4.00 m in length. At the other end of the bar sits another 5.00 kg ball, unattached to the bar. The dropped ball sticks to the bar after the collision. How high will the other ball go after the collision?

10.86 A uniform, 0.0300-kg rod of length 0.400 m rotates in a horizontal plane about a fixed axis through its center and perpendicular to the rod. Two small rings, each with mass 0.0200 kg, are mounted so that they can slide along the rod. They are initially held by catches at positions 0.0500 m on each side of the center of the rod, and the system is rotating at 30.0 rev/min. Without otherwise changing the system, the catches are released, and the rings slide outward along the rod and fly off at the ends. a) What is the angular speed of the system at the instant when the rings reach the ends

of the rod? b) What is the angular speed of the rod after the rings leave it?

10.87 A uniform rod of length L rests on a frictionless horizontal surface. The rod pivots about a fixed, frictionless axis at one end. The rod is initially at rest. A bullet traveling parallel to the horizontal surface and perpendicular to the rod with speed v strikes the rod at its center and becomes embedded in it. The mass of the bullet is one-fourth the mass of the rod. a) What is the final angular speed of the rod? b) What is the ratio of the kinetic energy of the system after the collision to the kinetic energy of the bullet before the collision?

10.88 The solid wood door of a gymnasium is 1.00 m wide and 2.00 m high, has total mass 35.0 kg, and is hinged along one side. The door is open and at rest when a stray basketball hits the center of the door head-on, applying an average force of 1500 N to the door for 8.00 ms. Find the angular speed of the door after the impact. (*Hint:* Integrating Eq. (10.32) yields $\Delta L_z = \int_{t_1}^{t_2}(\sum \tau_z)\,dt = (\sum \tau_z)_{av}\Delta t$. The quantity $\int_{t_1}^{t_2}(\sum \tau_z)\,dt$ is called the *angular impulse*.)

10.89 A target in a shooting gallery consists of a vertical square wooden board, 0.250 m on a side and with mass 0.750 kg, that pivots on a horizontal axis along its top edge. The board is struck face-on at its center by a bullet with mass 1.90 g that is traveling at 360 m/s and that remains embedded in the board. a) What is the angular speed of the board just after the bullet's impact? b) What maximum height above the equilibrium position does the center of the board reach before starting to swing down again? c) What minimum bullet speed would be required for the board to swing all the way over after impact?

10.90 Neutron Star Glitches. Occasionally, a rotating neutron star (Exercise 10.38) undergoes a sudden and unexpected speedup called a *glitch*. One explanation is that a glitch occurs when the crust of the neutron star settles slightly, decreasing the moment of inertia about the rotation axis. A neutron star with angular speed $\omega_0 = 70.4$ rad/s underwent such a glitch in October 1975 that increased its angular speed to $\omega = \omega_0 + \Delta\omega$, where $\Delta\omega/\omega_0 = 2.01 \times 10^{-6}$. If the radius of the neutron star before the glitch was 11 km, by how much did its radius decrease in the starquake? Assume that the neutron star is a uniform sphere.

10.91 Disks A and B are mounted on a shaft SS and may be connected or disconnected by a clutch C (Fig. 10.59). Disk A is made of a lighter material than disk B so that the moment of inertia of disk A about the shaft is one-third that of disk B. The moments of inertia of the shaft and clutch are negligible. With the clutch disconnected, A is brought up to an angular speed ω_0. The accelerating torque is then removed from A, and A is coupled to disk B by the clutch. (You can ignore bearing friction.) It is found that 2400 J of

Figure 10.59 Problem 10.91.

thermal energy is developed in the clutch when the connection is made. What was the original kinetic energy of disk *A?*

10.92 A small block with mass 0.250 kg is attached to a string passing through a hole in a frictionless, horizontal surface (see Fig. 10.47). The block is originally revolving in a circle with a radius of 0.800 m about the hole with a tangential speed of 4.00 m/s. The string is then pulled slowly from below, shortening the radius of the circle in which the block revolves. The breaking strength of the string is 30.0 N. What is the radius of the circle when the string breaks?

10.93 A horizontal, plywood disk with mass 7.00 kg and diameter 1.00 m pivots on frictionless bearings about a vertical axis through its center. You attach a circular model-railroad track of negligible mass and average diameter 0.95 m to the disk. A 1.20-kg, battery-driven model train rests on the tracks. To demonstrate conservation of angular momentum, you switch on the train's engine. The train moves counterclockwise, soon attaining a constant speed of 0.600 m/s relative to the tracks. Find the magnitude and direction of the angular velocity of the disk relative to the earth.

10.94 One particle of mass m moves at a constant speed v in a circle of radius R that is a distance R above the xz-plane. Another particle of mass m moves the same way at the same speed in another circle of radius R that is a distance R below the xz-plane. The two particles are one-half a revolution apart, so that when one is at (x, R, z), the other is at $(-x, -R, -z)$. Thus their center of mass is at the origin but their rotation axis (the y-axis) is not an axis of symmetry. a) Sketch these two particles at the instant they are at $(R, R, 0)$ and $(-R, -R, 0)$, and show their position, velocity, and angular momentum vectors with respect to the origin. b) Show that at any time the two particles have equal angular momentum. c) What angle do $\vec{\omega}$ (the angular velocity vector of the system of two particles) and the total angular momentum of the system make with each other? d) Show that the y-component of the system's total angular momentum is constant and equal to $L_y = 2mvR$. e) What is the y-component of the total torque acting on the system? f) Find the magnitude of the net force acting on each particle and the magnitude of the total torque acting on the system. g) Show, using your sketch from part (a), the direction of the net torque on the system and that this torque rotates parallel to the xz-plane.

10.95 In a physics laboratory you do the following ballistic pendulum experiment. You shoot a ball of mass m horizontally from a spring gun with a speed v. The ball is immediately caught a distance r below a frictionless pivot by a pivoted catcher assembly of mass M. The moment of inertia of this assembly about its rotation axis through the pivot is I. The distance r is much greater than the radius of the ball. a) Use conservation of angular momentum to show that the angular speed of the ball and catcher just after the ball is caught is $\omega = mvr/(mr^2 + I)$. b) After the ball is caught, the center of mass of the ball-catcher assembly system swings up with a maximum height increase h. Use conservation of energy to show that $\omega = \sqrt{2(M + m)gh/(mr^2 + I)}$. c) Your lab partner says that linear momentum is conserved in the collision and derives the expression $mv = (m + M)V$, where V is the speed of the ball immediately after the collision. She then uses conservation of

energy to derive that $V = \sqrt{2gh}$, so that $mv = (m + M)\sqrt{2gh}$. Use the results of parts (a) and (b) to show that this equation is satisfied only for the special case when r is given by $I = Mr^2$.

10.96 A 55-kg runner runs around the edge of a horizontal turntable mounted on a vertical, frictionless axis through its center. The runner's velocity relative to the earth has magnitude 2.8 m/s. The turntable is rotating in the opposite direction with an angular velocity of magnitude 0.20 rad/s relative to the earth. The radius of the turntable is 3.0 m, and its moment of inertia about the axis of rotation is 80 kg·m². Find the final angular velocity of the system if the runner comes to rest relative to the turntable. (You can treat the runner as a particle.)

10.97 A Falling Bicycle. The moment of inertia of the front wheel of a bicycle about its axle is 0.085 kg·m², its radius is 0.33 m, and the forward speed of the bicycle is 6.00 m/s. With what angular velocity must the front wheel be turned about a vertical axis to counteract the capsizing torque due to a mass 50.0 kg located 0.040 m horizontally to the right or left of the line of contact of wheels and ground? (Bicycle riders: Compare your own experience and decide whether your answer seems reasonable.)

10.98 Center of Percussion. A baseball bat rests on a frictionless, horizontal surface. The bat has a length of 0.900 m, a mass of 0.800 kg, and its center of mass is 0.600 m from the handle end of the bat (Fig. 10.60). The moment of inertia of the bat about its center of mass is 0.0530 kg·m². The bat is struck by a baseball traveling perpendicular to the bat. The impact applies an impulse $J = \int_{t_1}^{t_2} F\, dt$ at a point a distance x from the handle end of the bat. What must x be so that the handle end of the bat remains at rest as the bat begins to move? (*Hint:* Consider the motion of the center of mass and the rotation about the center of mass. Find x so that these two motions combine to give $v = 0$ for the end of the bat just after the collision. Also, note that integration of Eq. (10.32) gives $\Delta L = \int_{t_1}^{t_2}(\sum \tau)\, dt$ (Problem 10.88).) The point on the bat you have located is called the *center of percussion*. Hitting a pitched ball at the center of percussion of the bat minimizes the "sting" the batter experiences on the hands.

Figure 10.60 Problem 10.98.

10.99 Consider a gyroscope with an axis that is not horizontal, but is inclined from the horizontal by an angle β. Show that the precession angular frequency does not depend on the value of β, but is given by Eq. (10.36).

Challenge Problems

10.100 A uniform ball of radius R rolls without slipping between two rails such that the horizontal distance is d between the two contact points of the rails to the ball. a) In a sketch, show that at any instant $v_{cm} = \omega \sqrt{R^2 - d^2/4}$. Discuss this expression in the limits $d = 0$ and $d = 2R$. b) For a uniform ball starting from rest and descending a vertical distance h while rolling without slipping down a ramp, $v_{cm} = \sqrt{10gh/7}$. Replacing the ramp with the two rails, show that

$$v_{cm} = \sqrt{\frac{10gh}{5 + 2/(1 - d^2/4R^2)}}$$

In each case, the work done by friction has been ignored. c) Which speed in part (b) is smaller? Why? Answer in terms of how the loss of potential energy is shared between the gain in translational and rotational kinetic energies. d) For which value of the ratio d/R do the two expressions for the speed in part (b) differ by 5.0%? By 0.50%?

10.101 When an object is rolling without slipping, the rolling friction force is much less than the friction force when the object is sliding; a silver dollar will roll on its edge much farther than it will slide on its flat side (see Section 5.3). When an object is rolling without slipping on a horizontal surface, we can approximate the friction force to be zero, so that a_x and α_z are approximately zero and v_x and ω_z are approximately constant. Rolling without slipping means $v_x = r\omega_z$ and $a_x = r\alpha_z$. If an object is set into motion on a surface *without* these equalities, sliding (kinetic) friction will act on the object as it slips until rolling without slipping is established. A solid cylinder with mass M and radius R, rotating with angular speed ω_0 about an axis through its center, is set on a horizontal surface for which the kinetic friction coefficient is μ_k. a) Draw a free-body diagram for the cylinder on the surface. Think carefully about the direction of the kinetic friction force on the cylinder. Calculate the accelerations a_x of the center of mass and α_z of rotation about the center of mass. b) The cylinder is initially slipping completely, so initially $\omega_z = \omega_0$ but $v_x = 0$. Rolling without slipping sets in when $v_x = R\omega_z$. Calculate the *distance* the cylinder rolls before slipping stops. c) Calculate the work done by the friction force on the cylinder as it moves from where it was set down to where it begins to roll without slipping.

10.102 A demonstration gyroscope wheel is constructed by removing the tire from a bicycle wheel 0.650 m in diameter, wrapping lead wire around the rim, and taping it in place. The shaft projects 0.200 m at each side of the wheel, and a woman holds the ends of the shaft in her hands. The mass of the system is 8.00 kg; its entire mass may be assumed to be located at its rim. The shaft is horizontal and the wheel is spinning about the shaft at 5.00 rev/s. Find the magnitude and direction of the force each hand exerts on the shaft a) when the shaft is at rest; b) when the shaft is rotating in a horizontal plane about its center at 0.050 rev/s; c) when the shaft is rotating in a horizontal plane about its center at 0.300 rev/s. d) At what rate must the shaft rotate in order that it may be supported at one end only?

10.103 A block with mass m is revolving with linear speed v_1 in a circle of radius r_1 on a frictionless horizontal surface (see Fig. 10.47). The string is slowly pulled from below until the radius of the circle in which the block is revolving is reduced to r_2. a) Calculate the tension T in the string as a function of r, the distance of the block from the hole. Your answer will be in terms of the initial velocity v_1 and the radius r_1. b) Use $W = \int_{r_1}^{r_2} \vec{T}(r) \cdot d\vec{r}$ to calculate the work done by $\vec{T}$ when r changes from r_1 to r_2. c) Compare the results of part (b) to the change in the kinetic energy of the block.

11 EQUILIBRIUM AND ELASTICITY

Flamingos often stand with one foot tucked up against the body, which keeps that foot warm and conserves body heat. To remain in equilibrium, a standing flamingo must keep its center of gravity—the point where the flamingo's weight can be considered to act—over the foot that touches the ground.

? **Flamingos have broad, webbed feet. How does this enhance the bird's stability when standing on one leg?**

We've devoted a good deal of effort to understanding why and how bodies accelerate in response to the forces that act on them. But very often we're interested in making sure that bodies *don't* accelerate. Any building, from the Petronas Towers in Malaysia (the world's tallest) to the humblest shed, must be designed so that it won't topple over. Similar concerns occur with a suspension bridge, a ladder leaning against a wall, or a crane hoisting a bucket full of concrete.

A body that can be modeled as a *particle* is in equilibrium whenever the vector sum of the forces acting on it is zero. But for the situations we've just described, that condition isn't enough. If forces act at different points on an extended body, an additional requirement must be satisfied to ensure that the body has no tendency to *rotate:* the sum of the *torques* about any point must be zero. This requirement is based on the principles of rotational dynamics developed in Chapter 10. We can compute the torque due to the weight of a body using the concept of center of mass from Section 8.5 and the related concept of center of gravity, which we introduce in this chapter.

Rigid bodies don't bend, stretch, or squash when forces act on them. But the rigid body is an idealization; all real materials are *elastic* and do deform to some extent. Elastic properties of materials are tremendously important. You want the wings of an airplane to be able to bend a little, but you'd rather not have them break off. The steel frame of an earthquake-resistant building has to be able to flex, but not too much. Many of the necessities of everyday life, from rubber bands to suspension bridges, depend on the elastic properties of materials. In this chapter we'll introduce the concepts of *stress, strain,* and *elastic modulus* and a simple principle called *Hooke's law* that helps us predict what deformations will occur when forces are applied to a real (not perfectly rigid) body.

11.1 | Conditions For Equilibrium

We learned in Sections 4.2 and 5.1 that a particle is in equilibrium—that is, the particle does not accelerate—in an inertial frame of reference if the vector sum of all the forces acting on the particle is zero, $\sum \vec{F} = 0$. The equivalent statement for an *extended* body is that the center of mass of the body has zero acceleration if the vector sum of all external forces acting on the body is zero, as discussed in Section 8.5. This is often called the **first condition for equilibrium.** In vector and component forms,

$$\sum \vec{F} = 0$$

$$\sum F_x = 0 \qquad \sum F_y = 0 \qquad \sum F_z = 0 \qquad (11.1)$$

$$\text{(first condition for equilibrium)}$$

where the sum includes *external* forces only.

A second condition for an extended body to be in equilibrium is that the body must have no tendency to *rotate.* This condition is based on the dynamics of rotational motion in exactly the same way that the first condition is based on Newton's first law. A rigid body that, in an inertial frame, is not rotating about a certain point has zero angular momentum about that point, $\vec{L} = 0$. If it is not to start rotating about that point, the rate of change of angular momentum $d\vec{L}/dt$ must *also* be zero. From the discussion in Section 10.5, particularly Eq. (10.32), this means that the sum of torques $\sum \vec{\tau}$ due to all the external forces acting on the body must be zero. A rigid body in equilibrium can't have any tendency to start rotating about *any* point, so the sum of external torques must be zero about any point. This is the **second condition for equilibrium:**

$$\sum \vec{\tau} = 0 \quad \text{about any point} \qquad \text{(second condition for equilibrium)} \quad (11.2)$$

The sum of the torques due to all external forces acting on the body, with respect to any specified point, must be zero.

In this chapter we will apply the first and second conditions for equilibrium to situations in which a rigid body is at rest (no translation or rotation). Such a body is said to be in **static equilibrium.** But the same conditions apply to a rigid body in uniform *translational* motion (without rotation), such as an airplane in flight with constant speed, direction, and altitude. Such a body is in equilibrium but is not static.

Test Your Understanding

Which of the following situations satisfy both the first and second conditions for equilibrium: (a) a seagull gliding at a constant angle below the horizontal and at a constant speed; (b) an automobile crankshaft turning at an increasing angular speed in the engine of a parked car; (c) a thrown baseball that does not rotate as it sails through the air?

11.2 | Center of Gravity

In most equilibrium problems, one of the forces acting on the body is its weight. We need to be able to calculate the *torque* of this force. The weight doesn't act at a single point; it is distributed over the entire body. But we can always calculate the torque due to the body's weight by assuming that the entire force of gravity (weight) is concentrated at a point called the **center of gravity** (abbreviated "cg"). The acceleration due to gravity $\vec{g}$ decreases with altitude; but if we can ignore this variation over the vertical dimension of the body, then the body's center of gravity is identical with its *center of mass,* which we defined in Section 8.5. We stated this result without proof in Section 10.2, and now we'll prove it.

First let's review the definition of the center of mass. For a collection of particles with masses $m_1, m_2, \ldots$ and coordinates $(x_1, y_1, z_1), (x_2, y_2, z_2), \ldots$, the coordinates $x_{cm}, y_{cm},$ and z_{cm} of the center of mass are given by

$$x_{cm} = \frac{m_1 x_1 + m_2 x_2 + m_3 x_3 + \cdots}{m_1 + m_2 + m_3 + \cdots} = \frac{\sum_i m_i x_i}{\sum_i m_i}$$

$$y_{cm} = \frac{m_1 y_1 + m_2 y_2 + m_3 y_3 + \cdots}{m_1 + m_2 + m_3 + \cdots} = \frac{\sum_i m_i y_i}{\sum_i m_i}$$

$$z_{cm} = \frac{m_1 z_1 + m_2 z_2 + m_3 z_3 + \cdots}{m_1 + m_2 + m_3 + \cdots} = \frac{\sum_i m_i z_i}{\sum_i m_i} \quad \text{(center of mass)} \quad (11.3)$$

Also, $x_{cm}, y_{cm},$ and z_{cm} are the components of the position vector $\vec{r}_{cm}$ of the center of mass, so Eqs. (11.3) are equivalent to the vector equation

$$\vec{r}_{cm} = \frac{m_1 \vec{r}_1 + m_2 \vec{r}_2 + \cdots}{m_1 + m_2 + \cdots} = \frac{\sum_i m_i \vec{r}_i}{\sum_i m_i} \quad (11.4)$$

Now let's consider the gravitational torque on a body of arbitrary shape (Fig. 11.1). We assume that the acceleration due to gravity $\vec{g}$ has the same magnitude and direction at every point in the body. Every particle in the body experiences a gravitational force, and the total weight of the body is the vector sum of a large number of parallel forces. A typical particle has mass m_i and weight $\vec{w}_i = m_i \vec{g}$. If $\vec{r}_i$ is the position vector of this particle with respect to an arbitrary origin O, the torque vector $\vec{\tau}_i$ of the weight $\vec{w}_i$ with respect to O is, from Eq. (10.3),

$$\vec{\tau}_i = \vec{r}_i \times \vec{w}_i = \vec{r}_i \times m_i \vec{g}$$

The *total* torque due to the gravitational forces on all the particles is

$$\vec{\tau} = \sum_i \vec{\tau}_i = \vec{r}_1 \times m_1 \vec{g} + \vec{r}_2 \times m_2 \vec{g} + \cdots$$

$$= (m_1 \vec{r}_1 + m_2 \vec{r}_2 + \cdots) \times \vec{g}$$

$$= \left(\sum_i m_i \vec{r}_i \right) \times \vec{g}$$

When we multiply and divide this by the total mass of the body,

$$M = m_1 + m_2 + \cdots = \sum_i m_i$$

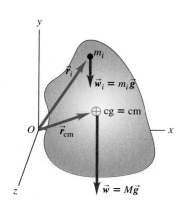

11.1 The gravitational torque about any point can be found by assuming that all the weight of the body acts at its center of gravity (cg). The cg is identical to the center of mass (cm) if $\vec{g}$ is the same at all points on the body.

we get

$$\vec{\tau} = \frac{m_1\vec{r}_1 + m_2\vec{r}_2 + \cdots}{m_1 + m_2 + \cdots} \times M\vec{g} = \frac{\sum_i m_i\vec{r}_i}{\sum_i m_i} \times M\vec{g}$$

The fraction in this equation is just the position vector $\vec{r}_{cm}$ of the center of mass, with components x_{cm}, y_{cm}, and z_{cm}, as given by Eq. (11.4), and $M\vec{g}$ is equal to the total weight $\vec{w}$ of the body. Thus

$$\vec{\tau} = \vec{r}_{cm} \times M\vec{g} = \vec{r}_{cm} \times \vec{w} \qquad (11.5)$$

The total gravitational torque, given by Eq. (11.5), is the same as though the total weight $\vec{w}$ were acting on the position $\vec{r}_{cm}$ of the center of mass, which we also call the *center of gravity*. **If $\vec{g}$ has the same value at all points on a body, its center of gravity is identical to its center of mass.** Note, however, that the center of mass is defined independently of any gravitational effect.

While the value of $\vec{g}$ does vary somewhat with elevation, the variation is extremely slight (Fig. 11.2). Hence we will assume throughout this chapter that the center of gravity and center of mass are identical unless explicitly stated otherwise.

11.2 The acceleration due to gravity at the bottom of the 452-m-tall Petronas Towers in Malaysia is only 0.014% greater than at the top. The center of gravity of the towers is only about 2 cm below the center of mass.

Finding and Using the Center of Gravity

We can often use symmetry considerations to locate the center of gravity of a body, just as we did for center of mass. The center of gravity of a homogeneous sphere, cube, circular sheet, or rectangular plate is at its geometric center. The center of gravity of a right circular cylinder or cone is on its axis of symmetry.

For a body of more complex shape, we can sometimes locate the center of gravity by thinking of the body as being made of symmetrical pieces. For example, we could approximate the human body as a collection of solid cylinders, with a sphere for the head. Then we can compute the coordinates of the center of gravity of the combination from Eqs. (11.3), letting m_1, m_2, ... be the masses of the individual pieces and (x_1, y_1, z_1), (x_2, y_2, z_2), ... be the coordinates of their centers of gravity.

When a body acted on by gravity is supported or suspended at a single point, the center of gravity is always at or directly above or below the point of suspension. If it were anywhere else, the weight would have a torque with respect to the point of suspension, and the body could not be in rotational equilibrium. This fact can be used to determine experimentally the location of the center of gravity of an irregular body. To see how this works, hold one corner of a piece of paper gently between your thumb and forefinger so that the paper hangs below your hand. Draw a line on the paper that extends vertically downward from the point of suspension; the center of gravity must lie somewhere along this line. Then hold the paper by a different corner and draw a line vertically downward from the new point of suspension. The point where the lines intersect is the paper's center of gravity. (Try this with a coffee cup suspended first from the upper part of its handle, then from the lower part of its handle.)

Using the same reasoning, we can see that a body supported at several points must have its center of gravity somewhere within the area bounded by the sup-

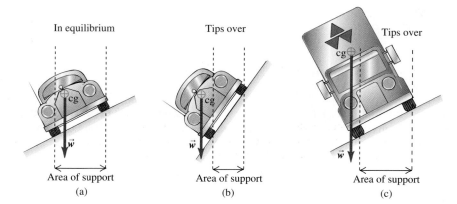

11.3 In (a) the center of gravity is within the area bounded by the supports, and the car is in equilibrium. The car in (b) and the truck in (c) will tip over because their centers of gravity lie outside the area of support.

7.2   A Tilted Beam: Torques and Equilibrium

7.3   Arm Levers

ports. This explains why a car can drive on a straight but slanted road if the slant angle is relatively small (Fig. 11.3a) but will tip over if the angle is too great (Fig. 11.3b). The truck in Fig. 11.3c has a higher center of gravity than the car and will tip over on a shallower incline; when a truck overturns on a highway and blocks traffic for hours, it's the high center of gravity that's to blame.

The lower the center of gravity and the larger the area of support, the more difficult it is to overturn a body. Four-legged animals such as deer and horses have a large area of support bounded by their legs; hence they are naturally stable and need only small feet or hooves. Animals that walk erect on two legs, such as humans and birds, need relatively large feet to give them a reasonable area of support. If a two-legged animal holds its body approximately horizontal, like a chicken or the dinosaur *Tyrannosaurus rex,* it must perform a delicate balancing act as it walks to keep its center of gravity over the foot that is on the ground. A chicken does this by moving its head; *T. rex* probably did it by moving its massive tail.

Example 11.1   **Walking the plank**

A uniform wooden plank of length $L = 6.0$ m and mass $M = 90$ kg rests on top of two sawhorses separated by $D = 1.5$ m, located equal distances from the center of the plank (Fig. 11.4). Your cousin Throckmorton tries to stand on the right-hand end of the plank. If the plank is to remain at rest, how massive can Throckmorton be?

11.4 Throckmorton and the plank, just barely in equilibrium.

SOLUTION

IDENTIFY: If the system of plank and Throckmorton is just in balance, the center of gravity of this system will be directly over the right-hand sawhorse (just barely within the area bounded by the two supports). The target variable is Throcky's mass.

SET UP: We take the origin at C, the geometrical center and center of gravity of the uniform plank, and take the positive x-axis to point horizontally to the right. Then the x-coordinates of the centers of gravity of the plank (mass M) and Throcky (unknown mass m) are $x_P = 0$ and $x_T = l/2 = 3.0$ m, respectively. We will use Eqs. (11.3) to locate the center of gravity of the system of plank and Throcky.

EXECUTE: From the first of Eqs. (11.3),

$$x_{cg} = \frac{M(0) + m(L/2)}{M + m} = \frac{m}{M + m}\frac{L}{2}$$

Setting this equal to $D/2$, the x-coordinate of the right-hand sawhorse, we have

$$\frac{m}{M+m}\frac{L}{2} = \frac{D}{2}$$
$$mL = (M+m)D$$
$$m = M\frac{D}{L-D} = (90 \text{ kg})\frac{1.5 \text{ m}}{6.0 \text{ m} - 1.5 \text{ m}}$$
$$= 30 \text{ kg}$$

EVALUATE: To check our result, let's repeat the calculation with a different choice of origin. Now we take the origin to be at S, the position of the right-hand sawhorse, so that $x_{cg} = 0$. The centers of gravity of the plank and Throcky are now at $x_P = -D/2$ and $x_T = (L/2) - (D/2)$ respectively, so

$$x_{cg} = \frac{M(-D/2) + m[(L/2) - (D/2)]}{M+m} = 0$$

$$m = \frac{MD/2}{(L/2) - (D/2)} = M\frac{D}{L-D} = 30 \text{ kg}$$

The mass doesn't depend on our arbitrary choice of origin.

A 60-kg child could only stand halfway between the right-hand sawhorse and the end of the plank. Can you see why?

Test Your Understanding

You are constructing a mobile with two wooden fish on one side of a lightweight rod and a counterweight on the other side (Fig. 11.5). What must be the mass m of the counterweight if the rod is to remain horizontal?

11.5 What is the unknown mass m?

11.3 | Solving Rigid-Body Equilibrium Problems

There are just two key principles of rigid-body equilibrium: the vector sum of the forces on the body must be zero, and the sum of torques about any point must be zero. To keep things simple, we'll restrict our attention to situations in which we can treat all forces as acting in a single plane, which we'll call the xy-plane. Then we can ignore the condition $\Sigma F_z = 0$ in Eq. (11.1), and in Eq. (11.2) we need consider only the z-components of torque (perpendicular to the plane). The first and second conditions for equilibrium are then

$$\Sigma F_x = 0 \quad \text{and} \quad \Sigma F_y = 0$$
(first condition for equilibrium, forces in xy-plane)

$$\Sigma \tau_z = 0 \tag{11.6}$$
(second condition for equilibrium, forces in xy-plane)

CAUTION Although the choice of reference point for calculating torques in $\Sigma \tau_z$ is arbitrary, once you choose a point you must use the *same* point to calculate *all* of the torques on a body. It helps to pick the point so as to simplify the calculations as much as possible.

Activ
ONLINE
Physics

7.4 Two Painters on a Beam

7.5 Lecturing from a Beam

The challenge is to apply these simple principles to specific problems. The following strategy is very similar to the suggestions in Section 5.2 for equilibrium of a particle. You should compare it with the strategy for rotational dynamics problems given in Section 10.2.

Problem-Solving
Strategy   **Equilibrium of a Rigid Body**

IDENTIFY *the relevant concepts:* The first and second conditions for equilibrium are useful whenever there is a rigid body that is not rotating and not accelerating in space.

SET UP *the problem using the following steps:*
1. Draw a sketch of the physical situation, including dimensions, and select the body in equilibrium to be analyzed.

2. Draw a free-body diagram showing the forces acting *on* the selected body and no others. *Do not* include forces exerted *by* this body on other bodies. Be careful to show correctly the point at which each force acts; this is crucial for correct torque calculations. You can't represent a rigid body as a point.
3. Choose coordinate axes and specify a positive sense of rotation for torques. Represent forces in terms of their components with respect to the axes you have chosen; when you do this, cross out the original force so that you don't include it twice.
4. In choosing a point to compute torques, note that if a force has a line of action that goes *through* a particular point, the torque of the force with respect to that point is zero. You can often eliminate unknown forces or components from the torque equation by a clever choice of point for your calculation. The body doesn't actually have to be pivoted about an axis through the chosen point.

EXECUTE *the solution as follows:*

1. Write equations expressing the equilibrium conditions. Remember that $\Sigma F_x = 0$, $\Sigma F_y = 0$, and $\Sigma \tau_z = 0$ are

always separate equations; *never* add x- and y-components in a single equation. Also remember that when a force is represented in terms of its components, you can compute the torque of that force by finding the torque of each component separately, each with its appropriate lever arm and sign, and adding the results. This is often easier than determining the lever arm of the original force.

2. You always need as many equations as you have unknowns. Depending on the number of unknowns, you may need to compute torques with respect to two or more axes to obtain enough equations. Often, there are several equally good sets of force and torque equations for a particular problem; there is usually no single "right" combination of equations. When you have as many independent equations as unknowns, you can solve the equations simultaneously.

EVALUATE *your answer:* A useful way to check your results is to rewrite the second condition for equilibrium, $\Sigma \tau_z = 0$, using a different choice of origin. If you've done everything correctly, you'll get the same answers using this new choice of origin as you did with your original choice.

Example 11.2

Weight distribution for a car

An auto magazine reports that a certain sports car (Fig. 11.6a) has 53% of its weight on the front wheels and 47% on its rear wheels, with a 2.46-m wheelbase. This means that the total normal force on the front wheels is $0.53w$ and that on the rear wheels is $0.47w$, where w is the total weight. The wheelbase is the distance between front and rear axles. How far in front of the rear axle is the car's center of gravity?

SOLUTION

IDENTIFY: We can use the two conditions for equilibrium since the car is assumed to be at rest. The conditions also apply when the car is driving in a straight line at constant speed, since the net force and net torque on the car are also zero in that situation. The target variable is the coordinate of the car's center of gravity.

SET UP: Figure 11.6 shows a free-body diagram for the car, along with the x- and y-axes and our convention that counterclockwise torques are positive. We have drawn the weight w as acting at the center of gravity, and the distance we want is L_{cg}. It is the lever arm of the weight with respect to the rear axle R, so it is reasonable to take torques with respect to R. Note that the torque due to the weight is negative because it tends to cause a clockwise rotation about R. The torque due to the upward normal force at the front wheels is positive because it tends to cause a counterclockwise rotation about R.

EXECUTE: You can see from Fig. 11.6b that the first condition for equilibrium is satisfied: $\Sigma F_x = 0$ because there aren't any x-components of force and $\Sigma F_y = 0$ because $0.47w + 0.53w + (-w) = 0$. The force equation doesn't involve the target vari-

11.6 (a) The car at rest on a horizontal surface. (b) Free-body diagram for the car.

able L_{cg}, so we must solve for it using the torque equation for point R:

$$\sum \tau_R = 0.47w(0) - wL_{cg} + 0.53w(2.46 \text{ m}) = 0$$
$$L_{cg} = 1.30 \text{ m}$$

EVALUATE: Note that the cg is between the two supports, as it must be (see Section 11.2). You can check the numerical result for the cg position by writing the torque equation about the front axle F.

You'll find that the cg is 1.16 m behind the front axle, or $(2.46 \text{ m}) - (1.16 \text{ m}) = 1.30 \text{ m}$ in front of the rear axle.

Can you show that if f is the fraction of the weight on the front wheels and d is the wheelbase, then the center of gravity is a distance fd in front of the rear wheels? In terms of the center of gravity, why do owners of rear-wheel-drive vehicles put bags of sand in their trunks to improve traction on snow and ice? Would this strategy help with a front-wheel-drive car? With a four-wheel-drive car?

Example 11.3	**A heroic rescue**

Sir Lancelot is trying to rescue the Lady Elayne from Castle Von Doom by climbing a uniform ladder that is 5.0 m long and weighs 180 N. Lancelot, who weighs 800 N, stops a third of the way up the ladder (Fig. 11.7a). The bottom of the ladder rests on a horizontal stone ledge and leans across the moat in equilibrium against a vertical wall that is frictionless because of a thick layer of moss. The ladder makes an angle of 53.1° with the horizontal, conveniently forming a 3–4–5 right triangle. a) Find the normal and friction forces on the ladder at its base. b) Find the minimum coefficient of static friction needed to prevent slipping at the base. c) Find the magnitude and direction of the contact force on the ladder at the base.

SOLUTION

IDENTIFY: The system of ladder and Lancelot is stationary, so we can use the two conditions for equilibrium. These conditions by themselves are enough to solve part (a). In part (b), we also need the relationship given in Section 5.3 between the static friction force, the coefficient of static friction, and the normal force. The contact force asked for in part (c) is the vector sum of the normal force and the friction force acting at the base of the ladder, which we find in part (a).

SET UP: Figure 11.7b shows the free-body diagram for the system of the ladder and Lancelot. We choose the x- and y-directions as shown and take counterclockwise torques to be positive. The ladder is described as "uniform," so its center of gravity is at its geometrical center. Lancelot's weight acts at a point on the ladder one-third of the way from the base toward the wall.

The frictionless wall exerts only a normal force n_1 at the top of the ladder. The forces at the base are the upward normal force n_2 and the static friction force f_s, which must point to the right to prevent slipping; the magnitudes n_2 and f_s are the target variables in part (a). From Eq. (5.6), these magnitudes are related by $f_s \leq \mu_s n_2$, where μ_s is the coefficient of static friction (the target variable in part (b)).

EXECUTE: a) From Eqs. (11.6), the first condition for equilibrium gives

$$\sum F_x = f_s + (-n_1) = 0$$
$$\sum F_y = n_2 + (-800 \text{ N}) + (-180 \text{ N}) = 0$$

(a)

(b)

(c)

11.7 (a) Sir Lancelot pauses a third of the way up the ladder, fearing it will slip. (b) Free-body diagram for the system of Sir Lancelot and the ladder. (c) The contact force at B is the superposition of the normal force and the static friction force.

These are two equations for the three unknowns n_1, n_2, and f_s. The first equation tells us that the two horizontal forces must be equal and opposite, and the second equation gives

$$n_2 = 980 \text{ N}$$

The ground pushes up with a force of 980 N to balance the total (downward) weight $(800 \text{ N} + 180 \text{ N})$.

We don't yet have enough equations, but now we can use the second condition for equilibrium. We can take torques about any point we choose. The smart choice is the point that will give us the fewest terms and fewest unknowns in the torque equation. With this in mind, we choose point B, at the base of the ladder. The two forces n_2 and f_s have no torque about that point. From Fig. 11.7b we see that the lever arm for the ladder's weight is 1.5 m, the lever arm for Lancelot's weight is 1.0 m, and the lever arm for n_1 is 4.0 m. The torque equation for point B is

$$\sum \tau_B = n_1(4.0 \text{ m}) - (180 \text{ N})(1.5 \text{ m}) - (800 \text{ N})(1.0 \text{ m})$$
$$+ n_2(0) + f_s(0) = 0$$

Solving for n_1, we get $n_1 = 268 \text{ N}$. We now substitute this back into the $\sum F_x = 0$ equation to get

$$f_s = 268 \text{ N}$$

b) The static friction force f_s cannot exceed $\mu_s n_2$, so the *minimum* coefficient of static friction to prevent slipping is

$$(\mu_s)_{\text{min}} = \frac{f_s}{n_2} = \frac{268 \text{ N}}{980 \text{ N}} = 0.27$$

c) The components of the contact force $\vec{F}_B$ at the base are the static friction force f_s and the normal force n_2, so

$$\vec{F}_B = f_s \hat{\imath} + n_2 \hat{\jmath} = (268 \text{ N})\hat{\imath} + (980 \text{ N})\hat{\jmath}$$

The magnitude and direction of $\vec{F}_B$ (Fig. 11.7c) are then

$$F_B = \sqrt{(268 \text{ N})^2 + (980 \text{ N})^2} = 1020 \text{ N}$$
$$\theta = \arctan \frac{980 \text{ N}}{268 \text{ N}} = 75°$$

EVALUATE: As Fig. 11.7c shows, the contact force $\vec{F}_B$ is *not* directed along the length of the ladder. You may be surprised by this, but there's really no good reason why the two directions should be the same. Can you show that if $\vec{F}_B$ were directed along the ladder, there would be a net counterclockwise torque with respect to the top of the ladder, and equilibrium would be impossible?

Here are a few final comments. First, as Lancelot climbs higher on the ladder, the lever arm and torque of his weight about B increase; this increases the values of n_1, f_s, and $(\mu_s)_{\text{min}}$. At the top, his lever arm would be nearly 3 m, giving a minimum coefficient of static friction of nearly 0.7. The value of μ_s would not be this large for a present-day aluminum ladder on a wood floor, which is why such ladders are usually equipped with nonslip rubber pads. Lancelot's medieval ladder is unlikely to have this feature, and the ladder is likely to slip as he climbs.

Second, a larger ladder angle would decrease the lever arms with respect to B of the weights of the ladder and Lancelot and increase the lever arm of n_1, all of which would decrease the required friction force. The R. D. Werner Ladder Co. recommends that its ladders be used at an angle of 75°. (Why not 90°?)

Finally, if we had assumed friction on the wall as well as on the floor, the problem would be impossible to solve by using the equilibrium conditions alone. (Try it!) Such a problem is said to be *statically indeterminate.* The difficulty is that it's no longer adequate to treat the body as being perfectly rigid. Another simple example of such a problem is a four-legged table; there is no way to use the equilibrium conditions alone to find the force on each separate leg. In a sense, four legs are one too many; three, properly located, are sufficient for stability.

<table>
<tr><td>Example
11.4</td><td>**Another rescue attempt**</td></tr>
</table>

After Lancelot falls into the moat, Sir Gawain tosses a grappling hook through an open upstairs window. In Fig. 11.8a he is resting partway up the rope. Gawain weighs 700 N, his body makes an angle of 60° with the wall, and his center of gravity is 0.85 m from his feet. The rope force acts 1.30 m from his feet, and the rope makes an angle of 20° with the vertical. Find the tension in the rope and the forces exerted on his feet by the moss-free region of the wall.

SOLUTION

IDENTIFY and SET UP: We use the conditions for equilibrium as in the previous example. Figure 11.8b shows the free-body diagram for Gawain. The forces on his feet include a (horizontal) normal component n and a (vertical) frictional component f. The rope tension is T. The angles needed to find the torques are shown.

EXECUTE: Because two unknown forces (n and f) act at point B at Gawain's feet, we take torques about B, obtaining a torque equation in which T is the only unknown. The easiest way to find the torque of each force is to use $\tau = rF \sin \phi$, where ϕ is the angle between the position vector $\vec{r}$ (from B to the point of application of the force) and the force vector $\vec{F}$. Taking counterclockwise torques as positive, we obtain the torque equation

$$\sum \tau_B = (1.30 \text{ m})T(\sin 140°)$$
$$- (0.85 \text{ m})(700 \text{ N})(\sin 60°) + (0)n + (0)f = 0$$

We solve this equation for T; the result is

$$T = 617 \text{ N}$$

Note that the rope tension is *less than* Gawain's weight.

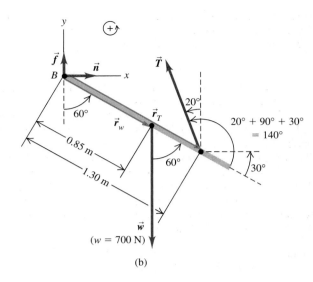

(a)   (b)

11.8 (a) Gawain pausing to catch his breath. (b) Free-body diagram for Gawain, considered as a rod.

To find the components of force n and f at Gawain's feet, we use the first equilibrium condition. From Fig. 11.8b, T_x is negative and T_y is positive, so $T_x = -T\sin 20°$ and $T_y = T\cos 20°$. Then

$$\sum F_x = n + (-T\sin 20°) = 0$$
$$\sum F_y = f + T\cos 20° + (-700\ \text{N}) = 0$$

When we substitute the value of T into these equations, we get

$$n = 211\ \text{N} \quad \text{and} \quad f = 120\ \text{N}$$

EVALUATE: As a check, note that the sum of the vertical friction force ($f = 120\ \text{N}$) and the vertical component of the tension ($T\cos 20° = 580\ \text{N}$) equals Gawain's 700-N weight.

| Example 11.5 | **Equilibrium and pumping iron** |

Figure 11.9a shows a human arm lifting a dumbbell, and Fig. 11.9b is a free-body diagram for the forearm. The forearm is in equilibrium under the action of the weight w of the dumbbell, the tension T in the tendon connected to the biceps muscle, and the force E exerted on the forearm by the upper arm at the elbow joint. For clarity the point A where the tendon is attached is drawn farther away from the elbow than its actual position. The weight w and the angle θ are given; we want to find the tendon tension and the two components of force at the elbow (three unknown scalar quantities in all). We neglect the weight of the forearm itself.

(a)

(b)

11.9 (a) Building up the biceps. (b) Free-body diagram for the forearm. The weight of the forearm is neglected, and the distance D is greatly exaggerated for clarity.

SOLUTION

IDENTIFY and SET UP: The system is at rest, so once again we use the conditions for equilibrium. As shown in Fig. 11.9b, we represent the tendon force in terms of its components T_x and T_y, using the given angle θ and the unknown magnitude T:

$$T_x = T \cos \theta \qquad T_y = T \sin \theta$$

We also represent the force at the elbow in terms of its components E_x and E_y. We'll guess that the directions of these components are as shown in Fig. 11.9b; there's no need to agonize over this guess, since the results for E_x and E_y will tell us the actual directions. Our target variables are the magnitude T of the tendon tension and the components E_x and E_y of the force at the elbow.

EXECUTE: The simplest way to find the tension T is to take torques about the elbow joint. The resulting torque equation does not contain E_x, E_y, or T_x because the lines of action of all these forces pass through this point. The torque equation is then simply

$$\sum \tau_E = Lw - DT_y = 0$$

From this we find

$$T_y = \frac{Lw}{D} \quad \text{and} \quad T = \frac{Lw}{D \sin \theta}$$

To find E_x and E_y, we use the first conditions for equilibrium, $\sum F_x = 0$ and $\sum F_y = 0$:

$$\sum F_x = T_x + (-E_x) = 0$$

$$E_x = T_x = T \cos \theta = \frac{Lw}{D \sin \theta} \cos \theta = \frac{Lw}{D} \cot \theta$$

$$= \frac{Lw}{D} \frac{D}{h} = \frac{Lw}{h}$$

$$\sum F_y = T_y + E_y + (-w) = 0$$

$$E_y = w - \frac{Lw}{D} = -\frac{(L - D)w}{D}$$

The negative sign shows that our guess for the direction of E_y, shown in Fig. 11.9b, was wrong; it is actually vertically *downward*.

EVALUATE: We can check our results by finding E_x and E_y in a different way that uses two more torque equations. We take torques about the tendon attach point, A:

$$\sum \tau_A = (L - D)w + DE_y = 0 \quad \text{and} \quad E_y = -\frac{(L - D)w}{D}$$

Finally, we take torques about point B in the figure:

$$\sum \tau_B = Lw - hE_x = 0 \quad \text{and} \quad E_x = \frac{Lw}{h}$$

We chose points A and B because the tendon tension T has zero torque about either of these points. (Can you see why from Fig. 11.9b?) Notice how much we have simplified these calculations by choosing the point for calculating torques so as to eliminate one or more of the unknown quantities.

In our alternative determination of E_x and E_y, we didn't explicitly use the first condition for equilibrium (that the vector sum of the forces is zero). As a check, compute $\sum F_x$ and $\sum F_y$ to verify that they really *are* zero. Consistency checks are always a good idea!

As a specific example, suppose $w = 200$ N, $D = 0.050$ m, $L = 0.30$ m, and $\theta = 80°$. Then from $\tan \theta = h/D$, we find

$$h = D \tan \theta = (0.050 \text{ m})(5.67) = 0.28 \text{ m}$$

From the previous general results we find

$$T = \frac{Lw}{D \sin \theta} = \frac{(0.30 \text{ m})(200 \text{ N})}{(0.050 \text{ m})(0.98)} = 1220 \text{ N}$$

$$E_y = -\frac{(L - D)w}{D} = -\frac{(0.30 \text{ m} - 0.050 \text{ m})(200 \text{ N})}{0.050 \text{ m}}$$

$$= -1000 \text{ N}$$

$$E_x = \frac{Lw}{h} = \frac{(0.30 \text{ m})(200 \text{ N})}{0.28 \text{ m}} = 210 \text{ N}$$

The magnitude of the force at the elbow is

$$E = \sqrt{E_x^2 + E_y^2} = 1020 \text{ N}$$

In view of the magnitudes of our results, neglecting the weight of the forearm itself, which may be 20 N or so, will cause only relatively small errors in our results.

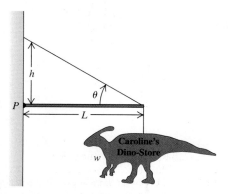

11.10 What is the tension in the diagonal wire and the force exerted by the hinge at P?

Test Your Understanding

A metal advertising sign (weight w) for a specialty shop is suspended from the end of a horizontal rod of length L and negligible mass (Fig. 11.10). The rod is supported by a cable at an angle θ from the horizontal and by a hinge at point P. Find the tension in the cable and the components of force exerted on the rod by the hinge at P.

11.4 | Stress, Strain, and Elastic Moduli

The rigid body is a useful idealized model, but the stretching, squeezing, and twisting of real bodies when forces are applied are often too important to ignore. Figure 11.11 shows three examples. We want to study the relation between the forces and deformations for each case.

(a) Tensile stress

(b) Bulk stress

(c) Shear stress

11.11 Three types of stress. (a) A cable under *tensile stress,* being stretched by forces acting at its ends. (b) A dolphin under *bulk stress,* being squeezed from all sides by forces due to water pressure. (c) A ribbon under *shear stress,* being deformed and eventually cut by forces exerted by the scissors.

For each kind of deformation we will introduce a quantity called **stress** that characterizes the strength of the forces causing the deformation, on a "force per unit area" basis. Another quantity, **strain,** describes the resulting deformation. When the stress and strain are small enough, we often find that the two are directly proportional, and we call the proportionality constant an **elastic modulus.** The harder you pull on something, the more it stretches; and the more you squeeze it, the more it compresses. The general pattern that emerges can be formulated as

$$\frac{\text{Stress}}{\text{Strain}} = \text{Elastic modulus} \quad (\text{Hooke's law}) \qquad (11.7)$$

The proportionality of stress and strain (under certain conditions) is called **Hooke's law,** after Robert Hooke (1635–1703), a contemporary of Newton. We used one form of Hooke's law in Sections 6.3 and 7.2: the elongation of an ideal spring is proportional to the stretching force. Remember that Hooke's law is not really a general law but an experimental finding that is valid only over a limited range. The last section of this chapter discusses what this limited range is.

Tensile and Compressive Stress and Strain

The simplest elastic behavior to understand is the stretching of a bar, rod, or wire when its ends are pulled (Fig. 11.11a). Figure 11.12 shows an object that initially has uniform cross-sectional area A and length l_0. We then apply forces of equal magnitude $F_\perp$ but opposite directions at the ends (this ensures that the object has no tendency to move left or right). We say that the object is in **tension.** We've already talked a lot about tension in ropes and strings; it's the same concept here. The subscript $\perp$ is a reminder that the forces act perpendicular to the cross section.

We define the **tensile stress** at the cross section as the ratio of the force $F_\perp$ to the cross-sectional area A:

$$\text{Tensile stress} = \frac{F_\perp}{A} \qquad (11.8)$$

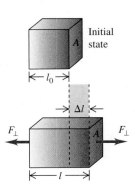

Tensile stress $= \dfrac{F_\perp}{A}$

Tensile strain $= \dfrac{\Delta l}{l_0}$

11.12 An object in tension. The net force on the object is zero, but the object deforms. The tensile stress (the ratio of the force to the cross-sectional area) produces a tensile strain (the elongation divided by the initial length). The elongation Δl is exaggerated for clarity.

This is a *scalar* quantity because $F_\perp$ is the *magnitude* of the force. The SI unit of stress is the **pascal** (abbreviated Pa, and named for the 17th-century French scientist and philosopher Blaise Pascal). Equation (11.8) shows that one pascal equals one newton per square meter (N/m^2):

$$1 \text{ pascal} = 1 \text{ Pa} = 1 \text{ N/m}^2$$

In the British system the logical unit of stress would be the pound per square foot, but the pound per square inch (lb/in.² or psi) is more commonly used. The conversion factors are

$$1 \text{ psi} = 6895 \text{ Pa} \quad \text{and} \quad 1 \text{ Pa} = 1.450 \times 10^{-4} \text{ psi}$$

The units of stress are the same as those of *pressure,* which we will encounter often in later chapters. Air pressure in automobile tires is typically around $3 \times 10^5 \text{ Pa} = 300 \text{ kPa}$, and steel cables are commonly required to withstand tensile stresses of the order of 10^8 Pa.

The object shown in Fig. 11.12 stretches to a length $l = l_0 + \Delta l$ when under tension. The elongation Δl does not occur only at the ends; every part of the bar stretches in the same proportion. The **tensile strain** of the object is equal to the fractional change in length, which is the ratio of the elongation Δl to the original length l_0:

$$\text{Tensile strain} = \frac{l - l_0}{l_0} = \frac{\Delta l}{l_0} \tag{11.9}$$

Tensile strain is stretch per unit length. It is a ratio of two lengths, always measured in the same units, and so is a pure (dimensionless) number with no units.

Experiment shows that for a sufficiently small tensile stress, stress and strain are proportional, as in Eq. (11.7). The corresponding elastic modulus is called **Young's modulus,** denoted by Y:

$$Y = \frac{\text{Tensile stress}}{\text{Tensile strain}} = \frac{F_\perp/A}{\Delta l/l_0} = \frac{F_\perp}{A}\frac{l_0}{\Delta l} \quad (\text{Young's modulus}) \tag{11.10}$$

Since strain is a pure number, the units of Young's modulus are the same as those of stress: force per unit area. Some typical values are listed in Table 11.1. (This table also gives values of two other elastic moduli that we will discuss later in this chapter.) A material with a large value of Y is relatively unstretchable; a large stress is required for a given strain. For example, the value of Y for cast steel $(2 \times 10^{11} \text{ Pa})$ is much larger than that for rubber $(5 \times 10^8 \text{ Pa})$.

When the forces on the ends of a bar are pushes rather than pulls (Fig. 11.13), the bar is in **compression** and the stress is a **compressive stress.** The compres-

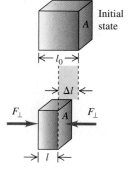

$$\text{Compressive stress} = \frac{F_\perp}{A}$$

$$\text{Compressive strain} = \frac{\Delta l}{l_0}$$

11.13 An object in compression. The compressive stress and compressive strain are defined in the same way as tensile stress and strain (see Fig. 11.12), except that Δl now denotes the distance that the object contracts.

Table 11.1 Approximate Elastic Moduli

Material	Young's Modulus, Y (Pa)	Bulk Modulus, B (Pa)	Shear Modulus, S (Pa)
Aluminum	7.0×10^{10}	7.5×10^{10}	2.5×10^{10}
Brass	9.0×10^{10}	6.0×10^{10}	3.5×10^{10}
Copper	11×10^{10}	14×10^{10}	4.4×10^{10}
Crown glass	6.0×10^{10}	5.0×10^{10}	2.5×10^{10}
Iron	21×10^{10}	16×10^{10}	7.7×10^{10}
Lead	1.6×10^{10}	4.1×10^{10}	0.6×10^{10}
Nickel	21×10^{10}	17×10^{10}	7.8×10^{10}
Steel	20×10^{10}	16×10^{10}	7.5×10^{10}

sive strain of an object in compression is defined in the same way as tensile strain, but Δl has the opposite direction. Hooke's law and Eq. (11.10) are valid for compression as well as tension if the compressive stress is not too great. For many materials, Young's modulus has the same value for both tensile and compressive stresses; composite materials such as concrete are an exception.

In many situations, bodies can experience both tensile and compressive stresses at the same time. As an example, a horizontal beam supported at each end sags under its own weight. As a result, the top of the beam is under compression, while the bottom of the beam is under tension (Fig. 11.14a). To minimize the stress and hence the bending strain, the top and bottom of the beam are given a large cross-sectional area. There is neither compression nor tension along the centerline of the beam, so this part can have a small cross section; this helps to keep the weight of the bar to a minimum and further helps to reduce the stress. The result is an I-beam of the familiar shape used in building construction (Fig. 11.14b).

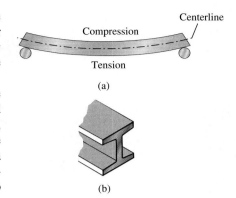

11.14 (a) A beam supported at both ends is under both compression and tension. (b) The cross-sectional shape of an I-beam minimizes both stress and weight.

Example 11.6 **Tensile stress and strain**

A steel rod 2.0 m long has a cross-sectional area of 0.30 cm². The rod is now hung by one end from a support structure and a 550-kg milling machine is hung from the rod's lower end. Determine the stress, the strain, and the elongation of the rod.

SOLUTION

IDENTIFY and SET UP: We use the definitions of stress, strain, and Young's modulus given by Eqs. (11.8), (11.9), and (11.10), respectively, and the value of Y for steel from Table 11.1.

EXECUTE: We find

$$\text{Stress} = \frac{F_\perp}{A} = \frac{(550 \text{ kg})(9.8 \text{ m/s}^2)}{3.0 \times 10^{-5} \text{ m}^2} = 1.8 \times 10^8 \text{ Pa}$$

$$\text{Strain} = \frac{\Delta l}{l_0} = \frac{\text{Stress}}{Y} = \frac{1.8 \times 10^8 \text{ Pa}}{20 \times 10^{10} \text{ Pa}} = 9.0 \times 10^{-4}$$

$$\text{Elongation} = \Delta l = (\text{Strain}) \times l_0 = (9.0 \times 10^{-4})(2.0 \text{ m})$$
$$= 0.0018 \text{ m} = 1.8 \text{ mm}$$

EVALUATE: The small size of this elongation, which results from a load of over half a ton, is a testament to the stiffness of steel.

Bulk Stress and Strain

When a dolphin plunges deep into the ocean, the water exerts nearly uniform pressure everywhere on its surface and squeezes it to a slightly smaller volume (Fig. 11.11b). This is a different situation from the tensile and compressive stresses and strains we have discussed. The stress is now a uniform pressure on all sides, and the resulting deformation is a volume change. We use the terms **bulk stress** (or **volume stress**) and **bulk strain** (or **volume strain**) to describe these quantities.

If an object is immersed in a fluid (liquid or gas) at rest, the fluid exerts a force on any part of the object's surface; this force is *perpendicular* to the surface. (If we tried to make the fluid exert a force parallel to the surface, the fluid would slip sideways to counteract the effort.) The force $F_\perp$ per unit area that the fluid exerts on the surface of an immersed object is called the **pressure** p in the fluid:

$$p = \frac{F_\perp}{A} \qquad (\text{pressure in a fluid}) \qquad (11.11)$$

The pressure in a fluid increases with depth. For example, the pressure of the air is about 21% greater at sea level than in Denver (at an elevation of 1.6 km, or 1.0 mi). If an immersed object is relatively small, however, we can ignore pressure differences due to depth for the purpose of calculating bulk stress. Hence we will treat the pressure as having the same value at all points on an immersed object's surface.

Pressure has the same units as stress; commonly used units include 1 Pa $(=1\ \mathrm{N/m^2})$ and 1 lb/in.^{2} (1 psi). Also in common use is the **atmosphere,** abbreviated atm. One atmosphere is the approximate average pressure of the earth's atmosphere at sea level:

$$1\ \mathrm{atmosphere} = 1\ \mathrm{atm} = 1.013 \times 10^5\ \mathrm{Pa} = 14.7\ \mathrm{lb/in.^2}$$

CAUTION Unlike force, pressure has no intrinsic direction: the pressure on the surface of an immersed object is the same no matter how the surface is oriented. Hence pressure is a *scalar* quantity, not a vector quantity.

Pressure plays the role of stress in a volume deformation. The corresponding strain is the fractional change in volume (Fig. 11.15), that is, the ratio of the volume change ΔV to the original volume V_0:

$$\mathrm{Bulk\ (volume)\ strain} = \frac{\Delta V}{V_0} \qquad (11.12)$$

Volume strain is the change in volume per unit volume. Like tensile or compressive strain, it is a pure number, without units.

When Hooke's law is obeyed, an increase in pressure (bulk stress) produces a *proportional* bulk strain (fractional change in volume). The corresponding elastic modulus (ratio of stress to strain) is called the **bulk modulus,** denoted by B. When the pressure on a body changes by a small amount Δp, from p_0 to $p_0 + \Delta p$, and the resulting bulk strain is $\Delta V/V_0$, Hooke's law takes the form

$$B = \frac{\mathrm{Bulk\ stress}}{\mathrm{Bulk\ strain}} = -\frac{\Delta p}{\Delta V/V_0} \qquad \mathrm{(bulk\ modulus)} \qquad (11.13)$$

We include a minus sign in this equation because an *increase* of pressure always causes a *decrease* in volume. In other words, if Δp is positive, ΔV is negative. The bulk modulus B itself is a positive quantity.

For small pressure changes in a solid or a liquid, we consider B to be constant. The bulk modulus of a *gas,* however, depends on the initial pressure p_0. Table 11.1 includes values of the bulk modulus for several solid materials. Its units, force per unit area, are the same as those of pressure (and of tensile or compressive stress).

The reciprocal of the bulk modulus is called the **compressibility** and is denoted by k. From Eq. (11.13),

$$k = \frac{1}{B} = -\frac{\Delta V/V_0}{\Delta p} = -\frac{1}{V_0}\frac{\Delta V}{\Delta p} \qquad \mathrm{(compressibility)} \qquad (11.14)$$

Compressibility is the fractional decrease in volume, $-\Delta V/V_0$, per unit increase Δp in pressure. The units of compressibility are those of *reciprocal pressure,* Pa^{-1} or atm^{-1}.

Values of compressibility k for several liquids are listed in Table 11.2. From this table the compressibility of water is 46.4×10^{-6} atm^{-1}. This means that for

Pressure $= p_0$

Volume V_0 Initial state

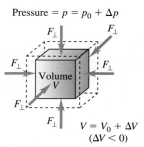

Pressure $= p = p_0 + \Delta p$

$V = V_0 + \Delta V$
$(\Delta V < 0)$

Bulk stress $= \Delta p$

Bulk strain $= \dfrac{\Delta V}{V_0}$

11.15 An object under bulk stress. Without the stress, the cube has volume V_0; when the stress is applied, the cube has a smaller volume V. The volume change ΔV is exaggerated for clarity.

Table 11.2 **Compressibilities of Liquids**

Liquid	Compressibility, k	
	Pa^{-1}	atm^{-1}
Carbon disulfide	93×10^{-11}	94×10^{-6}
Ethyl alcohol	110×10^{-11}	111×10^{-6}
Glycerine	21×10^{-11}	21×10^{-6}
Mercury	3.7×10^{-11}	3.8×10^{-6}
Water	45.8×10^{-11}	46.4×10^{-6}

each atmosphere increase in pressure, the volume of water decreases by 46.4 parts per million. Materials with small bulk modulus and large compressibility are easier to compress.

| Example 11.7 | **Bulk stress and strain** |

A hydraulic press contains 0.25 m³ (250 L) of oil. Find the decrease in volume of the oil when it is subjected to a pressure increase $\Delta p = 1.6 \times 10^7$ Pa (about 160 atm or 2300 psi). The bulk modulus of the oil is $B = 5.0 \times 10^9$ Pa (about 5.0×10^4 atm), and its compressibility is $k = 1/B = 20 \times 10^{-6}$ atm^{-1}.

SOLUTION

IDENTIFY and SET UP: We are given both the bulk modulus and the compressibility, so we can use either Eq. (11.13) or Eq. (11.14) to find the volume change ΔV.

EXECUTE: Solving Eq. (11.13) for ΔV, we find

$$\Delta V = -\frac{V_0 \Delta p}{B} = -\frac{(0.25 \text{ m}^3)(1.6 \times 10^7 \text{ Pa})}{5.0 \times 10^9 \text{ Pa}}$$

$$= -8.0 \times 10^{-4} \text{ m}^3 = -0.80 \text{ L}$$

Alternatively, we can use Eq. (11.14). Solving for ΔV and using the approximate unit conversions given above, we get

$$\Delta V = -kV_0 \Delta p = -(20 \times 10^{-6} \text{ atm}^{-1})(0.25 \text{ m}^3)(160 \text{ atm})$$

$$= -8.0 \times 10^{-4} \text{ m}^3$$

EVALUATE: We get the same result for ΔV with either approach, as we should. Note that ΔV is negative, indicating that the volume decreases when the pressure increases. Note that even though the pressure increase is very large, the *fractional* change in volume is very small:

$$\Delta V/V_0 = (-8.0 \times 10^{-4} \text{ m}^3)/(0.25 \text{ m}^3) = -0.0032, \text{ or } -0.32\%$$

Shear Stress and Strain

The third kind of stress-strain situation is called *shear*. The ribbon in Fig. 11.11c is under **shear stress:** one part of the ribbon is being pushed up while an adjacent part is being pushed down, producing a deformation of the ribbon. Figure 11.16 shows a body being deformed by a shear stress. In the figure, forces of equal magnitude but opposite direction act *tangent* to the surfaces of opposite ends of the object. We define the shear stress as the force $F_\parallel$ acting tangent to the surface, divided by the area A on which it acts:

$$\text{Shear stress} = \frac{F_\parallel}{A} \qquad (11.15)$$

Shear stress, like the other two types of stress, is a force per unit area.

Figure 11.16 shows that one face of the object under shear stress is displaced by a distance x relative to the opposite face. We define **shear strain** as the ratio of the displacement x to the transverse dimension h:

$$\text{Shear strain} = \frac{x}{h} \qquad (11.16)$$

In real-life situations, x is nearly always much smaller than h. Like all strains, shear strain is a dimensionless number; it is a ratio of two lengths.

If the forces are small enough that Hooke's law is obeyed, the shear strain is *proportional* to the shear stress. The corresponding elastic modulus (ratio of shear stress to shear strain) is called the **shear modulus,** denoted by S:

$$S = \frac{\text{Shear stress}}{\text{Shear strain}} = \frac{F_\parallel/A}{x/h} = \frac{F_\parallel}{A} \frac{h}{x} \qquad (\text{shear modulus}) \qquad (11.17)$$

with x and h defined as in Fig. 11.16.

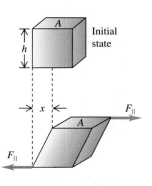

$$\text{Shear stress} = \frac{F_\parallel}{A}$$

$$\text{Shear strain} = \frac{x}{h}$$

11.16 An object under shear stress. Forces are applied tangent to opposite surfaces of the object (in contrast to the situation in Fig. 11.12, in which the forces act perpendicular to the surfaces). The deformation x is exaggerated for clarity.

Several values of shear modulus are given in Table 11.1. For a given material, S is usually one-third to one-half as large as Young's modulus Y for tensile stress. Keep in mind that the concepts of shear stress, shear strain, and shear modulus apply to *solid* materials only. The reason is that the shear forces in Fig. 11.16 are required to deform the solid block, and the block tends to return to its original shape if the shear forces are removed. By contrast, gases and liquids do not have a definite shape.

Example 11.8	**Shear stress and strain**

Suppose the object in Fig. 11.16 is the brass base plate of an outdoor sculpture; it experiences shear forces as a result of an earthquake. The frame is 0.80 m square and 0.50 cm thick. How large a force must be exerted on each of its edges if the displacement x (see Fig. 11.16b) is 0.16 mm?

SOLUTION

IDENTIFY and SET UP: We first find the shear strain using Eq. (11.16), then determine the shear stress using Eq. (11.17). We can then solve for the target variable $F_\parallel$ using Eq. (11.15). The values of all the other quantities are given, including the shear modulus of brass (from Table 11.1, $S = 3.5 \times 10^{10}$ Pa). Note that h in Fig. 11.16 represents the 0.80-m length of each side of the square plate, while the area A is the product of the 0.80-m length and the 0.50-cm thickness.

EXECUTE: The shear strain is

$$\text{Shear strain} = \frac{x}{h} = \frac{1.6 \times 10^{-4}\,\text{m}}{0.80\,\text{m}} = 2.0 \times 10^{-4}$$

From Eq. (11.17) the shear stress equals the shear strain multiplied by the shear modulus S,

$$\text{Stress} = (\text{Shear strain}) \times S$$
$$= (2.0 \times 10^{-4})(3.5 \times 10^{10}\,\text{Pa}) = 7.0 \times 10^6\,\text{Pa}$$

From Eq. (11.15), the force at each edge is the shear stress multiplied by the area of the edge:

$$F_\parallel = (\text{Shear stress}) \times A$$
$$= (7.0 \times 10^6\,\text{Pa})(0.80\,\text{m})(0.0050\,\text{m}) = 2.8 \times 10^4\,\text{N}$$

EVALUATE: The required force is more than three tons! Brass has a large shear modulus, which means that it's intrinsically difficult to deform. Furthermore, the plate is relatively thick (0.50 cm), so the area A is relatively large and a large force $F_\parallel$ is needed to provide the necessary stress $F_\parallel/A$.

Test Your Understanding

Which has greater tensile strain: a copper rod of cross-sectional area 0.500 cm^2 and length 1.00 m that is elongated by 2.00×10^{-2} mm, or a steel rod of the same cross-sectional area but 0.100 m in length and elongated by 2.00×10^{-3} mm? Which rod is under greater tensile stress?

11.5 | Elasticity and Plasticity

Hooke's law—the proportionality of stress and strain in elastic deformations—has a limited range of validity. In the preceding section we used phrases such as "provided that the forces are small enough that Hooke's law is obeyed." Just what *are* the limitations of Hooke's law? We know that if you pull, squeeze, or twist *anything* hard enough, it will bend or break. Can we be more precise than that?

Let's look at tensile stress and strain again. Suppose we plot a graph of stress as a function of strain. If Hooke's law is obeyed, the graph is a straight line with a slope equal to Young's modulus. Figure 11.17 shows a typical stress-strain graph for a metal such as copper or soft iron. The strain is shown as the *percent* elongation; the horizontal scale is not uniform beyond the first portion of the curve, up to a strain of less than 1%. The first portion is a straight line, indicating Hooke's law behavior with stress directly proportional to strain. This straight-line portion ends at point *a*; the stress at this point is called the *proportional limit.*

From *a* to *b*, stress and strain are no longer proportional, and Hooke's law is *not* obeyed. If the load is gradually removed, starting at any point between *O* and *b*, the curve is retraced until the material returns to its original length. The deformation is *reversible,* and the forces are conservative; the energy put into the material to cause the deformation is recovered when the stress is removed. In region *Ob* we say that the material shows *elastic behavior.* Point *b*, the end of this region, is called the *yield point;* the stress at the yield point is called the *elastic limit.*

When we increase the stress beyond point *b*, the strain continues to increase. But now when we remove the load at some point beyond *b*, say *c*, the material does not come back to its original length. Instead, it follows the red line in Fig. 11.17. The length at zero stress is now greater than the original length; the material has undergone an irreversible deformation and has acquired what we call a *permanent set.* Further increase of load beyond *c* produces a large increase in strain for a relatively small increase in stress, until a point *d* is reached at which *fracture* takes place. The behavior of the material from *b* to *d* is called *plastic flow* or *plastic deformation.* A plastic deformation is irreversible; when the stress is removed, the material does not return to its original state.

For some materials a large amount of plastic deformation takes place between the elastic limit and the fracture point. Such a material is said to be *ductile.* But if fracture occurs soon after the elastic limit is passed, the material is said to be *brittle.* A soft iron wire that can have considerable permanent stretch without breaking is ductile, while a steel piano string that breaks soon after its elastic limit is reached is brittle.

Figure 11.18 shows a stress-strain curve for vulcanized rubber that has been stretched by over seven times its original length. The stress is not proportional to the strain, but the behavior is elastic because when the load is removed, the material returns to its original length. However, the material follows different curves for increasing and decreasing stress. This is called *elastic hysteresis.* The work done by the material when it returns to its original shape is less than the work required to deform it; there are nonconservative forces associated with internal friction. Rubber with large elastic hysteresis is very useful for absorbing vibrations, such as in engine mounts and shock-absorber bushings for cars.

The stress required to cause actual fracture of a material is called the *breaking stress,* the *ultimate strength,* or (for tensile stress) the *tensile strength.* Two materials, such as two types of steel, may have very similar elastic constants but vastly different breaking stresses. Table 11.3 gives a few typical values of breaking stress for several materials in tension.

The conversion factor 6.9×10^8 Pa = 100,000 psi will help put these numbers in perspective. For example, if the breaking stress of a particular steel is 6.9×10^8 Pa, then a bar with 1-in.^{2} cross section has a breaking strength of 100,000 lb.

11.17 Typical stress-strain diagram for a ductile metal under tension.

11.18 Typical stress-strain diagram for vulcanized rubber. The curves are different for increasing and decreasing stress, a phenomenon called elastic hysteresis.

Test Your Understanding

While parking your car on a crowded street, you accidentally back into a steel post. You pull forward until the car no longer touches the post and then get out to inspect the damage. What does your rear bumper look like if the strain in the impact was (a) less than at the proportional limit; (b) greater than at the proportional limit, but less than at the yield point; (c) greater than at the yield point, but less than at the fracture point; or (d) greater than at the fracture point?

Table 11.3 Approximate Breaking Stresses of Materials

Material	Breaking Stress (Pa or N/m^2)
Aluminum	2.2×10^8
Brass	4.7×10^8
Glass	10×10^8
Iron	3.0×10^8
Phosphor bronze	5.6×10^8
Steel	$5–20 \times 10^8$

SUMMARY

For a rigid body to be in equilibrium, two conditions must be satisfied. First, the vector sum of forces must be zero. Second, the sum of torques about any point must be zero. To compute the torque due to a force, find the torque of each force component by using its appropriate lever arm and sign, and then add these values. (See Examples 11.2 through 11.5)

$$\sum F_x = 0 \quad \sum F_y = 0 \quad \sum F_z = 0 \tag{11.1}$$

$$\sum \vec{\tau} = 0 \text{ about } any \text{ point} \tag{11.2}$$

The torque due to the weight of a body can be obtained by assuming the entire weight to be concentrated at the center of gravity. If $\vec{g}$ has the same value at all points, the coordinates of the center of gravity are the same as those of the center of mass. (See Example 11.1)

$$\vec{r}_{cm} = \frac{m_1\vec{r}_1 + m_2\vec{r}_2 + m_3\vec{r}_3 + \cdots}{m_1 + m_2 + m_3 + \cdots}$$

$$= \frac{\sum_i m_i\vec{r}_i}{\sum_i m_i} \tag{11.4}$$

Hooke's law states that in elastic deformations, stress (force per unit area) is proportional to strain (fractional deformation). The proportionality constant is called the elastic modulus.

$$\frac{\text{Stress}}{\text{Strain}} = \text{Elastic modulus} \tag{11.7}$$

Tensile stress is tensile force per unit area, $F_\perp/A$. Tensile strain is fractional change in length, $\Delta l/l_0$. The elastic modulus is called Young's modulus Y. Compressive stress and strain are defined in the same way as tensile stress and strain. (See Example 11.6)

$$Y = \frac{\text{Tensile stress}}{\text{Tensile strain}} = \frac{F_\perp/A}{\Delta l/l_0} = \frac{F_\perp}{A} \frac{l_0}{\Delta l} \tag{11.10}$$

Pressure in a fluid is force per unit area. Bulk stress is pressure change, Δp, and bulk strain is fractional volume change, $\Delta V/V_0$. The elastic modulus is called the bulk modulus, B. Compressibility, k, is the reciprocal of bulk modulus: $k = 1/B$. (See Example 11.7)

$$p = \frac{F_\perp}{A} \tag{11.11}$$

$$B = \frac{\text{Bulk stress}}{\text{Bulk strain}} = -\frac{\Delta p}{\Delta V/V_0} \tag{11.13}$$

Shear stress is force per unit area, $F_\parallel/A$, for a force applied tangent to a surface. Shear strain is the displacement x of one side divided by the transverse dimension h. The elastic modulus is called the shear modulus, S. (See Example 11.8)

$$S = \frac{\text{Shear stress}}{\text{Shear strain}} = \frac{F_\parallel/A}{x/h} = \frac{F_\parallel}{A}\frac{h}{x}$$
$$(11.17)$$

The proportional limit is the maximum stress for which stress and strain are proportional. Beyond the proportional limit, Hooke's law is not valid. The elastic limit is the stress beyond which irreversible deformation occurs. The breaking stress, or ultimate strength, is the stress at which the material breaks.

Key Terms

Your Notes

Answer to Chapter Opening Question

To remain in equilibrium, the flamingo's center of gravity must be over the area of support. The broader the foot, the larger the area of support and the more latitude there is in the location of the center of gravity.

Answers to Test Your Understanding Questions

Section 11.1 Situation (a) satisfies both equilibrium conditions; the seagull has zero acceleration (so $\sum \vec{F} = 0$) and no tendency to start rotating (so $\sum \vec{\tau} = 0$). Situation (b) satisfies the first condition, because the crankshaft as a whole does not accelerate through space, but does not satisfy the second condition; the crankshaft has an angular acceleration, so $\sum \vec{\tau}$ is not zero. Situation (c) satisfies the second condition (there is no tendency to rotate), but not the first one; the baseball accelerates in its flight (due to gravity), so $\sum \vec{F}$ is not zero.

Section 11.2 In equilibrium, the center of gravity must be at the point of support. Choose this point to be the origin, so $x_{cg} = 0$. With the positive x-direction to the right, the coordinates of the three objects are $x = -10$ cm for the counterweight, $x = 15$ cm for the first fish, and $x = 40$ cm for the second fish. (We treat the rod as being massless.) From the x-equation in Eqs. (11.3),

$$x_{cg} = 0$$
$$= \frac{m(-10 \text{ cm}) + (0.10 \text{ kg})(15 \text{ cm}) + (0.10 \text{ kg})(40 \text{ cm})}{m + (0.10 \text{ kg}) + (0.10 \text{ kg})}$$

so

$$m = \frac{(0.10 \text{ kg})(15 \text{ cm}) + (0.10 \text{ kg})(40 \text{ cm})}{10 \text{ cm}} = 0.55 \text{ kg}$$

Section 11.3 This is the same situation as described in Example 11.5, with the rod replacing the forearm, the hinge replacing the elbow, and the cable replacing the tendon. The cable attachment point is at the end of the rod, so the distance D in Fig. 11.9b is equal to L. From Example 11.5, the tension is

$$T = \frac{Lw}{L \sin \theta} = \frac{w}{\sin \theta}$$

and the horizontal and vertical forces exerted by the hinge are

$$E_x = \frac{Lw}{h} = \frac{w}{\tan \theta} \quad \text{and} \quad E_y = -\frac{(L - L)}{L} w = 0$$

In this situation, the hinge exerts no vertical force.

Section 11.4 The tensile strain $\Delta l/l_0$ is the same for both rods. The copper rod has ten times the elongation Δl of the steel rod, but it also has ten times the original length l_0. The stress is equal to the product of Young's modulus Y and the strain. From Table 11.1, steel has a larger value of Y, so a greater stress is required to produce the same strain.

Section 11.5 In both cases (a) and (b), the bumper will have sprung back to its original shape (although the paint may be scratched). In case (c), the bumper will have a permanent dent or deformation. In case (d), the bumper will be torn or broken.

Discussion Questions

Q11.1 Does a rigid body in uniform rotation about a fixed axis satisfy the first and second conditions for equilibrium? Why or why not? An alternative definition of equilibrium states: "For a rigid body to be in equilibrium, every part of the body must be in equilibrium." On the basis of this definition, is a rigid body in uniform rotation about a fixed axis in equilibrium? Explain.

Q11.2 Is it possible for the net force on an object to be zero but the object not to be in equilibrium? Is it possible for an object to be in equilibrium if the net force is not zero? Explain.

Q11.3 Car tires are sometimes "balanced" on a machine that pivots the tire and wheel about the center. Weights are placed around the wheel rim until it does not tip from the horizontal plane. Discuss this procedure in terms of the center of gravity.

Q11.4 Does the center of gravity of a solid body always lie within the material of the body? If not, give a counterexample.

Q11.5 In Section 11.2 we always assumed that the value of g was the same at all points on the body. This is *not* a good approximation if the dimensions of the body are great enough, because the value of g decreases with altitude. If this is taken into account, will the center of gravity of a long, vertical rod be above, below, or at its center of mass? Explain how this can be used to keep the long axis of an orbiting spacecraft pointed toward the earth. (This would be useful for a weather satellite that must always keep its camera lens trained on the earth.) The moon is not exactly spherical, but is somewhat elongated. Explain why this same effect is responsible for keeping the same face of the moon pointed toward the earth at all times.

Q11.6 You are balancing a wrench by suspending it at a single point. Is the equilibrium stable, unstable, or neutral if the point is above, at, or below the wrench's center of gravity? In each case give the reasoning behind your answer. (For rotation, a rigid body is in *stable* equilibrium if a small rotation of the body produces a torque that tends to return the body to equilibrium; it is in *unstable* equilibrium if a small rotation produces a torque that tends to take the body farther from equilibrium; and it is in *neutral* equilibrium if a small rotation produces no torque.)

Q11.7 You can probably stand flatfooted on the floor, and then rise up and balance on your tiptoes. Why are you unable do it if your toes are touching the wall of your room? (Try it!)

Q11.8 You freely pivot a horseshoe from a horizontal nail through one of its nail holes. You then hang a long string with a weight at its bottom from the same nail, so that the string hangs vertically in front of the horseshoe without touching it. How do you know that the horseshoe's center of gravity is along the line behind the string? How can you locate the center of gravity by repeating the process at another nail hole? Will the center of gravity be within the solid material of the horseshoe?

Q11.9 A stunt driver drives a car in a straight line at constant speed while balanced on the two right-hand wheels. Where is the car's center of gravity?

Q11.10 When four-legged animals walk, they always have three of their legs on the ground at any instant. Explain why, using the concept of center of gravity.

Q11.11 Why must a water skier moving with constant velocity lean backwards? What determines how far back she must lean? Draw a free-body diagram for the water skier to justify your answers.

Q11.12 In pioneer days, when a Conestoga wagon was stuck in the mud, people would grasp the wheel spokes and try to turn the wheels, rather than simply pushing the wagon. Why?

Q11.13 A physical therapist says a physicist-patient has weak stomach muscles because he can't do a partial sit-up without his feet being held down. The physicist, who has short, slender legs and a long, heavy, upper torso, correctly disagrees. Explain, using the concept of center of gravity.

Q11.14 Why is it easier to hold a 10-kg dumbbell in your hand at your side than it is to hold it with your arm extended horizontally?

Q11.15 Figure 11.3 shows that an object at rest on an incline will tip over if its center of gravity is not within the area of support. Explain why this result does *not* depend on the value of the coefficient of static friction between the object and the surface of the incline. Draw a free-body diagram as part of your explanation.

Q11.16 In Example 11.3 (Section 11.3), explain why n_1 is negative in the $\Sigma F_x = 0$ equation, but gives a positive torque in the $\Sigma \tau_B = 0$ equation.

Q11.17 Why is a tapered water glass with a narrow base easier to tip over than one with straight sides? Does it matter whether the glass is full or empty?

Q11.18 When a tall, heavy refrigerator is pushed across a rough floor, what determines whether it slides or tips?

Q11.19 Climbing ropes used by mountaineers are usually made of nylon. Would a steel cable of equal strength be just as good? What advantages and disadvantages would it have in comparison to nylon?

Q11.20 Why is concrete with steel reinforcing rods embedded in it stronger than plain concrete?

Q11.21 Electric power lines are sometimes made by using wires with a steel core and a copper jacket or with strands of copper and steel twisted together. Why?

Q11.22 Compare the mechanical properties of a steel cable, made by twisting many thin wires together, with those of a solid steel rod of the same diameter. What advantages does each have?

Q11.23 The material in human bones and elephant bones is essentially the same, but an elephant has much thicker legs. Explain why, in terms of breaking stress.

Q11.24 There is a small but appreciable amount of elastic hysteresis in the large tendon at the back of a horse's leg. Explain how this can cause damage to the tendon if a horse runs too hard for too long a time.

Q11.25 When rubber mounting blocks are used to absorb machine vibrations through elastic hysteresis, as mentioned in Section 11.5, what becomes of the energy associated with the vibrations?

Exercises

Section 11.2 Center of Gravity

11.1 A ball with radius $r_1 = 0.080$ m and mass 1.00 kg is attached by a light rod 0.400 m in length to a second ball with radius $r_2 = 0.100$ m and mass 2.00 kg (Fig. 11.19). Where is the center of gravity of this system?

Figure 11.19 Exercise 11.1.

11.2 Suppose the rod in Exercise 11.1 is uniform with mass 1.50 kg. Where is the system's center of gravity? How does your result compare to that for Exercise 11.1?

11.3 For the plank of Example 11.1 (Section 11.2), show that a 60-kg child could stand no farther to the right than halfway between the right-hand sawhorse and the end of the plank without the plank tipping over.

Section 11.3 Solving Rigid-Body Equilibrium Problems

11.4 A uniform 300-N trapdoor in a ceiling is hinged at one side. Find the net upward force needed to begin to open it and the total force exerted on the door by the hinges, a) if the upward force is applied at the center; b) if the upward force is applied at the center of the edge opposite the hinges.

11.5 Raising a Ladder. A ladder carried by a fire truck is 20.0 m long. The ladder weighs 2800 N and its center of gravity is at its center. The ladder is pivoted at one end (A) about a pin (Fig. 11.20); you can ignore the friction torque at the pin. The ladder is raised into position by a force applied by a hydraulic piston at C. Point C is 8.0 m from A, and the force $\vec{F}$ exerted by the piston makes an angle of 40° with the ladder. What magnitude must $\vec{F}$ have to just lift the ladder off the support bracket at B?

Figure 11.20 Exercise 11.5.

11.6 Two people are carrying a uniform wooden board that is 3.00 m long and weighs 160 N. If one person applies an upward force equal to 60 N at one end, at what point does the other person lift?

11.7 Two people carry a heavy electric motor by placing it on a light board 2.00 m long. One person lifts at one end with a force of

400 N, and the other lifts the opposite end with a force of 600 N. What is the weight of the motor, and where along the board is its center of gravity located?

11.8 Suppose the board in Exercise 11.7 is not light but weighs 200 N, with its center of gravity at its center. The two people each exert the same forces as before. What is the weight of the motor in this case, and where is its center of gravity located?

11.9 Show that the torques about point B due to T_x and T_y in Example 11.5 (Section 11.3) sum to zero. Do this for the general case, rather than for the specific numerical values given at the end of the example.

11.10 A uniform ladder 5.0 m long rests against a frictionless, vertical wall with its lower end 3.0 m from the wall. The ladder weighs 160 N. The coefficient of static friction between the foot of the ladder and the ground is 0.40. A man weighing 740 N climbs slowly up the ladder. a) What is the maximum frictional force that the ground can exert on the ladder at its lower end? b) What is the actual frictional force when the man has climbed 1.0 m along the ladder? c) How far along the ladder can the man climb before the ladder starts to slip?

11.11 A diving board 3.00 m long is supported at a point 1.00 m from the end, and a diver weighing 500 N stands at the free end (Fig. 11.21). The diving board is of uniform cross section and weighs 280 N. Find a) the force at the support point and b) the force at the end that is held down.

Figure 11.21 Exercise 11.11.

11.12 A uniform, aluminum beam 9.00 m long, weighing 300 N, rests symmetrically on two supports 5.00 m apart (Fig. 11.22). A boy weighing 600 N starts at point A and walks toward the right. a) In the same diagram construct two graphs showing the upward forces F_A and F_B exerted on the beam at points A and B, as functions of the coordinate x of the boy. Let 1 cm = 100 N vertically, and 1 cm = 1.00 m horizontally. b) From your diagram, how far beyond point B can the boy walk before the beam tips? c) How far from the right end of the beam should support B be placed so that the boy can walk just to the end of the beam without causing it to tip?

Figure 11.22 Exercise 11.12.

11.13 Find the tension T in each cable and the magnitude and direction of the force exerted on the strut by the pivot in each of the arrangements in Fig. 11.23. In each case let w be the weight of the suspended crate full of priceless art objects. The strut is uniform and also has weight w.

(a)   (b)

Figure 11.23 Exercise 11.13.

11.14 The horizontal beam in Fig. 11.24 weighs 150 N, and its center of gravity is at its center. Find a) the tension in the cable and b) the horizontal and vertical components of the force exerted on the beam at the wall.

11.15 A door 1.00 m wide and 2.00 m high weighs 280 N and is supported by two hinges, one 0.50 m from the top and the other 0.50 m from the bottom. Each hinge supports half the total weight of the door. Assuming that the door's center of gravity is at its center, find the horizontal components of force exerted on the door by each hinge.

Figure 11.24 Exercise 11.14.

11.16 Suppose that you can lift no more than 450 N (about 100 lb) unaided. a) How much can you lift using a 2.0-m long wheelbarrow? Assume that the wheelbarrow weighs 80 N and that its center of gravity is 0.70 m from the wheel. Assume that the load carried in the wheelbarrow is also 0.70 m from the wheel. b) Where does the force come from to enable you to lift more than 450 N when you use the wheelbarrow?

11.17 You take your dog Clea to the vet, and the doctor decides he must locate the little beast's center of gravity. It would be awkward to hang the pooch from the ceiling, so the vet must devise another method. He places Clea's front feet on one scale and her hind feet on another. The front scale reads 157 N while the rear scale reads 89 N. The vet next measures Clea and finds that her rear feet are 0.95 m behind her front feet. How much does Clea weigh and where is her center of gravity?

11.18 The boom in Fig. 11.25 weighs 2600 N. The boom is

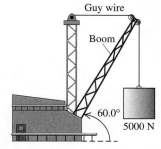

Figure 11.25 Exercise 11.18.

attached with a frictionless pivot at its lower end. It is not uniform; the distance of its center of gravity from the pivot is 35% of its length. a) Find the tension in the guy wire and the horizontal and vertical components of the force exerted on the boom at its lower end. b) Does the line of action of the force exerted on the boom at its lower end lie along the boom?

11.19 A 3.00-m-long, 240-N uniform rod at the zoo is held in a horizontal position by two ropes at its ends (Fig. 11.26). The left rope makes an angle of 150° with the rod and the right rope makes an angle θ with the horizontal. A 90-N howler monkey (*Alouatta seniculus*) hangs motionless 0.50 m from the right end of the rod as he carefully studies you. Calculate the tensions in the two ropes and the angle θ.

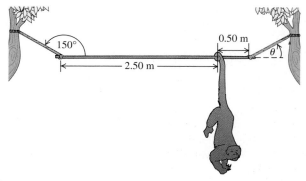

Figure 11.26 Exercise 11.19.

11.20 A non-uniform beam 4.50 m long and weighing 1.00 kN makes an angle of 25.0° with the horizontal. It is held in position by a frictionless pivot at its upper-right end and by a cable 3.00 m farther down the beam and perpendicular to it. The center of gravity of the beam is 2.00 m down the beam from the pivot. Lighting equipment exerts a 5.00-kN downward force on the lower-left end of the beam. Find the tension T in the cable and the horizontal and vertical components of the force exerted on the beam by the pivot.

11.21 A Couple. Two forces equal in magnitude and opposite in direction, acting on an object at two different points, form what is called a *couple*. Two antiparallel forces with equal magnitudes $F_1 = F_2 = 8.00 \, N$ are applied to a rod as shown in Fig. 11.27. a) What should the distance l between the forces be if they are to provide a net torque of 6.40 N·m about the left-hand end of the rod? b) Is the sense of this torque clockwise or counterclockwise? c) Repeat parts (a) and (b) for a pivot at the point on the rod where $\vec{F}_2$ is applied.

Figure 11.27 Exercise 11.21.

Section 11.4 Stress, Strain, and Elastic Moduli

11.22 A relaxed biceps muscle requires a force of 25.0 N for an elongation of 3.0 cm; the same muscle under maximum tension requires a force of 500 N for the same elongation. Find Young's modulus for the muscle tissue under each of these conditions if the muscle is assumed to be a uniform cylinder with length 0.200 m and cross-sectional area 50.0 cm².

11.23 A circular steel wire 2.00 m long must stretch no more than 0.25 cm when a tensile force of 400 N is applied to each end of the wire. What minimum diameter is required for the wire?

11.24 Two round rods, one steel and the other copper, are joined end to end. Each rod is 0.750 m long and 1.50 cm in diameter. The combination is subjected to a tensile force with magnitude 4000 N. For each rod, what is a) the strain? b) the elongation?

11.25 A metal rod that is 4.00 m long and 0.50 cm² in cross-sectional area is found to stretch 0.20 cm under a tension of 5000 N. What is Young's modulus for this metal?

11.26 Stress on a Mountaineer's Rope. A nylon rope used by mountaineers elongates 1.10 m under the weight of an 65.0-kg climber. If the rope is 45.0 m in length and 7.0 mm in diameter, what is Young's modulus for this material?

11.27 In constructing a large mobile, an artist hangs an aluminum sphere of mass 6.0 kg from a vertical steel wire 0.50 m long and 2.5×10^{-3} cm² in cross-sectional area. On the bottom of the sphere he attaches a similar steel wire, from which he hangs a brass cube of mass 10.0 kg. For each wire, compute a) the tensile strain and b) the elongation.

11.28 A vertical solid steel post 25 cm in diameter and 2.50 m long is required to support a load of 8000 kg. You can ignore the weight of the post. What is a) the stress in the post? b) the strain in the post? c) the change in the post's length when the load is applied?

11.29 Outside a house 1.0 km from ground zero of a 100-kiloton nuclear bomb explosion, the pressure will rapidly rise to as high as 2.8 atm while the pressure inside the house remains 1.0 atm. If the area of the front of the house is 50 m², what is the resulting net force exerted by the air on the front of the house?

11.30 A solid gold bar is pulled up from the hold of the sunken RMS *Titanic*. a) What happens to its volume as it goes from the pressure at the ship to the lower pressure at the ocean's surface? b) The pressure difference is proportional to the depth. How many times greater would the volume change have been had the ship been twice as deep? c) The bulk modulus of lead is one-fourth that of gold. Find the ratio of the volume change of a solid lead bar to that of a gold bar of equal volume for the same pressure change.

11.31 A petite young woman distributes her 500 N weight equally over the heels of her high-heeled shoes. Each heel has an area of 0.750 cm². a) What is the pressure exerted on the floor by each heel? b) With the same pressure, how much weight could be supported by two flat-bottomed sandals, each of area 200 cm²?

11.32 In the Challenger Deep of the Marianas Trench, the depth of seawater is 10.9 km and the pressure is 1.16×10^8 Pa (about 1.15×10^3 atm). a) If a cubic meter of water is taken from the surface to this depth, what is the change in its volume? (Normal atmospheric pressure is about 1.0×10^5 Pa. Assume that k for

seawater is the same as the freshwater value given in Table 11.2.)
b) What is the density of seawater at this depth? (At the surface, seawater has a density of 1.03×10^3 kg/m^3.)

11.33 A specimen of oil having an initial volume of 600 cm^3 is subjected to a pressure increase of 3.6×10^6 Pa, and the volume is found to decrease by 0.45 cm^3. What is the bulk modulus of the material? The compressibility?

11.34 A square steel plate is 10.0 cm on a side and 0.500 cm thick. a) Find the shear strain that results if a force of magnitude 9.0×10^5 N is applied to each of the four sides, parallel to the side. b) Find the displacement x in centimeters.

11.35 Shear forces are applied to a rectangular solid. The same forces are applied to another rectangular solid of the same material, but with three times each edge length. In each case the forces are small enough that Hooke's law is obeyed. What is the ratio of the shear strain for the larger object to that of the smaller object?

11.36 Two strips of brass, each 3.00 cm wide and 45.0 cm long, are placed with their ends overlapping by 1.00 cm. The overlapping ends are then riveted together by four rivets, each with diameter 0.250 cm. Tests show that when a tension of at least 1.20×10^4 N is applied to the ends of the riveted strip, the rivets fail and shear apart. What is the shearing stress on each rivet when it fails? Assume that each rivet carries one quarter of the load.

Section 11.5 Elasticity and Plasticity

11.37 In a materials testing laboratory, a metal wire made from a new alloy is found to break when a tensile force of 90.8 N is applied perpendicular to each end. If the diameter of the wire is 1.84 mm, what is the breaking stress of the alloy?

11.38 A 4.0-m-long steel wire has a cross-sectional area of 0.050 cm^2. Its proportional limit has a value of 0.0016 times its Young's modulus (Table 11.1). Its breaking stress has a value of 0.0065 times its Young's modulus. The wire is fastened at its upper end and hangs vertically. a) How great a weight can be hung from the wire without exceeding the proportional limit? b) How much will the wire stretch under this load? c) What is the maximum weight that the wire can support?

11.39 A steel cable with cross-sectional area 3.00 cm^2 has an elastic limit of 2.40×10^8 Pa. Find the maximum upward acceleration that can be given a 1200-kg elevator supported by the cable if the stress is not to exceed one-third of the elastic limit.

11.40 A brass wire is to withstand a tensile force of 350 N without breaking. What minimum diameter must the wire have?

Problems

11.41 Center of Gravity of a Car. For an automobile with weight w, the front wheels support a fraction f of the weight, so the normal force exerted at the front wheels is fw, and at the rear wheels the normal force is $(1 - f)w$. The distance between the front and rear axles is d. a) Show that the center of gravity of the car is a distance fd in front of the rear wheels. b) Show that the general result derived in part (a) reproduces the numerical answer of Example 11.2 (Section 11.3).

11.42 Sir Lancelot rides slowly out of the castle at Camelot and onto the 12.0-m-long drawbridge that passes over the moat (Fig. 11.28). Unbeknownst to him, his enemies have partially severed the vertical cable holding up the front end of the bridge so that it will break under a tension of 5.80×10^3 N. The bridge has mass 200 kg and its center of gravity is at its center. Lancelot, his lance, his armor, and his horse together have a combined mass of 600 kg. Will the cable break before Lancelot reaches the end of the drawbridge? If so, how far from the castle end of the bridge will the center of gravity of the horse plus rider be when the cable breaks?

Figure 11.28 Problem 11.42.

11.43 Three vertical forces act on an airplane when it is flying at a constant altitude and with a constant velocity. These are the weight of the airplane, an aerodynamic force on the wing of the airplane, and an aerodynamic force on the airplane's horizontal tail. (The aerodynamic forces are exerted by the surrounding air, and are reactions to the forces that the wing and tail exert on the air as the airplane flies through it.) For a particular light airplane with a weight of 6700 N, the center of gravity is 0.30 m in front of the point where the wing's vertical aerodynamic force acts and 3.66 m in front of the point where the tail's vertical aerodynamic force acts. Determine the magnitude and direction (upward or downward) of each of the two vertical aerodynamic forces.

11.44 A pickup truck has a wheelbase of 3.00 m. Ordinarily, 10,780 N rests on the front wheels and 8820 N on the rear wheels when the truck is parked on a level road. a) A box weighing 3600 N is now placed on the tailgate, 1.00 m behind the rear axle. How much total weight now rests on the front wheels? On the rear wheels? b) How much weight would need to be placed on the tailgate to make the front wheels come off the ground?

11.45 A uniform seesaw is 4.00 m long and weighs 240 N. The seesaw is pivoted about a point on its bottom surface and the location of this pivot along the length of the seesaw can be adjusted as far as 0.20 m from the center of the seesaw. Little Karen, who weighs 420 N, has her center of gravity over the right end, with the seesaw set so she exerts the most torque about the pivot. The seesaw balances horizontally when Elwood has his center of gravity 0.15 m from the left end. Both have their feet off the ground. a) What does Elwood weigh? b) Is this equilibrium stable, unstable, or neutral? (*Hint:* See Question Q11.6.)

11.46 A piece of steel will not balance with the tip of one end just barely resting on the edge of a table (point O in Fig. 11.29). But suppose we fasten on two other pieces as shown in Fig. 11.29. All pieces have the same square cross section and density. You can ignore the weight of the fasteners. The system can now balance if we find the right length *l* of the third piece. a) What is the length *l* for which the system will balance with the first and third pieces horizontal, as shown in the figure? b) How would your answer change if the pieces were made of copper rather than steel? Explain.

Figure 11.29 Problem 11.46.

11.47 Several external forces $\vec{F}_i$ are applied to a rigid body. With respect to an origin *O*, force $\vec{F}_1$ is applied at the point located at $\vec{r}_1$, force $\vec{F}_2$ at $\vec{r}_2$, and so on. Point *P* is located at $\vec{r}_P$. If the body is in translational equilibrium, so that $\Sigma \vec{F}_i = \mathbf{0}$, prove that the sum of the torques about point *P* equals the sum of torques about point *O*. (A special case of this result is when $\Sigma \vec{\tau}_O = \mathbf{0}$; for a body in translational equilibrium, if the resultant torque about one point is zero, the resultant torque about any point is zero.)

11.48 A claw hammer is used to pull a nail out of a board (Fig. 11.30). The nail is at an angle of 60° to the board, and a force $\vec{F}_1$ of magnitude 500 N applied to the nail is required to pull it from the board. The hammer head contacts the board at point *A*, which is 0.080 m from where the nail enters the board. A horizontal force $\vec{F}_2$ is applied to the hammer handle at a distance of 0.300 m above the board. What magnitude of force $\vec{F}_2$ is required to apply the required 500-N force (F_1) to the nail? (You can ignore the weight of the hammer.)

11.49 End *A* of the bar *AB* in Fig. 11.31 rests on a frictionless horizontal surface, and end *B* is hinged. A horizontal force $\vec{F}$ of magnitude 120 N is exerted on end *A*. You can ignore the weight of the bar. What are the horizontal and vertical components of

Figure 11.30 Problem 11.48.

Figure 11.31 Problem 11.49.

the force exerted by the bar on the hinge at *B*?

11.50 A single additional force is to be applied to the bar in Fig. 11.32 to maintain it in equilibrium in the position shown. You can ignore the weight of the bar. a) What are the horizontal and vertical components of the required force? b) What is the angle the force must make with the bar? c) What is the magnitude of the required force? d) Where should the force be applied?

Figure 11.32 Problem 11.50.

11.51 Your dog Nikita is 0.90 m long (nose to hindlegs). Her forelegs are 0.18 m horizontally from her nose, her center of gravity is 0.28 m horizontally from her hindlegs, and she weighs 190 N. a) How much force does a level floor exert on each of Nikita's front feet, and on each of her hind feet? b) If Nikita picks up a 25-N bone and holds it in her mouth (directly under her nose), what is the force exerted by the floor on each of her front feet and each of her hind feet?

11.52 **A Truck on a Drawbridge.** A loaded cement mixer drives onto an old drawbridge, where it stalls with its center of gravity three-quarters of the way across the span. The truck driver radios for help, sets the hand-brake, and waits. Meanwhile, a boat approaches, so the drawbridge is raised by means of a cable attached to the end opposite the hinge (Fig. 11.33). The drawbridge span is 40.0 m long and has a mass of 12,000 kg; its center of gravity is at its midpoint. The cement mixer, with driver, has mass 30,000 kg. When the drawbridge has been raised to an angle of 30° above the horizontal, the cable makes an angle of 70° with the surface of the bridge. a) What is the tension *T* in the cable when the drawbridge is held in this position? b) What are the horizontal and vertical components of the force the hinge exerts on the span?

Figure 11.33 Problem 11.52.

11.53 **Couples.** a) A force $\vec{F}_1$ is applied to a rod at a distance *x* from some point *P* on the rod. A second force $\vec{F}_2$ is applied at a distance *x* + *l* from point *P*. The forces are equal in magnitude and opposite in direction, and hence, form a couple (Exercise 11.21). The angle between the rod and the direction of each force is ϕ.

Derive an expression for the net torque about P produced by these two forces and show that it is independent of x. This shows that a couple produces the same torque about any pivot point. b) Two forces with magnitude $F_1 = F_2 = 14.0$ N are applied to a rod as shown in Fig. 11.34, with $\phi = 37°$. Calculate the net torque about point O due to these two forces by calculating the torque due to each separate force. Calculate the net torque about point P due to these two forces by calculating the torque due to each separate force. Compare your results to the general result derived in part (a).

Figure 11.34 Problem 11.53.

11.54 A non-uniform fire escape ladder is 6.0 m long when extended to the icy alley below. It is held at the top by a frictionless pivot, and there is negligible frictional force from the icy surface at the bottom. The ladder weighs 250 N, and its center of gravity is 2.0 m along the ladder from its bottom. A mother and child of total weight 750 N are on the ladder 1.5 m from the pivot. The ladder makes an angle θ with the horizontal. Find the magnitude and direction of a) the force exerted by the icy alley on the ladder and b) the force exerted by the ladder on the pivot. c) Do your answers in parts (a) and (b) depend on the angle θ?

11.55 A uniform strut of mass m makes an angle θ with the horizontal. It is supported by a frictionless pivot located at $\frac{1}{3}$ its length from its lower-left end and a horizontal rope at its upper-right end. A cable and package of total weight w hang from its upper-right end. a) Find the vertical and horizontal components V and H of the pivot's force on the strut as well as the tension T in the rope. b) If the maximum safe tension in the rope is 700 N and the mass of the strut is 20.0 kg, find the maximum safe weight of the cable and package when the strut makes an angle of 55.0° with the horizontal. c) For what angle θ can no weight be safely suspended from the right end of the strut?

11.56 You are asked to design the decorative mobile shown in Fig. 11.35. The strings and rods have negligible weight, and the rods are to hang horizontally. a) Draw a free-body diagram for each rod. b) Find the weights of the balls A, B, and C. Find the tensions in the strings S_1, S_2, and S_3. c) What can you say about the horizontal location of the mobile's center of gravity? Explain.

Figure 11.35 Problem 11.56.

11.57 A uniform 7.5-m-long beam weighing 9000 N is hinged to a wall and supported by a thin cable attached 1.5 m from the free end of the beam. The cable runs between the beam and the wall and makes a 40° angle with the beam. What is the tension in the cable when the beam is at an angle of 30° above the horizontal?

11.58 A uniform drawbridge must be held at a 37° angle above the horizontal to allow ships to pass underneath. The drawbridge weighs 45,000 N and is 14.0 m long. A cable is connected 3.5 m from the hinge where the bridge pivots (measured along the bridge) and pulls horizontally on the bridge to hold it in place. a) What is the tension in the cable? b) Find the magnitude and direction of the force the hinge exerts on the bridge.

11.59 A uniform, 250-kg beam is supported using a cable connected to the ceiling, as shown in Fig. 11.36. The lower end of the beam rests on the floor. a) What is the tension in the cable? b) What is the minimum coefficient of static friction between the beam and the floor required for the beam to remain in this position?

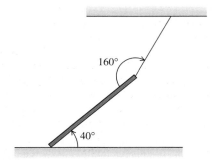

Figure 11.36 Problem 11.59.

11.60 a) In Fig. 11.37 a 6.00-m-long, uniform beam is hanging from a point 1.00 m to the right of its center. The beam weighs 140 N and makes an angle of 30.0° with the vertical. At the right-hand end of the beam a 100-N weight is hung; an unknown weight w hangs at the other end. If the system is in equilibrium, what is w? You can ignore the thickness of the beam. b) If the beam makes, instead, an angle of 45.0° with the vertical, what is w?

Figure 11.37 Problem 11.60.

11.61 A uniform, horizontal flagpole 5.00 m in length and with a weight of 200 N is hinged to a vertical wall at one end. A 600-N stuntwoman hangs from its other end. The flagpole is supported by a guy wire running from its outer end to a point on the wall directly above the pole. a) If the tension in this wire is not to exceed 1000 N, what is the minimum height above the pole at which it may be fastened to the wall? b) If the flagpole remains horizontal, by how many newtons would the tension be increased if the wire were fastened 0.50 m below this point?

11.62 A holiday decoration consists of two shiny glass spheres with masses 0.0240 kg and 0.0360 kg suspended, as shown in Fig. 11.38, from a uniform rod with mass 0.120 kg and length 1.00 m. The rod is suspended from the ceiling by a vertical cord at each end, so that it is horizontal. Calculate the tension in each of the cords A through F.

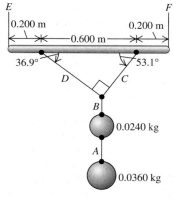

Figure 11.38 Problem 11.62.

11.63 The sign for Lew's Bait Shop and Sushi Restaurant is a uniform rectangle, 1.50 m wide and 0.800 m tall, weighing 300 N. It is supported by a wire at its upper-right corner and a pivot at its lower-left corner. a) Late one night, Lew's competitor from across the street cuts the wire. The sign stays in place because the pivot has rusted tight. After the wire has been cut, what are the magnitude and direction of the force and of the torque (about an axis at the pivot) exerted by the pivot on the sign? b) When the sign was brand new, the pivot was frictionless and the wire was at the angle that minimized the tension in it. What was the tension in the wire and the magnitude and direction of the force exerted on the sign by the pivot?

11.64 When you stretch a wire, rope, or rubber band, it gets thinner as well as longer. When Hooke's law holds, the fractional decrease in width is proportional to the tensile strain. If w_0 is the original width and Δw is the change in width, then $\Delta w/w_0 = -\sigma \Delta l/l_0$, where the minus sign reminds us that width decreases when length increases. The dimensionless constant, σ, different for different materials, is called *Poisson's ratio*. a) If the steel rod of Example 11.6 (Section 11.4) has a circular cross section and a Poisson's ratio of 0.23, what is its change in diameter when the milling machine is hung from it? b) A cylinder made of nickel (Poisson's ratio = 0.42) has radius 2.0 cm. What tensile force $F_\perp$ must be applied perpendicular to each end of the cylinder to cause its radius to decrease by 0.10 mm? Assume that the breaking stress and proportional limit for the metal are extremely large and are not exceeded.

11.65 A circular disk 0.500 m in diameter, pivoted about a horizontal axis through its center, has a cord wrapped around its rim.

Figure 11.39 Problem 11.65.

The cord passes over a frictionless pulley P and is attached to an object that weighs 240 N. A uniform rod 2.00 m long is fastened to the disk, with one end at the center of the disk. The apparatus is in equilibrium, with the rod horizontal (Fig. 11.39). a) What is the weight of the rod? b) What is the new equilibrium direction of the rod when a second object weighing 20.0 N is suspended from the other end of the rod, as shown by the broken line? That is, what angle does the rod then make with the horizontal?

11.66 One end of a uniform meter stick is placed against a vertical wall (Fig. 11.40). The other end is held by a light-weight cord that makes an angle θ with the stick. The coefficient of static friction between the end of the meter stick and the wall is 0.40. a) What is the maximum value the angle θ can have if the stick is to remain in equilibrium? b) Let the angle θ be 15°. A

Figure 11.40 Problem 11.66.

block of the same weight as the meter stick is suspended from the stick, as shown, at a distance x from the wall. What is the minimum value of x for which the stick will remain in equilibrium? c) When $\theta = 15°$, how large must the coefficient of static friction be so that the block can be attached 10 cm from the left end of the stick without causing it to slip?

11.67 Two friends are carrying a 200-kg crate up a flight of stairs. The crate is 1.25 m long and 0.50 m high, and its center of gravity is at its center. The stairs make a 45.0° angle with respect to the floor. The crate also is carried at a 45.0° angle, so that its bottom side is parallel to the slope of the stairs (Fig. 11.41). If the force each person applies is vertical, what is the magnitude of each of these forces? Is it best to be the person above or below on the stairs?

Figure 11.41 Problem 11.67.

11.68 Forearm. In the human arm, the forearm and hand pivot about the elbow joint. Consider a simplified model in which the biceps muscle is attached to the forearm 3.80 cm from the elbow joint. Assume that the person's hand and forearm together weigh 15.0 N and that their center of gravity is 15.0 cm from the elbow (not quite half-way to the hand). The forearm is held horizontally at a right angle to the upper arm, with the biceps muscle exerting its force perpendicular to the forearm. a) Draw a free-body diagram for the forearm and find the force exerted by the biceps when the hand is empty. b) Now the person holds a 80.0-N weight in his hand, with the forearm still horizontal. Assume that the center of gravity of this weight is 33.0 cm from the elbow. Construct a free-body diagram for the forearm and find the force now exerted by the biceps. Explain why the biceps muscle needs to be very strong. c) Under the conditions of part (b), find the magnitude and direction of the force which the elbow joint exerts on the forearm. d) While holding the 80.0-N weight, the person raises their forearm until it is at an angle of 53.0° above the horizontal. If the biceps muscle continues to exert its force perpendicular to the forearm, what is this force when the forearm is in this position? Has the force increased or decreased from its value in part (b)? Explain why this is so and test your answer by actually doing this with your own arm.

11.69 Refer to the discussion of holding a dumbbell in Example 11.5. The maximum weight that can be held in this way is limited by the maximum allowable tendon tension T (determined by the strength of the tendons) and by the distance D from the elbow to where the tendon attaches to the forearm. a) Let T_{max} represent the maximum value of the tendon tension. Use the results of Example 11.5 to express w_{max} (the maximum weight that can be held) in terms of T_{max}, L, D, and h. Your expression should *not* include the angle θ. b) The tendons of different primates are attached to the forearm at different values of D. Calculate the derivative of w_{max} with respect to D, and determine whether the derivative is positive or negative. c) A chimpanzee tendon is attached to the forearm at a point farther from the elbow than for humans. Use this to explain why chimpanzees have stronger arms than humans. (The disadvantage is that chimpanzees have less flexible arms than do humans.)

11.70 A uniform 90.0-N table is 3.6 m long, 1.0 m high, and 1.2 m wide. A 1500-N weight is placed 0.50 m from one end of the table, a distance of 0.60 m from each of the two legs at that end. Draw a free-body diagram for the table and find the force that each of the four legs exerts on the floor.

11.71 Flying Buttress. a) A symmetric building has a roof sloping upward at 35.0° above the horizontal on each side. If each side of the uniform roof weighs 10,000 N, find the horizontal force that this roof exerts at the top of the wall, which tends to push out the walls. Which type of building would be more in danger of collapsing, one with tall walls or one with short walls? Explain. b) As you saw in part (a), tall walls are in danger of collapsing from the weight of the roof. This problem plagued the ancient builders of large structures. A solution used in the great Gothic cathedrals during the 1200s was the flying buttress, a stone support running between the walls and the ground, that helped to hold in the walls. A Gothic church has a uniform roof weighing a total of 20,000 N and rising at 40° above the horizontal at each wall. The walls are 40 m tall, and a flying buttress meets each wall 10 m below the base of the roof. What horizontal force must this flying buttress apply to the wall?

11.72 You are trying to raise a bicycle wheel of mass m and radius R up over a curb of height h. To do this, you apply a horizontal force $\vec{F}$. What is the least magnitude of the force $\vec{F}$ that will succeed in raising the wheel onto the curb when the force is applied a) at the center of the

Figure 11.42 Problem 11.72.

wheel (Fig. 11.42)? b) at the top of the wheel? c) In which case is less force required?

11.73 The Farmyard Gate. A gate 4.00 m wide and 2.00 m high weighs 500 N. Its center of gravity is at its center, and it is hinged at A and B. To relieve the strain on the top hinge, a wire CD is connected as shown in Fig. 11.43. The tension in CD is increased until the horizontal force at hinge A is zero. a) What is the tension in the wire CD? b) What is the magnitude of the horizontal component of the force at hinge B? c) What is the combined vertical force exerted by hinges A and B?

Figure 11.43 Problem 11.73.

11.74 If you put a uniform block at the edge of a table, the center of the block must be over the table for the block not to fall off. a) If you stack two identical blocks at the table edge, the center of the top block must be over the bottom block, and the center of gravity of the two blocks together must be over the table. In terms of the length L of each block, what is the maximum overhang possible (Fig. 11.44)? b) Repeat part (a) for three identical blocks and for four identical blocks. c) Is it possible to make a stack of blocks such that the uppermost block is not directly over the table at all? How many blocks would it take to do this? (Try this with your friends using copies of this book.)

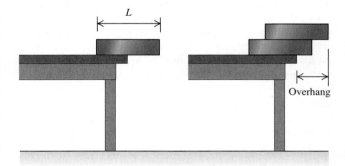

Figure 11.44 Problem 11.74.

11.75 Two uniform 75.0-g marbles 2.00 cm in diameter are stacked as shown in Fig. 11.45 in a container that is 3.00 cm wide. a) Find the force that the container exerts on the marbles at the points of contact A, B, and C. b) What force does each marble exert on the other?

11.76 Two identical, uniform beams weighing 260 N each are connected at one end by a frictionless hinge. A light horizontal crossbar attached at the mid-

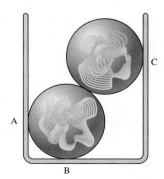

Figure 11.45 Problem 11.75.

points of the beams maintains an angle of 53.0° between the beams. The beams are suspended from the ceiling by vertical wires such that they form a "V," as shown in Fig. 11.46. a) What force does the crossbar exert on each beam? b) Is the crossbar under tension or compression? c) What force (magnitude and direction) does the hinge at point A exert on each beam?

Figure 11.46 Problem 11.76.

11.77 An engineer is designing a conveyor system for loading hay bales into a wagon (Fig. 11.47). Each bale is 0.25 m wide, 0.50 m high, and 0.80 m long (the dimension perpendicular to the plane of the figure), with mass 30.0 kg. The center of gravity of each bale is at its geometrical center. The coefficient of static friction

Figure 11.47 Problem 11.77.

between a bale and the conveyor belt is 0.60, and the belt moves with constant speed. a) The angle β of the conveyor is slowly increased. At some critical angle a bale will tip (if it doesn't slip first), and at some different critical angle it will slip (if it doesn't tip first). Find the two critical angles and determine which happens at the smaller angle. b) Would the outcome of part (a) be different if the coefficient of friction were 0.40?

11.78 The hay bale of Problem 11.77 is dragged along a horizontal surface with constant speed by a force $\vec{F}$ (Fig. 11.48). The coefficient of kinetic friction is 0.35. a) Find the magnitude of the force $\vec{F}$. b) Find the value of h at which the bale just begins to tip.

Figure 11.48 Problem 11.78.

11.79 A garage door is mounted on an overhead rail (Fig. 11.49). The wheels at A and B have rusted so that they do not roll, but rather slide along the track. The coefficient of kinetic friction is 0.52. The distance between the wheels is 2.00 m, and each is 0.50 m from the vertical sides of the door. The door is uniform and weighs 950 N. It is pushed to the left at constant speed by a horizontal force $\vec{F}$. a) If the distance h is 1.60 m, what is the vertical component of the force exerted on each wheel by the track? b) Find the maximum value h can have without causing one wheel to leave the track.

Figure 11.49 Problem 11.79.

11.80 A horizontal boom is supported at its left end by a frictionless pivot. It is held in place by a cable attached to the right-hand end of the boom. A chain and crate of total weight w hang from somewhere along the boom. The boom's weight w_b cannot be ignored and the boom may or may not be uniform. a) Show that the tension in the cable is the same whether the cable makes an angle θ or an angle $180° - \theta$ with the horizontal, and that the horizontal force component exerted on the boom by the pivot has equal magnitude but opposite direction for the two angles. b) Show that the cable cannot be horizontal. c) Show that the tension in the cable is a minimum when the cable is vertical, pulling upward on the right end of the boom. d) Show that when the cable is vertical, the force exerted by the pivot on the boom is vertical.

11.81 Prior to being placed in its hole, a 5700-N, 9.0-m-long, uniform utility pole makes some nonzero angle with the vertical. A vertical cable attached 2.0 m below its upper end holds it in place while its lower end rests on the ground. a) Find the tension in the cable and the magnitude and direction of the force exerted by the ground on the pole. b) Why don't we need the angle the pole makes with the vertical, as long as it is not zero?

11.82 Hooke's law for a tensile stress can be written as $F_x = kx$, where x is the object's change in length and k is the force constant. a) What is the force constant of a rod of length l_0, cross-sectional area A, and Young's modulus Y? b) In terms of l_0, A, and Y, how much work is required to stretch the object a distance x?

11.83 A 12.0-kg mass, fastened to the end of an aluminum wire with an unstretched length of 0.50 m, is whirled in a vertical circle with a constant angular speed of 120 rev/min. The cross-sectional area of the wire is 0.014 cm². Calculate the elongation of the wire when the mass is a) at the lowest point of the path and b) at the highest point of its path.

11.84 A metal wire 3.50 m long and 0.70 mm in diameter was given the following test. A load weighing 20 N was originally hung from the wire to keep it taut. The position of the lower end of the wire was read on a scale as load was added.

Added Load (N)	Scale Reading (cm)
0	3.02
10	3.07
20	3.12
30	3.17
40	3.22
50	3.27
60	3.32
70	4.27

a) Graph these values, plotting the increase in length horizontally and the added load vertically. b) Calculate the value of Young's modulus. c) The proportional limit occurred at a scale reading of 3.34 cm. What was the stress at this point?

11.85 A 1.05-m-long rod of negligible weight is supported at its ends by wires A and B of equal length (Fig. 11.50). The cross-sectional area of A is 2.00 mm² and that of B is 4.00 mm². Young's modulus for wire A is 1.80×10^{11} Pa; that for B is 1.20×10^{11} Pa. At what point along the rod should a weight w be suspended to produce a) equal stresses in A and B? b) equal strains in A and B?

Figure 11.50 Problem 11.85.

11.86 An amusement park ride consists of airplane-shaped cars attached to steel rods (Fig. 11.51). Each rod has a length of 15.0 m and a cross-sectional area of 8.00 cm². a) How much is the rod stretched when the ride is at rest? (Assume that each car plus two people seated in it has a total weight of 1900 N.) b) When operating, the ride has a maximum angular speed of 8.0 rev/min. How much is the rod stretched then?

Figure 11.51 Problem 11.86.

11.87 A brass rod with a length of 1.40 m and a cross-sectional area of 2.00 cm² is fastened end-to-end to a nickel rod with length L and cross-sectional area 1.00 cm². The compound rod is subjected to equal and opposite pulls of magnitude 4.00×10^4 N at its ends. a) Find the length L of the nickel rod if the elongations of the two rods are equal. b) What is the stress in each rod? c) What is the strain in each rod?

11.88 Stress on the Shin Bone. Compressive strength of our bones is important in everyday life. Young's modulus for bone is about 1.4×10^{10} Pa. Bone can take only about a 1.0% change in its length before fracturing. a) What is the maximum force that can be applied to a bone whose minimum cross-sectional area is 3.0 cm²? (This is approximately the cross-sectional area of a tibia, or shin bone, at its narrowest point.) b) Estimate the maximum height from which a 70-kg man could jump and not fracture the tibia. Take the time between when he first touches the floor and when he has stopped to be 0.030 s, and assume that the stress is distributed equally between his legs.

11.89 You hang a floodlamp from the end of a vertical steel wire. The floodlamp stretches the wire 0.18 mm and the stress is proportional to the strain. How much would it have stretched a) if the wire were twice as long? b) if the wire had the same length but twice the diameter? c) for a copper wire of the original length and diameter?

11.90 A moonshiner produces pure ethanol (ethyl alcohol) late at night and stores it in a stainless steel tank in the form of a cylinder 0.300 m in diameter with a tight-fitting piston at the top. The total volume of the tank is 250 L (0.250 m^3). In an attempt to squeeze a little more into the tank, the moonshiner piles 1420 kg of lead bricks on top of the piston. What additional volume of ethanol can the moonshiner squeeze into the tank? (Assume that the wall of the tank is perfectly rigid.)

11.91 A bar with cross-sectional area A is subjected to equal and opposite tensile forces $\vec{F}$ at its ends. Consider a plane through the bar making an angle θ with a plane at right angles to the bar (Fig. 11.52). a) What is the tensile (normal) stress at this plane in terms of F, A, and θ? b) What is the shear (tangential) stress at the plane in terms of F, A, and θ?

Figure 11.52 Problem 11.91.

c) For what value of θ is the tensile stress a maximum? d) For what value of θ is the shear stress a maximum?

11.92 A horizontal, uniform, copper rod has an original length l_0, a cross-sectional area A, a Young's modulus Y, and a mass m. It is supported by a frictionless pivot at its right end and by a cable at its left end. Both pivot and cable are attached so that they exert their forces uniformly over the rod's cross section. The cable makes an angle θ with the rod and compresses it. a) Find the stress exerted by the cable and pivot on the rod. b) Find the change in length of the rod due to this stress. c) The mass of the rod equals $\rho A l_0$, where ρ is the density. Show that the answers to parts (a) and (b) are independent of the cross-sectional area of the rod. d) The density of copper is 8900 kg/m³. Take Y for compression as given for copper in Table 11.1. Find the stress and change in length for an original length of 1.8 m and an angle of 30°. e) By how much would you multiply the answers of part (d) if the rod were twice as long?

Challenge Problems

11.93 A bookcase weighing 1500 N rests on a horizontal surface for which the coefficient of static friction is $\mu_s = 0.40$. The bookcase is 1.80 m tall and 2.00 m wide; its center of gravity is at its geometrical center. The bookcase rests on four short legs that are each 0.10 m from the edge of the bookcase. A person pulls on a rope attached to an upper corner of the bookcase with a force $\vec{F}$ that makes an angle θ with the bookcase (Fig. 11.53). a) If $\theta = 90°$, so $\vec{F}$ is horizontal, show that as F is increased from zero, the bookcase will start to slide before it tips, and calculate the magnitude of $\vec{F}$ that will start the bookcase sliding. b) If $\theta = 0°$, so $\vec{F}$ is vertical, show that the bookcase will tip over rather than slide, and calculate the magnitude of $\vec{F}$ that will cause the bookcase to start to tip. c) Calculate as a function of θ the magnitude of $\vec{F}$ that will cause the bookcase to start to slide and the magnitude that will cause it to start to tip. What is the smallest value that θ can have so that the bookcase will still start to slide before it starts to tip?

Figure 11.53 Challenge Problem 11.93.

11.94 Knocking Over a Post. One end of a post weighing 400 N and with height h rests on a rough horizontal surface with $\mu_s = 0.30$. The upper end is held by a rope fastened to the surface and making an angle of 36.9° with the post (Fig. 11.54). A horizontal force $\vec{F}$ is exerted on the post as shown. a) If the force $\vec{F}$ is applied at the midpoint of the post, what is the largest value it can have without causing the post to slip? b) How large can the force be without causing the post to slip if its point of application is $\frac{6}{10}$ of the way from the ground to the top of the post? c) Show that if the point of application of the force is too high, the post cannot be made to slip, no matter how great the force. Find the critical height for the point of application.

Figure 11.54 Challenge Problem 11.94.

11.95 Minimizing the Tension. A heavy horizontal girder of length L has several objects suspended from it. It is supported by a frictionless pivot at its left end and a cable of negligible weight that is attached to an I-beam at a point a distance h directly above the girder's center. Where should the other end of the cable be attached to the girder so that the cable's tension is a minimum? (*Hint:* In evaluating and presenting your answer, don't forget that the maximum distance of the point of attachment from the pivot is the length L of the beam.)

11.96 Two ladders, 4.00 m and 3.00 m long, are hinged at point A and tied together by a horizontal rope 0.90 m above the floor (Fig. 11.55). The ladders weigh 480 N and 360 N, respectively, and the center of gravity of each is at its center. Assume that the floor is freshly waxed and frictionless. a) Find the upward force at the bottom of each ladder. b) Find the tension in the rope. c) Find the magnitude of the force one ladder exerts on the other at point A. d) If an 800-N painter stands at point A, find the tension in the horizontal rope.

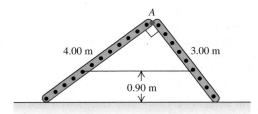

Figure 11.55 Challenge Problem 11.96.

11.97 A device for measuring compressibility consists of a cylinder filled with oil and fitted with a piston at one end. A block of sodium is immersed in the oil and a force is applied to the piston. Assume that the piston and walls of the cylinder are perfectly rigid and that there is no friction and no oil leak. Compute the compressibility of the sodium in terms of the applied force F, the piston displacement x, the piston area A, the initial volume of the oil V_O, the initial volume of the sodium V_S, and the compressibility of the oil k_O.

11.98 Bulk Modulus of an Ideal Gas. The equation of state (the equation relating pressure, volume, and temperature) for an ideal gas is $pV = nRT$, where n and R are constants. a) Show that if the gas is compressed while the temperature T is held constant, the bulk modulus is equal to the pressure. b) When an ideal gas is compressed without the transfer of any heat into or out of it, the pressure and volume are related by $pV^\gamma = $ constant, where γ is a constant having different values for different gases. Show that, in this case, the bulk modulus is given by $B = \gamma p$.

11.99 An angler hangs a 4.50-kg fish from a vertical steel wire 1.50 m long and 5.00×10^{-3} cm^2 in cross-sectional area. The upper end of the wire is securely fastened to a support. a) Calculate the amount the wire is stretched by the hanging fish. The angler now applies a force $\vec{F}$ to the fish, pulling it very slowly downward by 0.500 mm from its equilibrium position. For this downward motion, calculate b) the work done by gravity; c) the work done by the force $\vec{F}$; d) the work done by the force the wire exerts on the fish; and e) the change in the elastic potential energy (the potential energy associated with the tensile stress in the wire). Compare the answers in parts (d) and (e). (*Hint:* See Problem 11.82.)

12 GRAVITATION

These two immense galaxies—each containing billions of stars and located 114 million light years from earth—have been undergoing a slow-motion collision for tens of millions of years. This "collision" is actually a gravitational interaction between the two galaxies. (The individual stars of the two galaxies are so widely separated that they are unlikely to actually collide with each other.)

? **These two galaxies together have a mass more than 10^{11} times larger than the sun's mass. Why, then, isn't the earth pulled away from the sun and toward these galaxies?**

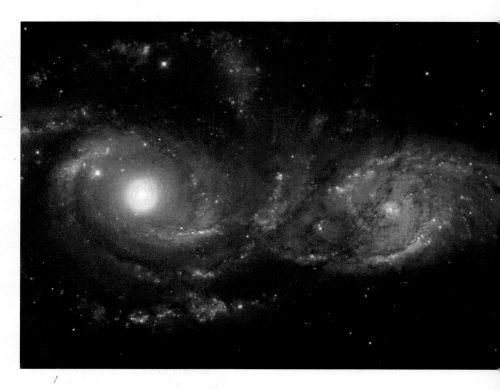

Some of the earliest investigations in physical science started with questions that people asked about the night sky. Why doesn't the moon fall to earth? Why do the planets move across the sky? And why doesn't the earth fly off into space rather than remaining in orbit around the sun? The study of gravitation provides the answers for these and many related questions.

As we remarked in Chapter 5, gravitation is one of the four classes of interactions found in nature, and it was the earliest of the four to be studied extensively. Newton discovered in the 17th century that the same interaction that makes an apple fall out of a tree also keeps the planets in their orbits around the sun. This was the beginning of *celestial mechanics,* the study of the dynamics of objects in space. Today, our knowledge of celestial mechanics allows us to determine how to put a satellite into any desired orbit around the earth or to choose just the right trajectory to send a spacecraft to another planet.

In this chapter you will learn the basic law that governs gravitational interactions. This law is *universal:* gravity acts in the same fundamental way between the earth and your body, between the sun and a planet, and between a planet and one of its moons. We'll apply the law of gravitation to phenomena such as the variation of weight with altitude, the orbits of satellites around the earth, and the orbits of planets around the sun.

12.1 | Newton's Law of Gravitation

The example of gravitational attraction that's probably most familiar to you is your *weight,* the force that attracts you toward the earth. During his study of the motions of the planets and of the moon, Newton discovered the fundamental character of the gravitational attraction between *any* two bodies. Along with his three laws of motion, Newton published the **law of gravitation** in 1687. It may be stated as follows:

> **Every particle of matter in the universe attracts every other particle with a force that is directly proportional to the product of the masses of the particles and inversely proportional to the square of the distance between them.**

Translating this into an equation, we have

$$F_g = \frac{Gm_1m_2}{r^2} \quad \text{(law of gravitation)} \quad (12.1)$$

where F_g is the magnitude of the gravitational force on either particle, m_1 and m_2 are their masses, r is the distance between them (Fig. 12.1), and G is a fundamental physical constant called the **gravitational constant.** The numerical value of G depends on the system of units used.

Equation (12.1) tells us that the gravitational force between two particles decreases with increasing distance r: if the distance is doubled, the force is only one-fourth as great, and so on. Although many of the stars in the night sky are far more massive than the sun, they are so far away that their gravitational force on the earth is negligibly small.

> **CAUTION** Because the symbols g and G are almost the same, it's common to confuse the two gravitational quantities that these symbols represent. Lowercase g is the acceleration due to gravity, which relates the weight w of a body to its mass m: $w = mg$. The value of g is different at different locations on the earth's surface and on the surfaces of different planets. By contrast, capital G relates the gravitational force between any two bodies to their masses and the distance between them. We call G a *universal* constant because it has the same value for any two bodies, no matter where in space they are located. In the next section we'll see how the values of g and G are related, but remember that they're two very different quantities!

Gravitational forces always act along the line joining the two particles, and they form an action-reaction pair. Even when the masses of the particles are different, the two interaction forces have equal magnitude (Fig. 12.1). The attractive force that your body exerts on the earth has the same magnitude as the force that it exerts on you. When you fall from a diving board into a swimming pool, the entire earth rises up to meet you! (Why don't you notice this? The earth's mass is greater than yours by a factor of about 10^{23}, so its acceleration is only 10^{-23} as great as yours.)

We have stated the law of gravitation in terms of the interaction between two *particles.* It turns out that the gravitational interaction of any two bodies having *spherically symmetric* mass distributions (such as solid spheres or spherical shells) is the same as though we concentrated all the mass of each at its center, as

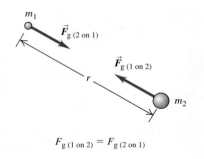

$$F_{g\,(1\,\text{on}\,2)} = F_{g\,(2\,\text{on}\,1)}$$

12.1 Even if two particles have very different masses, they exert equally strong forces of gravitational attraction on each other.

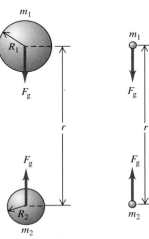

(a) Two spherically symmetric masses m_1 and m_2

(b) Two particles with masses m_1 and m_2

12.2 The gravitational effect *outside* any spherically symmetric mass distribution is the same as though all of the mass were concentrated at its center.

12.3 Thanks to the planet Jupiter's immense mass of 1.90×10^{27} kg, the mutual gravitational attraction of its atoms has pulled the planet into a nearly spherical shape. This has not happened for Amalthea (one of Jupiter's moons, shown in close-up). With a relatively tiny mass of 7.17×10^{18} kg, Amalthea has only weak mutual gravitation and so has an irregular shape.

in Fig. 12.2. Thus, if we model the earth as a spherically symmetric body with mass m_E, the force it exerts on a particle or a spherically symmetric body with mass m, at a distance r between centers, is

$$F_g = \frac{Gm_E m}{r^2} \qquad (12.2)$$

provided that the body lies outside the earth. A force of the same magnitude is exerted *on* the earth by the body. (We will prove these statements in Section 12.6.)

At points *inside* the earth the situation is different. If we could drill a hole to the center of the earth and measure the gravitational force on a body at various depths, we would find that toward the center of the earth the force *decreases,* rather than increasing as $1/r^2$. As the body enters the interior of the earth (or other spherical body), some of the earth's mass is on the side of the body opposite from the center and pulls in the opposite direction. Exactly at the center, the earth's gravitational force on the body is zero.

Spherically symmetric bodies are an important case because moons, planets, and stars all tend to be spherical. Since all particles in a body gravitationally attract each other, the particles tend to move to minimize the distance between them. As a result, the body naturally tends to assume a spherical shape, just as a lump of clay forms into a sphere if you squeeze it with equal forces on all sides. This effect is greatly reduced in celestial bodies of low mass since the gravitational attraction is less, and these bodies tend *not* to be spherical (Fig. 12.3).

Determining the Value of G

To determine the value of the gravitational constant G, we have to *measure* the gravitational force between two bodies of known masses m_1 and m_2 at a known distance r. The force is extremely small for bodies that are small enough to be brought into the laboratory, but it can be measured with an instrument called a *torsion balance,* which Sir Henry Cavendish used in 1798 to determine G.

A modern version of the Cavendish torsion balance is shown in Fig. 12.4. A light, rigid rod shaped like an inverted T is supported by a very thin, vertical quartz fiber. Two small spheres, each of mass m_1, are mounted at the ends of the horizontal arms of the T. When we bring two large spheres, each of mass m_2, to the positions shown, the attractive gravitational forces twist the T through a small angle. To

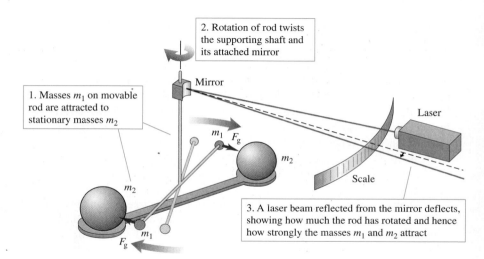

2. Rotation of rod twists the supporting shaft and its attached mirror

Mirror

1. Masses m_1 on movable rod are attracted to stationary masses m_2

Laser

Scale

3. A laser beam reflected from the mirror deflects, showing how much the rod has rotated and hence how strongly the masses m_1 and m_2 attract

12.4 The principle of the Cavendish balance, used for determining the value of G. The angle of deflection has been exaggerated here for clarity.

measure this angle, we shine a beam of light on a mirror fastened to the T. The reflected beam strikes a scale, and as the T twists, the reflected beam moves along the scale.

After calibrating the Cavendish balance, we can measure gravitational forces and thus determine G. The presently accepted value (in SI units) is

$$G = 6.673(10) \times 10^{-11} \, \text{N} \cdot \text{m}^2/\text{kg}^2$$

To three significant figures, $G = 6.67 \times 10^{-11} \, \text{N} \cdot \text{m}^2/\text{kg}^2$. Because $1 \, \text{N} = 1 \, \text{kg} \cdot \text{m}/\text{s}^2$, the units of G can also be expressed (in fundamental SI units) as $\text{m}^3/(\text{kg} \cdot \text{s}^2)$.

Gravitational forces combine vectorially. If each of two masses exerts a force on a third, the *total* force on the third mass is the vector sum of the individual forces of the first two. Example 12.3 makes use of this property, which is often called *superposition of forces.*

Example 12.1 Calculating gravitational force

The mass m_1 of one of the small spheres of a Cavendish balance is 0.0100 kg, the mass m_2 of one of the large spheres is 0.500 kg, and the center-to-center distance between each large sphere and the nearer small one is 0.0500 m. Find the gravitational force F_g on each sphere due to the nearest other sphere.

SOLUTION

IDENTIFY and SET UP: We will use the law of gravitation, Eq. (12.1), to determine F_g.

EXECUTE: Each sphere experiences the *same* magnitude of force from the other sphere, even though the masses may be very different (as they are in this case). The magnitude of each force is

$$F_g = \frac{(6.67 \times 10^{-11} \, \text{N} \cdot \text{m}^2/\text{kg}^2)(0.0100 \, \text{kg})(0.500 \, \text{kg})}{(0.0500 \, \text{m})^2}$$
$$= 1.33 \times 10^{-10} \, \text{N}$$

EVALUATE: This is a very small force, which is what we expect. After all, we don't experience noticeable gravitational pulls from ordinary low-mass objects in our environment. It takes a truly massive object such as the earth to exert a substantial gravitational force.

Example 12.2 Acceleration due to gravitational attraction

Suppose one large sphere and one small sphere are detached from the apparatus in Example 12.1 and placed 0.0500 m (between centers) from each other at a point in space far removed from all other bodies. What is the magnitude of the acceleration of each, relative to an inertial system?

SOLUTION

IDENTIFY and SET UP: We found the magnitude of the force on each sphere in Example 12.1. (The system of two spheres is so distant from other bodies that we can neglect any other forces.) To determine the magnitude of each sphere's acceleration, we'll use Newton's second law.

EXECUTE: The acceleration a_1 of the smaller sphere has magnitude

$$a_1 = \frac{F_g}{m_1} = \frac{1.33 \times 10^{-10} \, \text{N}}{0.0100 \, \text{kg}} = 1.33 \times 10^{-8} \, \text{m}/\text{s}^2$$

The acceleration a_2 of the larger sphere has magnitude

$$a_2 = \frac{F_g}{m_2} = \frac{1.33 \times 10^{-10} \, \text{N}}{0.500 \, \text{kg}} = 2.66 \times 10^{-10} \, \text{m}/\text{s}^2$$

EVALUATE: Although the forces on the bodies are equal in magnitude, the magnitudes of the two accelerations are *not* equal. Also, the accelerations are not constant; the gravitational forces increase as the spheres start to move toward each other.

Example 12.3	**Superposition of gravitational forces**

Many stars in the sky are actually systems of two or more stars that remain together due to their mutual gravitational attraction. Figure 12.5 shows a three-star system at an instant when the stars are at the vertices of a 45° right triangle. (For comparison, the mass of the sun—a rather ordinary star—is 1.99×10^{30} kg and the earth–sun distance is 1.50×10^{11} m.) We assume that the stars are spherical so that we can replace each star by a point mass at its center, as in Fig. 12.2. Find the magnitude and direction of the total gravitational force exerted on the small star by the two large ones.

SOLUTION

IDENTIFY: We use the principle of superposition: the total force on the small star is the vector sum of the forces due to each large star.

SET UP: We assume that the stars are spheres so that we can use the law of gravitation for each force. We first calculate the magnitude of each force using Eq. (12.1), then compute the vector sum using components along the axes shown in Fig. 12.5.

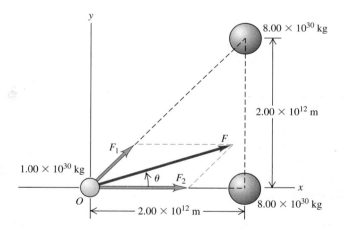

12.5 The total gravitational force on the smaller star is the vector sum of the forces exerted on it by the two larger stars.

EXECUTE: The magnitude F_1 of the force on the small star due to the upper large one is

$$F_1 = \frac{\left[\begin{array}{c} (6.67 \times 10^{-11} \text{ N} \cdot \text{m}^2/\text{kg}^2) \\ \times (8.00 \times 10^{30} \text{ kg})(1.00 \times 10^{30} \text{ kg}) \end{array} \right]}{(2.00 \times 10^{12} \text{ m})^2 + (2.00 \times 10^{12} \text{ m})^2}$$

$$= 6.67 \times 10^{25} \text{ N}$$

The magnitude F_2 of the force due to the lower large star is

$$F_2 = \frac{\left[\begin{array}{c} (6.67 \times 10^{-11} \text{ N} \cdot \text{m}^2/\text{kg}^2) \\ \times (8.00 \times 10^{30} \text{ kg})(1.00 \times 10^{30} \text{ kg}) \end{array} \right]}{(2.00 \times 10^{12} \text{ m})^2}$$

$$= 1.33 \times 10^{26} \text{ N}$$

The x- and y-components of these forces are

$$F_{1x} = (6.67 \times 10^{25} \text{ N})(\cos 45°) = 4.72 \times 10^{25} \text{ N}$$
$$F_{1y} = (6.67 \times 10^{25} \text{ N})(\sin 45°) = 4.72 \times 10^{25} \text{ N}$$
$$F_{2x} = 1.33 \times 10^{26} \text{ N}$$
$$F_{2y} = 0$$

The components of the total force on the small star are

$$F_x = F_{1x} + F_{2x} = 1.81 \times 10^{26} \text{ N}$$
$$F_y = F_{1y} + F_{2y} = 4.72 \times 10^{25} \text{ N}$$

The magnitude of this force is

$$F = \sqrt{F_x^2 + F_y^2} = \sqrt{(1.81 \times 10^{26} \text{ N})^2 + (4.72 \times 10^{25} \text{ N})^2}$$
$$= 1.87 \times 10^{26} \text{ N}$$

and its direction relative to the x-axis is

$$\theta = \arctan \frac{F_y}{F_x} = \arctan \frac{4.72 \times 10^{25} \text{ N}}{1.81 \times 10^{26} \text{ N}} = 14.6°$$

EVALUATE: While the total force on the small star is tremendous, the magnitude of the resulting acceleration is not: $a = F/m = (1.87 \times 10^{26} \text{ N})/(1.00 \times 10^{30} \text{ kg}) = 1.87 \times 10^{-4} \text{ m/s}^2$.

Can you show that the total force on the small star is *not* directed toward the center of mass of the two large stars? (See Problem 12.48.)

Comparing Examples 12.1 and 12.3 shows that gravitational forces are negligible between ordinary household-sized objects, but very substantial between objects that are the size of stars. Indeed, gravitation is *the* most important force on the scale of planets, stars, and galaxies (like those shown in the photograph that opens this chapter). It is responsible for holding our earth together and for keeping the planets in orbit about the sun. The mutual gravitational attraction between different parts of the sun compresses material at the sun's core to very high densities and temperatures, making it possible for nuclear reactions to take place there. These reactions generate the sun's energy output, which makes it possible for life to exist on earth and for you to read these words.

The gravitational force is so important on the cosmic scale because it acts *at a distance,* without any direct contact between bodies. Electric and magnetic forces have this same remarkable property, but they are less important on astronomical scales because large accumulations of matter are electrically neutral—that is, they contain equal amounts of positive and negative charge. As a result, the electric and magnetic forces between stars or planets are very small or zero. The strong and weak interactions that we discussed in Section 5.5 also act at a distance, but their influence is negligible at distances much greater than the diameter of an atomic nucleus (about 10^{-14} m).

A useful way to describe forces that act at a distance is in terms of a *field.* One body sets up a disturbance or field at all points in space, and the force that acts on a second body at a particular point is its response to the first body's field at that point. There is a field associated with each force that acts at a distance, and so we refer to gravitational fields, electric fields, magnetic fields, and so on. We won't need the field concept for our study of gravitation in this chapter, so we won't discuss it further here. But in later chapters we'll find that the field description is an extraordinarily powerful tool for describing electric and magnetic interactions.

Test Your Understanding

Calculate the magnitude of the gravitational force between two 900-kg automobiles when they are 3.0 m apart. Do gravitational forces play an appreciable role in automobile collisions?

12.2 | Weight

We defined the *weight* of a body in Section 4.4 as the attractive gravitational force exerted on it by the earth. We can now broaden our definition. **The weight of a body is the total gravitational force exerted on the body by all other bodies in the universe.** When the body is near the surface of the earth, we can neglect all other gravitational forces and consider the weight as just the earth's gravitational attraction. At the surface of the *moon* we consider a body's weight to be the gravitational attraction of the moon, and so on.

If we again model the earth as a spherically symmetric body with radius R_E and mass m_E, the weight w of a small body of mass m at the earth's surface (distance R_E from its center) is

$$w = F_g = \frac{Gm_E m}{R_E^2} \quad \text{(weight of a body of mass } m \text{ at the earth's surface)} \quad (12.3)$$

But we also know from Section 4.4 that the weight w of a body is the force that causes the acceleration g of free fall, so by Newton's second law, $w = mg$. Equating this with Eq. (12.3) and dividing by m, we find

$$g = \frac{Gm_E}{R_E^2} \quad \text{(acceleration due to gravity at the earth's surface)} \quad (12.4)$$

12.6 In an airliner at high altitude, you are farther from the center of the earth than when on the ground and you actually weigh less. The effect is quite small, though measurable. Can you show that at an altitude of 10 km above the surface, you weigh just 0.3% less than you do on the ground?

The acceleration due to gravity g is independent of the mass m of the body because m doesn't appear in this equation. We already knew that, but we can now see how it follows from the law of gravitation.

We can *measure* all the quantities in Eq. (12.4) except for m_E, so this relation allows us to compute the mass of the earth. Solving Eq. (12.4) for m_E and using $R_E = 6380$ km $= 6.38 \times 10^6$ m and $g = 9.80$ m/s^2, we find

$$m_E = \frac{gR_E^2}{G} = 5.98 \times 10^{24} \text{ kg}$$

very close to the currently accepted value of 5.974×10^{24} kg. Once Cavendish had measured G, he computed the mass of the earth in just this way.

At a point above the earth's surface a distance r from the center of the earth (a distance $r - R_E$ above the surface), the weight of a body is given by Eq. (12.3) with R_E replaced by r:

$$w = F_g = \frac{Gm_E m}{r^2} \tag{12.5}$$

The weight of a body decreases inversely with the square of its distance from the earth's center (Fig. 12.6). Figure 12.7 shows how the weight varies with height above the earth for an astronaut who weighs 700 N at the earth's surface.

The *apparent* weight of a body on earth differs slightly from the earth's gravitational force because the earth rotates and is therefore not precisely an inertial frame of reference. We have ignored this effect in the above discussion and have assumed that the earth *is* an inertial system. We will return to the effect of the earth's rotation in Section 12.7.

In our discussion of weight, we've used the fact that the earth is an approximately spherically symmetric distribution of mass. But this does *not* mean that the earth is uniform. To demonstrate that it cannot be uniform, let's first calculate the average *density,* or mass per unit volume, of the earth. If we assume a spherical earth, the volume is

$$V_E = \frac{4}{3}\pi R_E^3 = \frac{4}{3}\pi (6.38 \times 10^6 \text{ m})^3 = 1.09 \times 10^{21} \text{ m}^3$$

12.7 An astronaut weighing 700 N at the earth's surface experiences less gravitational attraction when above the surface. The relevant distance r is from the astronaut to the *center* of the earth (*not* from the astronaut to the earth's surface).

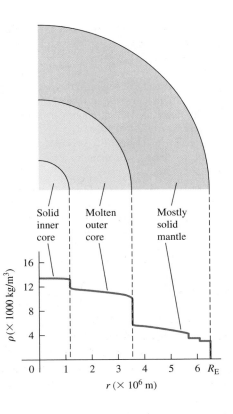

12.8 The density of the earth decreases with increasing distance from its center.

The average density ρ (the Greek letter "rho") of the earth is the total mass divided by the total volume:

$$\rho = \frac{m_E}{V_E} = \frac{5.97 \times 10^{24} \text{ kg}}{1.09 \times 10^{21} \text{ m}^3}$$

$$= 5500 \text{ kg/m}^3 = 5.5 \text{ g/cm}^3$$

(For comparison, the density of water is $1000 \text{ kg/m}^3 = 1.00 \text{ g/cm}^3$.) If the earth were uniform, we would expect the density of individual rocks near the earth's surface to have this same value. In fact, the density of surface rocks is substantially lower, ranging from about $2000 \text{ kg/m}^3 = 2 \text{ g/cm}^3$ for sedimentary rocks to about $3300 \text{ kg/m}^3 = 3.3 \text{ g/cm}^3$ for basalt. So the earth *cannot* be uniform, and the interior of the earth must be much more dense than the surface in order that the *average* density be $5500 \text{ kg/m}^3 = 5.5 \text{ g/cm}^3$. According to geophysical models of the earth's interior, the maximum density at the center is about $13,000 \text{ kg/m}^3 = 13 \text{ g/cm}^3$. Figure 12.8 is a graph of density as a function of distance from the center.

| Example 12.4 | **Gravity on Mars** |

You're involved in the design of a mission carrying humans to the surface of the planet Mars, which has a radius $r_M = 3.40 \times 10^6$ m and a mass $m_M = 6.42 \times 10^{23}$ kg. The earth weight of the Mars lander is 39,200 N. Calculate its weight F_g and the acceleration g_M due to the gravity of Mars: a) 6.0×10^6 m above the surface of Mars (the distance at which the moon Phobos orbits Mars); b) at the surface of Mars. Neglect the gravitational effects of the (very small) moons of Mars.

SOLUTION

IDENTIFY and SET UP: We find the weight F_g using Eq. (12.5) with m_E (the mass of the earth) replaced with m_M (the mass of Mars). Note that the value of G is the same everywhere in the universe; it is a fundamental physical constant. We then find the acceleration g_M using $F_g = mg_M$, where m is the mass of the lander. We're not given the value of this mass, but we can determine it from the lander's weight on earth.

EXECUTE: a) The distance r from the *center* of Mars is

$$r = (6.0 \times 10^6 \text{ m}) + (3.40 \times 10^6 \text{ m}) = 9.4 \times 10^6 \text{ m}$$

The mass m of the lander is its earth weight w divided by the acceleration of gravity g on earth:

$$m = \frac{w}{g} = \frac{39,200 \text{ N}}{9.8 \text{ m/s}^2} = 4000 \text{ kg}$$

The mass is the same whether the lander is on the earth, on Mars, or in between. From Eq. (12.5),

$$F_g = \frac{Gm_M m}{r^2}$$

$$= \frac{(6.67 \times 10^{-11} \text{ N} \cdot \text{m}^2/\text{kg}^2)(6.42 \times 10^{23} \text{ kg})(4000 \text{ kg})}{(9.4 \times 10^6 \text{ m})^2}$$

$$= 1940 \text{ N}$$

The acceleration due to the gravity of Mars at this point is

$$g_M = \frac{F_g}{m} = \frac{1940 \text{ N}}{4000 \text{ kg}} = 0.48 \text{ m/s}^2$$

This is also the acceleration experienced by Phobos in its orbit, 6.0×10^6 m above the surface of Mars.

b) To find F_g and g_M at the surface, we repeat the above calculations, replacing $r = 9.4 \times 10^6$ m with $R_M = 3.40 \times 10^6$ m. Alternatively, because F_g and g_M are inversely proportional to $1/r^2$ (at any point outside the planet), we can multiply the results of part (a) by the factor

$$\left(\frac{9.4 \times 10^6 \text{ m}}{3.40 \times 10^6 \text{ m}} \right)^2$$

You should use both methods to show that at the surface $F_g = 15{,}000$ N and $g_M = 3.7$ m/s².

EVALUATE: The results for part (b) show that an object's weight and the acceleration due to gravity are roughly 40% as large on the surface of Mars as they are on the earth's surface. Science-fiction films and stories set on Mars commonly describe the planet's lower temperatures and thinner atmosphere, but they seldom focus on the experience of being in a low-gravity environment.

Test Your Understanding

The mass of Mars is only 11% as large as the earth's mass. Why, then, isn't the surface gravity on Mars 11% as great as on earth?

12.3 | Gravitational Potential Energy

When we first developed the concept of gravitational potential energy in Section 7.1, we assumed that the gravitational force on a body is constant in magnitude and direction. This led to the expression $U = mgy$. But we now know that the earth's gravitational force on a body of mass m at any point outside the earth is given more generally by Eq. (12.2), $F_g = Gm_E m/r^2$, where m_E is the mass of the earth and r is the distance of the body from the earth's center. For problems in which r changes enough that the gravitational force can't be considered constant, we need a more general expression for gravitational potential energy.

To find this expression, we follow the same basic sequence of steps as in Section 7.1. We consider a body of mass m outside the earth, and first compute the work W_{grav} done by the gravitational force when the body moves directly away from or toward the center of the earth from $r = r_1$ to $r = r_2$, as in Fig. 12.9. This work is given by

$$W_{grav} = \int_{r_1}^{r_2} F_r \, dr \tag{12.6}$$

where F_r is the radial component of the gravitational force $\vec{F}$, that is, the component in the direction *outward* from the center of the earth. Because $\vec{F}$ points directly *inward* toward the center of the earth, F_r is negative. It differs from Eq. (12.2), the magnitude of the gravitational force, by a minus sign:

$$F_r = -\frac{Gm_E m}{r^2} \qquad (12.7)$$

Substituting Eq. (12.7) into Eq. (12.6), we see that W_{grav} is given by

$$W_{\text{grav}} = -Gm_E m \int_{r_1}^{r_2} \frac{dr}{r^2} = \frac{Gm_E m}{r_2} - \frac{Gm_E m}{r_1} \qquad (12.8)$$

The path doesn't have to be a straight line; it could also be a curve like the one in Fig. 12.9. By an argument similar to that in Section 7.1, this work depends only on the initial and final values of r, not on the path taken. This also proves that the gravitational force is always *conservative*.

We now define the corresponding potential energy U so that $W_{\text{grav}} = U_1 - U_2$, as in Eq. (7.3). Comparing this with Eq. (12.8), we see that the appropriate definition for **gravitational potential energy** is

$$U = -\frac{Gm_E m}{r} \qquad \text{(gravitational potential energy)} \qquad (12.9)$$

Figure 12.10 shows how the gravitational potential energy depends on the distance r between the body of mass m and the center of the earth. When the body moves away from the earth, r increases, the gravitational force does negative work, and U increases (i.e., becomes less negative). When the body "falls" toward earth, r decreases, the gravitational work is positive, and the potential energy decreases (i.e., becomes more negative).

You may be troubled by Eq. (12.9) because it states that gravitational potential energy is always negative. But in fact you've seen negative values of U before. In using the formula $U = mgy$ in Section 7.1, we found that U was negative whenever the body of mass m was at a value of y below the arbitrary height we chose to be $y = 0$—that is, whenever the body and the earth were closer together than some certain arbitrary distance. (See, for instance, Example 7.2 in Section 7.1.) In defining U by Eq. (12.9), we have chosen U to be zero when the body of mass m is infinitely far away from the earth $(r = \infty)$. As the body moves toward the earth, gravitational potential energy decreases and so becomes negative. If we wanted, we could make $U = 0$ at the surface of the earth, where $r = R_E$, by simply adding the quantity $Gm_E m/R_E$ to Eq. (12.9). This would make U positive when $r > R_E$, but at the price of making the expression for U more complicated. This added term would not affect the *difference* in U between any two points, which is the only physically significant quantity; that is why we omit this term and use Eq. (12.9) for potential energy.

CAUTION Be careful not to confuse the expressions for gravitational force, Eq. (12.7), and gravitational potential energy, Eq. (12.9). The force F_r is proportional to $1/r^2$, while potential energy U is proportional to $1/r$.

Armed with Eq. (12.9), we can now use general energy relations for problems in which the $1/r^2$ behavior of the earth's gravitational force has to be included. If

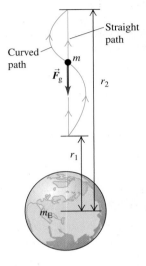

Work done by $\vec{F}_g$ does not depend on the path (gravitational force is conservative)

12.9 When a body moves from radial coordinate r_1 to r_2, the work done on it by the gravitational force is the same whether the body follows a straight path or any curved path.

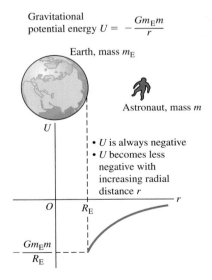

Gravitational potential energy $U = -\dfrac{Gm_E m}{r}$

Earth, mass m_E

Astronaut, mass m

• U is always negative
• U becomes less negative with increasing radial distance r

12.10 A graph of the gravitational potential energy U for the system of the earth (mass m_E) and an astronaut (mass m) versus the astronaut's distance r from the center of the earth.

the gravitational force on the body is the only force that does work, the total mechanical energy of the system is constant, or *conserved*. In the following example we'll use this principle to calculate **escape speed,** the speed required for a body to escape completely from a planet.

Example 12.5	"From the earth to the moon"

In Jules Verne's 1865 story with this title, three men were shot to the moon in a shell fired from a giant cannon sunk in the earth in Florida. a) Find the muzzle speed needed to shoot the shell straight up to a height above the earth equal to the earth's radius. (b) Find the *escape speed*—that is, the muzzle speed that would allow the shell to escape from the earth completely. Neglect air resistance, the earth's rotation, and the gravitational pull of the moon. The earth's radius is $R_E = 6380$ km $= 6.38 \times 10^6$ m, and its mass is $m_E = 5.97 \times 10^{24}$ kg (see Appendix F).

SOLUTION

IDENTIFY: Once the shell leaves the muzzle of the cannon, only the (conservative) gravitational force does work and mechanical energy is conserved. We use this fact to find the shell's speed as it leaves the muzzle in part (a), for which the shell comes to a halt at a distance of two earth radii from the planet's center, and in part (b), for which the shell comes to a halt at an infinite distance from earth.

SET UP: The equation to use in both parts is the statement of energy conservation, $K_1 + U_1 = K_2 + U_2$, where the potential energy U is given by Eq. (12.9). Point 1 is where the shell leaves the cannon with speed v_1 (the target variable). At this point the distance from the center of the earth is $r_1 = R_E$, the radius of the earth. Point 2 is where the shell reaches its maximum height; in part (a) it is at $r_2 = 2R_E$ (Fig. 12.11), and in part (b) it is infinitely far from the earth at $r_2 = \infty$. In either case the shell is at rest at point 2, so $v_2 = 0$. Let m be the mass of the shell (with passengers).

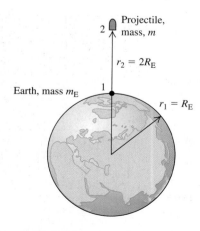

12.11 A projectile fired from the surface of the earth to a height equal to the earth's radius.

EXECUTE: a) We can determine v_1 from the energy-conservation equation

$$K_1 + U_1 = K_2 + U_2$$

$$\frac{1}{2}mv_1^2 + \left(-\frac{Gm_Em}{R_E}\right) = 0 + \left(-\frac{Gm_Em}{2R_E}\right)$$

Rearranging this, we find that

$$v_1 = \sqrt{\frac{Gm_E}{R_E}}$$

$$= \sqrt{\frac{(6.67 \times 10^{-11}\,\text{N}\cdot\text{m}^2/\text{kg}^2)(5.97 \times 10^{24}\,\text{kg})}{6.38 \times 10^6\,\text{m}}}$$

$$= 7900\,\text{m/s} \; (=28{,}400\,\text{km/h} = 17{,}700\,\text{mi/h})$$

b) We want the shell barely to be able to "reach" $r_2 = \infty$, with no kinetic energy left over. Hence $K_2 = 0$ and $U_2 = 0$ (the potential energy goes to zero at infinity; see Fig. 12.10). The total energy is therefore zero, and when the shell is fired its positive kinetic energy K_1 and negative potential energy U_1 must also add to zero:

$$\frac{1}{2}mv_1^2 + \left(-\frac{Gm_Em}{R_E}\right) = 0$$

$$v_1 = \sqrt{\frac{2Gm_E}{R_E}}$$

$$= \sqrt{\frac{2(6.67 \times 10^{-11}\,\text{N}\cdot\text{m}^2/\text{kg}^2)(5.97 \times 10^{24}\,\text{kg})}{6.38 \times 10^6\,\text{m}}}$$

$$= 1.12 \times 10^4\,\text{m/s} \; (=40{,}200\,\text{km/h} = 25{,}000\,\text{mi/h})$$

EVALUATE: This result does not depend on the mass of the shell, nor does it depend on the direction in which the shell is launched. Modern spacecraft launched from Florida must attain essentially this same speed to escape the earth. A spacecraft on the ground at Cape Canaveral is already moving at 410 m/s to the east because of the earth's rotation; by launching to the east, the spacecraft takes advantage of this "free" contribution toward escape speed.

Generalizing our result, the initial speed v_1 needed for a body to escape from the surface of a spherical mass M with radius R (ignoring air resistance) is

$$v_1 = \sqrt{\frac{2GM}{R}} \qquad \text{(escape speed)}$$

You can use this result to compute the escape speed for other bodies. You will find 5.02×10^3 m/s for Mars, 5.95×10^4 m/s for Jupiter, and 6.18×10^5 m/s for the sun.

More on Gravitational Force and Potential Energy

The relationship between the gravitational force on a body, given by Eq. (12.7), and its gravitational potential energy, given by Eq. (12.9), can be expressed in a different way by using the methods of Section 7.4. Equations (7.17) and (7.18) show that the component of force in a given direction equals the negative of the derivative of U with respect to the corresponding coordinate. For motion along the x-axis,

$$F_x = -\frac{dU}{dx} \tag{12.10}$$

The gravitational force has a component only in the radial direction, so we replace x by r in Eq. (12.10). We find

$$F_r = -\frac{dU}{dr} = -\frac{d}{dr}\left(-\frac{Gm_E m}{r}\right) = -\frac{Gm_E m}{r^2} \tag{12.11}$$

This agrees with the F_r we started with in Eq. (12.7). As we remarked before, F_r is negative, showing that the force points opposite to the direction of increasing r.

As a final note, let's show that when we are close to the earth's surface, Eq. (12.9) reduces to the familiar $U = mgy$ from Chapter 7. We first rewrite Eq. (12.8) as

$$W_{grav} = Gm_E m \frac{r_1 - r_2}{r_1 r_2}$$

If the body stays close to the earth, then in the denominator we may replace r_1 and r_2 by R_E, the earth's radius, so

$$W_{grav} = Gm_E m \frac{r_1 - r_2}{R_E^2}$$

According to Eq. (12.4), $g = Gm_E / R_E^2$, so

$$W_{grav} = mg(r_1 - r_2)$$

If we replace the r's by y's, this is just Eq. (7.1) for the work done by a constant gravitational force. In Section 7.1 we used this equation to derive Eq. (7.2), $U = mgy$, so we may consider this expression for gravitational potential energy to be a special case of the more general Eq. (12.9).

Test Your Understanding

Is it possible for a planet to have the same surface gravity (that is, the same value of g at the surface) as the earth and yet have a greater escape speed?

12.4 | The Motion of Satellites

Artificial satellites orbiting the earth are a familiar fact of contemporary life (Fig. 12.12). But how do they stay in orbit, and what determines the properties of their orbits? We can use Newton's laws and the law of gravitation to provide the answers. We'll see in the next section that the motion of planets can be analyzed in the same way.

To begin, think back to the discussion of projectile motion in Section 3.3. In Example 3.7 a motorcycle rider rides horizontally off the edge of a cliff,

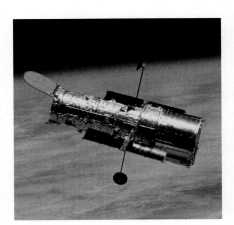

12.12 With a length of 13.2 m and a mass of 11,000 kg, the Hubble Space Telescope is among the largest satellites placed in orbit.

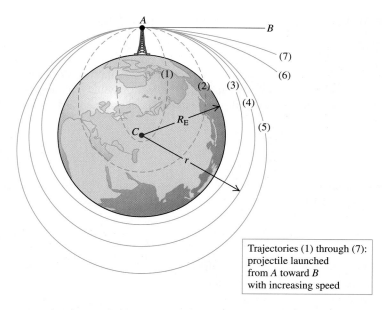

Trajectories (1) through (7): projectile launched from *A* toward *B* with increasing speed

12.13 Trajectories of a projectile launched from point *A* in the direction *AB* with different initial speeds. Orbits (1) and (2) would be completed as shown if the earth were a point mass at *C*. (This illustration is based on one in Isaac Newton's *Principia*.)

launching himself into a parabolic path that ends on the flat ground at the base of the cliff. If he survives and repeats the experiment with increased launch speed, he will land farther from the starting point. We can imagine him launching himself with great enough speed that the earth's curvature becomes significant. As he falls, the earth curves away beneath him. If he is going fast enough, and if his launch point is high enough that he clears the mountain tops, he may be able to go right on around the earth without ever landing.

Figure 12.13 shows a variation on this theme. We launch a projectile from point *A* in the direction *AB*, tangent to the earth's surface. Trajectories (1) to (7) show the effect of increasing the initial speed. In trajectories (3) through (5) the projectile misses the earth and becomes a satellite. If there is no retarding force, the projectile's speed when it returns to point *A* is the same as its initial speed and it repeats its motion indefinitely.

Trajectories (1) through (5) close on themselves and are called **closed orbits.** All closed orbits are ellipses or segments of ellipses; trajectory (4) is a circle, a special case of an ellipse. Trajectories (6) and (7) are **open orbits.** For these paths the projectile never returns to its starting point but travels ever farther away from the earth.

A circular orbit like trajectory (4) in Fig. 12.13 is the simplest case. It is also an important case, since many artificial satellites have nearly circular orbits and the orbits of the planets around the sun are also fairly circular. The only force acting on a satellite in circular orbit around the earth is the earth's gravitational attraction, which is directed toward the center of the earth and hence toward the center of the orbit (Fig. 12.14). As we discussed in Section 5.4, this means that the satellite is in *uniform* circular motion and its speed is constant. The satellite isn't falling *toward* the earth; rather, it's constantly falling *around* the earth. In a circu-

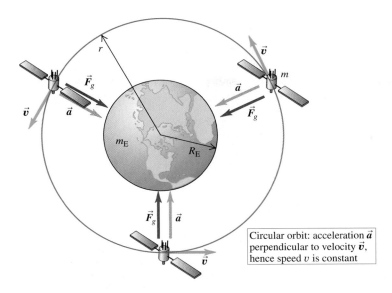

Circular orbit: acceleration $\vec{a}$ perpendicular to velocity $\vec{v}$, hence speed v is constant

12.14 The force $\vec{F}_g$ due to the earth's gravitational attraction provides the centripetal acceleration that keeps a satellite in orbit. Compare to Fig. 5.29.

lar orbit the speed is just right to keep the distance from the satellite to the center of the earth constant.

How can we find the constant speed v of a satellite in a circular orbit? The radius of the orbit is r, measured from the *center* of the earth; the acceleration of the satellite has magnitude $a_{\text{rad}} = v^2/r$ and is always directed toward the center of the circle. By the law of gravitation, the net force (gravitational force) on the satellite of mass m has magnitude $F_g = Gm_{\text{E}}m/r^2$ and is in the same direction as the acceleration. Newton's second law ($\sum \vec{F} = m\vec{a}$) then tells us that

$$\frac{Gm_{\text{E}}m}{r^2} = \frac{mv^2}{r}$$

Solving this for v, we find

4.6   Satellites Orbit

$$v = \sqrt{\frac{Gm_{\text{E}}}{r}} \qquad (\text{circular orbit}) \qquad (12.12)$$

This relation shows that we can't choose the orbit radius r and the speed v independently; for a given radius r, the speed v for a circular orbit is determined.

Equation (12.12) also shows that the satellite's motion does not depend on its mass m because m doesn't appear in the equation. If we could cut a satellite in half without changing its speed, each half would continue on with the original motion. An astronaut on board a space shuttle is herself a satellite of the earth, held by the earth's gravitational attraction in the same orbit as the shuttle. The astronaut has the same velocity and acceleration as the shuttle, so nothing is pushing her against the floor or walls of the shuttle. She is in a state of *apparent weightlessness*, as in a freely falling elevator; see the discussion following Example 5.10 in Section 5.2. (*True* weightlessness would occur only if the astronaut were infinitely far from any other masses, so that the gravitational force on her would be zero.) Indeed, every part of her body is apparently weightless; she feels nothing pushing her stomach against her intestines or her head against her shoulders (Fig. 12.15).

12.15 These space shuttle astronauts are in a state of apparent weightlessness. Which are right side up and which are upside down?

(a)

(b)

12.16 (a) The International Space Station orbits at an altitude of about 400 km, or about 6800 km from the center of the earth. Its orbital speed is about 7.7 km/s and its orbital period is about 93 minutes. (b) The moon orbits the earth at a much larger distance of about 384,000 km. As a result, it has a slow orbital speed (about 1.0 km/s) and a long orbital period (about 27.3 days).

12.17 The rings of Saturn are made up of countless numbers of icy particles, each of which is an independent satellite of Saturn.

Apparent weightlessness is not just a feature of circular orbits; it occurs whenever gravity is the only force acting on a spacecraft. Hence it occurs for orbits of any shape, including open orbits such as (6) and (7) in Fig. 12.13.

We can derive a relation between the radius r of a circular orbit and the period T, the time for one revolution. The speed v is the distance $2\pi r$ traveled in one revolution, divided by the period:

$$v = \frac{2\pi r}{T} \tag{12.13}$$

To get an expression for T, we solve Eq. (12.13) for T and substitute v from Eq. (12.12):

$$T = \frac{2\pi r}{v} = 2\pi r \sqrt{\frac{r}{Gm_E}} = \frac{2\pi r^{3/2}}{\sqrt{Gm_E}} \quad \text{(circular orbit)} \tag{12.14}$$

Equations (12.12) and (12.14) show that larger orbits correspond to slower speeds and longer periods (Fig. 12.16).

It's interesting to compare Eq. (12.12) to the calculation of escape speed in Example 12.5. We see that the escape speed from a spherical body with radius R is just $\sqrt{2}$ times as large as the speed of a satellite in a circular orbit at that radius. If our spacecraft is in circular orbit around *any* planet, we have to multiply our speed by a factor of $\sqrt{2}$ to escape to infinity, regardless of the planet's mass.

Since the speed v in a circular orbit is determined by Eq. (12.12) for a given orbit radius r, the total mechanical energy $E = K + U$ is determined as well. Using Eqs. (12.9) and (12.12), we have

$$E = K + U = \frac{1}{2}mv^2 + \left(-\frac{Gm_Em}{r}\right) = \frac{1}{2}m\left(\frac{Gm_E}{r}\right) - \frac{Gm_Em}{r}$$

$$E = -\frac{Gm_Em}{2r} \quad \text{(circular orbit)} \tag{12.15}$$

The total mechanical energy in a circular orbit is negative and equal to one-half the potential energy. Increasing the orbit radius r means increasing the mechanical energy (that is, making E less negative). If the satellite is in a relatively low orbit that encounters the outer fringes of earth's atmosphere, mechanical energy decreases due to negative work done by the force of air resistance; as a result, the orbit radius decreases until the satellite hits the ground or burns up in the atmosphere.

We have talked mostly about earth satellites, but we can apply the same analysis to the circular motion of *any* body under its gravitational attraction to a stationary body. Other examples include the earth's moon, the moons of other planets, and the rings of Saturn. If Saturn's rings were a rotating rigid body, all parts of the rings would have the same orbital period T, and by Eq. (12.13) the outer regions of the rings (large r) would have a greater speed v than the inner regions (small r). But since the rings are actually composed of many individual orbiting particles, Eqs. (12.12) and (12.14) tell us that particles in the outer regions of the rings move at a slower speed and have a longer period than do particles in the inner regions (Fig. 12.17).

Example 12.6	**A satellite orbit**

Suppose you want to place a 1000-kg weather satellite into a circular orbit 300 km above the earth's surface. a) What speed, period, and radial acceleration must it have? b) How much work has to be done to place this satellite in orbit? c) How much additional work would have to be done to make this satellite escape the earth? The earth's radius is $R_E = 6380$ km and its mass is $m_E = 5.97 \times 10^{24}$ kg.

SOLUTION

IDENTIFY: The satellite is in a circular orbit, so we can use the equations derived in this section.

SET UP: In part (a), given the satellite's altitude we first find the radius of its orbit. We then calculate the speed v and period T using Eqs. (12.12) and (12.14). The acceleration in a circular orbit is given by the familiar formula from Chapter 3, $a_{rad} = v^2/r$. For the processes described in parts (b) and (c), the work required is the difference between the initial and final mechanical energy, which for a circular orbit is given by Eq. (12.15).

EXECUTE: a) The radius of the satellite's orbit is

$$r = 6380 \text{ km} + 300 \text{ km} = 6680 \text{ km} = 6.68 \times 10^6 \text{ m}$$

From Eq. (12.12),

$$v = \sqrt{\frac{Gm_E}{r}} = \sqrt{\frac{(6.67 \times 10^{-11} \text{ N} \cdot \text{m}^2/\text{kg}^2)(5.97 \times 10^{24} \text{ kg})}{6.68 \times 10^6 \text{ m}}}$$

$$= 7720 \text{ m/s}$$

From Eq. (12.14),

$$T = \frac{2\pi r}{v} = \frac{2\pi (6.68 \times 10^6 \text{ m})}{7720 \text{ m/s}}$$

$$= 5440 \text{ s} = 90.6 \text{ min}$$

The radial acceleration is

$$a_{rad} = \frac{v^2}{r} = \frac{(7720 \text{ m/s})^2}{6.68 \times 10^6 \text{ m}}$$

$$= 8.92 \text{ m/s}^2$$

This is the value of g at a height of 300 km above the earth's surface; it is somewhat less than the value of g at the surface.

b) The work required is the difference between E_2, the total mechanical energy when the satellite is in orbit, and E_1, the original mechanical energy when the satellite was at rest on the launch pad back on earth. In orbit, if we use Eq. (12.15), the energy is

$$E_2 = -\frac{Gm_E m}{2r}$$

$$= -\frac{(6.67 \times 10^{-11} \text{ N} \cdot \text{m}^2/\text{kg}^2)(5.97 \times 10^{24} \text{ kg})(1000 \text{ kg})}{2(6.68 \times 10^6 \text{ m})}$$

$$= -2.99 \times 10^{10} \text{ J}$$

At rest on the earth's surface, the energy is purely potential:

$$E_1 = K_1 + U_1 = 0 + \left(-\frac{Gm_E m}{R_E}\right)$$

$$= -\frac{(6.67 \times 10^{-11} \text{ N} \cdot \text{m}^2/\text{kg}^2)(5.97 \times 10^{24} \text{ kg})(1000 \text{ kg})}{6.68 \times 10^6 \text{ m}}$$

$$= -6.25 \times 10^{10} \text{ J}$$

and so

$$W_{required} = E_2 - E_1 = -2.99 \times 10^{10} \text{ J} - (-6.25 \times 10^{10} \text{ J})$$

$$= 3.26 \times 10^{10} \text{ J}$$

c) We saw in part (b) of Example 12.5 that for a satellite to escape to infinity, the total mechanical energy must be zero. The total mechanical energy in the circular orbit is $E_2 = -2.99 \times 10^{10}$ J; to increase this to zero, an amount of work equal to 2.99×10^{10} J would have to be done. This extra energy could be supplied by rocket engines attached to the satellite.

EVALUATE: In part (b) we ignored the satellite's initial kinetic energy (while it was still on the launch pad) due to the rotation of the earth. You should check to see how much difference this makes (see Example 12.5 for useful data).

Test Your Understanding

A space shuttle and a satellite are in the same circular orbit but on opposite sides of the earth. Explain how the shuttle could rendezvous with the satellite by maneuvering into a circular orbit with a slightly smaller radius. Could this rendezvous also be achieved by maneuvering into a circular orbit of slightly *larger* radius?

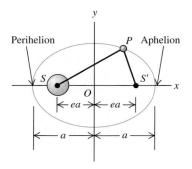

12.18 Geometry of an ellipse. The sum of the distances SP and $S'P$ is the same for every point on the curve. The sizes of the sun (S) and planet (P) are exaggerated for clarity.

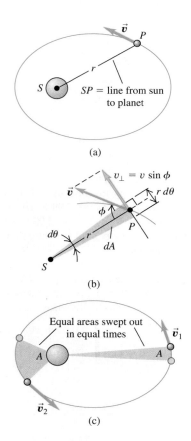

12.19 (a) The planet (P) moves about the sun (S) in an elliptical orbit. (b) In a time dt the line SP sweeps out an area $dA = \frac{1}{2}(r\,d\theta)r = \frac{1}{2}r^2\,d\theta$. (c) The planet's speed varies so that the line SP sweeps out the same area A in a given time t regardless of the planet's position in its orbit.

12.5 | Kepler's Laws and the Motion of Planets

The name *planet* comes from a Greek word meaning "wanderer," and indeed the planets continually change their positions in the sky relative to the background of stars. One of the great intellectual accomplishments of the 16th and 17th centuries was the threefold realization that the earth is also a planet, that all planets orbit the sun, and that the apparent motions of the planets as seen from the earth can be used to precisely determine the orbits of the planets.

The first and second of these ideas were published by Nicolaus Copernicus in Poland in 1543. The determination of planetary orbits was carried out between 1601 and 1619 by the German astronomer and mathematician Johannes Kepler, using a voluminous set of precise data on apparent planetary motions compiled by his mentor, the Danish astronomer Tycho Brahe. By trial and error, Kepler discovered three empirical laws that accurately described the motions of the planets:

1. **Each planet moves in an elliptical orbit, with the sun at one focus of the ellipse.**

2. **A line from the sun to a given planet sweeps out equal areas in equal times.**

3. **The periods of the planets are proportional to the $\frac{3}{2}$ powers of the major axis lengths of their orbits.**

Kepler did not know *why* the planets moved in this way. Three generations later, when Newton turned his attention to the motion of the planets, he discovered that each of Kepler's laws can be *derived;* they are consequences of Newton's laws of motion and the law of gravitation. Let's see how each of Kepler's laws arises.

First consider the elliptical orbits described in Kepler's first law. Figure 12.18 shows the geometry of the ellipse. The longest dimension is the *major axis,* with half-length a; this half-length is called the **semi-major axis.** The sum of the distances from S to P and from S' to P is the same for all points on the curve. S and S' are the *foci* (plural of focus). The sun is at S, and the planet is at P; we think of them both as points because the size of each is very small in comparison to the distance between them. There is nothing at the other focus S'.

The distance of each focus from the center of the ellipse is ea, where e is a dimensionless number between 0 and 1 called the **eccentricity.** If $e = 0$, the ellipse is a circle. The actual orbits of the planets are fairly circular; their eccentricities range from 0.007 for Venus to 0.248 for Pluto. (The earth's orbit has $e = 0.017$.) The point in the planet's orbit closest to the sun is the *perihelion,* and the point most distant from the sun is the *aphelion.*

Newton was able to show that for a body acted on by an attractive force proportional to $1/r^2$, the only possible closed orbits are a circle or an ellipse; he also showed that open orbits (trajectories (6) and (7) in Fig. 12.13) must be parabolas or hyperbolas. These results can be derived by a straightforward application of Newton's laws and the law of gravitation, together with a lot more differential equations than we're ready for.

Kepler's second law is shown in Fig. 12.19. In a small time interval dt, the line from the sun S to the planet P turns through an angle $d\theta$. The area swept out is the colored triangle with height r, base length $r\,d\theta$, and area $dA = \frac{1}{2}r^2\,d\theta$ (Fig. 12.19b). The rate at which area is swept out, dA/dt, is called the *sector velocity:*

$$\frac{dA}{dt} = \frac{1}{2}r^2\frac{d\theta}{dt} \qquad (12.16)$$

The essence of Kepler's second law is that the sector velocity has the same value at all points in the orbit. When the planet is close to the sun, r is small and $d\theta/dt$ is large; when the planet is far from the sun, r is large and $d\theta/dt$ is small.

To see how Kepler's second law follows from Newton's laws, we express dA/dt in terms of the velocity vector $\vec{v}$ of the planet P. The component of $\vec{v}$ perpendicular to the radial line is $v_\perp = v\sin\phi$. From Fig. 12.19b the displacement along the direction of $v_\perp$ during time dt is $r\,d\theta$, so we also have $v_\perp = r\,d\theta/dt$. Using this relation in Eq. (12.16), we find

$$\frac{dA}{dt} = \frac{1}{2}rv\sin\phi \qquad (\text{sector velocity}) \qquad (12.17)$$

Now $rv\sin\phi$ is the magnitude of the vector product $\vec{r}\times\vec{v}$, which in turn is $1/m$ times the angular momentum $\vec{L} = \vec{r}\times m\vec{v}$ of the planet with respect to the sun. So we have

$$\frac{dA}{dt} = \frac{1}{2m}\left|\vec{r}\times m\vec{v}\right| = \frac{L}{2m} \qquad (12.18)$$

Thus Kepler's second law, that sector velocity is constant, means that angular momentum is constant!

It is easy to see why the angular momentum of the planet *must* be constant. According to Eq. (10.29), the rate of change of $\vec{L}$ equals the torque of the gravitational force $\vec{F}$ acting on the planet:

$$\frac{d\vec{L}}{dt} = \vec{\tau} = \vec{r}\times\vec{F}$$

In our situation, $\vec{r}$ is the vector from the sun to the planet, and the force $\vec{F}$ is directed from the planet to the sun. So these vectors always lie along the same line, and their vector product $\vec{r}\times\vec{F}$ is zero. Hence $d\vec{L}/dt = 0$. This conclusion does not depend on the $1/r^2$ behavior of the force; angular momentum is conserved for *any* force that acts always along the line joining the particle to a fixed point. Such a force is called a *central force.* (Kepler's first and third laws are valid *only* for a $1/r^2$ force.)

Conservation of angular momentum also explains why the orbit lies in a plane. The vector $\vec{L} = \vec{r}\times m\vec{v}$ is always perpendicular to the plane of the vectors $\vec{r}$ and $\vec{v}$; since $\vec{L}$ is constant in magnitude *and* direction, $\vec{r}$ and $\vec{v}$ always lie in the same plane, which is just the plane of the planet's orbit.

We have already derived Kepler's third law for the particular case of circular orbits. Equation (12.14) shows that the period of a satellite or planet in a circular orbit is proportional to the $\frac{3}{2}$ power of the orbit radius. Newton was able to show that this same relationship holds for an elliptical orbit, with the orbit radius r replaced by the semi-major axis a:

$$T = \frac{2\pi a^{3/2}}{\sqrt{Gm_S}} \qquad (\text{elliptical orbit around the sun}) \qquad (12.19)$$

Since the planet orbits the sun, not the earth, we have replaced the earth's mass m_E in Eq. (12.14) with the sun's mass m_S. Note that the period does not depend on the eccentricity e. An asteroid in an elongated elliptical orbit with semi-major axis a will have the same orbital period as a planet in a circular orbit of radius a. The key difference is that the asteroid moves at different speeds at different points in its elliptical orbit (Fig. 12.19c), while the planet's speed is constant around its circular orbit.

Conceptual Example 12.7 — Orbital speeds

At what point in an elliptical orbit (Fig. 12.18) does a planet have the greatest speed?

SOLUTION

Mechanical energy is conserved as the planet moves around its orbit. The planet's kinetic energy $K = \frac{1}{2}mv^2$ is maximum when the potential energy $U = -Gm_S m/r$ is minimum (that is, most nega-

tive; see Fig. 12.10), which occurs when r is a minimum. Hence the speed v is maximum at perihelion.

Your intuition about falling bodies is helpful here. As the planet falls inward toward the sun, it picks up speed, and its speed is maximum when closest to the sun. By the same reasoning, the planet slows down as it moves away from the sun, and its speed is minimum at aphelion.

Example 12.8 — Kepler's third law

The asteroid Pallas has an orbital period of 4.62 years and an orbital eccentricity of 0.233. Find the semi-major axis of its orbit.

SOLUTION

IDENTIFY and SET UP: Asteroids orbit the sun, so we can use Kepler's third law in the form of Eq. (12.19) to determine the semi-major axis a from the given period T. Note that we don't need the value of the eccentricity.

EXECUTE: From Eq. (12.19), $a^{3/2} = (\sqrt{Gm_S}\,T)/2\pi$. To solve for a, raise this expression to the $\frac{2}{3}$ power:

$$a = \left(\frac{Gm_S T^2}{4\pi^2}\right)^{1/3}$$

Since $G = 6.67 \times 10^{-11}$ N·m²/kg² and $m_S = 1.99 \times 10^{30}$ kg (the mass of the sun from Appendix F) are given in SI units, we must

express the period T in seconds rather than years using a conversion factor from Appendix E: $T = (4.62\ \text{yr})(3.156 \times 10^7\ \text{s/yr}) = 1.46 \times 10^8$ s. Using this value, we find $a = 4.15 \times 10^{11}$ m. (Plug in the numbers yourself to check.)

EVALUATE: Our result is intermediate between the semi-major axes of Mars and Jupiter (see Appendix F). Indeed, most known asteroids orbit in an "asteroid belt" between the orbits of these two planets.

As a historical note, Pallas wasn't discovered until 1802, almost two centuries after Kepler's publication of his third law. While Kepler deduced his three laws from the motions of the five planets (other than the earth) known in his time, these laws have proven to apply equally well to all of the planets, asteroids, and comets subsequently discovered to be orbiting the sun.

Example 12.9 — Comet Halley

Comet Halley moves in an elongated elliptical orbit around the sun (Fig. 12.20). At perihelion, the comet is 8.75×10^7 km from the sun; at aphelion it is 5.26×10^9 km from the sun. Find the semi-major axis, eccentricity, and period of the orbit.

SOLUTION

IDENTIFY and SET UP: Given the perihelion and aphelion distances, we can find the semi-major axis and eccentricity from Fig. 12.18. Once we know the semi-major axis, we can find the orbital period from Eq. (12.19).

EXECUTE: From Fig. 12.18 the length of the major axis equals the sum of the comet–sun distance at perihelion plus the comet–sun distance at aphelion. The length of the major axis is $2a$, so

$$a = \frac{8.75 \times 10^7\ \text{km} + 5.26 \times 10^9\ \text{km}}{2} = 2.67 \times 10^9\ \text{km}$$

Further inspection of Fig. 12.18 shows that the comet–sun distance at perihelion is given by

$$a - ea = a(1 - e)$$

(a)

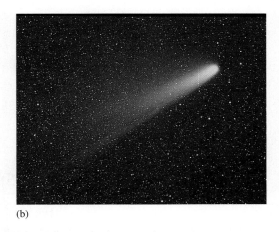

(b)

12.20 (a) The orbit of Comet Halley. (b) Comet Halley as it appeared in 1986. At the heart of the comet is an icy body, called the nucleus, that is about 10 km across. When the comet's orbit carries it close to the sun, the heat of sunlight causes the nucleus to partially evaporate. the evaporated material forms the tail, which can be tens of millions of kilometers long.

Since we are given that this distance is 8.75×10^7 km, the eccentricity is

$$e = 1 - \frac{8.75 \times 10^7 \text{ km}}{a} = 1 - \frac{8.75 \times 10^7 \text{ km}}{2.67 \times 10^9 \text{ km}} = 0.967$$

The period is given by Eq. (12.19):

$$T = \frac{2\pi a^{3/2}}{\sqrt{Gm_S}} = \frac{2\pi (2.67 \times 10^{12} \text{ m})^{3/2}}{\sqrt{(6.67 \times 10^{-11} \text{ N} \cdot \text{m}^2/\text{kg}^2)(1.99 \times 10^{30} \text{ kg})}}$$
$$= 2.38 \times 10^9 \text{ s} = 75.5 \text{ years}$$

EVALUATE: The eccentricity is very close to 1, so the comet has a very elongated orbit (see Fig. 12.20a). Comet Halley was at perihelion in early 1986; it will next reach perihelion one period later, in 2061.

We have assumed that as a planet or comet orbits the sun, the sun remains absolutely stationary. Of course, this can't be correct; because the sun exerts a gravitational force on the planet, the planet exerts a gravitational force on the sun of the same magnitude but opposite direction. In fact, *both* the sun and the planet orbit around their common center of mass (Fig. 12.21). We've made only a small error by ignoring this effect, however; the sun's mass is about 750 times the total mass of all the planets combined, so the center of mass of the solar system is not far from the center of the sun. Remarkably, astronomers have used this effect to detect the presence of planets orbiting other stars. Sensitive telescopes are able to detect the apparent "wobble" of a star as it orbits the common center of mass of the star and an unseen companion planet. (The planets are too faint to observe directly.) By analyzing the details of these "wobbles," astronomers have discovered planets in orbit around dozens of other stars.

Newton's analysis of planetary motions is used on a daily basis by modern-day astronomers. But the most remarkable result of Newton's work is that the motions of bodies in the heavens obey the *same* laws of motion as do bodies on the earth. This *Newtonian synthesis,* as it has come to be called, is one of the great unifying principles of science. It has had profound effects on the way that humanity looks

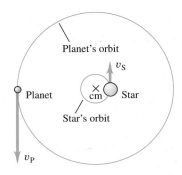

12.21 A star and its planet both orbit about their common center of mass. The star is much more massive than the planet and is much closer to the center of mass (far closer than shown here). The planet and star are on opposite sides of the center of mass at all times.

at the universe—not as a realm of impenetrable mystery, but as a direct extension of our everyday world, subject to scientific study and calculation.

Test Your Understanding

The orbit of Comet X has a semi-major axis that is four times larger than the semi-major axis of Comet Y. What is the ratio of the two comets' orbital periods?

*12.6 | Spherical Mass Distributions

We have used without proof the statement that the gravitational interaction between two spherically symmetric mass distributions is the same as though all the mass of each were concentrated at its center. Now we're ready to prove this statement. Newton searched for a proof for several years, and he delayed publication of the law of gravitation until he found one.

Here's our program. Rather than starting with two spherically symmetric masses, we'll tackle the simpler problem of a point mass m interacting with a thin spherical shell with total mass M. We will show that when m is outside the sphere, the *potential energy* associated with this gravitational interaction is the same as though M were all concentrated at the center of the sphere. According to Eq. (12.10), the force is the negative derivative of the potential energy, so the *force* on m is also the same as for a point mass M. Any spherically symmetric mass distribution can be thought of as being made up of many concentric spherical shells, so our result will also hold for *any* spherically symmetric M.

We start by considering a ring on the surface of the shell (Fig. 12.22a), centered on the line from the center of the shell to m. The reason we do this is that all of the particles that make up the ring are the same distance s from the point mass m. From Eq. (12.9) the potential energy of interaction between the earth (mass m_E) and a point mass m, separated by a distance r, is $U = -Gm_Em/r$. By changing notation in this expression, we see that in the situation shown in Fig. 12.22a, the potential energy of interaction between the point mass m and a particle of mass m_i within the ring is given by

$$U_i = -\frac{Gmm_i}{s}$$

To find the potential energy of interaction between m and the entire ring of mass $dM = \sum_i m_i$, we sum this expression for U_i over all particles in the ring. Calling this potential energy dU, we find

$$dU = \sum_i U_i = \sum_i \left(-\frac{Gmm_i}{s}\right) = -\frac{Gm}{s}\sum_i m_i = -\frac{Gm\,dM}{s} \quad (12.20)$$

To proceed, we need to know the mass dM of the ring. We can find this with the aid of a little geometry. The radius of the shell is R, so in terms of the angle ϕ shown in the figure, the radius of the ring is $R\sin\phi$, and its circumference is $2\pi R\sin\phi$. The width of the ring is $R\,d\phi$, and its area dA is approximately equal to its width times its circumference:

$$dA = 2\pi R^2 \sin\phi\,d\phi$$

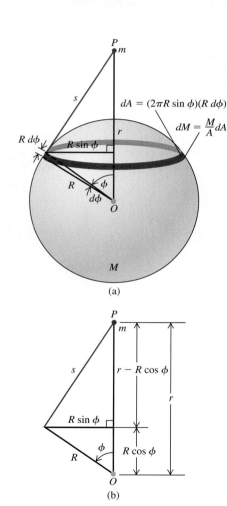

12.22 (a) To calculate the gravitational effects *outside* a spherical shell, all the mass M can be considered concentrated at the center. (b) The distance s is the hypotenuse of a right triangle with sides $(r - R\cos\phi)$ and $R\sin\phi$.

The ratio of the ring mass dM to the total mass M of the shell is equal to the ratio of the area dA of the ring to the total area $A = 4\pi R^2$ of the shell:

$$\frac{dM}{M} = \frac{2\pi R^2 \sin \phi \, d\phi}{4\pi R^2} = \frac{1}{2} \sin \phi \, d\phi \qquad (12.21)$$

Now we solve Eq. (12.21) for dM and substitute the result into Eq. (12.20) to find the potential energy of interaction between the point mass m and the ring:

$$dU = -\frac{GMm \sin \phi \, d\phi}{2s} \qquad (12.22)$$

The total potential energy of interaction between the point mass and the *shell* is the integral of Eq. (12.22) over the whole sphere as ϕ varies from 0 to π (*not* 2π!) and s varies from $r - R$ to $r + R$. To carry out the integration, we have to express the integrand in terms of a single variable; we choose s. To express ϕ and $d\phi$ in terms of s, we have to do a little more geometry. Figure 12.22b shows that s is the hypotenuse of a right triangle with sides $(r - R \cos \phi)$ and $R \sin \phi$, so the Pythagorean theorem gives

$$s^2 = (r - R \cos \phi)^2 + (R \sin \phi)^2$$
$$= r^2 - 2rR \cos \phi + R^2 \qquad (12.23)$$

We take differentials of both sides:

$$2s \, ds = 2rR \sin \phi \, d\phi$$

Next we divide this by $2rR$ and substitute the result into Eq. (12.22):

$$dU = -\frac{GMm}{2s} \frac{s \, ds}{rR} = -\frac{GMm}{2rR} ds \qquad (12.24)$$

We can now integrate Eq. (12.24), recalling that s varies from $r - R$ to $r + R$:

$$U = -\frac{GMm}{2rR} \int_{r-R}^{r+R} ds = -\frac{GMm}{2rR}[(r + R) - (r - R)] \qquad (12.25)$$

Finally, we have

$$U = -\frac{GMm}{r} \qquad (\text{point mass } m \text{ outside spherical shell } M) \qquad (12.26)$$

This is equal to the potential energy of two point masses m and M at a distance r. So we have proved that the gravitational potential energy of the spherical shell M and the point mass m at any distance r is the same as though they were point masses. Because the force is given by $F_r = -dU/dr$, the force is also the same.

Any spherically symmetric mass distribution can be thought of as a combination of concentric spherical shells. Because of the principle of superposition of forces, what is true of one shell is also true of the combination. So we have proved half of what we set out to prove, that the gravitational interaction between any spherically symmetric mass distribution and a point mass is the same as though all the mass of the spherically symmetric distribution were concentrated at its center.

The other half is to prove that *two* spherically symmetric mass distributions interact as though they were both points. That's easier. In Fig. 12.22 the forces the two bodies exert on each other are an action-reaction pair, and they obey Newton's third law. So we have also proved that the force that m exerts *on* the sphere

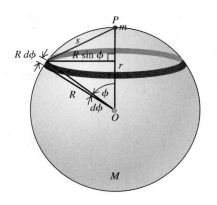

12.23 When a point mass m is *inside* a uniform spherical shell of mass M, the potential energy is the same no matter where inside the shell the point mass is located. The force from the masses' mutual gravitational interaction is zero.

M is the same as though M were a point. But now if we replace m with a spherically symmetric mass distribution centered at m's location, the resulting gravitational force on any part of M is the same as before, and so is the total force. This completes our proof.

A Point Mass Inside a Spherical Shell

We assumed at the beginning that the point mass m was outside the spherical shell, so our proof is valid only when m is outside a spherically symmetric mass distribution. When m is *inside* a spherical shell, the geometry is as shown in Fig. 12.23. The entire analysis goes just as before; Eqs. (12.20) through (12.24) are still valid. But when we get to Eq. (12.25), the limits of integration have to be changed to $R - r$ and $R + r$. We then have

$$U = -\frac{GMm}{2rR} \int_{R-r}^{R+r} ds = -\frac{GMm}{2rR}[(R+r) - (R-r)] \quad (12.27)$$

and the final result is

$$U = -\frac{GMm}{R} \qquad \text{(point mass } m \text{ inside spherical shell } M\text{)} \quad (12.28)$$

Compare this result to Eq. (12.26): instead of having r, the distance between m and the center of M, in the denominator, we have R, the radius of the shell. This means that U in Eq. (12.28) doesn't depend on r and thus has the same value everywhere inside the shell. When m moves around inside the shell, no work is done on it, so the force on m at any point inside the shell must be zero.

More generally, at any point in the interior of any spherically symmetric mass distribution (not necessarily a shell), at a distance r from its center, the gravitational force on a point mass m is the same as though we removed all the mass at points farther than r from the center and concentrated all the remaining mass at the center.

Example 12.10	**"Journey to the center of the earth"**

Suppose we drill a hole through the earth (radius R_E, mass m_E) along a diameter and drop a mail pouch (mass m) down the hole. Derive an expression for the gravitational force on the pouch as a function of its distance r from the center. Assume that the density of the earth is uniform (not a very realistic model; see Fig. 12.8).

SOLUTION

IDENTIFY and SET UP: According to the statements above, the gravitational force at distance r from the center is determined only by the mass M within a sphere of radius r (Fig. 12.24).

EXECUTE: With uniform density the mass is proportional to the volume of the sphere, which is $\frac{4}{3}\pi r^3$ for this sphere and $\frac{4}{3}\pi R_E^3$ for the entire earth. So we have

$$\frac{M}{m_E} = \frac{\frac{4}{3}\pi r^3}{\frac{4}{3}\pi R_E^3} = \frac{r^3}{R_E^3}$$

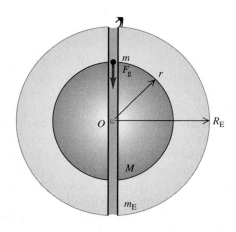

12.24 A hole through the center of the earth (assumed to be uniform). When an object is a distance r from the center, only the mass inside a sphere of radius r exerts a net gravitational force on it.

The magnitude of the gravitational force on m is given by

$$F_g = \frac{GMm}{r^2} = \frac{Gm_E m}{r^2}\frac{r^3}{R_E^3} = \frac{Gm_E m}{R_E^3}r$$

EVALUATE: At points inside the earth, F_g is *directly proportional* to the distance r from the center, rather than proportional to $1/r^2$ as it

is outside the sphere. Right at the surface, where $r = R_E$, the above expression gives $F_g = Gm_E m/R_E^2$, as we should expect. In the next chapter we'll learn how to compute the time it would take for the mail pouch to emerge on the other side of the earth. Meanwhile, keep in mind that the assumption of uniform density of the earth is, like the hole, pure fantasy. We discussed the variation of density in the earth in Section 12.2.

Test Your Understanding

In the classic 1913 science-fiction novel *At the Earth's Core* by Edgar Rice Burroughs, explorers discover that the earth is a hollow sphere and that an entire civilization lives on the inside of the sphere. Would it be possible to stand and walk around on the inner surface of a hollow planet?

*12.7 | Apparent Weight and the Earth's Rotation

Because the earth rotates on its axis, it is not precisely an inertial frame of reference. For this reason the apparent weight of a body on earth is not precisely equal to the earth's gravitational attraction, which we will call the **true weight** $\vec{w}_0$ of the body. Figure 12.25 is a cutaway view of the earth, showing three observers. Each one holds a spring scale with a body of mass m hanging from it. Each scale applies a tension force $\vec{F}$ to the body hanging from it, and the reading on each scale is the magnitude F of this force. If the observers are unaware of the earth's rotation, each one *thinks* that the scale reading equals the weight of the body because he

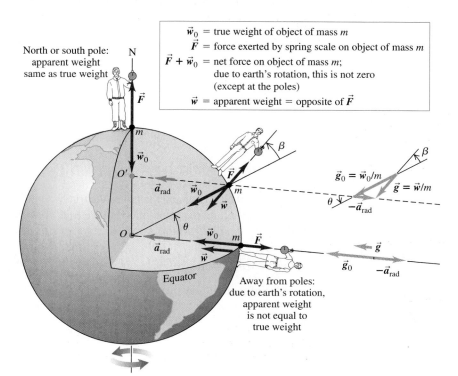

$\vec{w}_0$ = true weight of object of mass m
$\vec{F}$ = force exerted by spring scale on object of mass m
$\vec{F} + \vec{w}_0$ = net force on object of mass m; due to earth's rotation, this is not zero (except at the poles)
$\vec{w}$ = apparent weight = opposite of $\vec{F}$

North or south pole: apparent weight same as true weight

Equator

Away from poles: due to earth's rotation, apparent weight is not equal to true weight

12.25 Except at the poles, the reading for an object being weighed on a scale (the *apparent weight*) is less than the gravitational force of attraction on the object (the *true weight*). The reason is that a net force is needed to provide a centripetal acceleration as the object rotates with the earth. For clarity, the illustration greatly exaggerates the angle β between the true and apparent weight vectors.

thinks the body on his spring scale is in equilibrium. So each observer thinks that the tension $\vec{F}$ must be opposed by an equal and opposite force $\vec{w}$, which we call the **apparent weight.** But if the bodies are rotating with the earth, they are *not* precisely in equilibrium. Our problem is to find the relation between the apparent weight $\vec{w}$ and the true weight $\vec{w}_0$.

If we assume that the earth is spherically symmetric, then the true weight $\vec{w}_0$ has magnitude $Gm_{\mathrm{E}}m/R_{\mathrm{E}}^2$, where m_{E} and R_{E} are the mass and radius of the earth. This value is the same for all points on the earth's surface. If the center of the earth can be taken as the origin of an inertial coordinate system, then the body at the north pole really *is* in equilibrium in an inertial system, and the reading on that observer's spring scale is equal to w_0. But the body at the equator is moving in a circle of radius R_{E} with speed v, and there must be a net inward force equal to the mass times the centripetal acceleration:

$$w_0 - F = \frac{mv^2}{R_{\mathrm{E}}}$$

So the magnitude of the apparent weight (equal to the magnitude of F) is

$$w = w_0 - \frac{mv^2}{R_{\mathrm{E}}} \qquad (\text{at the equator}) \qquad (12.29)$$

If the earth were not rotating, the body when released would have a free-fall acceleration $g_0 = w_0/m$. Since the earth *is* rotating, the falling body's actual acceleration relative to the observer at the equator is $g = w/m$. Dividing Eq. (12.29) by m and using these relations, we find

$$g = g_0 - \frac{v^2}{R_{\mathrm{E}}} \qquad (\text{at the equator})$$

To evaluate v^2/R_{E}, we note that in 86,164 s a point on the equator moves a distance equal to the earth's circumference, $2\pi R_{\mathrm{E}} = 2\pi(6.38 \times 10^6 \text{ m})$. (The solar day, 86,400 s, is $\frac{1}{365}$ longer than this because in one day the earth also completes $\frac{1}{365}$ of its orbit around the sun.) Thus we find

$$v = \frac{2\pi(6.38 \times 10^6 \text{ m})}{86,164 \text{ s}} = 465 \text{ m/s}$$

$$\frac{v^2}{R_{\mathrm{E}}} = \frac{(465 \text{ m/s})^2}{6.38 \times 10^6 \text{ m}} = 0.0339 \text{ m/s}^2$$

So for a spherically symmetric earth the acceleration due to gravity should be about 0.03 m/s² less at the equator than at the poles.

At locations intermediate between the equator and the poles, the true weight $\vec{w}_0$ and the centripetal acceleration are not along the same line, and we need to write a vector equation corresponding to Eq. (12.29). From Fig. 12.25 we see that the appropriate equation is

$$\vec{w} = \vec{w}_0 - m\vec{a}_{\mathrm{rad}} = m\vec{g}_0 - m\vec{a}_{\mathrm{rad}} \qquad (12.30)$$

The difference in the magnitudes of g and g_0 lies between zero and 0.0339 m/s². As shown in Fig. 12.25, the *direction* of the apparent weight differs from the direction toward the center of the earth by a small angle β, which is 0.1° or less.

Table 12.1 gives the values of g at several locations, showing variations with latitude. There are also small additional variations due to the lack of perfect spherical symmetry of the earth, local variations in density, and differences in elevation.

Table 12.1 Variations of g with Latitude and Elevation

Station	North Latitude	Elevation (m)	g (m/s^2)
Canal Zone	9°	0	9.78243
Jamaica	18°	0	9.78591
Bermuda	32°	0	9.79806
Denver	40°	1638	9.79609
Pittsburgh, Pa.	40.5°	235	9.80118
Cambridge, Mass.	42°	0	9.80398
Greenland	70°	0	9.82534

12.26 This orbiting astronaut is acted on by the earth's gravity, but *feels* weightless because his acceleration is equal to $\vec{g}$.

Our discussion of apparent weight can also be applied to the phenomenon of apparent weightlessness in orbiting spacecraft, which we described in Section 12.4. Bodies in an orbiting spacecraft are *not* weightless; the earth's gravitational attraction continues to act on them just as though they were at rest relative to the earth. The apparent weight of a body in a spacecraft is again given by Eq. (12.30):

$$\vec{w} = \vec{w}_0 - m\vec{a}_{rad} = m\vec{g}_0 - m\vec{a}_{rad}$$

But for a spacecraft in orbit, as well as any body inside the spacecraft, the acceleration $\vec{a}_{rad}$ toward the earth's center is equal to the value of the acceleration of gravity $\vec{g}_0$ at the position of the spacecraft. Hence

$$\vec{g}_0 = \vec{a}_{rad}$$

and the apparent weight is

$$\vec{w} = 0$$

This is what we mean when we say that an astronaut or other body in the spacecraft is apparently weightless. Note that we didn't make any assumptions about the shape of the orbit; as we mentioned in Section 12.4, an astronaut will be apparently weightless no matter what the orbit (Fig. 12.26).

Test Your Understanding

What would the speed v of a point on the equator have to be for an object there to have an apparent weight of *zero?*

12.8 | Black Holes

The concept of a black hole is one of the most interesting and startling products of modern gravitational theory, yet the basic idea can be understood on the basis of Newtonian principles. Think first about the properties of our own sun. Its mass $M = 1.99 \times 10^{30}$ kg and radius $R = 6.96 \times 10^8$ m are much larger than those of any planet, but compared to other stars, our sun is not exceptionally massive.

What's the average *density* ρ of the sun? You can find it the same way we found the average density of the earth in Section 12.2:

$$\rho = \frac{M}{V} = \frac{M}{\frac{4}{3}\pi R^3} = \frac{1.99 \times 10^{30} \text{ kg}}{\frac{4}{3}\pi (6.96 \times 10^8 \text{ m})^3}$$

$$= 1410 \text{ kg/m}^3$$

The sun's temperatures range from 5800 K (about 5500°C, or 10,000°F) at the surface up to 1.5×10^7 K (about 2.7×10^7 °F) in the interior, so it surely contains no solids or liquids. Yet gravitational attraction pulls the sun's gas atoms together until the sun is, on average, 41% denser than water and about 1200 times as dense as the air we breathe.

Now think about the escape speed for a body at the surface of the sun. In Example 12.5 (Section 12.3) we found that the escape speed from the surface of a spherical mass M with radius R is $v = \sqrt{2GM/R}$. We can relate this to the average density. Substituting $M = \rho V = \rho\left(\frac{4}{3}\pi R^3\right)$ into the expression for escape speed gives

$$v = \sqrt{\frac{2GM}{R}} = \sqrt{\frac{8\pi G\rho}{3}}R \qquad (12.31)$$

Using either form of this equation, you can show that the escape speed for a body at the surface of our sun is $v = 6.18 \times 10^5$ m/s (about 2.2 million km/h, or 1.4 million mi/h). This value, roughly 1/500 the speed of light, is independent of the mass of the escaping body; it depends only on the mass and radius (or average density and radius) of the sun.

Now consider various stars with the same average density ρ and different radii R. Equation (12.31) shows that for a given value of density ρ, the escape speed v is directly proportional to R. In 1783 the Rev. John Mitchell, an amateur astronomer, noted that if a body with the same average density as the sun had about 500 times the radius of the sun, the magnitude of its escape velocity would be greater than the speed of light c. With his statement that "all light emitted from such a body would be made to return toward it," Mitchell became the first person to suggest the existence of what we now call a **black hole.**

The first expression for escape speed in Eq. (12.31) also suggests that a body of mass M will act as a black hole if its radius R is less than or equal to a certain critical radius. How can we determine this critical radius? You might think that you can find the answer by simply setting $v = c$ in Eq. (12.31). As a matter of fact, this does give the correct result, but only because of two compensating errors. The kinetic energy of light is *not* $mc^2/2$, and the gravitational potential energy near a black hole is *not* given by Eq. (12.9). In 1916, Karl Schwarzschild used Einstein's general theory of relativity (in part a generalization and extension of Newtonian gravitation theory) to derive an expression for the critical radius R_S, now called the **Schwarzschild radius.** The result turns out to be the same as though we had set $v = c$ in Eq. (12.31), so

$$c = \sqrt{\frac{2GM}{R_S}}$$

Solving for the Schwarzschild radius R_S, we find

$$R_S = \frac{2GM}{c^2} \qquad (\text{Schwarzschild radius}) \qquad (12.32)$$

If a spherical, nonrotating body with mass M has a radius less than R_S, then *nothing* (not even light) can escape from the surface of the body, and the body is a black hole (Fig. 12.27). In this case, any other body within a distance R_S of the center of the black hole is trapped by the gravitational attraction of the black hole and cannot escape from it.

The surface of the sphere with radius R_S surrounding a black hole is called the **event horizon:** since light can't escape from within that sphere, we can't see events occurring inside. All that an observer outside the event horizon can know about a black hole is its mass (from its gravitational effects on other bodies), its electric charge (from the electric forces it exerts on other charged bodies), and its

(a)

(b)

12.27 (a) When the radius R of a body is greater than the Schwarzschild radius R_S, light can escape from the surface of the body. As it travels away, it is "red-shifted" to longer wavelengths. (b) When the body is within the event horizon (radius R_S), it is a black hole with an escape speed greater than the speed of light. In that case we can obtain very little information about it.

angular momentum (because a rotating black hole tends to drag space—and everything in that space—around with it). All other information about the body is irretrievably lost when it collapses inside its event horizon.

| Example 12.11 | **Black hole calculations** |

Astrophysical theory suggests that a burned-out star will collapse under its own gravity to form a black hole when its mass is at least three solar masses. If it does, what is the radius of its event horizon?

SOLUTION

IDENTIFY and SET UP: The radius in question is R_S as given by Eq. (12.32). "Three solar masses" means $M = 3(1.99 \times 10^{30} \text{ kg}) = 6.0 \times 10^{30}$ kg.

EXECUTE: From Eq. (12.32),

$$R_S = \frac{2GM}{c^2} = \frac{2(6.67 \times 10^{-11} \text{ N·m}^2/\text{kg}^2)(6.0 \times 10^{30} \text{ kg})}{(3.00 \times 10^8 \text{ m/s})^2}$$

$$= 8.9 \times 10^3 \text{ m} = 8.9 \text{ km}$$

or less than 6 miles.

EVALUATE: If the radius of such an object is just equal to the Schwarzschild radius, the average density has the incredibly large value

$$\rho = \frac{M}{\frac{4}{3}\pi R^3} = \frac{6.0 \times 10^{30} \text{ kg}}{\frac{4}{3}\pi(8.9 \times 10^3 \text{ m})^3}$$

$$= 2.0 \times 10^{18} \text{ kg/m}^3$$

This is about 10^{15} times as great as the density of familiar matter on earth and is comparable to the densities of atomic nuclei. In fact, once the body collapses to a radius of R_S, nothing can prevent it from collapsing further. All of the mass ends up being crushed down to a single point called a *singularity* at the center of the event horizon. This point has zero volume and so has *infinite* density.

At points far from a black hole, its gravitational effects are the same as those of any normal body with the same mass. If the sun collapsed to form a black hole, the orbits of the planets would be unaffected. But things get dramatically different close to the black hole. If you decided to become a martyr for science and jump into a black hole, those you left behind would notice several odd effects as you moved toward the event horizon, most of them associated with effects of general relativity. If you carried a radio transmitter to send back your comments on what was happening, they would have to retune their receiver continuously to lower and lower frequencies, an effect called the *gravitational red shift.* Consistent with this shift, they would observe that your clocks (electronic or biological) would appear to run more and more slowly, an effect called *time dilation.* In fact, during their lifetimes they would never see you make it to the event horizon. In your frame of reference, you would make it to the event horizon in a rather short time but in a rather disquieting way. As you fell feet first into the black hole, the gravitational pull on your feet would be greater than that on your head, which would be slightly farther away from the black hole. The *differences* in gravitational force on different parts of your body would be great enough to stretch you along the direction toward the black hole and compress you perpendicular to it. These effects (called *tidal forces*) would rip you to atoms, and then rip your atoms apart, before you reached the event horizon.

If light cannot escape from a black hole and if black holes are as small as Example 12.11 suggests, how can we know that such things exist? The answer is that any gas or dust near the black hole tends to be pulled into an *accretion disk* that swirls around and into the black hole, rather like a whirlpool (Fig. 12.28). Friction within the accretion disk's material causes it to lose mechanical energy and spiral into the black hole; as it moves inward, it is compressed together. This causes heating of the material, just as air compressed in a bicycle pump gets hotter. Temperatures in excess of 10^6 K can occur in the accretion disk, so hot that the disk emits not just vis-

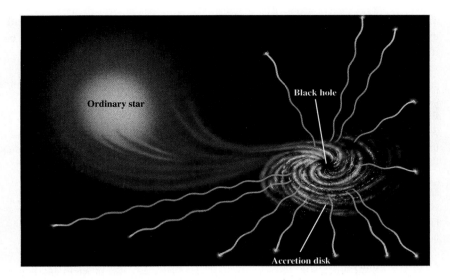

12.28 A binary star system in which an ordinary star and a black hole orbit each other. Matter is pulled from the ordinary star to form an accretion disk around the black hole. The gas in the accretion disk is compressed and heated to such high temperatures that it becomes an intense source of x rays.

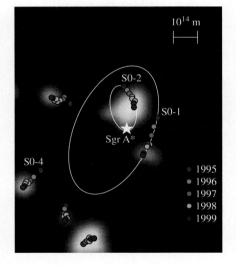

12.29 This false-color image shows the motions of stars at the center of our galaxy over a five-year period. Analyzing these orbits using Kepler's third law indicates that the stars are moving about an unseen object that is some 2.6×10^6 times the mass of the sun. The scale bar indicates a length of 10^{14} m (670 times the distance from the earth to the sun) at the distance of the galactic center.

ible light (as do bodies that are "red-hot" or "white-hot") but x rays. Astronomers look for these x rays (emitted *before* the material crosses the event horizon) to signal the presence of a black hole. Several promising candidates have been found, and astronomers now express considerable confidence in the existence of black holes.

Black holes in binary star systems like the one depicted in Fig. 12.28 have masses a few times greater than the sun's mass. There is also mounting evidence for the existence of much larger *supermassive black holes*. One example is thought to lie at the center of our Milky Way galaxy, some 26,000 light years from earth in the direction of the constellation Sagittarius. High-resolution images of the galactic center reveal stars moving at speeds greater than 1500 km/s about an unseen object that lies at the position of a source of radio waves called Sgr A* (Fig. 12.29). By analyzing these motions, astronomers can infer the period T and semi-major axis a of each star's orbit. The mass m_X of the unseen object can then be calculated using Kepler's third law in the form given in Eq. (12.19), with the mass of the sun m_S replaced by m_X:

$$T = \frac{2\pi a^{3/2}}{\sqrt{Gm_X}} \quad \text{so} \quad m_X = \frac{4\pi^2 a^3}{GT^2}$$

The conclusion is that the mysterious dark object at the galactic center has a mass of 5.2×10^{36} kg, or 2.6 *million* times the mass of the sun. Yet observations with radio telescopes show that it has a radius of no more than about 10^{11} m, comparable to the distance from the earth to the sun. These observations suggest that this massive, compact object is a black hole with a Schwarzschild radius of 7.8×10^9 m. In the next several years astronomers hope to improve the resolution of their observations and actually be able to see the event horizon of this black hole.

Other lines of research suggest that even larger black holes, in excess of 10^9 solar masses, lie at the centers of other galaxies. Observational and theoretical studies of black holes of all sizes continue to be an exciting area of research in both physics and astronomy.

Test Your Understanding

If the sun somehow collapsed to form a black hole, what effect would this event have on the orbit of the earth?

SUMMARY

Newton's law of gravitation states that any two bodies with masses m_1 and m_2, a distance r apart, attract each other with forces inversely proportional to r^2. These forces form an action-reaction pair and obey Newton's third law. When two or more bodies exert gravitational forces on a particular body, the total gravitational force on that individual body is the vector sum of the forces exerted by the other bodies. (See Examples 12.1 through 12.3)

$$F_g = \frac{Gm_1m_2}{r^2} \quad (12.1)$$

The weight w of a body is the total gravitational force exerted on it by all other bodies in the universe. Near the surface of the earth (mass m_E and radius R_E), it is essentially equal to the gravitational force of the earth alone. (See Example 12.4)

$$w = F_g = \frac{Gm_Em}{R_E^2} \quad (12.3)$$
(weight at earth's surface)

$$g = \frac{Gm_E}{R_E^2} \quad (12.4)$$
(acceleration due to gravity at earth's surface)

The gravitational potential energy U of two masses m and m_E separated by a distance r is inversely proportional to r. The potential energy is never positive; it is zero only when the two bodies are infinitely far apart. (See Example 12.5)

$$U = -\frac{Gm_Em}{r} \quad (12.9)$$

When a satellite moves in a circular orbit, the centripetal acceleration is provided by the gravitational attraction of the earth. (See Example 12.6)

$$v = \sqrt{\frac{Gm_E}{r}} \text{ (speed in circular orbit)} \quad (12.12)$$

$$T = \frac{2\pi r}{v} = 2\pi r\sqrt{\frac{r}{Gm_E}} = \frac{2\pi r^{3/2}}{\sqrt{Gm_E}} \quad (12.14)$$
(period in circular orbit)

Kepler's three laws describe characteristics of the elliptical orbits of planets around the sun or satellites around a planet. (See Examples 12.7 through 12.9)

The gravitational interaction of any spherically symmetric mass distribution, at points outside the distribution, is the same as if all the mass were concentrated at the center. (See Example 12.10)

Two spherically symmetric masses m_1 and m_2

Two particles with masses m_1 and m_2

If a nonrotating spherical mass distribution with total mass M has a radius less than its Schwarzschild radius R_S, it is called a black hole. The gravitational interaction prevents anything, including light, from escaping from within a sphere with radius R_S. (See Example 12.11)

$$R_S = \frac{2GM}{c^2} \qquad (12.32)$$
(Schwarzschild radius)

Key Terms

law of gravitation, 437
gravitational constant, 437
gravitational potential energy, 445
escape speed, 446
closed orbit, 448

open orbit, 448
semi-major axis, 452
eccentricity, 452
true weight, 459

apparent weight, 460
black hole, 462
Schwarzschild radius, 462
event horizon, 462

Your Notes

Answer to Chapter Opening Question

The gravitational force that an object of mass m_1 exerts on the earth (mass m_2) is inversely proportional to the distance r to that object: $F = Gm_1m_2/r^2$. While the galaxies are 10^{11} times more massive than the sun, they are 7.2×10^{12} times more distant from the earth and their gravitational force on the earth is only $(10^{11})/(7.2 \times 10^{12})^2 = 1.9 \times 10^{-15}$ times as great as the sun's gravitational force. Hence the galaxies' gravitational effect on the earth is negligible.

Answers to Test Your Understanding Questions

Section 12.1 From Eq. (12.1), the magnitude of the force on either car is

$$F_g = (6.67 \times 10^{-11}\ \text{N·m}^2/\text{kg}^2)(900\ \text{kg})(900\ \text{kg})/(3.0\ \text{m})^2$$
$$= 6.0 \times 10^{-6}\ \text{N}$$

You can exert a force more than 100,000 times greater than this with your little finger. Gravitational forces are negligible in automobile collisions.

Section 12.2 The acceleration due to gravity g_M at the surface of Mars depends on both the planet's mass m_M and its radius R_M: $g_M = Gm_M/R_M^2$. The radius of Mars is 53% as great as that of the earth, and its smaller radius makes the value of g_M larger than what you would expect based on the mass alone.

Section 12.3 This is indeed possible, because the two quantities depend in different ways on the planet's mass m_P and radius R_P: the value of g at the surface is Gm_P/R_P^2, while the escape speed is $\sqrt{2Gm_P/R_P}$. For the planet Saturn, for example, m_P is about 100 times the earth's mass and R_P is about 10 times the earth's radius. The value of g is different than on earth by a factor $(100)/(10)^2 = 1$ (i.e., it is the same as on earth), while the escape speed is greater by a factor $\sqrt{100/10} = 3.2$.

It may help to remember that the surface gravity tells you about conditions right next to the planet's surface, while the escape speed (which tells you how fast you must travel to escape to infinity) depends on conditions at *all* points between the planet's surface and infinity. Because Saturn has so much more mass than the earth, its gravitational effects are appreciable at much greater distances and its escape speed is higher.

Section 12.4 In a smaller-radius orbit, the shuttle has a faster speed and will eventually catch up with the satellite. The shuttle could then use its thrusters to maneuver into the same orbit as the satellite to complete the rendezvous. The shuttle could also enter a larger-radius orbit; then the satellite would have the greater speed and would eventually catch up with the shuttle.

Section 12.5 Equation (12.9) shows that the orbital period is proportional to the $\frac{3}{2}$ power of the semi-major axis. Hence the orbital period of Comet X is longer than that of Comet Y by a factor of $4^{3/2} = 8$.

Section 12.6 Our analysis shows that there is *zero* gravitational force inside a hollow spherical shell. Hence visitors to the interior of a hollow planet would find themselves quite weightless, and they could not stand or walk on the planet's inner surface.

Section 12.7 We want $g = g_0 - v^2/R_E = 0$. Hence v^2 must equal g_0R_E, or

$$v = \sqrt{g_0R_E} = \sqrt{(9.80\ \text{m/s}^2)(6.38 \times 10^6\ \text{m})} = 7910\ \text{m/s}$$

For this to happen, the earth would have to spin $(7910\ \text{m/s})/(465\ \text{m/s}) = 17.0$ times faster than it does now!

Section 12.8 There would be no effect. If the sun collapsed into a black hole (which, according to our understanding of stars, it cannot do), it would have the same mass but a much smaller radius. Because the gravitational attraction of the sun on the earth does not depend on the sun's radius, the earth's orbit would be unaffected.

Discussion Questions

Q12.1 A student wrote: "The only reason an apple falls downward to meet the earth instead of the earth rising upward to meet the apple is that the earth is much more massive and so exerts a much greater pull." Please comment.

Q12.2 To use the Cavendish balance (Fig. 12.4) to determine G, you must first know the masses m_1 and m_2. How could you determine these masses *without* weighing them?

Q12.3 If all planets had the same average density, how would the acceleration due to gravity at the surface of a planet depend on its radius?

Q12.4 Is a pound of butter on the earth the same amount as a pound of butter on Mars? What about a kilogram of butter? Explain.

Q12.5 Example 12.2 (Section 12.1) shows that the acceleration of each sphere caused by the gravitational force is inversely proportional to the mass of that sphere. So why does the force of gravity give all masses the same acceleration when they are dropped near the surface of the earth?

Q12.6 When will you attract the sun more: today at noon, or tonight at midnight? Explain.

Q12.7 Since the moon is constantly attracted toward the earth by the gravitational interaction, why doesn't it crash into the earth?

Q12.8 Imagine that the sun somehow became twice as massive as it is now. What effect would this have on your weight on the earth (as measured by standing on a scale)? Explain your answer.

Q12.9 The sun pulls on the moon with a force that is more than twice the magnitude of the force with which the earth attracts the moon. Why, then, doesn't the sun take the moon away from the earth?

Q12.10 As defined in Chapter 7, gravitational potential energy is $U = mgy$ and is positive for a body of mass m above the earth's surface (which is at $y = 0$). But in this chapter, gravitational potential energy is $U = -Gm_E m/r^2$, which is *negative* for a body of mass m above the earth's surface (which is at $r = R_E$). How can you reconcile these seemingly incompatible descriptions of gravitational potential energy?

Q12.11 A planet is moving at constant speed in a circular orbit around a star. In one complete orbit, what is the net amount of work done on the planet by the star's gravitational force: positive, negative, or zero? What if the planet's orbit is an ellipse, so that the speed is not constant? Explain your answers.

Q12.12 Does the escape speed for an object at the earth's surface depend on the direction in which it is launched? Explain. Does your answer depend on whether or not you include the effects of air resistance?

Q12.13 If a projectile is fired straight up from the earth's surface, what would happen if the total mechanical energy (kinetic plus potential) is a) less than zero? b) greater than zero? In each case, ignore air resistance and the gravitational effects of the sun, the moon, and the other planets.

Q12.14 Discuss whether the following statement is correct: "In the absence of air resistance, the trajectory of a projectile thrown near the earth's surface is an *ellipse,* not a parabola."

Q12.15 The earth is closer to the sun in November than in May. In which month does it move faster in its orbit? Explain.

Q12.16 A communications firm wants to place a satellite in orbit so that it is always directly above the earth's 45th parallel (latitude 45° north). This means that the plane of the orbit will not pass through the center of the earth. Is such an orbit possible? Why or why not?

Q12.17 At what point in an elliptical orbit is the acceleration maximum? At what point is it minimum? Justify your answers.

Q12.18 Which takes more fuel, a voyage from the earth to the moon or from the moon to the earth? Explain.

Q12.19 What would Kepler's third law be for circular orbits if an amendment to Newton's law of gravitation made the gravitational force inversely proportional to r^3? Would this change affect Kepler's other two laws? Explain.

Q12.20 In the elliptical orbit of Comet Halley shown in Fig. 12.20a, the sun's gravity is responsible for making the comet fall inward from aphelion to perihelion. But what is responsible for making the comet move from perihelion back outward to aphelion?

Q12.21 Many people believe that orbiting astronauts feel weightless because they are "beyond the pull of the earth's gravity." How far from the earth would a spacecraft have to travel to be truly beyond the earth's gravitational influence? If a spacecraft were really unaffected by the earth's gravity, would it remain in orbit? Explain. What is the real reason why astronauts in orbit feel weightless?

Q12.22 As part of their training before going into orbit, astronauts ride in an airliner that is flown along the same parabolic trajectory as a freely falling projectile. Explain why this gives the same experience of apparent weightlessness as being in orbit.

Exercises

Section 12.1 Newton's Law of Gravitation

12.1 What is the ratio of the gravitational pull of the sun on the moon to that of the earth on the moon? (Assume the distance of the moon from the sun can be approximated by the distance of the earth from the sun.) Use the data in Appendix F. Is it more accurate to say that the moon orbits the earth, or that the moon orbits the sun?

12.2 A 2150-kg satellite used in a cellular telephone network is in a circular orbit at a height of 780 km above the surface of the earth. What is the gravitational force on the satellite? What fraction is this of its weight at the surface of the earth?

12.3 A set of two uniform spheres with masses m_1 and m_2 have their centers separated by a distance r_{12}. They exert a gravitational force of magnitude F_{12} on each other. What magnitude of gravitational force does a second set of uniform spheres exert on one another if they have masses nm_1 and nm_2 and a center-to-center separation of nr_{12}, where n is any positive number?

12.4 Two uniform spheres, each with mass M and radius R, touch one another. What is the magnitude of their gravitational force of attraction?

12.5 An interplanetary spaceship passes through the point in space where the gravitational forces from the sun and the earth on the ship exactly cancel. a) How far from the center of the earth is it? Use the data in Appendix F. b) What, if anything, happens to the spaceship when it passes through the point described in part (a)? Explain.

12.6 a) In Fig. 12.30 what are the magnitude and direction of the net gravitational force exerted on the 0.100-kg uniform sphere by the other two uniform spheres? The centers of all three spheres are on the same line. b) According to Newton's third law, does the 0.100-kg sphere exert forces of the same magnitude as your answer to part (a), but opposite direction, on *each* of the other two spheres?

Figure 12.30 Exercise 12.6.

12.7 The typical adult human has a mass of about 70 kg. What force does a full moon exert on such a human when it is directly overhead with its center 378,000 km away?

12.8 The sun has a mass 333,000 times that of the earth. For a person on earth, the average distance to the center of the sun is 23,500 times the distance to the center of the earth. In magnitude, what is the ratio of the sun's gravitational force on you to the earth's gravitational force on you?

12.9 Calculate the magnitude and direction of the net gravitational force on the moon due to the earth and the sun when the moon is in each of the positions shown in Fig. 12.31. (Note that the figure is *not* drawn to scale. Assume that the sun is in the plane of the earth-

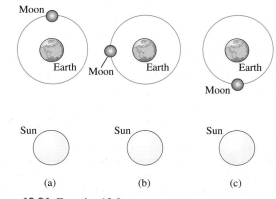

Figure 12.31 Exercise 12.9

moon orbit, even though this is not actually the case.) Use the data in Appendix F.

12.10 Four identical masses of 800 kg each are placed at the corners of a square whose side length is 10.0 cm. What is the net gravitational force (magnitude and direction) on one of the masses, due to the other three?

12.11 Three identical masses of 500 kg each are placed on the *x*-axis. One mass is at $x = -10$ cm, one is at the origin, and one is at $x = 40$ cm. What is the net gravitational force (magnitude and direction) on the mass at the origin, due to the other two masses?

12.12 Determine the magnitude and direction of the gravitational force on a particle of mass *m* that is midway between two spherically symmetric objects, one of mass m_1 and the other of mass m_2, where $m_2 > m_1$. The centers of the two spherically symmetric objects are a distance *d* apart.

12.13 Two uniform spheres, each of mass 0.260 kg, are fixed at points *A* and *B* (Fig. 12.32). Find the magnitude and direction of the initial acceleration of a uniform sphere with mass 0.010 kg if released from rest at point *P* and acted on only by forces of gravitational attraction of the spheres at *A* and *B*.

Figure 12.32 Exercise 12.13.

Section 12.2 Weight

12.14 Use the mass and radius of the planet Pluto given in Appendix F to calculate the acceleration due to gravity at the surface of Pluto.

12.15 At what distance above the surface of the earth is the acceleration due to the earth's gravity 0.980 m/s^2 if the acceleration due to gravity at the surface has magnitude 9.80 m/s^2?

12.16 The mass of Venus is 81.5% that of the earth, and its radius is 94.9% that of the earth. a) Compute the acceleration due to gravity on the surface of Venus from these data. b) What is the weight of a 5.00-kg rock on the surface of Venus?

12.17 Titania, the largest moon of the planet Uranus, has $\frac{1}{8}$ the radius of the earth and $\frac{1}{1700}$ the mass of the earth. a) What is the acceleration due to gravity at the surface of Titania? b) What is the average density of Titania? (This is less than the density of rock, which is one piece of evidence that Titania is made primarily of ice.)

12.18 Rhea, one of Saturn's moons, has a radius of 765 km and an acceleration due to gravity of 0.278 m/s^2 at its surface. Calculate its mass and average density.

12.19 Calculate the earth's gravity force on a 75-kg astronaut who is repairing the Hubble Space Telescope 600 km above the earth's surface, and then compare this value with his weight at the earth's surface. In view of your result, explain why we say astronauts are weightless when they orbit the earth in a satellite such as a space shuttle.

12.20 Neutron stars, such as the one at the center of the Crab Nebula, have about the same mass as our sun but a *much* smaller diameter. If you weigh 675 N on the earth, what would you weigh at the surface of a neutron star that has the same mass as our sun and a diameter of 20 km?

12.21 An experiment using the Cavendish balance to measure the gravitational constant *G* found that a uniform 0.400-kg sphere attracts another uniform 0.00300-kg sphere with a force of 8.00×10^{-10} N, when the distance between the centers of the spheres is 0.0100 m. The acceleration due to gravity at the earth's surface is 9.80 m/s^2, and the radius of the earth is 6380 km. Compute the mass of the earth from these data.

12.22 In Example 12.4 (Section 12.2) we ignored the gravitational effects of the moons of Mars. The larger of the two moons, Phobos, has a radius of approximately 12 km and an average density of 2000 kg/m^3. a) Calculate the gravitational force of Phobos on the Mars lander of Example 12.4 if the Mars lander were on the surface of Phobos. b) Discuss whether the approximation made in Example 12.4 was appropriate.

Section 12.3 Gravitational Potential Energy

12.23 The asteroid Dactyl, discovered in 1993, has a radius of only about 700 m and a mass of about 3.6×10^{12} kg. Use the results of Example 12.5 (Section 12.3) to calculate the escape speed for an object at the surface of Dactyl. Could a person reach this speed just by walking?

12.24 When in orbit, a communication satellite attracts the earth with a force of 19.0 kN and the earth-satellite gravitational potential energy (relative to zero at infinite separation) is -1.39×10^{11} J. a) Find the satellite's altitude above the earth's surface. b) Find the mass of the satellite.

12.25 Use the results of Example 12.5 (Section 12.3) to calculate the escape speed for a spacecraft a) from the surface of Mars; b) from the surface of Jupiter. Use the data in Appendix F. c) Why is the escape speed for a spacecraft independent of the spacecraft's mass?

12.26 Ten days after it was launched toward Mars in December 1998, the *Mars Climate Orbiter* spacecraft (mass 629 kg) was 2.87×10^6 km from the earth and traveling at 1.20×10^4 km/h relative to the earth. At this time, what were a) the spacecraft's kinetic energy relative to the earth and b) the potential energy of the earth-spacecraft system?

Section 12.4 The Motion of Satellites

12.27 An earth satellite moves in a circular orbit with an orbital speed of 6200 m/s. a) Find the time of one revolution. b) Find the radial acceleration of the satellite in its orbit.

12.28 What is the period of revolution of a satellite with mass *m* that orbits the earth in a circular path of radius 7880 km (about 1500 km above the surface of the earth)?

12.29 For a satellite to be in a circular orbit 780 km above the surface of the earth, what must its orbital speed be?

12.30 Assume that the earth's orbit around the sun is circular. Use the earth's orbital radius and orbital period given in Appendix F to calculate the mass of the sun.

12.31 Deimos, a moon of Mars, is about 12 km in diameter with mass 2.0×10^{15} kg. Suppose you are stranded alone on Deimos and want to play a one-person game of baseball. You would be the pitcher, and you would be the batter! With what speed would you have to throw a baseball so that it would go into orbit and return to you so you could hit it? Do you think you could actually throw it at this speed?

Section 12.5 Kepler's Laws and the Motion of Planets

12.32 Planet Vulcan. Suppose that a planet were discovered between the sun and Mercury, with a circular orbit of radius equal to $\frac{2}{3}$ of the average orbit radius of Mercury. (Such a planet was once postulated, in part to explain the precession of Mercury's orbit. It was even given the name Vulcan, although we now have no evidence that it actually exists. Mercury's precession has been explained by general relativity.) What would be the orbital period of such a planet?

12.33 The star Rho[1] Cancri is 57 light years from the earth and has a mass 0.85 times that of our sun. A planet has been detected in a circular orbit around Rho[1] Cancri with an orbital radius equal to 0.11 times the radius of the earth's orbit around the sun. What are a) the orbital speed and b) the orbital period of the planet of Rho[1] Cancri?

12.34 Venus orbits the sun in a nearly circular orbit. Use the data in Appendix F for the orbital radius and period of Venus to calculate the mass of the sun.

12.35 a) Use Fig. 12.18 to show that the sun-planet distance at perihelion is $(1 - e)a$, the sun-planet distance at aphelion is $(1 + e)a$, and therefore that the sum of these two distances is $2a$. b) Pluto is usually called the outermost planet, but at perihelion in 1989, it was almost 100 million km closer to the sun than Neptune. The semi-major axes of the orbits of Pluto and Neptune are 5.92×10^{12} m and 4.50×10^{12} m, respectively, and the eccentricities are 0.248 and 0.010. Find Pluto's closest distance and Neptune's farthest distance from the sun. c) How many years after being at perihelion in 1989 will Pluto again be at perihelion?

12.36 The star 70 Virginis is 59 light years from the earth and has a mass of 1.9×10^{30} kg. a) A large planet of mass 1.3×10^{28} kg is known to orbit this star. The planet is attracted to the star by a force of 3.3×10^{26} N, when their centers are separated by a distance equal to the semi-major axis of the planet's orbit. Calculate this distance (in km). b) What is the period (in days) of this planet's orbit? c) Other as yet undiscovered planets may orbit 70 Virginis. If a second planet is in a circular orbit around 70 Virginis with an orbital period equal to eight times the value found in part (b), what is the radius (in km) of the second planet's orbit?

12.37 The *Helios B* spacecraft had a speed of 71 km/s when it was 4.3×10^7 km from the sun. a) Prove that it was not in a circular orbit about the sun. b) Prove that its orbit about the sun was closed and therefore elliptical.

*Section 12.6 Spherical Mass Distributions

*12.38 A thin, uniform rod has length L and mass M. A small uniform sphere of mass m is placed a distance x from one end of the rod, along the axis of the rod (Fig. 12.33). a) Calculate the gravitational potential energy of the rod-sphere system. Take the potential energy to be zero when the rod and sphere are infinitely far apart. Show that your answer reduces to the expected result when x is

Figure 12.33 Exercise 12.38, Problem 12.83.

much larger than L. (*Hint:* Use the power series expansion for $\ln(1 + x)$ given in Appendix B.) b) Use $F_x = -dU/dx$ to find the magnitude and direction of the gravitational force exerted on the sphere by the rod. Show that your answer reduces to the expected result when x is much larger than L.

*12.39 Consider the ring-shaped body of Fig. 12.34. A particle with mass m is placed a distance x from the center of the ring, along the line through the center of the ring and perpendicular to its plane. a) Calculate the gravitational potential energy U of this system. Take the potential energy to be zero when the two objects are far apart. b) Show that your answer to part (a) reduces to the expected result when x is much larger than the radius a of the ring. c) Use $F_x = -dU/dx$ to find the magnitude and direction of the force on the particle. d) Show that your answer to part (c) reduces to the expected result when x is much larger than a. e) What are the values of U and F_x when $x = 0$? Explain why these results make sense.

Figure 12.34 Exercise 12.39, Problem 12.82.

Section 12.7 Apparent Weight and the Earth's Rotation

*12.40 The weight of a person at the equator, as determined by a spring balance, is 690 N. By how much does this differ from the true force of gravitational attraction at the same point? Assume that the earth is spherically symmetric.

*12.41 The acceleration due to gravity at the north pole of Neptune is approximately 10.7 m/s^2. Neptune has mass 1.0×10^{26} kg and radius 2.5×10^4 km and rotates once around its axis in about 16 h. a) What is the gravitational force on a 5.0-kg object at the north pole of Neptune? b) What is the apparent weight of this same object at Neptune's equator? (Note that Neptune's "surface" is gaseous, not solid, so it is impossible to stand on it.)

Section 12.8 Black Holes

12.42 a) Show that a black hole attracts an object with mass m with a force of $mc^2R_S/(2r^2)$, where r is the distance between the object and the center of the black hole. b) Calculate the magnitude of the gravitational force exerted by a black hole of Schwarzschild radius 14.0 mm on a 5.00-kg mass 3000 km from it. c) What is the mass of this black hole?

12.43 At the Galaxy's Core. Astronomers have observed a small, massive object at the center of our Milky Way galaxy (Section 12.8). A ring of material orbits this massive object; the ring has a diameter of about 15 light years and an orbital speed of about 200 km/s. a) Determine the mass of the massive object at the center of the Milky Way galaxy. Give your answer both in kilograms and in solar masses (one solar mass is the mass of the sun). b) Observations of stars, as well as theories of the structure of stars, suggest that it is impossible for a single star to have a mass of more than about 50 solar masses. Can this massive object be a single, ordinary star? c) Many astronomers believe that the massive object at the center of the Milky Way galaxy is a black hole. If so, what must the

Schwarzschild radius of this black hole be? Would a black hole of this size fit inside the earth's orbit around the sun?

12.44 Derive the relation $R_S = (3.0 \text{ km/solar mass})M$, where M is in solar masses. (One solar mass is the mass of the sun.) How accurate is this relation?

12.45 To what fraction of its current radius would the earth have to be compressed to become a black hole?

Problems

12.46 Three uniform spheres are fixed at positions shown in Fig. 12.35. a) What is the magnitude and direction of the force on a 0.0150-kg particle placed at P? b) If the spheres are in deep outer space and a 0.0150-kg particle is released from rest 300 m from the origin along a line 45° below the $-x$-axis, what will the particle's speed be when it reaches the origin?

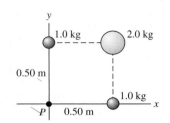

Figure 12.35 Problem 12.46.

12.47 A uniform sphere with mass 60.0 kg is held with its center at the origin and a second uniform sphere with mass 80.0 kg is held with its center at the point $x = 0$, $y = 3.00$ m. a) What are the magnitude and direction of the net gravitational force due to these objects on a third uniform sphere with mass 0.500 kg placed at the point $x = 4.00$ m, $y = 0$? b) Where, other than infinitely far away, could the third sphere be placed such that the net gravitational force acting on it from the other two spheres is equal to zero?

12.48 a) Show that the gravitational force on the small star due to the two large stars in Example 12.3 (Section 12.1) is *not* directed toward the point midway between the two large masses. b) Consider the two large stars as comprising a single, rigid body, as if they were joined by a rod of negligible mass. Calculate the torque exerted by the small star on the rigid body for a pivot at its center of mass. c) Explain how the result in part (b) shows that the center of mass does not coincide with the center of gravity. Why is this the case in this situation?

12.49 At a certain instant, the earth, the moon, and a stationary 1250-kg spacecraft lie at the vertices of an equilateral triangle whose sides are 3.84×10^5 km in length. a) Find the magnitude and direction of the net gravitational force exerted on the spacecraft by the earth and moon. State the direction as an angle measured from a line connecting the earth and the spacecraft. In a sketch, show the earth, the moon, the spacecraft, and the force vector. b) What is the minimum amount of work that you would have to do to move the spacecraft to a point far from the earth and moon? You can ignore any gravitational effects due to the other planets or the sun.

12.50 An experiment is performed in deep space with two uniform spheres, one with mass 25.0 kg and the other with mass 100.0 kg. They have equal radii, $r = 0.20$ m. The spheres are released from rest with their centers 40.0 m apart. They accelerate toward each other because of their mutual gravitational attraction. You can ignore

all gravitational forces other than that between the two spheres. a) Explain why linear momentum is conserved. b) When their centers are 20.0 m apart, find i) the speed of each sphere and ii) the magnitude of the relative velocity with which one sphere is approaching the other. c) How far from the initial position of the center of the 25.0-kg sphere do the surfaces of the two spheres collide?

12.51 Assume that the moon orbits the earth in a circular orbit. From the observed orbital period of 27.3 days, calculate the distance of the moon from the center of the earth. Assume that the moon's motion is determined solely by the gravitational force exerted on it by the earth, and use the mass of the earth given in Appendix F.

12.52 Gravitational Field. The *gravitational field* $\vec{g}$ produced by an object is defined as the gravitational force $\vec{F}_g$ that the object exerts on a small test particle with mass m divided by m. That is, $\vec{g} = \vec{F}_g/m$. What are the magnitude and direction of the gravitational field produced by a 20.0-kg spherically symmetric object at a point 1.50 m directly above the center of the object?

12.53 Geosynchronous Satellites. Many satellites are moving in a circle in the earth's equatorial plane. They are at such a height above the earth's surface that they always remain above the same point. a) Find the altitude of these satellites above the earth's surface. (Such an orbit is said to be *geosynchronous*.) b) Explain, with a sketch, why the radio signals from these satellites cannot directly reach receivers on earth that are north of 81.3° N latitude.

12.54 A landing craft with mass 12,500 kg is in a circular orbit 5.75×10^5 m above the surface of a planet. The period of the orbit is 5800 s. The astronauts in the lander measure the diameter of the planet to be 9.60×10^6 m. The lander sets down at the north pole of the planet. What is the weight of a 85.6-kg astronaut as he steps out onto the planet's surface?

12.55 What is the escape speed from a 300-km-diameter asteroid with a density of 2500 kg/m³?

12.56 a) Asteroids have average densities of about 2500 kg/m³ and radii from 470 km down to less than a kilometer. Assuming that the asteroid has a spherically symmetric mass distribution, estimate the radius of the largest asteroid from which you could escape simply by jumping off. (*Hint*: You can estimate your jump speed by relating it to the maximum height that you can jump on earth.) b) Europa, one of Jupiter's four large moons, has a radius of 1570 km. The acceleration due to gravity at its surface is 1.33 m/s². Calculate its average density.

12.57 a) Suppose you are at the earth's equator and observe a satellite passing directly overhead and moving from west to east in the sky. Exactly 12.0 hours later, you again observe this satellite to be directly overhead. How far above the earth's surface is the satellite's orbit? b) You observe another satellite directly overhead and traveling east to west. This satellite is again overhead in 12.0 hours. How far is this satellite's orbit above the surface of the earth?

12.58 Planet X rotates in the same manner as the earth, around an axis through its north and south poles, and is perfectly spherical. An astronaut who weighs 943.0 N on the earth weighs 915.0 N at the north pole of Planet X and only 850.0 N at its equator. The distance from the north pole to the equator is 18,850 km, measured along the surface of Planet X. a) How long is the day on Planet X? b) If a

45,000 kg satellite is placed in a circular orbit 2000 km above the surface of Planet X, what will be its orbital period?

12.59 There are two equations from which a change in the gravitational potential energy U of the system of a mass m and the earth can be calculated. One is $U = mgy$ (Eq. 7.2). The other is $U = -Gm_{\mathrm{E}}m/r$ (Eq. 12.9). As shown in Section 12.3, the first equation is correct only if the gravitational force is a constant over the change in height Δy. The second is always correct. Actually, the gravitational force is never exactly constant over any change in height, but if the variation is small, we can ignore it. Consider the difference in U between a mass at the earth's surface and a distance h above it using both equations, and find the value of h for which Eq. (7.2) is in error by 1%. Express this value of h as a fraction of the earth's radius, and also obtain a numerical value for it.

12.60 Your starship, the *Aimless Wanderer,* lands on the mysterious planet Mongo. As chief scientist-engineer, you make the following measurements: a 2.50-kg stone thrown upward from the ground at 12.0 m/s returns to the ground in 8.00 s; the circumference of Mongo at the equator is 2.00×10^5 km; and there is no appreciable atmosphere on Mongo. The starship commander, Captain Confusion, asks for the following information: a) What is the mass of Mongo? b) If the *Aimless Wanderer* goes into a circular orbit 30,000 km above the surface of Mongo, how many hours will it take the ship to complete one orbit?

12.61 Calculate the percent difference between your weight in Sacramento, near sea level, and at the top of Mount Everest, which is 8800 m above sea level.

12.62 In Example 12.5 (Section 12.3) we ignored the gravitational effects of the moon on a spacecraft en route from the earth to the moon. In fact, we must include the gravitational potential energy due to the moon as well. For this problem, you can ignore the motion of the earth and moon. a) If the moon has radius R_{M} and the distance between the centers of the earth and the moon is R_{EM}, find the total gravitational potential energy of the particle-earth and particle-moon systems when a particle with mass m is between the earth and the moon, and a distance r from the center of the earth. Take the gravitational potential energy to be zero when the objects are far from each other. b) There is a point along a line between the earth and the moon where the net gravitational force is zero. Use the expression derived in part (a) and numerical values from Appendix F to find the distance of this point from the center of the earth. With what speed must a spacecraft be launched from the surface of the earth just barely to reach this point? c) If a spacecraft were launched from the earth's surface toward the moon with an initial speed of 11.2 km/s, with what speed would it impact the moon?

12.63 An unmanned spacecraft is in a circular orbit around the moon, observing the lunar surface from an altitude of 50.0 km (see Appendix F). To the dismay of scientists on earth, an electrical fault causes an on-board thruster to fire, decreasing the speed of the spacecraft by 20.0 m/s. If nothing is done to correct its orbit, with what speed (in km/h) will the spacecraft crash into the lunar surface?

***12.64** What would be the length of a day (that is, the time required for one rotation of the earth on its axis) if the rate of rotation of the earth were such that $g = 0$ at the equator?

12.65 Falling Hammer. A hammer with mass m is dropped from rest from a height h above the earth's surface. This height is not necessarily small compared with the radius R_{E} of the earth. If you ignore air resistance, derive an expression for the speed v of the hammer when it reaches the surface of the earth. Your expression should involve h, R_{E}, and m_{E}, the mass of the earth.

12.66 a) Calculate how much work is required to launch a spacecraft of mass m from the surface of the earth (mass m_{E}, radius R_{E}) and place it in a circular *low earth orbit*, that is, an orbit whose altitude above the earth's surface is much less than R_{E}. (As an example, the International Space Station is in low earth orbit at an altitude of about 400 km, much less than $R_{\mathrm{E}} = 6380$ km.) You can ignore the kinetic energy that the spacecraft has on the ground due to the earth's rotation. b) Calculate the minimum amount of additional work required to move the spacecraft from low earth orbit to a very great distance from the earth. You can ignore the gravitational effects of the sun, the moon, and the other planets. c) Justify the statement that "in terms of energy, low earth orbit is halfway to the edge of the universe."

12.67 A spacecraft is to be launched from the surface of the earth so that it will escape from the solar system altogether. a) Find the speed relative to the center of the earth with which the spacecraft must be launched. Take into consideration the gravitational effects of both the earth and the sun, and include the effects of the earth's orbital speed, but ignore air resistance. b) The rotation of the earth can help this spacecraft achieve escape speed. Find the speed that the spacecraft must have relative to the earth's *surface* if it is launched from Florida at the point shown in Fig. 12.36. The rotation and orbital motions of the earth are in the same direction. The launch facilities in Florida are 28.5° north of the equator. c) The European Space Agency (ESA) uses launch facilities in French Guiana (immediately north of Brazil), 5.15° north of the equator. What speed relative to the earth's surface would a spacecraft need to escape the solar system if launched from French Guiana?

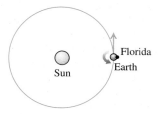

Figure 12.36 Problem 12.67

12.68 Gravitational Potential. We can define a useful quantity called the *gravitational potential* of the earth (denoted by ϕ) by requiring that the product of its value at any point, and the mass of a particle placed at that point, gives us the gravitational potential energy of that earth-particle system. (We will use an analogous quantity, the electric potential V, in the discussion of electrical interactions in Chapter 23.) a) Find the SI units for the gravitational potential and show they are equivalent to m^2/s^2. b) What is the equation for the gravitational potential at any point outside the earth, if it is taken to be zero infinitely far from the earth? c) What is the increase in gravitational potential between the earth's surface and the International Space Station's altitude of 400 km? d) Use the result of part (c) to calculate the amount of work that must be done against the earth's gravitational force to move a 15,000-kg payload from the surface of the earth to an altitude of 400 km.

12.69 Kirkwood Gaps. Hundreds of thousands of asteroids orbit the sun within the *asteroid belt*, which extends from about 3×10^8 km to about 5×10^8 km from the sun. a) Find the orbital period (in years) of i) an asteroid at the inside of the belt and ii) an asteroid at the outside of the belt. Assume circular orbits. b) In 1867 the American astronomer Daniel Kirkwood pointed out that several gaps exist in the asteroid belt where relatively few asteroids are found. It is now understood that these *Kirkwood gaps* are caused by the gravitational attraction of Jupiter, the largest planet, which orbits the sun once every 11.86 years. As an example, if an asteroid has an orbital period half that of Jupiter, or 5.93 years, on every other orbit this asteroid would be at its closest to Jupiter and feel a strong attraction toward the planet. This attraction, acting over and over on successive orbits, could sweep asteroids out of the Kirkwood gap. Use this hypothesis to determine the orbital radius for this Kirkwood gap. c) One of several other Kirkwood gaps appears at a distance from the sun where the orbital period is 0.400 that of Jupiter. Explain why this happens, and find the orbital radius for this Kirkwood gap.

12.70 If a satellite is in a sufficiently low orbit, it will encounter air drag from the earth's atmosphere. Since air drag does negative work (the force of air drag is directed opposite the motion), the mechanical energy will decrease. According to Eq. (12.15), if E decreases (becomes more negative), the radius r of the orbit will decrease. If air drag is relatively small, the satellite can be considered to be in a circular orbit of continually decreasing radius. a) According to Eq. (12.12), if the radius of a satellite's circular orbit decreases, the satellite's orbital speed v *increases*. How can you reconcile this with the statement that the mechanical energy *decreases*? (*Hint:* Is air drag the only force that does work on the satellite as the orbital radius decreases?) b) Due to air drag, the radius of a satellite's circular orbit decreases from r to $r - \Delta r$, where the positive quantity Δr is much less than r. The mass of the satellite is m. Show that the increase in orbital speed is $\Delta v = +(\Delta r/2)\sqrt{Gm_E/r^3}$; that the change in kinetic energy is $\Delta K = +(Gm_E m/2r^2)\Delta r$; that the change in gravitational potential energy is $\Delta U = -2\Delta K = -(Gm_E m/r^2)\Delta r$; and that the amount of work done by the force of air drag is $W = -(Gm_E m/2r^2)\Delta r$. Interpret these results in light of your comments in part (a). c) A satellite with mass 3000 kg is initially in a circular orbit 300 km above the earth's surface. Due to air drag, the satellite's altitude decreases to 250 km. Calculate the initial orbital speed; the increase in orbital speed; the initial mechanical energy; the change in kinetic energy; the change in gravitational potential energy; the change in mechanical energy; and the work done by the force of air drag. d) Eventually a satellite will descend to a low enough altitude in the atmosphere that the satellite burns up and the debris falls to the earth. What becomes of the initial mechanical energy?

12.71 Binary Star—Equal Masses. Two identical stars with mass M orbit around their center of mass. Each orbit is circular and has radius R, so that the two stars are always on opposite sides of the circle. a) Find the gravitational force of one star on the other. b) Find the orbital speed of each star and the period of the orbit. c) How much energy would be required to separate the two stars to infinity?

12.72 Binary Star—Different Masses. Two stars, with mass M_1 and M_2, are in circular orbits around their center of mass. The star with mass M_1 has an orbit of radius R_1; the star with mass M_2 has an orbit of radius R_2. a) Show that the ratio of the orbital radii of the two stars equals the reciprocal of the ratio of their masses, that is, $R_1/R_2 = M_2/M_1$. b) Explain why the two stars have the same orbital period, and show that the period T is given by $T = 2\pi(R_1 + R_2)^{3/2}/\sqrt{G(M_1 + M_2)}$. c) The two stars in a certain binary star system move in circular orbits. The first star, Alpha, has an orbital speed of 36.0 km/s. The second star, Beta, has an orbital speed of 12.0 km/s. The orbital period is 137 d. What are the masses of each of the two stars? d) One of the best candidates for a black hole is found in the binary system called A0620-0090. The two objects in the binary system are an orange star, V616 Monocerotis, and a compact object believed to be a black hole (Fig. 12.21). The orbital period of A0620-0090 is 7.75 hours, the mass of V616 Monocerotis is estimated to be 0.67 times the mass of the sun, and the mass of the black hole is estimated to be 3.8 times the mass of the sun. Assuming that the orbits are circular, find the radius of each object's orbit and the orbital speed of each object. Compare these answers to the orbital radius and orbital speed of the earth in its orbit around the sun.

12.73 Comets travel around the sun in elliptical orbits with large eccentricities. If a comet has speed 2.0×10^4 m/s when at a distance of 2.5×10^{11} m from the center of the sun, what is its speed when at a distance of 5.0×10^{10} m?

12.74 As Mars orbits the sun in its elliptical orbit, its distance of closest approach to the center of the sun (at perihelion) is 2.067×10^{11} m, and its maximum distance from the center of the sun (at aphelion) is 2.492×10^{11} m. If the orbital speed of Mars at aphelion is 2.198×10^4 m/s, what is its orbital speed at perihelion? (You can ignore the influence of the other planets.)

12.75 Consider a spacecraft in an elliptical orbit around the earth. At the low point, or perigee of its orbit, it is 400 km above the earth's surface; at the high point, or apogee, it is 4000 km above the earth's surface. a) What is the period of the spacecraft's orbit? b) Using conservation of angular momentum, find the ratio of the spacecraft's speed at perigee to its speed at apogee. c) Using conservation of energy, find the speed at perigee and the speed at apogee. d) It is necessary to have the spacecraft escape from the earth completely. If the spacecraft's rockets are fired at perigee, by how much would the speed have to be increased to achieve this? What if the rockets were fired at apogee? Which point in the orbit is more efficient to use?

12.76 The planet Uranus has a radius of 25,560 km and a surface acceleration due to gravity of 11.1 m/s^2 at its poles. Its moon Miranda (discovered by Kuiper in 1948) is in a circular orbit about Uranus at an altitude of 104,000 km above the planet's surface. Miranda has a mass of 6.6×10^{19} kg and a radius of 235 km. a) Calculate the mass of Uranus from the above data. b) Calculate the magnitude of Miranda's acceleration due to its orbital motion about Uranus. c) Calculate the acceleration due to Miranda's gravity at the surface of Miranda. d) Do the answers to parts (b) and (c) mean that an object released 1 m above Miranda's surface on the side toward Uranus will fall *up* relative to Miranda? Explain.

12.77 A 3000-kg spacecraft is in a circular orbit 2000 km above the surface of Mars. How much work must the spacecraft engines perform to move the spacecraft to a circular orbit that is 4000 km above the surface?

12.78 In Example 12.9 (Section 12.5) the orbital period of Comet Halley was calculated to be 75.5 years. Many other comets have orbits that extend much farther from the sun, and hence have much longer periods. a) At aphelion, a typical long-period comet is about 8×10^{12} km from the sun; at perihelion, it passes inside the earth's orbit. Estimate the orbital period of such a comet. Give your answer in years. b) Estimate the speed of a typical long-period comet at perihelion. (*Hint:* At aphelion, the comet is moving very slowly.) c) The massive nucleus of a comet contains about 10^{15} kg of material. If the earth were struck by a long-period comet, estimate the kinetic energy the comet would have just before impact. (*Hint:* The motion of the comet is due primarily to the gravitational attraction of the sun, not the earth. Why?) Compare this to the energy released in a very large volcanic eruption, about 6×10^{18} J, and to the energy that would be released by burning all the fossil fuel on earth, about 2×10^{23} J. An impact of this kind apparently occurred 65 million years ago in the Yucatan, and is implicated in the demise of the dinosaurs as well as many other ancient species.

12.79 One of the brightest comets of the twentieth century was Comet Hyakutake, which passed close to the sun in early 1996. The orbital period of this comet is estimated to be about 30,000 years. Find the semi-major axis of this comet's orbit. Compare to the average sun-Pluto distance and to the distance to Alpha Centauri, the nearest star to the sun, which is 4.3 light years distant.

12.80 Planets are not uniform inside. Normally, they are densest at the center and have decreasing density outward toward the surface. Model a spherically symmetric planet, with the same radius as the earth, as having a density that decreases linearly with distance from the center. Let the density be 15.0×10^3 kg/m³ at the center and 2.0×10^3 kg/m³ at the surface. What is the acceleration due to gravity at the surface of this planet?

12.81 A uniform wire with mass M and length L is bent into a semicircle. Find the magnitude and direction of the gravitational force this wire exerts on a point with mass m placed at the center of curvature of the semicircle.

*12.82 An object in the shape of a thin ring has radius a and mass M. A uniform sphere with mass m and radius R is placed with its center at a distance x to the right of the center of the ring, along a line through the center of the ring, and perpendicular to its plane (Fig. 12.34). What is the gravitational force that the sphere exerts on the ring-shaped object? Show that your result reduces to the expected result when x is much larger than a.

*12.83 A thin, uniform rod has length L and mass M. Calculate the magnitude of the gravitational force the rod exerts on a particle with mass m that is at a point along the axis of the rod a distance x from one end (Fig. 12.33). Show that your result reduces to the expected result when x is much larger than L.

*12.84 A shaft is drilled from the surface to the center of the earth (Fig. 12.24). As in Example 12.10 (Section 12.6), make the rather unrealistic assumption that the density of the earth is uniform. With this approximation, the gravitational force on an object with mass m, that is inside the earth at a distance r from the center, has mag-

nitude $F_g = Gm_{\mathrm{E}}mr/R_{\mathrm{E}}^3$ (as shown in Example 12.10) and points toward the center of the earth. a) Derive an expression for the gravitational potential energy $U(r)$ of the object-and-earth system as a function of the object's distance from the center of the earth. Take the potential energy to be zero when the object is at the center of the earth. b) If an object is released in the shaft at the earth's surface, what speed will it have when it reaches the center of the earth?

Challenge Problems

12.85 a) When an object is in a circular orbit of radius r around the earth (mass m_{E}), the period of the orbit is T [given by Eq. (12.14)] and the orbital speed is v, given by Eq. (12.12). Show that when the object is moved into a circular orbit of slightly larger radius $r + \Delta r$, where $\Delta r \ll r$, its new period is $T + \Delta T$ and its new orbital speed is $v - \Delta v$, where Δr, ΔT, and Δv are all positive quantities and

$$\Delta T = \frac{3\pi \Delta r}{v} \quad \text{and} \quad \Delta v = \frac{\pi \Delta r}{T}$$

(*Hint:* Use the expression $(1 + x)^n \approx 1 + nx$, valid for $|x| \ll 1$.) b) The International Space Station (ISS) is in a nearly circular orbit at an altitude of 398.00 km above the surface of the earth. A maintenance crew is about to arrive on the space shuttle that is also in a circular orbit in the same orbital plane as the ISS, but with an altitude of 398.10 km. The crew has come to remove a faulty 125-m electrical cable, one end of which is attached to the ISS and the other end of which is floating free in space. The plan is for the shuttle to snag the free end just at the moment that the shuttle, the ISS, and the center of the earth all lie along the same line. The cable will then break free from the ISS when it becomes taut. How long after the free end is caught by the space shuttle will it detach from the ISS? Give your answer in minutes. c) If the shuttle misses catching the cable, show that the crew must wait a time $t \approx T^2/\Delta T$ before they have a second chance. Find the numerical value of t and explain whether it would be worth the wait.

12.86 Interplanetary Navigation. The most efficient way to send a spacecraft from the earth to another planet is by using a *Hohmann transfer orbit* (Fig. 12.37). If the orbits of the departure and desti-

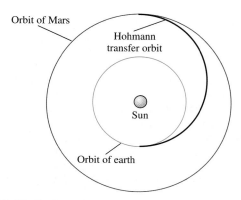

Figure 12.37 Challenge Problem 12.86.

nation planets are circular, the Hohmann transfer orbit is an elliptical orbit whose perihelion and aphelion are tangent to the orbits of the two planets. The rockets are fired briefly at the departure planet to put the spacecraft into the transfer orbit; the spacecraft then coasts until it reaches the destination planet. The rockets are then fired again to put the spacecraft into the same orbit about the sun as the destination planet. a) For a flight from earth to Mars? in what direction must the rockets be fired at the earth and at Mars: in the direction of motion, or opposite the direction of motion? What about from a flight from Mars to the earth? b) How long does a one-way trip from the the earth to Mars take, between the firings of the rockets? c) To reach Mars from the earth, the launch must be timed so that Mars will be at the right spot when the spacecraft reaches Mars's orbit around the sun. At launch, what must the angle between a sun-Mars line and a sun-earth line be? Use data from Appendix F.

12.87 Tidal Forces Near a Black Hole. An astronaut inside a spacecraft, which protects her from harmful radiation, is orbiting a black hole at a distance of 120 km from its center. The black hole is 5.00 times the mass of the sun and has a Schwarzschild radius of 15.0 km. The astronaut is positioned inside the spaceship such that one of her 0.030-kg ears is 6.0 cm farther from the black hole than the center of mass of the spacecraft and the other ear is 6.0 cm closer. a) What is the tension between her ears? Would the astronaut find it difficult to keep from being torn apart by the gravitational forces? (Since her whole body orbits with the same angular velocity, one ear is moving too slowly for the radius of its orbit and the other is moving too fast. Hence her head must exert forces on her ears to keep them in their orbits.) b) Is the center of gravity of her head at the same point as the center of mass? Explain.

*12.88 Mass M is distributed uniformly over a disk of radius a. Find the gravitational force (magnitude and direction) between this disk-shaped mass and a particle with mass m located a distance x above the center of the disk (Fig. 12.38). Does your result reduce to the correct expression as x becomes very large? (*Hint:* Divide the disk into infinitesimally thin

Figure 12.38 Challenge Problem 12.88.

concentric rings, use the expression derived in Problem 12.70 for the gravitational force due to each ring, and integrate to find the total force.)

*12.89 Mass M is distributed uniformly along a line of length $2L$. A particle with mass m is at a point that is a distance a above the center of the line on its perpendicular bisector (point P in Fig. 12.39). For the gravitational force that the line exerts on the particle, calculate the components perpendicular and parallel to the line. Does your result reduce to the correct expression as a becomes very large?

Figure 12.39 Challenge Problem 12.89

13 PERIODIC MOTION

Pendulum clocks have been keeping track of time since the mid-1600s. The underlying principle is that the time for a complete back-and-forth swing depends hardly at all on the size, or amplitude, of the swing. Hence a pendulum clock keeps proper track of time even as the driving mechanism winds down and the pendulum's oscillations die away due to friction.

? **Suppose you doubled the mass of a clock's pendulum (including the rod and the weight at the end) while keeping its dimensions the same. Would the clock run fast or slow?**

Many kinds of motion repeat themselves over and over: the vibration of a quartz crystal in a watch, the swinging pendulum of a grandfather clock, the sound vibrations produced by a clarinet or an organ pipe, and the back-and-forth motion of the pistons in a car engine. We call this **periodic motion** or **oscillation,** and it is the subject of this chapter. Understanding periodic motion will be essential for our later study of waves, sound, alternating electric currents, and light.

A body that undergoes periodic motion always has a stable equilibrium position. When it is moved away from this position and released, a force or torque comes into play to pull it back toward equilibrium. But by the time it gets there, it has picked up some kinetic energy, so it overshoots, stopping somewhere on the other side, and is again pulled back toward equilibrium. Picture a ball rolling back and forth in a round bowl or a pendulum that swings back and forth past its straight-down position.

In this chapter we will concentrate on two simple examples of systems that can undergo periodic motions: spring-mass systems and pendulums. We will also study why oscillations often tend to die out with time and why some oscillations can build up to greater and greater displacements from equilibrium when periodically varying forces act.

13.1 | Describing Oscillation

One of the simplest systems that can have periodic motion is shown in Fig. 13.1a. A body with mass m moves on a frictionless horizontal guide system, such as a linear air track, so it moves only along the x-axis. The body is attached to a spring of negligible mass that can be either stretched or compressed. The left end of the spring is held fixed and the right end is attached to the body. The spring force is the only horizontal force acting on the body; the vertical normal and gravitational forces always add to zero. The quantities x, v_x, a_x, and F_x refer to the x-components of the position, velocity, acceleration, and force vectors, respectively. Like all components, they can be positive, negative, or zero.

It's simplest to define our coordinate system so that the origin O is at the equilibrium position, where the spring is neither stretched nor compressed. Then x is the x-component of the **displacement** of the body from equilibrium and is also the change in length of the spring. The x-component of acceleration a_x is given by $a_x = F_x/m$.

Figure 13.1b shows free-body diagrams for three different positions of the body. Whenever the body is displaced from its equilibrium position, the spring force tends to restore it to the equilibrium position. We call a force with this character a **restoring force.** Oscillation can occur only when there is a restoring force tending to return the system to equilibrium.

Let's analyze how oscillation occurs in this system. If we displace the body to the right to $x = A$ and then let go, the net force and the acceleration are to the left. The speed increases as the body approaches the equilibrium position O. When the body is at O, the net force acting on it is zero, but because of its motion it *overshoots* the equilibrium position. On the other side of the equilibrium position the velocity is to the left but the acceleration is to the right; the speed decreases until the body comes to a stop. We will show later that with an ideal spring, the stopping point is at $x = -A$. The body then accelerates to the right, overshoots equilibrium again, and stops at the starting point $x = A$, ready to repeat the whole process. The body is oscillating! If there is no friction or other force to remove mechanical energy from the system, this motion repeats itself forever; the restoring force perpetually draws the body back toward the equilibrium position, only to have the body overshoot time after time.

In different situations the force may depend on the displacement x from equilibrium in different ways. But oscillation *always* occurs if the force is a *restoring* force that tends to return the system to equilibrium.

Here are some terms that we'll use in discussing periodic motions of all kinds:

The **amplitude** of the motion, denoted by A, is the maximum magnitude of displacement from equilibrium—that is, the maximum value of $|x|$. It is always positive. If the spring in Fig. 13.1 is an ideal one, the total overall range of the motion is $2A$. The SI unit of A is the meter. A complete vibration, or **cycle,** is one complete round trip, say from A to $-A$ and back to A, or from O to A, back through O to $-A$, and back to O. Note that motion from one side to the other (say, $-A$ to A) is a half-cycle, not a whole cycle.

The **period,** T, is the time for one cycle. It is always positive. The SI unit is the second, but it is sometimes expressed as "seconds per cycle."

The **frequency,** f, is the number of cycles in a unit of time. It is always positive. The SI unit of frequency is the hertz:

$$1 \text{ hertz} = 1 \text{ Hz} = 1 \text{ cycle/s} = 1 \text{ s}^{-1}$$

Displacement to the left, restoring force to the right

Zero displacement, zero restoring force

Displacement to the right, restoring force to the left

(a)

(b)

13.1 Model for periodic motion. (a) At the equilibrium position the spring exerts zero force. When the body is displaced from equilibrium, the spring exerts a restoring force back toward the equilibrium position. (b) Free-body diagrams for the three positions.

This unit is named in honor of the German physicist Heinrich Hertz (1857–1894), a pioneer in investigating electromagnetic waves.

The **angular frequency,** ω, is 2π times the frequency:

$$\omega = 2\pi f$$

We'll learn shortly why ω is a useful quantity. It represents the rate of change of an angular quantity (not necessarily related to a rotational motion) that is always measured in radians, so its units are rad/s. Since f is in cycle/s, we may regard the number 2π as having units rad/cycle.

From the definitions of period T and frequency f we see that each is the reciprocal of the other:

$$f = \frac{1}{T} \qquad T = \frac{1}{f} \qquad \text{(relationships between frequency and period)} \qquad (13.1)$$

Also, from the definition of ω,

$$\omega = 2\pi f = \frac{2\pi}{T} \qquad \text{(angular frequency)} \qquad (13.2)$$

Example 13.1	**Period, frequency, and angular frequency**

An ultrasonic transducer (a kind of loudspeaker) used for medical diagnosis oscillates at a frequency of 6.7 MHz = 6.7×10^6 Hz. How much time does each oscillation take, and what is the angular frequency?

SOLUTION

IDENTIFY and SET UP: Our target variables are the period T and the angular frequency ω. We are given the frequency f, so we can find these variables using Eqs. (13.1) and (13.2), respectively.

EXECUTE: From Eqs. (13.1) and (13.2),

$$T = \frac{1}{f} = \frac{1}{6.7 \times 10^6 \text{ Hz}} = 1.5 \times 10^{-7} \text{ s} = 0.15 \ \mu\text{s}$$

$$\omega = 2\pi f = 2\pi (6.7 \times 10^6 \text{ Hz})$$
$$= (2\pi \text{ rad/cycle})(6.7 \times 10^6 \text{ cycle/s})$$
$$= 4.2 \times 10^7 \text{ rad/s}$$

EVALUATE: This is a very rapid vibration, with large f and ω and small T. A slow vibration has small f and ω and large T.

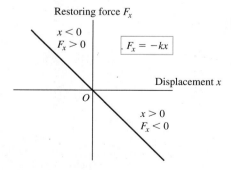

13.2 The restoring force of an idealized spring is directly proportional to the displacement. This is Hooke's law, $F_x = -kx$. Oscillation with a Hooke's law restoring force is called simple harmonic motion.

Test Your Understanding

A boat at anchor bobs up and down with the waves. The boat moves 6.0 cm above and 6.0 cm below its equilibrium position, and makes one complete up-and-down cycle every 5.00 s. What are the amplitude, period, frequency, and angular frequency of the motion?

13.2 | Simple Harmonic Motion

The very simplest kind of oscillation occurs when the restoring force F_x is *directly proportional* to the displacement from equilibrium x. This happens if the spring in Fig. 13.1 is an ideal one that obeys Hooke's law. The constant of proportionality between F_x and x is the force constant k. (You may want to review Hooke's law and the definition of the force constant in Section 6.3.) On either side of the equilibrium position, F_x and x always have opposite signs. In Section 6.3 we represented the force acting *on* a stretched ideal spring as $F_x = kx$. The x-component of force the spring exerts *on the body* is the negative of this, so the x-component of force F_x on the body is

$$F_x = -kx \quad \text{(restoring force exerted by an ideal spring)} \quad (13.3)$$

This equation gives the correct magnitude and sign of the force, whether x is positive, negative, or zero (Fig. 13.2). The force constant k is always positive and has units of N/m (a useful alternative set of units is kg/s^2). We are assuming that there is no friction, so Eq. (13.3) gives the *net* force on the body.

When the restoring force is directly proportional to the displacement from equilibrium, as given by Eq. (13.3), the oscillation is called **simple harmonic motion,** *abbreviated* **SHM.** The acceleration $a_x = d^2x/dt^2 = F_x/m$ of a body in SHM is given by

$$a_x = \frac{d^2x}{dt^2} = -\frac{k}{m}x \quad \text{(simple harmonic motion)} \quad (13.4)$$

The minus sign means the acceleration and displacement always have opposite signs. This acceleration is *not* constant, so don't even think of using the constant-acceleration equations from Chapter 2. We'll see shortly how to solve this equation to find the displacement x as a function of time. A body that undergoes simple harmonic motion is called a **harmonic oscillator.**

Why is simple harmonic motion important? Keep in mind that not all periodic motions are simple harmonic; in periodic motion in general, the restoring force depends on displacement in a more complicated way than in Eq. (13.3). But in many systems the restoring force is *approximately* proportional to displacement if the displacement is sufficiently small (Fig. 13.3). That is, if the amplitude is small enough, the oscillations of such systems are approximately simple harmonic and therefore approximately described by Eq. (13.4). Thus we can use SHM as an approximate model for many different periodic motions, such as the vibration of the quartz crystal in a watch, the motion of a tuning fork, the electric current in an alternating-current circuit, and the vibrations of atoms in molecules and solids.

Equations of Simple Harmonic Motion

To explore the properties of simple harmonic motion, we must express the displacement x of the oscillating body as a function of time, $x(t)$. The second derivative of this function, d^2x/dt^2, must be equal to $(-k/m)$ times the function itself, as required by Eq. (13.4). As we mentioned above, the formulas for constant acceleration from Section 2.4 are no help because the acceleration changes constantly as the displacement x changes. Instead, we'll find $x(t)$ by noticing a striking similarity between SHM and another form of motion that we've already studied in detail.

Figure 13.4 shows a top view of a horizontal disk of radius A with a ball attached to its rim at point Q. The disk rotates with constant angular speed ω (measured in rad/s), so the ball moves in uniform circular motion. A horizontal light beam shines on the rotating disk and casts a shadow of the ball on a screen. The shadow at point P oscillates back and forth as the ball moves in a circle. We then arrange a body attached to an ideal spring, like the combination shown in Fig. 13.1, so that the body oscillates parallel to the shadow. We will prove that the motion of the body

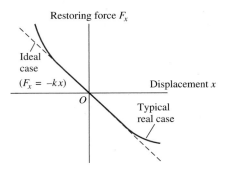

13.3 In most real oscillations the restoring force is not directly proportional to the displacement. However, $F_x = -kx$ is often a good approximation to the force if the displacement x is sufficiently small.

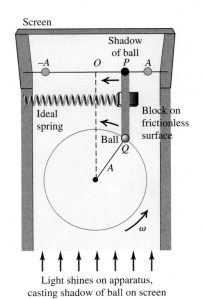

13.4 The ball at point Q rotates counterclockwise in uniform circular motion. Its shadow at point P moves in simple harmonic motion, exactly like a body oscillating on an ideal spring.

and the motion of the ball's shadow are *identical* if the amplitude of the body's oscillation is equal to the disk radius A, and if the angular frequency $2\pi f$ of the oscillating body is equal to the angular speed ω of the rotating disk. That is, *simple harmonic motion is the projection of uniform circular motion onto a diameter.*

We can verify this remarkable statement by finding the acceleration of the shadow at P and comparing it to the acceleration of a body undergoing SHM, given by Eq. (13.4). The circle in which the ball moves so that its projection matches the motion of the oscillating body is called the **circle of reference;** we will call the point Q the *reference point*. We take the circle of reference to lie in the xy-plane, with the origin O at the center of the circle (Fig. 13.5a). At time t the vector OQ from the origin to the reference point Q makes an angle θ with the positive x-axis. As the point Q moves around the reference circle with constant angular speed ω, the vector OQ rotates with the same angular speed. Such a rotating vector is called a **phasor.** (This term was in use long before the invention of the similarly named Star Trek stun gun. The phasor method for analyzing oscillations is useful in many areas of physics. We'll use phasors when we study alternating-current circuits in Chapter 31 and the interference of light in Chapters 35 and 36.)

The x-component of the phasor at time t is just the x-coordinate of the point Q:

$$x = A \cos \theta \tag{13.5}$$

This is also the x-coordinate of the shadow P, which is the *projection* of Q onto the x-axis. Hence the acceleration of the shadow P along the x-axis is equal to the x-component of the acceleration vector of the reference point Q (Fig. 13.5c). Since point Q is in uniform circular motion, its acceleration vector $\vec{a}_Q$ is always directed toward O. Furthermore, the magnitude of $\vec{a}_Q$ is constant and given by the angular speed squared times the radius of the circle (see Section 9.3):

$$a_Q = \omega^2 A \tag{13.6}$$

Figure 13.5c shows that the x-component of $\vec{a}_Q$ is $a_x = -a_Q \cos \theta$. Combining this with Eqs. (13.5) and (13.6), we get that the acceleration of point P is

$$a_x = -a_Q \cos \theta = -\omega^2 A \cos \theta \tag{13.7}$$

or

$$a_x = -\omega^2 x \tag{13.8}$$

The acceleration of the point P is directly proportional to the displacement x and always has the opposite sign. These are precisely the hallmarks of simple harmonic motion.

Equation (13.8) is *exactly* the same as Eq. (13.4) for the acceleration of a harmonic oscillator, provided that the angular speed ω of the reference point Q is related to the force constant k and mass m of the oscillating body by

$$\omega^2 = \frac{k}{m} \quad \text{or} \quad \omega = \sqrt{\frac{k}{m}} \tag{13.9}$$

We have been using the same symbol ω for the angular *speed* of the reference point Q and the angular *frequency* of the oscillating point P. The reason is that these quantities are equal! If point Q makes one complete revolution in time T, then point P goes through one complete cycle of oscillation in the same time; hence T is the period of the oscillation. During time T the point Q moves through 2π radians, so its angular speed is $\omega = 2\pi/T$. But this is just the same as Eq. (13.2) for the angular frequency of the point P, which verifies our statement about

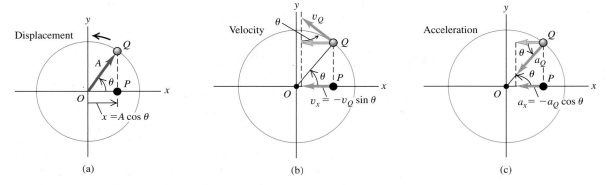

13.5 (a) The x-coordinate of the ball's shadow P (Fig. 13.4) changes as the ball Q moves in uniform circular motion. (b), (c) The x-velocity and x-acceleration of P are the x-components of the velocity and acceleration vectors, respectively, of Q.

the two interpretations of ω. This is why we introduced angular frequency in Section 13.1; this quantity makes the connection between oscillation and circular motion. So we re-interpret Eq. (13.9) as an expression for the angular frequency of simple harmonic motion for a body of mass m, acted on by a restoring force with force constant k:

$$\omega = \sqrt{\frac{k}{m}} \qquad \text{(simple harmonic motion)} \qquad (13.10)$$

When you start a body oscillating in SHM, the value of ω is not yours to choose; it is predetermined by the values of k and m. The units of k are N/m or kg/s², so k/m is in $(\text{kg/s}^2)/\text{kg} = \text{s}^{-2}$. When we take the square root in Eq. (13.10), we get s^{-1}, or more properly rad/s because this is an *angular* frequency (recall that a radian is not a true unit).

According to Eqs. (13.1) and (13.2), the frequency f and period T are

$$f = \frac{\omega}{2\pi} = \frac{1}{2\pi}\sqrt{\frac{k}{m}} \qquad \text{(simple harmonic motion)} \qquad (13.11)$$

$$T = \frac{1}{f} = \frac{2\pi}{\omega} = 2\pi\sqrt{\frac{m}{k}} \qquad \text{(simple harmonic motion)} \qquad (13.12)$$

We see from Eq. (13.12) that a larger mass m, with its greater inertia, will have less acceleration, move more slowly, and take a longer time for a complete cycle (Fig. 13.6). In contrast, a stiffer spring (one with a larger force constant k) exerts a greater force at a given deformation x, causing greater acceleration, higher speeds, and a shorter time T per cycle.

> **CAUTION** You can run into trouble if you don't make the distinction between frequency f and angular frequency $\omega = 2\pi f$. Frequency tells you how many cycles of oscillation occur per second, while angular frequency tells you how many radians per second this corresponds to on the reference circle. In solving problems, pay careful attention to whether the goal is to find f or ω.

Tines with large mass m: low frequency $f = 128$ Hz

Tines with small mass m: high frequency $f = 4096$ Hz

13.6 The greater the mass m in a tuning fork's tines, the lower the frequency of oscillation $f = (2\pi)\sqrt{k/m}$ and the lower the pitch of the sound that the tuning fork produces.

Equations (13.11) and (13.12) show that the period and frequency of simple harmonic motion are completely determined by the mass m and the force constant k. *In simple harmonic motion the period and frequency do not depend on the amplitude A.* For given values of m and k the time of one complete oscillation is the same whether the amplitude is large or small. Equation (13.3) shows why we should expect this. Larger A means that the body reaches larger values of $|x|$ and is subjected to larger restoring forces. This increases the average speed of the body over a complete cycle; this exactly compensates for having to travel a larger distance, so the same total time is involved.

The oscillations of a tuning fork are essentially simple harmonic motion, which means that it always vibrates with the same frequency, independent of amplitude. This is why a tuning fork can be used as a standard for musical pitch. If it were not for this characteristic of simple harmonic motion, it would be impossible to make familiar types of mechanical and electronic clocks run accurately or to play most musical instruments in tune. If you encounter an oscillating body with a period that *does* depend on the amplitude, the oscillation is *not* simple harmonic motion.

Example 13.2	Angular frequency, frequency, and period in SHM

A spring is mounted horizontally, with its left end held stationary. By attaching a spring balance to the free end and pulling toward the right (Fig. 13.7), we determine that the stretching force is proportional to the displacement and that a force of 6.0 N causes a displacement of 0.030 m. We remove the spring balance and attach a 0.50-kg body to the end, pull it a distance of 0.020 m, release it, and watch it oscillate. a) Find the force constant of the spring. b) Find the angular frequency, frequency, and period of the oscillation.

SOLUTION

IDENTIFY: Because the spring force (equal in magnitude to the stretching force) is proportional to the displacement, the motion is simple harmonic.

SET UP: We find the value of the force constant k using Hooke's law, Eq. (13.3), and the values of ω, f, and T using Eqs. (13.10), (13.11), and (13.12), respectively.

EXECUTE: a) When $x = 0.030$ m, the force the spring exerts on the spring balance is $F_x = -6.0$ N. From Eq. (13.3),

$$k = -\frac{F_x}{x} = -\frac{-6.0 \text{ N}}{0.030 \text{ m}} = 200 \text{ N/m} = 200 \text{ kg/s}^2$$

b) Using $m = 0.50$ kg in Eq. (13.10), we find

$$\omega = \sqrt{\frac{k}{m}} = \sqrt{\frac{200 \text{ kg/s}^2}{0.50 \text{ kg}}} = 20 \text{ rad/s}$$

The frequency f is

$$f = \frac{\omega}{2\pi} = \frac{20 \text{ rad/s}}{2\pi \text{ rad/cycle}} = 3.2 \text{ cycle/s} = 3.2 \text{ Hz}$$

The period T is the reciprocal of the frequency f:

$$T = \frac{1}{f} = \frac{1}{3.2 \text{ cycle/s}} = 0.31 \text{ s}$$

A period is usually stated in "seconds" rather than "seconds per cycle."

EVALUATE: The amplitude of the oscillation is 0.020 m, the distance to the right that we pulled the body attached to the spring before releasing it. We didn't need to use this information to find the angular frequency, frequency, or period because in SHM, none of these quantities depend on the amplitude.

13.7 The force exerted *on* the spring is (shown by the vector F) has x-component $F_x = +6.0$ N. The force exerted *by* the spring has x-component $F_x = -6.0$ N.

Displacement, Velocity, and Acceleration in SHM

We still need to find the displacement x as a function of time for a harmonic oscillator. Equation (13.4) for a body in simple harmonic motion along the x-axis is identical to Eq. (13.8) for the x-coordinate of the reference point in uniform circular motion with constant angular speed $\omega = \sqrt{k/m}$. It follows that Eq. (13.5), $x = A \cos \theta$, describes the coordinate x for both of these situations. If at $t = 0$ the phasor OQ makes an angle ϕ ("phi") with the positive x-axis, then at any later time t this angle is $\theta = \omega t + \phi$. We substitute this into Eq. (13.5) to obtain

$$x = A \cos (\omega t + \phi) \qquad \text{(displacement in SHM)} \qquad (13.13)$$

13.8 Graph of x versus t [Eq. (13.13)] for simple harmonic motion. The case shown has $\phi = 0$.

where $\omega = \sqrt{k/m}$. Figure. 13.8 shows a graph of Eq. (13.13) for the particular case $\phi = 0$. The displacement x is a periodic function of time, as expected for SHM. We could also have written Eq. (13.13) in terms of a sine function rather than a cosine by using the identity $\cos \alpha = \sin (\alpha + \pi/2)$. *In simple harmonic motion the position is a periodic, sinusoidal function of time.* There are many other periodic functions, but none so smooth and simple as a sine or cosine function.

The value of the cosine function is always between -1 and 1, so in Eq. (13.13), x is always between $-A$ and A. This confirms that A is the amplitude of the motion. Figure 13.9a shows the graph of x versus t for various values of A.

The period T is the time for one complete cycle of oscillation. The cosine function repeats itself whenever the quantity in parentheses in Eq. (13.13) increases by 2π radians. Thus, if we start at time $t = 0$, the time T to complete one cycle is given by

$$\omega T = \sqrt{\frac{k}{m}} \, T = 2\pi \quad \text{or} \quad T = 2\pi \sqrt{\frac{m}{k}}$$

which is just Eq. (13.12). Changing either m or k changes the period of oscillation, as shown in Figs. 13.9b and 13.9c.

The constant ϕ in Eq. (13.13) is called the **phase angle.** It tells us at what point in the cycle the motion was at $t = 0$ (equivalent to where around the circle the point Q was at $t = 0$). We denote the position at $t = 0$ by x_0. Putting $t = 0$ and $x = x_0$ in Eq. (13.13), we get

$$x_0 = A \cos \phi \qquad (13.14)$$

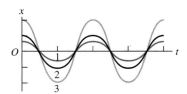

(a) Increasing A; same k and m

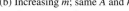

(b) Increasing m; same A and k

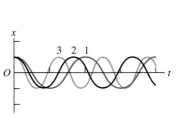

(c) Increasing k; same A and m

13.9 Variations of simple harmonic motion. All cases shown have $\phi = 0$. (a) Amplitude A increases from curve 1 to 2 to 3. Changing A alone has no effect on the period. (b) Mass m increases from 1 to 2 to 3. Increasing m alone increases the period. (c) Force constant k increases from 1 to 2 to 3. Increasing k alone decreases the period.

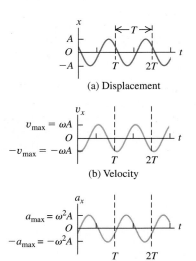

13.10 These three curves show SHM with the same period and amplitude but with different phase angles ϕ.

If $\phi = 0$, then $x_0 = A \cos 0 = A$, and the body starts at its maximum positive displacement. If $\phi = \pi$, then $x_0 = A \cos \pi = -A$, and the particle starts at its maximum *negative* displacement. If $\phi = \pi/2$, then $x_0 = A \cos (\pi/2) = 0$, and the particle is initially at the origin. Figure 13.10 shows the displacement x versus time for different phase angles.

We find the velocity v_x and acceleration a_x as functions of time for a harmonic oscillator by taking derivatives of Eq. (13.13) with respect to time:

$$v_x = \frac{dx}{dt} = -\omega A \sin (\omega t + \phi) \qquad \text{(velocity in SHM)} \qquad (13.15)$$

$$a_x = \frac{dv_x}{dt} = \frac{d^2 x}{dt^2} = -\omega^2 A \cos (\omega t + \phi) \qquad \text{(acceleration in SHM)} \qquad (13.16)$$

The velocity v_x oscillates between $v_{\text{max}} = +\omega A$ and $-v_{\text{max}} = -\omega A$, and the acceleration a_x oscillates between $a_{\text{max}} = +\omega^2 A$ and $-a_{\text{max}} = -\omega^2 A$ (Fig. 13.11). Comparing Eq. (13.16) with Eq. (13.13) and recalling that $\omega^2 = k/m$ from Eq. (13.9), we see that

$$a_x = -\omega^2 x = -\frac{k}{m} x$$

which is just Eq. (13.4) for simple harmonic motion. This confirms that Eq. (13.13) for x as a function of time is correct.

We actually derived Eq. (13.16) earlier in a geometrical way by taking the x-component of the acceleration vector of the reference point Q. This was done in Fig. 13.5c and Eq. (13.7) (recall that $\theta = \omega t + \phi$). In the same way, we could have derived Eq. (13.15) by taking the x-component of the velocity vector of Q, as shown in Fig. 13.5b. We'll leave the details for you to work out (see Problem 13.79).

Note that the sinusoidal graph of displacement versus time (Fig. 13.11a) is shifted by one-quarter period from the graph of velocity versus time (Fig. 13.11b) and by one-half period from the graph of acceleration versus time (Fig. 13.11c). When the body is passing through the equilibrium position so that the displacement is zero, the velocity equals either v_{max} or $-v_{\text{max}}$ (depending on which way the body is moving) and the acceleration is zero. When the body is at either its maximum positive displacement, $x = +A$, or its maximum negative displacement, $x = -A$, the velocity is zero and the body is instantaneously at rest. At these points, the restoring force $F_x = -kx$ and the acceleration of the body have their maximum magnitudes. At $x = +A$ the acceleration is negative and equal to $-a_{\text{max}}$. At $x = -A$ the acceleration is positive: $a_x = +a_{\text{max}}$.

If we are given the initial position x_0 and initial velocity v_{0x} for the oscillating body, we can determine the amplitude A and the phase angle ϕ. Here's how to do it. The initial velocity v_{0x} is the velocity at time $t = 0$; putting $v_x = v_{0x}$ and $t = 0$ in Eq. (13.15), we find

$$v_{0x} = -\omega A \sin \phi \qquad (13.17)$$

To find ϕ, divide Eq. (13.17) by Eq. (13.14). This eliminates A and gives an equation that we can solve for ϕ:

$$\frac{v_{0x}}{x_0} = \frac{-\omega A \sin \phi}{A \cos \phi} = -\omega \tan \phi$$

$$\phi = \arctan \left(-\frac{v_{0x}}{\omega x_0} \right) \qquad \text{(phase angle in SHM)} \qquad (13.18)$$

13.11 (a) Graph of x versus t for SHM. In this graph, $\phi = \pi/3$. (b) Graph of v_x versus t for the same motion. The v_x-t graph is shifted by $\frac{1}{4}$ cycle from the x-t graph. (c) Graph of a_x versus t for the same motion. The x-t graph is shifted by $\frac{1}{4}$ cycle from the v_x-t graph and by $\frac{1}{2}$ cycle from the a_x-t graph.

It is also easy to find the amplitude A if we are given x_0 and v_{0x}. We'll sketch the derivation, and you can fill in the details. Square Eq. (13.14); divide Eq. (13.17) by ω, square it, and add to the square of Eq. (13.14). The right side will be $A^2(\sin^2 \phi + \cos^2 \phi)$, which is equal to A^2. The final result is

$$A = \sqrt{x_0^2 + \frac{v_{0x}^2}{\omega^2}} \qquad \text{(amplitude in SHM)} \qquad (13.19)$$

Note that when the body has both an initial displacement x_0 and a nonzero initial velocity v_{0x}, the amplitude A is *not* equal to the initial displacement. That's reasonable; if you start the body at a positive x_0 but give it a positive velocity v_{0x}, it will go *farther* than x_0 before it turns and comes back.

9.1   Position Graphs and Equations

9.2   Describing Vibrational Motion

9.5   Ape Drops Tarzan

Problem-Solving Strategy	**Simple Harmonic Motion I**

IDENTIFY *the relevant concepts:* An oscillating system undergoes simple harmonic motion (SHM) *only* if the restoring force is directly proportional to the displacement. Be certain that this is the case for the problem at hand before attempting to use any of the results of this section. As always, identify the target variables.

SET UP *the problem using the following steps:*
1. Identify the known and unknown quantities, and determine which are the target variables.
2. It's useful to distinguish between two kinds of quantities. *Basic properties* of the system include the mass m and force constant k. (In some problems, m, k, or both can be determined from other information.) They also include quantities derived from m and k, such as the period T, frequency f, and angular frequency ω. *Properties of the motion* describe how the system behaves when it is set into motion in a particular way. They include the amplitude A, maximum velocity $v_{\max}$, and phase angle ϕ, as well as the values of x, v_x, and a_x at a particular time.
3. If necessary, define an x-axis as in Fig. 13.6.

EXECUTE *the solution as follows:*
1. Use the equations given in Sections 13.1 and 13.2 to solve for the target variables.
2. If you need to calculate the phase angle, be certain to express it in radians. The quantity ωt in Eq. (13.13) is naturally in radians, so ϕ must be as well.
3. If you need to find the values of x, v_x, and a_x at various times, use Eqs. (13.11), (13.15), and (13.16). If the initial position x_0 and initial velocity v_{0x} are both given, you can determine the phase angle and amplitude from Eqs. (13.18) and (13.19). If the body is given an initial positive displacement x_0 but zero initial velocity $(v_{0x} = 0)$, then the amplitude is $A = x_0$ and the phase angle is $\phi = 0$. If it has an initial positive velocity v_{0x} but no initial displacement $(x_0 = 0)$, the amplitude is $A = v_{0x}/\omega$ and the phase angle is $\phi = -\pi/2$.

EVALUATE *your answer:* Check your results to make sure that they're consistent. As an example, suppose you've used the initial position and velocity to find general expressions for x and v_x at time t. If you substitute $t = 0$ into these expressions, you should get back the correct values of x_0 and v_{0x}.

Example 13.3	**Describing SHM**

Let's return to the system of mass and horizontal spring we considered in Example 13.2, with $k = 200$ N/m and $m = 0.50$ kg. This time we give the body an initial displacement of $+0.015$ m and an initial velocity of $+0.40$ m/s. a) Find the period, amplitude, and phase angle of the motion. b) Write equations for the displacement, velocity, and acceleration as functions of time.

SOLUTION

IDENTIFY: As in Example 13.2, the oscillations are SHM and we may use the expressions developed in this section.

SET UP: We are given the values of k, m, x_0, and v_{0x}. From them, we calculate the target variables T, A, and ϕ and the expressions for x, v_x, and a_x as functions of time.

EXECUTE: a) The period is exactly the same as in Example 13.2, $T = 0.31$ s. In simple harmonic motion the period does not depend on the amplitude, only on the values of k and m.

In Example 13.2 we found that $\omega = 20$ rad/s. So from Eq. (13.19),

$$A = \sqrt{x_0^2 + \frac{v_{0x}^2}{\omega^2}}$$

$$= \sqrt{(0.015 \text{ m})^2 + \frac{(0.40 \text{ m/s})^2}{(20 \text{ rad/s})^2}}$$

$$= 0.025 \text{ m}$$

To find the phase angle ϕ, we use Eq. (13.18):

$$\phi = \arctan\left(\frac{-v_{0x}}{\omega x_0}\right)$$

$$= \arctan\left(\frac{-0.40 \text{ m/s}}{(20 \text{ rad/s})(0.015 \text{ m})}\right) = -53° = -0.93 \text{ rad}$$

b) The displacement, velocity, and acceleration at any time are given by Eqs. (13.13), (13.15), and (13.16), respectively. Substituting the values, we get

$$x = (0.025 \text{ m}) \cos\left[(20 \text{ rad/s})t - 0.93 \text{ rad}\right]$$
$$v_x = -(0.50 \text{ m/s}) \sin\left[(20 \text{ rad/s})t - 0.93 \text{ rad}\right]$$
$$a_x = -(10 \text{ m/s}^2) \cos\left[(20 \text{ rad/s})t - 0.93 \text{ rad}\right]$$

The velocity varies sinusoidally between -0.50 m/s and $+0.50$ m/s, and the acceleration varies sinusoidally between -10 m/s^2 and $+10$ m/s^2.

EVALUATE: You can check the results for x and v_x as functions of time by substituting $t = 0$ and evaluating the result. You should get $x = x_0 = 0.015$ m and $v_x = v_{0x} = 0.40$ m/s. Do you?

Test Your Understanding

The position of a certain object in SHM is given as a function of time by $x = (0.050 \text{ m}) \cos\left[(290 \text{ rad/s})t + (2.5 \text{ rad})\right]$. What are the amplitude, period, phase angle, and initial position for this motion?

13.3 | Energy in Simple Harmonic Motion

Act|v
ONLINE
Phys|cs

9.3   Vibrational Energy
9.4   Two Ways to Weigh Young Tarzan
9.6   Releasing a Vibrating Skier I
9.7   Releasing a Vibrating Skier II

We can learn even more about simple harmonic motion by using energy considerations. Take another look at the body oscillating on the end of a spring in Fig. 13.1. We've already noted that the spring force is the only horizontal force on the body. The force exerted by an ideal spring is a conservative force, and the vertical forces do no work, so the total mechanical energy of the system is *conserved*. We also assume that the mass of the spring itself is negligible.

The kinetic energy of the body is $K = \frac{1}{2}mv^2$ and the potential energy of the spring is $U = \frac{1}{2}kx^2$, just as in Section 7.2. (It would be helpful to review that section.) There are no nonconservative forces that do work, so the total mechanical energy $E = K + U$ is conserved:

$$E = \frac{1}{2}mv_x^2 + \frac{1}{2}kx^2 = \text{constant} \tag{13.20}$$

(Since the motion is one-dimensional, $v^2 = v_x^2$.)

The total mechanical energy E is also directly related to the amplitude A of the motion. When the body reaches the point $x = A$, its maximum displacement from equilibrium, it momentarily stops as it turns back toward the equilibrium position. That is, when $x = A$ (or $-A$), $v_x = 0$. At this point the energy is entirely potential, and $E = \frac{1}{2}kA^2$. Because E is constant, this quantity equals E at any other point. Combining this expression with Eq. (13.20), we get

$$E = \frac{1}{2}mv_x^2 + \frac{1}{2}kx^2 = \frac{1}{2}kA^2 = \text{constant} \tag{13.21}$$

(total mechanical energy in SHM)

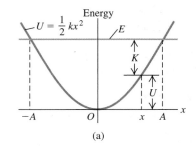

13.12 In SHM the total mechanical energy E is constant. As the body oscillates, the energy continually transforms back and forth between potential energy U and kinetic energy K.

We can verify this equation by substituting x and v_x from Eqs. (13.13) and (13.15) and using $\omega^2 = k/m$ from Eq. (13.9):

$$E = \frac{1}{2}mv_x^2 + \frac{1}{2}kx^2 = \frac{1}{2}m[-\omega A \sin(\omega t + \phi)]^2 + \frac{1}{2}k[A \cos(\omega t + \phi)]^2$$

$$= \frac{1}{2}kA^2 \sin^2(\omega t + \phi) + \frac{1}{2}kA^2 \cos^2(\omega t + \phi)$$

$$= \frac{1}{2}kA^2$$

(Recall that $\sin^2 \alpha + \cos^2 \alpha = 1$.) Hence our expressions for displacement and velocity in SHM are consistent with energy conservation, as they must be.

We can use Eq. (13.21) to solve for the velocity v_x of the body at a given displacement x:

$$v_x = \pm\sqrt{\frac{k}{m}}\sqrt{A^2 - x^2} \qquad (13.22)$$

The $\pm$ sign means that at a given value of x the body can be moving in either direction. For example, when $x = \pm A/2$,

$$v_x = \pm\sqrt{\frac{k}{m}}\sqrt{A^2 - \left(\pm\frac{A}{2}\right)^2} = \pm\sqrt{\frac{3}{4}}\sqrt{\frac{k}{m}}A$$

Equation (13.22) also shows that the *maximum* speed v_{max} occurs at $x = 0$. Using Eq. (13.10), $\omega = \sqrt{k/m}$, we find that

$$v_{max} = \sqrt{\frac{k}{m}}A = \omega A \qquad (13.23)$$

This agrees with Eq. (13.15), which showed that v_x oscillates between $-\omega A$ and $+\omega A$.

Figure 13.12 shows the energy quantities E, K, and U at $x = 0$, $x = \pm A/2$, and $x = \pm A$. Fig. 13.13 is a graphical display of Eq. (13.21); energy (kinetic, potential, and total) is plotted vertically and the coordinate x is plotted horizontally. The parabolic curve in Fig. 13.13a represents the potential energy $U = \frac{1}{2}kx^2$. The horizontal line represents the total mechanical energy E, which is constant and does not vary with x. This line intersects the potential-energy curve at $x = -A$ and $x = A$, where the energy is entirely potential. At any value of x between $-A$ and A, the vertical distance from the x-axis to the parabola is U; since $E = K + U$, the remaining vertical distance up to the horizontal line is K. Figure 13.13b shows

Activ Physics
ONLINE

9.8   One- and Two-Spring Vibrating Systems

9.9   Vibro-Ride

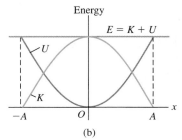

13.13 Kinetic energy K, potential energy U, and total mechanical energy E as functions of position for SHM. At each value of x the sum of the values of K and U equals the constant value of E.

both K and U as functions of x. As the body oscillates between $-A$ and A, the energy is continuously transformed from potential to kinetic and back again.

Figure 13.13a shows the connection between the amplitude A and the corresponding total mechanical energy $E = \frac{1}{2}kA^2$. If we tried to make x greater than A (or less than $-A$), U would be greater than E, and K would have to be negative. But K can never be negative, so x can't be greater than A or less than $-A$.

Problem-Solving Strategy

Simple Harmonic Motion II

The energy equation, Eq. (13.21), is a useful alternative relation between velocity and position, especially when energy quantities are also required. If the problem involves a relation among position, velocity, and acceleration without reference to time, it is usually easier to use Eq. (13.4) (from Newton's second law) or (13.21) (from energy conservation) than to use the general expressions for x, v_x, and a_x as functions of time [Eqs. (13.13),

(13.15), and (13.16), respectively]. Because the energy equation involves x^2 and v_x^2, it cannot tell you the sign of x or of v_x; you have to infer the sign from the situation. For instance, if the body is moving from the equilibrium position toward the point of greatest positive displacement, then x is positive and v_x is positive.

Example 13.4

Velocity, acceleration, and energy in SHM

In the oscillation described in Example 13.2, $k = 200$ N/m, $m = 0.50$ kg, and the oscillating mass is released from rest at $x = 0.020$ m. a) Find the maximum and minimum velocities attained by the oscillating body. b) Compute the maximum acceleration. c) Determine the velocity and acceleration when the body has moved halfway to the center from its original position. d) Find the total energy, potential energy, and kinetic energy at this position.

IDENTIFY: Note that the problem refers to the motion at various *positions* in the motion, not at specified *times*. This is a hint that we can use the energy relations found in this section to solve for the target variables.

SET UP: Figure 13.14 shows the choice of x-axis. For any position x we use Eqs. (13.22) and (13.4) to find the velocity v_x and acceleration a_x, respectively. Given the velocity and position, we use Eq. (13.21) to find the energy quantities K, U, and E.

> **CAUTION** The velocity of the body is *not* constant, so these images of the body at equally spaced positions are *not* equally spaced in time.

EXECUTE: a) The velocity v_x at any displacement x is given by Eq. (13.22):

$$v_x = \pm\sqrt{\frac{k}{m}}\sqrt{A^2 - x^2}$$

The maximum velocity occurs when the body is moving to the right through the equilibrium position, where $x = 0$:

$$v_x = v_{max} = \sqrt{\frac{k}{m}}A = \sqrt{\frac{200\ \text{N/m}}{0.50\ \text{kg}}}(0.020\ \text{m}) = 0.40\ \text{m/s}$$

The minimum (i.e., most negative) velocity occurs when the body is moving to the left through $x = 0$; its value is $-v_{max} = -0.40$ m/s.

b) From Eq. (13.4),

$$a_x = -\frac{k}{m}x$$

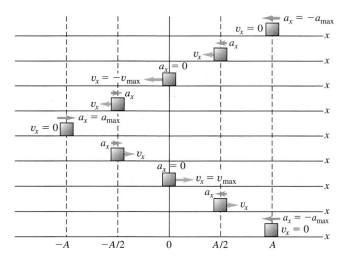

13.14 A body is attached to a spring, pulled a distance A from the equilibrium position and released. This figure shows the velocity and acceleration at nine points in the motion.

The maximum (most positive) acceleration occurs at the most negative value of x, $x = -A$; therefore

$$a_{max} = -\frac{k}{m}(-A) = -\frac{200 \text{ N/m}}{0.50 \text{ kg}}(-0.020 \text{ m}) = 8.0 \text{ m/s}^2$$

The minimum (most negative) acceleration is -8.0 m/s^2, occurring at $x = +A = +0.020 \text{ m}$.

c) At a point halfway to the center from the initial position, $x = A/2 = 0.010 \text{ m}$. From Eq. (13.22),

$$v_x = -\sqrt{\frac{200 \text{ N/m}}{0.50 \text{ kg}}}\sqrt{(0.020 \text{ m})^2 - (0.010 \text{ m})^2} = -0.35 \text{ m/s}$$

We choose the negative square root because the body is moving from $x = A$ toward $x = 0$. From Eq. (13.4),

$$a_x = -\frac{200 \text{ N/m}}{0.50 \text{ kg}}(0.010 \text{ m}) = -4.0 \text{ m/s}^2$$

At this point the velocity and the acceleration have the same sign, so the speed is increasing. The conditions at $x = 0$, $\pm A/2$, and $\pm A$ are shown in Fig. 13.14.

d) The total energy has the same value at all points during the motion:

$$E = \frac{1}{2}kA^2 = \frac{1}{2}(200 \text{ N/m})(0.020 \text{ m})^2 = 0.040 \text{ J}$$

The potential energy is

$$U = \frac{1}{2}kx^2 = \frac{1}{2}(200 \text{ N/m})(0.010 \text{ m})^2 = 0.010 \text{ J}$$

and the kinetic energy is

$$K = \frac{1}{2}mv_x^2 = \frac{1}{2}(0.50 \text{ kg})(-0.35 \text{ m/s})^2 = 0.030 \text{ J}$$

EVALUATE: At the point $x = A/2$, the energy is one-fourth potential energy and three-fourths kinetic energy. You can check this result by inspecting Fig. 13.13b.

Example 13.5	**Energy and momentum in SHM**

A block with mass M attached to a horizontal spring with force constant k is moving with simple harmonic motion having amplitude A_1. At the instant when the block passes through its equilibrium position, a lump of putty with mass m is dropped vertically onto the block from a very small height and sticks to it (Fig. 13.15a). a) Find the new amplitude and period. b) Repeat part (a) for the case in which the putty is dropped on the block when it is at one end of its path (Fig. 13.15b).

SOLUTION

IDENTIFY: The problem involves the motion at a given position, not a given time, so we can use energy methods to solve it. Before the putty lands on the block, the mechanical energy of the oscillating block and spring is constant. When the putty lands on the block, it's a completely inelastic collision (Section 8.3); the horizontal component of momentum is conserved, but kinetic energy decreases. Once the collision ends, the mechanical energy remains constant at its new value. We use this principle to find the new amplitude (which is related to the total energy of the system). We find the new period using the relationship between period and mass.

SET UP: Figure 13.15 shows the choice of coordinates. The target variables in each part are the amplitude A_2 and the period T_2 after the collision. In each part we consider what happens before, during, and after the collision.

EXECUTE: a) Before the collision the total mechanical energy of the block and spring is $E_1 = \frac{1}{2}kA_1^2$. Since the block is at the equilibrium position, $U = 0$, and the energy is purely kinetic. If we let v_1 be the speed of the block at the equilibrium position, we have

$$E_1 = \frac{1}{2}Mv_1^2 = \frac{1}{2}kA_1^2 \quad \text{so} \quad v_1 = \sqrt{\frac{k}{M}}A_1$$

During the collision the x-component of momentum of the system of block and putty is conserved. (Why?) Just before the collision

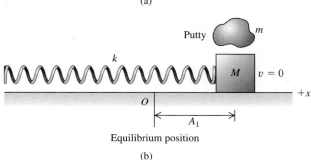

13.15 (a) Dropping a lump of putty on an oscillating block as it passes through equilibrium. (b) Dropping a lump of putty on the block at $x = A_1$.

this component is the sum of Mv_1 (for the block) and zero (for the putty). Just after the collision the block and putty move together with speed v_2, and their combined x-component of momentum is $(M + m)v_2$. From conservation of momentum,

$$Mv_1 + 0 = (M + m)v_2 \quad \text{so} \quad v_2 = \frac{M}{M + m}v_1$$

The collision lasts a very short time, so just after the collision the block and putty are still at the equilibrium position. The energy is still purely kinetic, but is *less* than before the collision:

$$E_2 = \frac{1}{2}(M + m)v_2^2 = \frac{1}{2}\frac{M^2}{M + m}v_1^2 = \frac{M}{M + m}\left(\frac{1}{2}Mv_1^2\right)$$

$$= \left(\frac{M}{M + m}\right)E_1$$

Since E_2 equals $\frac{1}{2}kA_2^2$, where A_2 is the amplitude after the collision, we have

$$\frac{1}{2}kA_2^2 = \left(\frac{M}{M + m}\right)\frac{1}{2}kA_1^2$$

$$A_2 = A_1\sqrt{\frac{M}{M + m}}$$

The larger the putty mass m, the smaller the final amplitude.

Finding the period of oscillation after the collision is the easy part. Using Eq. (13.12), we have

$$T_2 = 2\pi\sqrt{\frac{M + m}{k}}$$

b) When the putty drops on the block, the block is instantaneously at rest; all the mechanical energy is stored in the spring as potential energy. The x-component of momentum is zero both before and after the collision: the block has zero kinetic energy just before the collision, and the block and putty have zero kinetic energy just after the collision. So in this case, adding the extra mass of the putty has *no effect* on the mechanical energy. That is,

$$E_2 = E_1 = \frac{1}{2}kA_1^2$$

and the amplitude after the collision is unchanged $(A_2 = A_1)$. The period still changes when the putty is added, though; its value doesn't depend on how the mass is added, only on what the total mass is. So T_2 is the same as we found in part (a), $T_2 = 2\pi\sqrt{(M + m)/k}$.

EVALUATE: Why is energy lost in part (a) but not in part (b)? The difference is that in part (a) the putty slides against the moving block during the collision, and energy is dissipated by kinetic friction.

Test Your Understanding

To double the total energy for an oscillating mass-spring system, by what factor must the amplitude increase? What effect does this change have on the frequency? Assume SHM.

13.4 | Applications of Simple Harmonic Motion

So far, we've looked at a grand total of *one* situation in which simple harmonic motion (SHM) occurs: a body attached to an ideal horizontal spring. But SHM can occur in any system in which there is a restoring force that is directly proportional to the displacement from equilibrium, as given by Eq. (13.3), $F_x = -kx$. The restoring force will originate in different ways in different situations, so the force constant k has to be found for each case by examining the net force on the system. Once this is done, it's straightforward to find the angular frequency ω, frequency f, and period T; we just substitute the value of k into Eqs. (13.10), (13.11), and (13.12), respectively. Let's use these ideas to examine several examples of simple harmonic motion.

Vertical SHM

Suppose we hang a spring with force constant k (Fig. 13.16a) and suspend from it a body with mass m. Oscillations will now be vertical; will they still be SHM? In Fig. 13.16b the body hangs at rest, in equilibrium. In this position the spring is stretched an amount Δl just great enough that the spring's upward vertical force $k\Delta l$ on the body balances its weight mg:

$$k\Delta l = mg$$

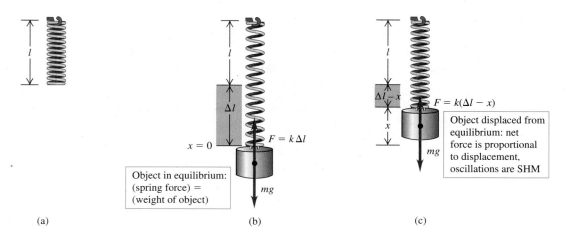

13.16 (a) A hanging spring. (b) A body suspended from the spring. When the spring is stretched just far enough that the upward spring force equals the object's weight, the object is in equilibrium. (c) If the spring obeys Hooke's law, the object's motion will be simple harmonic.

Take $x = 0$ to be this equilibrium position and take the positive x-direction to be upward. When the body is a distance x *above* its equilibrium position (Fig. 13.16c), the extension of the spring is $\Delta l - x$. The upward force it exerts on the body is then $k(\Delta l - x)$, and the net x-component of force on the body is

$$F_{\text{net}} = k(\Delta l - x) + (-mg) = -kx$$

that is, a net downward force of magnitude kx. Similarly, when the body is *below* the equilibrium position, there is a net upward force with magnitude kx. In either case there is a restoring force with magnitude kx. If the body is set in vertical motion, it oscillates in SHM with the same angular frequency as though it were horizontal, $\omega = \sqrt{k/m}$. So vertical SHM doesn't differ in any essential way from horizontal SHM. The only real change is that the equilibrium position $x = 0$ no longer corresponds to the point at which the spring is unstretched. The same ideas hold if a body with weight mg is placed atop a compressible spring (Fig. 13.17) and compresses it a distance Δl.

13.17 If the weight mg compresses the spring a distance Δl, the force constant is $k = mg/\Delta l$ and the angular frequency for vertical SHM is $\omega = \sqrt{k/m}$—the same as if the body were suspended from the spring (Fig. 13.16).

Example 13.6	**Vertical SHM in an old car**

The shock absorbers in an old car with mass 1000 kg are completely worn out. When a 980-N person climbs slowly into the car to its center of gravity, the car sinks 2.8 cm. When the car, with the person aboard, hits a bump, the car starts oscillating up and down in SHM. Model the car and person as a single body on a single spring, and find the period and frequency of the oscillation.

SOLUTION

IDENTIFY and SET UP: The situation is like that shown in Fig. 13.17. The compression of the spring when the extra weight is added tells us the force constant, which we can use to find the period and frequency (the target variables).

EXECUTE: When the force increases by 980 N, the spring compresses an additional 0.028 m, and the coordinate x of the car changes by -0.028 m. Hence the effective force constant (including the effect of the entire suspension) is

$$k = -\frac{F_x}{x} = -\frac{980\ \text{N}}{-0.028\ \text{m}} = 3.5 \times 10^4\ \text{kg/s}^2$$

The person's mass is $w/g = (980 \text{ N})/(9.8 \text{ m/s}^2) = 100$ kg. The *total* oscillating mass is $m = 1000 \text{ kg} + 100 \text{ kg} = 1100$ kg. The period T is

$$T = 2\pi\sqrt{\frac{m}{k}} = 2\pi\sqrt{\frac{1100 \text{ kg}}{3.5 \times 10^4 \text{ kg/s}^2}} = 1.11 \text{ s}$$

and the frequency is

$$f = \frac{1}{T} = \frac{1}{1.11 \text{ s}} = 0.90 \text{ Hz}$$

EVALUATE: A persistent oscillation with a period of about one second makes for a very unpleasant ride. The purpose of shock absorbers is to make such oscillations die out (see Section 13.7).

Balance wheel   Spring

τ_z θ

Spring torque τ_z opposes angular displacement

13.18 The balance wheel of a mechanical watch. The restoring torque exerted by the spring is proportional to the angular displacement θ, so the motion is angular SHM.

Angular SHM

Figure 13.18 shows the balance wheel of a mechanical watch. The wheel has a moment of inertia I about its axis. A coil spring exerts a restoring torque τ_z that is proportional to the angular displacement θ from the equilibrium position. We write $\tau_z = -\kappa\theta$, where κ (the Greek letter "kappa") is a constant called the *torsion constant*. Using the rotational analog of Newton's second law for a rigid body, $\sum\tau_z = I\alpha_z = I \, d^2\theta/dt^2$, the equation of motion is

$$-\kappa\theta = I\alpha \quad \text{or} \quad \frac{d^2\theta}{dt^2} = -\frac{\kappa}{I}\theta$$

The form of this equation is exactly the same as Eq. (13.4) for the acceleration in simple harmonic motion, with x replaced by θ and k/m replaced by κ/I. So we are dealing with a form of *angular* simple harmonic motion. The angular frequency ω and frequency f are given by Eqs. (13.10) and (13.11), respectively, with the same replacement:

$$\omega = \sqrt{\frac{\kappa}{I}} \quad \text{and} \quad f = \frac{1}{2\pi}\sqrt{\frac{\kappa}{I}} \qquad \text{(angular SHM)} \qquad (13.24)$$

The motion is described by the function

$$\theta = \Theta \cos(\omega t + \phi)$$

where Θ (capital "theta") plays the role of an angular amplitude.

It's a good thing that the motion of a balance wheel *is* simple harmonic. If it weren't, the frequency might depend on the amplitude, and the watch would run too fast or too slow as the spring ran down.

*Vibrations of Molecules

The following discussion of the vibrations of molecules uses the binomial theorem. If you aren't familiar with this theorem, you should read about it in the appropriate section of your math textbook.

When two atoms are separated from each other by a few atomic diameters, they can exert attractive forces on each other. But if the atoms are so close to each other that their electron shells overlap, the forces between the atoms are repulsive. Between these limits, there can be an equilibrium separation distance at which two atoms form a *molecule*. If these atoms are displaced slightly from equilibrium, they will oscillate. Let's see whether these oscillations can be simple harmonic.

As an example, we'll consider one type of interaction between atoms called the *van der Waals interaction*. Our immediate task here is to study oscillations, so we

13.19 (a) Two atoms with centers separated by r. (b) Potential energy U in the van der Waals interaction as a function of r. (c) Force F_r on the right-hand atom as function of r.

won't go into the details of how this interaction arises. Let the center of one atom be at the origin and let the center of the other atom be a distance r away (Fig. 13.19a); the equilibrium distance between centers is $r = R_0$. Experiment shows that the van der Waals interaction can be described by the potential energy function

$$U = U_0\left[\left(\frac{R_0}{r}\right)^{12} - 2\left(\frac{R_0}{r}\right)^6\right] \tag{13.25}$$

where U_0 is a positive constant with units of joules. When the two atoms are very far apart, $U = 0$; when they are separated by the equilibrium distance $r = R_0$, $U = -U_0$. The force on the second atom is the negative derivative of Eq. (13.25),

$$F_r = -\frac{dU}{dr} = U_0\left[\frac{12R_0^{12}}{r^{13}} - 2\frac{6R_0^6}{r^7}\right] = 12\frac{U_0}{R_0}\left[\left(\frac{R_0}{r}\right)^{13} - \left(\frac{R_0}{r}\right)^7\right] \tag{13.26}$$

The potential energy and force are plotted in Figs. 13.19b and 13.19c, respectively. The force is positive for $r < R_0$ and negative for $r > R_0$, so it is a *restoring* force.

To study small-amplitude oscillations around the equilibrium separation $r = R_0$, we introduce the quantity x to represent the displacement from equilibrium:

$$x = r - R_0 \quad \text{so} \quad r = R_0 + x$$

In terms of x, the force F_r in Eq. (13.26) becomes

$$F_r = 12\frac{U_0}{R_0}\left[\left(\frac{R_0}{R_0 + x}\right)^{13} - \left(\frac{R_0}{R_0 + x}\right)^7\right] = 12\frac{U_0}{R_0}\left[\frac{1}{(1 + x/R_0)^{13}} - \frac{1}{(1 + x/R_0)^7}\right] \tag{13.27}$$

This looks nothing like Hooke's law, $F_x = -kx$, so we might be tempted to conclude that molecular oscillations cannot be SHM. But let us restrict ourselves to *small-amplitude* oscillations so that the absolute value of the displacement x is small in comparison to R_0 and the absolute value of the ratio x/R_0 is much less than 1. We can then simplify Eq. (13.27) by using the *binomial theorem*:

$$(1 + u)^n = 1 + nu + \frac{n(n - 1)}{2!}u^2 + \frac{n(n - 1)(n - 2)}{3!}u^3 + \cdots \tag{13.28}$$

If $|u|$ is much less than 1, each successive term in Eq. (13.28) is much smaller than the one it follows, and we can safely approximate $(1 + u)^n$ by just the first two terms. In Eq. (13.27), u is replaced by x/R_0 and n equals -13 or -7, so

$$\frac{1}{(1 + x/R_0)^{13}} = (1 + x/R_0)^{-13} \approx 1 + (-13)\frac{x}{R_0}$$

$$\frac{1}{(1 + x/R_0)^7} = (1 + x/R_0)^{-7} \approx 1 + (-7)\frac{x}{R_0}$$

$$F_r \approx 12\frac{U_0}{R_0}\left[\left(1 + (-13)\frac{x}{R_0}\right) - \left(1 + (-7)\frac{x}{R_0}\right)\right] = -\left(\frac{72U_0}{R_0^2}\right)x \quad (13.29)$$

This is just Hooke's law, with force constant $k = 72U_0/R_0^2$. (Note that k has the correct units, J/m^2 or N/m.) So oscillations of molecules bound by the van der Waals interaction can be simple harmonic motion, provided that the amplitude is small in comparison to R_0 so that the approximation $|x/R_0| \ll 1$ used in the derivation of Eq. (13.29) is valid.

You can also show that the potential energy U, Eq. (13.25), can be written as $U \approx \frac{1}{2}kx^2 + C$, where $C = -U_0$ and k is again equal to $72U_0/R_0^2$. Adding a constant to the potential energy has no effect on the physics, so the system of two atoms is fundamentally no different from a mass attached to a horizontal spring for which $U = \frac{1}{2}kx^2$. The proof is left as an exercise.

Example 13.7   Molecular vibration

Two argon atoms can form a weakly bound molecule, Ar_2, held together by a van der Waals interaction with $U_0 = 1.68 \times 10^{-21}$ J and $R_0 = 3.82 \times 10^{-10}$ m. Find the frequency for small oscillations of one of the atoms about its equilibrium position.

SOLUTION

IDENTIFY and SET UP: Because the oscillations are small, we can use Eq. (13.11) to obtain the frequency of simple harmonic motion. The force constant is given by Eq. (13.29).

EXECUTE: The force constant is

$$k = \frac{72U_0}{R_0^2} = \frac{72(1.68 \times 10^{-21}\text{ J})}{(3.82 \times 10^{-10}\text{ m})^2} = 0.829 \text{ J/m}^2 = 0.829 \text{ N/m}$$

This is comparable to the force constant of a loose, floppy toy spring like a Slinky™.

From the periodic table of the elements (Appendix D), the average atomic mass of argon is $(39.948 \text{ u})(1.66 \times 10^{-27} \text{ kg/1 u}) = 6.63 \times 10^{-26}$ kg. If one of the argon atoms is fixed and the other atom oscillates, the frequency of oscillation is

$$f = \frac{1}{2\pi}\sqrt{\frac{k}{m}} = \frac{1}{2\pi}\sqrt{\frac{0.829 \text{ N/m}}{6.63 \times 10^{-26} \text{ kg}}} = 5.63 \times 10^{11} \text{ Hz}$$

The oscillating mass is very small, so even a loose spring causes very rapid oscillations.

EVALUATE: Our answer for f isn't quite right, however. If there is no net external force acting on the molecule, the center of mass of the molecule (located halfway between the two atoms) doesn't accelerate. To ensure this, *both* atoms must oscillate with the same amplitude in opposite directions. It turns out that we can account for this by replacing m with $m/2$ in the expression for f. (See Problem 13.81.) This makes f larger by a factor of $\sqrt{2}$, so $f = \sqrt{2}(5.63 \times 10^{11}\text{ Hz}) = 7.96 \times 10^{11}$ Hz. An additional complication is that on the atomic scale we must use *quantum mechanics,* not Newtonian mechanics, to describe oscillation and other motions; happily, the frequency has the same value in quantum mechanics.

Test Your Understanding

Suppose one of the argon atoms in an Ar_2 molecule (Example 13.7) is displaced from equilibrium by 1.00×10^{-11} m and released. What is the magnitude of the atom's initial acceleration?

13.5 | The Simple Pendulum

A **simple pendulum** is an idealized model consisting of a point mass suspended by a massless, unstretchable string. When the point mass is pulled to one side of its straight-down equilibrium position and released, it oscillates about the equilibrium position. Familiar situations such as a wrecking ball on a crane's cable, the plumb bob on a surveyor's transit, and a child on a swing can be modeled as simple pendulums.

The path of the point mass (sometimes called a pendulum bob) is not a straight line but the arc of a circle with radius L equal to the length of the string (Fig. 13.20). We use as our coordinate the distance x measured along the arc. If the motion is simple harmonic, the restoring force must be directly proportional to x or (because $x = L\theta$) to θ. Is it?

In Fig. 13.20 we represent the forces on the mass in terms of tangential and radial components. The restoring force F_θ is the tangential component of the net force:

$$F_\theta = -mg \sin \theta \qquad (13.30)$$

The restoring force is provided by gravity; the tension T merely acts to make the point mass move in an arc. The restoring force is proportional *not* to θ but to $\sin \theta$, so the motion is *not* simple harmonic. However, if the angle θ is small, $\sin \theta$ is very nearly equal to θ in radians (Fig. 13.21). For example, when $\theta = 0.1$ rad (about 6°), $\sin \theta = 0.0998$, a difference of only 0.2%. With this approximation, Eq. (13.30) becomes

$$F_\theta = -mg\theta = -mg\frac{x}{L} \quad \text{or}$$

$$F_\theta = -\frac{mg}{L}x \qquad (13.31)$$

The restoring force is then proportional to the coordinate *for small displacements,* and the force constant is $k = mg/L$. From Eq. (13.10) the angular frequency ω of a simple pendulum with small amplitude is

$$\omega = \sqrt{\frac{k}{m}} = \sqrt{\frac{mg/L}{m}} = \sqrt{\frac{g}{L}} \qquad (13.32)$$
$$\text{(simple pendulum, small amplitude)}$$

The corresponding frequency and period relations are

$$f = \frac{\omega}{2\pi} = \frac{1}{2\pi}\sqrt{\frac{g}{L}} \quad \text{(simple pendulum, small amplitude)} \quad (13.33)$$

$$T = \frac{2\pi}{\omega} = \frac{1}{f} = 2\pi\sqrt{\frac{L}{g}} \quad \text{(simple pendulum, small amplitude)} \quad (13.34)$$

Note that these expressions do not involve the *mass* of the particle. This is because the restoring force, a component of the particle's weight, is proportional to m. Thus the mass appears on *both* sides of $\Sigma \vec{F} = m\vec{a}$ and cancels out. (This is the same physics that explains why bodies of different masses fall with the same acceleration in a vacuum.) For small oscillations the period of a pendulum for a given value of g is determined entirely by its length.

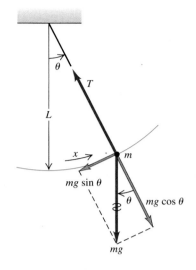

13.20 The dynamics of a simple pendulum.

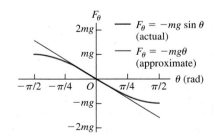

13.21 For small angular displacements θ, the restoring force $F_\theta = -mg \sin \theta$ on a simple pendulum is approximately equal to $-mg\theta$; that is, it is approximately proportional to the displacement θ. Hence for small angles the oscillations are simple harmonic.

The dependence on L and g in Eqs. (13.32) through (13.34) is just what we should expect. *A long pendulum has a longer period than a shorter one.* Increasing g increases the restoring force, causing the frequency to increase and the period to decrease.

We emphasize again that the motion of a pendulum is only *approximately* simple harmonic. When the amplitude is not small, the departures from simple harmonic motion can be substantial. But how small is "small"? The period can be expressed by an infinite series; when the maximum angular displacement is Θ, the period T is given by

$$T = 2\pi\sqrt{\frac{L}{g}}\left(1 + \frac{1^2}{2^2}\sin^2\frac{\Theta}{2} + \frac{1^2 \cdot 3^2}{2^2 \cdot 4^2}\sin^4\frac{\Theta}{2} + \cdots\right) \qquad (13.35)$$

We can compute the period to any desired degree of precision by taking enough terms in the series. We invite you to check that when $\Theta = 15°$ (on either side of the central position), the true period is longer than that given by the approximate Eq. (13.34) by less than 0.5%.

The usefulness of the pendulum as a timekeeper depends on the period being *very nearly* independent of amplitude, provided that the amplitude is small. Thus, as a pendulum clock runs down and the amplitude of the swings decreases a little, the clock still keeps very nearly correct time.

Example 13.8	**A simple pendulum**

Find the period and frequency of a simple pendulum 1.000 m long at a location where $g = 9.800$ m/s^2.

SOLUTION

IDENTIFY and SET UP: We use Eq. (13.34) to determine the period T of the pendulum from its length, and Eq. (13.1) to find the frequency f from T.

EXECUTE: From Eqs. (13.34) and (13.1),

$$T = 2\pi\sqrt{\frac{L}{g}} = 2\pi\sqrt{\frac{1.000 \text{ m}}{9.800 \text{ m/s}^2}} = 2.007 \text{ s}$$

$$f = \frac{1}{T} = 0.4983 \text{ Hz}$$

EVALUATE: The period is almost exactly 2 s. In fact, when the metric system was first established, the second was defined as half the period of a one-meter pendulum. This wasn't a very good standard for time, however, because the value of g varies from place to place. We discussed more modern time standards in Section 1.3.

Test Your Understanding

Deposits of ore or oil affect the local value of g because their density differs from that of their surroundings. Suppose a 1.000-m simple pendulum has a period of exactly 2.000 s at a certain location. What is the local value of g?

13.6 | The Physical Pendulum

A **physical pendulum** is any *real* pendulum, using a body of finite size, as contrasted to the idealized model of the *simple* pendulum with all the mass concentrated at a single point. For small oscillations, analyzing the motion of a real, physical pendulum is almost as easy as for a simple pendulum. Figure 13.22 shows a body of irregular shape pivoted so that it can turn without friction about an axis through point O. In the equilibrium position the center of gravity is directly below the pivot; in the position shown in the figure, the body is displaced from equilibrium by an angle θ, which we use as a coordinate for the system. The distance from O to the center of gravity is d, the moment of inertia of the body about the axis of rotation through O is I, and the total mass is m. When the body is displaced as shown, the weight mg causes a restoring torque

$$\tau_z = -(mg)(d\sin\theta) \tag{13.36}$$

The negative sign shows that the restoring torque is clockwise when the displacement is counterclockwise, and vice versa.

When the body is released, it oscillates about its equilibrium position. The motion is not simple harmonic because the torque τ_z is proportional to $\sin\theta$ rather than to θ itself. However, if θ is small, we can again approximate $\sin\theta$ by θ in radians, and the motion is *approximately* simple harmonic. With this approximation,

$$\tau_z = -(mgd)\theta$$

The equation of motion is $\Sigma\tau_z = I\alpha_z$, so

$$-(mgd)\theta = I\alpha_z = I\frac{d^2\theta}{dt^2}$$

$$\frac{d^2\theta}{dt^2} = -\frac{mgd}{I}\theta \tag{13.37}$$

Comparing this with Eq. (13.4), we see that the role of (k/m) for the spring-mass system is played here by the quantity (mgd/I). Thus the angular frequency is given by

$$\omega = \sqrt{\frac{mgd}{I}} \qquad \text{(physical pendulum, small amplitude)} \qquad (13.38)$$

The frequency f is $1/2\pi$ times this, and the period T is

$$T = 2\pi\sqrt{\frac{I}{mgd}} \qquad \text{(physical pendulum, small amplitude)} \qquad (13.39)$$

Equation (13.39) is the basis of a common method for experimentally determining the moment of inertia of a body with a complicated shape. The center of gravity of the body is first located by balancing. The body is then suspended so that it is free to oscillate about an axis, and the period T of small-amplitude oscillations is measured. By using Eq. (13.39), the moment of inertia I of the body about this axis can then be calculated from T, the body's mass m, and the distance d from the axis to the center of gravity (see Exercise 13.50). Biomechanics researchers use this method to find the moments of inertia of an animal's limbs. This information is important for analyzing how an animal walks, as we'll see in the second of the two following examples.

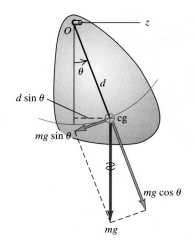

13.22 Dynamics of a physical pendulum.

9.12 Physical Pendulum

<table>
<tr><td>Example
13.9</td><td>**Physical pendulum versus simple pendulum**</td></tr>
</table>

Suppose the body in Fig. 13.22 is a uniform rod with length L, pivoted at one end. Find the period of its motion.

SOLUTION

IDENTIFY and SET UP: We use Table 9.2 (Section 9.4) to find the moment of inertia of the rod, then substitute this value into Eq. (13.39) to determine the period of oscillation.

EXECUTE: From Table 9.2, the moment of inertia of a uniform rod about an axis through one end is $I = \frac{1}{3}ML^2$. The distance from the pivot to the center of gravity is $d = L/2$. From Eq. (13.39),

$$T = 2\pi\sqrt{\frac{\frac{1}{3}ML^2}{MgL/2}} = 2\pi\sqrt{\frac{2L}{3g}}$$

EVALUATE: If the rod is a meter stick $(L = 1.00\text{ m})$ and $g = 9.80\text{ m/s}^2$, then

$$T = 2\pi\sqrt{\frac{2(1.00\text{ m})}{3(9.80\text{ m/s}^2)}} = 1.64\text{ s}$$

The period is smaller by a factor of $\sqrt{2/3} = 0.816$ than the period of a simple pendulum with the same length, calculated in Example 13.8 (Section 13.5).

<table>
<tr><td>Example
13.10</td><td>*Tyrannosaurus rex* and the physical pendulum</td></tr>
</table>

All walking animals, including humans, have a natural walking pace, a number of steps per minute that is more comfortable than a faster or slower pace. Suppose this natural pace is equal to the period of the leg, viewed as a uniform rod pivoted at the hip joint. a) How does the natural walking pace depend on the length L of the leg, measured from hip to foot? b) Fossil evidence shows that *Tyrannosaurus rex*, a two-legged dinosaur that lived about 65 million years ago at the end of the Cretaceous period, had a leg length $L = 3.1$ m and a stride length (the distance from one footprint to the next print of the same foot) $S = 4.0$ m (Fig. 13.23). Estimate the walking speed of *Tyrannosaurus rex*.

SOLUTION

IDENTIFY and SET UP: We treat the leg as a physical pendulum, with a period of oscillation as found in Example 13.9. Our target variables are a) the relationship between the walking pace and the leg length and b) the walking speed of *T. rex*.

EXECUTE: a) From Example 13.9 the period of oscillation of the leg is $T = 2\pi\sqrt{2L/3g}$, which is proportional to $\sqrt{L}$. Each period (a complete back-and-forth swing of the leg) corresponds to *two* steps, so the walking pace in steps per unit time is just twice the oscilla-

13.23 The walking speed of *Tyrannosaurus rex* can be estimated from its leg length L and its stride length S.

tion frequency $f = 1/T$. Hence the walking pace is proportional to $1/\sqrt{L}$. Animals with short legs (small values of L), such as mice or Chihuahuas, have rapid walking paces; humans, giraffes, and other animals with long legs (large values of L) walk at slower paces.

b) According to our model for the natural walking pace, the elapsed time for one stride of a walking *Tyrannosaurus rex* is

$$T = 2\pi\sqrt{\frac{2L}{3g}} = 2\pi\sqrt{\frac{2(3.1 \text{ m})}{3(9.8 \text{ m/s}^2)}} = 2.9 \text{ s}$$

The distance moved during this time is the stride length S, so the walking speed is

$$v = \frac{S}{T} = \frac{4.0 \text{ m}}{2.9 \text{ s}} = 1.4 \text{ m/s} = 5.0 \text{ km/h} = 3.1 \text{ mi/h}$$

This is about the same as a typical human walking speed!

EVALUATE: Our estimate must be somewhat in error because a uniform rod isn't a very good model for a leg. The legs of many animals, including *T. rex* as well as humans, are tapered; there is a lot more mass between the knee and the hip than between the knee and the foot. Thus the center of mass is less than $L/2$ from the hip; a reasonable guess would be about $L/4$. The moment of inertia is *considerably* less than $ML^2/3$, probably somewhere around $ML^2/15$. Try these numbers out with the analysis of Example 13.9; you'll get a shorter oscillation period and an even faster walking speed for *T. rex*.

Test Your Understanding

Suppose the body in Fig. 13.22 is a uniform rod of mass m and length L pivoted around its midpoint. The moment of inertia for this pivot point is $I = \frac{1}{12}mL^2$. What is the oscillation period? Interpret your result.

13.7 | Damped Oscillations

The idealized oscillating systems we have discussed so far are frictionless. There are no nonconservative forces, the total mechanical energy is constant, and a system set into motion continues oscillating forever with no decrease in amplitude.

Real-world systems always have some dissipative forces, however, and oscillations do die out with time unless we provide some means for replacing the dissipated mechanical energy (Fig 13.24). A mechanical pendulum clock continues to run because potential energy stored in the spring or a hanging weight system replaces the mechanical energy lost due to friction in the pivot and the gears. But eventually the spring runs down or the weights reach the bottom of their travel. Then no more energy is available, and the pendulum swings decrease in amplitude and stop.

The decrease in amplitude caused by dissipative forces is called **damping,** and the corresponding motion is called **damped oscillation.** The simplest case to analyze in detail is a simple harmonic oscillator with a frictional damping force that is directly proportional to the *velocity* of the oscillating body. This behavior occurs in friction involving viscous fluid flow, such as in shock absorbers or sliding between oil-lubricated surfaces. We then have an additional force on the body due to friction, $F_x = -bv_x$, where $v_x = dx/dt$ is the velocity and b is a constant that describes the strength of the damping force. The negative sign shows that the force is always opposite in direction to the velocity. The *net* force on the body is then

$$\sum F_x = -kx - bv_x \qquad (13.40)$$

and Newton's second law for the system is

$$-kx - bv_x = ma_x \quad \text{or} \quad -kx - b\frac{dx}{dt} = m\frac{d^2x}{dt^2} \qquad (13.41)$$

13.24 A swinging bell left to itself will eventually stop oscillating due to damping forces (air resistance and friction at the point of suspension).

Equation (13.41) is a differential equation for x; it would be the same as Eq. (13.4), the equation for the acceleration in SHM, except for the added term $-b\,dx/dt$. Solving this equation is a straightforward problem in differential equations, but we won't go into the details here. If the damping force is relatively small, the motion is described by

$$x = Ae^{-(b/2m)t} \cos\left(\omega't + \phi\right) \quad \text{(oscillator with little damping)} \quad (13.42)$$

The angular frequency of oscillation ω' is given by

$$\omega' = \sqrt{\frac{k}{m} - \frac{b^2}{4m^2}} \quad \text{(oscillator with little damping)} \quad (13.43)$$

You can verify that Eq. (13.42) is a solution of Eq. (13.41) by calculating the first and second derivatives of x, substituting them into Eq. (13.41), and checking whether the left and right sides are equal. This is a straightforward but slightly tedious procedure.

The motion described by Eq. (13.42) differs from the undamped case in two ways. First, the amplitude $Ae^{-(b/2m)t}$ is not constant but decreases with time because of the decreasing exponential factor $e^{-(b/2m)t}$. Figure 13.25 is a graph of Eq. (13.42) for the case $\phi = 0$; it shows that the larger the value of b, the more quickly the amplitude decreases.

Second, the angular frequency ω', given by Eq. (13.43), is no longer equal to $\omega = \sqrt{k/m}$ but is somewhat smaller. It becomes zero when b becomes so large that

$$\frac{k}{m} - \frac{b^2}{4m^2} = 0 \quad \text{or} \quad b = 2\sqrt{km} \quad (13.44)$$

When Eq. (13.44) is satisfied, the condition is called **critical damping.** The system no longer oscillates but returns to its equilibrium position without oscillation when it is displaced and released.

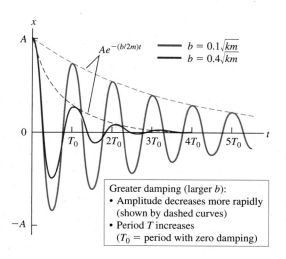

13.25 Graph of displacement versus time for an oscillator with little damping [Eq. (13.42)] with phase angle $\phi = 0$. Graphs are shown for two values of the damping constant b.

If b is greater than $2\sqrt{km}$, the condition is called **overdamping.** Again there is no oscillation, but the system returns to equilibrium more slowly than with critical damping. For the overdamped case the solutions of Eq. (13.41) have the form

$$x = C_1 e^{-a_1 t} + C_2 e^{-a_2 t}$$

where C_1 and C_2 are constants that depend on the initial conditions and a_1 and a_2 are constants determined by m, k, and b.

When b is less than the critical value, as in Eq. (13.42), the condition is called **underdamping.** The system oscillates with steadily decreasing amplitude.

In a vibrating tuning fork or guitar string, it is usually desirable to have as little damping as possible. By contrast, damping plays a beneficial role in the oscillations of an automobile's suspension system. The shock absorbers provide a velocity-dependent damping force so that when the car goes over a bump, it doesn't continue bouncing forever (Fig. 13.26). For optimal passenger comfort, the system should be critically damped or slightly underdamped. As the shocks get old and worn, the value of b decreases and the bouncing persists longer. Not only is this nauseating, but it is also bad for steering because the front wheels have less positive contact with the ground. Thus damping is an advantage in this system. Too much damping would be counterproductive, however; if b is excessively large, the system is overdamped and the suspension returns to equilibrium more slowly. If the suspension is overdamped and the car hits a second bump just after the first one, the springs in the suspension will still be compressed somewhat from the first bump and will be unable to fully absorb the impact.

In damped oscillations the damping force is nonconservative; the mechanical energy of the system is not constant but decreases continuously, approaching zero after a long time. To derive an expression for the rate of change of energy, we first write an expression for the total mechanical energy E at any instant:

$$E = \frac{1}{2} m v_x^2 + \frac{1}{2} k x^2$$

To find the rate of change of this quantity, we take its time derivative:

$$\frac{dE}{dt} = m v_x \frac{dv_x}{dt} + k x \frac{dx}{dt}$$

But $dv_x/dt = a_x$ and $dx/dt = v_x$, so

$$\frac{dE}{dt} = v_x(m a_x + k x)$$

From Eq. (13.41), $m a_x + k x = -b\,dx/dt = -b v_x$, so

$$\frac{dE}{dt} = v_x(-b v_x) = -b v_x^2 \qquad \text{(damped oscillations)} \qquad (13.45)$$

The right side of Eq. (13.45) is always negative, whether v is positive or negative. This shows that E continuously decreases, though not at a uniform rate. The term $-b v_x^2 = (-b v_x) v_x$ (force times velocity) is the rate at which the damping force does (negative) work on the system (that is, the damping *power*). This equals the rate of change of the total mechanical energy of the system.

Similar behavior occurs in electric circuits containing inductance, capacitance, and resistance. There is a natural frequency of oscillation, and the resistance plays the role of the damping constant b. In such circuits it is usually desirable to

Upper cylinder attached to frame of car: remains relatively stationary

Piston

Viscous fluid

Pushed up

Pushed down

Lower cylinder attached to axle and wheel: moves up and down

13.26 An automobile shock absorber. The viscous fluid causes a damping force that depends on the relative velocity of the two ends of the unit. This helps control wheel bounce and jounce.

minimize damping, but damping can never be prevented completely. We will study these circuits in detail in Chapters 30 and 31.

Test Your Understanding

Consider again the mass-spring combination in Example 13.2 (Section 13.2), with $k = 200$ N/m and $m = 0.50$ kg. With no damping, this system has an angular frequency $\omega = 20$ rad/s. What is the angular frequency if there is damping with $b = 10$ kg/s? For what value of b is there critical damping?

13.8 | Forced Oscillations and Resonance

A damped oscillator left to itself will eventually stop moving altogether. But we can maintain a constant-amplitude oscillation by applying a force that varies with time in a periodic or cyclic way, with a definite period and frequency. As an example, consider your cousin Throckmorton on a playground swing. You can keep him swinging with constant amplitude by giving him a little push once each cycle. We call this additional force a **driving force.**

If we apply a periodically varying driving force with angular frequency ω_d to a damped harmonic oscillator, the motion that results is called a **forced oscillation** or a *driven oscillation.* It is different from the motion that occurs when the system is simply displaced from equilibrium and then left alone, in which case the system oscillates with a **natural angular frequency** ω' determined by m, k, and b, as in Eq. (13.43). In a forced oscillation, however, the angular frequency with which the mass oscillates is equal to the driving angular frequency ω_d. This does *not* have to be equal to the angular frequency ω' with which the system would oscillate without a driving force. If you grab the ropes of Throckmorton's swing, you can force the swing to oscillate with any frequency you like.

Suppose we force the oscillator to vibrate with an angular frequency ω_d that is nearly *equal* to the angular frequency ω' it would have with no driving force. What happens? The oscillator is naturally disposed to oscillate at $\omega = \omega'$, so we expect the amplitude of the resulting oscillation to be larger than when the two frequencies are very different. Detailed analysis and experiment shows that this is just what happens. The easiest case to analyze is a *sinusoidally* varying force, say, $F(t) = F_{max} \cos \omega_d t$. If we vary the frequency ω_d of the driving force, the amplitude of the resulting forced oscillation varies in an interesting way (Fig. 13.27). When there is very little damping (small b), the amplitude goes through a sharp peak as the driving angular frequency ω_d nears the natural oscillation angular frequency ω'. When the damping is increased (larger b), the peak becomes broader and smaller in height and shifts toward lower frequencies.

We could work out an expression that shows how the amplitude A of the forced oscillation depends on the frequency of a sinusoidal driving force, with maximum value F_{max}. That would involve more differential equations than we're ready for, but here is the result:

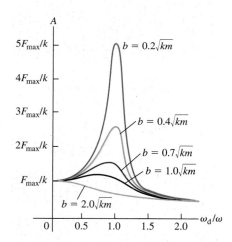

13.27 Graph of the amplitude A of forced oscillation as a function of the angular frequency ω_d of the driving force. The horizontal axis shows the ratio of ω_d to the angular frequency $\omega = \sqrt{k/m}$ of an undamped oscillator. Each curve has a different value of the damping constant b.

$$A = \frac{F_{max}}{\sqrt{(k - m\omega_d^2)^2 + b^2\omega_d^2}} \quad \text{(amplitude of a driven oscillator)} \quad (13.46)$$

When $k - m\omega_d^2 = 0$, the first term under the radical is zero, so A has a maximum near $\omega_d = \sqrt{k/m}$. The height of the curve at this point is proportional to $1/b$; the less damping, the higher the peak. At the low-frequency extreme, when $\omega_d = 0$, we get $A = F_{max}/k$. This corresponds to a *constant* force F_{max} and a constant displacement $A = F_{max}/k$ from equilibrium, as we might expect.

The fact that there is an amplitude peak at driving frequencies close to the natural frequency of the system is called **resonance.** Physics is full of examples of resonance; building up the oscillations of a child on a swing by pushing with a frequency equal to the swing's natural frequency is one. *A* vibrating rattle in a car that occurs only at a certain engine speed or wheel-rotation speed is an all-too-familiar example. Inexpensive loudspeakers often have an annoying boom or buzz when a musical note happens to coincide with the resonant frequency of the speaker cone or the speaker housing. *A* tuned circuit in a radio or television receiver responds strongly to waves having frequencies near its resonant frequency, and this fact is used to select a particular station and reject the others. We will study resonance in electric circuits in detail in Chapter 31.

Resonance in mechanical systems can be destructive. A company of soldiers once destroyed a bridge by marching across it in step; the frequency of their steps was close to a natural vibration frequency of the bridge, and the resulting oscillation had large enough amplitude to tear the bridge apart. Ever since, marching soldiers have been ordered to break step before crossing a bridge. Some years ago, vibrations of the engines of a particular airplane had just the right frequency to resonate with the natural frequencies of its wings. Large oscillations built up, and occasionally the wings fell off.

Nearly everyone has seen the film of the collapse of the Tacoma Narrows suspension bridge in 1940 (Fig. 13.28). This is usually cited as an example of resonance driven by the wind, but there's some doubt as to whether it should be called that. The wind didn't have to vary *periodically* with a frequency close to a natural frequency of the bridge. The air flow past the bridge was turbulent, and vortices were formed in the air with a regular frequency that depended on the flow speed. It is conceivable that this frequency may have coincided with a natural frequency of the bridge. But the cause may well have been something more subtle called a *self-excited oscillation,* in which the aerodynamic forces caused by a *steady* wind blowing on the bridge tended to displace it farther from equilibrium at times when it was already moving away from equilibrium. It is as though we had a damping force such as the $-bv_x$ term in Eq. (13.40) but with the sign reversed. Instead of draining mechanical energy away from the system, this anti-damping force pumps energy into the system, building up the oscillations to destructive amplitudes. The approximate differential equation is Eq. (13.41) with the sign of the *b* term reversed, and the oscillating solution is Eq. (13.42) with a *positive* sign in the exponent. You can see that we're headed for trouble. Engineers have learned how to stabilize suspension bridges, both structurally and aerodynamically, to prevent such disasters.

Test Your Understanding

Earlier we described how a certain airplane had an unwanted resonance between vibrations of the engines and vibrations of the wings. This problem was fixed by stiffening the wing structure. Use the ideas of resonance to explain how this solution worked.

13.28 The Tacoma Narrows Bridge collapsed four months and six days after it was opened for traffic. The main span was 2800 ft long and 39 ft wide, with 8-ft-high steel stiffening girders on both sides. The maximum amplitude of the torsional vibrations was 35°; the frequency was about 0.2 Hz.

SUMMARY

Periodic motion is motion that repeats itself in a definite cycle. It occurs whenever a body has a stable equilibrium position and a restoring force that acts when it is displaced from equilibrium. Period T is the time for one cycle. Frequency f is the number of cycles per unit time. Angular frequency ω is 2π times the frequency. (See Example 13.1)

$$f = \frac{1}{T} \qquad T = \frac{1}{f} \qquad (13.1)$$

$$\omega = 2\pi f = \frac{2\pi}{T} \qquad (13.2)$$

If the net force is a restoring force F_x that is directly proportional to the displacement x, the motion is called simple harmonic motion (SHM). In many cases this condition is satisfied if the displacement from equilibrium is small.

$$F_x = -kx \qquad (13.3)$$

$$a_x = \frac{F_x}{m} = -\frac{k}{m}x \qquad (13.4)$$

The circle of reference construction uses a rotating vector called a phasor, having a length equal to the amplitude of the motion. Its projection on the horizontal axis represents the actual motion of a body in simple harmonic motion.

The angular frequency, frequency, and period in SHM do not depend on the amplitude, but only on the mass m and force constant k. (See Examples 13.2, 13.6, and 13.7)

$$\omega = \sqrt{\frac{k}{m}} \qquad (13.10)$$

$$f = \frac{\omega}{2\pi} = \frac{1}{2\pi}\sqrt{\frac{k}{m}} \qquad (13.11)$$

$$T = \frac{1}{f} = 2\pi\sqrt{\frac{m}{k}} \qquad (13.12)$$

In SHM, the displacement, velocity, and acceleration are sinusoidal functions of time. The angular frequency is $\omega = \sqrt{k/m}$; the amplitude A and phase angle ϕ are determined by the initial position and velocity of the body. (See Example 13.3)

$$x = A\cos(\omega t + \phi) \qquad (13.13)$$

Energy is conserved in SHM. The total energy can be expressed in terms of the force constant k and amplitude A. (See Examples 13.4 and 13.5)

$$E = \frac{1}{2}mv_x^2 + \frac{1}{2}kx^2 = \frac{1}{2}kA^2 = \text{constant} \qquad (13.21)$$

In angular simple harmonic motion, the frequency and angular frequency are related to the moment of inertia I and the torsion constant κ.

$$\omega = \sqrt{\frac{\kappa}{I}} \quad \text{and}$$

$$f = \frac{1}{2\pi}\sqrt{\frac{\kappa}{I}} \qquad (13.24)$$

Balance wheel   Spring

Spring torque τ_z opposes angular displacement

A simple pendulum consists of a point mass m at the end of a massless string of length L. Its motion is approximately simple harmonic for sufficiently small amplitude; the angular frequency, frequency, and period then depend only on g and L, not on the mass or amplitude. (See Examples 13.8)

$$\omega = \sqrt{\frac{g}{L}} \qquad (13.32)$$

$$f = \frac{\omega}{2\pi} = \frac{1}{2\pi}\sqrt{\frac{g}{L}} \qquad (13.33)$$

$$T = \frac{2\pi}{\omega} = \frac{1}{f} = 2\pi\sqrt{\frac{L}{g}} \qquad (13.34)$$

A physical pendulum is a body suspended from an axis of rotation a distance d from its center of gravity. If the moment of inertia about the axis of rotation is I, the angular frequency and period for small-amplitude oscillations are independent of amplitude. (See Examples 13.9 and 13.10)

$$\omega = \sqrt{\frac{mgd}{I}} \qquad (13.38)$$

$$T = 2\pi\sqrt{\frac{I}{mgd}} \qquad (13.39)$$

When a damping force $F_x = -bv_x$ proportional to velocity is added to a simple harmonic oscillator, the motion is called a damped oscillation. For a relatively small damping force, the motion is sinusoidal with a decaying amplitude and an angular frequency ω' that is lower than it would be without damping. This situation occurs when $b < 2\sqrt{km}$, a condition called underdamping. When $b = 2\sqrt{km}$, the system is critically damped and no longer oscillates. When $b > 2\sqrt{km}$, the system is overdamped.

$$x = Ae^{-(b/2m)t}\cos\omega't \quad (13.42)$$

$$\omega' = \sqrt{\frac{k}{m} - \frac{b^2}{4m^2}} \qquad (13.43)$$

When a sinusoidally varying driving force is added to a damped harmonic oscillator, the resulting motion is called a forced oscillation. The amplitude is a function of driving frequency ω_d and reaches a peak at a driving frequency close to the natural frequency of the system. This behavior is called resonance.

$$A = \frac{F_{max}}{\sqrt{(k - m\omega_d^2)^2 + b^2\omega_d^2}} \qquad (13.46)$$

Key Terms

periodic motion (oscillation), 476
displacement, 477
restoring force, 477
amplitude, 477
cycle, 477
period, 477

frequency, 477
angular frequency, 478
simple harmonic motion (SHM), 479
harmonic oscillator, 479
circle of reference, 480
phasor, 480

phase angle, 483
simple pendulum, 495
physical pendulum, 496
damping, 499
damped oscillation, 499
critical damping, 500
overdamping, 501

underdamping, 501
driving force, 502
forced oscillation, 502
natural angular frequency, 502
resonance, 503

Your Notes

Answer to Chapter Opening Question

Neither—the clock would still keep time correctly. If its rod has negligible mass, then the pendulum is a simple pendulum and its period is independent of the mass [Eq. (13.34)]. If the rod's mass is included, the pendulum is a physical pendulum. Doubling its mass m also doubles its moment of inertia I, so the ratio I/m is unchanged and the period $T = 2\pi\sqrt{I/mgd}$ [Eq. (13.39)] remains the same.

Answers to Test Your Understanding Questions

Section 13.1 The amplitude is $A = 6.0$ cm (the maximum displacement from equilibrium). The period is $T = 5.00$ s, the frequency is $f = 1/T = 1/(5.00 \text{ s}) = 0.200$ Hz, and the angular frequency is $\omega = 2\pi f = 2\pi(0.200 \text{ Hz}) = 1.26$ rad/s.

Section 13.2 Comparing with Eq. (13.13), $x = A \cos(\omega t + \phi)$, we see that the amplitude $A = 0.050$ m, the phase angle $\phi = 2.5$ rad, and the angular frequency $\omega = 290$ rad/s. From Eq. (13.2), the period is $T = 2\pi/\omega = (2\pi \text{ rad})/(290 \text{ rad/s}) = 0.022$ s. At $t = 0$ the position is $x_0 = (0.050 \text{ m}) \cos(2.5 \text{ rad}) = -0.040$ m.

Section 13.3 To increase the energy $E = \frac{1}{2}kA^2$ by a factor of 2, the amplitude A must increase by a factor of $\sqrt{2}$. Because the motion is SHM, changing the amplitude has no effect on the frequency.

Section 13.4 From Eq. (13.4), $a_x = -(k/m)x$. With $k = 0.829$ N/m, $m = 6.63 \times 10^{-26}$ kg, and an initial displacement of 1.00×10^{-11} m, the magnitude of the acceleration is $[(0.829 \text{ N/m})/(6.63 \times 10^{-26} \text{ kg})] \times (1.00 \times 10^{-11} \text{ m}) = 1.25 \times 10^{14}$ m/s^2. Microscopic objects can undergo truly immense accelerations.

Section 13.5 From Eq. (13.34), $g = 4\pi^2 L/T^2 = 4\pi^2(1.000 \text{ m})/(2.000 \text{ s})^2 = 9.870$ m/s^2.

Section 13.6 The distance from the pivot to the center of gravity (the center of the rod) is $d = 0$. Hence the period $T = 2\pi\sqrt{I/mgd}$ is infinite. In other words, when it is displaced, the rod takes an infinite amount of time to move back. The rod is in equilibrium for *any* value of the angle θ in Fig. 13.22, so there is no restoring torque and no tendency for the rod to move.

Section 13.7 From Eq. (13.43), the angular frequency with $b = 10$ kg/s is

$$\omega' = \sqrt{\frac{k}{m} - \frac{b^2}{4m^2}}$$

$$= \sqrt{\frac{200 \text{ N/m}}{0.50 \text{ kg}} - \frac{(10 \text{ kg/s})^2}{4(0.50 \text{ kg})^2}} = 17 \text{ rad/s}$$

which is less than with zero damping. Critical damping requires

$$b = 2\sqrt{km} = 2\sqrt{(200 \text{ N/m})(0.50 \text{ kg})} = 20 \text{ kg/s}$$

Section 13.8 The stiffened wing had a higher force constant k and hence a higher natural frequency. As a result, its natural frequency no longer matched the driving frequency of the engines, and resonance no longer occurred.

Discussion Questions

Q13.1 An object is moving with SHM of amplitude A on the end of a spring. If the amplitude is doubled, what happens to the total distance the object travels in one period? What happens to the period? What happens to the maximum speed of the object? Discuss how these answers are related.

Q13.2 Think of several examples in everyday life of motion that are at least, approximately simple harmonic. In what respects does each differ from SHM?

Q13.3 Does a tuning fork or similar tuning instrument undergo simple harmonic motion? Why is this a crucial question for musicians?

Q13.4 A highly elastic "superball" bouncing on a hard floor has a motion that is approximately periodic. In what ways is the motion similar to SHM? In what ways is it different?

Q13.5 If a uniform spring is cut in half, what is the force constant of each half? Justify your answer. How would the frequency of SHM using a half-spring differ from that using the same mass and the entire spring?

Q13.6 The analysis of SHM in this chapter ignored the mass of the spring. How does the spring's mass change the characteristics of the motion?

Q13.7 Two identical gliders on an air track are connected by an ideal spring. Could such a system undergo SHM? Explain. How would the period compare with that of a single glider attached to a spring whose other end is rigidly attached to a stationary object? Explain.

Q13.8 You are captured by aliens, taken into their ship, and drugged. You awake some time later and find yourself locked in a small room with no windows. All they have left you with is your digital watch, your school ring, and your long silver-chain necklace. Explain how you can determine whether or not you are still on earth or if you have been transported to Mars.

Q13.9 The system shown in Fig. 13.16 is mounted in an elevator, which accelerates upward with constant acceleration. Does the period increase, decrease, or remain the same? Justify your answer.

Q13.10 If a pendulum has a period of 2.5 s on earth, what would be its period in a space station orbiting the earth? If a mass hung from a vertical spring has a period of 5.0 s on earth, what would be its period in the space station? Justify each of your answers.

Q13.11 A pendulum is mounted in an elevator that accelerates upward with constant acceleration. Does the period increase, decrease, or remain the same? Explain your answer.

Q13.12 What should you do to the length of the string of a simple pendulum to a) double its frequency? b) double its period? c) double its angular frequency?

Q13.13 If a pendulum clock is taken to a mountaintop, does it gain or lose time, assuming it is correct at a lower elevation? Explain your answer.

Q13.14 When the amplitude of a simple pendulum increases, should its period increase or decrease? Give a qualitative argument; do not rely on Eq. (13.35). Is your argument also valid for a physical pendulum?

Q13.15 Why do short dogs (like Chihuahuas) walk with quicker strides than do tall dogs (like Great Danes)?

Q13.16 At what point in the motion of a simple pendulum is the string tension greatest? Least? In each case give the reasoning behind your answer.

Q13.17 Could a standard of time be based on the period of a certain standard pendulum? What advantages and disadvantages would such a standard have compared to the actual present-day standard discussed in Section 1.3?

Q13.18 The frequency with which dogs pant is the natural frequency of their respiratory system. Why do they choose this frequency?

Q13.19 In designing structures in an earthquake-prone region, how should the natural frequencies of oscillation of a structure relate to typical earthquake frequencies? Why? Should the structure have a large or small amount of damping?

Exercises

Section 13.1 Describing Oscillation

13.1 A piano string sounds a middle A by vibrating primarily at 220 Hz. a) Calculate its period and angular frequency. b) Calculate the period and angular frequency for a soprano singing a "high A," two octaves up, which is four times the frequency of the piano string.

13.2 If an object on a horizontal frictionless surface is attached to a spring, displaced, and then released, it will oscillate. If it is displaced 0.120 m from its equilibrium position and released with zero initial speed, then after 0.800 s its displacement is found to be 0.120 m on the opposite side, and it has passed the equilibrium position once during this interval. Find a) the amplitude; b) the period; c) the frequency.

13.3 The tip of a tuning fork goes through 440 complete vibrations in 0.500 s. Find the angular frequency and the period of the motion.

13.4 The displacement of an oscillating object as a function of time is shown in Fig. 13.29. What is a) the frequency; b) the amplitude; c) the period?

Figure 13.29 Exercise 13.4.

Section 13.2 Simple Harmonic Motion

13.5 A machine part is undergoing SHM with a frequency of 5.00 Hz and amplitude 1.80 cm. How long does it take the part to go from $x = 0$ to $x = -1.80$ cm?

13.6 In a physics lab, you attach a 0.200-kg air-track glider to the end of an ideal spring of negligible mass and start it oscillating. The elapsed time from when the glider first moves through the equilibrium point to the second time it moves through that point is 2.60 s. Find the spring's force constant.

13.7 A body of unknown mass is attached to an ideal spring with force constant 120 N/m. It is found to vibrate with a frequency of 6.00 Hz. Find a) the period; b) the angular frequency; c) the mass of the body.

13.8 A harmonic oscillator is made by using a 0.600-kg frictionless block and an ideal spring of unknown force constant. The oscillator is found to have a period of 0.150 s. Find the force constant of the spring.

13.9 A harmonic oscillator has mass 0.500 kg and an ideal spring with force constant 140 N/m. Find a) the period; b) the frequency; c) the angular frequency.

13.10 Substitute the following equations, in which A, ω, and β are constant, into Eq. (13.4) to see if they describe SHM. If so, what must ω equal? a) $x = A \sin(\omega t + \beta)$. b) $x = A\omega t^2 + \beta$. c) $x = Ae^{i(\omega t + \beta)}$, where $i = \sqrt{-1}$.

13.11 Jerk. A guitar string vibrates at a frequency of 440 Hz. A point at its center moves in SHM with an amplitude of 3.0 mm and a phase angle of zero. a) Write an equation for the position of the center of the string as a function of time. b) What are the maximum values of the magnitudes of the velocity and acceleration of the center of the string? c) The derivative of the acceleration with respect to time is a quantity called the *jerk*. Write an equation for the jerk of the center of the string as a function of time, and find the maximum value of the magnitude of the jerk.

13.12 A 2.00-kg frictionless block is attached to an ideal spring with force constant 300 N/m. At $t = 0$ the spring is neither stretched nor compressed and the block is moving in the negative direction at 12.0 m/s. Find a) the amplitude and b) the phase angle. c) Write an equation for the position as a function of time.

13.13 Repeat Exercise 13.12, but assume that at $t = 0$ the block has velocity -4.00 m/s and displacement $+0.200$ m.

13.14 The point of the needle of a sewing machine moves in SHM along the x-axis with a frequency of 2.5 Hz. At $t = 0$ its position and velocity components are $+1.1$ cm and -15 cm/s. a) Find the acceleration component of the needle at $t = 0$. b) Write equations giving the position, velocity, and acceleration components of the point as a function of time.

13.15 An object is undergoing simple harmonic motion with period 1.200 s and amplitude 0.600 m. At $t = 0$ the object is at $x = 0$. How far is the object from the equilibrium position when $t = 0.480$ s?

13.16 A 42.5-kg chair is attached to a spring and allowed to oscillate. When the chair is empty the chair takes 1.30 s to make one complete vibration. But with a person sitting in it, with her feet off the floor, the chair now takes 2.54 s for one cycle. What is the mass of the person?

13.17 A 0.400-kg object undergoing SHM has $a_x = -2.70$ m/s^2, when $x = 0.300$ m. What is the time for one oscillation?

13.18 A 0.500-kg mass on a spring has velocity as a function of time given by $v_x(t) = (3.60 \text{ cm/s}) \sin[(4.71 \text{ s}^{-1})t - \pi/2]$. What is a) the period? b) the amplitude? c) the maximum acceleration of the mass?

13.19 A 1.50-kg mass on a spring has displacement as a function of time given by the equation $x(t) = (7.40 \text{ cm}) \cos[(4.16 \text{ s}^{-1})t - 2.42]$. Find a) the time for one complete vibration; b) the force constant of the spring; c) the maximum speed of the mass; d) the maximum force on the mass; e) the position, speed, and acceleration of the mass at $t = 1.00$ s, and the force on the mass at that time.

13.20 An object is undergoing simple harmonic motion with period 0.300 s and amplitude 6.00 cm. At $t = 0$ the object is instantaneously at rest at $x = 6.00$ cm. Calculate the time it takes the object to go from $x = 6.00$ cm to $x = -1.50$ cm.

Section 13.3 Energy in Simple Harmonic Motion

13.21 Derive Eq. (13.19) a) from Eqs. (13.14) and (13.17); b) from conservation of energy, Eq. (13.21).

13.22 A tuning fork labeled 392 Hz has the tip of each of its two prongs vibrating with an amplitude of 0.600 mm. a) What is the maximum speed of the tip of a prong? b) A housefly (*Musca domestica*) with mass 0.0270 g is holding on to the tip of one of the prongs. As the prong vibrates, what is the fly's maximum kinetic energy? Assume that the fly's mass has a negligible effect on the frequency of oscillation.

13.23 A harmonic oscillator has angular frequency ω and amplitude A. a) What are the magnitudes of the displacement and velocity when the elastic potential energy is equal to the kinetic energy? (Assume that $U = 0$ at equilibrium.) b) How often does this occur in each cycle? What is the time between occurrences? c) At an instant when the displacement is equal to $A/2$, what fraction of the total energy of the system is kinetic and what fraction is potential?

13.24 A 0.500-kg glider, attached to the end of an ideal spring with force constant $k = 450$ N/m, undergoes simple harmonic motion with an amplitude 0.040 m. Compute a) the maximum speed of the glider; b) the speed of the glider when it is at $x = -0.015$ m; c) the magnitude of the maximum acceleration of the glider; d) the acceleration of the glider at $x = -0.015$ m; e) the total mechanical energy of the glider at any point in its motion.

13.25 A cheerleader waves her pom-pom in SHM with an amplitude of 18.0 cm and a frequency of 0.850 Hz. Find a) the maximum magnitude of the acceleration and of the velocity; b) the acceleration and speed when the pom-pom's coordinate is $x = +9.0$ cm; c) the time required to move from the equilibrium position directly to a point 12.0 cm away. d) Which of the quantities asked for in parts (a), (b), and (c) can be found using the energy approach used in Section 13.3, and which cannot? Explain.

13.26 For the situation described in part (a) of Example 13.5, what should be the value of the putty mass m so that the amplitude after the collision is one-half the original amplitude? For this value of m, what fraction of the original mechanical energy is converted into heat?

13.27 A 0.150-kg toy is undergoing SHM on the end of a horizontal spring with force constant $k = 300$ N/m. When the object is 0.0120 m from its equilibrium position, it is observed to have a speed of 0.300 m/s. What is a) the total energy of the object at any point of its motion? b) the amplitude of the motion? c) the maximum speed attained by the object during its motion?

13.28 You are watching an object that is moving in SHM. When the object is displaced 0.600 m to the right of its equilibrium position, it has a velocity of 2.20 m/s to the right and an acceleration of 8.40 m/s² to the left. How much farther from this point will the object move before it stops momentarily and then starts to move back to the left?

13.29 An object moves in SHM with a period of 0.500 s. The object's maximum acceleration is 6.40 m/s². What is its maximum speed?

Section 13.4 Applications of Simple Harmonic Motion

13.30 The scale of a spring balance reading from zero to 200 N is 12.5 cm long. A fish hanging from the bottom of the spring oscillates vertically at 2.60 Hz. What is the mass of the fish? You can ignore the mass of the spring.

13.31 A proud deep-sea fisherman hangs a 65.0-kg fish from an ideal spring having negligible mass. The fish stretches the spring 0.120 m. a) What is the force constant of the spring? b) What is the period of oscillation of the fish if it is pulled down and released?

13.32 A thrill-seeking cat with mass 4.00 kg is attached by a harness to an ideal spring of negligible mass and oscillates vertically in SHM. The amplitude is 0.050 m, and at the highest point of the motion the spring has its natural unstretched length. Calculate the elastic potential energy of the spring (take it to be zero for the unstretched spring), the kinetic energy of the cat, the gravitational potential energy of the system relative to the lowest point of the motion, and the sum of these three energies when the cat is a) at its highest point; b) at its lowest point; c) at its equilibrium position.

13.33 A 2.00-kg block of Cheddar cheese hangs from an ideal spring of negligible mass. When displaced from equilibrium and released, the cheese oscillates with a period of 0.400 s. How much is the spring stretched when the cheese hangs in equilibrium (at rest)?

13.34 Your boss at the Cut-Rate Cuckoo Clock Company asks you what would happen to the frequency of the angular SHM of the balance wheel if it had the same density and the same coil spring (thus the same torsion constant), but all the balance wheel dimensions were made one-third as great to save material. a) What do you tell her? b) By what factor would the torsion constant need to be changed to make the smaller balance wheel oscillate at the original frequency?

13.35 A certain alarm clock ticks four times each second, each tick representing half a period. The balance wheel consists of a thin rim with radius 0.55 cm, connected to the balance staff by thin spokes of negligible mass. The total mass of the balance wheel is 0.90 g. a) What is the moment of inertia of the balance wheel about its shaft? b) What is the torsion constant of the hairspring?

13.36 A thin metal disk with mass 2.00×10^{-3} kg and radius 2.20 cm is attached at its center to a long fiber (Fig. 13.30). The disk, when twisted and released, oscillates with a period of 1.00 s. Find the torsion constant of the fiber.

Figure 13.30 Exercise 13.36.

13.37 You want to find the moment of inertia of a complicated machine part about an axis through its center of mass. You suspend it from a wire along this axis. The wire has a torsion constant of 0.450 N·m/rad. You twist the part a small amount about this axis and let it go, timing 125 oscillations in 265 s. What is the moment of inertia you want to find?

13.38 The balance wheel of a watch vibrates with an angular amplitude Θ, angular frequency ω, and phase angle $\phi = 0$. a) Find expressions for the angular velocity $d\theta/dt$ and angular acceleration $d^2\theta/dt^2$ as functions of time. b) Find the balance wheel's angular velocity and angular acceleration when its angular displacement is Θ, and when its angular displacement is $\Theta/2$ and θ is decreasing. (*Hint:* Sketch a graph of θ versus t.)

*13.39 For the van der Waals interaction with potential energy function given by Eq. (13.25), show that when the magnitude of the displacement x from equilibrium $(r = R_0)$ is small, the potential energy can be written approximately as $U \approx \frac{1}{2}kx^2 - U_0$. [*Hint:* In Eq. (13.25), let $r = R_0 + x$ and $u = x/R_0$. Then approximate $(1 + u)^n$ by the first *three* terms in Eq. (13.28).] How does k in this equation compare with the force constant in Eq. (13.29) for the force?

*13.40 When displaced from equilibrium, the two hydrogen atoms in an H_2 molecule are acted on by a restoring force $F_x = -kx$ with $k = 580$ N/m. Calculate the oscillation frequency of the H_2 molecule. (*Hint:* The mass of a hydrogen atom is 1.008 atomic mass units, or 1 u; see Appendix E. As in Example 13.7 of Section 13.4, use $m/2$ instead of m in the expression for f.)

Section 13.5 The Simple Pendulum

13.41 A Pendulum on Mars. A certain simple pendulum has a period on the earth of 1.60 s. What is its period on the surface of Mars, where $g = 3.71$ m/s^2?

13.42 You pull a simple pendulum of length 0.240 m to the side through an angle of 3.50° and release it. a) How much time does it take the pendulum bob to reach its highest speed? b) How much time does it take if the pendulum is released at an angle of 1.75° instead of 3.50°?

13.43 An apple weighs 1.00 N. When you hang it from the end of a long spring of force constant 1.50 N/m and negligible mass, it bounces up and down in SHM. If you stop the bouncing and let the apple swing from side to side through a small angle, the frequency of this simple pendulum is half the bounce frequency. (Because the angle is small, the back and forth swings do not cause any appreciable change in the length of the spring.) What is the unstretched length of the spring (with the apple removed)?

13.44 A small sphere with mass m is attached to a massless rod of length L that is pivoted at the top, forming a simple pendulum. The pendulum is pulled to one side so that the rod is at an angle Θ from the vertical, and released from rest. a) In a diagram, show the pendulum just after it is released. On the diagram, draw vectors representing the *forces* acting on the small sphere and the *acceleration* of the sphere. Accuracy counts! At this point, what is the linear acceleration of the sphere? b) Repeat part (a) for the instant when the pendulum rod is at an angle $\Theta/2$ from the vertical. c) Repeat part (a) for the instant when the pendulum rod is vertical. At this point, what is the linear speed of the sphere?

13.45 After landing on an unfamiliar planet, a space explorer constructs a simple pendulum of length 50.0 cm. She finds that the pendulum makes 100 complete swings in 136 s. What is the value of g on this planet?

Section 13.6 The Physical Pendulum

13.46 We want to support a thin hoop by a horizontal nail and have the hoop make one complete small-angle oscillation each 2.0 s. What must the hoop's radius be?

13.47 Show that the expression for the period of a physical pendulum reduces to that of a simple pendulum if the physical pendulum consists of a particle with mass m on the end of a massless string of length L.

13.48 A 1.80-kg monkey wrench is pivoted 0.250 m from its center of mass and allowed to swing as a physical pendulum. The period for small-angle oscillations is 0.940 s. a) What is the moment of inertia of the wrench about an axis through the pivot? b) If the wrench is initially displaced 0.400 rad from its equilibrium position, what is the angular speed of the wrench as it passes through the equilibrium position?

13.49 A holiday ornament in the shape of a hollow sphere with mass $M = 0.015$ kg and radius $R = 0.050$ m is hung from a tree limb by a small loop of wire attached to the surface of the sphere. If the ornament is displaced a small distance and released, it swings back and forth as a physical pendulum. Calculate its period. (You can ignore friction at the pivot. The moment of inertia of the sphere about the pivot at the tree limb is $5MR^2/3$.)

13.50 A 1.80-kg connecting rod from a car engine is pivoted about a horizontal knife edge as shown in Fig. 13.31. The center of gravity of the rod was located by balancing and is 0.200 m from the pivot. When it is set into small amplitude oscillation, the rod makes 100 complete swings in 120 s. Calculate the moment of inertia of the rod about the rotation axis through the pivot.

Figure 13.31 Exercise 13.50.

$d = 0.200$ m

cg

Section 13.7 Damped Oscillations

13.51 An unhappy 0.300-kg mouse, moving on the end of a spring with force constant $k = 2.50$ N/m is acted on by a damping force $F_x = -bv_x$. a) If the constant b has the value 0.900 kg/s, what is the frequency of oscillation of the mouse? b) For what value of the constant b will the motion be critically damped?

13.52 A 50.0-g hard-boiled egg moves on the end of a spring with force constant $k = 25.0$ N/m. Its initial displacement is 0.300 m. A damping force $F_x = -bv_x$ acts on the egg, and the amplitude of the motion decreases to 0.100 m in 5.00 s. Calculate the magnitude of the damping constant b.

13.53 The motion of an underdamped oscillator is described by Eq. (13.42). Let the phase angle ϕ be zero. a) According to this equation, what is the value of x at $t = 0$? b) What are the magnitude and direction of the velocity at $t = 0$? What does the result tell you about the slope of the graph of x versus t near $t = 0$? c) Obtain an expression for the acceleration a_x at $t = 0$. For what value or range of values of the damping constant b (in terms of k and m) is the acceleration at $t = 0$ negative, zero, and positive? Discuss each case in terms of the shape of the graph of x versus t near $t = 0$.

Section 13.8 Forced Oscillations and Resonance

13.54 A sinusoidally varying driving force is applied to a damped harmonic oscillator of force constant k and mass m. If the damping constant has a value b_1, the amplitude is A_1 when the driving angular frequency equals $\sqrt{k/m}$. In terms of A_1, what is the amplitude for the same driving frequency and the same driving force amplitude F_{max}, if the damping constant is a) $3b_1$? b) $b_1/2$?

13.55 A sinusoidally varying driving force is applied to a damped harmonic oscillator. a) What are the units of the damping constant b? b) Show that the quantity $\sqrt{km}$ has the same units as b. c) In terms of F_{max} and k, what is the amplitude for $\omega_d = \sqrt{k/m}$ when i) $b = 0.2\sqrt{km}$? ii) $b = 0.4\sqrt{km}$? Compare your results to Fig. 13.27.

13.56 An experimental package and its support structure which are to be placed on board the International Space Station act as an underdamped spring-mass system with a force constant of 2.1×10^6 N/m and mass 108 kg. A NASA requirement is that resonance for forced oscillations not occur for any frequency below 35 Hz. Does this package meet the requirement?

Problems

13.57 SHM in a Car Engine.
The motion of the piston of an automobile engine (Fig. 13.32) is approximately simple harmonic. a) If the stroke of an engine (twice the amplitude) is 0.100 m and the engine runs at 3500 rev/min, compute the acceleration of the piston at the

Figure 13.32 Problem 13.57.

endpoint of its stroke. b) If the piston has mass 0.450 kg, what net force must be exerted on it at this point? c) What are the speed and kinetic energy of the piston at the midpoint of its stroke? d) What average power is required to accelerate the piston from rest to the speed found in part (c)? e) If the engine runs at 7000 rev/min, what are the answers to parts (b), (c), and (d)?

13.58 Four passengers with combined mass 250 kg compress the springs of a car with worn-out shock absorbers by 4.00 cm when they enter it. Model the car and passengers as a single body on a single ideal spring. If the loaded car has a period of vibration of 1.08 s, what is the period of vibration of the empty car?

13.59 A glider is oscillating in SHM on an air track with an amplitude A_1. You slow it so that its amplitude is halved. What happens to its a) period, frequency, and angular frequency? b) total mechanical energy? c) maximum speed? d) speed at $x = \pm A_1/4$? e) potential and kinetic energies at $x = \pm A_1/4$?

13.60 You hang an unknown weight from the end of a spring, moving your hand down slowly so that when the weight comes to equilibrium, it has stretched the spring a distance L. If the spring has negligible mass, prove that the weight can oscillate in SHM with the same period as a simple pendulum of length L.

13.61 A child with poor table manners is sliding his 250-g dinner plate back and forth in SHM with an amplitude of 0.100 m on a horizontal surface. At a point 0.060 m away from equilibrium, the

speed of the plate is 0.300 m/s. a) What is the period? b) What is the displacement when the speed is 0.160 m/s? c) In the center of the dinner plate is a 10.0-g carrot slice. If the carrot slice is just on the verge of slipping at the end point of the path, what is the coefficient of static friction between the carrot slice and the plate?

13.62 No Bull! While visiting friends at Cal State Chico, you pay a visit to the Crazy Horse Saloon. This fine establishment features a 200-kg mechanical bucking bull, that has a mechanism that makes it move vertically in simple harmonic motion. Whether the "bull" has a rider or not, it moves with the same amplitude (0.250 m) and frequency (1.50 Hz). After watching other saloon patrons hold on to the "bull" while riding, you (mass 75.0 kg) decide to ride it the macho way by *not* holding on. No one is terribly surprised when you come out of the saddle. Later, while waiting for your bruises and pride to heal, you pass the time by calculating the following: a) You leave the saddle when it is moving upward. What is the magnitude of the downward acceleration of the saddle when you lose contact with it? b) How high is the top surface of the saddle above its equilibrium position when you first become airborne? c) How fast are you moving upward when you leave the saddle? d) You are in free fall until you return to the saddle. Show that this happens 0.538 s later. e) What is your speed relative to the saddle at the instant you return?

13.63 A block with mass M rests on a frictionless surface and is connected to a horizontal spring of force constant k. The other end of the spring is attached to a wall (Fig. 13.33). A second block with mass m rests on top of the first block. The coefficient of static friction between the blocks is μ_s. Find the *maximum* amplitude of oscillation such that the top block will not slip on the bottom block.

Figure 13.33 Problem 13.63.

13.64 A 10.0-kg mass is traveling to the right with a speed of 2.00 m/s on a smooth, horizontal surface when it collides with and sticks to a second 10.0-kg mass that is initially at rest but is attached to a light spring with force constant 80.0 N/m. a) Find the frequency, amplitude, and period of the subsequent oscillations. b) How long does it take the system to return the first time to the position it had immediately after the collision?

13.65 A block with mass m_1 attached to a horizontal spring that has force constant k, is moving with SHM that has amplitude A_1 and period T_1. a) At the instant that the block passes through its equilibrium position, it suddenly splits into two identical pieces. One piece remains attached to the spring; the other is quickly pushed out of the way. In terms of A_1 and T_1, what are the new amplitude and period of the SHM after the block has split? b) Repeat part (a), for the situation where the block splits when it is at $x = A_1$. c) Compare your results to those of Example 13.5 (Section 13.3) and comment on the similarities and differences.

13.66 An object with mass 0.200 kg is acted on by an elastic restoring force with force constant 10.0 N/m. a) Graph elastic potential energy U as a function of displacement x over a range of x from -0.300 m to $+0.300$ m. On your graph, let 1 cm = 0.05 J vertically and 1 cm = 0.05 m horizontally. The object is set into oscillation with an initial potential energy of 0.140 J and an initial kinetic energy of 0.060 J. Answer the following questions by referring to the graph. b) What is the amplitude of oscillation? c) What is the potential energy when the displacement is one-half the amplitude? d) At what displacement are the kinetic and potential energies equal? e) What is the value of the phase angle ϕ if the initial velocity is positive and the initial displacement is negative?

13.67 Figure 13.16 shows a body with mass m hanging from a spring of force constant k. The positive x-axis is upward and $x = 0$ is the equilibrium position of the body. a) Show that when the body is at coordinate x, the elastic potential energy in the spring is $U_{el} = \frac{1}{2}k(\Delta l - x)^2$. b) Let $x = x_0$ be the coordinate at which the gravitational potential energy is zero. Show that the total potential energy is $U = \frac{1}{2}kx^2 + \frac{1}{2}k(\Delta l)^2 - mgx_0$. c) The expression for the total potential energy in part (b) is of the form $U = \frac{1}{2}kx^2 + C$, where the constant C equals $\frac{1}{2}k(\Delta l)^2 - mgx_0$. Explain why the behavior of the system does not depend on the value of this constant, so that vertical SHM is fundamentally no different than horizontal SHM, for which $U = \frac{1}{2}kx^2$.

13.68 A hanging wire is 1.80 m in length. When a 60.0-kg steel ball is suspended from the wire, the wire stretches by 2.00 mm. If the ball is pulled down a small additional distance and released, at what frequency will it vibrate? Assume that the stress on the wire is less than the proportional limit (see Section 11.5).

13.69 A 5.00-kg partridge is suspended from a pear tree by an ideal spring of negligible mass. When the partridge is pulled down 0.100 m below its equilibrium position and released, it vibrates with a period of 4.20 s. a) What is its speed as it passes through the equilibrium position? b) What is its acceleration when it is 0.050 m above the equilibrium position? c) When it is moving upward, how much time is required for it to move from a point 0.050 m below its equilibrium position to a point 0.050 m above it? d) The motion of the partridge is stopped, and then it is removed from the spring. How much does the spring shorten?

13.70 A 0.0200-kg bolt moves with SHM that has an amplitude of 0.240 m and a period of 1.500 s. The displacement of the bolt is $+0.240$ m when $t = 0$. Compute a) the displacement of the bolt when $t = 0.500$ s; b) the magnitude and direction of the force acting on the bolt when $t = 0.500$ s; c) the minimum time required for the bolt to move from its initial position to the point where $x = -0.180$ m; d) the speed of the bolt when $x = -0.180$ m.

13.71 SHM of a Butcher's Scale. A spring of negligible mass and force constant $k = 400$ N/m is hung vertically, and a 0.200-kg pan is suspended from its lower end. A butcher drops a 2.2-kg steak onto the pan from a height of 0.40 m. The steak makes a totally inelastic collision with the pan and sets the system into vertical simple harmonic motion. What is a) the speed of the pan and steak immediately after the collision? b) the amplitude of the subsequent motion? c) the period of that motion?

13.72 A uniform beam is suspended horizontally by two identical, vertical springs that are attached between the ceiling and each end of the beam. The beam has mass 225 kg and a 175-kg sack of gravel sits on the middle of it. The beam is oscillating in SHM, with an amplitude of 40.0 cm and a frequency of 0.600 cycles/s. a) The sack of gravel falls off the beam when the beam has its maximum upward displacement. What are the frequency and amplitude of the subsequent SHM of the beam? b) If the gravel instead falls off when the beam has its maximum speed, what are the frequency and amplitude of the subsequent SHM of the beam?

13.73 On the planet Newtonia, a simple pendulum having a bob with mass 1.25 kg and a length of 185.0 cm takes 1.42 s, when released from rest, to swing through an angle of 12.5°, where it again has zero speed. The circumference of Newtonia is measured to be 51,400 km. What is the mass of the planet Newtonia?

13.74 A 40.0-N force stretches a vertical spring 0.250 m. a) What mass must be suspended from the spring so that the system will oscillate with a period of 1.00 s? b) If the amplitude of the motion is 0.050 m and the period is that specified in part (a), where is the object and in what direction is it moving 0.35 s after it has passed the equilibrium position, moving downward? c) What force (magnitude and direction) does the spring exert on the object when it is 0.030 m below the equilibrium position, moving upward?

13.75 Don't Miss the Boat. While on a visit to Minnesota ("Land of 10,000 Lakes"), you sign up to take an excursion around one of the larger lakes. When you go to the dock where the 1500-kg boat is tied, you find that the boat is bobbing up and down in the waves, executing simple harmonic motion with amplitude 20 cm. The boat takes 3.5 s to make one complete up-and-down cycle. When the boat is at its highest point, its deck is at the same height as the stationary dock. As you watch the boat bob up and down, you (mass 60 kg) begin to feel a bit woozy, due in part to the previous night's dinner of lutefisk. As a result, you refuse to board the boat unless the level of the boat's deck is within 10 cm of the dock level. How much time do you have to board the boat comfortably during each cycle of up-and-down motion?

13.76 An interesting though highly impractical example of oscillation is the motion of an object dropped down a hole that extends from one side of the earth, through its center, to the other side. With the assumption (not very realistic) that the earth is a sphere of uniform density, prove that the motion is simple harmonic and find the period. *Note:* The gravitational force on the object as a function of the object's distance r from the center of the earth was derived in Example 12.10 (Section 12.6). The motion is simple harmonic if the acceleration a_x and the displacement from equilibrium x are related by Eq. (13.8), and the period is then $T = 2\pi/\omega$.]

13.77 Let t_1 be the time it takes a body undergoing SHM to move from $x = 0$ (when $t = 0$) to $x = A$. Obtain an equation for t_1 in the following way. In Eq. (13.22) replace v_x with dx/dt. Separate variables by writing all factors containing x on one side and all those containing t on the other side so that each side can be integrated. Integrate the equation with limits on t of 0 to t_1 and limits on x of 0 to A and thereby obtain an equation for t_1. How does t_1 compare to the period T?

13.78 For a certain oscillator the net force on the body with mass m is given by $F_x = -cx^3$. a) What is the potential energy function for this oscillator, if we take $U = 0$ at $x = 0$? b) One-quarter of a period is the time for the body to move from $x = 0$ to $x = A$. Calculate this time and hence the period. [*Hint:* Begin with

Eq. (13.20), modified to include the potential energy function you found in part (a), and solve for the velocity v_x as a function of x. Then replace v_x with dx/dt and follow the steps outlined in Problem 13.77. In the x-integral make the change of variable $u = x/A$. The resulting integral can be evaluated by numerical methods on a computer and has the value $\int_0^1 du/\sqrt{1 - u^4} = 1.31$.] c) According to the result you obtained in part (b), does the period depend on the amplitude A of the motion? Are the oscillations simple harmonic?

13.79 Consider the circle of reference shown in Fig. 13.5. The x-component of the velocity of Q is the velocity of P. Compute this component and show that the velocity of P is as given by Eq. (13.15).

13.80 To measure g in an unorthodox manner, a student places a ball bearing on the concave side of a lens (Fig. 13.34). She attaches the lens to a simple harmonic oscillator (actually a small stereo speaker) whose amplitude is A and whose frequency f can be varied. She can measure both A and f with a strobe light. a) If the ball bearing has mass m, find the normal force exerted by the lens on the ball bearing as a function of time. Your result should be in terms of A, f, m, g, and a phase angle ϕ. b) The frequency is slowly increased. When it reaches a value f_b, the ball is heard to bounce. What is g in terms of A and f_b?

Figure 13.34 Problem 13.80.

*13.81 Diatomic Molecule. Two identical atoms in a diatomic molecule vibrate as harmonic oscillators. However, their center of mass, midway between them, remains at rest. a) Show that at any instant, the momenta of the atoms relative to the center of mass are $\vec{p}$ and $-\vec{p}$. b) Show that the total kinetic energy K of the two atoms at any instant is the same as that of a single object with mass $m/2$ with a momentum of magnitude p. (Use $K = p^2/2m$.) This result shows why $m/2$ should be used in the expression for f in Example 13.7 (Section 13.4). c) If the atoms are not identical, but have masses m_1 and m_2, show that the result of part (a) still holds and the single object's mass in part (b) is $m_1 m_2/(m_1 + m_2)$. The quantity $m_1 m_2/(m_1 + m_2)$ is called the *reduced mass* of the system.

*13.82 An approximation for the potential energy of a KCl molecule is $U = A[(R_0^7/8r^8) - 1/r]$, where $R_0 = 2.67 \times 10^{-10}$ m and $A = 2.31 \times 10^{-28}$ J·m. Using this approximation: a) Show that the radial component of the force on each atom is $F_r = A[(R_0^7/r^9) - 1/r^2]$. b) Show that R_0 is the equilibrium separation. c) Find the minimum potential energy. d) Use $r = R_0 + x$ and the first two terms of the binomial theorem (Eq. 13.28) to show that $F_r \approx -(7A/R_0^3)x$, so that the molecule's force constant is $k = 7A/R_0^3$. e) With both the K and Cl atoms vibrating in opposite directions on opposite sides of the molecule's center of mass, $m_1 m_2/(m_1 + m_2) = 3.06 \times 10^{-26}$ kg is the mass to use in calculating the frequency (see Problem 13.81). Calculate the frequency of small-amplitude vibrations.

*13.83 Three positively charged particles are maintained in a straight line. The end particles are identical and are held fixed at a distance $2R_0$ apart. The potential energy for the force acting on the center particle can be written as $U = A[1/r - 1/(r - 2R_0)]$,

where A is a constant and r is the distance from the left-hand particle to the center particle. a) Show that the force on the center particle is $F_r = A[1/r^2 - 1/(r - 2R_0)^2]$. b) Show that $r = R_0$ at equilibrium. c) Use $r = R_0 + x$ and the first two terms of the binomial theorem (Eq. 13.28) to show that $F_r \approx -(4A/R_0^3)x$, so that the force constant is $k = 4A/R_0^3$. d) What is the frequency of small-amplitude oscillation of the center particle if it has mass m?

13.84 Many real springs are more easily stretched than compressed. We can represent this by using different spring constants for $x > 0$ and for $x < 0$. As an example, consider a spring that exerts the following restoring force:

$$F_x = \begin{cases} -kx & \text{for } x > 0 \\ -2kx & \text{for } x < 0 \end{cases}$$

A mass m on a frictionless, horizontal surface is attached to this spring, displaced to $x = A$ by stretching the spring and released. a) Find the period of the motion. Does the period depend on A? Are the oscillations simple harmonic? b) What is the most negative value of x that the mass m reaches? Is the oscillation symmetric about the point $x = 0$?

13.85 Two solid cylinders connected along their common axis by a short, light rod have radius R and total mass M, and rest on a horizontal tabletop. A spring with force constant k has one end attached to a clamp and the other end attached to a frictionless ring at the center of mass of the cylinders (Fig. 13.35). The cylinders are pulled to the left a distance x, which stretches the spring, and released. There is sufficient friction between the tabletop and the cylinders for the cylinders to roll without slipping as they move back and forth on the end of the spring. Show that the motion of the center of mass of the cylinders is simple harmonic, and calculate its period in terms of M and k. [Hint: The motion is simple harmonic if a_x and x are related by Eq. (13.8), and the period then is $T = 2\pi/\omega$. Apply $\sum \tau_z = I_{cm}\alpha_z$ and $\sum F_x = Ma_{cm-x}$ to the cylinders in order to relate a_{cm-x} and the displacement x of the cylinders from their equilibrium position.]

Figure 13.35 Problem 13.85.

13.86 In Fig. 13.36 the upper ball is released from rest, collides with the stationary lower ball, and sticks to it. The strings are both 50.0 cm long. The upper ball has mass 2.00 kg and it is initially 10.0 cm higher than the lower ball, which has mass 3.00 kg. Find the frequency and maximum angular displacement of the motion after the collision.

Figure 13.36 Problem 13.86.

13.87 T-rex. Model the leg of the T-rex in Example 13.10 (Section 13.6) as two uniform rods joined end-to-end. Let each rod have length 1.55 m. Let the lower rod have mass M and the upper rod mass $2M$. The composite object is pivoted about the top of the upper rod. Compute the oscillation period of this object for small amplitude oscillations. Compare your result to that of Example 13.10.

13.88 A slender, uniform, metal rod with mass M is pivoted without friction about an axis through its midpoint and perpendicular to the rod. A horizontal spring with force constant k is attached to the lower end of the rod, with the other end of the spring attached to a rigid support. If the rod is displaced by a small angle Θ from the vertical (Fig. 13.37) and released, show that it moves in angular SHM and calculate the period. (*Hint:* Assume that the angle Θ is small enough for the approximations $\sin \Theta \approx \Theta$ and $\cos \Theta \approx 1$ to be valid. The motion is simple harmonic if $d^2\theta/dt^2 = -\omega^2\theta$, and the period is then $T = 2\pi/\omega$.)

Figure 13.37 Problem 13.88.

13.89 The Silently Ringing Bell Problem. A large bell is hung from a wooden beam so it can swing back and forth with negligible friction. The center of mass of the bell is 0.60 m below the pivot, the bell has mass 34.0 kg, and the moment of inertia of the bell about an axis at the pivot is 18.0 kg·m². The clapper is a small, 1.8-kg mass attached to one end of a slender rod that has length L and negligible mass. The other end of the rod is attached to the inside of the bell so it can swing freely about the same axis as the bell. What should be the length L of the clapper rod for the bell to ring silently, that is, for the period of oscillation for the bell to equal that for the clapper?

13.90 Two identical, thin rods, each with mass m and length L, are joined at right angles to form an L-shaped object. This object is balanced on top of a sharp edge (Fig. 13.38). If the L-shaped object is deflected slightly, it oscillates. Find the frequency of oscillation.

Figure 13.38 Problem 13.90.

13.91 You want to construct a pendulum with a period of 4.00 s at a location where $g = 9.80$ m/s². a) What is the length of a *simple* pendulum having this period? b) Suppose the pendulum must be mounted in a case that is not over 0.50 m high. Can you devise a pendulum having a period of 4.00 s that will satisfy this requirement?

13.92 Show that $x(t)$ as given in Eq. (13.42) is a solution of Newton's second law for damped oscillations, Eq. (13.41), if ω' is as defined in Eq. (13.43).

13.93 A uniform rod of length L oscillates through small angles about a point a distance x from its center. a) Prove that its angular frequency is $\sqrt{gx/[(L^2/12) + x^2]}$. b) Show that its maximum

angular frequency occurs when $x = L/\sqrt{12}$. c) What is the length of the rod if the maximum angular frequency is 2π rad/s?

13.94 When Is It Simple? To make a pendulum, you hang a uniform solid sphere of radius R on the end of a light string. The distance from the pivot to the center of the sphere is L. a) Prove that the period of your pendulum is $T = T_{sp}\sqrt{1 + 2R^2/5L^2}$, where T_{sp} is the period of a simple pendulum of length L. b) At what value of L is T only 0.10% larger than T_{sp}? c) For a sphere of diameter 2.540 cm, what is the value of L in part (b)?

Challenge Problems

13.95 Two springs, each with unstretched length 0.200 m, but with different force constants k_1 and k_2, are attached to opposite ends of a block with mass m on a level, frictionless surface. The outer ends of the springs are now attached to two pins P_1 and P_2, 0.100 m from the original positions of the ends of the springs (Fig. 13.39). Let $k_1 = 2.00$ N/m, $k_2 = 6.00$ N/m, and $m = 0.100$ kg. a) Find the length of each spring when the block is in its new equilibrium position after the springs have been attached to the pins. b) Find the period of vibration of the block if it is slightly displaced from its new equilibrium position and released.

Figure 13.39 Challenge Problem 13.95.

13.96 The Effective Force Constant of Two Springs. Two springs with the same unstretched length, but different force constants k_1 and k_2 are attached to a block mass m on a level, frictionless surface. Calculate the effective force constant k_{eff} in each of the three cases (a), (b), and (c) depicted in Fig. 13.40. (The effective force constant is defined by $\Sigma F_x = -k_{\mathrm{eff}} x$.) d) An object

Figure 13.40 Challenge Problem 13.96.

with mass m, suspended from a uniform spring with a force constant k, vibrates with a frequency f_1. When the spring is cut in half and the same object is suspended from one of the halves, the frequency is f_2. What is the ratio f_2/f_1?

13.97 a) What is the change ΔT in the period of a simple pendulum when the acceleration of gravity g changes by Δg? (*Hint:* The new period $T + \Delta T$ is obtained by substituting $g + \Delta g$ for g:

$$T + \Delta T = 2\pi\sqrt{\frac{L}{g + \Delta g}}$$

To obtain an approximate expression, expand the factor $(g + \Delta g)^{-1/2}$ using the binomial theorem (Appendix B) and keeping only the first two terms:

$$(g + \Delta g)^{-1/2} = g^{-1/2} - \frac{1}{2}g^{-3/2}\Delta g + \cdots$$

The other terms contain higher powers of Δg and are very small if Δg is small.) Express your result as the *fractional* change in period $\Delta T/T$ in terms of the fractional change $\Delta g/g$. b) A pendulum clock keeps correct time at a point where $g = 9.8000$ m/s^2, but is found to lose 4.0 s each day at a higher elevation. Use the result of part (a) to find the approximate value of g at this new location.

13.98 A Spring with Mass. The preceding problems in this chapter have assumed that the springs had negligible mass. But of course no spring is completely massless. To find the effect of the spring's mass, consider a spring with mass M, equilibrium length L_0, and spring constant k. When stretched or compressed to a length L, the potential energy is $\frac{1}{2}kx^2$, where $x = L - L_0$. a) Consider a spring, as described above, that has one end fixed and the other end moving with speed v. Assume that the speed of points along the length of the spring varies linearly with distance l from the fixed end. Assume also that the mass M of the spring is distributed uniformly along the length of the spring. Calculate the kinetic energy of the spring in terms of M and v. (*Hint:* Divide the spring into pieces of length dl; find the speed of each piece in terms of l, v, and L; find the mass of each piece in terms of dl, M, and L; and integrate from 0 to L. The result is *not* $\frac{1}{2}Mv^2$, since not all of the spring moves with the same speed.) b) Take the time derivative of the conservation of energy equation, Eq. (13.21), for a mass m moving on the end of a *massless* spring. By comparing your results to Eq. (13.8), which defines ω, show that the angular frequency of oscillation is $\omega = \sqrt{k/m}$. c) Apply the procedure of part (b) to obtain the angular frequency of oscillation ω of the spring considered in part (a). If the *effective mass* M' of the spring is defined by $\omega = \sqrt{k/M'}$, what is M' in terms of M?

13.99 A uniform 1.00-m stick hangs from a horizontal axis at one end and oscillates as a physical pendulum. An object of small dimensions and with mass equal to that of the meter stick can be clamped to the stick at a distance y below the axis. Let T represent the period of the system with the body attached and T_0 the period of the meter stick alone. a) Find the ratio T/T_0. Evaluate your expression for y ranging from 0 to 1.0 m in steps of 0.1 m, and graph T/T_0 versus y. b) Is there any value of y, other than $y = 0$, for which

$T = T_0$? If so, find it and explain why the period is unchanged when y has this value.

13.100 You measure the period of a physical pendulum about one pivot point to be T. Then you find another pivot point on the opposite side of the center of mass that gives the same period. The two points are separated by a distance L. Use the parallel-axis theorem to show that $g = L(2\pi/T)^2$. (This result shows a way that you can measure g without knowing the mass or any moments of inertia of the physical pendulum.)

13.101 Resonance in a Mechanical System. A mass m is attached to one end of a massless spring with a force constant k and an unstretched length l_0. The other end of the spring is free to turn about a nail driven into a frictionless, horizontal surface (Fig. 13.41). The mass is made to revolve in a circle with an angular frequency of revolution ω'. a) Calculate the length l of the spring as a function of ω'. b) What happens to the result in part (a) when ω' approaches the natural frequency $\omega = \sqrt{k/m}$ of the mass-spring system? (If your result bothers you, remember

Figure 13.41 Challenge Problem 13.101.

that massless springs and frictionless surfaces don't exist as such, but are only approximate descriptions of real springs and surfaces. Also, Hooke's law is itself only an approximation of the way real springs behave; the greater the elongation of the spring, the greater the deviation from Hooke's law.)

***13.102 Vibration of a Covalently Bonded Molecule.** Many diatomic (two-atom) molecules are bound together by *covalent bonds* that are much stronger than the van der Waals interaction. Examples include H_2, O_2, and N_2. Experiment shows that for many such molecules, the interaction can be described by a force of the form

$$F_r = A\left[e^{-2b(r - R_0)} - e^{-b(r - R_0)}\right]$$

where A and b are positive constants, r is the center-to-center separation of the atoms, and R_0 is the equilibrium separation. For the hydrogen molecule (H_2), $A = 2.97 \times 10^{-8}$ N, $b = 1.95 \times 10^{10}$ m^{-1}, and $R_0 = 7.4 \times 10^{-11}$ m. Find the force constant for small oscillations around equilibrium. (*Hint:* Use the expansion for e^x given in Appendix B.) Compare your result to the value given in Exercise 13.40.

FLUID MECHANICS 14

This Olympic cyclist is riding a stationary bike in a wind tunnel. By observing the airflow around his body (aided by trails of smoke), scientists are able to determine which bicycle designs and cycling techniques will minimize air resistance and maximize performance.

? **At various places around the cyclist, air is forced through a narrow constriction (for example, between his upper arm and his body). Does this make the air slow down, speed up, or neither?**

Fluids play a vital role in many aspects of everyday life. We drink them, breathe them, swim in them. They circulate through our bodies and control our weather. Airplanes fly through them; ships float in them. A fluid is any substance that can flow; we use the term for both liquids and gases. We usually think of a gas as easily compressed and a liquid as nearly incompressible, although there are exceptional cases.

We begin our study with **fluid statics,** the study of fluids at rest in equilibrium situations. Like other equilibrium situations, it is based on Newton's first and third laws. We will explore the key concepts of density, pressure, and buoyancy. **Fluid dynamics,** the study of fluids in motion, is much more complex; indeed, it is one of the most complex branches of mechanics. Fortunately, we can analyze many important situations using simple idealized models and familiar principles such as Newton's laws and conservation of energy. Even so, we will barely scratch the surface of this broad and interesting topic.

14.1 | Density

An important property of any material is its **density,** defined as its mass per unit volume. A homogeneous material such as ice or iron has the same density throughout. We use the Greek letter ρ ("rho") for density. If a mass m of homogeneous material has volume V, the density ρ is

$$\rho = \frac{m}{V} \qquad (\text{definition of density}) \qquad (14.1)$$

The density of some materials varies from point to point within the material; some examples are the earth's atmosphere (which is less dense at high altitude) and oceans (which are denser at greater depths). For these materials, Eq. (14.1) describes the *average* density. In general, the density of a material depends on environmental factors such as temperature and pressure.

The SI unit of density is the kilogram per cubic meter (1 kg/m^3). The cgs unit, the gram per cubic centimeter (1 g/cm^3), is also widely used. The conversion factor

$$1 \text{ g/cm}^3 = 1000 \text{ kg/m}^3$$

is useful. Densities of several common substances at ordinary temperatures are given in Table 14.1 (see also Fig. 14.1). Note the wide range of magnitudes. The densest material found on earth is the metal osmium ($\rho = 22{,}500 \text{ kg/m}^3$), but its density pales by comparison to the densities of exotic astronomical objects such as white dwarf stars and neutron stars.

14.1 The price of gold is quoted by weight (say, in dollars per ounce). Because gold is one of the densest of the metals, a fortune in gold can be stored in a small volume.

Table 14.1 Densities of Some Common Substances

Material	Density (kg/m^3)*	Material	Density (kg/m^3)*
Air (1 atm, 20° C)	1.20	Iron, steel	7.8×10^3
Ethanol	0.81×10^3	Brass	8.6×10^3
Benzene	0.90×10^3	Copper	8.9×10^3
Ice	0.92×10^3	Silver	10.5×10^3
Water	1.00×10^3	Lead	11.3×10^3
Seawater	1.03×10^3	Mercury	13.6×10^3
Blood	1.06×10^3	Gold	19.3×10^3
Glycerin	1.26×10^3	Platinum	21.4×10^3
Concrete	2×10^3	White dwarf star	10^{10}
Aluminum	2.7×10^3	Neutron star	10^{18}

*To obtain the densities in grams per cubic centimeter, simply divide by 10^3.

The **specific gravity** of a material is the ratio of its density to the density of water at 4.0°C, 1000 kg/m³; it is a pure number without units. For example, the specific gravity of aluminum is 2.7. "Specific gravity" is a poor term, since it has nothing to do with gravity; "relative density" would have been better.

Measuring density is an important analytical technique. For example, we can determine the charge condition of a storage battery by measuring the density of its electrolyte, a sulfuric acid solution. As the battery discharges, the sulfuric acid (H_2SO_4) combines with lead in the battery plates to form insoluble lead sulfate ($PbSO_4$), decreasing the concentration of the solution. The density decreases from about $1.30 \times 10^3 \text{ kg/m}^3$ for a fully charged battery to $1.15 \times 10^3 \text{ kg/m}^3$ for a discharged battery.

Another automotive example is permanent-type antifreeze, which is usually a solution of ethylene glycol ($\rho = 1.12 \times 10^3 \text{ kg/m}^3$) and water. The freezing point of the solution depends on the glycol concentration, which can be determined by measuring the specific gravity. Such measurements are performed routinely by mechanics using a device called a hydrometer, which we'll discuss in Section 14.3.

Example 14.1	**The weight of a roomful of air**

Find the mass and weight of the air in a living room with a 4.0 m × 5.0 m floor and a ceiling 3.0 m high. What is the mass and weight of an equal volume of water?

SOLUTION

IDENTIFY: We assume that air is homogeneous, so that the density is the same throughout the room. (It is true that air is less dense at high elevations than near sea level. The variation in density over the 3.0-m height of the room, however, is negligibly small; see Section 14.2.)

SET UP: We use Eq. (14.1) to relate the mass (the target variable) to the volume (which we calculate from the dimensions of the room) and the density (from Table 14.1).

EXECUTE: The volume of the room is $V = (3.0 \text{ m})(4.0 \text{ m}) \times (5.0 \text{ m}) = 60 \text{ m}^3$. The mass m_{air} of air is given by Eq. (14.1):

$$m_{air} = \rho_{air}V = (1.2 \text{ kg/m}^3)(60 \text{ m}^3) = 72 \text{ kg}$$

The weight of the air is

$$w_{air} = m_{air}g = (72 \text{ kg})(9.8 \text{ m/s}^2) = 700 \text{ N} = 160 \text{ lb}$$

The mass of an equal volume of water is

$$m_{water} = \rho_{water}V = (1000 \text{ kg/m}^3)(60 \text{ m}^3) = 6.0 \times 10^4 \text{ kg}$$

The weight is

$$w_{water} = m_{water}\,g = (6.0 \times 10^4 \text{ kg})(9.8 \text{ m/s}^2)$$
$$= 5.9 \times 10^5 \text{ N} = 1.3 \times 10^5 \text{ lb} = 66 \text{ tons}$$

EVALUATE: A roomful of air weighs about the same as an average adult! Water is nearly a thousand times denser than air, and its mass and weight are larger by the same factor. Indeed, the weight of a roomful of water would certainly collapse the floor of an ordinary house.

Test Your Understanding

What volume of water would have the same mass as one cubic meter of platinum? If this water occupied a cube, what would be the length of each side?

14.2 | Pressure in a Fluid

When a fluid (either liquid or gas) is at rest, it exerts a force perpendicular to any surface in contact with it, such as a container wall or a body immersed in the fluid. This is the force that you feel pressing on your legs when you dangle them in a swimming pool. While the fluid as a whole is at rest, the molecules that make up the fluid are in motion; the force exerted by the fluid is due to molecules colliding with their surroundings.

If we think of an imaginary surface *within* the fluid, the fluid on the two sides of the surface exerts equal and opposite forces on the surface. (Otherwise, the surface would accelerate and the fluid would not remain at rest.) Consider a small surface of area dA centered on a point in the fluid; the normal force exerted by the fluid on each side is $dF_\perp$ (Fig. 14.2). We define the **pressure** p at that point as the normal force per unit area, that is, the ratio of $dF_\perp$ to dA:

$$p = \frac{dF_\perp}{dA} \quad \text{(definition of pressure)} \tag{14.2}$$

If the pressure is the same at all points of a finite plane surface with area A, then

$$p = \frac{F_\perp}{A} \tag{14.3}$$

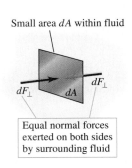

Small area dA within fluid

Equal normal forces exerted on both sides by surrounding fluid

14.2 The pressure acting on either side of a small area within a fluid is $p = dF_\perp/dA$.

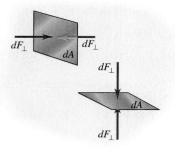

Pressure does not have a direction of its own: it can produce a force $dF_\perp = p\,dA$ in any direction

14.3 Pressure is a scalar (it has no intrinsic direction) and has units of newtons per square meter. By contrast, force is a vector with units of newtons.

where $F_\perp$ is the net normal force on one side of the surface. The SI unit of pressure is the **pascal**, where

$$1 \text{ pascal} = 1 \text{ Pa} = \text{N/m}^2$$

We introduced the pascal in Chapter 11. Two related units, used principally in meteorology, are the *bar*, equal to 10^5 Pa, and the *millibar*, equal to 100 Pa.

Atmospheric pressure p_a is the pressure of the earth's atmosphere, the pressure at the bottom of this sea of air in which we live. This pressure varies with weather changes and with elevation. Normal atmospheric pressure at sea level (an average value) is 1 *atmosphere* (atm), defined to be exactly 101,325 Pa. To four significant figures,

$$(p_a)_{av} = 1 \text{ atm} = 1.013 \times 10^5 \text{ Pa}$$
$$= 1.013 \text{ bar} = 1013 \text{ millibar} = 14.70 \text{ lb/in.}^2$$

CAUTION In everyday language the words "pressure" and "force" mean pretty much the same thing. In fluid mechanics, however, these words describe distinct quantities with different characteristics (Fig. 14.3). Fluid pressure acts perpendicular to any surface in the fluid, no matter how that surface is oriented. Hence pressure has no intrinsic direction of its own; it's a scalar. By contrast, force is a vector with a definite direction. Remember, too, that pressure is force per unit area.

Example 14.2   The force of air

In the room described in Example 14.1, what is the total downward force on the surface of the floor due to air pressure of 1.00 atm?

SOLUTION

IDENTIFY AND SET UP: The pressure is uniform, so we use Eq. (14.3) to determine the force $F_\perp$ from the pressure and area.

EXECUTE: The floor area is $A = (4.0 \text{ m})(5.0 \text{ m}) = 20 \text{ m}^2$, so from Eq. (14.3) the total downward force is

$$F_\perp = pA = (1.013 \times 10^5 \text{ N/m}^2)(20 \text{ m}^2)$$
$$= 2.0 \times 10^6 \text{ N} = 4.6 \times 10^5 \text{ lb} = 230 \text{ tons}$$

EVALUATE: As in Example 14.1, this is more than enough force to collapse the floor. So why doesn't it collapse? Because there is an upward force on the underside of the floor. If the house has a basement, this upward force is exerted by the air underneath the floor. In this case, if we neglect the thickness of the floor, the *net* force due to air pressure is zero.

Pressure, Depth, and Pascal's Law

If the weight of the fluid can be neglected, the pressure in a fluid is the same throughout its volume. We used that approximation in our discussion of bulk stress and strain in Section 11.4. But often the fluid's weight is *not* negligible. Atmospheric pressure is less at high altitude than at sea level, which is why an airplane cabin has to be pressurized when flying at 35,000 feet. When you dive into deep water, your ears tell you that the pressure increases rapidly with increasing depth below the surface.

We can derive a general relation between the pressure p at any point in a fluid at rest and the elevation y of the point. We'll assume that the density ρ and the

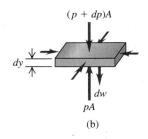

(a)   (b)

14.4 The forces on an element of fluid in equilibrium.

acceleration due to gravity g are the same throughout the fluid. If the fluid is in equilibrium, every volume element is in equilibrium. Consider a thin element of fluid, with thickness dy (Fig. 14.4). The bottom and top surfaces each have area A, and they are at elevations y and $y + dy$ above some reference level where $y = 0$. The volume of the fluid element is $dV = A\,dy$, its mass is $dm = \rho\,dV = \rho A\,dy$, and its weight is $dw = dm\,g = \rho g A\,dy$.

What are the other forces on this fluid element? Call the pressure at the bottom surface p; the total y-component of upward force on this surface is pA. The pressure at the top surface is $p + dp$, and the total y-component of (downward) force on the top surface is $-(p + dp)A$. The fluid element is in equilibrium, so the total y-component of force, including the weight and the forces at the bottom and top surfaces, must be zero:

$$\sum F_y = 0 \quad \text{so} \quad pA - (p + dp)A - \rho g A\,dy = 0$$

When we divide out the area A and rearrange, we get

$$\frac{dp}{dy} = -\rho g \tag{14.4}$$

This equation shows that when y increases, p decreases; that is, as we move upward in the fluid, pressure decreases, as we expect. If p_1 and p_2 are the pressures at elevations y_1 and y_2, respectively, and if ρ and g are constant, then

$$p_2 - p_1 = -\rho g(y_2 - y_1) \quad \text{(pressure in a fluid of uniform density)} \tag{14.5}$$

It's often convenient to express Eq. (14.5) in terms of the *depth* below the surface of a fluid (Fig. 14.5). Take point 1 at any level in the fluid and let p represent the pressure at this point. Take point 2 at the *surface* of the fluid, where the pressure is p_0 (subscript zero for zero depth). The depth of point 1 below the surface is $h = y_2 - y_1$, and Eq. (14.5) becomes

$$p_0 - p = -\rho g(y_2 - y_1) = -\rho g h$$

$$p = p_0 + \rho g h \quad \text{(pressure in a fluid of uniform density)} \tag{14.6}$$

The pressure p at a depth h is greater than the pressure p_0 at the surface by an amount $\rho g h$. Note that the pressure is the same at any two points at the same level in the fluid. The *shape* of the container does not matter (Fig. 14.6).

Equation (14.6) shows that if we increase the pressure p_0 at the top surface, possibly by using a piston that fits tightly inside the container to push down on

14.5 The pressure p at a depth h in a fluid is greater than the surface pressure p_0 by $\rho g h$.

14.6 The pressure at the top of each fluid column is equal to p_0, the pressure of the atmosphere. Pressure depends only on height, not the shape of the container, so each fluid column has the same height.

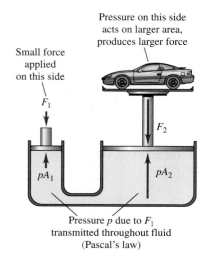

14.7 Principle of the hydraulic lift, an application of Pascal's law. The size of the fluid-filled container is exaggerated for clarity.

the fluid surface, the pressure p at any depth increases by exactly the same amount. This fact was recognized in 1653 by the French scientist Blaise Pascal (1623–1662) and is called **Pascal's law: Pressure applied to an enclosed fluid is transmitted undiminished to every portion of the fluid and the walls of the containing vessel.**

The hydraulic lift shown schematically in Fig. 14.7 illustrates Pascal's law. A piston with small cross-sectional area A_1 exerts a force F_1 on the surface of a liquid such as oil. The applied pressure $p = F_1/A_1$ is transmitted through the connecting pipe to a larger piston of area A_2. The applied pressure is the same in both cylinders, so

$$p = \frac{F_1}{A_1} = \frac{F_2}{A_2} \quad \text{and} \quad F_2 = \frac{A_2}{A_1} F_1 \qquad (14.7)$$

The hydraulic lift is a force-multiplying device with a multiplication factor equal to the ratio of the areas of the two pistons. Dentist's chairs, car lifts and jacks, many elevators, and hydraulic brakes all use this principle.

For gases the assumption that the density ρ is uniform is realistic only over short vertical distances. In a room with a ceiling height of 3.0 m filled with air of uniform density 1.2 kg/m³, the difference in pressure between floor and ceiling, given by Eq. (14.6), is

$$\rho g h = (1.2 \text{ kg/m}^3)(9.8 \text{ m/s}^2)(3.0 \text{ m}) = 35 \text{ Pa}$$

or about 0.00035 atm, a very small difference. But between sea level and the summit of Mount Everest (8882 m) the density of air changes by nearly a factor of 3, and in this case we cannot use Eq. (14.6). Liquids, by contrast, are nearly incompressible, and it is usually a very good approximation to regard their density as independent of pressure. A pressure of several hundred atmospheres will cause only a few percent increase in the density of most liquids.

Absolute Pressure, Gauge Pressure, and Pressure Gauges

If the pressure inside a car tire is equal to atmospheric pressure, the tire is flat. The pressure has to be *greater* than atmospheric to support the car, so the significant quantity is the *difference* between the inside and outside pressures. When we say that the pressure in a car tire is "32 pounds" (actually 32 lb/in.², equal to 220 kPa or 2.2×10^5 Pa), we mean that it is *greater* than atmospheric pressure (14.7 lb/in.² or 1.01×10^5 Pa) by this amount. The *total* pressure in the tire is then 47 lb/in.² or 320 kPa. The excess pressure above atmospheric pressure is usually called **gauge pressure,** and the total pressure is called **absolute pressure.** Engineers use the abbreviations psig and psia for "pounds per square inch gauge" and "pounds per square inch absolute," respectively. If the pressure is *less* than atmospheric, as in a partial vacuum, the gauge pressure is negative.

| Example 14.3 | **Finding absolute and gauge pressures** |

A solar water-heating system uses solar panels on the roof, 12.0 m above the storage tank. The water pressure at the level of the panels is one atmosphere. What is the absolute pressure in the tank? The gauge pressure?

SOLUTION

IDENTIFY: Water is nearly incompressible. (Imagine trying to compress a cylinder full of water with a piston—you couldn't do it!) Hence we can treat it as a fluid of uniform density.

SET UP: The level of the panels corresponds to point 2 in Fig. 14.5, and the level of the tank corresponds to point 1. Hence our target variable is p in Eq. (14.6); we are given $p_0 = 1$ atm $= 1.01 \times 10^{-5}$ Pa and $h = 12.0$ m.

EXECUTE: From Eq. (14.6), the absolute pressure is

$p = p_0 + \rho g h$

$\quad = (1.01 \times 10^5 \text{ Pa}) + (1000 \text{ kg/m}^3)(9.80 \text{ m/s}^2)(12.0 \text{ m})$

$\quad = 2.19 \times 10^5 \text{ Pa} = 2.16 \text{ atm} = 31.8 \text{ lb/in.}^2$

The gauge pressure is

$p - p_0 = (2.19 - 1.01) \times 10^5 \text{ Pa}$

$\qquad = 1.18 \times 10^5 \text{ Pa} = 1.16 \text{ atm} = 17.1 \text{ lb/in.}^2$

EVALUATE: If such a tank has a pressure gauge, it is usually calibrated to read gauge pressure rather than absolute pressure. As we have mentioned, the variation in *atmospheric* pressure over this height is negligibly small.

The simplest pressure gauge is the open-tube *manometer* (Fig. 14.8a). The U-shaped tube contains a liquid of density ρ, often mercury or water. One end of the tube is connected to the container where the pressure p is to be measured, and the other end is open to the atmosphere at pressure $p_0 = p_a$. The pressure at the bottom of the tube due to the fluid in the left column is $p + \rho g y_1$, and the pressure at the bottom due to the fluid in the right column is $p_a + \rho g y_2$. These pressures are measured at the same point, so they must be equal:

$$p + \rho g y_1 = p_a + \rho g y_2$$
$$p - p_a = \rho g(y_2 - y_1) = \rho g h \qquad (14.8)$$

In Eq. (14.8), p is the *absolute pressure,* and the difference $p - p_a$ between absolute and atmospheric pressure is the gauge pressure. Thus the gauge pressure is proportional to the difference in height $(y_2 - y_1)$ of the liquid columns.

Another common pressure gauge is the **mercury barometer.** It consists of a long glass tube, closed at one end, that has been filled with mercury and then inverted in a dish of mercury (Fig. 14.8b). The space above the mercury column contains only mercury vapor; its pressure is negligibly small, so the pressure p_0 at the top of the mercury column is practically zero. From Eq. (14.6),

$$p_a = p = 0 + \rho g(y_2 - y_1) = \rho g h \qquad (14.9)$$

Thus the mercury barometer reads the atmospheric pressure p_a directly from the height of the mercury column.

In many applications, pressures are still commonly described in terms of the height of the corresponding mercury column, as so many "inches of mercury" or "millimeters of mercury" (abbreviated mm Hg). A pressure of 1 mm Hg is called *one torr,* after Evangelista Torricelli, inventor of the mercury barometer. But these units depend on the density of mercury, which varies with temperature, and on the value of g, which varies with location, so the pascal is the preferred unit of pressure.

One common type of blood-pressure gauge, called a *sphygmomanometer,* uses a mercury-filled manometer. Blood-pressure readings, such as 130/80, refer to the maximum and minimum gauge pressures in the arteries, measured in mm Hg or torr. Blood pressure varies with height; the standard reference point is the upper arm, level with the heart.

Many types of pressure gauges use a flexible sealed vessel (Fig. 14.9). A change in the pressure either inside or outside the vessel causes a change in its dimensions. This change is detected optically, electrically, or mechanically.

14.8 Two types of pressure gauge. (a) The open-tube manometer. (b) The mercury barometer.

(a)   (b)

14.9 (a) A Bourdon pressure gauge. When the pressure inside the spiral metal tube increases, the tube straightens out a little, deflecting the attached pointer. (b) A Bourdon-type pressure gauge used on a compressed gas tank.

Example 14.4	**A tale of two fluids**

A manometer tube is partially filled with water. Oil (which does not mix with water, and has a lower density than water) is then poured into the left arm of the tube until the oil–water interface is at the midpoint of the tube (Fig. 14.10). Both arms of the tube are open to the air. Find a relationship between the heights h_{oil} and h_{water}.

SOLUTION

IDENTIFY: The relationship between pressure and depth in a fluid applies only to fluids of uniform density. Hence we can't write a single equation for the oil and the water together. We can, however, write a pressure–depth relation for each fluid separately, noting that both fluid columns have the same pressure at the bottom (where they are in contact and in equilibrium, so the pressures must be the same) and at the top (where both are in contact with and in equilibrium with the air).

SET UP: Let p_0 be atmospheric pressure and let p be the pressure at the bottom of the tube. The densities of the two fluids are ρ_{water} and ρ_{oil} (which is less than ρ_{water}). We use Eq. (14.6) for each fluid.

EXECUTE: For the two fluids, Eq. (14.6) becomes

$$p = p_0 + \rho_{water}gh_{water}$$
$$p = p_0 + \rho_{oil}gh_{oil}$$

Since the pressure p at the bottom of the tube is the same for both fluids, we set these two expressions equal to each other and solve for h_{oil} in terms of h_{water}. You can show that the result is

$$h_{oil} = \frac{\rho_{water}}{\rho_{oil}}h_{water}$$

EVALUATE: Since oil is less dense than water, the ratio ρ_{water}/ρ_{oil} is greater than unity and h_{oil} is greater than h_{water} (as shown in Fig. 14.10). This result makes sense: a greater height of low-density oil is needed to produce the same pressure p at the bottom of the tube.

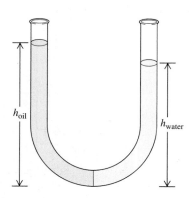

14.10 A U-shaped tube containing oil (on the left) and water (on the right). What is the relationship between the heights of the two fluid columns?

Mercury is less dense at high temperatures than at low temperatures. Suppose you move a mercury barometer from the inside of a tightly sealed refrigerator to outdoors on a hot summer day. You find that the column of mercury remains at the same height in the tube. How does the air pressure outdoors compare to the air pressure inside the refrigerator? (Ignore the even smaller change in the dimensions of the glass tube due to the temperature change.)

14.3 | Buoyancy

Buoyancy is a familiar phenomenon: a body immersed in water seems to weigh less than when it is in air. When the body is less dense than the fluid, it floats. The human body usually floats in water, and a helium-filled balloon floats in air.

Archimedes's principle states: **When a body is completely or partially immersed in a fluid, the fluid exerts an upward force on the body equal to the weight of the fluid displaced by the body.** To prove this principle, we consider an arbitrary portion of fluid at rest. In Fig. 14.11a the irregular outline is the surface bounding this portion of fluid. The arrows represent the forces exerted on the boundary surface by the surrounding fluid.

The entire fluid is in equilibrium, so the sum of all the y-components of force on this portion of fluid is zero. Hence the sum of the y-components of the *surface* forces must be an upward force equal in magnitude to the weight mg of the fluid inside the surface. Also, the sum of the torques on the portion of fluid must be zero, so the line of action of the resultant y-component of surface force must pass through the center of gravity of this portion of fluid.

Now we remove the fluid inside the surface and replace it with a solid body having exactly the same shape (Fig. 14.11b). The pressure at every point is exactly the same as before. So the total upward force exerted on the body by the fluid is also the same, again equal in magnitude to the weight mg of the fluid displaced to make way for the body. We call this upward force the **buoyant force** on the solid body. The line of action of the buoyant force again passes through the center of gravity of the displaced fluid (which doesn't necessarily coincide with the center of gravity of the body).

(a)

(b)

14.11 Archimedes's principle. (a) An element of a fluid in equilibrium. The buoyant force of the surrounding fluid equals the weight of the element. (b) If the fluid element is replaced by a body of identical shape, the body experiences the same buoyant force as in (a). This force is equal to the weight of the displaced fluid.

When a balloon floats in equilibrium in air, its weight (including the gas inside it) must be the same as the weight of the air displaced by the balloon. When a submerged submarine is in equilibrium, its weight must equal the weight of the water it displaces. A body whose average density is less than that of a liquid can float partially submerged at the free upper surface of the liquid. The greater the density of the liquid, the less of the body is submerged. When you swim in seawater (density 1030 kg/m^3), your body floats higher than in fresh water (1000 kg/m^3). Unlikely as it may seem, lead floats in mercury. Very flat-surfaced "float glass" is made by floating molten glass on molten tin and then letting it cool.

Another familiar example is the hydrometer, used to measure the density of liquids (Fig. 14.12a). The calibrated float sinks into the fluid until the weight of the fluid it displaces is exactly equal to its own weight. The hydrometer floats *higher* in denser liquids than in less dense liquids. It is weighted at its bottom end so that the upright position is stable, and a scale in the top stem permits direct density readings. Figure 14.12b shows a type of hydrometer that is commonly used to measure the density of battery acid or antifreeze. The bottom of the large tube is immersed in the liquid; the bulb is squeezed to expel air and is then released, like a giant medicine dropper. The liquid rises into the outer tube, and the hydrometer floats in this sample of the liquid.

14.12 (a) A simple hydrometer. (b) Use of a hydrometer as a tester for battery acid or antifreeze.

Example 14.5	**Buoyancy**

A 15.0-kg solid gold statue is being raised from a sunken ship (Fig. 14.13a). What is the tension in the hoisting cable when the statue is at rest and a) completely immersed; b) out of the water?

SOLUTION

IDENTIFY: When the statue is immersed, it experiences an upward buoyant force equal in magnitude to the weight of fluid displaced.

14.13 (a) The completely immersed statue at rest. (b) Free-body diagram for the immersed statue.

To find the tension, we note that the statue is in equilibrium (it is at rest) and consider the three forces acting on it: weight, the buoyant force, and the tension in the cable.

SET UP: Figure 14.13b shows the free-body diagram for the statue in equilibrium. Our target variable is the tension T. We are given the weight mg, and we can calculate the buoyant force B by using Archimedes's principle. We do this for two cases: (a) when the statue is immersed in water and (b) when it is out of the water and immersed in air.

EXECUTE: a) To find the buoyant force, we first find the volume of the statue, using the density of gold from Table 14.1:

$$V = \frac{m}{\rho_{\text{gold}}} = \frac{15.0 \text{ kg}}{19.3 \times 10^3 \text{ kg/m}^3} = 7.77 \times 10^{-4} \text{ m}^3$$

Using Table 14.1 again, we find the weight of this volume of seawater:

$$w_{\text{sw}} = m_{\text{sw}}g = \rho_{\text{sw}}Vg$$
$$= (1.03 \times 10^3 \text{ kg/m}^3)(7.77 \times 10^{-4} \text{ m}^3)(9.80 \text{ m/s}^2)$$
$$= 7.84 \text{ N}$$

This equals the buoyant force B.

The statue is at rest, so the net external force acting on it is zero. From Fig. 14.13b,

$$\sum F_y = B + T + (-mg) = 0$$
$$T = mg - B = (15.0 \text{ kg})(9.80 \text{ m/s}^2) - 7.84 \text{ N}$$
$$= 147 \text{ N} - 7.84 \text{ N} = 139 \text{ N}$$

If a spring scale is attached to the upper end of the cable, it will indicate 7.84 N less than if the statue were not immersed in seawater. Hence the submerged statue seems to weigh 139 N, about 5% less than its actual weight of 147 N.

b) The density of air is about 1.2 kg/m^3, so the buoyant force of air on the statue is

$$B = \rho_{\text{air}} V g = (1.2 \text{ kg/m}^3)(7.77 \times 10^{-4} \text{ m}^3)(9.80 \text{ m/s}^2)$$
$$= 9.1 \times 10^{-3} \text{ N}$$

This is only 62 parts per million of the statue's actual weight. This effect is not within the precision of our data, and we ignore it. Thus the tension in the cable with the statue in air is equal to the statue's weight, 147 N.

EVALUATE: Note that the buoyant force is proportional to the density of the *fluid,* not the density of the statue. The denser the fluid, the greater the buoyant force and the smaller the cable tension. If the fluid had the same density as the statue, the buoyant force would be equal to the statue's weight and the tension would be zero (the cable would go slack). If the fluid were denser than the statue, the tension would be *negative:* the buoyant force would be greater than the statue's weight, and a downward force would be required to keep the statue from rising upward.

Conceptual Example 14.6	A weighty question

You place a container of seawater on a scale and note the reading on the scale. You now suspend the statue of Example 14.5 in the water (Fig. 14.14). How does the scale reading change?

SOLUTION

Consider the water, the statue, and the container together as a system; the total weight of the system does not depend on whether the statue is immersed. The total supporting force, including the tension T and the upward force F of the scale on the container (equal to the scale reading), is the same in both cases. But we saw in Example 14.5 that T decreases by 7.84 N when the statue is immersed, so the scale reading F must *increase* by 7.84 N. An alternative viewpoint is that the water exerts an upward buoyant force of 7.84 N on the statue, so the statue must exert an equal downward force on the water, making the scale reading 7.84 N greater than the weight of water and container.

14.14 How does the scale reading change when the statue is immersed in water?

Surface Tension

An object less dense than water, such as an air-filled beach ball, floats with part of its volume below the surface. Conversely, a paper clip can rest *atop* a water surface even though its density is several times that of water. This is an example of **surface tension:** the surface of the liquid behaves like a membrane under tension (Fig. 14.15). Surface tension arises because the molecules of the liquid exert attractive forces on each other. There is zero net force on a molecule inside the volume of the liquid, but a surface molecule is drawn into the volume (Fig. 14.16). Thus the liquid tends to minimize its surface area, just as a stretched membrane does.

14.15 The surface of the water acts like a membrane under tension, allowing this water strider to literally "walk on water."

14.16 Each molecule in a liquid is attracted to its fellow molecules. A molecule of the surface is attracted into the bulk liquid, which tends to reduce the liquid's surface area.

14.17 Surface tension makes it difficult to force water through small crevices. The required water pressure p can be reduced by using hot, soapy water, which has less surface tension.

14.18 A flow tube bounded by flow lines. In steady flow, fluid cannot cross the walls of a flow tube.

Surface tension explains why freely falling raindrops are spherical (*not* teardrop-shaped): a sphere has a smaller surface area for its volume than any other shape. It also explains why hot, soapy water is used for washing. To wash clothing thoroughly, water must be forced through the tiny spaces between the fibers (Fig. 14.17). To do so requires increasing the surface area of the water, which is difficult to achieve because of surface tension. The job is made easier by increasing the temperature of the water and adding soap, both of which decrease the surface tension.

Surface tension is important for a millimeter-sized water drop, which has a relatively large surface area for its volume. (A sphere of radius r has surface area $4\pi r^2$ and volume $(4\pi/3)r^3$. The ratio of surface area to volume is $3/r$, which increases with decreasing radius.) For large quantities of liquid, however, the ratio of surface area to volume is relatively small, and surface tension is negligible compared to pressure forces. For the remainder of this chapter, we will consider only fluids in bulk and hence will ignore the effects of surface tension.

Test Your Understanding

An object with uniform density floats in water with one-third of its volume above the surface. How does the density of the object compare to the density of water?

14.4 | Fluid Flow

We are now ready to consider *motion* of a fluid. Fluid flow can be extremely complex, as shown by the currents in river rapids or the swirling flames of a campfire. But some situations can be represented by relatively simple idealized models. An **ideal fluid** is a fluid that is *incompressible* (that is, its density cannot change) and has no internal friction (called **viscosity**). Liquids are approximately incompressible in most situations, and we may also treat a gas as incompressible if the pressure differences from one region to another are not too great. Internal friction in a fluid causes shear stresses when two adjacent layers of fluid move relative to each other, as when fluid flows inside a tube or around an obstacle. In some cases we can neglect these shear forces in comparison with forces arising from gravitation and pressure differences.

The path of an individual particle in a moving fluid is called a **flow line.** If the overall flow pattern does not change with time, the flow is called **steady flow.** In steady flow, every element passing through a given point follows the same flow line. In this case the "map" of the fluid velocities at various points in space remains constant, although the velocity of a particular particle may change in both magnitude and direction during its motion. A **streamline** is a curve whose tangent at any point is in the direction of the fluid velocity at that point. When the flow pattern changes with time, the streamlines do not coincide with the flow lines. We will consider only steady-flow situations, for which flow lines and streamlines are identical.

The flow lines passing through the edge of an imaginary element of area, such as the area A in Fig. 14.18, form a tube called a **flow tube.** From the definition of a flow line, in steady flow no fluid can cross the side walls of a flow tube; the fluids in different flow tubes cannot mix.

14.19 Laminar flow around obstacles of different shapes.

Figure 14.19 shows patterns of fluid flow from left to right around a number of obstacles. The photographs were made by injecting dye into water flowing between two closely spaced glass plates. These patterns are typical of **laminar flow,** in which adjacent layers of fluid slide smoothly past each other and the flow is steady. (A *lamina* is a thin sheet.) At sufficiently high flow rates, or when boundary surfaces cause abrupt changes in velocity, the flow can become irregular and chaotic. This is called **turbulent flow** (Fig. 14.20). In turbulent flow there is no steady-state pattern; the flow pattern continuously changes.

The Continuity Equation

The mass of a moving fluid doesn't change as it flows. This leads to an important quantitative relationship called the **continuity equation.** Consider a portion of a flow tube between two stationary cross sections with areas A_1 and A_2 (Fig. 14.21). The fluid speeds at these sections are v_1 and v_2, respectively. No fluid flows in or out across the sides of the tube because the fluid velocity is tangent to the wall at every point on the wall. During a small time interval dt, the fluid at A_1 moves a distance $v_1\, dt$, so a cylinder of fluid with height $v_1\, dt$ and volume $dV_1 = A_1 v_1\, dt$ flows into the tube across A_1. During this same interval, a cylinder of volume $dV_2 = A_2 v_2\, dt$ flows out of the tube across A_2.

Let's first consider the case of an incompressible fluid so that the density ρ has the same value at all points. The mass dm_1 flowing into the tube across A_1 in time dt is $dm_1 = \rho A_1 v_1\, dt$. Similarly, the mass dm_2 that flows out across A_2 in the same time is $dm_2 = \rho A_2 v_2\, dt$. In steady flow the total mass in the tube is constant, so $dm_1 = dm_2$ and

$$\rho A_1 v_1\, dt = \rho A_2 v_2\, dt \qquad \text{or}$$

$$A_1 v_1 = A_2 v_2 \qquad \text{(continuity equation, incompressible fluid)} \quad (14.10)$$

The product Av is the *volume flow rate* dV/dt, the rate at which volume crosses a section of the tube:

$$\frac{dV}{dt} = Av \qquad \text{(volume flow rate)} \qquad (14.11)$$

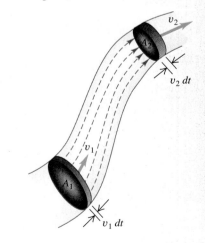

14.20 The flow of smoke rising from these incense sticks is laminar up until a certain point, then becomes turbulent.

14.21 A flow tube with changing cross-sectional area. If the fluid is incompressible, the product Av has the same value at all points along the tube.

The *mass* flow rate is the mass flow per unit time through a cross section. This is equal to the density ρ times the volume flow rate dV/dt.

Equation (14.10) shows that the volume flow rate has the same value at all points along any flow tube. When the cross section of a flow tube decreases, the speed increases, and vice versa. The deep part of a river has larger cross section and slower current than the shallow part, but the volume flow rates are the same in both. This is the essence of the familiar maxim, "Still waters run deep." The stream of water from a faucet narrows as it gains speed during its fall, but dV/dt is the same everywhere along the stream. If a water pipe with 2-cm diameter is connected to a pipe with 1-cm diameter, the flow speed is four times as great in the 1-cm part as in the 2-cm part.

We can generalize Eq. (14.10) for the case in which the fluid is *not* incompressible. If ρ_1 and ρ_2 are the densities at sections 1 and 2, then

$$\rho_1 A_1 v_1 = \rho_2 A_2 v_2 \qquad \text{(continuity equation, compressible fluid)} \qquad (14.12)$$

We leave the details for an exercise. If the fluid is incompressible so that ρ_1 and ρ_2 are always equal, Eq. (14.12) reduces to Eq. (14.10).

Example 14.7	Incompressible fluid flow

As part of a lubricating system for heavy machinery, oil of density 850 kg/m³ is pumped through a cylindrical pipe of diameter 8.0 cm at a rate of 9.5 liters per second. a) What is the speed of the oil? What is the mass flow rate? b) If the pipe diameter is reduced to 4.0 cm, what are the new values of the speed and volume flow rate? Assume that the oil is incompressible.

SOLUTION

IDENTIFY AND SET UP: We use the definition of volume flow rate, Eq. (14.11), to determine the speed v_1 in the 8.0-cm-diameter section. The mass flow rate is the product of the density and the volume flow rate. The continuity equation for incompressible flow, Eq. (14.10), allows us to find the speed v_2 in the 4.0-cm-diameter section.

EXECUTE: a) The volume flow rate dV/dt equals the product $A_1 v_1$, where A_1 is the cross-sectional area of the pipe of diameter 8.0 cm and radius 4.0 cm. Hence

$$v_1 = \frac{dV/dt}{A_1} = \frac{(9.5 \text{ L/s})(10^{-3} \text{ m}^3/\text{L})}{\pi(4.0 \times 10^{-2} \text{ m})^2} = 1.9 \text{ m/s}$$

The mass flow rate is $\rho \, dV/dt = (850 \text{ kg/m}^3)(9.5 \times 10^{-3} \text{ m}^3/\text{s}) = 8.1 \text{ kg/s}$.

b) Since the oil is incompressible, the volume flow rate has the same value (9.5 L/s) in both sections of pipe. From Eq. (14.10),

$$v_2 = \frac{A_1}{A_2} v_1 = \frac{\pi(4.0 \times 10^{-2} \text{ m})^2}{\pi(2.0 \times 10^{-2} \text{ m})^2}(1.9 \text{ m/s}) = 7.6 \text{ m/s}$$

EVALUATE: The second section of pipe has one-half the diameter and one-fourth the cross-sectional area of the first section. Hence the speed must be four times greater in the second section, which is just what our result shows $(v_2 = 4v_1)$.

Test Your Understanding

Air in the atmosphere is nearly incompressible. Use this fact to explain why particularly fast-moving winds are found in mountain passes.

14.5 | Bernoulli's Equation

According to the continuity equation, the speed of fluid flow can vary along the paths of the fluid. The pressure can also vary; it depends on height as in the static situation (Section 14.2), and it also depends on the speed of flow. We can derive an important relationship called *Bernoulli's equation* that relates the pressure, flow speed, and height for flow of an ideal, incompressible fluid. Bernoulli's equation is an essential tool in analyzing plumbing systems, hydroelectric generating stations, and the flight of airplanes.

The dependence of pressure on speed follows from the continuity equation, Eq. (14.10). When an incompressible fluid flows along a flow tube with varying cross section, its speed *must* change, and so an element of fluid must have an acceleration. If the tube is horizontal, the force that causes this acceleration has to be applied by the surrounding fluid. This means that the pressure *must* be different in regions of different cross section; if it were the same everywhere, the net force on every fluid element would be zero. When a horizontal flow tube narrows and a fluid element speeds up, it must be moving toward a region of lower pressure in order to have a net forward force to accelerate it. If the elevation also changes, this causes an additional pressure difference.

To derive Bernoulli's equation, we apply the work-energy theorem to the fluid in a section of a flow tube. In Fig. 14.22 we consider the element of fluid that at some initial time lies between the two cross sections a and c. The speeds at the lower and upper ends are v_1 and v_2. In a small time interval dt, the fluid that is initially at a moves to b, a distance $ds_1 = v_1\,dt$, and the fluid that is initially at c moves to d, a distance $ds_2 = v_2\,dt$. The cross-sectional areas at the two ends are A_1 and A_2, as shown. The fluid is incompressible; hence by the continuity equation, Eq. (14.10), the volume of fluid dV passing *any* cross section during time dt is the same. That is, $dV = A_1\,ds_1 = A_2\,ds_2$.

Let's compute the *work* done on this fluid element during dt. We assume that there is negligible internal friction in the fluid (i.e., no viscosity), so the only nongravitational forces that do work on the fluid element are due to the pressure of the surrounding fluid. The pressures at the two ends are p_1 and p_2; the force on the cross section at a is $p_1 A_1$, and the force at c is $p_2 A_2$. The net work dW done on the element by the surrounding fluid during this displacement is therefore

$$dW = p_1 A_1\,ds_1 - p_2 A_2\,ds_2 = (p_1 - p_2)dV \qquad (14.13)$$

The second term has a negative sign because the force at c opposes the displacement of the fluid.

The work dW is due to forces other than the conservative force of gravity, so it equals the change in the total mechanical energy (kinetic energy plus gravitational potential energy) associated with the fluid element. The mechanical energy for the fluid between sections b and c does not change. At the beginning of dt the fluid between a and b has volume $A_1\,ds_1$, mass $\rho A_1\,ds_1$, and kinetic energy $\frac{1}{2}\rho(A_1\,ds_1)v_1^2$. At the end of dt the fluid between c and d has kinetic energy $\frac{1}{2}\rho(A_2\,ds_2)v_2^2$. The net change in kinetic energy dK during time dt is

$$dK = \frac{1}{2}\rho\,dV(v_2^2 - v_1^2) \qquad (14.14)$$

What about the change in gravitational potential energy? At the beginning of dt the potential energy for the mass between a and b is $dm\,gy_1 = \rho\,dV\,gy_1$. At the end of dt the potential energy for the mass between c and d is $dm\,gy_2 = \rho\,dV\,gy_2$. The net change in potential energy dU during dt is

$$dU = \rho\,dV\,g(y_2 - y_1) \qquad (14.15)$$

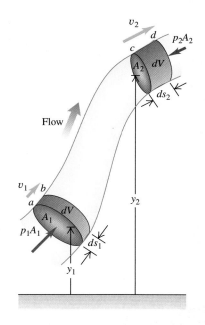

14.22 The net work done on a fluid element by the pressure of the surrounding fluid equals the change in the kinetic energy plus the change in the gravitational potential energy.

Combining Eqs. (14.13), (14.14), and (14.15) in the energy equation $dW = dK + dU$, we obtain

$$(p_1 - p_2)dV = \frac{1}{2}\rho\, dV(v_2^2 - v_1^2) + \rho\, dV g(y_2 - y_1)$$

$$p_1 - p_2 = \frac{1}{2}\rho(v_2^2 - v_1^2) + \rho g(y_2 - y_1) \qquad (14.16)$$

This is **Bernoulli's equation.** It states that the work done on a unit volume of fluid by the surrounding fluid is equal to the sum of the changes in kinetic and potential energies per unit volume that occur during the flow. We may also interpret Eq. (14.16) in terms of pressures. The first term on the right is the pressure difference associated with the change of speed of the fluid. The second term on the right is the additional pressure difference caused by the weight of the fluid and the difference in elevation of the two ends.

We can also express Eq. (14.16) in a more convenient form as

$$p_1 + \rho g y_1 + \frac{1}{2}\rho v_1^2 = p_2 + \rho g y_2 + \frac{1}{2}\rho v_2^2 \quad \text{(Bernoulli's equation)} \qquad (14.17)$$

The subscripts 1 and 2 refer to *any* two points along the flow tube, so we can also write

$$p + \rho g y + \frac{1}{2}\rho v^2 = \text{constant} \qquad (14.18)$$

Note that when the fluid is *not* moving (so $v_1 = v_2 = 0$), Eq. (14.17) reduces to the pressure relation we derived for a fluid at rest, Eq. (14.5).

CAUTION We stress again that Bernoulli's equation is valid only for incompressible, steady flow of a fluid with no internal friction (no viscosity). It's a simple equation that's easy to use; don't let this tempt you into using it in situations in which it doesn't apply!

Problem-Solving Strategy   **Bernoulli's Equation**

Bernoulli's equation is derived from the work-energy theorem, so it isn't surprising that much of the problem-solving strategy suggested in Section 7.1 is applicable here.

IDENTIFY *the relevant concepts:* First ensure that the fluid flow is steady and that the fluid is incompressible and has no internal friction. This case is an idealization, but it holds up surprisingly well for fluids flowing through sufficiently large pipes and for flows within bulk fluids (e.g., air flowing around an airplane or water flowing around a fish).

SET UP *the problem using the following steps:*
1. Always begin by identifying clearly the points 1 and 2 referred to in Bernoulli's equation.
2. Define your coordinate system, particularly the level at which $y = 0$.
3. Make lists of the unknown and known quantities in Eq. (14.17). The variables are p_1, p_2, v_1, v_2, y_1, and y_2, and the constants are ρ and g. Decide which unknowns are your target variables.

EXECUTE *the solution as follows:* Write Bernoulli's equation and solve for the unknowns. In some problems you will need to use the continuity equation, Eq. (14.10), to get a relation between the two speeds in terms of cross-sectional areas of pipes or containers. Or perhaps you will know both speeds and need to determine one of the areas. You may also need to use Eq. (14.11) to find the volume flow rate.

EVALUATE *your answer:* As always, verify that the results make physical sense. Double-check that you have used consistent units. In SI units, pressure is in pascals, density in kilograms per cubic meter, and speed in meters per second. Also note that the pressures must be either all absolute pressures or all gauge pressures.

Example 14.8   Water pressure in the home

Water enters a house through a pipe with an inside diameter of 2.0 cm at an absolute pressure of 4.0×10^5 Pa (about 4 atm). A 1.0-cm-diameter pipe leads to the second-floor bathroom 5.0 m above (Fig. 14.23). When the flow speed at the inlet pipe is 1.5 m/s, find the flow speed, pressure, and volume flow rate in the bathroom.

SOLUTION

IDENTIFY AND SET UP: Let points 1 and 2 be at the inlet pipe and at the bathroom, respectively. We are given the speed v_1 and pressure p_1 at the inlet pipe, and the pipe diameters at points 1 and 2 (from which we calculate the areas A_1 and A_2). We take $y_1 = 0$ (at the inlet) and $y_2 = 5.0$ m (at the bathroom). Our first two target variables are the speed v_2 and pressure p_2. Since we have more than one unknown, we use both Bernoulli's equation and the continuity equation. Once we find v_2, we calculate the volume flow rate $v_2 A_2$ at point 2.

EXECUTE: The speed v_2 at the bathroom is obtained from the continuity equation, Eq. (14.10):

$$v_2 = \frac{A_1}{A_2}v_1 = \frac{\pi (1.0 \text{ cm})^2}{\pi (0.50 \text{ cm})^2}(1.5 \text{ m/s}) = 6.0 \text{ m/s}$$

We are given p_1 and v_1, and can find p_2 from Bernoulli's equation:

$$p_2 = p_1 - \frac{1}{2}\rho(v_2^2 - v_1^2) - \rho g(y_2 - y_1) = 4.0 \times 10^5 \text{ Pa}$$

$$- \frac{1}{2}(1.0 \times 10^3 \text{ kg/m}^3)(36 \text{ m}^2/\text{s}^2 - 2.25 \text{ m}^2/\text{s}^2)$$

$$- (1.0 \times 10^3 \text{ kg/m}^3)(9.8 \text{ m/s}^2)(5.0 \text{ m})$$

$$= 4.0 \times 10^5 \text{ Pa} - 0.17 \times 10^5 \text{ Pa} - 0.49 \times 10^5 \text{ Pa}$$

$$= 3.3 \times 10^5 \text{ Pa} = 3.3 \text{ atm} = 48 \text{ lb/in.}^2$$

The volume flow rate is

$$\frac{dV}{dt} = A_2 v_2 = \pi (0.50 \times 10^{-2} \text{ m})^2 (6.0 \text{ m/s})$$

$$= 4.7 \times 10^{-4} \text{ m}^3/\text{s} = 0.47 \text{ L/s}$$

EVALUATE: This is a reasonable flow rate for a bathroom faucet or shower. Note that after the water is turned off, the term $\frac{1}{2}\rho(v_2^2 - v_1^2)$ in the equation for pressure vanishes, and the pressure rises to 3.5×10^5 Pa.

14.23 What is the water pressure in the second-story bathroom of this house?

Example 14.9   Speed of efflux

Figure 14.24 shows a gasoline storage tank with cross-sectional area A_1, filled to a depth h. The space above the gasoline contains air at pressure p_0, and the gasoline flows out through a short pipe

with area A_2. Derive expressions for the flow speed in the pipe and the volume flow rate.

14.24 Calculating the speed of efflux for gasoline flowing out the bottom of a storage tank.

IDENTIFY: We consider the entire volume of moving liquid as a single flow tube, so we can use Bernoulli's principle.

SET UP: Points 1 and 2 in Fig. 14.24 are at the surface of the gasoline and at the short exit pipe, respectively. At point 1 the pressure is p_0 and at point 2 it is atmospheric pressure p_a. We take $y = 0$ at the exit pipe, so $y_1 = h$ and $y_2 = 0$. Because A_1 is very much larger than A_2, the upper surface of the gasoline will drop very slowly and we can regard v_1 as essentially equal to zero. We find the target variable v_2 from Eq. (14.17) and the volume flow rate from Eq. (14.11).

EXECUTE: Applying Bernoulli's equation to points 1 and 2 and taking $y = 0$ at the bottom of the tank, we find

$$p_0 + \frac{1}{2}\rho v_1^2 + \rho g h = p_a + \frac{1}{2}\rho v_2^2$$

$$v_2^2 = v_1^2 + 2\frac{p_0 - p_a}{\rho} + 2gh$$

Using $v_1 = 0$, we find

$$v_2^2 = 2\frac{p_0 - p_a}{\rho} + 2gh$$

From Eq. (14.11), the volume flow rate is $dV/dt = v_2 A_2$.

EVALUATE: The speed v_2, sometimes called the *speed of efflux*, depends on both the pressure difference $(p_0 - p_a)$ and the height h of the liquid level in the tank. If the top of the tank is vented to the atmosphere, there is no excess pressure: $p_0 = p_a$ and $p_0 - p_a = 0$. In that case,

$$v_2 = \sqrt{2gh}$$

That is, the speed of efflux from an opening at a distance h below the top surface of the liquid is the *same* as the speed a body would acquire in falling freely through a height h. This result is called *Torricelli's theorem*. It is valid not only for an opening in the bottom of a container, but also for a hole in a side wall at a depth h below the surface. In this case the volume flow rate is

$$\frac{dV}{dt} = A_2\sqrt{2gh}$$

Example 14.10

The Venturi meter

Figure 14.25 shows a *Venturi meter,* used to measure flow speed in a pipe. The narrow part of the pipe is called the *throat*. Derive an expression for the flow speed v_1 in terms of the cross-sectional areas A_1 and A_2 and the difference in height h of the liquid levels in the two vertical tubes.

IDENTIFY AND SET UP: We apply Bernoulli's equation to the wide (point 1) and narrow (point 2) parts of the pipe. The difference in height between the two vertical tubes tells us the pressure difference between points 1 and 2.

14.25 The Venturi meter.

EXECUTE: The two points are at the same vertical coordinate $(y_1 = y_2)$, so Eq. (14.17) says

$$p_1 + \frac{1}{2}\rho v_1^2 = p_2 + \frac{1}{2}\rho v_2^2$$

From the continuity equation, $v_2 = (A_1/A_2)v_1$. Substituting this and rearranging, we get

$$p_1 - p_2 = \frac{1}{2}\rho v_1^2\left(\frac{A_1^2}{A_2^2} - 1\right)$$

The pressure difference $p_1 - p_2$ is also equal to $\rho g h$, where h is the difference in liquid level in the two tubes. Combining this with the above result and solving for v_1, we get

$$v_1 = \sqrt{\frac{2gh}{(A_1/A_2)^2 - 1}}$$

EVALUATE: Because A_1 is greater than A_2, v_2 is greater than v_1 and the pressure p_2 in the throat is *less* than p_1. A net force to the right accelerates the fluid as it enters the throat, and a net force to the left slows it as it leaves.

| Conceptual Example 14.11 | **Lift on an airplane wing** |

Figure 14.26 shows flow lines around a cross section of an airplane wing. The flow lines crowd together above the wing, corresponding to increased flow speed and reduced pressure in this region, just as in the Venturi throat. The upward force on the underside of the wing is greater than the downward force on the top side; there is a net upward force, or *lift*. Lift is not simply due to the impulse of air striking the underside of the wing; in fact, it turns out that the reduced pressure on the upper wing surface makes the greatest contribution to the lift. (This highly simplified discussion ignores the formation of vortices; a more complete discussion would take these into account.)

We can also understand the lift force on the basis of momentum changes. Figure 14.26 shows that there is a net *downward* change in the vertical component of momentum of the air flowing past the wing, corresponding to the downward force the wing exerts on the air. The reaction force *on* the wing is *upward,* as we concluded above.

A similar flow pattern and lift force are found in the vicinity of any humped object in a wind (see the airflow over the cyclist's back in the photo that opens this chapter). In a sufficiently strong wind, the lift force on the top of an open umbrella can collapse the umbrella upward. A lift force also acts on a car driving at high speed due to air moving over the car's curved upper surface. Such lift can reduce traction on the car's tires, which is why many cars are equipped with an aerodynamic "spoiler" at the car's tail. The spoiler is shaped like an upside-down wing and provides a downward force on the rear wheels.

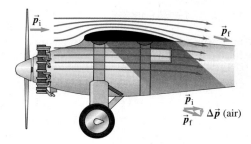

14.26 Flow lines around an airplane wing. The momentum of a parcel of air (relative to the wing) is $\vec{p}_i$ before encountering the wing and $\vec{p}_f$ afterward.

Test Your Understanding

It's not surprising that a wind blowing directly on an open door can make the door slam shut. Use Bernoulli's principle to explain how a wind blowing *across* a doorway (that is, parallel to the opening) can make the door close. (Assume that the door opens inward.)

*14.6 | Viscosity and Turbulence

In our discussion of fluid flow we assumed that the fluid had no internal friction and that the flow was laminar. While these assumptions are often quite valid, in many important physical situations the effects of viscosity (internal friction) and turbulence (nonlaminar flow) are extremely important. Let's take a brief look at some of these situations.

Viscosity

Viscosity is internal friction in a fluid. Viscous forces oppose the motion of one portion of a fluid relative to another. Viscosity is the reason why it takes effort to paddle a canoe through calm water, but it is also the reason why the paddle works. Viscous effects are important in the flow of fluids in pipes, the flow of blood, the lubrication of engine parts, and many other situations.

Fluids that flow readily, such as water or gasoline, have smaller viscosities than do "thick" liquids such as honey or motor oil. Viscosities of all fluids are strongly temperature dependent, increasing for gases and decreasing for liquids as the temperature increases (Fig. 14.27). An important goal in the design of oils for engine lubrication is to *reduce* the temperature variation of viscosity as much as possible.

14.27 Lava is an example of a viscous fluid. The viscosity decreases with increasing temperature: the hotter the lava, the more easily it can flow.

A viscous fluid always tends to cling to a solid surface in contact with it. There is always a thin *boundary layer* of fluid near the surface, in which the fluid is nearly at rest with respect to the surface. That's why dust particles can cling to a fan blade even when it is rotating rapidly, and why you can't get all the dirt off your car by just squirting a hose at it.

Viscosity has important effects on the flow of liquids through pipes, including the flow of blood in the circulatory system. First think about a fluid with zero viscosity so that we can apply Bernoulli's equation, Eq. (14.17). If the two ends of a long cylindrical pipe are at the same height ($y_1 = y_2$) and the flow speed is the same at both ends (so $v_1 = v_2$), Bernoulli's equation tells us that the pressure is the same at both ends of the pipe. But this result simply isn't true if we take viscosity into account. To see why, consider Fig. 14.28, which shows the flow-speed profile for laminar flow of a viscous fluid in a long cylindrical pipe. Due to viscosity, the speed is *zero* at the pipe walls (to which the fluid clings) and is greatest at the center of the pipe. The motion is like a lot of concentric tubes sliding relative to one another, with the central tube moving fastest and the outermost tube at rest. Viscous forces between the tubes oppose this sliding, so to keep the flow going we must apply a greater pressure at the back of the flow than at the front. That's why you have to keep squeezing a tube of toothpaste or a packet of ketchup (both viscous fluids) to keep the fluid coming out of its container. Your fingers provide a pressure at the back of the flow that is far greater than the atmospheric pressure at the front of the flow.

The pressure difference required to sustain a given volume flow rate through a cylindrical pipe of length L and radius R turns out to be proportional to L/R^4. If we decrease R by one-half, the required pressure increases by $2^4 = 16$; decreasing R by a factor of 0.90 (a 10% reduction) increases the required pressure difference by a factor of $(1/0.90)^4 = 1.52$ (a 52% increase). This simple relationship explains the connection between a high-cholesterol diet (which tends to narrow the arteries) and high blood pressure. Due to the R^4 dependence, even a small narrowing of the arteries can result in substantially elevated blood pressure and added strain on the heart muscle.

14.28 Velocity profile for a viscous fluid in a cylindrical pipe.

Turbulence

When the speed of a flowing fluid exceeds a certain critical value, the flow is no longer laminar. Instead, the flow pattern becomes extremely irregular and complex, and it changes continuously with time; there is no steady-state pattern. This irregular, chaotic flow is called **turbulence.** Figure 14.20 shows the contrast between laminar and turbulent flow for smoke rising in air. Bernoulli's equation is *not* applicable to regions where there is turbulence, because the flow is not steady.

Whether a flow is laminar or turbulent depends in part on the fluid's viscosity. The greater the viscosity, the greater the tendency for the fluid to flow in sheets or lamina and the more likely the flow is to be laminar. (When we discussed Bernoulli's equation in Section 14.5, we assumed that the flow was laminar and that the fluid had zero viscosity. In fact, a *little* viscosity is needed to ensure that the flow is laminar.)

For a fluid of a given viscosity, flow speed is a determining factor. A flow pattern that is stable at low speeds suddenly becomes unstable when a critical speed is reached. Irregularities in the flow pattern can be caused by roughness in the pipe wall, variations in the density of the fluid, and many other factors. At low flow speeds, these disturbances damp out; the flow pattern is *stable* and tends to maintain its laminar nature. When the critical speed is reached, however the flow pattern becomes unstable. The disturbances no longer damp out, but grow until they destroy the entire laminar-flow pattern.

Normal blood flow in the human aorta is laminar, but a small disturbance such as a heart pathology can cause the flow to become turbulent. Turbulence makes noise, which is why listening to blood flow with a stethoscope is a useful diagnostic technique.

The curve ball

Does a curve ball *really* curve? Yes, it certainly does, and the reason is turbulence. Figure 14.29a shows a ball moving through the air from left to right. To an observer moving with the center of the ball, the air stream appears to move from right to left, as shown by the flow lines in the figure. Because of the large speeds that are ordinarily involved (near 160 km/h, or 100 mi/h), there is a region of *turbulent* flow behind the ball.

Figure 14.29b shows a *spinning* ball with "top spin." Layers of air near the ball's surface are pulled around in the direction of spin by friction between the ball and air and by the air's internal friction (viscosity). The speed of air relative to the ball's surface becomes smaller at the top of the ball than at the bottom, and turbulence occurs farther forward on the top side than on the bottom. This asymmetry causes a pressure difference; the average pressure at the top of the ball is now greater than that at the bottom. The net force

deflects the ball downward, as shown in Fig. 14.29c. This is why "top spin" is used in tennis to keep a very fast serve in the court (Fig. 14.29d). In a baseball curve pitch, the ball spins about a nearly vertical axis, and the actual deflection is sideways. In that case, Fig. 14.29c is a *top* view of the situation. A curve ball thrown by a left-handed pitcher curves *toward* a right-handed batter, making it harder to hit (Fig. 14.29e).

A similar effect occurs with golf balls, which always have "back spin" from impact with the slanted face of the golf club. The resulting pressure difference between the top and bottom of the ball causes a lift force that keeps the ball in the air considerably longer than would be possible without spin. A well-hit drive appears from the tee to "float" or even curve *upward* during the initial portion of its flight. This is a real effect, not an illusion. The dimples on the ball play an essential role; the viscosity of air gives an undimpled ball a much shorter trajectory than a dimpled one with the same initial velocity and spin. Figure 14.30 shows the backspin of a golf ball just after it is struck by a club.

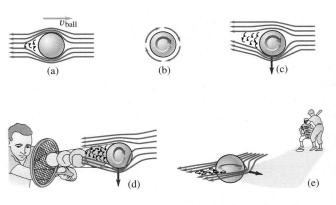

14.29 Motion of the air from right to left relative to the ball corresponds to motion of a ball through still air from left to right. (a) A nonspinning ball has a symmetric region of turbulence behind it. (b) A spinning ball drags along layers of air near its surface. (c) The resulting asymmetric region of turbulence and the deflection of the air stream by the spinning ball. The force shown is exerted on the ball by the air stream; it pushes the ball in the direction of the tangential velocity of the front of the ball. The force can (d) push a tennis ball downward or (e) make a baseball curve.

14.30 Stroboscopic photograph of a golf ball being struck by a club. The picture was taken at 1000 flashes per second. The ball rotates about once in eight pictures, corresponding to an angular speed of 125 rev/s, or 7500 rpm.

Test Your Understanding

How much more thumb pressure must a nurse use to administer an injection with a hypodermic needle of inside diameter 0.40 mm compared to one with inside diameter 0.55 mm? Assume that the two needles have the same length and that the volume flow rate is the same in both cases.

SUMMARY

Density is mass per unit volume. If a mass m of homogeneous material has volume V, its density ρ is the ratio m/V. Specific gravity is the ratio of the density of a material to the density of water. (See Example 14.1)

$$\rho = \frac{m}{V} \qquad (14.1)$$

Pressure is normal force per unit area. Pascal's law states that pressure applied to the surface of an enclosed fluid is transmitted undiminished to every portion of the fluid. Absolute pressure is the total pressure in a fluid; gauge pressure is the difference between absolute pressure and atmospheric pressure. The SI unit of pressure is the pascal (Pa): 1 Pa $= 1$ N/m^2. (See Example 14.2)

$$p = \frac{dF_\perp}{dA} \qquad (14.2)$$

Small area dA within fluid

$dF_\perp$ dA $dF_\perp$

Equal normal forces exerted on both sides by surrounding fluid

The pressure difference between points 1 and 2 in a static fluid of uniform density ρ (an incompressible fluid) is proportional to the difference between the elevations y_1 and y_2. If the pressure at the surface of an incompressible liquid at rest is ρ_0, the pressure at a depth h is greater by an amount $\rho g h$. (See Examples 14.3 and 14.4)

$$p_2 - p_1 = -\rho g(y_2 - y_1)$$
(pressure in a fluid of uniform density)
$$(14.5)$$

$$p = p_0 + \rho g h$$
(pressure in a fluid of uniform density)
$$(14.6)$$

Fluid, density ρ

$p_2 = p_0$

$y_2 - y_1 = h$

$p_1 = p$ y_2

1   y_1

Archimedes's principle states that when a body is immersed in a fluid, the fluid exerts an upward buoyant force on the body equal to the weight of the fluid that the body displaces. (See Examples 14.5 and 14.6)

Fluid replaced by body of same shape: experiences same bouyant force

$dF_\perp$

B cg $dF_\perp$

w_{body}

$dF_\perp$

An ideal fluid is incompressible and has no viscosity (no internal friction). A flow line is the path of a fluid particle; a streamline is a curve tangent at each point to the velocity vector at that point. A flow tube is a tube bounded at its sides by flow lines. In laminar flow, layers of fluid slide smoothly past each other. In turbulent flow, there is great disorder and a constantly changing flow pattern.

A Flow lines

Flow tube

Conservation of mass in an incompressible fluid is expressed by the continuity equation, which relates the flow speeds v_1 and v_2 for two cross sections A_1 and A_2 in a flow tube. The product Av equals the volume flow rate, dV/dt, the rate at which volume crosses a section of the tube. (See Example 14.7)

$$A_1 v_1 = A_2 v_2$$
(continuity equation, incompressible fluid)
$$(14.10)$$

$$\frac{dV}{dt} = Av$$
(volume flow rate)
$$(14.11)$$

v_2

$v_2\,dt$

v_1

A_1

$v_1\,dt$

Bernoulli's equation relates the pressure p, flow speed v, and elevation y for any two points, assuming steady flow in an ideal fluid. (See Examples 14.8 through 14.11)

$$p_1 + \rho g y_1 + \frac{1}{2}\rho v_1^2 = p_2 + \rho g y_2 + \frac{1}{2}\rho v_2^2$$

(Bernoulli's equation)   (14.17)

Key Terms

fluid statics, 515
fluid dynamics, 515
density, 515
specific gravity, 516
pressure, 517
pascal, 518
atmospheric pressure, 518
Pascal's law, 520
gauge pressure, 520

absolute pressure, 520
mercury barometer, 521
buoyancy, 523
Archimedes's principle, 523
buoyant force, 523
surface tension, 525
ideal fluid, 526
viscosity, 526
flow line, 526

steady flow, 526
streamline, 526
flow tube, 526
laminar flow, 527
turbulent flow, 527
continuity equation, 527
Bernoulli's equation, 530
viscosity, 533
turbulence, 534

Your Notes

Answer to Chapter Opening Question

Air maintains nearly the same density when passing through such a constriction, so the equation of continuity for an incompressible fluid, Eq. (14.10), applies. A constriction corresponds to a reduced cross-sectional area A, so the speed v must increase.

Answers to Test Your Understanding Questions

Section 14.1 From Table 14.1, the density of platinum is 21.4 times that of water (21.4×10^3 kg/m^3 versus 1.00×10^3 kg/m^3). From Eq. (14.1), volume and density are inversely proportional to each other, so the same mass of water has 21.4 times the volume of the platinum, or 21.4 m^3. The length of each side of the cube would be $\sqrt[3]{21.4 \text{ m}^3} = 2.78$ m.

Section 14.2 From Eq. (14.9), the outside pressure is equal to the product $\rho g h$. The density ρ decreases while the height h of the mercury column remains the same, so the pressure must be lower outdoors than inside the refrigerator.

Section 14.3 The object displaces two-thirds of its volume V, so the upward buoyant force is $B = \frac{2}{3}\rho_{\text{water}}Vg$. The object is in equilibrium, so B is equal to the weight of the object $\rho_{\text{object}}Vg$. Hence $\rho_{\text{object}} = \frac{2}{3}\rho_{\text{water}}$. This is an example of a general rule: if an object floats in a liquid with a fraction x of its volume submerged, the object's average density is x times the density of the liquid.

Section 14.4 Because air is nearly incompressible, the volume flow rate of air is essentially constant. When air blows through a constriction such as a mountain pass, the speed increases to keep the volume flow rate the same.

Section 14.5 From Bernoulli's equation, an increase in the flow speed v corresponds to a decrease in the air pressure p. The reduced air pressure on the "outdoor" side of the door makes the door swing toward that side, closing it.

Section 14.6 The required pressure is proportional to $1/R^4$. With the smaller-diameter needle, the pressure is greater by a factor of $[(0.55 \text{ mm})/(0.40 \text{ mm})]^4 = 3.6$.

Discussion Questions

Q14.1 If the weight of a roomful of water is so great (see Example 14.1 in Section 14.1), why don't the first floors of houses built over basements or crawl spaces all collapse when flooded to the ceiling?

Q14.2 A rubber hose is attached to a funnel, and the free end is bent around to point upward. When water is poured into the funnel, it rises in the hose to the same level as in the funnel, even though the funnel has a lot more water in it than the hose. Why?

Q14.3 Comparing Examples 14.1 and 14.2 in Sections 14.1 and 14.2, it seems that 700 N of air is exerting a downward force of 2.0×10^6 N on the floor. How is this possible?

Q14.4 Equation (14.7) shows that an area ratio of 100 to 1 can give 100 times more output force than input force. Doesn't this violate conservation of energy? Explain.

Q14.5 You have probably noticed that the lower the tire pressure, the greater the contact area between the tire and the pavement. Why?

Q14.6 In hot-air ballooning, a large balloon is filled with air heated by a gas burner at the bottom. Why must the air be heated? How does the balloonist control ascent and descent?

Q14.7 In describing the size of a large ship, one uses such expressions as "it displaces 20,000 tons." What does this mean? Can the weight of the ship be obtained from this information?

Q14.8 You drop a solid sphere of aluminum in a bucket of water that sits on the ground. The buoyant force equals the weight of water displaced; this is less than the weight of the sphere, so the sphere sinks to the bottom. If you take the bucket with you on an elevator that accelerates upward, the apparent weight of the water increases and the buoyant force on the sphere increases. Could the acceleration of the elevator be great enough to make the sphere pop up out of the water? Explain.

Q14.9 A rigid, lighter-than-air dirigible filled with helium cannot continue to rise indefinitely. Why not? What determines the maximum height it can attain?

Q14.10 Air pressure decreases with increasing altitude. So why is air near the surface not continuously drawn upward toward the lower-pressure regions above?

Q14.11 The purity of gold can be tested by weighing it in air and in water. How? Do you think you could get away with making a fake gold brick by gold-plating some cheaper material?

Q14.12 During the Great Mississippi Flood of 1993, the levees in St. Louis tended to rupture first at the bottom. Why?

Q14.13 A cargo ship travels from the Atlantic Ocean (salt water) to Lake Ontario (fresh water) via the St. Lawrence River. The ship rides several centimeters lower in the water in Lake Ontario than it did in the ocean. Explain why.

Q14.14 A submarine is more compressible than water. Why, then, can a submarine completely surrounded by water only be in unstable equilibrium?

Q14.15 An old question is "Which weighs more, a pound of feathers or a pound of lead?" If the weight in pounds is the gravitational force, will a pound of feathers balance a pound of lead on opposite pans of an equal-arm balance? Explain, taking into account buoyant forces.

Q14.16 Suppose the door of a room makes an airtight but frictionless fit in its frame. Do you think you could open the door if the air pressure on one side were standard atmospheric pressure and that on the other side differed from standard by 1%? Explain.

Q14.17 A balloon is less compressible than air. Why, then, is there a height at which a helium balloon is in stable equilibrium?

Q14.18 A piece of iron is glued to the top of a block of wood. When the block is placed in a bucket of water with the iron on top, the block floats. The block is now turned over so that the iron is submerged beneath the wood. Does the block float or sink? Does the water level in the bucket rise, drop, or stay the same? Explain your answers.

Q14.19 You take an empty glass jar and push it into a tank of water with the open mouth of the jar downward, so that the air inside the jar is trapped and cannot get out. If you push the jar deeper into the water, does the buoyant force on the jar stay the same? If not, does it increase or decrease? Explain your answer.

Q14.20 You are floating in a canoe in the middle of a swimming pool. Your friend is at the edge of the pool, carefully noting the level of the water on the side of the pool. You have a bowling ball with you in the canoe. If you carefully drop the bowling ball over the side of the canoe and it sinks to the bottom of the pool, does the water level in the pool rise or fall?

Q14.21 You are floating in a canoe in the middle of a swimming pool. A large bird flies up and lights on your shoulder. Does the water level in the pool rise or fall?

Q14.22 You are swimming in a swimming pool and then climb onto an inflatable plastic raft that is floating in the pool. If you are totally out of the water when you are floating on the raft, when you climb onto the raft does the water level in the pool rise or fall?

Q14.23 An ice cube floats in a glass of water. When the ice melts, will the water level in the glass rise, fall, or remain unchanged? Explain.

Q14.24 You are told, "Bernoulli's equation tells us where there is higher fluid speed, there is lower fluid pressure, and vice versa." Is this statement always true, even for an idealized fluid? Explain.

Q14.25 If the velocity at each point in space in steady-state fluid flow is constant, how can a fluid particle accelerate?

Q14.26 In a store-window vacuum cleaner display, a table-tennis ball is suspended in midair in a jet of air blown from the outlet hose of a tank-type vacuum cleaner. The ball bounces around a little but always moves back toward the center of the jet, even if the jet is tilted from the vertical. How does this behavior illustrate Bernoulli's equation?

Q14.27 A tornado consists of a rapidly whirling air vortex. Why is the pressure always much lower in the center than at the outside? How does this condition account for the destructive power of a tornado?

Q14.28 Airports at high elevations have longer runways for take-offs and landings than do airports at sea level. One reason is that aircraft engines develop less power in the thin air well above sea level. What is another reason?

Q14.29 When a smooth-flowing stream of water comes out of a faucet, it narrows as it falls. Explain why this happens.

Exercises

Section 14.1 Density

14.1 On a part-time job, you are asked to bring a cylindrical iron rod of length 85.8 cm and diameter 2.85 cm from a storage room to a machinist. Will you need a cart? (To answer, calculate the weight of the rod.)

14.2 The moon has radius 1740 km and mass 7.35×10^{22} kg. What is its average density?

14.3 You purchase a rectangular piece of metal that has dimensions $5.0 \times 15.0 \times 30.0$ mm and mass 0.0158 kg. The seller tells you that the metal is gold. To check this, you compute the average density of the piece. What value do you get? Were you cheated?

14.4 A kidnapper demands a 40.0-kg cube of platinum as a ransom. What is the length of a side?

14.5 A uniform, lead sphere and a uniform aluminum sphere have the same mass. What is the ratio of the radius of the aluminum sphere to the radius of the lead sphere?

14.6 a) What is the average density of the sun? b) What is the average density of a neutron star that has the same mass as the sun but a radius of only 20.0 km?

Section 14.2 Pressure in a Fluid

14.7 At what depth in seawater is the gauge pressure 1.00×10^5 Pa?

14.8 In intravenous feeding, a needle is inserted in a vein in the patient's arm and a tube leads from the needle to a reservoir of fluid (density 1050 kg/m^3) located at height h above the arm. The top of the reservoir is open to the air. If the gauge pressure inside the vein is 5980 Pa, what is the minimum value of h that allows fluid to enter the vein? Assume the needle diameter is large enough that you can ignore the viscosity (Section 14.6) of the fluid.

14.9 A barrel contains a 0.120-m layer of oil floating on water that is 0.250 m deep. The density of the oil is 600 kg/m^3 a) What is the gauge pressure at the oil-water interface? b) What is the gauge pressure at the bottom of the barrel?

14.10 An empty sport utility vehicle weighs 16.5 kN. Each of its tires has a gauge pressure of 205 kPa (29.7 lb/in.^{2}). a) What is the total contact area of the four tires with the pavement? (Assume that the tire walls are flexible so that the pressure exerted by the tire on the pavement equals the air pressure inside the tire.) b) For the same tire pressure, what is that area when the vehicle is loaded with 9.1 kN of passengers and cargo?

14.11 You are designing a diving bell to withstand the pressure of seawater at a depth of 250 m. a) What is the gauge pressure at this depth? (You can ignore changes in the density of the water with depth.) b) At this depth, what is the net force due to the water outside and the air inside the bell on a circular glass window 30.0 cm in diameter if the pressure inside the diving bell equals the pressure at the surface of the water? (You can ignore the small variation of pressure over the surface of the window.)

14.12 What gauge pressure must a pump produce to pump water from the bottom of the Grand Canyon (elevation 730 m) to Indian Gardens (elevation 1370 m)? Express your results in pascals and in atmospheres.

14.13 The liquid in the open-tube manometer in Fig. 14.8a is mercury, $y_1 = 3.00$ cm, and $y_2 = 7.00$ cm. Atmospheric pressure is 980 millibars. a) What is the absolute pressure at the bottom of the U-shaped tube? b) What is the absolute pressure in the open tube at a depth of 4.00 cm below the free surface? c) What is the absolute pressure of the gas in the tank? d) What is the gauge pressure of the gas in pascals?

14.14 There is a maximum depth at which a diver can breathe through a snorkel tube (Fig. 14.31) because as the depth increases, so does the pressure difference, tending to collapse the diver's lungs. Since the snorkel connects the air in the lungs to the atmosphere at the surface, the pressure inside the lungs is atmospheric pressure. What is the external-internal pressure difference when the diver's lungs are at a depth of 6.1 m (about 20 ft)? Assume that the diver is in fresh water. (A scuba diver breathing from compressed air tanks can operate at greater depths than can a snorkeler, since the pressure of the air inside the scuba diver's lungs increases to match the external pressure of the water.)

14.15 A tall cylinder with a cross-sectional area 12.0 cm² is partially filled with mercury; the surface of the mercury is 5.00 cm above the bottom of the cylinder. Water is slowly poured in on top of the mercury and the two fluids don't mix. What volume of water must be added to double the gauge pressure at the bottom of the cylinder?

14.16 A closed container is partially filled with water. Initially, the air above the water is at atmospheric pressure $(1.01 \times 10^5 \text{ Pa})$ and the gauge pressure at the bottom of the water is 2500 Pa. Then additional air is pumped in, increasing the pressure of the air above the water by 1500 Pa. a) What then is the gauge pressure at the bottom of the water? b) By how much must the water level in the container be reduced, by drawing some water out through a valve at the bottom of the container, to return the gauge pressure at the bottom of the water to its original value of 2500 Pa? The pressure of the air above the water is maintained at 1500 Pa above atmospheric pressure.

14.17 An electrical short cuts off all power to a submersible when it is 30 m below the surface of the ocean. The crew must push out a hatch of area 0.75 m² and weight 300 N on the bottom to escape. If the pressure inside is 1.0 atm, what downward force must they exert on the hatch to open it?

14.18 You are assigned the design of a cylindrical, pressurized water tank for a future colony on Mars, where the acceleration due to gravity is 3.71 m/s². The pressure at the surface of the water will be 130 kPa and the depth of the water will be 14.2 m. The pressure of the air in the building outside the tank will be 93 kPa. Find the net downward force on the tank's flat bottom, of area 2.00 m², exerted by the water and air inside the tank and the air outside the tank.

14.19 A tapered, pressurized tank for a rocket contains 0.250 m³ of kerosene, with mass 205 kg. The pressure at the top of the kerosene is 2.01×10^5 Pa. The kerosene exerts a force of 16.4 kN on the tank's bottom, which has an area of 0.0700 m². Find the depth of the kerosene.

14.20 The piston of a hydraulic automobile lift is 0.30 m in diameter. What gauge pressure, in pascals, is required to lift a car with a mass of 1200 kg? Also express this pressure in atmospheres.

Section 14.3 Buoyancy

14.21 A slab of ice floats on a freshwater lake. What minimum volume must the slab have for a 45.0-kg woman to be able to stand on it without getting her feet wet?

14.22 An ore sample weighs 17.50 N in air. When the sample is suspended by a light cord and totally immersed in water, the tension in the cord is 11.20 N. Find the total volume and the density of the sample.

p_a

6.1 m

Figure 14.31 Exercise 14.14.

14.23 An object of average density ρ floats at the surface of a fluid of density ρ_{fluid}. a) How must the two densities be related? b) In view of the answer to part (a), how can steel ships float in water? c) In terms of ρ and ρ_{fluid}, what fraction of the object is submerged and what fraction is above the fluid? Check that your answers give the correct limiting behavior as $\rho \rightarrow \rho_{\text{fluid}}$ and as $\rho \rightarrow 0$. d) While on board your yacht, your cousin Throckmorton cuts a rectangular piece (dimensions 5.0 × 4.0 × 3.0 cm) out of a life preserver and throws it into the ocean. The piece has a mass of 42 g. As it floats in the ocean, what percentage of its volume is above the surface?

14.24 A hollow, plastic sphere is held below the surface of a freshwater lake by a cord anchored to the bottom of the lake. The sphere has a volume of 0.650 m³ and the tension in the cord is 900 N. a) Calculate the buoyant force exerted by the water on the sphere. b) What is the mass of the sphere? c) The cord breaks and the sphere rises to the surface. When the sphere comes to rest, what fraction of its volume will be submerged?

14.25 A cubical block of wood, 10.0 cm on a side, floats at the interface between oil and water with its lower surface 1.50 cm below the interface (Fig. 14.32). The density of the oil is 790 kg/m³. a) What is the gauge pressure at the upper face of the block? b) What is the gauge pressure at the lower face of the block? c) What is the mass and density of the block?

Figure 14.32 Exercise 14.25.

14.26 A solid, aluminum ingot weighs 89 N in air. a) What is its volume? b) The ingot is suspended from a rope and totally immersed in water. What is the tension in the rope (the *apparent* weight of the ingot in water)?

14.27 Two cubical blocks of identical size and shape are suspended by strings and completely immersed in a swimming pool. Block A is aluminum; the top of block A is 0.5 m below the surface of the water. Block B is brass; the top of block B is 1.5 m below the surface of the water. For each of the following, is the quantity larger for block A, larger for B, or the same for both: a) the water pressure at the top of the block; b) the buoyant force exerted on the block by the water; c) the tension in the string attached to the block.

14.28 A rock is suspended by a light string. When the rock is in air, the tension in the string is 39.2 N. When the rock is totally immersed in water, the tension is 28.4 N. When the rock is totally immersed in an unknown liquid, the tension is 18.6 N. What is the density of the unknown liquid?

Section 14.4 Fluid Flow

14.29 A shower head has 20 circular openings, each with radius 1.0 mm. The shower head is connected to a pipe with radius 0.80 cm. If the speed of water in the pipe is 3.0 m/s, what is its speed as it exits the shower-head openings?

14.30 Water is flowing in a pipe with a varying cross-sectional area, and at all points the water completely fills the pipe. At point 1 the cross-sectional area of the pipe is 0.070 m², and the magnitude of the fluid velocity is 3.50 m/s. a) What is the fluid speed at points in the pipe where the cross-sectional area is i) 0.105 m^2? ii) 0.047 m^2? b) Calculate the volume of water discharged from the open end of the pipe in 1.00 hour.

14.31 Water is flowing in a circular pipe with varying cross-sectional area, and at all points the water completely fills the pipe. a) At one point in the pipe the radius is 0.150 m. What is the speed of the water at this point if the volume flow rate in the pipe is 1.20 m^3/s? b) At a second point in the pipe the water speed is 3.80 m/s. What is the radius of the pipe at this point?

14.32 a) Derive Eq. (14.12). b) If the density increases by 1.50% from point 1 to point 2, what happens to the volume flow rate?

Section 14.5 Bernoulli's Equation

14.33 A sealed tank containing seawater to a height of 11.0 m also contains air above the water at a gauge pressure of 3.00 atm. Water flows out from the bottom through a small hole. Calculate the efflux speed of the water.

14.34 A small circular hole 6.00 mm in diameter is cut in the side of a large water tank, 14.0 m below the water level in the tank. The top of the tank is open to the air. Find a) the speed of efflux; b) the volume discharged per unit time.

14.35 What gauge pressure is required in the city water mains for a stream from a fire hose connected to the mains to reach a vertical height of 15.0 m? (Assume that the mains have a much larger diameter than the fire hose.)

14.36 At one point in a pipeline the water's speed is 3.00 m/s and the gauge pressure is 5.00×10^4 Pa. Find the gauge pressure at a second point in the line, 11.0 m lower than the first, if the pipe diameter at the second point is twice that at the first.

14.37 **Lift on an Airplane.** Air streams horizontally past a small airplane's wings such that the speed is 70.0 m/s over the top surface and 60.0 m/s past the bottom surface. If the plane has mass 1340 kg and a wing area of 16.2 m^2, what is the net vertical force (including the effects of gravity) on the airplane? The density of the air is 1.20 kg/m^3.

14.38 A soft drink (mostly water) flows in a pipe at a beverage plant with a mass flow rate that would fill 220 0.355-L cans per minute. At point 2 in the pipe, the gauge pressure is 152 kPa and the cross-sectional area is 8.00 cm^2. At point 1, 1.35 m above point 2, the cross-sectional area is 2.00 cm^2. Find the a) mass flow rate; b) volume flow rate; c) flow speeds at points 1 and 2; d) gauge pressure at point 1.

14.39 Water discharges from a horizontal, cylindrical pipe at the rate of 465 cm^3/s. At a point in the pipe where the radius is 2.05 cm, the absolute pressure is 1.60×10^5 Pa. What is the pipe radius at a constriction if the pressure there is reduced to 1.20×10^5 Pa?

14.40 At a certain point in a horizontal pipeline, the water's speed is 2.50 m/s, and the gauge pressure is 1.80×10^4 Pa. Find the gauge pressure at a second point in the line if the cross-sectional area at the second point is twice that at the first.

14.41 A golf course sprinkler system discharges water from a horizontal pipe at the rate of 7200 cm^3/s. At one point in the pipe, where the radius is 4.00 cm, the water's absolute pressure is 2.40×10^5 Pa. At a second point in the pipe, the water passes through a constriction where the radius is 2.00 cm. What is the water's absolute pressure as it flows through this constriction?

Problems

14.42 In a lecture demonstration, a professor pulls apart two hemispherical, steel shells (diameter D) with ease using their attached handles. She then places them together, pumps out the air to an absolute pressure of p, and hands them to a bodybuilder in the back row to pull apart. a) If atmospheric pressure is p_0, how much force must the bodybuilder exert on each shell? b) Evaluate your answer for the case $p = 0.025$ atm, $D = 10.0$ cm.

14.43 The deepest point known in any of the earth's oceans is in the Marianas Trench, 10.92 km deep. a) Assuming water to be incompressible, what is the pressure at this depth? Use the density of seawater. b) The actual pressure is 1.16×10^8 Pa; your calculated value will be less because the density actually varies with depth. Using the compressibility of water and the actual pressure, find the density of the water at the bottom of the Marianas Trench. What is the percent change in the density of the water?

14.44 A swimming pool measures 5.0 m long $\times$ 4.0 m wide $\times$ 3.0 m deep. Compute the force exerted by the water against a) the bottom; b) either end. (*Hint:* Calculate the force on a thin, horizontal strip at a depth h, and integrate this over the end of the pool.) Do not include the force due to air pressure.

14.45 The upper edge of a gate in a dam runs along the water surface. The gate is 2.00 m high and 4.00 m wide and is hinged along a horizontal line through its center (Fig. 14.33). Calculate the torque about the hinge arising from the force due to the water. (*Hint:* Use a procedure

Figure 14.33 Problem 14.45.

similar to that used in Problem 14.44; calculate the torque of a thin, horizontal strip at a depth h and integrate this over the gate.)

14.46 **Force and Torque on a Dam.** A dam has the shape of a rectangular solid. The side facing the lake has area A and height H. The surface of the freshwater lake behind the dam is at the top of the dam. a) Show that the net horizontal force exerted by the water on the dam equals $\frac{1}{2}\rho gHA$, that is, the average gauge pressure across the face of the dam times the area (see Problem 14.44). b) Show that the torque exerted by the water about an axis along the bottom of the dam is $\rho gH^2A/6$. c) How do the force and torque depend on the size of the lake?

14.47 An astronaut is standing at the north pole of a newly discovered, spherically symmetric planet of radius R. In his hands he holds a container full of a liquid with mass m and volume V. At the surface of the liquid, the pressure is p_0; at a depth d below the surface, the pressure has a greater value p. From this information, determine the mass of the planet.

14.48 To find the density at a given point within a material, consider a small volume dV centered on that point. If the mass within that volume is dm, the density at the point is $\rho = dm/dV$. Consider a cylindrical rod with mass M, radius R, and length L with a density that is proportional to the square of the distance from one end, $\rho = Cx^2$. a) Show that $C = 3M/\pi R^2L^3$. b) Show that the average density, given by Eq. (14.1), is $\frac{1}{3}$ of the density at the $x = L$ end.

14.49 The earth does not have a uniform density; it is most dense at its center and least dense at its surface. An approximation of its density is $\rho(r) = A - Br$, where $A = 12{,}700$ kg/m^3 and $B = 1.50 \times 10^{-3}$ kg/m^4. Use $R = 6.37 \times 10^6$ m for the radius of the earth approximated as a sphere. a) Geological evidence indicates that the densities are 13,100 kg/m^3 and 2,400 kg/m^3 at the earth's center and surface. What values does the linear approximation model give for the densities at these two locations? b) Imagine dividing the earth into concentric, spherical shells. Each shell has radius r, thickness dr, volume $dV = 4\pi r^2 dr$, and mass $dm = \rho(r)\, dV$. By integrating from $r = 0$ to $r = R$, show that the mass of the earth in this model is $M = \frac{4}{3}\pi R^3 (A - \frac{3}{4}BR)$. c) Show that the given values of A and B give the correct mass of the earth to within 0.4%. d) We saw in Section 12.6 that a uniform spherical shell gives no contribution to g inside it. Show that $g(r) = \frac{4}{3}\pi Gr(A - \frac{3}{4}Br)$ inside the earth in this model. e) Verify that the expression of part (d) gives $g = 0$ at the center of the earth and $g = 9.85$ m/s^2 at the surface. f) Show that in this model g does *not* decrease uniformly with depth but rather has a maximum of $4\pi GA^2/9B = 10.01$ m/s^2 at $r = 2A/3B = 5640$ km.

14.50 In Example 12.10 (Section 12.6) we saw that inside a planet of uniform density (not a very realistic assumption for the earth) the acceleration due to gravity increases uniformly with distance from the center of the planet. That is, $g(r) = g_s r/R$, where g_s is the acceleration due to gravity at the surface, r is the distance from the center of the planet, and R is the radius of the planet. The interior of the planet can be treated approximately as an incompressible fluid of density ρ. a) Replace the height y in Eq. (14.4) with the radial coordinate r and integrate to find the pressure inside a uniform planet as a function of r. Let the pressure at the surface be zero. (This means ignoring the pressure of the planet's atmosphere.) b) Using this model, calculate the pressure at the center of the earth. (Use a value of ρ equal to the average density of the earth, calculated from the mass and radius given in Appendix F.) c) Geologists estimate the pressure at the center of the earth to be approximately 4×10^{11} Pa. Does this agree with your calculation for the pressure at $r = 0$? What might account for any differences?

14.51 A U-shaped tube open to the air at both ends contains some mercury. A quantity of water is carefully poured into the left arm of the U-shaped tube until the vertical height of the water column is 15.0 cm (Fig. 14.34). a) What is the gauge pressure at the water-mercury interface? b) Calculate the vertical distance h from the top of the mercury in the right-hand arm of the tube to the top of the water in the left-hand arm.

Figure 14.34 Problem 14.51.

14.52 The Great Molasses Flood. On the afternoon of January 15, 1919, an unusually warm day in Boston, a 27.4-m-high, 27.4-m-diameter cylindrical metal tank used for storing molasses ruptured. Molasses flooded into the streets in a 9-m-deep stream, killing pedestrians and horses, and knocking down buildings. The molasses had a density of 1600 kg/m^3. If the tank was full before the accident, what was the total outward force the molasses exerted on its sides? (*Hint:* Consider the outward force on a circular ring of the tank wall of width dy and at a depth y below the surface. Integrate to find the total outward force. Assume that before the tank ruptured, the pressure at the surface of the molasses was equal to the air pressure outside the tank.)

14.53 An open barge has the dimensions shown in Fig. 14.35. If the barge is made out of 4.0-cm-thick steel plate on each of its four sides and its bottom, what mass of coal can the barge carry in fresh water without sinking? Is there enough room in the barge to hold this amount of coal? (The density of coal is about 1500 kg/m^3.)

Figure 14.35 Problem 14.53.

14.54 A hot-air balloon has a volume of 2200 m^3. The balloon fabric (the envelope) weighs 900 N. The basket with gear and full propane tanks weighs 1700 N. If the balloon can barely lift an additional 3200 N of passengers, breakfast, and champagne when the outside air density is 1.23 kg/m^3, what is the average density of the heated gases in the envelope?

14.55 Advertisements for a certain small car claim that it floats in water. a) If the car's mass is 900 kg and its interior volume is 3.0 m^3, what fraction of the car is immersed when it floats? You can ignore the volume of steel and other materials. b) Water gradually leaks in and displaces the air in the car. What fraction of the interior volume is filled with water when the car sinks?

14.56 A single ice cube with mass 9.70 g floats in a glass completely full of 420 cm^3 of water. You can ignore the water's surface tension and its variation in density with temperature (as long as it remains a liquid). a) What volume of water does the ice cube displace? b) When the ice cube has completely melted, has any water overflowed? If so, how much? If not, explain why this is so. c) Suppose the water in the glass had been very salty water of density 1050 kg/m^3. What volume of salt water would the 9.70-g ice cube displace? d) Redo part (b) for the freshwater ice cube in the salty water.

14.57 A piece of wood is 0.600 m long, 0.250 m wide, and 0.080 m thick. Its density is 600 kg/m^3. What volume of lead must be fastened underneath it to sink the wood in calm water so that its top is just even with the water level? What is the mass of this volume of lead?

14.58 A hydrometer consists of a spherical bulb and a cylindrical stem with a cross-sectional area of 0.400 cm^2 (Fig. 14.12a). The total volume of bulb and stem is 13.2 cm^3. When immersed in water, the hydrometer floats with 8.00 cm of the stem above the water surface. When immersed in an organic fluid, 3.20 cm of the stem is above the surface. Find the density of the organic fluid. (*Note:* This illustrates the precision of such a hydrometer. Relatively small density differences give rise to relatively large differences in hydrometer readings.)

14.59 The densities of air, helium, and hydrogen (at $p = 1.0$ atm and $T = 20°C$) are 1.20 kg/m^3, 0.166 kg/m^3, and 0.0899 kg/m^3, respectively. a) What is the volume in cubic meters displaced by a hydrogen-filled airship that has a total "lift" of 120 kN? (The "lift" is the amount by which the buoyant force exceeds the weight of the gas that fills the airship.) b) What would be the "lift" if helium were used instead of hydrogen? In view of your answer, why is helium used in modern airships like advertising blimps?

14.60 SHM of a Floating Object. An object with height h, mass M, and a uniform cross-sectional area A floats upright in a liquid with density ρ. a) Calculate the vertical distance from the surface of the liquid to the bottom of the floating object at equilibrium. b) A downward force with magnitude F is applied to the top of the object. At the new equilibrium position, how much farther below the surface of the liquid is the bottom of the object than it was in part (a)? (Assume that some of the object remains above the surface of the liquid.) c) Your result in part (b) shows that if the force is suddenly removed, the object will oscillate up and down in SHM. Calculate the period of this motion in terms of the density ρ of the liquid, the mass M, and cross-sectional area A of the object. You can ignore the damping due to fluid friction (Section 13.7).

14.61 A 950-kg cylindrical can buoy floats vertically in salt water. The diameter of the buoy is 0.900 m. a) Calculate the additional distance the buoy will sink when a 70.0-kg man stands on top. (Use the expression derived in part (b) of Problem 14.60.) b) Calculate the period of the resulting vertical SHM when the man dives off. (Use the expression derived in part (c) of Problem 14.60, and as in that problem, you can ignore the damping due to fluid friction.)

14.62 In seawater, a life preserver with a volume of 0.0400 m^3 will support a 75.0-kg person (average density 980 kg/m^3) with 20% of the person's volume above water when the life preserver is fully submerged. What is the density of the material composing the life preserver?

14.63 A block of balsa wood placed in one scale pan of an equal-arm balance is exactly balanced by a 0.0950-kg brass mass in the other scale pan. Find the true mass of the balsa wood if its density is 150 kg/m^3. Explain why it is accurate to ignore the buoyancy in air of the brass but *not* the buoyancy in air of the balsa wood.

14.64 Block A in Fig. 14.36 hangs by a cord from spring balance D and is submerged in a liquid C contained in beaker B. The mass of the beaker is 1.00 kg; the mass of the liquid is 1.80 kg. Balance D reads 3.50 kg and balance E reads 7.50 kg. The volume of block A is 3.80×10^{-3} m^3. a) What is the density of the liquid? b) What will each balance read if block A is pulled up out of the liquid?

Figure 14.36 Problem 14.64.

14.65 A hunk of aluminum is completely covered with a gold shell to form an ingot of weight 45.0 N. When you suspend the ingot from a spring balance and submerge the ingot in water, the balance reads 39.0 N. What is the weight of the gold in the shell?

14.66 A plastic ball has radius 12.0 cm and floats in water with 16.0% of its volume submerged. a) What force must you apply to the ball to hold it at rest totally below the surface of the water? b) If you let go of the ball, what is its acceleration the instant you release it?

14.67 The weight of a king's solid crown is w. When suspended by a light rope and completely immersed in water, the tension in the rope (the crown's apparent weight) is fw. a) Prove that the crown's relative density (specific gravity) is $1/(1 - f)$. Discuss the meaning of the limits as f approaches 0 and 1. b) If the crown is solid gold and weighs 12.9 N in air, what is its apparent weight when completely immersed in water? c) Repeat part (b) if the crown is solid lead with a very thin gold plating, but still has a weight in air of 12.9 N.

14.68 A piece of steel has a weight w, an apparent weight (see Problem 14.67) w_{water} when completely immersed in water, and an apparent weight w_{fluid} when completely immersed in an unknown fluid. a) Prove that the fluid's density relative to water (specific gravity) is $(w - w_{fluid})/(w - w_{water})$. b) Is this result reasonable for the three cases of w_{fluid} greater than, equal to, or less than w_{water}? c) The apparent weight of the piece of steel in water of density 1000 kg/m^3 is 87.2% of its weight. What percentage of its weight will its apparent weight be in formic acid (density 1220 kg/m^3)?

14.69 You cast some metal of density ρ_m in a mold, but you are worried that there might be cavities within the casting. You measure the weight of the casting to be w, and the buoyant force when it is completely surrounded by water to be B. a) Show that $V_0 = B/(\rho_{water}g) - w/(\rho_m g)$ is the total volume of any enclosed cavities. b) If your metal is copper, the casting's weight is 156 N, and the buoyant force is 20 N, what is the total volume of any enclosed cavities in your casting? What fraction is this of the total volume of the casting?

14.70 A cubical block of wood 0.100 m on a side and with a density of 550 kg/m^3 floats in a jar of water. Oil with a density of 750 kg/m^3 is poured on the water until the top of the oil layer is 0.035 m below the top of the block. a) How deep is the oil layer? b) What is the gauge pressure at the block's lower face?

14.71 Dropping Anchor. An iron anchor with mass 35.0 kg and density 7860 kg/m^3 lies on the deck of a small barge that has vertical sides and floats in a freshwater river. The area of the bottom of the barge is 8.00 m^2. The anchor is thrown overboard but is suspended above the bottom of the river by a rope; the mass and volume of the rope are small enough to ignore. After the anchor is overboard and the barge has finally stopped bobbing up and down, has the barge risen or sunk down in the water? By what vertical distance?

14.72 Assume that crude oil from a supertanker has density 750 kg/m^3. The tanker runs aground on a sandbar. To refloat the tanker, its oil cargo is pumped out into steel barrels, each of which has a mass of 15.0 kg when empty and holds 0.120 m^3 of oil. You can ignore the volume occupied by the steel from which the barrel is made. a) If a salvage worker accidentally drops a filled, sealed barrel overboard, will it float or sink in the seawater? b) If the barrel floats, what fraction of its volume will be above the water surface? If it sinks, what minimum tension would have to be exerted by a rope to haul the barrel up from the ocean floor? c) Repeat parts (a) and (b) if the density of the oil is 910 kg/m³ and the mass of each empty barrel is 32.0 kg.

14.73 A cubical block of density ρ_B and with sides of length L floats in a liquid of greater density ρ_L. a) What fraction of the block's volume is above the surface of the liquid? b) The liquid is denser than water (density ρ_W) and does not mix with it. If water is poured on the surface of the liquid, how deep must the water layer be so that the water surface just rises to the top of the block? Express your answer in terms of L, ρ_B, ρ_L, and ρ_W. c) Find the depth of the water layer in part (b) if the liquid is mercury, the block is made of iron, and the side length is 10.0 cm.

14.74 A barge is in a rectangular lock on a freshwater river. The lock is 60.0 m long and 20.0 m wide, and the steel doors on each end are closed. With the barge floating in the lock, a 2.50×10^6 N load of scrap metal is put onto the barge. The metal has density 9000 kg/m³. a) When the load of scrap metal, initially on the bank, is placed onto the barge, what vertical distance does the water in the lock rise? b) The scrap metal is now pushed overboard into the water. Does the water level in the lock rise, fall, or remain the same? If it rises or falls, by what vertical distance does it change?

14.75 A U-shaped tube with a horizontal portion of length l (Fig. 14.37) contains a liquid. What is the difference in height between the liquid columns in the vertical arms a) if the tube has an acceleration a toward the right? b) if the tube is mounted on a horizontal turntable rotating with an angular speed ω with

Figure 14.37 Problem 14.75.

one of the vertical arms on the axis of rotation? c) Explain why the difference in height does not depend on the density of the liquid or on the cross-sectional area of the tube. Would it be the same if the vertical tubes did not have equal cross-sectional areas? Would it be the same if the horizontal portion were tapered from one end to the other? Explain.

14.76 A cylindrical container of an incompressible liquid with density ρ rotates with constant angular speed ω about its axis of symmetry, which we take to be the y-axis (Fig. 14.38). a) Show that the pressure at a given height within the fluid increases in the radial direction (outward from the axis of rotation) according to $\partial p / \partial r = \rho \omega^2 r$. b) Integrate this partial differen-

Figure 14.38 Problem 14.76.

tial equation to find the pressure as a function of distance from the axis of rotation along a horizontal line at $y = 0$. c) Combine the result of part (b) with Eq. (14.5) to show that the surface of the rotating liquid has a *parabolic* shape, that is, the height of the liquid is given by $h(r) = \omega^2 r^2 / 2g$. (This technique is used for making parabolic telescope mirrors; liquid glass is rotated and allowed to solidify while rotating.)

14.77 An incompressible fluid with density ρ is in a horizontal test tube of inner cross-sectional area A. The test tube spins in a horizontal circle in an ultracentrifuge at an angular speed ω. Gravitational forces are negligible. Consider a volume element of the fluid of area A and thickness dr' a distance r' from the rotation axis. The pressure on its inner surface is p and on its outer surface is $p + dp$. a) Apply Newton's second law to the volume element to show $dp = \rho \omega^2 r' \, dr'$. b) If the surface of the fluid is at a radius r_0 where the pressure is p_0, show that the pressure p at a distance $r \geq r_0$ is $p = p_0 + \rho \omega^2 (r^2 - r_0^2)/2$. c) An object of volume V and density ρ_{ob} has its center of mass at a distance R_{cmob} from the axis. Show that the net horizontal force on the object is $\rho V \omega^2 R_{cm}$, where R_{cm} is the distance from the axis to the center of mass of the displaced fluid. d) Explain why the object will move inward if $\rho R_{cm} > \rho_{ob} R_{cmob}$ and outward if $\rho R_{cm} < \rho_{ob} R_{cmob}$. e) For small objects of uniform density, $R_{cm} = R_{cmob}$. What happens to a mixture of small objects of this kind with different densities in an ultracentrifuge?

14.78 Untethered helium balloons, floating in a car that has all the windows rolled up and outside air vents closed, move in the direction of the car's acceleration, but loose balloons filled with air move in the opposite direction. To show why, consider only the horizontal forces acting on the balloons. Let a be the magnitude of the car's forward acceleration. Consider a horizontal tube of air with a cross-sectional area A that extends from the windshield, where $x = 0$ and $p = p_0$, back along the x-axis. Now consider a volume element of thickness dx in this tube. The pressure on its front surface is p and the pressure on its rear surface is $p + dp$. Assume the air has a constant density ρ. a) Apply Newton's second law to the volume element to show that $dp = \rho a \, dx$. b) Integrate the result of part (a) to find the pressure at the front surface in terms of a and x. c) To show that considering ρ constant is reasonable, calculate the pressure difference in atm for a distance as long as 2.5 m and a large acceleration of 5.0 m/s². d) Show that the net horizontal force on a balloon of volume V is ρVa. e) For negligible friction forces, show that the acceleration of the balloon (average density ρ_{bal}) is $(\rho / \rho_{bal})a$, so that the acceleration relative to the car is $a_{rel} = [(\rho / \rho_{bal}) - 1]a$. f) Use the expression for a_{rel} in part (e) to explain the movement of the balloons.

14.79 A cubical block of wood 0.30 m on a side is weighted so that its center of gravity is at the point shown in Fig. 14.39a and it floats in water with $\frac{1}{2}$ its volume submerged. If the block is "heeled" at an angle of 45.0° as in Fig. 14.39b, compute the net torque about a horizontal axis perpendicular to the block and passing through its geometrical center.

Figure 14.39 Problem 14.79.

14.80 Water stands at a depth H in a large, open tank whose side walls are vertical (Fig. 14.40). A hole is made in one of the walls at a depth h below the water surface. a) At what distance R from the foot of the wall does the emerging stream strike the floor? b) How far above the bottom of the tank could a second hole be cut so that the stream emerging from it could have the same range as for the first hole?

Figure 14.40 Problem 14.80.

14.81 A cylindrical bucket, open at the top, is 25.0 cm high and 10.0 cm in diameter. A circular hole with a cross-sectional area 1.50 cm^2 is cut in the center of the bottom of the bucket. Water flows into the bucket from a tube above it at the rate of 2.40×10^{-4} m^3/s. How high will the water in the bucket rise?

14.82 Water flows steadily from an open tank as in Fig. 14.41. The elevation of point 1 is 10.0 m, and the elevation of points 2 and 3 is 2.00 m. The cross-sectional area at point 2 is 0.0480 m^2; at point 3 it is 0.0160 m^2. The area of the tank is very large compared with the cross-sectional area of the pipe. Assuming that Bernoulli's equation applies, compute a) the discharge rate in cubic meters per second; b) the gauge pressure at point 2.

Figure 14.41 Problem 14.82.

14.83 Modern airplane design calls for a lift, due to the net force of the moving air on the wing, of about 2000 N per square meter of wing area. Assume that air flows past the wing of an aircraft with streamline flow. If the speed of flow past the lower wing surface is 120 m/s, what is the required speed over the upper surface to give a lift of 2000 N/m^2? The density of the air is 1.20 kg/m^3.

14.84 In 1993 the radius of Hurricane Emily was about 350 km. The wind speed near the center ("eye") of the hurricane, whose radius was about 30 km, reached about 200 km/h. As air swirled in from the rim of the hurricane toward the eye, its angular momentum remained roughly constant. a) Estimate the wind speed at the rim of the hurricane. b) Estimate the pressure difference at the

earth's surface between the eye and the rim. (*Hint:* See Table 14.1.). Where is the pressure greater? c) If the kinetic energy of the swirling air in the eye could be converted completely to gravitational potential energy, how high would the air go? d) In fact the air in the eye is lifted to heights of several kilometers. How can you reconcile this with your answer to part (c)?

14.85 Two very large open tanks A and F (Fig. 14.42) contain the same liquid. A horizontal pipe BCD, having a constriction at C and open to the air at D, leads out of the bottom of tank A, and a vertical pipe E opens into the constriction at C and dips into the liquid in tank F. Assume streamline flow and no viscosity. If the cross-sectional area at C is one half that at D and if D is a distance h_1 below the level of the liquid in A, to what height h_2 will liquid rise in pipe E? Express your answer in terms of h_1.

Figure 14.42 Problem 14.85.

14.86 The horizontal pipe, shown in Fig. 14.43, has a cross-sectional area of 40.0 cm^2 at the wider portions and 10.0 cm^2 at the constriction. Water is flowing in the pipe, and the discharge from the pipe is 6.00×10^{-3} m^3/s $(6.00\ \text{L/s})$. Find a) the flow speeds at the wide and the narrow portions; b) the pressure difference

Figure 14.43 Problem 14.86.

between these portions; c) the difference in height between the mercury columns in the U-shaped tube.

14.87 A liquid flowing from a vertical pipe has a very definite shape as it flows from the pipe. To get the equation for this shape, assume that the liquid is in free fall once it leaves the pipe. Just as it leaves the pipe, the liquid has speed v_0 and the radius of the stream of liquid is r_0. a) Find an equation for the speed of the liquid as a function of the distance y it has fallen. Combining this with the equation of continuity, find an expression for the radius of the stream as a function of y. b) If water flows out of a vertical pipe at a speed of 1.20 m/s, how far below the outlet will the radius be $\frac{1}{2}$ the original radius of the stream?

Challenge Problems

14.88 A rock with mass $m = 3.00$ kg is suspended from the roof of an elevator by a light cord. The rock is totally immersed in a bucket of water that sits on the floor of the elevator, but the rock doesn't touch the bottom or sides of the bucket. a) When the elevator is at rest, the tension in the cord is 21.0 N. Calculate the volume of the rock. b) Derive an expression for the tension in the cord when the elevator is accelerating *upward* with an acceleration of magnitude a. Calculate the tension when $a = 2.50$ m/s^2 upward. c) Derive an expression for the tension in the cord when the elevator is accelerating *downward* with an acceleration of magnitude a. Calculate the tension when $a = 2.50$ m/s^2 downward. d) What is the tension when the elevator is in free fall with a downward acceleration equal to g?

14.89 Suppose a piece of styrofoam, $\rho = 180$ kg/m^3, is held completely submerged in water (Fig. 14.44). a) What is the tension in the cord? Find this using Archimedes's principle. b) Use $p = p_0 + \rho g h$ to calculate directly the force exerted by the water on the two sloped sides and the bottom of the styrofoam; then show that the vector sum of these forces is the buoyant force.

Figure 14.44 Challenge Problem 14.89.

14.90 A large tank with diameter D, open to the air, contains water to a height H. A small hole with diameter $d (d \ll D)$ is made at the base of the tank. Ignoring any effects of viscosity, calculate the time it takes for the tank to drain completely.

14.91 A *siphon*, as shown in Fig. 14.45, is a convenient device for removing liquids from containers. To establish the

Figure 14.45 Challenge Problem 14.91.

flow, the tube must be initially filled with fluid. Let the fluid have density ρ and let the atmospheric pressure be p_a. Assume that the cross-sectional area of the tube is the same at all points along it. a) If the lower end of the siphon is at a distance h below the surface of the liquid in the container, what is the speed of the fluid as it flows out the lower end of the siphon? (Assume that the container has a very large diameter, and ignore any effects of viscosity.) b) A curious feature of a siphon is that the fluid initially flows "uphill." What is the greatest height H that the high point of the tube can have if flow is still to occur?

14.92 The following is quoted from a letter. *It is the practice of carpenters hereabouts, when laying out and leveling up the foundations of relatively long buildings, to use a garden hose filled with water, with glass tubes 10 to 12 inches long thrust into the ends of the hose. The theory is that water, seeking a common level, will be the same height in both the tubes and thus effect a level. Now the question rises as to what happens if a bubble of air is left in the hose. Our greybeards contend the air will not affect the reading from one end to the other. Others say that it will cause important inaccuracies.* Can you give a relatively simple answer to this question, together with an explanation? Fig. 14.46 gives a rough sketch of the situation that caused the dispute.

Figure 14.46 Challenge Problem 14.92.

MECHANICAL WAVES

When an earthquake strikes, the news of the event travels through the body of the earth in the form of seismic waves. By studying such waves, geophysicists learn about the internal structure of the earth and where future earthquake activity is likely to occur.

? **Some seismic waves are more energetic and destructive than others. Which aspects of a seismic wave determine how much power is carried by the wave?**

Ripples on a pond, musical sounds, seismic tremors triggered by an earthquake—all these are *wave* phenomena. Waves can occur whenever a system is disturbed from equilibrium and when the disturbance can travel, or *propagate,* from one region of the system to another. As a wave propagates, it carries energy. The energy in light waves from the sun warms the surface of our planet; the energy in seismic waves can crack our planet's crust.

This chapter and the next are about mechanical waves—waves that travel within some material called a *medium*. (Chapter 16 is concerned with sound, an important type of mechanical wave.) We'll begin this chapter by deriving the basic equations for describing waves, including the important special case of *sinusoidal* waves in which the wave pattern is a repeating sine or cosine function. To help us understand waves in general, we'll look at the simple case of waves that travel on a stretched string or rope.

Waves on a string play an important role in music. When a musician strums a guitar or bows a violin, she makes waves that travel in opposite directions along the instrument's strings. What happens when these oppositely directed waves overlap is called *interference*. We'll discover that sinusoidal waves can occur on a guitar or violin string only for certain special frequencies, called *normal-mode frequencies,* determined by the properties of the string. The normal-mode frequencies of a stringed instrument determine the pitch of the musical sounds that the instrument produces. (In the next chapter we'll find that interference also helps explain the pitches of *wind* instruments such as flutes and pipe organs.)

Not all waves are mechanical in nature. *Electromagnetic* waves—including light, radio waves, infrared and ultraviolet radiation, and x-rays—can propagate even in empty space, where there is *no* medium. We'll explore these and other nonmechanical waves in later chapters.

15.1 | Types of Mechanical Waves

A **mechanical wave** is a disturbance that travels through some material or substance called the **medium** for the wave. As the wave travels through the medium, the particles that make up the medium undergo displacements of various kinds, depending on the nature of the wave.

Figure 15.1 shows three varieties of mechanical waves. In Fig. 15.1a the medium is a string or rope under tension. If we give the left end a small upward shake or wiggle, the wiggle travels along the length of the string. Successive sections of string go through the same motion that we gave to the end, but at successively later times. Because the displacements of the medium are perpendicular or *transverse* to the direction of travel of the wave along the medium, this is called a **transverse wave.**

(a) Transverse waves on a string

(b) Longitudinal waves in a fluid

(c) Combined transverse and longitudinal waves on the surface of a liquid

Particles of medium

Particles of medium

Surface particles of medium

15.1 Three ways to make a wave that moves to the right. (a) The hand moves the string up and then returns, producing a transverse wave. (b) The piston moves to the right, compressing the gas or liquid, then returns, producing a longitudinal wave. (c) The board moves to the right then returns, producing a combination of longitudinal and transverse waves.

In Fig. 15.1b the medium is a liquid or gas in a tube with a rigid wall at the right end and a movable piston at the left end. If we give the piston a single back-and-forth motion, displacement and pressure fluctuations travel down the length of the medium. This time the motions of the particles of the medium are back and forth along the *same* direction that the wave travels. We call this a **longitudinal wave.**

In Fig. 15.1c the medium is water in a channel, such as an irrigation ditch or canal. When we move the flat board at the left end forward and back once, a wave disturbance travels down the length of the channel. In this case the displacements of the water have *both* longitudinal and transverse components.

Each of these systems has an equilibrium state. For the stretched string it is the state in which the system is at rest, stretched out along a straight line. For the fluid in a tube it is a state in which the fluid is at rest with uniform pressure, and for the water in a trough it is a smooth, level water surface. In each case the wave motion is a disturbance from the equilibrium state that travels from one region of the medium to another. And in each case there are forces that tend to restore the system to its equilibrium position when it is displaced, just as the force of gravity tends to pull a pendulum toward its straight-down equilibrium position when it is displaced.

These examples have three things in common. First, in each case the disturbance travels or *propagates* with a definite speed through the medium. This speed is called the speed of propagation, or simply the **wave speed.** It is determined in each case by the mechanical properties of the medium. We will use the symbol v for wave speed. (The wave speed is *not* the same as the speed with which particles move when they are disturbed by the wave. We'll return to this point in Section 15.3.) Second, the medium itself does not travel through space; its individual particles undergo back-and-forth or up-and-down motions around their equilibrium positions. The overall pattern of the wave disturbance is what travels. Third, to set any of these systems into motion, we have to put in energy by doing mechanical work on the system. The wave motion transports this energy from one region of the medium to another. *Waves transport energy, but not matter, from one region to another.*

10.1 Properties of Mechanical Waves

15.2 | Periodic Waves

The transverse wave on a stretched string in Fig. 15.1a is an example of a *wave pulse.* The hand shakes the string up and down just once, exerting a transverse force on it as it does so. The result is a single "wiggle," or pulse, that travels along the length of the string. The tension in the string restores its straight-line shape once the pulse has passed.

A more interesting situation develops when we give the free end of the string a repetitive, or *periodic,* motion. (You may want to review the discussion of periodic motion in Chapter 13 before going ahead.) Then each particle in the string also undergoes periodic motion as the wave propagates, and we have a **periodic wave.**

In particular, suppose we move the string up and down with *simple harmonic motion* (SHM) with amplitude A, frequency f, angular frequency $\omega = 2\pi f$, and period $T = 1/f = 2\pi/\omega$. Figure 15.2 shows one way to do this. The wave that results is a symmetrical sequence of *crests* and *troughs.* As we will see, periodic waves with simple harmonic motion are particularly easy to analyze; we call them **sinusoidal waves.** It also turns out that *any* periodic wave can be represented as a combination of sinusoidal waves. So this particular kind of wave motion is worth special attention.

In Fig. 15.2 the wave that advances along the string is a *continuous succession* of transverse sinusoidal disturbances. Figure 15.3 shows the shape of a part of the string near the left end at time intervals of $\frac{1}{8}$ of a period, for a total time of one period. The wave shape advances steadily toward the right, as indicated by the

$t = 0$

All particles on string oscillate in SHM
with same amplitude and period

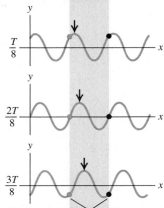

$\dfrac{T}{8}$

$\dfrac{2T}{8}$

$\dfrac{3T}{8}$

Two particles one wavelength apart
oscillate in phase with each other

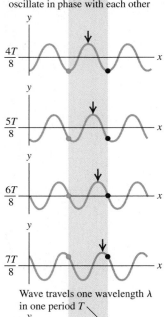

$\dfrac{4T}{8}$

$\dfrac{5T}{8}$

$\dfrac{6T}{8}$

$\dfrac{7T}{8}$

Wave travels one wavelength λ
in one period T

T

15.3 A sinusoidal transverse wave traveling to the right along a string. The shape of
the string is shown at intervals of $\frac{1}{8}$ of a
period; the vertical scale is exaggerated.
The blue and red dots represent particles in
the string whose x-coordinates differ by the
wavelength λ.

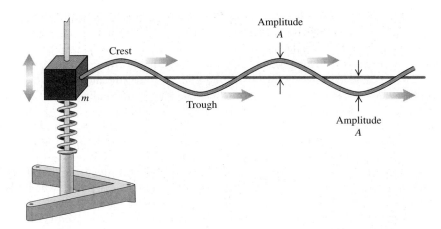

15.2 A block of mass m attached to a spring undergoes simple harmonic motion, producing a sinusoidal wave that travels to the right on the string. The amplitude of the wave is
the same as the amplitude of the spring's oscillation. (In a real-life system a driving force
would have to be applied to the block to replace the energy carried away by the wave.)

short red arrow pointing to a particular wave crest. As the wave moves, any point
on the string (the blue dot, for example) oscillates up and down about its equilibrium position with simple harmonic motion. *When a sinusoidal wave passes
through a medium, every particle in the medium undergoes simple harmonic
motion with the same frequency.*

> **CAUTION** Be very careful to distinguish between the motion of the *trans-
> verse wave* along the string and the motion of a *particle* of the string. The wave
> moves with constant speed v *along* the length of the string, while the motion
> of the particle is simple harmonic and *transverse* (perpendicular) to the length
> of the string.

For a periodic wave, the shape of the string at any instant is a repeating pattern. The length of one complete wave pattern is the distance from one crest to
the next, or from one trough to the next, or from any point to the corresponding
point on the next repetition of the wave shape. We call this distance the **wavelength** of the wave, denoted by λ (the Greek letter "lambda"). The wave pattern
travels with constant speed v and advances a distance of one wavelength λ in a
time interval of one period T. So the wave speed v is given by $v = \lambda/T$, or,
because $f = 1/T$,

$$v = \lambda f \qquad \text{(periodic wave)} \qquad (15.1)$$

The speed of propagation equals the product of wavelength and frequency. The
frequency is a property of the *entire* periodic wave because all points on the string
oscillate with the same frequency f.

Waves on a string propagate in just one dimension (in Fig. 15.3, along the
x-axis). But the ideas of frequency, wavelength, and amplitude apply equally well
to waves that propagate in two or three dimensions. Figure 15.4 shows a wave
propagating in two dimensions on the surface of a tank of water. As with waves on
a string, the wavelength is the distance from one crest to the next and the amplitude is the height of a crest.

15.4 A series of drops falling into water produces a periodic wave that spreads radially outward. The wave crests and troughs are concentric circles. The wavelength λ is the radial distance between adjacent crests or adjacent troughs.

In many important situations including waves on a string, the wave speed v is determined entirely by the mechanical properties of the medium. In this case, increasing f causes λ to decrease so that the product $v = \lambda f$ remains the same, and waves of *all* frequencies propagate with the same wave speed. In this chapter we will consider *only* waves of this kind. (In later chapters we will study the propagation of light waves in matter for which the wave speed depends on frequency; this turns out to be the reason why prisms break white light into a spectrum and why raindrops create a rainbow.)

To understand the mechanics of a periodic *longitudinal* wave, we consider a long tube filled with a fluid, with a piston at the left end as in Fig. 15.1b. If we push the piston in, we compress the fluid near the piston, increasing the pressure in this region. This region then pushes against the neighboring region of fluid, and so on, and a wave pulse moves along the tube.

Now suppose we move the piston back and forth with simple harmonic motion, along a line parallel to the axis of the tube (Fig. 15.5). This motion forms regions in the fluid where the pressure and density are greater or less than the equilibrium values. We call a region of increased density a *compression;* a region of reduced density is a *rarefaction*. Figure 15.5 shows compressions as darkly shaded areas and rarefactions as lightly shaded areas. The short red arrow in Fig. 15.5 shows the position of one particular compression; the compressions and rarefactions move to the right with constant speed v.

The motion of a single particle of the medium, such as the one shown by a blue dot in Fig. 15.5, is simple harmonic motion parallel to the direction of wave propagation. The wavelength is the distance from one compression to the next or from one rarefaction to the next. The fundamental equation $v = \lambda f$ holds for longitudinal waves as well as for transverse waves, and indeed for *all* types of periodic waves. Just as for transverse waves, in this chapter and the next we will consider only situations in which the speed of longitudinal waves does not depend on the frequency.

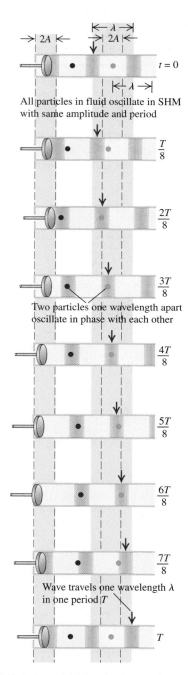

15.5 A sinusoidal longitudinal wave traveling to the right in a fluid. The wave has the same amplitude and period as the oscillation of the piston. The blue and red dots represent particles in the fluid whose x-coordinates differ by the wavelength λ. In equilibrium the particle shown by the red dot is $\frac{1}{2}$ wavelength from the piston; its oscillations are one half-cycle out of phase with the piston.

Within figure 15.5, the labels read:

All particles in fluid oscillate in SHM with same amplitude and period

$t = 0$

$\dfrac{T}{8}$

$\dfrac{2T}{8}$

$\dfrac{3T}{8}$

Two particles one wavelength apart oscillate in phase with each other

$\dfrac{4T}{8}$

$\dfrac{5T}{8}$

$\dfrac{6T}{8}$

$\dfrac{7T}{8}$

Wave travels one wavelength λ in one period T

T

Example 15.1	**Wavelength of a musical sound**

Sound waves are longitudinal waves in air. The speed of sound depends on temperature; at 20°C it is 344 m/s (1130 ft/s). What is the wavelength of a sound wave in air at 20°C if the frequency is 262 Hz (the approximate frequency of middle C on a piano)?

SOLUTION

IDENTIFY AND SET UP: The target variable is the wavelength λ. The wave speed $v = 344$ m/s and frequency $f = 262$ Hz are given, so we can use the relationship in Eq. (15.1) between λ, v, and f for periodic waves.

EXECUTE: We solve Eq. (15.1) for the target variable λ:

$$\lambda = \frac{v}{f} = \frac{344 \text{ m/s}}{262 \text{ Hz}} = \frac{344 \text{ m/s}}{262 \text{ s}^{-1}} = 1.31 \text{ m}$$

Note that the units of frequency are either hertz (Hz) or inverse seconds (s^{-1}).

EVALUATE: What happens to the wavelength if the frequency changes? The speed of sound waves is unaffected by changes in frequency, so the relationship $\lambda = v/f$ tells us that wavelength will change in inverse proportion to the frequency. As an example, the "high C" sung by coloratura sopranos is two octaves above middle C. Each octave corresponds to a factor of two in frequency, so the frequency of high C is four times that of middle C: $f = 4(262 \text{ Hz}) = 1048$ Hz. Hence the *wavelength* of high C is *one-fourth* as large: $\lambda = (1.31 \text{ m})/4 = 0.328$ m.

Test Your Understanding

If you double the wavelength of a wave on a string, what happens to the speed of the wave? What happens to the frequency of the wave?

15.3 | Mathematical Description of a Wave

Many characteristics of periodic waves can be described by using the concepts of wave speed, amplitude period, frequency, and wavelength. Often, though, we need a more detailed description of the positions and motions of individual particles of the medium at particular times during wave propagation. For this description we need the concept of a *wave function,* a function that describes the position of any particle in the medium at any time. We will concentrate on *sinusoidal* waves, in which each particle undergoes simple harmonic motion about its equilibrium position.

As a specific example, let's look at waves on a stretched string. If we ignore the sag of the string due to gravity, the equilibrium position of the string is along a straight line. We take this to be the *x*-axis of a coordinate system. Waves on a string are *transverse;* during wave motion a particle with equilibrium position *x* is displaced some distance *y* in the direction perpendicular to the *x*-axis. The value of *y* depends on which particle we are talking about (that is, *y* depends on *x*) and also on the time *t* when we look at it. Thus *y* is a *function* of both *x* and *t*; $y = y(x, t)$. We call $y(x, t)$ the **wave function** that describes the wave. If we know this function for a particular wave motion, we can use it to find the displacement (from equilibrium) of any particle at any time. From this we can find the velocity and acceleration of any particle, the shape of the string, and anything else we want to know about the behavior of the string at any time.

Wave Function for a Sinusoidal Wave

Let's see how to determine the form of the wave function for a sinusoidal wave. Suppose a sinusoidal wave travels from left to right (the direction of increasing *x*) along the string, as in Fig. 15.3. Every particle of the string oscillates with simple harmonic motion with the same amplitude and frequency. But the oscillations of particles at different points on the string are *not* all in step with each other. The particle marked by the blue dot in Fig. 15.3 is at its maximum positive value of y at $t = 0$ and returns to $y = 0$ at $t = 2T/8$; these same events occur for a particle at the *center* of the colored band at $t = 4T/8$ and $t = 6T/8$, exactly one half-period later. For any two particles of the string, the motion of the particle on the right (in terms of the wave, the "downstream" particle) lags behind the motion of the particle on the left by an amount proportional to the distance between the particles.

Hence the cyclic motions of various points on the string are out of step with each other by various fractions of a cycle. We call these differences *phase differences,* and we say that the *phase* of the motion is different for different points. For example, if one point has its maximum positive displacement at the same time that another has its maximum negative displacement, the two are a half-cycle out of phase. (This is the case for the blue dot in Fig. 15.3 and a point at the center of the colored band.)

Suppose that the displacement of a particle at the left end of the string $(x = 0)$, where the wave originates, is given by

$$y(x = 0, t) = A \cos \omega t = A \cos 2\pi ft \qquad (15.2)$$

That is, the particle oscillates in simple harmonic motion with amplitude A, frequency f, and angular frequency $\omega = 2\pi f$. The notation $y(x = 0, t)$ reminds us that the motion of this particle is a special case of the wave function $y(x, t)$ that describes the entire wave. At $t = 0$ the particle at $x = 0$ is at its maximum positive displacement $(y = A)$ and is instantaneously at rest (because the value of y is a maximum).

The wave disturbance travels from $x = 0$ to some point x to the right of the origin in an amount of time given by x/v, where v is the wave speed. So the motion of point x at time t is the same as the motion of point $x = 0$ at the earlier time $t - x/v$. Hence we can find the displacement of point x at time t by simply replacing t in Eq. (15.2) by $(t - x/v)$. When we do that, we find the following expression for the wave function:

$$y(x, t) = A \cos\left[\omega\left(t - \frac{x}{v}\right)\right]$$

Because $\cos(-\theta) = \cos\theta$, we can rewrite the wave function as

$$y(x, t) = A \cos\left[\omega\left(\frac{x}{v} - t\right)\right] = A \cos 2\pi f\left(\frac{x}{v} - t\right) \qquad (15.3)$$

(sinusoidal wave moving in $+x$-direction)

The displacement $y(x, t)$ is a function of both the location x of the point and the time t. We could make Eq. (15.3) more general by allowing for different values of the phase angle, as we did for simple harmonic motion in Section 13.2, but for now we omit this.

We can rewrite the wave function given by Eq. (15.3) in several different but useful forms. We can express it in terms of the period $T = 1/f$ and the wavelength $\lambda = v/f$:

$$y(x, t) = A \cos 2\pi\left(\frac{x}{\lambda} - \frac{t}{T}\right) \qquad (15.4)$$

(sinusoidal wave moving in $+x$-direction)

We get another convenient form of the wave function if we define a quantity k, called the **wave number:**

$$k = \frac{2\pi}{\lambda} \quad (\text{wave number}) \tag{15.5}$$

Substituting $\lambda = 2\pi/k$ and $f = \omega/2\pi$ into the wavelength-frequency relation $v = \lambda f$ gives

$$\omega = vk \quad (\text{periodic wave}) \tag{15.6}$$

We can then rewrite Eq. (15.4) as

$$y(x, t) = A\cos(kx - \omega t) \tag{15.7}$$

$$(\text{sinusoidal wave moving in } +x\text{-direction})$$

Which of these various forms for the wave function $y(x, t)$ we use in any specific problem is a matter of convenience. Note that ω has units rad/s, so for unit consistency in Eqs. (15.6) and (15.7) the wave number k must have the units rad/m. (Some physicists define the wave number as $1/\lambda$ rather than $2\pi/\lambda$. When reading other texts, be sure to determine how this term is defined.)

The wave function $y(x, t)$ is graphed as a function of x for a specific time t in Fig. 15.6a. This graph gives the displacement y of a particle from its equilibrium position as a function of the coordinate x of the particle. If the wave is a transverse wave on a string, the graph in Fig. 15.6a represents the shape of the string at that instant, like a flash photograph of the string. In particular, at time $t = 0$,

$$y(x, t = 0) = A\cos kx = A\cos 2\pi\frac{x}{\lambda}$$

Figure 15.6b is a graph of the wave function versus time t for a specific coordinate x. This graph gives the displacement y of the particle at that coordinate as a function of time—that is, it describes the motion of that particle. In particular, at the position $x = 0$,

$$y(x = 0, t) = A\cos(-\omega t) = A\cos \omega t = A\cos 2\pi\frac{t}{T}$$

This is consistent with our original statement about the motion at $x = 0$, Eq. (15.2).

CAUTION Although they may look the same at first glance, Figs. 15.6a and 15.6b are *not* identical. Figure 15.6a is a picture of the shape of the string at $t = 0$, while Fig. 15.6b is a graph of the displacement y of a particle at $x = 0$ as a function of time.

15.6 (a) The graph of $y(x, t)$ versus coordinate x for a specific time t shows the shape of the wave at that time. (b) The graph of $y(x, t)$ versus time t at a specific coordinate x depicts the motion of a particle at that coordinate. The vertical scale is exaggerated in both (a) and (b).

Wave displacement versus coordinate x at time $t = 0$

Wavelength λ

(a)

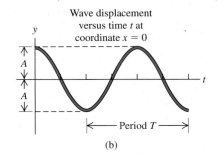

Wave displacement versus time t at coordinate $x = 0$

Period T

(b)

We can modify Eqs. (15.3) through (15.7) to represent a wave traveling in the *negative x*-direction. In this case the displacement of point x at time t is the same as the motion of point $x = 0$ at the *later* time $(t + x/v)$, so in Eq. (15.2) we replace t by $(t + x/v)$. For a wave traveling in the negative x-direction,

$$y(x, t) = A \cos 2\pi f\left(\frac{x}{v} + t\right) = A \cos 2\pi\left(\frac{x}{\lambda} + \frac{t}{T}\right) = A \cos(kx + \omega t)$$

(sinusoidal wave moving in $-x$-direction)   (15.8)

In the expression $y(x, t) = A \cos(kx \pm \omega t)$ for a wave traveling in the $-x$ or $+x$-direction, the quantity $(kx \pm \omega t)$ is called the **phase.** It plays the role of an angular quantity (always measured in radians) in Eq. (15.7) or (15.8), and its value for any values of x and t determines what part of the sinusoidal cycle is occurring at a particular point and time. For a crest (where $y = A$ and the cosine function has the value 1), the phase could be 0, 2π, 4π, and so on; for a trough (where $y = -A$ and the cosine has the value -1) it could be π, 3π, 5π, and so on.

The wave speed is the speed with which we have to move along with the wave to keep alongside a point of a given phase, such as a particular crest of a wave on a string. For a wave traveling in the $+x$-direction, that means $kx - \omega t = $ constant. Taking the derivative with respect to t, we find $k \, dx/dt = \omega$, or

$$\frac{dx}{dt} = \frac{\omega}{k}$$

Comparing this with Eq. (15.6), we see that dx/dt is equal to the speed v of the wave. Because of this relationship, v is sometimes called the *phase velocity* of the wave. (*Phase speed* would be a better term.)

Problem-Solving Strategy — Mechanical Waves

IDENTIFY *the relevant concepts:* Wave problems fall into two broad categories. *Kinematics* problems are concerned with describing wave motion; they involve wave speed v, wavelength λ (or wave number k), frequency f (or angular frequency ω), and amplitude A. They may also involve the position, velocity, and acceleration of individual particles in the medium. *Dynamics* problems also use concepts from Newton's laws such as force and mass. As an example, later in this chapter we'll encounter problems that involve the relation of wave speed to the mechanical properties of the wave medium. We'll get into these relations in the next section and in Chapter 16.

As always, make sure that you identify the target variable(s) for the problem. In some cases it will be the wavelength, frequency, or wave speed; in other cases you'll be asked to find an expression for the wave function.

SET UP *the problem using the following steps:*

1. Make a list of the quantities whose values are given. To help you visualize the situation, you'll find it useful to sketch graphs of y versus x (like Fig. 15.6a) and of y versus t (like Fig. 15.6b). Label your graphs with the values of the known quantities.

2. Decide which equations you'll need to use. If any two of v, f, and λ are given, you'll need to use Eq. (15.1) $(v = \lambda f)$ to find the third quantity (see Example 15.1). If the problem involves the angular frequency ω and/or the wave number k, you'll need to use the definitions of those quantities and Eq. (15.6) $(\omega = vk)$. You may also need the various forms of the wave function given in Eqs. (15.3), (15.4), and (15.7).

3. If the wave speed isn't given, and you don't have enough information to determine it using $v = \lambda f$, you may be able to find v using the relationship between v and the mechanical properties of the system. (In the next section we'll develop this relation for waves on a string.)

EXECUTE *the solution as follows:*

Solve for the unknown quantities using the equations you've selected. In some problems all you need to do is find the value of one of the wave variables.

If you're asked to determine the wave function, you need to know A and any two of v, λ, and f (or v, k, and ω). Once you have this information, you can use it in Eq. (15.3), (15.4), or (15.7) to get the specific wave function for the problem at hand. Once you have that, you can find the value of y at any point (value of x) and at any time by substituting into the wave function.

EVALUATE *your answer:* Look at your results with a critical eye. Check to see whether the values of v, f, and λ (or v, ω, and k) agree with the relationships given in Eq. (15.1) or (15.6). If you've calculated the wave function, check one or more special cases for which you can guess what the results ought to be.

Example 15.2	**Wave on a clothesline**

Your cousin Throckmorton is playing with the clothesline. He unties one end, holds it taut, and wiggles the end up and down sinusoidally with frequency 2.00 Hz and amplitude 0.075 m. The wave speed is $v = 12.0$ m/s. At time $t = 0$ the end has maximum positive displacement and is instantaneously at rest. Assume no wave bounces back from the far end to muddle up the pattern. a) Find the amplitude, angular frequency, period, wavelength, and wave number of the wave. b) Write a wave function describing the wave. c) Write equations for the displacement as a function of time of Throckmorton's end of the clothesline and of a point 3.00 m from his end.

SOLUTION

IDENTIFY: This is a kinematics problem about the motion of the clothesline. Since Throcky moves his hand in a sinusoidal way, he produces a sinusoidal wave that propagates down the clothesline. Hence we can use all of the expressions we've developed in this section. Our target variables in part (a) are amplitude A, angular frequency ω, period T, wavelength λ, and wave number k, so we need to use the equations that relate these quantities. In parts (b) and (c) our target "variables" are actually expressions for displacement; to find these, we use the general equations for the wave function of a sinusoidal wave.

SET UP: A photograph of the clothesline at time $t = 0$ would look just like Fig. 15.6a, with the maximum displacement at $x = 0$ (the end that Throcky has in his hand). We take the positive x-direction to be the direction in which the wave propagates, so we can use Eqs. (15.4) and (15.7) to describe the displacement of the clothesline as a function of position x and time t. We also use the relationships $f = 1/T$, $\omega = 2\pi f$, $k = 2\pi/\lambda$, $v = \lambda f$, and $\omega = vk$.

EXECUTE: a) The amplitude A of the wave is just the amplitude of the motion of the end of the clothesline, $A = 0.075$ m. Similarly, the wave frequency is $f = 2.00$ Hz, the same as the frequency of the end of the clothesline. The angular frequency is

$$\omega = 2\pi f = (2\pi \text{ rad/cycle})(2.00 \text{ cycles/s}) = 4.00\pi \text{ rad/s}$$
$$= 12.6 \text{ rad/s}$$

The period is $T = 1/f = 0.500$ s. We get the wavelength from Eq. (15.1):

$$\lambda = \frac{v}{f} = \frac{12.0 \text{ m/s}}{2.00 \text{ s}^{-1}} = 6.00 \text{ m}$$

We find the wave number from Eq. (15.5) or (15.6):

$$k = \frac{2\pi}{\lambda} = \frac{2\pi \text{ rad}}{6.00 \text{ m}} = 1.05 \text{ rad/m} \quad \text{or}$$

$$k = \frac{\omega}{v} = \frac{4.00\pi \text{ rad/s}}{12.0 \text{ m/s}} = 1.05 \text{ rad/m}$$

b) Since we found the values of A, T, and λ in part (a), we can write down the wave function using Eq. (15.4):

$$y(x, t) = A \cos 2\pi\left(\frac{x}{\lambda} - \frac{t}{T}\right)$$

$$= (0.075 \text{ m})\cos 2\pi\left(\frac{x}{6.00 \text{ m}} - \frac{t}{0.500 \text{ s}}\right)$$

$$= (0.075 \text{ m})\cos[(1.05 \text{ rad/m})x - (12.6 \text{ rad/s})t]$$

We can also get this same equation from Eq. (15.7) by using the values of ω and k we obtained in part (a).

c) With our choice of the positive x-direction, the two points in question are at $x = 0$ and $x = +3.00$ m. For each point, we can find the displacement as a function of time by substituting these values of x into the wave function we found in part (b):

$$y(x = 0, t) = (0.075 \text{ m})\cos 2\pi\left(\frac{0}{6.00 \text{ m}} - \frac{t}{0.500 \text{ s}}\right)$$

$$= (0.075 \text{ m})\cos(12.6 \text{ rad/s})t$$

$$y(x = +3.00 \text{ m}, t) = (0.075 \text{ m})\cos 2\pi\left(\frac{3.00 \text{ m}}{6.00 \text{ m}} - \frac{t}{0.500 \text{ s}}\right)$$

$$= (0.075 \text{ m})\cos[\pi - (12.6 \text{ rad/s})t]$$

$$= -(0.075 \text{ m})\cos(12.6 \text{ rad/s})t$$

EVALUATE: In part (b), the quantity $(1.05 \text{ rad/m})x - (12.6 \text{ rad/s})t$ is the *phase* of a point x on the string at time t. The phases of the two points in part (c) differ by π because these points are separated by one half-wavelength ($\lambda/2 = (6.00 \text{ m})/2 = 3.00$ m). Both points oscillate in SHM with the same frequency and amplitude, but their oscillations are one half-cycle out of phase. Thus, while a graph of y versus t for the point at $x = 0$ is a cosine curve (like Fig. 15.6b), a graph of y versus t for the point $x = 3.00$ m is a *negative* cosine (the same as a cosine curve shifted by one half-cycle).

Using the expression for $y(x = 0, t)$ in part (c), can you show that the end of the string at $x = 0$ is instantaneously at rest at $t = 0$, just as we stated at the beginning of this example? (*Hint:* calculate the velocity at this point by taking the derivative of y with respect to t.)

Particle Velocity and Acceleration in a Sinusoidal Wave

From the wave function we can get an expression for the transverse velocity of any *particle* in a transverse wave. We call this v_y to distinguish it from the wave propagation speed v. To find the transverse velocity v_y at a particular point x, we take the derivative of the wave function $y(x, t)$ with respect to t, keeping x constant. If the wave function is

$$y(x, t) = A \cos(kx - \omega t)$$

then

$$v_y(x, t) = \frac{\partial y(x, t)}{\partial t} = \omega A \sin(kx - \omega t) \tag{15.9}$$

The ∂ in this expression is a modified d, used to remind us that $y(x, t)$ is a function of *two* variables and that we are allowing only one (t) to vary. The other (x) is constant because we are looking at a particular point on the string. This derivative is called a *partial derivative.* If you haven't reached this point yet in your study of calculus, don't fret; it's a simple idea.

Equation (15.9) shows that the transverse velocity of a particle varies with time, as we expect for simple harmonic motion. The maximum particle speed is ωA; this can be greater than, less than, or equal to the wave speed v, depending on the amplitude and frequency of the wave.

The *acceleration* of any particle is the *second* partial derivative of $y(x, t)$ with respect to t:

$$a_y(x, t) = \frac{\partial^2 y(x, t)}{\partial t^2} = -\omega^2 A \cos(kx - \omega t) = -\omega^2 y(x, t) \tag{15.10}$$

The acceleration of a particle equals $-\omega^2$ times its displacement, which is the result we obtained in Section 13.2 for simple harmonic motion.

We can also compute partial derivatives of $y(x, t)$ with respect to x, holding t constant. This corresponds to studying the shape of the string at one instant of time, like a flash photo. The first derivative $\partial y(x, t)/\partial x$ is the *slope* of the string at any point. The second partial derivative with respect to x is the *curvature* of the string:

$$\frac{\partial^2 y(x, t)}{\partial x^2} = -k^2 A \cos(kx - \omega t) = -k^2 y(x, t) \tag{15.11}$$

From Eqs. (15.10) and (15.11) and the relation $\omega = vk$ we see that

$$\frac{\partial^2 y(x, t)/\partial t^2}{\partial^2 y(x, t)/\partial x^2} = \frac{\omega^2}{k^2} = v^2 \quad \text{and}$$

$$\frac{\partial^2 y(x, t)}{\partial x^2} = \frac{1}{v^2} \frac{\partial^2 y(x, t)}{\partial t^2} \quad \text{(wave equation)} \tag{15.12}$$

We've derived Eq. (15.12) for a wave traveling in the positive *x*-direction. You can use the same steps to show that the wave function for a sinusoidal wave propagating in the *negative x*-direction, $y(x, t) = A \cos (kx + \omega t)$, also satisfies this equation.

Equation (15.12), called the **wave equation,** is one of the most important equations in all of physics. Whenever it occurs, we know that a disturbance can propagate as a wave along the *x*-axis with wave speed *v*. The disturbance need not be a sinusoidal wave; we'll see in the next section that *any* wave on a string obeys Eq. (15.12), whether the wave is periodic or not (see also Problems 15.54 and 15.59). In Chapter 32 we will find that electric and magnetic fields satisfy the wave equation; the wave speed will turn out to be the speed of light, which will lead us to the conclusion that light is an electromagnetic wave.

Figure 15.7a shows the velocity v_y and acceleration a_y, given by Eqs. (15.9) and (15.10), for several points on a string as a sinusoidal wave passes along it. Note that at points where the string has an upward curvature $(\partial^2 y/\partial x^2 > 0)$, the acceleration of that point is positive $(a_y = \partial^2 y/\partial t^2 > 0)$; this follows from the wave equation, Eq. (15.12). For the same reason the acceleration is negative $(a_y = \partial^2 y/\partial t^2 < 0)$ at points where the string has a downward curvature $(\partial^2 y/\partial x^2 < 0)$, and the acceleration is zero $(a_y = \partial^2 y/\partial t^2 = 0)$ at points of inflection where the curvature is zero $(\partial^2 y/\partial x^2 = 0)$. We emphasize again that v_y and a_y are the *transverse* velocity and acceleration of points on the string; these points move along the *y*-direction, not along the propagation direction of the wave. Figure 15.7b shows the transverse motions of several points on the string.

The concept of wave function is equally useful with *longitudinal* waves, and everything we have said about wave functions can be adapted to this case. The quantity *y* still measures the displacement of a particle of the medium from its equilibrium position; the difference is that for a longitudinal wave, this displacement is *parallel* to the *x*-axis instead of perpendicular to it. We'll discuss longitudinal waves in detail in Chapter 16.

Acceleration a_y at each point is proportional to displacement *y* at that point

Acceleration is upward where string curves upward, downward where string curves downward

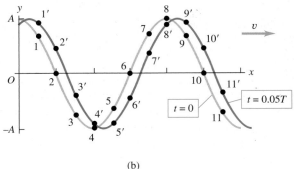

(a)   (b)

15.7 (a) Another view of the wave at $t = 0$ in Fig. 15.6a. The vectors show the transverse velocity v_y and transverse acceleration a_y at several points on the string. (b) The two curves show the wave at $t = 0$ and $t = 0.05T$. During this interval, a particle at point 1 is displaced to point 1′, a particle at point 2 is displaced to point 2′, and so on.

The wave function for a sinusoidal wave is

$$y(x, t) = (2.50 \text{ mm}) \cos \left[(2.00 \text{ rad/m})x + (3.20 \text{ rad/s})t \right]$$

In which direction is this wave propagating? What is its wave number? Its angular frequency? Its amplitude?

15.4 | Speed of a Transverse Wave

One of the key properties of any wave is the wave speed. Light waves in air have a much greater speed of propagation than do sound waves in air (3.00×10^8 m/s versus 344 m/s); that's why you see the flash from a bolt of lightning before you hear the clap of thunder. In this section we'll see what determines the speed of propagation of one particular kind of wave: transverse waves on a string. The speed of these waves is important to understand in its own right because it is an essential part of analyzing stringed musical instruments, as we'll discuss later in this chapter. Furthermore, the speeds of many kinds of mechanical waves turn out to have the same basic mathematical expression as does the speed of waves on a string.

The physical quantities that determine the speed of transverse waves on a string are the *tension* in the string and its *mass per unit length* (also called *linear mass density*). We might guess that increasing the tension should increase the restoring forces that tend to straighten the string when it is disturbed, thus increasing the wave speed. We might also guess that increasing the mass should make the motion more sluggish and decrease the speed. Both these guesses turn out to be right. We'll develop the exact relationship between wave speed, tension, and mass per unit length by two different methods. The first is simple in concept and considers a specific wave shape; the second is more general but also more formal. Choose whichever you like better.

Wave Speed on a String: First Method

We consider a perfectly flexible string (Fig. 15.8). In the equilibrium position the tension is F, and the linear mass density (mass per unit length) is μ. (When portions of the string are displaced from equilibrium, the mass per unit length decreases a little, and the tension increases a little.) We ignore the weight of the string so that when the string is at rest in the equilibrium position, the string forms a perfectly straight line as in Fig. 15.8a.

Starting at time $t = 0$, we apply a constant upward force F_y at the left end of the string. We might expect that the end would move with constant acceleration; that would happen if the force were applied to a *point* mass. But here the effect of the force F_y is to set successively more and more mass in motion. The wave travels with constant speed v, so the division point P between moving and nonmoving portions moves with the same constant speed v (Fig. 15.8b).

Figure 15.8b shows that all particles in the moving portion of the string move upward with constant *velocity* v_y, not constant acceleration. To see why this is so, we note that the *impulse* of the force F_y up to time t is $F_y t$. According to the impulse-momentum theorem (Section 8.1), the impulse is equal to the change in the total transverse component of momentum $(mv_y - 0)$ of the moving part of

10.2 Speed of Waves on a String

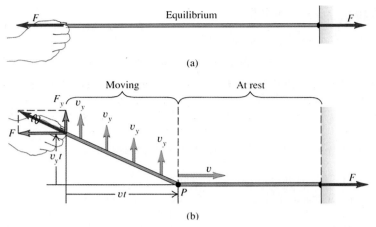

15.8 Propagation of a transverse wave on a string. (a) String in equilibrium; (b) part of the string in motion.

the string. Because the system started with *no* transverse momentum, this is equal to the total momentum at time t:

$$F_y t = m v_y$$

The total momentum thus must increase proportionately with time. But since the division point P moves with constant speed, the length of string that is in motion and hence the total mass m in motion are also proportional to the time t that the force has been acting. So the *change* of momentum must be associated entirely with the increasing amount of mass in motion, not with an increasing velocity of an individual mass element. That is, $m v_y$ changes because m, not v_y, changes.

At time t, the left end of the string has moved up a distance $v_y t$, and the boundary point P has advanced a distance vt. The total force at the left end of the string has components F and F_y. Why F? There is no motion in the direction along the length of the string, so there is no unbalanced horizontal force. Therefore F, the magnitude of the horizontal component, does not change when the string is displaced. In the displaced position the tension is $(F^2 + F_y^2)^{1/2}$ (greater than F), and the string stretches somewhat.

To derive an expression for the wave speed v, we again apply the impulse-momentum theorem to the portion of the string in motion at time t, that is, the portion to the left of P in Fig. 15.8b. The transverse *impulse* (transverse force times time) is equal to the change of transverse *momentum* of the moving portion (mass times transverse component of velocity). The impulse of the transverse force F_y in time t is $F_y t$. In Fig. 15.8b the right triangle whose vertex is at P, with sides $v_y t$ and vt, is similar to the right triangle whose vertex is at the position of the hand, with sides F_y and F. Hence

$$\frac{F_y}{F} = \frac{v_y t}{vt} \qquad F_y = F\frac{v_y}{v}$$

and

$$\text{Transverse impulse} = F_y t = F\frac{v_y}{v}t$$

The mass of the moving portion of the string is the product of the mass per unit length μ and the length vt, or μvt. The transverse momentum is the product of this mass and the transverse velocity v_y:

$$\text{Transverse momentum} = (\mu vt)v_y$$

We note again that the momentum increases with time *not* because mass is moving faster, as was usually the case in Chapter 8, but because *more mass* is brought into motion. But the impulse of the force F_y is still equal to the total change in momentum of the system. Applying this relation, we obtain

$$F\frac{v_y}{v}t = \mu vt v_y$$

Solving this for v, we find

$$v = \sqrt{\frac{F}{\mu}} \qquad (\text{speed of a transverse wave on a string}) \qquad (15.13)$$

Equation (15.13) confirms our prediction that the wave speed v should increase when the tension F increases but decrease when the mass per unit length μ increases (Fig. 15.9).

Note that v_y does not appear in Eq. (15.13); thus the wave speed doesn't depend on v_y. Our calculation considered only a very special kind of pulse, but we can consider *any* shape of wave disturbance as a series of pulses with different values of v_y. So even though we derived Eq. (15.13) for a special case, it is valid for *any* transverse wave motion on a string, including the sinusoidal and other periodic waves we discussed in Section 15.3. Note also that the wave speed doesn't depend on the amplitude or frequency of the wave, in accordance with our assumptions in Section 15.3.

Wave Speed on a String: Second Method

Here is an alternative derivation of Eq. (15.13). If you aren't comfortable with partial derivatives, it can be omitted. We apply Newton's second law, $\sum \vec{F} = m\vec{a}$, to a small segment of string whose length in the equilibrium position is Δx (Fig. 15.10). The mass of the segment is $m = \mu\,\Delta x$; the forces at the ends are represented in terms of their x- and y-components. The x-components have equal magnitude F and add to zero because the motion is transverse and there is no component of acceleration in the x-direction. To obtain F_{1y} and F_{2y}, we note that the ratio F_{1y}/F is

15.9 These cables have a relatively large amount of mass per unit length (μ) and a low tension (F). If the cables are disturbed—say, by a bird landing on them—transverse waves will travel along them at a slow speed $v = \sqrt{F/\mu}$.

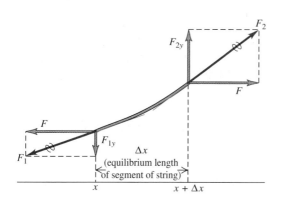

15.10 Free-body diagram for a segment of string. The force at each end of the string is tangent to the string at the point of application.

equal in magnitude to the *slope* of the string at point x and that F_{2y}/F is equal to the slope at point $x + \Delta x$. Taking proper account of signs, we find

$$\frac{F_{1y}}{F} = -\left(\frac{\partial y}{\partial x}\right)_x \qquad \frac{F_{2y}}{F} = \left(\frac{\partial y}{\partial x}\right)_{x+\Delta x} \qquad (15.14)$$

The notation reminds us that the derivatives are evaluated at points x and $x + \Delta x$, respectively. From Eq. (15.14) we find that the net y-component of force is

$$F_y = F_{1y} + F_{2y} = F\left[\left(\frac{\partial y}{\partial x}\right)_{x+\Delta x} - \left(\frac{\partial y}{\partial x}\right)_x\right] \qquad (15.15)$$

We now equate F_y from Eq. (15.15) to the mass $\mu\,\Delta x$ times the y-component of acceleration $\partial^2 y/\partial t^2$. We obtain

$$F\left[\left(\frac{\partial y}{\partial x}\right)_{x+\Delta x} - \left(\frac{\partial y}{\partial x}\right)_x\right] = \mu\,\Delta x\frac{\partial^2 y}{\partial t^2} \qquad (15.16)$$

or, dividing by $F\,\Delta x$,

$$\frac{\left(\frac{\partial y}{\partial x}\right)_{x+\Delta x} - \left(\frac{\partial y}{\partial x}\right)_x}{\Delta x} = \frac{\mu}{F}\frac{\partial^2 y}{\partial t^2} \qquad (15.17)$$

We now take the limit as $\Delta x \to 0$. In this limit, the left side of Eq. (15.17) becomes the derivative of $\partial y/\partial x$ with respect to x (at constant t), that is, the *second* (partial) derivative of y with respect to x:

$$\frac{\partial^2 y}{\partial x^2} = \frac{\mu}{F}\frac{\partial^2 y}{\partial t^2} \qquad (15.18)$$

Now, finally, comes the punch line of our story. Equation (15.18) has exactly the same form as the *wave equation,* Eq. (15.12), that we derived at the end of Section 15.3. That equation and Eq. (15.18) describe the very same wave motion, so they must be identical. Comparing the two equations, we see that for this to be so, we must have

$$v = \sqrt{\frac{F}{\mu}} \qquad (15.19)$$

which is the same expression as Eq. (15.13).

In going through this derivation, we didn't make any special assumptions about the shape of the wave. Since our derivation led us to rediscover Eq. (15.12), the wave equation, we conclude that the wave equation is valid for waves on a string that have *any* shape.

Equation (15.13) or (15.19) gives the wave speed only for the special case of mechanical waves on a stretched string or rope. Remarkably, it turns out that for many types of mechanical waves, including waves on a string, the expression for wave speed has the same general form:

$$v = \sqrt{\frac{(\text{restoring force returning the system to equilibrium})}{(\text{inertia resisting the return to equilibrium})}}$$

To interpret this expression, let's look at the now-familiar case of waves on a string. The tension F in the string plays the role of the restoring force; it tends to bring the string back to its undisturbed, equilibrium configuration. The mass of the string—or, more properly, the linear mass density μ—provides the inertia that prevents the string from returning instantaneously to equilibrium. Hence we have $v = \sqrt{F/\mu}$ for the speed of waves on a string.

In Chapter 16 we'll see a similar expression for the speed of sound waves in a gas. Roughly speaking, the gas pressure provides the force that tends to return the gas to its undisturbed state when a sound wave passes through. The inertia is provided by the density, or mass per unit volume, of the gas.

Example 15.3	**Varying the speed of a wave**

In Example 15.2 the linear mass density of the clothesline is 0.250 kg/m. a) How much tension does Throcky have to apply to produce the observed wave speed of 12.0 m/s? b) If the tension is increased to four times the value in (a) but the frequency is still 2.00 Hz, what will be the wavelength of the wave on the clothesline?

SOLUTION

IDENTIFY: The target variable is the tension in part (a) and the wavelength in part (b). Both parts of the problem involve *dynamics*, that is, the relationship between the wave speed and the properties of the clothesline itself (tension and linear mass density). Part (b) also involves *kinematics*, since we'll need the relationship between the wave speed, wavelength, and frequency.

SET UP: In part (a) we're given the linear mass density μ and the wave speed v. Hence we can find the tension F using Eq. (15.13), $v = \sqrt{F/\mu}$. In part (b) we use this same relationship to find the new wave speed with the new value of tension, then find the new wavelength λ using Eq. (15.1), $v = \lambda f$.

EXECUTE: (a) Solving Eq. (15.13) for F, we find

$$F = \mu v^2 = (0.250 \text{ kg/m})(12.0 \text{ m/s})^2 = 36.0 \text{ kg-m/s}^2$$
$$= 36.0 \text{ N} = 8.09 \text{ lb}$$

which is well within Throcky's capabilities.

b) Equation (15.13) says that the wave speed is proportional to the square root of the tension F. Furthermore, Eq. (15.1) tells us that $\lambda = v/f$, so that the wavelength is linearly proportional to the wave speed (for constant frequency f, which is the case in this problem). So if we increase the value of F by a factor of 4, the wave speed and the wavelength both increase by a factor of $\sqrt{4} = 2$. Hence, using the old wavelength value from Example 15.2 will give the new wavelength

$$\lambda = 2(6.00 \text{ m}) = 12.0 \text{ m}$$

EVALUATE: We can check our answer to part (b) by using the new tension $F = 4(36.0 \text{ N})$ in the expression $v = \sqrt{F/\mu}$ to find the new wave speed, then calculating the new wavelength using $\lambda = v/f$:

$$v = \sqrt{\frac{144 \text{ N}}{0.250 \text{ kg/m}}} = 24.0 \text{ m/s} \quad \text{so}$$
$$\lambda = \frac{24.0 \text{ m/s}}{2.00 \text{ s}^{-1}} = 12.0 \text{ m}$$

Example 15.4	**Calculating wave speed**

One end of a nylon rope is tied to a stationary support at the top of a vertical mine shaft 80.0 m deep (Fig. 15.11). The rope is stretched taut by a box of mineral samples with mass 20.0 kg attached at the lower end. The mass of the rope is 2.00 kg. The geologist at the bottom of the mine signals to his colleague at the top by jerking the rope sideways. a) What is the speed of a transverse wave on the rope? b) If a point on the rope is given a transverse simple harmonic motion with a frequency of 2.00 Hz, how many cycles of the wave are there in the rope's length?

SOLUTION

IDENTIFY: This example is similar to the previous one; it involves both dynamics [in part (a)] and kinematics [in part (b)].

In part (a) the target variable is the wave speed. The new feature is that the tension is provided by the weight of the box of samples. In fact, the weight of the rope itself contributes to the tension, which means that the tension is different at the top and bottom of the rope. We'll ignore this effect here since the weight of the rope is small compared to the weight of the samples.

$m_{rope} = 2.00$ kg

80.0 m

$m_{samples} = 20.0$ kg

15.11 Sending signals along a vertical rope using transverse waves.

Note that the target variable in part (b) is actually the number of wavelengths that fit into the length of the rope.

SET UP: As in Example 15.3, we use the relationship $v = \sqrt{F/\mu}$ in part (a). If we neglect the weight of the rope itself, the tension F is just equal to the weight of the box. In part (b) we use the equation

$v = f\lambda$ to find the wavelength, which we then compare to the 80.0-m length of the rope.

EXECUTE: a) The tension in the rope (due to the sample box) is

$$F = m_{samples}g = (20.0 \text{ kg})(9.80 \text{ m/s}^2) = 196 \text{ N}$$

and the mass per unit length of the rope is

$$\mu = \frac{m_{rope}}{L} = \frac{2.00 \text{ kg}}{80.0 \text{ m}} = 0.0250 \text{ kg/m}$$

Hence, from Eq. (15.13), the wave speed is

$$v = \sqrt{\frac{F}{\mu}} = \sqrt{\frac{196 \text{ N}}{0.0250 \text{ kg/m}}} = 88.5 \text{ m/s}$$

b) From Eq. (15.1),

$$\lambda = \frac{v}{f} = \frac{88.5 \text{ m/s}}{2.00 \text{ s}^{-1}} = 44.3 \text{ m}$$

The length of the rope is 80.0 m, so the number of wave cycles in the rope is

$$\frac{80.0 \text{ m/s}}{44.3 \text{ m/cycle}} = 1.81 \text{ cycles}$$

EVALUATE: If we account for the weight of the rope, the tension is greater at the top of the rope than at the bottom. Hence the wave speed increases and the wavelength increases as the wave travels up the rope. Can you verify that the wave speed at the top is 92.9 m/s?

Test Your Understanding

The six strings of a guitar are of the same length and are under nearly the same tension, but have different thicknesses. On which string do waves travel the fastest?

15.5 | Energy in Wave Motion

Every wave motion has *energy* associated with it. The energy we receive from sunlight and the destructive effects of ocean surf and earthquakes bear this out. To produce any of the wave motions we have discussed in this chapter, we have to apply a force to a portion of the wave medium; the point where the force is applied moves, so we do *work* on the system. As the wave propagates, each portion of the medium exerts a force and does work on the adjoining portion. In this way a wave can transport energy from one region of space to another.

As an example of energy considerations in wave motion, let's look again at transverse waves on a string. How is energy transferred from one portion of string to another? Picture a wave traveling from left to right (the positive *x*-direction) on the string, and consider a particular point *a* on the string (Fig. 15.12a). The string to the left of *a* exerts a force on the string to the right of it, and vice versa. In Fig. 15.12b the string to the left of *a* has been removed, and the force it exerts at *a* is represented

Slope $= \dfrac{\delta y}{\delta x}$

(a)   (b)

15.12 (a) Point *a* on a string carrying a wave from left to right. (b) The components of the force exerted on the right part of the string by the left part at point *a*.

by the components F and F_y, as we did in Figs. 15.8 and 15.10. We note again that F_y/F is equal to the negative of the *slope* of the string at a, which is also given by $\partial y/\partial x$. Putting these together, we have

$$F_y(x, t) = -F\frac{\partial y(x, t)}{\partial x} \tag{15.20}$$

We need the negative sign because F_y is negative when the slope is positive. We write the vertical force as $F_y(x, t)$ as a reminder that its value may be different at different points along the string and at different times.

When point a moves in the y-direction, the force F_y does *work* on this point and therefore transfers energy into the part of the string to the right of a. The corresponding power P (rate of doing work) at the point a is the transverse force $F_y(x, t)$ at a times the transverse velocity $v_y(x, t) = \partial y(x, t)/\partial t$ of that point:

$$P(x, t) = F_y(x, t)v_y(x, t) = -F\frac{\partial y(x, t)}{\partial x}\frac{\partial y(x, t)}{\partial t} \tag{15.21}$$

This power is the *instantaneous* rate at which energy is transferred along the string. Its value depends on the position x on the string and on the time t. Note that energy is being transferred only at points where the string has a nonzero slope ($\partial y/\partial x$ is nonzero), so that there is a transverse component of the tension force, and where the string has a nonzero transverse velocity ($\partial y/\partial t$ is nonzero) so that the transverse force can do work.

Equation (15.21) is valid for *any* wave on a string, sinusoidal or not. For a sinusoidal wave with wave function given by Eq. (15.7), we have

$$y(x, t) = A \cos(kx - \omega t)$$

$$\frac{\partial y(x, t)}{\partial x} = -kA \sin(kx - \omega t)$$

$$\frac{\partial y(x, t)}{\partial t} = \omega A \sin(kx - \omega t)$$

$$P(x, t) = Fk\omega A^2 \sin^2(kx - \omega t) \tag{15.22}$$

By using the relations $\omega = vk$ and $v^2 = F/\mu$, we can also express Eq. (15.22) in the alternative form

$$P(x, t) = \sqrt{\mu F}\,\omega^2 A^2 \sin^2(kx - \omega t) \tag{15.23}$$

The $\sin^2$ function is never negative, so the instantaneous power in a sinusoidal wave is either positive (so that energy flows in the positive x-direction) or zero (at points where there is no energy transfer). Energy is never transferred in the direction opposite to the direction of wave propagation (Fig. 15.13).

The maximum value of the instantaneous power $P(x, t)$ occurs when the $\sin^2$ function has the value unity:

$$P_{max} = \sqrt{\mu F}\,\omega^2 A^2 \tag{15.24}$$

To obtain the *average* power from Eq. (15.23), we note that the *average* value of the $\sin^2$ function, averaged over any whole number of cycles, is $\frac{1}{2}$. Hence the average power is

$$P_{av} = \frac{1}{2}\sqrt{\mu F}\,\omega^2 A^2 \quad \text{(average power, sinusoidal wave on a string)} \tag{15.25}$$

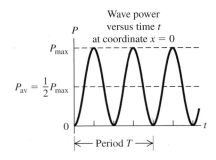

15.13 The instantaneous power $P(x, t)$ in a sinusoidal wave as given by Eq. (15.23), shown as a function of time at coordinate $x = 0$. The power is never negative, which means that energy never flows opposite to the direction of wave propagation.

The average power is just one-half of the maximum instantaneous power (see Fig. 15.13).

The average rate of energy transfer is proportional to the square of the amplitude and to the square of the frequency. This proportionality is a general result for mechanical waves of all types, including seismic waves (see the photo that opens this chapter). For a mechanical wave, the rate of energy transfer quadruples if the frequency is doubled (for the same amplitude) or if the amplitude is doubled (for the same frequency).

Electromagnetic waves turn out to be a bit different. While the average rate of energy transfer in an electromagnetic wave is proportional to the square of the amplitude, just as for mechanical waves, it is independent of the value of ω.

| Example 15.5 | **Power in a wave** |

a) In Examples 15.2 and 15.3, at what maximum rate does Throcky put energy into the rope? That is, what is his maximum instantaneous power? b) What is his average power? c) As Throcky tires, the amplitude decreases. What is the average power when the amplitude has dropped to 7.50 mm?

SOLUTION

IDENTIFY: Our target variable in (a) is the *maximum instantaneous* power, while the target variable in (b) and (c) is the *average* power. As we've seen, these two quantities have different values for a sinusoidal wave. We'll be able to calculate the values of both quantities because we know all the other properties of the wave from Example 15.2.

SET UP: For part (a) we use Eq. (15.24), while for parts (b) and (c) we use Eq. (15.25).

EXECUTE: a) The maximum instantaneous power is

$$P_{max} = \sqrt{\mu F}\,\omega^2 A^2$$
$$= \sqrt{(0.250 \text{ kg/m})(36.0 \text{ N})}\,(4.00\pi \text{ rad/s})^2 (0.075 \text{ m})^2$$
$$= 2.66 \text{ W}$$

b) From Eqs. (15.24) and (15.25), the average power is one-half of the maximum instantaneous power, so

$$P_{av} = \tfrac{1}{2}(2.66 \text{ W}) = 1.33 \text{ W}$$

c) The new amplitude is $\frac{1}{10}$ of the value we used in parts (a) and (b). The average power is proportional to the *square* of the amplitude, so now the average power is

$$P_{av} = \left(\frac{1}{10}\right)^2 (1.33 \text{ W}) = 0.0133 \text{ W} = 13.3 \text{ mW}$$

EVALUATE: The *maximum* instantaneous power in part (a) occurs when the quantity $\sin^2(kx - \omega t)$ in Eq. (15.23) is equal to 1. At any given value of x, this happens twice per period of the wave—once when the sine function is equal to $+1$, once when it's equal to -1. The *minimum* instantaneous power is zero; this occurs when $\sin(kx - \omega t) = 0$, which also happens twice per period.

Wave Intensity

Waves on a string carry energy in just one dimension of space (along the direction of the string). But other types of waves, including sound waves in air and seismic waves in the body of the earth, carry energy across all three dimensions of space. For waves that travel in three dimensions, we define the **intensity** (denoted by I) to be *the time average rate at which energy is transported by the wave, per unit area,* across a surface perpendicular to the direction of propagation. That is, intensity I is average power per unit area. It is usually measured in watts per square meter (W/m^2).

If waves spread out equally in all directions from a source, the intensity at a distance r from the source is inversely proportional to r^2. (Fig. 15.14). This follows directly from energy conservation. If the power output of the source is

P, then the average intensity I_1 through a sphere with radius r_1 and surface area $4\pi r_1^2$ is

$$I_1 = \frac{P}{4\pi r_1^2}$$

The average intensity I_2 through a sphere with a different radius r_2 is given by a similar expression. If no energy is absorbed between the two spheres, the power P must be the same for both, and

$$4\pi r_1^2 I_1 = 4\pi r_2^2 I_2$$

$$\frac{I_1}{I_2} = \frac{r_2^2}{r_1^2} \quad \text{(inverse-square law for intensity)} \quad (15.26)$$

The intensity I at any distance r is therefore inversely proportional to r^2. This relationship is called the *inverse-square law* for intensity.

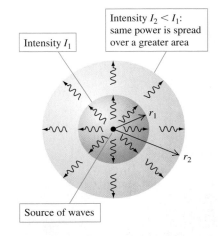

15.14 The greater the distance from a wave source, the greater the area over which the wave power is distributed and the smaller the wave intensity.

Example 15.6	**The inverse-square law**

A tornado warning siren on top of a tall pole radiates sound waves uniformly in all directions. At a distance of 15.0 m the intensity of the sound is 0.250 W/m². At what distance from the siren is the intensity 0.010 W/m²?

SOLUTION

IDENTIFY: Because waves spread out equally in all directions, we can use the inverse-square law. Our target variable is a distance from the wave source.

SET UP: The relation to use is Eq. (15.26). We are given the distance $r_1 = 15.0$ m at which the intensity is $I_1 = 0.250$ W/m², and we want to find the distance r_2 at which the intensity is $I_2 = 0.010$ W/m².

EXECUTE: We solve Eq. (15.26) for r_2:

$$r_2 = r_1\sqrt{\frac{I_1}{I_2}} = (15.0 \text{ m})\sqrt{\frac{0.250 \text{ W/m}^2}{0.010 \text{ W/m}^2}} = 75.0 \text{ m}$$

EVALUATE: As a check on our answer, note that r_2 is five times greater than r_1. By the inverse-square law, the intensity I_2 should be $1/5^2 = 1/25$ as great as I_1, and indeed it is.

By using the inverse-square law we've assumed that the sound waves travel in straight lines away from the siren. A more realistic solution of this problem would account for the reflection of sound waves from the ground. Such a solution is beyond our scope, however.

Test Your Understanding

In Example 15.5, suppose the wave amplitude remained equal to 0.075 m but the frequency increased from 2.00 Hz to 4.00 Hz. What effect would this have on the average power?

15.6 | Wave Interference, Boundary Conditions, and Superposition

Up to this point we've been discussing waves that propagate continuously in the same direction. But when a wave strikes the boundaries of its medium, all or part of the wave is *reflected*. When you yell at a building wall or a cliff face some distance away, the sound wave is reflected from the rigid surface and you hear an echo. When you flip the end of a rope whose far end is tied to a rigid support, a pulse travels the length of the rope and is reflected back to you. In both cases, the initial and reflected waves overlap in the same region of the medium. This overlapping of waves is called **interference.** (In general, the term *interference* refers to what happens when two or more waves pass through the same region at the same time.)

As a simple example of wave reflections and the role of the boundary of a wave medium, let's look again at transverse waves on a stretched string. What happens when a wave pulse or a sinusoidal wave arrives at the *end* of the string?

If the end is fastened to a rigid support, it is a *fixed* end that cannot move. The arriving wave exerts a force on the support; the reaction to this force, exerted *by* the support *on* the string, "kicks back" on the string and sets up a *reflected* pulse or wave traveling in the reverse direction. Figure 15.15 is a series of photographs showing the reflection of a pulse at the fixed end of a long coiled spring. The reflected pulse moves in the opposite direction from the initial, or *incident,* pulse, and its displacement is also opposite. This situation is illustrated for a wave pulse on a string in Fig. 15.16a.

15.15 A series of images of a wave pulse, equally spaced in time from top to bottom. The pulse starts at the left in the top image, travels to the right, and is reflected from the fixed end at the right.

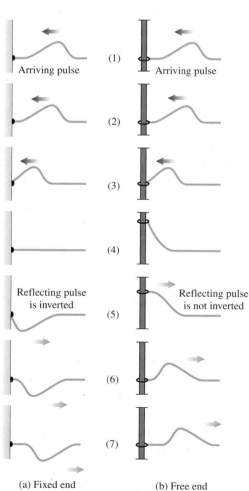

(a) Fixed end   (b) Free end

15.16 Reflection of a wave pulse (a) at a fixed end of a string and (b) at a free end. Time increases from top to bottom in each figure.

The opposite situation from an end that is held stationary is a *free* end, one that is perfectly free to move in the direction perpendicular to the length of the string. For example, the string might be tied to a light ring that slides on a frictionless rod perpendicular to the rope, as in Fig. 15.16b. The ring and rod maintain the tension but exert no transverse force. When a wave arrives at this free end, the ring slides along the rod. The ring reaches a maximum displacement, and both it and the string come momentarily to rest, as in drawing (4) in Fig. 15.16b. But the string is now stretched, giving increased tension, so the free end of the string is pulled back down, and again a reflected pulse is produced (drawing (7)). As for a fixed end, the reflected pulse moves in the opposite direction from the initial pulse, but now the direction of the displacement is the same as for the initial pulse. The conditions at the end of the string, such as a rigid support or the complete absence of transverse force, are called **boundary conditions.**

The formation of the reflected pulse is similar to the overlap of two pulses traveling in opposite directions. Figure 15.17 shows two pulses with the same shape, one inverted with respect to the other, traveling in opposite directions. As the pulses overlap and pass each other, the total displacement of the string is the *algebraic sum* of the displacements at that point in the individual pulses. Because these two pulses have the same shape, the total displacement at point O in the middle of the figure is zero at all times. Thus the motion of the right half of the string would be the same if we cut the string at point O, threw away the left side, and held the end at O fixed. The two pulses on the right side then correspond to the incident and reflected pulses, combining so that the total displacement at O is *always* zero. For this to occur, the reflected pulse must be inverted relative to the incident pulse.

Figure 15.18 shows two pulses with the same shape, traveling in opposite directions but *not* inverted relative to each other. The displacement at point O in the middle of the figure is not zero, but the slope of the string at this point is always zero. According to Eq. (15.20), this corresponds to the absence of any transverse force at this point. In this case the motion of the right half of the string would be the same as if we cut the string at point O and anchored the end with a frictionless sliding ring (Fig. 15.16b) that maintains tension without exerting any transverse force. In other words, this situation corresponds to reflection of a pulse at a free end of a string at point O. In this case the reflected pulse is *not* inverted.

The Principle of Superposition

Combining the displacements of the separate pulses at each point to obtain the actual displacement is an example of the **principle of superposition:** when two waves overlap, the actual displacement of any point on the string at any time is obtained by adding the displacement the point would have if only the first wave were present and the displacement it would have if only the second wave were present. In other words, the wave function $y(x, t)$ that describes the resulting motion in this situation is obtained by *adding* the two wave functions for the two separate waves:

$$y(x, t) = y_1(x, t) + y_2(x, t) \quad \text{(principle of superposition)} \quad (15.27)$$

Mathematically, this additive property of wave functions follows from the form of the wave equation, Eq. (15.12) or (15.18), which every physically possible wave function must satisfy. Specifically, the wave equation is *linear;* that is, it contains the function $y(x, t)$ only to the first power (there are no terms involving

15.17 Overlap of two wave pulses—one right side up, one inverted—traveling in opposite directions. Time increases from top to bottom.

15.18 Overlap of two wave pulses—both right side up—traveling in opposite directions. Time increases from top to bottom. Compare to Fig. 15.17.

$y(x, t)^2$, $y(x, t)^{1/2}$, etc.). As a result, if any two functions $y_1(x, t)$ and $y_2(x, t)$ satisfy the wave equation separately, their sum $y_1(x, t) + y_2(x, t)$ also satisfies it and is therefore a physically possible motion. Because this principle depends on the linearity of the wave equation and the corresponding linear-combination property of its solutions, it is also called the *principle of linear superposition.* For some physical systems, such as a medium that does not obey Hooke's law, the wave equation is *not* linear; this principle does not hold for such systems.

The principle of superposition is of central importance in all types of waves. When a friend talks to you while you are listening to music, you can distinguish the sound of speech and the sound of music from each other. This is precisely because the total sound wave reaching your ears is the algebraic sum of the wave produced by your friend's voice and the wave produced by the speakers of your stereo. If two sound waves did *not* combine in this simple linear way, the sound you would hear in this situation would be a hopeless jumble. Superposition also applies to electromagnetic waves (such as light) and many other types of waves.

15.19 Two wave pulses with different shapes.

Test Your Understanding

Two wave pulses with different shapes are traveling in opposite directions along a string (Fig. 15.19). Make a series of sketches like Fig. 15.18 showing the shape of the string as the two pulses approach, overlap, and then pass each other.

15.7 | Standing Waves on a String

We have talked about the reflection of a wave *pulse* on a string when it arrives at a boundary point (either a fixed end or a free end). Now let's look at what happens when a *sinusoidal* wave is reflected by a fixed end of a string. We'll again approach the problem by considering the superposition of two waves propagating through the string, one representing the original or incident wave and the other representing the wave reflected at the fixed end.

Figure 15.20 shows a string that is fixed at its left-hand end. Its right-hand end is moved up and down in simple harmonic motion to produce a wave that travels to the left; the wave reflected from the fixed end travels to the right. The resulting motion when the two waves combine no longer looks like two waves traveling in opposite directions. The string appears to be subdivided into a number of segments, as in the time-exposure photographs of Figs. 15.20a, 15.20b, 15.20c, and 15.20d. Figure 15.20e shows two instantaneous shapes of the string in Fig. 15.20b. Let's compare this behavior with the waves we studied in Sections 15.1 through 15.5. In a wave that travels along the string, the amplitude is constant and the wave pattern moves with a speed equal to the wave speed. Here, instead, the wave pattern remains in the same position along the string and its amplitude fluctuates. There are particular points called **nodes** (labeled N in Fig. 15.20e) that never move at all. Midway between the nodes are points called **antinodes** (labeled A in Fig. 15.20e) where the amplitude of motion is greatest. Because the wave pattern doesn't appear to be moving in either direction along the string, it is called a **standing wave.** (To emphasize the difference, a wave that *does* move along the string is called a **traveling wave.**)

The principle of superposition explains how the incident and reflected wave combine to form a standing wave. In Fig. 15.21 the red curves show a wave traveling to the left. The blue curves show a wave traveling to the right with the same propagation speed, wavelength, and amplitude. The waves are shown at nine

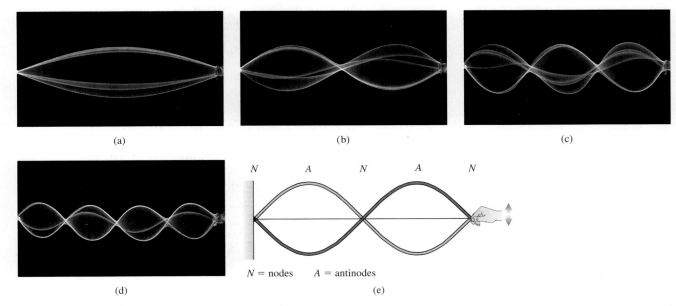

(a)   (b)   (c)

(d)   (e)

N = nodes A = antinodes

15.20 (a)–(d) Time exposures of standing waves in a stretched string. From (a) to (d), the frequency of oscillation of the right-hand end increases and the wavelength of the standing wave decreases. (e) The extremes of the motion of the standing wave in (b), with nodes at the center and at the ends. The right-hand end of the string moves very little compared to the antinodes and so is essentially a node.

instants, one sixteenth of a period apart. At each point along the string, we add the displacements (the values of y) for the two separate waves; the result is the total wave on the string, shown in black.

At certain instants, such as $t = 4T/16$, the two wave patterns are exactly in phase with each other, and the shape of the string is a sine curve with twice the amplitude of either individual wave. At other instants, such as $t = 8T/16$, the two waves are exactly out of phase with each other, and the total wave at that instant is zero. The resultant displacement is *always* zero at those places marked N at the bottom of Fig. 15.21. These are the *nodes*. At a node the displacements of the two waves in red and blue are always equal and opposite and cancel each other out. This cancellation is called **destructive interference.** Midway between the nodes are the points of *greatest* amplitude or *antinodes,* marked A. At the antinodes the displacements of the two waves in red and blue are always identical, giving a large resultant displacement; this phenomenon is called **constructive interference.** We can see from the figure that the distance between successive nodes or between successive antinodes is one half-wavelength, or $\lambda/2$.

We can derive a wave function for the standing wave of Fig. 15.21 by adding the wave functions $y_1(x, t)$ and $y_2(x, t)$ for two waves with equal amplitude, period, and wavelength traveling in opposite directions. Here $y_1(x, t)$ (the red curves in Fig. 15.21) represents an incoming, or *incident,* wave traveling to the left along the $+x$-axis, arriving at the point $x = 0$ and being reflected; $y_2(x, t)$ (the blue curves in Fig. 15.21) represents the *reflected* wave traveling to the right from $x = 0$. We noted in Section 15.6 that the wave reflected from a fixed end of a string is inverted, so we give a negative sign to one of the waves:

$$y_1(x, t) = -A\cos(kx + \omega t) \qquad \text{(incident wave traveling to the left)}$$

$$y_2(x, t) = A\cos(kx - \omega t) \qquad \text{(reflected wave traveling to the right)}$$

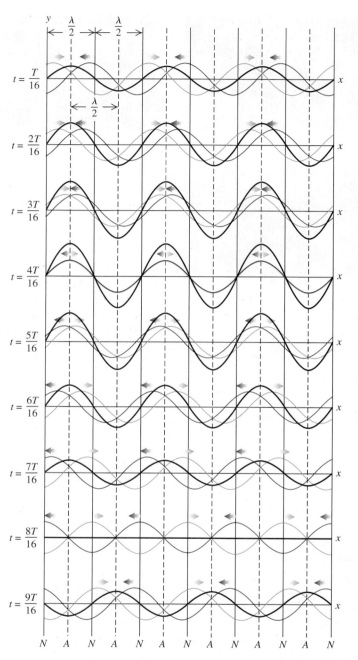

15.21 Formation of a standing wave. A wave traveling to the left (red curves) combines with a wave traveling to the right (blue curves) to form a standing wave (black curves). The horizontal x-axis in each part shows the equilibrium position of the string.

Note also that the change in sign corresponds to a shift in *phase* of 180° or π radians. At $x = 0$ the motion from the reflected wave is $A \cos \omega t$; and the motion from the incident wave is $-A \cos \omega t$, which we can also write as $A \cos(\omega t + \pi)$. From Eq. (15.27), the wave function for the standing wave is the sum of the individual wave functions:

$$y(x, t) = y_1(x, t) + y_2(x, t) = A\left[-\cos(kx + \omega t) + \cos(kx - \omega t)\right]$$

We can rewrite each of the cosine terms by using the identities for the cosine of the sum and difference of two angles: $\cos(a \pm b) = \cos a \cos b \mp \sin a \sin b$. Applying these and combining terms, we obtain the wave function for the standing wave:

$$y(x, t) = y_1(x, t) + y_2(x, t) = (2A \sin kx)\sin \omega t$$

or

$$y(x, t) = (A_{SW} \sin kx)\sin \omega t \qquad (15.28)$$
$$\text{(standing wave on a string, fixed end at } x = 0)$$

The standing wave amplitude A_{SW} is twice the amplitude A of either of the original traveling waves:

$$A_{SW} = 2A$$

Equation (15.28) has two factors: a function of x and a function of t. The factor $A_{SW} \sin kx$ shows that at each instant the shape of the string is a sine curve. But unlike a wave traveling along a string, the wave shape stays in the same position, oscillating up and down as described by the $\sin \omega t$ factor. This behavior is shown graphically by the black curves in Fig. 15.21. Each point in the string still undergoes simple harmonic motion, but all the points between any successive pair of nodes oscillate *in phase*. This is in contrast to the phase differences between oscillations of adjacent points that we see with a wave traveling in one direction.

We can use Eq. (15.28) to find the positions of the nodes; these are the points for which $\sin kx = 0$, so the displacement is *always* zero. This occurs when $kx = 0, \pi, 2\pi, 3\pi, \ldots$, or, using $k = 2\pi/\lambda$,

$$x = 0, \frac{\pi}{k}, \frac{2\pi}{k}, \frac{3\pi}{k}, \ldots$$
$$= 0, \frac{\lambda}{2}, \frac{2\lambda}{2}, \frac{3\lambda}{2}, \ldots \qquad (15.29)$$

$$\text{(nodes of a standing wave on a string, fixed end at } x = 0)$$

In particular, there is a node at $x = 0$, as there should be, since this point is a fixed end of the string.

A standing wave, unlike a traveling wave, *does not* transfer energy from one end to the other. The two waves that form it would individually carry equal amounts of power in opposite directions. There is a local flow of energy from each node to the adjacent antinodes and back, but the *average* rate of energy transfer is zero at every point. If you evaluate the wave power given by Eq. (15.21) using the wave function of Eq. (15.28), you will find that the average power is zero (see Challenge Problem 15.82).

Problem-Solving Strategy	**Standing Waves**

IDENTIFY *the relevant concepts:* As with traveling waves, it's useful to distinguish between the purely kinematic quantities, such as wave speed v, wavelength λ, and frequency f, and the dynamic quantities involving the properties of the medium, such as F and μ for transverse waves on a string. Once you decide what the target variable is, try to determine whether the problem is only kinematic in nature or whether the properties of the medium are also involved.

SET UP *the problem using the following steps:*

1. In visualizing nodes and antinodes in standing waves, it is always helpful to draw diagrams. For a string you can draw the shape at one instant and label the nodes N and antinodes A. The distance between two adjacent nodes or two adjacent antinodes is always $\lambda/2$, and the distance between a node and the adjacent antinode is always $\lambda/4$.

2. Decide which equations you'll need to use. The wave function for the standing wave is almost always useful (like Eq. (15.28)).

3. You can compute the wave speed if you know either λ and f (or, equivalently, $k = 2\pi/\lambda$ and $\omega = 2\pi f$) or the properties of the medium (for a string, F and μ).

EXECUTE *the solution as follows:*

Solve for the unknown quantities using the equations you've selected. Once you have the wave function, you can find the value of the displacement y at any point in the wave medium (value of x) and at any time. You can find the velocity of a particle in the wave medium by taking the partial derivative of y with respect to time. To find the acceleration of such a particle, take the second partial derivative of y with respect to time.

EVALUATE *your answer:* Compare your numerical answers with your diagram. Check that the wave function is compatible with the boundary conditions (for example, the displacement should be zero at a fixed end).

Example 15.7	**Standing waves on a guitar string**

One of the strings of a guitar lies along the x-axis when in equilibrium. The end of the string at $x = 0$ (the bridge of the guitar) is tied down. An incident sinusoidal wave, corresponding to the red curves in Fig. 15.21, travels along the string in the $-x$-direction at 143 m/s with an amplitude of 0.750 mm and a frequency of 440 Hz. This wave is reflected from the fixed end at $x = 0$, and the superposition of the incident traveling wave and the reflected traveling wave forms a standing wave. a) Find the equation giving the displacement of a point on the string as a function of position and time. b) Locate the points on the string that don't move at all. c) Find the amplitude, maximum transverse velocity, and maximum transverse acceleration at points of maximum oscillation.

SOLUTION

IDENTIFY: This is a *kinematics* problem in which we are asked to describe the motion of the string (see the Problem-Solving Strategy in Section 15.3). The target variables are the wave function of the standing wave (part (a)), the locations of the points that don't move, or *nodes* (part (b)), and the maximum values of displacement y, transverse velocity v_y, and transverse acceleration a_y. (Waves on a string are transverse waves, so *transverse* means "in the direction of the displacement," that is, in the y-direction.) To find these quantities we use the expression that we derived in this section for a standing wave on a string with a fixed end, as well as other relations from Sections 15.2 and 15.3.

SET UP: Since there is a fixed end at $x = 0$, we may use Eqs. (15.28) and (15.29) to describe this standing wave. We also use the relationships among ω, k, f, λ, and the wave speed v.

EXECUTE: a) To use Eq. (15.28) we need the values of A_{SW}, ω, and k. The amplitude of the incident wave is $A = 0.750$ mm $= 7.50 \times 10^{-4}$ m; the reflected wave has the same amplitude, and the standing wave amplitude is $A_{SW} = 2A = 1.50 \times 10^{-3}$ m. The angular frequency ω and wave number k are

$$\omega = 2\pi f = (2\pi \text{ rad})(440 \text{ s}^{-1}) = 2760 \text{ rad/s}$$

$$k = \frac{\omega}{v} = \frac{2760 \text{ rad/s}}{143 \text{ m/s}} = 19.3 \text{ rad/m}$$

Then Eq. (15.28) gives

$$y(x, t) = (A_{SW} \sin kx) \sin \omega t$$
$$= [(1.50 \times 10^{-3} \text{ m}) \sin (19.3 \text{ rad/m})x] \sin (2760 \text{ rad/s})t$$

b) The positions of the nodes are given by Eq. (15.29): $x = 0, \lambda/2, \lambda, 3\lambda/2, \ldots$ The wavelength is

$$\lambda = \frac{v}{f} = \frac{143 \text{ m/s}}{440 \text{ Hz}} = 0.325 \text{ m}$$

so the nodes are at the following distances from the fixed end:

$$x = 0, 0.163 \text{ m}, 0.325 \text{ m}, 0.488 \text{ m}, \ldots$$

c) From the expression in part (a) for $y(x, t)$, we see that the maximum displacement from equilibrium is 1.50×10^{-3} m $= 1.50$ mm, which is just twice the amplitude of the incident wave. This maximum occurs at the *antinodes*, which are midway between adjacent nodes (that is, at $x = 0.081$ m, 0.244 m, 0.406 m, $\ldots$).

For a particle on the string at any point x, the transverse (y-) velocity is

$$v_y(x, t) = \frac{\partial y(x, t)}{\partial t}$$
$$= [(1.50 \times 10^{-3} \text{ m}) \sin (19.3 \text{ rad/m})x]$$
$$\times [(2760 \text{ rad/s}) \cos (2760 \text{ rad/s})t]$$
$$= [(4.15 \text{ m/s}) \sin (19.3 \text{ rad/m})x] \cos (2760 \text{ rad/s})t$$

At an antinode, $\sin(19.3 \text{ rad/m})x = \pm 1$ and the transverse velocity varies in value between 4.15 m/s and -4.15 m/s. As is always the case in simple harmonic motion, the maximum velocity occurs when the particle is passing through the equilibrium position ($y = 0$).

The transverse acceleration $a_y(x, t)$ is the first partial derivative of $v_y(x, t)$ with respect to time (that is, the *second* partial derivative of $y(x, t)$ with respect to time). We leave the calculation to you; the result is

$$a_y(x, t) = \frac{\partial v_y(x, t)}{\partial t} = \frac{\partial^2 y(x, t)}{\partial t^2}$$

$$= [(-1.15 \times 10^4 \text{ m/s}^2) \sin(19.3 \text{ rad/m})x]$$
$$\times \sin(2760 \text{ rad/s})t$$

At the antinodes, the transverse acceleration varies in value between $+1.15 \times 10^4$ m/s^2 and -1.15×10^4 m/s^2.

EVALUATE: The maximum transverse velocity at an antinode is quite respectable (about 15 km/h, or 9.3 mi/h). But the maximum transverse acceleration is tremendous, 1170 times the acceleration due to gravity! Guitar strings are made of sturdy stuff to be able to withstand such acceleration.

Guitar strings are actually tied down at *both* ends. We'll see the consequences of this in the next section.

Test Your Understanding

Suppose the frequency of the standing wave in Example 15.7 were doubled from 440 Hz to 880 Hz. Would all of the nodes for $f = 440$ Hz also be nodes for $f = 880$ Hz? If so, would there be additional nodes for $f = 880$ Hz? If not, which nodes are absent for $f = 880$ Hz?

15.8 | Normal Modes of a String

When we described standing waves on a string rigidly held at one end, as in Fig. 15.20, we made no assumptions about the length of the string or about what was happening at the other end. Let's now consider a string of a definite length L, rigidly held at *both* ends. Such strings are found in many musical instruments, including pianos, violins, and guitars. When a guitar string is plucked, a wave is produced in the string; this wave is reflected and re-reflected from the ends of the string, making a standing wave. This standing wave on the string in turn produces a sound wave in the air, with a frequency determined by the properties of the string. This is what makes stringed instruments so useful in making music.

To understand these properties of standing waves on a string fixed at both ends, let's first examine what happens when we set up a sinusoidal wave on such a string. The standing wave that results must have a node at *both* ends of the string. We saw in the preceding section that adjacent nodes are one half-wavelength $(\lambda/2)$ apart, so the length of the string must be $\lambda/2$, or $2(\lambda/2)$, or $3(\lambda/2)$, or in general some integer number of half-wavelengths:

$$L = n\frac{\lambda}{2} \quad (n = 1, 2, 3, \dots) \quad \text{(string fixed at both ends)} \quad (15.30)$$

That is, if a string with length L is fixed at both ends, a standing wave can exist only if its wavelength satisfies Eq. (15.30).

Solving this equation for λ and labeling the possible values of λ as λ_n, we find

$$\lambda_n = \frac{2L}{n} \quad (n = 1, 2, 3, \dots) \quad \text{(string fixed at both ends)} \quad (15.31)$$

Waves can exist on the string if the wavelength is *not* equal to one of these values, but there cannot be a steady wave pattern with nodes and antinodes, and the total wave cannot be a standing wave. Equation (15.31) is illustrated by the standing waves shown in Figs. 15.20a, 15.20b, 15.20c, and 15.20d; these represent $n = 1, 2, 3,$ and 4, respectively.

Activ
Physics
ONLINE

10.4 Standing Waves on Strings

10.5 Tuning a Stringed Instrument: Standing Waves

10.6 String Mass and Standing Waves

Corresponding to the series of possible standing-wave wavelengths λ_n is a series of possible standing-wave frequencies f_n, each related to its corresponding wavelength by $f_n = v/\lambda_n$. The smallest frequency f_1 corresponds to the largest wavelength (the $n = 1$ case), $\lambda_1 = 2L$:

$$f_1 = \frac{v}{2L} \qquad \text{(string fixed at both ends)} \qquad (15.32)$$

This is called the **fundamental frequency.** The other standing-wave frequencies are $f_2 = 2v/2L$, $f_3 = 3v/2L$, and so on. These are all integer multiples of the fundamental frequency f_1, such as $2f_1$, $3f_1$, $4f_1$, and so on, and we can express *all* the frequencies as

$$f_n = n\frac{v}{2L} = nf_1 \qquad (n = 1, 2, 3, \dots) \qquad \text{(string fixed at both ends)}$$

$$(15.33)$$

These frequencies are called **harmonics,** and the series is called a **harmonic series.** Musicians sometimes call f_2, f_3, and so on **overtones;** f_2 is the second harmonic or the first overtone, f_3 is the third harmonic or the second overtone, and so on. The first harmonic is the same as the fundamental frequency (Fig. 15.22).

For a string with fixed ends at $x = 0$ and $x = L$, the wave function $y(x, t)$ of the nth standing wave is given by Eq. (15.28) (which satisfies the condition that there is a node at $x = 0$), with $\omega = \omega_n = 2\pi f_n$ and $k = k_n = 2\pi/\lambda_n$:

$$y_n(x, t) = A_{\text{SW}} \sin k_n x \sin \omega_n t \qquad (15.34)$$

You can easily show that this wave function has nodes at both $x = 0$ and $x = L$, as it must.

A **normal mode** of an oscillating system is a motion in which all particles of the system move sinusoidally with the same frequency. For a system made up of a string of length L fixed at both ends, each of the wavelengths given by Eq. (15.31) corresponds to a possible normal-mode pattern and frequency. There are infinitely many normal modes, each with its characteristic frequency and vibration pattern. Figure 15.23 shows the first four normal-mode patterns and their associated frequencies and wave-

15.22 Each string of a violin naturally oscillates at one or more of its harmonic frequencies, producing sound waves in the air with the same frequencies.

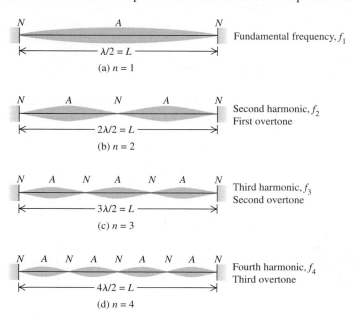

15.23 The first four normal modes of a string fixed at both ends. (Compare these to the photographs in Fig. 15.20.)

lengths; these correspond to Eq. (15.34) with $n = 1, 2, 3,$ and 4. By contrast, a harmonic oscillator, which has only one oscillating particle, has only one normal mode and one characteristic frequency. The string fixed at both ends has infinitely many normal modes because it is made up of a very large (effectively infinite) number of particles. More complicated oscillating systems also have infinite numbers of normal modes, though with more complex normal-mode patterns than a string (Fig. 15.24).

If we could displace a string so that its shape is the same as one of the normal-mode patterns and then release it, it would vibrate with the frequency of that mode. Such a vibrating string would displace the surrounding air with the same frequency, producing a traveling sinusoidal sound wave that your ears would perceive as a pure tone. But when a string is struck (as in a piano) or plucked (as is done to guitar strings), the shape of the displaced string is *not* as simple as one of the patterns in Fig. 15.23. The fundamental as well as many overtones are present in the resulting vibration. This motion is therefore a combination or *superposition* of many normal modes. Several simple-harmonic motions of different frequencies are present simultaneously, and the displacement of any point on the string is the sum (or superposition) of displacements associated with the individual modes. The sound produced by the vibrating string is likewise a superposition of traveling sinusoidal sound waves, which you perceive as a rich, complex tone with the fundamental frequency f_1. The standing wave on the string and the traveling sound wave in the air have similar **harmonic content** (the extent to which frequencies higher than the fundamental are present). The harmonic content depends on how the string is initially set into motion. If you pluck the strings of an acoustic guitar in the normal location over the sound hole, the sound that you hear has a different harmonic content than if you pluck the strings next to the fixed end on the guitar body.

It is possible to represent every possible motion of the string as some superposition of normal-mode motions. Finding this representation for a given vibration pattern is called *harmonic analysis*. The sum of sinusoidal functions that represents a complex wave is called a *Fourier series*. Figure 15.25 shows how a standing wave that is produced by plucking a guitar string of length L at a point $L/4$ from one end can be represented as a combination of sinusoidal functions.

As we have seen, the fundamental frequency of a vibrating string is $f_1 = v/2L$. The speed v of waves on the string is determined by Eq. (15.13), $v = \sqrt{F/\mu}$. Combining these, we find

$$f_1 = \frac{1}{2L}\sqrt{\frac{F}{\mu}} \quad \text{(string fixed at both ends)} \quad (15.35)$$

15.24 Astronomers have discovered that the sun oscillates in several different normal modes. In this computer simulation of one such mode, blue denotes places where solar material is moving outward and red denotes where it is moving inward.

Activ
ONLINE
Physics

10.10 Complex Waves: Fourier Analysis

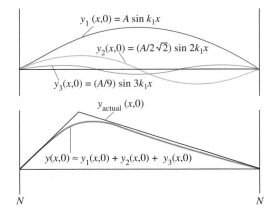

15.25 When a guitar string is plucked (pulled into a triangular shape) and released, a standing wave results. The standing wave is well represented (except at the sharp maximum point) by the sum of just three sinusoidal functions. Including additional sinusoidal functions further improves the representation.

15.26 Comparing the range of a concert grand piano to those of a violin (red bar), a viola (yellow bar), a cello (green bar), and a string bass (blue bar). In all cases, longer strings provide bass notes and shorter strings produce treble notes.

This is also the fundamental frequency of the sound wave created in the surrounding air by the vibrating string. Familiar musical instruments show how f_1 depends on the properties of the string. The inverse dependence of frequency on length L is illustrated by the long strings of the bass (low-frequency) section of the piano or the bass viol compared with the shorter strings on the treble section of the piano or the violin (Fig. 15.26). The pitch of a violin or guitar is usually varied by pressing a string against the fingerboard with the fingers to change the length L of the vibrating portion of the string. Increasing the tension F increases the wave speed v and thus increases the frequency (and the pitch). All string instruments are "tuned" to the correct frequencies by varying the tension; you tighten the string to raise the pitch. Finally, increasing the mass per unit length μ decreases the wave speed and thus the frequency. The lower notes on a steel guitar are produced by thicker strings, and one reason for winding the bass strings of a piano with wire is to obtain the desired low frequency without resorting to a string that is inconveniently long.

Wind instruments such as saxophones and trombones also have normal modes. As for stringed instruments, the frequencies of these normal modes determine the pitch of the musical tones that these instruments produce. We'll discuss these instruments and many other aspects of sound in Chapter 16.

Example 15.8	**A giant bass viol**

In an effort to get your name in the *Guinness Book of World Records*, you set out to build a bass viol with strings that have a length of 5.00 m between fixed points. One string has a linear mass density of 40.0 g/m and a fundamental frequency of 20.0 Hz (the lowest frequency that the human ear can hear). Calculate (a) the tension of this string, (b) the frequency and wavelength on the string of the second harmonic, and (c) the frequency and wavelength on the string of the second overtone.

SOLUTION

IDENTIFY: The target variable in (a) is the string tension; we find this from the expression for the fundamental frequency of the string, which involves the tension. In parts (b) and (c) the target variables are the frequency and wavelength of different harmonics. We determine these from the given length of the string and fundamental frequency.

SET UP: For part (a), the equation to use is Eq. (15.35); it involves the known values of f_1, L, and μ as well as the target variable F. We solve parts (b) and (c) using Eqs. (15.31) and (15.33).

EXECUTE: a) We solve Eq. (15.35) for the string tension F:

$$F = 4\mu L^2 f_1^2 = 4(40.0 \times 10^{-3}\ \text{kg/m})(5.00\ \text{m})^2(20.0\ \text{s}^{-1})^2$$
$$= 1600\ \text{N} = 360\ \text{lb}$$

b) The second harmonic is denoted by $n = 2$. From Eq. (15.33), the second harmonic frequency is

$$f_2 = 2f_1 = 2(20.0\ \text{Hz}) = 40.0\ \text{Hz}$$

From Eq. (15.31), the wavelength on the string of the second harmonic is

$$\lambda_2 = \frac{2L}{2} = 5.00\ \text{m}$$

c) The second overtone is the "second tone over" (above) the fundamental, that is, $n = 3$. Its frequency and wavelength are

$$f_3 = 3f_1 = 3(20.0\ \text{Hz}) = 60.0\ \text{Hz}$$
$$\lambda_3 = \frac{2L}{3} = 3.33\ \text{m}$$

EVALUATE: The tension in part (a) is a bit larger than in a real bass viol, for which the string tension is typically a few hundred newtons. The wavelengths in parts (b) and (c) are equal to the length of the string and two-thirds the length of the string respectively; these results agree with the drawings of standing waves in Fig. 15.23.

Example 15.9

From waves on a string to sound waves in air

What are the frequency and wavelength of the sound waves produced in the air when the string in the previous example is vibrating at its fundamental frequency? The speed of sound in air at 20°C is 344 m/s.

SOLUTION

IDENTIFY: Our target variables are f and λ for the sound wave produced by the bass viol, *not* for the standing wave on the string. However, when the string vibrates at a particular frequency, the surrounding air is forced to vibrate at the same frequency. So the frequency of the sound wave is the same as that of the standing wave on the string. However, the relationship $\lambda = v/f$ shows that the *wavelength* of the sound wave is typically different from the wavelength of the standing wave on the string, because the two waves have different speeds.

SET UP: The only equation we need is $v = \lambda f$. We apply this to both the standing wave on the string (speed v_{string}) and the traveling sound wave (speed v_{sound}).

EXECUTE: The sound wave frequency is the same as the standing wave fundamental frequency: $f = f_1 = 20.0$ Hz. The wavelength of the sound wave is

$$\lambda_{1\,(\text{sound})} = \frac{v_{\text{sound}}}{f_1} = \frac{344\ \text{m/s}}{20.0\ \text{Hz}} = 17.2\ \text{m}$$

EVALUATE: Note that $\lambda_{1\,(\text{sound})}$ is greater than the wavelength of the standing wave on the string, $\lambda_{1\,(\text{string})} = 2L = 2(5.00\ \text{m}) = 10.0$ m. Fundamentally this is because the speed of sound is greater than the speed of waves on the string, $v_{\text{string}} = \lambda_{1\,(\text{string})}f_1 = (10.0\ \text{m})(20.0\ \text{Hz}) = 200$ m/s. Hence, for *any* normal mode on this string, the sound wave that is produced has the same frequency as the wave on the string but a wavelength that is greater by a factor of $v_{\text{sound}}/v_{\text{string}} = (344\ \text{m/s})/(200\ \text{m/s}) = 1.72$.

Test Your Understanding

While a guitar string is vibrating, you gently touch the midpoint of the string to ensure that the string does not vibrate at that point. Which normal modes *cannot* be present on the string while you are touching it in this way?

SUMMARY

A wave is any disturbance from an equilibrium condition that propagates from one region to another. A mechanical wave always travels within some material called the medium. The wave disturbance propagates at the wave speed v, which depends on the type of wave and the properties of the medium.

In a periodic wave, the motion of each point of the medium is periodic. A sinusoidal wave is a special periodic wave in which each point moves in simple harmonic motion. For any periodic wave, the frequency f is the number of cycles per unit time, the period T is the time for one cycle, the wavelength λ is the distance over which the wave pattern repeats, and the amplitude A is the maximum displacement of a particle in the medium. The product of λ and f equals the wave speed. (See Example 15.1)

$$v = \lambda f \qquad (15.1)$$

The wave function $y(x, t)$ describes the displacements of individual particles in the medium. Equations (15.3), (15.4), and (15.7) give the wave equation for a sinusoidal wave traveling in the $+x$-direction. If the wave is moving in the $-x$-direction, the minus signs in the cosine functions are replaced by plus signs. (See Example 15.2)

$$y(x, t) = A \cos\left[\omega\left(\frac{x}{v} - t\right)\right]$$

$$= A \cos 2\pi f\left(\frac{x}{v} - t\right) \qquad (15.3)$$

$$y(x, t) = A \cos 2\pi\left(\frac{x}{\lambda} - \frac{t}{T}\right) \qquad (15.4)$$

$$y(x, t) = A \cos(kx - \omega t) \qquad (15.7)$$

where $k = 2\pi/\lambda$ and $\omega = 2\pi f = vk$

The wave function obeys a partial differential equation called the wave equation.

$$\frac{\partial^2 y(x, t)}{\partial x^2} = \frac{1}{v^2}\frac{\partial^2 y(x, t)}{\partial t^2} \qquad (15.12)$$

The speed of transverse waves on a string depends on the tension F and mass per unit length μ. (See Examples 15.3 and 15.4)

$$v = \sqrt{\frac{F}{\mu}} \qquad \text{(waves on a string)} \qquad (15.13)$$

Wave motion conveys energy from one region to another. For a sinusoidal mechanical wave, the average power P_{av} is proportional to the square of the wave amplitude and the square of the frequency. For waves that spread out in three dimensions, the wave intensity I is inversely proportional to the distance from the source. (See Examples 15.5 and 15.6)

$$P_{av} = \frac{1}{2}\sqrt{\mu F}\,\omega^2 A^2 \qquad (15.25)$$

(average power, sinusoidal wave)

$$\frac{I_1}{I_2} = \frac{r_2^2}{r_1^2} \qquad (15.26)$$

(inverse-square law for intensity)

A wave that reaches a boundary of the medium in which it propagates is reflected. The principle of superposition states that the total wave displacement at any point where two or more waves overlap is the sum of the displacements of the individual waves.

$$y(x, t) = y_1(x, t) + y_2(x, t)$$
(principle of superposition) (15.27)

When a sinusoidal wave is reflected from a fixed or free end of a stretched string, the incident and reflected waves combine to form a standing sinusoidal wave with nodes and antinodes. Adjacent nodes are spaced a distance $\lambda/2$ apart, as are adjacent antinodes. (See Example 15.7)

$$y(x, t) = (A_{SW} \sin kx) \sin \omega t$$
(standing wave on a string, fixed end at $x = 0$) (15.28)

$N = $ nodes $A = $ antinodes

When both ends of a string with length L are held fixed, standing waves can occur only when L is an integer multiple of $\lambda/2$. Each frequency and its associated vibration pattern is called a normal mode. The lowest frequency f_1 is called the fundamental frequency. (See Examples 15.8 and 15.9)

$$f_n = n\frac{v}{2L} = nf_1 \quad (n = 1, 2, 3, \ldots)$$
(15.33)

$$f_1 = \frac{1}{2L}\sqrt{\frac{F}{\mu}}$$
(15.35)

(string fixed at both ends)

Key Terms

Your Notes

Answer to Chapter Opening Question

The power of a mechanical wave depends on its frequency and amplitude [see Eq. (15.25)].

Answers to Test Your Understanding Questions

Section 15.2 The speed of waves on a string doesn't change at all when the wavelength is changed. The relation $v = \lambda f$ [Eq. (15.1)] tells us that $f = v/\lambda$, so doubling the wavelength λ will reduce the frequency f by a factor of 1/2.

Section 15.3 The plus sign in the cosine function tells us that the wave is propagating in the negative x-direction. The wave has wave number $k = 2.00 \text{ m}^{-1}$, angular frequency $\omega = 3.20 \text{ rad/s}$, and amplitude $A = 2.50 \text{ mm}$.

Section 15.4 The relation $v = \sqrt{F/\mu}$ [Eq. (15.13)] says that the wave speed is greatest on the string with the smallest linear mass density. This is the thinnest string, which has the smallest amount of mass m and hence the smallest linear mass density $\mu = m/L$ (all strings are the same length).

Section 15.5 The average power is proportional to the square of the frequency [see Eq. (15.25)]. Hence doubling the frequency means that the average power increases by a factor of $2^2 = 4$.

Section 15.6

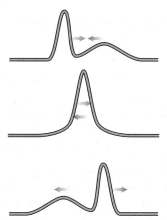

Section 15.7 Doubling the frequency makes the wavelength half as large. Hence the spacing between nodes (equal to $\lambda/2$) is also half as large. There are nodes at all of the previous positions, but there is also a new node between every pair of old nodes.

Section 15.8 When you touch the string at its center, you are demanding that there be a node at the center. Hence only standing waves with a node at $x = L/2$ are allowed. From Figure 15.23 you can see that the normal modes $n = 1, 3, 5, \ldots$ cannot be present.

Discussion Questions

Q15.1 A common pastime at sporting events is "doing the wave," in which the spectators in one section raise and lower their arms, signaling the spectators in the next section to do the same, and so

on. In the language of this chapter, is this truly a wave? Why or why not?

Q15.2 Explain why it is necessary to include the factor 2π in Equation (15.4). What unit does 2π have? Could you correctly replace this factor with 360°?

Q15.3 What kinds of energy are associated with waves on a stretched string? How could you detect such energy experimentally?

Q15.4 The amplitude of a wave decreases gradually as the wave travels down a long, stretched string. What happens to the energy of the wave when this happens?

Q15.5 For the wave motions discussed in this chapter, does the speed of propagation depend on the amplitude? How can you tell?

Q15.6 The speed of ocean waves depends on the depth of the water; the deeper the water, the faster the wave travels. Use this to explain why ocean waves crest and "break" as the wave nears the shore.

Q15.7 Is it possible to have a longitudinal wave on a stretched string? Why or why not? Is it possible to have a transverse wave on a steel rod? Again, why or why not? If your answer is yes in either case, explain how you would create such a wave.

Q15.8 An echo is sound reflected from a distant object, such as a wall or a cliff. Explain how you can determine how far away the object is by timing the echo.

Q15.9 Why do you see lightning before you hear the thunder? A familiar rule of thumb is to start counting slowly, once per second, when you see the lightning; when you hear the thunder, divide the number you have reached by 3 to obtain your distance from the lightning in kilometers (or by 5 to obtain your distance in miles). Why does this work, or does it?

Q15.10 For transverse waves on a string, is the wave speed the same as the maximum speed of any part of the string, or is there a different relation between these two speeds? Explain your reasoning.

Q15.11 Children make toy telephones by sticking each end of a long string through a hole in the bottom of a paper cup and knotting it so it will not pull out. When the spring is pulled taut, sound can be transmitted from one cup to the other. How does this work? Why is the transmitted sound louder than the sound traveling through air for the same distance?

Q15.12 The four strings on a violin have different thicknesses, but are all under approximately the same tension. Do waves travel faster on the thick strings or the thin strings? Why? How does the fundamental vibration frequency compare for the thick versus the thin strings?

Q15.13 When a rock is thrown into a pond and the resulting ripples spread in ever-widening circles, the amplitude decreases with increasing distance from the center. Why?

Q15.14 Two strings of different mass per unit length μ_1 and μ_2 are tied together and stretched with a tension F. A wave travels along the string and passes the discontinuity in μ. Which of the following wave properties will be the same on both sides of the discontinuity and which ones will change: Speed of the wave? Frequency? Wavelength? Explain the physical reasoning behind each of your answers.

Q15.15 A long rope with mass m is suspended from the ceiling and hangs vertically. A wave pulse is produced at the lower end of the rope and the pulse travels up the rope. Does the speed of the wave pulse change as it moves up the rope, and if so, does it increase or decrease?

Q15.16 In a transverse wave on a string, the motion of the string is perpendicular to the length of the string. How, then, is it possible for energy to move along the length of the string?

Q15.17 Justify the following statement: "A wave transfers energy fastest where the particles of the medium are moving fastest." In particular, explain how can this can be true for a transverse wave, in which the particles of the medium move in a direction perpendicular to that of wave propagation.

Q15.18 Energy can be transferred along a string by wave motion. However, in a standing wave on a string, no energy can ever be transferred past a node. Why not?

Q15.19 Can a standing wave be produced on a string by superposing two waves traveling in opposite directions with the same frequency but different amplitudes? Why or why not? Can a standing wave be produced by superposing two waves traveling in opposite directions with different frequencies but the same amplitude? Why or why not?

Q15.20 If you stretch a rubber band and pluck it, you hear a (somewhat) musical tone. How does the frequency of this tone change as you stretch the rubber band further? (Try it!) Does this agree with Eq. (15.35) for a string fixed at both ends? Explain.

Q15.21 A musical interval of an *octave* corresponds to a factor of two in frequency. By what factor must the tension in a guitar or violin string be increased to raise its pitch one octave? To raise it two octaves? Explain your reasoning. Is there any danger in attempting these changes in pitch?

Q15.22 By touching a string lightly at its center while bowing, a violinist can produce a note exactly one octave above the note to which the string is tuned, that is, a note with exactly twice the frequency. Why is this possible?

Q15.23 As we discussed in Section 15.1, water waves are a combination of longitudinal and transverse waves. Defend the following statement: "When water waves hit a vertical wall, the wall is a node of the longitudinal displacement but an antinode of the transverse displacement."

Q15.24 Violins are short instruments, while cellos and basses are long. In terms of the frequency of the waves they produce, explain why this is so.

Q15.25 What is the purpose of the frets on a guitar? In terms of the frequency of the vibration of the strings, explain their use.

Exercises

Section 15.2 Periodic Waves

15.1 A fisherman notices that his boat is moving up and down periodically, owing to waves on the surface of the water. It takes 2.5 s for the boat to travel from its highest point to its lowest, a total distance of 0.62 m. The fisherman sees that the wave crests are spaced 6.0 m apart. a) How fast are the waves traveling? b) What is the amplitude of each wave? c) If the total vertical distance traveled by the boat were 0.30 m, but the other data remained the same, how would the answers to parts (a) and (b) be affected? d) Would you

expect that the motion of the boat would be purely vertical? Why or why not?

15.2 Ultrasound Imaging. Ultrasound is the name given to frequencies above the human range of hearing, which is about 20,000 Hz. Waves above this frequency can be used to penetrate the body and to produce images by reflecting from surfaces. In a typical ultrasound scan, the waves travel with a speed of 1500 m/s. For a good detailed image, the wavelength should be no more than 1.0 mm. What frequency is required?

15.3 The speed of sound in air at 20°C is 344 m/s. a) What is the wavelength of a sound wave with a frequency of 784 Hz, corresponding to the note G_5 on a piano? b) What is the frequency of a sound wave with a wavelength of 0.0655 mm? (This frequency is too high for you to hear.)

15.4 The speed of radio waves in vacuum (equal to the speed of light) is 3.00×10^8 m/s. Find the wavelength for a) an AM radio station with frequency 540 kHz; b) an FM radio station with frequency 104.5 MHz.

15.5 Audible Wavelengths. Provided that the amplitude is sufficiently great, the human ear can respond to longitudinal waves over a range of frequencies from about 20.0 Hz to about 20,000 Hz. Compute the wavelengths corresponding to these frequencies a) for waves in air $(v = 344$ m/s$)$; b) for waves in water $(v = 1480$ m/s$)$.

Section 15.3 Mathematical Description of a Wave

15.6 A certain transverse wave is described by

$$y(x, t) = (6.50 \text{ mm}) \cos 2\pi \left(\frac{x}{28.0 \text{ cm}} - \frac{t}{0.0360 \text{ s}} \right)$$

Determine the wave's a) amplitude; b) wavelength; c) frequency; d) speed of propagation; e) direction of propagation.

15.7 Transverse waves on a string have wave speed 8.00 m/s, amplitude 0.0700 m, and wavelength 0.320 m. The waves travel in the $-x$-direction, and at $t = 0$ the $x = 0$ end of the string has its maximum upward displacement. a) Find the frequency, period, and wave number of these waves. b) Write a wave function describing the wave. c) Find the transverse displacement of a particle at $x = 0.360$ m at time $t = 0.150$ s. d) How much time must elapse from the instant in part (c) until the particle at $x = 0.360$ m next has maximum upward displacement?

15.8 Redraw Figs. 15.7a and 15.7b for the case in which the wave is propagating to the *left*.

15.9 Show that the following wave functions satisfy the wave equation, Eq. (15.12): a) $y(x, t) = A \cos(kx + \omega t)$; b) $y(x, t) = A \sin(kx + \omega t)$. c) In what directions are these waves traveling? How can you tell? d) For the wave of part (b), write the equations for the transverse velocity and transverse acceleration of a particle at point x.

15.10 a) For a wave described by $y(x, t) = A \cos(kx - \omega t)$, graph y, v_y, and a_y as functions of x for time $t = 0$. b) Consider the following points on the string: i) $x = 0$; ii) $x = \pi/4k$; iii) $x = \pi/2k$; iv) $x = 3\pi/4k$; v) $x = \pi/k$; vi) $x = 5\pi/4k$; vii) $x = 3\pi/2k$; and viii) $x = 7\pi/4k$. For a particle at each of these points at $t = 0$, describe in words whether the particle is moving and in what direction, and whether the particle is speeding up, slowing down, or instantaneously not accelerating.

15.11 A sinusoidal wave is propagating along a stretched string that lies along the x-axis. The displacement of the string as a function of time is graphed in Fig. 15.27 for particles at $x = 0$ and at $x = 0.0900$ m. a) What is the amplitude of the wave? b) What is the period of the wave? c) You are told that the two points $x = 0$ and $x = 0.0900$ m are within one wavelength of each other. If the wave is moving in the $+x$-direction, determine the wavelength and the wave speed. d) If instead the wave is moving in the $-x$-direction, determine the wavelength and the wave speed. e) Would it be possible to determine definitively the wavelength in parts (c) and (d) if you were not told that the two points were within one wavelength of each other? Why or why not?

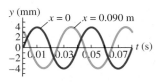

Figure 15.27 Exercise 15.11.

15.12 Speed of Propagation vs. Particle Speed. a) Show that Eq. (15.3) may be written as

$$y(x, t) = A \cos\left[\frac{2\pi}{\lambda}(x - vt)\right]$$

b) Use $y(x, t)$ to find an expression for the transverse velocity v_y of a particle in the string on which the wave travels. c) Find the maximum speed of a particle of the string. Under what circumstances is this equal to the propagation speed v? Less than v? Greater than v?

15.13 A transverse wave on a string has amplitude 0.300 cm, wavelength 12.0 cm, and speed 6.00 cm/s. It is represented by $y(x, t)$ as given in Exercise 15.12. a) At time $t = 0$, compute y at 1.5-cm intervals of x (that is, at $x = 0$, $x = 1.5$ cm, $x = 3.0$ cm, and so on) from $x = 0$ to $x = 12.0$ cm. Graph the results. This is the shape of the string at time $t = 0$. b) Repeat the calculations for the same values of x at times $t = 0.400$ s and $t = 0.800$ s. Graph the shape of the string at these instants. In what direction is the wave traveling?

Section 15.4 Speed of a Transverse Wave

15.14 With what tension must a rope with length 2.50 m and mass 0.120 kg be stretched for transverse waves of frequency 40.0 Hz to have a wavelength of 0.750 m?

15.15 One end of a horizontal rope is attached to a prong of an electrically driven tuning fork that vibrates at 120 Hz. The other end passes over a pulley and supports a 1.50-kg mass. The linear mass density of the rope is 0.0550 kg/m. a) What is the speed of a transverse wave on the rope? b) What is the wavelength? c) How would your answers to parts (a) and (b) change if the mass were increased to 3.00 kg?

15.16 A cowgirl ties one end of a 10.0-m-long rope to a fence post and pulls on the other end so the rope is stretched horizontally with a tension of 140 N. The mass of the rope is 0.800 kg. a) What is the speed of transverse waves on the rope? b) If the cowgirl moves the free end up and down with a frequency of 1.20 Hz, what is the wavelength of the transverse waves on the rope? c) The cowgirl pulls harder so that the tension is doubled to 280 N.

With what frequency must she move the free end of the rope up and down to produce transverse waves of the same wavelength as in part (a)?

15.17 One end of a 14.0-m-long rubber tube, with total mass 0.800 kg, is fastened to a fixed support. A cord attached to the other end passes over a pulley and supports an object with a mass of 7.50 kg. The tube is struck a transverse blow at one end. Find the time required for the pulse to reach the other end.

15.18 Weighty Rope. If in Example 15.4 (Section 15.4) we do *not* neglect the weight of the rope, what is the wave speed a) at the bottom of the rope? b) at the middle of the rope? c) at the top of the rope?

15.19 A simple, harmonic oscillator at the point $x = 0$ generates a wave on a rope. The oscillator operates at a frequency of 40.0 Hz and with an amplitude of 3.00 cm. The rope has a linear mass density of 50.0 g/m and is stretched with a tension of 5.00 N. a) Determine the speed of the wave. b) Find the wavelength. c) Write the wave function $y(x, t)$ for the wave. Assume that the oscillator has its maximum upward displacement at time $t = 0$. d) Find the maximum transverse acceleration of points on the rope. e) In the discussion of transverse waves in this chapter, the force of gravity was ignored. Is that a reasonable approximation for this wave? Explain.

Section 15.5 Energy in Wave Motion

15.20 A piano wire with mass 3.00 g and length 80.0 cm is stretched with a tension of 25.0 N. A wave with frequency 120.0 Hz and amplitude 1.6 mm travels along the wire. a) Calculate the average power carried by the wave. b) What happens to the average power if the wave amplitude is halved?

15.21 Redraw Fig. 15.13 for a position on the string at $x = \lambda/4$. Explain any differences from Fig. 15.13.

15.22 Threshold of Pain. You are investigating the report of a UFO landing in an isolated portion of New Mexico, and encounter a strange object that is radiating sound waves uniformly in all directions. Assume that the sound comes from a point source and that you can ignore reflections. You are slowly walking toward the source. When you are 7.5 m from it, you measure its intensity to be 0.11 W/m². An intensity of 1.0 W/m² is often used as the "threshold of pain." How much closer to the source can you move before the sound intensity reaches this threshold?

15.23 Energy Output. By measurement you determine that sound waves are spreading out equally in all directions from a point source and that the intensity is 0.026 W/m² at a distance of 4.3 m from the source. a) What is the intensity at a distance of 3.1 m from the source? b) How much sound energy does the source emit in one hour if its power output remains constant?

15.24 A fellow student of mathematical bent tells you that the wave function of a traveling wave on a thin rope is $y(x, t) = 2.30$ mm $\cos[(6.98 \text{ rad/m})x + (742 \text{ rad/s})t]$. Being more practical-minded, you measure the rope to have a length of 1.35 m and a mass of 3.38 grams. You are then asked to determine the following: a) amplitude; b) frequency; c) wavelength; d) wave speed; e) direction the wave is traveling; f) tension in the rope; g) average power transmitted by the wave.

15.25 What is the total power output of the siren in Example 15.6?

Section 15.6 Wave Interference, Boundary Conditions, and Superposition

15.26 Reflection. A wave pulse on a string has the dimensions shown in Fig. 15.28 at $t = 0$. The wave speed is 40 cm/s. a) If point O is a fixed end, draw the total wave on the string at $t = 15$ ms, 20 ms, 25 ms, 30 ms, 35 ms, 40 ms, and 45 ms. b) Repeat part (a) for the case in which point O is a free end.

Figure 15.28 Exercise 15.26.

15.27 Reflection. A wave pulse on a string has the dimensions shown in Fig. 15.29 at $t = 0$. The wave speed is 5.0 m/s. a) If point O is a fixed end, draw the total wave on the string at $t = 1.0$ ms, 2.0 ms, 3.0 ms, 4.0 ms, 5.0 ms, 6.0 ms, and 7.0 ms. b) Repeat part (a) for the case in which point O is a free end.

Figure 15.29 Exercise 15.27.

15.28 Interference of Triangular Pulses. Two triangular wave pulses are traveling toward each other on a stretched string as shown in Fig. 15.30. Each pulse is identical to the other and travels at 2.00 cm/s. The leading edges of the pulses are 1.00 cm apart at $t = 0$. Sketch the shape of the string at $t = 0.250$ s, $t = 0.500$ s, $t = 0.750$ s, $t = 1.000$ s, and $t = 1.250$ s.

Figure 15.30 Exercise 15.28.

15.29 Interference of Rectangular Pulses. Figure 15.31 shows two rectangular wave pulses on a stretched string traveling toward each other. Each pulse is traveling with a speed of 1.00 mm/s and has the height and width shown in the figure. If the leading edges of the pulses are 8.00 mm apart at $t = 0$, sketch the shape of the string at $t = 4.00$ s, $t = 6.00$ s, and $t = 10.0$ s.

Figure 15.31 Exercise 15.29.

15.30 Two traveling waves move on a string that has a fixed end at $x = 0$. They are identical, expect for opposite velocities. Each has an amplitude of 2.46 mm, a period of 3.65 ms, and a speed of 111 m/s. Write the wave function of the resulting standing wave.

Section 15.7 Standing Waves on a String

15.31 Standing waves on a wire are described by Eq. (15.28), with $A_{SW} = 2.50$ mm, $\omega = 942$ rad/s, and $k = 0.750\pi$ rad/m. The left-hand end of the wire is at $x = 0$. At what distances from the left-hand end are a) the nodes of the standing wave? b) the antinodes of the standing wave?

15.32 Wave Equation and Standing Waves. a) Prove by direct substitution that $y(x, t) = (A_{SW} \sin kx) \sin \omega t$ is a solution of the wave equation Eq. (15.12), for $v = \omega/k$. b) Explain why the relation $v = \omega/k$ for *traveling* waves also applies to *standing* waves.

15.33 Adjacent antinodes of a standing wave on a string are 15.0 cm apart. A particle at an antinode oscillates in simple harmonic motion with amplitude 0.850 cm and period 0.0750 s. The string lies along the $+x$-axis and is fixed at $x = 0$. a) Find the displacement of a point on the string as a function of position and time. b) Find the speed of propagation of a transverse wave in the string. c) Find the amplitude at a point 3.0 cm to the right of an antinode.

15.34 Give the details of the derivation of Eq. (15.28) from $y_1(x, t) + y_2(x, t) = A[-\cos(kx + \omega t) + \cos(kx - \omega t)]$.

15.35 Let $y_1(x, t) = A \cos(k_1 x - \omega_1 t)$ and $y_2(x, t) = A \cos(k_2 x - \omega_2 t)$ be two solutions to the wave equation, Eq. (15.12), for the same v. Show that $y(x, t) = y_1(x, t) + y_2(x, t)$ is also a solution to the wave equation.

Section 15.8 Normal Modes of a String

15.36 A piano tuner stretches a steel piano wire with a tension of 800 N. The steel wire is 0.400 m long and has a mass of 3.00 g. a) What is the frequency of its fundamental mode of vibration? b) What is the number of the highest harmonic that could be heard by a person who is capable of hearing frequencies up to 10,000 Hz?

15.37 A wire with mass 40.0 g is stretched so that its ends are tied down at points 80.0 cm apart. The wire vibrates in its fundamental mode with frequency 60.0 Hz and with an amplitude at the antinodes of 0.300 cm. a) What is the speed of propagation of transverse waves in the wire? b) Compute the tension in the wire. c) Find the maximum transverse velocity and acceleration of particles in the wire.

15.38 Waves on a Stick. A flexible stick 2.0 m long is not fixed in any way and is free to vibrate. Make clear drawings of this stick vibrating in its first three harmonics and then use your drawings to find the wavelengths of each of these harmonics.

15.39 A thin, taut string tied at both ends and oscillating in its third harmonic has its shape described by the equation $y(x, t) = (5.60$ cm$)\sin[(0.0340$ rad/cm$)x]\sin[(50.0$ rad/s$)t]$, where the origin is at the left end of the string, the x-axis is along the string and the y-axis is perpendicular to the string. a) Draw a sketch that shows the standing wave pattern. b) Find the amplitude of the two traveling waves that make up this standing wave. c) What is the length of the string? d) Find the wavelength, frequency, period, and speed of the traveling waves. e) Find the maximum transverse speed of a point on the string. f) What would be the equation $y(x, t)$ for this string if it were vibrating in its eighth harmonic?

15.40 The wave function of a standing wave is $y(x, t) =$ 4.44 mm $\sin[(32.5 \text{ rad/m}) x] \sin[(754 \text{ rad/s}) t]$. For the two traveling waves that make up this standing wave, find the a) amplitude; b) wavelength; c) frequency; d) wave speed; e) wave functions. f) From the information given, can you determine which harmonic this is? Explain.

15.41 Consider again the rope and traveling wave of Exercise 15.24. Assume that the ends of the rope are held fixed and that there is both this traveling wave and the reflected wave traveling in the opposite direction. a) What is the wavefunction $y(x, t)$ for the standing wave that is produced? b) In which harmonic is the standing wave oscillating? c) What is the frequency of the fundamental oscillation?

15.42 In deriving Eq. (15.29) for a string with a fixed end at $x = 0$, we found that nodes occur at positions x that satisfy $kx = n\pi$, where $n = 0, 1, 2, 3, \ldots$ Apply this condition to a string with fixed ends at $x = 0$ and $x = L$ to rederive Eq. (15.31) for the possible standing-wave wavelengths.

15.43 Playing a Cello. The portion of a cello string between the bridge and upper end of the finger board (that part of the string that is free to vibrate) is 60.0 cm long, and this length of the string has mass 2.00 g. The string sounds an A_4 note (440 Hz) when played. a) Where must the cellist put a finger (what distance x from bridge) to play a D_5 note (587 Hz)? (See Fig. 15.32.) For both the A_4 and D_5 notes, the string vibrates in its fundamental mode. b) Without retuning, is it possible to play a G_4 note (392 Hz) on this string? Why or why not?

Figure 15.32 Exercise 15.43.

15.44 a) A horizontal string tied at both ends is vibrating in its fundamental mode. The traveling waves have speed v, frequency f, amplitude A, and wavelength λ. Calculate the maximum transverse velocity and maximum transverse acceleration of points located at i) $x = \lambda/2$, ii) $x = \lambda/4$, and iii) $x = \lambda/8$ from the left-hand end of the string. b) At each of the points in part (a), what is the amplitude of the motion? c) At each of the points in part (a), how much time does it take the string to go from its largest upward displacement to its largest downward displacement?

15.45 A 1.50-m-long rope is stretched between two supports with a tension that makes the speed of transverse waves 48.0 m/s. What are the wavelength and frequency of a) the fundamental? b) the second overtone? c) the fourth harmonic?

15.46 Guitar String. One of the 63.5-cm-long strings of an ordinary guitar is tuned to produce the note B_3 (frequency 245 Hz) when vibrating in its fundamental mode. a) Find the speed of transverse waves on this string. b) If the tension in this string is increased by 1.0%, what will be the new fundamental frequency of the string? c) If the speed of sound in the surrounding air is 344 m/s, find the frequency and wavelength of the sound wave produced in the air by

the vibration of the B_3 string. How do these compare to the frequency and wavelength of the standing wave on the string?

Problems

15.47 A transverse sine wave with an amplitude of 2.50 mm and a wavelength of 1.80 m travels from left to right along a long, horizontal stretched string with a speed of 36.0 m/s. Take the origin at the left end of the undisturbed string. At time $t = 0$ the left end of the string has its maximum upward displacement. a) What are the frequency, angular frequency, and wave number of the wave? b) What is the function $y(x, t)$ that describes the wave? c) What is $y(t)$ for a particle at the left end of the string? d) What is $y(t)$ for a particle 1.35 m to the right of the origin? e) What is the maximum magnitude of transverse velocity of any particle of the string? f) Find the transverse displacement and the transverse velocity of a particle 1.35 m to the right of the origin at time $t = 0.0625$ s.

15.48 A transverse wave on a rope is given by

$$y(x, t) = (0.750 \text{ cm}) \cos \pi[(0.400 \text{ cm}^{-1})x + (250 \text{ s}^{-1})t]$$

a) Find the amplitude, period, frequency, wavelength, and speed of propagation. b) Sketch the shape of the rope at the following values of t: 0, 0.0005 s, and 0.0010 s. c) Is the wave traveling in the $+x$- or $-x$-direction? d) The mass per unit length of the rope is 0.0500 kg/m. Find the tension. e) Find the average power of this wave.

15.49 Three pieces of string, each of length L, are joined together end-to-end, to make a combined string of length $3L$. The first piece of string has mass per unit length μ_1, the second piece has mass per unit length $\mu_2 = 4\mu_1$, and the third piece has mass per unit length $\mu_3 = \mu_1/4$. a) If the combined string is under tension F, how much time does it take a transverse wave to travel the entire length $3L$? Give your answer in terms of L, F, and μ_1. b) Does your answer to part (a) depend on the order in which the three pieces are joined together? Explain.

15.50 Low-Cost Remote Control. Your roommates have lost the TV remote control, and no amount of searching can find it. Rather than buy a new one, you build a low-cost replacement. You attach one end of a small lever mechanism to the TV's channel-changing button. You plan to attach the other end of the lever to a 3.0-m long string that will run from the TV to the couch. When you pull the string tight and pluck your end of the string, a wave will travel down the string and trigger the lever, changing the channel. Your design assumes that you will disturb the string vertically by only 5.0 mm when you pluck it (so that using your "remote" will not require undue exertion) and that your wave will take only 0.20 s to travel horizontally along the string from your hand to the lever (to give quick channel changes). Unfortunately, your roommates have also taken most of your supply of string. All you can find around the apartment are two different pieces of 1.5-m long string; one piece has a mass of 90 g, while the other has a mass of just 10 g. You tie the two pieces together to make a combined 3.0-m long string and attach one end of the combined string to the lever mechanism. You then take the other end in hand and head for the couch. When you stretch the combined string to make it taut, how hard do you have to pull?

15.51 Ant Joy Ride. You place your pet ant Klyde (mass m) on top of a horizontal, stretched rope, where he holds on tightly. The rope has mass M and length L and is under tension F. You start a sinusoidal transverse wave of wavelength λ and amplitude A propagating along the rope. The motion of the rope is in a vertical plane. Klyde's mass is so small that his presence has no effect on the propagation of the wave. a) What is Klyde's top speed as he oscillates up and down? b) Klyde enjoys the ride and begs for more. You decide to double his top speed by changing the tension while keeping the wavelength and amplitude the same. Should the tension be increased or decreased, and by what factor?

15.52 Weightless Ant. An ant with mass m is standing peacefully on top of a horizontal, stretched rope. The rope has mass per unit length μ and is under tension F. Without warning, Throckmorton starts a sinusoidal transverse wave of wavelength λ propagating along the rope. The motion of the rope is in a vertical plane. What minimum wave amplitude will make the ant become momentarily weightless? Assume that m is so small that the presence of the ant has no effect on the propagation of the wave.

15.53 When a transverse sinusoidal wave is present on a string, the particles of the string undergo SHM. This is the same motion as that of a mass m attached to an ideal spring of force constant k', for which the angular frequency of oscillation was found in Chapter 13 to be $\omega = \sqrt{k'/m}$. Consider a string with tension F and mass per unit length μ, along which is propagating a sinusoidal wave with amplitude A and wavelength λ. a) Find the "force constant" k' of the restoring force that acts on a short segment of the string of length Δx (where $\Delta x \ll \lambda$). b) How does the "force constant" calculated in part (b) depend on F, μ, A, and λ? Explain the physical reasons why this should be so.

15.54 Motion of a Wave Pulse. The sinusoidal waves described in Section 15.2 can be produced on a string by continually oscillating one end of the string. If instead the end of the string is given a single shake, a *wave pulse* propagates down the string. A particular wave pulse is described by the function

$$y(x, t) = \frac{A^3}{A^2 + (x - vt)^2}$$

where $A = 1.00$ cm and $v = 20.0$ m/s. a) Sketch the pulse as a function of x at $t = 0$. How far along the string does the pulse extend? b) Sketch the pulse as a function of x at $t = 0.001$ s. c) At the point $x = 4.50$ cm, at what time t is the displacement maximum? At which two times is the displacement at $x = 4.50$ cm equal to half its maximum value? d) Show that $y(x, t)$ satisfies the wave equation Eq. (15.12).

15.55 A Non-Sinusoidal Wave. The shape of a wave on a string at a specific instant is shown in Fig. 15.33. The wave is propagating to the right, in the $+x$-direction. a) Determine the direction of the transverse *velocity* of each of the six labeled points on the string. If the velocity is zero, state it as such.

Fig. 15.33 Exercise 15.55.

Explain your reasoning. b) Determine the direction of the transverse *acceleration* of each of the six labeled points on the string. Explain your reasoning. c) How would your answers be affected if the wave was propagating to the left, in the $-x$-direction?

15.56 A continuous succession of sinusoidal wave pulses are produced at one end of a very long string and travel along the length of the string. The wave has frequency 40.0 Hz, amplitude 5.00 mm, and wavelength 0.600 m. a) How long does it take the wave to travel a distance of 8.00 m along the length of the string? b) How long does it take a point on the string to travel a distance of 8.00 m, once the wave train has reached the point and set it into motion? c) In parts (a) and (b), how does the time change if the amplitude is doubled?

15.57 Two-Dimensional Waves. A stretched string lies along the x-axis. The string is displaced along both the y- and z-direction, so that the transverse displacement of the string is given by

$$y(x, t) = A \cos(kx - \omega t) \qquad z(x, t) = A \sin(kx - \omega t)$$

a) Draw a graph of z versus y for a particle on the string at $x = 0$. This shows the trajectory of the particle as seen by an observer on the $+x$-axis looking back toward $x = 0$. Indicate the position of the particle at $t = 0$, $t = \pi/2\omega$, $t = \pi/\omega$, and $t = 3\pi/2\omega$. b) Find the velocity vector of a particle at an *arbitrary* position x on the string. Show that this represents the tangential velocity of a particle moving in a circle of radius A with angular velocity ω, and show that the speed of the particle is constant (i.e. the particle is in uniform circular motion). (See Problem 3.75.) c) Find the acceleration vector of the particle in part (b). Show that the acceleration is always directed toward the center of the circle and that its magnitude is $a = \omega^2 A$. Explain these results in terms of uniform circular motion. Suppose that the displacement of the string was instead given by

$$y(x, t) = A \cos(kx - \omega t) \qquad z(x, t) = -A \sin(kx - \omega t)$$

Describe how the motion of a particle at x would be different from the motion described in part (a).

15.58 Locating Lightning by Radio. While doing her physics homework, a student in the Boston area is in her dorm room listening to the radio broadcast of a Red Sox–Yankees baseball game at Boston's Fenway Park. In the bottom of the fourth inning a thunderstorm moving from west to east makes its presence known in three ways: (1) The student sees a lightning flash (and hears the electromagnetic pulse on her radio receiver); (2) 3.00 s later, she hears the thunder over the radio. (3) 4.43 s after the lightning flash, the thunder rattles her window. By a previous careful measurement she knows that she is 1.12 km due north of the broadcast booth at the ballpark. The speed of sound is 344 m/s. Where did the lightning flash occur in relation to the ballpark?

15.59 Waves of Arbitrary Shape. a) Explain why *any* wave described by a function of the form $y(x, t) = f(x - vt)$ moves in the $+x$-direction with speed v. b) Show that $y(x, t) = f(x - vt)$ satisfies the wave equation, no matter what the functional form of f. To do this, write $y(x, t) = f(u)$, where $u = x - vt$. Then, to take partial derivatives of $y(x, t)$, use the chain rule:

$$\frac{\partial y(x, t)}{\partial t} = \frac{df(u)}{du}\frac{\partial u}{\partial t} = \frac{df(u)}{du}(-v)$$

$$\frac{\partial y(x, t)}{\partial x} = \frac{df(u)}{du}\frac{\partial u}{\partial x} = \frac{df(u)}{du}$$

c) A wave pulse is described by the function $y(x, t) = De^{-(Bx - Ct)^2}$, where B, C, and D are all positive constants. What is the speed of this wave?

15.60 Equation (15.7) for a sinusoidal wave can be made more general by including a phase angle ϕ, where $0 \le \phi \le 2\pi$ (in radians) Then the wave function $y(x, t)$ becomes

$$y(x, t) = A \cos(kx - \omega t + \phi)$$

a) Sketch the wave as a function of x at $t = 0$ for $\phi = 0$, $\phi = \pi/4$, $\phi = \pi/2$, $\phi = 3\pi/4$, and $\phi = 3\pi/2$. b) Calculate the transverse velocity $v_y = \partial y/\partial t$. c) At $t = 0$, a particle on the string at $x = 0$ has displacement $y = A/\sqrt{2}$. Is this enough information to determine the value of ϕ? In addition, if you are told that a particle at $x = 0$ is moving toward $y = 0$ at $t = 0$, what is the value of ϕ? d) Explain in general what you must know about the wave's behavior at a given instant to determine the value of ϕ.

15.61 a) Show that Eq. (15.25) can also be written as $P_{av} = \frac{1}{2}Fk\omega A^2$, where k is the wave number of the wave. b) If the tension F in the string is quadrupled while the amplitude A is kept the same, how must k and ω each change to keep the average power constant? (*Hint:* Recall Eq. (15.6).)

15.62 Energy in a Triangular Pulse. A triangular wave pulse on a taut string travels in the positive x-direction with speed v. The tension in the string is F and the linear mass density of the string is μ. At $t = 0$, the shape of the pulse is given by

$$y(x, 0) = \begin{cases} 0 & \text{if } x < -L \\ h(L + x)/L & \text{for } -L < x < 0 \\ h(L - x)/L & \text{for } 0 < x < L \\ 0 & \text{for } x > L \end{cases}$$

a) Draw the pulse at $t = 0$. b) Determine the wave function $y(x, t)$ at all times t. c) Find the instantaneous power in the wave. Show that the power is zero except for $-L < (x - vt) < L$ and that in this interval the power is constant. Find the value of this constant power.

15.63 A sinusoidal transverse wave travels on a string. The string has length 8.00 m and mass 6.00 grams. The wave speed is 30.0 m/s and the wavelength is 0.200 m. a) If the wave is to have an average power of 50.0 W, what must be the amplitude of the wave? b) For this same string, if the amplitude and wavelength are the same as in part (a), what is the average power for the wave if the tension is increased such that the wave speed is doubled?

15.64 Instantaneous Power in a Wave. a) Graph $y(x, t)$ as given by Eq. (15.7) as a function of x for a given time t (say, $t = 0$). On the same axes, make a graph of the instantaneous power $P(x, t)$ as given by Eq. (15.23). b) Explain the connection between the slope of the graph of $y(x, t)$ versus x and the value of $P(x, t)$. In particular, explain what is happening at points where $P = 0$, where there is no instantaneous energy transfer. c) The quantity $P(x, t)$ always has the same sign. What does this imply about the direction of energy flow? d) Consider a wave moving in the $-x$-direction, for which

$y(x, t) = A \cos(kx + \omega t)$. Calculate $P(x, t)$ for this wave, and make a graph of $y(x, t)$ and $P(x, t)$ as functions of x for a given time t (say, $t = 0$). What differences arise from reversing the direction of the wave?

15.65 A metal wire, with density ρ and Young's modulus Y, is stretched between rigid supports. At temperature T, the speed of a transverse wave is found to be v_1. When the temperature is increased to $T + \Delta T$, the speed decreases to $v_2 < v_1$. Determine the coefficient of linear expansion of the wire.

15.66 A vibrating string 50.0 cm long is under a tension of 1.00 N. The results from five successive stroboscopic pictures are shown in Fig. 15.34. The strobe rate is set at 5000 flashes per minute and observations reveal that the maximum displacement occurred at flashes 1 and 5 with no other maxima in between. a) Find the period, frequency, and wavelength for the traveling waves on this string. b) In what normal mode (harmonic) is the string vibrating? c) What is the speed of the traveling waves on the string? d) How fast is point P moving when the string is in: (i) position 1? (ii) position 3? e) What is the mass of this string?

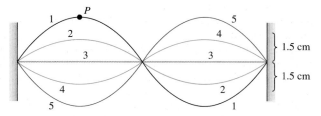

Figure 15.34 Problem 15.66.

15.67 Clothesline Nodes. Cousin Throckmorton is once again playing with the clothesline in Example 15.2. One end of the clothesline is attached to a vertical post. Throcky holds the other end loosely in his hand, so that the speed of waves on the clothesline is a relatively slow 0.720 m/s. He finds several frequencies at which he can oscillate his end of the clothesline so that a light clothespin 45.0 cm from the post doesn't move. What are these frequencies?

15.68 A guitar string is vibrating in its fundamental mode, with nodes at each end. The length of the segment of the string that is free to vibrate is 0.386 m. The maximum transverse acceleration of a point at the middle of the segment is 8.40×10^3 m/s² and the maximum transverse velocity is 3.80 m/s. a) What is the amplitude of this standing wave? b) What is the wave speed for the transverse traveling waves on this string?

15.69 As shown in Exercise 15.32, a standing wave given by Eq. (15.28) satisfies the wave equation Eq. (15.12). a) Show that a standing wave given by Eq. (15.28) also satisfies the equation

$$\frac{\partial^2 y(x, t)}{\partial t^2} = -\omega^2 y(x, t)$$

Interpret this equation in terms of what you know about simple harmonic motion. b) Does a traveling wave given by $y(x, t) = A \cos(kx - \omega t)$ also satisfy the equation in part (a)? Interpret this result.

15.70 a) The red and blue waves in Fig. 15.17 combine so that the displacement of the string at O is always zero. To show this mathematically for a wave of arbitrary shape, consider a wave moving to the right along the string in Fig. 15.17 (shown in blue) that, at time T, is given by $y_1(x, T) = f(x)$, where f is some function of x. (The form of $f(x)$ determines the shape of the wave.) If the point O corresponds to $x = 0$, explain why, at time T, the wave moving to the left in Fig. 15.17 (shown in red) is given by the function $y_2(x, T) = -f(-x)$. b) Show that the total wave function $y(x, T) = y_1(x, T) + y_2(x, T)$ is zero at O, independent of the form of the function $f(x)$. c) The red and blue waves in Fig. 15.18 combine so that the slope of the string at O is always zero. To show this mathematically for a wave of arbitrary shape, again let the wave moving to the right in Fig. 15.18 (shown in blue) be given by $y_1(x, T) = f(x)$ at time T. Explain why the wave moving to the left (shown in red) at this same time T is given by $y_2(x, T) = f(-x)$. d) Show that the total wave function $y(x, T) = y_1(x, T) + y_2(x, T)$ has zero slope at O, independent of the form of the function $f(x)$, as long as $f(x)$ has a finite first derivative.

15.71 A string that lies along the $+x$-axis has a free end at $x = 0$. a) By using steps similar to those used to derive Eq. (15.28), show that an incident traveling wave $y_1(x, t) = A \cos(kx + \omega t)$ gives rise to a standing wave $y(x, t) = 2A \cos \omega t \cos kx$. b) Show that the standing wave has an antinode at its free end $(x = 0)$. c) Find the maximum displacement, maximum speed, and maximum acceleration of the free end of the string.

15.72 A string with both ends held fixed is vibrating in its third harmonic. The waves have a speed of 192 m/s and a frequency of 240 Hz. The amplitude of the standing wave at an antinode is 0.400 cm. a) Calculate the amplitude at points on the string a distance of i) 40.0 cm; ii) 20.0 cm; and iii) 10.0 cm from the left-hand end of the string. b) At each point in part (a), how much time does it take the string to go from its largest upward displacement to its largest downward displacement? c) Calculate the maximum transverse velocity and the maximum transverse acceleration of the string at each of the points in part (a).

15.73 Pit and Plank. A wooden plank is placed over a pit that is 5.00 m wide. A physics student stands in the middle of the plank and begins to jump up and down such that she jumps upward from the plank two times each second. The plank oscillates with a large amplitude, with maximum amplitude at its center. a) What is the speed of transverse waves on the plank? b) At what rate does the student have to jump to produce large-amplitude oscillations if she is standing 1.25 m from the edge of the pit? (*Note:* The transverse standing waves of the plank have nodes at the two ends that rest on the ground on either side of the pit.)

15.74 Holding Up under Stress. A string or rope will break apart if it is placed under too much tensile stress [Eq. (11.8)]. Thicker ropes can withstand more tension without breaking because the thicker the rope, the greater the cross-sectional area and the smaller the stress. One type of steel has density 7800 kg/m³ and will break if the tensile stress exceeds 7.0×10^8 N/m². You want to make a guitar string from 4.0 g of this type of steel. In use, the guitar string must be able to withstand a tension of 900 N without breaking.

Your job is the following: a) Determine the maximum length and minimum radius the string can have. b) Determine the highest possible fundamental frequency of standing waves on this string, if the entire length of the string is free to vibrate.

15.75 Combining Standing Waves. A guitar string of length L is plucked in such a way that the total wave produced is the sum of the fundamental and the second harmonic. That is, the standing wave is given by

$$y(x, t) = y_1(x, t) + y_2(x, t)$$

where

$$y_1(x, t) = C \sin \omega_1 t \sin k_1 x$$
$$y_2(x, t) = C \sin \omega_2 t \sin k_2 x$$

with $\omega_1 = vk_1$ and $\omega_2 = vk_2$. a) At what values of x are the nodes of y_1? b) At what values of x are the nodes of y_2? c) Graph the total wave at $t = 0$, $t = \frac{1}{8}f_1$, $t = \frac{1}{4}f_1$, $t = \frac{3}{8}f_1$, and $t = \frac{1}{2}f_1$. d) Does the sum of the two standing waves y_1 and y_2 produce a standing wave? Explain.

15.76 A massive aluminum sculpture is hung from a steel wire. The fundamental frequency for transverse standing waves on the wire is 200 Hz. The sculpture is then immersed in water so that $\frac{1}{3}$ of its volume is submerged. a) What is the new fundamental frequency? b) Why is it a good approximation to treat the wire as being fixed at both ends?

15.77 Tuning a Cello. A cellist tunes the C-string of her instrument to a fundamental frequency of 65.4 Hz. The vibrating portion of the string is 0.600 m long and has a mass of 14.4 g. a) With what tension must she stretch it? b) What percent increase in tension is needed to increase the frequency from 65.4 Hz to 73.4 Hz, corresponding to a rise in pitch from C to D?

Challenge Problems

15.78 Longitudinal Waves on a Spring. A long spring such as a Slinky™ is often used to demonstrate longitudinal waves. a) Show that if a spring that obeys Hooke's law has mass m, length L, and force constant k', the speed of longitudinal waves on the spring is $v = L\sqrt{k'/m}$. b) Evaluate v for a spring with $m = 0.250$ kg, $L = 2.00$ m, and $k' = 1.50$ N/m.

15.79 a) Show that for a wave on a string, the kinetic energy *per unit length of string* is

$$u_k(x, t) = \frac{1}{2}\mu v_y^2(x, t) = \frac{1}{2}\mu\left(\frac{\partial y(x, t)}{\partial t}\right)^2$$

where μ is the mass per unit length. b) Calculate $u_k(x, t)$ for a sinusoidal wave given by Eq. (15.7). c) There is also elastic potential energy in the string, associated with the work required to deform and stretch the string. Consider a short segment of string at position x that has unstretched length Δx, as in Fig. 15.10. Ignoring the (small) curvature of the segment, its slope is $\partial y(x, t)/\partial x$. Assume that the displacement of the string from equilibrium is small, so that $\partial y/\partial x$ has a magnitude much less than unity. Show that the stretched length of the segment is approximately

$$\Delta x \left[1 + \frac{1}{2}\left(\frac{\partial y(x,t)}{\partial x}\right)^2 \right]$$

(*Hint:* Use the relationship $\sqrt{1+u} \approx 1 + \frac{1}{2}u$, valid for $|u| \ll 1$.)
d) The potential energy stored in the segment equals the work done by the string tension F (which acts along the string) to stretch the segment from its unstretched length Δx to the length calculated in part (c). Calculate this work and show that the potential energy *per unit length of string* is

$$u_p(x,t) = \frac{1}{2}F\left(\frac{\partial y(x,t)}{\partial x}\right)^2$$

e) Calculate $u_p(x,t)$ for a sinusoidal wave given by Eq. (15.7). f) Show that $u_k(x,t) = u_p(x,t)$ for all x and t. g) Show $y(x,t)$, $u_k(x,t)$, and $u_p(x,t)$ as functions of x for $t = 0$ in one graph with all three functions on the same axes. Explain why u_k and u_p are maximum where y is zero, and vice versa. h) Show that the instantaneous power in the wave, given by Eq. (15.22), is equal to the total energy per unit length multiplied by the wave speed v. Explain why this result is reasonable.

15.80 A deep-sea diver is suspended beneath the surface of Loch Ness by a 100-m long cable that is attached to a boat on the surface (Fig. 15.35). The diver and his suit have a total mass of 120 kg and a volume of 0.0800 m^3. The cable has a diameter of 2.00 cm and a linear mass density of $\mu = 1.10$ kg/m. The diver thinks he sees something moving in the murky depths and jerks the end of the cable back and forth to send transverse waves up the cable as a signal to his companions in the boat.

100 m

x

$m = 120$ kg

Figure 15.35 Challenge Problem 15.80.

a) What is the tension in the cable at its lower end, where it is attached to the diver? Do not forget to include the buoyant force that the water (density 1000 kg/m^3) exerts on him. b) Calculate the tension in the cable a distance x above the diver. The buoyant force on the cable must be included in your calculation. c) The speed of transverse waves on the cable is given by $v = \sqrt{F/\mu}$ (Eq. 15.13) . The speed therefore varies along the cable, since the tension is not constant. (This expression neglects the damping force that the water exerts on the moving cable.) Integrate to find the time required for the first signal to reach the surface.

15.81 A uniform rope with length L and mass m is held at one end and whirled in a horizontal circle with angular velocity ω. You can ignore the force of gravity on the rope. Find the time required for a transverse wave to travel from one end of the rope to the other.

15.82 Instantaneous Power in a Standing Wave. From Eq. (15.21), the instantaneous rate at which a wave transmits energy along a string (instantaneous power) is

$$P(x,t) = -F\frac{\partial y(x,t)}{\partial x}\frac{\partial y(x,t)}{\partial t}$$

where F is the tension. a) Evaluate $P(x,t)$ for a standing wave of the form given by Eq. (15.28). b) Show that for all values of x, the *average* power P_{av} carried by the standing wave is zero. (Equation (15.25) does not apply here. Can you see why?) c) For a standing wave given by Eq. (15.28), graph $P(x,t)$ and the displacement $y(x,t)$ as functions of x for $t = 0$, $t = \pi/4\omega$, $t = \pi/2\omega$, and $t = 3\pi/4\omega$. (Positive $P(x,t)$ means energy is flowing in the $+x$-direction; negative $P(x,t)$ means the flow is in the $-x$-direction.) d) The *kinetic* energy per unit length of string is greatest where the string has the greatest transverse speed, and the *potential* energy per unit length of string is greatest where the string has the steepest slope (because there the string is stretched the most). (See Challenge Problem 15.79.) Using these ideas, discuss the flow of energy along the string.

15.83 Out of Tune. The B-string of a guitar is made of steel (density 7800 kg/m^3), is 63.5 cm long, and has diameter 0.406 mm. The fundamental frequency is $f = 247.0$ Hz. a) Find the string tension. b) If the tension F is changed by a small amount ΔF, the frequency f changes by a small amount Δf. Show that

$$\frac{\Delta f}{f} = \frac{1}{2}\frac{\Delta f}{f}$$

c) The string is tuned as in part (a) when its temperature is 18.5°C. Strenuous playing can make the temperature of the string rise, changing its vibration frequency. Find Δf if the temperature of the string rises to 29.5°C. The steel string has a Young's modulus of 2.00 × 10^{11} Pa and a coefficient of linear expansion of 1.20 × 10^{-5} (C°)$^{-1}$. Assume that the temperature of the body of the guitar remains constant. Will the vibration frequency rise or fall?

SOUND AND HEARING 16

A human voice and a trumpet are both examples of wind instruments. In both cases, a standing sound wave is set up in the air within a long pipe (the singer's vocal tract or the metal tube of the trumpet) in response to a vibration at one end (the singer's vocal folds or the trumpeter's lips). The character of the sound is determined by the properties of the standing wave.

? **What is the physical difference between musical sound and noise?**

Of all the mechanical waves that occur in nature, the most important in our everyday lives are longitudinal waves in a medium, usually air, called *sound* waves. The reason is that the human ear is tremendously sensitive and can detect sound waves even of very low intensity. Besides their use in spoken communication, our ears allow us to pick up a myriad of cues about our environment, from the welcome sound of a meal being prepared to the warning sound of an approaching car. The ability to hear an unseen nocturnal predator was essential to the survival of our ancestors, so it is no exaggeration to say that we humans owe our existence to our highly evolved sense of hearing.

Up to this point we have described mechanical waves primarily in terms of displacement; however, a description of sound waves in terms of *pressure* fluctuations is often more appropriate, largely because the ear is primarily sensitive to changes in pressure. We'll study the relations among displacement, pressure fluctuation, and intensity and the connections between these quantities and human sound perception.

When a source of sound or a listener moves through the air, the listener may hear a different frequency than the one emitted by the source. This is the Doppler effect, which has important applications in medicine and technology.

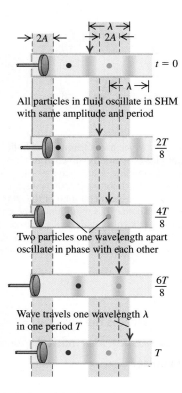

16.1 A sinusoidal longitudinal wave traveling to the right in a fluid. The blue and red dots represent two particles in the fluid that are one-half wavelength apart, and the magenta arrow tracks the motion of a wave compression. (Compare to Fig. 15.5.)

16.1 | Sound Waves

The most general definition of **sound** is that it is a longitudinal wave in a medium. Our main concern in this chapter is with sound waves in air, but sound can travel through any gas, liquid, or solid. You may be all too familiar with the propagation of sound through a solid if your neighbor's stereo speakers are right next to your wall.

The simplest sound waves are sinusoidal waves, which have definite frequency, amplitude, and wavelength. The human ear is sensitive to waves in the frequency range from about 20 to 20,000 Hz, called the **audible range,** but we also use the term *sound* for similar waves with frequencies above (**ultrasonic**) and below (**infrasonic**) the range of human hearing.

Sound waves usually travel out in all directions from the source of sound, with an amplitude that depends on the direction and distance from the source. We'll return to this point in the next section. For now, we concentrate on the idealized case of a sound wave that propagates in the positive x-direction only. As we discussed in Section 15.3, such a wave is described by a wave function $y(x, t)$, which gives the instantaneous displacement y of a particle in the medium at position x at time t. If the wave is sinusoidal, we can express it using Eq. (15.7):

$$y(x, t) = A \cos(kx - \omega t) \tag{16.1}$$
(sound wave propagating in the $+x$-direction)

Remember that in a longitudinal wave the displacements are *parallel* to the direction of travel of the wave, so distances x and y are measured parallel to each other, not perpendicular as in a transverse wave. The amplitude A is the maximum displacement of a particle in the medium from its equilibrium position (Fig. 16.1). It is also called the **displacement amplitude.**

Sound waves may also be described in terms of variations of *pressure* at various points. In a sinusoidal sound wave in air, the pressure fluctuates above and below atmospheric pressure p_a in a sinusoidal variation with the same frequency as the motions of the air particles. The human ear operates by sensing such pressure variations. A sound wave entering the ear canal exerts a fluctuating pressure on one side of the eardrum; the air on the other side of the eardrum, vented to the outside by the Eustachian tube, is at atmospheric pressure. The pressure difference on the two sides of the eardrum sets it into motion. Microphones and similar devices also usually sense pressure differences, not displacements, so it is very useful to develop a relation between these two descriptions.

Let $p(x, t)$ be the instantaneous pressure fluctuation in a sound wave at any point x at time t. That is, $p(x, t)$ is the amount by which the pressure *differs* from normal atmospheric pressure p_a. Think of $p(x, t)$ as the *gauge pressure* defined in Section 14.2; it can be either positive or negative. The *absolute* pressure at a point is then $p_a + p(x, t)$.

To see the connection between the pressure fluctuation $p(x, t)$ and the displacement $y(x, t)$ in a sound wave propagating in the $+x$-direction, consider an imaginary cylinder of air with cross-sectional area S and axis along the direction of propagation (Fig. 16.2). When no sound wave is present, the cylinder has length Δx and volume $V = S \Delta x$, as shown by the shaded volume in Fig. 16.2. When a wave is present, at time t the end of the cylinder that is initially at x is displaced by $y_1 = y(x, t)$, and the end that is initially at $x + \Delta x$ is displaced by $y_2 = y(x + \Delta x, t)$; this is shown by the red lines. If $y_2 > y_1$, as shown in Fig. 16.2, the cylinder's volume has increased, which causes a decrease in pressure. If $y_2 < y_1$, the cylinder's volume has decreased and the pressure has increased. If

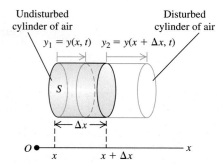

16.2 A cylindrical volume of air with cross-sectional area S and undisturbed length Δx. As a sound wave propagates along the x-axis, the left and right ends undergo different displacements y_1 and y_2. The resulting change in volume is $S(y_2 - y_1)$.

$y_2 = y_1$, the cylinder is simply shifted to the left or right; there is no volume change and no pressure fluctuation. The pressure fluctuation depends on the *difference* between the displacement at neighboring points in the medium.

Quantitatively, the change in volume ΔV of the cylinder is

$$\Delta V = S(y_2 - y_1) = S[y(x + \Delta x, t) - y(x, t)]$$

In the limit as $\Delta x \to 0$, the fractional change in volume dV/V is

$$\frac{dV}{V} = \lim_{\Delta x \to 0} \frac{S[y(x + \Delta x, t) - y(x, t)]}{S\Delta x} = \frac{\partial y(x, t)}{\partial x} \qquad (16.2)$$

The fractional volume change is related to the pressure fluctuation by the bulk modulus B, which by definition [Eq. (11.13)] is $B = -p(x, t)/(dV/V)$ (see Section 11.4). Solving for $p(x, t)$, we have

$$p(x, t) = -B\frac{\partial y(x, t)}{\partial x} \qquad (16.3)$$

The negative sign arises because when $\partial y(x, t)/\partial x$ is positive, the displacement is greater at $x + \Delta x$ than at x, corresponding to an increase in volume and a *decrease* in pressure.

When we evaluate $\partial y(x, t)/\partial x$ for the sinusoidal wave of Eq. (16.1), we find

$$p(x, t) = BkA \sin(kx - \omega t) \qquad (16.4)$$

Figure 16.3 shows $y(x, t)$ and $p(x, t)$ for a sinusoidal sound wave at $t = 0$. It also shows how individual particles of the wave are displaced at this time. While $y(x, t)$ and $p(x, t)$ describe the same wave, these two functions are one-quarter cycle out of phase: at any time, the displacement is greatest where the pressure fluctuation is zero, and vice versa. In particular, note that the compressions (points of greatest pressure and density) and rarefactions (points of lowest pressure and density) are points of *zero* displacement.

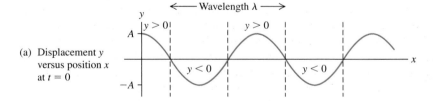

(a) Displacement y versus position x at $t = 0$

(b) Undisplaced particles
Displaced particles at $t = 0$

Particles pile up: **Compression**   Particles pulled apart: **Rarefaction**   Particles pile up: **Compression**   Particles pulled apart: **Rarefaction**

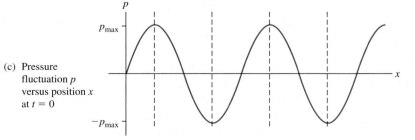

(c) Pressure fluctuation p versus position x at $t = 0$

16.3 Three ways to describe a sound wave. (a) A graph of displacement y versus position x [Eq. (16.1)] at $t = 0$. (b) A cartoon showing the displacement of individual particles in the gas at $t = 0$. Particles are displaced to the right where $y > 0$ and to the left where $y < 0$. Points where particles pile up are compressions (dark shading), and points where they are pulled apart are rarefactions (light shading). (c) A graph of pressure fluctuation p versus position x [Eq. (16.4)] at $t = 0$. Note that p is most positive at a compression and most negative at a rarefaction.

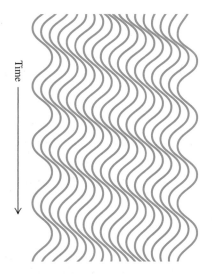

16.4 Diagram for illustrating longitudinal traveling waves.

CAUTION Keep in mind that the graphs in Fig. 16.3 show the wave at only *one* instant of time. Because the wave is propagating in the +x-direction, as time goes by the wave patterns in the functions $y(x, t)$ and $p(x, t)$ move to the right at the wave speed $v = \omega/k$. Hence the positions of the compressions and rarefactions also move to the right at this same speed. The particles, by contrast, simply oscillate back and forth in simple harmonic motion as shown in Fig. 16.1.

Visualizing the relationship between particle motion and wave motion is not as easy for longitudinal waves as it is for transverse waves on a string. Figure 16.4 will help you understand these motions. To use this figure, tape two index cards edge-to-edge with a gap of 1 mm or so between them, forming a thin slit. Place the cards over the top of the figure with the slit horizontal, and move them downward with constant speed. The portions of the sine curves that are visible through the slit correspond to a row of particles in a medium in which a longitudinal sinusoidal wave is traveling. Each particle undergoes simple harmonic motion about its equilibrium position, with delays or phase shifts that increase continuously along the slit. The regions of maximum compression and expansion move from left to right with constant speed. Moving the card upward simulates a wave traveling from right to left.

Equation (16.4) shows that the quantity BkA represents the maximum pressure fluctuation. We call this the **pressure amplitude,** denoted by p_{max}:

$$p_{max} = BkA \qquad \text{(sinusoidal sound wave)} \qquad (16.5)$$

The pressure amplitude is directly proportional to the displacement amplitude A, as we might expect, and it also depends on wavelength. Waves of shorter wavelength λ (larger wave number $k = 2\pi/\lambda$) have greater pressure variations for a given amplitude because the maxima and minima are squeezed closer together. A medium with a large value of bulk modulus B requires a relatively large pressure amplitude for a given displacement amplitude because large B means a less compressible medium; that is, greater pressure change is required for a given volume change.

Example 16.1 | **Amplitude of a sound wave in air**

In a sinusoidal sound wave of moderate loudness the maximum pressure variations are of the order of 3.0×10^{-2} Pa above and below atmospheric pressure p_a (nominally 1.013×10^5 Pa at sea level). Find the corresponding maximum displacement if the frequency is 1000 Hz. In air at normal atmospheric pressure and density, the speed of sound is 344 m/s and the bulk modulus is 1.42×10^5 Pa.

SOLUTION

IDENTIFY and SET UP: We are given the pressure amplitude p_{max}, wave speed v, frequency f, and bulk modulus B. Our target variable is the displacement A, which is related to p_{max} by Eq. (16.5). We also use the relationship $\omega = vk$ [Eq. (15.6)] to determine the wave number k from v and the angular frequency $\omega = 2\pi f$.

EXECUTE: From Eq. (16.5), the maximum displacement is $A = p_{max}/Bk$. From Eq. (15.6), the wave number is

$$k = \frac{\omega}{v} = \frac{2\pi f}{v} = \frac{(2\pi \text{ rad})(1000 \text{ Hz})}{344 \text{ m/s}} = 18.3 \text{ rad/m}$$

Then

$$A = \frac{p_{max}}{Bk} = \frac{3.0 \times 10^{-2} \text{ Pa}}{(1.42 \times 10^5 \text{ Pa})(18.3 \text{ rad/m})}$$
$$= 1.2 \times 10^{-8} \text{ m}$$

EVALUATE: This displacement amplitude is only about $\frac{1}{100}$ the size of a human cell. Remember that the ear actually senses pressure fluctuations; it detects these minuscule displacements only indirectly.

| Example 16.2 | **Amplitude of a sound wave in the inner ear** |

When a sound wave enters the ear, it sets the eardrum into oscillation, which in turn causes oscillation of the three tiny bones in the middle ear called the *ossicles* (Fig. 16.5). This oscillation is finally transmitted to the fluid-filled inner ear; the motion of the fluid disturbs hair cells within the inner ear, which transmit nerve impulses to the brain with the information that a sound is present. The moving part of the eardrum has an area of about 43 mm², and the area of the stirrup (the smallest of the ossicles) where it connects to the inner ear is about 3.2 mm². For the sound in the previous example, determine a) the pressure amplitude and b) the displacement amplitude of the wave in the fluid of the inner ear. The speed of sound in this fluid is about 1500 m/s.

SOLUTION

IDENTIFY: Although the sound wave is now traveling in fluid (mostly water) rather than gas, the same principles and relationships among the properties of the wave apply.

SET UP: We can safely neglect the mass of the ossicles (about 58 mg = 5.8×10^{-5} kg), so the force exerted by the ossicles on the fluid in the inner ear is the same as the force exerted on the eardrum and ossicles by the sound wave in air. (We used this same idea in Chapters 4 and 5 when we said that the tension is the same at either end of a massless rope.) Hence the pressure amplitude p_{max} is greater in the inner ear than in the outside air because the same force is exerted on a smaller area (the area of the stirrup versus the area of the eardrum).

Given the pressure amplitude in the inner ear, we find the displacement amplitude using Eq. (16.5). The values of B and k are different than in the air. To determine k, note that the wave in the inner ear has the same angular frequency ω as the wave in the air because the air, eardrum, ossicles, and inner-ear fluid all oscillate together. But because the wave speed v is greater in the inner ear

than in the air (1500 m/s versus 344 m/s), the wave number $k = \omega/v$ is greater as well.

EXECUTE: a) Using the area of the eardrum and pressure amplitude found in Example 16.1, the maximum force exerted by the sound wave in air on the eardrum is $F_{max} = p_{max\,(air)} S_{eardrum}$. Hence the pressure amplitude in the inner ear fluid is

$$p_{max\,(inner\,ear)} = \frac{F_{max}}{S_{stirrup}} = \frac{p_{max\,(air)} S_{eardrum}}{S_{stirrup}}$$

$$= \frac{(3.0 \times 10^{-2}\,\text{Pa})(43\,\text{mm}^2)}{3.2\,\text{mm}^2} = 0.40\,\text{Pa}$$

b) To find the maximum displacement, we again use the relationship $A = p_{max}/Bk$ as in Example 16.1. The fluid in the inner ear is mostly water, which has a much greater bulk modulus than air because water is much more difficult to compress. From Table 11.2 the compressibility of water (unfortunately also called k) equals $45.8 \times 10^{-11}\,\text{Pa}^{-1}$, so $B_{fluid} = 1/(45.8 \times 10^{-11}\,\text{Pa}^{-1}) = 2.18 \times 10^9\,\text{Pa}$.

We determine the value of the wave number k using the value of ω from Example 16.1 and $v = 1500$ m/s for the inner-ear fluid. Hence

$$k_{inner\,ear} = \frac{\omega}{v_{inner\,ear}} = \frac{6283\,\text{rad/s}}{1500\,\text{m/s}} = 4.2\,\text{rad/m}$$

When we put everything together, the maximum displacement of the fluid in the inner ear is

$$A_{inner\,ear} = \frac{p_{max\,(inner\,ear)}}{B_{fluid}\,k_{inner\,ear}} = \frac{0.40\,\text{Pa}}{(2.18 \times 10^9\,\text{Pa})(4.2\,\text{rad/m})}$$

$$= 4.4 \times 10^{-11}\,\text{m}$$

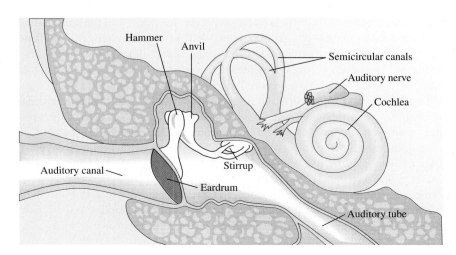

16.5 The anatomy of the human ear. The middle ear is the size of a small marble; the ossicles (hammer, anvil, and stirrup) are the smallest bones in the human body.

EVALUATE: The result in part (a) shows that the effect of the ossicles is to increase the pressure amplitude in the inner ear by a factor of $(43 \text{ mm}^2)/(3.2 \text{ mm}^2) = 13$. This amplification factor helps give the human ear its great sensitivity.

The displacement amplitude in the inner ear is even smaller than in the air. What really matters in the inner ear, however, is the *pressure* amplitude, since the pressure variations within the fluid cause the forces that set the hair cells into motion.

Perception of Sound Waves

The physical characteristics of a sound wave are directly related to the perception of that sound by a listener. For a given frequency, the greater the pressure amplitude of a sinusoidal sound wave, the greater the perceived **loudness.** The relationship between pressure amplitude and loudness is not a simple one, and it varies from one person to another. One important factor is that the ear is not equally sensitive to all frequencies in the audible range. A sound at one frequency may seem louder than one of equal pressure amplitude at a different frequency. At 1000 Hz the minimum pressure amplitude that can be perceived with normal hearing is about 3×10^{-5} Pa; to produce the same loudness at 200 Hz or 15,000 Hz requires about 3×10^{-4} Pa. Perceived loudness also depends on the health of the ear. A loss of sensitivity at the high-frequency end usually happens naturally with age but can be further aggravated by excessive noise levels. Studies have shown that many young rock musicians have suffered permanent ear damage and have hearing that is typical of persons 65 years of age. Portable stereo headsets used at high volume pose similar threats to hearing. Be careful!

The frequency of a sound wave is the primary factor in determining the **pitch** of a sound, the quality that lets us classify the sound as "high" or "low." The higher the frequency of a sound (within the audible range), the higher the pitch that a listener will perceive. Pressure amplitude also plays a role in determining pitch. When a listener compares two sinusoidal sound waves with the same frequency but different pressure amplitudes, the one with the greater pressure amplitude is usually perceived as louder but also as slightly lower in pitch.

Musical sounds have wave functions that are more complicated than a simple sine function. The pressure fluctuation in the sound wave produced by a clarinet is shown in Fig. 16.6a. The pattern is so complex because the column of air in a wind instrument like a clarinet vibrates at a fundamental frequency and at many harmonics at the same time. (In Section 15.8, we described this same behavior for a string that has been plucked, bowed, or struck. We'll examine the physics of wind instruments in Section 16.5.) The sound wave produced in the surrounding air has a similar amount of each harmonic, that is, a similar *harmonic content.* Figure 16.6b shows the harmonic content of the sound of a clarinet. The mathematical process of translating a pressure-time graph like Fig. 16.6a into a graph of harmonic content like Fig. 16.6b is called *Fourier analysis.*

Two tones produced by different instruments might have the same fundamental frequency (and thus the same pitch) but sound different because of the presence of different amounts of the various harmonics. The difference is called *tone color, quality,* or **timbre** and is often described in subjective terms such as reedy, golden, round, mellow, and tinny. A tone that is rich in harmonics, like the clarinet tone in Figs. 16.6a and 16.6b usually sounds thin and "stringy" or "reedy," while a tone containing mostly a fundamental, like the alto recorder tone in Figs. 16.6c and 16.6d is more mellow and flutelike. The same principle applies to the human voice, which is another example of a wind instrument; the vowels "a" and "e" sound different because of differences in harmonic content.

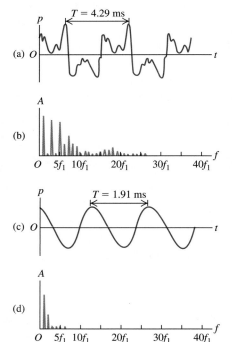

16.6 (a) Graph of pressure fluctuation versus time from a clarinet with fundamental frequency $f_1 = 233$ Hz. (b) Harmonic content of the sound of the clarinet. (c) Graph of pressure fluctuation versus time from an alto recorder with fundamental frequency $f_1 = 523$ Hz. (d) Harmonic content of the sound of the alto recorder. (From R.E. Berg and D.G. Stork, *The Physics of Sound,* Prentice-Hall, 1982.)

Another factor in determining tone quality is the behavior at the beginning (*attack*) and end (*decay*) of a tone. A piano tone begins with a thump and then dies away gradually. A harpsichord tone, in addition to having different harmonic content, begins much more quickly with a click, and the higher harmonics begin before the lower ones. When the key is released, the sound also dies away much more rapidly with a harpsichord than with a piano. Similar effects are present in other musical instruments. With wind and string instruments the player has considerable control over the attack and decay of the tone, and these characteristics help to define the unique characteristics of each instrument.

Unlike the tones made by musical instruments or the vowels in human speech, **noise** is a combination of *all* frequencies, not just frequencies that are integer multiples of a fundamental frequency. (An extreme case is "white noise," which contains equal amounts of all frequencies across the audible range.) Examples include the sound of the wind and the hissing sound you make in saying the consonant "s."

10.10 Complex Waves: Fourier Analysis

Test Your Understanding

You use an electronic signal generator to produce a sinusoidal sound wave in air. You then increase the frequency of the wave while keeping the pressure amplitude constant. What effect does this have on the displacement amplitude of the sound wave?

16.2 | Speed of Sound Waves

We found in Section 15.4 that the speed v of a transverse wave on a string depends on the string tension F and the linear mass density μ: $v = \sqrt{F/\mu}$. What, we may ask, is the corresponding expression for the speed of sound waves in a gas or liquid? On what properties of the medium does the speed depend?

We can make an educated guess about these questions by remembering a claim that we made in Section 15.4: For mechanical waves in general, the expression for the wave speed is of the form

$$v = \sqrt{\frac{(\text{restoring force returning the system to equilibrium})}{(\text{inertia resisting the return to equilibrium})}}$$

A sound wave in a bulk fluid causes compressions and rarefactions of the fluid, so the restoring-force term in the above expression must be related to how easy or difficult it is to compress the fluid. This is precisely what the bulk modulus B of the medium tells us. According to Newton's second law, inertia is related to mass. The "massiveness" of a bulk fluid is described by its density, or mass per unit volume, ρ. (The corresponding quantity for a string is the mass per unit length, μ.) Hence we expect that the speed of sound waves should be of the form $v = \sqrt{B/\rho}$.

To check our guess, we'll derive the speed of sound (longitudinal) waves in a fluid in a pipe. This is a situation of some importance, since all musical wind instruments are fundamentally pipes in which a longitudinal wave (sound) propagates in a fluid (air). Human speech works on the same principle; sound waves propagate in your vocal tract, which is basically an air-filled pipe connected to the lungs at one end (your larynx) and to the outside air at the other end (your mouth). The steps in our derivation are completely parallel to those we used in Section 15.4 to find the speed of transverse waves, so you'll find it useful to review that section.

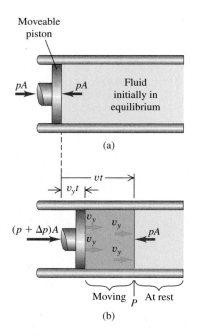

16.7 A sound wave propagating in a fluid confined to a tube. (a) Fluid in equilibrium. (b) A time t after the piston begins moving to the right at speed v_y, the fluid between the piston and point P is in motion. The speed of sound waves is v.

Figure 16.7 shows a fluid (either liquid or gas) with density ρ in a pipe with cross-sectional area A. In the equilibrium state, the fluid is under a uniform pressure p. In Fig. 16.7a the fluid is at rest. We take the x-axis along the length of the pipe. This is also the direction in which we make a longitudinal wave propagate, so the displacement y is also measured along the pipe, just as in Section 16.1 (see Fig. 16.2).

At time $t = 0$ we start the piston at the left end moving toward the right with constant speed v_y. This initiates a wave motion that travels to the right along the length of the pipe, in which successive sections of fluid begin to move and become compressed at successively later times.

Figure 16.7b shows the fluid at time t. All portions of fluid to the left of point P are moving to the right with speed v_y, and all portions to the right of P are still at rest. The boundary between the moving and stationary portions travels to the right with a speed equal to the speed of propagation or wave speed v. At time t the piston has moved a distance $v_y t$, and the boundary has advanced a distance vt. As with a transverse disturbance in a string, we can compute the speed of propagation from the impulse-momentum theorem.

The quantity of fluid set in motion in time t is the amount that originally occupied a section of the cylinder with length vt, cross-sectional area A, and volume vtA. The mass of this fluid is ρvtA, and its longitudinal momentum (that is, momentum along the length of the pipe) is

$$\text{Longitudinal momentum} = (\rho vtA)v_y$$

Next we compute the increase of pressure, Δp, in the moving fluid. The original volume of the moving fluid, Avt, has decreased by an amount $Av_y t$. From the definition of the bulk modulus B, Eq. (11.13) in Section 11.5,

$$B = \frac{-\text{Pressure change}}{\text{Fractional volume change}} = \frac{-\Delta p}{-Av_y t/Avt}$$

$$\Delta p = B\frac{v_y}{v}$$

The pressure in the moving fluid is $p + \Delta p$ and the force exerted on it by the piston is $(p + \Delta p)A$. The net force on the moving fluid (see Fig. 16.7b) is ΔpA, and the longitudinal impulse is

$$\text{Longitudinal impulse} = \Delta pAt = B\frac{v_y}{v}At$$

Because the fluid was at rest at time $t = 0$, the change in momentum up to time t is equal to the momentum at that time. Applying the impulse-momentum theorem, we find

$$B\frac{v_y}{v}At = \rho vtAv_y \tag{16.6}$$

When we solve this expression for v, we get

$$v = \sqrt{\frac{B}{\rho}} \qquad \text{(speed of a longitudinal wave in a fluid)} \tag{16.7}$$

which agrees with our educated guess. Thus the speed of propagation of a longitudinal pulse in a fluid depends only on the bulk modulus B and the density ρ of the medium.

While we derived Eq. (16.7) for waves in a pipe, it also applies to longitudinal waves in a bulk fluid. Thus the speed of sound waves traveling in air or water is determined by this equation.

When a longitudinal wave propagates in a *solid* rod or bar, the situation is somewhat different. The rod expands sideways slightly when it is compressed longitudinally, while a fluid in a pipe with constant cross section cannot move sideways. Using the same kind of reasoning that led us to Eq. (16.7), we can show that the speed of a longitudinal pulse in the rod is given by

$$v = \sqrt{\frac{Y}{\rho}} \quad \text{(speed of a longitudinal wave in a solid rod)} \quad (16.8)$$

where Y is Young's modulus, defined in Section 11.4.

CAUTION Equation (16.8) applies only to a rod or bar whose sides are free to bulge and shrink a little as the wave travels. It does not apply to longitudinal waves in a *bulk* solid, since in these materials, sideways motion in any element of material is prevented by the surrounding material. The speed of longitudinal waves in a bulk solid depends on the density, the bulk modulus, and the *shear* modulus; a full discussion is beyond the scope of this book.

As with the derivation for a transverse wave on a string, Eqs. (16.7) and (16.8) are valid for sinusoidal and other periodic waves, not just for the special case discussed here.

Table 16.1 lists the speed of sound in several bulk materials. Sound waves travel more slowly in lead than in aluminum or steel because lead has a lower bulk modulus and shear modulus and a higher density.

Table 16.1 Speed of Sound in Various Bulk Materials

Material	Speed of Sound (m/s)
Gases	
Air (20°C)	344
Helium (20°C)	999
Hydrogen (20°C)	1330
Liquids	
Liquid helium (4 K)	211
Mercury (20°C)	1451
Water (0°C)	1402
Water (20°C)	1482
Water (100°C)	1543
Solids	
Aluminum	6420
Lead	1960
Steel	5941

Example 16.3

Wavelength of sonar waves

A ship uses a sonar system to detect underwater objects (Fig. 16.8). The system emits underwater sound waves and measures the time interval for the reflected wave (echo) to return to the detector. Determine the speed of sound waves in water using Eq. (16.7) and find the wavelength of a 262-Hz wave.

SOLUTION

IDENTIFY and SET UP: To use Eq. (16.7), we find the bulk modulus of water from the compressibility (Table 11.2) and the density ($\rho = 1.00 \times 10^3 \text{ kg/m}^3$). Given the speed and the frequency $f = 262$ Hz, we find the wavelength from the relationship $v = f\lambda$.

EXECUTE: From Table 11.2 we find that the compressibility of water, which is the reciprocal of the bulk modulus, is $k = 45.8 \times 10^{-11} \text{ Pa}^{-1}$. Thus $B = (1/45.8) \times 10^{11}$ Pa. We obtain

$$v = \sqrt{\frac{B}{\rho}} = \sqrt{\frac{(1/45.8) \times 10^{11} \text{ Pa}}{1.00 \times 10^3 \text{ kg/m}^3}}$$

$$= 1480 \text{ m/s}$$

The wavelength is given by

$$\lambda = \frac{v}{f} = \frac{1480 \text{ m/s}}{262 \text{ s}^{-1}} = 5.65 \text{ m}$$

16.8 A sonar system uses underwater sound waves to detect and locate submerged objects.

EVALUATE: The calculated value of v agrees well with the experimental value in Table 16.1. Though water is far denser than air (ρ is larger), it is also far more incompressible (B is larger) and the speed $v = \sqrt{B/\rho}$ turns out to be more than four times the speed of sound in air at ordinary temperatures.

We found in Example 15.1 (Section 15.2) that a wave of this frequency in air has a wavelength of 1.31 m. The speed of sound in water is greater than in air, so the wavelength $\lambda = v/f$ must be greater as well; this is just as we calculated.

Dolphins emit high-frequency sound waves (typically 100,000 Hz) and use the echoes for guidance and for hunting. The corresponding wavelength in water is 1.48 cm. With this high-frequency "sonar" system they can sense objects that are roughly as small as the wavelength (but not much smaller). *Ultrasonic imaging* is a medical technique that uses exactly the same physical principle; sound waves of very high frequency and very short wavelength, called *ultrasound,* are scanned over the human body, and the "echoes" from interior organs are used to create an image. With ultrasound of frequency 5 MHz $= 5 \times 10^6$ Hz, the wavelength in water (the primary constituent of the body) is 0.3 mm, and features as small as this can be discerned in the image. Ultrasound is used for the study of heart-valve action, detection of tumors, and prenatal examinations (Fig. 16.9). Ultrasound is more sensitive than x rays in distinguishing various kinds of tissues and does not have the radiation hazards associated with x rays.

Example 16.4	**Speed of a longitudinal wave**

What is the speed of longitudinal waves in a lead rod?

SOLUTION

IDENTIFY: Waves of this kind are made by clamping the rod in place and striking one end face-on with a hammer. The amplitude of the resulting waves is so small as to be nearly invisible to the naked eye. But the question is about the *speed* of the waves, which does not depend on the amplitude. Note that we cannot simply use the value for lead from Table 16.1, as that value refers to the speed of sound in a *bulk* material, not a rod.

SET UP: Equation (16.8) applies to this situation. We find the values of Young's modulus Y and the density ρ from Tables 11.1 and 14.1, respectively.

EXECUTE: We find $Y = 1.6 \times 10^{10}$ Pa and $\rho = 11.3 \times 10^3$ kg/m^3 (that is, 11.3 times the density of water), so

$$v = \sqrt{\frac{Y}{\rho}} = \sqrt{\frac{1.6 \times 10^{10} \text{ Pa}}{11.3 \times 10^3 \text{ kg/m}^3}} = 1.2 \times 10^3 \text{ m/s}$$

EVALUATE: This is more than three times the speed of sound in air, but slower than the speed of sound in *bulk* lead (see Table 16.1). The reason is that for lead the bulk modulus is greater than Young's modulus.

Speed of Sound in Gases

10.3 Speed of Sound in a Gas

Most of the sound waves that we encounter on a daily basis propagate in air. To use Eq. (16.7) to find the speed of sound waves in air, we must keep in mind that the bulk modulus of a gas depends on the pressure of the gas: the greater the pressure applied to a gas to compress it, the more it resists further compression and hence the greater the bulk modulus. (That's why specific values of the bulk modulus for gases are not given in Table 11.1.) The expression for the bulk modulus of a gas for use in Eq. (16.7) is

$$B = \gamma p_0 \qquad (16.9)$$

where p_0 is the equilibrium pressure of the gas. The quantity γ (the Greek letter "gamma") is called the *ratio of heat capacities.* It is a dimensionless number that characterizes the thermal properties of the gas. (We'll learn more about this quantity in Chapter 19.) As an example, the ratio of heat capacities for air is $\gamma = 1.40$. At normal atmospheric pressure $p_0 = 1.01 \times 10^5$ Pa, so $B = (1.40)(1.013 \times 10^5$ Pa$) = 1.42 \times 10^5$ Pa. This value is miniscule compared to the bulk modulus of a typical solid (Table 11.1), which is approximately 10^{10} to 10^{11} Pa. This shouldn't be surprising: it's simply a statement that air is far easier to compress than steel.

The density ρ of a gas also depends on the pressure, which in turn depends on the temperature. It turns out that the ratio B/ρ for a given type of ideal gas does *not* depend on the pressure at all, only the temperature. From Eq. (16.7), this means that the speed of sound in a gas is fundamentally a function of temperature T:

$$v = \sqrt{\frac{\gamma RT}{M}} \quad \text{(speed of sound in an ideal gas)} \quad (16.10)$$

This expression incorporates several quantities that you may recognize from your study of ideal gases in chemistry and which we will study in Chapters 17, 18, and 19. The temperature T is the *absolute* temperature in kelvins (K), equal to the Celsius temperature plus 273.15; thus 20.00°C corresponds to $T = 293.15$ K. The quantity M is the *molar mass*, or mass per mole of the substance of which the gas is composed. The *gas constant R* has the same value for all gases. The current best numerical value of R is

$$R = 8.314472(15) \text{ J/mol} \cdot \text{K}$$

which for practical calculations we can write as 8.314 J/mol · K.

For any particular gas, γ, R, and M are constants, and the wave speed is proportional to the square root of the absolute temperature. We will see in Chapter 18 that Eq. (16.10) is almost identical to the expression for the average speed of molecules in an ideal gas. This shows that sound speeds and molecular speeds are closely related; exploring that relationship in detail would be beyond our scope.

16.9 This three-dimensional image of a fetus in the womb was made using a sequence of ultrasound scans. Each individual scan reveals a two-dimensional "slice" through the fetus; a large number of these slices were then combined digitally to yield this remarkable view.

Example 16.5	**Speed of sound in air**

Compute the speed of sound waves in air at room temperature $(T = 20°C)$ and find the range of wavelengths in air to which the human ear (which can hear frequencies in the range of 20–20,000 Hz) is sensitive. The mean molar mass for air (a mixture of principally nitrogen and oxygen) is 28.8×10^{-3} kg/mol and the ratio of heat capacities is $\gamma = 1.40$.

SOLUTION

IDENTIFY and SET UP: We use Eq. (16.10) to find the sound speed and the relation $v = f\lambda$ to determine the wavelength that corresponds to each frequency.

EXECUTE: At $T = 20°C = 293$ K we find

$$v = \sqrt{\frac{\gamma RT}{M}}$$

$$= \sqrt{\frac{(1.40)(8.314 \text{ J/mol} \cdot \text{K})(293 \text{ K})}{28.8 \times 10^{-3} \text{ kg/mol}}} = 344 \text{ m/s}$$

Using this value of v and the expression $\lambda = v/f$, we find that at 20°C a 20-Hz note corresponds to a wavelength of 17 m and a 20,000-Hz note corresponds to a wavelength of 1.7 cm.

EVALUATE: Our calculated value of v agrees with the measured speed of sound at this temperature to within 0.3%.

It's interesting to note that bats can hear much higher frequencies. Like dolphins, bats use high-frequency sound waves for navigation. A typical frequency is 100 kHz; the corresponding wavelength in air at 20°C is about 3.4 mm, small enough for them to detect the flying insects they eat.

In this discussion we have ignored the *molecular* nature of a gas and have treated it as a continuous medium. A gas is actually composed of molecules in random motion, separated by distances that are large in comparison with their diameters. The vibrations that constitute a wave in a gas are superposed on the random thermal motion. At atmospheric pressure, a molecule travels an average distance of about 10^{-7} m between collisions, while the displacement amplitude of a faint sound may be only 10^{-9} m. We can think of a gas with a sound wave passing through as being comparable to a swarm of bees; the swarm as a whole oscillates slightly while individual insects move about through the swarm, apparently at random.

Test Your Understanding

Calculate the speed of longitudinal waves along a steel railroad track. Explain why you can hear the sound of an approaching train from listening to the rails before you can hear it in the air.

16.3 | Sound Intensity

Traveling sound waves, like all other traveling waves, transfer energy from one region of space to another. We saw in Section 15.5 that a useful way to describe the energy carried by a sound wave is through the *wave intensity I,* equal to the time average rate at which energy is transported per unit area across a surface perpendicular to the direction of propagation. In particular, we will express the intensity of a sound wave in terms of the displacement amplitude A or pressure amplitude p_{max}.

For simplicity, let us consider a sound wave propagating in the $+x$-direction so that we can use our expressions for Section 16.1 for the displacement $y(x, t)$ and pressure fluctuation $p(x, t)$—Eqs. (16.1) and (16.4), respectively. In Section 6.4 we saw that power equals the product of force and velocity [see Eq. (6.18)]. So the power per unit area in this sound wave equals the product of $p(x, t)$ (force per unit area) and the *particle* velocity $v_y(x, t)$. The particle velocity $v_y(x, t)$ is the velocity at time t of that portion of the wave medium at coordinate x.) Using Eqs. (16.1) and (16.4), we find

$$v_y(x, t) = \frac{\partial y(x, t)}{\partial t} = \omega A \sin(kx - \omega t)$$

$$p(x, t)v_y(x, t) = [BkA \sin(kx - \omega t)][\omega A \sin(kx - \omega t)]$$

$$= B\omega k A^2 \sin^2(kx - \omega t)$$

CAUTION Remember that the velocity of the wave as a whole is *not* the same as the particle velocity. While the wave continues to move in the direction of propagation, individual particles in the wave medium merely slosh back and forth, as shown in Fig. (16.1). Furthermore, the maximum speed of a particle of the medium can be very different from the wave speed.

The intensity is, by definition, the time average value of $p(x, t)v_y(x, t)$. For any value of x the average value of the function $\sin^2(kx - \omega t)$ over one period $T = 2\pi/\omega$ is 1/2, so

$$I = \frac{1}{2}B\omega k A^2 \tag{16.11}$$

By using the relations $\omega = vk$ and $v^2 = B/\rho$, we can transform Eq. (16.11) into the form

$$I = \frac{1}{2}\sqrt{\rho B}\,\omega^2 A^2 \qquad \text{(intensity of a sinusoidal sound wave)} \qquad (16.12)$$

This equation shows why in a stereo system, a low-frequency woofer has to vibrate with much larger amplitude than a high-frequency tweeter to produce the same sound intensity.

It is usually more useful to express I in terms of the pressure amplitude p_{max}. Using Eq. (16.5) and the relation $\omega = vk$, we find

$$I = \frac{\omega p_{max}{}^2}{2Bk} = \frac{v p_{max}{}^2}{2B} \qquad (16.13)$$

By using the wave speed relation $v^2 = B/\rho$, we can also write Eq. (16.13) in the alternative forms

$$I = \frac{p_{max}{}^2}{2\rho v} = \frac{p_{max}{}^2}{2\sqrt{\rho B}} \qquad \text{(intensity of a sinusoidal sound wave)} \qquad (16.14)$$

You should verify these expressions (see Exercise 16.16). Comparison of Eqs. (16.12) and (16.14) shows that sinusoidal sound waves of the same intensity but different frequency have different displacement amplitudes A but the *same* pressure amplitude p_{max}. This is another reason why it is usually more convenient to describe a sound wave in terms of pressure fluctuations, not displacement.

The *total* average power carried across a surface by a sound wave equals the product of the intensity at the surface and the surface area if the intensity over the surface is uniform. The average total sound power emitted by a person speaking in an ordinary conversational tone is about 10^{-5} W, while a loud shout corresponds to about 3×10^{-2} W. If all the residents of New York City were to talk at the same time, the total sound power would be about 100 W, equivalent to the electric power requirement of a medium-sized light bulb. On the other hand, the power required to fill a large auditorium or stadium with loud sound is considerable (see Example 16.9.)

If the sound source emits waves in all directions equally, the intensity decreases with increasing distance r from the source according to the inverse-square law: the intensity is proportional to $1/r^2$. We discussed this law and its consequences in Section 15.5.

The inverse-square relationship does not apply indoors because sound energy can also reach a listener by reflection from the walls and ceiling. Indeed, part of the architect's job in designing an auditorium is to tailor these reflections so that the intensity is as nearly constant as possible over the entire auditorium.

Problem-Solving Strategy **Sound Intensity**

IDENTIFY *the relevant concepts:* The relationships between intensity and amplitude of a sound wave are rather straightforward. Quite a few other quantities are involved in these rela- tionships, however, so it's particularly important to decide which is your target variable.

SET UP *the problem using the following steps:*

1. Sort the various physical quantities into categories. The amplitude is described by A or p_{max}, and the frequency f can be determined from ω, k, or λ. These quantities are related through the wave speed v, which in turn is determined by the properties of the medium: B and ρ for a liquid; γ, T, and M for a gas.
2. Determine which quantities are given and which are the unknown target variables. Then start looking for relationships that take you where you want to go.

EXECUTE *the solution as follows:* Use the equations you've selected to solve for the target variables. Be certain that all of the quantities are expressed in the correct units. In particular, if temperature is used to calculate the speed of sound in a gas, make sure that it is expressed in kelvins (Celsius temperature plus 273.15).

EVALUATE *your answer:* There are multiple relationships among the quantities that describe a wave. Try using an alternative one to check your results.

Example 16.6 Intensity of a sound wave in air

Find the intensity of the sound wave in Example 16.1, with $p_{max} = 3.0 \times 10^{-2}$ Pa. Assume the temperature is 20°C so that the density of air is $\rho = 1.20$ kg/m³ and the speed of sound is $v = 344$ m/s.

SOLUTION

IDENTIFY and SET UP: We are given the pressure amplitude p_{max}, density ρ, and wave speed v, and want to find the intensity I. This is done most easily with Eq. (16.14).

EXECUTE: From Eq. (16.14),

$$I = \frac{p_{max}^2}{2\rho v} = \frac{(3.0 \times 10^{-2}\,\text{Pa})^2}{2(1.20\,\text{kg/m}^3)(344\,\text{m/s})}$$
$$= 1.1 \times 10^{-6}\,\text{J/(s·m}^2) = 1.1 \times 10^{-6}\,\text{W/m}^2$$

EVALUATE: This seems like a very low intensity, but in fact it is well within the range of sound intensities encountered on a daily basis. A very loud sound wave at the threshold of pain has a pressure amplitude of about 30 Pa and an intensity of about 1 W/m². The pressure amplitude of the faintest sound wave that can be heard is about 3×10^{-5} Pa, and the corresponding intensity is about 10^{-12} W/m². (Try these values of p_{max} in Eq. (16.14) to check that the corresponding intensities are as we have stated.)

Example 16.7 Same intensity, different frequencies

A 20-Hz sound wave has the same intensity as the 1000-Hz sound wave in Examples 16.1 and 16.6. What are the displacement amplitude and pressure amplitude of the 20-Hz sound wave?

SOLUTION

IDENTIFY and SET UP: Given the intensity, we can use Eq. (16.12) to determine the displacement amplitude A. Note that we are not given the value of B. However, ρ and B depend only on the properties of the medium, not the amplitude or frequency, so their values will cancel out if we equate the intensities at 20 Hz and at 1000 Hz. We can also use Eq. (16.14) to find the pressure amplitude at 20 Hz.

EXECUTE:

Inspection of Eq. (16.12) shows that if a wave in a given medium (same ρ and B) has the same intensity I at two different frequencies, then the product ωA must have the same value for both frequencies. From Example 16.1, $A = 1.2 \times 10^{-8}$ m at 1000 Hz, so

$$(20\,\text{Hz})A_{20} = (1000\,\text{Hz})(1.2 \times 10^{-8}\,\text{m})$$
$$A_{20} = 6.0 \times 10^{-7}\,\text{m} = 0.60\,\mu\text{m}$$

Do you understand why we didn't have to convert the frequencies to angular frequencies?

Since the intensity is the same for both frequencies, Eq. (16.14) shows that the *pressure* amplitude p_{max} must also be the same for both. Hence $p_{max} = 3.0 \times 10^{-2}$ Pa for $f = 20$ Hz.

EVALUATE: Our result reinforces the idea that pressure amplitude offers a more convenient description of a sound wave than displacement amplitude. Note also that using Eq. (16.5) and $k = \omega/v$, we get $p_{max} = BkA = (B/v)\omega A$; the bulk modulus B and wave speed v depend only on the medium, so we again conclude that the product ωA must have the same value for both frequencies.

| Example 16.8 | **A high-altitude sound wave** |

At an altitude of 11,000 m, near the cruising altitude of jetliners (Fig. 16.10), the atmosphere is cold and thin: the temperature is $-57°C$, the pressure is 2.26×10^4 Pa, and the density is 0.364 kg/m^3. What would be the intensity of a 1000-Hz sound wave with the same displacement amplitude as the wave at sea level in Examples 16.1 and 16.6?

16.10 How do sound waves differ in the upper atmosphere?

SOLUTION

IDENTIFY and SET UP: We can use Eq. (16.12) to determine the intensity from the density, bulk modulus, frequency, and amplitude. We are not given the value of B, but we can calculate it using Eq. (16.9).

EXECUTE: Using Eq. (16.9) and the value $\gamma = 1.40$ for air, we find $B = \gamma p_0 = (1.40)(2.26 \times 10^4 \text{ Pa}) = 3.16 \times 10^4$ Pa. Then, from Eq. (16.12),

$$I = \frac{1}{2}\sqrt{(0.364 \text{ kg/m}^3)(2.26 \times 10^4 \text{ Pa})}$$
$$\times (2\pi)^2(1000 \text{ Hz})^2(1.2 \times 10^{-8} \text{ m})^2$$
$$= 3.1 \times 10^{-7} \text{ W/m}^2$$

EVALUATE: This intensity is only 28% as great as that of a wave with the same amplitude at sea level (3.1×10^{-7} W/m^2 versus 1.1×10^{-6} W/m^2). Hence a loudspeaker oscillating with a certain amplitude would produce a substantially less intense sound at high altitude. At even higher altitude, the atmosphere fades into the near-vacuum of interplanetary space and the density and pressure (and hence bulk modulus) decrease to zero. The intensity likewise decreases, and outer space is nearly silent.

| Example 16.9 | **"Play it loud!"** |

For an outdoor concert we want the sound intensity at a distance of 20 m from the speaker array to be 1 W/m^2. Assuming that the sound waves have the same intensity in all directions, what acoustic power output is needed from the speaker array?

SOLUTION

IDENTIFY: This example uses the definition of intensity as power per unit area. Here the total power is the target variable, and the area in question is a hemisphere centered on the speaker array.

SET UP: We make the assumptions that the speakers are near ground level and that the acoustic power is spread uniformly over a

hemisphere 20 m in radius (that is, we assume that none of the acoustic power is directed into the ground). The surface area of this hemisphere is equal to $(1/2)(4\pi)(20 \text{ m})^2$, or about 2500 m^2. The required power is the product of this area and the intensity.

EXECUTE: The speaker array power is

$$(1 \text{ W/m}^2)(2500 \text{ m}^2) = 2500 \text{ W} = 2.5 \text{ kW}$$

EVALUATE: The electrical power input to the speaker would need to be considerably larger because the efficiency of such devices is not very high (typically a few percent for ordinary speakers, and up to 25% for horn-type speakers).

The Decibel Scale

Because the ear is sensitive over a broad range of intensities, a *logarithmic* intensity scale is usually used. The **sound intensity level** β of a sound wave is defined by the equation

$$\beta = (10 \text{ dB}) \log \frac{I}{I_0} \qquad \text{(definition of sound intensity level)} \qquad (16.15)$$

In this equation, I_0 is a reference intensity, chosen to be 10^{-12} W/m^2, approximately the threshold of human hearing at 1000 Hz. Recall that "log" means the logarithm to base 10. Sound intensity levels are expressed in **decibels,** abbreviated dB. A decibel is $\frac{1}{10}$ of a *bel,* a unit named for Alexander Graham Bell (the inventor of the telephone). The bel is inconveniently large for most purposes, and the decibel is the usual unit of sound intensity level.

If the intensity of a sound wave equals I_0 or 10^{-12} W/m^2, its sound intensity level is 0 dB. An intensity of 1 W/m^2 corresponds to 120 dB. Table 16.2 gives the sound intensity levels in decibels of several familiar sounds. You can use Eq. (16.15) to check the value of sound intensity level β given for each intensity in the table.

Table 16.2 Sound Intensity Levels from Various Sources (Representative Values)

Source or Description of Sound	Sound Intensity Level, β (dB)	Intensity, I (W/m^2)
Military jet aircraft 30 m away	140	10^2
Threshold of pain	120	1
Riveter	95	3.2×10^{-3}
Elevated train	90	10^{-3}
Busy street traffic	70	10^{-5}
Ordinary conversation	65	3.2×10^{-6}
Quiet automobile	50	10^{-7}
Quiet radio in home	40	10^{-8}
Average whisper	20	10^{-10}
Rustle of leaves	10	10^{-11}
Threshold of hearing at 1000 Hz	0	10^{-12}

Because the ear is not equally sensitive to all frequencies in the audible range, some sound-level meters weight the various frequencies unequally. One such scheme leads to the so-called dBA scale; this scale deemphasizes the low and very high frequencies, where the ear is less sensitive than at midrange frequencies.

Example 16.10

Temporary deafness

A ten-minute exposure to 120-dB sound will typically shift your threshold of hearing at 1000 Hz from 0 dB up to 28 dB for a while. Ten years of exposure to 92-dB sound will cause a *permanent* shift up to 28 dB. What intensities correspond to 28 dB and 92 dB?

SOLUTION

IDENTIFY and SET UP: We solve Eq. (16.15) to find the intensity I (the target variable) for each value of the sound intensity level β.

EXECUTE: We rearrange Eq. (16.15) by dividing both sides by 10 dB and then using the relationship $10^{\log x} = x$:

$$I = I_0 10^{\beta/(10\text{ dB})}$$

When $\beta = 28$ dB,

$$I = (10^{-12}\text{ W/m}^2)10^{(28\text{ dB}/10\text{ dB})}$$
$$= (10^{-12}\text{ W/m}^2)10^{2.8} = 6.3 \times 10^{-10}\text{ W/m}^2$$

Similarly, for $\beta = 92$ dB,

$$I = (10^{-12}\text{ W/m}^2)10^{(92\text{ dB}/10\text{ dB})} = 1.6 \times 10^{-3}\text{ W/m}^2$$

EVALUATE: If your answers are a factor of 10 too large, you may have entered 10×10^{-12} in your calculator instead of 1×10^{-12}. Be careful!

<table>
<tr><td>Example
16.11</td><td colspan="2">**A bird sings in a meadow**</td></tr>
</table>

Consider an idealized model with a bird (treated as a point source) emitting constant sound power, with intensity inversely proportional to the square of the distance from the bird. By how many decibels does the sound intensity level drop when you move twice as far away from the bird?

SOLUTION

IDENTIFY: Because the decibel scale is logarithmic, the *difference* between two sound intensity levels (the target variable) corresponds to the *ratio* of the corresponding intensities. The ratio of the intensities comes from the inverse-square law.

SET UP: We label the two points 1 and 2 (see Fig. 16.11). We use Eq. (16.15), the definition of sound intensity level, twice (once at each point). We use Eq. (15.26), the statement of the inverse-square law, to relate the intensities at the two points.

EXECUTE: The difference in sound intensity level, $\beta_2 - \beta_1$, is given by

$$\beta_2 - \beta_1 = (10\ \text{dB})\left(\log\frac{I_2}{I_0} - \log\frac{I_1}{I_0}\right)$$
$$= (10\ \text{dB})[(\log I_2 - \log I_0) - (\log I_1 - \log I_0)]$$
$$= (10\ \text{dB})\log\frac{I_2}{I_1}$$

Now we use the reciprocal of Eq. (15.26); $I_2/I_1 = r_1^2/r_2^2$, so

$$\beta_2 - \beta_1 = (10\ \text{dB})\log\frac{r_1^2}{r_2^2} = (10\ \text{dB})\log\frac{r_1^2}{(2r_1)^2}$$
$$= (10\ \text{dB})\log\frac{1}{4} = -6.0\ \text{dB}$$

A decrease in intensity of a factor of 4 corresponds to a 6-dB decrease in sound intensity level.

EVALUATE: Our result is negative, which tells us (correctly) that the sound intensity level is less at point 2 than at point 1. The 6-dB difference doesn't depend on the value of the sound intensity level at point 1. If point 1 is relatively close to the bird so that $\beta_1 = 56$ dB, then at a point twice as far away, $\beta_2 = 50$ dB; if point 1 is more distant from the bird so that $\beta_1 = 28$ dB, then at a point twice as far away, $\beta_2 = 22$ dB.

It's interesting to note that the perceived *loudness* of a sound is not directly proportional to its intensity. As an example, most people usually interpret an increase of 8 to 10 dB in sound intensity level (corresponding to an intensity that increases by a factor of 6 to 10) as a doubling of loudness.

16.11 When you double your distance from a point source of sound, by how much does the sound intensity level decrease?

Test Your Understanding

You double the intensity of a sound wave in air while leaving the frequency unchanged. (The pressure, density, and temperature of the air remain unchanged as well.) What effect does this have on the displacement amplitude, pressure amplitude, bulk modulus, sound speed, and sound intensity level?

16.4 | Standing Sound Waves and Normal Modes

When longitudinal (sound) waves propagate in a fluid in a pipe with finite length, the waves are reflected from the ends in the same way that transverse waves on a string are reflected at its ends. The superposition of the waves traveling in opposite directions again forms a standing wave. Just as for transverse standing waves on a string (Section 15.7), standing sound waves (normal modes) in a pipe can be used to create sound waves in the surrounding air. This is the operating principle of the human voice as well as many musical instruments, including woodwinds, brasses, and pipe organs.

Transverse waves on a string, including standing waves, are usually described only in terms of the displacement of the string. But, as we have seen, sound waves in a fluid may be described either in terms of the displacement of the fluid or in terms of the pressure variation in the fluid. To avoid confusion, we'll use the terms **displacement node** and **displacement antinode** to refer to points where particles of the fluid have zero displacement and maximum displacement, respectively.

We can demonstrate standing sound waves in a column of gas using an apparatus called Kundt's tube (Fig. 16.12). A horizontal glass tube a meter or so long is closed at one end and has a flexible diaphragm at the other end that can transmit vibrations. A nearby loudspeaker is driven by an audio oscillator and amplifier; this produces sound waves that force the diaphragm to vibrate sinusoidally with a frequency that we can vary. The sound waves within the tube are reflected at the other, closed end of the tube. We spread a small amount of light powder uniformly along the bottom of the tube. As we vary the frequency of the sound, we pass through frequencies at which the amplitude of the standing waves becomes large enough for the powder to be swept along the tube at those points where the gas is in motion. The powder therefore collects at the displacement nodes (where the gas is not moving). Adjacent nodes are separated by a distance equal to $\lambda/2$, and we can measure this distance. Given the wavelength, we can use this experiment to determine the wave speed: We read the frequency f from the

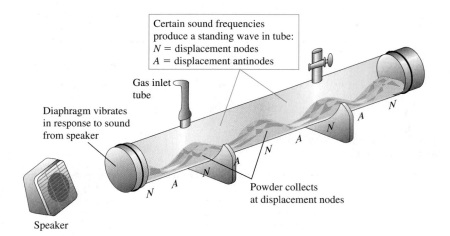

16.12 Demonstrating standing sound waves using a Kundt's tube. The blue shading represents the density of the gas at an instant when the gas pressure at the displacement nodes is a maximum or a minimum.

oscillator dial, and we can then calculate the speed v of the waves from the relation $v = \lambda f$.

Figure 16.13 will help you to visualize standing sound waves; it is analogous to Fig. 16.4 for longitudinal traveling waves. Again tape two index cards together edge-to-edge, with a gap of a millimeter or two forming a thin slit. Place the card over the diagram with the slit horizontal and move it vertically with constant velocity. The portions of the sine curves that appear in the slit correspond to the oscillations of the particles in a longitudinal standing wave. Each particle moves with longitudinal simple harmonic motion about its equilibrium position.

Figure 16.14 is an enlarged version of a portion of Fig. 16.13 centered on a displacement node. Note that particles on opposite sides of a displacement node vibrate in opposite phase. When these particles approach each other, the gas between them is compressed and the pressure rises; when they recede from each other, there is an expansion and the pressure drops. Hence at a displacement *node* the gas undergoes the maximum amount of compression and expansion, and the variations in pressure and density above and below the average have their maximum value. By contrast, particles on opposite sides of a displacement *antinode* vibrate *in phase;* the distance between the particles is nearly constant, and there is *no* variation in pressure or density at a displacement antinode.

We use the term **pressure node** to describe a point in a standing sound wave at which the pressure and density do not vary and the term **pressure antinode** to describe a point at which the variations in pressure and density are greatest. Using these terms, we can summarize our observations about standing sound waves as follows: **a pressure node is always a displacement antinode, and a pressure antinode is always a displacement node.** Figure 16.12 depicts a standing sound wave at an instant at which the pressure variations are greatest; the blue shading shows that the density and pressure of the gas have their maximum and minimum values at the displacement nodes (labeled N).

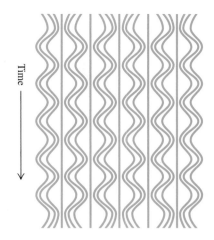

16.13 Diagram for illustrating longitudinal standing waves. The straight vertical lines mark the locations of displacement nodes and pressure antinodes.

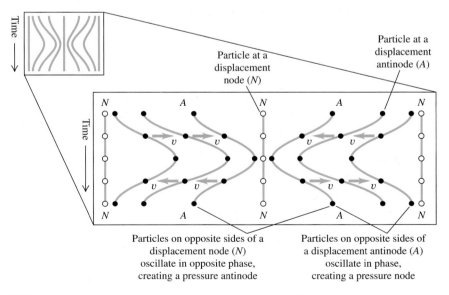

16.14 In a standing longitudinal wave, a displacement node N is a pressure antinode (a point where the pressure fluctuates the most) and a displacement antinode A is a pressure node (a point where the pressure does not fluctuate at all).

When reflection takes place at a *closed* end of a pipe (an end with a rigid barrier or plug), the displacement of the particles at this end must always be zero, analogous to a fixed end of a string. Thus a closed end of a pipe is a displacement node and a pressure antinode; the particles do not move, but the pressure variations are maximum. An *open* end of a pipe is a pressure node because it is open to the atmosphere, where the pressure is constant. Because of this, an open end is always a displacement *antinode,* in analogy to a free end of a string; the particles oscillate with maximum amplitude, but the pressure does not vary. (Strictly speaking, the pressure node actually occurs somewhat beyond an open end of a pipe. But if the diameter of the pipe is small in comparison to the wavelength, which is true for most musical instruments, this effect can safely be neglected.) Thus longitudinal waves in a column of fluid are reflected at the closed and open ends of a pipe in the same way that transverse waves in a string are reflected at fixed and free ends, respectively.

Conceptual Example 16.12	**The sound of silence**

A directional loudspeaker aims a sound wave of wavelength λ at a wall (Fig. 16.15). At what distances from the wall could you stand and hear no sound at all?

SOLUTION

Your ear detects *pressure* variations in the air; increases or decreases in the pressure outside your eardrum cause it to move

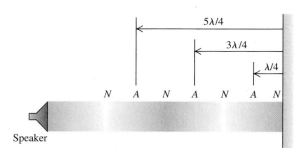

16.15 When a sound wave is directed at a wall, it interferes with the reflected wave to create a standing wave. The *N*'s and *A*'s are *displacement* nodes and antinodes.

slightly in or out, a motion that generates an electrical signal that is sent to the brain. (If you've ever had trouble getting your ears to "pop" on a drive up into the mountains or on an airline flight, you're familiar with just how sensitive your ears are to pressure changes.) Hence you will hear no sound if your ear is at a pressure node, which is a displacement antinode. The wall is a displacement node; the distance from a node to an adjacent antinode is $\lambda/4$, and the distance from one antinode to the next is $\lambda/2$ (Fig. 16.15). Hence the distances d from the wall at which no sound will be heard are

$$d = \lambda/4$$
(first displacement antinode and pressure node.)

$$d = \lambda/4 + \lambda/2 = 3\lambda/4$$
(second displacement antiode and pressure node)

$$d = 3\lambda/4 + \lambda/2 = 5\lambda/4$$
(third displacement antinode and pressure node)

and so on. If the loudspeaker is not highly directional, this effect is hard to notice because of multiple reflections of sound waves from the floor, ceiling, and other points on the walls.

Organ Pipes and Wind Instruments

The most important application of standing sound waves is the production of musical tones by wind instruments. Organ pipes are one of the simplest examples (Fig. 16.16). Air is supplied by a blower, at a gauge pressure typically of the order of 10^3 Pa (10^{-2} atm), to the bottom end of the pipe (Fig. 16.17). A stream of air emerges from the narrow opening at the edge of the horizontal surface and is directed against the top edge of the opening, which is called the *mouth* of the pipe. The column of air in the pipe is set into vibration, and there is a series of possible normal modes, just as with the stretched string. The mouth always acts as an open end; thus it is a pressure node and a displacement antinode. The other end of the pipe (at the top in Fig. 16.17) may be either open or closed.

In Fig. 16.18, both ends of the pipe are open, so both ends are pressure nodes and displacement antinodes. An organ pipe that is open at both ends is called an *open pipe*. The fundamental frequency f_1 corresponds to a standing-wave pattern with a displacement antinode at each end and a displacement node in the middle (Fig. 16.18a). The distance between adjacent antinodes is always equal to one half-wavelength, and in this case that is equal to the length L of the pipe; $\lambda/2 = L$. The corresponding frequency, obtained from the relation $f = v/\lambda$, is

$$f_1 = \frac{v}{2L} \qquad \text{(open pipe)} \qquad (16.16)$$

Figures 16.18b and 16.18c show the second and third harmonics (first and second overtones); their vibration patterns have two and three displacement nodes, respectively. For these, a half-wavelength is equal to $L/2$ and $L/3$, respectively, and the frequencies are twice and three times the fundamental, respectively. That is, $f_2 = 2f_1$ and $f_3 = 3f_1$. For *every* normal mode of an open pipe the length L

16.16 Organ pipes of different sizes produce tones with different frequencies.

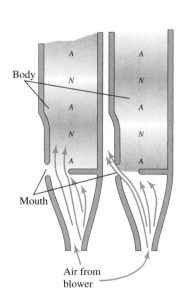

16.17 Cross sections of an organ pipe at two instants one half-period apart. Vibrations from the turbulent air flow set up standing waves in the pipe. The N's and A's are *displacement* nodes and antinodes; as the blue shading shows, these are points of maximum pressure variation and zero pressure variation, respectively.

16.18 A cross section of an open pipe showing the first three normal modes. The shading indicates the pressure variations. The red curves are graphs of the displacement along the pipe axis at two instants separated in time by one half-period. The N's and A's are the *displacement* nodes and antinodes; interchange these to show the *pressure* nodes and antinodes.

$(a) f_1 = \dfrac{v}{4L}$

$(b) f_3 = 3\dfrac{v}{4L} = 3f_1$

$(c) f_5 = 5\dfrac{v}{4L} = 5f_1$

16.19 A cross section of a stopped pipe showing the first three normal modes as well as the *displacement* nodes and antinodes. Only odd harmonics are possible.

must be an integer number of half-wavelengths, and the possible wavelengths λ_n are given by

$$ L = n\frac{\lambda_n}{2} \quad \text{or} \quad \lambda_n = \frac{2L}{n} \quad (n = 1, 2, 3, \dots) \quad \text{(open pipe)} \quad (16.17) $$

The corresponding frequencies f_n are given by $f_n = v/\lambda_n$, so all the normal-mode frequencies for a pipe that is open at both ends are given by

$$ f_n = \frac{nv}{2L} \quad (n = 1, 2, 3, \dots) \quad \text{(open pipe)} \quad (16.18) $$

The value $n = 1$ corresponds to the fundamental frequency, $n = 2$ to the second harmonic (or first overtone), and so on. Alternatively, we can say

$$ f_n = nf_1 \quad (n = 1, 2, 3, \dots) \quad \text{(open pipe)} \quad (16.19) $$

with f_1 given by Eq. (16.16).

Figure 16.19 shows a pipe that is open at the left end but closed at the right end. This is called a *stopped pipe*. The left (open) end is a displacement antinode (pressure node), but the right (closed) end is a displacement node (pressure antinode). The distance between a node and the adjacent antinode is always one quarter-wavelength. Figure 16.19a shows the lowest-frequency mode; the length of the pipe is a quarter-wavelength ($L = \lambda_1/4$). The fundamental frequency is $f_1 = v/\lambda_1$, or

$$ f_1 = \frac{v}{4L} \quad \text{(stopped pipe)} \quad (16.20) $$

This is one-half the fundamental frequency for an *open* pipe of the same length. In musical language, the *pitch* of a closed pipe is one octave lower (a factor of 2 in frequency) than that of an open pipe of the same length. Figure 16.19b shows the next mode, for which the length of the pipe is *three-quarters* of a wavelength, corresponding to a frequency $3f_1$. For Fig. 16.19c, $L = 5\lambda/4$ and the frequency is $5f_1$. The possible wavelengths are given by

$$ L = n\frac{\lambda_n}{4} \quad \text{or} \quad \lambda_n = \frac{4L}{n} \quad (n = 1, 3, 5, \dots) \quad \text{(stopped pipe)} \quad (16.21) $$

The normal-mode frequencies are given by $f_n = v/\lambda_n$, or

$$ f_n = \frac{nv}{4L} \quad (n = 1, 3, 5, \dots) \quad \text{(stopped pipe)} \quad (16.22) $$

or

$$ f_n = nf_1 \quad (n = 1, 3, 5, \dots) \quad \text{(stopped pipe)} \quad (16.23) $$

with f_1 given by Eq. (16.20). We see that the second, fourth, and all *even* harmonics are missing. In a pipe that is closed at one end, the fundamental frequency is $f_1 = v/4L$, and only the odd harmonics in the series ($3f_1, 5f_1, \dots$) are possible.

A final possibility is a pipe that is closed at *both* ends, with displacement nodes and pressure antinodes at both ends. This wouldn't be of much use as a musical instrument because there would be no way for the vibrations to get out of the pipe.

Example 16.13	A tale of two pipes

On a day when the speed of sound is 345 m/s, the fundamental frequency of a stopped organ pipe is 220 Hz. a) How long is this stopped pipe? b) The second *overtone* of this pipe has the same wavelength as the third *harmonic* of an open pipe. How long is the open pipe?

SOLUTION

IDENTIFY and SET UP: Since this is a stopped pipe (open at one end and closed at the other), the normal-mode frequencies are given by Eq. (16.22). We use this to determine the length L from the frequency in part (a). In part (b), we must make a comparison to an open pipe, for which the frequencies are given by Eq. (16.18).

EXECUTE: a) For a stopped pipe, $f_1 = v/4L$, so the length of the stopped pipe is

$$L_{stopped} = \frac{v}{4f_1} = \frac{345 \text{ m/s}}{4(220 \text{ s}^{-1})} = 0.392 \text{ m}$$

b) The frequency of the first overtone of a stopped pipe is $f_3 = 3f_1$, and the frequency of the second overtone is $f_5 = 5f_1$:

$$f_5 = 5f_1 = 5(220 \text{ Hz}) = 1100 \text{ Hz}$$

If the wavelengths are the same, the frequencies are the same, so the frequency of the third harmonic of the open pipe is also 1100 Hz. The third harmonic of an open pipe is at $3f_1 = 3(v/2L)$. If this equals 1100 Hz, then

$$1100 \text{ Hz} = 3\left(\frac{345 \text{ m/s}}{2L_{open}}\right) \quad \text{and} \quad L_{open} = 0.470 \text{ m}$$

EVALUATE: The stopped pipe is 0.392 m long and has a fundamental frequency of 220 Hz; the open pipe is longer, 0.470 m, but has a higher fundamental frequency of $(1100 \text{ Hz})/3 = 367$ Hz. If this seems like a contradiction, you should again compare Figs. 16.18a and 16.19a.

In an organ pipe in actual use, several modes are always present at once; the motion of the air is a superposition of these modes. This situation is analogous to a string that is struck or plucked, as in Fig. 15.25. Just as for a vibrating string, a complex standing wave in the pipe produces a traveling sound wave in the surrounding air with a harmonic content similar to that of the standing wave. A very narrow pipe produces a sound wave rich in higher harmonics, which we hear as a thin and "stringy" tone; a fatter pipe produces mostly the fundamental mode, heard as a softer, more flute-like tone. The harmonic content also depends on the shape of the pipe's mouth.

We have talked about organ pipes, but this discussion is also applicable to other wind instruments. The flute and the recorder are directly analogous. The most significant difference is that those instruments have holes along the pipe. Opening and closing the holes with the fingers changes the effective length L of the air column and thus changes the pitch. Any individual organ pipe, by comparison, can play only a single note. The flute and recorder behave as *open* pipes, while the clarinet acts as a *stopped* pipe (closed at the reed end, open at the bell).

Equations (16.18) and (16.22) show that the frequencies of any wind instrument are proportional to the speed of sound v in the air column inside the instrument. As Eq. (16.10) shows, v depends on temperature; it increases when temperature increases. Thus the pitch of all wind instruments rises with increasing temperature. An organ that has some of its pipes at one temperature and others at a different temperature is bound to sound out of tune.

Test Your Understanding

If you connect a hose to one end of a metal pipe and blow compressed air into it, the pipe produces a musical tone. If instead you blow compressed helium into the pipe at the same pressure and temperature, the pipe produces a higher tone. Why?

16.5 | Resonance

Many mechanical systems have normal modes of oscillation. As we have seen, these include columns of air (as in an organ pipe) and stretched strings (as in a guitar; see Section 15.8). In each mode, every particle of the system oscillates with simple harmonic motion at the same frequency as the mode. Air columns and stretched strings have an infinite series of normal modes, but the basic concept is closely related to the simple harmonic oscillator, discussed in Chapter 13, which has only a single normal mode (that is, only one frequency at which it oscillates after being disturbed).

Suppose we apply a periodically varying force to a system that can oscillate. The system is then forced to oscillate with a frequency equal to the frequency of the applied force (called the *driving frequency*). This motion is called a *forced oscillation*. We talked about forced oscillations of the harmonic oscillator in Section 13.8, and we suggest that you review that discussion. In particular, we described the phenomenon of mechanical **resonance.** A simple example of resonance is pushing Cousin Throckmorton on a swing. The swing is a pendulum; it has only a single normal mode, with a frequency determined by its length. If we push the swing periodically with this frequency, we can build up the amplitude of the motion. But if we push with a very different frequency, the swing hardly moves at all.

Resonance also occurs when a periodically varying force is applied to a system with many normal modes. An example is shown in Fig. 16.20a. An open organ pipe is placed next to a loudspeaker that is driven by an amplifier and emits pure sinusoidal sound waves of frequency f, which can be varied by adjusting the amplifier. The air in the pipe is forced to vibrate with the same frequency f as the *driving force* provided by the loudspeaker. In general the amplitude of this motion is relatively small, and the air inside the pipe will not move in any of the normal-mode patterns shown in Fig. 16.18. But if the frequency f of the force is close to one of the normal-mode frequencies, the air in the pipe moves in the normal-mode

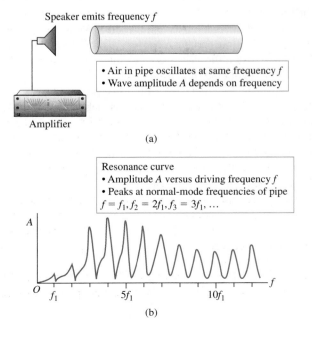

16.20 (a) The air in an open pipe is forced to oscillate at the same frequency as the sinusoidal sound waves coming from the loudspeaker. (b) The resonance curve of the open pipe graphs the amplitude of the standing sound wave in the pipe as a function of the driving frequency.

pattern for that frequency, and the amplitude can become quite large. Figure 16.20b shows the amplitude of oscillation of the air in the pipe as a function of the driving frequency f. The shape of this graph is called the **resonance curve** of the pipe; it has peaks where f equals the normal-mode frequencies of the pipe. The detailed shape of the resonance curve depends on the geometry of the pipe.

If the frequency of the force is precisely *equal* to a normal-mode frequency, the system is in resonance, and the amplitude of the forced oscillation is maximum. If there were no friction or other energy-dissipating mechanism, a driving force at a normal-mode frequency would continue to add energy to the system, and the amplitude would increase indefinitely. In such an idealized case the peaks in the resonance curve of Fig. 16.20b would be infinitely high. But in any real system there is always some dissipation of energy, or damping, as we discussed in Section 13.8; the amplitude of oscillation in resonance may be large, but it cannot be infinite.

The "sound of the ocean" you hear when you put your ear next to a large seashell is due to resonance. The noise of the outside air moving past the seashell is a mixture of sound waves of almost all audible frequencies, which forces the air inside the seashell to oscillate. The seashell behaves like an organ pipe, with a set of normal-mode frequencies; hence the inside air oscillates most strongly at those frequencies, producing the seashell's characteristic sound. To hear a similar phenomenon, uncap a full bottle of your favorite beverage and blow across the open top. The noise is provided by your breath blowing across the top, and the "organ pipe" is the column of air inside the bottle above the surface of the liquid. If you take a drink and repeat the experiment, you will hear a lower tone because the "pipe" is longer and the normal-mode frequencies are lower.

Resonance also occurs when a stretched string is forced to oscillate (Section 15.8). Suppose that one end of a stretched string is held fixed while the other is given a transverse sinusoidal motion with small amplitude, setting up standing waves. If the frequency of the driving mechanism is *not* equal to one of the normal-mode frequencies of the string, the amplitude at the antinodes is fairly small. However, if the frequency is equal to any one of the normal-mode frequencies, the string is in resonance, and the amplitude at the antinodes is very much larger than that at the driven end. The driven end is not precisely a node, but it lies much closer to a node than to an antinode when the string is in resonance. The photographs in Fig. 15.20 were made this way, with the left end of the string fixed and the right end oscillating vertically with small amplitude; large-amplitude standing waves resulted when the frequency of oscillation of the right end was equal to the fundamental frequency or to one of the first three overtones.

It is easy to demonstrate resonance with a piano. Push down the damper pedal (the right-hand pedal) so that the dampers are lifted and the strings are free to vibrate, and then sing a steady tone into the piano. When you stop singing, the piano seems to continue to sing the same note. The sound waves from your voice excite vibrations in the strings that have natural frequencies close to the frequencies (fundamental and harmonics) present in the note you sang.

A more spectacular example is a singer breaking a wine glass with her amplified voice. A good-quality wine glass has normal-mode frequencies that you can hear by tapping it. If the singer emits a loud note with a frequency corresponding exactly to one of these normal-mode frequencies, large-amplitude oscillations can build up and break the glass (Fig. 16.21).

Resonance is a very important concept, not only in mechanical systems but in all areas of physics. In Chapter 31 we will see examples of resonance in electric circuits.

16.21 The frequency of the sound from this trumpet exactly matches one of the normal-mode frequencies of the goblet. The resonant vibrations of the goblet have such large amplitude that the goblet tears itself apart.

<table>
<tr><td>Example
16.14</td><td>**An organ-guitar duet**</td></tr>
</table>

A stopped organ pipe is sounded near a guitar, causing one of the strings to vibrate with large amplitude. We vary the tension of the string until we find the maximum amplitude. The string is 80% as long as the stopped pipe. If both the pipe and the string vibrate at their fundamental frequency, calculate the ratio of the wave speed on the string to the speed of sound in air.

SOLUTION

IDENTIFY: The large response of the string is an example of resonance. It occurs because the organ pipe and the guitar string have the same fundamental frequency.

SET UP: Letting the subscripts a and s stand for the air in the pipe and the string, respectively, the condition for resonance is $f_{1a} = f_{1s}$. Equation (16.20) gives the fundamental frequency for a stopped pipe, while the fundamental frequency for a guitar string held at both ends is given by Eq. (15.32). These expressions involve the wave speed in air (v_a) and on the string (v_s) and the lengths of the

pipe and string; we are given that $L_s = 0.80L_a$, and our target variable is the ratio v_s/v_a.

EXECUTE: From Eqs. (16.20) and (15.32), $f_{1a} = v_a/4L_a$ and $f_{1s} = v_s/2L_s$. Setting these equal to each other, we find

$$\frac{v_a}{4L_a} = \frac{v_s}{2L_s}$$

Substituting $L_s = 0.80 L_a$ and rearranging, we get

$$\frac{v_s}{v_a} = 0.40$$

EVALUATE: As an example, if the speed of sound in air is 345 m/s, the wave speed on the string is $(0.40)(345 \text{ m/s}) = 138$ m/s. Note that while the standing waves in the pipe and on the string have the same frequency, they have different *wavelengths* $\lambda = v/f$ because the two media have different wave speeds v. Which standing wave has the greater wavelength?

Test Your Understanding

Suppose the open pipe in Fig. 16.20 is 60.0 cm long. If the speed of sound in air is 345 m/s, what frequency emitted by the speaker produces the greatest response in the pipe?

16.6 | Interference of Waves

Wave phenomena that occur when two or more waves overlap in the same region of space are grouped under the heading *interference*. As we have seen, standing waves are a simple example of an interference effect: two waves traveling in opposite directions in a medium combine to produce a standing wave pattern with nodes and antinodes that do not move.

Figure 16.22 shows an example of another type of interference that involves waves that spread out in space. Two speakers, driven in phase by the same amplifier, emit identical sinusoidal sound waves with the same constant frequency. We place a microphone at point P in the figure, equidistant from the speakers. Wave crests emitted from the two speakers at the same time travel equal distances and arrive at point P at the same time; hence the waves arrive in phase, and there is constructive interference. The total wave amplitude at P is twice the amplitude from each individual wave, and we can measure this combined amplitude with the microphone.

Now let's move the microphone to point Q, where the distances from the two speakers to the microphone differ by a half-wavelength. Then the two waves arrive a half-cycle out of step, or *out of phase;* a positive crest from one speaker arrives at the same time as a negative crest from the other. Destructive interference takes place, and the amplitude measured by the microphone is much *smaller*

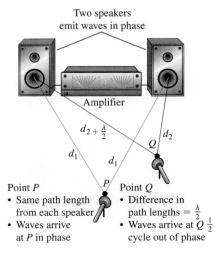

16.22 Two speakers driven by the same amplifier. Constructive interference occurs at point P and destructive interference occurs at point Q.

than when only one speaker is present. If the amplitudes from the two speakers are equal, the two waves cancel each other out completely at point Q, and the total amplitude there is zero.

CAUTION Although this situation bears some resemblance to standing waves in a pipe, the total wave in Fig. 16.22 is a *traveling* wave, not a standing wave. To see why, recall that in a standing wave there is no net flow of energy in any direction. By contrast, in Fig. 16.22 there is an overall flow of energy from the speakers into the surrounding air; this is characteristic of a traveling wave. The interference between the waves from the two speakers simply causes the energy flow to be *channeled* into certain directions (for example, toward P) and away from other directions (for example, away from Q). You can see another difference between Fig. 16.22 and a standing wave by considering a point, such as Q, where destructive interference occurs. Such a point is *both* a displacement node *and* a pressure node because there is no wave at all at this point. Compare this to a standing wave, in which a pressure node is a displacement antinode and vice versa.

Constructive interference occurs wherever the distances traveled by the two waves differ by a whole number of wavelengths, $0, \lambda, 2\lambda, 3\lambda, \ldots$; in all these cases the waves arrive at the microphone in phase (Fig. 16.23a). If the distances from the two speakers to the microphone differ by any half-integer number of wavelengths, $\lambda/2, 3\lambda/2, 5\lambda/2, \ldots$, the waves arrive at the microphone out of phase and there will be destructive interference (Fig. 16.23b). In this case, little or no sound energy flows toward the microphone directly in front of the speakers. The energy is instead directed to the sides, where constructive interference occurs.

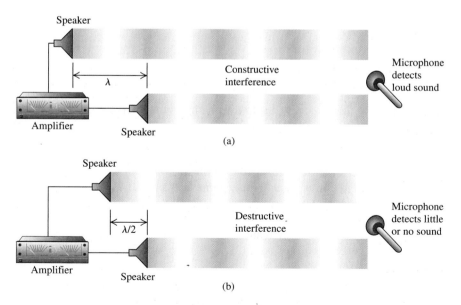

16.23 Two speakers driven by the same amplifier, emitting waves in phase. Only the waves directed toward the microphone are shown, and they are separated for clarity. (a) Constructive interference occurs when the path difference is $0, \lambda, 2\lambda, 3\lambda, \ldots$. (b) Destructive interference occurs when the path difference is $\lambda/2, 3\lambda/2, 5\lambda/2, \ldots$

<table>
<tr><td>

Example
16.15
</td><td>

Loudspeaker interference
</td></tr>
</table>

Two small loudspeakers, A and B (Fig. 16.24), are driven by the same amplifier and emit pure sinusoidal waves in phase. If the speed of sound is 350 m/s, a) for what frequencies does constructive interference occur at point P? b) for what frequencies does destructive interference occur at point P?

SOLUTION

IDENTIFY: The nature of the interference at P depends on the difference in path lengths from points A and B to P and how this difference compares to the wavelength.

SET UP: We calculate the path lengths from A to P and from B to P using the Pythagorean theorem. Constructive interference occurs when the difference in path lengths equals a whole number of wavelengths, while destructive interference occurs when the path

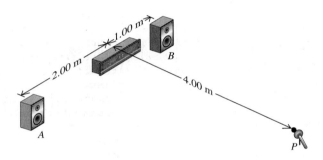

16.24 What sort of interference occurs at P?

length difference is a half-integer number of wavelengths. To find the corresponding frequencies, we use the relationship $v = f\lambda$.

EXECUTE: The distance from speaker A to point P is $[(2.00\text{ m})^2 + (4.00\text{ m})^2]^{1/2} = 4.47$ m, and the distance from speaker B to point P is $[(1.00\text{ m})^2 + (4.00\text{ m})^2]^{1/2} = 4.12$ m. The path difference is $d = 4.47\text{ m} - 4.12\text{ m} = 0.35$ m. a) Constructive interference occurs when the path difference is $d = 0, \lambda, 2\lambda, \ldots$ or $d = 0, v/f, 2v/f, \ldots = nv/f$. So the possible frequencies are

$$f_n = \frac{nv}{d} = n\frac{350\text{ m/s}}{0.35\text{ m}} \qquad (n = 1, 2, 3, \ldots)$$
$$= 1000\text{ Hz}, 2000\text{ Hz}, 3000\text{ Hz}, \ldots$$

b) Destructive interference occurs when the path difference is $d = \lambda/2, 3\lambda/2, 5\lambda/2, \ldots$ or $d = v/2f, 3v/2f, 5v/2f, \ldots$. The possible frequencies are

$$f_n = \frac{nv}{2d} = n\frac{350\text{ m/s}}{2(0.35\text{ m})} \qquad (n = 1, 2, 3, \ldots)$$
$$= 500\text{ Hz}, 1500\text{ Hz}, 2500\text{ Hz}, \ldots$$

EVALUATE: As we increase the frequency, the sound at point P alternates between large and small amplitudes; the maxima and minima occur at the frequencies we have found. It can be hard to notice this effect in an ordinary room because of multiple reflections from the walls, floor, and ceiling. Such an experiment is best done either outdoors or in an anechoic chamber, which has walls that absorb almost all sound and thereby eliminate reflections.

Experiments closely analogous to the one in Example 16.15, but using light, have provided both strong evidence for the wave nature of light and a means of measuring its wavelengths. We will discuss these experiments in detail in Chapter 35.

Interference effects are used to control noise from very loud sound sources such as gas-turbine power plants or jet engine test cells. The idea is to use additional sound sources that in some regions of space interfere destructively with the unwanted sound and cancel it out. Microphones in the controlled area feed signals back to the sound sources, which are continuously adjusted for optimum cancellation of noise in the controlled area.

Test Your Understanding

Suppose the frequency from the two loudspeakers in Fig. 16.22 is 250 Hz, so that there is neither constructive nor destructive interference at P. What will the microphone detect if it is moved toward the amplifier along the line in Fig. 16.24 labeled 4.00 m?

16.7 | Beats

In Section 16.6 we talked about *interference* effects that occur when two different waves with the same frequency overlap in the same region of space. Now let's look at what happens when we have two waves with equal amplitude but slightly different frequencies. This occurs, for example, when two tuning forks with slightly different frequencies are sounded together, or when two organ pipes that are supposed to have exactly the same frequency are slightly "out of tune."

Consider a particular point in space where the two waves overlap. The displacements of the individual waves at this point are plotted as functions of time in Fig. 16.25a. The total length of the time axis represents one second, and the frequencies are 16 Hz (blue graph) and 18 Hz (red graph). Applying the principle of superposition, we add the two displacements at each instant of time to find the total displacement at that time. The result is the graph of Fig. 16.25b. At certain times the two waves are in phase; their maxima coincide and their amplitudes add. But because of their slightly different frequencies, the two waves cannot be in phase at all times. Indeed, at certain times (like $t = 0.50$ s in Fig. 16.25) the two waves are exactly *out* of phase. The two waves then cancel each other, and the total amplitude is zero.

The resultant wave in Fig. 16.25b looks like a single sinusoidal wave with a varying amplitude that goes from a maximum to zero and back. In this example the amplitude goes through two maxima and two minima in one second, so the frequency of this amplitude variation is 2 Hz. The amplitude variation causes variations of loudness called **beats,** and the frequency with which the loudness varies is called the **beat frequency.** In this example the beat frequency is the *difference* of the two frequencies. If the beat frequency is a few hertz, we hear it as a waver or pulsation in the tone.

We can prove that the beat frequency is *always* the difference of the two frequencies f_a and f_b. Suppose f_a is larger than f_b; the corresponding periods are T_a and T_b, with $T_a < T_b$. If the two waves start out in phase at time $t = 0$, they are again in phase when the first wave has gone through exactly one more cycle than the second. This happens at a value of t equal to T_{beat}, the *period* of the beat. Let n be the number of cycles of the first wave in time T_{beat}; then the number of cycles of the second wave in the same time is $(n - 1)$, and we have the relations

$$T_{\text{beat}} = nT_a \quad \text{and} \quad T_{\text{beat}} = (n - 1)T_b$$

Eliminating n between these two equations, we find

$$T_{\text{beat}} = \frac{T_a T_b}{T_b - T_a}$$

Act**i**v
ONLINE
Phys**i**cs

10.7 Beats and Beat Frequency

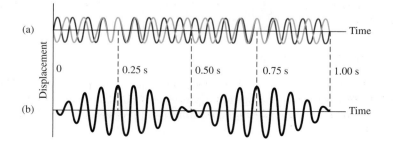

16.25 Beats are fluctuations in amplitude produced by two sound waves of slightly different frequency, 16 Hz and 18 Hz. (a) Individual waves. (b) Resultant wave formed by superposition of the two waves. The beat frequency is 18 Hz − 16 Hz = 2 Hz.

The reciprocal of the beat period is the beat *frequency,* $f_{\text{beat}} = 1/T_{\text{beat}}$, so

$$f_{\text{beat}} = \frac{T_b - T_a}{T_a T_b} = \frac{1}{T_a} - \frac{1}{T_b}$$

and finally

$$f_{\text{beat}} = f_a - f_b \qquad (\text{beat frequency}) \qquad (16.24)$$

As claimed, the beat frequency is the difference of the two frequencies. In using Eq. (16.24), remember that f_a is the higher frequency.

An alternative way to derive Eq. (16.24) is to write functions to describe the curves in Fig. 16.25a and then add them. Suppose that at a certain position the two waves are given by $y_a(t) = A \sin 2\pi f_a t$ and $y_b(t) = -A \sin 2\pi f_b t$. We use the trigonometric identity

$$\sin a - \sin b = 2 \sin \frac{1}{2}(a - b) \cos \frac{1}{2}(a + b)$$

We can then express the total wave $y(t) = y_a(t) + y_b(t)$ as

$$y_a(t) + y_b(t) = \left[2A \sin \frac{1}{2}(2\pi)(f_a - f_b)t \right] \cos \frac{1}{2}(2\pi)(f_a + f_b)t$$

The amplitude factor (the quantity in brackets) varies slowly with frequency $\frac{1}{2}(f_a - f_b)$. The cosine factor varies with a frequency equal to the *average* frequency $\frac{1}{2}(f_a + f_b)$. The *square* of the amplitude factor, which is proportional to the intensity that the ear hears, goes through two maxima and two minima per cycle. So the beat frequency f_{beat} that is heard is twice the quantity $\frac{1}{2}(f_a - f_b)$, or just $f_a - f_b$, in agreement with Eq. (16.24).

Beats between two tones can be heard up to a beat frequency of about 6 or 7 Hz. Two piano strings or two organ pipes differing in frequency by 2 or 3 Hz sound wavery and "out of tune," although some organ stops contain two sets of pipes deliberately tuned to beat frequencies of about 1 to 2 Hz for a gently undulating effect. Listening for beats is an important technique in tuning all musical instruments.

At frequency differences greater than about 6 or 7 Hz, we no longer hear individual beats, and the sensation merges into one of *consonance* or *dissonance,* depending on the frequency ratio of the two tones. In some cases the ear perceives a tone called a *difference tone,* with a pitch equal to the beat frequency of the two tones. For example, if you listen to a whistle that produces sounds at 1800 Hz and 1900 Hz when blown, you will hear not only these tones but also a much lower 100-Hz tone.

The engines on multiengine propeller aircraft have to be synchronized so that the sounds don't cause annoying beats, which are heard as loud throbbing sounds (Fig. 16.26). On some planes this is done electronically; on others the pilot does it by ear, just like tuning a piano.

Test Your Understanding

16.26 If the two propellers on this airplane are not precisely synchronized, the pilots, passengers, and listeners on the ground will hear beats.

One tuning fork vibrates at 440 Hz, while a second tuning fork vibrates at an unknown frequency. When both tuning forks are sounded simultaneously, you hear a tone that rises and falls in intensity three times per second. What do you conclude about the frequency of the second tuning fork?

16.8 | The Doppler Effect

You've probably noticed that when a car approaches you with its horn sounding, the pitch seems to drop as the car passes. This phenomenon, first described by the 19th-century Austrian scientist Christian Doppler, is called the **Doppler effect.** When a source of sound and a listener are in motion relative to each other, the frequency of the sound heard by the listener is not the same as the source frequency. A similar effect occurs for light and radio waves; we'll return to this later in this section.

To analyze the Doppler effect for sound, we'll work out a relation between the frequency shift and the velocities of source and listener relative to the medium (usually air) through which the sound waves propagate. To keep things simple, we consider only the special case in which the velocities of both source and listener lie along the line joining them. Let v_S and v_L be the velocity components along this line for the source and the listener, respectively, relative to the medium. We choose the positive direction for both v_S and v_L to be the direction from the listener L to the source S. The speed of sound relative to the medium, v, is always considered positive.

Moving Listener

Let's think first about a listener L moving with velocity v_L toward a stationary source S (Fig. 16.27). The source emits a sound wave with frequency f_S and wavelength $\lambda = v/f_S$. The figure shows several wave crests, separated by equal distances λ. The wave crests approaching the moving listener have a speed of propagation *relative to the listener* of $(v + v_L)$. So the frequency f_L with which the crests arrive at the listener's position (that is, the frequency the listener hears) is

$$f_L = \frac{v + v_L}{\lambda} = \frac{v + v_L}{v/f_S} \qquad (16.25)$$

or

$$f_L = \left(\frac{v + v_L}{v}\right)f_S = \left(1 + \frac{v_L}{v}\right)f_S \qquad (16.26)$$

(moving listener, stationary source)

So a listener moving toward a source $(v_L > 0)$, as in Fig. 16.27, hears a higher frequency (higher pitch) than does a stationary listener. A listener moving away from the source $(v_L < 0)$ hears a lower frequency (lower pitch).

16.27 A listener moving toward a stationary source hears a frequency that is higher than the source frequency because the relative speed of listener and wave is greater than the wave speed v.

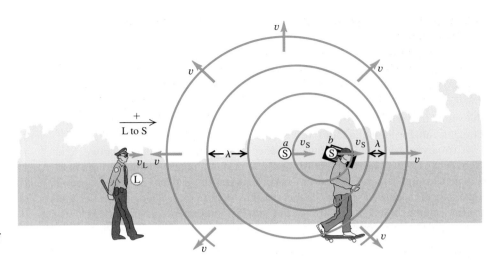

16.28 Wave crests emitted by a moving source are crowded together in front of the source (to the right of this source) and stretched out behind it (to the left of this source).

16.29 The Doppler effect explains why the siren on a fire engine or ambulance has a high pitch $(f_L > f_S)$ when it is approaching you $(v_S < 0)$ and a low pitch $(f_L < f_S)$ when it is moving away $(v_S > 0)$.

Moving Source and Moving Listener

Now suppose the source is also moving, with velocity v_S (Fig. 16.28). The wave speed relative to the wave medium (air) is still v; it is determined by the properties of the medium and is not changed by the motion of the source. But the wavelength is no longer equal to v/f_S. Here's why. The time for emission of one cycle of the wave is the period $T = 1/f_S$. During this time, the wave travels a distance $vT = v/f_S$ and the source moves a distance $v_S T = v_S/f_S$. The wavelength is the distance between successive wave crests, and this is determined by the *relative* displacement of source and wave. As Fig. 16.28 shows, this is different in front of and behind the source. In the region to the right of the source in Fig. 16.28 (that is, in front of the source), the wavelength is

$$\lambda = \frac{v}{f_S} - \frac{v_S}{f_S} = \frac{v - v_S}{f_S} \qquad (16.27)$$

(wavelength in front of a moving source)

In the region to the left of the source (that is, behind the source), it is

$$\lambda = \frac{v + v_S}{f_S} \qquad (\text{wavelength behind a moving source}) \qquad (16.28)$$

The waves in front of and behind the source are compressed and stretched out, respectively, by the motion of the source (Fig. 16.29).

To find the frequency heard by the listener behind the source, we substitute Eq. (16.28) into the first form of Eq. (16.25):

$$f_L = \frac{v + v_L}{\lambda} = \frac{v + v_L}{(v + v_S)/f_S}$$

$$f_L = \frac{v + v_L}{v + v_S} f_S \qquad (16.29)$$

(Doppler effect, moving source and moving listener)

This expresses the frequency f_L heard by the listener in terms of the frequency f_S of the source.

Equation (16.29) includes all possibilities for motion of source and listener (relative to the medium) along the line joining them. If the listener happens to be at rest in the medium, v_L is zero. When both source and listener are at rest or have the same velocity relative to the medium, then $v_L = v_S$ and $f_L = f_S$. Whenever the direction of the source or listener velocity is opposite to the direction from the listener toward the source (which we have defined as positive), the corresponding velocity to be used in Eq. (16.29) is negative.

Problem-Solving Strategy

Doppler Effect

IDENTIFY *the relevant concepts:* The Doppler effect is relevant whenever the source of waves, the wave detector (listener), or both are in motion.

SET UP *the problem using the following steps:*
1. Establish a coordinate system. Define the positive direction to be the direction from the listener to the source, and make sure you know the signs of all relevant velocities. A velocity in the direction from the listener toward the source is positive; a velocity in the opposite direction is negative. Also, the velocities must all be measured relative to the air in which the sound is traveling.
2. Use consistent notation to identify the various quantities: subscript S for source, L for listener.
3. Determine which unknown quantities are your target variables.

EXECUTE *the solution:*
1. Use Eq. (16.29) to relate the frequencies at the source and the listener, the sound speed, and the velocities of the source and the listener. If the source is moving, you can find the wavelength measured by the listener using Eq. (16.27) or (16.28).
2. When a wave is reflected from a surface, either stationary or moving, the analysis can be carried out in two steps. In the first, the surface plays the role of listener; the frequency with which the wave crests arrive at the surface is f_L. Then think of the surface as a new source, emitting waves with this same frequency f_L. Finally, determine what frequency is heard by a listener detecting this new wave.

EVALUATE *your answer:* Ask whether your final result makes sense. If the source and the listener are moving toward each other, $f_L > f_S$; if they are moving apart, $f_L < f_S$. If the source and the listener have no relative motion, $f_L = f_S$.

Example 16.16

Doppler effect I: Wavelengths

A police siren emits a sinusoidal wave with frequency $f_S = 300$ Hz. The speed of sound is 340 m/s. a) Find the wavelength of the waves if the siren is at rest in the air. b) If the siren is moving at 30 m/s (108 km/h, or 67 mi/h), find the wavelengths of the waves ahead of and behind the source.

SOLUTION

IDENTIFY: The Doppler effect is not involved in part (a), since neither the source nor the listener is moving. In part (b), the source is in motion and we must invoke the Doppler effect.

SET UP: We use the relationship $v = \lambda f$ to determine the wavelength when the police siren is at rest. When it is in motion, we find

the wavelength on either side of the siren using Eqs. (16.27) and (16.28).

16.30 Wavelengths ahead of and behind the police siren when the siren is moving through the air at 30 m/s.

EXECUTE: a) When the source is at rest,

$$\lambda = \frac{v}{f_S} = \frac{340 \text{ m/s}}{300 \text{ Hz}} = 1.13 \text{ m}$$

b) The situation is shown in Fig. 16.30. From Eq. (16.27), in front of the siren,

$$\lambda = \frac{v - v_S}{f_S} = \frac{340 \text{ m/s} - 30 \text{ m/s}}{300 \text{ Hz}} = 1.03 \text{ m}$$

From Eq. (16.28), behind the siren,

$$\lambda = \frac{v + v_S}{f_S} = \frac{340 \text{ m/s} + 30 \text{ m/s}}{300 \text{ Hz}} = 1.23 \text{ m}$$

EVALUATE: The wavelength is less in front of the siren and greater behind the siren, as it should be.

Example
16.17   **Doppler effect II: Frequencies**

If a listener L is at rest and the siren in Example 16.16 is moving away from L at 30 m/s (Fig. 16.31), what frequency does the listener hear?

SOLUTION

IDENTIFY and SET UP: Our target variable is the listener's frequency f_L. We know $f_S = 300$ Hz from Example 16.16, and we have $v_L = 0$ and $v_S = 30$ m/s. (The source velocity v_S is positive because the siren is moving in the same direction as the direction from listener to source.)

EXECUTE: From Eq. (16.29),

$$f_L = \frac{v}{v + v_S} f_S = \frac{340 \text{ m/s}}{340 \text{ m/s} + 30 \text{ m/s}} (300 \text{ Hz}) = 276 \text{ Hz}$$

EVALUATE: The source and listener are moving apart, so the frequency f_L heard by the listener is less than the frequency f_S emitted by the source.

Here's an alternative approach we can use to check our result. From Example 16.16, the wavelength behind the source (which is where the listener in Fig. 16.31 is located) is 1.23 m, so

$$f_L = \frac{v}{\lambda} = \frac{340 \text{ m/s}}{1.23 \text{ m}} = 276 \text{ Hz}$$

Even though the source is moving, the wave speed v relative to the stationary listener is unchanged.

16.31 The listener is at rest, and the siren moves away from the listener at 30 m/s.

Example
16.18   **Doppler effect III: A moving listener**

If the siren is at rest and the listener is moving toward the left at 30 m/s (Fig. 16.32), what frequency does the listener hear?

16.32 The siren is at rest, and the listener moves away from it at 30 m/s.

SOLUTION

IDENTIFY and SET UP: The key difference between this example and Example 16.17 is that the source is at rest (so $v_S = 0$) and the listener is in motion. The positive direction (from listener to source) is still from left to right, so $v_L = -30$ m/s.

EXECUTE: From Eq. (16.29),

$$f_L = \frac{v + v_L}{v} f_S = \frac{340 \text{ m/s} + (-30 \text{ m/s})}{340 \text{ m/s}} (300 \text{ Hz}) = 274 \text{ Hz}$$

EVALUATE: Again the frequency heard by the listener is less than the source frequency. Note that the *relative velocity* of source and listener is the same as in the previous example, but the Doppler shift is different because the velocities relative to the *air* are different.

| Example 16.19 | **Doppler effect IV: Moving source, moving listener** |

If the siren is moving away from the listener with a speed of 45 m/s relative to the air and the listener is moving toward the siren with a speed of 15 m/s relative to the air (Fig. 16.33), what frequency does the listener hear?

SOLUTION

IDENTIFY and SET UP: Now *both* the listener and the source are in motion, with $v_L = 15$ m/s and $v_S = 45$ m/s. (Both velocities are positive because both velocity vectors point in the direction from listener to source.)

EXECUTE: Once again using Eq. (16.29), we find

$$f_L = \frac{v + v_L}{v + v_S} f_S = \frac{340 \text{ m/s} + 15 \text{ m/s}}{340 \text{ m/s} + 45 \text{ m/s}} (300 \text{ Hz})$$
$$= 277 \text{ Hz}$$

EVALUATE: The frequency heard by the listener is again less than the source frequency, but the value is different than in the previous two examples, even though the source and listener move away from each other at 30 m/s in all three cases. The *sign* of the Doppler shift of frequency (that is, whether f_L is less than or greater than f_S)

depends on how the source and the listener are moving relative to each other; to determine the *value* of the Doppler shift of frequency, you must know the velocities of source and listener relative to the air.

16.33 The velocities relative to the air when the siren is moving away from the listener with a relative velocity of 30 m/s.

| Example 16.20 | **Doppler effect V: A double Doppler shift** |

The police car with its 300-Hz siren is moving toward a warehouse at 30 m/s, intending to crash through the door. What frequency does the driver of the police car hear reflected from the warehouse?

SOLUTION

IDENTIFY: In this situation there are *two* Doppler shifts, as shown in Fig. 16.34. In the first shift, the warehouse is the stationary "listener." The frequency of sound reaching the warehouse, which we call f_W, is greater than 300 Hz because the source is approaching. In the second shift, the warehouse acts as a source of sound with frequency f_W, and the listener is the driver of the police car; she hears a frequency greater than f_W because she is approaching the source.

SET UP: To determine f_W, we use Eq. (16.29) with f_L replaced by f_W. For this part of the problem, $v_L = v_W = 0$ (the warehouse is at rest) and $v_S = -30$ m/s (the siren is moving in the negative direction from source to listener).

To determine the frequency heard by the driver, which is our target variable, we again use Eq. (16.29) but now with f_S replaced by f_W. For this second part of the problem, $v_S = 0$ because the stationary warehouse is the source and the velocity of the listener (the driver) is $v_L = +30$ m/s. (The listener's velocity is positive because it is in the direction from listener to source.)

(a)

(b)

16.34 (a) The police car with siren (source) moves at 30 m/s toward the stationary warehouse (listener). (b) The sound wave is reflected from the stationary warehouse (source) toward the police car (listener).

EXECUTE: The frequency reaching the warehouse is

$$f_W = \frac{v}{v + v_S} f_S = \frac{340 \text{ m/s}}{340 \text{ m/s} + (-30 \text{ m/s})} (300 \text{ Hz}) = 329 \text{ Hz}$$

Then the frequency heard by the driver is

$$f_L = \frac{v + v_L}{v} f_W = \frac{340 \text{ m/s} + 30 \text{ m/s}}{340 \text{ m/s}} (329 \text{ Hz}) = 358 \text{ Hz}$$

EVALUATE: Because there are two Doppler shifts, the reflected sound heard by the driver has an even higher frequency than the sound heard by a stationary listener in the warehouse.

Doppler Effect for Electromagnetic Waves

In the Doppler effect for sound, the velocities v_L and v_S are always measured relative to the *air* or whatever medium we are considering. There is also a Doppler effect for *electromagnetic* waves in empty space, such as light waves or radio waves. In this case there is no medium that we can use as a reference to measure velocities, and all that matters is the *relative* velocity of source and receiver. (By contrast, the Doppler effect for sound does not depend simply on this relative velocity, as discussed in Example 16.19.)

To derive the expression for the Doppler frequency shift for light, we have to use the special theory of relativity. We will discuss this in Chapter 37, but for now we quote the result without derivation. The wave speed is the speed of light, usually denoted by c, and it is the same for both source and receiver. In the frame of reference in which the receiver is at rest, the source is moving away from the receiver with velocity v. (If the source is *approaching* the receiver, v is negative.) The source frequency is again f_S. The frequency f_R measured by the receiver R (the frequency of arrival of the waves at the receiver) is then given by

$$f_R = \sqrt{\frac{c - v}{c + v}} f_S \qquad \text{(Doppler effect for light)} \qquad (16.30)$$

When v is positive, the source is moving directly *away* from the receiver and f_R is always less than f_S; when v is negative, the source is moving directly *toward* the receiver and f_R is *greater* than f_S. The qualitative effect is the same as for sound, but the quantitative relationship is different.

A familiar application of the Doppler effect for radio waves is the radar device mounted on the side window of a police car to check other cars' speeds. The electromagnetic wave emitted by the device is reflected from a moving car, which acts as a moving source, and the wave reflected back to the device is Doppler-shifted in frequency. The transmitted and reflected signals are combined to produce beats, and the speed can be computed from the frequency of the beats. Similar techniques ("Doppler radar") are used to measure wind velocities in the atmosphere.

The Doppler effect is also used to track satellites and other space vehicles. In Fig. 16.35 a satellite emits a radio signal with constant frequency f_S. As the satellite orbits past, it first approaches and then moves away from the receiver; the frequency f_R of the signal received on earth changes from a value greater than f_S to a value less than f_S as the satellite passes overhead.

The Doppler effect for electromagnetic waves, including visible light, is important in astronomy. Astronomers compare wavelengths of light from distant stars to those emitted by the same elements on earth. For example, in a binary star system, in which two stars orbit about their common center of mass, the light is

16.35 Change of velocity component along the line of sight of a satellite passing a tracking station. The frequency received at the tracking station changes from high to low as the satellite passes overhead.

Doppler-shifted to higher frequencies when a star is moving toward an observer on earth and to lower frequencies when it's moving away. Measurements of the frequency shifts reveal information about the orbits and masses of the stars that comprise the binary system.

Light from most galaxies is shifted toward the longer-wavelength or red end of the visible spectrum, an effect called the *red shift*. This is often described as a Doppler shift resulting from motion of these galaxies away from us. However, from the point of view of the general theory of relativity, it is something much more fundamental: it is associated with the expansion of space itself. Distant galaxies have large red shifts because their light has been in transit for a long time and has shared in the expansion of all the space through which it moved. Extrapolating this expansion backward to over 10^{10} years ago leads to the "big bang" picture. From this point of view, the big bang was not an explosion in space but the initial rapid expansion of space itself.

Test Your Understanding

You are at an outdoor concert with a wind blowing at 10 m/s from the performers toward you. What is the value of v_S? Of v_L? Is the sound you hear Doppler-shifted? If so, is it shifted to lower or higher frequencies?

*16.9 | Shock Waves

You may have experienced "sonic booms" caused by an airplane flying overhead faster than the speed of sound. We can see qualitatively why this happens from Fig. 16.36. Let v_S denote the *speed* of the airplane relative to the air, so that it is always positive. The motion of the airplane through the air produces sound; if v_S is less than the speed of sound v, the waves in front of the airplane are crowded together with a wavelength given by Eq. (16.27):

$$\lambda = \frac{v - v_S}{f_S}$$

As the speed v_S of the airplane approaches the speed of sound v, the wavelength approaches zero and the wave crests pile up on each other (Fig. 16.36a). The airplane must exert a large force to compress the air in front of it; by Newton's third law, the air exerts an equally large force back on the airplane. Hence there is a large increase in aerodynamic drag (air resistance) as the airplane approaches the speed of sound, a phenomenon known as the "sound barrier."

When v_S is greater in magnitude than v, the source of sound is **supersonic,** and Eqs. (16.27) and (16.29) for the Doppler effect no longer describe the sound wave in front of the source. Figure 16.36b shows a cross section of what happens. As the airplane moves, it displaces the surrounding air and produces sound. A series of wave crests is emitted from the nose of the airplane; each spreads out in a circle centered at the position of the airplane when it emitted the crest. After a time t the crest emitted from point S_1 has spread to a circle with radius vt, and the airplane has moved a greater distance v_St to position S_2. You can see that the circular crests interfere constructively at points along the green line that makes an angle α with the direction of the airplane velocity, leading to a very-large-amplitude wave crest along this line. This large-amplitude crest is called a **shock wave** (Fig. 16.36c).

(a)

(b)

(c)

16.36 (a) As the speed of the source of sound S approaches the speed of sound, the wave crests begin to pile up in front of S. (b) A shock wave forms when the speed of the source is greater than the speed of sound. (c) Photograph of shock waves produced by a T-38 jet aircraft moving at 1.1 times the speed of sound. Separate shock waves are generated by the nose, wings, and tail. The angles of these waves vary because the air is accelerated and decelerated as it moves relative to the airplane, so the relative speed of the airplane and air is different at different points on the airplane.

16.37 The first supersonic airplane, the Bell X-1, was shaped much like a .50-caliber bullet—which was known to be able to travel faster than sound.

From the right triangle in Fig. 16.36b we can see that the angle α is given by

$$\sin \alpha = \frac{vt}{v_S t} = \frac{v}{v_S} \qquad (\text{shock wave}) \qquad (16.31)$$

In this relation, v_S is the *speed* of the source (the magnitude of its velocity) relative to the air and is always positive. The ratio v_S/v is called the **Mach number.** It is greater than unity for all supersonic speeds, and $\sin \alpha$ in Eq. (16.31) is the reciprocal of the Mach number. The first person to break the sound barrier was Capt. Chuck Yeager of the U.S. Air Force, flying the Bell X-1 at Mach 1.06 on October 14, 1947 (Fig. 16.37).

The actual situation is three-dimensional; the shock wave forms a *cone* around the direction of motion of the source. If the source (possibly a supersonic jet airplane or a rifle bullet) moves with constant velocity, the angle α is constant, and the shock-wave cone moves along with the source. It's the arrival of this shock wave that causes the sonic boom you hear after a supersonic airplane has passed by. The larger the airplane, the stronger the sonic boom; the shock wave produced at ground level by the (now retired) Concorde supersonic airliner flying at 12,000 m (40,000 ft) causes a sudden jump in air pressure of about 20 Pa. In front of the shock-wave cone, there is no sound. Inside the cone a stationary listener hears the Doppler-shifted sound of the airplane moving away.

CAUTION We emphasize that a shock wave is produced *continuously* by any object that moves through the air at supersonic speed, not only at the instant that it "breaks the sound barrier." The sound waves that combine to form the shock wave, as in Fig. 16.36b, are created by the motion of the object itself, not by any sound source that the object may carry. The cracking noises of a bullet and of the tip of a circus whip are due to their supersonic motion. A supersonic jet airplane may have very loud engines, but these do not cause the shock wave. Indeed, a space shuttle makes a very loud sonic boom when coming in for a landing; its engines are out of fuel at this point, so it is a supersonic glider.

Shock waves have applications outside of aviation. They are used to break up kidney stones and gallstones without invasive surgery, using a technique with the impressive name *extracorporeal shock-wave lithotripsy*. A shock wave produced outside the body is focused by a reflector or acoustic lens so that as much of it as possible converges on the stone. When the resulting stresses in the stone exceed its tensile strength, it breaks into small pieces and can be eliminated. This technique requires accurate determination of the location of the stone, which may be done using ultrasonic imaging techniques (see the discussion of ultrasonic imaging on page 600).

Example 16.21 | **Sonic boom of the Concorde**

The Concorde is flying at Mach 1.75 at an altitude of 8000 m, where the speed of sound is 320 m/s. How long after the plane passes directly overhead will you hear the sonic boom?

SOLUTION

IDENTIFY: The shock wave forms a cone trailing backward from the airplane, so the problem is really asking for how much time elapses from when the Concorde flies overhead to when the shock wave reaches you.

SET UP: Figure 16.38 shows the situation just as the shock wave reaches you at point L. A time t (our target variable) has elapsed since the Concorde passed overhead, during which the airliner flying at speed v_S has traveled a distance $v_S t$. We use trigonometry to solve for t.

EXECUTE: From Eq. (16.31) the angle α of the shock cone is

$$\alpha = \arcsin \frac{1}{1.75} = 34.8°$$

The speed of the plane is the speed of sound multiplied by the Mach number:

$$v_S = (1.75)(320 \text{ m/s}) = 560 \text{ m/s}$$

From Fig. 16.38 we have

$$\tan \alpha = \frac{8000 \text{ m}}{v_S t}$$

$$t = \frac{8000 \text{ m}}{(560 \text{ m/s})(\tan 34.8°)} = 20.5 \text{ s}$$

v_S = Mach 1.75

v_S

8000 m

Shock
wave

L

α

$v_S t$

EVALUATE: You hear the boom 20.5 s after the Concorde passes overhead, and at that time it has traveled $(560 \text{ m/s})(20.5 \text{ s}) = 11.5$ km past the straight-overhead point.

In this calculation we assumed that the speed of sound is the same at all altitudes, so $\alpha = \arcsin v/v_S$ is a constant and the shock wave forms a perfect cone. In fact, the speed of sound decreases with increasing altitude. How would this affect the result?

16.38 You hear a sonic boom when the shock wave reaches you at L (*not* just when the plane breaks the sound barrier). A listener to the right of L has not yet heard the sonic boom but will shortly; a listener to the left of L has already heard the sonic boom and now hears the Doppler-shifted sound of the airplane.

Test Your Understanding

Large meteors have been known to produce sonic booms as they descend at supersonic speeds through the earth's atmosphere. If the shock wave from such a meteor has angle $\alpha = 4.0°$, what is the Mach number of the meteor?

SUMMARY

Sound consists of longitudinal waves in a medium. A sinusoidal sound wave is characterized by its frequency f and wavelength λ (or angular frequency ω and wave number k) and by its displacement amplitude A. The pressure amplitude $p_{\max}$ is directly proportional to the displacement amplitude; the proportionality constant is the product of the wave number and the bulk modulus B of the wave medium. (See Examples 16.1 and 16.2)

$$p_{\max} = BkA \qquad (16.5)$$
(sinusoidal sound wave)

All particles in fluid oscillate in SHM with same amplitude and period

---

The loudness of a sound depends on its amplitude and frequency, while the pitch depends primarily on its frequency. The tone quality or timbre depends on the harmonic content and the attack and decay characteristics.

The speed of a longitudinal (sound) wave in a fluid depends on the bulk modulus B and density ρ. If the fluid is an ideal gas, the speed can be expressed in terms of the temperature T, molar mass M, and ratio of heat capacities γ of the gas. The speed of longitudinal waves in a solid rod depends on the density and Young's modulus Y. (See Examples 16.3 through 16.5)

$$v = \sqrt{\frac{B}{\rho}} \quad \text{(longitudinal wave in a fluid)} \qquad (16.7)$$

$$v = \sqrt{\frac{\gamma RT}{M}} \quad \text{(sound wave in an ideal gas)} \qquad (16.7)$$

$$v = \sqrt{\frac{Y}{\rho}} \quad \text{(longitudinal wave in a solid rod)} \qquad (16.8)$$

Moving   At rest

The intensity I of a sound wave is the time average rate at which energy is transported by the wave, per unit area. For a sinusoidal wave, the intensity can be expressed in terms of the displacement amplitude A or the pressure amplitude $p_{\max}$. (See Examples 16.6 through 16.9)

$$I = \frac{1}{2}\sqrt{\rho B}\,\omega^2 A^2 = \frac{p_{\max}^2}{2\rho v}$$

$$= \frac{p_{\max}^2}{2\sqrt{\rho B}} \qquad (16.12),\ (16.14)$$

(intensity of a sinusoidal sound wave)

The sound intensity level β of a sound wave is a logarithmic measure of its intensity. It is measured relative to I_0, an arbitrary intensity defined to be 10^{-12} W/m^2. Sound intensity levels are expressed in decibels (dB). (See Examples 16.10 and 16.11)

$$\beta = (10 \text{ dB}) \log \frac{I}{I_0}$$

(definition of sound intensity level)   (16.15)

Standing sound waves can be set up in a pipe or tube. A closed end is a displacement node and a pressure antinode; an open end is a displacement antinode and a pressure node. For a pipe of length L open at both ends, the normal-mode frequencies are integer multiples of the sound speed divided by $2L$. For a stopped pipe (one that is open at only one end), the normal-mode frequencies are the odd multiples of the sound speed divided by $4L$. (See Examples 16.12 and 16.13)

$$f_n = \frac{nv}{2L} \quad (n = 1, 2, 3, \ldots) \qquad (16.18)$$
(open pipe)

$$f_n = \frac{nv}{4L} \quad (n = 1, 3, 5, \ldots) \qquad (16.22)$$
(stopped pipe)

A system with normal modes of oscillation can be driven to oscillate at any frequency. A maximum response, or resonance, occurs if the driving frequency is close to one of the normal-mode frequencies of the system. (See Example 16.14)

When two or more waves overlap in the same region of space, the resulting effects are called interference. The resulting amplitude can be either larger or smaller than the amplitude of each individual wave, depending on whether the waves are in phase (constructive interference) or out of phase (destructive interference). (See Example 16.15)

Beats are heard when two tones with slightly different frequencies f_a and f_b are sounded together. The beat frequency f_{beat} is the difference between f_a and f_b.

$$f_{beat} = f_a - f_b \quad (16.24)$$
(beat frequency)

The Doppler effect for sound is the frequency shift that occurs when there is motion of a source of sound, a listener, or both, relative to the medium. The source and listener frequencies f_S and f_L are related by the source and listener velocities v_S and v_L relative to the medium and to the speed of sound v. (See Examples 16.16 through 16.20)

$$f_L = \frac{v + v_L}{v + v_S} f_S \quad (16.29)$$
(Doppler effect, moving source and moving listener)

A sound source moving with a speed v_S greater than the speed of sound v creates a shock wave. The wave front is a cone with angle α. (See Example 16.21)

$$\sin \alpha = \frac{v}{v_S} \quad (16.31)$$
(shock wave)

Key Terms

Your Notes

Answer to Chapter Opening Question

Both musical sounds and noise are made up of a combination of sinusoidal sound waves. The difference is that the frequencies of the sine waves in musical sound are all integer multiples of a fundamental frequency, while *all* frequencies are present in noise.

Answers to Test Your Understanding Questions

Section 16.1 From Eq. (16.5), the displacement amplitude is $A = p_{max}/Bk$. The pressure amplitude p_{max} and bulk modulus B remain the same, but as the frequency f increases the wave number $k = \omega/v = 2\pi f/v$ increases as well. Since A is inversely proportional to k, the displacement amplitude decreases. In other words, at higher frequency a smaller maximum displacement is required to produce the same maximum pressure fluctuation.

Section 16.2 The steel rails are long rods, so the expression to use is Eq. (16.8). Young's modulus for steel (from Table 11.1) is $Y = 20 \times 10^{10}$ Pa and the density (from Table 14.1) is $\rho = 7.8 \times 10^3$ kg/m^3. Hence

$$v = \sqrt{\frac{Y}{\rho}} = \sqrt{\frac{20 \times 10^{10}\ \text{Pa}}{7.8 \times 10^3\ \text{kg/m}^3}} = 5.1 \times 10^3\ \text{m/s}$$

This is almost 15 times faster than the speed of sound in air, which is why the train can be heard first through the rails.

Section 16.3 Equations (16.9) and (16.10) show that the bulk modulus B and sound speed v remain the same because the physical properties of the air are unchanged. From Eqs. (16.12) and (16.14), the intensity is proportional to the square of the displacement amplitude or the square of the pressure amplitude. Hence doubling the intensity means that A and p_{max} both increase by a factor of $\sqrt{2}$. Example 16.11 shows that *multiplying* the intensity by a factor of $2(I_2/I_1 = 2)$ corresponds to *adding* to the sound intensity level by $(10\ \text{dB})\log(I_2/I_1) = (10\ \text{dB})\log 2 = 3.0$ dB.

Section 16.4 Helium is less dense and has a lower molar mass than air, so sound travels faster in helium than in air. The normal-mode frequencies for a pipe are proportional to the sound speed v, so the frequency and hence the pitch increase when the air in the pipe is replaced with helium.

Section 16.5 The highest peak in the resonance curve of Fig. 16.23b is at $f = 4f_1$, where $f_1 = v/2L$ is the fundamental frequency of the open pipe. Hence the frequency that produces the greatest response is $4v/2L = 2v/L = 2(345\ \text{m/s})/(0.600\ \text{m}) = 1150$ Hz.

Section 16.6 If $f = 250$ Hz, the wavelength is $\lambda = v/f = (350\ \text{m/s})/(250\ \text{Hz}) = 1.4$ m. As the microphone is moved toward the amplifier, the path difference increases from 0.35 m at P to 1.00 m at the amplifier. At a certain point along the way, the path difference will be equal to $0.70\ \text{m} = \lambda/2$ and destructive interference will occur. So as the microphone is moved inward, the sound intensity decreases to a minimum at the destructive-interference point, and then increases again as the microphone moves past that point.

Section 16.7 The beat frequency is 3 Hz, so the difference between the two tuning fork frequencies is also 3 Hz. Hence the second tuning fork vibrates at a frequency of either 443 Hz or 437 Hz. You can distinguish between the two possibilities by comparing the pitches of the two tuning forks sounded one at a time: the frequency is 437 Hz if the second tuning fork has a lower pitch and 443 Hz if it has a higher pitch.

Section 16.8 The air (the medium for sound waves) is moving from the source toward the listener. Hence, relative to the air, both the source and the listener are moving in the direction from listener to source. So both velocities are positive and $v_S = v_L = +10$ m/s. The equality of these two velocities means that the numerator and the denominator in Eq. (16.29) are the same, so $f_L = f_S$ and there is *no* Doppler shift.

Section 16.9 From Eq. (16.31), $\sin\alpha$ is equal to the ratio of the speed of sound to the speed of the source, which in turn is the reciprocal of the Mach number: $\sin\alpha = v/v_S = 1/M$. Hence $M = 1/(\sin 4.0°) = 14$; that is, the meteor is traveling at 14 times the speed of sound.

Discussion Questions

Q16.1 When sound travels from air into water, does the frequency of the wave change? The speed? The wavelength? Explain your reasoning.

Q16.2 The hero of a western movie listens for an oncoming train by putting his ear to the track. Why does this method give early warning of the approach of a train?

Q16.3 Does the pitch (or frequency) of an organ pipe increase or decrease with increasing temperature? Explain.

Q16.4 In most modern wind instruments the pitch is changed by using keys or valves to change the length of the vibrating air column. The bugle, however, has no valves or keys, yet it can play many notes. How is this possible? Are there restrictions on what notes a bugle can play?

Q16.5 Symphonic musicians always "warm up" their wind instruments by blowing them before a performance. What purpose does this serve?

Q16.6 When you inhale helium, your voice becomes high and squeaky. Why? (*Warning:* Inhaling too much helium can cause unconsciousness or death.)

Q16.7 Lane dividers on highways sometimes have regularly spaced ridges or ripples. When the tires of a moving car roll along such a divider, a musical note is produced. Why? Explain how this phenomenon could be used to measure the car's speed.

Q16.8 The tone quality of an acoustic guitar is different when the strings are plucked near the bridge (the lower end of the strings) than when they are plucked near the sound hole (close to the center of the strings). Why?

Q16.9 Which has a more direct influence on the loudness of a sound wave: the *displacement* amplitude or the *pressure* amplitude? Explain your reasoning.

Q16.10 If the pressure amplitude of a sound wave is halved, by what factor does the intensity of the wave decrease? By what factor must the pressure amplitude of a sound wave be increased in order to increase the intensity by a factor of 16? Explain.

Q16.11 Defend the following statement: "*Adding* a certain number of decibels to the sound intensity level corresponds to *multiplying* the intensity by a corresponding factor."

Q16.12 A small fraction of the energy in a sound wave is absorbed by the air through which the sound passes. How does this modify the inverse-square relationship between intensity and distance from the source? Explain your reasoning.

Q16.13 An organist in a cathedral plays a loud chord and then releases the keys. The sound persists for a few seconds and gradually dies away. Why does it persist? What happens to the sound energy when the sound dies away?

Q16.14 Two vibrating tuning forks have identical frequencies, but one is stationary and the other is mounted at the rim of a rotating platform. What does a listener hear? Explain.

Q16.15 A large church has part of the organ in the front of the church and part in the back. A person walking rapidly down the aisle while both segments are playing at once reports that the two segments sound out of tune. Why?

Q16.16 A sound source and a listener are both at rest on the earth, but a strong wind is blowing from the source toward the listener. Is there a Doppler effect? Why or why not?

Q16.17 Can you think of circumstances in which a Doppler effect would be observed for surface waves in water? For elastic waves propagating in a body of water deep below the surface? If so, describe the circumstances and explain your reasoning. If not, explain why not.

*Q16.18 Does an aircraft make a sonic boom only at the instant its speed exceeds Mach 1? Explain your reasoning.

*Q16.19 If you are riding in a supersonic aircraft, what do you hear? Explain your reasoning. In particular, do you hear a continuous sonic boom? Why or why not?

*Q16.20 The speed of sound is not the same at all altitudes, but decreases with increasing altitude. How does this affect the calculation in Example 16.21 (Section 16.9)? Is the shape of the shock wave still a cone? Explain your reasoning.

*Q16.21 A jet airplane is flying at a constant altitude at a steady speed v_S greater than the speed of sound. Describe what is being heard by observers at points A, B, and C at the instant shown in Fig. 16.39, when the shock wave has just reached point B. Explain your reasoning.

Figure 16.39 Question Q16.21.

Exercises

Unless indicated otherwise, assume the speed of sound in air to be $v = 344$ m/s.

Section 16.1 Sound Waves

16.1 Example 16.1 (Section 16.1) showed that for sound waves in air with frequency 1000 Hz, a displacement amplitude of 1.2×10^{-8} m produces a pressure amplitude of 3.0×10^{-2} Pa. a) What is the wavelength of these waves? b) For 1000-Hz waves in air, what displacement amplitude would be needed for the pressure amplitude to be at the pain threshold, which is 30 Pa? c) For what wavelength and frequency will waves with a displacement amplitude of 1.2×10^{-8} m produce a pressure amplitude of 1.5×10^{-3} Pa?

16.2 Example 16.1 (Section 16.1) showed that for sound waves in air with frequency 1000 Hz, a displacement amplitude of 1.2×10^{-8} m produces a pressure amplitude of 3.0×10^{-2} Pa. Water at 20°C has a bulk modulus of 2.2×10^9 Pa, and the speed of sound in water at this temperature is 1480 m/s. For 1000-Hz sound waves in 20°C water, what displacement amplitude is produced if the pressure amplitude is 3.0×10^{-2} Pa? Explain why your answer is much less than 1.2×10^{-8} Pa.

16.3 Consider a sound wave in air that has displacement amplitude 0.0200 mm. Calculate the pressure amplitude for frequencies of a) 150 Hz; b) 1500 Hz; c) 15,000 Hz. In each case compare the results to the pain threshold, which is 30 Pa.

16.4 Find the pressure amplitude of the wave traveling in the air at an altitude of 11,000 m that is described in Example 16.8 (Section 16.3). How does your answer compare to the pressure amplitude of the wave at sea level that has this same displacement amplitude (Example 16.1 in Section 16.2)? What does this tell you about the variation of pressure amplitude with altitude, if the frequency and displacement amplitude of the wave is kept fixed?

Section 16.2 Speed of Sound Waves

16.5 a) In a liquid with density 1300 kg/m^3, longitudinal waves with frequency 400 Hz are found to have wavelength 8.00 m. Calculate the bulk modulus of the liquid. b) A metal bar with a length of 1.50 m has density 6400 kg/m^3. Longitudinal sound waves take 3.90×10^{-4} s to travel from one end of the bar to the other. What is Young's modulus for this metal?

16.6 A major earthquake centered on Loma Prieta, California, near San Francisco, occurred at 5:04 P.M. local time on October 17, 1989 (in UTC, Coordinated Universal Time, 0h 4m 15s on October 18, 1989). The primary seismic waves (*P* waves) from such an earthquake are longitudinal waves that travel through the earth's crust. *P* waves were detected at Caracas, Venezuela, at 0h 13m 54s UTC; at Kevo, Finland, at 0h 15m 35s UTC; and at Vienna, Austria, at 0h 17m 02s UTC. The distances the *P* waves traveled from Loma Prieta were 6280 km to Caracas, 8690 km to Kevo, and 9650 km to Vienna. a) Use the arrival times to calculate the average speed of the *P* waves that traveled to these three cities. How can you account for any differences between the average speeds? b) The average density of the earth's crust is about 3.3 g/cm^3. Use this value to calculate the bulk modulus of the earth's crust along the path traveled by the *P* waves to each of the three cities. How do your answers compare to the bulk moduli in Table 11.1?

16.7 A submerged scuba diver hears the sound of a boat horn directly above her on the surface of the lake. At the same time, a friend on dry land 22.0 m from the boat also hears the horn (Fig. 16.40). The horn is 1.2 m above the surface of the water. What is the distance (labeled by "?" in Fig. 16.40) from the horn to the diver? Both air and water are at 20°C.

16.8 At a temperature of 27.0°C, what is the speed of longitudinal waves in a) hydrogen (molar mass 2.02 g/mol)? b) helium (molar mass 4.00 g/mol)? c) argon (molar mass

Figure 16.40 Exercise 16.7

39.9 g/mol)? See Table 19.1 for values of γ. d) Compare your answers for parts (a), (b) and (c) with the speed in air at the same temperature.

16.9 A jet airliner is cruising at high altitude at a speed of 850 km/h (about 530 mi/h). This is equal to 0.85 times the speed of sound at that altitude (also called "Mach 0.85"). a) What is the air temperature at this altitude? b) With this data, can you determine the air pressure at this altitude? Explain.

16.10 The speed of sound in air at 20°C was found to be 344 m/s. What is the change in speed for a 1.0 C° change in air temperature?

16.11 An 80.0-m-long brass rod is struck at one end. A person at the other end hears two sounds as a result of two longitudinal waves, one traveling in the metal rod and the other traveling in the air. What is the time interval between the two sounds? The speed of sound in air is 344 m/s; relevant information about brass can be found in Table 11.1 and Table 14.1.

16.12 What is the difference between the speed of longitudinal waves in air at 27.0°C and their speed at −13.0°C?

16.13 What must be the stress (F/A) in a stretched wire of a material whose Young's modulus is Y for the speed of longitudinal waves to equal 30 times the speed of transverse waves?

Section 16.3 Sound Intensity

16.14 A longitudinal wave of frequency 220 Hz travels down a copper rod of radius 8.00 mm. The average power in the wave is 6.50 μW. (See Tables 11.1 and 14.1 for the necessary data about copper.) a) Find the wavelength of the wave. b) Find the amplitude A of the wave. c) Find the maximum longitudinal velocity of a particle in the rod.

16.15 **Longitudinal Waves in Different Fluids.** a) A longitudinal wave propagating in a water-filled pipe has intensity 3.00×10^{-6} W/m^2 and frequency 3400 Hz. Find the amplitude A and wavelength λ of the wave. Water has density 1000 kg/m^3 and bulk modulus 2.18×10^9 Pa. b) If the pipe is filled with air at pressure 1.00×10^5 Pa and density 1.20 kg/m^3, what will be the amplitude A and wavelength λ of a longitudinal wave with the same intensity and frequency as in part (a)? c) In which fluid is the amplitude larger, water or air? What is the ratio of the two amplitudes? Why is this ratio so different from 1.00?

16.16 Derive Eq. (16.14) from the equations that precede it.

16.17 A sound wave in air at 20°C has a frequency of 150 Hz and a displacement amplitude of 5.00×10^{-3} mm. For this sound wave calculate the a) pressure amplitude (in Pa); b) intensity (in W/m^2); c) sound intensity level (in decibels).

16.18 a) What is the sound intensity level in a car when the sound intensity is 0.500 μW/m^2? b) What is the sound intensity level in

the air near a jackhammer when the pressure amplitude of the sound is 0.150 Pa and the temperature is 20.0°C?

16.19 For a person with normal hearing, the faintest sound that can be heard at a frequency of 400 Hz has a pressure amplitude of about 6.0×10^{-5} Pa. Calculate the corresponding intensity and sound intensity level at 20°C.

16.20 The intensity due to a number of independent sound sources is the sum of the individual intensities. a) When four quadruplets cry simultaneously, how many decibels greater is the sound intensity level than when a single one cries? b) To increase the sound intensity level again by the same number of decibels as in part (a), how many more crying babies are required?

16.21 A baby's mouth is 30 cm from her father's ear and 1.50 m from her mother's ear. What is the difference between the sound intensity levels heard by the father and by the mother?

16.22 In October 1991, the Sacramento City Council adopted a law to reduce the allowed sound intensity level of the much despised leaf blowers from their current level of about 95 dB to 70 dB by May 1992. With the new law, what is the ratio of the new allowed intensity to the previously allowed intensity?

16.23 a) By what factor must the sound intensity be increased to increase the sound intensity level by 13.0 dB? b) Explain why you don't need to know the original sound intensity.

Section 16.4 Standing Sound Waves and Normal Modes

16.24 The fundamental frequency of an open pipe is 594 Hz. What is the fundamental frequency if one end is plugged?

16.25 Standing sound waves are produced in a pipe that is 1.20 m long. For the fundamental and first two overtones, determine the locations along the pipe (measured from the left end) of the displacement nodes and the pressure nodes if a) the pipe is open at both ends; b) the pipe is closed at the left end and open at the right end.

16.26 Find the fundamental frequency and the frequency of the first three overtones of a pipe 45.0 cm long a) if the pipe is open at both ends; b) if the pipe is closed at one end. Use $v = 344$ m/s. c) For each of the above cases, what is the number of the highest harmonic that may be heard by a person who can hear frequencies from 20 Hz to 20,000 Hz?

16.27 The human vocal tract is a pipe that extends about 17 cm from the lips to the vocal folds (also called "vocal cords") near the middle of your throat. The vocal folds behave rather like the reed of a clarinet, and the vocal tract acts as a stopped pipe. Estimate the first three standing-wave frequencies of the vocal tract. Use $v = 344$ m/s. (The answers are only an estimate, since the position of lips and tongue affect the motion of air in the vocal tract.)

16.28 A certain pipe produces a fundamental frequency of 262 Hz in air. a) If the pipe is filled with helium at the same temperature, what fundamental frequency does it produce? (The molar mass of air is 28.8 g/mol, and the molar mass of helium is 4.00 g/mol.) b) Does your answer to part (a) depend on whether the pipe is open or stopped? Why or why not?

Section 16.5 Resonance

16.29 You blow across the open mouth of an empty test tube and produce the fundamental standing wave of the air column inside the test tube. The speed of sound in air is 344 m/s and the test tube acts as a stopped pipe. a) If the length of the air column in the test tube is 14.0 cm, what is the frequency of this standing wave? b) What is the frequency of the fundamental standing wave in the air column if the test tube is half-filled with water?

16.30 The D note (588 Hz) from a trumpet causes a guitar string to vibrate in its second overtone with large amplitude. The vibrating portion of the guitar string has length 63.5 cm. What is the speed of transverse waves on the guitar string?

Section 16.6 Interference of Waves

16.31 Two loudspeakers, A and B (Fig. 16.41), are driven by the same amplifier and emit sinusoidal waves in phase. Speaker B is 2.00 m to the right of speaker A. Consider point Q along the extension of the line connecting the speakers, 1.00 m to the right of speaker B. Both speakers emit sound waves that travel directly from the speaker to point Q. a) What is the lowest frequency for which *constructive* interference occurs at point Q? b) What is the lowest frequency for which *destructive* interference occurs at point Q?

Figure 16.41 Exercises 16.31 and 16.32.

16.32 Two loudspeakers, A and B (Fig. 16.41), are driven by the same amplifier and emit sinusoidal waves in phase. Speaker B is 2.00 m to the right of speaker A. The frequency of the sound waves produced by the loudspeakers is 206 Hz. Consider point P between the speakers and along the line connecting them, a distance x to the right of speaker A. Both speakers emit sound waves that travel directly from the speaker to point P. a) For what values of x will *destructive* interference occur at point P? b) For what values of x will *constructive* interference occur at point P? c) Interference effects like those in parts (a) and (b) are almost never a factor in listening to home stereo equipment. Why not?

16.33 Two loudspeakers, A and B, are driven by the same amplifier and emit sinusoidal waves in phase. Speaker B is 12.0 m to the right of speaker A. The frequency of the waves emitted by each speaker is 688 Hz. You are standing between the speakers, along the line connecting them and are at a point of constructive interference. How far must you walk toward speaker B to move to a point of destructive interference?

16.34 Two loudspeakers, A and B, are driven by the same amplifier and emit sinusoidal waves in phase. The frequency of the waves emitted by each speaker is 172 Hz. You are 8.00 m from A. What is the closest you can be to B and be at a point of destructive interference?

16.35 Two loudspeakers, A and B, are driven by the same amplifier and emit sinusoidal waves in phase. The frequency of the waves emitted by each speaker is 860 Hz. Point P is 12.0 m from A and 13.4 m from B. Is the interference at P constructive or destructive? Give the reasoning behind your answer.

Section 16.7 Beats

16.36 Two identical violin strings, when in tune and stretched with the same tension, have a fundamental frequency of 440.0 Hz. One of the strings is retuned by adjusting its tension. When this is done, 1.5 beats per second are heard when both strings are plucked simultaneously. a) What are the possible fundamental frequencies of the retuned string? b) By what fractional amount was the string tension changed if it was i) increased; ii) decreased?

16.37 Two sinusoidal sound waves with frequencies 108 Hz and 112 Hz arrive at your ear simultaneously. Each wave has an amplitude of 1.5×10^{-8} m when it reaches your ear. a) Describe in detail what you hear. b) What is the maximum amplitude of the total sound wave? What is the minimum amplitude?

Section 16.8 The Doppler Effect

16.38 On the planet Arrakis a male ornithoid is flying toward his mate at 25.0 m/s while singing at a frequency of 1200 Hz. If the stationary female hears a tone of 1240 Hz, what is the speed of sound in the atmosphere of Arrakis?

16.39 In Example 16.20 (Section 16.8), suppose the police car is moving away from the warehouse at 20 m/s. What frequency does the driver of the police car hear reflected from the warehouse?

16.40 Two train whistles, A and B, each have a frequency of 392 Hz. A is stationary and B is moving toward the right (away from A) at a speed of 35.0 m/s. A listener is between the two whistles and is moving toward the right with a speed of 15.0 m/s (Fig. 16.42). No wind is blowing. a) What is the frequency from A as heard by the listener? b) What is the frequency from B as heard by the listener? c) What is the beat frequency detected by the listener?

Figure 16.42 Exercise 16.40.

16.41 A railroad train is traveling at 25.0 m/s in still air. The frequency of the note emitted by the locomotive whistle is 400 Hz. What is the wavelength of the sound waves a) in front of the locomotive? b) behind the locomotive? What is the frequency of the sound heard by a stationary listener c) in front of the locomotive? d) behind the locomotive?

16.42 A swimming duck paddles the water with its feet once every 1.6 s, producing surface waves with this period. The duck is moving at constant speed in a pond where the speed of surface waves is 0.32 m/s, and the crests of the waves ahead of the duck are spaced 0.12 m apart. a) What is the duck's speed? b) How far apart are the crests behind the duck?

16.43 **Moving Source vs. Moving Listener.** a) A sound source producing 1.00 kHz waves moves toward a stationary listener at $\frac{1}{2}$ the speed of sound. What frequency will the listener hear? b) Suppose instead that the source is stationary and the listener moves toward the source at $\frac{1}{2}$ the speed of sound. What frequency does the listener hear? How does your answer compare to that in part (a)? Explain on physical grounds why the two answers differ.

16.44 A car alarm is emitting sound waves of frequency 520 Hz. You are on a motorcycle, traveling directly away from the car. How fast must you be traveling if you detect a frequency of 490 Hz?

16.45 A railroad train is traveling at 30.0 m/s in still air. The frequency of the note emitted by the train whistle is 262 Hz. What frequency is heard by a passenger on a train moving in the opposite direction to the first at 18.0 m/s and a) approaching the first? b) receding from the first?

*Section 16.9 Shock Waves
*16.46 A jet plane flies overhead at Mach 1.70 and at a constant altitude of 950 m. a) What is the angle α of the shock-wave cone? b) How much time after the plane passes directly overhead do you hear the sonic boom? Neglect the variation of the speed of sound with altitude.

Problems

16.47 a) Defend the following statement: "In a sinusoidal sound wave, the pressure variation given by Eq. (16.4) is greatest where the displacement given by Eq. (16.1) is zero." b) For a sinusoidal sound wave given by Eq. (16.1) with amplitude $A = 10.0 \mu m$ and wavelength $\lambda = 0.250$ m, graph the displacement y and pressure fluctuation p as functions of x at time $t = 0$. Show at least two wavelengths of the wave on your graphs. c) The displacement y in a nonsinusoidal sound wave is shown in Fig. 16.43 as a function of x for $t = 0$. Draw a graph showing the pressure fluc-

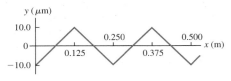

Figure 16.43 Problem 16.47.

tuation p in this wave as a function of x at $t = 0$. This sound wave has the same 10.0-μm amplitude as the wave in part (b). Does it have the same pressure amplitude? Why or why not? d) Is the statement in part (a) necessarily true if the sound wave is *not* sinusoidal? Explain your reasoning.

16.48 Many airports have noise ordinances restricting the maximum sound intensity that an aircraft may produce when it takes off. At one California airport the maximum allowable sound intensity level is 98.5 dB as measured by a microphone at the end of the 1740-m-long runway. A certain airliner produces a sound intensity level of 100.0 dB on the ground when it flies over at an altitude of 100 m. On takeoff, this airliner rolls for 1200 m along the runway before leaving the ground, at which point it climbs at a 15° angle. Does this airliner violate the noise ordinance? You can ignore any effects due to reflection of the sound waves from the ground.

16.49 The sound from a trumpet radiates uniformly in all directions in 20°C air. At a distance of 5.00 m from the trumpet the sound intensity level is 52.0 dB. The frequency is 587 Hz. a) What is the pressure amplitude at this distance? b) What is the displacement amplitude? c) At what distance is the sound intensity level 30.0 dB?

16.50 A sound wave entering a window of area A has a sound intensity level of β. a) How much acoustic power enters the window? b) Find the value of the acoustic power for the case of a 1.20-m² window and 55.0-dB sound.

16.51 A person is playing a small flute 10.75 cm long, open at one end and closed at the other, near a taut string having a fundamental frequency of 600.0 Hz. If the speed of sound is 344.0 m/s, for which harmonics of the flute will the string resonate? In each case, which harmonic of the string is in resonance?

16.52 A New Musical Instrument. You have designed a new musical instrument of very simple construction. Your design consists of a metal tube with length L and diameter $L/10$. You have stretched a string of mass per unit length μ across the open end of the tube. The other end of the tube is closed. To produce the musical effect you're looking for, you want the frequency of the third-harmonic standing wave on the string to be the same as the fundamental frequency for sound waves in the air column in the tube. The speed of sound waves in this air column is v_s. a) What must be the tension of the string to produce the desired effect? b) What happens to the sound produced by the instrument if the tension is changed to twice the value calculated in part (a)? c) For the tension calculated in part (a), what other harmonics of the string, if any, are in resonance with standing waves in the air column?

16.53 An organ pipe has two successive harmonics with frequencies 1372 and 1764 Hz. a) Is this an open or a stopped pipe? Explain. b) What two harmonics are these? c) What is the length of the pipe?

16.54 Longitudinal Standing Waves in a Solid. Longitudinal standing waves can be produced in a solid rod by holding it at some point between the fingers of one hand and stroking it with the other hand. The rod oscillates with antinodes at both ends. a) Why are the ends antinodes and not nodes? b) The fundamental frequency can be obtained by stroking the rod while it is held at its center. Explain why this is the *only* place to hold the rod to obtain the fundamental. c) Calculate the fundamental frequency of a steel rod of length 1.50 m (see Table 16.1). d) What is the next possible standing-wave frequency of this rod? Where should the rod be held to excite a standing wave of this frequency?

16.55 a) Determine the first three normal-mode frequencies for a pipe of length L that is closed at *both* ends. Explain your reasoning. b) Use the results of part (a) to estimate the normal-mode frequencies of a shower stall. Explain the connection between these frequencies and the observation that your singing voice probably sounds better in the shower, especially when you sing at certain frequencies.

16.56 One type of steel has a density of 7.8×10^3 kg/m³ and a breaking stress of 7.0×10^8 N/m². A cylindrical guitar string is to be made of 4.00 g of this steel. a) What are the length and radius of the longest and thinnest string that can be placed under a tension of 900 N without breaking? b) What is the highest fundamental frequency that this string could have?

16.57 A long tube contains air at a pressure of 1.00 atm and a temperature of 77.0°C. The tube is open at one end and closed at the other by a movable piston. A tuning fork near the open end is vibrating with a frequency of 500 Hz. Resonance is produced when the piston is at distances 18.0, 55.5, and 93.0 cm from the open end. a) From these measurements, what is the speed of sound in air at 77.0°C? b) From the result of part (a), what is the value of γ? c) These data show that a displacement antinode is slightly outside of the open end of the tube. How far outside is it?

16.58 The auditory canal of the human ear (Fig. 16.5) extends about 2.5 cm from the outside ear to the eardrum. a) Explain why the human ear is especially sensitive to sounds at frequencies around 3500 Hz. Use $v = 344$ m/s. b) Would you expect the ear to be especially sensitive to frequencies around 7000 Hz? Around 10,500 Hz? Why or why not?

16.59 A platinum wire (density 21.4 g/cm^3) is 225 μm in diameter and 0.450 m long. One end of the wire is attached to the ceiling, while a 420-g mass is attached to the other end so that the wire hangs vertically under tension. If a vibrating tuning fork of just the right frequency is held next to the wire, the wire begins to vibrate as well. a) What tuning-fork frequencies will cause this to happen? You may assume that the bottom end of the wire (to which the mass is attached) is essentially stationary, and that the tension in the wire is essentially constant along its length. b) Justify the assumptions made in part (a).

16.60 The frequency of the note F$_4$ is 349 Hz. a) If an organ pipe is open at one end and closed at the other, what length must it have for its fundamental mode to produce this note at 20.0°C? b) At what air temperature will the frequency be 370 Hz, corresponding to a rise in pitch from F to F-sharp? (Ignore the change in length of the pipe due to the temperature change.)

16.61 A standing wave with a frequency of 1100 Hz in a column of methane (CH_4) at 20.0°C produces nodes that are 0.200 m apart. What is the value of γ for methane? (The molar mass of methane is 16.0 g/mol.)

16.62 Two identical loudspeakers are located at points A and B, 2.00 m apart. The loudspeakers are driven by the same amplifier and produce sound waves with a frequency of 784 Hz. Take the speed of sound in air to be 344 m/s. A small microphone is moved out from point B along a line perpendicular to the line connecting A and B (line BC in Fig. 16.44). a) At what distances from B will there be *destructive* interference? b) At what distances from B will there be *constructive* interference? c) If the frequency is made low enough, there will be no positions along the line BC at which destructive interference occurs. How low must the frequency be for this to be the case?

Figure 16.44 Problem 16.62.

16.63 Soap Opera. A man marries a great Wagnerian soprano but alas, he discovers he cannot stand Wagnerian opera. In order to save his eardrums, the unhappy man decides he must silence his lark-like wife for good. His plan is to tie her to the front of his car and send car and soprano speeding towards a brick wall. This soprano, however is quite shrewd, having studied physics in her student days at the music conservatory. She realizes that this wall has a resonant frequency of 600 Hz, which means that if a continuous sound wave of this frequency hits the wall, it will fall down, and she will be saved to sing again. The car is heading toward the wall at a high speed of 30 m/s. a) At what frequency must the soprano sing so that the wall will crumble? b) What frequency will the soprano hear reflected from the wall just before it crumbles?

16.64 A bat flies toward a wall, emitting a steady sound of frequency 2.00 kHz. This bat hears its own sound plus the sound reflected by the wall. How fast should the bat fly in order to hear a beat frequency of 10.0 Hz?

16.65 A small sphere of radius R is arranged to pulsate so that its radius varies in simple harmonic motion between a minimum of $R - \Delta R$ and a maximum of $R + \Delta R$ with frequency f. This produces sound waves in the surrounding air of density ρ and bulk modulus B. a) Find the intensity of sound waves at the surface of the sphere. (The amplitude of oscillation of the sphere is the same as that of the air at the surface of the sphere.) b) Find the total acoustic power radiated by the sphere. c) At a distance $d \gg R$ from the center of the sphere, find the amplitude, pressure amplitude, and intensity of the sound wave.

16.66 A 2.00-MHz sound wave travels through a pregnant woman's abdomen and is reflected from the fetal heart wall of her unborn baby. The heart wall is moving toward the sound receiver as the heart beats. The reflected sound is then mixed with the transmitted sound, and 85 beats per second are detected. The speed of sound in body tissue is 1500 m/s. Calculate the speed of the fetal heart wall at the instant this measurement is made.

16.67 The sound source of a ship's sonar system operates at a frequency of 22.0 kHz. The speed of sound in water (assumed to be at a uniform 20°C) is 1482 m/s. a) What is the wavelength of the waves emitted by the source? b) What is the difference in frequency between the directly radiated waves and the waves reflected from a whale traveling directly toward the ship at 4.95 m/s? The ship is at rest in the water.

16.68 A police siren of frequency f_{siren} is attached to a vibrating platform. The platform and siren oscillate up and down in simple harmonic motion with amplitude A_p and frequency f_p. a) Find the maximum and minimum sound frequencies that you would hear at a position directly above the siren. b) At what point in the motion of the platform is the maximum frequency heard? The minimum frequency? Explain.

16.69 Horseshoe bats (genus *Rhinolophus*) emit sounds from their nostrils, then listen to the frequency of the sound reflected from their prey to determine the prey's speed. (The "horseshoe" that gives the bat its name is a depression around the nostrils that acts like a focusing mirror, so that the bat emits sound in a narrow beam like a flashlight.) A *Rhinolophus* flying at speed v_{bat} emits sound of frequency f_{bat}; the sound it hears reflected from an insect flying toward it has a higher frequency f_{refl}. a) Show that the speed of the insect is

$$v_{insect} = v \left[\frac{f_{refl}(v - v_{bat}) - f_{bat}(v + v_{bat})}{f_{refl}(v - v_{bat}) + f_{bat}(v + v_{bat})} \right]$$

where v is the speed of sound. b) If $f_{bat} = 80.7$ kHz, $f_{refl} = 83.5$ kHz, and $v_{bat} = 3.9$ m/s, calculate the speed of the insect.

16.70 A sound wave with frequency f_0 and wavelength λ_0 travels horizontally toward the right. It strikes and is reflected from a large, rigid, vertical plane surface, perpendicular to the direction of propagation of the wave and moving toward the left with a speed v_1. a) How many positive wave crests strike the surface in a time interval t? b) At the end of this time interval, how far to the left of the surface is the wave that was reflected at the beginning of the time interval? c) What is the wavelength of the reflected waves in terms of λ_0? d) What is the frequency of the reflected waves in terms of f_0? Is your result consistent with the assertion made in the

Problem-Solving Strategy in Section 16.8? e) A listener is at rest at the left of the moving surface. How many beats per second does she detect as a result of the combined effect of the incident and reflected waves?

16.71 a) Show that Eq. (16.30) can be written as

$$f_R = f_S\left(1 - \frac{v}{c}\right)^{1/2}\left(1 + \frac{v}{c}\right)^{-1/2}$$

b) Use the binomial theorem to show that if $v \ll c$, this is approximately equal to

$$f_R = f_S\left(1 - \frac{v}{c}\right)$$

c) A pilotless reconnaissance aircraft emits a radio signal with a frequency of 243 MHz. It is flying directly toward a test engineer on the ground. The engineer detects beats between the received signal and a local signal also of frequency 243 MHz. The beat frequency is 46.0 Hz. What is the speed of the aircraft? (Radio waves travel at the speed of light, $c = 3.00 \times 10^8$ m/s.)

16.72 The gas cloud known as the Crab Nebula (Fig. 9.34) can be seen with even a small telescope. It is the remnant of a *supernova*, a cataclysmic explosion of a star. The explosion was seen on the earth on July 4, 1054 A.D. The streamers in Fig. 9.34 glow with the characteristic red color of heated hydrogen gas. In a laboratory on the earth, heated hydrogen produces red light with frequency 4.568×10^{14} Hz; the red light received from streamers in the Crab Nebula pointed toward the earth has frequency 4.586×10^{14} Hz. a) Estimate the speed with which the outer edges of the Crab Nebula are expanding. Assume that the speed of the center of the nebula relative to the earth is negligible. (You may use the formulas derived in Problem 16.71. The speed of light is 3.00×10^8 m/s.) b) Assuming that the expansion speed has been constant since the supernova explosion, estimate the diameter of the Crab Nebula. Give your answer in meters and in light years. c) The angular diameter of the Crab Nebula as seen from earth is about 5 arc minutes (1 arc minute $= \frac{1}{60}$ of a degree). Estimate the distance (in light years) to the Crab Nebula, and estimate the year in which the supernova explosion actually took place.

16.73 A radar installation used for monitoring thunderstorms emits radio waves at a frequency of 2800 MHz. A line of thunderstorms is approaching the installation at 42.0 km/h. a) In the frame of reference of the thunderstorms, is the frequency of the radio waves greater than or less than 2800 MHz? Why? By what amount does the frequency differ from 2800 MHz? (You may use the formulas derived in Problem 16.71. Radio waves travel at the speed of light, $c = 3.00 \times 10^8$ m/s.) b) Radio waves are reflected from the water drops in the thunderstorms, and the reflected waves are detected back at the installation. As measured by the receiver at the installation, is the frequency of these reflected waves greater than or less than 2800 MHz? Why? By what amount does the frequency of the reflected waves differ from 2800 MHz?

16.74 A woman stands at rest in front of a large, smooth wall. She holds a vibrating tuning fork of frequency f_0 directly in front of her (between her and the wall). a) She now runs toward the wall with a speed v_W. She detects beats due to the interference between the sound waves reaching her directly from the fork and those reaching her after being reflected from the wall. How many beats per second will she detect? (*Note:* If the beat frequency is too large, the woman may have to use some instrumentation other than her ears to detect and count the beats.) b) If she instead runs away from the wall, holding the tuning fork at her back so it is between her and the wall, how many beats per second will she detect?

*16.75 On a clear day you see a jet plane flying overhead. From the apparent size of the plane, you determine that it is flying at a constant altitude h. You hear the sonic boom at time T after the plane passes directly overhead. Show that if the speed of sound v is the same at all altitudes, the speed of the plane is

$$v_S = \frac{hv}{\sqrt{h^2 - v^2 T^2}}$$

(*Hint:* Trigonometric identities will be useful.)

Challenge Problems

16.76 Figure 16.45 shows the pressure fluctuation p of a nonsinusoidal sound wave as a function of x for $t = 0$. The wave

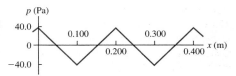

Figure 16.45 Challenge Problem 16.76.

is traveling in the $+x$-direction. a) Graph the pressure fluctuation p as a function of t for $x = 0$. Show at least two cycles of oscillation. b) Graph the displacement y in this sound wave as a function of x at $t = 0$. At $x = 0$, the displacement at $t = 0$ is zero. Show at least two wavelengths of the wave. c) Graph the displacement y as a function of t for $x = 0$. Show at least two cycles of oscillation. d) Calculate the maximum velocity and the maximum acceleration of an element of the air through which this sound wave is traveling. e) Describe how the cone of a loudspeaker must move as a function of time to produce the sound wave in this problem.

16.77 Two loudspeakers, A and B, radiate sound uniformly in all directions in air at 20°C. The acoustic power output from A is 8.00×10^{-4} W, and from B it is 6.00×10^{-5} W. Both loudspeakers are vibrating in phase at a frequency of 172 Hz. a) Determine the difference in phase of the two signals at a point C along the line joining A and B, 3.00 m from B and 4.00 m from A (Fig. 16.46). b) Determine the intensity and sound intensity level at C from speaker A if speaker B is turned off and the intensity and sound intensity level at point C from speaker B if speaker A is turned off. c) With both speakers on, what are the intensity and sound intensity level at C?

Fig. 16.46 Challenge Problem 16.77.

17

TEMPERATURE AND HEAT

Perhaps the highest-temperature material you will ever see is the sun's outer atmosphere, or corona. At a temperature of about 2,000,000°C (3,600,000°F), the corona glows with a light that is literally unearthly. But because the corona is also very thin, its light is rather faint. You can only see the corona during a total solar eclipse when the sun's disk is covered by the moon, as shown here.

? **Is it accurate to say that the corona contains heat?**

Whether it's a sweltering summer day or a frozen midwinter night, your body needs to be kept at a nearly constant temperature. It has effective temperature-control mechanisms, but sometimes it needs help. On a hot day you wear less clothing to improve heat transfer from your body to the air and for better cooling by evaporation of perspiration. You probably drink cold beverages possibly with ice in them, and sit near a fan or in an air-conditioned room. On a cold day you wear more clothes or stay indoors where it's warm. When you're outside, you keep active and drink hot liquids to stay warm. The concepts in this chapter will help you understand the basic physics of keeping warm or cool.

The terms *temperature* and *heat* are often used interchangeably in everyday language. In physics, however, these two terms have very different meanings. In this chapter we'll define temperature in terms of how it's measured and see how temperature changes affect the dimensions of objects. We'll see that heat refers to energy transfer caused by temperature differences and learn how to calculate and control such energy transfers.

Our emphasis in this chapter is on the concepts of temperature and heat as they relate to *macroscopic* objects such as cylinders of gas, ice cubes, and the human body. In Chapter 18 we'll look at these same concepts from a *microscopic* viewpoint in terms of the behavior of individual atoms and molecules. These two chapters lay the groundwork for the subject of **thermodynamics,** the study of energy transformations involving heat, mechanical work, and other aspects of energy and how these transformations relate to the properties of matter. Thermodynamics forms an indispensable part of the foundation of physics, chemistry, and the life sciences, and its applications turn up in such places as car engines, refrigerators, biochemical processes, and the structure of stars. We'll explore the key ideas of thermodynamics in Chapters 19 and 20.

17.1 | Temperature and Thermal Equilibrium

The concept of **temperature** is rooted in qualitative ideas of "hot" and "cold" based on our sense of touch. A body that feels hot usually has a higher temperature than a similar body that feels cold. That's pretty vague, and the senses can be deceived. But many properties of matter that we can *measure* depend on temperature. The length of a metal rod, steam pressure in a boiler, the ability of a wire to conduct an electric current, and the color of a very hot glowing object—all these depend on temperature.

Temperature is also related to the kinetic energies of the molecules of a material. In general this relationship is fairly complex, so it's not a good place to start in *defining* temperature. In Chapter 18 we will look at the relationship between temperature and the energy of molecular motion for an ideal gas. It is important to understand, however, that temperature and heat can be defined independently of any detailed molecular picture. In this section we'll develop a *macroscopic* definition of temperature.

To use temperature as a measure of hotness or coldness, we need to construct a temperature scale. To do this, we can use any measurable property of a system that varies with its "hotness" or "coldness." Figure 17.1a shows a familiar system that is used to measure temperature. When the system becomes hotter, the colored liquid (usually mercury or ethanol) expands and rises in the tube, and the value of L increases. Another simple system is a quantity of gas in a constant-volume container (Fig. 17.1b). The pressure p, measured by the gauge, increases or decreases as the gas becomes hotter or colder. A third example is the electrical resistance R of a conducting wire, which also varies when the wire becomes hotter or colder. Each of these properties gives us a number (L, p, or R) that varies with hotness and coldness, so each property can be used to make a **thermometer.**

To measure the temperature of a body, you place the thermometer in contact with the body. If you want to know the temperature of a cup of hot coffee, you stick the thermometer in the coffee; as the two interact, the thermometer becomes hotter and the coffee cools off a little. After the thermometer settles down to a steady value, you read the temperature. The system has reached an *equilibrium* condition, in which the interaction between the thermometer and the coffee causes no further change in the system. We call this a state of **thermal equilibrium.**

If two systems are separated by an insulating material or **insulator** such as wood, plastic foam, or fiberglass, they influence each other more slowly. Camping coolers are made with insulating materials to delay the ice and cold food inside from warming up and attaining thermal equilibrium with the hot summer air outside. An *ideal insulator* is a material that permits no interaction at all between the two systems. It prevents the systems from attaining thermal equilibrium if they aren't in thermal equilibrium at the start. An ideal insulator is just that, an idealization; real insulators, like those in camping coolers, aren't ideal, so the contents of the cooler will warm up eventually.

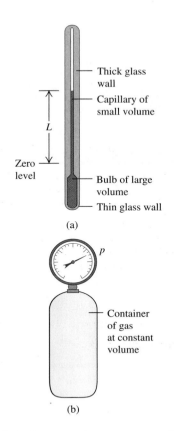

17.1 (a) A system whose temperature is specified by the value of length L. (b) A system whose temperature is given by the value of the pressure p.

(a) If systems *A* and *B* are each in thermal equilibrium with system *C* ...

(b) ... then systems *A* and *B* are in thermal equilibrium with each other

17.2 The zeroth law of thermodynamics. Blue slabs represent thermal insulators, and orange slabs represent thermal conductors.

We can discover an important property of thermal equilibrium by considering three systems, *A*, *B*, and *C*, that initially are not in thermal equilibrium (Fig. 17.2). We surround them with an ideal insulating box so that they cannot interact with anything except each other. We separate systems *A* and *B* with an ideal insulating wall (the green slab in Fig. 17.2a), but we let system *C* interact with both systems *A* and *B*. This interaction is shown in the figure by a yellow slab representing a thermal **conductor,** a material that *permits* thermal interactions through it. We wait until thermal equilibrium is attained; then *A* and *B* are each in thermal equilibrium with *C*. But are they in thermal equilibrium *with each other?*

To find out, we separate system *C* from systems *A* and *B* with an ideal insulating wall (Fig. 17.2b), and then we replace the insulating wall between *A* and *B* with a *conducting* wall that lets *A* and *B* interact. What happens? Experiment shows that *nothing* happens; there are no additional changes to *A* or *B*. We conclude that **if *C* is initially in thermal equilibrium with both *A* and *B*, then *A* and *B* are also in thermal equilibrium with each other.** This result is called the **zeroth law of thermodynamics.** (The importance of this law was recognized only after the first, second, and third laws of thermodynamics had been named. Since it is fundamental to all of them, the name "zeroth" seemed appropriate.)

Now suppose system *C* is a thermometer, such as the tube-and-liquid system of Fig. 17.1a. In Fig. 17.2a the thermometer *C* is in contact with both *A* and *B*. In thermal equilibrium, when the thermometer reading reaches a stable value, the thermometer measures the temperature of both *A* and *B*; hence *A* and *B* both have the *same* temperature. Experiment shows that thermal equilibrium isn't affected by adding or removing insulators, so the reading of thermometer *C* wouldn't change if it were in contact only with *A* or only with *B*. We conclude that **two systems are in thermal equilibrium if and only if they have the same temperature.** This is what makes a thermometer useful; a thermometer actually measures *its own* temperature, but when a thermometer is in thermal equilibrium with another body, the temperatures must be equal. When the temperatures of two systems are different, they *cannot* be in thermal equilibrium.

Test Your Understanding

Why does a nurse taking your temperature wait for the thermometer reading to stop changing? What object is it whose temperature the nurse is reading?

17.2 | Thermometers and Temperature Scales

To make the liquid-in-tube device shown in Fig. 17.1a into a useful thermometer, we need to mark a scale on the tube wall with numbers on it. These numbers are arbitrary, and historically many different schemes have been used. Suppose we label the thermometer's liquid level at the freezing temperature of pure water "zero" and the level at the boiling temperature "100," and divide the distance between these two points into 100 equal intervals called *degrees*. The result is the **Celsius temperature scale** (formerly called the *centigrade* scale in English-speaking countries). The Celsius temperature for a state colder than freezing water is a negative number. The Celsius scale is used, both in everyday life and in science and industry, almost everywhere in the world.

Another common type of thermometer uses a *bimetallic strip,* made by bonding strips of two different metals together (Fig. 17.3a). When the temperature of the composite strip increases, one metal expands more than the other and the strip bends. This strip is usually formed into a spiral, with the outer end anchored to the thermometer case and the inner end attached to a pointer (Fig. 17.3c). The pointer rotates in response to temperature changes.

In a *resistance thermometer* the changing electrical resistance of a coil of fine wire, a carbon cylinder, or a germanium crystal is measured. Because resistance can be measured very precisely, resistance thermometers are usually more precise than most other types.

Some thermometers don't require physical contact with the object they're measuring. An example is an ear thermometer (Fig. 17.4). Such a thermometer uses a device called a *thermopile* to measure the amount of infrared radiation emitted by the eardrum, which indicates the eardrum's temperature. (We'll see in Section 17.7 that *all* objects emit electromagnetic radiation as a consequence of their temperature.) The advantage of this technique is that it doesn't require touching the fragile and easily damaged eardrum.

In the **Fahrenheit temperature scale,** still used in everyday life in the United States, the freezing temperature of water is 32°F (thirty-two degrees Fahrenheit) and the boiling temperature is 212°F, both at standard atmospheric pressure. There are 180 degrees between freezing and boiling, compared to 100 on the Celsius scale, so one Fahrenheit degree represents only $\frac{100}{180}$, or $\frac{5}{9}$, as great a temperature change as one Celsius degree.

To convert temperatures from Celsius to Fahrenheit, note that a Celsius temperature T_C is the number of Celsius degrees above freezing; the number of Fahrenheit degrees above freezing is $\frac{9}{5}$ of this. But freezing on the Fahrenheit scale is at 32°F, so to obtain the actual Fahrenheit temperature T_F, multiply the Celsius value by $\frac{9}{5}$ and then add 32°. Symbolically,

$$T_F = \frac{9}{5}T_C + 32° \qquad (17.1)$$

To convert Fahrenheit to Celsius, solve this equation for T_C:

$$T_C = \frac{5}{9}(T_F - 32°) \qquad (17.2)$$

In words, subtract 32° to get the number of Fahrenheit degrees above freezing, and then multiply by $\frac{5}{9}$ to obtain the number of Celsius degrees above freezing, that is, the Celsius temperature.

We don't recommend memorizing Eqs. (17.1) and (17.2). Instead, try to understand the reasoning that led to them so that you can derive them on the spot when you need them, checking your reasoning with the relation 100°C = 212°F.

It is useful to distinguish between an actual temperature and a temperature *interval* (a difference or change in temperature). An actual temperature of 20° is stated as 20°C (twenty degrees Celsius), and a temperature *interval* of 10° is 10 C° (ten Celsius degrees). A beaker of water heated from 20°C to 30°C has a temperature change of 10 C°.

Test Your Understanding

Find the average Fahrenheit temperature on the planet Venus (average Celsius temperature 460°C), and find the temperature at which the Fahrenheit and Celsius scales agree.

17.3 (a) A bimetallic strip. (b) The strip bends when its temperature is raised. (c) A bimetallic strip used in a thermometer.

17.4 An ear thermometer measures infrared radiation from the eardrum, which is located far enough inside the head that it gives an excellent indication of the body's internal temperature.

17.3 | Gas Thermometers and the Kelvin Scale

When we calibrate two thermometers, such as a liquid-in-tube system and a resistance thermometer, so that they agree at 0°C and 100°C, they may not agree exactly at intermediate temperatures. Any temperature scale defined in this way always depends somewhat on the specific properties of the material used. Ideally, we would like to define a temperature scale that *doesn't* depend on the properties of a particular material. To establish a truly material-independent scale, we first need to develop some principles of thermodynamics. We'll return to this fundamental problem in Chapter 20. Here we'll discuss a thermometer that comes close to the ideal, the *gas thermometer.*

The principle of a gas thermometer is that the pressure of a gas at constant volume increases with temperature. A quantity of gas is placed in a constant-volume container (Fig. 17.5a), and its pressure is measured by one of the devices described in Section 14.2. To calibrate a constant-volume gas thermometer, we measure the pressure at two temperatures, say 0°C and 100°C, plot these points on a graph, and draw a straight line between them. Then we can read from the graph the temperature corresponding to any other pressure. Figure 17.5b shows the results of three such experiments, each using a different type and quantity of gas.

By extrapolating this graph, we see that there is a hypothetical temperature, −273.15°C, at which the absolute pressure of the gas would become zero. We might expect that this temperature would be different for different gases, but it turns out to be the *same* for many different gases (at least in the limit of very low gas density). We can't actually observe this zero-pressure condition. Gases liquefy and solidify at very low temperatures, and the proportionality of pressure to temperature no longer holds.

We use this extrapolated zero-pressure temperature as the basis for a temperature scale with its zero at this temperature. This is the **Kelvin temperature scale,**

17.5 (a) A constant-volume gas thermometer. (b) Graphs of absolute pressure versus temperature for a constant-volume low-density gas thermometer. The three graphs are for experiments with different types and quantities of gas: the greater the amount of gas, the higher the graph. The dashed lines are extrapolations of the data to low temperature.

Same experiment with different amounts of different gases

−273.15 −200 −100   0   100   200 T (°C)

0   100   200   300   400   500 T (K)

In each case, straight-line extrapolation predicts that pressure would become 0 at −273.15 °C

(a)

(b)

named for the British physicist Lord Kelvin (1824–1907). The units are the same size as those on the Celsius scale, but the zero is shifted so that 0 K $= -273.15°$C and 273.15 K $= 0°$C; that is,

$$T_K = T_C + 273.15 \qquad (17.3)$$

This scale is shown in Fig. 17.5b. A common room temperature, $20°$C ($= 68°$F), is $20 + 273.15$, or about 293 K.

> **CAUTION** In SI nomenclature, "degree" is not used with the Kelvin scale; the temperature mentioned above is read "293 kelvins," not "degrees Kelvin" (Fig. 17.6). We capitalize Kelvin when it refers to the temperature scale; however, the *unit* of temperature is the *kelvin,* which is not capitalized (but is nonetheless abbreviated as a capital K).

17.6 Kelvin temperatures are measured in kelvins (K), *not* "degrees Kelvin."

Example 17.1	**Body temperature**

You place a small piece of melting ice in your mouth. Eventually, the water all converts from ice at $T_1 = 32.00°$F to body temperature, $T_2 = 98.60°$F. Express these temperatures as $°$C and K, and find $\Delta T = T_2 - T_1$ in both cases.

SOLUTION

IDENTIFY and SET UP: We convert Fahrenheit to Celsius temperatures using Eq. (17.2), and Celsius to Kelvin temperatures using Eq. (17.3).

EXECUTE: First we find the Celsius temperatures. We know that $T_1 = 32.00°$F $= 0.00°$C, and $98.60°$F is $98.60 - 32.00 = 66.60$ F$°$ above freezing; multiply this by $(5$ C$°/9$ F$°)$ to find 37.00 C$°$ above freezing, or $T_2 = 37.00°$C.

To get the Kelvin temperatures, we just add 273.15 to each Celsius temperature: $T_1 = 273.15$ K and $T_2 = 310.15$ K. "Normal" body temperature is $37.0°$C, but if your doctor says that your temperature is 310 K, don't be alarmed.

The temperature *difference* $\Delta T = T_2 - T_1$ is 37.00 C$° = 37.00$ K.

EVALUATE: The Celsius and Kelvin scales have different zero points but the same size degrees. Therefore any temperature difference is the *same* on the Celsius and Kelvin scales but not the same on the Fahrenheit scale.

The Celsius scale has two fixed points, the normal freezing and boiling temperatures of water. But we can define the Kelvin scale using a gas thermometer with only a single reference temperature. We define the ratio of any two temperatures T_1 and T_2 on the Kelvin scale as the ratio of the corresponding gas-thermometer pressures p_1 and p_2:

$$\frac{T_2}{T_1} = \frac{p_2}{p_1} \qquad \left(\text{constant-volume gas thermometer, } T \text{ in kelvins}\right) \qquad (17.4)$$

The pressure p is directly proportional to the Kelvin temperature, as shown in Fig. 17.5b. To complete the definition of T, we need only specify the Kelvin temperature of a single specific state. For reasons of precision and reproducibility the state

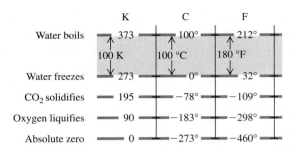

17.7 Relations among Kelvin, Celsius, and Fahrenheit temperature scales. Temperatures have been rounded off to the nearest degree.

chosen is the *triple point* of water. This is the unique combination of temperature and pressure at which solid water (ice), liquid water, and water vapor can all coexist. It occurs at a temperature of 0.01°C and a water-vapor pressure of 610 Pa (about 0.006 atm). (This is the pressure of the *water;* it has nothing to do directly with the gas pressure in the *thermometer.*) The triple-point temperature T_{triple} of water is *defined* to have the value $T_{\text{triple}} = 273.16$ K, corresponding to 0.01°C. From Eq. (17.4), if p_{triple} is the pressure in a gas thermometer at temperature T_{triple} and p is the pressure at some other temperature T, then T is given on the Kelvin scale by

$$T = T_{\text{triple}}\frac{p}{p_{\text{triple}}} = (273.16 \text{ K})\frac{p}{p_{\text{triple}}} \qquad (17.5)$$

Low-pressure gas thermometers using various gases are found to agree very closely, but they are large, bulky, and very slow to come to thermal equilibrium. They are used principally to establish high-precision standards and to calibrate other thermometers.

The relationships among the three temperature scales we have discussed are shown graphically in Fig. 17.7. The Kelvin scale is called an **absolute temperature scale,** and its zero point ($T = 0$ K $= -273.15$°C, the temperature at which $p = 0$ in Eq. (17.5)) is called **absolute zero.** At absolute zero a system of molecules (such as a quantity of a gas, a liquid, or a solid) has its *minimum* possible total energy (kinetic plus potential); because of quantum effects, however, it is *not* correct to say that all molecular motion ceases at absolute zero. To define more completely what we mean by absolute zero, we need to use the thermodynamic principles developed in the next several chapters. We will return to this concept in Chapter 20.

Test Your Understanding

The temperature of the solar corona (see the photograph that opens this chapter) is 2.0×10^7 °C, and the temperature at which helium becomes a liquid at standard pressure is -268.93°C. Express these temperatures in kelvins, and explain why the Kelvin scale is usually employed for very high and very low temperatures.

17.4 | Thermal Expansion

Most materials expand when their temperatures increase. Rising temperatures make the liquid expand in a liquid–in–tube thermometer (Fig. 17.1a) and bend bimetallic strips (Fig. 17.3). The decks of bridges need special joints and supports to allow for expansion. A completely filled and tightly capped bottle of water cracks when it is heated, but you can loosen a metal jar lid by running hot water over it. These are all examples of *thermal expansion.*

Linear Expansion

Suppose a rod of material has a length L_0 at some initial temperature T_0. When the temperature changes by ΔT, the length changes by ΔL. Experiments show that if ΔT is not too large (say, less than 100 C° or so), ΔL is *directly proportional* to ΔT. If two rods made of the same material have the same temperature change, but one is twice as long as the other, then the *change* in its length is also twice as great. Therefore ΔL must also be proportional to L_0. Introducing a proportionality constant α (which is different for different materials), we may express these relations in an equation:

$$\Delta L = \alpha L_0 \, \Delta T \qquad (\text{linear thermal expansion}) \qquad (17.6)$$

If a body has length L_0 at temperature T_0, then its length L at a temperature $T = T_0 + \Delta T$ is

$$L = L_0 + \Delta L = L_0 + \alpha L_0 \, \Delta T = L_0(1 + \alpha \, \Delta T) \qquad (17.7)$$

The constant α, which describes the thermal expansion properties of a particular material, is called the **coefficient of linear expansion.** The units of α are K^{-1} or $(C°)^{-1}$. (Remember that a temperature *interval* is the same in the Kelvin and Celsius scales.) For many materials, every linear dimension changes according to Eq. (17.6) or (17.7). Thus L could be the thickness of a rod, the side length of a square sheet, or the diameter of a hole. Some materials, such as wood or single crystals, expand differently in different directions. We won't consider this complication.

We can understand thermal expansion qualitatively on a molecular basis. Picture the interatomic forces in a solid as springs, as in Fig. 17.8. (We explored the analogy between spring forces and interatomic forces in Section 13.4.) Each atom vibrates about its equilibrium position. When the temperature increases, the energy and amplitude of the vibration also increase. The interatomic spring forces are not symmetrical about the equilibrium position; they usually behave like a spring that is easier to stretch than to compress. As a result, when the amplitude of vibration increases, the *average* distance between atoms also increases. As the atoms get farther apart, every dimension increases.

> **CAUTION** If a solid object has a hole in it, what happens to the size of the hole when the temperature of the object increases? A common misconception is that if the object expands, the hole will shrink because material expands into the hole. But the truth of the matter is that if the object expands, the hole will expand too (Fig. 17.9); as we stated above, *every* linear dimension of an object changes in the same way when the temperature changes. If you're not convinced, think of the atoms in Fig. 17.8a as outlining a cubical hole. When the object expands, the atoms move apart and the hole

Average distance between atoms

(a)

$U(x)$ x = distance between atoms
● = average distance
between atoms

O ——————— x

E_3
E_2
E_1

As energy increases from E_1 to E_2 to E_3, average distance between atoms increases

(b)

17.8 (a) A model of the forces between neighboring atoms in a solid. The "springs" are easier to stretch than to compress. (b) A graph of the "spring" potential energy U versus distance x between neighboring atoms (compare Fig. 13.19a). The graph is not symmetrical: as the energy increases and the atoms oscillate with greater amplitude, the *average* distance increases.

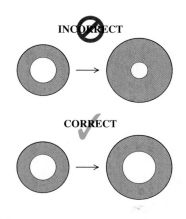

INCORRECT

CORRECT

17.9 When an object undergoes thermal expansion, any holes in the object expand as well.

17.10 When this SR-71 aircraft is sitting on the ground, its wing panels fit together so loosely that fuel leaks out of the wings onto the ground. But once it is in flight at over three times the speed of sound, air friction heats the panels so much that they expand to make a perfect fit. (In-flight refueling makes up for the lost fuel.)

increases in size. The only situation in which a "hole" will fill in due to thermal expansion is when two separate objects expand and close the gap between them (Fig. 17.10).

The direct proportionality expressed by Eq. (17.6) is not exact; it is *approximately* correct only for sufficiently small temperature changes. For a given material, α varies somewhat with the initial temperature T_0 and the size of the temperature interval. We'll ignore this complication here, however. Average values of α for several materials are listed in Table 17.1. Within the precision of these values we don't need to worry whether T_0 is 0°C or 20°C or some other temperature. Note that typical values of α are very small; even for a temperature change of 100 C°, the fractional length change $\Delta L/L_0$ is only of the order of 1/1000 for the metals in the table.

Volume Expansion

Increasing temperature usually causes increases in *volume* for both solid and liquid materials. Just as with linear expansion, experiments show that if the temperature change ΔT is not too great (less than 100 C° or so), the increase in volume ΔV is approximately proportional to both the temperature change ΔT and the initial volume V_0:

$$\Delta V = \beta V_0 \Delta T \quad \text{(volume thermal expansion)} \quad (17.8)$$

The constant β characterizes the volume expansion properties of a particular material; it is called the **coefficient of volume expansion.** The units of β are K^{-1} or $(C°)^{-1}$. As with linear expansion, β varies somewhat with temperature, and Eq. (17.8) is an approximate relation that is valid only for small temperature changes. For many substances, β decreases at low temperatures. Several values of β in the neighborhood of room temperature are listed in Table 17.2. Note that the values for liquids are generally much larger than those for solids.

For solid materials there is a simple relation between the volume expansion coefficient β and the linear expansion coefficient α. To derive this relation, we consider a cube of material with side length L and volume $V = L^3$. At the initial temperature the values are L_0 and V_0. When the temperature increases by dT, the side length increases by dL and the volume increases by an amount dV given by

$$dV = \frac{dV}{dL} dL = 3L^2 \, dL$$

Now we replace L and V by the initial values L_0 and V_0. From Eq. (17.6), dL is

$$dL = \alpha L_0 \, dT$$

since $V_0 = L_0^3$, this means that dV can also be expressed as

$$dV = 3L_0^2 \alpha L_0 \, dT = 3\alpha V_0 \, dT$$

This is consistent with the infinitesimal form of Eq. (17.8), $dV = \beta V_0 \, dT$, only if

$$\beta = 3\alpha \quad (17.9)$$

You should check this relation for some of the materials listed in Tables 17.1 and 17.2.

Table 17.1 Coefficients of Linear Expansion

Material	$\alpha\left[\mathrm{K}^{-1} \text{ or } (\mathrm{C}°)^{-1}\right]$
Aluminum	2.4×10^{-5}
Brass	2.0×10^{-5}
Copper	1.7×10^{-5}
Glass	$0.4–0.9 \times 10^{-5}$
Invar (nickel–iron alloy)	0.09×10^{-5}
Quartz (fused)	0.04×10^{-5}
Steel	1.2×10^{-5}

Table 17.2 Coefficients of Volume Expansion

Solids	$\beta\left[\mathrm{K}^{-1} \text{ or } (\mathrm{C}°)^{-1}\right]$	Liquids	$\beta\left[\mathrm{K}^{-1} \text{ or } (\mathrm{C}°)^{-1}\right]$
Aluminum	7.2×10^{-5}	Ethanol	75×10^{-5}
Brass	6.0×10^{-5}	Carbon disulfide	115×10^{-5}
Copper	5.1×10^{-5}	Glycerin	49×10^{-5}
Glass	$1.2–2.7 \times 10^{-5}$	Mercury	18×10^{-5}
Invar	0.27×10^{-5}		
Quartz (fused)	0.12×10^{-5}		
Steel	3.6×10^{-5}		

Problem-Solving Strategy — Thermal Expansion

IDENTIFY *the relevant concepts:* Decide whether the problem involves changes in length (linear thermal expansion) or in volume (volume thermal expansion).

SET UP *the problem using the following steps:*
1. Choose Eq. (17.6) for linear expansion and Eq. (17.8) for volume expansion.
2. Identify which quantities in Eq. (17.6) or (17.8) are known and which are the unknown target variables.

EXECUTE *the solution as follows:*
1. Solve for the target variables. Often you will be given two temperatures and asked to compute ΔT. Or you may be given an initial temperature T_0 and asked to find a final temperature corresponding to a given length or volume change. In this case, plan to find ΔT first; then the final temperature is $T_0 + \Delta T$.
2. Unit consistency is crucial, as always. L_0 and ΔL (or V_0 and ΔV) must have the same units, and if you use a value of α or β in K^{-1} or $(\mathrm{C}°)^{-1}$, then ΔT must be in kelvins or Celsius degrees ($\mathrm{C}°$). But you can use K and $\mathrm{C}°$ interchangeably.

EVALUATE *your answer:* Check whether your results make sense. Remember that the sizes of holes in a material expand with temperature just the same way as any other linear dimension, and the volume of a hole (such as the volume of a container) expands the same way as the corresponding solid shape.

Example 17.2 — Length change due to temperature change I

A surveyor uses a steel measuring tape that is exactly 50.000 m long at a temperature of 20°C. What is its length on a hot summer day when the temperature is 35°C?

SOLUTION

IDENTIFY and SET UP: This problem concerns linear expansion, so we use Eq. (17.6). We are given $L = 50.000$ m, $T_0 = 20°C$, and $T = 35°C$, and the value of α is found from Table 17.1. The target variable is the new length L.

EXECUTE: The temperature change is $\Delta T = T - T_0 = 15\,\mathrm{C}°$, so from Eq. (17.6) the change in length ΔL and the final length $L = L_0 + \Delta L$ are

$$\Delta L = \alpha L_0 \Delta T = (1.2 \times 10^{-5}\,\mathrm{K}^{-1})(50\,\mathrm{m})(15\,\mathrm{K})$$
$$= 9.0 \times 10^{-3}\,\mathrm{m} = 9.0\,\mathrm{mm}$$
$$L = L_0 + \Delta L = 50.000\,\mathrm{m} + 0.009\,\mathrm{m} = 50.009\,\mathrm{m}$$

Thus the length at 35°C is 50.009 m.

EVALUATE: Note that L_0 is given to five significant figures but that we need only two of them to compute ΔL. Note also that ΔL is proportional to the initial length L_0: a 5.0-m tape would expand by 0.90 mm, and a 0.50-m (50-cm) tape would expand by a mere 0.090 mm.

This example shows that metals expand very little under moderate temperature changes. Even a metal baking pan in a 200°C (392°F) oven is only slightly larger than it is at room temperature.

| Example 17.3 | **Length change due to temperature change II** |

In Example 17.2 the surveyor uses the measuring tape to measure a distance when the temperature is 35°C; the value that she reads off the tape is 35.794 m. What is the actual distance? Assume that the tape is calibrated for use at 20°C.

SOLUTION

IDENTIFY and SET UP: As we saw in Example 17.2, at 35°C the tape has expanded slightly. The distance between two successive meter marks is slightly more than one meter, so the scale underestimates the actual distance. Hence the actual distance is *larger* than the distance read off the tape by a factor equal to the ratio of the tape's length L at 35°C to its length L_0 at 20°C.

EXECUTE: The ratio L/L_0 is $(50.009 \text{ m})/(50.000 \text{ m})$, so the true distance is

$$\frac{50.009 \text{ m}}{50.000 \text{ m}} (35.794 \text{ m}) = 35.800 \text{ m}$$

EVALUATE: Although the difference of 0.008 m = 8 mm between the scale reading and the actual distance seems small, it can be important in precision work.

| Example 17.4 | **Volume change due to temperature change** |

A glass flask with volume 200 cm³ is filled to the brim with mercury at 20°C. How much mercury overflows when the temperature of the system is raised to 100°C? The coefficient of linear expansion of the glass is 0.40×10^{-5} K⁻¹.

SOLUTION

IDENTIFY: This problem involves the volume expansion of the glass and of the mercury. The amount of overflow depends on the *difference* between the volume changes for these two materials.

SET UP: The amount of overflow is equal to the difference between the values of ΔV for mercury and for glass, both given by Eq. (17.8). For the mercury to overflow, its coefficient of volume expansion β must be larger than that for glass. The value for mercury is $\beta_{mercury} = 18 \times 10^{-5}$ K⁻¹ from Table 17.2, and we find the value of β for this type of glass from Eq. (17.9), $\beta = 3\alpha$.

EXECUTE: The coefficient of volume expansion for the glass is

$$\beta_{glass} = 3\alpha_{glass} = 3(0.40 \times 10^{-5} \text{ K}^{-1}) = 1.2 \times 10^{-5} \text{ K}^{-1}$$

The increase in volume of the glass flask is

$$\begin{aligned}
\Delta V_{glass} &= \beta_{glass} V_0 \Delta T \\
&= (1.2 \times 10^{-5} \text{ K}^{-1})(200 \text{ cm}^3)(100°C - 20°C) \\
&= 0.19 \text{ cm}^3
\end{aligned}$$

The increase in volume of the mercury is

$$\begin{aligned}
\Delta V_{mercury} &= \beta_{mercury} V_0 \Delta T \\
&= (18 \times 10^{-5} \text{ K}^{-1})(200 \text{ cm}^3)(100°C - 20°C) \\
&= 2.9 \text{ cm}^3
\end{aligned}$$

The volume of mercury that overflows is

$$\Delta V_{mercury} - \Delta V_{glass} = 2.9 \text{ cm}^3 - 0.19 \text{ cm}^3 = 2.7 \text{ cm}^3$$

EVALUATE: This is basically how a mercury-in-glass thermometer works, except that instead of letting the mercury overflow and run all over the place, the thermometer has it rise inside a sealed tube as T increases.

As Tables 17.1 and 17.2 show, glass has smaller coefficients of expansion α and β than do most metals. This is why you can use hot water to loosen a metal lid on a glass jar; the metal expands more than the glass does.

Thermal Expansion of Water

Water, in the temperature range from 0°C to 4°C, *decreases* in volume with increasing temperature. In this range its coefficient of volume expansion is *negative*. Above 4°C, water expands when heated (Fig. 17.11). Hence water has its greatest density at 4°C. Water also expands when it freezes, which is why ice humps up in the middle of the compartments in an ice cube tray. By contrast, most materials contract when they freeze.

This anomalous behavior of water has an important effect on plant and animal life in lakes. A lake cools from the surface down; above 4°C, the cooled water at the surface flows to the bottom because of its greater density. But when the surface temperature drops below 4°C, the water near the surface is less dense than the warmer water below. Hence the downward flow ceases, and the water near the surface remains colder than that at the bottom. As the surface freezes, the ice floats because it is less dense than water. The water at the bottom remains at 4°C until nearly the entire lake is frozen. If water behaved like most substances, contracting continuously on cooling and freezing, lakes would freeze from the bottom up. Circulation due to density differences would continuously carry warmer water to the surface for efficient cooling, and lakes would freeze solid much more easily. This would destroy all plant and animal life that cannot withstand freezing. If water did not have this special property, the evolution of life would have taken a very different course.

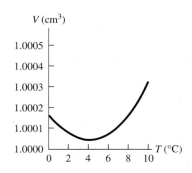

17.11 The volume of one gram of water in the temperature range from 0°C to 10°C. By 100°C the volume has increased to 1.034 cm³. If the coefficient of volume expansion were constant, the curve would be a straight line.

Thermal Stress

If we clamp the ends of a rod rigidly to prevent expansion or contraction and then change the temperature, tensile or compressive stresses called **thermal stresses** develop. The rod would like to expand or contract, but the clamps won't let it. The resulting stresses may become large enough to strain the rod irreversibly or even break it. (You may want to review the discussion of stress and strain in Section 11.4).

Engineers must account for thermal stress when designing structures. Concrete highways and bridge decks usually have gaps between sections, filled with a flexible material or bridged by interlocking teeth (Fig. 17.12), to permit expansion and contraction of the concrete. Long steam pipes have expansion joints or U-shaped sections to prevent buckling or stretching with temperature changes. If one end of a steel bridge is rigidly fastened to its abutment, the other end usually rests on rollers.

To calculate the thermal stress in a clamped rod, we compute the amount the rod *would* expand (or contract) if not held and then find the stress needed to compress (or stretch) it back to its original length. Suppose that a rod with length L_0 and cross-sectional area A is held at constant length while the temperature is reduced (negative ΔT), causing a tensile stress. The fractional change in length if the rod were free to contract would be

$$\left(\frac{\Delta L}{L_0}\right)_{\text{thermal}} = \alpha \, \Delta T \qquad (17.10)$$

Both ΔL and ΔT are negative. The tension must increase by an amount F that is just enough to produce an equal and opposite fractional change in length $\left(\Delta L/L_0\right)_{\text{tension}}$. From the definition of Young's modulus, Eq. (11.10),

$$Y = \frac{F/A}{\Delta L/L_0} \quad \text{so} \quad \left(\frac{\Delta L}{L_0}\right)_{\text{tension}} = \frac{F}{AY} \qquad (17.11)$$

If the length is to be constant, the *total* fractional change in length must be zero. From Eqs. (17.10) and (17.11), this means that

$$\left(\frac{\Delta L}{L_0}\right)_{\text{thermal}} + \left(\frac{\Delta L}{L_0}\right)_{\text{tension}} = \alpha \, \Delta T + \frac{F}{AY} = 0$$

17.12 The interlocking teeth of an expansion joint on a bridge. These joints are needed to accommodate changes in length that result from thermal expansion.

Solving for the tensile stress F/A required to keep the rod's length constant, we find

$$\frac{F}{A} = -Y\alpha\,\Delta T \qquad \text{(thermal stress)} \qquad (17.12)$$

For a decrease in temperature, ΔT is negative, so F and F/A are positive; this means that a *tensile* force and stress are needed to maintain the length. If ΔT is positive, F and F/A are negative, and the required force and stress are *compressive*.

If there are temperature differences within a body, non-uniform expansion or contraction will result and thermal stresses can be induced. You can break a glass bowl by pouring very hot water into it; the thermal stress between the hot and cold parts of the bowl exceeds the breaking stress of the glass, causing cracks. The same phenomenon makes ice cubes crack when dropped into warm water. Heat-resistant glasses such as Pyrex™ have exceptionally low expansion coefficients and high strength.

Example 17.5	**Thermal stress**

An aluminum cylinder 10 cm long, with a cross-sectional area of 20 cm², is to be used as a spacer between two steel walls. At 17.2°C it just slips in between the walls. When it warms to 22.3°C, calculate the stress in the cylinder and the total force it exerts on each wall, assuming that the walls are perfectly rigid and a constant distance apart.

SOLUTION

IDENTIFY and SET UP: We use Eq. (17.12) to relate the stress (our target variable) to the temperature change. The relevant values of Young's modulus Y and the coefficient of linear expansion α are those for aluminum, the material of which the cylinder is made; we find these values from Tables 11.1 and 17.1, respectively.

EXECUTE: For aluminum, $Y = 7.0 \times 10^{10}$ Pa and $\alpha = 2.4 \times 10^{-5}$ K^{-1}. The temperature change is $\Delta T = 22.3°C - 17.2°C = 5.1$ C° $= 5.1$ K. The stress is F/A; from Eq. (17.12),

$$\frac{F}{A} = -Y\alpha\,\Delta T = -(0.70 \times 10^{11}\text{ Pa})(2.4 \times 10^{-5}\text{ K}^{-1})(5.1\text{ K})$$
$$= -8.6 \times 10^{6}\text{ Pa (or } -1200\text{ lb/in.}^2)$$

The negative sign indicates that compressive rather than tensile stress is needed to keep the cylinder's length constant. This stress is independent of the length and cross-sectional area of the cylinder. The total force F is the cross-sectional area times the stress:

$$F = A\left(\frac{F}{A}\right) = (20 \times 10^{-4}\text{ m}^2)(-8.6 \times 10^{6}\text{ Pa})$$
$$= -1.7 \times 10^{4}\text{ N}$$

or nearly two tons. The negative sign indicates compression.

EVALUATE: The stress on the cylinder and the force it exerts on each wall are immense. This points out the importance of accounting for such thermal stresses in engineering.

Test Your Understanding

In the bimetallic strip shown in Fig. 17.3, metal 1 is copper. What are two different materials that could be used for metal 2?

17.5 | Quantity of Heat

When you put a cold spoon into a cup of hot coffee, the spoon warms up and the coffee cools down as they approach thermal equilibrium. The interaction that causes these temperature changes is fundamentally a transfer of *energy* from one substance to another. Energy transfer that takes place solely because of a temperature difference is called *heat flow* or *heat transfer,* and energy transferred in this way is called **heat.**

An understanding of the relation between heat and other forms of energy emerged gradually during the 18th and 19th centuries. Sir James Joule (1818–1889) studied how water can be warmed by vigorous stirring with a paddle wheel (Fig. 17.13a). The paddle wheel adds energy to the water by doing *work* on it, and Joule found that *the temperature rise is directly proportional to the amount of work done.* The same temperature change can also be caused by putting the water in contact with some hotter body (Fig. 17.13b); hence this interaction must also involve an energy exchange. We will explore the relation between heat and mechanical energy in greater detail in Chapters 19 and 20.

CAUTION It is absolutely essential for you to keep clearly in mind the distinction between *temperature* and *heat.* Temperature depends on the physical state of a material and is a quantitative description of its hotness or coldness. In physics the term "heat" always refers to energy in transit from one body or system to another because of a temperature difference, never to the amount of energy contained within a particular system. We can change the temperature of a body by adding heat to it or taking heat away, or by adding or subtracting energy in other ways, such as mechanical work (Fig. 17.13a). If we cut a body in half, each half has the same temperature as the whole; but to raise the temperature of each half by a given interval, we add *half* as much heat as for the whole.

(a)

(b)

17.13 The same temperature change of the same system may be accomplished by (a) doing work on it or (b) adding heat to it.

We can define a *unit* of quantity of heat based on temperature changes of some specific material. The **calorie** (abbreviated cal) is defined as *the amount of heat required to raise the temperature of one gram of water from* 14.5°C *to* 15.5°C. The kilocalorie (kcal), equal to 1000 cal, is also used; a food-value calorie is actually a kilocalorie. A corresponding unit of heat using Fahrenheit degrees and British units is the **British thermal unit,** or Btu. One Btu is the quantity of heat required to raise the temperature of one pound (weight) of water 1 F° from 63°F to 64°F.

Because heat is energy in transit, there must be a definite relation between these units and the familiar mechanical energy units such as the joule. Experiments similar in concept to Joule's have shown that

$$1 \text{ cal} = 4.186 \text{ J}$$

$$1 \text{ kcal} = 1000 \text{ cal} = 4186 \text{ J}$$

$$1 \text{ Btu} = 778 \text{ ft} \cdot \text{lb} = 252 \text{ cal} = 1055 \text{ J}$$

The calorie is not a fundamental SI unit. The International Committee on Weights and Measures recommends using the joule as the basic unit of energy in all forms, including heat. We will follow that recommendation in this book.

Specific Heat

We use the symbol Q for quantity of heat. When it is associated with an infinitesimal temperature change dT, we call it dQ. The quantity of heat Q required to increase the temperature of a mass m of a certain material from T_1 to T_2 is found to be approximately proportional to the temperature change $\Delta T = T_2 - T_1$. It is also proportional to the mass m of material. When you're heating water to make

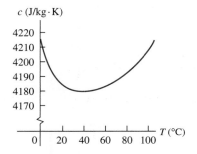

17.14 Specific heat of water as a function of temperature. The value of c varies by less than 1% between 0°C and 100°C.

tea, you need twice as much heat for two cups as for one if the temperature interval is the same. The quantity of heat needed also depends on the nature of the material; raising the temperature of one kilogram of water by 1 C° requires 4190 J of heat, but only 910 J is needed to raise the temperature of one kilogram of aluminum by 1 C°.

Putting all these relationships together, we have

$$Q = mc\,\Delta T \qquad (\text{heat required for temperature change of mass } m) \quad (17.13)$$

where c is a quantity, different for different materials, called the **specific heat** of the material. For an infinitesimal temperature change dT and corresponding quantity of heat dQ,

$$dQ = mc\,dT \tag{17.14}$$

$$c = \frac{1}{m}\frac{dQ}{dT} \qquad (\text{specific heat}) \tag{17.15}$$

In Eqs. (17.13), (17.14), and (17.15), Q (or dQ) and ΔT (or dT) can be either positive or negative. When they are positive, heat enters the body and its temperature increases; when they are negative, heat leaves the body and its temperature decreases.

CAUTION Remember that dQ does not represent a change in the amount of heat *contained* in a body; this is a meaningless concept. Heat is always energy *in transit* as a result of a temperature difference. There is no such thing as "the amount of heat in a body."

The specific heat of water is approximately

$$4190 \text{ J/kg}\cdot\text{K} \qquad 1 \text{ cal/g}\cdot\text{C}° \qquad \text{or } 1 \text{ Btu/lb}\cdot\text{F}°$$

The specific heat of a material always depends somewhat on the initial temperature and the temperature interval. Figure 17.14 shows this variation for water. In the problems and examples in this chapter we will usually ignore this small variation.

Example 17.6	**Feed a cold, starve a fever**

During a bout with the flu an 80·kg man ran a fever of 39.0°C (102.2°F) instead of the normal body temperature of 37.0°C (98.6°F). Assuming that the human body is mostly water, how much heat is required to raise his temperature by that amount?

SOLUTION

IDENTIFY: This problem uses the relationship between heat (the target variable), mass, specific heat, and temperature change.

SET UP: We are given the values of $m = 80$ kg, $c = 4190$ J/kg·K (for water), and $\Delta T = 39.0°\text{C} - 37.0°\text{C} = 2.0$ C° $= 2.0$ K. We use Eq. (17.13) to determine the required heat.

EXECUTE: From Eq. (17.13),

$$Q = mc\,\Delta T = (80 \text{ kg})(4190 \text{ J/kg}\cdot\text{K})(2.0 \text{ K}) = 6.7 \times 10^5 \text{ J}$$

EVALUATE: This corresponds to 160 kcal, or 160 food-value calories. In fact, the specific heat of the human body is more nearly equal to 3480 J/kg·K, about 83% that of water. The difference is due to the presence of protein, fat, and minerals, which have lower specific heats. With this value of c, the required heat is 5.6×10^5 J $= 133$ kcal. Either result shows us that were it not for the body's temperature-regulating systems, taking in energy in the form of food would produce measurable changes in body temperature. (In the case of a person with the flu, the elevated temperature results from the body's extra activity in fighting infection.)

Example
17.7

Overheating electronics

You are designing an electronic circuit element made of 23 mg of silicon. The electric current through it adds energy at the rate of 7.4 mW = 7.4×10^{-3} J/s. If your design doesn't allow any heat transfer out of the element, at what rate does its temperature increase? The specific heat of silicon is 705 J/kg·K.

SOLUTION

IDENTIFY and SET UP: The target variable is the *rate* of temperature change. From Eq. (17.14), the temperature change ΔT in kelvins is proportional to the heat transferred in joules, so the rate of temperature change in K/s is proportional to the rate of heat transfer in J/s.

EXECUTE: In one second, $Q = (7.4 \times 10^{-3} \text{ J/s})(1 \text{ s}) = 7.4 \times 10^{-3}$ J. From Eq. (17.13), $Q = mc \, \Delta T$, the temperature change in one second is

$$\Delta T = \frac{Q}{mc} = \frac{7.4 \times 10^{-3} \text{ J}}{(23 \times 10^{-6} \text{ kg})(705 \text{ J/kg·K})} = 0.46 \text{ K}$$

Alternatively, we can divide both sides of Eq. (17.14) by dt and rearrange:

$$\frac{dT}{dt} = \frac{dQ/dt}{mc}$$

$$= \frac{7.4 \times 10^{-3} \text{ J/s}}{(23 \times 10^{-6} \text{ kg})(705 \text{ J/kg·K})} = 0.46 \text{ K/s}$$

EVALUATE: At this rate of temperature rise (27 K every minute), the circuit element would soon self-destruct. Heat transfer is an important design consideration in electronic circuit elements.

Molar Heat Capacity

Sometimes it's more convenient to describe a quantity of substance in terms of the number of *moles n* rather than the *mass m* of material. Recall from your study of chemistry that a mole of any pure substance always contains the same number of molecules. (We will discuss this point in more detail in Chapter 18.) The *molar mass* of any substance, denoted by M, is the mass per mole. (The quantity M is sometimes called *molecular weight,* but *molar mass* is preferable; the quantity depends on the mass of a molecule, not its weight.) For example, the molar mass of water is 18.0 g/mol = 18.0×10^{-3} kg/mol; one mole of water has a mass of 18.0 g = 0.0180 kg. The total mass m of material is equal to the mass per mole M times the number of moles n:

$$m = nM \qquad (17.16)$$

Replacing the mass m in Eq. (17.13) by the product nM, we find

$$Q = nMc \, \Delta T \qquad (17.17)$$

The product Mc is called the **molar heat capacity** (or *molar specific heat*) and is denoted by C (capitalized). With this notation we rewrite Eq. (17.17) as

$$Q = nC \, \Delta T \qquad (\text{heat required for temperature change of } n \text{ moles}) \quad (17.18)$$

Comparing to Eq. (17.15), we can express the molar heat capacity C (heat per mole per temperature change) in terms of the specific heat c (heat per mass per temperature change) and the molar mass M (mass per mole):

$$C = \frac{1}{n}\frac{dQ}{dT} = Mc \qquad (\text{molar heat capacity}) \qquad (17.19)$$

For example, the molar heat capacity of water is

$$C = Mc = (0.0180 \text{ kg/mol})(4190 \text{ J/kg·K}) = 75.4 \text{ J/mol·K}$$

Table 17.3 Approximate Specific Heats and Molar Heat Capacities (Constant Pressure)

Substance	Specific Heat, c ($J/kg \cdot K$)	M (kg/mol)	Molar Heat Capacity, C ($J/mol \cdot K$)
Aluminum	910	0.0270	24.6
Beryllium	1970	0.00901	17.7
Copper	390	0.0635	24.8
Ethanol	2428	0.0461	111.9
Ethylene glycol	2386	0.0620	148.0
Ice (near 0°C)	2100	0.0180	37.8
Iron	470	0.0559	26.3
Lead	130	0.207	26.9
Marble ($CaCO_3$)	879	0.100	87.9
Mercury	138	0.201	27.7
Salt (NaCl)	879	0.0585	51.4
Silver	234	0.108	25.3
Water (liquid)	4190	0.0180	75.4

17.15 Water has a much higher specific heat than the glass or metals used to make cookware. This helps explain why it takes several minutes to boil water on a stove, even though the pot or kettle reaches a high temperature very quickly.

Values of specific heat and molar heat capacity for several substances are given in Table 17.3. Note the remarkably large specific heat for water (Fig. 17.15).

CAUTION The term *heat capacity* is unfortunate because it gives the erroneous impression that a body *contains* a certain amount of heat. Remember, heat is energy in transit to or from a body, not the energy residing in the body.

Precise measurements of specific heats and molar heat capacities require great experimental skill. Usually, a measured quantity of energy is supplied by an electric current in a heater wire wound around the specimen. The temperature change ΔT is measured with a resistance thermometer or thermocouple embedded in the specimen. This sounds simple, but great care is needed to avoid or compensate for unwanted heat transfer between the sample and its surroundings. Measurements for solid materials are usually made at constant atmospheric pressure; the corresponding values are called the *specific heat* and *molar heat capacity at constant pressure,* denoted by c_p and C_p. For a gas it is usually easier to keep the substance in a container with constant *volume;* the corresponding values are called the *specific heat* and *molar heat capacity at constant volume,* denoted by c_V and C_V. For a given substance, C_V and C_p are different. If the system can expand while heat is added, there is additional energy exchange through the performance of *work* by the system on its surroundings. If the volume is constant, the system does no work. For gases the difference between C_p and C_V is substantial. We will study heat capacities of gases in detail in Section 19.7.

The last column of Table 17.3 shows something interesting. The molar heat capacities for most elemental solids are about the same, about 25 $J/mol \cdot K$. This correlation, named the *rule of Dulong and Petit* (for its discoverers), forms the basis for a very important idea. The number of atoms in one mole is the same for all elemental substances. This means that on a *per atom* basis, about the same amount of heat is required to raise the temperature of each of these elements by a given amount, even though the *masses* of the atoms are very different. The heat required for a given temperature increase depends only on *how many* atoms the sample contains, not on the mass of an individual atom. We will see the reason why the rule of Dulong and Petit works so well when we study the molecular basis of heat capacities in greater detail in Chapter 18.

Test Your Understanding

Suppose you want to use the apparatus shown in Fig. 17.13a to heat water to make a cup of tea. Through what distance would a 1.00-kg block have to fall to raise the temperature of 0.250 kg of water from 20.0°C to 90.0°C? Assume that all of the potential energy that is lost as the block falls goes into raising the temperature of the water.

17.6 | Calorimetry and Phase Changes

Calorimetry means "measuring heat." We have discussed the energy transfer (heat) involved in temperature changes. Heat is also involved in *phase changes,* such as the melting of ice or boiling of water. Once we understand these additional heat relationships, we can analyze a variety of problems involving quantity of heat.

Phase Changes

We use the term **phase** to describe a specific state of matter, such as a solid, liquid, or gas. The compound H_2O exists in the *solid phase* as ice, in the *liquid phase* as water, and in the *gaseous phase* as steam. (These are also referred to as **states of matter:** the solid state, the liquid state, and the gaseous state.) A transition from one phase to another is called a **phase change** or *phase transition*. For any given pressure a phase change takes place at a definite temperature, usually accompanied by absorption or emission of heat and a change of volume and density.

A familiar example of a phase change is the melting of ice. When we add heat to ice at 0°C and normal atmospheric pressure, the temperature of the ice *does not* increase. Instead, some of it melts to form liquid water. If we add the heat slowly, to maintain the system very close to thermal equilibrium, the temperature remains at 0°C until all the ice is melted (Fig. 17.16). The effect of adding heat to this system is not to raise its temperature but to change its *phase* from solid to liquid.

To change 1 kg of ice at 0°C to 1 kg of liquid water at 0°C and normal atmospheric pressure requires 3.34×10^5 J of heat. The heat required per unit mass is called the **heat of fusion** (or sometimes *latent heat of fusion*), denoted by L_f. For water at normal atmospheric pressure the heat of fusion is

$$L_f = 3.34 \times 10^5 \text{ J/kg} = 79.6 \text{ cal/g} = 143 \text{ Btu/lb}$$

More generally, to melt a mass m of material that has a heat of fusion L_f requires a quantity of heat Q given by

$$Q = mL_f$$

This process is *reversible*. To freeze liquid water to ice at 0°C, we have to *remove* heat; the magnitude is the same, but in this case, Q is negative because heat is removed rather than added. To cover both possibilities and to include other kinds of phase changes, we write

$$Q = \pm mL \quad \text{(heat transfer in a phase change)} \quad (17.20)$$

17.16 The surrounding air is at room temperature, but this ice–water mixture remains at 0°C until all of the ice has melted and the phase change is complete.

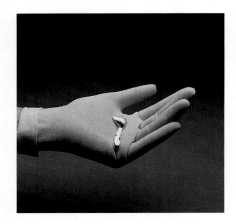

17.17 The metal gallium, shown here melting in a person's hand, is one of the few elements that melt in the vicinity of room temperature. Its melting temperature is 29.8°C, and its heat of fusion is 8.04×10^4 J/kg.

The plus sign (heat entering) is used when the material melts; the minus sign (heat leaving) is used when it freezes. The heat of fusion is different for different materials, and it also varies somewhat with pressure.

For any given material at any given pressure, the freezing temperature is the same as the melting temperature. At this unique temperature the liquid and solid phases (liquid water and ice, for example) can coexist in a condition called **phase equilibrium.**

We can go through this whole story again for *boiling* or *evaporation,* a phase transition between liquid and gaseous phases. The corresponding heat (per unit mass) is called the **heat of vaporization** L_v. At normal atmospheric pressure the heat of vaporization L_v for water is

$$L_v = 2.256 \times 10^6 \text{ J/kg} = 539 \text{ cal/g} = 970 \text{ Btu/lb}$$

That is, it takes 2.256×10^6 J to change 1 kg of liquid water at 100°C to 1 kg of water vapor at 100°C. By comparison, to raise the temperature of a kilogram of water from 0°C to 100°C requires $Q = mc\,\Delta T = (1.00 \text{ kg})(4190 \text{ J/kg} \cdot \text{C}°) \times (100 \text{ C}°) = 4.19 \times 10^5$ J, less than one-fifth as much heat as is required for vaporization at 100°C. This agrees with everyday kitchen experience; a pot of water may reach boiling temperature in a few minutes, but it takes a much longer time to completely evaporate all the water away.

Like melting, boiling is a reversible transition. When heat is removed from a gas at the boiling temperature, the gas returns to the liquid phase, or *condenses,* giving up to its surroundings the same quantity of heat (heat of vaporization) that was needed to vaporize it. At a given pressure the boiling and condensation temperatures are always the same; at this temperature the liquid and gaseous phases can coexist in phase equilibrium.

Both L_v and the boiling temperature of a material depend on pressure. Water boils at a lower temperature (about 95°C) in Denver than in Pittsburgh because Denver is at higher elevation and the average atmospheric pressure is less. The heat of vaporization is somewhat greater at this lower pressure, about 2.27×10^6 J/kg.

Table 17.4 lists heats of fusion and vaporization for several materials and their melting and boiling temperatures at normal atmospheric pressure. Very few ele-

Table 17.4 Heats of Fusion and Vaporization

Substance	Normal Melting Point K	Normal Melting Point °C	Heat of Fusion, L_f (J/kg)	Normal Boiling Point K	Normal Boiling Point °C	Heat of Vaporization, L_v (J/kg)
Helium	*	*	*	4.216	−268.93	20.9×10^3
Hydrogen	13.84	−259.31	58.6×10^3	20.26	−252.89	452×10^3
Nitrogen	63.18	−209.97	25.5×10^3	77.34	−195.8	201×10^3
Oxygen	54.36	−218.79	13.8×10^3	90.18	−183.0	213×10^3
Ethanol	159	−114	104.2×10^3	351	78	854×10^3
Mercury	234	−39	11.8×10^3	630	357	272×10^3
Water	273.15	0.00	334×10^3	373.15	100.00	2256×10^3
Sulfur	392	119	38.1×10^3	717.75	444.60	326×10^3
Lead	600.5	327.3	24.5×10^3	2023	1750	871×10^3
Antimony	903.65	630.50	165×10^3	1713	1440	561×10^3
Silver	1233.95	960.80	88.3×10^3	2466	2193	2336×10^3
Gold	1336.15	1063.00	64.5×10^3	2933	2660	1578×10^3
Copper	1356	1083	134×10^3	1460	1187	5069×10^3

*A pressure in excess of 25 atmospheres is required to make helium solidify. At 1 atmosphere pressure, helium remains a liquid down to absolute zero.

17.18 Graph of temperature versus time for a specimen of water initially in the solid phase (ice). Heat is added to the specimen at a constant rate. The temperature remains constant during each change of phase, provided that the pressure remains constant.

ments have melting temperatures in the vicinity of ordinary room temperatures; one of the few is the metal gallium, shown in Fig. 17.17.

Figure 17.18 shows how the temperature varies when we add heat continuously to a specimen of ice with an initial temperature below 0°C (point *a*). The temperature rises until we reach the melting point (point *b*). As more heat is added, the temperature remains constant until all the ice has melted (point *c*). Then the temperature starts to rise again until the boiling temperature is reached (point *d*). At that point the temperature again is constant until all the water is transformed into the vapor phase (point *e*). If the rate of heat input is constant, the slope of the line for the solid phase (ice) has a steeper slope than does the line for the liquid phase (water). Do you see why? (See Table 17.3.)

A substance can sometimes change directly from the solid to the gaseous phase. This process is called *sublimation,* and the solid is said to *sublime.* The corresponding heat is called the *heat of sublimation,* L_s. Liquid carbon dioxide cannot exist at a pressure lower than about 5×10^5 Pa (about 5 atm), and "dry ice" (solid carbon dioxide) sublimes at atmospheric pressure. Sublimation of water from frozen food causes freezer burn. The reverse process, a phase change from gas to solid, occurs when frost forms on cold bodies such as refrigerator cooling coils.

Very pure water can be cooled several degrees below the freezing temperature without freezing; the resulting unstable state is described as *supercooled.* When a small ice crystal is dropped in or the water is agitated, it crystallizes within a second or less. Supercooled water *vapor* condenses quickly into fog droplets when a disturbance, such as dust particles or ionizing radiation, is introduced. This principle is used in "seeding" clouds, which often contain supercooled water vapor, to cause condensation and rain.

A liquid can sometimes be *superheated* above its normal boiling temperature. Any small disturbance such as agitation causes local boiling with bubble formation.

Steam heating systems for buildings use a boiling–condensing process to transfer heat from the furnace to the radiators. Each kilogram of water that is turned to steam in the boiler absorbs over 2×10^6 J (the heat of vaporization L_v of water) from the boiler and gives it up when it condenses in the radiators. Boiling–condensing processes are also used in refrigerators, air conditioners, and heat pumps. We will discuss these systems in Chapter 20.

The temperature-control mechanisms of many warm-blooded animals make use of heat of vaporization, removing heat from the body by using it to evaporate

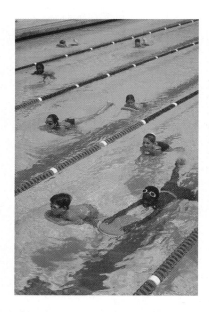

17.19 The water may be warm and it may be a hot day, but these children will feel cold when they first step out of the swimming pool. That's because as water evaporates from their skin, it removes the heat of vaporization from their bodies. To stay warm, they will need to dry off immediately.

water from the tongue (panting) or from the skin (sweating). Evaporative cooling enables humans to maintain normal body temperature in hot, dry desert climates where the air temperature may reach 55°C (about 130°F). The skin temperature may be as much as 30°C cooler than the surrounding air. Under these conditions a normal person may perspire several liters per day, and this lost water must be replaced. Old-time desert rats (such as one of the authors) state that in the desert, any canteen that holds less than a gallon should be viewed as a toy! Evaporative cooling also explains why you feel cold when you first step out of a swimming pool (Fig. 17.19).

Evaporative cooling is also used to cool buildings in hot, dry climates and to condense and recirculate "used" steam in coal-fired or nuclear-powered electric-generating plants. That's what goes on in the large, tapered concrete towers that you see at such plants.

Chemical reactions such as combustion are analogous to phase changes in that they involve definite quantities of heat. Complete combustion of one gram of gasoline produces about 46,000 J or about 11,000 cal, so the **heat of combustion** L_c of gasoline is

$$L_c = 46,000 \text{ J/g} = 4.6 \times 10^7 \text{ J/kg}$$

Energy values of foods are defined similarly; the U.S. unit of food energy, although called a calorie, is a *kilocalorie,* equal to 1000 cal or 4186 J. When we say that a gram of peanut butter "contains 6 calories," we mean that 6 kcal of heat (6,000 cal or 25,000 J) is released when the carbon and hydrogen atoms in the peanut butter react with oxygen (with the help of enzymes) and are completely converted to CO_2 and H_2O. Not all of this energy is directly useful for mechanical work. We will study the *efficiency* of energy utilization in Chapter 20.

Heat Calculations

Let's look at some examples of calorimetry calculations (calculations with heat). The basic principle is very simple: when heat flow occurs between two bodies that are isolated from their surroundings, the amount of heat lost by one body must equal the amount gained by the other. Heat is energy in transit, so this principle is really just conservation of energy. Calorimetry, dealing entirely with one conserved quantity, is in many ways the simplest of all physical theories!

Calorimetry Problems

IDENTIFY *the relevant concepts:* The basic principle underlying calorimetry calculations (calculations with heat) is very simple: when heat flow occurs between two bodies that are isolated from their surroundings, the amount of heat lost by one body must equal the amount gained by the other body. Heat is energy in transit, so this principle is really just conservation of energy. Calorimetry, dealing entirely with one conserved quantity, is in many ways the simplest of all physical theories!

SET UP *the problem using the following steps:*

1. Identify which objects exchange heat. To avoid confusion with algebraic signs, take each quantity of heat *added* to a body as *positive* and each quantity *leaving* a body as *negative.* When several bodies interact, the *algebraic sum* of the quantities of heat transferred to all the bodies must be zero.

2. Each object will undergo a temperature change with no phase change, a phase change at constant temperature, or both. Use Eq. (17.13) to describe temperature changes and Eq. (17.20) to describe phase changes.

3. Consult Table 17.3 for values of the specific heat or molar heat capacity and Table 17.4 for heats of fusion or vaporization.

4. Be certain to identify which quantities are known and which are the unknown target variables.

EXECUTE *the solution as follows:*

1. Solve Eq. (17.13) and/or Eq. (17.20) for the target variables. Often you will need to find an unknown temperature. Represent it by an algebraic symbol such as T. Then if a body has an initial temperature of 20°C and an unknown final temperature T, the temperature change for the body is $\Delta T = T_{\text{final}} - T_{\text{initial}} = T - 20°C$ (*not* 20°C $- T$).

2. In problems where a phase change takes place, as when ice melts, you may not know in advance whether *all* the material undergoes a phase change or only part of it. You can always assume one or the other, and if the resulting calculation gives an absurd result (such as a final temperature higher or lower than *any* of the initial temperatures), you know the initial assumption was wrong. Back up and try again!

EVALUATE *your answer:* A common error is to use the wrong algebraic sign for either a Q or ΔT term. Double-check your calculations, and make sure that the final results are physically sensible.

Example 17.8	A temperature change with no phase change

A geologist working in the field drinks her morning coffee out of an aluminum cup. The cup has a mass of 0.120 kg and is initially at 20.0°C when she pours in 0.300 kg of coffee initially at 70.0°C. What is the final temperature after the coffee and the cup attain thermal equilibrium? (Assume that coffee has the same specific heat as water and that there is no heat exchange with the surroundings.)

SOLUTION

IDENTIFY and SET UP: No phase changes occur in this situation, so we need use only Eq. (17.13). The two objects we must consider are the cup and the coffee, and the target variable is their common final temperature.

EXECUTE: By using Table 17.3, the (negative) heat gained by the coffee is

$$Q_{\text{coffee}} = m_{\text{coffee}} c_{\text{water}} \, \Delta T_{\text{coffee}}$$
$$= (0.300 \text{ kg})(4190 \text{ J/kg·K})(T - 70.0°C)$$

The (positive) heat gained by the aluminum cup is

$$Q_{\text{aluminum}} = m_{\text{aluminum}} c_{\text{aluminum}} \, \Delta T_{\text{aluminum}}$$
$$= (0.120 \text{ kg})(910 \text{ J/kg·K})(T - 20.0°C)$$

We equate the sum of these two quantities of heat to zero, obtaining an algebraic equation for T:

$$Q_{\text{coffee}} + Q_{\text{aluminum}} = 0 \quad \text{or}$$
$$(0.300 \text{ kg})(4190 \text{ J/kg·K})(T - 70.0°C)$$
$$+ (0.120 \text{ kg})(910 \text{ J/kg·K})(T - 20.0°C) = 0$$

Solution of this equation gives $T = 66.0°C$.

EVALUATE: The final temperature is much closer to the initial temperature of the coffee than to that of the cup; water has a much larger specific heat than aluminum, and we have more than twice as much mass of water. We can also find the quantities of heat by substituting this value for T back into the original equations. We find that $Q_{\text{coffee}} = -5.0 \times 10^3$ J and $Q_{\text{aluminum}} = +5.0 \times 10^3$ J; Q_{coffee} is negative, which means that the coffee loses heat.

Example 17.9	Changes in both temperature and phase

A physics student wants to cool 0.25 kg of Diet Omni-Cola (mostly water), initially at 25°C, by adding ice initially at −20°C. How much ice should she add so that the final temperature will be 0°C with all the ice melted if the heat capacity of the container may be neglected?

SOLUTION

IDENTIFY and SET UP: The ice and the Omni-Cola are the objects that exchange heat. The Omni-Cola undergoes a temperature change only, while the ice undergoes both a temperature change and a phase change from solid to liquid. The target variable is the mass of ice, m_{ice}.

EXECUTE: The Omni-Cola loses heat, so the heat added to it is negative:

$$Q_{\text{Omni}} = m_{\text{Omni}} c_{\text{water}} \, \Delta T_{\text{Omni}}$$
$$= (0.25 \text{ kg})(4190 \text{ J/kg·K})(0°C - 25°C)$$
$$= -26,000 \text{ J}$$

From Table 17.3, the specific heat of ice (not the same as for liquid water) is 2.1×10^3 J/kg·K. Let the mass of ice be m_{ice}; then the heat Q_1 needed to warm it from −20°C to 0°C is

$$Q_1 = m_{ice}c_{ice} \, \Delta T_{ice}$$
$$= m_{ice}(2.1 \times 10^3 \text{ J/kg} \cdot \text{K})[0°C - (-20°C)]$$
$$= m_{ice}(4.2 \times 10^4 \text{ J/kg})$$

From Eq. (17.20) the additional heat Q_2 needed to melt this mass of ice is the mass times the heat of fusion. Using Table 17.4, we find

$$Q_2 = m_{ice}L_f$$
$$= m_{ice}(3.34 \times 10^5 \text{ J/kg})$$

The sum of these three quantities must equal zero:

$$Q_{Omni} + Q_1 + Q_2 = -26,000 \text{ J} + m_{ice}(42,000 \text{ J/kg})$$
$$+ m_{ice}(334,000 \text{ J/kg}) = 0$$

Solving this for m_{ice}, we get $m_{ice} = 0.069$ kg $= 69$ g.

EVALUATE: This mass of ice corresponds to three or four medium-size ice cubes, which seems reasonable for the quantity of Omni-Cola in this problem.

Example 17.10   What's cooking?

A heavy copper pot of mass 2.0 kg (including the copper lid) is at a temperature of 150°C. You pour 0.10 kg of water at 25°C into the pot, then quickly close the lid of the pot so that no steam can escape. Find the final temperature of the pot and its contents, and determine the phase (liquid or gas) of the water. Assume that no heat is lost to the surroundings.

SOLUTION

IDENTIFY and SET UP: The two objects that exchange heat are the water and the pot. Note that there are three conceivable outcomes in this situation. One, none of the water boils, and the final temperature is less than 100°C; two, a portion of the water boils, giving a mixture of water and steam at 100°C; or three, all the water boils, giving 0.10 kg of steam at a temperature of 100°C or greater.

EXECUTE: The simplest case to calculate is the first possibility, so let's try that first. Let the common final temperature of the liquid water and the copper pot be T. Since we are assuming that no phase changes take place, the sum of the quantities of heat added to the two materials is

$$Q_{water} + Q_{copper} = m_{water}c_{water}(T - 25°C)$$
$$+ m_{copper}c_{copper}(T - 150°C)$$
$$= (0.10 \text{ kg})(4190 \text{ J/kg} \cdot \text{K})(T - 25°C)$$
$$+ (2.0 \text{ kg})(390 \text{ J/kg} \cdot \text{K})(T - 150°C)$$
$$= 0$$

Solving this for T, we find $T = 106°C$. But this is above the boiling point of water, which contradicts our assumption that none of the water boils! So this assumption can't be correct; at least some of the water undergoes a phase change.

If we try the second possibility, in which the final temperature is 100°C, we have to find the fraction of water that changes to the gaseous phase. Let this fraction be x; the (positive) amount of heat needed to vaporize this water is $(xm_{water})L_v$. Setting the final temperature T equal to 100°C, we have

$$Q_{water} = m_{water}c_{water}(100°C - 25°C) + xm_{water}L_v$$
$$= (0.10 \text{ kg})(4190 \text{ J/kg} \cdot \text{K})(75 \text{ K})$$
$$+ x(0.10 \text{ kg})(2.256 \times 10^6 \text{ J/kg})$$
$$= 3.14 \times 10^4 \text{ J} + x(2.256 \times 10^5 \text{ J})$$
$$Q_{copper} = m_{copper}c_{copper}(100°C - 150°C)$$
$$= (2.0 \text{ kg})(390 \text{ J/kg} \cdot \text{K})(-50 \text{ K}) = -3.90 \times 10^4 \text{ J}$$

The requirement that the sum of all the quantities of heat be zero then gives

$$Q_{water} + Q_{copper} = 3.14 \times 10^4 \text{ J} + x(2.256 \times 10^5 \text{ J})$$
$$-3.90 \times 10^4 \text{ J} = 0$$
$$x = \frac{3.90 \times 10^4 \text{ J} - 3.14 \times 10^4 \text{ J}}{2.256 \times 10^5 \text{ J}} = 0.034$$

This makes sense, and we conclude that the final temperature of the water and copper is 100°C. Of the original 0.10 kg of water, $0.034(0.10 \text{ kg}) = 0.0034$ kg $= 3.4$ g has been converted to steam at 100°C.

EVALUATE: Had x turned out to be greater than 1, we would have again had a contradiction (the fraction of water that vaporized can't be greater than 1). In this case the third possibility would have been the correct description, all the water would have vaporized, and the final temperature would have been greater than 100°C. Can you show that this would have been the case if we had originally poured less than 15 g of 25°C water into the pot?

| Example 17.11 | Combustion, temperature change, and phase change |

In a particular gasoline camp stove, 30% of the energy released in burning the fuel actually goes to heating the water in the pot on the stove. If we heat 1.00 L (1.00 kg) of water from 20°C to 100°C and boil 0.25 kg of it away, how much gasoline do we burn in the process?

SOLUTION

IDENTIFY and SET UP: In this problem, Eqs. (17.13) and (17.20) apply to the water, all of which undergoes a temperature change and part of which also undergoes a phase change from liquid to gas. This requires a certain amount of heat, which we use to determine the amount of gasoline that must be burned (the target variable).

EXECUTE: The heat required to raise the temperature of the water from 20°C to 100°C is

$$Q_1 = mc\,\Delta T = (1.00\text{ kg})(4190\text{ J/kg}\cdot\text{K})(80\text{ K})$$
$$= 3.35 \times 10^5\text{ J}$$

To boil 0.25 kg of water at 100°C requires

$$Q_2 = mL_\text{v} = (0.25\text{ kg})(2.256 \times 10^6\text{ J/kg}) = 5.64 \times 10^5\text{ J}$$

The total energy needed is the sum of these, or 8.99×10^5 J. This is only 0.30 of the total heat of combustion, so that energy is $(8.99 \times 10^5\text{ J})/0.30 = 3.00 \times 10^6$ J. As we mentioned earlier, one gram of gasoline releases 46,000 J, so the mass of gasoline required is

$$\frac{3.00 \times 10^6\text{ J}}{46,000\text{ J/g}} = 65\text{ g}$$

or a volume of about 0.09 L of gasoline.

EVALUATE: This result is a testament to the tremendous amount of energy that can be released by burning even a small quantity of gasoline. Note that most of the heat delivered was used to boil away 0.25 L of water. Can you show that another 123 g of gasoline would be required to boil away the remaining water?

Test Your Understanding

You take a block of ice at 0°C and add heat to it at a steady rate. After a time t, the block of ice has been converted completely to steam at 100°C. What is the temperature of the (ice? water? steam?) at time $t/2$, and what is its phase?

17.7 | Mechanisms of Heat Transfer

We have talked about *conductors* and *insulators,* materials that permit or prevent heat transfer between bodies. Now let's look in more detail at *rates* of energy transfer. In the kitchen you use an aluminum pot for good heat transfer from the stove to whatever you're cooking, but your refrigerator is insulated with a material that *prevents* heat from flowing into the food inside the refrigerator. How do we describe the difference between these two materials?

The three mechanisms of heat transfer are conduction, convection, and radiation. *Conduction* occurs within a body or between two bodies in contact. *Convection* depends on motion of mass from one region of space to another. *Radiation* is heat transfer by electromagnetic radiation, such as sunshine, with no need for matter to be present in the space between bodies.

Conduction

If you hold one end of a copper rod and place the other end in a flame, the end you are holding gets hotter and hotter, even though it is not in direct contact with the flame. Heat reaches the cooler end by **conduction** through the material. On the atomic level, the atoms in the hotter regions have more kinetic energy, on the average, than

17.20 Steady-state heat flow due to conduction in a uniform rod.

their cooler neighbors. They jostle their neighbors, giving them some of their energy. The neighbors jostle *their* neighbors, and so on through the material. The atoms themselves do not move from one region of material to another, but their energy does.

Most metals also use another, more effective mechanism to conduct heat. Within the metal, some electrons can leave their parent atoms and wander through the crystal lattice. These "free" electrons can rapidly carry energy from the hotter to the cooler regions of the metal, so metals are generally good conductors of heat. A metal rod at 20°C feels colder than a piece of wood at 20°C because heat can flow more easily from your hand into the metal. The presence of "free" electrons also causes most metals to be good electrical conductors.

Heat transfer occurs only between regions that are at different temperatures, and the direction of heat flow is always from higher to lower temperature. Figure 17.20 shows a rod of conducting material with cross-sectional area A and length L. The left end of the rod is kept at a temperature T_H and the right end at a lower temperature T_C, so heat flows from left to right. The sides of the rod are covered by an ideal insulator, so no heat transfer occurs at the sides.

When a quantity of heat dQ is transferred through the rod in a time dt, the rate of heat flow is dQ/dt. We call this rate the **heat current,** denoted by H. That is, $H = dQ/dt$. Experiments show that the heat current is proportional to the cross-sectional area A of the rod and to the temperature difference $(T_H - T_C)$ and is inversely proportional to the rod length L. Introducing a proportionality constant k called the **thermal conductivity** of the material, we have

$$H = \frac{dQ}{dt} = kA\frac{T_H - T_C}{L} \qquad \text{(heat current in conduction)} \quad (17.21)$$

The quantity $(T_H - T_C)/L$ is the temperature difference *per unit length;* it is called the magnitude of the **temperature gradient.** The numerical value of k depends on the material of the rod. Materials with large k are good conductors of heat; materials with small k are poor conductors or insulators. Equation (17.21) also gives the heat current through a slab or through *any* homogeneous body with uniform cross section A perpendicular to the direction of flow; L is the length of the heat-flow path.

The units of heat current H are units of energy per time, or power; the SI unit of heat current is the watt $(1\ \text{W} = 1\ \text{J/s})$. We can find the units of k by solving Eq. (17.21) for k; you can show that the SI units are $\text{W/m} \cdot \text{K}$. Some numerical values of k are given in Table 17.5.

The thermal conductivity of "dead" (that is, nonmoving) air is very small. A wool sweater keeps you warm because it traps air between the fibers. In fact, many insulating materials such as Styrofoam and fiberglass are mostly dead air. Figure 17.21 shows a ceramic material with very unusual thermal properties, including very small conductivity.

If the temperature varies in a non-uniform way along the length of the conducting rod, we introduce a coordinate x along the length and generalize the temperature gradient to be dT/dx. The corresponding generalization of Eq. (17.21) is

$$H = \frac{dQ}{dt} = -kA\frac{dT}{dx} \qquad (17.22)$$

Table 17.5 Thermal Conductivities

Substance	$k\ (\text{W/m} \cdot \text{K})$
Metals	
Aluminum	205.0
Brass	109.0
Copper	385.0
Lead	34.7
Mercury	8.3
Silver	406.0
Steel	50.2
Various solids (representative values)	
Brick, insulating	0.15
Brick, red	0.6
Concrete	0.8
Cork	0.04
Felt	0.04
Fiberglass	0.04
Glass	0.8
Ice	1.6
Rock wool	0.04
Styrofoam	0.01
Wood	0.12–0.04
Gases	
Air	0.024
Argon	0.016
Helium	0.14
Hydrogen	0.14
Oxygen	0.023

The negative sign shows that heat always flows in the direction of *decreasing* temperature.

For thermal insulation in buildings, engineers use the concept of **thermal resistance,** denoted by R. The thermal resistance R of a slab of material with area A is defined so that the heat current H through the slab is

$$H = \frac{A(T_H - T_C)}{R} \tag{17.23}$$

where T_H and T_C are the temperatures on the two sides of the slab. Comparing this with Eq. (17.21), we see that R is given by

$$R = \frac{L}{k} \tag{17.24}$$

where L is the thickness of the slab. The SI unit of R is $1\ m^2 \cdot K/W$. In the units used for commercial insulating materials in the United States, H is expressed in Btu/h, A is in ft^2, and $T_H - T_C$ in F°. (1 Btu/h $= 0.293$ W.) The units of R are then $ft^2 \cdot F° \cdot h/Btu$, though values of R are usually quoted without units; a 6-inch-thick layer of fiberglass has an R value of 19 (that is, $R = 19\ ft^2 \cdot F° \cdot h/Btu$), a two-inch-thick slab of polyurethane foam has a value of 12, and so on. Doubling the thickness doubles the R-value. Common practice in new construction in severe northern climates is to specify R values of around 30 for exterior walls and ceilings. When the insulating material is in layers, such as a plastered wall, fiberglass insulation, and wood exterior siding, the R values are additive. Do you see why? (See Problem 17.110.)

17.21 This protective tile, developed for use in the space shuttle, has extraordinary thermal properties. The extremely small thermal conductivity and small heat capacity of the material make it possible to hold the tile by its edges, even though its temperature is high enough to emit the light for this photograph.

Problem-Solving Strategy	**Heat Conduction**

IDENTIFY *the relevant concepts:* The concept of heat conduction comes into play whenever two objects at different temperature are placed in contact.

SET UP *the problem using the following steps:*
1. Identify the direction of heat flow in the problem (from hot to cold). In Eq. (17.21), L is always measured along this direction, and A is always an area perpendicular to this direction. Often when a box or other container has an irregular shape but uniform wall thickness, you can approximate it as a flat slab with the same thickness and total wall area.
2. Identify the target variable.

EXECUTE *the solution as follows:*
1. If heat flows through a single object, use Eq. (17.21) to solve for the target variable.
2. In some problems the heat flows through two different materials in succession. The temperature at the interface between the two materials is then intermediate between T_H and T_C; represent it by a symbol such as T. The temperature differences for the two materials are then $(T_H - T)$ and $(T - T_C)$. In steady-state heat flow, the same heat has to pass through both materials in succession, so the heat current H must be *the same* in both materials.
3. If there are two *parallel* heat-flow paths, so that some heat flows through each, then the total H is the sum of the quantities H_1 and H_2 for the separate paths. An example is heat flow from inside to outside a house, both through the glass in a window and through the surrounding frame. In this case the temperature difference is the same for the two paths, but L, A, and k may be different for the two paths.
4. As always, it is essential to use a consistent set of units. If you use a value of k expressed in W/m·K, don't use distances in centimeters, heat in calories, or T in degrees Fahrenheit!

EVALUATE *your answer:* As always, ask yourself whether the results are physically reasonable.

| Example 17.12 | **Conduction through a picnic cooler** |

A Styrofoam box used to keep drinks cold at a picnic has total wall area (including the lid) of 0.80 m^2 and wall thickness 2.0 cm. It is filled with ice, water, and cans of Omni-Cola at 0°C. What is the rate of heat flow into the box if the temperature of the outside wall is 30°C? How much ice melts in one day?

SOLUTION

IDENTIFY and SET UP: The first target variable is the heat current H. The second is the amount of ice melted, which depends on the heat current (heat per unit time), the elapsed time, and the heat of fusion.

EXECUTE: We assume that the total heat flow is approximately the same as it would be through a flat slab of area 0.80 m^2 and thickness 2.0 cm = 0.020 m (Fig. 17.22). We find k from Table 17.5. From Eq. (17.21) the heat current (rate of heat flow) is

$$H = kA\frac{T_H - T_C}{L} = (0.010 \text{ W/m} \cdot \text{K})(0.80 \text{ m}^2)\frac{30°C - 0°C}{0.020 \text{ m}}$$
$$= 12 \text{ W} = 12 \text{ J/s}$$

The total heat flow Q in one day (86,400 s) is

$$Q = Ht = (12 \text{ J/s})(86,400 \text{ s}) = 1.04 \times 10^6 \text{ J}$$

The heat of fusion of ice is 3.34×10^5 J/kg, so the quantity of ice melted by this quantity of heat is

$$m = \frac{Q}{L_f}$$
$$= \frac{1.04 \times 10^6 \text{ J}}{3.34 \times 10^5 \text{ J/kg}} = 3.1 \text{ kg}$$

EVALUATE: The low heat current is a result of the low thermal conductivity of Styrofoam. A substantial amount of heat flows in 24 hours, but a relatively small amount of ice melts because the heat of fusion is high.

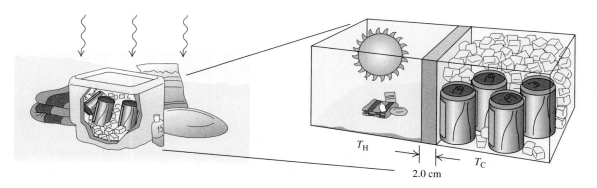

17.22 Conduction of heat. We can approximate heat flow through the walls of a picnic cooler by heat flow through a single flat slab of Styrofoam.

| Example 17.13 | **Conduction through two bars I** |

A steel bar 10.0 cm long is welded end-to-end to a copper bar 20.0 cm long (Fig. 17.23). Both bars are insulated perfectly on their sides. Each bar has a square cross section, 2.00 cm on a side. The free end of the steel bar is maintained at 100°C by placing it in contact with steam, and the free end of the copper bar is maintained at 0°C by placing it in contact with ice. Find the temperature at the junction of the two bars and the total rate of heat flow.

SOLUTION

IDENTIFY and SET UP: As we discussed in the Problem-Solving Strategy, the heat currents in the two bars must be equal; this idea is the key to the solution. We write Eq. (17.21) twice, once for each bar, and set the heat currents H_{steel} and H_{copper} equal to each other. Both expressions for the heat current involve the temperature T at the junction, which is one of our target variables.

17.23 Heat flow along two metal bars, one of steel and one of copper, connected end-to-end.

EXECUTE: Setting the two heat currents equal,

$$H_{steel} = \frac{k_{steel}A(100°C - T)}{L_{steel}} = H_{copper} = \frac{k_{copper}A(T - 0°C)}{L_{copper}}$$

The areas A are equal and may be divided out. Substituting $L_{steel} = 0.100$ m, $L_{copper} = 0.200$ m, and numerical values of k from Table 17.5, we find

$$\frac{(50.2 \text{ W/m} \cdot \text{K})(100°C - T)}{0.100 \text{ m}} = \frac{(385 \text{ W/m} \cdot \text{K})(T - 0°C)}{0.200 \text{ m}}$$

Rearranging and solving for T, we find

$$T = 20.7°C$$

We can find the total heat current by substituting this value for T back into either of the above expressions:

$$H_{steel} = \frac{(50.2 \text{ W/m} \cdot \text{K})(0.0200 \text{ m})^2(100°C - 20.7°C)}{0.100 \text{ m}}$$

$$= 15.9 \text{ W}$$

or

$$H_{copper} = \frac{(385 \text{ W/m} \cdot \text{K})(0.0200 \text{ m})^2(20.7°C)}{0.200 \text{ m}} = 15.9 \text{ W}$$

EVALUATE: Even though the steel bar is shorter, the temperature drop across it is much greater than across the copper bar (from 100°C to 20.7°C in the steel versus from 20.7°C to 0°C in the copper). This difference arises because steel is a much poorer conductor than copper.

Example 17.14	**Conduction through two bars II**

In Example 17.13, suppose the two bars are separated. One end of each bar is maintained at 100°C and the other end of each bar is maintained at 0°C (Fig. 17.24). What is the *total* rate of heat flow in the two bars?

17.24 Heat flow along two metal bars, one of steel and one of copper, parallel to and separated from each other.

SOLUTION

IDENTIFY and SET UP: In this case the bars are in parallel rather than in series. The total heat current is now the *sum* of the currents in the two bars, and for each bar, $T_H - T_C = 100°C - 0°C = 100$ K.

EXECUTE: We write the heat currents for the two rods individually, then add them to get the total heat current:

$$H = H_{steel} + H_{copper} = \frac{k_{steel}A(T_H - T_C)}{L_{steel}} + \frac{k_{copper}A(T_H - T_C)}{L_{copper}}$$

$$= \frac{(50.2 \text{ W/m} \cdot \text{K})(0.0200 \text{ m})^2(100 \text{ K})}{0.100 \text{ m}}$$

$$+ \frac{(385 \text{ W/m} \cdot \text{K})(0.0200 \text{ m})^2(100 \text{ K})}{0.200 \text{ m}}$$

$$= 20.1 \text{ W} + 77.0 \text{ W} = 97.1 \text{ W}$$

EVALUATE: The heat flow in the copper bar is much greater than that in the steel bar, even though it is longer, because the thermal conductivity of copper is much larger. The total heat flow is much greater than in Example 17.13, partly because the total cross section for heat flow is greater and partly because the full 100-K temperature difference appears across each bar.

17.25 A heating element in the tip of this submerged tube warms the surrounding water, producing a complex pattern of free convection.

Convection

Convection is the transfer of heat by mass motion of a fluid from one region of space to another. Familiar examples include hot-air and hot-water home heating systems, the cooling system of an automobile engine, and the flow of blood in the body. If the fluid is circulated by a blower or pump, the process is called *forced convection;* if the flow is caused by differences in density due to thermal expansion, such as hot air rising, the process is called *natural convection* or *free convection* (Fig. 17.25).

Free convection in the atmosphere plays a dominant role in determining the daily weather, and convection in the oceans is an important global heat-transfer mechanism. On a smaller scale, soaring hawks and glider pilots make use of thermal updrafts from the warm earth. The most important mechanism for heat transfer within the human body (needed to maintain nearly constant temperature in various environments) is *forced* convection of blood, with the heart serving as the pump.

Convective heat transfer is a very complex process, and there is no simple equation to describe it. Here are a few experimental facts:

1. The heat current due to convection is directly proportional to the surface area. This is the reason for the large surface areas of radiators and cooling fins.
2. The viscosity of fluids slows natural convection near a stationary surface, giving a surface film that on a vertical surface typically has about the same insulating value as 1.3 cm of plywood $(R$ value $= 0.7)$. Forced convection decreases the thickness of this film, increasing the rate of heat transfer. This is the reason for the "wind-chill factor"; you get cold faster in a cold wind than in still air with the same temperature.
3. The heat current due to convection is found to be approximately proportional to the $\frac{5}{4}$ power of the temperature difference between the surface and the main body of fluid.

Radiation

Radiation is the transfer of heat by electromagnetic waves such as visible light, infrared, and ultraviolet radiation. Everyone has felt the warmth of the sun's radiation and the intense heat from a charcoal grill or the glowing coals in a fireplace. Most of the heat from these very hot bodies reaches you not by conduction or convection in the intervening air but by *radiation*. This heat transfer would occur even if there were nothing but vacuum between you and the source of heat.

Every body, even at ordinary temperatures, emits energy in the form of electromagnetic radiation. At ordinary temperatures, say 20°C, nearly all the energy is carried by infrared waves with wavelengths much longer than those of visible light (see Figs. 17.4 and 17.26). As the temperature rises, the wavelengths shift to shorter values. At 800°C a body emits enough visible radiation to be self-luminous and appears "red-hot," although even at this temperature most of the energy is carried by infrared waves. At 3000°C, the temperature of an incandescent lamp filament, the radiation contains enough visible light so the body appears "white-hot."

The rate of energy radiation from a surface is proportional to the surface area A. The rate increases very rapidly with temperature, depending on the fourth power of the absolute (Kelvin) temperature. The rate also depends on the nature of the surface; this dependence is described by a quantity e called the **emissivity.** A dimensionless number between 0 and 1, it represents the ratio of the rate of radiation from a particular surface to the rate of radiation from an equal area of an ideal radiating surface at the same temperature. Emissivity also depends some-

17.26 This false-color infrared photograph reveals radiation emitted by various parts of the man's body. The strongest emission (colored red) comes from the warmest areas, while there is very little emission from the bottle of cold beverage.

what on temperature. Thus the heat current $H = dQ/dt$ due to radiation from a surface area A with emissivity e at absolute temperature T can be expressed as

$$H = Ae\sigma T^4 \qquad \text{(heat current in radiation)} \qquad (17.25)$$

where σ is a fundamental physical constant called the **Stefan-Boltzmann constant.** This relation is called the **Stefan-Boltzmann law** in honor of its late-19th-century discoverers. The current best numerical value of σ is

$$\sigma = 5.670400(40) \times 10^{-8} \text{ W/m}^2 \cdot \text{K}^4$$

We invite you to check unit consistency in Eq. (17.25). Emissivity (e) is often larger for dark surfaces than for light ones. The emissivity of a smooth copper surface is about 0.3, but e for a dull black surface can be close to unity.

Example 17.15	**Heat transfer by radiation**

A thin square steel plate, 10 cm on a side, is heated in a black-smith's forge to a temperature of 800°C. If the emissivity is 0.60, what is the total rate of radiation of energy?

SOLUTION

IDENTIFY and SET UP: We use Eq. (17.25). The target variable is H, the rate of emission of energy. All of the other quantities are given.

EXECUTE: The total surface area, including both sides, is $2(0.10 \text{ m})^2 = 0.020 \text{ m}^2$. We must convert the temperature to the Kelvin scale; 800°C = 1073 K. Then Eq. (17.25) gives

$$\begin{aligned} H &= Ae\sigma T^4 \\ &= (0.020 \text{ m}^2)(0.60)(5.67 \times 10^{-8} \text{ W/m}^2 \cdot \text{K}^4)(1073 \text{ K})^4 \\ &= 900 \text{ W} \end{aligned}$$

EVALUATE: A blacksmith standing nearby will easily feel heat being radiated from this plate.

While a body at absolute temperature T is radiating, its surroundings at temperature T_s are also radiating, and the body *absorbs* some of this radiation. If it is in thermal equilibrium with its surroundings, $T = T_s$ and the rates of radiation and absorption must be equal. For this to be true, the rate of absorption must be given in general by $H = Ae\sigma T_s^4$. Then the *net* rate of radiation from a body at temperature T with surroundings at temperature T_s is

$$H_{\text{net}} = Ae\sigma T^4 - Ae\sigma T_s^4 = Ae\sigma(T^4 - T_s^4) \qquad (17.26)$$

In this equation a positive value of H means a net heat flow *out of* the body. Equation (17.26) shows that for radiation, as for conduction and convection, the heat current depends on the temperature *difference* between two bodies.

Example 17.16	**Radiation from the human body**

If the total surface area of the human body is 1.20 m^2 and the surface temperature is 30°C = 303 K, find the total rate of radiation of energy from the body. If the surroundings are at a temperature of 20°C, what is the *net* rate of heat loss from the body by radiation? The emissivity of the body is very close to unity, irrespective of skin pigmentation.

SOLUTION

IDENTIFY and SET UP: The rate of radiation of energy from the body is given by Eq. (17.25), and the net rate of heat loss is given by Eq. (17.26).

EXECUTE: Taking $e = 1$ in Eq. (17.25), we find that the body radiates at a rate

$H = Ae\sigma T^4$

$\quad = (1.20 \text{ m}^2)(1)(5.67 \times 10^{-8} \text{ W/m}^2 \cdot \text{K}^4)(303 \text{ K})^4$

$\quad = 574 \text{ W}$

This loss is partly offset by *absorption* of radiation, which depends on the temperature of the surroundings. The *net* rate of radiative energy transfer is given by Eq. (17.26):

$H_{\text{net}} = Ae\sigma(T^4 - T_s^4)$

$\quad = (1.20 \text{ m}^2)(1)(5.67 \times 10^{-8} \text{ W/m}^2 \cdot \text{K}^4)$

$\quad \quad \times [(303 \text{ K})^4 - (293 \text{ K})^4] = 72 \text{ W}$

EVALUATE: The value of H_{net} is positive because the body is losing heat to its colder surroundings.

Heat transfer by radiation is important in some surprising places. A premature baby in an incubator can be cooled dangerously by radiation if the walls of the incubator happen to be cold, even when the *air* in the incubator is warm. Some incubators regulate the air temperature by measuring the baby's skin temperature.

A body that is a good absorber must also be a good emitter. An ideal radiator, with an emissivity of unity, is also an ideal absorber, absorbing *all* of the radiation that strikes it. Such an ideal surface is called an ideal black body or simply a **blackbody.** Conversely, an ideal *reflector*, which absorbs *no* radiation at all, is also a very ineffective radiator.

This is the reason for the silver coatings on vacuum ("Thermos") bottles, invented by Sir James Dewar (1842–1923). A vacuum bottle has double glass walls. The air is pumped out of the spaces between the walls; this eliminates nearly all heat transfer by conduction and convection. The silver coating on the walls reflects most of the radiation from the contents back into the container, and the wall itself is a very poor emitter. Thus a vacuum bottle can keep coffee or soup hot for several hours. The Dewar flask, used to store very cold liquefied gases, is exactly the same in principle.

Test Your Understanding

The ear thermometer in Fig. 17.4 measures the radiation emitted by the eardrum. By what percentage does the radiation rate increase if the eardrum's temperature increases from 37.00°C to 37.10°C?

SUMMARY

A thermometer measures temperature. Two bodies in thermal equilibrium must have the same temperature. A conducting material between two bodies permits them to interact and come to thermal equilibrium; an insulating material prevents or impedes this interaction.

If A and B are each in thermal equilibrium with C... Then A and B are in thermal equilibrium with each other

---

The Celsius and Fahrenheit temperature scales are based on the freezing temperature $(0°C = 32°F)$ and boiling temperature $(100°C = 212°F)$. One Celsius degree equals $\frac{9}{5}$ Fahrenheit degrees. (See Example 17.1)

$$T_F = \frac{9}{5}T_C + 32° \qquad (17.1)$$

$$T_C = \frac{5}{9}(T_F - 32°) \qquad (17.2)$$

The Kelvin scale has its zero at the extrapolated zero-pressure temperature for a gas thermometer, $-273.15°C = 0$ K. In the gas-thermometer scale, the ratio of two temperatures T_1 and T_2 is defined to be equal to the ratio of the two corresponding gas-thermometer pressures p_1 and p_2. The triple-point temperature of water $(0.01°C)$ is defined to be 273.16 K.

$$T_K = T_C + 273.15 \qquad (17.3)$$

$$\frac{T_2}{T_1} = \frac{p_2}{p_1} \qquad (17.4)$$

In each case, straight-line extrapolation predicts that pressure would become 0 at -273.15 °C

Same experiment with different amounts of different gases

A temperature change ΔT causes a change in any linear dimension L_0 of a solid body. The change ΔL is approximately proportional to L_0 and ΔT. Similarly, a temperature change causes a change ΔV in the volume V_0 of any solid or liquid material that is approximately proportional to V_0 and ΔT. The quantities α and β are the coefficients of linear expansion and volume expansion. For solids, $\beta = 3\alpha$. (See Examples 17.2 through 17.4)

$$\Delta L = \alpha L_0 \Delta T \quad (17.6)$$
$$\Delta V = \beta V_0 \Delta T \quad (17.8)$$

INCORRECT

CORRECT

When a material is cooled or heated and held so it cannot contract or expand, it is under a tensile stress F/A. (See Example 17.5)

$$\frac{F}{A} = -Y\alpha \Delta T \qquad (17.12)$$

Heat is energy in transit from one body to another as a result of a temperature difference. The quantity of heat Q required to raise the temperature of a quantity of material by a small amount ΔT is proportional to ΔT. This proportionality can be expressed either in terms of the mass m and specific heat capacity c or in terms of the number of moles n and the molar heat capacity $C = Mc$. Here M is the molar mass and $m = nM$. (See Examples 17.6 and 17.7)

$$Q = mc\,\Delta T \qquad (17.13)$$

$$Q = nC\,\Delta T \qquad (17.18)$$

To change a mass m of a material to a different phase at the same temperature (such as liquid to solid or liquid to vapor) requires the addition or subtraction of a quantity of heat. The amount of heat is equal to the product of m and L, the heat of fusion, vaporization, or sublimation.

$$Q = \pm mL \qquad (17.20)$$

When heat is added to a body, the corresponding Q is positive; when it is removed, Q is negative. The basic principle of calorimetry comes from conservation of energy. In an isolated system whose parts interact by heat exchange, the algebraic sum of the Q's for all parts of the system must be zero. (See Examples 17.8 through 17.11)

Conduction is the transfer of energy of molecular motion within bulk materials without bulk motion of the materials. The heat current H or conduction depends on the area A through which the heat flows, the length L of the heat-flow path, the temperature difference $(T_H - T_C)$, and the thermal conductivity k of the material.
(See Examples 17.2 through 17.4)

$$H = \frac{dQ}{dt} = kA\frac{T_H - T_C}{L}$$
$$(17.21)$$

Convection is a complex heat transfer process that involves mass motion from one region to another. It depends on surface area, orientation, and the temperature difference between a body and its surroundings.

Radiation is energy transfer through electromagnetic radiation. The radiation heat current H depends on the surface area A, the emissivity e of the surface (a pure number between 0 and 1), and the Kelvin temperature T. It involves a fundamental constant σ called the Stefan-Boltzmann constant. When a body at temperature T is surrounded by material at temperature T_s, the *net* heat current H_{net} from the body to its surroundings depends on both T and T_s.
(See Examples 17.5 and 17.16)

$$H = Ae\sigma T^4$$
$$(17.25)$$
$$H_{net} = Ae\sigma(T^4 - T_s^4)$$
$$(17.26)$$

Key Terms

Your Notes

Answer to Chapter Opening Question

No. By "heat" we mean energy that is in transit from one body to another as a result of temperature difference between the bodies. Bodies do not *contain* heat.

Answers to Test Your Understanding Questions

Section 17.1 The nurse is waiting for the thermometer and your body to come into thermal equilibrium. The reading actually shows the reading of the thermometer; if the thermometer is in thermal equilibrium with your body, it also indicates body temperature.

Section 17.2 From Eq. (17.1), on Venus $T_F = \frac{9}{5}T_C + 32° = \frac{9}{5}(460°) + 32° = 860°F$. The two scales agree when $T_F = T_C$; again using Eq. (17.1), this means $T_C = \frac{9}{5}T_C + 32°$. Solving, $\frac{4}{5}T_C = -32°$ and $T_C = \frac{5}{4}(-32°C) = -40°C$. Hence $-40°F$ is the same temperature as $-40°C$.

Section 17.3 The Kelvin temperature of the corona is $2.0 \times 10^7 + 273.15 = 2.0 \times 10^7$ K. (Note that the Celsius temperature is given to only two significant figures.) The Kelvin temperature of the boiling point of helium is $-268.93 + 273.15 = 4.22$ K. The Kelvin scale is convenient at high temperatures where the difference between 0 K and 0°C is negligible. At low temperatures, using the Kelvin scale gives smaller, more easily interpreted values.

Section 17.4 Metal 2 must expand more than metal 1 when heated and so must have a larger coefficient of linear expansion α. From Table 17.1, two metals with larger values of α than copper are aluminum and brass.

Section 17.5 The temperature change is $\Delta T = 70$ K, so the required amount of heat is $Q = m_{water}c\,\Delta T = (1.00 \text{ kg})(4190 \text{ J/kg}\cdot\text{K})(70 \text{ K}) = 2.93 \times 10^5$ J. If the block falls a distance h, the amount of potential energy lost is $m_{block}gh$, so $h = Q/m_{block}g = (2.93 \times 10^5 \text{ J})/(1.00 \text{ kg})(9.80 \text{ m/s}^2) = 2.99 \times 10^4$ m, or 29.9 kilometers! Because the specific heat of water is so high, this is a *very* impractical way to make tea.

Section 17.6 In time t the system goes from point b to point e in Fig. 17.18. According to this figure, at time $t/2$ (halfway along the horizontal axis from b to e), the system is at 100°C and is still boiling; that is, it is a mixture of liquid and gas. This says that most of the heat added goes into boiling the water.

Section 17.7 From Eq. (17.25), the radiation rate (heat current) is proportional to the fourth power of the Kelvin temperature. The temperature increases from $(37.00 + 273.15)$ K $= 310.15$ K to $(37.10 + 273.15)$ K $= 310.25$ K, so the radiation rate increases by a factor $[(310.25 \text{ K})/(310.15 \text{ K})]^4 = 1.0013$. Hence the percentage increase is 0.13%, which is well within the sensitivity of an ear thermometer.

Discussion Questions

Q17.1 A thermometer is laid out in direct sunlight. Does it measure the temperature of the air, or of the sun, or of something else? Explain.

Q17.2 Does it make sense to say that one body is twice as hot as another? Why or why not?

Q17.3 Many automobile engines have cast-iron cylinders and aluminum pistons. What kinds of problems could occur if the engine gets too hot? (The coefficient of volume expansion of cast iron is approximately the same as that of steel.)

Q17.4 Why do frozen water pipes burst? Would a mercury thermometer break if the temperature went below the freezing temperature of mercury? Why or why not?

Q17.5 Two bodies made of the same material have the same external dimensions and appearance, but one is solid and the other is hollow. When their temperature is increased, is the overall volume expansion the same or different? Why?

Q17.6 The inside of an oven is at a temperature of 200°C (392°F). You can put your hand in the oven without injury as long as you don't touch anything. But since the air inside the oven is also at 200°C, why isn't your hand burned just the same?

Q17.7 A newspaper article about the weather states that "the temperature of a body measures how much heat the body contains." Is this description correct? Why or why not?

Q17.8 To raise the temperature of an object, must you add heat to it? If you add heat to an object, must you raise its temperature? Explain.

Q17.9 A student asserts that a suitable unit for specific heat capacity is $1 \text{ m}^2/\text{s}^2\cdot\text{C}°$. Is she correct? Why or why not?

Q17.10 In some household air conditioners used in dry climates, air is cooled by blowing it through a water-soaked filter, evaporating some of the water. How does this cool the air? Would such a system work well in a high-humidity climate? Why or why not?

Q17.11 The units of specific heat capacity c are J/kg$\cdot$K, but the units of heat of fusion L_f or heat of vaporization L_v are simply J/kg. Why do the units of L_f and L_v not include a factor of $(\text{K})^{-1}$ to account for a temperature change?

Q17.12 If you set an ice cube on your kitchen table, it will stay frozen longer if it is wrapped in a wet paper towel. Why?

Q17.13 Why does food cook faster in a pressure cooker than in an open pot of boiling water?

Q17.14 Desert travelers sometimes keep water in a canvas bag. Some water seeps through the bag and evaporates. How does this cool the water inside?

Q17.15 When you first step out of the shower, you feel cold. But as soon as you are dry you feel warmer, even though the room temperature does not change. Why?

Q17.16 The climate of regions adjacent to large bodies of water (like the Pacific and Atlantic coasts) is usually more moderate than that of regions far from large bodies of water (like the prairies). Why?

Q17.17 When water is placed in ice-cube trays in a freezer, why does the water not freeze all at once when the temperature has reached 0°C? In fact the water freezes first in a layer adjacent to the sides of the tray. Why?

Q17.18 Before giving you an injection, a physician swabs your arm with isopropyl alcohol at room temperature. Why does this make your arm feel cold? (*Hint:* The reason is *not* the fear of the injection! The boiling point of isopropyl alcohol is 82.4°C.)

Q17.19 A cold block of metal feels colder than a block of wood at the same temperature. Why? A *hot* block of metal feels hotter than a block of wood at the same temperature. Again, why? Is there any temperature at which the two blocks feel equally hot or cold? What temperature is this?

Q17.20 A person pours a cup of hot coffee, intending to drink it five minutes later. To keep the coffee as hot as possible, should she put cream in it now, or wait until just before she drinks it? Explain.

Q17.21 When a freshly baked apple pie has just been removed from the oven, the crust and filling are both at the same temperature. Yet if you sample the pie, the filling will burn your tongue but the crust will not. Why is there a difference? (*Hint:* The filling is moist while the crust is dry.)

Q17.22 Old-time kitchen lore suggests that things cook better (evenly and without burning) in heavy cast-iron pots. What desirable characteristics do such pots have?

Q17.23 Coastal land has a higher temperature than the ocean during the day, but a lower temperature at night. Explain why. (*Hint:* The specific heat capacity of soil is only 0.2–0.8 times as great as that of water.)

Q17.24 It is well known that a potato bakes faster if a large nail is stuck through it. Why? Is an aluminum nail better than a steel one? Why or why not? (*Note:* Don't try this in a microwave oven!) There is also a gadget on the market to hasten the roasting of meat, consisting of a hollow metal tube containing a wick and some water; this is claimed to be much better than a solid metal rod. How does this work?

Q17.25 Glider pilots in the Midwest know that thermal updrafts are likely to occur in the vicinity of freshly plowed fields. Why?

Q17.26 Some folks claim that ice cubes freeze faster if the trays are filled with hot water, because hot water cools off faster than cold water. What do you think?

Q17.27 We're lucky that the earth isn't in thermal equilibrium with the sun (which has a surface temperature of 5800 K). But why are they not in thermal equilibrium?

Q17.28 When energy shortages occur, magazines sometimes urge us to keep our homes at a constant temperature day and night to conserve fuel. They argue that by turning down the heat at night, the walls, ceilings, etc. cool off and must be reheated in the morning. So, by keeping the temperature constant, these parts of the house would not cool off and would not have to be reheated. Does this argument make sense? Would we really save energy by following this advice?

Exercises

Section 17.2 Thermometers and Temperature Scales

17.1 Convert the following Celsius temperatures to Fahrenheit: a) $-62.8°C$, the lowest temperature ever recorded in North America (February 3, 1947, Snag, Yukon); b) $56.7°C$, the highest temperature ever recorded in the United States (July 10, 1913, Death Valley, California); c) $31.1°C$, the world's highest average annual temperature (Lugh Ferrandi, Somalia).

17.2 Find the Celsius temperatures corresponding to a) a winter night in Seattle ($41.0°F$); b) a hot summer day in Palm Springs ($107.0°F$); c) a cold winter day in northern Manitoba ($-18.0°F$).

17.3 You work in a materials testing lab and your boss tells you to increase the temperature of a sample by $40.0 C°$. The only thermometer you can find at your workbench reads in °F. If the initial temperature of the sample is $68.2°F$, what is its temperature in °F when the desired temperature increase has been achieved?

17.4 a) On January 22, 1943, the temperature in Spearfish, South Dakota, rose from $-4.0°F$ to $45.0°F$ in just two minutes. What was the temperature change in Celsius degrees? b) The temperature in Browning, Montana, was $44.0°F$ on January 23, 1916. The next day the temperature plummeted to $-56.0°F$. What was the temperature change in Celsius degrees?

17.5 a) You feel sick and are told that you have a temperature of $40.2°C$. What is your temperature in °F? Should you be concerned? b) The morning weather report in Sydney gives a current temperature of $12°C$. What is this temperature in °F?

17.6 A "blue norther" passes through Lubbock, Texas, one September afternoon and the temperature drops $11.8 C°$ in one hour. What is this temperature change in F°?

17.7 Two beakers of water, A and B, initially are at the same temperature. The temperature of the water in beaker A is increased $10 F°$ and the temperature of the water in beaker B is increased $10 K$. After these temperature changes, which beaker of water has the higher temperature? Explain.

17.8 You put a bottle of soft drink in a refrigerator and leave it until its temperature has dropped $10.0 K$. What is its temperature change in a) F°; b) C°?

Section 17.3 Gas Thermometers and the Kelvin Scale

17.9 Convert the following record-setting temperatures to the Kelvin scale: a) the lowest temperature recorded in the 48 contiguous states ($-70.0°F$ at Rogers Pass, Montana, on January 20, 1954); b) Australia's highest temperature ($127.0°F$ at Cloncurry, Queensland, on January 16, 1889); c) the lowest temperature recorded in the northern hemisphere ($-90.0°F$ at Verkhoyansk, Siberia, in 1892).

17.10 Convert the following Kelvin temperatures to the Celsius and Fahrenheit scales: a) the midday temperature at the surface of the moon ($400 K$); b) the temperature at the tops of the clouds in the atmosphere of Saturn ($95 K$); c) the temperature at the center of the sun ($1.55 \times 10^7 K$).

17.11 The normal boiling point of liquid neon is $-245.92°C$. What is this temperature on the Kelvin scale?

17.12 The ratio of the pressure of a gas at the melting point of platinum and its pressure at the triple point of water, when the gas is kept at constant volume, is found to be 7.476. What is the Celsius temperature of the melting point of platinum?

17.13 A gas thermometer registers an absolute pressure corresponding to 325 mm of mercury when in contact with water at the triple point. What pressure does it read when in contact with water at the normal boiling point?

17.14 Like the Kelvin scale, the *Rankine scale* is an absolute temperature scale: absolute zero is zero degrees Rankine ($0°R$). However, the units of this scale are the same size as those on the Fahrenheit scale rather than the Celsius scale. What is the numerical value of the triple-point temperature of water on the Rankine scale?

17.15 A Constant-Volume Gas Thermometer. An experimenter using a gas thermometer found the pressure at the triple point of water (0.01°C) to be 4.80×10^4 Pa and the pressure at the normal boiling point (100°C) to be 6.50×10^4 Pa. a) Assuming that the pressure varies linearly with temperature, use these two data points to find the Celsius temperature at which the gas pressure would be zero (that is, find the Celsius temperature of absolute zero). b) Does the gas in this thermometer obey Eq. (17.4) precisely? If that equation were precisely obeyed, and the pressure at 100°C were 6.50×10^4 Pa, what pressure would the experimenter have measured at 0.01°C? (As we will learn in Section 18.1, Eq. (17.4) is accurate only for gases at very low density.)

Section 17.4 Thermal Expansion

17.16 Air Friction and Thermal Expansion. The supersonic airliner Concorde is 62.1 m long when sitting on the ground on a typical (15°C) day. It is primarily made of aluminum. In flight at twice the speed of sound, friction with the air warms the Concorde's skin and causes the aircraft to lengthen by 25 cm. (The passenger cabin is on rollers, and the airplane expands around the passengers.) What is the temperature of the Concorde's skin in flight?

17.17 The Humber Bridge in England has the world's longest single span, 1410 m in length. Calculate the change in length of the steel deck of the span when the temperature increases from −5.0°C to 18.0°C.

17.18 Ensuring a Tight Fit. Aluminum rivets used in airplane construction are made slightly larger than the rivet holes and cooled by "dry ice" (solid CO_2) before being driven. If the diameter of a hole is 4.500 mm, what should be the diameter of a rivet at 23.0°C, if its diameter is to equal that of the hole when the rivet is cooled to −78.0°C, the temperature of dry ice? Assume that the expansion coefficient remains constant at the value given in Table 17.1.

17.19 A U.S. one-cent coin has a diameter of 1.9000 cm at 20.0°C. The coin is made of a metal alloy (mostly zinc) for which the coefficient of linear expansion is 2.6×10^{-5} K^{-1}. What would its diameter be a) on a hot day in Death Valley (48.0°C)? b) on a cold night in the mountains of Greenland (−53.0°C)?

17.20 The pendulum shaft of a clock is made of brass. What is the fractional change in length of the shaft when it is cooled from 19.50°C to 5.00°C?

17.21 A metal rod is 40.125 cm long at 20.0°C and 40.148 cm long at 45.0°C. Calculate the average coefficient of linear expansion of the rod for this temperature range.

17.22 A copper cylinder is initially at 20.0°C. At what temperature will its volume be 0.150% larger than it is at 20.0°C?

17.23 An underground tank with a capacity of 1700 L (1.70 m^3) is filled with ethanol that has an initial temperature of 19.0°C. After the ethanol has cooled off to the temperature of the tank and ground, which is 10.0°C, how much air space will there be above the ethanol in the tank? (Assume that the volume of the tank doesn't change.)

17.24 A steel tank is completely filled with 2.80 m³ of ethanol when both the tank and the ethanol are at a temperature of 32.0°C. When the tank and its contents have cooled to 18.0°C, what additional volume of ethanol can be put into the tank?

17.25 A glass flask whose volume is 1000.00 cm³ at 0.0°C is completely filled with mercury at this temperature. When flask and mercury are warmed to 55.0°C, 8.95 cm³ of mercury overflow. If the coefficient of volume expansion of mercury is 18.0×10^{-5} K^{-1}, compute the coefficient of volume expansion of the glass.

17.26 a) If an area measured on the surface of a solid body is A_0 at some initial temperature and then changes by ΔA when the temperature changes by ΔT, show that

$$\Delta A = (2\alpha)A_0\Delta T$$

where α is the coefficient of linear expansion. b) A circular sheet of aluminum is 55.0 cm in diameter at 15.0°C. By how much does the area of one side of the sheet change when the temperature increases to 27.5°C?

17.27 A machinist bores a hole of diameter 1.350 cm in a steel plate at a temperature of 25°C. What is the cross-sectional area of the hole a) at 25°C; b) when the temperature of the plate is increased to 175°C? Assume that the coefficient of linear expansion remains constant over this temperature range. (*Hint:* See Exercise 17.26.)

17.28 As a new mechanical engineer for Engines Inc., you have been assigned to design brass pistons to slide inside steel cylinders. The engines in which these pistons will be used will operate between 20°C and 150°C. Assume that the coefficients of expansion are constant over this temperature range. a) If the piston just fits inside the chamber at 20°C, will the engines be able to run at higher temperatures? Explain. b) If the cylindrical pistons are 25.000 cm in diameter at 20°C, what should be the minimum diameter of the cylinders at that temperature so the pistons will operate at 150°C?

17.29 The markings on an aluminum ruler and a brass ruler are perfectly aligned at 0°C. How far apart will the 20.0 cm marks be on the two rulers at 100°C, if precise alignment of the left-hand ends of the rulers is maintained?

17.30 A brass rod is 185 cm long and 1.60 cm in diameter. What force must be applied to each end of the rod to prevent it from contracting when it is cooled from 120°C to 10°C?

17.31 a) A wire that is 1.50 m long at 20°C is found to increase in length by 1.9 cm when warmed to 420°C. Compute its average coefficient of linear expansion for this temperature range. b) The wire is stretched just taut (zero tension) at 420°C. Find the stress in the wire if it is cooled to 20°C without being allowed to contract. Young's modulus for the wire is 2.0×10^{11} Pa.

17.32 Steel train rails are laid in 12.0-m-long segments placed end-to-end. The rails are laid on a winter day when their temperature is −2.0°C. a) How much space must be left between adjacent rails if they are just to touch on a summer day when their temperature is 33.0°C? b) If the rails are originally laid in contact, what is the stress in them on a summer day when their temperature is 33.0°C?

Section 17.5 Quantity of Heat

17.33 Heat Loss During Breathing. In very cold weather a significant mechanism for heat loss by the human body is energy expended in warming the air taken into the lungs with each breath. a) On a cold winter day when the temperature is −20°C, what is the amount of heat needed to warm to body temperature (37°C) the

0.50 L of air exchanged with each breath? Assume that the specific heat capacity of air is 1020 J/kg·K and that 1.0 L of air has mass 1.3×10^{-3} kg. b) How much heat is lost per hour if the respiration rate is 20 breaths per minute?

17.34 While running, a 70-kg student generates thermal energy at a rate of 1200 W. To maintain a constant body temperature of 37°C, this energy must be removed by perspiration or other mechanisms. If these mechanisms failed and the heat could not flow out of the student's body, for what amount of time could a student run before irreversible body damage occurs? (Protein structures in the body are irreversibly damaged if body temperature rises to 44°C or above. The specific heat capacity of a typical human body is 3480 J/kg·K, slightly less than that of water. The difference is due to the presence of protein, fat, and minerals, which have lower specific heat capacities.)

17.35 While painting the top of an antenna 225 m in height, a worker accidentally lets a 1.00-L water bottle fall from his lunchbox. The bottle lands in some bushes at ground level and does not break. If a quantity of heat equal to the magnitude of the change in mechanical energy of the water goes into the water, what is its increase in temperature?

17.36 A crate of fruit with mass 35.0 kg and specific heat capacity 3650 J/kg·K slides down a ramp inclined at 36.9° below the horizontal. The ramp is 8.00 m long. a) If the crate was at rest at the top of the incline and has a speed of 2.50 m/s at the bottom, how much work was done on the crate by friction? b) If an amount of heat equal to the magnitude of the work done by friction goes into the crate of fruit and the fruit reaches a uniform final temperature, what is its temperature change?

17.37 An engineer is working on a new engine design. One of the moving parts contains 1.60 kg of aluminum and 0.30 kg of iron and is designed to operate at 210°C. How much heat is required to raise its temperature from 20° to 210°C?

17.38 A nail driven into a board increases in temperature. If we assume that 60% of the kinetic energy delivered by a 1.80-kg hammer with a speed of 7.80 m/s is transformed into heat that flows into the nail and does not flow out, what is the temperature increase of an 8.00-g aluminum nail after it is struck ten times?

17.39 An aluminum tea kettle with mass 1.50 kg and containing 1.80 kg of water is placed on a stove. If no heat is lost to the surroundings, how much heat must be added to raise the temperature from 20.0°C to 85.0°C?

17.40 In an effort to stay awake for an all-night study session, a student makes a cup of coffee by first placing a 200-W electric immersion heater in 0.320 kg of water. a) How much heat must be added to the water to raise its temperature from 20.0°C to 80.0°C? b) How much time is required? Assume that all of the heater's power goes into heating the water.

17.41 A technician measures the specific heat capacity of an unidentified liquid by immersing an electrical resistor in it. Electrical energy is converted to heat transferred to the liquid for 120 s at a constant rate of 65.0 W. The mass of the liquid is 0.780 kg, and its temperature increases from 18.55°C to 22.54°C. a) Find the average specific heat capacity of the liquid in this temperature range. Assume that negligible heat is transferred to the container that holds the liquid and that no heat is lost to the surroundings. b) Sup-

pose that in this experiment heat transfer from the liquid to the container or surroundings cannot be ignored. Is the result calculated in part (a) an *overestimate* or an *underestimate* of the average specific heat capacity? Explain.

17.42 You are given a sample of metal and are asked to determine its specific heat capacity. You weigh the sample and find that its weight is 28.4 N. You carefully add 1.25×10^4 J of heat energy to the sample and find that its temperature rises 18.0 C°. What is the sample's specific heat capacity?

17.43 You add 8950 J of heat to 3.00 moles of iron. a) What is the temperature increase of the iron? b) If this same amount of heat is added to 3.00 kg of iron, what is the iron's temperature increase? c) How do the results of parts (a) and (b) compare? Explain.

Section 17.6 Calorimetry and Phase Changes

17.44 As a physicist, you put heat into a 500-g solid sample at the rate of 10.0 kJ/min, while recording its temperature as a function of time. You plot your data and obtain the graph shown in Fig. 17.27. a) What is the latent heat of fusion for this solid? b) What are the specific heats of the liquid and solid states of the material?

Figure 17.27 Exercise 17.44.

17.45 A 500-g chunk of an unknown metal, which has been in boiling water for several minutes, is quickly dropped into an insulating Styrofoam beaker containing 1.00 kg of water at room temperature (20.0°C). After waiting and gently stirring for 5.00 minutes, you observe that the water's temperature has reached a constant value of 22.0°C. a) Assuming that the Styrofoam absorbs a negligibly small amount of heat and that no heat was lost to the surroundings, what is the specific heat of the metal? b) Which is more useful for storing heat, this metal or an equal weight of water? Explain. c) What if the heat absorbed by the Styrofoam actually is not negligible. How would the specific heat you calculated in part (a) be in error? Would it be too large, too small, or still correct? Explain.

17.46 Before going in for his annual physical, a 70.0-kg man whose body temperature is 37.0°C consumes an entire 0.355-L can of a soft drink (mostly water) at 12.0°C. a) What will be his body temperature after equilibrium is attained? Ignore any heating by the man's metabolism. The specific heat capacity of the man's body is 3480 J/kg·K. b) Is the change in his body temperature great enough to be measured by a medical thermometer?

17.47 In the situation described in Exercise 17.46, the man's metabolism will eventually return the temperature of his body (and of the soft drink that he consumed) to 37.0°C. If his body releases energy at a rate of 7.00×10^3 kJ/day (the *basal metabolic rate,* or BMR), how long does this take? Assume that all of the released energy goes into raising the temperature.

17.48 An ice cube tray of negligible mass contains 0.350 kg of water at 18.0°C. How much heat must be removed to cool the water to 0.0°C and freeze it? Express your answer in joules, calories, and Btu.

17.49 How much heat is required to convert 12.0 g of ice at -10.0°C to steam at 100.0°C? Express your answer in joules, calories, and Btu.

17.50 An open container holds 0.550 kg of ice at -15.0°C. The mass of the container can be ignored. Heat is supplied to the container at the constant rate of 800 J/min for 500 min. a) After how many minutes does the ice *start* to melt? b) After how many minutes, from the time when the heating is first started, does the temperature begin to rise above 0°C? c) Plot a curve showing the temperature as a function of the elapsed time.

17.51 The capacity of commercial air conditioners is sometimes expressed in "tons," the number of tons of ice $\left(1 \text{ ton} = 2000 \text{ lb}\right)$ that can be frozen from water at 0°C in 24 h by the unit. Express the capacity of a 2-ton air conditioner in Btu/h and in watts.

17.52 Steam Burns vs. Water Burns. What is the amount of heat input to your skin when it receives the heat released a) by 25.0 g of steam initially at 100.0°C, when it is cooled to skin temperature (34.0°C)? b) by 25.0 g of water initially at 100.0°C, when it is cooled to 34.0°C? c) What does this tell you about the relative severity of steam and hot water burns?

17.53 What must the initial speed of a lead bullet be at a temperature of 25°C so that the heat developed when it is brought to rest will be just sufficient to melt it? Assume that all the initial mechanical energy of the bullet is converted to heat and that no heat flows from the bullet to its surroundings. (Typical rifles have muzzle speeds that exceed the speed of sound in air, which is 347 m/s at 25°C.)

17.54 Evaporation of sweat is an important mechanism for temperature control in some warm-blooded animals. a) What mass of water must evaporate from the skin of a 70.0-kg man to cool his body 1.00 C°? The heat of vaporization of water at body temperature (37°C) is 2.42×10^6 J/kg·K. The specific heat capacity of a typical human body is 3480 J/kg·K (see Exercise 17.34). b) What volume of water must the man drink to replenish the evaporated water? Compare to the volume of a soft-drink can $\left(355 \text{ cm}^3\right)$.

17.55 "The Ship of the Desert." Camels require very little water because they are able to tolerate relatively large changes in their body temperature. While humans keep their body temperatures constant to within one or two Celsius degrees, a dehydrated camel permits its body temperature to drop to 34.0°C overnight and rise to 40.0°C during the day. To see how effective this mechanism is for saving water, calculate how many liters of water a 400-kg camel would have to drink if it attempted to keep its body temperature at a constant 34.0°C by evaporation of sweat during the day (12 hours) instead of rising to 40.0°C. (The specific heat capacity of a camel or other mammal is about the same as that of a typical human, 3480 J/kg·K. The heat of vaporization of water at 34°C is 2.42×10^6 J/kg.)

17.56 In a physics lab experiment a student immersed 200 one-cent coins (each having a mass of 3.00 g) in boiling water. After they reached thermal equilibrium, she fished them out and dropped them into 0.240 kg of water at 20.0°C in an insulated container of negligible mass. What was the final temperature of the coins? (One-cent coins are made of a metal alloy—mostly zinc—with a specific heat capacity of 390 J/kg·K.)

17.57 A copper pot with mass 0.500 kg contains 0.170 kg of water at a temperature of 20.0°C. A 0.250-kg block of iron at 85.0°C is dropped into the pot. Find the final temperature, assuming no heat loss to the surroundings.

17.58 A laboratory technician drops a 0.0850-kg sample of unknown material, at a temperature of 100.0°C, into a calorimeter. The calorimeter can, initially at 19.0°C, is made of 0.150 kg of copper and contains 0.200 kg of water. The final temperature of the calorimeter can is 26.1°C. Compute the specific heat capacity of the sample.

17.59 An insulated beaker with negligible mass contains 0.250 kg of water at a temperature of 75.0°C. How many kilograms of ice at a temperature of -20.0°C must be dropped in the water so that the final temperature of the system will be 30.0°C?

17.60 A glass vial containing a 16.0-g sample of an enzyme is cooled in an ice bath. The bath contains water and 0.120 kg of ice. The sample has specific heat capacity 2250 J/kg·K; the glass vial has mass 6.0 g and specific heat capacity 2800 J/kg·K. How much ice melts in cooling the enzyme sample from room temperature (19.5°C) to the temperature of the ice bath?

17.61 A 4.00-kg silver ingot is taken from a furnace, where its temperature is 750°C, and placed on a large block of ice at 0°C. Assuming that all the heat given up by the silver is used to melt the ice, how much ice is melted?

17.62 A copper calorimeter can with mass 0.100 kg contains 0.160 kg of water and 0.018 kg of ice in thermal equilibrium at atmospheric pressure. If 0.750 kg of lead at a temperature of 255°C is dropped into the calorimeter can, what is the final temperature? Assume that no heat is lost to the surroundings.

17.63 A vessel whose walls are thermally insulated contains 2.40 kg of water and 0.450 kg of ice, all at a temperature of 0.0°C. The outlet of a tube leading from a boiler in which water is boiling at atmospheric pressure is inserted into the water. How many grams of steam must condense inside the vessel (also at atmospheric pressure) to raise the temperature of the system to 28.0°C? You can ignore the heat transferred to the container.

Section 17.7 Mechanisms of Heat Transfer

17.64 Use Eq. (17.21) to show that the SI units of thermal conductivity are W/m·K.

17.65 Suppose that the rod in Fig. 17.20 is made of copper, is 45.0 cm long, and has a cross-sectional area of 1.25 cm². Let $T_H = 100.0$°C and $T_C = 0.0$°C. a) What is the final steady-state temperature gradient along the rod? b) What is the heat current in the rod in the final steady state? c) What is the final steady-state temperature at a point in the rod 12.0 cm from its left end?

17.66 One end of an insulated metal rod is maintained at 100°C, and the other end is maintained at 0°C by an ice-water mixture. The rod is 60.0 cm long and has a cross-sectional area of 1.25 cm^2. The heat conducted by the rod melts 8.50 g of ice in 10.0 min. Find the thermal conductivity k of the metal.

17.67 A carpenter builds an exterior house wall with a layer of wood 3.0 cm thick on the outside and a layer of Styrofoam insulation 2.2 cm thick on the inside wall surface. The wood has $k = 0.080$ W/m·K and the Styrofoam has $k = 0.010$ W/m·K. The interior surface temperature is 19.0°C, and the exterior surface temperature is -10.0°C. a) What is the temperature at the plane where the wood meets the Styrofoam? b) What is the rate of heat flow per square meter through this wall?

17.68 An electric kitchen range has a total wall area of 1.40 m^2 and is insulated with a layer of fiberglass 4.0 cm thick. The inside surface of the fiberglass has a temperature of 175°C and its outside surface is at 35°C. The fiberglass has a thermal conductivity of 0.040 W/m·K. a) What is the heat current through the insulation, assuming it may be treated as a flat slab with an area of 1.40 m^2? b) What electric-power input to the heating element is required to maintain this temperature?

17.69 The ceiling of a room has an area of 125 ft^2. The ceiling is insulated to an R value of 30 (in units of ft^2·F°·h/Btu). The surface in the room is maintained at 69°F, and the surface in the attic has a temperature of 35°F. What is the heat flow through the ceiling into the attic in 5.0 h? Express your answer in Btu and in joules.

17.70 A long rod, insulated to prevent heat loss along its sides, is in perfect thermal contact with boiling water (at atmospheric pressure) at one end and with an ice-water mixture at the other (Fig. 17.28). The rod consists of 1.00-m section of copper (one end in steam) joined end-to-end to a length L_2 of steel (one end in ice). Both sections of the rod have cross-sectional areas of 4.00 cm^2. The temperature of the copper-steel junction is 65.0°C after a steady state has been set up. a) How much heat per second flows from the steam bath to the ice-water mixture? b) What is the length L_2 of the steel section?

Figure 17.28 Exercise 17.70.

17.71 A pot with a steel bottom 8.50 mm thick rests on a hot stove. The area of the bottom of the pot is 0.150 m^2. The water inside the pot is at 100.0°C, and 0.390 kg are evaporated every 3.00 min. Find the temperature of the lower surface of the pot, which is in contact with the stove.

17.72 You are asked to design a cylindrical steel rod 50.0 cm long, with a circular cross-section, that will conduct 150 J/s from a furnace at 400°C to a container of boiling water under one atmosphere. What must the rod's diameter be?

17.73 A rod 1.300 m long consists of a 0.800-m length of aluminum joined end-to-end to a 0.500-m length of brass. The free end

of the aluminum section is maintained at 150.0°C and the free end of the brass piece is maintained at 20.0°C. No heat is lost through the sides of the rod. At a steady state, what is the temperature T of the point where the two metals are joined?

17.74 What is the rate of energy radiation per unit area of a blackbody at a temperature of a) 273 K? b) 2730 K?

17.75 What is the net rate of heat loss by radiation in Example 17.16 (Section 17.7) if the temperature of the surroundings is 5.0°C?

17.76 The emissivity of tungsten is 0.35. A tungsten sphere with radius 1.50 cm is suspended within a large evacuated enclosure whose walls are at 290 K. What power input is required to maintain the sphere at a temperature of 3000 K if heat conduction along the supports is neglected?

17.77 Size of a Light-Bulb Filament. The operating temperature of a tungsten filament in an incandescent light bulb is 2450 K, and its emissivity is 0.35. Find the surface area of the filament of a 150-W bulb if all the electrical energy consumed by the bulb is radiated by the filament as electromagnetic waves. (Only a fraction of the radiation appears as visible light.)

17.78 The Sizes of Stars. The hot glowing surfaces of stars emit energy in the form of electromagnetic radiation. It is a good approximation to assume $e = 1$ for these surfaces. Find the radii of the following stars (assumed to be spherical): a) Rigel, the bright blue star in the constellation Orion, that radiates energy at a rate of 2.7×10^{32} W and has surface temperature 11,000 K; b) Procyon B (only visible using a telescope), that radiates energy at a rate of 2.1×10^{23} W and has surface temperature 10,000 K. c) Compare your answers to the radius of the earth, the radius of the sun, and the distance between the earth and the sun. (Rigel is an example of a *supergiant* star, and Procyon B is an example of a *white dwarf* star.)

Problems

17.79 You propose a new temperature scale, with temperatures given in °M. You define 0.0°M to be the normal melting point of mercury and 100.0° to be the normal boiling point of mercury. a) What is the normal boiling point of water in °M? b) A temperature change of 10.0 M° corresponds to how many C°?

17.80 Suppose that a steel hoop could be constructed to fit just around the earth's equator at a temperature of 20.0°C. What would be the thickness of space between the hoop and the earth if the temperature of the hoop were increased by 0.50 C°?

17.81 At an absolute temperature T_0, a cube has sides of length L_0 and has density ρ_0. The cube is made of a material with coefficient of volume expansion β. a) Show that if the temperature increases to $T_0 + \Delta T$, the density of the cube becomes, approximately,

$$\rho \approx \rho_0(1 - \beta\Delta T)$$

(*Hint:* Use the expression $(1 + x)^n \approx 1 + nx$, valid for $|x| \ll 1$.) Explain why this approximate result is only valid if ΔT is much less than $1/\beta$, and explain why you would expect this to be the case in most situations. b) A copper cube has sides of length 1.25 cm at 20.0°C. Find the change in its volume and density when its temperature is increased to 70.0°C.

17.82 A 250-kg weight is hanging from the ceiling by a thin copper wire. In its fundamental mode, this wire vibrates at the frequency of concert A (440 Hz). You then increase the temperature of the wire by 40 C°. a) By how much will the fundamental frequency change? Will it increase or decrease? b) By what percent will the speed of a wave on the wire change? c) By what percent will the wavelength of the fundamental standing wave change? Will it increase or decrease?

17.83 You are making pesto for your pasta and have a cylindrical measuring cup 10.0 cm high made of ordinary glass ($\beta = 2.7 \times 10^{-5}(C°)^{-1}$) that is filled with olive oil ($\beta = 6.8 \times 10^{-4}(C°)^{-1}$) to a height of 1.00 mm below the top of the cup. Initially, the cup and oil are at room temperature (22.0°C). You get a phone call and forget about the olive oil, which you inadvertently leave on the hot stove. The cup and oil heat up slowly, and have a common temperature. At what temperature will the olive oil start to spill out of the cup?

17.84 Use Fig. 17.11 to find the approximate coefficient of volume expansion of water at 2.0°C and at 8°C.

17.85 As in Problem 17.90, a steel rod and an aluminum rod are placed end-to-end between rigid supports. But now 0.20 cm has been cut from each rod, so the steel rod is 0.348 m long, the aluminum rod is 0.248 m long and there is a 0.40 cm gap between them. Initially, each rod has a temperature of 20.0°C. If both rods are either heated or cooled to the same final temperature, at what temperature will the gap just become closed?

17.86 At a temperature of 20.0°C the volume of a certain glass flask, up to a reference mark on the long stem of the flask, is exactly 100 cm³. The flask is filled to this point with a liquid whose coefficient of volume expansion is 8.00×10^{-4} K⁻¹, with both flask and liquid at 20.0°C. The coefficient of volume expansion of the glass is 2.00×10^{-5} K⁻¹. The cross-sectional area of the stem is 50.0 mm² and can be considered constant. a) Explain why it is a good approximation to ignore the change in the cross-sectional area of the stem. b) How far will the liquid rise or fall in the stem when the temperature is raised to 50.0°C?

17.87 A metal rod that is 30.0 cm long expands by 0.0650 cm when its temperature is raised from 0°C to 100°C. A rod of a different metal and of the same length expands by 0.0350 cm for the same rise in temperature. A third rod, also 30.0 cm long, is made up of pieces of each of the above metals placed end-to-end and expands 0.0580 cm between 0°C and 100°C. Find the length of each portion of the composite bar.

17.88 On a cool (4°C) Saturday morning, a pilot fills the fuel tanks of her Pitts S-2C (a two-seat aerobatic airplane) to their full capacity of 106.0 L. Before flying on Sunday morning, when the temperature is again 4°C, she checks the fuel level and finds only 103.4 L of gasoline in the tanks. She realizes that it was hot on Saturday afternoon, and that thermal expansion of the gasoline caused the missing fuel to empty out of the tank's vent. a) What was the maximum temperature (in °C) reached by the fuel and the tank on Saturday afternoon? The coefficient of volume expansion of gasoline is 9.5×10^{-4} K⁻¹, and the tank is made of aluminum. b) In order to have the maximum amount of fuel available for flight, when should the pilot have filled the fuel tanks?

17.89 a) Equation (17.12) gives the stress required to keep the length of a rod constant as its temperature changes. Show that if the

length is permitted to change by an amount ΔL when its temperature changes by ΔT, the stress is equal to

$$\frac{F}{A} = Y\left(\frac{\Delta L}{L_0} - \alpha\Delta T\right)$$

where F is the tension on the rod, L_0 is the original length of the rod, A its cross-sectional area, α its coefficient of linear expansion, and Y its Young's modulus. b) A heavy brass bar has projections at its ends, as in Fig. 17.29. Two fine steel wires, fastened between the projections, are just taut (zero tension) when the whole system is at 20°C. What is the tensile stress in the steel wires when the temperature of the system is raised to 140°C? Make any simplifying assumptions you think are justified, but state what they are.

Figure 17.29 Problem 17.89.

17.90 A steel rod with a length of 0.350 m and an aluminum rod with a length of 0.250 m, both with the same diameter, are placed end-to-end between rigid supports with no initial stress in the rods. The temperature of the rods is now raised by 60.0 C°. What is the stress in each rod? (*Hint:* The length of the combined rod remains the same, but the lengths of the individual rods do not. See Problem 17.89.)

17.91 A steel ring with a 2.5000-in. inside diameter at 20.0°C is to be warmed and slipped over a brass shaft with a 2.5020-in. outside diameter at 20.0°C. a) To what temperature should the ring be warmed? b) If the ring and the shaft together are cooled by some means such as liquid air, at what temperature will the ring just slip off the shaft?

17.92 Bulk Stress Due to a Temperature Increase. a) Prove that, if an object under pressure is raised in temperature but not allowed to expand, the increase in pressure is

$$\Delta p = B\beta\Delta T$$

where the bulk modulus B and the average coefficient of volume expansion β are both assumed positive and constant. b) What pressure is necessary to prevent a steel block from expanding when its temperature is increased from 20.0°C to 35.0°C?

17.93 A liquid is enclosed in a metal cylinder that is provided with a piston of the same metal. The system is originally at a pressure of 1.00 atm (1.013×10^5 Pa) and at a temperature of 30.0°C. The piston is forced down until the pressure on the liquid is increased by 50.0 atm, and then clamped in this position. Find the new temperature at which the pressure of the liquid is again 1.00 atm. Assume that the cylinder is sufficiently strong so that its volume is not altered by changes in pressure, but only by changes in temperature. Use the result derived in Problem 17.92. (*Hint:* See Section 11.4.)

Compressibility of liquid: $k = 8.50 \times 10^{-10}$ Pa⁻¹

Coefficient of volume expansion of liquid: $\beta = 4.80 \times 10^{-4}$ K⁻¹

Coefficient of volume expansion of metal: $\beta = 3.90 \times 10^{-5}$ K⁻¹

17.94 A thirsty mechanic cools a 2.00-L bottle of a soft drink (mostly water) by pouring it into a large aluminum mug with mass

0.257 kg and adding 0.120 kg of ice initially at $-15.0°C$. If the soft drink and mug are initially at 20.0°C, what is the final temperature of the system, assuming no heat losses?

17.95 Spacecraft Reentry. A spacecraft made of aluminum circles the earth at a speed of 7700 m/s. a) Find the ratio of its kinetic energy to the energy required to raise its temperature from 0°C to 600°C. (The melting point of aluminum is 660°C. Assume a constant specific heat capacity of 910 J/kg·K.) b) Discuss the bearing of your answer on the problem of the reentry of a manned space vehicle in the earth's atmosphere.

17.96 A capstan is a rotating drum or cylinder over which a rope or cord slides in order to provide a great amplification of the rope's tension while keeping both ends free (Fig. 17.30). Since the added tension in the rope is due to friction, the capstan generates thermal energy. a) If the difference in tension between the two ends of the rope is 520 N and the capstan has a diameter of 10.0 cm and turns once in 0.900 s, find the rate at which thermal energy is generated. Why does the number of turns not matter? b) If the capstan is made of iron and has mass 6.00 kg, at what rate does its temperature rise? Assume that the temperature in the capstan is uniform and that all the thermal energy generated flows into it.

Figure 17.30 Problem 17.96.

17.97 Debye's T^3 Law. At very low temperatures the molar heat capacity of rock salt varies with temperature according to Debye's T^3 law:

$$C = k\frac{T^3}{\Theta^3}$$

where $k = 1940$ J/mol·K and $\Theta = 281$ K. a) How much heat is required to raise the temperature of 1.50 mol of rock salt from 10.0 K to 40.0 K? (Hint: Use Eq. (17.18) in the form $dQ = nC\,dT$ and integrate.) b) What is the average molar heat capacity in this range? c) What is the true molar heat capacity at 40.0 K?

17.98 A person of mass 70.0 kg is sitting in the bathtub. The bathtub is 190 cm by 80 cm; before the person got in, the water was 10 cm deep. The water is at a temperature of 37.0°C. Suppose that the water were to cool down spontaneously to form ice at 0.0°C, and that all the energy released was used to launch the hapless bather vertically into the air. How high would the bather go? (As

you will see in Chapter 20, this event is allowed by energy conservation but is prohibited by the second law of thermodynamics.)

17.99 Hot Air in a Physics Lecture. a) A typical student listening attentively to a physics lecture has a heat output of 100 W. How much heat energy does a class of 90 physics students release into a lecture hall over the course of a 50-min lecture? b) Assume that all the heat energy in part (a) is transferred to the 3200 m³ of air in the room. The air has specific heat capacity 1020 J/kg·K and density 1.20 kg/m³. If none of the heat escapes and the air conditioning system is off, how much will the temperature of the air in the room rise during the 50-min lecture? c) If the class is taking an exam, the heat output per student rises to 280 W. What is the temperature rise during 50 min in this case?

17.100 The molar heat capacity of a certain substance varies with temperature according to the empirical equation

$$C = 29.5 \text{ J/mol·K} + (8.20 \times 10^{-3} \text{ J/mol·K}^2)T$$

How much heat is necessary to change the temperature of 3.00 mol of this substance from 27°C to 227°C? (Hint: Use Eq. (17.18) in the form $dQ = nC\,dT$ and integrate.)

17.101 a) An ice cube with mass 0.075 kg is taken from a freezer, where the cube's temperature was $-10.0°C$, and dropped into a glass of water at 0.0°C. If no heat is gained or lost from outside, how much water will freeze onto the cube? b) Is it possible to add so much ice to the glass that all of the water freezes? Explain.

17.102 Hot Water versus Steam Heating. In a household hot-water heating system, water is delivered to the radiators at 70.0°C (158.0°F) and leaves at 28.0°C (82.4°F). The system is to be replaced by a steam system in which steam at atmospheric pressure condenses in the radiators and the condensed steam leaves the radiators at 35.0°C (95.0°F). How many kilograms of steam will supply the same heat as was supplied by 1.00 kg of hot water in the first system?

17.103 A copper calorimeter can with mass 0.446 kg contains 0.0950 kg of ice. The system is initially at 0.0°C. a) If 0.0350 kg of steam at 100.0°C and 1.00 atm pressure is added to the can, what is the final temperature of the calorimeter can and its contents? b) At the final temperature, how many kilograms are there of ice, how many of liquid water, and how many of steam?

17.104 In a container of negligible mass, 0.140 kg of ice initially at $-15.0°C$ is added to 0.190 kg of water that has a temperature of 35.0°C. a) If no heat is lost to the surroundings, what is the final temperature of the system? b) At the final temperature, how many kilograms are there of ice and how many of liquid water?

17.105 In a container of negligible mass, 0.0400 kg of steam at 100°C and atmospheric pressure is added to 0.200 kg of water at 50.0°C. a) If no heat is lost to the surroundings, what is the final temperature of the system? b) At the final temperature, how many kilograms are there of steam and how many of liquid water?

17.106 A tube leads from a 0.150-kg calorimeter to a flask in which water is boiling under atmospheric pressure. The calorimeter has specific heat capacity 420 J/kg·K, and it originally contains 0.340 kg of water at 15.0°C. Steam is allowed to condense in the calorimeter at atmospheric pressure until the temperature of the calorimeter and contents reaches 71.0°C, at which point the total mass of the calorimeter and its contents is found to be 0.525 kg. Compute the heat of vaporization of water from these data.

17.107 In a container of negligible mass, 0.150 kg of ice at 0°C and 0.0950 kg of steam at 100°C are added to 0.200 kg of water that has a temperature of 50.0°C. a) If no heat is lost to the surroundings and the pressure in the container is a constant 1.00 atm, what is the final temperature of the system? b) At the final temperature, how many kilograms are there of ice, how many of liquid water, and how many of steam? c) Repeat parts (a) and (b) if 0.350 kg of ice at 0°C and 0.012 kg of steam at 100°C are added to 0.200 kg of water that has a temperature of 40.0°C.

17.108 One experimental method of measuring an insulating material's thermal conductivity is to construct a box of the material and measure the power input to an electric heater inside the box that maintains the interior at a measured temperature above the outside surface. Suppose that in such an apparatus a power input of 180 W is required to keep the interior surface of the box 65.0 C° (about 120 F°) above the temperature of the outer surface. The total area of the box is 2.18 m², and the wall thickness is 3.9 cm. Find the thermal conductivity of the material in SI units.

17.109 Effect of a Window in a Door. A carpenter builds a solid wood door with dimensions 2.00 m × 0.95 m × 5.0 cm. Its thermal conductivity is $k = 0.120$ W/m·K. The air films on the inner and outer surfaces of the door have the same combined thermal resistance as an additional 1.8 cm thickness of solid wood. The inside air temperature is 20.0°C and the outside air temperature is −8.0°C. a) What is the rate of heat flow through the door? b) By what factor is the heat flow increased if a window 0.50 m on a side is inserted in the door? The glass is 0.45 cm thick and the glass has a thermal conductivity of 0.80 W/m·K. The air films on the two sides of the glass have a total thermal resistance that is the same as an additional 12.0 cm of glass.

17.110 A wood ceiling with thermal resistance R_1 is covered with a layer of insulation with thermal resistance R_2. Prove that the effective thermal resistance of the combination is $R = R_1 + R_2$.

17.111 Compute the ratio of the rate of heat loss through a single-pane window with area 0.15 m² to that for a double-pane window with the same area. The glass of a single pane is 4.2 mm thick, and the air space between the two panes of the double-pane window is 7.0 mm thick. The glass has thermal conductivity 0.80 W/m·K. The air films on the room and outdoor surfaces of either window have a combined thermal resistance of 0.15 m²·K/W.

17.112 Rods of copper, brass, and steel are welded together to form a Y-shaped figure. The cross-sectional area of each rod is 2.00 cm². The free end of the copper rod is maintained at 100.0°C and the free ends of the brass and steel rods at 0.0°C. Assume that there is no heat loss from the surfaces of the rods. The lengths of the rods are as follows: copper, 13.0 cm; brass, 18.0 cm; steel, 24.0 cm. a) What is the temperature of the junction point? b) What is the heat current in each of the three rods?

17.113 Time Needed for a Lake to Freeze Over. a) When the air temperature is below 0°C, the water at the surface of a lake freezes to form an ice sheet. Why doesn't freezing occur throughout the entire volume of the lake? b) Show that the thickness of the ice sheet formed on the surface of a lake is proportional to the square root of the time if the heat of fusion of the water freezing on the underside of the ice sheet is conducted through the sheet. c) Assuming that the upper surface of the ice sheet is at −10°C and that the bottom surface is at 0°C, calculate the time it will take to form an ice sheet 25 cm thick. d) If the lake in part (c) is uniformly 40 m deep, how long would it take to freeze all the water in the lake? Is this likely to occur?

17.114 A rod is initially at a uniform temperature of 0°C throughout. One end is kept at 0°C, and the other is brought into contact with a steam bath at 100°C. The surface of the rod is insulated so that heat can flow only lengthwise along the rod. The cross-sectional area of the rod is 2.50 cm², its length is 120 cm, its thermal conductivity is 380 W/m·K, its density is 1.00×10^4 kg/m³, and its specific heat capacity is 520 J/kg·K. Consider a short cylindrical element of the rod 1.00 cm in length. a) If the temperature gradient at the cooler end of this element is 140 C°/m, how many joules of heat energy flow across this end per second? b) If the average temperature of the element is increasing at the rate of 0.250 C°/s, what is the temperature gradient at the other end of the element?

17.115 If the solar radiation energy incident per second on the frozen surface of a lake is 600 W/m² and 70% of this energy is absorbed by the ice, how much time will it take for a 2.50-cm-thick layer of ice to melt? The ice and the water beneath it are at a temperature of 0°C.

17.116 The rate at which radiant energy from the sun reaches the earth's upper atmosphere is about 1.50 kW/m². The distance from the earth to the sun is 1.50×10^{11} m, and the radius of the sun is 6.96×10^8 m. a) What is the rate of radiation of energy per unit area from the sun's surface? b) If the sun radiates as an ideal blackbody, what is the temperature of its surface?

17.117 A Thermos for Liquid Helium. A physicist uses a cylindrical metal can 0.250 m high and 0.090 m in diameter to store liquid helium at 4.22 K; at that temperature the heat of vaporization of helium is 2.09×10^4 J/kg. Completely surrounding the metal can are walls maintained at the temperature of liquid nitrogen, 77.3 K, with vacuum between the can and the surrounding walls. How much helium is lost per hour? The emissivity of the metal can is 0.200. The only heat transfer between the metal can and the surrounding walls is by radiation.

17.118 Thermal Expansion of an Ideal Gas. a) The pressure p, volume V, number of moles n, and Kelvin temperature T of an ideal gas are related by the equation $pV = nRT$, where R is a constant. Prove that the coefficient of volume expansion for an ideal gas is equal to the reciprocal of the Kelvin temperature if the expansion occurs at constant pressure. b) Compare the coefficients of volume expansion of copper and air at a temperature of 20°C. Assume that air may be treated as an ideal gas and that the pressure remains constant.

17.119 An engineer is developing an electric water heater to provide a continuous supply of hot water. One trial design is shown in

Fig. 17.31. Water is flowing at the rate of 0.500 kg/min, the inlet thermometer registers 18.0°C, the voltmeter reads 120 V, and the ammeter reads 15.0 A (corresponding to a power input of (120 V) $\times (15.0 \text{ A}) = 1800 \text{ W}$). a) When a steady state is finally reached, what is the reading of the outlet thermometer? b) Why is it unnecessary to take into account the heat capacity mc of the apparatus itself?

Figure 17.31 Problem 17.119.

17.120 Food Intake of a Hamster. The energy output of an animal engaged in an activity is called the basal metabolic rate (BMR) and is a measure of the conversion of food energy into other forms of energy. A simple calorimeter to measure the BMR consists of an insulated box with a thermometer to measure the temperature of the air. The air has density 1.20 kg/m³ and specific heat capacity 1020 J/kg·K. A 50.0-g hamster is placed in a calorimeter that contains 0.0500 m³ of air at room temperature. a) When the hamster is running in a wheel, the temperature of the air in the calorimeter rises 1.60 C° per hour. How much heat does the running hamster generate in an hour? Assume that all this heat goes into the air in the calorimeter. You can ignore the heat that goes into the walls of the box and into the thermometer, and assume that no heat is lost to the surroundings. b) Assuming that the hamster converts seed into heat with an efficiency of 10% and that hamster seed has a food energy value of 24 J/g, how many grams of seed must the hamster eat per hour to supply this energy?

Challenge Problems

17.121 a) A spherical shell has inner and outer radii a and b, respectively, and the temperatures at the inner and outer surfaces are T_2 and T_1. The thermal conductivity of the material of which the shell is made is k. Derive an equation for the total heat current through the shell. b) Derive an equation for the temperature variation within the shell in part (a); that is, calculate T as a function of r, the distance from the center of the shell. c) A hollow cylinder has length L, inner radius a, and outer radius b, and the temperatures at the inner and outer surfaces are T_2 and T_1. (The cylinder could represent an insulated hot-water pipe, for example.) The thermal conductivity of the material of which the cylinder is made is k. Derive an equation for the total heat current through the walls of the cylinder. d) For the cylinder of part (c), derive an equation for the temperature variation inside the cylinder walls. e) For the spherical shell of part (a) and the hollow cylinder of part (c), show that the equation for the total heat current in each case reduces to Eq. (17.21) for linear heat flow when the shell or cylinder is very thin.

17.122 A steam pipe with a radius of 2.00 cm, carrying steam at 140°C, is surrounded by a cylindrical jacket with inner and outer radii 2.00 cm and 4.00 cm and made of a type of cork with thermal conductivity 4.00×10^{-2} W/m·K. This in turn is surrounded by a cylindrical jacket made of a brand of Styrofoam with thermal conductivity 1.00×10^{-2} W/m·K and having inner and outer radii 4.00 cm and 6.00 cm (Fig. 17.32). The outer surface of the Styrofoam is in contact with air at 15°C. Assume that this outer surface has a temperature of 15°C. a) What is the temperature at a radius of 4.00 cm, where the two insulating layers meet? b) What is the total rate of transfer of heat out of a 2.00-m length of pipe? (*Hint:* Use the expression derived in part (c) of Challenge Problem 17.121.)

Figure 17.32 Challenge Problem 17.122.

17.123 Suppose that both ends of the rod in Fig. 17.20 are kept at a temperature of 0°C, and that the initial temperature distribution along the rod is given by $T = (100°\text{C}) \sin \pi x/L$, where x is measured from the left end of the rod. Let the rod be copper, with length $L = 0.100$ m and cross-sectional area 1.00 cm². a) Show the initial temperature distribution in a diagram. b) What is the final temperature distribution after a very long time has elapsed? c) Sketch curves that you think would represent the temperature distribution at intermediate times. d) What is the initial temperature gradient at the ends of the rod? e) What is the initial heat current from the ends of the rod into the bodies making contact with its ends? f) What is the initial heat current at the center of the rod? Explain. What is the heat current at this point at any later time? g) What is the value of the *thermal diffusivity* $k/\rho c$ for copper, and in what unit is it expressed? (Here k is the thermal conductivity, $\rho = 8.9 \times 10^3$ kg/m³ is the density, and c is the specific heat capacity.) h) What is the initial time rate of change of temperature at the center of the rod? i) How much time would be required for the center of the rod to reach its final temperature if the temperature continued to decrease at this rate? (This time is called the *relaxation time* of the rod.) j) From the graphs in part (c), would you expect the magnitude of the rate of temperature change at the midpoint to remain constant, increase, or decrease as a function of time? k) What is the initial rate of change of temperature at a point in the rod 2.5 cm from its left end?

17.124 Temperature Change in a Clock. A pendulum clock is designed to tick off one second on each side-to-side swing of the pendulum (two ticks per complete period). a) Will a pendulum clock gain time in hot weather and lose it in cold, or the reverse? Explain your reasoning. b) A particular pendulum clock keeps correct time at 20.0°C. The pendulum shaft is of steel, and its mass can be ignored compared with that of the bob. What is the fractional change in length of the shaft when it is cooled to 10.0°C? c) How many seconds per day will the clock gain or lose at 10.0°C? d) How closely must the temperature be controlled if the clock is not to gain or lose more than 1.00 s a day? Does the answer depend on the period of the pendulum?

17.125 A solid cylindrical copper rod 0.200 m long has one end maintained at a temperature of 20.00 K. The other end is blackened and exposed to thermal radiation from surrounding walls at 500 K. The sides of the rod are insulated, so no energy is lost or gained except at the ends of the rod. When equilibrium is reached, what is the temperature of the blackened end? (*Hint:* Since copper is a very good conductor of heat at low temperature, with $k = 1670$ W/m·K at 20 K, the temperature of the blackened end is only slightly greater than 20.00 K.)

17.126 A Walk in the Sun. Consider a poor lost soul walking at 5 km/h on a hot day in the desert, wearing only a bathing suit. This person's skin temperature tends to rise due to four mechanisms: i) energy is generated by metabolic reactions in the body at a rate of 280 W, and almost all of this energy is converted to heat that flows to the skin; ii) heat is delivered to the skin by convection from the outside air at a rate equal to $k'A_{skin}(T_{air} - T_{skin})$, where k' is 54 J/h·C°·m², the exposed skin area A_{skin} is 1.5 m², the air temperature T_{air} is 47°C, and the skin temperature T_{skin} is 36°C; iii) the skin absorbs radiant energy from the sun at a rate of 1400 W/m²; iv) the skin absorbs radiant energy from the environment, which has temperature 47°C. a) Calculate the net rate (in watts) at that the person's skin is heated by all four of these mechanisms. Assume that the emissivity of the skin is $e = 1$ and that the skin temperature is initially 36°C. Which mechanism is the most important? b) At what rate (in L/h) must perspiration evaporate from this person's skin to maintain a constant skin temperature? (The heat of vaporization of water at 36°C is 2.42×10^6 J/kg.) c) Suppose instead the person is protected by light-colored clothing ($e \approx 0$) so that the exposed skin area is only 0.45 m². What rate of perspiration is required now? Discuss the usefulness of the traditional clothing worn by desert peoples.

18 THERMAL PROPERTIES OF MATTER

What is the fundamental difference between the warm air of summer and the cold air of winter? The answer is kinetic energy: the higher the temperature of a gas, the greater the average kinetic energy of the gas molecules. At a given temperature, gas molecules with different masses (such as oxygen and nitrogen) have different average speeds but the same average kinetic energy.

? **How much faster do air molecules move on a hot day in the desert (+40°C) than on a frigid arctic night (−40°C)?**

The kitchen is a great place to learn about how the properties of matter depend on temperature. When you boil water in a tea kettle, the increase in temperature produces steam that whistles out of the spout at high pressure. If you forget to poke holes in a potato before baking it, the high-pressure steam produced inside the potato can cause it to explode messily. Water vapor in the air can condense into droplets of liquid on the sides of a glass of ice water; if the glass is just out of the freezer, frost will form on the sides as water vapor changes to a solid.

All of these examples show the interrelations between the large-scale or *macroscopic* properties of a substance, such as pressure, volume, temperature, and mass of substance. But we can also describe a substance using a *microscopic* perspective. This means investigating small-scale quantities such as the masses, speeds, kinetic energies, and momenta of the individual molecules that make up a substance.

The macroscopic and microscopic descriptions are intimately related. For example, the (microscopic) collision forces that occur when air molecules strike a solid surface (such as your skin) cause (macroscopic) atmospheric pressure. Standard atmospheric pressure is 1.01×10^5 Pa; to produce this pressure, 10^{32} molecules strike your skin every day with an average speed of over 1700 km/h (1000 mi/h)!

In this chapter we'll begin our study of the thermal properties of matter by looking at some macroscopic aspects of matter in general. We'll pay special attention to the *ideal gas,* one of the simplest types of matter to understand. Using our knowledge of momentum and kinetic energy, we'll relate the macroscopic properties of an ideal gas to the microscopic behavior of its individual molecules. We'll also use microscopic ideas to understand the heat capacities of both gases and solids. Finally, we'll take a look at the various phases of matter—gas, liquid, and solid—and the conditions under which each occurs.

18.1 | Equations of State

The conditions in which a particular material exists are described by physical quantities such as pressure, volume, temperature, and amount of substance. For example, a tank of oxygen in a welding outfit has a pressure gauge and a label stating its volume. We could add a thermometer and place the tank on a scale to determine its mass. These variables describe the *state* of the material and are called **state variables.**

The volume V of a substance is usually determined by its pressure p, temperature T, and amount of substance, described by the mass m or number of moles n. Ordinarily, we can't change one of these variables without causing a change in another. When the tank of oxygen gets hotter, the pressure increases. If the tank gets too hot, it explodes; this happens occasionally with overheated steam boilers.

In a few cases the relationship among p, V, T, and m (or n) is simple enough that we can express it as an equation called the **equation of state.** When it's too complicated for that, we can use graphs or numerical tables. Even then, the relation among the variables still exists; we call it an equation of state even when we don't know the actual equation.

Here's a simple (though approximate) equation of state for a solid material. The temperature coefficient of volume expansion β (Section 17.4) is the fractional volume change $\Delta V / V_0$ per unit temperature change, and the compressibility k (Section 11.4) is the negative of the fractional volume change $\Delta V / V_0$ per unit pressure change. If a certain amount of material has volume V_0 when the pressure is p_0 and the temperature is T_0, the volume V at slightly differing pressure p and temperature T is approximately

$$V = V_0[1 + \beta(T - T_0) - k(p - p_0)] \tag{18.1}$$

(There is a negative sign in front of the term $k(p - p_0)$ because an *increase* in pressure causes a *decrease* in the volume.) Equation (18.1) is called an *equation of state* for the material.

The Ideal Gas Equation

Another simple equation of state is the one for an *ideal gas.* Figure 18.1 shows an experimental setup to study the behavior of a gas. The cylinder has a movable piston to vary the volume, the temperature can be varied by heating, and we can pump any desired amount of any gas into the cylinder. We then measure the pressure, volume, temperature, and amount of gas. Note that *pressure* refers both to the force per unit area exerted by the cylinder on the gas and to the force per unit area exerted by the gas on the cylinder; by Newton's third law, these must be equal.

It is usually easiest to describe the amount of gas in terms of the number of moles n, rather than the mass. We did this when we defined molar heat capacity in Section 17.5; you may want to review that section. The **molar mass M** of a

18.1 A hypothetical setup for studying the behavior of gases. By (1) heating the gas, (2) varying the volume with a movable piston, and (3) adding more gas, we can control the gas pressure p, volume V, temperature T, and number of moles n.

compound (sometimes called *molecular weight*) is the mass per mole, and the total mass m_{tot} of a given quantity of that compound is the number of moles n times the mass per mole M:

$$m_{tot} = nM \qquad \text{(total mass, number of moles, and molar mass)} \quad (18.2)$$

We are calling the total mass m_{tot} because later in the chapter we will use m for the mass of one molecule.

Measurements of the behavior of various gases lead to several conclusions. First, the volume V is proportional to the number of moles n. If we double the number of moles, keeping pressure and temperature constant, the volume doubles.

Second, the volume varies *inversely* with the absolute pressure p. If we double the pressure while holding the temperature T and number of moles n constant, the gas compresses to one-half of its initial volume. In other words, $pV = $ constant when n and T are constant.

Third, the pressure is proportional to the *absolute* temperature. If we double the absolute temperature, keeping the volume and number of moles constant, the pressure doubles. In other words, $p = $ (constant)T when n and V are constant.

These three relationships can be combined neatly into a single equation, called the **ideal-gas equation:**

$$pV = nRT \qquad \text{(ideal-gas equation)} \qquad (18.3)$$

8.4   State Variables and Ideal Gas Law

where R is a proportionality constant. An **ideal gas** is one for which Eq. (18.3) holds precisely for *all* pressures and temperatures. This is an idealized model; it works best at very low pressures and high temperatures, when the gas molecules are far apart and in rapid motion. It is reasonably good (within a few percent) at moderate pressures (such as a few atmospheres) and at temperatures well above those at which the gas liquefies.

We might expect that the constant R in the ideal-gas equation would have different values for different gases, but it turns out to have the same value for *all* gases, at least at sufficiently high temperature and low pressure. It is called the **gas constant** (or *ideal-gas constant*). The numerical value of R depends on the units of p, V, and T. In SI units, in which the unit of p is Pa $(1 \text{ Pa} = 1 \text{ N/m}^2)$ and the unit of V is m^3, the current best numerical value of R is

$$R = 8.314472(15) \text{ J/mol} \cdot \text{K}$$

or $R = 8.314$ J/mol$\cdot$K to four significant digits. Note that the units of pressure times volume are the same as units of work or energy (for example, N/m^2 times m^3); that's why R has units of energy per mole per unit of absolute temperature. In chemical calculations, volumes are often expressed in liters (L) and pressures in atmospheres (atm). In this system, to four significant digits,

$$R = 0.08206 \frac{\text{L} \cdot \text{atm}}{\text{mol} \cdot \text{K}}$$

We can express the ideal-gas equation, Eq. (18.3), in terms of the mass m_{tot} of gas, using $m_{tot} = nM$ from Eq. (18.2):

$$pV = \frac{m_{tot}}{M}RT \qquad (18.4)$$

From this we can get an expression for the density $\rho = m_{tot}/V$ of the gas:

$$\rho = \frac{pM}{RT} \qquad (18.5)$$

CAUTION When using Eq. (18.5), be certain that you distinguish between the Greek letter ρ ("rho") for density and the letter p for pressure.

For a *constant mass* (or constant number of moles) of an ideal gas the product nR is constant, so the quantity pV/T is also constant. If the subscripts 1 and 2 refer to any two states of the same mass of a gas, then

$$\frac{p_1 V_1}{T_1} = \frac{p_2 V_2}{T_2} = \text{constant} \qquad (\text{ideal gas, constant mass}) \qquad (18.6)$$

Notice that you don't need the value of R to use this equation.

The proportionality of pressure to absolute temperature is familiar; in fact, in Chapter 17 we *defined* a temperature scale in terms of pressure in a constant-volume gas thermometer. That may make it seem that the pressure–temperature relation in the ideal-gas equation, Eq. (18.3), is just a result of the way we define temperature. But the equation also tells us what happens when we change the volume or the amount of substance. Also, the gas-thermometer scale turns out to correspond closely to a temperature scale that we will define in Chapter 20 that doesn't depend on the properties of any particular material. For now, consider this equation as being based on this genuinely material-independent temperature scale, even though we haven't defined it yet.

Problem-Solving Strategy

Ideal Gases

IDENTIFY *the relevant concepts:* Unless the problem explicitly states otherwise, you can use the ideal-gas equation for any situation in which you need to find the state (pressure, volume, temperature, and/or number of moles) of a gas.

SET UP *the problem using the following steps:*
1. Identify the target variables.
2. In some problems you will be concerned with only one state of the system, in which case Eq. (18.3) is the relationship to use. Some of the quantities in this equation will be known; others will be unknown. Make a list of what you know and what you have to find. For example, $p = 1.0 \times 10^6$ Pa, $V = 4\,\text{m}^3$, $T = ?$, $n = 2$ mol, or something comparable.
3. In other problems you will compare two different states of the same amount of gas. Decide which is state 1 and which is state 2, and make a list of the quantities for each: p_1, p_2, V_1, V_2, T_1, T_2. If all but one of these quantities are known, you can use Eq. (18.6). Otherwise, use Eq. (18.3).

For example, if p_1, V_1, and n are given, you can't use Eq. (18.6) because you don't know T_1.
4. Some problems involve the density ρ (mass per volume) rather than the number of moles n and the volume V. In this case it's most convenient to use Eq. (18.5), $\rho = pM/RT$.

EXECUTE *the solution as follows:*
1. As always, use a consistent set of units. If you need the value of the gas constant R, choose its units and then convert the units of the other quantities accordingly. You may have to convert atmospheres to pascals or liters to cubic meters ($1\,\text{m}^3 = 10^3\,\text{L} = 10^6\,\text{cm}^3$). Sometimes the problem statement will make one system of units clearly more convenient than others. Decide on your system and stick to it.
2. Don't forget that T must always be an *absolute* temperature. If you are given temperatures in °C, be sure to convert to Kelvin temperature by adding 273.15 (to three

significant figures, 273). Likewise, p is always the absolute pressure, never the gauge pressure.

3. You may sometimes have to convert between mass m_{tot} and number of moles n. The relationship is $m_{tot} = Mn$, where M is the molecular mass. Here's a tricky point: If you use Eq. (18.4), you *must* use the same mass units for m_{tot} and M. So if M is in grams per mole (the usual units for molecular mass), then m_{tot} must also be in grams. If you want to use m_{tot} in kilograms, then you must convert M to kg/mol. For example, the molecular mass of oxygen is 32 g/mol or 32×10^{-3} kg/mol. Be careful!

4. Once you have taken care of steps 1–3, solve for the target variables.

EVALUATE *your answer:* Look carefully at your results and see whether they make physical sense. For example, we'll find in Example 18.1 that a mole of gas at one atmosphere pressure and 0°C occupies a volume of 22.4 liters. If you do a calculation of the amount of air inside a soccer ball (radius 11 cm, volume 5.6 L) and get a fantastically large answer like $n = 5000$ moles, you probably converted units incorrectly or made an algebraic error.

Example 18.1	**Volume of a gas at STP**

The condition called **standard temperature and pressure** (STP) for a gas is defined to be a temperature of 0°C = 273.15 K and a pressure of 1 atm = 1.013×10^5 Pa. If you want to keep a mole of an ideal gas in your room at STP, how big a container do you need?

SOLUTION

IDENTIFY and SET UP: We are asked about the properties of a single state of the system, so we use Eq. (18.3).

EXECUTE: From Eq. (18.3), using R in J/mol·K,

$$V = \frac{nRT}{p} = \frac{(1 \text{ mol})(8.314 \text{ J/mol·K})(273.15 \text{ K})}{1.013 \times 10^5 \text{ Pa}}$$
$$= 0.0224 \text{ m}^3 = 22.4 \text{ L}$$

EVALUATE: You may be familiar with this result from your study of chemistry. Note that 22.4 L is almost exactly the volume of three basketballs. A cube 0.282 m on a side would also do the job.

Example 18.2	**Compressing gas in an automobile engine**

In an automobile engine, a mixture of air and gasoline is compressed in the cylinders before being ignited. A typical engine has a compression ratio of 9.00 to 1; this means that the gas in the cylinders is compressed to 1/(9.00) of its original volume (Fig. 18.2). The initial pressure is 1.00 atm and the initial temperature is 27°C. If the pressure after compression is 21.7 atm, find the temperature of the compressed gas.

SOLUTION

IDENTIFY and SET UP: Let state 1 be the uncompressed gas and let state 2 be the fully compressed gas. Then $p_1 = 1.00$ atm, $p_2 = 21.7$ atm, and $V_1 = 9.00 V_2$. Converting temperature to the Kelvin scale by adding 273, $T_1 = 300$ K; the final temperature T_2 is the target variable. The intake and exhaust valves at the top of the cylinder in Fig. 18.2 stay closed during the compression, so the number of moles of gas n is constant. Hence we can use Eq. (18.6):

EXECUTE: Solving Eq. (18.6) for the temperature T_2 of the compressed gas,

$$T_2 = T_1 \frac{p_2 V_2}{p_1 V_1} = (300 \text{ K}) \frac{(21.7 \text{ atm}) V_2}{(1.00 \text{ atm})(9.00 V_2)}$$
$$= 723 \text{ K} = 450°C$$

18.2 Cutaway of an automobile engine. While the air–gasoline mixture is being compressed prior to ignition, the intake and exhaust valves are both in the closed (up) position.

We didn't need to know the values of V_1 or V_2, only their ratio.

EVALUATE: Note that T_2 is the temperature of the air–gasoline mixture *before* the mixture is ignited; when burning starts, the temperature becomes higher still.

Example	
18.3	**Mass of air in a scuba tank**

A typical tank used for scuba diving has a volume of 11.0 L (about 0.4 ft^3) and a gauge pressure, when full, of 2.10×10^7 Pa (about 3000 psig). The "empty" tank contains 11.0 L of air at 21°C and 1 atm $(1.013 \times 10^5 \text{ Pa})$. When the tank is filled with hot air from a compressor, the temperature is 42°C and the gauge pressure is 2.10×10^7 Pa. What mass of air was added? (Air is a mixture of gases, about 78% nitrogen, 21% oxygen, and 1% miscellaneous; its average molar mass is 28.8 g/mol $= 28.8 \times 10^{-3}$ kg/mol.)

SOLUTION

IDENTIFY and SET UP: Our target variable is the *difference* between the mass present at the beginning (state 1) and at the end (state 2). We are given the molar mass of air, so we can use Eq. (18.2) to find the target variable if we know the number of moles present in states 1 and 2. We determine n_1 and n_2 by applying Eq. (18.3) to each state individually.

EXECUTE: We must remember to convert the temperatures to the Kelvin scale by adding 273 and to convert the pressure to absolute by adding 1.013×10^5 Pa. From Eq. (18.3), the number of moles n_1 in the "empty" tank is

$$n_1 = \frac{p_1 V_1}{R T_1} = \frac{(1.013 \times 10^5 \text{ Pa})(11.0 \times 10^{-3} \text{ m}^3)}{(8.315 \text{ J/mol} \cdot \text{K})(294 \text{ K})} = 0.46 \text{ mol}$$

The volume of the metal tank is hardly affected by the increased pressure, so $V_1 = V_2$. The number of moles in the full tank is

$$n_2 = \frac{p_2 V_2}{R T_2} = \frac{(21.1 \times 10^6 \text{ Pa})(11.0 \times 10^{-3} \text{ m}^3)}{(8.315 \text{ J/mol} \cdot \text{K})(315 \text{ K})} = 88.6 \text{ mol}$$

We added $n_2 - n_1 = 88.6 \text{ mol} - 0.46 \text{ mol} = 88.1$ mol to the tank. From Eq. (18.2), the added mass is $M(n_2 - n_1) = (88.1 \text{ mol})(28.8 \times 10^{-3} \text{ kg/mol}) = 2.54$ kg.

EVALUATE: The added mass is not insubstantial: you could certainly use a scale to determine whether the tank was empty or full.

Could this problem have been solved in the same way as Example 18.2? The volume is constant, so $p/nT = R/V$ is constant and $p_1/n_1 T_1 = p_2/n_2 T_2$; this can be solved for n_2/n_1, the ratio of the final and initial number of moles. But we need the *difference* of these two numbers, not the ratio, so this equation by itself isn't enough to solve the problem.

Example	
18.4	**Variation of atmospheric pressure with elevation**

Find the variation of atmospheric pressure with elevation in the earth's atmosphere, assuming that the temperature is 0°C at all elevations. Ignore the variation of g with elevation.

SOLUTION

IDENTIFY: As the elevation increases, both the atmospheric pressure and the density decrease. Hence we have *two* unknown functions of elevation; to solve for them, we need two separate relationships. One of these is the ideal-gas equation, which we can write in terms of pressure and density; the other is the relationship between pressure and density in a fluid in equilibrium, discussed in Section 14-2.

SET UP: In Section 14-2, we found the general equation $dp/dy = -\rho g$, Eq. (14.4), for the variation of pressure p with elevation y as a function of density ρ. Equation (18.5), $\rho = pM/RT$, states the ideal-gas equation in terms of density. We are told to assume that g and T are the same at all elevations; we also assume that the atmosphere has the same chemical composition, and hence the same molar mass M, at all heights. We then combine the two expressions and solve for $p(y)$.

EXECUTE: We substitute $\rho = pM/RT$ into $dp/dy = -\rho g$, separate variables, and integrate, letting p_1 be the pressure at elevation y_1 and p_2 be the pressure at y_2:

$$\frac{dp}{dy} = -\frac{pM}{RT}g$$

$$\int_{p_1}^{p_2} \frac{dp}{p} = -\frac{Mg}{RT} \int_{y_1}^{y_2} dy$$

$$\ln\frac{p_2}{p_1} = -\frac{Mg}{RT}(y_2 - y_1)$$

$$\frac{p_2}{p_1} = e^{-Mg(y_2 - y_1)/RT}$$

Now let $y_1 = 0$ be at sea level and let the pressure at that point be $p_0 = 1.013 \times 10^5$ Pa. Then our final expression for the pressure p at any height y is

$$p = p_0 e^{-Mgy/RT}$$

EVALUATE: According to our calculation, the pressure decreases exponentially with elevation. The graph in Fig. 18.3 shows that the slope dp/dy becomes less negative with greater elevation. That result makes sense, since $dp/dy = -\rho g$ and the density also decreases with elevation. At the summit of Mount Everest, where $y = 8863$ m,

$$\frac{Mgy}{RT} = \frac{(28.8 \times 10^{-3}\ \text{kg/mol})(9.80\ \text{m/s}^2)(8863\ \text{m})}{(8.315\ \text{J/mol} \cdot \text{K})(273\ \text{K})} = 1.10$$

$$p = (1.013 \times 10^5\ \text{Pa})e^{-1.10} = 0.337 \times 10^5\ \text{Pa}$$
$$= 0.33\ \text{atm}$$

The assumption of constant temperature isn't realistic, and g decreases a little with increasing elevation (see Challenge Problem 18.88). Even so, this example shows why mountaineers need to carry oxygen on Mount Everest.

The ability of the human body to absorb oxygen from the atmosphere depends critically on atmospheric pressure. Absorption drops sharply when the pressure is less than about 0.65×10^5 Pa, corresponding to an elevation above sea level of about 4700 m (15,000 ft). There is no permanent human habitation on earth above 6000 m (about 20,000 ft), although survival for short periods of

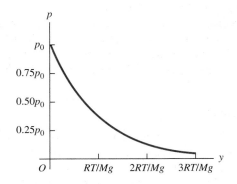

18.3 The variation of atmospheric pressure p with elevation y, assuming a constant temperature T.

time is possible at higher elevations. Jet airplanes, which typically fly at altitudes of 8000 to 12,000 m, *must* have pressurized cabins for passenger comfort and health.

The Van Der Waals Equation

The ideal-gas equation, Eq. (18.3), can be obtained from a simple molecular model that ignores the volumes of the molecules themselves and the attractive forces between them. We'll examine that model in Section 18.3. Meanwhile, we mention another equation of state, the **van der Waals equation,** that makes approximate corrections for these two omissions. This equation was developed by the 19th-century Dutch physicist J. D. van der Waals; the interaction between atoms that we discussed in Section 13.4 was named the *van der Waals interaction* after him. The van der Waals equation is

$$\left(p + \frac{an^2}{V^2}\right)(V - nb) = nRT \tag{18.7}$$

The constants a and b are empirical constants, different for different gases. Roughly speaking, b represents the volume of a mole of molecules; the total volume of the molecules is then nb, and the net volume available for the molecules to move around in is $V - nb$. The constant a depends on the attractive intermolecular forces, which reduce the pressure of the gas for given values of n, V, and T by *pulling* the molecules together as they *push* on the walls of the container. The decrease in pressure is proportional to the number of molecules per unit volume in a layer near the wall (which are exerting the pressure on the wall) and is also proportional to the number per unit volume in the next layer beyond the wall (which are doing the attracting). Hence the decrease in pressure due to intermolecular forces is proportional to n^2/V^2.

When n/V is small (that is, when the gas is *dilute*), the average distance between molecules is large, the corrections in the van der Waals equation become insignificant, and Eq. (18.7) reduces to the ideal-gas equation. As an example, for carbon dioxide gas (CO_2) the constants in the van der Waals equation are

$a = 0.364 \, \text{J} \cdot \text{m}^3/\text{mol}^2$ and $b = 4.27 \times 10^{-5} \, \text{m}^3/\text{mol}$. We found in Example 18.1 that one mole of an ideal gas at $T = 0°\text{C} = 273.15 \, \text{K}$ and $p = 1 \, \text{atm} = 1.013 \times 10^5 \, \text{Pa}$ occupies a volume $V = 0.0224 \, \text{m}^3$; according to Eq. (18.7), one mole of CO_2 occupying this volume at this temperature would be at a pressure 532 Pa less than 1 atm, a difference of only 0.5% from the ideal-gas value.

pV-Diagrams

We could in principle represent the $p\text{-}V\text{-}T$ relationship graphically as a *surface* in a three-dimensional space with coordinates p, V, and T. This representation sometimes helps in grasping the overall behavior of the substance, but ordinary two-dimensional graphs are usually more convenient. One of the most useful of these is a set of graphs of pressure as a function of volume, each for a particular constant temperature. Such a diagram is called a ***pV*-diagram.** Each curve, representing behavior at a specific temperature, is called an **isotherm,** or a *pV-isotherm.*

Figure 18.4 shows pV-isotherms for a constant amount of an ideal gas. The highest temperature is T_4; the lowest is T_1. This is a graphical representation of the ideal-gas equation of state. We can read off the volume V corresponding to any given pressure p and temperature T in the range shown.

Figure 18.5 shows a pV-diagram for a material that *does not* obey the ideal-gas equation. At temperatures below T_c the isotherms develop flat regions in which we can compress the material without an increase in pressure. Observation of the gas shows that it is *condensing* from the vapor (gas) to the liquid phase. The flat parts of the isotherms in the shaded area of Fig. 18.5 represent conditions of liquid-vapor *phase equilibrium.* As the volume decreases, more and more material goes from vapor to liquid, but the pressure does not change. (To keep the temperature constant during condensation, we have to remove the heat of vaporization, discussed in Section 17.6.)

When we compress such a gas at a constant temperature T_2 in Fig. 18.5, it is vapor until point a is reached. Then it begins to liquefy; as the volume decreases further, more material liquefies, and *both* the pressure and the temperature remain constant. At point b, all the material is in the liquid state. After this, any further compression results in a very rapid rise of pressure, because liquids are in general much less compressible than gases. At a lower constant temperature T_1, similar behavior occurs, but the condensation begins at lower pressure and greater volume than at the constant temperature T_2. At temperatures greater than T_c, *no* phase transition occurs as the material is compressed; at the highest temperatures, such as T_4, the curves resemble the ideal-gas curves of Fig. 18.4. We call T_c the *critical temperature* for this material. In Section 18.6 we'll discuss what happens to the phase of the gas above the critical temperature.

We will use pV-diagrams often in the next two chapters. We will show that the *area* under a pV-curve (whether or not it is an isotherm) represents the *work* done by the system during a volume change. This work, in turn, is directly related to heat transfer and changes in the *internal energy* of the system, which we'll get to in Chapter 19.

Test Your Understanding

Use the ideal-gas equation to explain why it's important to make holes in the crust of an apple pie before baking it.

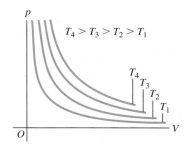

18.4 Isotherms, or constant-temperature curves, for a constant amount of an ideal gas. For each curve, the product $pV = nRT$ is constant, so p is proportional to $1/V$; the proportionality constant increases with increasing T.

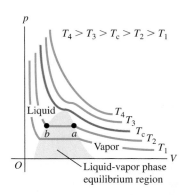

18.5 A pV-diagram for a non-ideal gas, showing isotherms for temperatures above and below the critical temperature T_c. The liquid-vapor equilibrium region is shown as a green shaded area. At still lower temperatures the material might undergo phase transitions from liquid to solid or from gas to solid; these are not shown in this diagram.

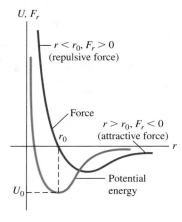

18.6 The force between two molecules (red curve) is zero at a separation $r = r_0$, where the potential energy (blue curve) is a minimum. The force is attractive when the separation is greater than r_0 and repulsive when the separation is less than r_0. (Compare this figure to Fig. 13.19.)

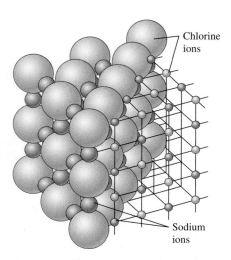

18.7 Schematic representation of the cubic crystal structure of sodium chloride.

18.2 | Molecular Properties of Matter

We have studied several properties of matter in bulk, including elasticity, density, surface tension, heat capacities, and equations of state, with only passing references to molecular structure. Now we want to look in more detail at the relation of bulk behavior to microscopic structure. We begin with a general discussion of the molecular structure of matter. Then in the next two sections we develop the kinetic-molecular model of an ideal gas, obtaining from this molecular model the equation of state and an expression for heat capacity.

All familiar matter is made up of **molecules.** For any specific chemical compound, all the molecules are identical. The smallest molecules contain one atom each and are of the order of 10^{-10} m in size; the largest contain many atoms and are at least 10,000 times that large. In gases the molecules move nearly independently; in liquids and solids they are held together by intermolecular forces that are electrical in nature, arising from interactions of the electrically charged particles that make up the molecules. Gravitational forces between molecules are negligible in comparison with electrical forces.

The interaction of two *point* electric charges is described by a force (repulsive for like charges, attractive for unlike charges) with a magnitude proportional to $1/r^2$, where r is the distance between the points. We will study this relationship, called *Coulomb's law,* in Chapter 21. Molecules are *not* point charges but complex structures containing both positive and negative charge, and their interactions are more complex. The force between molecules in a gas varies with the distance r between molecules somewhat as shown in Fig. 18.6, where a positive F_r corresponds to a repulsive force and a negative F_r to an attractive force. When molecules are far apart, the intermolecular forces are very small and usually attractive. As a gas is compressed and its molecules are brought closer together, the attractive forces increase. The intermolecular force becomes zero at an equilibrium spacing r_0, corresponding roughly to the spacing between molecules in the liquid and solid states. In liquids and solids, relatively large pressures are needed to compress the substance appreciably. This shows that at molecular distances slightly *less* than the equilibrium spacing, the forces become *repulsive* and relatively large.

Figure 18.6 also shows the potential energy as a function of r. This function has a *minimum* at r_0, where the force is zero. The two curves are related by $F_r(r) = -dU/dr$, as we showed in Section 7.4. Such a potential energy function is often called a **potential well.** A molecule at rest at a distance r_0 from a second molecule would need an additional energy $|U_0|$, the "depth" of the potential well, to "escape" to an indefinitely large value of r.

Molecules are always in motion; their kinetic energies usually increase with temperature. At very low temperatures the average kinetic energy of a molecule may be much *less* than the depth of the potential well. The molecules then condense into the liquid or solid phase with average intermolecular spacings of about r_0. But at higher temperatures the average kinetic energy becomes larger than the depth $|U_0|$ of the potential well. Molecules can then escape the intermolecular force and become free to move independently, as in the gaseous phase of matter.

In *solids,* molecules vibrate about more or less fixed points. In a crystalline solid these points are arranged in a recurring *crystal lattice.* Figure 18.7 shows the cubic crystal structure of sodium chloride (ordinary salt). A scanning tunneling microscope image of individual silicon atoms on the surface of a crystal is shown in Fig. 18.8.

The vibration of molecules in a solid about their equilibrium positions may be nearly simple harmonic if the potential well is approximately parabolic in shape at distances close to r_0. (We discussed this kind of simple harmonic motion in Section 13.4.) But if the potential-energy curve rises more gradually for $r > r_0$ than for $r < r_0$, as in Fig. 18.6, the average position shifts to larger r with increasing amplitude. As we pointed out in Section 17.4, this is the basis of thermal expansion.

In a *liquid,* the intermolecular distances are usually only slightly greater than in the solid phase of the same substance, but the molecules have much greater freedom of movement. Liquids show regularity of structure only in the immediate neighborhood of a few molecules. This is called *short-range order,* in contrast with the *long-range order* of a solid crystal.

The molecules of a *gas* are usually widely separated and so have only very small attractive forces. A gas molecule moves in a straight line until it collides with another molecule or with a wall of the container. In molecular terms, an *ideal gas* is a gas whose molecules exert *no* attractive forces on each other and therefore have no *potential* energy.

At low temperatures, most common substances are in the solid phase. As the temperature rises, a substance melts and then vaporizes. From a molecular point of view, these transitions are in the direction of increasing molecular kinetic energy. Thus temperature and molecular kinetic energy are closely related.

We have used the mole as a measure of quantity of substance. One **mole** of any pure chemical element or compound contains a definite number of molecules, the same number for all elements and compounds. The official SI definition is that **one mole is the amount of substance that contains as many elementary entities as there are atoms in 0.012 kilogram of carbon-12.** In our discussion, the "elementary entities" are molecules. (In a monatomic substance such as carbon or helium, each molecule is a single atom, but we'll still call it a molecule here.) Note that atoms of a given element may occur in any of several isotopes, which are chemically identical but have different atomic masses; "carbon-12" refers to a specific isotope of carbon.

The number of molecules in a mole is called **Avogadro's number,** denoted by N_A. The current best numerical value of N_A is

$$N_A = 6.02214199(47) \times 10^{23} \text{ molecules/mol} \quad (\text{Avogadro's number})$$

The *molar mass* M of a compound is the mass of one mole. It is equal to the mass m of a single molecule multiplied by Avogadro's number.

$$M = N_A m \quad (18.8)$$
(molar mass, Avogadro's number, and mass of a molecule)

When the molecule consists of a single atom, the term *atomic mass* is often used instead of molar mass or molecular weight.

18.8 A scanning tunneling microscope image of the surface of a silicon crystal. The area shown is only 9.0 nm (9.0×10^{-9} m) tall. Each blue "bead" is an individual silicon atom; you can clearly see how these atoms are arranged in a (nearly) perfect array of hexagons.

Example 18.5	**Atomic and molecular mass**

Find the mass of a single hydrogen atom and the mass of an oxygen molecule.

SOLUTION

IDENTIFY and SET UP: We use Eq. (18.8) in the form $m = M/N_A$ and the values of the molar masses from the periodic table of the elements (Appendix D).

EXECUTE: The mass per mole of atomic hydrogen (that is, the atomic mass) is 1.008 g/mol. Therefore the mass m_H of a single hydrogen atom is

$$m_H = \frac{1.008 \text{ g/mol}}{6.022 \times 10^{23} \text{ molecules/mol}} = 1.674 \times 10^{-24} \text{ g/molecule}$$

From Appendix D, the atomic mass of oxygen is 16.0 g/mol, so the molar mass of oxygen, which has diatomic (two-atom) molecules, is 32.0 g/mol. The mass of a single molecule of O_2 is

$$m_{O_2} = \frac{32.0 \text{ g/mol}}{6.022 \times 10^{23} \text{ molecules/mol}} = 53.1 \times 10^{-24} \text{ g/molecule}$$

EVALUATE: We note that the values in Appendix D are for the *average* atomic masses of a natural sample of each element. Such a sample may contain several different isotopes of the element, each with a different atomic mass. Natural samples of hydrogen and oxygen are almost entirely made up of just one isotope; this is not the case for all elements, however.

Test Your Understanding

Suppose you could adjust the value of r_0 for the molecules of a certain chemical compound (Fig. 18.6) by turning a dial. If you doubled the value of r_0, how would this change the density of the solid form of this compound?

18.3 | Kinetic-Molecular Model of an Ideal Gas

The goal of any molecular theory of matter is to understand the *macroscopic* properties of matter in terms of its atomic or molecular structure and behavior. Such theories are of tremendous practical importance; once we have this understanding, we can design materials to have specific desired properties. Such analysis has led to the development of high-strength steels, glasses with special optical properties, semiconductor materials for electronic devices, and countless other materials essential to contemporary technology.

In this and the following sections we will consider a simple molecular model of an ideal gas. This *kinetic-molecular model* represents the gas as a large number of particles bouncing around in a closed container. In this section we use the kinetic-molecular model to understand how the ideal-gas equation of state, Eq. (18.3), is related to Newton's laws. In the following section we use the kinetic-molecular model to predict the molar heat capacity of an ideal gas. We'll go on to elaborate the model to include "particles" that are not points but have a finite size. We will be able to see why polyatomic gases have larger molar heat capacities than monatomic gases.

The following discussion of the kinetic-molecular model has several steps, and you may need to go over it several times to grasp how it all goes together. Don't get discouraged!

Here are the assumptions of our model:

1. A container with volume V contains a very large number N of identical molecules, each with mass m.
2. The molecules behave as point particles; their size is small in comparison to the average distance between particles and to the dimensions of the container.
3. The molecules are in constant motion; they obey Newton's laws of motion. Each molecule collides occasionally with a wall of the container. These collisions are perfectly elastic.
4. The container walls are perfectly rigid and infinitely massive and do not move.

Act|v
ONLINE
Phys|cs

8.1   Characteristics of a Gas

CAUTION Make sure you don't confuse *N*, the number of *molecules* in the gas, with *n*, the number of *moles*. The number of molecules is equal to the number of moles multiplied by Avogadro's number: $N = nN_A$.

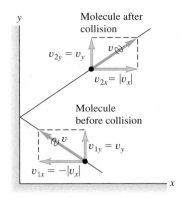

During collisions the molecules exert *forces* on the walls of the container; this is the origin of the *pressure* that the gas exerts. In a typical collision (Fig. 18.9) the velocity component parallel to the wall is unchanged, and the component perpendicular to the wall reverses direction but does not change in magnitude.

Our program is first to determine the *number* of collisions that occur per unit time for a certain area *A* of wall. Then we find the total momentum change associated with these collisions and the force needed to cause this momentum change. From this we can determine the pressure, which is force per unit area, and compare the result to the ideal-gas equation. We'll find a direct connection between the temperature of the gas and the kinetic energy of the gas molecules.

To begin, we will assume that all molecules in the gas have the same *magnitude* of x-velocity, $|v_x|$. This isn't right, but making this temporary assumption helps to clarify the basic ideas. We will show later that this assumption isn't really necessary.

As shown in Fig. 18.9, for each collision the x-component of velocity changes from $-|v_x|$ to $+|v_x|$. So the x-component of momentum changes from $-m|v_x|$ to $+m|v_x|$, and the *change* in the x-component of momentum is $m|v_x| - (-m|v_x|) = 2m|v_x|$.

If a molecule is going to collide with a given wall area *A* during a small time interval *dt*, then at the beginning of *dt* it must be within a distance $|v_x| \, dt$ from the wall (Fig. 18.10) and it must be headed toward the wall. So the number of molecules that collide with *A* during *dt* is equal to the number of molecules within a cylinder with base area *A* and length $|v_x| \, dt$ that have their x-velocity aimed toward the wall. The volume of such a cylinder is $A|v_x| \, dt$. Assuming that the number of molecules per unit volume (N/V) is uniform, the *number* of molecules in this cylinder is $(N/V)(A|v_x| \, dt)$. On the average, half of these molecules are moving toward the wall and half are moving away from it. So the number of collisions with *A* during *dt* is

$$\frac{1}{2}\left(\frac{N}{V}\right)(A|v_x|dt)$$

For the system of all molecules in the gas, the total momentum change dP_x during *dt* is the *number* of collisions multiplied by $2m|v_x|$:

$$dP_x = \frac{1}{2}\left(\frac{N}{V}\right)(A|v_x|dt)(2m|v_x|) = \frac{NAmv_x^2 \, dt}{V} \tag{18.9}$$

(We are using capital *P* for total momentum and small *p* for pressure. Be careful!) We wrote v_x^2 rather than $|v_x|^2$ in the final expression because the square of the absolute value of a number is equal to the square of that number. The *rate* of change of momentum component P_x is

$$\frac{dP_x}{dt} = \frac{NAmv_x^2}{V} \tag{18.10}$$

According to Newton's second law, this rate of change of momentum equals the force exerted by the wall area *A* on the gas molecules. From Newton's *third* law

18.9 Elastic collision of a molecule with an idealized container wall. The velocity component parallel to the wall does not change; the velocity component perpendicular to the wall reverses direction. The speed *v* does not change.

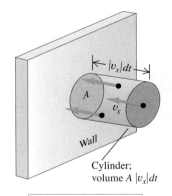

18.10 For a molecule to strike the wall in area *A* during a time interval *dt*, the molecule must be headed for the wall and be within the shaded cylinder of height $|v_x|dt$ at the beginning of the interval.

this is equal and opposite to the force exerted *on* the wall *by* the molecules. Pressure p is the magnitude of the force exerted on the wall per unit area, and we obtain

$$p = \frac{F}{A} = \frac{Nmv_x^2}{V} \qquad (18.11)$$

The pressure exerted by the gas depends on the number of molecules per volume (N/V), the mass m per molecule, and the speed of the molecules.

We mentioned that $|v_x|$ is really *not* the same for all the molecules. But we could have sorted the molecules into groups having the same $|v_x|$ within each group, then added up the resulting contributions to the pressure. The net effect of all this is just to replace v_x^2 in Eq. (18.11) by the *average* value of v_x^2, which we denote by $(v_x^2)_{av}$. Furthermore, $(v_x^2)_{av}$ is related simply to the *speeds* of the molecules. The speed v of any molecule is related to the velocity components v_x, v_y, and v_z by

$$v^2 = v_x^2 + v_y^2 + v_z^2$$

We can average this relation over all molecules:

$$(v^2)_{av} = (v_x^2)_{av} + (v_y^2)_{av} + (v_z^2)_{av}$$

But there is no real difference in our model between the *x*-, *y*-, and *z*-directions. (Molecular speeds are very fast in a typical gas, so the effects of gravity are negligibly small.) It follows that $(v_x^2)_{av}$, $(v_y^2)_{av}$, and $(v_z^2)_{av}$ must all be *equal*. Hence $(v^2)_{av}$ is equal to $3(v_x^2)_{av}$, and

$$(v_x^2)_{av} = \frac{1}{3}(v^2)_{av}$$

so Eq. (18.11) becomes

$$pV = \frac{1}{3}Nm(v^2)_{av} = \frac{2}{3}N\left[\frac{1}{2}m(v^2)_{av}\right] \qquad (18.12)$$

We notice that $\frac{1}{2}m(v^2)_{av}$ is the average translational kinetic energy of a single molecule. The product of this and the total number of molecules N equals the total random kinetic energy K_{tr} of translational motion of all the molecules. (The notation K_{tr} reminds us that this energy is associated with *translational* motion. There may be additional energies associated with rotational and vibrational motion of molecules.) The product pV equals two-thirds of the total translational kinetic energy:

$$pV = \frac{2}{3}K_{tr} \qquad (18.13)$$

Now we compare this with the ideal-gas equation,

$$pV = nRT$$

which is based on experimental studies of gas behavior. For the two equations to agree, we must have

$$K_{tr} = \frac{3}{2}nRT \qquad (18.14)$$

(average translational kinetic energy of n moles of ideal gas)

This remarkably simple result shows that K_{tr} is *directly proportional* to the absolute temperature T. We will use this important result several times in the following discussion.

The average translational kinetic energy of a single molecule is the total translational kinetic energy K_{tr} of all molecules divided by the number of molecules, N:

$$\frac{K_{tr}}{N} = \frac{1}{2}m(v^2)_{av} = \frac{3nRT}{2N}$$

Also, the total number of molecules N is the number of moles n multiplied by Avogadro's number N_A, so

$$N = nN_A \qquad \frac{n}{N} = \frac{1}{N_A}$$

and

$$\frac{K_{tr}}{N} = \frac{1}{2}m(v^2)_{av} = \frac{3}{2}\left(\frac{R}{N_A}\right)T \qquad (18.15)$$

The ratio R/N_A occurs frequently in molecular theory. It is called the **Boltzmann constant,** k:

$$k = \frac{R}{N_A} = \frac{8.314 \text{ J/mol} \cdot \text{K}}{6.022 \times 10^{23} \text{ molecules/mol}}$$

$$= 1.381 \times 10^{-23} \text{ J/molecule} \cdot \text{K}$$

(The current best numerical value of k is $1.3806503(24) \times 10^{-23}$ J/molecule $\cdot$ K.) In terms of k we can rewrite Eq. (18.15) as

$$\frac{1}{2}m(v^2)_{av} = \frac{3}{2}kT \qquad (18.16)$$

(average translational kinetic energy of a gas molecule)

This shows that the average translational kinetic energy *per molecule* depends only on the temperature, not on the pressure, volume, or kind of molecule. We can obtain the average translational kinetic energy *per mole* by multiplying Eq. (18.16) by Avogadro's number and using the relation $M = N_A m$:

$$N_A \frac{1}{2}m(v^2)_{av} = \frac{1}{2}M(v^2)_{av} = \frac{3}{2}RT \qquad (18.17)$$

(average translational kinetic energy per mole of gas)

The translational kinetic energy of a mole of ideal-gas molecules depends only on T.

Finally, it is sometimes convenient to rewrite the ideal-gas equation on a molecular basis. We use $N = N_A n$ and $R = N_A k$ to obtain the alternative form of the ideal-gas equation:

$$pV = NkT \qquad (18.18)$$

This shows that we can think of the Boltzmann constant k as a gas constant on a "per-molecule" basis instead of the usual "per-mole" basis for R.

Molecular Speeds

From Eqs. (18.16) and (18.17) we can obtain expressions for the square root of $(v^2)_{av}$, called the **root-mean-square speed** (or **rms speed**) v_{rms}:

$$v_{rms} = \sqrt{(v^2)_{av}} = \sqrt{\frac{3kT}{m}} = \sqrt{\frac{3RT}{M}} \qquad (18.19)$$

(root-mean-square speed of a gas molecule)

It might seem more natural to characterize molecular speeds by their *average* value rather than by v_{rms}, but we see that v_{rms} evolves more directly from Eqs. (18.16) and (18.17). To compute the rms speed, we square each molecular speed, add, divide by the number of molecules, and take the square root; v_{rms} is the *root* of the *mean* of the *squares*. Example 18.7 illustrates this procedure.

Equations (18.16) and (18.19) show that at a given temperature T, gas molecules of different mass m have the same average kinetic energy but different root-mean-square speeds. On average, the nitrogen molecules ($M = 28$ g/mol) in the air around you are moving faster than are the oxygen molecules ($M = 32$ g/mol). Hydrogen molecules ($M = 2$ g/mol) are fastest of all; this is why there is hardly any hydrogen in the earth's atmosphere, despite its being the most common element in the universe (Fig. 18.11). A sizable fraction of any H_2 molecules in the atmosphere would have speeds greater than the earth's "escape speed" of 1.12×10^4 m/s (calculated in Example 12.5, Section 12.3) and would escape into space. The heavier, slower-moving gases cannot escape so easily, which is why they dominate our atmosphere.

The assumption that individual molecules undergo perfectly elastic collisions with the container wall is actually a little too simple. More detailed investigation has shown that in most cases, molecules actually adhere to the wall for a short time and then leave again with speeds that are characteristic of the temperature *of the wall*. However, the gas and the wall are ordinarily in thermal equilibrium and have the same temperature. So there is no net energy transfer between gas and wall, and this discovery does not alter the validity of our conclusions.

18.11 The relationship between molecular speeds and molecular masses explains why hydrogen—the most common element in the universe—is only a trace constituent of our atmosphere (only 0.00005% hydrogen by volume).

Problem-Solving Strategy

Kinetic-Molecular Theory

IDENTIFY *the relevant concepts:* Use the results of the kinetic-molecular model whenever you are asked to relate macroscopic properties of a gas, such as temperature and pressure, to microscopic properties, such as molecular speeds.

SET UP *the problem using the following steps:*
1. Identify which variables are known and which are the unknown target variables.
2. Choose the equation(s) to be used from among Eqs. (18.14), (18.16), and (18.19).

EXECUTE *the solution as follows:* As you solve for the target variable, be on your guard for inconsistency in units. Special caution is needed in the following places:

1. The usual units for molar mass M are grams per mole; the molar mass of oxygen (O_2) is 32 g/mol, for example. These units are often omitted in tables. In equations such as Eq. (18.19), when you use SI units you *must* express M in kilograms per mole by multiplying the table value by $(1 \text{ kg}/10^3 \text{ g})$. Thus in SI units, M for oxygen is 32×10^{-3} kg/mol.

2. Are you working on a "per molecule" basis or a "per mole" basis? Remember that m is the mass of a single molecule and M is the mass of a mole of molecules; N is the number of molecules and n is the number of moles; k is the gas constant per molecule and R is the gas constant per mole. Although N, the number of molecules, is in one sense a dimensionless number, you can do a complete unit

check if you think of N as having units of "molecules"; then m has units of "mass per molecule," and k has units of "joules per molecule per kelvin."

3. Remember that T is always *absolute* (Kelvin) temperature.

EVALUATE *your answer:* Are your answers reasonable? Keep in mind that typical molecular speeds at room temperature are several hundred meters per second. If your answer seems dramatically different, recheck your calculations.

Example 18.6 Calculating molecular kinetic energy and v_{rms}

a) What is the average translational kinetic energy of a molecule of an ideal gas at a temperature of 27°C? b) What is the total random translational kinetic energy of the molecules in one mole of this gas? c) What is the root-mean-square speed of oxygen molecules at this temperature?

SOLUTION

IDENTIFY and SET UP: We are given temperature $T = 27°C$ and number of moles $n = 1$ mol, and the molecular mass m is that for oxygen. We use Eq. (18.16) to determine the average kinetic energy of a molecule, Eq. (18.14) to find the total molecular kinetic energy, and Eq. (18.19) to find the rms speed of a molecule.

EXECUTE: a) To use Eq. (18.16), we first convert the temperature to the Kelvin scale: 27°C = 300 K. Then

$$\frac{1}{2}m(v^2)_{av} = \frac{3}{2}kT = \frac{3}{2}(1.38 \times 10^{-23} \text{ J/K})(300 \text{ K})$$
$$= 6.21 \times 10^{-21} \text{ J}$$

This answer does not depend on the mass of the molecule.
b) From Eq. (18.14), the total translational kinetic energy of a mole of molecules is

$$K_{tr} = \frac{3}{2}nRT = \frac{3}{2}(1 \text{ mol})(8.315 \text{ J/mol} \cdot \text{K})(300 \text{ K})$$
$$= 3740 \text{ J}$$

This is about the same kinetic energy as that of a sprinter in a 100-m dash.
c) From Example 18.5 (Section 18.2), the mass of an oxygen molecule is

$$m_{O_2} = (53.1 \times 10^{-24} \text{ g})(1 \text{ kg}/10^3 \text{ g}) = 5.31 \times 10^{-26} \text{ kg}$$

From Eq. (18.19),

$$v_{rms} = \sqrt{\frac{3kT}{m}} = \sqrt{\frac{3(1.38 \times 10^{-23} \text{ J/K})(300 \text{ K})}{5.31 \times 10^{-26} \text{ kg}}}$$
$$= 484 \text{ m/s}$$

This is 1740 km/h, or 1080 mi/h! Alternatively,

$$v_{rms} = \sqrt{\frac{3RT}{M}} = \sqrt{\frac{3(8.314 \text{ J/mol} \cdot \text{K})(300 \text{ K})}{32.0 \times 10^{-3} \text{ kg/mol}}}$$
$$= 484 \text{ m/s}$$

EVALUATE: We can check our result in part (b) by noting that the translational kinetic energy per mole must be equal to the average translational kinetic energy per molecule from part (a) multiplied by Avogadro's number N_A: $K_{tr} = (6.022 \times 10^{23} \text{ molecules})(6.21 \times 10^{-21} \text{ J/molecule}) = 3740 \text{ J}$.

In part (c), note that when we use Eq. (18.19) with R in SI units, we must express M in *kilograms* per mole, not grams per mole. In this example we use $M = 32.0 \times 10^{-3} \text{ kg/mol}$, *not* 32.0 g/mol.

Example 18.7 Rms and average speeds

Five gas molecules chosen at random are found to have speeds of 500, 600, 700, 800, and 900 m/s. Find the rms speed. Is it the same as the *average* speed?

SOLUTION

IDENTIFY and SET UP: To solve this problem, we use the definition of the root-mean-square value: square each value, find the average (mean) of the squares, and then take the square root of the result.

EXECUTE: The average value of v^2 for the five molecules is

$$(v^2)_{av} = \frac{\left[\begin{array}{c}(500 \text{ m/s})^2 + (600 \text{ m/s})^2 + (700 \text{ m/s})^2 \\ + (800 \text{ m/s})^2 + (900 \text{ m/s})^2\end{array}\right]}{5}$$
$$= 5.10 \times 10^5 \text{ m}^2/\text{s}^2$$

The square root of this is v_{rms}:

$$v_{rms} = 714 \text{ m/s}$$

The *average* speed v_{av} is given by

$$v_{av} = \frac{500 \text{ m/s} + 600 \text{ m/s} + 700 \text{ m/s} + 800 \text{ m/s} + 900 \text{ m/s}}{5}$$
$$= 700 \text{ m/s}$$

EVALUATE: We see that in general, v_{rms} and v_{av} are *not* the same. Roughly speaking, v_{rms} gives greater weight to the larger speeds than does v_{av}.

Example
18.8   **Volume of a gas at STP, revisited**

Find the number of molecules and the number of moles in one cubic meter of air at atmospheric pressure and 0°C.

SOLUTION

IDENTIFY and SET UP: We are given the volume V, the pressure p, and the temperature T. We find the first target variable, the number of molecules N, using Eq. (18.18); the number of moles is N divided by Avogadro's number.

EXECUTE: From Eq. (18.18),

$$N = \frac{pV}{kT} = \frac{(1.013 \times 10^5 \text{ Pa})(1 \text{ m}^3)}{(1.38 \times 10^{-23} \text{ J/K})(273 \text{ K})}.$$

$$= 2.69 \times 10^{25} \text{ molecules}$$

The number of moles n is

$$n = \frac{N}{N_A} = \frac{2.69 \times 10^{25} \text{ molecules}}{6.022 \times 10^{23} \text{ molecules/mol}} = 44.7 \text{ mol}$$

EVALUATE: To check our result, note that the total volume is 1.00 m³, so the volume of one mole is

$$\frac{1.00 \text{ m}^3}{44.7 \text{ mol}} = 0.0224 \text{ m}^3/\text{mol} = 22.4 \text{ L/mol}$$

This is the same result that we obtained in Example 18.1 (Section 18.1) using only macroscopic quantities.

Collisions Between Molecules

We have ignored the possibility that two gas molecules might collide. If they are really points, they *never* collide. But consider a more realistic model in which the molecules are rigid spheres with radius r. How often do they collide with other molecules? How far do they travel, on average, between collisions? We can get approximate answers from the following rather primitive model.

Consider N spherical molecules with radius r in a volume V. Suppose only one molecule is moving. When it collides with another molecule, the distance between centers is $2r$. Suppose we draw a cylinder with radius $2r$, with its axis parallel to the velocity of the molecule (Fig. 18.12). The moving molecule collides with any other molecule whose center is inside this cylinder. In a short time dt a molecule with speed v travels a distance $v\,dt$; during this time it collides with any molecule that is in the cylindrical volume of radius $2r$ and length $v\,dt$. The volume of the cylinder is $4\pi r^2 v\,dt$. There are N/V molecules per unit volume, so the number dN with centers in this cylinder is

$$dN = 4\pi r^2 v\,dt\,N/V$$

Thus the number of collisions *per unit time* is

$$\frac{dN}{dt} = \frac{4\pi r^2 v N}{V}$$

This result assumes that only one molecule is moving. The analysis is quite a bit more involved when all the molecules move at once. It turns out that in this case the collisions are more frequent, and the above equation has to be multiplied by a factor of $\sqrt{2}$:

$$\frac{dN}{dt} = \frac{4\pi\sqrt{2}r^2 v N}{V}$$

The average time t_{mean} between collisions, called the *mean free time,* is the reciprocal of this expression:

$$t_{\text{mean}} = \frac{V}{4\pi\sqrt{2}r^2 v N} \tag{18.20}$$

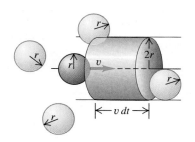

18.12 In a time dt a molecule with radius r will collide with any other molecule within a cylindrical volume of radius $2r$ and length $v\,dt$.

The average distance traveled between collisions is called the **mean free path,** denoted by λ (the Greek letter "lambda"). In our simple model, this is just the molecule's speed v multiplied by t_{mean}:

$$\lambda = vt_{\text{mean}} = \frac{V}{4\pi\sqrt{2}r^2N} \quad (\text{mean free path of a gas molecule}) \quad (18.21)$$

The mean free path is inversely proportional to the number of molecules per unit volume (N/V) and inversely proportional to the cross-sectional area πr^2 of a molecule; the more molecules there are and the larger the size of a molecule, the shorter the mean distance between collisions (Fig. 18.13). Note that the mean free path *does not* depend on the speed of the molecule.

We can express Eq. (18.21) in terms of macroscopic properties of the gas, using the ideal-gas equation in the form of Eq. (18.18), $pV = NkT$. We find

$$\lambda = \frac{kT}{4\pi\sqrt{2}r^2p} \quad (18.22)$$

If the temperature is increased at constant pressure, the gas expands, the average distance between molecules increases, and λ increases. If the pressure is increased at constant temperature, the gas compresses and λ decreases.

18.13 If you try to walk through a crowd, your mean free path—the distance you can travel on average without running into another person—depends on how large the people are and how closely they are spaced.

Example 18.9	**Calculating mean free path**

a) Estimate the mean free path of a molecule of air at 27°C and 1 atm. Model the molecules as spheres with radius $r = 2.0 \times 10^{-10}$ m. b) Estimate the mean free time of an oxygen molecule with $v = v_{\text{rms}}$.

SOLUTION

IDENTIFY and SET UP: We use Eq. (18.21) to determine the mean free path λ. To find the mean free time t_{mean} we could use Eq. (18.20), but it's more convenient to use the basic relationship $\lambda = vt_{\text{mean}}$ in Eq. (18.21). For the speed v we use the root-mean-square speed for oxygen calculated in Example 18.6.

EXECUTE: a) From Eq. (18.22),

$$\lambda = \frac{kT}{4\pi\sqrt{2}r^2p}$$

$$= \frac{(1.38 \times 10^{-23}\ \text{J/K})(300\ \text{K})}{4\pi\sqrt{2}(2.0 \times 10^{-10}\ \text{m})^2(1.01 \times 10^5\ \text{Pa})}$$

$$= 5.8 \times 10^{-8}\ \text{m}$$

The molecule doesn't get very far between collisions, but the distance is still several hundred times the radius of the molecule. To get a mean free path of one meter, the pressure must be about 5.8×10^{-8} atm. Pressures this low are found 100 km or so above the earth's surface, at the outer fringe of our atmosphere.

b) From Example 18.6, for oxygen at 27°C the root-mean-square speed is $v_{\text{rms}} = 484$ m/s, so the mean free time for a molecule with this speed is

$$t_{\text{mean}} = \frac{\lambda}{v} = \frac{5.8 \times 10^{-8}\ \text{m}}{484\ \text{m/s}} = 1.2 \times 10^{-10}\ \text{s}$$

This molecule undergoes about 10^{10} collisions per second!

EVALUATE: Note that the mean free *path* calculated in part (a) doesn't depend on the molecule's speed, but the mean free *time* does. Slower molecules have a longer average time interval t_{mean} between collisions than do fast ones, but the average *distance* λ between collisions is the same no matter what the molecule's speed.

Test Your Understanding

What is the root-mean-square speed of hydrogen atoms in the sun's corona, shown in the photo that opens Chapter 17? (The corona's temperature is so high and its density so low that hydrogen atoms cannot form into molecules.) Compare your result to the sun's escape speed of 6.18×10^5 m/s.

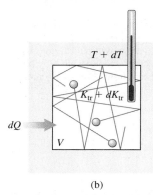

18.14 (a) A fixed volume V of a monatomic ideal gas. (b) When an amount of heat dQ is added to the gas, the total translational kinetic energy increases by $dK_{tr} = dQ$ and the temperature increases by $dT = dQ/nC_V$.

18.4 | Heat Capacities

When we introduced the concept of heat capacity in Section 17.5, we talked about ways to *measure* the specific heat or molar heat capacity of a particular material. Now we'll see how these numbers can be *predicted* on theoretical grounds. That's a significant step forward.

Heat Capacities of Gases

The basis of our analysis is that heat is *energy* in transit. When we add heat to a substance, we are increasing its molecular energy. In this discussion we will keep the volume of the gas constant so that we don't have to worry about energy transfer through mechanical work. If we were to let the gas expand, it would do work by pushing on moving walls of its container, and this additional energy transfer would have to be included in our calculations. We'll return to this more general case in Chapter 19. For now, with the volume held constant, we are concerned with C_V, the molar heat capacity *at constant volume.*

In the simple kinetic-molecular model of Section 18.3 the molecular energy consists only of the translational kinetic energy K_{tr} of the pointlike molecules. This energy is directly proportional to the absolute temperature T, as shown by Eq. (18.14), $K_{tr} = \frac{3}{2}nRT$. When the temperature changes by a small amount dT, the corresponding change in kinetic energy is

$$dK_{tr} = \frac{3}{2}nR\,dT \qquad (18.23)$$

From the definition of molar heat capacity at constant volume, C_V (Section 17.5), we also have

$$dQ = nC_V\,dT \qquad (18.24)$$

where dQ is the heat input needed for a temperature change dT. Now if K_{tr} represents the total molecular energy, as we have assumed, then dQ and dK_{tr} must be *equal* (Fig. 18.14). Equating the expressions given by Eqs. (18.23) and (18.24), we get

$$nC_V\,dT = \frac{3}{2}nR\,dT$$

$$C_V = \frac{3}{2}R \qquad \text{(ideal gas of point particles)} \qquad (18.25)$$

This surprisingly simple result says that the molar heat capacity (at constant volume) of *every* gas whose molecules can be represented as points is equal to $3R/2$.

To see whether this makes sense, let's first check the units. The gas constant *does* have units of energy per mole per kelvin, the correct units for a molar heat capacity. But more important is whether Eq. (18.25) agrees with *measured* values of molar heat capacities. In SI units, Eq. (18.25) gives

$$C_V = \frac{3}{2}(8.314\ \text{J/mol·K}) = 12.47\ \text{J/mol·K}$$

For comparison, Table 18.1 gives measured values of C_V for several gases. We see that for *monatomic* gases our prediction is right on the money, but that it is way off for diatomic and polyatomic gases.

Table 18.1 Molar Heat Capacities of Gases

Type of Gas	Gas	C_V (J/mol·K)
Monatomic	He	12.47
	Ar	12.47
Diatomic	H_2	20.42
	N_2	20.76
	O_2	21.10
	CO	20.85
Polyatomic	CO_2	28.46
	SO_2	31.39
	H_2S	25.95

This comparison tells us that our point-molecule model is good enough for monatomic gases but that for diatomic and polyatomic molecules we need something more sophisticated. For example, we can picture a diatomic molecule as *two* point masses, like a little elastic dumbbell, with an interaction force between the atoms of the kind shown in Fig. 18.6. Such a molecule can have additional kinetic energy associated with *rotation* about axes through its center of mass. The atoms may also have a back-and-forth *vibrating* motion along the line joining them, with additional kinetic and potential energies. These possibilities are shown in Fig. 18.15.

When heat flows into a *monatomic* gas at constant volume, *all* of the added energy goes into an increase in random *translational* molecular kinetic energy. Equation (18.23) shows that this gives rise to an increase in temperature. But when the temperature is increased by the same amount in a *diatomic* or *polyatomic* gas, additional heat is needed to supply the increased rotational and vibrational energies. Thus polyatomic gases have *larger* molar heat capacities than monatomic gases, as Table 18.1 shows.

But how do we know how much energy is associated with each additional kind of motion of a complex molecule, compared to the translational kinetic energy? The new principle that we need is called the principle of **equipartition of energy.** It can be derived from sophisticated statistical-mechanics considerations; that derivation is beyond our scope, and we will treat the principle as an axiom.

The principle of equipartition of energy states that each velocity component (either linear or angular) has, on average, an associated kinetic energy per molecule of $\frac{1}{2}kT$, or one-half the product of the Boltzmann constant and the absolute temperature. The number of velocity components needed to describe the motion of a molecule completely is called the number of **degrees of freedom.** For a monatomic gas, there are three degrees of freedom (for the velocity components v_x, v_y, and v_z); this gives a total average kinetic energy per molecule of $3(\frac{1}{2}kT)$, consistent with Eq. (18.16).

For a *diatomic* molecule there are two possible axes of rotation, perpendicular to each other and to the molecule's axis. (We don't include rotation about the molecule's own axis because in ordinary collisions there is no way for this rotational motion to change.) If we assign five degrees of freedom to a diatomic molecule, the average total kinetic energy per molecule is $\frac{5}{2}kT$ instead of $\frac{3}{2}kT$. The total kinetic energy of n moles is $K_{\text{tot}} = nN_A(\frac{5}{2}kT) = \frac{5}{2}n(kN_A)T = \frac{5}{2}nRT$, and the molar heat capacity (at constant volume) is

$$C_V = \frac{5}{2}R \quad \text{(diatomic gas, including rotation)} \quad (18.26)$$

In SI units,

$$C_V = \frac{5}{2}(8.314 \text{ J/mol} \cdot \text{K}) = 20.79 \text{ J/mol} \cdot \text{K}$$

This agrees within a few percent with the measured values for diatomic gases given in Table 18.1.

Vibrational motion can also contribute to the heat capacities of gases. Molecular bonds are not rigid; they can stretch and bend, and the resulting vibrations lead to additional degrees of freedom and additional energies. For most diatomic gases, however, vibrational motion does *not* contribute appreciably to heat capacity. The reason for this is a little subtle and involves some concepts of quantum mechanics.

(a) Translational motion

(b) Rotational motion

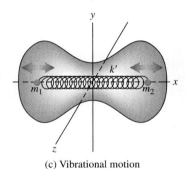

(c) Vibrational motion

18.15 A diatomic molecule. Almost all the mass of each atom is in its tiny nucleus. (a) The center of mass has three independent velocity components. (b) The molecule has two independent axes of rotation through its center of mass. (c) The atoms and "spring" have additional kinetic and potential energies of vibration.

18.16 Experimental values of C_V for hydrogen gas (H_2). The temperature is plotted on a logarithmic scale. Appreciable rotational motion begins to occur above 50 K, and above 600 K the molecule begins to appreciably increase its vibrational motion.

Briefly, vibrational energy can change only in finite steps. If the energy change of the first step is much larger than the energy possessed by most molecules, then nearly all the molecules remain in the minimum-energy state of motion. In that case, changing the temperature does not change their average vibrational energy appreciably, and the vibrational degrees of freedom are said to be "frozen out." In more complex molecules the gaps between permitted energy levels are sometimes much smaller, and then vibration *does* contribute to heat capacity. The rotational energy of a molecule also changes by finite steps, but they are usually much smaller; the "freezing out" of rotational degrees of freedom occurs only in rare instances, such as for the hydrogen molecule below about 100 K.

In Table 18.1 the large values of C_V for some polyatomic molecules show the contributions of vibrational energy. In addition, a molecule with three or more atoms that are not in a straight line has three, not two, rotational degrees of freedom.

From this discussion we expect heat capacities to be temperature-dependent, generally increasing with increasing temperature. Figure 18.16 is a graph of the temperature dependence of C_V for hydrogen gas (H_2), showing the temperatures at which the rotational and vibrational energies begin to contribute to the heat capacity.

Heat Capacities of Solids

We can carry out a similar heat-capacity analysis for a crystalline solid. Consider a crystal consisting of N identical atoms (a *monatomic solid*). Each atom is bound to an equilibrium position by interatomic forces. The elasticity of solid materials shows us that these forces must permit stretching and bending of the bonds. We can think of a crystal as an array of atoms connected by little springs (Fig. 18.17). Each atom can *vibrate* about its equilibrium position.

Each atom has three degrees of freedom, corresponding to its three components of velocity. According to the equipartition principle, each atom has an average kinetic energy of $\frac{1}{2}kT$ for each degree of freedom. In addition, each atom has *potential* energy associated with the elastic deformation. For a simple harmonic oscillator (discussed in Chapter 13) it is not hard to show that the average kinetic energy of an atom is *equal* to its average potential energy. In our model of a crystal, each atom is essentially a three-dimensional harmonic oscillator; it can be shown that the equality of average kinetic and potential energies also holds here, provided that the "spring" forces obey Hooke's law.

18.17 To visualize the forces between neighboring atoms in a crystal, envision every atom as being attached to its neighbors by springs.

Thus we expect each atom to have an average kinetic energy $\frac{3}{2}kT$ and an average potential energy $\frac{3}{2}kT$, or an average total energy $3kT$ per atom. If the crystal contains N atoms or n moles, its total energy is

$$E_{\text{tot}} = 3NkT = 3nRT \qquad (18.27)$$

From this we conclude that the molar heat capacity of a crystal should be

$$C_V = 3R \qquad \text{(ideal monatomic solid)} \qquad (18.28)$$

In SI units,

$$C_V = (3)(8.314\ \text{J/mol}\cdot\text{K}) = 24.9\ \text{J/mol}\cdot\text{K}$$

This is the **rule of Dulong and Petit,** which we encountered as an *empirical* finding in Section 17.5: elemental solids all have molar heat capacities of about 25 J/mol·K. Now we have *derived* this rule from kinetic theory. The agreement is only approximate, to be sure, but considering the very simple nature of our model, it is quite significant.

At low temperatures, the heat capacities of most solids *decrease* with decreasing temperature (Fig. 18.18) for the same reason that vibrational degrees of freedom of molecules are frozen out at low temperatures. At very low temperatures the quantity kT is much *smaller* than the smallest energy step the vibrating atoms can take. Hence most of the atoms remain in their lowest energy states because the next higher energy level is out of reach. The average vibrational energy per atom is then *less* than $3kT$, and the heat capacity per molecule is *less* than $3k$. At higher temperatures when kT is *large* in comparison to the minimum energy step, the equipartition principle holds, and the total heat capacity is $3k$ per molecule or $3R$ per mole as the Dulong and Petit relation predicts. Quantitative understanding of the temperature variation of heat capacities was one of the triumphs of quantum mechanics during its initial development in the 1920s.

Test Your Understanding

A cylinder with a fixed volume contains hydrogen gas (H_2) at 25 K. You then add heat to the gas at a constant rate until its temperature reaches 500 K. Does the temperature of the gas increase at a constant rate? Why or why not? If not, does the temperature increase most rapidly near the beginning or near the end of this process?

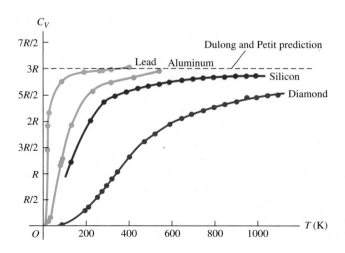

18.18 Experimental values of C_V for lead, aluminum, silicon, and diamond. At high temperatures, C_V for each solid approaches about $3R$, in agreement with the rule of Dulong and Petit. At low temperatures, C_V is much less than $3R$.

*18.5 | Molecular Speeds

As we mentioned in Section 18.3, the molecules in a gas don't all have the same speed. Figure 18.19 shows one experimental scheme for measuring the distribution of molecular speeds. A substance is vaporized in a hot oven; molecules of the vapor escape through an aperture in the oven wall and into a vacuum chamber. A series of slits blocks all molecules except those in a narrow beam, which is aimed at a pair of rotating disks. A molecule passing through the slit in the first disk is blocked by the second disk unless it arrives just as the slit in the second disk is lined up with the beam. The disks function as a speed selector that passes only molecules within a certain narrow speed range. This range can be varied by changing the disk rotation speed, and we can measure how many molecules lie within each of various speed ranges.

To describe the results of such measurements, we define a function $f(v)$ called a *distribution function*. If we observe a total of N molecules, the number dN having speeds in the range between v and $v + dv$ is given by

$$dN = Nf(v)\,dv \tag{18.29}$$

We can also say that the *probability* that a randomly chosen molecule will have a speed in the interval v to $v + dv$ is $f(v)\,dv$. Hence $f(v)$ is the probability per unit speed *interval;* it is *not* equal to the probability that a molecule has speed exactly equal to v. Since a probability is a pure number, $f(v)$ has units of reciprocal speed (s/m).

Figure 18.20a shows distribution functions for several different temperatures. At each temperature the height of the curve for any value of v is proportional to the number of molecules with speeds near v. The peak of the curve represents the *most probable speed* v_{mp} for the corresponding temperature. As the temperature increases, the average molecular kinetic energy increases, and so the peak of $f(v)$ shifts to higher and higher speeds.

Figure 18.20b shows that the area under a curve between any two values of v represents the fraction of all the molecules having speeds in that range. Every molecule must have *some* value of v, so the integral of $f(v)$ over all v must be unity for any T.

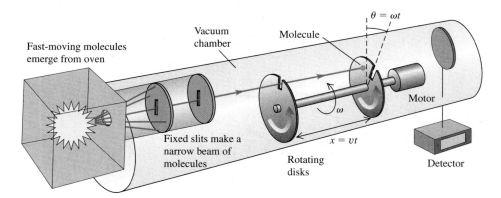

18.19 A molecule with a speed v passes through the slit in the first rotating disk. When the molecule reaches the second rotating disk, the disks have rotated through the offset angle θ. If $v = \omega x/\theta$, the molecule passes through the slit in the second rotating disk and reaches the detector.

If we know $f(v)$, we can calculate the most probable speed v_{mp}, the average speed v_{av}, and the rms speed v_{rms}. To find v_{mp}, we simply find the point where $df/dv = 0$; this gives the value of the speed where the curve has its peak. To find v_{av}, we take the number $Nf(v)\,dv$ having speeds in each interval dv, multiply each number by the corresponding speed v, add all these products (by integrating over all v from zero to infinity), and finally divide by N. That is,

$$v_{\text{av}} = \int_0^{\infty} vf(v)\,dv \qquad (18.30)$$

The rms speed is obtained similarly; the average of v^2 is given by

$$(v^2)_{\text{av}} = \int_0^{\infty} v^2 f(v)\,dv \qquad (18.31)$$

and v_{rms} is the square root of this.

The function $f(v)$ describing the actual distribution of molecular speeds is called the **Maxwell-Boltzmann distribution.** It can be derived from statistical-mechanics considerations, but that derivation is beyond our scope. Here is the result:

$$f(v) = 4\pi \left(\frac{m}{2\pi kT}\right)^{3/2} v^2 e^{-mv^2/2kT}$$

(Maxwell-Boltzmann distribution) $\qquad (18.32)$

We can also express this function in terms of the translational kinetic energy of a molecule, which we denote by ϵ. That is, $\epsilon = \frac{1}{2}mv^2$. We invite you to verify that when this is substituted into Eq. (18.32), the result is

$$f(v) = \frac{8\pi}{m} \left(\frac{m}{2\pi kT}\right)^{3/2} \epsilon e^{-\epsilon/kT} \qquad (18.33)$$

This form shows that the exponent in the Maxwell-Boltzmann distribution function is $-\epsilon/kT$ and that the shape of the curve is determined by the relative magnitude of ϵ and kT at any point. We leave it as an exercise to prove that the *peak* of each curve occurs where $\epsilon = kT$, corresponding to a most probable speed v_{mp} given by

$$v_{\text{mp}} = \sqrt{\frac{2kT}{m}} \qquad (18.34)$$

To find the average speed, we substitute Eq. (18.32) into Eq. (18.30) and carry out the integration, making a change of variable $v^2 = x$ and then integrating by parts. We leave the details as an exercise; the result is

$$v_{\text{av}} = \sqrt{\frac{8kT}{\pi m}} \qquad (18.35)$$

Finally, to find the rms speed, we substitute Eq. (18.32) into Eq. (18.31). Evaluating the resulting integral takes some mathematical acrobatics, but we can find it in a table of integrals. The result is

$$v_{\text{rms}} = \sqrt{\frac{3kT}{m}} \qquad (18.36)$$

This result agrees with Eq. (18.19); it *must* agree if the Maxwell-Boltzmann distribution is to be consistent with the equipartition theorem and our other kinetic-theory calculations.

18.20 (a) Curves of the Maxwell-Boltzmann distribution function $f(v)$ for various temperatures. (b)The shaded areas under the curve represent the fractions of molecules within certain speed ranges. The most probable speed v_{mp} for a given temperature is at the peak of the curve.

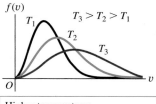

8.2   Maxwell-Boltzmann Distribution—
      Conceptual Analysis

8.3   Maxwell-Boltzmann Distribution—
      Quantitative Analysis

Table 18.2 Fractions of Molecules in an Ideal Gas with Speeds Less Than Various Multiples of v/v_{rms}

v/v_{rms}	Fraction
0.20	0.011
0.40	0.077
0.60	0.218
0.80	0.411
1.00	0.608
1.20	0.771
1.40	0.882
1.60	0.947
1.80	0.979
2.00	0.993

Table 18.2 shows the fraction of all the molecules in an ideal gas that have speeds *less than* various multiples of v_{rms}. These numbers were obtained by numerical integration; they are the same for all ideal gases.

The distribution of molecular speeds in liquids is similar, although not identical, to that for gases. We can understand the vapor pressure of a liquid and the phenomenon of boiling on this basis. Suppose a molecule must have a speed at least as great as v_A in Fig. 18.20b to escape from the surface of a liquid into the adjacent vapor. The number of such molecules, represented by the area under the "tail" of each curve (to the right of v_A), increases rapidly with temperature. Thus the rate at which molecules can escape is strongly temperature-dependent. This process is balanced by another one in which molecules in the vapor phase collide inelastically with the surface and are trapped back into the liquid phase. The number of molecules suffering this fate per unit time is proportional to the pressure in the vapor phase. Phase equilibrium between liquid and vapor occurs when these two competing processes proceed at exactly the same rate. So if the molecular speed distributions are known for various temperatures, we can make a theoretical prediction of vapor pressure as a function of temperature. When liquid evaporates, it's the high-speed molecules that escape from the surface. The ones that are left have less energy on average; this gives us a molecular view of evaporative cooling.

Rates of chemical reactions are often strongly temperature-dependent, and the reason is contained in the Maxwell-Boltzmann distribution. When two reacting molecules collide, the reaction can occur only when the molecules are close enough for the electric-charge distributions of their electrons to interact strongly. This requires a minimum energy, called the *activation energy,* and thus a certain minimum molecular speed. Figure 18.20a shows that the number of molecules in the high-speed tail of the curve increases rapidly with temperature. Thus we expect the rate of any reaction that depends on an activation energy to increase rapidly with temperature. Similarly, many plant growth processes have strongly temperature-dependent rates, as can be seen by the rapid and diverse growth in tropical rain forests.

Test Your Understanding

The Maxwell-Boltzmann distribution curve has a long "tail" that extends to infinity. Use this fact to explain why v_{av} is greater than v_{mp}.

18.6 | Phases of Matter

We've talked a lot about ideal gases in the last few sections. An ideal gas is the simplest system to analyze from a molecular viewpoint because we ignore the interactions between molecules. But those interactions are the very thing that makes matter condense into the liquid and solid phases under some conditions. So it's not surprising that theoretical analysis of liquid and solid structure and behavior is a lot more complicated than that for gases. We won't try to go far here with a microscopic picture, but we can talk in general about phases of matter, phase equilibrium, and phase transitions.

In Section 17.6 we learned that each phase is stable only in certain ranges of temperature and pressure. A transition from one phase to another ordinarily takes place under conditions of **phase equilibrium** between the two phases, and for a given pressure this occurs at only one specific temperature. We can represent these conditions on a graph with axes p and T, called a **phase diagram;** Fig. 18.21 shows an example. Each point on the diagram represents a pair of values of p and

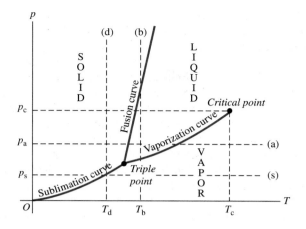

18.21 A typical pT phase diagram, showing regions of temperature and pressure at which the various phases exist and where phase changes occur.

T. Only a single phase can exist at each point, except for points on the solid lines, where two phases can coexist in phase equilibrium.

These lines separate the diagram into solid, liquid, and vapor regions. For example, the fusion curve separates the solid and liquid areas and represents possible conditions of solid-liquid phase equilibrium. Similarly, the vaporization curve separates the liquid and vapor areas, and the sublimation curve separates the solid and vapor areas. The three curves meet at the **triple point,** the only condition under which all three phases can coexist (Fig. 18.22). In Section 17.3 we used the triple-point temperature of water to define the Kelvin temperature scale. Triple-point data for several substances are given in Table 18.3.

Table 18.3 Triple-Point Data

Substance	Temperature (K)	Pressure (Pa)
Hydrogen	13.80	0.0704×10^5
Deuterium	18.63	0.171×10^5
Neon	24.56	0.432×10^5
Nitrogen	63.18	0.125×10^5
Oxygen	54.36	0.00152×10^5
Ammonia	195.40	0.0607×10^5
Carbon dioxide	216.55	5.17×10^5
Sulfur dioxide	197.68	0.00167×10^5
Water	273.16	0.00610×10^5

If we add heat to a substance at a constant pressure p_a, it goes through a series of states represented by the horizontal line (a) in Figure 18.21. The melting and boiling temperatures at this pressure are the temperatures at which the line intersects the fusion and vaporization curves, respectively. When the pressure is p_s, constant-pressure heating transforms a substance from solid directly to vapor. This process is called *sublimation;* the intersection of line (s) with the sublimation curve gives the temperature T_s at which it occurs for a pressure p_s. At any pressure less than the triple-point pressure, no liquid phase is possible. The triple-point pressure for carbon dioxide is 5.1 atm. At normal atmospheric pressure, solid carbon dioxide ("dry ice") undergoes sublimation; there is no liquid phase at this pressure.

Line (b) in Fig. 18.21 represents compression at a constant temperature T_b. The material passes from vapor to liquid and then to solid at the points where line (b) crosses the vaporization curve and fusion curve, respectively. Line (d) shows

18.22 Atmospheric pressure on earth is higher than the triple-point pressure of water (see line (a) in Fig. 18.21). Depending on the temperature, water can exist as a vapor (in the atmosphere), as a liquid (in the ocean), or a solid (like the iceberg shown here).

constant-temperature compression at a lower temperature T_d; the material passes from vapor to solid at the point where line (d) crosses the sublimation curve.

We saw in the pV-diagram of Fig. 18.5 that a liquid-vapor phase transition occurs only when the temperature and pressure are less than those at the point lying at the top of the green shaded area labeled "Liquid-vapor phase equilibrium region." This point corresponds to the end point at the top of the vaporization curve in Fig. 18.21. It is called the **critical point,** and the corresponding values of p and T are called the critical pressure and temperature, p_c and T_c. A gas at a pressure *above* the critical pressure does not separate into two phases when it is cooled at constant pressure (along a horizontal line above the critical point in Fig. 18.21). Instead, its properties change gradually and continuously from those we ordinarily associate with a gas (low density, large compressibility) to those of a liquid (high density, small compressibility) *without a phase transition.*

If this stretches credibility, think about liquid-phase transitions at successively higher points on the vaporization curve. As we approach the critical point, the *differences* in physical properties (such as density, bulk modulus, and viscosity) between the liquid and vapor phases become smaller and smaller. Exactly *at* the critical point they all become zero, and at this point the distinction between liquid and vapor disappears. The heat of vaporization also grows smaller and smaller as we approach the critical point, and it too becomes zero at the critical point.

For nearly all familiar materials the critical pressures are much greater than atmospheric pressure, so we don't observe this behavior in everyday life. For example, the critical point for water is at 647.4 K and 221.2×10^5 Pa (about 218 atm or 3210 psi). But high-pressure steam boilers in electric generating plants regularly run at pressures and temperatures well above the critical point.

Many substances can exist in more than one solid phase. A familiar example is carbon, which exists as noncrystalline soot and crystalline graphite and diamond. Water is another example; at least eight types of ice, differing in crystal structure and physical properties, have been observed at very high pressures.

We remarked in Section 18.1 that the equation of state of any material can be represented graphically as a surface in a three-dimensional space with coordinates p, V, and T. Such a surface is seldom useful in representing detailed quantitative information, but it can add to our general understanding of the behavior of materials at various temperatures and pressure. Figure 18.23 shows a typical pVT-surface. The light lines represent pV-isotherms; projecting them onto the pV-plane would give a diagram similar to Fig. 18.5. The pV-isotherms represent contour lines on the pVT-surface, just as contour lines on a topographic map represent the elevation (the third dimension) at each point. The projections of the edges of the surface onto the pT-plane gives the pT-phase diagram of Fig. 18.21.

Line *abcdef* represents constant-pressure heating, with melting along *bc* and vaporization along *de*. Note the volume changes that occur as T increases along this line. Line *ghjklm* corresponds to an isothermal (constant temperature) compression, with liquefaction along *hj* and solidification along *kl*. Between these, segments *gh* and *jk* represent isothermal compression with increase in pressure; the pressure increases are much greater in the liquid region *jk* and the solid region *lm* than in the vapor region *gh*. Finally, *nopq* represents isothermal solidification directly from the vapor phase; this is the process involved in growth of crystals directly from vapor, as in the formation of snowflakes or frost and in the fabrication of some solid-state electronic devices. These three lines on the pVT-surface are worth careful study.

For contrast, Fig. 18.24 shows the much simpler pVT-surface for a substance that obeys the ideal-gas equation of state under all conditions. The projections of

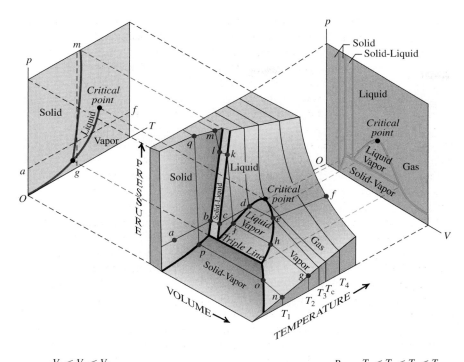

18.23 A *pVT*-surface for a substance that expands on melting. Projections of the boundaries on the surface on the *pT*- and *pV*-planes are also shown.

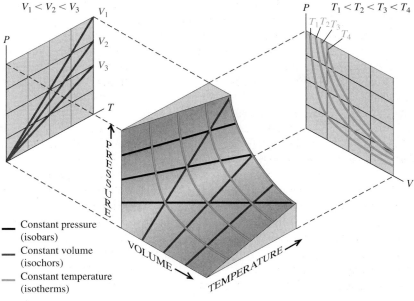

18.24 A *pVT*-surface for an ideal gas. At the left, each red line corresponds to a certain constant volume; at the right, each blue line corresponds to a certain constant temperature.

the constant-temperature curves onto the *pV*-plane correspond to the curves of Fig. 18.4, and the projections of the constant-volume curves onto the *pT*-plane show the direct proportionality of pressure to absolute temperature.

Test Your Understanding

The average atmospheric pressure on Mars is 6.0×10^2 Pa. Could there be lakes or rivers on Mars today? What about in the past, when the atmospheric pressure is thought to have been substantially greater than today?

SUMMARY

The pressure p, volume V, and absolute temperature T of a given quantity of a substance are called state variables. They are related by an equation of state. This relation pertains only to equilibrium states, in which p and T are uniform throughout the system. The ideal-gas equation of state relates p, V, T and the number of moles n through a constant R that is the same for all gases. (See Examples 18.1 through 18.4)

$$pV = nRT \quad (18.3)$$

A pV-diagram is a set of graphs, called isotherms, each showing pressure as a function of volume for a constant temperature.

The molar mass M of a pure substance is the mass per mole. The mass m_{tot} of a quantity of substance equals M multiplied by the number of moles n. Avogadro's number N_A is the number of molecules in a mole. The mass m of an individual molecule is M divided by N_A. (See Example 18.5)

$$m_{tot} = nM \quad (18.2)$$

$$M = N_A m \quad (18.8)$$

Chlorine ions

Sodium ions

In an ideal gas, the total translational kinetic energy of the gas as a whole (K_{tr}) and the average translational kinetic energy per molecule $[\frac{1}{2}m(v^2)_{av}]$ are proportional to the absolute temperature T. The root-mean-square speed of molecules in an ideal gas is proportional to the square root of T. These expressions involve the Boltzmann constant $k = R/N_A$. (See Examples 18.6 through 18.8)

$$K_{tr} = \frac{3}{2}nRT \quad (18.14)$$

$$\frac{1}{2}m(v^2)_{av} = \frac{3}{2}kT \quad (18.16)$$

$$v_{rms} = \sqrt{(v^2)_{av}} = \sqrt{\frac{3kT}{m}} = \sqrt{\frac{3RT}{M}} \quad (18.19)$$

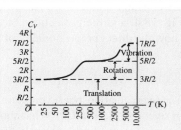

Molecule after collision

$v_{2y} = v_y$

$v_{2x} = |v_x|$

Molecule before collision

$v_{1y} = v_y$

$v_{1x} = -|v_x|$

The mean free path λ of molecules in an ideal gas depends on the number of molecules per volume (N/V) and the molecular radius r. (See Example 18.9)

$$\lambda = vt_{mean} = \frac{V}{4\pi\sqrt{2}r^2 N} \quad (18.21)$$

$v\,dt$

The molar heat capacity C_V at constant volume can be expressed as a simple multiple of the gas constant R for certain idealized cases: an ideal monatomic gas [Eq. (18.25)]; an ideal diatomic gas including rotation energy [Eq. (18.26)]; and an ideal monatomic solid [Eq. (18.28)]. Many real systems are approximated well by these idealizations.

$$C_V = \frac{3}{2}R \quad (18.25)$$

$$C_V = \frac{5}{2}R \quad (18.26)$$

$$C_V = 3R \quad (18.28)$$

The speeds of molecules in an ideal gas are distributed according to the Maxwell-Boltzmann distribution $f(v)$. The quantity $f(v)\,dv$ describes what fraction of the molecules have speeds between v and $v + dv$.

$$f(v) = 4\pi\left(\frac{m}{2\pi kT}\right)^{3/2} v^2 e^{-mv^2/2kT} \qquad (18.32)$$

Ordinary matter exists in the solid, liquid, and gas phases. A phase diagram shows conditions under which two phases can coexist in phase equilibrium. All three phases can coexist at the triple point. The vaporization curve ends at the critical point, above which the distinction between the liquid and gas phases disappears.

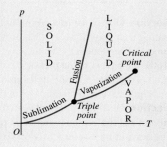

Key Terms

Your Notes

Answer to Chapter Opening Question

From Eq. (18.19), the root-mean-square speed of a gas molecule is proportional to the square root of the absolute temperature T. The temperature range we're considering is from $(-40 + 273.15)$ K $= 233$ K to $(40 + 273.15)$ K $= 313$ K. Hence the speeds increase by a factor of $\sqrt{(313 \text{ K})/(233 \text{ K})} = 1.16$; that is, there is a 16% increase. While 40°C feels far warmer than −40°C, the difference in molecular speeds is relatively minor.

Answers to Test Your Understanding Questions

Section 18.1 If there were no holes in the crust, the steam emanating from the filling as it bakes would be confined to a fixed volume V. The amount of steam n and the temperature T both increase as the pie bakes, so the pressure $p = nRT/V$ increases as well, until the crust bursts. Cutting holes in the crust allows steam to escape and the pressure to remain within limits.

Section 18.2 The value of r_0 determines the equilibrium separation of the molecules in the solid phase, so doubling r_0 means that the separation doubles as well. Hence a solid cube of this compound might grow from 1 cm on a side to 2 cm on a side. The volume would then be $2^3 = 8$ times larger, and the density (mass divided by volume) would be $\frac{1}{8}$ as great.

Section 18.3 The temperature of the corona is 2,000,000°C, or to two significant figures 2.0×10^6 K (the difference between Celsius and Kelvin temperatures is irrelevant). Hydrogen has a molar mass (from Appendix D) of $M = 1.008$ g/mol $= 1.008 \times 10^{-3}$ kg/mol, so

$$v_{\text{rms}} = \sqrt{\frac{3RT}{M}} = \sqrt{\frac{3(8.314 \text{ J/mol·K})(2.0 \times 10^6 \text{ K})}{1.008 \times 10^{-3} \text{ kg/mol}}}$$

$$= 2.2 \times 10^5 \text{ m/s}$$

This is less than the escape speed, so not all of the hydrogen in the corona can escape into space. But remember that not all the atoms move at the rms speed: some are moving fast enough to escape, and they help make up a continuous flow of particles called the *solar wind*.

Section 18.4 Adding a small amount of heat dQ to the gas changes the temperature by dT, where $dQ = nC_V\,dT$ from Eq. (18.24). Figure 18.16 shows that C_V for H_2 varies with temperature between 25 K and 500 K, so a given amount of heat gives rise to different amounts of temperature change during the process. Hence the temperature will *not* increase at a constant rate. The temperature change $dT = dQ/nC_V$ is inversely proportional to C_V, so the temperature increases most rapidly at the beginning of the process when the temperature is lowest and C_V is smallest (see Fig. 18.16).

Section 18.5 If the curve were symmetrical like a bell curve, the average value would be at the peak of the curve and v_{av} would be equal to v_{mp} (which, by definition, is at the peak of the curve). The curve is not symmetrical, however, and its long tail shifts the average value to the right of the peak. Hence $v_{\text{av}} > v_{\text{mp}}$.

Section 18.6 The triple-point pressure of water from Table 18.3 is 6.10×10^2 Pa. The present-day pressure on Mars is just less than this value, corresponding to the line labeled p_s in Fig. 18.21. Hence liquid water cannot exist on the present-day Martian surface, and there are no rivers or lakes. Planetary scientists conclude that liquid water could have and almost certainly did exist on Mars in the past, when the atmosphere was thicker.

Discussion Questions

Q18.1 Section 18.1 states that ordinarily, pressure, volume, and temperature cannot change individually without one affecting the others. Yet when a liquid evaporates, its volume changes, even though its pressure and temperature are constant. Is this inconsistent? Why or why not?

Q18.2 In the ideal-gas equation, could an equivalent Celsius temperature be used instead of the Kelvin one if an appropriate numerical value of the constant R is used? Why or why not?

Q18.3 On a chilly morning you can "see your breath." Can you really? What are you actually seeing? Does this phenomenon depend on the temperature of the air, the humidity, or both? Explain.

Q18.4 When a car is driven some distance, the air pressure in the tires increases. Why? Should you let out some air to reduce the pressure? Why or why not?

Q18.5 The coolant in an automobile radiator is kept at a pressure higher than atmospheric pressure. Why is this desirable? The radiator cap will release coolant when the gauge pressure of the coolant reaches a certain value, typically 15 lb/in.^{2} or so. Why not just seal the system completely?

Q18.6 Unwrapped food placed in a freezer experiences dehydration, known as "freezer burn." Why?

Q18.7 "Freeze-drying" food involves the same process as "freezer burn," referred to in the previous question. For freeze-drying, the food is usually frozen first, then placed in a vacuum chamber and irradiated with infrared radiation. What is the purpose of the vacuum? The radiation? What advantages might freeze-drying have in comparison to ordinary drying?

Q18.8 A group of students drove from their university (near sea level) up into the mountains for a skiing weekend. Upon arriving at the slopes, they discovered that the bags of potato chips they had brought for snacks had all burst open. What caused this to happen?

Q18.9 How does evaporation of perspiration from your skin cool your body?

Q18.10 Helium is a mixture of two isotopes, one having molar mass 3 g/mol and the other 4 g/mol. In a sample of helium gas which atoms move faster, on the average? Explain why.

Q18.11 Which has more atoms, a kilogram of hydrogen or a kilogram of lead? Which has more mass? Explain.

Q18.12 Use the concepts of the kinetic-molecular model to explain: a) why the pressure of a gas in a rigid container increases as heat is added to the gas; b) why the pressure of a gas increases as we compress it, even if we do not change its temperature.

Q18.13 The proportion of various gases in the earth's atmosphere changes somewhat with altitude. Would you expect the proportion of oxygen at high altitude to be greater or less than at sea level compared to the proportion of nitrogen? Why?

Q18.14 Comment on the following statement: *When two gases are mixed, if they are to be in thermal equilibrium they must have the same average molecular speed.* Is the statement correct? Why or why not?

Q18.15 The kinetic-molecular model contains a hidden assumption about the temperature of the container walls. What is this assumption? What would happen if this assumption were not valid?

Q18.16 The temperature of an ideal gas is directly proportional to the average kinetic energy of its molecules. If a container of ideal gas is moving past you at 2000 m/s, is the temperature of the gas higher than if the container was at rest? Defend your answer.

Q18.17 If the pressure of an ideal monatomic gas is increased while the number of moles is kept constant, what happens to the average translational kinetic energy of one atom of the gas? Is it possible to change *both* the volume and pressure of an ideal gas and keep the average translational kinetic energy of the atoms constant? Explain.

Q18.18 In deriving the ideal-gas equation from the kinetic-molecular model, we ignored potential energy due to the earth's gravity. Is this omission justified? Why or why not?

Q18.19 The derivation of the ideal-gas equation included the assumption that the number of molecules is very large, so that we could compute the average force due to many collisions. However, the ideal-gas equation holds accurately only at low pressures, where the molecules are few and far between. Is this inconsistent? Why or why not?

Q18.20 A gas storage tank has a small leak. The pressure in the tank drops more quickly if the gas is hydrogen or helium than if it is oxygen. Why?

Q18.21 Consider two specimens of ideal gas at the same temperature. Specimen A has the same total mass as specimen B, but the molecules in specimen A have greater molar mass than they do in specimen B. In which specimen is the total kinetic energy of the gas greater? Does your answer depend on the molecular structure of the gases? Why or why not?

Q18.22 The temperature of an ideal monatomic gas is increased from 25°C to 50°C. Does the average translational kinetic energy of each gas atom double? Explain. If your answer is no, what would the final temperature be if the average translational kinetic energy is doubled?

Q18.23 If the root-mean-square speed of the atoms of an ideal gas is to be doubled, by what factor must the Kelvin temperature of the gas be increased? Explain.

Q18.24 Some elements that form solid crystals have molar heat capacities *greater* than $3R$. What could account for this?

*Q18.25 In a gas that contains N molecules, would it be accurate to say that the number of molecules with speed v is equal to $f(v)$? Or would it be accurate to say that this number is given by $Nf(v)$? Explain your answers.

*Q18.26 Fig. 18.19 shows a velocity selector. The two disks are a distance x apart, the slits in the two disks are at an angle θ to each other, and the disks rotate together with angular speed ω. Explain why a molecule that passes through the first slit of the selector will only be able to pass through the second slit if its speed is $v = \omega x / \theta$.

*Q18.27 Imagine a special air filter placed in a window of a house. The tiny holes in the filter only allow air molecules moving faster than a certain speed to exit the house, and only allow air molecules moving slower than that speed to enter the house from outside. What effect would this filter have on the temperature inside the house? (It turns out that the second law of thermodynamics—which we will discuss in Chapter 20—tells us that such a wonderful air filter would be quite impossible to make.)

Q18.28 A beaker of water at room temperature is placed in an enclosure, and the air pressure in the enclosure is slowly reduced. When the air pressure is reduced sufficiently, the water begins to boil. The temperature of the water does not rise when it boils; in fact, the temperature *drops* slightly. Explain these phenomena.

Q18.29 Ice is slippery to walk on, and especially slippery if you wear ice skates. What does this tell you about how the melting temperature of ice depends on pressure? Explain.

Q18.30 Hydrothermal vents are openings in the ocean floor that discharge very hot water. The water emerging from one such vent off the Oregon coast, 2400 m below the surface, has a temperature of 279°C. Despite its high temperature, the water doesn't boil. Why not?

Q18.31 The dark areas on the moon's surface are called *maria,* Latin for "seas," and were once thought to be bodies of water. In fact the maria are not "seas" at all, but plains of solidified lava. Given that there is no atmosphere on the moon, how can you explain the absence of liquid water on the moon's surface?

Q18.32 In addition to the normal cooking directions printed on the back of a box of rice, there are also "high altitude directions." The only difference is that the "high altitude directions" suggest increasing the cooking time and using a greater volume of boiling water in which to cook the rice. Why should the directions depend on the altitude in this way?

Exercises

Section 18.1 Equations of State

18.1 A 20.0-L tank contains 0.225 kg of helium at 18.0°C. The molar mass of helium is 4.00 g/mol. a) How many moles of helium are in the tank? b) What is the pressure in the tank, in pascals and in atmospheres?

18.2 Helium gas with a volume of 2.60 L, under a pressure of 1.30 atm and at a temperature of 41.0°C, is warmed until both pressure and volume are doubled. a) What is the final temperature? b) How many grams of helium are there? The molar mass of helium is 4.00 g/mol.

18.3 A cylindrical tank has a tight-fitting piston that allows the volume of the tank to be changed. The tank originally contains 0.110 m³ of air at a pressure of 3.40 atm. The piston is slowly pulled out until the volume of the gas is increased to 0.390 m³. If the temperature remains constant, what is the final value of the pressure?

18.4 A 3.00-L tank contains air at 3.00 atm and 20.0°C. The tank is sealed and cooled until the pressure is 1.00 atm. a) What is the temperature then in degrees Celsius? Assume that the volume of the tank is constant. b) If the temperature is kept at the value found in part (a) and the gas is compressed, what is the volume when the pressure again becomes 3.00 atm?

18.5 a) Use the ideal gas law to estimate the number of air molecules in your physics lab room, assuming all the air is N_2. b) Calculate the particle density in the lab (that is, the number of molecules per cubic centimeter).

18.6 You have several identical balloons. You experimentally determine that a balloon will break if its volume exceeds 0.900 L. The pressure of the gas inside the balloon equals air pressure (1.00 atm). a) If the air inside the balloon is at a constant temperature of 22.0°C and behaves as an ideal gas, what mass of air can you blow into one of the balloons before it bursts? b) Repeat part (a) if the gas is helium rather than air.

18.7 A Jaguar XK8 convertible has an eight-cylinder engine. At the beginning of its compression stroke, one of the cylinders contains 499 cm^3 of air at atmospheric pressure (1.01×10^5 Pa) and a temperature of 27.0°C. At the end of the stroke the air has been compressed to a volume of 46.2 cm^3 and the gauge pressure has increased to 2.72×10^6 Pa. Compute the final temperature.

18.8 A welder using a tank of volume 0.0750 m^3 fills it with oxygen (molar mass = 32.0 g/mol) at a gauge pressure of 3.00×10^5 Pa and temperature of 37.0°C. The tank has a small leak, and in time some of the oxygen leaks out. On a day when the temperature is 22.0°C, the gauge pressure of the oxygen in the tank is 1.80×10^5 Pa. Find a) the initial mass of oxygen; b) the mass of oxygen that has leaked out.

18.9 A large cylindrical tank contains 0.750 m^3 of nitrogen gas at 27°C and 1.50×10^5 Pa (absolute pressure). The tank has a tight-fitting piston that allows the volume to be changed. What will be the pressure if the volume is decreased to 0.480 m^3 and the temperature is increased to 157°C?

18.10 A room with dimensions 7.00 m × 8.00 m × 2.50 m is filled with pure oxygen at 22.0°C and 1.00 atm. The molar mass of oxygen is 32.0 g/mol. a) How many moles of oxygen are required? b) What is the mass of this oxygen, in kilograms?

18.11 The gas inside a balloon will always have a pressure nearly equal to atmospheric pressure, since that is the pressure applied to the outside of the balloon. You fill a balloon with helium (a nearly ideal gas) to a volume of 0.600 L at a temperature of 19.0°C. What is the volume of the balloon if you cool it to the boiling point of liquid nitrogen (77.3 K)?

18.12 Deviations from the Ideal-Gas Equation. For carbon dioxide gas (CO_2), the constants in the van der Waals equation are $a = 0.364$ J·m^3/mol^2 and $b = 4.27 \times 10^{-5}$ m^3/mol. a) If 1.00 mol of CO_2 gas at 350 K is confined to a volume of 400 cm^3, find the pressure of the gas using the ideal-gas equation and the van der Waals equation. b) Which equation gives a lower pressure? Why? What is the percentage difference of the van der Waals result from the ideal-gas equation result? c) The gas is kept at the same temperature as it expands to a volume of 4000 cm^3. Repeat the calculations of parts (a) and (b). d) Explain how your calculations show that the van der Waals equation is equivalent to the ideal-gas equation if n/V is small.

18.13 The total lung volume for a typical physics student is 6.00 L. A physics student fills her lungs with air at an absolute pressure of 1.00 atm. Then, holding her breath, she compresses her chest cavity, decreasing her lung volume to 5.70 L. What is the pressure of the air in her lungs then? Assume that the temperature of the air remains constant.

18.14 A diver observes a bubble of air rising from the bottom of a lake (where the absolute pressure is 3.50 atm) to the surface (where the pressure is 1.00 atm). The temperature at the bottom is 4.0°C, and the temperature at the surface is 23.0°C. a) What is the ratio of the volume of the bubble as it reaches the surface to its volume at the bottom? b) Would it be safe for the diver to hold his breath while ascending from the bottom of the lake to the surface? Why or why not?

18.15 A metal tank with volume 3.10 L will burst if the absolute pressure of the gas it contains exceeds 100 atm. a) If 11.0 mol of an ideal gas is put into the tank at a temperature of 23.0°C, to what temperature can the gas be warmed before the tank ruptures? You can ignore the thermal expansion of the tank. b) Based on your answer to part (a), is it reasonable to ignore the thermal expansion of the tank? Explain.

18.16 Three moles of ideal gas are in a rigid cubical box with sides of length 0.200 m. a) What is the force that the gas exerts on each of the six sides of the box when the gas temperature is 20.0°C? b) What is the force when the temperature of the gas is increased to 100.0°C?

18.17 With the assumptions of Example 18.4 (Section 18.1), at what altitude above sea level is air pressure 90% of the pressure at sea level?

18.18 Make the same assumptions as in Example 18.4 (Section 18.1). How does the percent decrease in air pressure in going from sea level to an altitude of 100 m compare to that when going from sea level to an altitude of 1000 m? If your second answer is not ten times your first answer, explain why.

18.19 With the assumptions of Example 18.4 (Section 18.1), how does the density of air at sea level compare to that at an altitude of 100 m above sea level?

18.20 With the assumption (not very realistic) that the air temperature is a uniform 0°C, what is the atmospheric pressure at an altitude of 5000 m? This is roughly the maximum altitude usually attained by aircraft without pressurized cabins.

18.21 At an altitude of 11.000 m (a typical cruising altitude for a jet airliner), the air temperature is −56.5°C and the air density is 0.364 kg/m^3. What is the pressure of the atmosphere at that altitude? (Note that the temperature at this altitude is not the same as at the surface of the earth, so that the calculation of Example 18.4 (Section 18.1) doesn't apply.)

Section 18.2 Molecular Properties of Matter

18.22 A large organic molecule has a mass of 1.41×10^{-21} kg. What is the molar mass of this compound?

18.23 What is the volume of 3.00 moles of copper?

18.24 Modern vacuum pumps make it easy to attain pressures of the order of 10^{-13} atm in the laboratory. At a pressure of 9.00×10^{-14} atm and an ordinary temperature (say $T = 300$ K), how many molecules are present in a volume of 1.00 cm^3?

18.25 The Lagoon Nebula (Fig. 18.25) is a cloud of hydrogen gas located 3900 light years from the earth. The cloud is about 45 light years in diameter, and glows because of its high temperature of 7500 K. (The gas is raised to this temperature by the stars that lie within the nebula). The cloud is also very thin; there are only 80 molecules per cubic centimeter. a) Find the gas pressure (in atmospheres) in the Lagoon Nebula. Compare to the laboratory pressure referred to in Exercise 18.24. b) Science fiction films sometimes show starships being buffeted by turbulence as they fly through gas clouds such as the Lagoon Nebula. Does this seem realistic? Why or why not?

Figure 18.25 Exercise 18.25.

18.26 In a gas at standard conditions, what is the length of the side of a cube that contains a number of molecules equal to the population of the earth (about 6×10^9)?

18.27 How many moles are there in a 1.00-kg bottle of water? How many molecules? The molar mass of water is 18.0 g/mol.

18.28 Consider an ideal gas at 27°C and 1.00 atm pressure. Imagine the molecules to be, on average, uniformly spaced, with each molecule at the center of a small cube. a) What is the length of an edge of each small cube if adjacent cubes touch but don't overlap? b) How does this distance compare with the diameter of a molecule?

18.29 Consider 5.00 mol of liquid water. a) What volume is occupied by this amount of water? The molar mass of water is 18.0 g/mol. b) Imagine the molecules to be, on average, uniformly spaced, with each molecule at the center of a small cube. What is the length of an edge of each small cube if adjacent cubes touch but don't overlap? c) How does this distance compare with the diameter of a molecule?

Section 18.3 Kinetic-Molecular Model of an Ideal Gas

18.30 A flask contains a mixture of neon (Ne), krypton (Kr), and radon (Rn) gases. Compare a) the average kinetic energies of the three types of atoms; b) the root-mean-square speeds. (*Hint:* the periodic table in Appendix D shows the molar mass (in g/mol) of each element under the chemical symbol for that element.)

18.31 Gaseous Diffusion of Uranium. a) A process called *gaseous diffusion* is often used to separate isotopes of uranium, that is, atoms of the elements that have different masses, such as ^{235}U and ^{238}U. The only gaseous compound of uranium at ordinary temperatures is uranium hexafluoride, UF_6. Speculate on how $^{235}UF_6$ and $^{238}UF_6$ molecules might be separated by diffusion. b) The molar masses for $^{235}UF_6$ and $^{238}UF_6$ molecules are 0.349 kg/mol and 0.352 kg/mol, respectively. If uranium hexafluoride acts as an ideal gas, what is the ratio of the root-mean-square speed of $^{235}UF_6$ molecules to that of $^{238}UF_6$ molecules if the temperature is uniform?

18.32 The ideas of average and root-mean-square value can be applied to any distribution. A class of 150 students had the following scores on a 100-point quiz:

Score	Number of Students
10	11
20	12
30	24
40	15
50	19
60	10
70	12
80	20
90	17
100	10

a) Find the average score for the class. b) Find the root-mean-square score for the class.

18.33 We have two equal size boxes, A and B. Each box contains gas that behaves as an ideal gas. We insert a thermometer into each box and find that the gas in box A is at a temperature of 50°C while that in B is at 10°C. This is all we know about the gas in the boxes. Which of the following statements *must* be true? Which *could* be true? a) The pressure in A is higher than in B. b) There are more molecules in A than in B. c) A and B cannot contain the same type of gas. d) The molecules in A have more average kinetic energy per molecule than those in B. e) The molecules in A are moving faster than those in B. Explain the reasoning behind your answers.

18.34 We have two equal size boxes, A and B. Each box contains gas that behaves as an ideal gas. We put a pressure gauge into each box and find that the gauge reads 0.200 atm in box A, but only 0.040 atm in box B. This is all we know about the gas in the boxes. Which of the following statements *must* be true? Which *could* be true? a) There are more molecules in A than in B. b) The molecules in A are moving faster than those in B. c) The temperature in A is higher than in B. d) The molecules in A are heavier than those in B. e) The molecules in A have more average kinetic energy per molecule than those in B. Explain the reasoning behind your answers.

18.35 a) A deuteron, ^2_1H, is the nucleus of a hydrogen isotope and consists of one proton and one neutron. The plasma of deuterons in a nuclear fusion reactor must be heated to about 300 million K. What is the rms speed of the deuterons? Is this a significant fraction of the speed of light ($c = 3.0 \times 10^8$ m/s)? b) What would the temperature of the plasma be if the deuterons had an rms speed equal to $0.10c$?

18.36 Initially, the translational rms speed of an atom of a monatomic ideal gas is 250 m/s. The pressure and volume of the gas are each doubled while the number of moles of the gas is kept constant. What is the final translational rms speed of the atoms?

18.37 a) Oxygen (O_2) has a molar mass of 32.0 g/mol. What is the average translational kinetic energy of an oxygen molecule at a temperature of 300 K? b) What is the average value of the square of its speed? c) What is the root-mean-square speed? d) What is the momentum of an oxygen molecule traveling at this speed? e) Suppose an oxygen molecule traveling at this speed bounces back and forth between opposite sides of a cubical vessel 0.10 m on a side.

What is the average force it exerts on one of the walls of the container? (Assume that the molecule's velocity is perpendicular to the two sides that it strikes.) f) What is the average force per unit area? g) How many oxygen molecules traveling at this speed are necessary to produce an average pressure of 1 atm? h) Compute the number of oxygen molecules that are actually contained in a vessel of this size at 300 K and atmospheric pressure. i) Your answer for part (h) should be three times as large as the answer for part (g). Where does this discrepancy arise?

18.38 Calculate the mean free path of air molecules at a pressure of 3.50×10^{-13} atm and a temperature of 300 K. (This pressure is readily attainable in the laboratory; see Exercise 18.24.) As in Example 18.9, model the air molecules as spheres of radius 2.0×10^{-10} m.

18.39 At what temperature is the root-mean-square speed of nitrogen molecules equal to the root-mean-square speed of hydrogen molecules at 20.0°C? (*Hint:* The periodic table in Appendix D shows the molar mass (in g/mol) of each element under the chemical symbol for that element. The molar mass of H_2 is twice the molar mass of hydrogen atoms, and similarly for N_2.)

18.40 Smoke particles in the air typically have masses of the order of 10^{-16} kg. The Brownian motion (rapid, irregular movement) of these particles, resulting from collisions with air molecules, can be observed with a microscope. a) Find the root-mean-square speed of Brownian motion for a particle with a mass of 3.00×10^{-16} kg in air at 300 K. b) Would the root-mean-square speed be different if the particle were in hydrogen gas at the same temperature? Explain.

Section 18.4 Heat Capacities

18.41 a) Calculate the specific heat capacity at constant volume of water vapor, assuming the nonlinear triatomic molecule has three translational and three rotational degrees of freedom and that vibrational motion does not contribute. The molar mass of water is 18.0 g/mol. b) The actual specific heat capacity of water vapor at low pressures is about 2000 J/kg·K. Compare this with your calculation and comment on the actual role of vibrational motion.

18.42 a) The specific heat of ice at constant volume is 833 J/kg·K at −180°C, 1640 J/kg·K at −60°C, and 2060 J/kg·K at −5.0°C. Calculate C_V, the molar heat capacity at constant volume for ice, at each of these temperatures. The molar mass of H_2O is 18.0 g/mol. b) Why do you suppose the value of C_V increases with increasing temperature? c) Do the values that you calculate approach the value of $3R$ (given in the rule of Dulong and Petit) as the temperature increases? Speculate about why this should be so.

18.43 a) How much heat does it take to increase the temperature of 2.50 mol of a diatomic ideal gas by 30.0 K near room temperature if the gas is held at constant volume? b) What is the answer to the question in part (a) if the gas is monatomic rather than diatomic?

18.44 a) Compute the specific heat capacity at constant volume of nitrogen (N_2) gas, and compare with the specific heat capacity of liquid water. The molar mass of N_2 is 28.0 g/mol. b) You warm 1.00 kg of water at a constant volume of 1.00 L from 20.0°C to 30.0°C in a kettle. For the same amount of heat, how many kilograms of 20.0°C air would you be able to warm to 30.0°C? What volume (in liters) would this air occupy at 20.0°C and a pressure of 1.00 atm? Make the simplifying assumption that air is 100% N_2.

***18.45** For a gas of nitrogen molecules (N_2), what must the temperature be if 94.7% of all the molecules have speeds less than a) 1500 m/s; b) 1000 m/s; c) 500 m/s? Use Table 18.2. The molar mass of N_2 is 28.0 g/mol.

***18.46** Derive Eq. (18.33) from Eq. (18.32).

***18.47** Prove that $f(v)$ as given by Eq. (18.33) is maximum for $\epsilon = kT$. Use this result to obtain Eq. (18.34).

***18.48** For diatomic carbon dioxide gas $(CO_2$, molar mass = 44.0 g/mol$)$ at $T = 300$ K, calculate a) the most probable speed v_{mp}; b) the average speed v_{av}; c) the root-mean-square speed v_{rms}.

Section 18.6 Phases of Matter

18.49 Puffy cumulus clouds, which are made of water droplets, occur at lower altitudes in the atmosphere. Wispy cirrus clouds, which are made of ice crystals, occur only at higher altitudes. Find the altitude y (measured from sea level) above which only cirrus clouds can occur. On a typical day and at altitudes less than 11 km, the temperature at an altitude y is given by $T = T_0 - \alpha y$, where $T_0 = 15.0$°C and $\alpha = 6.0$ C°/1000 m.

18.50 Solid water (ice) is slowly warmed from a very low temperature. a) What minimum external pressure p_1 must be applied to the solid if a melting phase transition is to be observed? Describe the sequence of phase transitions that occur if the applied pressure p is such that $p < p_1$. b) Above a certain maximum pressure p_2, no boiling transition is observed. What is this pressure? Describe the sequence of phase transitions that occur if $p_1 < p < p_2$.

18.51 A physicist places a piece of ice at 0.00°C and a beaker of water at 0.00°C inside a glass box and closes the lid of the box. All the air is then removed from the box. If the ice, water, and beaker are all maintained at a temperature of 0.00°C, describe the final equilibrium state inside the box.

18.52 The atmosphere of the planet Mars is 95.3% carbon dioxide (CO_2) and about 0.03% water vapor. The atmospheric pressure is only 600 Pa, and the surface temperature varies from −30°C to −100°C. The polar ice caps contain both CO_2 ice and water ice. Could there be *liquid* CO_2 on the surface of Mars? Could there be liquid water? Why or why not?

Problems

18.53 a) Use Eq. (18.1) to estimate the change in the volume of a solid steel sphere of volume 11 L when the temperature and pressure increase from 21°C and 1.013×10^5 Pa to 42°C and 2.10×10^7 Pa. (*Hint:* Consult Chapters 11 and 17 to determine the values of β and k.) b) In Example 18.3 the change in volume of an 11-L steel scuba tank was ignored. Was this a good approximation? Explain.

18.54 At an absolute pressure of 2.00×10^{-13} atm, a partial vacuum easily obtained in laboratories (see Exercise 18.24), calculate the mass of nitrogen present in a volume of 3000 cm³ if the temperature of the gas is 22.0°C. The molar mass of nitrogen (N_2) is 28.0 g/mol.

18.55 A cylinder 1.00 m tall with inside diameter 0.120 m is used to hold propane gas (molar mass 44.1 g/mol) for use in a barbecue. It is initially filled with gas until the gauge pressure is 1.30×10^6 Pa and the temperature is 22.0°C. The temperature of the gas remains constant as it is partially emptied out of the tank, until the gauge pressure is 2.50×10^5 Pa. Calculate the mass of propane that has been used.

18.56 During a test dive in 1939, prior to being accepted by the U.S. Navy, the submarine *Squalus* sank at a point where the depth of water was 73.0 m. The temperature at the surface was 27.0°C and at the bottom it was 7.0°C. The density of seawater is 1030 kg/m³. a) A diving bell was used to rescue 33 trapped crewmen from the *Squalus*. The diving bell was in the form of a circular cylinder 2.30 m high, open at the bottom and closed at the top. When the diving bell was lowered to the bottom of the sea, to what height did water rise within the diving bell? (*Hint:* You may ignore the relatively small variation in water pressure between the bottom of the bell and the surface of the water within the bell.) b) At what gauge pressure must compressed air have been supplied to the bell while on the bottom to expel all the water from it?

18.57 A glassblower makes a barometer using a tube 0.900 m long and with a cross-sectional area of 0.620 cm². Mercury stands in this tube to a height of 0.750 m. The room temperature is 20.0°C. A small amount of air is accidentally introduced into the evacuated space above the mercury, and the column drops to a height of 0.690 m. How many grams of air were introduced? The molar mass of air is 28.8 g/mol.

18.58 A hot-air balloon stays aloft because hot air at atmospheric pressure is less dense than cooler air at the same pressure; the calculation of the buoyant force is discussed in Chapter 14. If the volume of the balloon is 500 m³ and the surrounding air is at 15.0°C, what must the temperature of the air in the balloon be for it to lift a total load of 290 kg (in addition to the mass of the hot air)? The density of air at 15.0°C and atmospheric pressure is 1.23 kg/m³.

18.59 An automobile tire has a volume of 0.0150 m³ on a cold day when the temperature of the air in the tire is 5.0°C and atmospheric pressure is 1.02 atm. Under these conditions the gauge pressure is measured to be 1.70 atm (about 25 lb/in.²). After the car is driven on the highway for 30 min, the temperature of the air in the tires has risen to 45.0°C and the volume to 0.0159 m³. What then is the gauge pressure?

18.60 A flask with a volume of 1.50 L, provided with a stopcock, contains ethane gas (C_2H_6) at 300 K and atmospheric pressure $(1.013 \times 10^5$ Pa). The molar mass of ethane is 30.1 g/mol. The system is warmed to a temperature of 380 K, with the stopcock open to the atmosphere. The stopcock is then closed, and the flask cooled to its original temperature. a) What is the final pressure of the ethane in the flask? b) How many grams of ethane remain in the flask?

18.61 A balloon whose volume is 750 m³ is to be filled with hydrogen at atmospheric pressure $(1.01 \times 10^5$ Pa). a) If the hydrogen is stored in cylinders with volumes of 1.90 m³ at a gauge pressure of 1.20×10^6 Pa, how many cylinders are required? Assume that the temperature of the hydrogen remains constant. b) What is the total weight (in addition to the weight of the gas) that can be supported by the balloon if the gas in the balloon and the surrounding air are both at 15.0°C? The molar mass of hydrogen

(H_2) is 2.02 g/mol. The density of air at 15.0°C and atmospheric pressure is 1.23 kg/m³. See Chapter 14 for a discussion of buoyancy. c) What weight could be supported if the balloon were filled with helium (with a molar mass of 4.00 g/mol) instead of hydrogen, again at 15.0°C?

18.62 A vertical cylindrical tank 0.900 m high has its top end closed by a tightly fitting frictionless piston of negligible weight. The air inside the cylinder is at an absolute pressure of 1.00 atm. The piston is depressed by pouring mercury on it slowly (Fig. 18.26). How far will the piston descend before mercury spills over the top of the cylinder? The temperature of the air is kept constant.

Figure 18.26 Problem 18.62.

18.63 A large tank of water has a hose connected to it, as shown in Fig. 18.27. The tank is sealed at the top and has compressed air between the water surface and the top. When the water height h has the value 3.50 m, the absolute pressure p of the compressed air is 4.20×10^5 Pa.

Figure 18.27 Problem 18.63.

Assume that the air above the water expands at constant temperature, and take the atmospheric pressure to be 1.00×10^5 Pa. a) What is the speed with which water flows out of the hose when $h = 3.50$ m? b) As water flows out of the tank, h decreases. Calculate the speed of flow for $h = 3.00$ m and for $h = 2.00$ m. c) At what value of h does the flow stop?

18.64 In one hour, an average person at rest consumes 14.5 L of oxygen at a pressure of 1.00 atm and a temperature of 20.0°C. a) Express this rate of oxygen consumption in terms of the number of molecules per second. b) A person at rest inhales and exhales 0.5 L of air with each breath. The inhaled air is 21.0% oxygen, and the exhaled air is 16.3% oxygen. How many breaths per minute will satisfy this person's oxygen requirements? c) Repeat part (b) for a resting person at an elevation of 3000 m, where the pressure is 0.72 atm and the temperature is 0.0°C. Assume that the oxygen percentages and volume per inhalation are the same as stated in part (b). (To maintain its functions, the body still requires the same number of oxygen molecules per second as at sea level.) Explain why some people report "shortness of breath" at such elevations.

18.65 How Many Atoms Are You? Estimate the number of atoms in the body of a 50-kg physics student. Note that the human body is mostly water, which has molar mass 18.0 g/mol, and that each water molecule contains three atoms.

18.66 The size of an oxygen molecule is about 2.0×10^{-10} m. Make a rough estimate of the pressure at which the finite volume of the molecules should cause noticeable deviations from ideal-gas behavior at ordinary temperatures $(T = 300$ K).

18.67 Successive Approximations and the van der Waals Equation. In the ideal-gas equation, the number of moles per volume n/V is simply equal to p/RT. In the van der Waals equation, solving for n/V in terms of the pressure p and temperature T is somewhat more involved. a) Show that the van der Waals equation can be written as

$$\frac{n}{V} = \left(\frac{p + an^2/V^2}{RT}\right)\left(1 - \frac{bn}{V}\right)$$

b) The van der Waals parameters for hydrogen sulfide gas (H_2S) are $a = 0.448 \; J \cdot m^3/mol^2$ and $b = 4.29 \times 10^{-5} \; m^3/mol$. Determine the number of moles per volume of H_2S gas at $127°C$ and an absolute pressure of 9.80×10^5 Pa as follows: i) Calculate a first approximation using the ideal-gas equation, $n/V = p/RT$. ii) Substitute this approximation for n/V into the right-hand side of the equation in part (a). The result is a new, improved approximation for n/V. iii) Substitute the new approximation for n/V into the right-hand side of the equation in (a). The result is a further-improved approximation for n/V. iv) Repeat step (iii) until successive approximations agree to the desired level of accuracy (in this case, to three significant figures). c) Compare your final result in part (b) to the result p/RT obtained using the ideal-gas equation. Which result gives a larger value of n/V? Why?

18.68 a) Compute the increase in gravitational potential energy for a nitrogen molecule (molar mass 28.0 g/mol) for an increase in elevation of 400 m near the earth's surface. b) At what temperature is this equal to the average kinetic energy of a nitrogen molecule? c) Is it possible that a nitrogen molecule near sea level where $T = 15.0°C$ could rise to an altitude of 400 m? Is it likely that it could do so without hitting any other molecules along the way? Explain.

18.69 The Lennard-Jones Potential. A commonly used potential-energy function for the interaction of two molecules (Fig. 18.6) is the Lennard-Jones 6-12 potential

$$U(r) = U_0\left[\left(\frac{R_0}{r}\right)^{12} - 2\left(\frac{R_0}{r}\right)^6\right]$$

where r is the distance between the centers of the molecules and U_0 and R_0 are positive constants. The corresponding force $F(r)$ is given in Eq. (13.26). a) Graph $U(r)$ and $F(r)$ versus r. b) Let r_1 be the value of r at which $U(r) = 0$, and let r_2 be the value of r at which $F(r) = 0$. Show the locations of r_1 and r_2 on your graphs of $U(r)$ and $F(r)$. Which of these values represents the equilibrium separation between the molecules? c) Find the values of r_1 and r_2 in terms of R_0, and find the ratio r_1/r_2. d) If the molecules are located a distance r_2 apart (as calculated in part (c)), how much work must be done to pull them apart so that $r \to \infty$?

18.70 a) What is the total random translational kinetic energy of 5.00 L of hydrogen gas, with pressure 1.01×10^5 Pa and temperature 300 K? b) If the tank containing the gas is moved with a speed of 30.0 m/s, by what percentage is the total kinetic energy of the gas increased? The molar mass of hydrogen (H_2) is 2.016 g/mol.

18.71 The speed of propagation of a sound wave in air at $27°C$ is about 350 m/s. Calculate, for comparison, a) v_{rms} for nitrogen molecules; b) the rms value of v_x at this temperature. The molar mass of nitrogen (N_2) is 28.0 g/mol.

18.72 The surface of the sun has a temperature of about 5800 K and consists largely of hydrogen atoms. a) Find the rms speed of a

hydrogen atom at this temperature. (The mass of a single hydrogen atom is 1.67×10^{-27} kg.) b) The escape speed for a particle to leave the gravitational influence of the sun is given by $(2GM/R)^{1/2}$, where M is the sun's mass, R its radius, and G the gravitational constant (Example 12.5 of Section 12.3). Use the data in Appendix F to calculate this escape speed. c) Can appreciable quantities of hydrogen escape from the sun? Can *any* hydrogen escape? Explain.

18.73 a) Show that a projectile with mass m can "escape" from the surface of a planet if it is launched vertically upward with a kinetic energy greater than mgR_p, where g is the acceleration due to gravity at the planet's surface and R_p is the planet's radius. Ignore air resistance. (See Problem 18.72.) b) If the planet in question is the earth, at what temperature does the average translational kinetic energy of a nitrogen molecule (molar mass = 28.0 g/mol) equal that required to escape? What about a hydrogen molecule (molar mass = 2.02 g/mol)? c) Repeat part (b) for the moon, for which $g = 1.63 \; m/s^2$ and $R_p = 1740$ km. d) While the earth and the moon have similar average surface temperatures, the moon has essentially no atmosphere. Use your results from parts (b) and (c) to explain why.

18.74 Planetary Atmospheres. a) The temperature near the top of Jupiter's multicolored cloud layer is about 140 K. The temperature at the top of the earth's troposphere, at an altitude of about 20 km, is about 220 K. Calculate the rms speed of hydrogen molecules in both these environments. Give your answers in m/s and as a fraction of the escape speed from the respective planet (see Problem 18.72). b) Hydrogen gas (H_2) is a rare element in the earth's atmosphere. In the atmosphere of Jupiter, by contrast, 89% of all molecules are H_2. Explain why, using your results from part (a). c) Suppose an astronomer claims to have discovered an oxygen (O_2) atmosphere on the asteroid Ceres. How likely is this? Ceres has a mass equal to 0.014 times the mass of the moon, a density of 2400 kg/m³, and a surface temperature of about 200 K.

18.75 a) For what mass of molecule or particle is v_{rms} equal to 1.00 mm/s at 300 K? b) If the particle is an ice crystal, how many molecules does it contain? The molar mass of water is 18.0 g/mol. c) Calculate the diameter of the particle if it is a spherical piece of ice. Would it be visible to the naked eye?

18.76 In describing the heat capacities of solids in Section 18.4, we stated that the potential energy $U = \frac{1}{2}kx^2$ of a harmonic oscillator averaged over one period of the motion is equal to the kinetic energy $K = \frac{1}{2}mv^2$ averaged over one period. Prove this result using Eqs. (13.13) and (13.15) for the position and velocity of a simple harmonic oscillator. For simplicity, assume that the initial position and velocity make the phase angle ϕ equal to zero. (*Hint:* Use the trigonometric identities $\cos^2(\theta) = [1 + \cos(2\theta)]/2$ and $\sin^2(\theta) = [1 - \cos(2\theta)]/2$. What is the average value of $\cos(2\omega t)$ averaged over one period?)

18.77 It is possible to make crystalline solids that are only one layer of atoms thick. Such "two-dimensional" crystals can be created by depositing atoms on a very flat surface. a) If the atoms in such a two-dimensional crystal can only move within the plane of the crystal, what will be its molar heat capacity near room temperature? Give your answer as a multiple of R and in J/mol·K. b) At very low temperatures, will the molar heat capacity of a two-dimensional crystal be greater than, less, or equal to the result you found in part (a)? Explain why.

18.78 a) Calculate the total *rotational* kinetic energy of the molecules in 1.00 mol of a diatomic gas at 300 K. b) Calculate the moment of inertia of an oxygen molecule (O_2) for rotation about either the y- or z-axis shown in Fig. 18.15. Treat the molecule as two massive points (representing the oxygen atoms) separated by a distance of 1.21×10^{-10} m. The molar mass of oxygen *atoms* is 16.0 g/mol. c) Find the rms angular velocity of rotation of an oxygen molecule about either the y- or z-axis shown in Fig. 18.15. How does your answer compare to the angular velocity of a typical piece of rapidly rotating machinery $(10{,}000$ rev/min$)$?

18.79 For each polyatomic gas in Table 18.1, compute the value of the molar heat capacity at constant volume, C_V, on the assumption that there is no vibrational energy. Compare with the measured values in the table, and compute the fraction of the total heat capacity that is due to vibration for each of the three gases. (*Note:* CO_2 is linear; SO_2 and H_2S are not. Recall that a linear polyatomic molecule has two rotational degrees of freedom, and a nonlinear molecule has three.)

*18.80 a) Show that $\int_0^\infty f(v)\, dv = 1$, where $f(v)$ is the Maxwell-Boltzmann distribution of Eq. (18.32). b) In terms of the physical definition of $f(v)$, explain why the integral in part (a) *must* have this value.

*18.81 Calculate the integral in Eq. (18.31), $\int_0^\infty v^2 f(v)\, dv$, and compare this result to $(v^2)_{\text{av}}$ as given by Eq. (18.16). (*Hint:* You may use the tabulated integral

$$\int_0^\infty x^{2n} e^{-\alpha x^2}\, dx = \frac{1 \cdot 3 \cdot 5 \cdots (2n-1)}{2^{n+1} \alpha^n} \sqrt{\frac{\pi}{\alpha}}$$

where n is a positive integer and α is a positive constant.)

*18.82 Calculate the integral in Eq. (18.30), $\int_0^\infty v f(v)\, dv$, and compare this result to v_{av} as given by Eq. (18.35). (*Hint:* Make the change of variable $v^2 = x$ and use the tabulated integral

$$\int_0^\infty x^n e^{-\alpha x}\, dx = \frac{n!}{\alpha^{n+1}}$$

where n is a positive integer and α is a positive constant.)

*18.83 a) Explain why in a gas of N molecules, the number of molecules having speeds in the *finite* interval v to $v + \Delta v$ is $\Delta N = N \int_v^{v+\Delta v} f(v)\, dv$. b) If Δv is small, then $f(v)$ is approximately constant over the interval and $\Delta N \approx N f(v)\, \Delta v$. For oxygen gas $(O_2$, molar mass $= 32.0$ g/mol$)$ at $T = 300$ K, use this approximation to calculate the number of molecules with speeds within $\Delta v = 20$ m/s of v_{mp}. Express your answer as a multiple of N. c) Repeat part (b) for speeds within $\Delta v = 20$ m/s of $7v_{\text{mp}}$. d) Repeat parts (b) and (c) for a temperature of 600 K. e) Repeat parts (b) and (c) for a temperature of 150 K. f) What do your results tell you about the shape of the distribution as a function of temperature? Do your conclusions agree with what is shown in Fig. 18.20?

18.84 The *vapor pressure* is the pressure of the vapor phase of a substance when it is in equilibrium with the solid or liquid phase of the substance. The *relative humidity* is the partial pressure of water vapor in the air divided by the vapor pressure of water at that same temperature, expressed as a percentage. The air is saturated when the humidity is 100%. a) The vapor pressure of water at 20.0°C is 2.34×10^3 Pa. If the air temperature is 20.0°C and the relative humidity is 60%, what is the partial pressure of water vapor in the atmosphere (that is, the pressure due to water vapor alone)? b) Under the conditions of part (a), what is the mass of water in 1.00 m³ of air? (The molar mass of water is 18.0 g/mol. Assume that water vapor can be treated as an ideal gas.)

18.85 The Dew Point. The vapor pressure of water (Problem 18.84) decreases as the temperature decreases. If the amount of water vapor in the air is kept constant as the air is cooled, a temperature is reached, called the *dew point,* at which the partial pressure and vapor pressure coincide and the vapor is saturated. If the air is cooled further, vapor condenses to liquid until the partial pressure again equals the vapor pressure at that temperature. The temperature in a room is 30.0°C. A meteorologist cools a metal can by gradually adding cold water. When the can temperature reaches 16.0°C, water droplets form on its outside surface. What is the relative humidity of the 30.0°C air in the room? The table below lists the vapor pressure of water at various temperatures:

Temperature (°C)	Vapor Pressure (Pa)
10.0	1.23×10^3
12.0	1.40×10^3
14.0	1.60×10^3
16.0	1.81×10^3
18.0	2.06×10^3
20.0	2.34×10^3
22.0	2.65×10^3
24.0	2.99×10^3
26.0	3.36×10^3
28.0	3.78×10^3
30.0	4.25×10^3

18.86 Altitude at Which Clouds Form. On a spring day in the midwestern United States, the air temperature at the surface is 28.0°C. Puffy cumulus clouds form at an altitude where the air temperature equals the dew point (Problem 18.85). If the air temperature decreases with altitude at a rate of 0.6 C°/100 m, at approximately what height above the ground will clouds form if the relative humidity at the surface is a) 35% b) 80%? (*Hint:* Use the table in Problem 18.85.)

Challenge Problems

18.87 Dark Nebulae and the Interstellar Medium. The dark area in Fig. 18.28 that appears devoid of stars is a *dark nebula,* a cold gas cloud in interstellar space that contains enough material to block out light from the stars behind it. A typical dark nebula is about 20 light years in diameter and contains about 50 hydrogen atoms per cubic centimeter (monatomic hydrogen, *not* H_2) at a temperature of about 20 K. (A light year is the distance light travels in vacuum in one year and is equal to 9.46×10^{15} m.) a) Estimate the mean free path for a hydrogen atom in a dark nebula. The radius of a hydrogen atom is 5.0×10^{-11} m. b) Estimate the rms speed of a hydrogen

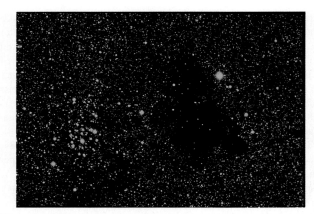

Figure 18.28 Challenge Problem 18.87.

atom and the mean free time (the average time between collisions for a given atom). Based on this result, do you think that atomic collisions, such as those leading to H_2 molecule formation, are very important in determining the composition of the nebula? c) Estimate the pressure inside a dark nebula. d) Compare the rms speed of a hydrogen atom to the escape speed at the surface of the nebula (assumed spherical). If the space around the nebula were a vacuum, would such a cloud be stable or would it tend to evaporate? e) The stability of dark nebulae is explained by the presence of the *interstellar medium* (ISM), an even thinner gas that permeates space and into which the dark nebulae are embedded. Show that for dark nebulae to be in equilibrium with the ISM, the numbers of atoms per volume (N/V) and the temperatures (T) of dark nebulae and the ISM must be related by

$$\frac{(N/V)_{\text{nebula}}}{(N/V)_{\text{ISM}}} = \frac{T_{\text{ISM}}}{T_{\text{nebula}}}$$

f) In the vicinity of the sun, the ISM contains about 1 hydrogen atom per 200 cm³. Estimate the temperature of the ISM in the vicinity of the sun. Compare to the temperature of the sun's surface, about 5800 K. Would a spacecraft coasting through interstellar space burn up? Why or why not?

18.88 In the *troposphere*, the part of the atmosphere that extends from the surface to an altitude of about 11 km, the temperature is not uniform but decreases with increasing elevation. a) Show that if the temperature variation is approximated by the linear relation

$$T = T_0 - \alpha y$$

where T_0 is the temperature at the earth's surface and T is the temperature at height y, the pressure p at height y is given by

$$\ln\left(\frac{p}{p_0}\right) = \frac{Mg}{R\alpha} \ln\left(\frac{T_0 - \alpha y}{T_0}\right)$$

where p_0 is the pressure at the earth's surface and M is the molar mass for air. The coefficient α is called the lapse rate of temperature. It varies with atmospheric conditions, but an average value is about 0.6 C°/100 m. b) Show that the above result reduces to the result of Example 18.4 (Section 18.1) in the limit that $\alpha \to 0$.

c) With $\alpha = 0.6$ C°/100 m, calculate p for $y = 8863$ m and compare your answer to the result of Example 18.4. Take $T_0 = 288$ K and $p_0 = 1.00$ atm.

18.89 Van der Waals Equation and Critical Points. a) In pV-diagrams the slope $\partial p/\partial V$ along an isotherm is never positive. Explain why. b) Regions where $\partial p/\partial V = 0$ represent equilibrium between two phases; volume can change with no change in pressure, as when water boils at atmospheric pressure. We can use this to determine the temperature, pressure, and volume per mole at the critical point using the equation of state $p = p(V, T, n)$. If $T > T_c$, then $p(V)$ has no maximum along an isotherm, but if $T < T_c$, then $p(V)$ has a maximum. Explain how this leads to the following condition for determining the critical point:

$$\frac{\partial p}{\partial V} = 0 \quad \text{and} \quad \frac{\partial^2 p}{\partial V^2} = 0 \quad \text{at the critical point}$$

c) Solve the van der Waals equation (Eq. 18.7) for p. That is, find $p(V, T, n)$. Find $\partial p/\partial V$ and $\partial^2 p/\partial V^2$. Set these equal to zero to obtain two equations for V, T, and n. d) Simultaneous solution of the two equations obtained in part (c) gives the temperature and volume per mole at the critical point, T_c and $(V/n)_c$. Find these constants in terms of a and b. (*Hint:* Divide one equation by the other to eliminate T.) e) Substitute these values into the equation of state to find p_c, the pressure at the critical point. f) Use the results from parts (d) and (e) to find the ratio $RT_c/p_c(V/n)_c$. This should not contain either a or b, and so should have the same value for all gases. g) Compute the ratio $RT_c/p_c(V/n)_c$ for the gases H_2, N_2, and H_2O using the critical point data given below.

Gas	T_c (K)	p_c (Pa)	$(V/n)_c$ (m³/mol)
H_2	33.3	13.0×10^5	65.0×10^{-6}
N_2	126.2	33.9×10^5	90.1×10^{-6}
H_2O	647.4	221.2×10^5	56.0×10^{-6}

h) Discuss how well the results of part (g) compare to the prediction of part (f) based on the van der Waals equation. What do you conclude about the accuracy of the van der Waals equation as a description of the behavior of gases near the critical point?

*18.90 In Example 18.7 (Section 18.3) we saw that $v_{\text{rms}} > v_{\text{av}}$. It is not difficult to show that this is *always* the case. (The only exception is when the particles have the same speed, in which case $v_{\text{rms}} = v_{\text{av}}$.) a) For two particles with speeds v_1 and v_2, show that $v_{\text{rms}} \geq v_{\text{av}}$, regardless of the numerical values of v_1 and v_2. Then show that $v_{\text{rms}} > v_{\text{av}}$ if $v_1 \neq v_2$. b) Suppose that for a collection of N particles you know that $v_{\text{rms}} > v_{\text{av}}$. Another particle, with speed u, is added to the collection of particles. If the new rms and average speeds are denoted as v'_{rms} and v'_{av}, show that

$$v'_{\text{rms}} = \sqrt{\frac{N v_{\text{rms}}^2 + u^2}{N + 1}} \quad \text{and} \quad v'_{\text{av}} = \frac{N v_{\text{av}} + u}{N + 1}$$

c) Use the expressions in part (b) to show that $v'_{\text{rms}} > v'_{\text{av}}$ regardless of the numerical value of u. d) Explain why your results for (a) and (c) together show that $v_{\text{rms}} > v_{\text{av}}$ for any collection of particles if the particles do not all have the same speed.

THE FIRST LAW OF THERMODYNAMICS 19

This steam locomotive operates using the first law of thermodynamics. When heat is generated by burning wood or coal in the locomotive's engine, part of the energy goes into increasing the temperature of the engine's water (which boils and produces steam). The remainder is used by the expanding steam to do work and propel the locomotive forward.

? **Would it be possible for the steam to propel the locomotive by doing work as it *condenses*?**

E very time you drive a car, turn on an air conditioner, or cook a meal, you reap the practical benefits of *thermodynamics,* the study of relationships involving heat, mechanical work, and other aspects of energy and energy transfer. For example, in a car engine heat is generated by the chemical reaction of oxygen and vaporized gasoline in the engine's cylinders. The heated gas pushes on the pistons within the cylinders, doing mechanical work that is used to power the car. This is an example of a *thermodynamic process.*

The first law of thermodynamics, central to the understanding of such processes, is an extension of the principle of conservation of energy. It broadens this principle to include energy exchange by both heat transfer and mechanical work and introduces the concept of the *internal energy* of a system. Conservation of energy plays a vital role in every area of physical science, and the first law has extremely broad usefulness. To state energy relationships precisely, we need the concept of a *thermodynamic system.* We'll discuss *heat* and *work* as two means of transferring energy into or out of such a system.

19.1 | Thermodynamic Systems

We have studied energy transfer through mechanical work (Chapter 6) and through heat transfer (Chapters 17 and 18). Now we are ready to combine and generalize these principles.

19.1 The popcorn in the pot is a thermodynamic system. In the thermodynamic process shown here, heat is added to the system, and the system does work on its surroundings to lift the lid of the pot.

We always talk about energy transfer to or from some specific *system*. The system might be a mechanical device, a biological organism, or a specified quantity of material, such as the refrigerant in an air conditioner or steam expanding in a turbine. In general, a **thermodynamic system** is any collection of objects that is convenient to regard as a unit, and that may have the potential to exchange energy with its surroundings. A familiar example is a quantity of popcorn kernels in a pot with a lid. When the pot is placed on a stove, energy is added to the popcorn by conduction of heat. As the popcorn pops and expands, it does work as it exerts an upward force on the lid and moves it through a displacement (Fig. 19.1). The *state* of the popcorn changes in this process, since the volume, temperature, and pressure of the popcorn all change as it pops. A process such as this one, in which there are changes in the state of a thermodynamic system, is called a **thermodynamic process.**

In mechanics we used the concept of *system* on a regular basis in connection with free-body diagrams and conservation of energy and momentum. With *thermodynamic* systems, as with all others, it is essential to define clearly at the start exactly what is and is not included in the system. Only then can we describe unambiguously the energy transfers into and out of that system. For instance, in our popcorn example we defined the system to include the popcorn but not the pot, lid, or stove.

Thermodynamics has its roots in many practical problems other than popping popcorn (Fig. 19.2). The engine in an automobile and the jet engines in an airplane use the heat of combustion of their fuel to perform mechanical work in propelling the vehicle. Muscle tissue in living organisms metabolizes chemical energy in food and performs mechanical work on the organism's surroundings. A steam engine or steam turbine uses the heat of combustion of coal or other fuel to perform mechanical work such as driving an electric generator or pulling a train.

Signs For Heat and Work in Thermodynamics

We describe the energy relations in any thermodynamic process in terms of the quantity of heat Q added *to* the system and the work W done *by* the system. Both Q and W may be positive, negative, or zero (Fig. 19.3). A positive value of Q represents heat flow *into* the system, with a corresponding input of energy to it; neg-

19.2 (a) A rocket engine uses the heat of combustion of its fuel to do work propelling the launch vehicle. (b) Humans and other biological organisms are more complicated systems than we can analyze fully in this book, but the same basic principles of thermodynamics apply to them.

(a)

(b)

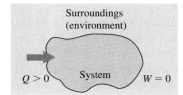

(a) When heat is added *to* the system, Q is positive

(b) When heat is transferred *out* of the system, Q is negative

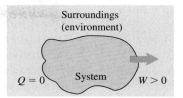

(c) When work is done *by* the system, W is positive

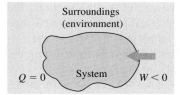

(d) When work is done *on* the system, W is negative

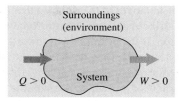

(e) Heat is added to the system and work is done by the system

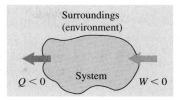

(f) Heat is transferred out of the system and work is done on the system

19.3 A thermodynamic system may exchange energy with its surroundings (environment) by means of heat [(a) and (b)], work. [(c) and (d)], or heat and work simultaneously [(e) and (f)]. Note the sign conventions for Q and W.

ative Q represents heat flow *out of* the system. A positive value of W represents work done *by* the system against its surroundings, such as work done by an expanding gas, and hence corresponds to energy *leaving* the system. Negative W, such as work done during compression of a gas in which work is done *on the gas* by its surroundings, represents energy *entering* the system. We will use these conventions consistently in the examples in this chapter and the next.

> **CAUTION** Note that our sign rule for work is *opposite* to the one we used in mechanics, in which we always spoke of the work done by the forces acting *on* a body. In thermodynamics it is usually more convenient to call W the work done *by* the system so that when a system expands, the pressure, volume change, and work are all positive. Take care to use the sign rules for work and heat consistently!

Test Your Understanding

In Example 17.8 (Section 17.6), what is the sign of Q for the coffee? For the aluminum cup? If a block slides along a horizontal surface with friction, what is the sign of W for the block?

19.2 | Work Done During Volume Changes

A simple but common example of a thermodynamic system is a quantity of gas enclosed in a cylinder with a movable piston. Internal-combustion engines, steam engines, and compressors in refrigerators and air conditioners all use some version of such a system. In the next several sections we will use the gas-in-cylinder system to explore several kinds of processes involving energy transformations.

We'll use a microscopic viewpoint, based on the kinetic and potential energies of individual molecules in a material, to develop intuition about thermodynamic quantities. But it is important to understand that the central principles of thermodynamics can be treated in a completely *macroscopic* way, without reference to microscopic models. Indeed, part of the great power and generality of thermodynamics is that it does *not* depend on details of the structure of matter.

(a) Piston moves away from molecule during collision: molecule loses kinetic energy, does positive work on piston

(b) Piston moves toward molecule during collision: molecule gains kinetic energy, does negative work on piston

19.4 (a) A molecule striking a piston (a) does positive work if the piston is moving away from the molecule and (b) does negative work if the piston is moving toward the molecule. Hence a gas does positive work when it expands as in (a) but does negative work when it compresses as in (b).

19.5 The infinitesimal work done by the system during the small expansion dx is $dW = pA\,dx$.

19.6 When fuel is burned in one of the cylinders of an automotive engine, the heated fuel–air mixture expands and does an amount of work given by Eq. (19.2).

First we consider the *work* done by the system during a volume change. When a gas expands, it pushes outward on its boundary surfaces as they move outward. Hence an expanding gas always does positive work. The same thing is true of any solid or fluid material that expands under pressure, such as the popcorn in Fig. 19.1.

We can understand the work done by a gas in a volume change by considering the molecules that make up the gas. When one such molecule collides with a stationary surface, it exerts a momentary force on the wall but does no work because the wall does not move. But if the surface is moving, such as a piston in a gasoline engine, the molecule *does* do work on the surface during the collision. If the piston in Fig. 19.4a moves to the right, so that the volume of the gas increases, the molecules that strike the piston exert a force through a distance and do *positive* work on the piston. If the piston moves toward the left as in Fig. 19.4b, so that the volume of the gas decreases, then positive work is done *on* the molecule during the collision. Hence the gas molecules do *negative* work on the piston.

Figure 19.5 shows a system whose volume can change (a gas, liquid, or solid) in a cylinder with a movable piston. Suppose that the cylinder has a cross-section area A and that the pressure exerted by the system at the piston face is p. The total force F exerted by the system on the piston is $F = pA$. When the piston moves out an infinitesimal distance dx, the work dW done by this force is

$$dW = F\,dx = pA\,dx$$

But

$$A\,dx = dV$$

where dV is the infinitesimal change of volume of the system. Thus we can express the work done by the system in this infinitesimal volume change as

$$dW = p\,dV \qquad (19.1)$$

In a finite change of volume from V_1 to V_2,

$$W = \int_{V_1}^{V_2} p\,dV \qquad \text{(work done in a volume change)} \qquad (19.2)$$

In general, the pressure of the system may vary during the volume change. For example, this is the case in the cylinders of an automobile engine as the pistons move back and forth (Fig. 19.6). To evaluate the integral in Eq. (19.2), we have to know how the pressure varies as a function of volume. We can represent this relationship as a graph of p as a function of V (a pV-diagram, described at the end of Section 18.1). Figure 19.7a shows a simple example. In this figure, Eq. (19.2) is represented graphically as the *area* under the curve of p versus V between the limits V_1 and V_2. (In Section 6.3 we used a similar interpretation of the work done by a force F as the area under the curve of F versus x between the limits x_1 and x_2.)

According to the rule we stated in Section 19.1, work is *positive* when a system *expands*. In an expansion from state 1 to state 2 in Fig. 19.7a the area under the curve and the work are positive. A *compression* from 1 to 2 in Fig. 19.7b gives a *negative* area; when a system is compressed, its volume decreases and it does *negative* work on its surroundings (see also Fig. 19.4b).

CAUTION When using Eq. (19.2), always remember that V_1 is the *initial* volume and V_2 is the *final* volume. That's why the labels 1 and 2 are reversed in Fig. 19.7b compared to Fig. 19.7a, even though both processes move between the same two thermodynamic states.

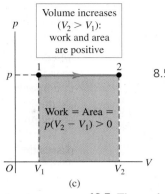

8.5 Work Done By a Gas

(a)   (b)   (c)

If the pressure p remains constant while the volume changes from V_1 to V_2 (Fig. 19.7c), the work done by the system is

$$W = p(V_2 - V_1)$$ (19.3)

(work done in a volume change at constant pressure)

In any process in which the volume is *constant,* the system does no work because there is no displacement.

19.7 The work done equals the area under the curve on a pV-diagram. (a) In an expansion, the work done is positive. (b) In a compression, the work done is negative. (c) In a constant-pressure process, the work done is easy to calculate because the area (in this case positive) is rectangular.

Example 19.1	**Isothermal expansion of an ideal gas**

An ideal gas undergoes an *isothermal* (constant-temperature) *expansion* at temperature T, during which its volume changes from V_1 to V_2. How much work does the gas do?

SOLUTION

IDENTIFY and SET UP: We are asked to calculate the work done due to a volume change, which means that we must use Eq. (19.2). To use this equation we need to know the pressure as a function of volume; for this we use the ideal-gas law, Eq. (18.3). Although it may be tempting to do so, we *cannot* use Eq. (19.3) because the temperature, not the pressure, is held constant.

EXECUTE: From Eq. (19.2),

$$W = \int_{V_1}^{V_2} p \, dV$$

From Eq. (18.3) the pressure p of n moles of ideal gas occupying volume V at absolute temperature T is

$$p = \frac{nRT}{V}$$

where R is the gas constant. We substitute this into the integral, take the constants n, R, and T outside, and evaluate the integral:

$$W = nRT \int_{V_1}^{V_2} \frac{dV}{V} = nRT \ln \frac{V_2}{V_1}$$

(ideal gas, isothermal process)

Also, when T is constant,

$$p_1 V_1 = p_2 V_2 \quad \text{or} \quad \frac{V_2}{V_1} = \frac{p_1}{p_2}$$

so the isothermal work may also be expressed as

$$W = nRT \ln \frac{p_1}{p_2} \quad \left(\text{ideal gas, isothermal process}\right)$$

EVALUATE: We check our result by noting that in an expansion $V_2 > V_1$ and the ratio V_2/V_1 is greater than 1. The logarithm of a number greater than 1 is positive, so $W > 0$, as it should be. As an additional check, look at our second expression for W: in an isothermal expansion the volume increases and the pressure drops, so $p_2 < p_1$, the ratio $p_1/p_2 > 1$, and $W = nRT \ln(p_1/p_2)$ is again positive.

These results also apply to an isothermal *compression* of a gas, for which $V_2 < V_1$ and $p_2 > p_1$.

Test Your Understanding

In an isothermal expansion of n moles of an ideal gas at temperature T, how many times larger than the initial volume must the final volume be if the gas does work nRT? If the gas does work $2nRT$?

(a)

(b)

(c)

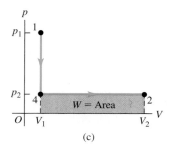

(d)

19.8 (a) Three different paths between state 1 and state 2. (b)–(d) The work done by the system during a transition between two states depends on the path chosen.

19.3 | Paths Between Thermodynamic States

We've seen that if a thermodynamic process involves a change in volume, the system undergoing the process does work (either positive or negative) on its surroundings. Heat also flows into or out of the system during the process if there is a temperature difference between the system and its surroundings. Let's now examine how the work done by and heat added to the system during a thermodynamic process depend on the details of how the process takes place.

When a thermodynamic system changes from an initial state to a final state, it passes through a series of intermediate states. We call this series of states a **path.** There are always infinitely many different possibilities for these intermediate states. When they are all equilibrium states, the path can be plotted on a pV-diagram (Fig. 19.8a). Point 1 represents an initial state with pressure p_1 and volume V_1, and point 2 represents a final state with pressure p_2 and volume V_2. To pass from state 1 to state 2, we could keep the pressure constant at p_1 while the system expands to volume V_2 (point 3 in Fig. 19.8b), then reduce the pressure to p_2 (probably by decreasing the temperature) while keeping the volume constant at V_2 (to point 2 on the diagram). The work done by the system during this process is the area under the line $1 \rightarrow 3$; no work is done during the constant-volume process $3 \rightarrow 2$. Or the system might traverse the path $1 \rightarrow 4 \rightarrow 2$ (Fig. 19.8c); in that case the work is the area under the line $4 \rightarrow 2$, since no work is done during the constant-volume process $1 \rightarrow 4$. The smooth curve from 1 to 2 is another possibility (Fig. 19.8d), and the work for this path is different from that for either of the other paths.

We conclude that *the work done by the system depends not only on the initial and final states, but also on the intermediate states, that is, on the path.* Furthermore, we can take the system through a series of states forming a closed loop, such as $1 \rightarrow 3 \rightarrow 2 \rightarrow 4 \rightarrow 1$. In this case the final state is the same as the initial state, but the total work done by the system is *not* zero. (In fact, it is represented on the graph by the area enclosed by the loop; can you prove that? See Exercise 19.8.) It follows that it doesn't make sense to talk about the amount of work *contained in* a system. In a particular state, a system may have definite values of the state coordinates p, V, and T, but it wouldn't make sense to say that it has a definite value of W.

Like work, the *heat* added to a thermodynamic system when it undergoes a change of state depends on the path from the initial state to the final state. Here's an example. Suppose we want to change the volume of a certain quantity of an ideal gas from 2.0 L to 5.0 L while keeping the temperature constant at $T = 300$ K. Figure 19.9 shows two different ways in which we can do this. In Fig. 19.9a the gas is contained in a cylinder with a piston, with an initial volume of 2.0 L. We let the gas expand slowly, supplying heat from the electric heater to keep the temperature at 300 K. After expanding in this slow, controlled, isothermal manner, the gas reaches its final volume of 5.0 L; it absorbs a definite amount of heat in the process.

Figure 19.9b shows a different process leading to the same final state. The container is surrounded by insulating walls and is divided by a thin, breakable partition into two compartments. The lower part has volume 2.0 L and the upper part has volume 3.0 L. In the lower compartment we place the same amount of the same gas as in Fig. 19.9a, again at $T = 300$ K. The initial state is the same as before. Now we break the partition; the gas undergoes a rapid, uncontrolled expansion, with no heat passing through the insulating walls. The final volume is 5.0 L, the same as in Fig. 19.9a. The gas does no work during this expansion because it doesn't push against anything that moves. This uncontrolled expansion

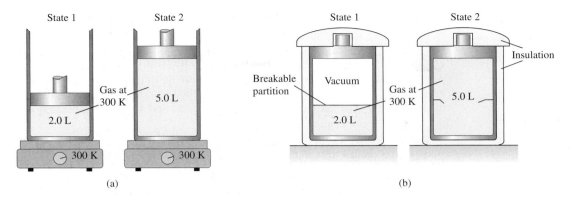

19.9 (a) Slow, controlled isothermal expansion of a gas from an initial state 1 to a final state 2 with the same temperature but lower pressure. (b) Rapid, uncontrolled expansion of the same gas starting at the same state 1 and ending at the same state 2.

of a gas into vacuum is called a **free expansion;** we will discuss it further in Section 19.6.

Experiments have shown that when an ideal gas undergoes a free expansion, there is no temperature change. Therefore the final state of the gas is the same as in Fig. 19.9a. The intermediate states (pressures and volumes) during the transition from state 1 to state 2 are entirely different in the two cases; Figs. 19.9a and 19.9b represent *two different paths* connecting the *same states* 1 and 2. For the path in Fig. 19.9b, *no* heat is transferred into the system, and the system does no work. Like work, *heat depends not only on the initial and final states but also on the path.*

Because of this path dependence, it would not make sense to say that a system "contains" a certain quantity of heat. To see this, suppose we assign an arbitrary value to "the heat in a body" in some reference state. Then presumably the "heat in the body" in some other state would equal the heat in the reference state plus the heat added when the body goes to the second state. But that's ambiguous, as we have just seen; the heat added depends on the *path* we take from the reference state to the second state. We are forced to conclude that there is *no* consistent way to define "heat in a body"; it is not a useful concept.

While it doesn't make sense to talk about "work in a body" or "heat in a body," it *does* make sense to speak of the amount of *internal energy* in a body. This important concept is our next topic.

Test Your Understanding

If there is one mole of gas in the systems shown in Fig. 19.9, how much work is done by the gas in the process shown in (a)? In the process shown in (b)?

19.4 | Internal Energy and the First Law of Thermodynamics

Internal energy is one of the most important concepts in thermodynamics. In Section 7.3, when we discussed energy changes for a body sliding with friction, we stated that warming a body increased its internal energy and that cooling the body decreased its internal energy. But what *is* internal energy? We can look at it in various ways; let's start with one based on the ideas of mechanics. Matter consists of atoms and molecules, and these are made up of particles having kinetic and potential energies. We *tentatively* define the **internal energy** of a system as the sum of the kinetic energies of all of its constituent particles, plus the sum of all the potential energies of interaction among these particles.

Note that internal energy does *not* include potential energy arising from the interaction between the system and its surroundings. If the system is a glass of water, placing it on a high shelf increases the gravitational potential energy arising from the interaction between the glass and the earth. But this has no effect on the interaction between the molecules of the water, and so the internal energy of the water does not change.

We use the symbol U for internal energy. (We used this same symbol in our study of mechanics to represent potential energy. You may have to remind yourself occasionally that U has a different meaning in thermodynamics.) During a change of state of the system the internal energy may change from an initial value U_1 to a final value U_2. We denote the change in internal energy as $\Delta U = U_2 - U_1$.

We know that heat transfer is energy transfer. When we add a quantity of heat Q to a system and it does no work during the process, the internal energy increases by an amount equal to Q; that is, $\Delta U = Q$. When a system does work W by expanding against its surroundings and no heat is added during the process, energy leaves the system and the internal energy decreases. That is, when W is positive, ΔU is negative, and vice versa. So $\Delta U = -W$. When *both* heat transfer and work occur, the *total* change in internal energy is

$$U_2 - U_1 = \Delta U = Q - W \quad \text{(first law of thermodynamics)} \quad (19.4)$$

We can rearrange this to the form

$$Q = \Delta U + W \quad (19.5)$$

The message of Eq. (19.5) is that in general, when heat Q is added to a system, some of this added energy remains within the system, changing its internal energy by an amount ΔU; the remainder leaves the system again as the system does work W against its surroundings. Because W and Q may be positive, negative, or zero, ΔU can be positive, negative, or zero for different processes (Fig. 19.10).

Equation (19.4) or (19.5) is the **first law of thermodynamics.** It is a generalization of the principle of conservation of energy to include energy transfer through heat as well as mechanical work. As you will see in later chapters, this principle can be extended to ever broader classes of phenomena by identifying additional forms of energy and energy transfer. In every situation in which it seems that the total energy in all known forms is not conserved, it has been possible to identify a new form of energy such that the total energy, including the new form, *is* conserved. There is energy associated with electric fields, with magnetic fields, and, according to the theory of relativity, even with mass itself.

At the beginning of this discussion we tentatively defined internal energy in terms of microscopic kinetic and potential energies. This has drawbacks, however. Actually *calculating* internal energy in this way for any real system would be hopelessly complicated. Furthermore, this definition isn't an *operational* one because it doesn't describe how to determine internal energy from physical quantities that we can measure directly.

So let's look at internal energy in another way. Starting over, we define the *change* in internal energy ΔU during any change of a system as the quantity given by Eq. (19.4), $\Delta U = Q - W$. This *is* an operational definition, because we can measure Q and W. It does not define U itself, only ΔU. This is not a shortcoming, because we can *define* the internal energy of a system to have a specified value in

(a) More heat added to system than system does work: internal energy of system increases

(b) More heat flows out of system than work is done: internal energy of system decreases

(c) Heat added to system equals work done by system: internal energy of system unchanged

19.10 In a thermodynamic process, the internal energy U of a system may (a) increase ($\Delta U > 0$), (b) decrease ($\Delta U < 0$), or (c) remain the same ($\Delta U = 0$).

some reference state, and then use Eq. (19.4) to define the internal energy in any other state. This is analogous to our treatment of potential energy in Chapter 7, in which we arbitrarily defined the potential energy of a mechanical system to be zero at a certain position.

This new definition trades one difficulty for another. If we define ΔU by Eq. (19.4), then when the system goes from state 1 to state 2 by two different paths, how do we know that ΔU is the same for the two paths? We have already seen that Q and W are, in general, *not* the same for different paths. If ΔU, which equals $Q - W$, is also path-dependent, then ΔU is ambiguous. If so, the concept of internal energy of a system is subject to the same criticism as the erroneous concept of quantity of heat in a system, as we discussed at the end of Section 19.3.

The only way to answer this question is through *experiment*. For various materials we measure Q and W for various changes of state and various paths to learn whether ΔU is or is not path-dependent. The results of many such investigations are clear and unambiguous: While Q and W depend on the path, $\Delta U = Q - W$ *is independent of path. The change in internal energy of a system during any thermodynamic process depends only on the initial and final states, not on the path leading from one to the other.*

Experiment, then, is the ultimate justification for believing that a thermodynamic system in a specific state has a unique internal energy that depends only on that state. An equivalent statement is that the internal energy U of a system is a function of the state coordinates p, V, and T (actually, any two of these, since the three variables are related by the equation of state).

To say that the first law of thermodynamics, given by Eq. (19.4) or (19.5), represents conservation of energy for thermodynamic processes is correct, as far as it goes. But an important *additional* aspect of the first law is the fact that internal energy depends only on the state of a system (Fig. 19.11). In changes of state, the change in internal energy is path-independent.

All this may seem a little abstract if you are satisfied to think of internal energy as microscopic mechanical energy. There's nothing wrong with that view, and we will make use of it at various times during our discussion. But in the interest of precise *operational* definitions, internal energy, like heat, can and must be defined in a way that is independent of the detailed microscopic structure of the material.

Two special cases of the first law of thermodynamics are worth mentioning. A process that eventually returns a system to its initial state is called a *cyclic* process. For such a process, the final state is the same as the initial state, and so the *total* internal energy change must be zero. Then

$$U_2 = U_1 \quad \text{and} \quad Q = W$$

If a net quantity of work W is done by the system during this process, an equal amount of energy must have flowed into the system as heat Q. But there is no reason why either Q or W individually has to be zero (Fig. 19.12).

Another special case occurs in an *isolated* system, one that does no work on its surroundings and has no heat flow to or from its surroundings. For any process taking place in an isolated system,

$$W = Q = 0$$

and therefore

$$U_2 = U_1 = \Delta U = 0$$

In other words, *the internal energy of an isolated system is constant.*

19.11 The internal energy of a cup of coffee just depends on its thermodynamic state—how much water and ground coffee it contains, and what its temperature is. It does not depend on the history of how the coffee was prepared—that is, the thermodynamic path that led to its current state.

Activ
Physics
ONLINE

8.6   Heat, Internal Energy, and First
Law of Thermodynamics

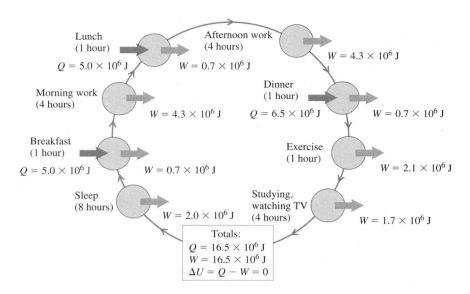

19.12 Every day, your body (a thermodynamic system) goes through a cyclic thermodynamic process like this one. Heat Q is added by metabolizing food, and your body does work W in breathing, walking, and other activities. If you return to the same state at the end of the day, $Q = W$ and the net change in your internal energy is zero.

The First Law of Thermodynamics

IDENTIFY *the relevant concepts:* The first law of thermodynamics is the statement of the law of conservation of energy in its most general form. You can apply it to *any* situation in which you are concerned with changes in the internal energy of a system, with heat flow into or out of a system, and/or with work done by or on a system.

SET UP *the problem using the following steps:*
1. Carefully define what the thermodynamic system is.
2. The first law of thermodynamics focuses on systems that go through thermodynamic processes. Some problems involve processes with more than one step, so make sure that you identify the initial and final state for each step.
3. Identify the known quantities and the target variables.
4. Check whether you have enough equations. The first law, $\Delta U = Q - W$, can be applied just once to each step in a thermodynamic process, so you will often need additional equations. These often include Eq. (19.2) for the work done in a volume change and the equation of state of the material that makes up the thermodynamic system (for an ideal gas, $pV = nRT$).

EXECUTE *the solution as follows:*
1. You shouldn't be surprised to be told that consistent units are essential. If p is in Pa and V in m³, then W is in joules. Otherwise, you may want to convert the pressure and volume units into units of Pa and m³. If a heat capacity is given in terms of calories, usually the simplest procedure is to convert it to joules. Be especially careful with moles. When you use $n = m_{tot}/M$ to convert between total mass

and number of moles, remember that if m_{tot} is in kilograms, M must be in *kilograms* per mole. The usual units for M are *grams* per mole; be careful!
2. The internal energy change ΔU in any thermodynamic process or series of processes is independent of the path, whether the substance is an ideal gas or not. This point is of the utmost importance in the problems in this chapter and the next. Sometimes you will be given enough information about one path between the given initial and final states to calculate ΔU for that path. Since ΔU is the same for every possible path between the same two states, you can then relate the various energy quantities for other paths.
3. When a process consists of several distinct steps, it often helps to make a chart showing Q, W, and ΔU for each step. Put these quantities for each step on a different line, and arrange them so the Q's, W's, and ΔU's form columns. Then you can apply the first law to each line; in addition, you can add each column and apply the first law to the sums. Do you see why?
4. Using steps (1)–(3), solve for the target variables.

EVALUATE *your answer:* Check your results for reasonableness. In particular, make sure that each of your answers has the correct algebraic sign. Remember that a positive Q means that heat flows *into* the system, and that a negative Q means that heat flows *out of* the system. A positive W means that work is done *by* the system on its environment, while a negative W means that work is done *on* the system by its environment.

| Example 19.2 | **Working off your dessert** |

You propose to eat a 900-calorie hot fudge sundae (with whipped cream) and then run up several flights of stairs to work off the energy you have taken in. How high do you have to climb? Assume that your mass is 60.0 kg.

SOLUTION

IDENTIFY: Eating the hot fudge sundae corresponds to a heat flow into your body, and running up stairs means that you do work. We can relate these quantities using the first law of thermodynamics.

SET UP: The system consists of your body. We are given that 900 food calories of heat flows into your body. The purpose of running up the stairs is to make sure that the final state of the system is the same as the initial state (no fatter, no leaner), so there is no net change in internal energy: $\Delta U = 0$. The work you must do to raise your mass m a height h is $W = mgh$; our target variable is h.

EXECUTE: Using the first law of thermodynamics, $\Delta U = 0 = Q - W$, so $W = Q$: the work that you do running up the stairs must

just equal the heat input from the sundae. Using $W = mgh$, the height that you climb is $h = Q/mg$. Before substituting values into this equation, we first convert units: one food-value calorie is 1 kcal = 1000 cal = 4190 J (to three significant figures), so

$$Q = 900 \text{ kcal} \, (4190 \text{ J/1 kcal}) = 3.77 \times 10^6 \text{ J}$$

Then

$$h = \frac{Q}{mg} = \frac{3.77 \times 10^6 \text{ J}}{(60.0 \text{ kg})(9.80 \text{ m/s}^2)}$$
$$= 6410 \text{ m} \quad (\text{about } 21{,}000 \text{ ft})$$

EVALUATE: Good luck! We have assumed 100% efficiency in the conversion of food energy into mechanical work; this isn't very realistic. We'll talk more about efficiency later.

| Example 19.3 | **A cyclic process** |

Figure 19.13 shows a pV-diagram for a *cyclic* process, one in which the initial and final states are the same. It starts at point a and proceeds counterclockwise in the pV-diagram to point b, then back to a, and the total work is $W = -500$ J. a) Why is the work negative? b) Find the change in internal energy and the heat added during this process.

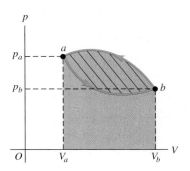

19.13 The net work done by the system in the process aba is -500 J. What would it have been if the process had proceeded clockwise in this pV-diagram?

SOLUTION

IDENTIFY and SET UP: The thermodynamic process here has two steps: $a \to b$ via the lower curve in Fig. 19.13 and $b \to a$ via the upper curve. But note that the questions in (a) and (b) are about the *entire* cyclic process $a \to b \to a$ (around the loop in Fig. 19.13).

EXECUTE: a) The work done equals the area under the curve, with the area taken as positive for increasing volume and negative for decreasing volume. The area under the lower curve from a to b is positive, but it is smaller than the absolute value of the negative area under the upper curve from b back to a. Therefore the net area (the area enclosed by the path, shown with red stripes) and the work are negative. In other words, 500 more joules of work is done *on* the system than *by* the system.

b) For this and any other cyclic process (in which the beginning and end points are the same), $\Delta U = 0$, so $Q = W = -500$ J. That is, 500 joules of heat must come *out of* the system.

EVALUATE: This example illustrates a general principle about pV-diagrams of cyclic processes: The total work is positive if the process goes around the cycle in a clockwise direction, and the total work is negative if the process goes around the cycle in a counterclockwise direction (as in Fig. 19.13).

<table>
<tr><td>Example 19.4</td><td>**Comparing thermodynamic processes**</td></tr>
</table>

A series of thermodynamic processes is shown in the pV-diagram of Fig. 19.14. In process ab, 150 J of heat is added to the system, and in process bd, 600 J of heat is added. Find a) the internal energy change in process ab; b) the internal energy change in process abd (shown in light blue); and c) the total heat added in process acd (shown in dark blue).

SOLUTION

IDENTIFY and SET UP: In each process we use $\Delta U = Q - W$. We are given $Q_{ab} = +150$ J and $Q_{bd} = +600$ J (both values are positive because heat is *added* to the system). Our target variables are (a) ΔU_{ab}, (b) ΔU_{abd}, and (c) Q_{acd}.

EXECUTE: a) No volume change occurs during process ab, so $W_{ab} = 0$ and $\Delta U_{ab} = Q_{ab} = 150$ J.

b) Process bd occurs at constant pressure, so the work done by the system during this expansion is

$$W_{bd} = p(V_2 - V_1)$$
$$= (8.0 \times 10^4 \text{ Pa})(5.0 \times 10^{-3} \text{ m}^3 - 2.0 \times 10^{-3} \text{ m}^3)$$
$$= 240 \text{ J}$$

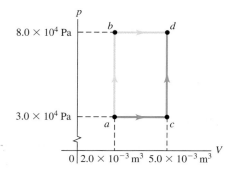

19.14 A pV-diagram showing the various thermodynamic processes.

The total work for process abd is

$$W_{abd} = W_{ab} + W_{bd} = 0 + 240 \text{ J} = 240 \text{ J}$$

and the total heat is

$$Q_{abd} = Q_{ab} + Q_{bd} = 150 \text{ J} + 600 \text{ J} = 750 \text{ J}$$

Applying Eq. (19.4) to abd, we find

$$\Delta U_{abd} = Q_{abd} - W_{abd} = 750 \text{ J} - 240 \text{ J} = 510 \text{ J}$$

c) Because ΔU is independent of path, the internal energy change is the same for path acd as for path abd; that is,

$$\Delta U_{acd} = \Delta U_{abd} = 510 \text{ J}$$

The total work for the path acd is

$$W_{acd} = W_{ac} + W_{cd} = p(V_2 - V_1) + 0$$
$$= (3.0 \times 10^4 \text{ Pa})(5.0 \times 10^{-3} \text{ m}^3 - 2.0 \times 10^{-3} \text{ m}^3)$$
$$= 90 \text{ J}$$

Now we apply Eq. (19.5) to process acd:

$$Q_{acd} = \Delta U_{acd} + W_{acd} = 510 \text{ J} + 90 \text{ J} = 600 \text{ J}$$

Here is a tabulation of the various quantities:

STEP	Q	W	$\Delta U = Q - W$	STEP	Q	W	$\Delta U = Q - W$
ab	150 J	0 J	150 J	ac	?	90 J	?
bd	600 J	240 J	360 J	cd	?	0 J	?
abd	750 J	240 J	510 J	acd	600 J	90 J	510 J

EVALUATE: We see that although ΔU is the same (510 J) for abd and acd, W (240 J versus 90 J) and Q (750 J versus 600 J) are quite different for the two processes.

Notice that we don't have enough information to find Q or ΔU for the processes ac and cd. We were nonetheless able to analyze the composite process acd by comparing it to the process abd, which has the same initial and final states and for which we have more complete information.

<table>
<tr><td>Example 19.5</td><td>**Thermodynamics of boiling water**</td></tr>
</table>

One gram of water (1 cm^3) becomes 1671 cm³ of steam when boiled at a constant pressure of 1 atm $(1.013 \times 10^5 \text{ Pa})$. The heat of vaporization at this pressure is $L_v = 2.256 \times 10^6$ J/kg. Compute a) the work done by the water when it vaporizes and b) its increase in internal energy.

SOLUTION

IDENTIFY and SET UP: The water is boiled at a constant pressure, so we can calculate the work W done by the water using Eq. (19.3). We can calculate the heat Q added to the water from the mass and the heat of vaporization, and can then find the internal energy change using $\Delta U = Q - W$.

EXECUTE: a) From Eq. (19.2), the work done by the vaporizing water is

$$W = p(V_2 - V_1)$$
$$= (1.013 \times 10^5 \, \text{Pa})(1671 \times 10^{-6} \, \text{m}^3 - 1 \times 10^{-6} \, \text{m}^3)$$
$$= 169 \, \text{J}$$

b) Using Eq. (17.20), the heat added to the water to vaporize it is

$$Q = mL_v = (10^{-3} \, \text{kg})(2.256 \times 10^6 \, \text{J/kg}) = 2256 \, \text{J}$$

From the first law of thermodynamics, Eq. (19.4), the change in internal energy is

$$\Delta U = Q - W = 2256 \, \text{J} - 169 \, \text{J} = 2087 \, \text{J}$$

EVALUATE: To vaporize one gram of water, we have to add 2256 J of heat. Most (2087 J) of this added energy remains in the system as an increase in internal energy. The remaining 169 J leaves the system again as it does work against the surroundings while expanding from liquid to vapor. The increase in internal energy is associated mostly with the intermolecular forces that hold the molecules together in the liquid state. These forces are attractive, so the associated potential energies are greater after work has been done to pull the molecules apart, forming the vapor state. It's like increasing gravitational potential energy by pulling an elevator farther from the center of the earth.

Infinitesimal Changes of State

In the preceding examples the initial and final states differ by a finite amount. Later we will consider *infinitesimal* changes of state in which a small amount of heat dQ is added to the system, the system does a small amount of work dW, and its internal energy changes by an amount dU. For such a process we state the first law in differential form as

$$dU = dQ - dW \tag{19.6}$$
$$\text{(first law of thermodynamics, infinitesimal process)}$$

For the systems we will discuss, the work dW is given by $dW = p \, dV$, so we can also state the first law as

$$dU = dQ - p \, dV \tag{19.7}$$

Test Your Understanding

Consider a thermodynamic process that goes directly from state a to state d in Example 19.4. In the pV-diagram of Fig. 19.14, this process is a straight line from a to d. What is the internal energy change for this process?

19.5 | Kinds of Thermodynamic Processes

In this section we describe four specific kinds of thermodynamic processes that occur often in practical situations. These can be summarized briefly as "no heat transfer" or *adiabatic,* "constant volume" or *isochoric,* "constant pressure" or *isobaric,* and "constant temperature" or *isothermal.* For some of these processes we can use a simplified form of the first law of thermodynamics.

Adiabatic Process

An **adiabatic process** (pronounced "ay-dee-ah-*bat*-ic") is defined as one with no heat transfer into or out of a system; $Q = 0$. We can prevent heat flow either by surrounding the system with thermally insulating material or by carrying out the process so quickly that there is not enough time for appreciable heat flow. From the first law we find that for every adiabatic process,

$$U_2 - U_1 = \Delta U = -W \qquad \text{(adiabatic process)} \qquad (19.8)$$

When a system expands adiabatically, W is positive (the system does work on its surroundings), so ΔU is negative and the internal energy decreases. When a system is *compressed* adiabatically, W is negative (work is done on the system by its surroundings) and U increases. In many (but not all) systems an increase of internal energy is accompanied by a rise in temperature.

The compression stroke in an internal-combustion engine is an approximately adiabatic process. The temperature rises as the air–fuel mixture in the cylinder is compressed. The expansion of the burned fuel during the power stroke is also an approximately adiabatic expansion with a drop in temperature. In Section 19.8 we'll consider adiabatic processes in an ideal gas.

Isochoric Process

An **isochoric process** (pronounced "eye-so-*kor*-ic") is a *constant-volume* process. When the volume of a thermodynamic system is constant, it does no work on its surroundings. Then $W = 0$ and

$$U_2 - U_1 = \Delta U = Q \qquad \text{(isochoric process)} \qquad (19.9)$$

In an isochoric process, all the energy added as heat remains in the system as an increase in internal energy. Heating a gas in a closed constant-volume container is an example of an isochoric process. The processes ab and cd in Example 19.4 are also examples of isochoric processes. (Note that there are types of work that do not involve a volume change. For example, we can do work on a fluid by stirring it. In some literature, "isochoric" is used to mean that no work of any kind is done.)

Isobaric Process

An **isobaric process** (pronounced "eye-so-*bear*-ic") is a *constant-pressure* process. In general, none of the three quantities ΔU, Q, and W is zero in an isobaric process, but calculating W is easy nonetheless. From Eq. (19.3),

$$W = p(V_2 - V_1) \qquad \text{(isobaric process)} \qquad (19.10)$$

Example 19.5 concerns an isobaric process, boiling water at constant pressure (Fig. 19.15).

Isothermal Process

An **isothermal process** is a *constant-temperature* process. For a process to be isothermal, any heat flow into or out of the system must occur slowly enough that thermal equilibrium is maintained. In general, none of the quantities ΔU, Q, or W is zero in an isothermal process.

In some special cases the internal energy of a system depends *only* on its temperature, not on its pressure or volume. The most familiar system having this special property is an ideal gas, as we'll discuss in the next section. For such systems, if the temperature is constant, the internal energy is also constant; $\Delta U = 0$ and $Q = W$. That is, any energy entering the system as heat Q must leave it again as work W done by the system. Example 19.1, involving an ideal gas, is an example

Act|v
Phys|cs
ONLINE

8.4   State Variables and Ideal Gas Law

19.15 Most cooking involves isobaric processes. That's because the air pressure above a saucepan or frying pan, or inside a microwave oven, remains essentially constant while the food is being heated. (But see Fig. 19.20 for an example of an *adiabatic* process in the kitchen.)

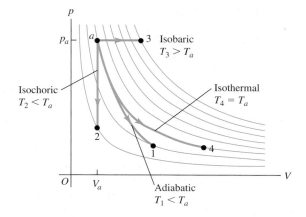

19.16 Four different processes for a constant amount of an ideal gas, all starting at state a. For the adiabatic process, $Q = 0$; for the isochoric process, $W = 0$; and for the isothermal process, $\Delta U = 0$. The temperature increases only during the isobaric expansion.

of an isothermal process in which U is also constant. For most systems other than ideal gases the internal energy depends on pressure as well as temperature, so U may vary even when T is constant.

Figure 19.16 shows a pV-diagram for each of these four processes for a constant amount of an ideal gas. The path followed in an adiabatic process (a to 1) is called an **adiabat.** A vertical line (constant volume) is an **isochor,** a horizontal line (constant pressure) is an **isobar,** and a curve of constant temperature (shown as light blue lines in Fig. 19.16) is an **isotherm.**

Test Your Understanding

Which of the processes in Fig. 19.8 are isochoric? Which are isobaric? Is it possible to tell if any of the processes are isothermal or adiabatic?

19.6 | Internal Energy of an Ideal Gas

We now show that for an ideal gas, the internal energy U depends only on temperature, not on pressure or volume. Let's think again about the free-expansion experiment described in Section 19.3. A thermally insulated container with rigid walls is divided into two compartments by a partition (Fig. 19.17). One compartment has a quantity of an ideal gas and the other is evacuated.

When the partition is removed or broken, the gas expands to fill both parts of the container. The gas does no work on its surroundings because the walls of the container don't move, and there is no heat flow through the insulation. So both Q and W are zero and the internal energy U is constant. This is true of any substance, whether it is an ideal gas or not.

Does the *temperature* change during a free expansion? Suppose it *does* change, while the internal energy stays the same. In that case we have to conclude that the internal energy depends on both the temperature and the volume or on both the temperature and the pressure, but certainly not on the temperature alone. But if T is constant during a free expansion, for which we know that U is constant even though both p and V change, then we have to conclude that U depends only on T, not on p or V.

Many experiments have shown that when a low-density gas undergoes a free expansion, its temperature *does not* change. Such a gas is essentially an ideal gas.

19.17 The partition is broken (or removed) to start the free expansion of gas into the vacuum region.

The conclusion is: **The internal energy of an ideal gas depends only on its temperature, not on its pressure or volume.** This property, in addition to the ideal-gas equation of state, is part of the ideal-gas model. Make sure that you understand that U depends only on T for an ideal gas, for we will make frequent use of this fact.

For non-ideal gases, some temperature change occurs during free expansions, even though the internal energy is constant. This shows that the internal energy cannot depend *only* on temperature; it must depend on pressure as well. From the microscopic viewpoint, in which internal energy U is the sum of the kinetic and potential energies for all the particles that make up the system, this is not surprising. Non-ideal gases usually have attractive intermolecular forces, and when molecules move farther apart, the associated potential energies increase. If the total internal energy is constant, the kinetic energies must decrease. Temperature is directly related to molecular *kinetic* energy, and for such a gas a free expansion is usually accompanied by a *drop* in temperature.

Test Your Understanding

Is the internal energy of a solid likely to be independent of its volume, as is the case for an ideal gas? Explain your reasoning. (*Hint:* See Fig. 18.17.)

19.7 | Heat Capacities of an Ideal Gas

We defined specific heat and molar heat capacity in Section 17.5. We also remarked at the end of that section that the specific heat or molar heat capacity of a substance depends on the conditions under which the heat is added. It is usually easiest to measure the heat capacity of a gas in a closed container under constant-volume conditions. The corresponding heat capacity is the **molar heat capacity at constant volume,** denoted by C_V. Heat capacity measurements for solids and liquids are usually carried out in the atmosphere under constant atmospheric pressure, and we call the corresponding heat capacity the **molar heat capacity at constant pressure,** C_p. If neither p nor V is constant, we have an infinite number of possible heat capacities.

Let's consider C_V and C_p for an ideal gas. To measure C_V, we raise the temperature of an ideal gas in a rigid container with constant volume (neglecting its thermal expansion). To measure C_p, we let the gas expand just enough to keep the pressure constant as the temperature rises.

Why should these two molar heat capacities be different? The answer lies in the first law of thermodynamics. In a constant-volume temperature increase, the system does no work, and the change in internal energy ΔU equals the heat added Q. In a constant-pressure temperature increase, on the other hand, the volume *must* increase; otherwise, the pressure (given by the ideal-gas equation of state $p = nRT/V$) could not remain constant. As the material expands, it does an amount of work W. According to the first law,

$$Q = \Delta U + W \tag{19.11}$$

For a given temperature increase, the internal energy change ΔU of an ideal gas has the same value no matter what the process (remember that the internal energy of an ideal gas depends only on temperature, not pressure or volume). Equation (19.11) then shows that the heat input for a constant-pressure process must be *greater* than that for a constant-volume process because additional energy must be supplied to account for the work W done during the expansion. So C_p is greater than C_V for an ideal gas. The pV-diagram in Fig. 19.18 shows this relationship. For air, C_p is 40% greater than C_V.

For a very few substances (one of which is water between 0°C and 4°C) the volume *decreases* during heating. In this case, W is negative, the heat input is *less* than in the constant-volume case, and C_p is *less* than C_V.

We can derive a simple relation between C_p and C_V for an ideal gas. First consider the constant-*volume* process. We place n moles of an ideal gas at temperature T in a constant-volume container. We place it in thermal contact with a hotter body; an infinitesimal quantity of heat dQ flows into the gas, and its temperature increases by an infinitesimal amount dT. By the definition of C_V, the molar heat capacity at constant volume,

$$dQ = nC_V \, dT \qquad (19.12)$$

The pressure increases during this process, but the gas does no work $(dW = 0)$ because the volume is constant. The first law in differential form, Eq. (19.6), is $dQ = dU + dW$. Since $dW = 0$, $dQ = dU$ and Eq. (19.12) can also be written as

$$dU = nC_V \, dT \qquad (19.13)$$

Now consider a constant-*pressure* process with the same temperature change dT. We place the same gas in a cylinder with a piston that we can allow to move just enough to maintain constant pressure. Again we bring the system into contact with a hotter body. As heat flows into the gas, it expands at constant pressure and does work. By the definition of C_p, the molar heat capacity at constant pressure, the amount of heat dQ entering the gas is

$$dQ = nC_p \, dT \qquad (19.14)$$

The work dW done by the gas in this constant-pressure process is

$$dW = p \, dV$$

We can also express dW in terms of the temperature change dT by using the ideal-gas equation of state, $pV = nRT$. Because p is constant, the change in V is proportional to the change in T:

$$dW = p \, dV = nR \, dT \qquad (19.15)$$

Now we substitute Eqs. (19.14) and (19.15) into the first law, $dQ = dU + dW$. We obtain

$$nC_p \, dT = dU + nR \, dT \qquad (19.16)$$

Now here comes the crux of the calculation. The internal energy change dU for the constant-pressure process is again given by Eq. (19.13), $dU = nC_V \, dT$, *even though now the volume is not constant.* Why is this so? Recall the discussion of Section 19.6; one of the special properties of an ideal gas is that its internal energy depends *only* on temperature. Thus the *change* in internal energy during any process must be determined only by the temperature change. If Eq. (19.13) is valid for an ideal gas for one particular kind of process, it must be valid for an ideal gas for *every* kind of process with the same dT. So we may replace dU in Eq. (19.16) by $nC_V \, dT$:

$$nC_p \, dT = nC_V \, dT + nR \, dT$$

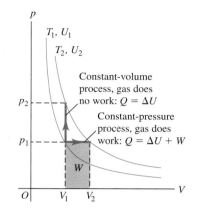

19.18 Raising the temperature of an ideal gas from T_1 to T_2 by a constant-volume or a constant-pressure process. For an ideal gas, U depends only on T, so ΔU is the same for both processes. But for the constant-pressure process, more heat Q must be added to both increase U and do work W. Hence $C_p > C_V$.

When we divide each term by the common factor $n\,dT$, we get

$$C_p = C_V + R \qquad \text{(molar heat capacities of an ideal gas)} \qquad (19.17)$$

As we predicted, the molar heat capacity of an ideal gas at constant pressure is *greater* than the molar heat capacity at constant volume; the difference is the gas constant R. (Of course, R must be expressed in the same units as C_p and C_V, such as J/mol·K.)

We have used the ideal-gas model to derive Eq. (19.17), but it turns out to be obeyed to within a few percent by many real gases at moderate pressures. Measured values of C_p and C_V are given in Table 19.1 for several real gases at low pressures; the difference in most cases is approximately $R = 8.314$ J/mol·K.

Table 19.1 Molar Heat Capacities of Gases at Low Pressure

Type of Gas	Gas	C_V (J/mol·K)	C_p (J/mol·K)	$C_p - C_V$ (J/mol·K)	$\gamma = C_p/C_V$
Monatomic	He	12.47	20.78	8.31	1.67
	Ar	12.47	20.78	8.31	1.67
Diatomic	H_2	20.42	28.74	8.32	1.41
	N_2	20.76	29.07	8.31	1.40
	O_2	20.85	29.17	8.31	1.40
	CO	20.85	29.16	8.31	1.40
Polyatomic	CO_2	28.46	36.94	8.48	1.30
	SO_2	31.39	40.37	8.98	1.29
	H_2S	25.95	34.60	8.65	1.33

The table also shows that the molar heat capacity of a gas is related to its molecular structure, as we discussed in Section 18.4. In fact, the first two columns of Table 19.1 are the same as Table 18.1.

The last column of Table 19.1 lists the values of the dimensionless **ratio of heat capacities,** C_p/C_V, denoted by the Greek letter γ ("gamma"):

$$\gamma = \frac{C_p}{C_V} \qquad \text{(ratio of heat capacities)} \qquad (19.18)$$

(This is sometimes called the "ratio of specific heats.") For gases, C_p is always greater than C_V and γ is always greater than unity. This quantity plays an important role in *adiabatic* processes for an ideal gas, which we will study in the next section.

We can use our kinetic-theory discussion of the molar heat capacity of an ideal gas (Section 18.4) to predict values of γ. As an example, an ideal monatomic gas has $C_V = \frac{3}{2}R$. From Eq. (19.17),

$$C_p = C_V + R = \frac{3}{2}R + R = \frac{5}{2}R$$

so

$$\gamma = \frac{C_p}{C_V} = \frac{\frac{5}{2}R}{\frac{3}{2}R} = \frac{5}{3} = 1.67$$

As Table 19.1 shows, this agrees well with values of γ computed from measured heat capacities. For most diatomic gases near room temperature, $C_V = \frac{5}{2}R$, $C_p = C_V + R = \frac{7}{2}R$, and

$$\gamma = \frac{C_p}{C_V} = \frac{\frac{7}{2}R}{\frac{5}{2}R} = \frac{7}{5} = 1.40$$

also in good agreement with measured values.

Here's a final reminder: For an ideal gas the internal energy change in *any* process is given by $\Delta U = nC_V\,\Delta T$, *whether the volume is constant or not.* This relation, which comes in handy in the following example, holds for other substances *only* when the volume is constant.

| Example 19.6 | **Cooling your room** |

A typical dorm room or bedroom contains about 2500 moles of air. Find the change in the internal energy of this much air when it is cooled from 23.9°C to 11.6°C at a constant pressure of 1.00 atm. Treat the air as an ideal gas with $\gamma = 1.400$.

SOLUTION

IDENTIFY and SET UP: This is a constant-pressure process. Your first impulse may be to find C_p and then calculate Q from $Q = nC_p\,\Delta T$; find the volume change and find the work done by the gas from $W = p\,\Delta V$; then finally use the first law to find ΔU. This would be perfectly correct, but there's a much easier way. For an ideal gas the internal-energy change is $\Delta U = nC_V\,\Delta T$ for *every* process, *whether the volume is constant or not.* So all we have to do is find C_V and use this expression for ΔU. We are given the number of moles and the initial and final temperatures.

EXECUTE: We are given the value of γ for air, so we use Eqs. (19.17) and (19.18) to determine C_V:

$$\gamma = \frac{C_p}{C_V} = \frac{C_V + R}{C_V} = 1 + \frac{R}{C_V}$$

$$C_V = \frac{R}{\gamma - 1} = \frac{8.314\ \text{J/mol}\cdot\text{K}}{1.400 - 1} = 20.79\ \text{J/mol}\cdot\text{K}$$

Then

$$\Delta U = nC_V\,\Delta T$$
$$= (2500\ \text{mol})(20.79\ \text{J/mol}\cdot\text{K})(11.6°\text{C} - 23.9°\text{C})$$
$$= -6.39 \times 10^5\ \text{J}$$

EVALUATE: A room air conditioner must extract this much internal energy from the air in your room and transfer it to the air outside. We'll discuss how this is done in Chapter 20.

Test Your Understanding

Would it be easier to cool your room (Example 19.6) if air were an ideal monatomic gas? If so, how much easier would it be?

19.8 | Adiabatic Processes for an Ideal Gas

An adiabatic process, defined in Section 19.5, is a process in which no heat transfer takes place between a system and its surroundings. Zero heat transfer is an idealization, but a process is approximately adiabatic if the system is well insulated or if the process takes place so quickly that there is not enough time for appreciable heat flow to occur.

In an adiabatic process, $Q = 0$, so from the first law, $\Delta U = -W$. An adiabatic process for an ideal gas is shown in the pV-diagram of Fig. 19.19. As the gas expands from volume V_a to V_b, it does positive work, so its internal energy decreases and its temperature drops. If point a, representing the initial state, lies on an isotherm at

8.11 Adiabatic Process

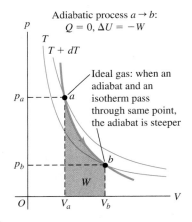

19.19 A pV-diagram of an adiabatic $(Q = 0)$ process for an ideal gas. As the gas expands from V_a to V_b, it does positive work W on its environment, its internal energy decreases $(\Delta U = -W < 0)$, and its temperature drops from $T + dT$ to T. (An adiabatic process is also shown in Fig. 19.16.)

temperature $T + dT$, then point b for the final state is on a different isotherm at a lower temperature T. For an ideal gas an adiabatic curve (adiabat) at any point is always *steeper* than the isotherm passing through the same point. For an adiabatic *compression* from V_b to V_a the situation is reversed and the temperature rises.

The air in the output pipes of air compressors used in gasoline stations, in paint-spraying equipment, and to fill scuba tanks is always warmer than the air entering the compressor; this is because the compression is rapid and hence approximately adiabatic. Adiabatic *cooling* occurs when you open a bottle of your favorite carbonated beverage. The gas just above the beverage surface expands rapidly in a nearly adiabatic process; the temperature of the gas drops so much that water vapor in the gas condenses, forming a miniature cloud.

CAUTION Keep in mind that when we talk about "adiabatic heating" and "adiabatic cooling," we really mean "raising the temperature" and "lowering the temperature," respectively. In an adiabatic process, the temperature change is due to work done by or on the system; there is *no* heat flow at all.

We can derive a relation between volume and temperature changes for an infinitesimal adiabatic process in an ideal gas. Equation (19.13) gives the internal energy change dU for *any* process for an ideal gas, adiabatic or not, so we have $dU = nC_V\,dT$. Also, the work done by the gas during the process is given by $dW = p\,dV$. Then, since $dU = -dW$ for an adiabatic process, we have

$$nC_V\,dT = -p\,dV \qquad (19.19)$$

To obtain a relation containing only the volume V and temperature T, we eliminate p using the ideal-gas equation in the form $p = nRT/V$. Substituting this into Eq. (19.19) and rearranging, we get

$$nC_V\,dT = -\frac{nRT}{V}\,dV$$

$$\frac{dT}{T} + \frac{R}{C_V}\frac{dV}{V} = 0$$

The coefficient R/C_V can be expressed in terms of $\gamma = C_p/C_V$. We have

$$\frac{R}{C_V} = \frac{C_p - C_V}{C_V} = \frac{C_p}{C_V} - 1 = \gamma - 1$$

$$\frac{dT}{T} + (\gamma - 1)\frac{dV}{V} = 0 \qquad (19.20)$$

Because γ is always greater than unity for a gas, $(\gamma - 1)$ is always positive. This means that in Eq. (19.20), dV and dT always have opposite signs. An adiabatic *expansion* of an ideal gas $(dV > 0)$ always occurs with a *drop* in temperature $(dT < 0)$, and an adiabatic *compression* $(dV < 0)$ always occurs with a *rise* in temperature $(dT > 0)$; this confirms our earlier prediction (Fig. 19.20).

For finite changes in temperature and volume we integrate Eq. (19.20), obtaining

$$\ln T + (\gamma - 1)\ln V = \text{constant}$$

$$\ln T + \ln V^{\gamma - 1} = \text{constant}$$

$$\ln(TV^{\gamma - 1}) = \text{constant}$$

19.20 Hot steam escapes from the top of this pressure cooker at high speed. Hence it has no time to exchange heat with its surroundings, and its expansion is nearly adiabatic. As the steam's volume increases, its temperature drops so much [see Eq. (19.22)] that it feels cool on this chef's hand.

and finally,

$$TV^{\gamma-1} = \text{constant} \qquad (19.21)$$

Thus for an initial state (T_1, V_1) and a final state (T_2, V_2),

$$T_1 V_1^{\gamma-1} = T_2 V_2^{\gamma-1} \qquad \text{(adiabatic process, ideal gas)} \qquad (19.22)$$

Because we have used the ideal-gas equation in our derivation of Eqs. (19.21) and (19.22), the T's must always be *absolute* (Kelvin) temperatures.

We can also convert Eq. (19.21) into a relation between pressure and volume by eliminating T, using the ideal-gas equation in the form $T = pV/nR$. Substituting this into Eq. (19.21), we find

$$\frac{pV}{nR} V^{\gamma-1} = \text{constant}$$

or, because n and R are constant,

$$pV^{\gamma} = \text{constant} \qquad (19.23)$$

For an initial state (p_1, V_1) and a final state (p_2, V_2), Eq. (19.23) becomes

$$p_1 V_1^{\gamma} = p_2 V_2^{\gamma} \qquad \text{(adiabatic process, ideal gas)} \qquad (19.24)$$

We can also calculate the *work* done by an ideal gas during an adiabatic process. We know that $Q = 0$ and $W = -\Delta U$ for *any* adiabatic process. For an ideal gas, $\Delta U = nC_V(T_2 - T_1)$. If the number of moles n and the initial and final temperatures T_1 and T_2 are known, we have simply

$$W = nC_V(T_1 - T_2) \qquad \text{(adiabatic process, ideal gas)} \qquad (19.25)$$

We may also use $pV = nRT$ in this equation to obtain

$$W = \frac{C_V}{R}(p_1 V_1 - p_2 V_2) = \frac{1}{\gamma - 1}(p_1 V_1 - p_2 V_2) \qquad (19.26)$$
$$\text{(adiabatic process, ideal gas)}$$

(We used the result $C_V = R/(\gamma - 1)$ from Example 19.6.) If the process is an expansion, the temperature drops, T_1 is greater than T_2, $p_1 V_1$ is greater than $p_2 V_2$, and the work is *positive*, as we should expect. If the process is a compression, the work is negative.

Throughout this analysis of adiabatic processes we have used the ideal-gas equation of state, which is valid only for *equilibrium* states. Strictly speaking, our results are valid only for a process that is fast enough to prevent appreciable heat exchange with the surroundings (so that $Q = 0$ and the process is adiabatic), yet slow enough that the system does not depart very much from thermal and mechanical equilibrium. Even when these conditions are not strictly satisfied, though, Eqs. (19.22), (19.24), and (19.26) give useful approximate results.

| Example 19.7 | **Adiabatic compression in a diesel engine** |

The compression ratio of a diesel engine is 15 to 1; this means that air in the cylinders is compressed to $\frac{1}{15}$ of its initial volume (Fig. 19.21). If the initial pressure is 1.01×10^5 Pa and the initial temperature is 27°C (300 K), find the final pressure and the temperature after compression. Air is mostly a mixture of diatomic oxygen and nitrogen; treat it as an ideal gas with $\gamma = 1.40$.

SOLUTION

IDENTIFY and SET UP: We are given the initial pressure $p_1 = 1.01 \times 10^5$ Pa and the initial temperature $T_1 = 300$ K, and we are told that the ratio of initial and final volumes is $V_1/V_2 = 15$. Since the gas is ideal and the process is adiabatic, we can find the final temperature T_2 using Eq. (19.22) and the final pressure p_2 using Eq. (19.24).

EXECUTE: From Eq. (19.22),

$$T_2 = T_1\left(\frac{V_1}{V_2}\right)^{\gamma-1} = (300\ \text{K})(15)^{0.40} = 886\ \text{K} = 613°\text{C}$$

From Eq. (19.24),

$$p_2 = p_1\left(\frac{V_1}{V_2}\right)^{\gamma} = (1.01 \times 10^5\ \text{Pa})(15)^{1.40}$$
$$= 44.8 \times 10^5\ \text{Pa} = 44\ \text{atm}$$

EVALUATE: If the compression had been isothermal, the final pressure would have been 15 atm, but because the temperature also increases during an adiabatic compression, the final pressure is much greater. When fuel is injected into the cylinders near the end of the compression stroke, the high temperature of the air attained during compression causes the fuel to ignite spontaneously without the need for spark plugs.

19.21 Adiabatic compression of air in a cylinder of a diesel engine.

| Example 19.8 | **Work done in an adiabatic process** |

In Example 19.7, how much work does the gas do during the compression if the initial volume of the cylinder is $1.00\ \text{L} = 1.00 \times 10^{-3}\ \text{m}^3$? Assume that C_V for air is $20.8\ \text{J/mol·K}$ and $\gamma = 1.40$.

SOLUTION

IDENTIFY and SET UP: We use Eq. (19.25) to determine the work done. We are given the initial temperature T_1 and we calculated the final temperature T_2 in Example 19.7. We are not given the number of moles n, but we can calculate it from the given information using the ideal-gas law $pV = nRT$.

EXECUTE: The number of moles is

$$n = \frac{p_1 V_1}{RT_1} = \frac{(1.01 \times 10^5\ \text{Pa})(1.00 \times 10^{-3}\ \text{m}^3)}{(8.314\ \text{J/mol·K})(300\ \text{K})}$$
$$= 0.0405\ \text{mol}$$

and Eq. (19.25) gives

$$W = nC_V(T_1 - T_2)$$
$$= (0.0405\ \text{mol})(20.8\ \text{J/mol·K})(300\ \text{K} - 886\ \text{K})$$
$$= -494\ \text{J}$$

EVALUATE: We can check our result using Eq. (19.26), the alternative expression for work done by an ideal gas in an adiabatic process:

$$W = \frac{1}{\gamma - 1}(p_1 V_1 - p_2 V_2)$$
$$= \frac{1}{1.40 - 1}\left[(1.01 \times 10^5\ \text{Pa})(1.00 \times 10^{-3}\ \text{m}^3) - (44.8 \times 10^5\ \text{Pa})\left(\frac{1.00 \times 10^{-3}\ \text{m}^3}{15}\right)\right]$$
$$= -494\ \text{J}$$

The work is negative because the gas is compressed.

Test Your Understanding

If you compress a certain quantity of a monatomic ideal gas to one-half of its initial volume, by what factor does the pressure change if the compression is done isothermally? If it is done adiabatically? Would these answers be different if the gas were diatomic?

SUMMARY

A thermodynamic system has the potential to exchange energy with its surroundings by heat transfer or by mechanical work. When a system at pressure p changes volume from V_1 to V_2, it does an amount of work W given by the integral of p with respect to volume. If the pressure is constant, the work done is equal to p times the change in volume. A negative value of W means that work is done on the system. (See Example 19.1)

$$W = \int_{V_1}^{V_2} p\, dV \qquad (19.2)$$

$$W = p(V_2 - V_1) \qquad (19.3)$$
(constant pressure only)

In any thermodynamic process, the heat added to the system and the work done by the system depend not only on the initial and final states, but also on the path (the series of intermediate states through which the system passes).

The first law of thermodynamics states that when heat Q is added to a system while it does work W, the internal energy U changes by an amount equal to $Q - W$. This law can also be expressed for an infinitesimal process. (See Examples 19.2, 19.3, and 19.5)

$$\Delta U = Q - W \qquad (19.4)$$

$$dU = dQ - dW \qquad (19.6)$$
(infinitesimal process)

The internal energy of any thermodynamic system depends only on its state. The change in internal energy in any process depends only on the initial and final states, not on the path. The internal energy of an isolated system is constant. (See Example 19.4)

Some important types of thermodynamic processes:

- Adiabatic process: No heat transfer in or out of a system; $Q = 0$.
- Isochoric process: Constant volume; $W = 0$.
- Isobaric process: Constant pressure; $W = p(V_2 - V_1)$.
- Isothermal process: Constant temperature.

The internal energy of an ideal gas depends only on its temperature, not its pressure or volume. For other substances the internal energy generally depends on both pressure and temperature.

The molar heat capacities C_V and C_p of an ideal gas differ by R, the ideal gas constant. The dimensionless ratio of heat capacities, C_p/C_V, is denoted by γ. (See Example 19.6)	$C_p = C_V + R$ (19.17) $\gamma = \dfrac{C_p}{C_V}$ (19.18)
For an adiabatic process for an ideal gas, the quantities $TV^{\gamma-1}$ and pV^{γ} are constant. The work done by an ideal gas during an adiabatic expansion can be expressed in terms of the initial and final values of temperature, or in terms of the initial and final values of pressure and volume. (See Examples 19.7 and 19.8)	$\begin{aligned} W &= nC_V(T_1 - T_2) \\ &= \dfrac{C_V}{R}(p_1 V_1 - p_2 V_2) \\ &= \dfrac{1}{\gamma - 1}(p_1 V_1 - p_2 V_2) \end{aligned}$ (19.25), (19.26)

Key Terms

thermodynamic system, 724
thermodynamic process, 724
path, 728
free expansion, 729
internal energy, 729
first law of thermodynamics, 730
adiabatic process, 735

isochoric process, 736
isobaric process, 736
isothermal process, 736
adiabat, 737
isochor, 737
isobar, 737

isotherm, 737
molar heat capacity at constant volume, 738
molar heat capacity at constant pressure, 738
ratio of heat capacities, 740

Your Notes

Answer to Chapter Opening Question

No. The work done by a gas as its volume changes from V_1 to V_2 is equal to the integral $\int p\, dV$ between those two volume limits. If the volume of the gas contracts, the final volume V_2 is less than the initial volume V_1 and the gas does negative work. Propelling the locomotive requires that the gas do positive work, so the gas doesn't contribute to propulsion while contracting.

Answers to Test Your Understanding Questions

Section 19.1 Heat flows out of the coffee, so $Q_{coffee} < 0$; heat flows into the aluminum cup, so $Q_{aluminum} > 0$. In mechanics, we would say that negative work is done *on* the block, since the surface exerts a force on the block that opposes the block's motion. But in thermodynamics we use the opposite convention and say that $W > 0$, which means that positive work is done *by* the block on the surface.

Section 19.2 We use the expression $W = nRT \ln(V_2/V_1)$. If $W = nRT$, then $\ln(V_2/V_1) = 1$ and $V_2/V_1 = e = 2.718$. If $W = 2nRT$, then $\ln(V_2/V_1) = 2$ and $V_2/V_1 = e^2 = 7.389$. Note that the expansion ratio V_2/V_1 must more than double in order to double the amount of work done.

Section 19.3 The situation in Fig. 19.9a is an isothermal expansion, so from Example 19.1 the gas does work $W = nRT \ln(V_2/V_1) = (1.00\ \text{mol})\ (8.314\ \text{J/mol} \cdot \text{K})\ (300\ \text{K})$ $\ln[(5.0\ \text{L}) / (2.0\ \text{L})] = 2.3 \times 10^3$ J. The gas does zero work in the free expansion shown in Fig. 19.9b.

Section 19.4 The internal energy change in a thermodynamic process depends only on the initial and final states, not on the particular path taken. Hence the internal energy change for the process ad is the same as for either of the processes abd or acd in Example 19.4: $\Delta U_{ad} = 510$ J.

Section 19.5 In a pV-diagram like those shown in Fig. 19.8, isochoric processes are represented by vertical lines (lines of constant volume) and isobaric processes are represented by horizontal lines (lines of constant pressure). Hence the processes $1 \rightarrow 4$ and $3 \rightarrow 2$ are isochoric and the processes $1 \rightarrow 3$ and $4 \rightarrow 2$ are isobaric. The process $1 \rightarrow 2$ in Fig. 19.8 is shown as a curved line, which superficially resembles the adiabatic and isothermal processes for an ideal gas in Fig. 19.16. Without more information we can't tell whether process $1 \rightarrow 2$ is isothermal, adiabatic, or neither.

Section 19.6 Using the model of a solid in Fig. 18.17, we can see that the internal energy of a solid *does* depend on its volume. Compressing the solid means compressing the "springs" between the atoms, thereby increasing their stored potential energy and hence the internal energy of the solid.

Section 19.7 The molar heat capacity of an ideal monatomic gas at constant volume is $C_V = \frac{3}{2}R = 12.47$ J/mol·K, which is $\frac{3}{5}$ as great as the actual value of C_V for air calculated in Example 19.6. Hence, for the same number of moles n and temperature change ΔT, the amount of energy that would have to be extracted would be only $\frac{3}{5}$ as much.

Section 19.8 In an isothermal process, $pV = $ constant. If the volume decreases to one-half of its initial value, the pressure must double. By contrast, in an adiabatic process $pV^\gamma = $ constant and the pressure increases by a factor of $2^\gamma = 2^{5/3} = 3.17$. (Because the final pressure is greater in the adiabatic case, more work is done in the adiabatic compression.) If the gas were diatomic, the isothermal result would be unaffected (the ideal-gas law doesn't change), but the adiabatic result would be different due to the different value of γ. Can you show that if $\gamma = 1.40$ for the diatomic gas, the pressure would increase by a factor of 2.64?

Discussion Questions

Q19.1 It is not correct to say that a body contains a certain amount of heat, yet a body can transfer heat to another body. How can a body give away something it does not have in the first place?

Q19.2 If you are told the initial and final states of a system and the associated change in internal energy, can you determine whether the internal energy change was due to work or to heat transfer? Explain.

Q19.3 Discuss the application of the first law of thermodynamics to a mountaineer who eats food, gets warm and perspires a lot during a climb, and does a lot of mechanical work in raising herself to the summit. What about the descent? The mountaineer also gets warm during the descent. Is the source of this energy the same as during the ascent?

Q19.4 When ice melts at 0°C, its volume decreases. Is the internal energy change greater than, less than, or equal to the heat added? How can you tell?

Q19.5 Example 19.5 (Section 19.4) describes a process that occurs at a constant temperature of 100°C. Why, then, can't we use in this example the expressions for W derived for an isothermal process in Example 19.1 (Section 19.2)?

Q19.6 Translated literally from the Greek, "isothermal" means "same heat." Give as many reasons as you can why such a literal translation is misleading.

Q19.7 Imagine a gas made up entirely of negatively charged electrons. Like charges repel, so the electrons exert repulsive forces on each other. Would you expect that the temperature of such a gas would rise, fall, or stay the same in a free expansion? Why?

Q19.8 There are a few materials that contract when their temperature is increased, such as water between 0°C and 4°C. Would you expect C_p for such materials to be greater or less than C_V? Why or why not?

Q19.9 In a constant-volume process, $dU = nC_V\, dT$. But in a constant-pressure process, it is *not* true that $dU = nC_p\, dT$. Why not?

Q19.10 When you blow on the back of your hand with your mouth wide open, your breath feels warm. But if you partially close your mouth to form an "o" and then blow on your hand, your breath feels cool. Why?

Q19.11 On a warm summer day, a large cylinder of compressed gas (propane or butane) was used to supply several large gas burners at a cookout. After a while, frost formed on the outside of the tank. Why?

Q19.12 In the carburetor of an aircraft or automobile engine, air flows through a relatively small aperture, then expands. In cool, foggy weather, ice sometimes forms in this aperture even though the outside air temperature is above freezing. Why?

Q19.13 For each of the following equations, does the equation apply only to isothermal, or adiabatic, or isobaric, or isochoric processes, or does it apply to all processes: $p_1V_1^\gamma = p_2V_2^\gamma$; $Q = nC_p\Delta T$; $Q = nC_V\Delta T$; $\Delta U = nC_V\Delta T$; $Q = \Delta U + W$; $W = p\Delta V$?

Q19.14 When you use a hand pump to inflate the tires of your bicycle, the pump gets warm after a while. Why? What happens to the temperature of the air in the pump as you compress it? Why does this happen? When you raise the pump handle to draw outside air into the pump, what happens to the temperature of the air taken in? Again, why does this happen?

Q19.15 The prevailing winds on the Hawaiian island of Kauai blow from the northeast. The winds cool as they go up the slope of Mt. Waialeale (elevation 1523 m), causing water vapor to condense and rain to fall. There is much more precipitation at the summit than at the base of the mountain. In fact, Mt. Waialeale is the rainiest spot on earth, averaging 11.7 m of rainfall a year. But what makes the winds cool?

Q19.16 Applying the same considerations as in Question 19.15, explain why the island of Niihau, a few kilometers to the southwest of Kauai, is almost a desert, and farms there need to be irrigated.

Q19.17 When a gas is compressed adiabatically into the surrounding air, its temperature rises even though there is no heat input to the gas. Where does the energy come from to raise the temperature?

Q19.18 When a gas expands adiabatically, it does work on its surroundings. But if there is no heat input to the gas, where does the energy come from to do the work?

Exercises

Section 19.2 Work Done During Volume Changes, and Section 19.3 Paths Between Thermodynamic States

19.1 Two moles of an ideal gas are heated at constant pressure from $T = 27°C$ to $107°C$. a) Draw a pV-diagram for this process. b) Calculate the work done by the gas.

19.2 Three moles of an ideal gas have an initial temperature of $127.0°C$. While the temperature is kept constant, the volume is increased until the pressure drops to 40.0% of its original value. a) Draw a pV-diagram for this process. b) Calculate the work done by the gas.

19.3 Five moles of an ideal gas are kept at a constant temperature of $53.0°C$ while the pressure of the gas is increased from 1.00 atm to 3.00 atm. a) Draw a pV-diagram for this process. b) Calculate the work done by the gas.

19.4 Six moles of an ideal gas are in a cylinder fitted at one end with a movable piston. The initial temperature of the gas is $27.0°C$ and the pressure is constant. As part of a machine design project, calculate the final temperature of the gas after it has done 1.75×10^3 J of work.

19.5 A metal cylinder with rigid walls contains 2.50 mol of oxygen gas. The gas is cooled until the pressure decreases to 30.0% of its original value. You can ignore the thermal contraction of the cylinder. a) Draw a pV-diagram for this process. b) Calculate the work done by the gas.

19.6 A gas under a constant pressure of 1.50×10^5 Pa and with an initial volume of 0.0900 m³ is cooled until its volume becomes 0.0600 m³. a) Draw a pV-diagram for this process. b) Calculate the work done by the gas.

19.7 A gas undergoes two processes. In the first, the volume remains constant at 0.200 m³ and the pressure increases from 2.00×10^5 Pa to 5.00×10^5 Pa. The second process is a compression to a volume of 0.120 m³ at a constant pressure of 5.00×10^5 Pa. a) In a pV-diagram, show both processes. b) Find the total work done by the gas during both processes.

19.8 Work Done in a Cyclic Process. a) In Fig. 19.8a, consider the closed loop $1 \rightarrow 3 \rightarrow 2 \rightarrow 4 \rightarrow 1$. This is a *cyclic* process in which the initial and final states are the same. Find the total work done by the system in this cyclic process, and show that it is equal to the area enclosed by the loop. b) How is the work done for the process in part (a) related to the work done if the loop is traversed in the opposite direction, $1 \rightarrow 4 \rightarrow 2 \rightarrow 3 \rightarrow 1$? Explain.

Section 19.4 Internal Energy and the First Law of Thermodynamics

19.9 In a certain chemical process, a lab technician supplies 254 J of heat to a system. At the same time, 73 J of work are done on the system by its surroundings. What is the increase in the internal energy of the system?

19.10 A gas in a cylinder expands from a volume of 0.110 m³ to 0.320 m³. Heat flows into the gas just rapidly enough to keep the pressure constant at 1.80×10^5 Pa during the expansion. The total heat added is 1.15×10^5 J. a) Find the work done by the gas. b) Find the change in internal energy of the gas. c) Does it matter whether or not the gas is ideal? Why or why not?

19.11 Five moles of an ideal monatomic gas with an initial temperature of $127°C$ expand and, in the process, absorb 1200 J of heat and do 2100 J of work. What is the final temperature of the gas?

19.12 The pressure of a system is decreased while the volume is kept constant. If heat flows into the system during this process, does the internal energy of the system increase or decrease? Explain your reasoning.

19.13 A gas in a cylinder is held at a constant pressure of 2.30×10^5 Pa and is cooled and compressed from 1.70 m³ to 1.20 m³. The internal energy of the gas decreases by 1.40×10^5 J. a) Find the work done by the gas. b) Find the absolute value $|Q|$ of the heat flow into or out of the gas, and state the direction of the heat flow. c) Does it matter whether or not the gas is ideal? Why or why not?

19.14 A system is taken from state a to state b along the three paths shown in Fig. 19.22. a) Along which path is the work done by the system the greatest? The least? b) If $U_b > U_a$, along which path is the absolute value $|Q|$ of the heat transfer the greatest? For this path, is heat absorbed or liberated by the system?

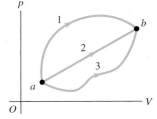

Figure 19.22 Exercise 19.14.

19.15 Doughnuts: Breakfast of Champions! A typical doughnut contains 2.0 g of protein, 17.0 g of carbohydrates, and 7.0 g of fat. The average food-energy values of these substances are

4.0 kcal/g for protein and carbohydrates and 9.0 kcal/g for fat. a) During heavy exercise, an average person uses energy at a rate of 510 kcal/h. How long would you have to exercise to "work off" one doughnut? b) If the energy in the doughnut could somehow be converted into the kinetic energy of your body as a whole, how fast could you move after eating the doughnut? Take your mass to be 60 kg, and express your answer in m/s and km/h.

19.16 A liquid is irregularly stirred in a well-insulated container and thereby undergoes a rise in temperature. Regard the liquid as the system. a) Has heat been transferred? How can you tell? b) Has work been done? How can you tell? Why is it important that the stirring is irregular? c) What is the sign of ΔU? How can you tell?

19.17 A system is taken around the cycle shown in Fig. 19.23, from state a to state b and then back to state a. The absolute value of the heat transfer during one cycle is 7200 J. a) Does the system absorb or liberate heat when it goes around the cycle in the direction shown in the figure? How can you tell? b) What is the work W done by the system in one cycle? c) If the system goes around the cycle in a counterclockwise direction, does it absorb or liberate heat in one cycle? What is the magnitude of the heat absorbed or liberated in one counterclockwise cycle?

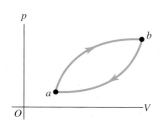

Figure 19.23 Exercise 19.17.

19.18 A thermodynamic system undergoes a cyclic process as shown in Fig. 19.24. The cycle consists of two closed loops, loop I and loop II. a) Over one complete cycle, does the system do positive or negative work? b) In each of loops I and II, is the net work done by the system positive or negative? c) Over one complete cycle, does heat flow into or out of the system? d) In each of loops I and II, does heat flow into or out of the system?

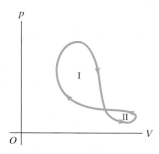

Figure 19.24 Exercise 19.18.

19.19 A student performs a combustion experiment by burning a mixture of fuel and oxygen in a constant-volume metal can surrounded by a water bath. During the experiment the temperature of the water is observed to rise. Regard the mixture of fuel and oxygen as the system. a) Has heat been transferred? How can you tell? b) Has work been done? How can you tell? c) What is the sign of ΔU? How can you tell?

19.20 Boiling Water at High Pressure. When water is boiled under a pressure of 2.00 atm, the heat of vaporization is 2.20×10^6 J/kg and the boiling point is 120°C. At this pressure, 1.00 kg of water has a volume of 1.00×10^{-3} m³, and 1.00 kg of steam has a volume of 0.824 m³. a) Compute the work done when 1.00 kg of steam is formed at this temperature. b) Compute the increase in internal energy of the water.

Section 19.5 Kinds of Thermodynamic Processes,
Section 19.6 Internal Energy of an Ideal Gas, and
Section 19.7 Heat Capacities of an Ideal Gas

19.21 In an experiment to simulate conditions within an automobile engine, 645 J of heat is transferred to 0.185 mol of air contained within a cylinder of volume 40.0 cm³. Initially the air is at a pressure of 3.00×10^6 Pa and a temperature of 780 K. a) If the volume of the cylinder is held fixed, what is the final temperature of the air? Assume that the air is essentially all nitrogen gas, and use the data in Table 19.1 even though the pressure is not low. Draw a pV-diagram for this process. b) Find the final temperature of the air if the volume of the cylinder is allowed to increase while the pressure remains constant. Draw a pV-diagram for this process.

19.22 A cylinder contains 0.0100 mol of helium at $T = 27.0$°C. a) How much heat is needed to raise the temperature to 67.0°C while keeping the volume constant? Draw a pV-diagram for this process. b) If instead the pressure of the helium is kept constant, how much heat is needed to raise the temperature from 27.0°C to 67.0°C? Draw a pV-diagram for this process. c) What accounts for the difference between your answers to parts (a) and (b)? In which case is more heat required? What becomes of the additional heat? d) If the gas is ideal, what is the change in its internal energy in part (a)? In part (b)? How do the two answers compare? Why?

19.23 Five moles of gas have their temperature increased from -10.0°C to $+20.0$°C. Calculate the heat that must be transferred into the gas if the gas is a) He at a constant pressure of 1.5 atm; b) Ar at a fixed volume of 8.2 m³; c) CO_2 at a constant pressure of 20,000 Pa.

19.24 When a quantity of monatomic ideal gas expands at a constant pressure of 4.00×10^4 Pa, the volume of the gas increases from 2.00×10^{-3} m³ to 8.00×10^{-3} m³. What is the change in the internal energy of the gas?

19.25 An ideal gas expands while the pressure is kept constant. During this process, does heat flow into the gas or out of the gas? Justify your answer.

19.26 Heat Q flows into a monatomic ideal gas, and the volume increases while the pressure is kept constant. What fraction of the heat energy is used to do the expansion work of the gas?

19.27 The temperature of 0.150 mol of an ideal gas is held constant at 77.0°C while its volume is reduced to 25.0% of its initial volume. The initial pressure of the gas is 1.25 atm. a) Determine the work done by the gas. b) What is the change in its internal energy? c) Does the gas exchange heat with its surroundings? If so, how much? Does the gas absorb or liberate heat?

19.28 During an isothermal compression of an ideal gas, 335 J of heat must be removed from the gas to maintain constant temperature. How much work is done by the gas during the process?

19.29 Propane gas (C_3H_8) behaves like an ideal gas with $\gamma = 1.127$. Determine the molar heat capacity at constant volume and the molar heat capacity at constant pressure.

19.30 A cylinder contains 0.250 mol of carbon dioxide (CO_2) gas at a temperature of 27.0°C. The cylinder is provided with a frictionless piston, which maintains a constant pressure of 1.00 atm on the gas. The gas is heated until its temperature increases to 127.0°C. Assume that the CO_2 may be treated as an ideal gas. a) Draw a *pV*-diagram for this process. b) How much work is done by the gas in this process? c) On what is this work done? d) What is the change in internal energy of the gas? e) How much heat was supplied to the gas? f) How much work would have been done if the pressure had been 0.50 atm?

19.31 Gaseous ethane (C_2H_6) has $\gamma = 1.220$ and may be treated as an ideal gas. a) If 2.40 mol of ethane is to be heated from 20.0°C to 25.0°C at a constant pressure of 1.00 atm, how much heat is required? b) What will be the change in internal energy of the ethane?

Section 19.8 Adiabatic Processes for an Ideal Gas

19.32 A monatomic ideal gas that is initially at a pressure of 1.50×10^5 Pa and with a volume of 0.0800 m³ is compressed adiabatically to a volume of 0.0400 m³. a) What is the final pressure? b) How much work is done by the gas? c) What is the ratio of the final temperature of the gas to its initial temperature? Is the gas heated or cooled by this compression?

19.33 The engine of a Ferrari F355 F1 sports car takes in air at 20.0°C and 1.00 atm and compresses it adiabatically to 0.0900 times the original volume. The air may be treated as an ideal gas with $\gamma = 1.40$. a) Draw a *pV*-diagram for this process. b) Find the final temperature and pressure.

19.34 Two moles of carbon monoxide (CO) start at a pressure of 1.2 atm and a volume of 30 liters. The gas is then compressed adiabatically to $\frac{1}{3}$ this volume. Assume that the gas may be treated as ideal. What is the change in the internal energy of the gas? Does the internal energy increase or decrease? Does the temperature of the gas increase or decrease during this process? Explain.

19.35 In an adiabatic process for an ideal gas the pressure decreases. In this process does the internal energy of the gas increase or decrease? Explain your reasoning.

19.36 An ideal gas initially at 4.00 atm and 350 K is permitted to expand adiabatically to 1.50 times its initial volume. Find the final pressure and temperature if the gas is a) monatomic; b) diatomic with $C_V = \frac{5}{2}R$.

19.37 During an adiabatic expansion the temperature of 0.450 mol of argon (Ar) drops from 50.0°C to 10.0°C. The argon may be treated as an ideal gas. a) Draw a *pV*-diagram for this process. b) How much work does the gas do? c) What is the change in internal energy of the gas?

19.38 A cylinder contains 0.100 mol of an ideal monatomic gas. Initially the gas is at a pressure of 1.00×10^5 Pa and occupies a volume of 2.50×10^{-3} m³. a) Find the initial temperature of the gas in kelvins. b) If the gas is allowed to expand to twice the initial volume, find the final temperature (in kelvins) and pressure of the gas if the expansion is i) isothermal; ii) isobaric; iii) adiabatic.

19.39 A quantity of sulfur dioxide gas (SO_2) occupies a volume of 5.00×10^{-3} m³ at a pressure of 1.10×10^5 Pa. The gas expands adiabatically to a volume of 1.00×10^{-2} m³. Assume that the gas may be treated as ideal. a) Find the final pressure of the gas. (*Hint:* See Table 19.1.) b) How much work does the gas do on its surroundings? c) What is the ratio of the final temperature of the gas to its initial temperature?

Problems

19.40 A quantity of air is taken from state *a* to state *b* along a path that is a straight line in the *pV*-diagram (Fig. 19.25). a) In this process, does the temperature of the gas increase, decrease, or stay the same? Explain. b) If $V_a = 0.0700$ m³, $V_b = 0.1100$ m³, $p_a = 1.00 \times 10^5$ Pa, and $p_b = 1.40 \times 10^5$ Pa, what is the work *W* done by the gas in this process? Assume that the gas may be treated as ideal.

Figure 19.25 Problem 19.40.

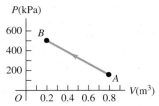
Figure 19.26 Problem 19.41.

19.41 Four moles of O_2 are taken from A to B by the process shown in the *pV*-diagram in Fig. 19.26. Assume the gas may be treated as ideal. What is the heat flow *Q* during this process? Does heat enter or leave the gas?

19.42 a) One-third of a mole of He gas is taken along the path *abc* that is shown as the solid line in Fig. 19.27. Assume the gas may be treated as ideal. How much heat is transferred into or out of the gas? b) If the gas instead went from state *a* to state *c* along the horizontal dashed line in Fig. 19.27, how much heat would be transferred into or out of the gas? c) How does *Q* in part (b) compare to *Q* in part (a)? Explain.

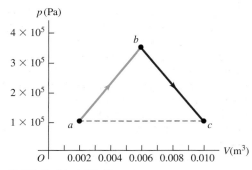
Figure 19.27 Problem 19.42.

19.43 When a system is taken from state *a* to state *b* in Fig. 19.28 along the path *acb*, 90.0 J of heat flows into the system and 60.0 J of work is done by the system. a) How much heat flows into the system along path *adb* if the work done by the system is 15.0 J? b) When the system is returned from *b* to *a* along the curved path,

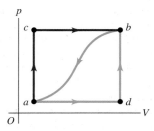
Figure 19.28 Problem 19.43.

the absolute value of the work done by the system is 35.0 J. Does the system absorb or liberate heat? How much heat? c) If $U_a = 0$ and $U_d = 8.0$ J, find the heat absorbed in the processes ad and db.

19.44 A thermodynamic system is taken from state a to state c in Fig. 19.29 along either path abc or path adc. Along path abc the work W done by the system is 450 J. Along path adc, W is 120 J. The internal energies of each of the four states shown in the figure are $U_a = 150$ J, $U_b = 240$ J, $U_c = 680$ J, and $U_d = 330$ J. Calculate the heat flow Q for each of the four processes ab, bc, ad, and dc. In each process, does the system absorb or liberate heat?

19.45 Fig. 19.29 shows four states of a thermodynamic system, a, b, c, and d. The volume of the system is V_a for both state a and state b, and is V_c for both state c and state d. The pressure of the system is p_a for both state

Figure 19.29 Problems 19.44 and 19.45.

a and state d, and is p_c for both state b and state c. The internal energies of the four states are U_a, U_b, U_c, and U_d. For each of the processes ab, bc, ad, and dc, find a) the work done by the system during the process and b) the heat flow into the system during the process. c) The system can be taken from state a to state c along either path abc or path adc. Find the net heat flow into the system and the net work done by the system for each path. For which path is the net heat flow greater? For which path is the net work greater? d) A friend tells you that the amounts of heat flow should be the same for both path abc and path adc, since the initial state (a) and final state (c) of the system are the same for both paths. How would you respond?

19.46 Two moles of N_2 gas undergo the cycle $abcd$. The pressure of the gas in each state is: $p_a = p_d = 6000$ Pa; $p_b = p_c = 2000$ Pa. The volume of the gas in each state is: $V_a = V_b = 2.00$ m^3; $V_c = V_d = 9.00$ m^3. The gas may be treated as ideal. a) Draw the pV-diagram for the cycle. b) Find the total work done on (or by) the gas in the complete cycle. Does the gas do work or is work done on it? c) Find the net heat put into (or taken out of) the gas in one cycle. Does heat go into the gas or out of the gas?

19.47 Two moles of an ideal monatomic gas go through the cycle abc. For the complete cycle, 800 J of heat flows out of the gas. Process ab is at constant pressure and process bc is at constant volume. States a and b have temperatures $T_a = 200$ K and $T_b = 300$ K. a) Sketch the pV-diagram for the cycle. b) What is the work W for the process ca?

19.48 Three moles of an ideal gas are taken around the cycle abc shown in Fig. 19.30. For this gas, $C_p = 29.1$ J/mol·K. Process ac is at constant pressure, process ba is at constant volume, and process cb is adiabatic. The temperature of the gas in states

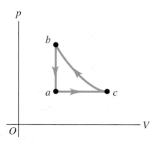

Figure 19.30 Problem 19.48.

a, c and b is $T_a = 300$ K, $T_c = 492$ K, and $T_b = 600$ K. Calculate the total work W for the cycle.

19.49 An ideal monatomic gas first has its pressure increased while its volume is held fixed (process ab). Its volume is then decreased at constant pressure until its temperature returns to the value it had in state a (process bc). The pressure for states b and c is 6.00×10^4 Pa; the volumes for these states are $V_b = 0.700$ m^3 and $V_c = 0.300$ m^3. a) Sketch the processes ab and bc on the same pV-diagram. b) What is the total heat flow for these two processes?

19.50 Nitrogen gas in an expandable container is cooled from 50.0°C to 10.0°C with the pressure held constant at 3.00×10^5 Pa. The total heat liberated by the gas is 2.50×10^4 J. Assume that the gas may be treated as ideal. a) Find the number of moles of gas. b) Find the change in internal energy of the gas. c) Find the work done by the gas. d) How much heat would be liberated by the gas for the same temperature change if the volume were constant?

19.51 In a certain process, 2.15×10^5 J of heat is liberated by a system, and at the same time the system contracts under a constant external pressure of 9.50×10^5 Pa. The internal energy of the system is the same at the beginning and end of the process. Find the change in volume of the system. (The system is *not* an ideal gas.)

19.52 A cylinder with a frictionless, movable piston like that shown in Fig. 19.5, contains a quantity of helium gas. Initially the gas is at a pressure of 1.00×10^5 Pa, has a temperature of 300 K, and occupies a volume of 1.50 L. The gas then undergoes two processes. In the first, the gas is heated and the piston is allowed to move to keep the temperature equal to 300 K. This continues until the pressure reaches 2.50×10^4 Pa. In the second process, the gas is compressed at constant pressure until it returns to its original volume of 1.50 L. Assume that the gas may be treated as ideal. a) In a pV-diagram, show both processes. b) Find the volume of the gas at the end of the first process, and find the pressure and temperature at the end of the second process. c) Find the total work done by the gas during both processes. d) What would you have to do to the gas to return it to its original pressure and temperature?

19.53 A Thermodynamic Process in a Liquid. A chemical engineer is studying the properties of liquid methanol (CH_3OH). She uses a steel cylinder with a cross-sectional area of 0.0200 m^2 and containing 1.20×10^{-2} m^3 of methanol. The cylinder is equipped with a tightly fitting piston that supports a load of 3.00×10^4 N. The temperature of the system is increased from 20.0°C to 50.0°C. For methanol, the coefficient of volume expansion is 1.20×10^{-3} K^{-1}, the density is 791 kg/m^3, and the specific heat capacity at constant pressure is $c_p = 2.51 \times 10^3$ J/kg·K. You can ignore the expansion of the steel cylinder. Find a) the increase in volume of the methanol; b) the mechanical work done by the methanol against the 3.00×10^4 N force; c) the amount of heat added to the methanol; d) the change in internal energy of the methanol. e) Based on your results, explain whether there is any substantial difference between the specific heat capacities c_p (at constant pressure) and c_V (at constant volume) for methanol under these conditions.

19.54 A Thermodynamic Process in a Solid. A cube of copper 2.00 cm on a side is suspended by a string. (The physical properties of copper are given in Tables 14.1, 17.2, and 17.3.) The cube is heated with a burner from 20.0°C to 90.0°C. The air surrounding the cube is at atmospheric pressure $(1.01 \times 10^5 \text{ Pa})$. Find a) the increase in volume of the cube; b) the mechanical work done by the cube to expand against the pressure of the surrounding air; c) the amount of heat added to the cube; d) the change in internal energy of the cube. e) Based on your results, explain whether there is any substantial difference between the specific heat capacities c_p (at constant pressure) and c_V (at constant volume) for copper under these conditions.

19.55 A Thermodynamic Process in an Insect. The African bombardier beetle *Stenaptinus insignis* can emit a jet of defensive spray from the moveable tip of its abdomen (Fig. 19.31). The beetle's body has reservoirs of two different chemicals; when the beetle is disturbed, these chemicals are combined in a reaction chamber, producing a compound that is warmed from 20°C to 100°C by the heat of reaction. The high pressure produced allows the compound to be sprayed out at speeds up to 19 m/s (68 km/h), scaring away predators of all kinds. (The beetle shown in the figure is tethered to a wire fastened to its back by wax. It is responding to having its left foreleg pinched by forceps. The beetle is 2 cm long.) Calculate the heat of reaction of the two chemicals (in J/kg). Assume that the specific heat capacity of the two chemicals and the spray is the same as that of water, 4.19×10^3 J/kg·K, and that the initial temperature of the chemicals is 20°C.

Figure 19.31 Problem 19.55.

19.56 A Compressed-Air Engine. You are designing an engine that runs on compressed air. Air enters the engine at a pressure of 1.60×10^6 Pa and leaves at a pressure of 2.80×10^5 Pa. What must the temperature of the compressed air be for there to be no possibility of frost forming in the exhaust ports of the engine? Assume that the expansion is adiabatic. (*Note:* Frost will form if the moist air is cooled below 0°C in the expansion.)

19.57 Chinook. During certain seasons strong winds called "chinooks" blow from the west across the eastern slopes of the Rockies and downhill into Denver and adjoining areas. Although the mountains are cool, the wind in Denver is very hot; within a few minutes after the chinook wind arrives, the temperature can climb 20 C° ("chinook" is a Native American word meaning "snow eater"). Similar winds occur in the Alps (called "foehns") and in southern California (called "Santa Anas"). a) Explain why the temperature of the chinook wind rises as it descends the slopes. Why is it important that the wind be fast-moving? b) Suppose a strong wind is blowing toward Denver (elevation 1630 m) from Grays Peak (80 km west of Denver, at an elevation of 4350 m), where the air pressure is 5.60×10^4 Pa and the air temperature is -15.0°C. The temperature and pressure in Denver before the wind arrives are 2.0°C and 8.12×10^4 Pa. By how many Celsius degrees will the temperature in Denver rise when the chinook arrives?

19.58 A certain ideal gas has molar heat capacity at constant volume C_V. A sample of this gas initially occupies a volume V_0 at pres-

sure p_0 and absolute temperature T_0. The gas expands isobarically to a volume $2V_0$, then expands further adiabatically to a final volume of $4V_0$. a) Draw a pV-diagram for this sequence of processes. b) Compute the total work done by the gas for this sequence of processes. c) Find the final temperature of the gas. d) Find the absolute value $|Q|$ of the total heat flow into or out of the gas for this sequence of processes, and state the direction of heat flow.

19.59 An air pump has a cylinder 0.250 m long with a movable piston. The pump is used to compress air from the atmosphere (at absolute pressure 1.01×10^5 Pa) into a very large tank at 4.20×10^5 Pa gauge pressure. (For air, $C_V = 20.8$ J/mol·K.) a) The piston begins the compression stroke at the open end of the cylinder. How far down the length of the cylinder has the piston moved when air first begins to flow from the cylinder into the tank? Assume that the compression is adiabatic. b) If the air is taken into the pump at 27.0°C, what is the temperature of the compressed air? c) How much work does the pump do in putting 20.0 mol of air into the tank?

19.60 Engine Turbochargers and Intercoolers. The power output of an automobile engine is directly proportional to the mass of air that can be forced into the volume of the engine's cylinders to react chemically with gasoline. Many cars have a *turbocharger* which compresses the air before it enters the engine, giving a greater mass of air per volume. This rapid, essentially adiabatic compression also heats the air. To compress it further, the air then passes through an *intercooler* in which the air exchanges heat with its surroundings at essentially constant pressure. The air is then drawn into the cylinders. In a typical installation, air is taken into the turbocharger at atmospheric pressure $(1.01 \times 10^5$ Pa$)$, density $\rho = 1.23$ kg/m^3, and temperature 15.0°C. It is compressed adiabatically to 1.45×10^5 Pa. In the intercooler, the air is cooled to the original temperature of 15.0°C at a constant pressure of 1.45×10^5 Pa. a) Draw a pV-diagram for this sequence of processes. b) If the volume of one of the engine's cylinders is 575 cm^3, what mass of air exiting from the intercooler will fill the cylinder at 1.45×10^5 Pa? Compared to the power output of an engine that takes in air at 1.01×10^5 Pa at 15.0°C, what percentage increase in power is obtained by using the turbocharger and intercooler? c) If the intercooler is not used, what mass of air exiting from the turbocharger will fill the cylinder at 1.45×10^5 Pa? Compared to the power output of an engine that takes in air at 1.01×10^5 Pa at 15.0°C, what percentage increase in power is obtained by using the turbocharger alone?

19.61 A monatomic ideal gas expands slowly to twice its original volume, doing 300 J of work in the process. Find the heat added to the gas and the change in internal energy of the gas if the process is a) isothermal; b) adiabatic; c) isobaric.

19.62 A cylinder with a piston contains 0.250 mol of oxygen at 2.40×10^5 Pa and 355 K. The oxygen may be treated as an ideal gas. The gas first expands isobarically to twice its original volume. It is then compressed isothermally back to its original volume, and finally it is cooled isochorically to its original pressure. a) Show the series of processes on a pV-diagram. b) Compute the temperature during the isothermal compression. c) Compute the maximum pressure. d) Compute the total work done by the piston on the gas during the series of processes.

19.63 Use the conditions and processes of Problem 19.62 to compute a) the work done by the gas, the heat added to it, and its internal-energy change during the initial expansion; b) the work done, the heat added, and the internal-energy change during the final cooling; c) the internal-energy change during the isothermal compression.

19.64 A cylinder with a piston contains 0.150 mol of nitrogen at 1.80×10^5 Pa and 300 K. The nitrogen may be treated as an ideal gas. The gas is first compressed isobarically to $\frac{1}{2}$ its original volume. It then expands adiabatically back to its original volume, and finally it is heated isochorically to its original pressure. a) Show the series of processes in a pV-diagram. b) Compute the temperatures at the beginning and end of the adiabatic expansion. c) Compute the minimum pressure.

19.65 Use the conditions and processes of Problem 19.64 to compute a) the work done by the gas, the heat added to it, and its internal-energy change during the initial compression; b) the work done by the gas, the heat added to it, and its internal-energy change during the adiabatic expansion; c) the work done, the heat added, and the internal-energy change during the final heating.

19.66 Comparing Thermodynamic Processes. In a cylinder, 1.20 mol of an ideal monatomic gas, initially at 3.60×10^5 Pa and 300 K, expands until its volume triples. Compute the work done by the gas if the expansion is a) isothermal; b) adiabatic; c) isobaric. d) Show each process in a pV-diagram. In which case is the absolute value of the work done by the gas greatest? Least? e) In which case is the absolute value of the heat transfer greatest? Least? f) In which case is the absolute value of the change in internal energy of the gas greatest? Least?

19.67 A flexible balloon contains 0.350 mol of hydrogen sulfide gas (H_2S). Initially the H_2S has a volume of 7.00×10^3 cm^3 and a temperature of 27.0°C. The H_2S first expands isobarically until the volume doubles. Then it expands adiabatically until the temperature returns to its initial value. Assume that the H_2S may be treated as an ideal gas. a) Draw a diagram of the process in the pV-plane. b) What is the total heat supplied to the H_2S in the process? c) What is the total change in the internal energy of the H_2S? d) What is the total work done by the H_2S? e) What is the final volume?

Challenge Problems

19.68 Oscillations of a Piston. A vertical cylinder of radius r contains a quantity of ideal gas, and is fitted with a piston with mass m that is free to move (Fig. 19.32). The piston and the walls of the cylin-

Figure 19.32 Challenge Problem 19.68.

der are frictionless and are made of a perfect thermal insulator. The outside air pressure is p_0. In equilibrium, the piston sits at a height h above the bottom of the cylinder. a) Find the absolute pressure of the gas trapped below the piston when in equilibrium. b) The piston is pulled up by a small distance and released. Find the net force acting on the piston when its base is a distance $h + y$ above the bottom of the cylinder, where y is much less than h. c) After the piston is displaced from equilibrium and released, it oscillates up and down. Find the frequency of these small oscillations. If the displacement is not small, are the oscillations simple harmonic? How can you tell?

19.69 The van der Waals equation of state, an approximate representation of the behavior of gases at high pressure, is given by Eq. (18.7):

$$\left(p + \frac{an^2}{V^2} \right) (V - nb) = nRT$$

where a and b are constants having different values for different gases. (In the special case of $a = b = 0$, this is the ideal-gas equation.) a) Calculate the work done by a gas with this equation of state in an isothermal expansion from V_1 to V_2. Show that your answer agrees with the ideal-gas result found in Example 19.1 (Section 19.2) when you set $a = b = 0$. b) For ethane gas (C_2H_6), $a = 0.554$ J·m^3/mol^2 and $b = 6.38 \times 10^{-5}$ m^3/mol. Calculate the work W done by 1.80 mol of ethane when it expands from 2.00×10^{-3} m^3 to 4.00×10^{-3} m^3 at a constant temperature of 300 K. Do the calculation using i) the van der Waals equation of state and ii) the ideal-gas equation of state. c) How large is the difference between the two results for W in part (b)? For which equation of state is W larger? Use the interpretation of the terms a and b given in Section 18.1 to explain why this should be so. Are the differences between the two equations of state important in this case?

20 THE SECOND LAW OF THERMODYNAMICS

Could heat spontaneously flow out of these fried eggs, thus "unfrying" them and making them cooler, and flow back into the hot frying pan? Such a process would conserve energy and is allowed by the first law of thermodynamics. But such a process is in fact impossible because it would violate the *second* law of thermodynamics: heat cannot by itself flow from one body into a hotter body.

? **Is it *ever* possible for heat to flow from a cold body into a hot one?**

Many thermodynamic processes proceed naturally in one direction but not the opposite. For example, heat by itself always flows from a hot body to a cooler body, never the reverse. Heat flow from a cool body to a hot body would not violate the first law of thermodynamics; energy would be conserved. But it doesn't happen in nature. Why not? As another example, note that it is easy to convert mechanical energy completely into heat; this happens every time we use a car's brakes to stop it. Going in the reverse direction, there are plenty of devices that convert heat *partially* into mechanical energy. (An automobile engine is an example.) But even the cleverest would-be inventors have never succeeded in building a machine that converts heat *completely* into mechanical energy. Again, why not?

The answer to both of these questions has to do with the *directions* of thermodynamic processes and is called the *second law of thermodynamics*. This law places fundamental limitations on the efficiency of an engine or a power plant. It also places limitations on the minimum energy input needed to operate a refrigerator. So the second law is directly relevant for many important practical problems.

We can also state the second law in terms of the concept of *entropy,* a quantitative measure of the degree of disorder or randomness of a system. The idea of entropy helps explain why ink mixed with water never spontaneously unmixes and why we never observe a host of other seemingly possible processes.

20.1 | Directions of Thermodynamic Processes

Thermodynamic processes that occur in nature are all **irreversible processes.** These are processes that proceed spontaneously in one direction but not the other (Fig. 20.1a). The flow of heat from a hot body to a cooler body is irreversible, as is the free expansion of a gas discussed in Sections 19.3 and 19.6. Sliding a book across a table converts mechanical energy into heat by friction; this process is irreversible, for no one has ever observed the reverse process (in which a book initially at rest on the table would spontaneously start moving and the table and book would cool down). Our main topic for this chapter is the *second law of thermodynamics,* which determines the preferred direction for such processes.

Despite this preferred direction for every natural process, we can think of a class of idealized processes that *would* be reversible. A system that undergoes such an idealized **reversible process** is always very close to being in thermodynamic equilibrium within itself and with its surroundings. Any change of state that takes place can then be reversed (made to go the other way) by making only an infinitesimal change in the conditions of the system. For example, heat flow between two bodies whose temperatures differ only infinitesimally can be reversed by making only a very small change in one temperature or the other (Fig. 20.1b).

Reversible processes are thus **equilibrium processes,** with the system always in thermodynamic equilibrium. Of course, if a system were *truly* in thermodynamic equilibrium, no change of state would take place. Heat would not flow into or out of a system with truly uniform temperature throughout, and a system that is truly in mechanical equilibrium would not expand and do work against its surroundings. A reversible process is an idealization that can never be precisely attained in the real world. But by making the temperature gradients and the pressure differences in the substance very small, we can keep the system very close to equilibrium states and make the process nearly reversible. That's why we call a reversible process a *quasi-equilibrium process.*

By contrast, heat flow with finite temperature difference, free expansion of a gas, and conversion of work to heat by friction are all *irreversible* processes; no small change in conditions could make any of them go the other way. They are also all *nonequilibrium* processes, in that the system is not in thermodynamic equilibrium at any point until the end of the process.

There is a relationship between the direction of a process and the *disorder* or *randomness* of the resulting state. For example, imagine a tedious sorting job, such as alphabetizing a thousand book titles written on file cards. Throw the alphabetized stack of cards into the air. Do they come down in alphabetical order? Alas, no: their tendency is to come down in a random or disordered state. In the free expansion of a gas discussed in Sections 19.3 and 19.6, the air is more disordered after it has expanded into the entire box than when it was confined in one side, just as your clothes are more disordered when scattered all over your floor than when confined to your closet.

Similarly, macroscopic kinetic energy is energy associated with organized, coordinated motions of many molecules, but heat transfer involves changes in energy of random, disordered molecular motion. Therefore conversion of mechanical energy into heat involves an increase of randomness or disorder.

In the following sections we will introduce the second law of thermodynamics by considering two broad classes of devices: *heat engines,* which are partly successful in converting heat into work, and *refrigerators,* which are partly successful in transporting heat from cooler to hotter bodies.

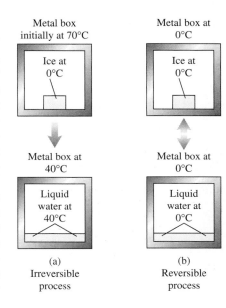

20.1 (a) A block of ice melts *irreversibly* when we place it in a hot metal box. Heat flows from the box into the ice and water, never the reverse. (b) We can melt ice *reversibly* in a metal box at 0°C. By raising or lowering the temperature of the box by an infinitesimal amount, we can make heat flow out of the box into the ice (thus melting it) or into the box from the water (thus refreezing it).

Your left and right hands are normally at the same temperature, just like the metal box and ice in Fig. 20.1b. Is rubbing your hands together to warm them a reversible process? Explain.

20.2 | Heat Engines

The essence of our technological society is the ability to use sources of energy other than muscle power. Sometimes, mechanical energy is directly available; water power is an example. But most of our energy comes from the burning of fossil fuels (coal, oil, and gas) and from nuclear reactions. They supply energy that is transferred as *heat*. This is directly useful for heating buildings, for cooking, and for chemical processing, but to operate a machine or propel a vehicle, we need *mechanical* energy.

Thus it's important to know how to take heat from a source and convert as much of it as possible into mechanical energy or work. This is what happens in gasoline engines in automobiles, jet engines in airplanes, steam turbines in electric power plants, and many other systems. Closely related processes occur in the animal kingdom; food energy is "burned" (that is, carbohydrates combine with oxygen to yield water, carbon dioxide, and energy) and partly converted to mechanical energy as an animal's muscles do work on its surroundings.

Any device that transforms heat partly into work or mechanical energy is called a **heat engine.** Usually, a quantity of matter inside the engine undergoes inflow and outflow of heat, expansion and compression, and sometimes change of phase. We call this matter the **working substance** of the engine. In internal-combustion engines the working substance is a mixture of air and fuel; in a steam turbine it is water (Fig. 20.2).

The simplest kind of engine to analyze is one in which the working substance undergoes a **cyclic process,** a sequence of processes that eventually leaves the substance in the same state in which it started. In a steam turbine the water is recycled and used over and over. Internal-combustion engines do not use the same air over and over, but we can still analyze them in terms of cyclic processes that approximate their actual operation.

All heat engines *absorb* heat from a source at a relatively high temperature, perform some mechanical work, and *discard* or *reject* some heat at a lower temperature. As far as the engine is concerned, the discarded heat is wasted. In internal-combustion engines the waste heat is that discarded in the hot exhaust gases and the cooling system; in a steam turbine it is the heat that must flow out of the used steam to condense and recycle the water.

When a system is carried through a cyclic process, its initial and final internal energies are equal. For any cyclic process, the first law of thermodynamics requires that

$$U_2 - U_1 = 0 = Q - W \quad \text{so} \quad Q = W$$

That is, the net heat flowing into the engine in a cyclic process equals the net work done by the engine.

When we analyze heat engines, it helps to think of two bodies with which the working substance of the engine can interact. One of these, called the *hot reservoir,* represents the heat source; it can give the working substance large amounts of heat at a constant temperature T_H without appreciably changing its own tem-

20.2 All motorized vehicles (except electric-powered vehicles) use heat engines for propulsion.

perature. The other body, called the *cold reservoir,* can absorb large amounts of discarded heat from the engine at a constant lower temperature T_C. In a steam-turbine system the flames and hot gases in the boiler are the hot reservoir, and the cold water and air used to condense and cool the used steam are the cold reservoir.

We denote the quantities of heat transferred from the hot and cold reservoirs as Q_H and Q_C, respectively. A quantity of heat Q is positive when heat is transferred *into* the working substance and is negative when heat leaves the working substance. Thus in a heat engine, Q_H is positive but Q_C is negative, representing heat *leaving* the working substance. This sign convention is consistent with the rules we stated in Section 19.1; we will continue to use those rules here. Frequently, it clarifies the relationships to state them in terms of the absolute values of the Q's and W's because absolute values are always positive. When we do this, our notation will show it explicitly.

We can represent the energy transformations in a heat engine by the *energy-flow diagram* of Fig. 20.3. The engine itself is represented by the circle. The amount of heat Q_H supplied to the engine by the hot reservoir is proportional to the width of the incoming "pipeline" at the top of the diagram. The width of the outgoing pipeline at the bottom is proportional to the magnitude $|Q_C|$ of the heat rejected in the exhaust. The branch line to the right represents the portion of the heat supplied that the engine converts to mechanical work, W.

When an engine repeats the same cycle over and over, Q_H and Q_C represent the quantities of heat absorbed and rejected by the engine *during one cycle*; Q_H is positive, and Q_C is negative. The *net* heat Q absorbed per cycle is

$$Q = Q_H + Q_C = |Q_H| - |Q_C| \qquad (20.1)$$

The useful output of the engine is the net work W done by the working substance. From the first law,

$$W = Q = Q_H + Q_C = |Q_H| - |Q_C| \qquad (20.2)$$

Ideally, we would like to convert *all* the heat Q_H into work; in that case we would have $Q_H = W$ and $Q_C = 0$. Experience shows that this is impossible; there is always some heat wasted, and Q_C *is never zero.* We define the **thermal efficiency** of an engine, denoted by e, as the quotient

$$e = \frac{W}{Q_H} \qquad (20.3)$$

The thermal efficiency e represents the fraction of Q_H that *is* converted to work. To put it another way, e is what you get divided by what you pay for. This is always less than unity, an all-too-familiar experience! In terms of the flow diagram of Fig. 20.3, the most efficient engine is the one for which the branch pipeline representing the work output is as wide as possible and the exhaust pipeline representing the heat thrown away is as narrow as possible.

When we substitute the two expressions for W given by Eq. (20.2) into Eq. (20.3), we get the following equivalent expressions for e:

$$e = \frac{W}{Q_H} = 1 + \frac{Q_C}{Q_H} = 1 - \left| \frac{Q_C}{Q_H} \right| \quad \text{(thermal efficiency of an engine)} \quad (20.4)$$

Note that e is a quotient of two energy quantities and thus is a pure number, without units. Of course, we must always express W, Q_H, and Q_C in the same units.

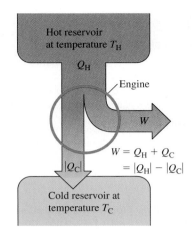

20.3 Schematic energy-flow diagram for a heat engine.

Hot reservoir at temperature T_H

Q_H

Engine

W

$|Q_C|$

$W = Q_H + Q_C$
$= |Q_H| - |Q_C|$

Cold reservoir at temperature T_C

Heat Engines

Problems involving heat engines are, first and foremost, problems in the first law of thermodynamics. Hence the Problem-Solving Strategy in Section 19.4 is equally useful throughout the present chapter, and we suggest that you reread it.

IDENTIFY *the relevant concepts:* A heat engine is any device that converts work partially to heat, as shown schematically in Fig. 20.3. We will see in Section 20.4 that a refrigerator is essentially a heat engine running in reverse, so many of the same concepts apply.

SET UP *the problem* as suggested in the Problem-Solving Strategy in Section 19.4. Equation (20.4) is useful in situations for which the thermal efficiency of the engine is relevant.

EXECUTE *the solution as follows:*
1. Be very careful with the sign conventions for W and the various Q's. W is positive when the system expands and does work; W is negative when the system is compressed. Each Q is positive if it represents heat entering the system and is negative if it represents heat leaving the system. When you know that a quantity is negative, such as Q_C in the above discussion, it sometimes helps to write it as $Q_C = -|Q_C|$.
2. Some problems deal with power rather than energy quantities. Power is work per unit time ($P = W/t$), and rate of heat transfer (heat current) H is heat transfer per unit time ($H = Q/t$). In such problems it helps to ask, "What is W or Q in one second (or one hour)?"
3. Keeping steps (1) and (2) in mind, solve for the target variables.

EVALUATE *your answer:* Use the first law of thermodynamics to check your results, paying particular attention to algebraic signs.

Analyzing a heat engine

A gasoline engine in a large truck takes in 10,000 J of heat and delivers 2000 J of mechanical work per cycle. The heat is obtained by burning gasoline with heat of combustion $L_c = 5.0 \times 10^4$ J/g. a) What is the thermal efficiency of this engine? b) How much heat is discarded in each cycle? c) How much gasoline is burned in each cycle? d) If the engine goes through 25 cycles per second, what is its power output in watts? In horsepower? e) How much gasoline is burned per second? Per hour?

SOLUTION

IDENTIFY and SET UP: The problem is about a heat engine, for which we are told the amount of work done per cycle ($W = 2000$ J) and the amount of heat taken in per cycle ($Q_H = 10,000$ J). Hence we use the first form of Eq. (20.4) to find the thermal efficiency. The first law of thermodynamics tells us the amount of heat rejected per cycle, and the heat of combustion tells us how much gasoline must be burned per cycle and hence per unit time.

EXECUTE: a) From the first expression in Eq. (20.4), the thermal efficiency is

$$e = \frac{W}{Q_H} = \frac{2000 \text{ J}}{10,000 \text{ J}} = 0.20 = 20\%$$

This is a fairly typical figure for cars and trucks if W includes only the work actually delivered to the wheels. b) From Eq. (20.2), $W = Q_H + Q_C$ so

$$Q_C = W - Q_H = 2000 \text{ J} - 10,000 \text{ J}$$
$$= -8000 \text{ J}$$

That is, 8000 J of heat leaves the engine during each cycle.
c) Let m be the mass of gasoline burned during each cycle. Then

$$Q_H = mL_c$$
$$m = \frac{Q_H}{L_c} = \frac{10,000 \text{ J}}{5.0 \times 10^4 \text{ J/g}} = 0.20 \text{ g}$$

d) The power P (rate of doing work) is the work per cycle multiplied by the number of cycles per second:

$$P = (2000 \text{ J/cycle})(25 \text{ cycles/s}) = 50,000 \text{ W} = 50 \text{ kW}$$

$$= (50,000 \text{ W})\frac{1 \text{ hp}}{746 \text{ W}} = 67 \text{ hp}$$

e) The mass of gasoline burned per second is the mass per cycle multiplied by the number of cycles per second:

$$(0.20 \text{ g/cycle})(25 \text{ cycles/s}) = 5.0 \text{ g/s}$$

The mass burned per hour is

$$(5.0 \text{ g/s})\frac{3600 \text{ s}}{1 \text{ h}} = 18,000 \text{ g/h} = 18 \text{ kg/h}$$

EVALUATE: We can check our result in part (e) by converting it to a more familiar quantity, the amount of fuel consumed per unit distance. The density of gasoline is about 0.70 g/cm^3, so this is about 25,700 cm^3, 25.7 L, or 6.8 gallons of gasoline per hour. If the truck is traveling at 55 mi/h (88 km/h), this represents fuel consumption of 8.1 miles/gallon (3.4 km/L). This is substantially worse fuel consumption than a passenger car, but fairly typical of large trucks.

Test Your Understanding

How much heat is rejected into the environment during a 24-hour period by 1000 trucks like those in Example 20.1 driving around a city?

20.3 | Internal-Combustion Engines

The gasoline engine, used in automobiles and many other types of machinery, is a familiar example of a heat engine. Let's look at its thermal efficiency. Figure 20.4 shows the operation of one type of gasoline engine. First a mixture of air and gasoline vapor flows into a cylinder through an open intake valve while the piston descends, increasing the volume of the cylinder from a minimum of V (when the piston is all the way up) to a maximum of rV (when it is all the way down). The quantity r is called the **compression ratio;** for present-day automobile engines its value is typically 8 to 10. At the end of this *intake stroke,* the intake valve closes and the mixture is compressed, approximately adiabatically, to volume V during the *compression stroke*. The mixture is then ignited by the spark plug, and the heated gas expands, approximately adiabatically, back to volume rV, pushing on the piston and doing work; this is the *power stroke*. Finally, the exhaust valve opens, and the combustion products are pushed out (during the *exhaust stroke*), leaving the cylinder ready for the next intake stroke.

The Otto Cycle

Figure 20.5 is a pV-diagram for an idealized model of the thermodynamic processes in a gasoline engine. This model is called the **Otto cycle.** At point a the gasoline–air mixture has entered the cylinder. The mixture is compressed adiabatically to point b and is then ignited. Heat Q_H is added to the system by the burn-

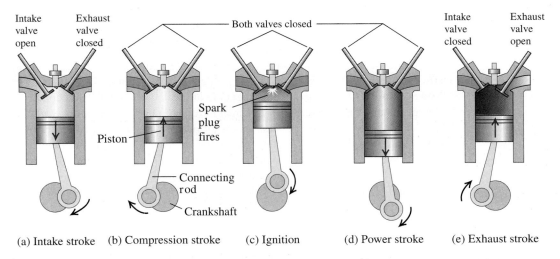

| Intake valve open | Exhaust valve closed | ───── Both valves closed ───── | | | Intake valve closed | Exhaust valve open |

Spark plug fires

Piston

Connecting rod

Crankshaft

(a) Intake stroke   (b) Compression stroke   (c) Ignition   (d) Power stroke   (e) Exhaust stroke

20.4 Cycle of a four-stroke internal-combustion engine. (a) *Intake stroke:* Piston moves down, causing a partial vacuum in cylinder; gasoline–air mixture flows through open intake valve into cylinder. (b) *Compression stroke:* Intake valve closes, and mixture is compressed as piston moves up. (c) *Ignition:* Spark plug ignites mixture. (d) *Power stroke:* Hot burned mixture pushes piston down, doing work. (e) *Exhaust stroke:* Exhaust valve opens and piston moves up, pushing burned mixture out of cylinder. Engine is now ready for next intake stroke, and the cycle repeats.

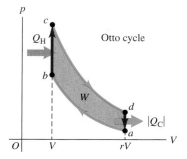

$a \rightarrow b$: Compression stroke
(adiabatic compression)
$b \rightarrow c$: Ignite fuel
(heating at constant volume)
$c \rightarrow d$: Power stroke
(adiabatic expansion)
$d \rightarrow a$: Reject heat to environment
(cooling at constant volume)

20.5 The pV-diagram for the Otto cycle, an idealized model of the thermodynamic processes in a gasoline engine.

ing gasoline along line bc, and the power stroke is the adiabatic expansion to d. The gas is cooled to the temperature of the outside air along line da; during this process, heat $|Q_C|$ is rejected. In practice, this gas leaves the engine as exhaust and does not enter the engine again. But since an equivalent amount of gasoline and air enters, we may consider the process to be cyclic.

We can calculate the efficiency of this idealized cycle. Processes bc and da are constant-volume, so the heats Q_H and Q_C are related simply to the temperatures:

$$Q_H = nC_V(T_c - T_b) > 0$$

$$Q_C = nC_V(T_a - T_d) < 0$$

The thermal efficiency is given by Eq. (20.4). Inserting the above expressions and cancelling out the common factor nC_V, we find

$$e = \frac{Q_H + Q_C}{Q_H} = \frac{T_c - T_b + T_a - T_d}{T_c - T_b} \qquad (20.5)$$

To simplify this further, we use the temperature-volume relation for adiabatic processes for an ideal gas, Eq. (19.22). For the two adiabatic processes ab and cd,

$$T_a(rV)^{\gamma-1} = T_b V^{\gamma-1} \quad \text{and} \quad T_d(rV)^{\gamma-1} = T_c V^{\gamma-1}$$

We divide each of these equations by the common factor $V^{\gamma-1}$ and substitute the resulting expressions for T_b and T_c back into Eq. (20.5). The result is

$$e = \frac{T_d r^{\gamma-1} - T_a r^{\gamma-1} + T_a - T_d}{T_d r^{\gamma-1} - T_a r^{\gamma-1}} = \frac{(T_d - T_a)(r^{\gamma-1} - 1)}{(T_d - T_a)r^{\gamma-1}}$$

Dividing out the common factor $(T_d - T_a)$, we get

$$e = 1 - \frac{1}{r^{\gamma-1}} \qquad (\text{thermal efficiency in Otto cycle}) \qquad (20.6)$$

The thermal efficiency given by Eq. (20.6) is always less than unity, even for this idealized model. With $r = 8$ and $\gamma = 1.4$ (the value for air) the theoretical efficiency is $e = 0.56$, or 56%. The efficiency can be increased by increasing r. However, this also increases the temperature at the end of the adiabatic compression of the air–fuel mixture. If the temperature is too high, the mixture explodes spontaneously during compression instead of burning evenly after the spark plug ignites it. This is called *pre-ignition* or *detonation;* it causes a knocking sound and can damage the engine. The octane rating of a gasoline is a measure of its anti-knock qualities. The maximum practical compression ratio for high-octane, or "premium," gasoline is about 10. Higher ratios can be used with more exotic fuels.

The Otto cycle, which we have just described, is a highly idealized model. It assumes that the mixture behaves as an ideal gas; it neglects friction, turbulence, loss of heat to cylinder walls, and many other effects that combine to reduce the efficiency of a real engine. Another source of inefficiency is incomplete combustion. A mixture of gasoline vapor with just enough air for complete combustion of the hydrocarbons to H_2O and CO_2 does not ignite readily. Reliable ignition requires a mixture that is "richer" in gasoline. The resulting incomplete combustion leads to CO and unburned hydrocarbons in the exhaust. The heat obtained from the gasoline is then less than the total heat of combustion; the difference is wasted, and the exhaust products contribute to air pollution. Efficiencies of real gasoline engines are typically around 35%.

The Diesel Cycle

The Diesel engine is similar in operation to the gasoline engine. The most important difference is that there is no fuel in the cylinder at the beginning of the compression stroke. A little before the beginning of the power stroke, the injectors start to inject fuel directly into the cylinder, just fast enough to keep the pressure approximately constant during the first part of the power stroke. Because of the high temperature developed during the adiabatic compression, the fuel ignites spontaneously as it is injected; no spark plugs are needed.

The idealized **Diesel cycle** is shown in Fig. 20.6. Starting at point a, air is compressed adiabatically to point b, heated at constant pressure to point c, expanded adiabatically to point d, and cooled at constant volume to point a. Because there is no fuel in the cylinder during most of the compression stroke, pre-ignition cannot occur, and the compression ratio r can be much higher than for a gasoline engine. This improves efficiency and ensures reliable ignition when the fuel is injected (because of the high temperature reached during the adiabatic compression). Values of r of 15 to 20 are typical; with these values and $\gamma = 1.4$, the theoretical efficiency of the idealized Diesel cycle is about 0.65 to 0.70. As with the Otto cycle, the efficiency of any actual engine is substantially less than this. While Diesel engines are very efficient, they must be built to much tighter tolerances than gasoline engines and the fuel-injection system requires careful maintenance.

$a \rightarrow b$: Compression stroke
(adiabatic compression)
$b \rightarrow c$: Ignite fuel
(heating at constant pressure)
$c \rightarrow d$: Power stroke
(adiabatic expansion)
$d \rightarrow a$: Reject heat to environment
(cooling at constant volume)

20.6 The pV-diagram for the idealized Diesel cycle.

Test Your Understanding

Use Fig. 20.5, *not* Eq. (20.6), to explain why the efficiency of an Otto-cycle engine increases with increasing compression ratio.

20.4 | Refrigerators

We can think of a **refrigerator** as a heat engine operating in reverse. A heat engine takes heat from a hot place and gives off heat to a colder place. A refrigerator does the opposite; it takes heat from a cold place (the inside of the refrigerator) and gives it off to a warmer place (usually the air in the room where the refrigerator is located). A heat engine has a net *output* of mechanical work; the refrigerator requires a net *input* of mechanical work. Using the sign conventions from Section 20.2, for a refrigerator Q_C is positive but both W and Q_H are negative; hence $|W| = -W$ and $|Q_H| = -Q_H$.

A flow diagram for a refrigerator is shown in Fig. 20.7. From the first law for a cyclic process,

$$Q_H + Q_C - W = 0 \quad \text{or} \quad -Q_H = Q_C - W$$

or, because both Q_H and W are negative,

$$|Q_H| = Q_C + |W| \tag{20.7}$$

Thus, as the diagram shows, the heat $|Q_H|$ leaving the working substance and given to the hot reservoir is always *greater* than the heat Q_C taken from the cold reservoir. Note that the absolute-value relation

$$|Q_H| = |Q_C| + |W| \tag{20.8}$$

is valid for both heat engines and refrigerators.

20.7 Schematic energy-flow diagram of a refrigerator.

From an economic point of view, the best refrigeration cycle is one that removes the greatest amount of heat $|Q_C|$ from the inside of the refrigerator for the least expenditure of mechanical work, $|W|$. The relevant ratio is therefore $|Q_C|/|W|$; the larger this ratio, the better the refrigerator. We call this ratio the **coefficient of performance,** denoted by K. From Eq. (20.8), $|W| = |Q_H| - |Q_C|$, so

$$K = \frac{|Q_C|}{|W|} = \frac{|Q_C|}{|Q_H| - |Q_C|} \tag{20.9}$$
(coefficient of performance of a refrigerator)

As always, we measure Q_H, Q_C, and W all in the same energy units; K is then a dimensionless number.

The principles of the common refrigeration cycle are shown schematically in Fig. 20.8a. The fluid "circuit" contains a refrigerant fluid (the working substance). The left side of the circuit (including the cooling coils inside the refrigerator) is at low temperature and low pressure; the right side (including the condenser coils outside the refrigerator) is at high temperature and high pressure. Ordinarily, both sides contain liquid and vapor in phase equilibrium.

The compressor takes in fluid, compresses it adiabatically, and delivers it to the condenser coil at high pressure. The fluid temperature is then higher than that of the air surrounding the condenser, so the refrigerant gives off heat $|Q_H|$ and partially condenses to liquid. The fluid then expands adiabatically into the evaporator at a rate controlled by the expansion valve. As the fluid expands, it cools considerably, enough that the fluid in the evaporator coil is colder than its surroundings. It absorbs heat $|Q_C|$ from its surroundings, cooling them and partially vaporizing. The fluid then enters the compressor to begin another cycle. The compressor, usually driven by an electric motor (Fig. 20.8b), requires energy input and does work $|W|$ *on* the working substance during each cycle.

20.8 (a) Principle of the mechanical refrigeration cycle. (b) How the key elements are arranged in a practical refrigerator.

Hot air

Fan

Blower

Condenser

Cool air

Expansion valve

Warm, outside air

Hot, humid air

Compressor

Evaporator

20.9 An air conditioner works on the same principle as a refrigerator.

An air conditioner operates on exactly the same principle. In this case the refrigerator box becomes a room or an entire building. The evaporator coils are inside, the condenser is outside, and fans circulate air through these (Fig. 20.9). In large installations the condenser coils are often cooled by water. For air conditioners the quantities of greatest practical importance are the *rate* of heat removal (the heat current H from the region being cooled) and the *power* input $P = W/t$ to the compressor. If heat $|Q_C|$ is removed in time t, then $H = |Q_C|/t$. Then we can express the coefficient of performance as

$$K = \frac{|Q_C|}{|W|} = \frac{Ht}{Pt} = \frac{H}{P}$$

Typical room air conditioners have heat removal rates H of 5000 to 10,000 Btu/h, or about 1500–3000 W, and require electric power input of about 600 to 1200 W. Typical coefficients of performance are about 3; the actual values depend on the inside and outside temperatures.

Unfortunately, K is often expressed commercially in mixed units, with H in Btu per hour and P in watts. In these units, H/P is called the **energy efficiency rating** (EER); the units, customarily omitted, are (Btu/h)/W. Because 1 W = 3.413 Btu/h, the EER is numerically 3.413 times as large as the dimensionless K. Room air conditioners typically have an EER of about 10.

A variation on this theme is the **heat pump,** used to heat buildings by cooling the outside air. It functions like a refrigerator turned inside out. The evaporator coils are outside, where they take heat from cold air, and the condenser coils are inside, where they give off heat to the warmer air. With proper design, the heat $|Q_H|$ delivered to the inside per cycle can be considerably greater than the work $|W|$ required to get it there.

Work is *always* needed to transfer heat from a colder to a hotter body. Heat flows spontaneously from hotter to colder, and to reverse this flow requires the addition of work from the outside. Experience shows that it is impossible to make a refrigerator that transports heat from a colder body to a hotter body without the addition of work. If no work were needed, the coefficient of performance would be infinite. We call such a device a *workless refrigerator;* it is a mythical beast, like the unicorn and the free lunch.

Test Your Understanding

Why can't you cool your house by leaving the refrigerator door open?

20.5 | The Second Law of Thermodynamics

Experimental evidence suggests strongly that it is *impossible* to build a heat engine that converts heat completely to work, that is, an engine with 100% thermal efficiency. This impossibility is the basis of one statement of the **second law of thermodynamics,** as follows:

> **It is impossible for any system to undergo a process in which it absorbs heat from a reservoir at a single temperature and converts the heat completely into mechanical work, with the system ending in the same state in which it began.**

We will call this the "engine" statement of the second law. (It is also known to physicists as the *Kelvin-Planck statement* of this law.)

The basis of the second law of thermodynamics lies in the difference between the nature of internal energy and that of macroscopic mechanical energy. In a moving body the molecules have random motion, but superimposed on this is a coordinated motion of every molecule in the direction of the body's velocity. The kinetic energy associated with this *coordinated* macroscopic motion is what we call the kinetic energy of the moving body. The kinetic and potential energies associated with the *random* motion constitute the internal energy.

When a body sliding on a surface comes to rest as a result of friction, the organized motion of the body is converted to random motion of molecules in the body and in the surface. Since we cannot control the motions of individual molecules, we cannot convert this random motion completely back to organized motion. We can convert *part* of it, and this is what a heat engine does.

If the second law were *not* true, we could power an automobile or run a power plant by cooling the surrounding air. Neither of these impossibilities violates the *first* law of thermodynamics. The second law, therefore, is not a deduction from the first but stands by itself as a separate law of nature. The first law denies the possibility of creating or destroying energy; the second law limits the *availability* of energy and the ways in which it can be used and converted.

Our analysis of refrigerators in Section 20.4 forms the basis for an alternative statement of the second law of thermodynamics. Heat flows spontaneously from hotter to colder bodies, never the reverse. A refrigerator does take heat from a colder to a hotter body, but its operation requires an input of mechanical energy or work. Generalizing this observation, we state:

> **It is impossible for any process to have as its sole result the transfer of heat from a cooler to a hotter body.**

We'll call this the "refrigerator" statement of the second law (it is also known as the *Clausius statement*). It may not seem to be very closely related to the "engine" statement. In fact, though, the two statements are completely equivalent. For example, if we could build a workless refrigerator, violating the second or "refrigerator" statement of the second law, we could use it in conjunction with a heat engine, pumping the heat rejected by the engine back to the hot reservoir to be

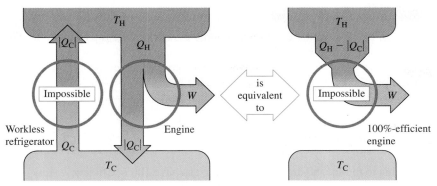

(a) If a workless refrigerator existed, it could be used to make a 100%-efficient engine

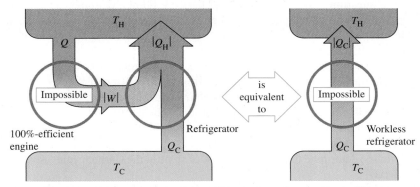

(b) If a 100%-efficient engine existed, it could be used to make a workless refrigerator

20.10 Energy-flow diagrams showing that the two forms of the second law are equivalent. (a) Using an impossible workless refrigerator and an ordinary heat engine to form a composite device that functions as an equally impossible 100%-efficient engine. (b) Using an impossible 100%-efficient engine and an ordinary refrigerator to form an equally impossible workless refrigerator.

reused. This composite machine (Fig. 20.10a) would violate the "engine" statement of the second law because its net effect would be to take a net quantity of heat $Q_H - |Q_C|$ from the hot reservoir and convert it completely to work W.

Alternatively, if we could make an engine with 100% thermal efficiency, in violation of the first statement, we could run it using heat from the hot reservoir and use the work output to drive a refrigerator that pumps heat from the cold reservoir to the hot (Fig. 20.10b). This composite device would violate the "refrigerator" statement because its net effect would be to take heat Q_C from the cold reservoir and deliver it to the hot reservoir without requiring any input of work. Thus any device that violates one form of the second law can be used to make a device that violates the other form. If violations of the first form are impossible, so are violations of the second!

The conversion of work to heat, as in friction or viscous fluid flow, and heat flow from hot to cold across a finite temperature gradient, are *irreversible* processes. The "engine" and "refrigerator" statements of the second law state that these processes can be only partially reversed. We could cite other examples. Gases always seep spontaneously through an opening from a region of high pressure to a region of low pressure; gases and miscible liquids left by themselves always tend to mix, not to unmix. The second law of thermodynamics is an expression of the inherent one-way aspect of these and many other irreversible processes. Energy conversion is an essential aspect of all plant and animal life and of human technology, so the second law of thermodynamics is of the utmost fundamental importance in the world we live in.

Test Your Understanding

You partially fill a test tube with water and put a cork in the top of the tube. You place a burner under the tube to boil the water, causing the cork to pop off the tube. Explain how this process converts heat into mechanical work, and explain why the process is not 100% efficient.

20.6 | The Carnot Cycle

8.14 Carnot Cycle

According to the second law, no heat engine can have 100% efficiency. How great an efficiency *can* an engine have, given two heat reservoirs at temperatures T_H and T_C? This question was answered in 1824 by the French engineer Sadi Carnot (1796-1832), who developed a hypothetical, idealized heat engine that has the maximum possible efficiency consistent with the second law. The cycle of this engine is called the **Carnot cycle.**

To understand the rationale of the Carnot cycle, we return to a recurrent theme in this chapter: *reversibility* and its relation to directions of thermodynamic processes. Conversion of work to heat is an irreversible process; the purpose of a heat engine is a *partial* reversal of this process, the conversion of heat to work with as great an efficiency as possible. For maximum heat-engine efficiency, therefore, *we must avoid all irreversible processes.* This requirement turns out to be enough to determine the basic sequence of steps in the Carnot cycle, as we will show next.

Heat flow through a finite temperature drop is an irreversible process. Therefore, during heat transfer in the Carnot cycle there must be *no* finite temperature difference. When the engine takes heat from the hot reservoir at temperature T_H, the working substance of the engine must also be at T_H; otherwise, irreversible heat flow would occur. Similarly, when the engine discards heat to the cold reservoir at T_C, the engine itself must be at T_C. That is, every process that involves heat transfer must be *isothermal* at either T_H or T_C.

Conversely, in any process in which the temperature of the working substance of the engine is intermediate between T_H and T_C, there must be *no* heat transfer between the engine and either reservoir because such heat transfer could not be reversible. Therefore any process in which the temperature T of the working substance changes must be *adiabatic.*

The bottom line is that every process in our idealized cycle must be either isothermal or adiabatic. In addition, thermal and mechanical equilibrium must be maintained at all times so that each process is completely reversible.

The Carnot Cycle

The Carnot cycle consists of two reversible isothermal and two reversible adiabatic processes. Figure 20.11 shows a Carnot cycle using as its working substance an ideal gas in a cylinder with a piston. It consists of the following steps:

1. The gas expands isothermally at temperature T_H, absorbing heat Q_H (*ab*).
2. It expands adiabatically until its temperature drops to T_C (*bc*).
3. It is compressed isothermally at T_C, rejecting heat $|Q_C|$ (*cd*).
4. It is compressed adiabatically back to its initial state at temperature T_H (*da*).

20.11 The Carnot cycle for an ideal gas. The light blue lines in the pV-diagram are isotherms (curves of constant temperature) and the dark blue lines are adiabats (curves of zero heat flow).

We can calculate the thermal efficiency e of a Carnot engine in the special case in which the working substance is an *ideal gas*. To carry out this calculation, we will first find the ratio Q_C/Q_H of the quantities of heat transferred in the two isothermal processes and then use Eq. (20.4) to find e.

For an ideal gas the internal energy U depends only on temperature and is thus constant in any isothermal process. For the isothermal expansion ab, $\Delta U_{ab} = 0$ and Q_H is equal to the work W_{ab} done by the gas during its isothermal expansion at temperature T_H. We calculated this work in Example 19.1 (Section 19.2); using that result, we have

$$Q_H = W_{ab} = nRT_H \ln \frac{V_b}{V_a} \tag{20.10}$$

Similarly,

$$Q_C = W_{cd} = nRT_C \ln \frac{V_d}{V_c} = -nRT_C \ln \frac{V_c}{V_d} \tag{20.11}$$

Because V_d is less than V_c, Q_C is negative $(Q_C = -|Q_C|)$; heat flows out of the gas during the isothermal compression at temperature T_C.

The ratio of the two quantities of heat is thus

$$\frac{Q_C}{Q_H} = -\left(\frac{T_C}{T_H}\right) \frac{\ln\,(V_c/V_d)}{\ln\,(V_b/V_a)} \tag{20.12}$$

This can be simplified further by use of the temperature-volume relation for an adiabatic process, Eq. (19.22). We find for the two adiabatic processes:

$$T_H V_b^{\gamma-1} = T_C V_c^{\gamma-1} \quad \text{and} \quad T_H V_a^{\gamma-1} = T_C V_d^{\gamma-1}$$

Dividing the first of these by the second, we find

$$\frac{V_b^{\gamma-1}}{V_a^{\gamma-1}} = \frac{V_c^{\gamma-1}}{V_d^{\gamma-1}} \quad \text{and} \quad \frac{V_b}{V_a} = \frac{V_c}{V_d}$$

Thus the two logarithms in Eq. (20.12) are equal, and that equation reduces to

$$\frac{Q_C}{Q_H} = -\frac{T_C}{T_H} \quad \text{or} \quad \frac{|Q_C|}{|Q_H|} = \frac{T_C}{T_H} \quad \text{(heat transfer in a Carnot engine)} \quad (20.13)$$

The ratio of the heat rejected at T_C to the heat absorbed at T_H is just equal to the ratio T_C/T_H. Then from Eq. (20.4) the efficiency of the Carnot engine is

$$e_{\text{Carnot}} = 1 - \frac{T_C}{T_H} = \frac{T_H - T_C}{T_H} \quad \text{(efficiency of a Carnot engine)} \quad (20.14)$$

This simple result says that the efficiency of a Carnot engine depends only on the temperatures of the two heat reservoirs. The efficiency is large when the temperature *difference* is large, and it is very small when the temperatures are nearly equal. The efficiency can never be exactly unity unless $T_C = 0$; we'll see later that this, too, is impossible.

CAUTION In all calculations involving the Carnot cycle, you must make sure that you use *absolute* (Kelvin) temperatures only. That's because Eqs. (20.10) through (20.14) come from the ideal-gas equation $pV = nRT$, in which T is absolute temperature.

Example 20.2   **Analyzing a Carnot engine I**

A Carnot engine takes 2000 J of heat from a reservoir at 500 K, does some work, and discards some heat to a reservoir at 350 K. How much work does it do, how much heat is discarded, and what is the efficiency?

SOLUTION

IDENTIFY and SET UP: For this Carnot engine we are given $Q_H = 2000$ J, the amount of heat absorbed, and the temperatures $T_H = 500$ K and $T_C = 350$ K of the hot and cold reservoirs, respectively. We find the amount of heat discarded using Eq. (20.13), then calculate the amount of work done using the first law of thermodynamics: the work done in a complete cycle is the sum of the heat absorbed and the (negative) heat discarded (see Eq. (20.2)). We find the efficiency from the two temperatures using Eq. (20.14).

EXECUTE: From Eq. (20.13) the heat Q_C discarded by the engine is

$$Q_C = -Q_H\frac{T_C}{T_H} = -(2000\text{ J})\frac{350\text{ K}}{500\text{ K}}$$
$$= -1400\text{ J}$$

Then from the first law, the work W done by the engine is

$$W = Q_H + Q_C = 2000\text{ J} + (-1400\text{ J})$$
$$= 600\text{ J}$$

From Eq. (20.14) the thermal efficiency is

$$e = 1 - \frac{T_C}{T_H} = 1 - \frac{350\text{ K}}{500\text{ K}} = 0.30 = 30\%$$

EVALUATE: The negative sign of Q_C is correct: it shows that heat flows *out* of the engine and into the cold reservoir. Note that we can check our result for e by using the basic definition of thermal efficiency,

$$e = \frac{W}{Q_H} = \frac{600\text{ J}}{2000\text{ J}} = 0.30 = 30\%$$

Analyzing a Carnot engine II

Suppose 0.200 mol of an ideal diatomic gas ($\gamma = 1.40$) undergoes a Carnot cycle with temperatures 227°C and 27°C. The initial pressure is $p_a = 10.0 \times 10^5$ Pa, and during the isothermal expansion at the higher temperature the volume doubles. a) Find the pressure and volume at each of points a, b, c, and d in the pV-diagram of Fig. 20.11. b) Find Q, W, and ΔU for each step and for the entire cycle. c) Determine the efficiency directly from the results of part (b), and compare it with the result from Eq. (20.14).

SOLUTION

IDENTIFY: This problem involves the properties of the Carnot cycle as well as those of an ideal gas.

SET UP: We are given the number of moles and the pressure and temperature at point a (which is at the higher of the two reservoir temperatures), so we can find the volume at a using the ideal-gas equation. We then find the pressure and volume at the other points from the equations given in this section in combination with the ideal-gas equation. Next, for each step in the cycle, we use Eqs. (20.10) and (20.11) to find the heat flow and work done and we use Eq. (19.13) to calculate the internal energy change. As in the preceding example, we find the efficiency using Eq. (20.14).

EXECUTE: a) We first remember to convert the Celsius temperatures to absolute temperatures: The higher temperature is $T_{\mathrm{H}} = (227 + 273.15)$ K $= 500$ K and the lower temperature is $T_{\mathrm{C}} = (27 + 273.15)$ K $= 300$ K. We then use the ideal-gas equation to find V_a:

$$V_a = \frac{nRT_{\mathrm{H}}}{p_a} = \frac{(0.200 \text{ mol})(8.314 \text{ J/mol} \cdot \text{K})(500 \text{ K})}{10.0 \times 10^5 \text{ Pa}}$$

$$= 8.31 \times 10^{-4} \text{ m}^3$$

The volume doubles during the isothermal expansion $a \rightarrow b$, so

$$V_b = 2V_a = 2(8.31 \times 10^{-4} \text{ m}^3)$$

$$= 16.6 \times 10^{-4} \text{ m}^3$$

Also, during the isothermal expansion $a \rightarrow b$, $p_a V_a = p_b V_b$, so

$$p_b = \frac{p_a V_a}{V_b} = 5.00 \times 10^5 \text{ Pa}$$

For the adiabatic expansion $b \rightarrow c$, $T_{\mathrm{H}} V_b^{\gamma-1} = T_{\mathrm{C}} V_c^{\gamma-1}$, so

$$V_c = V_b \left(\frac{T_{\mathrm{H}}}{T_{\mathrm{C}}}\right)^{1/(\gamma-1)} = (16.6 \times 10^{-4} \text{ m}^3)\left(\frac{500 \text{ K}}{300 \text{ K}}\right)^{2.5}$$

$$= 59.6 \times 10^{-4} \text{ m}^3$$

Using the ideal-gas equation again for point c, we find

$$p_c = \frac{nRT_{\mathrm{C}}}{V_c} = \frac{(0.200 \text{ mol})(8.314 \text{ J/mol} \cdot \text{K})(300 \text{ K})}{59.6 \times 10^{-4} \text{ m}^3}$$

$$= 0.837 \times 10^5 \text{ Pa}$$

For the adiabatic compression $d \rightarrow a$, $T_{\mathrm{C}} V_d^{\gamma-1} = T_{\mathrm{H}} V_a^{\gamma-1}$, and

$$V_d = V_a \left(\frac{T_{\mathrm{H}}}{T_{\mathrm{C}}}\right)^{1/(\gamma-1)} = (8.31 \times 10^{-4} \text{ m}^3)\left(\frac{500 \text{ K}}{300 \text{ K}}\right)^{2.5}$$

$$= 29.8 \times 10^{-4} \text{ m}^3$$

$$p_d = \frac{nRT_{\mathrm{C}}}{V_d} = \frac{(0.200 \text{ mol})(8.314 \text{ J/mol} \cdot \text{K})(300 \text{ K})}{29.8 \times 10^{-4} \text{ m}^3}$$

$$= 1.67 \times 10^5 \text{ Pa}$$

b) For the isothermal expansion $a \rightarrow b$, $\Delta U_{ab} = 0$. To find W_{ab} ($= Q_{\mathrm{H}}$), we use Eq. (20.10):

$$W_{ab} = Q_{\mathrm{H}} = nRT_{\mathrm{H}} \ln \frac{V_b}{V_a}$$

$$= (0.200 \text{ mol})(8.314 \text{ J/mol} \cdot \text{K})(500 \text{ K})(\ln 2) = 576 \text{ J}$$

For the adiabatic expansion $b \rightarrow c$, $Q_{bc} = 0$. From the first law of thermodynamics, $\Delta U_{bc} = Q_{bc} - W_{bc} = -W_{bc}$; hence the work W_{bc} done by the gas in this process equals the negative of the change in internal energy of the gas. From Eq. (19.13) we have $\Delta U = nC_V \Delta T$, where $\Delta T = T_{\mathrm{C}} - T_{\mathrm{H}}$ (final temperature minus initial temperature). Using $C_V = 20.8$ J/mol·K for an ideal diatomic gas, we find

$$W_{bc} = -\Delta U_{bc} = -nC_V(T_{\mathrm{C}} - T_{\mathrm{H}}) = nC_V(T_{\mathrm{H}} - T_{\mathrm{C}})$$

$$= (0.200 \text{ mol})(20.8 \text{ J/mol} \cdot \text{K})(500 \text{ K} - 300 \text{ K}) = 832 \text{ J}$$

For the isothermal compression $c \rightarrow d$, $\Delta U_{cd} = 0$; Eq. (20.11) gives

$$W_{cd} = Q_{\mathrm{C}} = nRT_{\mathrm{C}} \ln \frac{V_d}{V_c}$$

$$= (0.200 \text{ mol})(8.314 \text{ J/mol} \cdot \text{K})(300 \text{ K})\left(\ln \frac{29.8 \times 10^{-4} \text{ m}^3}{59.6 \times 10^{-4} \text{ m}^3}\right)$$

$$= -346 \text{ J}$$

For the adiabatic compression $d \rightarrow a$, $Q_{da} = 0$, and

$$W_{da} = -\Delta U_{da} = -nC_V(T_{\mathrm{H}} - T_{\mathrm{C}}) = nC_V(T_{\mathrm{C}} - T_{\mathrm{H}})$$

$$= (0.200 \text{ mol})(20.8 \text{ J/mol} \cdot \text{K})(300 \text{ K} - 500 \text{ K})$$

$$= -832 \text{ J}$$

We can tabulate the results as follows:

PROCESS	Q	W	ΔU
$a \rightarrow b$	576 J	576 J	0
$b \rightarrow c$	0	832 J	-832 J
$c \rightarrow d$	-346 J	-346 J	0
$d \rightarrow a$	0	-832 J	832 J
Total	230 J	230 J	0

c) From the table, $Q_H = 576$ J and the total work is 230 J. Thus

$$e = \frac{W}{Q_H} = \frac{230 \text{ J}}{576 \text{ J}} = 0.40 = 40\%$$

We can compare this with the result from Eq. (20.14):

$$e = \frac{T_H - T_C}{T_H} = \frac{500 \text{ K} - 300 \text{ K}}{500 \text{ K}} = 0.40 = 40\%$$

EVALUATE: In the table of results in part (b), note that for the entire cycle $Q = W$ and $\Delta U = 0$. These results are just what we would

expect: in a complete cycle, the net heat input is used to do work with zero net change in the internal energy of the system. Note also that the quantities of work in the two adiabatic processes are negatives of each other. Can you show from the analysis leading to Eq. (20.13) that this must *always* be the case in a Carnot cycle?

Note that the efficiency in this example is greater than that obtained in Example 20.2. That's because the ratio of the high and low temperatures is higher, (500 K)/(300 K) as compared to (500 K)/(350 K).

The Carnot Refrigerator

Because each step in the Carnot cycle is reversible, the *entire cycle* may be reversed, converting the engine into a refrigerator. The coefficient of performance of the Carnot refrigerator is obtained by combining the general definition of K, Eq. (20.9), with Eq. (20.13) for the Carnot cycle. We first rewrite Eq. (20.9) as

$$K = \frac{|Q_C|}{|Q_H| - |Q_C|} = \frac{|Q_C|/|Q_H|}{1 - |Q_C|/|Q_H|}$$

Then we substitute Eq. (20.13), $|Q_C|/|Q_H| = T_C/T_H$, into this expression. The result is

$$K_{\text{Carnot}} = \frac{T_C}{T_H - T_C} \qquad (20.15)$$

(coefficient of performance of a Carnot refrigerator)

When the temperature difference $T_H - T_C$ is small, K is much larger than unity; in this case a lot of heat can be "pumped" from the lower to the higher temperature with only a little expenditure of work. But the greater the temperature difference, the smaller the value of K and the more work is required to transfer a given quantity of heat.

Example 20.4	**Analyzing a Carnot refrigerator**

If the cycle described in Example 20.3 is run backward as a refrigerator, what is its coefficient of performance?

SOLUTION

IDENTIFY and SET UP: Equation (20.9) gives the coefficient of performance of *any* refrigerator in terms of the heat extracted from the cold reservoir per cycle and the work that must be done per cycle.

EXECUTE: In Example 20.3 we found that in one cycle the Carnot engine rejects heat $Q_C = -346$ J to the cold reservoir and does work $W = 230$ J. Hence, when run in reverse as a refrigerator, the system extracts heat $Q_C = +346$ J from the cold reservoir while requiring a work input of $W = -230$ J. From Eq. (20.9),

$$K = \frac{|Q_C|}{|W|} = \frac{346 \text{ J}}{230 \text{ J}} = 1.50$$

Because the cycle is a Carnot cycle, we may also use Eq. (20.15):

$$K = \frac{T_C}{T_H - T_C} = \frac{300 \text{ K}}{500 \text{ K} - 300 \text{ K}} = 1.50$$

EVALUATE: For a Carnot cycle, e and K depend only on the temperatures, as shown by Eqs. (20.14) and (20.15), and we don't need to calculate Q and W. For cycles containing irreversible processes, however, these two equations are not valid, and more detailed calculations are necessary.

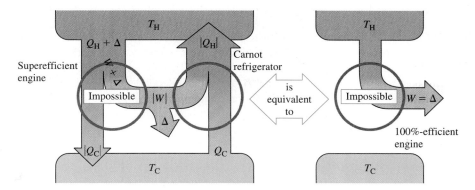

20.12 Proving that the Carnot engine has the highest possible efficiency. A "super-efficient" engine (more efficient than a Carnot engine) combined with a Carnot refrigerator could convert heat completely into work with no net heat transfer to the cold reservoir. This would violate the second law of thermodynamics.

The Carnot Cycle and the Second Law

We can prove that **no engine can be more efficient than a Carnot engine operating between the same two temperatures.** The key to the proof is the above observation that since each step in the Carnot cycle is reversible, the *entire cycle* may be reversed. Run backward, the engine becomes a refrigerator. Suppose we have an engine that is more efficient than a Carnot engine (Fig. 20.12). Let the Carnot engine, run backward as a refrigerator by negative work $-|W|$, take in heat Q_C from the cold reservoir and expel heat $|Q_H|$ to the hot reservoir. The superefficient engine expels heat $|Q_C|$, but to do this, it takes in a greater amount of heat $Q_H + \Delta$. Its work output is then $W + \Delta$, and the net effect of the two machines together is to take a quantity of heat Δ and convert it completely into work. This violates the engine statement of the second law. We could construct a similar argument that a superefficient engine could be used to violate the refrigerator statement of the second law. Note that we don't have to assume that the superefficient engine is reversible. In a similar way we can show that *no refrigerator can have a greater coefficient of performance than a Carnot refrigerator operating between the same two temperatures.*

Thus the statement that no engine can be more efficient than a Carnot engine is yet another equivalent statement of the second law of thermodynamics. It also follows directly that **all Carnot engines operating between the same two temperatures have the same efficiency, irrespective of the nature of the working substance.** Although we derived Eq. (20.14) for a Carnot engine using an ideal gas as its working substance, it is in fact valid for *any* Carnot engine, no matter what its working substance.

Equation (20.14), the expression for the efficiency of a Carnot engine, sets an upper limit to the efficiency of a real engine such as a steam turbine. To maximize this upper limit and the actual efficiency of the real engine, the designer must make the intake temperature T_H as high as possible and the exhaust temperature T_C as low as possible (Fig. 20.13).

The exhaust temperature cannot be lower than the lowest temperature available for cooling the exhaust. For a steam turbine at an electric power plant, T_C may be the temperature of river or lake water; then we want the boiler temperature T_H to be as high as possible. The vapor pressures of all liquids increase rapidly with temperature, so we are limited by the mechanical strength of the boiler. At 500°C the vapor pressure of water is about 240×10^5 Pa (235 atm); this is about the maximum practical pressure in large present-day steam boilers.

20.13 To maximize efficiency, the temperatures inside a jet engine are made as high as possible. Exotic ceramic materials are used that can withstand temperatures in excess of 1000°C without melting or becoming soft.

*The Kelvin Temperature Scale

In Chapter 17 we expressed the need for a temperature scale that doesn't depend on the properties of any particular material. We can now use the Carnot cycle to define such a scale. The thermal efficiency of a Carnot engine operating between two heat reservoirs at temperatures T_H and T_C is independent of the nature of the working substance and depends only on the temperatures. From Eq. (20.4), this thermal efficiency is

$$e = \frac{Q_H + Q_C}{Q_H} = 1 + \frac{Q_C}{Q_H}$$

Therefore the ratio Q_C/Q_H is the same for *all* Carnot engines operating between two given temperatures T_H and T_C.

Kelvin proposed that we *define* the ratio of the temperatures, T_C/T_H, to be equal to the magnitude of the ratio Q_C/Q_H of the quantities of heat absorbed and rejected:

$$\frac{T_C}{T_H} = \frac{|Q_C|}{|Q_H|} = -\frac{Q_C}{Q_H} \qquad \text{(definition of Kelvin temperature)} \qquad (20.16)$$

Equation (20.16) looks identical to Eq. (20.13), but there is a subtle and crucial difference. The temperatures in Eq. (20.13) are based on an ideal-gas thermometer, as defined in Section 17.3, while Eq. (20.16) *defines* a temperature scale based on the Carnot cycle and the second law of thermodynamics and is independent of the behavior of any particular substance. Thus the **Kelvin temperature scale** is truly *absolute.* To complete the definition of the Kelvin scale, we assign, as in Section 17.3, the arbitrary value of 273.16 K to the temperature of the triple point of water. When a substance is taken around a Carnot cycle, the ratio of the heats absorbed and rejected, $|Q_H|/|Q_C|$, is equal to the ratio of the temperatures of the reservoirs *as expressed on the gas-thermometer scale* defined in Section 17.3. Since the triple point of water is chosen to be 273.16 K in both scales, it follows that *the Kelvin and ideal-gas scales are identical.*

The zero point on the Kelvin scale is called **absolute zero.** Absolute zero can be interpreted on a molecular level; at absolute zero the system has its *minimum* possible total internal energy (kinetic plus potential). Because of quantum effects, however, it is *not* true that at $T = 0$, all molecular motion ceases. There are theoretical reasons for believing that absolute zero cannot be attained experimentally, although temperatures below 10^{-7} K have been achieved. The more closely we approach absolute zero, the more difficult it is to get closer. One statement of the *third law of thermodynamics* is that it is impossible to reach absolute zero in a finite number of thermodynamic steps.

Test Your Understanding

An inventor looking for financial support comes to you with an idea for a gasoline engine that runs on a novel type of thermodynamic cycle. His design is made entirely of copper and is air-cooled. He claims that the engine will be 85% efficient. Should you invest in this marvelous new engine? (*Hint:* See Table 17.4.)

20.7 | Entropy

The second law of thermodynamics, as we have stated it, is rather different in form from many familiar physical laws. It is not an equation or a quantitative relationship but rather a statement of *impossibility*. However, the second law *can* be stated as a quantitative relation with the concept of *entropy,* the subject of this section.

We have talked about several processes that proceed naturally in the direction of increasing disorder. Irreversible heat flow increases disorder because the molecules are initially sorted into hotter and cooler regions; this sorting is lost when the system comes to thermal equilibrium. Adding heat to a body increases its disorder because it increases average molecular speeds and therefore the randomness of molecular motion. Free expansion of a gas increases its disorder because the molecules have greater randomness of position after the expansion than before.

Entropy and Disorder

Entropy provides a *quantitative* measure of disorder. To introduce this concept, let's consider an infinitesimal isothermal expansion of an ideal gas. We add heat dQ and let the gas expand just enough to keep the temperature constant. Because the internal energy of an ideal gas depends only on its temperature, the internal energy is also constant; thus from the first law, the work dW done by the gas is equal to the heat dQ added. That is,

$$dQ = dW = p\,dV = \frac{nRT}{V}\,dV \quad \text{so} \quad \frac{dV}{V} = \frac{dQ}{nRT}$$

The gas is in a more disordered state after the expansion than before because the molecules are moving in a larger volume and have more randomness of position. Thus the fractional volume change dV/V is a measure of the increase in disorder, and the above equation shows that it is proportional to the quantity dQ/T. We introduce the symbol S for the entropy of the system, and we define the infinitesimal entropy change dS during an infinitesimal reversible process at absolute temperature T as

$$dS = \frac{dQ}{T} \qquad (\text{infinitesimal reversible process}) \qquad (20.17)$$

If a total amount of heat Q is added during a reversible isothermal process at absolute temperature T, the total entropy change $\Delta S = S_2 - S_1$ is given by

$$\Delta S = S_2 - S_1 = \frac{Q}{T} \qquad (\text{reversible isothermal process}) \qquad (20.18)$$

Entropy has units of energy divided by temperature; the SI unit of entropy is 1 J/K.

We can see how the quotient Q/T is related to the increase in disorder. Higher temperature means greater randomness of motion. If the substance is initially cold, with little molecular motion, adding heat Q causes a substantial fractional increase in molecular motion and randomness. But if the substance is already hot, the same quantity of heat adds relatively little to the greater molecular motion already present. So the quotient Q/T is an appropriate characterization of the increase in randomness or disorder when heat flows into a system.

| Example 20.5 | **Entropy change in melting** |

One kilogram of ice at 0°C is melted and converted to water at 0°C. Compute its change in entropy, assuming that the melting is done reversibly. The heat of fusion of water is $L_f = 3.34 \times 10^5$ J/kg.

SOLUTION

IDENTIFY: The melting occurs at a constant temperature of 0°C, so this is a reversible isothermal process.

SET UP: We are given the amount of heat added (in terms of the heat of fusion) and the temperature $T = 273$ K. (Note that in entropy calculations we must always use absolute, or Kelvin, temperatures.) We can then calculate the entropy change using Eq. (20.18).

EXECUTE: The heat needed to melt the ice is $Q = mL_f = 3.34 \times 10^5$ J. From Eq. (20.18) the increase in entropy of the system is

$$\Delta S = S_2 - S_1 = \frac{Q}{T} = \frac{3.34 \times 10^5 \text{ J}}{273 \text{ K}} = 1.22 \times 10^3 \text{ J/K}$$

EVALUATE: This increase corresponds to the increase in disorder when the water molecules go from the highly ordered state of a crystalline solid to the much more disordered state of a liquid (Fig. 20.14).

In any *isothermal* reversible process, the entropy change equals the heat transferred divided by the absolute temperature. When we refreeze the water, Q has the opposite sign, and the entropy change of the water is $\Delta S = -1.22 \times 10^3$ J/K. The water molecules rearrange themselves into a crystal to form ice, so disorder and entropy both decrease.

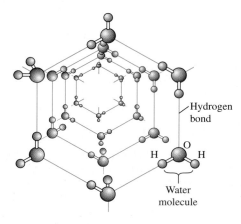

20.14 Water molecules are arranged in a regular, ordered way in an ice crystal. When the ice melts, the hydrogen bonds between molecules are broken, increasing the water's disorder and its entropy.

We can generalize the definition of entropy change to include *any* reversible process leading from one state to another, whether it is isothermal or not. We represent the process as a series of infinitesimal reversible steps. During a typical step, an infinitesimal quantity of heat dQ is added to the system at absolute temperature T. Then we sum (integrate) the quotients dQ/T for the entire process; that is,

$$\Delta S = \int_1^2 \frac{dQ}{T} \qquad \text{(entropy change in a reversible process)} \qquad (20.19)$$

The limits 1 and 2 refer to the initial and final states.

Because entropy is a measure of the disorder of a system in any specific state, it must depend only on the current state of the system, not on its past history. We will show later that this is indeed the case. When a system proceeds from an initial state with entropy S_1 to a final state with entropy S_2, the change in entropy $\Delta S = S_2 - S_1$ defined by Eq. (20.19) does not depend on the path leading from the initial to the final state but is the same for *all possible* processes leading from state 1 to state 2. Thus the entropy of a system must also have a definite value for any given state of the system. We recall that *internal energy*, introduced in Chapter 19, also has this property, although entropy and internal energy are very different quantities.

Since entropy is a function only of the state of a system, we can also compute entropy changes in *irreversible* (non-equilibrium) processes for which Eqs. (20.17) and (20.19) are not applicable. We simply invent a path connecting the given initial and final states that *does* consist entirely of reversible, equilibrium processes and compute the total entropy change for that path. It is not the actual path, but the entropy change must be the same as for the actual path.

As with internal energy, the above discussion does not tell us how to calculate entropy itself, but only the change in entropy in any given process. Just as with internal energy, we may arbitrarily assign a value to the entropy of a system in a specified reference state and then calculate the entropy of any other state with reference to this.

Example 20.6

Entropy change in a temperature change

One kilogram of water at 0°C is heated to 100°C. Compute its change in entropy.

SOLUTION

IDENTIFY and SET UP: In practice, the process described would be done irreversibly, perhaps by setting a pan of water on an electric range whose cooking surface is maintained at 100°C. But the entropy change of the water depends only on the initial and final states of the system, and is the same whether the process is reversible or irreversible. Hence we can imagine that the temperature of the water is increased reversibly in a series of infinitesimal steps, in each of which the temperature is raised by an infinitesimal amount dT. We then use Eq. (20.19) to integrate over all these steps and calculate the entropy change for the total process.

EXECUTE: From Eq. (17.14) the heat required to carry out each such infinitesimal step is $dQ = mc\,dT$. Substituting this into Eq. (20.19) and integrating, we find

$$\Delta S = S_2 - S_1 = \int_1^2 \frac{dQ}{T} = \int_{T_1}^{T_2} mc\,\frac{dT}{T} = mc\,\ln\frac{T_2}{T_1}$$

$$= (1.00\ \text{kg})(4190\ \text{J/kg}\cdot\text{K})\left(\ln\frac{373\ \text{K}}{273\ \text{K}}\right)$$

$$= 1.31 \times 10^3\ \text{J/K}$$

EVALUATE: The entropy change is positive, as it must be for a process in which the system absorbs heat.

In this calculation we assumed that the specific heat c doesn't depend on temperature. That's a pretty good approximation, since c for water increases by only 1% between 0°C and 100°C.

CAUTION In solving this problem you might be tempted to avoid doing an integral by using the simpler expression in Eq. (20.18), $\Delta S = Q/T$. This would be incorrect, however, because Eq. (20.18) is applicable only to *isothermal* processes and the initial and final temperatures in our example are *not* the same. If you persist and use $\Delta S = Q/T$ anyway, you would know what to use for the numerator $[Q = (4190\ \text{J/kg}\cdot\text{K})(1.00\ \text{kg})(100\ \text{K}) = 4.19 \times 10^5\ \text{J}]$ but you would be confused about whether to use $T = 273\ \text{K}$ (the initial temperature T_1), 373 K (the final temperature T_2), or perhaps 323 K (the average of T_1 and T_2). It doesn't matter, because none of these choices would give you the right answer! (Try it and see.) The *only* correct way to find the entropy change in a process with different initial and final temperatures is to use Eq. (20.19).

Conceptual Example 20.7

A reversible adiabatic process

A gas expands adiabatically and reversibly. What is its change in entropy?

SOLUTION

In an adiabatic process, no heat enters or leaves the system. Hence $dQ = 0$ and there is *no* change in entropy in this reversible process:

$\Delta S = 0$. Every *reversible* adiabatic process is a constant-entropy process. (For this reason, reversible adiabatic processes are also called *isentropic* processes.) The increase in disorder resulting from the gas occupying a greater volume is exactly balanced by the decrease in disorder associated with the lowered temperature and reduced molecular speeds.

| Example 20.8 | **Entropy change in a free expansion** |

A thermally insulated box is divided by a partition into two compartments, each having volume V (Fig. 20.15). Initially, one compartment contains n moles of an ideal gas at temperature T, and the other compartment is evacuated. We then break the partition, and the gas expands to fill both compartments. What is the entropy change in this free-expansion process?

SOLUTION

IDENTIFY: For this process, $Q = 0$, $W = 0$, $\Delta U = 0$, and therefore (because the system is an ideal gas) $\Delta T = 0$. We might think that the entropy change is zero because there is no heat exchange. But Eq. (20.19) can be used to calculate entropy changes for *reversible* processes only; this free expansion is *not* reversible, and there *is* an entropy change. The process is adiabatic because $Q = 0$, but it is not isentropic because $\Delta S \neq 0$. As we mentioned at the beginning of this section, entropy increases in a free expansion because the positions of the molecules are more random than before the expansion.

SET UP: To calculate ΔS, we recall that the entropy change depends only on the initial and final states. We can devise a *reversible* process having the same endpoints, use Eq. (20.19) to calculate its entropy change, and thus determine the entropy change in the original process. An appropriate reversible process in this case is an isothermal expansion from V to $2V$ at temperature T. The gas does work W during this substitute expansion, so an equal amount of heat Q must be supplied to keep the internal energy constant. We find the entropy change for this reversible isothermal process using Eq. (20.18); the entropy change for the free expansion will be the same.

EXECUTE: We found in Example 19.1 (Section 19.2) that the work done by n moles of ideal gas in an isothermal expansion from V_1 to V_2 is $W = nRT \ln (V_2/V_1)$. Using $V_1 = V$ and $V_2 = 2V$, we have

$$Q = W = nRT \ln \frac{2V}{V} = nRT \ln 2$$

Thus the entropy change is

$$\Delta S = \frac{Q}{T} = nR \ln 2$$

which is also the entropy change for the free expansion with the same initial and final states. For one mole,

$$\Delta S = (1 \text{ mol})(8.314 \text{ J/mol} \cdot \text{K})(\ln 2) = 5.76 \text{ J/K}$$

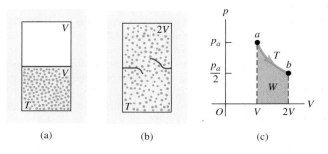

(a)   (b)   (c)

20.15 (a,b) Free expansion of an insulated ideal gas. (c) The free-expansion process doesn't pass through equilibrium states from a to b. However, the entropy change $S_b - S_a$ can be calculated by using the isothermal path shown or *any* reversible path from a to b.

EVALUATE: The entropy change is positive, as we predicted. The factor ($\ln 2$) in our answer is a result of the volume having increased by a factor of 2. Can you show that if the volume had increased in the free expansion from V to xV, where x is an arbitrary number, the entropy change would have been $\Delta S = nR \ln x$?

| Example 20.9 | **Entropy and the Carnot cycle** |

For the Carnot engine in Example 20.2 (Section 20.6), find the total entropy change in the engine during one cycle.

SOLUTION

IDENTIFY and SET UP: The Carnot cycle is made up of four reversible steps, shown in Fig. 20.11. We find the entropy change

ΔS for each step, then add the entropy changes to get the total ΔS for the cycle as a whole.

EXECUTE: There is no entropy change during the adiabatic expansion or adiabatic compression. During the isothermal expansion at $T_H = 500$ K the engine takes in 2000 J of heat, and its entropy change, from Eq. (20.18), is

$$\Delta S_H = \frac{Q_H}{T_H} = \frac{2000 \text{ J}}{500 \text{ K}} = 4.0 \text{ J/K}$$

During the isothermal compression at $T_C = 350$ K the engine gives off 1400 J of heat, and its entropy change is

$$\Delta S_C = \frac{Q_C}{T_C} = \frac{-1400 \text{ J}}{350 \text{ K}} = -4.0 \text{ J/K}$$

The total entropy change in the engine during one cycle is $\Delta S_{total} = \Delta S_H + \Delta S_C = 4.0 \text{ J/K} + (-4.0 \text{ J/K}) = 0$.

EVALUATE: The result $\Delta S_{total} = 0$ tells us that when the Carnot engine completes a cycle, it has the same entropy as it did at the beginning of the cycle. We'll explore this result in the following subsection.

What is the total entropy change of the engine's *environment* during this cycle? The hot (500 K) reservoir gives off 2000 J of heat during the reversible isothermal expansion, so its entropy change is $(-2000 \text{ J})/(500 \text{ K}) = -4.0$ J/K; the cold (350 K) reservoir absorbs 1400 J of heat during the reversible isothermal compression, so its entropy change is $(+1400 \text{ J})/(350 \text{ K}) = +4.0$ J/K. Thus each individual reservoir has an entropy change; however, the sum of these changes—that is, the total entropy change of the system's environment—is zero.

These results apply to the special case of the Carnot cycle, for which *all* of the processes are reversible. In this case we find that the total entropy change of the system and the environment together is zero. We will see that if the cycle includes irreversible processes (as is the case for the Otto cycle or Diesel cycle of Section 20.3), the total entropy change of the system and the environment *cannot* be zero, but rather must be positive.

Entropy in Cyclic Processes

Example 20.9 showed that the total entropy change for a cycle of a particular Carnot engine, which uses an ideal gas as its working substance, is zero. This result follows directly from Eq. (20.13), which we can rewrite as

$$\frac{Q_H}{T_H} + \frac{Q_C}{T_C} = 0 \qquad (20.20)$$

The quotient Q_H/T_H equals ΔS_H, the entropy change of the engine that occurs at $T = T_H$. Likewise, Q_C/T_C equals ΔS_C, the (negative) entropy change of the engine that occurs at $T = T_C$. Hence Eq. (20.20) says that $\Delta S_H + \Delta S_C = 0$; that is, there is zero net entropy change in one cycle.

What about Carnot engines that use a different working substance? According to the second law, *any* Carnot engine operating between given temperatures T_H and T_C has the same efficiency $e = 1 - T_C/T_H$ (Eq. (20.14)). Combining this expression for e with Eq. (20.4), $e = 1 + Q_C/Q_H$, just reproduces Eq. (20.20). So Eq. (20.20) is valid for any Carnot engine working between these temperatures, whether its working substance is an ideal gas or not. We conclude that *the total entropy change in one cycle of any Carnot engine is zero*.

This result can be generalized to show that the total entropy change during *any* reversible cyclic process is zero. A reversible cyclic process appears on a pV-diagram as a closed path (Fig. 20.16a). We can approximate such as path as closely as we like by a sequence of isothermal and adiabatic processes forming parts of

20.16 (a) A reversible cyclic process for an ideal gas is shown as a red closed path on a pV-diagram. Several ideal-gas isotherms are shown in blue. (b) We can approximate the path in (a) by a series of long, thin Carnot cycles; one cycle is highlighted in gold. The total entropy change is zero for each Carnot cycle and for the actual cyclic process. (c) The entropy change between points a and b is independent of the path.

(a)

(b)

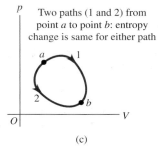

(c)

many long, thin Carnot cycles (Fig. 20.16b). The total entropy change for the full cycle is the sum of the entropy changes for each small Carnot cycle, each of which is zero. So **the total entropy change during *any* reversible cycle is zero:**

$$\oint \frac{dQ}{T} = 0 \quad \text{(reversible cyclic process)} \tag{20.21}$$

It follows that when a system undergoes a reversible process leading from any state *a* to any other state *b*, *the entropy change of the system is independent of the path* (Fig. 20.16c). If the entropy change for path 1 were different from the change for path 2, the system could be taken along path 1 and then backward along path 2 to the starting point, with a nonzero net change in entropy. This would violate the conclusion that the total entropy change in such a cyclic process must be zero. Because the entropy change in such processes is independent of path, we conclude that in any given state, the system has a definite value of entropy that depends only on the state, not on the processes that led to that state.

Entropy in Irreversible Processes

In an idealized, reversible process involving only equilibrium states, the total entropy change of the system and its surroundings is zero. But all *irreversible* processes involve an increase in entropy. Unlike energy, *entropy is not a conserved quantity.* The entropy of an isolated system *can* change, but as we shall see, it can never decrease. The free expansion of a gas, described in Example 20.8, is an irreversible process in an isolated system in which there is an entropy increase.

Example 20.10	**An irreversible process**

Suppose 1.00 kg of water at 100°C is placed in thermal contact with 1.00 kg of water at 0°C. What is the total change in entropy? Assume that the specific heat of water is constant at 4190 J/kg·K over this temperature range.

SOLUTION

IDENTIFY and SET UP: This process involves irreversible heat flow because of the temperature differences. We nonetheless calculate the entropy changes for the (initially) hot water and the (initially) cold water in the same way as in Example 20.6 by assuming that the process occurs reversibly. We calculate ΔS for each substance using Eq. (20.19) because the temperatures change in the process.

EXECUTE: The final temperature is 50°C = 323 K. The entropy change of the hot water is

$$\Delta S_{\text{hot}} = mc \int_{T_1}^{T_2} \frac{dT}{T} = (1.00 \text{ kg})(4190 \text{ J/kg·K}) \int_{373 \text{ K}}^{323 \text{ K}} \frac{dT}{T}$$

$$= (4190 \text{ J/K})\left(\ln \frac{323 \text{ K}}{373 \text{ K}}\right) = -603 \text{ J/K}$$

The entropy change of the cold water is

$$\Delta S_{\text{cold}} = (4190 \text{ J/K})\left(\ln \frac{323 \text{ K}}{273 \text{ K}}\right) = +705 \text{ J/K}$$

The *total* entropy change of the system is

$$\Delta S_{\text{total}} = \Delta S_{\text{hot}} + \Delta S_{\text{cold}} = (-603 \text{ J/K}) + 705 \text{ J/K} = +102 \text{ J/K}$$

EVALUATE: An irreversible heat flow in an isolated system is accompanied by an increase in entropy. We could have reached the same end state by simply mixing the two quantities of water. This, too, is an irreversible process; because the entropy depends only on the state of the system, the total entropy change would be the same, 102 J/K.

It's worth noting that the entropy of the system increases *continuously* as the two quantities of water come to equilibrium. For example, the first 4190 J of heat transferred cools the hot water to 99°C and warms the cold water to 1°C. The net change in entropy for this step is approximately

$$\Delta S = \frac{-4190 \text{ J}}{373 \text{ K}} + \frac{4190 \text{ J}}{273 \text{ K}} = +4.1 \text{ J/K}$$

Can you show in a similar way that the net entropy change is positive for *any* one-degree temperature change leading to the equilibrium condition?

Entropy and the Second Law

The results of Example 20.10 about the flow of heat from a higher to a lower temperature, or the mixing of substances at different temperatures, are characteristic of *all* natural (that is, irreversible) processes. When we include the entropy changes of all the systems taking part in the process, the increases in entropy are always greater than the decreases. In the special case of a *reversible* process, the increases and decreases are equal. Hence we can state the general principle: **When all systems taking part in a process are included, the entropy either remains constant or increases.** In other words, **no process is possible in which the total entropy decreases, when all systems taking part in the process are included.** This is an alternative statement of the second law of thermodynamics in terms of entropy. Thus it is equivalent to the "engine" and "refrigerator" statements discussed earlier. Figure 20.17 shows a specific example of this general principle.

The increase of entropy in every natural, irreversible process measures the increase of disorder or randomness in the universe associated with that process. Consider again the example of mixing hot and cold water (Example 20.10). We *might* have used the hot and cold water as the high- and low-temperature reservoirs of a heat engine. While removing heat from the hot water and giving heat to the cold water, we could have obtained some mechanical work. But once the hot and cold water have been mixed and have come to a uniform temperature, this opportunity to convert heat to mechanical work is lost irretrievably. The lukewarm water will never *unmix* itself and separate into hotter and colder portions. No decrease in *energy* occurs when the hot and cold water are mixed. What has been lost is not *energy,* but *opportunity,* the opportunity to convert part of the heat from the hot water into mechanical work. Hence when entropy increases, energy becomes less *available,* and the universe becomes more random or "run down."

Test Your Understanding

Suppose 2.00 kg of water at 50°C spontaneously changes temperature, so that half of the water cools to 0°C while the other half spontaneously warms to 100°C. (All of the water remains liquid, so it doesn't freeze or boil.) What would be the entropy change of the water? Is this process possible? Why or why not?

20.17 The mixing of colored ink and water starts from a state of relative order (low entropy) in which each fluid is separate and distinct from the other. The final state after mixing is more disordered (has greater entropy). Spontaneous unmixing of the ink and water, a process in which there would be a net decrease in entropy, is never observed.

*20.8 | Microscopic Interpretation of Entropy

We described in Section 19.4 how the internal energy of a system could be calculated, at least in principle, by adding up all the kinetic energies of its constituent particles and all the potential energies of interaction among the particles. This is called a *microscopic* calculation of the internal energy. We can also make a microscopic calculation of the entropy S of a system. Unlike energy, however, entropy is not something that belongs to each individual particle or pair of particles in the system. Rather, entropy is a measure of the disorder of the system as a whole. To see how to calculate entropy microscopically, we first have to introduce the idea of *macroscopic* and *microscopic states.*

Suppose you toss N identical coins on the floor, and half of them show heads and half show tails. This is a description of the large-scale or **macroscopic state** of the system of N coins. A description of the **microscopic state** of the system includes information about each individual coin: Coin 1 was heads, coin 2 was tails, coin 3 was tails, and so on. There can be many microscopic states that cor-

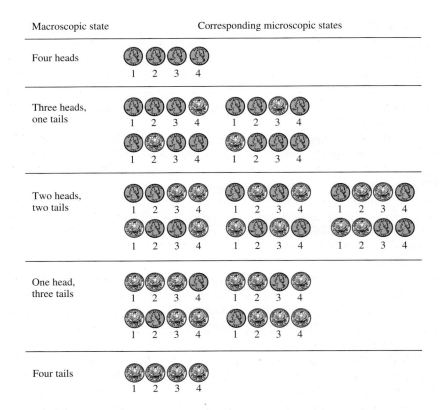

Macroscopic state	Corresponding microscopic states

20.18 All possible microscopic states of four coins. There can be several possible microscopic states for each macroscopic state.

respond to the same macroscopic description. For instance, with $N = 4$ coins there are six possible states in which half are heads and half are tails (Fig. 20.18). The number of microscopic states grows rapidly with increasing N; for $N = 100$ there are $2^{100} = 1.27 \times 10^{30}$ microscopic states, of which 1.01×10^{29} are half heads and half tails.

The least probable outcomes of the coin toss are the states that are either all heads or all tails. It is certainly possible that you could throw 100 heads in a row, but don't bet on it; the probability of doing this is only 1 in 1.27×10^{30}. The most probable outcome of tossing N coins is that half are heads and half are tails. The reason is that this *macroscopic* state has the greatest number of corresponding *microscopic* states, as shown in Fig. 20.18.

To make the connection to the concept of entropy, note that N coins that are all heads constitutes a completely ordered macroscopic state; the description "all heads" completely specifies the state of each one of the N coins. The same is true if the coins are all tails. But the macroscopic description "half heads, half tails" by itself tells you very little about the state (heads or tails) of each individual coin. We say that the system is *disordered* because we know so little about its microscopic state. Compared to the state "all heads" or "all tails," the state "half heads, half tails" has a much greater number of possible microscopic states, much greater disorder, and hence much greater entropy (which is a quantitative measure of disorder).

Now instead of N coins, consider a mole of an ideal gas containing Avogadro's number of molecules. The macroscopic state of this gas is given by its pressure p, volume V, and temperature T; a description of the microscopic state involves stating the position and velocity for each molecule in the gas. At a given pressure, volume, and temperature, the gas may be in any one of an astronomically large number of microscopic states, depending on the positions and velocities of its 6.02×10^{23} molecules. If the gas undergoes a free expansion into a greater volume, the range of possible positions increases, as does the number of possible microscopic states. The system becomes more disordered, and the entropy increases as calculated in Example 20.8 (Section 20.7).

We can draw the following general conclusion: *For any system, the most probable macroscopic state is the one with the greatest number of corresponding microscopic states, which is also the macroscopic state with the greatest disorder and the greatest entropy.*

Let w represent the number of possible microscopic states for a given macroscopic state. (For the four coins shown in Fig. 20.18 the state of four heads has $w = 1$, the state of three heads and one tails has $w = 4$, and so on.) Then the entropy S of a macroscopic state can be shown to be given by

$$S = k \ln w \quad \text{(microscopic expression for entropy)} \quad (20.22)$$

where $k = R/N_A$ is the Boltzmann constant (gas constant per molecule) introduced in Section 18.3. As Eq. (20.22) shows, increasing the number of possible microscopic states w increases the entropy S.

What matters in a thermodynamic process is not the absolute entropy S but the *difference* in entropy between the initial and final states. Hence an equally valid and useful definition would be $S = k \ln w + C$, where C is a constant, since C cancels in any calculation of an entropy difference between two states. But it's convenient to set this constant equal to zero and use Eq. (20.22). With this choice, since the smallest possible value of w is unity, the smallest possible value of S for any system is $k \ln 1 = 0$. Entropy can *never* be negative.

In practice, calculating w is a difficult task, so Eq. (20.22) is typically used only to calculate the absolute entropy S of certain special systems. But we can use this relation to calculate *differences* in entropy between one state and another. Consider a system that undergoes a thermodynamic process that takes it from macroscopic state 1, for which there are w_1 possible microscopic states, to macroscopic state 2, with w_2 associated microscopic states. The change in entropy in this process is

$$\Delta S = S_2 - S_1 = k \ln w_2 - k \ln w_1 = k \ln \frac{w_2}{w_1} \quad (20.23)$$

The *difference* in entropy between the two macroscopic states depends on the *ratio* of the numbers of possible microscopic states.

As the following example shows, using Eq. (20.23) to calculate a change in entropy from one macroscopic state to another gives the same results as considering a reversible process connecting those two states and using Eq. (20.19).

Example
20.11

A microscopic calculation of entropy change

Use Eq. (20.23) to calculate the entropy change in the free expansion of n moles of gas at temperature T described in Example 20.8.

SOLUTION

IDENTIFY and SET UP: The situation is shown in Figure 20.19. When the partition is broken, the velocities of the molecules are unaffected, since no work is done. But each molecule now has twice as much volume in which it can move and hence has twice the number of possible positions. This is all we need to calculate the entropy change using Eq. (20.23).

(a) Gas occupies volume V; number of microstates $= w_1$

(b) Gas occupies volume $2V$; number of microstates $= w_2 = 2^N w_1$

20.19 In a free expansion of N molecules in which the volume doubles, the number of possible microscopic states increases by 2^N.

EXECUTE: Let w_1 be the number of microscopic states of the system as a whole when the gas occupies volume V_1 (Fig. 20.19a). The number of molecules is $N = nN_A$, and each molecule has twice as many possible states after the partition is broken. Hence the number w_2 of microscopic states when the volume occupies volume $2V$ (Fig. 20.19b) is greater by a factor of 2^N; that is, $w_2 = 2^N w_1$. The change in entropy in this process is

$$\Delta S = k \ln \frac{w_2}{w_1} = k \ln \frac{2^N w_1}{w_1} = k \ln 2^N$$

$$= Nk \ln 2$$

Since $N = nN_A$ and $k = R/N_A$, this becomes

$$\Delta S = (nN_A)(R/N_A) \ln 2 = nR \ln 2$$

EVALUATE: We have found the same result as in Example 20.8, but without any reference to the thermodynamic path taken.

The relationship between entropy and the number of microscopic states gives us new insight into the entropy statement of the second law of thermodynamics, that the entropy of a closed system can never decrease. From Eq. (20.22) this means that a closed system can never spontaneously undergo a process that decreases the number of possible microscopic states.

An example of such a forbidden process would be if all of the air in your room spontaneously moved to one half of the room, leaving a vacuum in the other half. Such a "free compression" would be the reverse of the free expansion of Examples 20.8 and 20.11. This would decrease the number of possible microscopic states by a factor of 2^N. Strictly speaking, this process is not impossible! The probability of finding a given molecule in one half of the room is 1/2, so the probability of finding all of the molecules in one half of the room at once is $(1/2)^N$. (This is exactly the same as the probability of having a tossed coin come up heads N times in a row.) This probability is *not* zero. But lest you worry about suddenly finding yourself gasping for breath in the evacuated half of your room, consider that a typical room might hold 1000 moles of air, and so $N = 1000N_A = 6.02 \times 10^{26}$ molecules. The probability of all the molecules being in the same half of the room is therefore $(1/2)^{6.02 \times 10^{26}}$. Expressed as a decimal, this number has more than 10^{26} zeros to the right of the decimal point!

Because the probability of such a "free compression" taking place is so vanishingly small, it has almost certainly never occurred anywhere in the universe since the beginning of time. We conclude that for all practical purposes the second law of thermodynamics is never violated.

Test Your Understanding

A quantity of N molecules of an ideal gas initially occupies volume V. By what factor does the number of microscopic states of the gas increase if the gas expands to volume $2V$ in (a) a reversible isothermal process or (b) a reversible adiabatic process? Are the answers the same or different? Why?

SUMMARY

A reversible process is one whose direction can be reversed by an infinitesimal change in the conditions of the process, and in which the system is always in or very close to thermal equilibrium. All other thermodynamic processes are irreversible.

A heat engine takes heat Q_H from a source, converts part of it to work W, and discards the remainder $|Q_C|$ at a lower temperature. The thermal efficiency e of a heat engine measures how much of the absorbed heat is converted to work. (See Example 20.1)

$$e = \frac{W}{Q_H} = 1 + \frac{Q_C}{Q_H} = 1 - \left|\frac{Q_C}{Q_H}\right| \quad (20.4)$$

A gasoline engine operating on the Otto cycle has a theoretical maximum thermal efficiency e that depends on the compression ratio r and the ratio of heat capacities γ of the working substance.

$$e = 1 - \frac{1}{r^{\gamma - 1}} \quad (20.6)$$

A refrigerator takes heat Q_C from a colder place, has a work input $|W|$, and discards heat $|Q_H|$ at a warmer place. The effectiveness of the refrigerator is given by its coefficient of performance K.

$$K = \frac{|Q_C|}{|W|} = \frac{|Q_C|}{|Q_H| - |Q_C|} \quad (20.9)$$

The second law of thermodynamics describes the directionality of natural thermodynamic processes. It can be stated in several equivalent forms. The *engine* statement is that no cyclic process can convert heat completely into work. The *refrigerator* statement is that no cyclic process can transfer heat from a colder place to a hotter place with no input of mechanical work.

The Carnot cycle operates between two heat reservoirs at temperatures T_H and T_C and uses only reversible processes. Its thermal efficiency depends only on T_H and T_C. An additional equivalent statement of the second law is that no engine operating between the same two temperatures can be more efficient than a Carnot engine. (See Examples 20.2 and 20.3)

$$e_{\text{Carnot}} = 1 - \frac{T_C}{T_H} = \frac{T_H - T_C}{T_H} \quad (20.14)$$

A Carnot engine run backward is a Carnot refrigerator. Its coefficient of performance depends only on T_H and T_C. Another form of the second law states that no refrigerator operating between the same two temperatures can have a larger coefficient of performance than a Carnot refrigerator. (See Example 20.4)

$$K_{Carnot} = \frac{T_C}{T_H - T_C}$$

$$(20.15)$$

The Kelvin temperature scale is based on the efficiency of the Carnot cycle and is independent of the properties of any specific material. The zero point on the Kelvin scale is called absolute zero.

Entropy is a quantitative measure of the disorder of a system. The entropy change in any reversible process depends on the amount of heat flow and the absolute temperature T. Entropy depends only on the state of the system, and the change in entropy between given initial and final states is the same for all processes leading from one state to the other. This fact can be used to find the entropy change in an irreversible process. (See Examples 20.5 through 20.10)

$$\Delta S = \int_1^2 \frac{dQ}{T}$$

$$(20.19)$$

(reversible process)

An important statement of the second law of thermodynamics is that the entropy of an isolated system may increase but can never decrease. When a system interacts with its surroundings, the total entropy change of system and surroundings can never decrease. When the interaction involves only reversible processes, the total entropy is constant and $\Delta S = 0$; when there is any irreversible process, the total entropy increases and $\Delta S > 0$.

When a system is in a particular macroscopic state, the particles that make up the system may be in any of w possible microscopic states. The greater the number w, the greater the entropy. (See Example 20.11)

$$S = k \ln w \qquad (20.22)$$

N gas molecules in volume V; number of microstates = w

In volume $2V$; number of microstates = $2^N w$

Key Terms

Your Notes

Answer to Chapter Opening Question

Yes. That's what a refrigerator does: it makes heat flow from the cold interior of the refrigerator to the warm outside. The second law of thermodynamics says that heat cannot *spontaneously* flow from a cold body to a hot one. A refrigerator has a motor that does work on the system to *force* the heat to flow in that way.

Answers to Test Your Understanding Questions

Section 20.1 This process is irreversible. Like sliding a book across a table, rubbing your hands together uses friction to convert mechanical energy into heat. The (impossible) reverse process would involve your hands spontaneously getting colder, with the released energy forcing your hands to move rhythmically back and forth!

Section 20.2 Each truck's engine rejects 8000 J of heat per cycle and goes through 25 cycles per second. The total heat released in 24 hours is therefore

$$|Q_{C-\text{total}}| = (1000 \text{ trucks})\left(\frac{8000 \text{ J}}{\text{truck} \cdot \text{cycle}}\right)\left(\frac{25 \text{ cycles}}{\text{s}}\right)$$
$$\times \left(\frac{60 \text{ s}}{\text{min}}\right)\left(\frac{60 \text{ min}}{\text{h}}\right)(24 \text{ h})$$
$$= 1.7 \times 10^{13} \text{ J}$$

This is equivalent to 4.8×10^6 kilowatt-hours (a kilowatt-hour is a unit of energy used in electric power systems, equal to 10^3 watts of power delivered for one hour). A kilowatt-hour of electric energy costs a few cents, so this discarded heat is equivalent to approximately $100,000 of lost energy.

Section 20.3 Let Q_H, the heat input at volume V, be constant. Then a greater compression ratio r means a greater maximum volume rV, so the points d and a are farther to the right in the pV-diagram of Fig. 20.5 and the area enclosed by the loop $abcd$ is greater. The area within this loop is W, the work done by the engine, so a greater compression ratio means more work for the same input of heat and hence a greater efficiency $e = W/Q_H$.

Section 20.4 A refrigerator uses an input of work to transfer heat from one system (the refrigerator's interior) to another system (its exterior, which includes the house in which the refrigerator is installed). If the door is open, these two systems are really the *same* system and will eventually come to the same temperature. By the first law of thermodynamics, all of the work input to the refrigerator motor will be converted into heat and the temperature in your house will actually *increase*. To cool the house you need a system that will transfer heat from it to the outside world, such as an air conditioner or heat pump.

Section 20.5 The expanding steam does work to push the cork upward and pop it into the air. This process is not 100% efficient because only part of the heat absorbed from the burner goes into doing work; the remainder goes into raising the temperature of the water and the test tube.

Section 20.6 The efficiency can be no better than that of a Carnot engine running between the same two temperature limits, $e_{\text{Carnot}} = 1 - (T_C/T_H)$ [Eq. (20.14)]. The temperature T_C of the cold reservoir for this air-cooled engine is about 300 K (ambient temperature) and the temperature T_H of the hot reservoir cannot exceed the melting point of copper, 1356 K (see Table 17.4). Hence the maximum possible Carnot efficiency is $e = 1 - (300 \text{ K})/(1356 \text{ K}) = 0.78$, or 78%. The temperature of any real engine would be less than this, so it would be impossible for the inventor's engine to attain 85% efficiency. You should invest your money elsewhere.

Section 20.7 The process described is exactly the opposite of the process used in Example 20.10, so the net entropy change would be -102 J/K. This result violates the second law of thermodynamics, which states that the entropy of an isolated system cannot decrease. This remarkable process is quite impossible.

Section 20.8 (a) We saw in Example 20.8 (Section 20.7) that for an ideal gas, the entropy change in a free expansion is the same as in an isothermal expansion. From Eq. (20.23), this implies that the ratio of the number of microscopic states after and before the expansion, w_2/w_1, is also the same for these two cases. From Example 20.11, $w_2/w_1 = 2^N$, so the number of microscopic states increases by a factor 2^N. (b) In a reversible expansion the entropy change is $\Delta S = \int dQ/T = 0$; if the expansion is adiabatic there is no heat flow, so $\Delta S = 0$. From Eq. (20.23), $w_2/w_1 = 1$ and there is *no* change in the number of microscopic states. The difference is that in an adiabatic expansion the temperature drops and the molecules move more slowly, so they have fewer microscopic states available to them than in an isothermal expansion.

Discussion Questions

Q20.1 A pot is half-filled with water, and a lid is placed on the pot, forming a tight seal so that no water vapor can escape. The pot is heated on a stove, forming water vapor inside the pot. The heat is then turned off and the water vapor condenses back to liquid. Is this cycle reversible or irreversible? Why?

Q20.2 Give two examples of reversible processes and two examples of irreversible processes in purely mechanical systems, such as blocks sliding on planes, springs, pulleys, and strings. Explain what makes each process reversible or irreversible.

Q20.3 What irreversible processes occur in a gasoline engine? Why are they irreversible?

Q20.4 Suppose you try to cool the kitchen of your house by leaving the refrigerator door open. What happens? Why? Would the result be the same if you left open a picnic cooler full of ice? Explain the reason for any differences.

Q20.5 A member of the U.S. Congress proposed a scheme to produce energy as follows. Water molecules (H_2O) are to be broken apart to produce hydrogen and oxygen. The hydrogen is then burned (that is, combined with oxygen), releasing energy in the process. The only product of this combustion is water, so there is no pollution. In light of the second law of thermodynamics, what do you think of this energy-producing scheme?

Q20.6 Some critics of biological evolution claim that it violates the second law of thermodynamics, since evolution involves simple life forms developing into more complex and more highly ordered organisms. Explain why this is not a valid argument against evolution.

Q20.7 Is it a violation of the second law of thermodynamics to convert mechanical energy completely into heat? To convert heat completely into work? Explain your answers.

Q20.8 A growing plant creates a highly complex and organized structure out of simple materials such as air, water, and trace minerals. Does this violate the second law of thermodynamics? Why or why not? What is the plant's ultimate source of energy? Explain your reasoning.

Q20.9 Imagine a special air filter placed in a window of a house. The tiny holes in the filter only allow air molecules moving faster than a certain speed to exit the house, and only allows air molecules moving slower than that speed to enter the house from outside. Explain why such an air filter would cool the house, and why the second law of thermodynamics makes building such a filter an impossible task.

Q20.10 An electric motor has its shaft coupled to that of an electric generator. The motor drives the generator, and some current from the generator is used to run the motor. The excess current is used to light a home. What is wrong with this scheme?

Q20.11 When a wet cloth is hung up in a hot wind in the desert, it is cooled by evaporation to a temperature that may be 20 C° or so below that of the air. Discuss this process in light of the second law of thermodynamics.

Q20.12 If no real engine can be as efficient as a Carnot engine operating between the same two temperatures, what is the point of developing and using Eq. (20.14)?

Q20.13 Suppose you want to increase the efficiency of a heat engine. Would it be better to increase T_H or to decrease T_C by an equal amount? Why?

Q20.14 What would be the efficiency of a Carnot engine operating with $T_H = T_C$? What would be the efficiency if $T_C = 0$ K and T_H were any temperature above 0 K? Interpret your answers.

Q20.15 Real heat engines, like the gasoline engine in a car, always have some friction between their moving parts, although lubricants keep the friction to a minimum. Would a heat engine with completely frictionless parts be 100% efficient? Why or why not? Does the answer depend on whether or not the engine runs on the Carnot cycle? Again, why or why not?

Q20.16 Does a refrigerator full of food consume more power if the room temperature is 20°C than if it is 15°C? Or is the power consumption the same? Explain your reasoning.

Q20.17 Explain why each of the following processes are examples of increasing disorder or randomness: mixing hot and cold water; free expansion of a gas; irreversible heat flow; and developing heat by mechanical friction. Are entropy increases involved in all of these? Why or why not?

Q20.18 Are the earth and sun in thermal equilibrium? Are there entropy changes associated with the transmission of energy from the sun to the earth? Does radiation differ from other modes of heat transfer with respect to entropy changes? Explain your reasoning.

Q20.19 Discuss the entropy changes involved in the preparation and consumption of a hot fudge sundae.

Q20.20 If you run a movie film backwards, it is as if the direction of time is reversed. In the time-reversed movie, would you see processes that violate conservation of energy? Conservation of linear momentum? Would you see processes that violate the second law of thermodynamics? In each case, if law-breaking processes could occur, give some examples.

Exercises

Section 20.2 Heat Engines

20.1 A diesel engine performs 2200 J of mechanical work and discards 4300 J of heat each cycle. a) How much heat must be supplied to the engine in each cycle? b) What is the thermal efficiency of the engine?

20.2 An aircraft engine takes in 9000 J of heat and discards 6400 J each cycle. a) What is the mechanical work output of the engine during one cycle? b) What is the thermal efficiency of the engine?

20.3 A Gasoline Engine. A gasoline engine takes in 1.61×10^4 J of heat and delivers 3700 J of work per cycle. The heat is obtained by burning gasoline with a heat of combustion of 4.60×10^4 J/g. a) What is the thermal efficiency? b) How much heat is discarded in each cycle? c) What mass of fuel is burned in each cycle? d) If the engine goes through 60.0 cycles per second, what is its power output in kilowatts? In horsepower?

20.4 A gasoline engine has a power output of 180 kW (about 241 hp). Its thermal efficiency is 28.0%. a) How much heat must be supplied to the engine per second? b) How much heat is discarded by the engine per second?

20.5 A certain nuclear-power plant has a mechanical-power output (used to drive an electric generator) of 330 MW. Its rate of heat input from the nuclear reactor is 1300 MW. a) What is the thermal efficiency of the system? b) At what rate is heat discarded by the system?

Section 20.3 Internal-Combustion Engines

20.6 What compression ratio r must an Otto cycle have to achieve an ideal efficiency of 65.0% if $\gamma = 1.40$?

20.7 For an Otto cycle with $\gamma = 1.40$ and $r = 9.50$, the temperature of the gasoline-air mixture when it enters the cylinder is 22.0°C (point a of Fig. 20.5). a) What is the temperature at the end of the compression stroke (point b)? b) The initial pressure of the gasoline-air mixture (point a) is 8.50×10^4 Pa, slightly below atmospheric pressure. What is the pressure at the end of the compression stroke?

20.8 The Otto-cycle engine in a Mercedes-Benz SLK230 has a compression ratio of 8.8. a) What is the ideal efficiency of the engine? Use $\gamma = 1.40$. b) The engine in a Dodge Viper GT2 has a slightly higher compression ratio of 9.6. How much increase in the ideal efficiency results from this increase in the compression ratio?

Section 20.4 Refrigerators

20.9 A refrigerator has a coefficient of performance of 2.10. Each cycle it absorbs 3.40×10^4 J of heat from the cold reservoir. a) How much mechanical energy is required each cycle to operate the refrigerator? b) During each cycle, how much heat is discarded to the high-temperature reservoir?

20.10 Liquid refrigerant at a pressure of 1.34×10^5 Pa leaves the expansion valve of a refrigerator at $-23.0°C$. It then flows through the evaporation coils inside the refrigerator and leaves as vapor at the same pressure and at $-20.5°C$, the same temperature as the inside of the refrigerator. The boiling point of the refrigerant at this pressure is $-23.0°C$, the heat of vaporization is 1.60×10^5 J/kg, and the specific heat capacity of the vapor at constant pressure is 485 J/kg·K. The coefficient of performance of the refrigerator is $K = 2.8$. If 8.00 kg of refrigerant flows through the refrigerator each hour, find the electric power that must be supplied to the refrigerator.

20.11 A window air-conditioner unit absorbs 9.80×10^4 J of heat per minute from the room being cooled and in the same time period deposits 1.44×10^5 J of heat into the outside air. a) What is the power consumption of the unit in watts? b) What is the energy efficiency rating of the unit?

20.12 A freezer has a coefficient of performance of 2.40. The freezer is to convert 1.80 kg of water at $25.0°C$ to 1.80 kg of ice at $-5.0°C$ in one hour. a) What amount of heat must be removed from the water at $25.0°C$ to convert it to ice at $-5.0°C$? b) How much electrical energy is consumed by the freezer during this hour? c) How much wasted heat is rejected to the room in which the freezer sits?

Section 20.6 The Carnot Cycle

20.13 A Carnot engine whose high-temperature reservoir is at 620 K takes in 550 J of heat at this temperature in each cycle and gives up 335 J to the low-temperature reservoir. a) How much mechanical work does the engine perform during each cycle? b) What is the temperature of the low-temperature reservoir? c) What is the thermal efficiency of the cycle?

20.14 A Carnot engine is operated between two heat reservoirs at temperatures of 520 K and 300 K. a) If the engine receives 6.45 kJ of heat energy from the reservoir at 520 K in each cycle, how many joules per cycle does it reject to the reservoir at 300 K? b) How much mechanical work is performed by the engine during each cycle? c) What is the thermal efficiency of the engine?

20.15 An ice-making machine operates in a Carnot cycle. It takes heat from water at $0.0°C$ and rejects heat to a room at $24.0°C$. Suppose that 85.0 kg of water at $0.0°C$ are converted to ice at $0.0°C$. a) How much heat is rejected to the room? b) How much energy must be supplied to the device?

20.16 A Carnot refrigerator is operated between two heat reservoirs at temperatures of 320 K and 270 K. a) If in each cycle the refrigerator receives 415 J of heat energy from the reservoir at 270 K, how many joules of heat energy does it deliver to the reservoir at 320 K? b) If the refrigerator goes through 165 cycles each minute, what power input is required to operate the refrigerator? c) What is the coefficient of performance of the refrigerator?

20.17 A Carnot device extracts 5.00 kJ of heat from a body at $-10.0°C$. How much work is done if the device exhausts heat into the environment at a) $25.0°C$? b) $0.0°C$? c) $-25.0°C$? In each case, is the device acting as an engine or as a refrigerator?

20.18 The Kwik-Freez Appliance Co. wants you to design a food freezer that will keep the freezing compartment at $-5.0°C$ and will operate in a room at $20.0°C$. This freezer is to make 5.00 kg of ice

at $0.0°C$. starting with water at $20.0°C$. Find the least possible amount of electrical energy needed to make this ice and the smallest possible amount of heat expelled into the room.

20.19 An ideal Carnot engine operates between $500°C$ and $100°C$ with a heat input of 250 J per cycle. What minimum number of cycles are necessary for the engine to lift a 500-kg rock through a height of 100 m?

20.20 A Carnot heat engine has a thermal efficiency of 0.600 and the temperature of its hot reservoir is 800 K. If 3000 J of heat is rejected to the cold reservoir in one cycle, what is the work output of the engine during one cycle?

20.21 A Carnot heat engine uses a hot reservoir consisting of a large amount of boiling water and a cold reservoir consisting of a large tub of ice and water. In five minutes of operation of the engine, the heat rejected by the engine melts 0.0400 kg of ice. During this time, how much work W is performed by the engine?

20.22 An inventor claims to have developed an engine that in each cycle takes in 2.60×10^8 J of heat at a temperature of 400 K, does 42.0 kWh of mechanical work, and rejects heat at a temperature of 250 K. Would you advise investing money to put this engine on the market? Why or why not?

20.23 a) Show that the efficiency e of a Carnot engine and the coefficient of performance K of a Carnot refrigerator are related by $K = (1 - e)/e$. The engine and refrigerator operate between the same hot and cold reservoirs. b) What is K for the limiting values $e \rightarrow 1$ and $e \rightarrow 0$? Explain.

Section 20.7 Entropy

20.24 A sophomore with nothing better to do adds heat to 0.350 kg of ice at $0.0°C$ until it is all melted. a) What is the change in entropy of the water? b) The source of heat is a very massive body at a temperature of $25.0°C$. What is the change in entropy of this body? c) What is the total change in entropy of the water and the heat source?

20.25 You pour 100 g of water at $80.0°C$ into the ocean, which is at $20.0°C$, and wait about ten minutes. Treat the water you pour and the ocean as an isolated system. a) Is this process reversible or irreversible? Explain your reasoning using simple physical reasoning without resorting to any equations. b) Calculate the net entropy change of the system during this process. Explain whether or not this result is consistent with your answer to part (a).

20.26 A 15.0-kg block of ice at $0.0°C$ melts to liquid water at $0.0∞C$ inside a large room that has a temperature of $20.0°C$. Treat the ice and the room as an isolated system and assume that the room is large enough for its temperature change to be ignored. a) Is the melting of the ice reversible or irreversible? Explain your reasoning using simple physical reasoning without resorting to any equations. b) Calculate the net entropy change of the system during this process. Explain whether or not this result is consistent with your answer to part (a).

20.27 Calculate the entropy change that occurs when 1.00 kg of water at $20.0°C$ is mixed with 2.00 kg of water at $80.0°C$.

20.28 Three moles of an ideal gas undergo a reversible isothermal compression at $20.0°C$. During this compression, 1850 J of work is done on the gas. What is the change of entropy of the gas?

20.29 What is the change in entropy of 0.130 kg of helium gas at the normal boiling point of helium when it all condenses isothermally to 1.00 L of liquid helium? (*Hint:* See Table 17.4 in Section 17.6.)

20.30 a) Calculate the change in entropy when 1.00 kg of water at 100°C is vaporized and converted to steam at 100°C (see Table 17.4). b) Compare your answer to the change in entropy when 1.00 kg of ice is melted at 0°C, calculated in Example 20.5 (Section 20.7). Is the change in entropy greater for melting or for vaporization? Interpret your answer, using the idea that entropy is a measure of the randomness of a system.

20.31 a) Calculate the change in entropy when 1.00 mol of water (molecular mass 18.0 g/mol) at 100°C evaporates to form water vapor at 100°C. b) Repeat the calculation of part (a) for 1.00 mol of liquid nitrogen, 1.00 mol of silver, and 1.00 mol of mercury when each is vaporized at its normal boiling point. (See Table 17.4 for the heats of vaporization, and Appendix D for the molar masses. Note that the nitrogen molecule is N_2.) c) Your results in parts (a) and (b) should be in relatively close agreement. (This is called the *rule of Drepez and Trouton.*) Explain why this should be so, using the idea that entropy is a measure of the randomness of a system.

20.32 A block of copper with a mass of 3.50 kg, initially at 100.0°C, is dropped into 0.800 kg of water initially at 0.0°C. a) What is the final temperature of the system? b) What is the total change in entropy of the system?

20.33 Two moles of an ideal gas undergo a reversible isothermal expansion from 0.0280 m³ to 0.0420 m³ at a temperature of 25.0°C. What is the change in entropy of the gas?

*Section 20.8 Microscopic Interpretation of Entropy

***20.34** A box is separated by a partition into two parts of equal volume. The left side of the box contains 500 molecules of nitrogen gas; the right side contains 100 molecules of oxygen gas. The two gases are at the same temperature. The partition is punctured, and equilibrium is eventually attained. Assume that the volume of the box is large enough for each gas to undergo a free expansion and not change temperature. a) On average, how many molecules of each type will there be in either half of the box? b) What is the change in entropy of the system when the partition is punctured? c) What is the probability that the molecules will be found in the same distribution as they were before the partition was punctured, that is, 500 nitrogen molecules in the left half and 100 oxygen molecules in the right half?

***20.35** Two moles of an ideal gas occupy a volume V. The gas expands isothermally and reversibly to a volume $3V$. a) Is the velocity distribution changed by the isothermal expansion? Explain. b) Use Eq. (20.23) to calculate the change in entropy of the gas. c) Use Eq. (20.18) to calculate the change in entropy of the gas. Compare this result to that obtained in part (b).

***20.36** You toss four identical coins on the floor. Each coin has an equal probability of showing heads or tails. a) What is the probability of all four coins showing heads? Of all four showing tails? b) What is the probability of three coins showing heads and one tails? What is the probability of three coins showing tails and one heads? c) What is the probability of two coins showing heads and two showing tails? d) What is the sum of the five probabilities calculated in parts (a) through (c)? Explain.

Problems

20.37 You are designing a Carnot engine that has two moles of CO_2 as its working substance; the gas may be treated as ideal. The gas is to have a maximum temperature of 527°C and a maximum pressure of 5.00 atm. With a heat input of 400 J per cycle, you want 300 J of useful work. a) Find the temperature of the cold reservoir. b) For how many cycles must this engine run to melt completely a 10.0-kg block of ice originally at 0.0°C, using only the heat rejected by the engine?

20.38 A Carnot engine whose low-temperature reservoir is at −90.0°C has an efficiency of 40.0%. An engineer is assigned the problem of increasing this to 45.0%. a) By how many Celsius degrees must the temperature of the high-temperature reservoir be increased if the temperature of the low-temperature reservoir remains constant? b) By how many Celsius degrees must the temperature of the low-temperature reservoir be decreased if that of the high-temperature reservoir remains constant?

20.39 A heat engine takes 0.350 mol of a diatomic ideal gas around the cycle shown in the pV-diagram of Fig. 20.20. Process $1 \rightarrow 2$ is at constant volume, process $2 \rightarrow 3$ is adiabatic, and process $3 \rightarrow 1$ is at a constant pressure of 1.00 atm. The value of γ for this gas is 1.40. a) Find the pressure and volume at points 1, 2, and 3. b) Calculate Q, W, and ΔU for each of the three processes. c) Find the net work done by the gas in the cycle. d) Find the net heat flow into the engine in one cycle. e) What is the thermal efficiency of the engine? How does this compare to the efficiency of a Carnot-cycle engine operating between the same minimum and maximum temperatures T_1 and T_2?

Figure 20.20 Problem 20.39.

20.40 A heat engine operates with cycle *abcd*. The working substance of the engine is CO_2 gas; the gas can be treated as ideal. The maximum temperature of the gas during the cycle is 1000 K. The pressure and volume of the gas in each state is $p_a = p_d = 2.00 \times 10^5$ Pa; $p_b = p_c = 6.00 \times 10^5$ Pa; $V_a = V_b = 0.0100$ m³; $V_c = V_d = 0.0300$ m³. Calculate a) the number of moles of CO_2; b) the heat input in each cycle; c) the waste heat per cycle; d) the work performed by the engine in each cycle; e) the thermal efficiency of the engine.

20.41 Heat Pump. A heat pump is a heat engine run in reverse. In winter it pumps heat from the cold air outside into the warmer air inside the building, maintaining the building at a comfortable temperature. In summer, it pumps heat from the cooler air inside the building to the warmer air outside, acting as an air conditioner. a) If the outside temperature in winter is −5.0°C and the inside temperature is 17.0°C, how many joules of heat will the heat pump deliver to the inside for each joule of electrical energy used to run the unit, assuming an ideal Carnot cycle? b) Suppose you have the option of using electrical resistance heating rather than a heat pump. How much electrical energy would you need in order to deliver the same amount of heat to the inside of the house as in part (a)? Consider a Carnot heat pump delivering heat to the inside of a house to maintain it at 68°F. Show that the heat pump delivers less heat for each joule of electrical energy used to operate the unit as the outside temperature decreases. Notice that this behavior is opposite to the dependence of the efficiency of a Carnot heat engine on the difference in the reservoir temperatures. Explain why this is so.

20.42 A heat engine operates using the cycle shown in Fig. 20.21. The working substance is 2.00 moles of helium gas, which reaches a maximum temperature of 327°C. Assume the helium can be treated as an ideal gas. Process bc is isothermal. The pressure in states a and c is 1.00×10^5 Pa and the pressure in state b is 3.00×10^5 Pa. a) How much heat enters the gas and how much leaves the gas each cycle? b) How much work does the engine do each cycle and what is its efficiency? c) Compare this engine's efficiency with the maximum possible efficiency attainable with the hot and cold reservoirs used by this cycle.

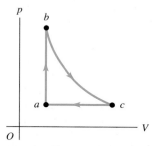

Figure 20.21 Problem 20.42.

20.43 As a budding mechanical engineer, you are called upon to design a Carnot engine that has 2.00 moles of a monatomic ideal gas as its working substance and that operates from a high-temperature reservoir at 500°C. The engine is to lift a 15.0-kg weight 2.00 m per cycle, using 500 J of heat input. The gas in the engine chamber can have a minimum volume of 5.00 L during the cycle. a) Draw a pV-diagram for this cycle. Show in your diagram where heat enters and leaves the gas. b) What must be the temperature of the cold reservoir? c) What is the thermal efficiency of the engine? d) How much heat energy does this engine waste per cycle? e) What is the maximum pressure that the gas chamber will have to withstand?

20.44 An experimental power plant at the Natural Energy Laboratory of Hawaii generates electricity from the temperature gradient of the ocean. The surface and deep-water temperatures are 27°C and 6°C, respectively. a) What is the maximum theoretical efficiency of this power plant? b) If the power plant is to produce 210 kW of power, at what rate must heat be extracted from the warm water? At what rate must heat be absorbed by the cold water? Assume the maximum theoretical efficiency. c) The cold water that enters the plant leaves it at a temperature of 10°C. What must be the flow rate of cold water through the system? Give your answer in kg/h and L/h.

20.45 What is the thermal efficiency of an engine that operates by taking n moles of diatomic ideal gas through the cycle $1 \rightarrow 2 \rightarrow 3 \rightarrow 4 \rightarrow 1$ shown in Fig. 20.22?

20.46 A cylinder contains oxygen at a pressure of 2.00 atm. The volume is 4.00 L, and the

Figure 20.22 Problem 20.45.

temperature is 300 K. Assume that the oxygen may be treated as an ideal gas. The oxygen is carried through the following processes:

(i) Heated at constant pressure from the initial state (state 1) to state 2, which has $T = 450$ K.
(ii) Cooled at constant volume to 250 K (state 3).
(iii) Compressed at constant temperature to a volume of 4.00 L (state 4).
(iv) Heated at constant volume to 300 K, which takes the system back to state 1.

a) Show these four processes in a pV-diagram, giving the numerical values of p and V in each of the four states. b) Calculate Q and W for each of the four processes. c) Calculate the net work done by the oxygen. d) What is the efficiency of this device as a heat engine? How does this compare to the efficiency of a Carnot-cycle engine operating between the same minimum and maximum temperatures of 250 K and 450 K?

20.47 Thermodynamic Processes for a Refrigerator. A refrigerator operates on the cycle shown in Fig. 20.23. The compression $(d \rightarrow a)$ and expansion $(b \rightarrow c)$ steps are adiabatic. The temperature, pressure, and volume of the coolant in each of the four states a, b, c, and d are given in the table below.

State	T (°C)	P (kPa)	V (m³)	U (kJ)	Percentage That Is Liquid
a	80	2305	0.0682	1969	0
b	80	2305	0.00946	1171	100
c	5	363	0.2202	1005	54
d	5	363	0.4513	1657	5

a) In each cycle, how much heat is taken from inside the refrigerator into the coolant while the coolant is in the evaporator? b) In each cycle, how much heat is exhausted from the coolant into the air outside the refrigerator while the coolant is in the condenser? c) In each cycle, how much work is done by the motor that operates the compressor? d) Calculate the coefficient of performance of the refrigerator.

20.48 A monatomic ideal gas is taken around the cycle shown in Fig. 20.24 in the direction

Figure 20.23 Problem 20.47.

Figure 20.24 Problem 20.48.

shown in the figure. The path for process $c \rightarrow a$ is a straight line in the pV-diagram. a) Calculate Q, W, and ΔU for each process $a \rightarrow b$, $b \rightarrow c$, and $c \rightarrow a$. b) What are Q, W, and ΔU for one complete cycle? c) What is the efficiency of the cycle?

20.49 A Stirling-cycle Engine. The *Stirling cycle* is similar to the Otto cycle, except that the compression and expansion of the gas is done at constant temperature, not adiabatically as in the Otto cycle. The Stirling cycle is used in *external* combustion engines, which means that the gas inside the cylinder is not used in the combustion process. Heat is supplied by burning fuel steadily outside the cylinder, instead of explosively inside the cylinder as in the Otto cycle. For this reason Stirling-cycle

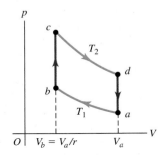

Figure 20.25 Problem 20.49.

engines are quieter than Otto-cycle engines, since there are no intake and exhaust valves (a major source of engine noise). While small Stirling engines are used for a variety of purposes, Stirling engines for automobiles have not been successful because they are larger, heavier, and more expensive than conventional automobile engines. In the cycle, the working fluid goes through the following sequence of steps (Fig. 20.25):

(i) Compressed isothermally at temperature T_1 from the initial state a to state b, with a compression ratio r.
(ii) Heated at constant volume to state c at temperature T_2.
(iii) Expanded isothermally at T_2 to state d.
(iv) Cooled at constant volume back to the initial state a.

Assume that the working fluid is n moles of an ideal gas (for which C_V is independent of temperature). a) Calculate Q, W, and ΔU for each of the processes $a \rightarrow b$, $b \rightarrow c$, $c \rightarrow d$, and $d \rightarrow a$. b) In the Stirling cycle, the heat transfers in the processes $b \rightarrow c$ and $d \rightarrow a$ do not involve external heat sources, but rather use *regeneration:* the same substance that transfers heat to the gas inside the cylinder in the process $b \rightarrow c$ also absorbs heat back from the gas in the process $d \rightarrow a$. Hence the heat transfers $Q_{b \rightarrow c}$ and $Q_{d \rightarrow a}$ do not play a role in determining the efficiency of the engine. Explain this last statement by comparing the expressions for $Q_{b \rightarrow c}$ and $Q_{d \rightarrow a}$ calculated in part (a). c) Calculate the efficiency of a Stirling-cycle engine in terms of the temperatures T_1 and T_2. How does this compare to the efficiency of a Carnot-cycle engine operating between these same two temperatures? (Historically, the Stirling cycle was devised before the Carnot cycle.) Does this result violate the second law of thermodynamics? Explain. Unfortunately, actual Stirling-cycle engines cannot achieve this efficiency, due to problems with the heat-transfer processes and pressure losses in the engine.

20.50 A Carnot engine operates between two heat reservoirs at temperatures T_H and T_C. An inventor proposes to increase the efficiency by running one engine between T_H and an intermediate temperature T' and a second engine between T' and T_C using as input the heat expelled by the first engine. Compute the efficiency of this composite system and compare it to that of the original engine.

20.51 The maximum power that can be extracted by a wind turbine from an air stream is approximately

$$P = kd^2v^3$$

where d is the blade diameter, v is the wind speed, and the constant $k = 0.5 \ \mathrm{W \cdot s^3/m^5}$. a) Explain the dependence of P on d and on v by considering a cylinder of air that passes over the turbine blades in time t (Fig. 20.26). This cylinder has diameter d, length $L = vt$, and density ρ. b) The Mod-5B wind turbine at Kahaku on the Hawaiian island of Oahu has a blade diameter of 97 m (slightly longer than a football field) and sits atop a 58-m tower. It can produce 3.2 MW of electric power. Assuming 25% efficiency, what wind speed is required to produce this amount of power? Give your answer in m/s and in km/h. c) Commercial wind turbines are commonly located in or downwind of mountain passes. Why?

Figure 20.26 Problem 20.51.

20.52 Fuel Economy and Automobile Performance. The Otto-cycle engine of a Volvo V70 has a compression ratio $r = 8.5$. The U.S. Environmental Protection Agency (EPA) fuel economy rating of this car is 25 miles per gallon on the highway (105 km/h, or 65 mi/h). Gasoline has a heat of combustion of 4.60×10^7 J/kg, and its density is 740 kg/m^3. a) At 105 km/h, what is the rate of gasoline consumption in L/h? b) What is the theoretical efficiency of the engine? Use $\gamma = 1.40$. c) How much power is the engine producing at 105 km/h? Assume that the engine is operating at its theoretical efficiency, and give your answer in watts and in horsepower. By comparison, the engine of the Volvo V70 has a maximum power of 236 hp. d) Due to heat losses and friction in the drive train, the actual efficiency is approximately 15%. Repeat part (c) using this information. What fraction of the maximum possible power is used for highway driving?

20.53 Automotive Thermodynamics. A Volkswagen Passat has a six-cylinder Otto-cycle engine with compression ratio $r = 10.6$. The diameter of each cylinder, called the *bore* of the engine, is 82.5 mm. The distance that the piston moves during the compression in Fig. 20.4, called the *stroke* of the engine, is 86.4 mm. The initial pressure of the air-fuel mixture (at point a in Fig. 20.5) is 8.50×10^4 Pa, and the initial temperature is 300 K (the same as the outside air). Assume that 200 J of heat is added to each cylinder in each cycle by the burning gasoline, and that the gas has $C_V = 20.5$ J/mol·K and $\gamma = 1.40$. a) Calculate the total work done in one cycle in each cylinder of the engine, and the heat released when the gas is cooled to the temperature of the outside air. b) Calculate the volume of the air-fuel mixture at point a in the cycle. c) Calculate the pressure, volume, and temperature of the gas at points b, c, and d in the cycle. In a pV-diagram, show the numerical values of p, V, and T for each of the four states. d) Compare the efficiency of this engine with the efficiency of a Carnot-cycle engine operating between the same maximum and minimum temperatures.

20.54 In a factory, an insulated iron bar 65.0 cm long having thermal conductivity 79.5 W/m·K and cross-sectional area 15.0 cm² conducts heat from a furnace at 250.0°C to a cool water reservoir at 40.0°C, a) By how much does the entropy of the factory change each second due to this process? b) Interpret your answer in terms of the reversibility or irreversibility of this process.

20.55 Unavailable Energy. The discussion of entropy and the second law that follows Example 20.10 (Section 20.7) says that the increase in entropy in an irreversible process is associated with energy becoming less available. Consider a Carnot cycle that uses a low temperature reservoir with Kelvin temperature T_c. This is a true reservoir, that is large enough not to change temperature when it accepts heat from the engine. Let the engine accept heat from an object of temperature T', where $T' > T_c$. The object is of finite size, so cools as heat is extracted from it. The engine continues to operate until $T' = T_c$. a) Show that the total magnitude of heat rejected to the low temperature reservoir is $T_c |\Delta S_h|$, where ΔS_h is the change in entropy of the high-temperature reservoir. b) Apply the result of part (a) to 1.00 kg of water initially at a temperature of 373 K as the heat source for the engine and $T_c = 273$ K. How much total mechanical work can be performed by the engine until it stops? c) Repeat part (b) for 2.00 kg of water at 323 K. d) Compare the amount of work that can be obtained from the energy in the water of Example 20.10 before and after it is mixed. Discuss whether your result shows that energy has become less available.

20.56 Recalculate the change in entropy between points a and b of Fig. 20.15c if the reversible path is a) an isobaric expansion to $2V$ followed by an isochoric process; b) an isochoric cooling to $p_a/2$ followed by an isobaric expansion.

20.57 A 0.0500-kg cube of ice at an initial temperature of -15.0°C is placed in 0.600 kg of water at $T = 45.0$°C in an insulated container of negligible mass. Calculate the change in entropy of the system.

20.58 a) For the Otto cycle shown in Fig. 20.5, calculate the changes in entropy of the gas in each of the constant-volume processes $b \rightarrow c$ and $d \rightarrow a$ in terms of the temperatures T_a, T_b, T_c, and T_d and the number of moles n and the heat capacity C_V of the gas. b) What is the total entropy change in the engine during one cycle? (*Hint:* Use the relation between T_a and T_b and between T_d and T_c.) c) The processes $b \rightarrow c$ and $d \rightarrow a$ occur irreversibly in a real Otto engine. Explain how can this be reconciled with your result in part (b).

20.59 A *TS*-Diagram. a) Graph a Carnot cycle, plotting Kelvin temperature vertically and entropy horizontally. This is called a temperature-entropy diagram, or *TS*-diagram. b) Show that the area under any curve representing a reversible path in a temperature-

entropy diagram represents the heat absorbed by the system. c) Derive from your diagram the expression for the thermal efficiency of a Carnot cycle. d) Draw a temperature-entropy diagram for the Stirling cycle, described in Problem 20.49. Use this diagram to relate the efficiency of the Carnot and Stirling cycles.

20.60 A physics student immerses one end of a copper rod in boiling water at 100°C and the other end in an ice-water mixture at 0°C. The sides of the rod are insulated. After steady-state conditions have been achieved in the rod, 0.160 kg of ice melts in a certain time interval. For this time interval find a) the entropy change of the boiling water; b) the entropy change of the ice water mixture; c) the entropy change of the copper rod; d) the total entropy change of the entire system.

20.61 To heat one cup of water (250 cm^3) to make coffee, you place an electric heating element in the cup. As the water temperature increases from 20°C to 65°C, the temperature of the heating element remains at a constant 120°C. Calculate the change in entropy of a) the water; b) the heating element; c) the system of water and heating element. (Make the same assumption about the specific heat of water as in Example 20.10 (Section 20.7), and ignore the heat that flows into the ceramic coffee cup itself.) d) Is this process reversible or irreversible? Explain.

20.62 An object of mass m_1, specific heat capacity c_1, and temperature T_1 is placed in contact with a second object of mass m_2, specific heat capacity c_2, and temperature $T_2 > T_1$. As a result, the temperature of the first object increases to T and the temperature of the second object decreases to T'. a) Show that the entropy increase of the system is

$$\Delta S = m_1 c_1 \ln \frac{T}{T_1} + m_2 c_2 \ln \frac{T'}{T_2}$$

and show that energy conservation requires that

$$m_1 c_1 (T - T_1) = m_2 c_2 (T_2 - T')$$

b) Show that the entropy change ΔS, considered as a function of T, is a *maximum* if $T = T'$, which is just the condition of thermodynamic equilibrium. c) Discuss the result of part (b) in terms of the idea of entropy as a measure of disorder.

Challenge Problem

20.63 Consider a Diesel cycle that starts (at point a in Fig. 20.6) with air at temperature T_a. The air may be treated as an ideal gas. a) If the temperature at point c is T_c, derive an expression for the efficiency of the cycle in terms of the compression ratio r. b) What is the efficiency if $T_a = 300$ K, $T_c = 950$ K, $\gamma = 1.40$, and $r = 21.0$?

APPENDIX A

The International System of Units

The Système International d'Unités, abbreviated SI, is the system developed by the General Conference on Weights and Measures and adopted by nearly all the industrial nations of the world. The following material is adapted from B. N. Taylor, ed., National Institute of Standards and Technology Spec. Pub. 811 (U.S. Govt. Printing Office, Washington, DC, 1995). See also **http://physics.nist.gov/cuu**

Quantity	Name of unit	Symbol	Equivalent units
	SI base units		
length	meter	m	
mass	kilogram	kg	
time	second	s	
electric current	ampere	A	
thermodynamic temperature	kelvin	K	
amount of substance	mole	mol	
luminous intensity	candela	cd	
	SI derived units		
area	square meter	m^2	
volume	cubic meter	m^3	
frequency	hertz	Hz	s^{-1}
mass density (density)	kilogram per cubic meter	kg/m^3	
speed, velocity	meter per second	m/s	
angular velocity	radian per second	rad/s	
acceleration	meter per second squared	m/s^2	
angular acceleration	radian per second squared	rad/s^2	
force	newton	N	$kg \cdot m/s^2$
pressure (mechanical stress)	pascal	Pa	N/m^2
kinematic viscosity	square meter per second	m^2/s	
dynamic viscosity	newton-second per square meter	$N \cdot s/m^2$	
work, energy, quantity of heat	joule	J	$N \cdot m$
power	watt	W	J/s
quantity of electricity	coulomb	C	$A \cdot s$
potential difference, electromotive force	volt	V	J/C, W/A
electric field strength	volt per meter	V/m	N/C
electric resistance	ohm	Ω	V/A
capacitance	farad	F	$A \cdot s/V$
magnetic flux	weber	Wb	$V \cdot s$
inductance	henry	H	$V \cdot s/A$
magnetic flux density	tesla	T	Wb/m^2
magnetic field strength	ampere per meter	A/m	
magnetomotive force	ampere	A	
luminous flux	lumen	lm	$cd \cdot sr$
luminance	candela per square meter	cd/m^2	
illuminance	lux	lx	lm/m^2
wave number	1 per meter	m^{-1}	
entropy	joule per kelvin	J/K	
specific heat capacity	joule per kilogram-kelvin	$J/kg \cdot K$	
thermal conductivity	watt per meter-kelvin	$W/m \cdot K$	

Quantity	Name of unit	Symbol	Equivalent units
radiant intensity	watt per steradian	W/sr	
activity (of a radioactive source)	becquerel	Bq	s^{-1}
radiation dose	gray	Gy	J/kg
radiation dose equivalent	sievert	Sv	J/kg
	SI supplementary units		
plane angle	radian	rad	
solid angle	steradian	sr	

DEFINITIONS OF SI UNITS

meter (m)   The *meter* is the length equal to the distance traveled by light, in vacuum, in a time of 1/299,792,458 second.

kilogram (kg)   The *kilogram* is the unit of mass; it is equal to the mass of the international prototype of the kilogram. (The international prototype of the kilogram is a particular cylinder of platinum-iridium alloy that is preserved in a vault at Sèvres, France, by the International Bureau of Weights and Measures.)

second (s)   The *second* is the duration of 9,192,631,770 periods of the radiation corresponding to the transition between the two hyperfine levels of the ground state of the cesium-133 atom.

ampere (A)   The *ampere* is that constant current that, if maintained in two straight parallel conductors of infinite length, of negligible circular cross section, and placed 1 meter apart in vacuum, would produce between these conductors a force equal to 2×10^{-7} newton per meter of length.

kelvin (K)   The *kelvin,* unit of thermodynamic temperature, is the fraction 1/273.16 of the thermodynamic temperature of the triple point of water.

ohm (Ω)   The *ohm* is the electric resistance between two points of a conductor when a constant difference of potential of 1 volt, applied between these two points, produces in this conductor a current of 1 ampere, this conductor not being the source of any electromotive force.

coulomb (C)   The *coulomb* is the quantity of electricity transported in 1 second by a current of 1 ampere.

candela (cd)   The *candela* is the luminous intensity, in a given direction, of a source that emits monochromatic radiation of frequency 540×10^{12} hertz and that has a radiant intensity in that direction of 1/683 watt per steradian.

mole (mol)   The *mole* is the amount of substance of a system that contains as many elementary entities as there are carbon atoms in 0.012 kg of carbon 12. The elementary entities must be specified and may be atoms, molecules, ions, electrons, other particles, or specified groups of such particles.

newton (N)   The *newton* is that force that gives to a mass of 1 kilogram an acceleration of 1 meter per second per second.

joule (J)   The *joule* is the work done when the point of application of a constant force of 1 newton is displaced a distance of 1 meter in the direction of the force.

watt (W)   The *watt* is the power that gives rise to the production of energy at the rate of 1 joule per second.

volt (V)   The *volt* is the difference of electric potential between two points of a conducting wire carrying a constant current of 1 ampere, when the power dissipated between these points is equal to 1 watt.

weber (Wb)   The *weber* is the magnetic flux that, linking a circuit of one turn, produces in it an electromotive force of 1 volt as it is reduced to zero at a uniform rate in 1 second.

lumen (lm)   The *lumen* is the luminous flux emitted in a solid angle of 1 steradian by a uniform point source having an intensity of 1 candela.

farad (F)   The *farad* is the capacitance of a capacitor between the plates of which there appears a difference of potential of 1 volt when it is charged by a quantity of electricity equal to 1 coulomb.

henry (H)   The *henry* is the inductance of a closed circuit in which an electromotive force of 1 volt is produced when the electric current in the circuit varies uniformly at a rate of 1 ampere per second.

radian (rad)   The *radian* is the plane angle between two radii of a circle that cut off on the circumference an arc equal in length to the radius.

steradian (sr)   The *steradian* is the solid angle that, having its vertex in the center of a sphere, cuts off an area of the surface of the sphere equal to that of a square with sides of length equal to the radius of the sphere.

SI Prefixes   The names of multiples and submultiples of SI units may be formed by application of the prefixes listed in Appendix F.

APPENDIX B

Useful Mathematical Relations

ALGEBRA

$$a^{-x} = \frac{1}{a^x} \qquad a^{(x+y)} = a^x a^y \qquad a^{(x-y)} = \frac{a^x}{a^y}$$

Logarithms: If $\log a = x$, then $a = 10^x$. $\log a + \log b = \log(ab)$ $\log a - \log b = \log(a/b)$ $\log(a^n) = n \log a$
If $\ln a = x$, then $a = e^x$. $\ln a + \ln b = \ln(ab)$ $\ln a - \ln b = \ln(a/b)$ $\ln(a^n) = n \ln a$

Quadratic formula: If $ax^2 + bx + c = 0,$ $x = \dfrac{-b \pm \sqrt{b^2 - 4ac}}{2a}.$

BINOMIAL THEOREM

$$(a+b)^n = a^n + na^{n-1}b + \frac{n(n-1)a^{n-2}b^2}{2!} + \frac{n(n-1)(n-2)a^{n-3}b^3}{3!} + \cdots$$

TRIGONOMETRY

In the right triangle ABC, $x^2 + y^2 = r^2$.

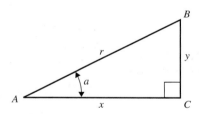

Definitions of the trigonometric functions: $\sin a = y/r$ $\cos a = x/r$ $\tan a = y/x$

Identities: $\sin^2 a + \cos^2 a = 1$ $\tan a = \dfrac{\sin a}{\cos a}$

$\sin 2a = 2 \sin a \cos a$ $\cos 2a = \cos^2 a - \sin^2 a = 2\cos^2 a - 1$
$= 1 - 2\sin^2 a$

$\sin \frac{1}{2}a = \sqrt{\dfrac{1 - \cos a}{2}}$ $\cos \frac{1}{2}a = \sqrt{\dfrac{1 + \cos a}{2}}$

$\sin(-a) = -\sin a$ $\sin(a \pm b) = \sin a \cos b \pm \cos a \sin b$
$\cos(-a) = \cos a$ $\cos(a \pm b) = \cos a \cos b \mp \sin a \sin b$
$\sin(a \pm \pi/2) = \pm \cos a$ $\sin a + \sin b = 2 \sin \frac{1}{2}(a+b) \cos \frac{1}{2}(a-b)$
$\cos(a \pm \pi/2) = \mp \sin a$ $\cos a + \cos b = 2 \cos \frac{1}{2}(a+b) \cos \frac{1}{2}(a-b)$

GEOMETRY

Circumference of circle of radius r: $C = 2\pi r$
Area of circle of radius r: $A = \pi r^2$
Volume of sphere of radius r: $V = 4\pi r^3/3$
Surface area of sphere of radius r: $A = 4\pi r^2$
Volume of cylinder of radius r and height h: $V = \pi r^2 h$

CALCULUS

Derivatives:

$$\frac{d}{dx}x^n = nx^{n-1}$$

$$\frac{d}{dx}\sin ax = a\cos ax$$

$$\frac{d}{dx}\cos ax = -a\sin ax$$

$$\frac{d}{dx}e^{ax} = ae^{ax}$$

$$\frac{d}{dx}\ln ax = \frac{1}{x}$$

$$\int \frac{dx}{\sqrt{a^2 - x^2}} = \arcsin\frac{x}{a}$$

$$\int \frac{dx}{\sqrt{x^2 + a^2}} = \ln\left(x + \sqrt{x^2 + a^2}\right)$$

$$\int \frac{dx}{x^2 + a^2} = \frac{1}{a}\arctan\frac{x}{a}$$

$$\int \frac{dx}{(x^2 + a^2)^{3/2}} = \frac{1}{a^2}\frac{x}{\sqrt{x^2 + a^2}}$$

$$\int \frac{x\,dx}{(x^2 + a^2)^{3/2}} = -\frac{1}{\sqrt{x^2 + a^2}}$$

Integrals:

Power series (convergent for range of x shown):

$$\int x^n\,dx = \frac{x^{n+1}}{n+1} \quad (n \neq -1)$$

$$(1 + x)^n = 1 + nx + \frac{n(n-1)x^2}{2!} + \frac{n(n-1)(n-2)}{3!}x^3 + \cdots \quad (|x| < 1)$$

$$\int \frac{dx}{x} = \ln x$$

$$\sin x = x - \frac{x^3}{3!} + \frac{x^5}{5!} - \frac{x^7}{7!} + \cdots \quad (\text{all } x)$$

$$\int \sin ax\,dx = -\frac{1}{a}\cos ax$$

$$\cos x = 1 - \frac{x^2}{2!} + \frac{x^4}{4!} - \frac{x^6}{6!} + \cdots \quad (\text{all } x)$$

$$\int \cos ax\,dx = \frac{1}{a}\sin ax$$

$$\tan x = x + \frac{x^3}{3} + \frac{2x^2}{15} + \frac{17x^7}{315} + \cdots \quad (|x| < \pi/2)$$

$$\int e^{ax}\,dx = \frac{1}{a}e^{ax}$$

$$e^x = 1 + x + \frac{x^2}{2!} + \frac{x^3}{3!} + \cdots \quad (\text{all } x)$$

$$\ln(1 + x) = x - \frac{x^2}{2} + \frac{x^3}{3} - \frac{x^4}{4} + \cdots \quad (|x| < 1)$$

APPENDIX C

The Greek Alphabet

Name	Capital	Lowercase	Name	Capital	Lowercase
Alpha	A	α	Nu	N	ν
Beta	B	β	Xi	Ξ	ξ
Gamma	Γ	γ	Omicron	O	o
Delta	Δ	δ	Pi	Π	π
Epsilon	E	ϵ	Rho	P	ρ
Zeta	Z	ζ	Sigma	Σ	σ
Eta	H	η	Tau	T	τ
Theta	Θ	θ	Upsilon	Υ	υ
Iota	I	ι	Phi	Φ	ϕ
Kappa	K	κ	Chi	X	χ
Lambda	Λ	λ	Psi	Ψ	ψ
Mu	M	μ	Omega	Ω	ω

APPENDIX D

Periodic Table of the Elements

Group	1	2	3	4	5	6	7	8	9	10	11	12	13	14	15	16	17	18
Period																		
1	1 **H** 1.008																	2 **He** 4.003
2	3 **Li** 6.941	4 **Be** 9.012											5 **B** 10.811	6 **C** 12.011	7 **N** 14.007	8 **O** 15.999	9 **F** 18.998	10 **Ne** 20.180
3	11 **Na** 22.990	12 **Mg** 24.305											13 **Al** 26.982	14 **Si** 28.086	15 **P** 30.974	16 **S** 32.065	17 **Cl** 35.453	18 **Ar** 39.948
4	19 **K** 39.098	20 **Ca** 40.078	21 **Sc** 44.956	22 **Ti** 47.867	23 **V** 50.942	24 **Cr** 51.996	25 **Mn** 54.938	26 **Fe** 55.845	27 **Co** 58.933	28 **Ni** 58.693	29 **Cu** 63.546	30 **Zn** 65.409	31 **Ga** 69.723	32 **Ge** 72.64	33 **As** 74.922	34 **Se** 78.96	35 **Br** 79.904	36 **Kr** 83.80
5	37 **Rb** 85.468	38 **Sr** 87.62	39 **Y** 88.906	40 **Zr** 91.224	41 **Nb** 92.906	42 **Mo** 95.94	43 **Tc** (98)	44 **Ru** 101.07	45 **Rh** 102.905	46 **Pd** 106.42	47 **Ag** 107.868	48 **Cd** 112.411	49 **In** 114.818	50 **Sn** 118.710	51 **Sb** 121.760	52 **Te** 127.60	53 **I** 126.904	54 **Xe** 131.293
6	55 **Cs** 132.905	56 **Ba** 137.327	71 **Lu** 174.967	72 **Hf** 178.49	73 **Ta** 180.948	74 **W** 183.84	75 **Re** 186.207	76 **Os** 190.23	77 **Ir** 192.217	78 **Pt** 195.078	79 **Au** 196.967	80 **Hg** 200.59	81 **Tl** 204.383	82 **Pb** 207.2	83 **Bi** 208.980	84 **Po** (209)	85 **At** (210)	86 **Rn** (222)
7	87 **Fr** (223)	88 **Ra** (226)	103 **Lr** (262)	104 **Rf** (261)	105 **Db** (262)	106 **Sg** (266)	107 **Bh** (264)	108 **Hs** (269)	109 **Mt** (268)	110 **Ds** (271)	111 **Uuu** (272)	112 **Uub** (285)	113 **Uut**	114 **Uuq** (289)	115 **Uup**	116 **Uuh**	117 **Uus**	118 **Uuo**

Lanthanoids	57 **La** 138.905	58 **Ce** 140.116	59 **Pr** 140.908	60 **Nd** 144.24	61 **Pm** (145)	62 **Sm** 150.36	63 **Eu** 151.964	64 **Gd** 157.25	65 **Tb** 158.925	66 **Dy** 162.500	67 **Ho** 164.930	68 **Er** 167.259	69 **Tm** 168.934	70 **Yb** 173.04
Actinoids	89 **Ac** (227)	90 **Th** (232)	91 **Pa** (231)	92 **U** (238)	93 **Np** (237)	94 **Pu** (244)	95 **Am** (243)	96 **Cm** (247)	97 **Bk** (247)	98 **Cf** (251)	99 **Es** (252)	100 **Fm** (257)	101 **Md** (258)	102 **No** (259)

For each element the average atomic mass of the mixture of isotopes occurring in nature is shown. For elements having no stable isotope, the approximate atomic mass of the longest-lived isotope is shown in parentheses. For elements that have been predicted but not yet detected, no atomic mass is given. All atomic masses are expressed in atomic mass units (1 u = 1.66053873(13) × 10^{-27} kg), equivalent to grams per mole (g/mol).

APPENDIX E

Unit Conversion Factors

LENGTH

1 m = 100 cm = 1000 mm = 10^6 μm = 10^9 nm
1 km = 1000 m = 0.6214 mi
1 m = 3.281 ft = 39.37 in.
1 cm = 0.3937 in.
1 in. = 2.540 cm
1 ft = 30.48 cm
1 yd = 91.44 cm
1 mi = 5280 ft = 1.609 km
1 Å = 10^{-10} m = 10^{-8} cm = 10^{-1} nm
1 nautical mile = 6080 ft
1 light year = 9.461×10^{15} m

AREA

1 cm^2 = 0.155 $in.^2$
1 m^2 = 10^4 cm^2 = 10.76 ft^2
1 $in.^2$ = 6.452 cm^2
1 ft^2 = 144 $in.^2$ = 0.0929 m^2

VOLUME

1 liter = 1000 cm^3 = 10^{-3} m^3 = 0.03531 ft^3 = 61.02 $in.^3$
1 ft^3 = 0.02832 m^3 = 28.32 liters = 7.477 gallons
1 gallon = 3.788 liters

TIME

1 min = 60 s
1 h = 3600 s
1 d = 86,400 s
1 y = 365.24 d = 3.156×10^7 s

ANGLE

1 rad = 57.30° = 180°/π
1° = 0.01745 rad = π/180 rad
1 revolution = 360° = 2π rad
1 rev/min (rpm) = 0.1047 rad/s

SPEED

1 m/s = 3.281 ft/s
1 ft/s = 0.3048 m/s
1 mi/min = 60 mi/h = 88 ft/s
1 km/h = 0.2778 m/s = 0.6214 mi/h
1 mi/h = 1.466 ft/s = 0.4470 m/s = 1.609 km/h
1 furlong/fortnight = 1.662×10^{-4} m/s

ACCELERATION

1 m/s^2 = 100 cm/s^2 = 3.281 ft/s^2
1 cm/s^2 = 0.01 m/s^2 = 0.03281 ft/s^2
1 ft/s^2 = 0.3048 m/s^2 = 30.48 cm/s^2
1 mi/h · s = 1.467 ft/s^2

MASS

1 kg = 10^3 g = 0.0685 slug
1 g = 6.85×10^{-5} slug
1 slug = 14.59 kg
1 u = 1.661×10^{-27} kg
1 kg has a weight of 2.205 lb when g = 9.80 m/s^2

FORCE

1 N = 10^5 dyn = 0.2248 lb
1 lb = 4.448 N = 4.448×10^5 dyn

PRESSURE

1 Pa = 1 N/m^2 = 1.450×10^{-4} $lb/in.^2$ = 0.209 lb/ft^2
1 bar = 10^5 Pa
1 $lb/in.^2$ = 6895 Pa
1 lb/ft^2 = 47.88 Pa
1 atm = 1.013×10^5 Pa = 1.013 bar
    = 14.7 $lb/in.^2$ = 2117 lb/ft^2
1 mm Hg = 1 torr = 133.3 Pa

ENERGY

1 J = 10^7 ergs = 0.239 cal
1 cal = 4.186 J (based on 15° calorie)
1 ft · lb = 1.356 J
1 Btu = 1055 J = 252 cal = 778 ft · lb
1 eV = 1.602×10^{-19} J
1 kWh = 3.600×10^6 J

MASS–ENERGY EQUIVALENCE

1 kg $\leftrightarrow$ 8.988×10^{16} J
1 u $\leftrightarrow$ 931.5 MeV
1 eV $\leftrightarrow$ 1.074×10^{-9} u

POWER

1 W = 1 J/s
1 hp = 746 W = 550 ft · lb/s
1 Btu/h = 0.293 W

APPENDIX F

Numerical Constants

FUNDAMENTAL PHYSICAL CONSTANTS*

Name	Symbol	Value
Speed of light	c	2.99792458×10^8 m/s
Magnitude of charge of electron	e	$1.602176462(63) \times 10^{-19}$ C
Gravitational constant	G	$6.673(10) \times 10^{-11}$ N $\cdot$ m^2/kg^2
Planck's constant	h	$6.62606876(52) \times 10^{-34}$ J $\cdot$ s
Boltzmann constant	k	$1.3806503(24) \times 10^{-23}$ J/K
Avogadro's number	N_A	$6.02214199(47) \times 10^{23}$ molecules/mol
Gas constant	R	$8.314472(15)$ J/mol $\cdot$ K
Mass of electron	m_e	$9.10938188(72) \times 10^{-31}$ kg
Mass of proton	m_p	$1.67262158(13) \times 10^{-27}$ kg
Mass of neutron	m_n	$1.67492716(13) \times 10^{-27}$ kg
Permeability of free space	μ_0	$4\pi \times 10^{-7}$ Wb/A $\cdot$ m
Permittivity of free space	$\epsilon_0 = 1/\mu_0 c^2$	$8.854187817 \ldots \times 10^{-12}$ C^2/N $\cdot$ m^2
	$1/4\pi\epsilon_0$	$8.987551787 \ldots \times 10^9$ N $\cdot$ m^2/C^2

OTHER USEFUL CONSTANTS*

Mechanical equivalent of heat		4.186 J/cal (15° calorie)
Standard atmospheric pressure	1 atm	1.01325×10^5 Pa
Absolute zero	0 K	-273.15°C
Electron volt	1 eV	$1.602176462(63) \times 10^{-19}$ J
Atomic mass unit	1 u	$1.66053873(13) \times 10^{-27}$ kg
Electron rest energy	$m_e c^2$	$0.510998902(21)$ MeV
Volume of ideal gas (0°C and 1 atm)		$22.413996(39)$ liter/mol
Acceleration due to gravity (standard)	g	9.80665 m/s^2

*Source: National Institute of Standards and Technology (**http://physics.nist.gov/cuu**). Numbers in parentheses show the uncertainty in the final digits of the main number; for example, the number 1.6454(21) means 1.6454 ± 0.0021. Values shown without uncertainties are exact.

ASTRONOMICAL DATA[†]

Body	Mass (kg)	Radius (m)	Orbit radius (m)	Orbit period
Sun	1.99×10^{30}	6.96×10^{8}	—	—
Moon	7.35×10^{22}	1.74×10^{6}	3.84×10^{8}	27.3 d
Mercury	3.30×10^{23}	2.44×10^{6}	5.79×10^{10}	88.0 d
Venus	4.87×10^{24}	6.05×10^{6}	1.08×10^{11}	224.7 d
Earth	5.97×10^{24}	6.38×10^{6}	1.50×10^{11}	365.3 d
Mars	6.42×10^{23}	3.40×10^{6}	2.28×10^{11}	687.0 d
Jupiter	1.90×10^{27}	6.91×10^{7}	7.78×10^{11}	11.86 y
Saturn	5.68×10^{26}	6.03×10^{7}	1.43×10^{12}	29.45 y
Uranus	8.68×10^{25}	2.56×10^{7}	2.87×10^{12}	84.02 y
Neptune	1.02×10^{26}	2.48×10^{7}	4.50×10^{12}	164.8 y
Pluto	1.31×10^{22}	1.15×10^{6}	5.91×10^{12}	247.9 y

[†]Source: NASA Jet Propulsion Laboratory Solar System Dynamics Group (**http://ssd.jpl.nasa.gov**), and P. Kenneth Seidelmann, ed., ***Explanatory Supplement to the Astronomical Almanac*** (University Science Books, Mill Valley, CA, 1992), pp. 704–706. For each body, "radius" is its radius at its equator and "orbit radius" is its average distance from the sun (for the planets) or from the earth (for the moon).

PREFIXES FOR POWERS OF 10

Power of ten	Prefix	Abbreviation	Pronunciation
10^{-24}	yocto-	y	*yoc*-toe
10^{-21}	zepto-	z	*zep*-toe
10^{-18}	atto-	a	*at*-toe
10^{-15}	femto-	f	*fem*-toe
10^{-12}	pico-	p	*pee*-koe
10^{-9}	nano-	n	*nan*-oe
10^{-6}	micro-	μ	*my*-crow
10^{-3}	milli-	m	*mil*-i
10^{-2}	centi-	c	*cen*-ti
10^{3}	kilo-	k	*kil*-oe
10^{6}	mega-	M	*meg*-a
10^{9}	giga-	G	*jig*-a or *gig*-a
10^{12}	tera-	T	*ter*-a
10^{15}	peta-	P	*pet*-a
10^{18}	exa-	E	*ex*-a
10^{21}	zetta-	Z	*zet*-a
10^{24}	yotta-	Y	*yot*-a

Examples:

1 femtometer = 1 fm = 10^{-15} m

1 picosecond = 1 ps = 10^{-12} s

1 nanocoulomb = 1 nC = 10^{-9} C

1 microkelvin = 1 μK = 10^{-6} K

1 millivolt = 1 mV = 10^{-3} V

1 kilopascal = 1 kPa = 10^{3} Pa

1 megawatt = 1 MW = 10^{6} W

1 gigahertz = 1 GHz = 10^{9} Hz

Answers to Odd-Numbered Problems

Chapter 1

1.1 1.61 km
1.3 3.34×10^3 ns
1.5 5.36 L
1.7 a) 2330 km/h b) 648 m/s
1.9 35.3 mi/gal
1.11 9.0 cm
1.13 a) $1.1 \times 10^{-3}\%$ b) no
1.15 a) 0.1% b) 0.008% c) 0.03%
1.17 a) 2.8 ± 0.3 cm^3 b) 170 ± 20
1.19 ten thousand
1.21 10^6
1.23 10^5 (with a full head of hair)
1.25 about 10^8
1.29 9×10^{14}; about $3.50 million per person
1.31 7.8 m, 38° north of east
1.33 144 m, 41° south of west
1.35 $A_x = 7.22$ m, $A_y = 9.58$ m; $B_x = 11.5$ m, $B_y = -9.64$ m; $C_x = -3.00$ m, $C_y = -5.20$ m
1.37 1190 N; 13.4° above forward direction
1.39 a) 11.1 m, 77.6° b) 11.1 m, 77.6° c) 28.5 m, 202.3° d) 28.5 m, 22.3°
1.41 5.06 km, 20.2° north of west
1.43 a) 2.48 cm, 18.3° b) 4.10 cm, 83.7° c) 4.10 cm, 263.7°
1.45 $\vec{A} = (7.2 \text{ m})\hat{\imath} + (9.6 \text{ m})\hat{\jmath}$,
$\vec{B} = (11.5 \text{ m})\hat{\imath} - (9.6 \text{ m})\hat{\jmath}$,
$\vec{C} = (-3.0 \text{ m})\hat{\imath} - (5.2 \text{ m})\hat{\jmath}$
1.47 a) $A = 5.00$, $B = 5.39$ b) $-1.00\hat{\imath} + 5.00\hat{\jmath}$ c) 5.10, 101.3°
1.51 a) 14.00 b) 58.7°
1.53 a) out of the page b) into the page
1.55 $-23.0\hat{k}$; 23.0
1.57 a) 640 acres/mi^2 b) 43,560 ft^2/acre c) 3.26×10^5 gal/acre-ft
1.59 a) 7.04×10^{-10} s b) 5.11×10^{12} cycle/h c) 2.1×10^{26} d) 4.6×10^4 s
1.61 a) 0.5 kg b) 6×10^{-14} g c) 0.3 g
1.63 about 1 dentist per 1,000 population
1.65 90.2 N, 256° counterclockwise from $+x$-axis
1.67 b) $A_x = 3.03$ cm, $A_y = 8.10$ cm c) 8.65 cm, 69.5° measured in the sense from the $+x$-axis toward the $+y$-axis
1.69 144 m, 41° south of west
1.71 b) 1.45 km
1.73 a) (87, 258) b) 136 pixels, 25° below straight left
1.75 380 km, 28.8° south of east
1.77 160 N, 13° below horizontal
1.79 B = 255 m, C = 70 m
1.83 b) 90°
1.85 a) $A = 5.39$, $B = 4.36$ b) $-5.00\hat{\imath} + 2.00\hat{\jmath} + 7.00\hat{k}$ c) 8.83; yes
1.89 54.7° b) 35.3°
1.91 $C_x = 8.0$, $C_y = 6.1$
1.93 b) 72.2
1.95 38.5 yd, 24.6° right of downfield
1.97 a) 76 ly b) 129°

Chapter 2

2.1 a) 197 m/s b) 169 m/s
2.3 1 h 10 min
2.5 286 s; fast: 1770 m; slow: 1570 m
2.7 a) 14.0 m/s b) 11.4 m/s c) case (a)
2.9 a) 12.0 m/s b) i) 0 m/s ii) 15.0 m/s iii) 12.0 m/s c) 13.3 s
2.11 a) (in m/s^2) 0, 1.0, 2.0, 2.0, 3.0, 1.5, 1.5, 0; no; yes b) 2.5 m/s^2, 1.5 m/s^2, 0
2.13 a) near 5 s b) 30 s to 40 s c) 0 d) -1.7 m/s^2
2.15 a) 50.0 cm, 2.00 cm/s, -0.125 cm/s^2 b) 16.0 s c) 32.0 s d) 6.20 s, 1.22 cm/s; 25.8 s, -1.22 cm/s; 36.4 s, -2.55 cm/s

2.17 a) about 10 m/s^2
2.21 a) 5.0 m/s b) 1.43 m/s^2
2.23 1.70 m
2.25 a) 1.7 m/s^2 b) 12 s c) 240 m
2.27 a) speeding up: 6.67 ft/s^2; slowing down: -26.5 ft/s^2 b) 90.4 mph; a_x not constant c) 3.32 s
2.31 a) 0, 6.3 m/s^2, -11.2 m/s^2 b) 100 m, 230 m, 320 m
2.33 a) 1.80×10^4 m/s b) 0.957 c) 6 h 11 min
2.35 b) 1 s, 3 s d) 2 s e) 3 s f) 1 s
2.37 b) $d/4$
2.39 a) 2.94 m/s b) 0.599 s
2.41 a) $t = \sqrt{2d/g}$ b) 0.190 s
2.43 a) 5.56 m/s, downward b) 9.80 m/s^2, downward c) 2.16 s d) 16.1 m/s
2.45 a) 25.6 m/s b) 31.6 m/s c) 15.2 m/s
2.47 a) 249 m/s^2 b) 25.4 c) 101 m d) no
2.49 a) 7.5 m b) 180 m c) 20 m
2.51 a) $x(t) = (0.250 \text{ m/s}^3)t^3 - (0.0100 \text{ m/s}^4)t^4$, $v_x(t) = (0.750 \text{ m/s}^3)t^2 - (0.0400 \text{ m/s}^4)t^3$ b) 39.1 m/s
2.53 a) 30.0 cm/s
2.55 b) 0.627 s, 1.60 s c) negative at 0.627 s, positive at 1.60 s d) 1.11 s e) 2.45 m f) 2.00 s, 0 s
2.57 a) 82 km/h b) 31 km/h
2.59 a) 3.5 m/s^2 b) 0 c) 1.5 m/s^2
2.61 brake pedal; no
2.63 a) 464 m/s b) 2.99×10^4 m/s c) 7.48
2.65 50.0 m
2.67 4.6 m/s^2
2.69 a) 6.17 s b) 24.8 m c) $v_{\text{truck}} = 13.0$ m/s, $v_{\text{auto}} = 21.0$ m/s
2.71 a) 17 m/s b) 1.6 s
2.73 a) 15.9 s b) 393 m c) 29.5 m/s
2.75 a) -4.00 m/s b) 12.0 m/s
2.77 28.6 m
2.79 a) no b) yes, 14.4 m/s; not physically attainable
2.81 a) 13.3 m b) 1.65 s
2.83 a) 7.59 m/s b) 5.14 m c) 1.60 s
2.85 a) 7.7 m/s b) 0.78 s c) 0.59 s d) 1.3 m
2.87 270 m
2.89 a) 20.5 m/s b) yes
2.91 a) 92.2 m/s b) 75.1 m/s
2.93 a) A b) 2.27 s, 5.73 s c) 1.00 s, 4.33 s d) 2.67 s
2.95 a) 9.55 s, 47.8 m b) 1.62 m/s d) 8.38 m/s e) no f) 3.69 m/s, 21.7 s, 80.0 m
2.97 8.18 m/s b) i) 0.411 m ii) 1.15 km c) 9.80 m/s d) 4.90 m/s

Chapter 3

3.1 a) $(v_{\text{av}})_x = 1.4$ m/s, $(v_{\text{av}})_y = -1.3$ m/s b) 1.9 m/s, $-43°$
3.3 a) 7.1 cm/s, 45° b) 5.0 cm/s, 90°; 7.1 cm/s, 45°; 11 cm/s, 27°
3.5 b) $(a_{\text{av}})_x = -8.67$ m/s^2, $(a_{\text{av}})_y = -2.33$ m/s^2 c) 8.98 m/s^2, 195°
3.7 b) $\vec{v} = \alpha\hat{\imath} - 2\beta t\hat{\jmath}$, $\vec{a} = -2\beta\hat{\jmath}$ c) $v = 5.4$ m/s, $-63°$; $a = 2.4$ m/s^2, $-90°$ d) speeding up and turning right
3.9 a) 0.600 m b) 0.385 m c) $v_x = 1.10$ m/s, $v_y = -3.43$ m/s; $v = 3.60$ m/s, 72.2° below the horizontal
3.11 2.86 m
3.13 a) 30.6 m/s b) 36.3 m/s
3.15 a) 1.08 s, 6.18 m, 4.51 m; 11.5 m, 5.74 m; 16.8 m, 4.51 m c) 11.7 m/s, $+24.8°$; 10.6 m/s, 0°; 11.7 m/s, $-24.8°$ d) parallel: -4.11 m/s^2, 0, 4.11 m/s^2; perpendicular: 8.90 m/s^2, 9.80 m/s^2, 8.90 m/s^2
3.17 a) 1.4 km b) 8.5 km

3.19 a) 0.682 s, 2.99 s b) 24.0 m/s, 11.3 m/s; 24.0 m/s, -11.3 m/s c) 30.0 m/s, $-36.9°$
3.21 a) 1.5 m b) -0.89 m/s
3.23 a) 13.6 m b) 34.6 m/s c) 103 m
3.25 a) 296 m b) 176 m c) 198 m d) horizontal: 15 m/s; vertical: 58.8 m/s e) horizontal: 15 m/s; vertical: 78.8 m/s
3.27 795 m
3.29 a) 0.034 m/s^2 = 0.0034g b) 1.4 h
3.31 a) 3.07 s b) 1.68 s
3.33 a) 3.50 m/s^2, upward b) 3.50 m/s^2, downward c) 12.6 s
3.35 b) no c) where the car is farthest from the center of the ellipse
3.37 a) 14 s b) 70 s
3.39 0.36 m/s, 38° west of south
3.41 a) 4.7 m/s, 25° south of east b) 190 s c) 380 m
3.43 b) -7.1 m/s, -42 m/s c) 43 m/s, 9.5° west of south
3.45 a) $t = 0$ b) $t = 0$ c) $t = 0$, $x = 0$, $y = 15.0$ m; $t = 5.21$ s, $x = 6.25$ m, $y = 1.44$ m d) 6.41 m/s, 5.21 s
3.47 a) 1.0 m/s^2 b) 5.4° c) 2.3 m/s d) 31 s
3.49 22 m/s
3.51 40 m/s
3.53 274 m
3.55 a) 42.8 m/s b) 42.0 m/s
3.57 $4\sqrt{2}D$
3.59 c) less than 45°
3.61 b) 15°, 75°
3.63 a) 17.8 m/s b) in river, 28.4 m from the near bank
3.65 a) 81.6 m b) in cart c) 245 m d) 53.1°
3.67 a) 49 m/s b) 50 m
3.69 a) 2000 m b) 2180 m
3.71 a) 2.23 m b) 3.84 m c) 8.65 m/s d) 3.09 m, 0.62 m
3.73 c) $(18/25)gT^2$ d) $(3/2)gT^2$
3.77 b) $v_x = R\omega(1 - \cos \omega t)$, $v_y = R\omega \sin \omega t$, $a_x = R\omega^2 \sin \omega t$, $a_y = R\omega^2 \cos \omega t$ c) $t = 0$, $2\pi/\omega$, $4\pi/\omega$, ...; $x = 0$, $2\pi R$, $4\pi R$, ...; $y = 0$; $a = R\omega^2$ in the $+y$-direction d) no
3.79 30 km
3.81 a) 44.7 km/h, 26.6° west of south b) 10.5° north of west
3.83 a) 0.782 s b) 7.67 m/s c) 5.17 m/s d) 1.04 m
3.85 7.39 m/s, 12.4° north of east
3.87 a) 80 m b) 1.6×10^{-3} c) overall effect is to reduce the radius
3.89 a) $\left(\dfrac{2v_0^2}{g}\right)[\tan(\theta + \phi) - \tan\theta]\dfrac{\cos^2(\theta + \phi)}{\cos\theta}$
b) $\dfrac{\pi}{4} - \dfrac{\theta}{2}$
3.91 $\Delta t = 0.5$ s: 9.589 m/s^2, 118.6°; $\Delta t = 0.1$ s: 9.983 m/s^2, 95.73°; $\Delta t = 0.05$ s: 9.996 m/s^2, 92.86°
3.93 a) 1.5 km/h b) 3.5 km/h

Chapter 4

4.1 a) 0 b) 90° c) 180°
4.3 7.1 N to the right, 7.1 N downward
4.5 494 N, 31.7°
4.7 2.2 m/s^2
4.9 16.0 kg
4.11 a) 3.13 m, 3.13 m/s b) 21.9 m, 6.25 m/s
4.13 a) $\Sigma\vec{F} = 0$
4.15 2.94×10^3 N
4.17 a) 4.49 kg b) 4.49 kg, 8.13 N
4.19 825 N, blocks
4.23 7.4×10^{-23} m/s^2
4.25 no
4.27 142 N

A9

4.31   1840 N, 135°

4.33   a) 17 N, 90° clockwise from +x-direction
b) 840 N

4.35   a) 4.8 m/s   b) 16 m/s²   c) 2360 N

4.37   b) 5.83 m/s²

4.39   a) 2.50 m/s²   b) 10.0 N   c) to the right; F
d) 25.0 N

4.41   a) 4.4 m   b) 300 m/s   c) i) 2.7×10^4 N
ii) 9.0×10^3 N

4.43   a) $4m\vec{a}$ b) $3m\vec{a}$ c) $2m\vec{a}$ d) $-m\vec{a}$
e) same magnitudes but opposite directions

4.45   a) 2.93 m/s²   b) 11.1 m/s²

4.47   b) 5.9×10^2 N   c) 2.2×10^3 N

4.49   3.53 m/s² upward   c) 120 N   d) 93.3 N

4.51   b) $2Mg/3$ c) 5/9

4.53 $(1.7 \times 10^4 \text{ N})\hat{\imath} - (3.4 \times 10^3)\,\hat{k}$

4.55 $(k_1 t/m)\hat{\imath} + (k_2 t^4/4m)\hat{\jmath}$

4.57 $\vec{r} = \left(\dfrac{k_1}{2m}t^2 + \dfrac{k_2 k_3}{120m^2}t^5\right)\hat{\imath} + \left(\dfrac{k_3}{6m}t^3\right)\hat{\jmath}$

$\vec{v} = \left(\dfrac{k_1}{m}t + \dfrac{k_2 k_3}{24m^2}t^4\right)\hat{\imath} + \left(\dfrac{k_3}{2m}t^2\right)\hat{\jmath}$

Chapter 5

5.1   a) 25.0 N   b) 50.0 N

5.3   a) 2.54×10^3 N   b) 1.01°

5.5   yes

5.7   a) 5.23×10^4 N   b) 3.36×10^4 N

5.9   523 N

5.11   a) 84.9 N   b) 60.0 N, 60.0 N

5.13   a) $w \sin \alpha$ b) $2w \sin \alpha$ c) $w \cos \alpha$ for each
block

5.15   b) 2.96 m/s²   c) 191 N; more than the bricks,
less than the counterweight

5.17   b) 2.50 m/s²   c) 1.37 kg   d) $T < mg$

5.19   a) 1.78 m/s², downward   b) 2.14 m/s², upward
c) yes

5.23   a) 22 N   b) 3.1 m

5.25   a) 7.06 N   b) 8.14 N   c) 1.17 N, 2.25 N

5.27   a) pulled down   b) 379 N

5.29   11 times farther

5.31   a) $\mu_k(m_A + m_B)g$ b) $\mu_k m_A g$

5.33   3.82 m/s²

5.35   a) 0.218 m/s   b) 11.7 N

5.37   a) $\mu_k mg/(\cos \theta - \mu_k \sin \theta)$
b) $1/(\tan \theta) = \mu_s$

5.39   b) 8.75 N   c) 30.8 N   d) 1.54 m/s²

5.41   a) 0.44 kg/m   b) 42 m/s

5.43   26.0 m/s

5.45   a) 75°   b) no

5.47   20.7°

5.49   a) 1.49 rev/min   b) 0.918 rev/min

5.51   a) 138 km/h   b) 3580 N

5.53   2.43 m/s

5.55   a) rope making 60° angle   b) 6400 N

5.57 $mg \tan \alpha$

5.59 $F[M + m(1 - x/L)]/(M + m)$

5.61   a) 12.0 N   b) 15.0 N

5.63   a) 16.9 N   b) 10.2 N

5.65   a) 25.3 s; 5.72×10^5 N   b) 4170 m

5.67   5.9 m/s

5.69   2.8 m/s²

5.71   a) plunge into hole   b) 16 m/s

5.73 $\mu_s/(1 + \mu_s)$

5.75   a) 66 N, northward   b) 59 N, southward

5.79 3.09×10^3 N

5.81   2.52 N

5.83   a) 12.9 kg   b) 47.2 in left-hand cord, 101 N in
right-hand cord

5.85 $a_1 = 2m_2 g/(4m_1 + m_2)$,
$a_2 = m_2 g/(4m_1 + m_2)$

5.87   1.46 m above the floor

5.89 g/μ_s

5.91   b) 0.450

5.93   a) $(F \cos \alpha)/(m_1 + m_2)$

5.95   a) 21 m/s   b) 8.5 m/s

5.97   a) 19 m/s   b) 50 m behind you; yes

5.99   a) move up   b) remains constant   c) remains
constant   d) stop

5.101   a) 6.00 m/s²   b) 3.80 m/s²   c) 7.36 m/s
d) 8.18 m/s   e) 7.78 m, 6.29 m/s, 1.38 m/s²
f) 3.14 s

5.103   1/3

5.105   a) $v_y(t) = v_t + (v_0 - v_t)\text{e}^{-kt/m}$ b) $v_y(t) = v_t$

5.107   a) 0.015; 0.36 N·s²/m²   b) 29 m/s   c) ratio is
$(\sin \beta - 0.015 \cos \beta)^{1/2}$

5.109   a) 120 N   b) 3.79 m/s

5.111   b) 0.28   c) no

5.113   a) right   b) 120 m

5.115   a) 81.1°   b) no   c) bead rides at bottom of loop
$(\beta = 0)$

5.119 $T_{\max} = 2\pi \sqrt{\dfrac{h \tan \beta}{g}}\sqrt{\left(\dfrac{\sin \beta + \mu_s \cos \beta}{\cos \beta - \mu_s \sin \beta}\right)}$

$T_{\min} = 2\pi \sqrt{\dfrac{h \tan \beta}{g}}\sqrt{\left(\dfrac{\sin \beta - \mu_s \cos \beta}{\cos \beta + \mu_s \sin \beta}\right)}$

5.121 $(M + m)g \tan \alpha$

5.123   a) $F = \dfrac{\mu_k w}{\cos \theta + \mu_k \sin \theta}$
c) $\theta = \tan^{-1} \mu_k = 14.0°$

5.125   a) $a_3 = g\dfrac{(-4m_1 m_2 + m_2 m_3 + m_3 m_1)}{(4m_1 m_2 + m_2 m_3 + m_3 m_1)}$
b) $a_B = -a_3$
c) $a_1 = g\dfrac{(4m_1 m_2 - 3m_2 m_3 + m_3 m_1)}{(4m_1 m_2 + m_2 m_3 + m_3 m_1)}$
d) $a_2 = g\dfrac{(4m_1 m_2 + m_2 m_3 - 3m_3 m_1)}{(4m_1 m_2 + m_2 m_3 + m_3 m_1)}$
e) $T_A = \frac{1}{2}T_C$
f) $T_C = \dfrac{8gm_1 m_2 m_3}{(4m_1 m_2 + m_2 m_3 + m_3 m_1)}$
g) $a_1 = a_2 = a_3 = a_B = 0$, $T_C = 2m_2 g$,
$T_A = m_2 g$; yes

5.127 $\cos^2 \beta$

Chapter 6

6.1   a) 3.60 J   b) -0.900 J   c) 2.70 J

6.3   300 J

6.5   a) 99.2 N   b) 387 J   c) -387 J   d) zero; zero
e) zero

6.7 2.6×10^9 J

6.9   a) (i) zero (ii) zero   b) (i) zero (ii) -25.1 J

6.11   a) 4.3×10^3 J   b) 40 km/h

6.13 $(2gh(1 + \mu_k/\tan \alpha))^{1/2}$

6.15   c) 17.1 m/s

6.17   74.2 J

6.19   a) $-(15/16)K_1$ b) no

6.21   16.8 cm

6.23   a) -28.4 J   b) 15.3 m/s   c) no

6.25   a) 4.96 m/s   b) $a = 1.43$ m/s², $v = 4.96$ m/s

6.27   a) $v_0^2/2\mu_k g$ b) 51.3 m

6.29   a) 48.0 N, 64.0 N   b) 0.360 J, 0.640 J

6.31   a) 2.8 m/s   b) 3.5 m/s

6.33   8.5 cm

6.35   a) 1.76   b) 0.67 m/s

6.37   a) 4.0 J   b) zero   c) -1.0 J   d) 3.0 J   e) -1.0 J

6.39   a) 2.83 m/s   b) 2.40 m/s

6.41   a) 5.65 cm   b) no; 0.57 J

6.43 3.6×10^5 J; 100 m/s

6.45   743 W = 0.995 hp

6.47   0.23

6.49   28

6.51 8.1×10^6 N

6.53   b) 9   c) 108 W

6.55   0.20 W

6.57   a) 532 J   b) -315 J   c) zero   d) -203 J
e) 14.7 J   f) 1.21 m/s

6.59   a) $1/\sin \alpha$ b) $W_{\text{in}} = W_{\text{out}}$

6.61   a) 2.59×10^{12} J   b) 4800 J

6.63   2.59 m

6.65   11,000 m/s; 1.2×10^{12} J

6.67   a) 4.00 N   b) 32.0 N   c) -15.0 J

6.69   a) 0.15 N   b) 9.4 N   c) 0.44 J

6.71   a) 2.56 m/s   b) 5.28 N   c) 19.7 J

6.73   a) -910 J   b) 3.17×10^3 J

6.75 1.0×10^5 N/m

6.77   1.1 m from where the spring is released

6.79   a) 1.02×10^4 N/m, 8.16 m

6.81   a) 0.600 m   b) 1.50 m/s

6.83   0.786

6.85   1.5 m

6.87   a) 1.10×10^5 J   b) 1.30×10^5 J   c) 3.99 kW

6.89   3.6 h

6.91 1.30×10^3 m³/s

6.93   a) 1.26×10^5 J   b) 1.46 W

6.95   a) 2.4 MW   b) 61 MW   c) 6.0 MW

6.97   a) 513 W   b) 355 W   c) 52.1 W

6.99   a) 358 N   b) 47.2 hp   c) 4.06 hp   d) 2.03%

6.101   a) $\frac{1}{6}Mv^2$ b) 6.1 m/s   c) 3.9 m/s
d) $K_{\text{ball}} = 0.40$ J, $K_{\text{spring}} = 0.60$ J

6.103   a) 2.0×10^5 J   b) 2.8×10^5 J   c) 2.8×10^5 J
d) 5 km/h

Chapter 7

7.1 3.45×10^6 J

7.5   a) 24.0 m/s   b) 24.0 m/s   c) part (b)

7.9   2.8 m/s

7.11 -5400 J

7.13   a) 880 J   b) -157 J   c) 471 J   d) 253 J
e) $a = 3.16$ m/s², $v = 7.11$ m/s, $\Delta K = 253$ J

7.15   a) 80.0 J   b) 5.00 J

7.17   a) 6.32 cm   b) 12 cm

7.19   1.72 m

7.21   a) 0.500 m/s   b) 0.500 m

7.23   a) 23.6 m/s   b) 69.2 m/s², upward

7.25   a) -1.2×10^2 J   b) $+1.2 \times 10^2$ J   c) zero
d) conservative

7.27   a) zero   b) -0.10 J   c) $+0.10$ J
d) conservative; $U(x) = \frac{1}{3}\alpha x^3$

7.29   a) -3.6 J   b) -3.6 J   c) -7.2 J
d) nonconservative

7.31   a) -59 J   b) -42 J   c) -59 J
d) nonconservative

7.33 $-6C_6/x^7$, attractive

7.35 $\vec{F} = -(2kx + k'y)\hat{\imath} - (k'x + 2ky)\hat{\jmath}$

7.37   a) $F(r) = (12a/r^{13}) - (6b/r^7)$ b) $(2a/b)^{1/6}$,
stable   c) $b^2/4a$ d) $a = 6.68 \times 10^{-138}$ J·m¹²,
$b = 6.41 \times 10^{-78}$ J·m⁶

7.39   a) 229 N   b) 823 J

7.41   a) no   b) yes, $150

7.43   0.41

7.45   a) 15.9 J   b) 4.0 J   c) 3.0 J

7.47   a) 20.0 m from left-hand edge of horizontal
section   b) -78.4 J

7.49   a) 22.2 m/s   b) 16.4 m   c) no

7.51   0.602 m

7.53   15.5 m/s

7.55   4.4 m/s

7.57   a) 919 N/m, 39.8 m   b) 17.0 m/s²

7.59   a) 7.00 m/s   b) 2.94 N

7.61   a) $mg(1 - h/d)$ b) 4.4×10^2 N
c) $\sqrt{2gh(1 - y/d)}$

7.63   48.2°

7.65   a) 0.392   b) -0.832 J

7.67   a) $U(x) = (30.0 \text{ N/m})x^2 + (6.00 \text{ N/m}^2)x^3$
b) 7.85 m/s

7.69   7.01 m/s

7.71   a) $m(g + a)^2/2gh$ b) $2gh/(g + a)$

7.73   119 J

7.75   a) 3.87 m/s   b) 0.10 m

7.77   a) 3.1 m/s   b) 1.4 m   c) no

7.79   a) 0.053 gallons   b) 19

7.81 3.3×10^{14} J = 9.2×10^7 kWh

7.83   a) -50.6 J   b) -67.5 J   c) nonconservative

7.85   a) no   b) $x_0 = F/k$ d) no   e) $3F/k$, $-F/k$
f) $v_{\max} = 2F/\sqrt{mk}$ at $x = x_0 = F/k$

7.87   b) $v(x) = \left[\dfrac{2\alpha}{mx_0^2}\left(\left(\dfrac{x_0}{x}\right) - \left(\dfrac{x_0}{x}\right)^2\right)\right]^{1/2}$
c) $x = 2x_0$, $v = \sqrt{\alpha/2mx_0^2}$ d) zero
e) $v(x) = \left[\dfrac{2\alpha}{mx_0^2}\left(\left(\dfrac{x_0}{x}\right) - \left(\dfrac{x_0}{x}\right)^2 - \dfrac{2}{9}\right)\right]^{1/2}$
f) first case: x_0, ∞; second case: $3x_0/2$, $3x_0$

Chapter 8

8.1 a) 1.20×10^5 kg·m/s b) i) 60.0 m/s
ii) 26.8 m/s
8.3 b) baseball, 0.525 c) woman, 0.643
8.5 0.256 kg·m/s, $-y$-direction
8.7 562 N, no
8.9 a) 10.8 m/s to the right b) 0.75 m/s to the left
8.11 a) $(18.8$ kg·m/s$)\hat{i}$
b) $(-3.55 \times 10^{-3}$ kg·m/s$)\hat{j}$
c) $(7.50 \times 10^3$ N$)\hat{i}$
d) $(13.0\hat{i} - 0.73\hat{j})$ kg·m/s,
$(89.3\hat{i} - 5.0\hat{j})$ m/s
8.13 $At_2 + (B/3)t_2^3$ b) $(A/m)t_2 + (B/3m)t_2^3$
8.15 a) 5.68×10^{-2} m/s b) 0.103 m/s
8.17 a) 4.66 m/s in the direction opposite the
defender's original motion b) -6580 J
8.19 a) 3.60 m/s b) 8.64 J
8.21 a) $v_B = (m_A/m_B)v_A$
8.23 2.11 m/s
8.25 525 N
8.27 a) $v_{A2} = 29.3$ m/s, $v_{B2} = 20.7$ m/s
b) $0.196 = 19.6\%$
8.29 a) 6.44 m/s, eastbound b) 2.49 m/s
c) -2.81×10^5 J, -1.38×10^5 J; part (a)
8.31 0.0300 J; converted to other forms of energy
8.33 35.3 km/h, 19.3° west of south
8.35 229 m/s
8.37 a) Sam: 9.67 m/s; Abigail: 2.26 m/s b) 640 J
8.39 0.150-kg glider: 3.2 m/s to the left; 0.300-kg
glider: 0.20 m/s to the left
8.43 a) $9.00m$ b) 3.00×10^6 m/s
8.45 7.42×10^8 m from the center of the sun;
outside
8.47 a) 0.30 kg b) $(2.0$ kg·m/s$)\hat{i}$ c) $(6.7$ m/s$)\hat{i}$
8.49 a) 1.25 kg b) $(1.5$ m/s$^3)t\hat{i}$ c) $(5.6$ N$)\hat{i}$
8.51 a) 80.0 N b) yes
8.53 75.0 kg
8.55 a) 0.442 b) 800 m/s c) 530 m/s
8.57 a) 7.2×10^{-66} b) 0.223
8.59 $(0.521\hat{i} + 7.81\hat{j})$ m/s
8.61 15
8.63 30° east of north, 30° east of south
8.65 a) 5.00 m/s, east b) 5.71 m/s, east
c) 3.78 m/s, east
8.67 a) Nash, 1.68 b) Packard, 0.933 c) F_P is
larger, 0.933 d) F_N is larger, 1.68
8.69 23.2 cm
8.71 a) 25.8 m/s, 35.5° b) no
8.73 $R/4$
8.75 65.5 m/s
8.77 a) 0.222 b) -291 J c) 0.784 J
8.79 b) $M = m$ c) zero
8.81 a) 9.35 m/s b) 3.29 m/s
8.83 b) $\frac{1}{2}Mv_{cm}^2$
8.85 a) 3.56 m/s b) 5.22 m/s c) 4.67 m/s
8.87 0.00544%
8.89 1.61×10^{-22} kg·m/s, to the left
8.91 A: 13.6 m/s; B: 6.34 m/s, 65.0°
8.93 a) $(L/2)\cos(\alpha/2)$, along axis from apex
b) $(L/3)$, along bisector from bottom c) $L/\sqrt{8}$
along bisector d) $L/\sqrt{12}$ from each side
8.95 0.400 m/s
8.97 a) 1.40 kg: 14.3 m/s; 0.28 kg: 71.6 m/s
b) 347 m
8.99 222 m/s, 1.01×10^3 m/s; $v_{Kr} = 1.5v_{Ba}$
8.101 a) zero b) 1 d) 0.87 m f) 0.089 m
8.103 a) yes b) no; kinetic energy decreases by
4.8×10^3 J
8.105 a) $1.37v_{ex}$ b) $1.18v_{ex}$ c) $2.38v_{ex}$ d) 2.94 km/s
8.107 a)

$$v = \begin{cases} (2400 \text{ m/s})\ln\left[\dfrac{1}{1 - t/(120 \text{ s})}\right], & 0 \le t \le 90 \text{ s} \\ 3.33 \text{ km/s}, & t \ge 90 \text{ s} \end{cases}$$

b) $(20$ m/s$^2)/[1 - t/(120$ s$)]$ for $0 \le t \le 90$ s
c) 6000 N vs. 735 N on earth
8.109 b) $2L/3$
8.111 a) $l^2\lambda g/32$ b) $l^2\lambda g/32$

Chapter 9

9.1 a) 0.600 rad $= 34.4°$ b) 6.27 cm c) 1.05 m
9.3 a) 42 rad/s^2 b) 74 rad/s
9.5 a) $\omega_z(t) = (0.400$ rad/s$) + (0.0360$ rad/s$^3)t^2$
b) 0.400 rad/s c) $\omega_z = 1.30$ rad/s,
$\omega_{\text{av-}z} = 0.700$ rad/s
9.7 a) $\alpha_z(t) = 2b - 6ct$ b) $b/3c$
9.9 a) 300 rpm b) 75.0 s; 312 rev
9.11 a) -1.25 rev/s^2, 23.3 rev b) 2.67 s
9.13 a) 24.0 s b) 68.8 rev
9.15 10.5 rad/s
9.17 9.00 rev
9.19 a) 540 rad b) 12.3 s c) -8.17 rad/s^2
9.21 a) 47.1 m/s b) 47.3 m/s
9.23 a) $a_{\text{rad}} = 18.0$ m/s^2 b) $v = 3.00$ m/s,
$a_{\text{rad}} = 18.0$ m/s^2
9.25 a) 0.180 m/s^2, 0, 0.180 m/s^2 b) 0.180 m/s^2,
0.377 m/s^2, 0.418 m/s^2 c) 0.180 m/s^2,
0.754 m/s^2, 0.775 m/s^2
9.27 10.7 cm; no
9.29 a) 0.831 m/s b) 109 m/s^2
9.31 a) 2.29 b) 1.51 c) 15.7 m/s,
1.06×10^3 m/s$^2 = 108g$
9.33 2.99 cm
9.35 $(M/12 + m/2)L^2$
9.37 a) 0.0640 kg·m^2 b) 0.0320 kg·m^2
c) 0.0320 kg·m^2
9.39 8.52 kg·m^2
9.41 0.193 kg·m^2
9.43 b) $K = \pi^2 I\omega^2/1800$
9.45 0.600 kg·m^2
9.47 7.35×10^4 J
9.49 a) $K = 2\pi^2 I/T^2$
b) $dK/dt = (-4\pi^2 I/T^3)(dT/dt)$ c) 70 J
d) -0.56 J/s
9.51 823 J
9.53 an axis parallel to a diameter and $(2/\sqrt{15})R$
from the center of the sphere
9.55 $\frac{1}{3}M(a^2 + b^2)$
9.57 $\frac{1}{3}M(L^2 - 3Lh + 3h^2)$
9.59 $\frac{1}{3}ML^2$
9.63 a) 840 rpm b) 75 mph c) 60 mph
9.65 a) $(6.40$ rad/s$^2)t - (1.50$ rad/s$^3)t^2$
b) 6.40 rad/s$^2 - (3.00$ rad/s$^3)t$
c) $\omega_{\text{max}} = 6.83$ rad/s at $t = 2.13$ s
9.67 a) 35.0 km/h $= 9.72$ m/s b) 8.51 J
c) 652 rad/s
9.69 a) 211 rev/min b) 800 W
9.71 a) 1.70 m/s b) 84.8 rad/s
9.73 b) 2.00 m/s^2 d) 0.208 kg·m^2
9.77 a) 7.36 m b) 327 m/s
9.79 a) 2.14×10^{29} J b) 2.66×10^{33} J
9.81 a) $Mb^2/6$ b) 182 J
9.83 a) -0.784 J b) 5.42 rad/s c) 5.42 m/s
d) particle speed $= 4.43$ m/s
9.85 $\sqrt{(2gd)(m_B - \mu_k m_A)/(m_A + m_B + I/R^2)}$
9.87 $\sqrt{(g/R)(1 - \cos\beta)}$
9.89 a) 2.25×10^{-3} kg·m^2 b) 3.40 m/s
c) 4.95 m/s
9.91 7.23 m
9.93 a) $(247/512)MR^2$ b) $(383/512)MR^2$
9.95 b) $\frac{1}{4}M(R_1^2 + R_2^2)$
9.97 a) $\frac{3}{5}MR^2$ b) larger
9.99 b) 5.97×10^{24} kg c) $0.334MR^2$
9.101 a) $s = r_0\theta + \beta\theta^2/2$

b) $\theta = \left(\dfrac{1}{\beta}\right)[\sqrt{r_0^2 + 2\beta vt} - r_0]$

c) $\omega_z = \dfrac{v}{\sqrt{r_0^2 + 2\beta vt}}$, $\alpha_z = -\dfrac{\beta v^2}{(r_0^2 + 2\beta vt)^{3/2}}$;
no d) $r_0 = 2.50$ cm, $\beta = 0.247$ μm/rad,
2.13×10^4 rev

Chapter 10

10.1 a) 40.0 N·m, out of the page b) 34.6 N·m, out
of the page c) 20.0 N·m, out of the page
d) 17.3 N·m, into the page e) zero f) zero
10.3 2.50 N·m, counterclockwise

10.5 c) $(-1.05$ N·m$)\hat{k}$
10.7 0.0524 N·m
10.9 1.2 m/s
10.11 a) $g(M + 3m)/(1 + 2m/M)$ b) less
c) no effect
10.13 0.482
10.15 7.47 N
10.17 a) 7.5 N in horizontal part, 18.2 N in hanging
part b) 0.0160 kg·m^2
10.19 a) 0.882 N b) 0.553 s c) 33.9 rad/s
10.21 a) 1/3 b) 2/7 c) 2/5 d) 5/13
10.23 a) 0.613 b) no, requires $\mu_s = 0.858$ c) no
slipping
10.25 11.7 m
10.27 a) 0.309 rad/s b) 100 J c) 6.67 W
10.29 a) 0.38 N·m b) 160 rad c) 59 J d) 59 J
10.31 b) 65.6 N
10.33 a) 358 N·m b) 1.79×10^3 N c) 83.8 m/s
10.35 a) 115 kg·m^2/s, into the page
b) 125 kg·m^2/s, out of the page
10.37 4.71×10^{-6} kg·m^2/s
10.39 b) 7.00 rad/s c) 1.03×10^{-2} J
d) 1.03×10^{-2} J
10.41 0.60 rev
10.43 a) 1.38 rad/s b) 1080 J before, 499 J after
10.45 a) 0.120 rad/s b) 3.20×10^{-4} J; work done by
bug
10.49 a) 36.8 min b) 1.10×10^5 N·m
10.51 5.4×10^{22} N·m
10.53 a) 0.955 kg·m^2 b) -0.0800 N·m c) 104 rev
10.55 a) 4500 W d) 2600 W
10.57 0.675 s
10.59 a) $L/4$ from the end with the clay
b) $(9g/8L)\sin\theta$ c) $(3g/2L)\sin\theta$
10.61 1200 N
10.63 a) 1.12 m/s^2 14.0 N
10.65 $a = F/2M$, $f = F/2$
10.67 a) 0.957 m
10.69 a) $(27R - 17r)/10$ b) $(5R - 3r)/2$
10.71 a) 1.76 N b) 123 rad/s^2 c) 9.80 m/s^2 d) T
would be unchanged, α and a would double
10.73 a) 23 m c) 16 m; 0.7 times smaller
10.75 a) 36.5 m, 28.0 m/s b) no; total energy is same
but less of it is rotational kinetic energy
10.77 a) 2.07 m/s b) 0.50 rev/s c) 1.50 rev/s
10.79 a) 9.34 m/s b) 18.7 m/s c) zero d) 5.60 m
10.81 3.4 m
10.83 $g/3$
10.85 1.87 m
10.87 a) $6v/19L$ b) $3/19$
10.89 a) 5.46 rad/s b) 3.17 cm c) 1.01×10^3 m/s
10.91 3200 J
10.93 0.30 rad/s, clockwise
10.97 12.7 rad/s
10.101 a) $a = +\mu_k g$, $\alpha = -2\mu_k g/R$ b) $\omega_0^2 R^2/18\mu_k g$
c) $-M\omega_0^2 R^2/6$
10.103 a) $mv_1^2 r_1^2/r^3$ b) $(mv_1^2/2)[(r_1/r_2)^2 - 1]$
c) same

Chapter 11

11.1 0.387 m from center of 1.00-kg sphere
11.5 5450 N
11.7 1000 N, 1.20 m from end where 400-N force is
applied
11.11 a) 1920 N b) 1140 N
11.13 a) $T = 2.60w$; $F_{\text{pivot}} = 3.28w$ at 37.6°
b) $T = 4.10w$; $F_{\text{pivot}} = 5.38w$ at 48.8°
11.15 140 N by each hinge
11.17 246 N; 0.34 m behind front feet
11.19 $T_{\text{left}} = 270$ N, $T_{\text{right}} = 304$ N, $\theta = 40°$
11.21 a) 0.800 m b) clockwise c) 0.800 m,
clockwise
11.23 1.4 mm
11.25 2.0×10^{11} Pa
11.27 a) upper: 3.1×10^{-3}; lower: 2.0×10^{-3}
b) upper: 1.6 mm; lower: 0.98 mm
11.29 9.1×10^6 N
11.31 a) 3.33×10^6 Pa b) 1.33×10^5 N
11.33 4.8×10^9 Pa, 2.1×10^{-10} Pa^{-1}
11.35 1/9

11.37 3.41×10^7 Pa

11.39 10.2 m/s^2

11.43 wing force: 7300 N upward; tail force: 600 N downward

11.45 a) 590 N b) neutral

11.49 120 N to the right, 160 N upward

11.51 a) fore: 37 N; hind: 58 N b) fore: 53 N; hind: 55 N

11.53 a) $\tau = Fl \sin \phi$ b) 12.6 N·m with either method of calculation

11.55 a) $V = mg + w$, $H = T = (w + mg/4)\cot\theta$ b) 9.50×10^2 N c) $4.00°$

11.57 7600 N

11.59 a) 2700 N b) 19

11.61 a) 4.90 m b) 60 N

11.63 a) 300 N upward, 225 N·m clockwise b) $T = 132$ N; $F_{pivot} = 194$ N, $71.2°$ above horizontal

11.65 a) 60.0 N b) 53.1°

11.67 person above: 590 N; person below: 1370 N; above

11.69 a) $w_{max} = T_{max}hD/L\sqrt{h^2 + D^2}$

11.71 a) 7140 N; tall walls b) 7900 N

11.73 a) 268 N b) 232 N c) 366 N

11.75 a) A: 0.424 N; B: 1.47 N; C: 0.424 N b) 0.848 N

11.77 a) tips at 27°, slips at 31°; the bale tips before it slips b) tips at 27°, slips at 22°; the bale slips before it tips

11.79 a) $F_A = 80$ N, $F_B = 870$ N b) 1.92 m

11.81 a) 3.7 kN, 2.0 kN vertically upward

11.83 a) 5.4 mm b) 4.2 mm

11.85 a) 0.70 m from wire A b) 0.45 m from wire B

11.87 a) 1.63 m; brass: 2.00×10^8 Pa; nickel: 4.00×10^8 Pa b) brass: 2.2×10^{-3}; nickel: 1.9×10^{-3}

11.89 a) 0.36 mm b) 0.045 mm c) 0.33 mm

11.91 a) $(F\cos^2\theta)/A$ b) $(F\sin 2\theta)/2A$ c) 0 d) 45°

11.93 a) 600 N b) 13.5 kN c) slide: $\mu_s w/(\sin\theta - \mu_s\cos\theta)$; tip: $w/[(1/9)\cos\theta + 2\sin\theta]$; 66°

11.95 the lesser of $h^2/L + L/2$ and L

11.97 $[(A^2x/F) - k_OV_O]/V_S$

11.99 a) 0.662 mm b) 2.20×10^{-2} J c) 8.33×10^{-3} J d) -3.04×10^{-2} J e) 3.04×10^{-2} J

Chapter 12

12.1 2.18

12.3 F_{12}

12.5 a) 2.59×10^8 m

12.7 2.4×10^{-3} N

12.9 a) 6.30×10^{20} N, toward sun b) 4.77×10^{20} N, 24.6° toward earth from sun c) 2.37×10^{20} N, toward sun

12.11 1.6×10^{-3} N, to the left

12.13 2.1×10^{-9} m/s^2, down

12.15 1.38×10^7 m

12.17 a) 0.37 m/s^2 b) 1700 kg/m^3

12.19 610 N; 83% of weight at surface

12.21 5.98×10^{24} kg

12.23 0.83 m/s; yes

12.25 a) 5.02×10^3 m/s b) 6.06×10^4 m/s

12.27 a) 175 min b) 3.71 m/s^2

12.29 7.46×10^3 m/s

12.31 4.7 m/s; yes

12.33 a) 8.3×10^4 m/s b) 1.3×10^6 s

12.35 a) 4.45×10^{12} m, 4.55×10^{12} m b) 248 y

12.39 a) $-GMm/\sqrt{a^2 + x^2}$ c) $GMmx/(a^2 + x^2)^{3/2}$ toward the center of the ring e) $-GMm/a$, zero

12.41 a) 53 N b) 52 N

12.43 a) 4.3×10^{37} kg $= 2.1 \times 10^7 M_{sun}$ b) no c) 6.3×10^{10} m; yes

12.45 1.39×10^{-9}

12.47 a) 2.20×10^{-10} N, 163° from $+x$-direction toward $+y$-direction b) $x = 0$, $y = 1.39$ m

12.49 a) 3.40 N, 0.61° b) 1.31×10^9 J

12.51 3.83×10^8 m

12.53 a) 3.58×10^7 m

12.55 1.8×10^2 m/s

12.57 a) 1.39×10^7 m b) 3.59×10^7 m

12.59 $0.01R_E = 64$ km

12.61 0.28%

12.63 6.06×10^3 km/h

12.65 $\sqrt{2Gm_Eh/(R_E^2 + hR_E)}$

12.67 a) 13.7 km/s b) 13.3 km/s c) 13.2 km/s

12.69 a) i) 2.8 y ii) 6.1 y b) 4.90×10^8 km c) 4.22×10^8 km

12.71 a) $GM^2/4R^2$ b) $\sqrt{GM/4R}$, $4\pi\sqrt{R^3/GM}$ c) $GM^2/4R$

12.73 6.8×10^4 m/s

12.75 a) 7.91×10^3 s b) 1.53 c) 8.43×10^3 m/s, 5.51×10^3 m/s d) 2.41×10^3 m/s, 3.26×10^3 m/s; perigee

12.77 3.22×10^9 J

12.79 1.4×10^{14} m

12.81 $2\pi GmM/L^2$, toward center of semicircle

12.83 $GMm/(x^2 + xL)$

12.85 b) 22 min c) 2.5×10^8 s $= 2900$ d $= 7.9$ y

12.87 a) 2.1 kN b) no

12.89 perpendicular: $GMm/(a\sqrt{L^2 + a^2})$; parallel: zero

Chapter 13

13.1 a) 4.54×10^{-3} s, 1.38×10^3 rad/s b) 1.14×10^{-3} s, 5.53×10^3 rad/s

13.3 5.53×10^3 rad/s, 1.14×10^{-3} s

13.5 0.0500 s

13.7 a) 0.167 s b) 37.7 rad/s c) 8.44×10^{-2} kg

13.9 a) 0.375 s b) 2.66 Hz c) 16.7 rad/s

13.11 a) $x = (3.0 \text{ mm})\cos[2\pi(440 \text{ Hz})t]$ b) 8.3 m/s, 2.3×10^4 m/s^2 c) jerk $= (6.3 \times 10^7 \text{ m/s}^3)\sin[2\pi(440 \text{ Hz})t]$, 6.3×10^7 m/s^3

13.13 a) 0.383 m b) 1.02 rad $= 58.5°$ c) $x = (0.383 \text{ m})\cos[(12.2 \text{ rad/s})t + 1.02 \text{ rad}]$

13.15 0.353 m

13.17 2.09 s

13.19 a) 1.51 s b) 26.0 N/m c) 0.308 m/s d) 1.92 N e) -0.0125 m; 0.303 m/s; 0.216 m/s^2

13.23 a) $A/\sqrt{2}$, $\omega A/\sqrt{2}$ b) four times; $\pi/2\omega$ c) 3/4 kinetic, 1/4 potential

13.25 a) 5.13 m/s^2, 0.961 m/s b) -2.57 m/s^2, 0.833 m/s c) 0.137 s d) v, a can be found with energy approach; t cannot

13.27 a) 2.84×10^{-2} J b) 0.0138 m c) 0.615 m/s

13.29 0.509 m/s

13.31 a) 5.31×10^3 N/m b) 0.695 s

13.33 3.97 cm

13.35 a) 2.7×10^{-8} kg·m^2 b) 4.3×10^{-6} N·m/rad

13.37 5.12×10^{-2} kg·m^2

13.41 2.60 s

13.43 2.00 m

13.45 10.7 m/s^2

13.49 0.58 s

13.51 a) 0.393 Hz b) 1.73 kg/s

13.53 a) A b) magnitude $= bA/2m$, in $-x$-direction; slope is negative c) $a(0) = A\left(\dfrac{b^2}{2m^2} - \dfrac{k}{m}\right)$; if $b < \sqrt{2mk}$, $a(0) < 0$; if $b = \sqrt{2mk}$, $a(0) = 0$; if $b > \sqrt{2mk}$, $a(0) > 0$

13.55 a) kg/s c) i) $5.0F_{max}/k$ ii) $2.5F_{max}/k$

13.57 a) 6.72×10^3 m/s^2 b) 3.02 kN c) 18.3 m/s, 75.6 J d) 17.6 kW e) 12.1 kN, 36.7 m/s, 302 J, 141 kW

13.59 a) all unchanged b) 1/4 as large c) halved d) $1/\sqrt{5}$ as great e) U: unchanged; K: 1/5 as large

13.61 a) 1.68 s b) 0.090 m c) 0.143

13.63 $\mu_s(m + M)g/k$

13.65 a) $A_1/\sqrt{2}$; $T_1/\sqrt{2}$ b) A_1; $T_1/\sqrt{2}$

13.69 a) 0.150 m/s b) 0.112 m/s^2, downward c) 0.700 s d) 4.38 m

13.71 a) 2.6 m/s b) 0.21 m c) 0.49 s

13.73 9.08×10^{24} kg

13.75 1.17 s

13.77 $t_1 = T/4$

13.83 d) $\sqrt{A/\pi^2 mR_0^3}$

13.85 $2\pi\sqrt{3M/2k}$

13.87 2.74 s; smaller

13.89 0.88 m

13.91 a) 3.97 m

13.93 c) 0.430 m

13.95 a) spring 1: 0.350 m; spring 2: 0.250 m b) 0.702 s

13.97 a) $\Delta T/T = -\frac{1}{2}(\Delta g/g)$ b) 9.7791 m/s^2

13.99 a) $T/T_0 = \sqrt{(L^2 + 3y^2)/(L^2 + 2yL)}$ b) $y = 2L/3$

13.101 a) $kl_0/(k - m\omega'^2)$

Chapter 14

14.1 $w = 41.8$ N; no

14.3 7.02×10^3 kg/m^3; yes

14.5 1.6

14.7 9.91 m

14.9 a) 706 Pa b) 3160 Pa

14.11 a) 2.52×10^6 Pa b) 1.78×10^5 Pa

14.13 a) 1.07×10^5 Pa b) 1.03×10^5 Pa c) 1.03×10^5 Pa d) 5.33×10^3 Pa

14.15 816 cm^3

14.17 2.3×10^5 N

14.19 4.14 m

14.21 0.562 m^3

14.23 a) $\rho < \rho_{fluid}$ c) submerged: ρ/ρ_{fluid}; above: $(\rho_{fluid} - \rho)/\rho_{fluid}$ d) 32%

14.25 a) 116 Pa b) 921 Pa c) 0.822 kg, 822 kg/m^3

14.27 a) larger for B b) same c) larger for B

14.29 9.6 m/s

14.31 a) 17.0 m/s b) 0.317 m

14.33 28.4 m/s

14.35 1.47×10^5 Pa

14.37 500 N downward

14.39 0.41 cm

14.41 2.25×10^5 Pa

14.43 a) 1.10×10^8 Pa b) 1080 kg/m^3, 5%

14.45 2.61×10^4 N·m

14.47 $(p - p_0)VR^2/Gmd$

14.49 a) 12,700 kg/m^3, 3140 kg/m^3

14.51 a) 1470 Pa b) 13.9 cm

14.53 9.8×10^6 kg, yes

14.55 a) 30% b) 70%

14.57 4.66×10^{-4} m^3, 5.27 kg

14.59 a) 1.10×10^4 m^3 b) 112 kN

14.61 a) 0.107 m b) 2.42 m

14.63 0.0958 kg

14.65 33.5 N

14.67 b) 12.2 N c) 11.8 N

14.69 b) 2.52×10^{-4} m^3, 0.124

14.71 risen by 5.57×10^{-4} m

14.73 a) $1 - \rho_B/\rho_L$ b) $(\rho_L - \rho_B)L/(\rho_L - \rho_w)$ c) 4.60 cm

14.75 a) la/g b) $\omega^2l^2/2g$

14.79 7.0 N·m

14.81 13.1 cm

14.83 133 m/s

14.85 $3h_1$

14.87 a) $r = r_0(1 + 2gy/v_0^2)^{-1/4}$ b) 1.10 m

14.89 a) 80.4 N

14.91 a) $\sqrt{2gh}$ b) $(p_a/\rho g) - h$

Chapter 15

15.1 a) 1.2 m/s b) 0.31 m c) a): unchanged; b): 0.15 m

15.3 a) 0.439 m b) 5.25×10^6 Hz

15.5 a) 17.2 m, 0.0172 m b) 74.0 m, 0.0740 m

15.7 a) $f = 25.0$ Hz, $T = 0.0400$ s, $k = 19.6$ rad/m b) $y(x, t)$ $= (0.0700 \text{ m})\cos 2\pi\left(\dfrac{x}{0.320 \text{ m}} + \dfrac{t}{0.0400 \text{ s}}\right)$ c) $+0.0495$ m d) 0.0050 s

15.9 c) $-x$-direction for both d) $v_y(x, t) = \omega A \cos(kx + \omega t)$, $a_y(x, t) = -\omega^2 A \sin(kx + \omega t)$

15.11 a) 4.0 mm b) 0.040 s c) 0.14 m, 3.6 m/s d) 0.24 m, 6.0 m/s e) no

15.13 b) $+x$-direction

15.15 a) 16.3 m/s b) 0.136 m c) both increase by $\sqrt{2}$

15.17 0.390 s

15.19 a) 10.0 m/s b) 0.250 m
c) $y(x, t) = (3.00 \text{ cm}) \cos[(8.00\pi \text{ rad/m})x - (80.0\pi \text{ rad/s})t]$ d) 1890 m/s^2 e) yes

15.23 a) 0.050 W/m^2 b) 2.2×10^4 J

15.25 707 W

15.31 a) $(1.33 \text{ m})n$, $n = 0, 1, 2, 3, \ldots$
b) $(1.33 \text{ m})(n + 1/2)$, $n = 0, 1, 2, 3, \ldots$

15.33 a) $y(x, t) = (0.850 \text{ cm}) \sin[2\pi x/(0.300 \text{ m})] \times \sin[2\pi t/(0.0750 \text{ s})]$ b) 4.00 m/s c) 0.688 cm

15.37 a) 96.0 m/s b) 461 N c) 1.13 m/s, 426 m/s^2

15.39 b) 2.80 cm c) 277 cm d) 185 cm, 0.126 s, 1470 cm/s e) 280 cm/s
f) $y(x, t) = (5.60 \text{ cm}) \sin[(0.0907 \text{ rad/cm})x] \times \sin([133 \text{ rad/s}]t)$

15.41 a) $y(x, t) = (4.60 \text{ mm}) \sin[(6.98 \text{ rad/m})x] \times \sin([742 \text{ rad/s}]t)$ b) 3rd c) 39.4 Hz

15.43 a) 45.0 cm b) no

15.45 a) 3.00 m, 16.0 Hz b) 1.00 m, 48.0 Hz
c) 0.750 m, 64.0 Hz

15.47 a) 20.0 Hz, 126 rad/s, 3.49 rad/m
b) $y(x, t) = (2.50 \text{ mm}) \times \cos[(3.49 \text{ rad/m})x - (126 \text{ rad/s})t]$
c) $y(t) = (2.50 \text{ mm}) \cos[(126 \text{ rad/s})t]$
d) $y(t) = (2.50 \text{ mm}) \times \cos[3\pi/2 - (126 \text{ rad/s})t]$ e) 0.314 m/s
f) −2.50 mm, 0

15.49 a) $(7L/2)\sqrt{\mu_1/F}$ b) no

15.51 a) $(2\pi A/\lambda)\sqrt{FL/M}$ b) increase by factor of 4

15.53 $4\pi^2 F\Delta x/\lambda^2$

15.55 a) 1, 0; 2, +; 3, −; 4, 0; 5, −; 6, + b) 1, −; 2, +; 3, −; 4, +; 5, −; 6; 0 c) (a): answers would reverse sign; (b): no change

15.59 c) C/B

15.61 b) k decreases by factor of $2\sqrt{2}$; ω decreases by factor of $\sqrt{2}$

15.63 a) 7.07 cm b) 400.0 W

15.65 $\alpha = (v_1{}^2 - v_2{}^2)\rho/Y\Delta T$

15.67 $n(0.800 \text{ Hz})$, $n = 1, 2, 3, \ldots$

15.69 c) yes

15.71 c) $2A$, $2A\omega$, $2A\omega^2$

15.73 a) 20.0 m/s b) four times per second

15.75 a) 0, L b) 0, $L/2$, L d) no

15.77 a) 148 N b) 26%

15.79 b) $(1/2)\mu\omega^2 A^2 \sin^2(kx - \omega t)$
e) $(1/2)Fk^2A^2 \sin^2(kx - \omega t)$

15.81 $\pi/\omega\sqrt{2}$

15.83 a) 99.4 N c) −4.25 Hz, falls

Chapter 16

16.1 a) 0.344 m b) 1.2×10^{-5} m c) 6.9 m, 50 Hz

16.3 a) 7.78 Pa b) 77.8 Pa c) 778 Pa

16.5 a) 1.33×10^{10} Pa b) 9.47×10^{10} Pa

16.7 90.8 m

16.9 a) −82°C b) no

16.11 0.208 s

16.13 $Y/900$

16.15 a) 9.44×10^{-11} m; 0.434 m
b) 5.66×10^{-9} m; 0.101 m c) air; $A_{\text{air}}/A_{\text{water}} = 60.0$

16.17 a) 1.94 Pa b) 4.58×10^{-3} W/m^2 c) 96.6 dB

16.19 4.36×10^{-12} W/m^2, 6.40 dB

16.21 14.0 dB

16.23 a) 20.0

16.25 a) *fundamental:* displacement node at 0.60 m, pressure nodes at 0 and 1.20 m; *first overtone:* displacement nodes at 0.30 m and 0.90 m, pressure nodes at 0, 0.60 m, and 1.20 m; *second overtone:* displacement nodes at 0.20 m, 0.60 m, and 1.00 m, pressure nodes at 0, 0.40 m, 0.80 m, and 1.20 m b) *fundamental:* displacement node at 0, pressure node at 1.20 m; *first overtone:* displacement nodes at 0 and 0.80 m, pressure node at 0.40 m and 1.20 m; *second overtone:* displacement nodes at 0, 0.48 m, and 0.96 m, pressure nodes at 0.24 m, 0.72 m, and 1.20 m

16.27 506 Hz, 1520 Hz, 2530 Hz

16.29 a) 614 Hz b) 1.23 kHz

16.31 a) 172 Hz b) 86 Hz

16.33 0.125 m

16.35 destructive

16.37 a) 4 beats/s b) 3.0×10^{-8} m, 0

16.39 267 Hz

16.41 a) 0.798 m b) 0.922 m c) 431 Hz d) 373 Hz

16.43 a) 2000 Hz b) 1500 Hz

16.45 a) 302 Hz b) 228 Hz

16.49 a) 0.0114 Pa b) 7.51×10^{-9} m c) 63.0 m

16.51 flute harmonic $3N$ ($N = 1, 3, \ldots$) resonates with string harmonic $4N$

16.53 a) stopped b) $n = 7$, $n = 9$ c) 43.9 cm

16.55 a) $v/2L$, v/L, $3v/2L$

16.57 a) 375 m/s b) 1.39 c) 0.8 cm

16.59 $n(77.3 \text{ Hz})$, $n = 1, 2, 3, \ldots$

16.61 1.27

16.63 a) 548 Hz b) 652 Hz

16.65 a) $I = 2\pi^2 \sqrt{\rho B} f^2 (\Delta R)^2$
b) $P = 8\pi^3 \sqrt{\rho B} f^2 R^2 (\Delta R)^2$
c) $A = (R/d)\Delta R$, $p_{\text{max}} = 2\pi \sqrt{\rho B} (Rf/d)\Delta R$, $I = 2\pi^2 \sqrt{\rho B} (fR/d)^2 (\Delta R)^2$

16.67 a) 6.74 cm b) 147 Hz

16.69 b) 2.0 m/s

16.71 c) 56.8 m/s

16.73 a) greater, 109 Hz b) greater, 218 Hz

16.77 a) 180° b) A: 3.98×10^{-6} W/m^2, 66.0 dB; B: 5.31×10^{-7} W/m^2, 57.2 dB
c) 1.60×10^{-6} W/m^2, 62.0 dB

Chapter 17

17.1 a) −81.0°F b) 134.1°F c) 88.0°F

17.3 140.2°F

17.5 a) 104.4°F, yes b) 54°F

17.7 B

17.9 a) 216.5 K b) 325.9 K c) 205.4 K

17.11 27.23 K

17.13 444 mm Hg

17.15 a) −282°C b) no, 4.76×10^4 Pa

17.17 0.39 m

17.19 a) 1.9014 cm b) 1.8964 cm

17.21 2.3×10^{-5} (C°)$^{-1}$

17.23 11 L

17.25 1.7×10^{-5} (C°)$^{-1}$

17.27 a) 1.431 cm^2 b) 1.437 cm^2

17.29 0.008 cm

17.31 a) 3.2×10^{-5} (C°)$^{-1}$ b) 2.5×10^9 Pa

17.33 a) 38 J b) 4.5×10^4 J

17.35 0.526 C°

17.37 3.03×10^5 J

17.39 5.79×10^5 J

17.41 a) 2.51×10^3 J/kg·K b) overestimate

17.43 a) 114 C° b) 6.35 C°

17.45 a) 215 J/kg·K b) water c) too small

17.47 8 min

17.49 3.64×10^4 J $= 8.69 \times 10^3$ cal $= 34.5$ Btu

17.51 2.39×10^4 Btu/h $= 7.01 \times 10^3$ W

17.53 357 m/s

17.55 3.45 L

17.57 27.5°C

17.59 0.0940 kg

17.61 2.10 kg

17.63 190 g

17.65 a) 222 K/m b) 10.7 W c) 73.3°C

17.67 a) −5.8°C b) 11 W/m^2

17.69 7.1×10^2 Btu $= 7.5 \times 10^5$ J

17.71 105.5°C

17.73 90.2°C

17.75 167 W

17.77 2.1 cm^2

17.79 a) 35.2 M° b) 39.6 C°

17.81 b) 5.0×10^{-3} cm^3, −23 kg/m^3

17.83 37.5°C

17.85 415°C

17.87 23.0 cm, 7.0 cm

17.89 b) 1.9×10^8 Pa

17.91 a) 87°C b) −80°C

17.93 20.2°C

17.95 a) 54.3

17.97 a) 83.6 J b) 1.86 J/mol·K c) 5.60 J/mol·K

17.99 a) 2.7×10^7 J b) 6.89 K c) 19.3 K

17.101 a) 4.7×10^{-3} kg b) yes

17.103 a) 86.1°C b) no ice, 0.130 kg liquid water, no steam

17.105 a) 100°C b) 0.0214 kg steam, 0.219 kg liquid water

17.107 a) 100°C b) no ice, 0.419 kg liquid water, 0.0264 kg steam c) 0°C, 0.154 kg ice, 0.408 kg liquid water, no steam

17.109 a) 94 W b) 1.3

17.111 2.9

17.113 a) 6.0×10^5 s (about 170 h) d) 1.5×10^{10} s (about 500 y); no

17.115 305 min

17.117 5.82×10^{-3} kg

17.119 a) 69.6°C

17.121 a) $H = 4\pi kab\Delta T/(b - a)$
b) $T(r) = \dfrac{[T_1 b(r - a) + T_2 a(b - r)]}{r(b - a)}$
c) $H = 2\pi kL\Delta T/\ln(b/a)$
d) $T(r) = \dfrac{[T_1 \ln(r/a) + T_2 \ln(b/r)]}{\ln(b/a)}$

17.123 b) uniform 0°C d) 3.14×10^3 C°/m
e) 121 W f) 0 g) 1.1×10^{-4} m^2/s
h) −11 C°/s i) 9.1 s j) decrease
k) −7.7 C°/s

17.125 20.42 K

Chapter 18

18.1 a) 56.2 mol b) 6.81×10^6 Pa $= 67.2$ atm

18.3 0.959 atm

18.5 a) 3×10^{27} molecules
b) 3×10^{19} molecules/cm^3

18.7 503.0°C

18.9 3.36×10^5 Pa

18.11 0.159 L

18.13 1.05 atm

18.15 a) 70.2°C b) yes

18.17 850 m

18.19 density at sea level 1.2% larger

18.21 2.28×10^4 Pa

18.23 21.4 cm^3

18.25 a) 8.2×10^{-17} atm b) no

18.27 55.6 mol, 3.35×10^{25} molecules

18.29 a) 9.00×10^{-5} m^3 b) 3.10×10^{-10} m
c) about the same

18.31 b) 1.004

18.33 a) could be true b) could be true c) not true
d) must be true e) could be true

18.35 a) 1.9×10^6 m/s; no, 0.64% of c
b) 7.3×10^{10} K

18.37 a) 6.21×10^{-21} J b) 2.34×10^5 m^2/s^2
c) 484 m/s d) 2.57×10^{-23} kg·m/s
e) 1.24×10^{-19} N f) 1.24×10^{-17} Pa
g) 8.15×10^{21} h) 2.45×10^{22}

18.39 3800°C

18.41 1380 J/kg·K

18.43 a) 1560 J b) 935 J

18.45 a) 987 K b) 438 K c) 110 K

18.49 2500 m

18.51 solid and vapor, with no liquid

18.53 a) 6.9×10^{-3} L b) yes

18.55 0.213 kg

18.57 0.00123 g

18.59 1.92 atm

18.61 a) 31 b) 8.41×10^3 N c) 7.80×10^3 N

18.63 a) 26.2 m/s b) 16.1 m/s, 5.44 m/s c) 1.74 m

18.65 5.0×10^{27}

18.67 a) 303 mol/m^3 c) van der Waals

18.69 b) r_2 c) $r_1 = R_0/2^{1/6}$, $r_2 = R_0$, $r_1/r_2 = 2^{-1/6}$
d) U_0

18.71 a) 517 m/s b) 299 m/s

18.73 b) 1.40×10^5 K, 1.01×10^4 K
c) 6.37×10^3 K, 4.59×10^2 K

18.75 a) 1.24×10^{-14} kg b) 4.16×10^{11}
c) 2.95×10^{-6} m, no

18.77 a) $2R$ b) less

18.79 CO_2: 20.79 J/mol·K, 0.270;
SO_2: 24.94 J/mol·K, 0.205;
H_2S: 24.94 J/mol·K, 0.039

18.83 b) 0.0420N c) $(2.94 \times 10^{-21})N$ d) 0.0297N, $(2.08 \times 10^{-21})N$ e) 0.0595N, $(4.15 \times 10^{-21})N$

18.85 42.6%

18.87 a) 4.5×10^{11} m b) 703 m/s, 6.4×10^8 s
(about 20 y), no c) 1.4×10^{-14} Pa d) about
650 m/s; evaporate f) 2×10^5 K; no
18.89 d) $T_c = 8a/27Rb$, $(V/n)_c = 3b$
e) $p_c = a/27b^2$ f) 8/3 g) 3.28, 3.44, 4.35

Chapter 19

19.1 b) 1330 J
19.3 b) -1.49×10^4 J
19.5 b) 0
19.7 b) 0 for first process, -4.00×10^4 J for second
process
19.9 327 J
19.11 113°C
19.13 a) -1.15×10^5 J b) 2.55×10^5 J, out of gas
c) no
19.15 a) 16.4 min b) 139 m/s = 501 km/h
19.17 a) absorbs b) 7200 J c) liberates 7200 J
19.19 a) yes b) no c) negative
19.21 a) 948 K b) 900 K
19.23 a) 3120 J b) 1870 J c) 5540 J
19.25 into gas
19.27 a) -605 J b) 0 c) yes, 605 J, liberates
19.29 $C_V = 65.5$ J/mol·K, $C_p = 73.8$ J/mol·K
19.31 a) 553 J b) 454 J
19.33 b) 495°C, 29.1 atm
19.35 decreases
19.37 b) 224 J c) -224 J
19.39 a) 4.50×10^4 Pa b) 345 J c) 0.818
19.41 -2.45×10^5 J; out of gas
19.43 a) 45.0 J b) liberates 65.0 J c) $Q_{ad} = 23.0$ J,
$Q_{db} = 22.0$ J
19.45 a) $W_{ab} = 0$, $W_{bc} = p_c(V_c - V_a)$,
$W_{ad} = p_a(V_c - V_a)$, $W_{dc} = 0$
b) $Q_{ab} = U_b - U_a$,
$Q_{bc} = U_c - U_b + p_c(V_c - V_a)$,
$Q_{ad} = U_d - U_a + p_a(V_c - V_a)$,
$Q_{dc} = U_c - U_d$
c) $Q_{abc} = U_c - U_a + p_c(V_c - V_a)$,
$W_{abc} = p_c(V_d - V_a)$;
$Q_{adc} = U_c - U_a + p_a(V_c - V_a)$,
$W_{adc} = p_a(V_d - V_a)$; abc, abc

19.47 -2460 J
19.49 b) -2.40×10^4 J
19.51 -0.226 m^3
19.53 a) 4.32×10^{-4} m^3 b) 648 J c) 7.15×10^5 J
d) 7.14×10^5 J e) no substantial difference
19.55 3.4×10^5 J/kg
19.57 b) 11.9 C°
19.59 a) 0.173 m b) 206°C c) 7.46×10^4 J
19.61 a) $Q = 300$ J, $\Delta U = 0$ b) $Q = 0$,
$\Delta U = -300$ J c) $Q = 750$ J, $\Delta U = 450$ J
19.63 a) $W = 738$ J, $Q = 2588$ J, $\Delta U = 1850$ J
b) $W = 0$, $Q = -1850$ J, $\Delta U = -1850$ J c) 0
19.65 a) $W = -187$ J, $Q = -654$ J, $\Delta U = -467$ J
b) $W = 113$ J, $Q = 0$, $\Delta U = -113$ J
c) $W = 0$, $Q = 580$ J, $\Delta U = 580$ J
19.67 b) 3.63×10^3 J c) 0 d) 3.63×10^3 J
e) 0.114 m^3
19.69 a) $W = nRT \ln[(V_2 - nb)/(V_1 - nb)] +$
$an^2(1/V_2 - 1/V_1)$ b) i) 2.80×10^3 J
ii) 3.11×10^3 J c) 310 J; larger for ideal gas

Chapter 20

20.1 a) 6500 J b) 0.34 = 34%
20.3 a) 0.23 = 23% b) 12,400 J c) 0.350 g
d) 222 kW = 298 hp
20.5 a) 0.25 = 25% b) 970 MW
20.7 a) 453.2°C b) 1.99×10^6 Pa
20.9 a) 1.62×10^4 J b) 5.02×10^4 J
20.11 a) 767 W b) 7.27
20.13 a) 215 J b) 378 K c) 0.391 = 39.1%
20.15 a) 3.09×10^7 J b) 2.49×10^6 J
20.17 a) -665 J, refrigerator b) -190 J, refrigerator
c) 285 J, engine
20.19 3790 cycles
20.21 4.89×10^3 J
20.23 b) $e \to 1$, $K = 0$; $e \to 0$, $K \to \infty$
20.25 a) irreversible b) 7.7 J/K
20.27 47.4 J/K
20.29 -644 J/K
20.31 a) 109 J/K b) N$_2$: 72.8 J/K; Ag: 102 J/K;
Hg: 86.6 J/K
20.33 6.74 J/K

20.35 a) no b) 18.3 J/K c) 18.3 J/K
20.37 a) -73°C b) 3.34×10^4 cycles
20.39 a) $p_1 = 1.00$ atm, $V_1 = 8.62 \times 10^{-3}$ m^3;
$p_2 = 2.00$ atm, $V_2 = 8.62 \times 10^{-3}$ m^3;
$p_3 = 1.00$ atm, $V_3 = 1.41 \times 10^{-2}$ m^3
b) $1 \to 2$: $Q = 2183$ J, $W = 0$, $\Delta U = 2183$ J;
$2 \to 3$: $Q = 0$, $W = 786$ J, $\Delta U = -786$ J;
$3 \to 1$: $Q = -1960$ J, $W = -559$ J,
$\Delta U = -1400$ J c) 227 J d) 227 J
e) $e = 10.4\%$, $e_{Carnot} = 50.0\%$
20.41 a) 13.2 J b) 13.2 J
20.43 b) 45°C c) 58.8% d) 206 J e) 2.57×10^6 Pa
20.45 10.5%
20.47 a) 7.36×10^5 J b) 9.33×10^5 J
c) 1.98×10^5 J d) 3.73
20.49 a) $a \to b$: $Q = -nRT_1 \ln r$, $W = -nRT_1 \ln r$,
$\Delta U = 0$; $b \to c$: $Q = nC_V(T_2 - T_1)$, $W = 0$,
$\Delta U = nC_V(T_2 - T_1)$; $c \to d$: $Q = nRT_2 \ln r$,
$W = nRT_2 \ln r$, $\Delta U = 0$; $d \to a$:
$Q = -nC_V(T_2 - T_1)$, $W = 0$,
$\Delta U = -nC_V(T_2 - T_1)$ c) $1 - T_1/T_2$, same, no
20.51 b) 14 m/s = 50 km/h
20.53 a) 122 J, 78 J b) 5.10×10^{-4} m^3
c) $p_b = 2.32 \times 10^6$ Pa, $V_b = 4.81 \times 10^{-5}$ m^3,
$T_b = 771$ K; $p_c = 4.00 \times 10^6$ Pa,
$V_c = 4.81 \times 10^{-5}$ m^3, $T_c = 1333$ K;
$p_d = 1.47 \times 10^5$ Pa, $V_d = 5.10 \times 10^{-4}$ m^3,
$T_d = 518$ K d) $e_{Carnot} = 77.5\%$, $e = 61.1\%$
20.55 b) 6.20×10^4 J c) 3.42×10^4 J d) before:
6.20×10^4 J; after 3.42×10^4 J
20.57 10.5 J/K
20.61 a) 150 J/K b) -120 J/K c) 30 J/K
d) irreversible
20.63 a) $e = 1 - \dfrac{1}{\gamma}\left[\dfrac{(T_c/T_a)^\gamma r^{\gamma(1-\gamma)} - 1}{(T_c/T_a) - r^{\gamma-1}}\right]$
b) 70.8%

PHOTO CREDITS

INDEX

For users of the three-volume edition, pages 1–791 (Chapters 1–20) are in Volume 1; pages 792–1444 (Chapters 21–37) are in Volume 2; and pages 1403–1714 (Chapters 37–44) are in Volume 3.

Pages 1445–1714 (Chapters 38–44) are not in the Standard Edition of this textbook.